Introduction to the Variational Formulation in Mechanics

Introduction to the Variational Formulation in Mechanics

Fundamentals and Applications

Edgardo O. Taroco, Pablo J. Blanco and Raúl A. Feijóo

HeMoLab - Hemodynamics Modeling Laboratory
LNCC/MCTIC - National Laboratory for Scientific Computing, Brazil

INCT-MACC - National Institute of Science and Technology in Medicine Assisted by Scientific Computing, Brazil

WILEY

This edition first published 2020

Registered Offices
John Wiley & Sons, Inc., 111 River Street, Hoboken, NJ 07030, USA
John Wiley & Sons Ltd, The Atrium, Southern Gate, Chichester, West Sussex, PO19 8SQ, UK

Editorial Office
John Wiley & Sons Ltd, The Atrium, Southern Gate, Chichester, West Sussex, PO19 8SQ, UK

For details of our global editorial offices, customer services, and more information about Wiley products visit us at www.wiley.com.

Wiley also publishes its books in a variety of electronic formats and by print-on-demand. Some content that appears in standard print versions of this book may not be available in other formats.

Library of Congress Cataloging-in-Publication data applied for

HB ISBN: 9781119600909

Cover Design by Wiley
Cover Images: © Sandipkumar Patel/Getty Images, © Pablo Javier Blanco

Set in 10/12pt Warnock by SPi Global, Pondicherry, India

Printed and bound by CPI Group (UK) Ltd, Croydon, CR0 4YY

10 9 8 7 6 5 4 3 2 1

To our families

Contents

Preface

This book was written intermittently over the period between 1980 and 2016 with an aim to provide students attending the courses organized by the authors, particularly for the graduate students at the National Laboratory for Scientific Computing (LNCC), with the foundational material of Mechanics using a variational tapestry. It is the result of the knowledge acquired and divulged by E.O.T and R.A.F. since the LNCC was established, which was initiated with the creation of the Laboratory of Computing (LAC) of the Brazilian Center for Research in Physics (CBPF) in 1977, through the foundation of the Laboratory for Scientific Computing (LCC) in 1980, its conversion into the category of a national laboratory (LNCC) in 1982 and the definitive move to the city of Petrópolis in 1998.

Part of the material presented here was used in various courses of theoretical and applied mechanics organized by E.O.T. and R.A.F. These course were

- 1st Course on Theoretical and Applied Mechanics: Theory of Shells and their Applications in Engineering (Module I – Basic Principles, July 5 to 30, 1982; Module II – Mechanical Models, January 3 to February 11, 1983; and Module III – Instability of Shells, July 4 to 30, 1983).
- 2nd Course on Theoretical and Applied Mechanics: Fundamentals of the Finite Element Method and its Applications in Engineering (Module I – Fundamentals of the Finite Element Method, July 2 to 27, 1984; Module II – Applications of the Finite Element Method in Solid Mechanics, January 7 to February 1, 1985, Module III – Modern Aspects of the Finite Element Method, July 1 to 26, 1985).
- 3rd Course on Theoretical and Applied Mechanics: Optimization: Fundamentals and Applications in Engineering (Module I – Optimization in the Modeling and Analysis of Engineering Problems, July 7 to August 1, 1986; Module II – Optimal Design: Foundations and Applications, July 6 to 31, 1987).

Unfortunately, and largely due to the long period of time that this document took to be finished, one of the authors, Prof. Edgardo O. Taroco, left us (passed away in January 2010). Nevertheless, we decided to keep his name among the authors as an acknowledgement to his deep contributions and in honor to his memory, as well as to the friendship and generosity that he always offered us throughout all these years. Therefore, all errors (of any kind) are the sole responsibility of P.J.B. and R.A.F.

Part of this material was used by R.A.F. in the course dictated during the Post-Doctoral Latin American Seminar on Continuum Mechanics and Microstructure, organized by the National Atomic Energy Commission in Argentina, sponsored by the Organization of American States (OAS), and held in Buenos Aires, July-August 1984. It was only after 1998 that the LNCC was moved to Petrópolis, and the LNCC Graduate Program was initiated. Then, we decided to consolidate the aforementioned texts into a monograph that would give emphasis to the formulation of the mechanics within a purely variational structure. More recently, since P.J.B. joined the LNCC in 2009, the idea of finishing this document, also including topics in other areas of physics as well as the extension of the variational framework to a multi-scale paradigm, resurfaced. Currently, this material is being used in several courses in the graduate program in Computational Modeling at LNCC, Brazil, and also in the graduate program in Mechanical Engineering at National University of Mar del Plata, Argentina.

Such variational framework used to present the mechanics was chosen not only for being one of our main areas of research, but also, and fundamentally, because we strongly believe that this way of looking into the roots of mechanics is the most suitable and convenient perspective to approach the mathematical modeling. In fact, and as it will become increasingly clear as we proceed, the foundational pillars upon which the whole modeling journey rests using this variational tapestry are the following

- The first pillar is related to the description of the kinematics, that is the formalization of the kinematical hypotheses which provides the definition of the generalized motion actions and admissible generalized strain rate actions for the model under study.
- The second pillar consists of the mathematical duality postulated between quantities related to such motion actions and generalized strain rates with, respectively, the external generalized forces and internal generalized stresses. In this way, forces and stresses are constructs fully shaped by linear continuous functionals whose arguments are kinematical entities. This aspect establishes a clear difference between the approach developed in this book and the procedure followed by most of the literature in the field of continuum mechanics, where forces and stresses are malleable entities introduced a priori, regardless of the kinematics defined for the physical system.
- The third pillar is the Principle of Virtual Power (or its generalization, the Principle of Multi-scale Virtual Power in the context of modeling problems with more than one scale). This principle allows to establish, for the physical system under study, the concept of mechanical equilibrium between external forces and internal stresses and, when proper constitutive relations are given, this principle characterizes the generalized displacement field for which the associated generalized stress state equilibrates the external forces.

These three pillars support the realm of so-called Method of Virtual Power (MVP), which establishes well-defined basic steps targeting a fully consistent modelling technique. Such variational structure was proposed, although with subtle modifications, by Prof. Paul Germain [114–118]. Particularly, Prof. Germain (lifetime member of the French Academy of Science) was invited by us in 1982 to teach the course Four Lectures on the Foundation of Shell Theory within Module I of the 1st Course on Theoretical and Applied Mechanics that we organized, precisely, to disseminate these concepts among students and professors from different Latin American countries. Another professor who greatly contributed to consolidate this approach in these courses was Prof. Giovanni Romano (Facoltà di Ingegneria dell'Università degli Studi di Napoli, Italy).

The organization of this book closely follows this spirit. In Part I we present the basic concepts of vector and tensor algebra and analysis, where the reader is introduced to the ubiquitous use of compact vector and tensor notation. This compact notation allows to go through the basic principles and concepts of the mechanics in a clear and concise way, without being obscured by the presence of indices, components and metric-related entities, which should be relegated to their specific role at the time of the calculation. Part II is dedicated to presenting the Method of Virtual Power (MVP), from the kinematics, through the duality and to the Principle of Virtual Power. This Part presents the variational groundwork which is used as a guiding theme in all that follows. The MVP is then applied to the most general case in the mechanics of deformable bodies, while its application to the case of hyperelastic materials, and materials which may experience creep and plasticity phenomena is also discussed. In Part III we present the application of the MVP to the modelling of structural components such as beams, plates and shells. For these components, it will be clear how forces and stresses emerge as natural outgrowth of the kinematical hypotheses. In Part IV, the application of the MVP to other problems from physics is addressed, including heat conduction, incompressible fluid flow, and high order continua. This will help the reader to illustrate the use of this unified theoretical framework in problems in which a purely variational approach is seldom encountered in the textbooks. Finally, in Part V, we expand this variational realm to embrace problems which require a multi-scale paradigm. This extended variational structure has been called the Method of Multi-scale Virtual Power (MMVP). Pursuing the same standard, the MMVP allows to provide a convenient and safe tool to substantiate multi-scale models of complex physical systems, allowing to consistently provide the groundwork on top of which multi-scale homogenization should take place. Finally, in the Appendices, we present various mathematical concepts and results to make the document self-contained.

Last but not least, we would like to thank all those who somehow contributed to our excursion towards this book. In particular we thank Gonzalo R. Feijóo for his contribution in the first drafting of some chapters of this book, and to Professors Enzo A. Dari (Bariloche Atomic Centre, Argentina), Sebastián Giusti (National Technological University, Faculty of Córdoba, Argentina), Santiago A. Urquiza (National University of Mar del Plata, Argentina), Pablo J. Sánchez (National Technological University, Regional Faculty of Santa Fe, Argentina), Alejandro Clausse (National University of Central Buenos Aires, Argentina) and Eduardo A. de Souza Neto (Zienkiewicz Center for Computational Engineering, Swansea University, United Kingdom) for their comments and discussions that definitely enriched us and, therefore, our work. We would also like to thank our Ph.D. students, particularly to Gonzalo D. Ares, Gonzalo D. Maso Talou, Carlos A. Bulant, Alonso M. Alvarez and Felipe Figueredo Rocha who, with their criticisms and observations, have also helped to improve this text.

Petrópolis, Brazil
March 2018

Pablo J. Blanco
Raúl A. Feijóo

Part I

Vector and Tensor Algebra and Analysis

The goal of the following two chapters, which comprise Part I of this book, is to provide the reader with the basic concepts and training in the area of tensor algebra (Chapter 1) and tensor analysis (Chapter 2) which will be omnipresent throughout this book. It is important to highlight here the use of compact notation (also called intrinsic notation) when invoking vector and tensors as well as algebraic and differential operations among them. That is, vectors and tensors are written independently from the adopted coordinate system. In contrast to the use of indicial notation, which puts in evidence the components of vectors and tensors in the given coordinate system, the compact notation appears as a clean and elegant form that allows the concepts to be presented without being obnubilated by the sometimes overwhelming presence of indexes.

At the end of these two chapters, some further reading material is suggested to complement and deepen the concepts presented here. However, we highlight that for the study of the rest of the book this reading is not mandatory. Moreover, the reader will also find in Appendices A and B a more detailed exposition of the topics addressed in what follows, in an attempt to make the book self-contained.

1

Vector and Tensor Algebra

1.1 Points and Vectors

Consider the three-dimensional Euclidean space, denoted by $\mathscr{E}$, whose geometry is built upon a set of primitive elements called points. Note that $\mathscr{E}$ is not a vector space in the sense of the algebra because the addition of points is a concept without meaning.

The difference between two points X and Y of space $\mathscr{E}$ is defined by

$$\mathbf{v} = X - Y, \tag{1.1}$$

where $\mathbf{v}$ is the vector whose origin is in Y and ends at X. All the vectors which can be determined through the differences between points belonging to $\mathscr{E}$ form the set V associated with $\mathscr{E}$. The set V is a (real) vector space, where the two basic operations inherent to the notion of vector space are defined, which are (i) the addition of vectors and (ii) the product of a vector by a real number.

The addition operation between a point $Y \in \mathscr{E}$ and the vector $\mathbf{v} \in V$ defines the point $X \in \mathscr{E}$ such that (1.1) is verified. This operation allows us to establish a biunivocal correspondence between points of $\mathscr{E}$ and vectors of V. In fact, we can arbitrarily pick a point O from $\mathscr{E}$, and then for each point $X \in \mathscr{E}$ there exists a unique vector $\mathbf{v} \in V$ such that $\mathbf{v} = X - O$.

Space V is called three-dimensional provided there exist in V sets of three vectors $\{\mathbf{e}_i\} = \{\mathbf{e}_1, \mathbf{e}_2, \mathbf{e}_3\}$ which are linearly independent[1] and can span the entire vector space V, that is, they can generate any vector $\mathbf{v} \in V$ through the linear combination $\mathbf{v} = \alpha_i \mathbf{e}_i$.[2] Any of these linearly independent sets of vectors $\{\mathbf{e}_i\}$ is called a basis for the (real) vector space V.

Beyond basic operations, multiplication by a real number, and addition of vectors, the vector space V associated with $\mathscr{E}$ is endowed with an inner product, also called the scalar product, operation between vectors of V. For $\mathbf{u}, \mathbf{v} \in V$, this operation is denoted by $\mathbf{u} \cdot \mathbf{v}$, which is geometrically defined by the product of the lengths of the vectors multiplied by the cosine of the angle between them. This operation satisfies the properties of the inner product in the sense of the algebra.

---

1 Vectors $\{\mathbf{e}_i\}$ are called linearly independent if $\alpha_i \mathbf{e}_i = \mathbf{0}$ implies $\alpha_i = 0$ for $i = 1, 2, 3$.

2 Throughout this book, unless stated otherwise, Einstein notation is adopted to shorten summation notations, resulting, for example, in the following lumped notation when indexes are repeated: $\mathbf{v} = \sum_{i=1}^{3} \alpha_i \mathbf{e}_i = \alpha_i \mathbf{e}_i$.

Introduction to the Variational Formulation in Mechanics: Fundamentals and Applications, First Edition.
Edgardo O. Taroco, Pablo J. Blanco and Raúl A. Feijóo.

Also, and through the introduction of this operation, the space V has a topological structure induced by the inner product through the definition of the norm operation for a vector $\mathbf{v} \in V$

$$\|\mathbf{v}\| = \sqrt{\mathbf{v} \cdot \mathbf{v}}. \tag{1.2}$$

Making use of these operations, vectors $\mathbf{u}, \mathbf{v} \in V$ are said to be orthogonal if $\mathbf{u} \cdot \mathbf{v} = 0$. Similarly, a basis $\{\mathbf{e}_i\}$ of V is called orthogonal if $\mathbf{e}_i \cdot \mathbf{e}_j = \|\mathbf{e}_i\|\|\mathbf{e}_j\|\delta_{ij}$ is verified, $i,j = 1,2,3$, where δ_{ij} is the Kronecker symbol.[3] Finally, a basis of V is called orthonormal if it is orthogonal and the norm of the vectors in the basis is unitary, that is, $\|\mathbf{e}_i\| = 1$, $i = 1,2,3$.

The matrix $[g_{ij}]$, $i,j \in \{1,2,3\}$, defined by

$$g_{ij} = \mathbf{e}_i \cdot \mathbf{e}_j, \qquad i,j = 1,2,3, \tag{1.3}$$

is not singular when the set $\{\mathbf{e}_i\}$ is a basis for V. In turn, from the definition it follows that $g_{ij} = g_{ji}$, that is, $[g_{ij}]$ is a symmetric matrix. Let us call

$$[g^{ij}] = [g_{ij}]^{-1}, \tag{1.4}$$

the inverse of matrix $[g_{ij}]$. From the definition we obtain

$$g^{ik}g_{kj} = g_{ik}g^{kj} = \delta_{ij}, \qquad i,j = 1,2,3. \tag{1.5}$$

With these results, we can introduce the dual basis $\{\mathbf{e}^i\}$ associated with $\{\mathbf{e}_i\}$ as the image of the linear transformation

$$\begin{aligned} [g^{ij}] &: V \to V, \\ \mathbf{e}_j &\mapsto \mathbf{e}^i = g^{ij}\mathbf{e}_j. \end{aligned} \tag{1.6}$$

It can be proved that this transformation produces a basis for V. Reciprocally, the application of $[g_{ij}]$ over $\{\mathbf{e}^i\}$ yields the original basis $\{\mathbf{e}_i\}$. In fact

$$g_{ij}\mathbf{e}^j = g_{ij}g^{jk}\mathbf{e}_k = \delta_{ik}\mathbf{e}_k = \mathbf{e}_i. \tag{1.7}$$

Another useful result is the following

$$\mathbf{e}^i \cdot \mathbf{e}_j = g^{ik}\mathbf{e}_k \cdot \mathbf{e}_j = g^{ik}g_{kj} = \delta_{ij}. \tag{1.8}$$

A coordinate system consists of a basis $\{\mathbf{e}_i\}$ for V, not necessarily orthogonal, and an arbitrary point O of $\mathscr{E}$ called the origin of the coordinate system. When the basis is orthonormal, the coordinate system is called Cartesian.

Observe that the notion of vector was introduced independently from the adopted basis, or, equivalently, from the coordinate system. When a basis $\{\mathbf{e}_i\}$, and then its dual basis $\{\mathbf{e}^i\}$, is chosen, then each vector $\mathbf{u} \in V$ can be associated with a triple of real numbers $\{u_1, u_2, u_3\}$ called components of $\mathbf{u}$ with respect to the basis $\{\mathbf{e}^i\}$, which are defined as follows

$$u_i = \mathbf{u} \cdot \mathbf{e}_i \qquad i = 1,2,3. \tag{1.9}$$

These are the components with respect to $\{\mathbf{e}^i\}$ because, as $\{\mathbf{e}^i\}$ is a basis, it results in $\mathbf{u} = \alpha_i\mathbf{e}^i$ and then

$$u_i = \mathbf{u} \cdot \mathbf{e}_i = \alpha_k\mathbf{e}^k \cdot \mathbf{e}_i = \alpha_k\delta_{ki} = \alpha_i. \tag{1.10}$$

---

3 The Kronecker symbol δ_{ij} is such that $\delta_{ii} = 1$ and $\delta_{ij} = 0$ for $i \neq j$.

In particular, given a basis $\{\mathbf{e}_i\}$, and its dual $\{\mathbf{e}^i\}$, the components of $\mathbf{u}$ with respect to these bases are usually named as follows

- Components of $\mathbf{u}$ with respect to $\{\mathbf{e}_i\}$ are contravariant components, and are defined by

$$u^i = \mathbf{u} \cdot \mathbf{e}^i \qquad i = 1, 2, 3. \tag{1.11}$$

  With these components the vector $\mathbf{u}$ can be represented through the linear combination $\mathbf{u} = u^i\mathbf{e}_i$.
- Components of $\mathbf{u}$ with respect to $\{\mathbf{e}^i\}$ are covariant components, and are defined by

$$u_i = \mathbf{u} \cdot \mathbf{e}_i \qquad i = 1, 2, 3. \tag{1.12}$$

  With these components the vector $\mathbf{u}$ can be represented through the linear combination $\mathbf{u} = u_i\mathbf{e}^i$.

If the basis is orthonormal, it is easy to show that it is identical to its dual basis, and therefore there is no distinction between covariant and contravariant components.

Likewise, given a coordinate system in $\mathscr{E}$, characterized by the basis $\{\mathbf{e}_i\}$ and the point $O \in \mathscr{E}$, we can define the coordinates of an arbitrary point $X \in \mathscr{E}$ as the covariant components of the vector $X - O$ from V, that is

$$X_i = (X - O) \cdot \mathbf{e}_i. \tag{1.13}$$

Thus, the same vector $\mathbf{u} \in V$, or the same point $X \in \mathscr{E}$, can be associated with different triples of components and representations depending upon the chosen coordinate system.[4]

To underline the difference between intrinsic and indicial notation, note that the inner product of vectors $\mathbf{u}, \mathbf{v} \in V$ as a function of their different components results in

$$\mathbf{u} \cdot \mathbf{v} = u^i v_i = u_i v^i = u_i g^{ij} v_j = u^i g_{ij} v^j, \tag{1.14}$$

that is, several expressions are possible for the same concept, which may delay the understanding progress.

Exercise 1.1 Verify the validity of identities given by (1.14).

4 As stated above, we will employ compact notation, which means that we highlight the entity in detriment of its components, in this case vector $\mathbf{u}$, or point X. Notation emphasizing components is also called indicial notation, and it should only be employed during the analysis of a given specific problem, specifically when calculations are required. In this way, we can present the concepts and operations in a clear and concise manner, without being confined to a certain coordinate system. This also helps to call the attention to the fact that, during calculations, we should choose that basis which makes the treatment of the problem as simple as possible. To sum up, indicial notation is not related to concepts, but to calculus.

1.2 Second-Order Tensors

We will employ the term tensor, or second-order tensor, as a synonym for the linear transformation between V and V.[5] Then, we have

$$\begin{aligned} \mathbf{T} :& V \to V, \\ & \mathbf{u} \mapsto \mathbf{v} = \mathbf{T}\mathbf{u}, \end{aligned} \tag{1.15}$$

verifying

$$\mathbf{T}(\mathbf{u} + \mathbf{v}) = \mathbf{T}\mathbf{u} + \mathbf{T}\mathbf{v} \qquad \forall \mathbf{u}, \mathbf{v} \in V, \tag{1.16}$$

$$\mathbf{T}(\alpha \mathbf{u}) = \alpha \mathbf{T}\mathbf{u} \qquad \forall \alpha \in \mathbb{R} \text{ and } \forall \mathbf{u} \in V. \tag{1.17}$$

Clearly, expressions (1.16) and (1.17) are equivalent to

$$\mathbf{T}(\alpha \mathbf{u} + \beta \mathbf{v}) = \alpha \mathbf{T}\mathbf{u} + \beta \mathbf{T}\mathbf{v} \qquad \forall \alpha, \beta \in \mathbb{R} \text{ and } \forall \mathbf{u}, \mathbf{v} \in V. \tag{1.18}$$

Taking α and β zero, we obtain an expression that will be frequently used hereafter,

$$\mathbf{T}\mathbf{0} = \mathbf{0}, \tag{1.19}$$

where $\mathbf{0}$ represents the null element in V.

The set of all second-order tensors, that is, the set of all linear transformations between V and V, will be called *Lin*

$$Lin = \{\mathbf{T};\ \mathbf{T} : V \to V, \text{linear}\}. \tag{1.20}$$

Introducing the addition and multiplication by real numbers in *Lin* defined by

$$(\mathbf{T} + \mathbf{S})\mathbf{u} = \mathbf{T}\mathbf{u} + \mathbf{S}\mathbf{u} \qquad \forall \mathbf{u} \in V, \tag{1.21}$$

$$(\alpha \mathbf{T})\mathbf{u} = \alpha(\mathbf{T}\mathbf{u}) \qquad \forall \mathbf{u} \in V, \tag{1.22}$$

turns *Lin* into a vector space, where the null tensor $\mathbf{O}$ transforms every vector $\mathbf{u} \in V$ into the null vector $\mathbf{0} \in V$, that is,

$$\mathbf{O}\mathbf{u} = \mathbf{0} \qquad \forall \mathbf{u} \in V. \tag{1.23}$$

The identity tensor is denoted by $\mathbf{I}$ and is defined by

$$\mathbf{I}\mathbf{u} = \mathbf{u} \qquad \forall \mathbf{u} \in V. \tag{1.24}$$

Given $\mathbf{T} \in Lin$, the set of vectors $\mathbf{v} \in V$ satisfying $\mathbf{T}\mathbf{v} = \mathbf{0}$ is denoted by $\mathcal{N}(\mathbf{T})$, that is,

$$\mathcal{N}(\mathbf{T}) = \{\mathbf{v} \in V;\ \mathbf{T}\mathbf{v} = \mathbf{0}\}. \tag{1.25}$$

It is easy to prove that $\mathcal{N}(\mathbf{T})$ is also a vector subspace of V, called the null space of $\mathbf{T}$ (also the kernel of $\mathbf{T}$).

Given $\mathbf{A}, \mathbf{B} \in Lin$, the composition of these tensors (composed transformation) is another tensor (linear transformation) $\mathbf{T} \in Lin$ such that

$$\mathbf{T}\mathbf{u} = (\mathbf{A}\mathbf{B})\mathbf{u} = \mathbf{A}\mathbf{B}\mathbf{u} \qquad \forall \mathbf{u} \in \mathbf{V}. \tag{1.26}$$

5 In this chapter we limit the presentation to second- and third-order tensors, and some associated operations because these are the most used elements along the book. However, in Appendices A and B the reader will find material related to linear transformations between vector spaces of possibly different dimensions, which naturally embraces the case of third- and also second-order tensors.

Since in general $\mathbf{AB} \neq \mathbf{BA}$, when the identity is verified we say that $\mathbf{A}$ and $\mathbf{B}$ are commutative.

From previous definitions, and given arbitrary $\mathbf{T}, \mathbf{S}, \mathbf{D} \in Lin$ and $\alpha \in \mathbb{R}$, then we have

$$\mathbf{T}(\mathbf{SD}) = (\mathbf{TS})\mathbf{D} = \mathbf{TSD}, \tag{1.27}$$

$$\mathbf{T}(\mathbf{S} + \mathbf{D}) = \mathbf{TS} + \mathbf{TD}, \tag{1.28}$$

$$\alpha(\mathbf{TS}) = \mathbf{T}(\alpha\mathbf{S}), \tag{1.29}$$

$$\mathbf{IT} = \mathbf{TI} = \mathbf{T}. \tag{1.30}$$

Indeed, for arbitrary $\mathbf{v} \in V$, it is

$$\mathbf{T}(\mathbf{SD})\mathbf{v} = \mathbf{T}(\mathbf{S}(\mathbf{Dv})) = (\mathbf{TS})(\mathbf{Dv}) = (\mathbf{TS})\mathbf{Dv}, \tag{1.31}$$

from which (1.27) is verified. Also, we have

$$\mathbf{T}(\mathbf{S} + \mathbf{D})\mathbf{v} = \mathbf{T}((\mathbf{S} + \mathbf{D})\mathbf{v}) = \mathbf{T}(\mathbf{Sv} + \mathbf{Dv}) = \mathbf{TSv} + \mathbf{TDv} = (\mathbf{TS} + \mathbf{TD})\mathbf{v}, \tag{1.32}$$

which verifies (1.28). Similarly,

$$\alpha(\mathbf{TS})\mathbf{v} = \alpha(\mathbf{T}(\mathbf{Sv})) = \mathbf{T}(\alpha\mathbf{Sv})) = \mathbf{T}(\alpha\mathbf{S})\mathbf{v}, \tag{1.33}$$

and (1.29) holds. Note finally that

$$\mathbf{ITv} = \mathbf{I}(\mathbf{Tv}) = \mathbf{Tv}, \tag{1.34}$$

and then we have proved (1.30).

The following notation will also be used

$$\mathbf{T}^n = \overbrace{\mathbf{TT}...\mathbf{T}}^{n} \qquad n \in \mathbf{N}, \tag{1.35}$$

with $\mathbf{T}^0 = \mathbf{I}$.

The transpose of a tensor $\mathbf{T}$ is the unique tensor $\mathbf{T}^T$ satisfying

$$\mathbf{Tu} \cdot \mathbf{v} = \mathbf{u} \cdot \mathbf{T}^T\mathbf{v} \qquad \forall \mathbf{u}, \mathbf{v} \in V. \tag{1.36}$$

Uniqueness is proved assuming that there are two tensor transposes for $\mathbf{T}$, denoted by $\mathbf{T}_1^T$ and $\mathbf{T}_2^T$, and which will be assumed to be different. From definition (1.36), each tensor satisfies

$$\mathbf{Tu} \cdot \mathbf{v} = \mathbf{u} \cdot \mathbf{T}_1^T\mathbf{v} \quad \forall \mathbf{u}, \mathbf{v} \in V, \tag{1.37}$$

$$\mathbf{Tu} \cdot \mathbf{v} = \mathbf{u} \cdot \mathbf{T}_2^T\mathbf{v} \quad \forall \mathbf{u}, \mathbf{v} \in V. \tag{1.38}$$

Subtracting both expressions yields

$$\mathbf{u} \cdot (\mathbf{T}_1^T\mathbf{v} - \mathbf{T}_2^T\mathbf{v}) = \mathbf{u} \cdot (\mathbf{T}_1^T - \mathbf{T}_2^T)\mathbf{v} = 0 \qquad \forall \mathbf{u}, \mathbf{v} \in V. \tag{1.39}$$

Recalling that $\mathbf{a} \cdot \mathbf{b} = 0$, $\forall \mathbf{a} \in V$ implies $\mathbf{b} = \mathbf{0}$, we have

$$(\mathbf{T}_1^T - \mathbf{T}_2^T)\mathbf{v} = \mathbf{0} \qquad \forall \mathbf{v} \in V, \tag{1.40}$$

and from the definition of the null tensor (1.23), we arrive at

$$\mathbf{T}_1^T = \mathbf{T}_2^T, \tag{1.41}$$

which contradicts the fact that both tensors were different. Therefore, there exists a unique transpose of a tensor.

Using the definition of the transpose of a tensor it is straightforward to conclude that, for arbitrary $\mathbf{A}, \mathbf{S} \in Lin$ and $\alpha \in \mathbb{R}$, we obtain

$$(\mathbf{S} + \mathbf{A})^T = \mathbf{S}^T + \mathbf{A}^T, \tag{1.42}$$

$$(\alpha \mathbf{S})^T = \alpha \mathbf{S}^T. \tag{1.43}$$

In fact, taking arbitrary $\mathbf{u}, \mathbf{v} \in V$, then

$$\begin{aligned}(\mathbf{S} + \mathbf{A})^T\mathbf{u} \cdot \mathbf{v} &= \mathbf{u} \cdot (\mathbf{S} + \mathbf{A})\mathbf{v} = \mathbf{u} \cdot \mathbf{S}\mathbf{v} + \mathbf{u} \cdot \mathbf{A}\mathbf{v} = \mathbf{S}^T\mathbf{u} \cdot \mathbf{v} + \mathbf{A}^T\mathbf{u} \cdot \mathbf{v} \\ &= (\mathbf{S}^T\mathbf{u} + \mathbf{A}^T\mathbf{u}) \cdot \mathbf{v} = (\mathbf{S}^T + \mathbf{A}^T)\mathbf{u} \cdot \mathbf{v},\end{aligned} \tag{1.44}$$

and so (1.42) is proved. Analogously,

$$(\alpha \mathbf{S})^T\mathbf{u} \cdot \mathbf{v} = \mathbf{u} \cdot (\alpha \mathbf{S})\mathbf{v} = \alpha \mathbf{u} \cdot \mathbf{S}\mathbf{v} = \alpha \mathbf{S}^T\mathbf{u} \cdot \mathbf{v}, \tag{1.45}$$

and so (1.43) is demonstrated.

As it is easy to see, the transpose operation is a linear transformation between *Lin* and *Lin*. In addition, given arbitrary $\mathbf{A}, \mathbf{S} \in Lin$, it is

$$(\mathbf{S}\mathbf{A})^T = \mathbf{A}^T\mathbf{S}^T, \tag{1.46}$$

$$(\mathbf{S}^T)^T = \mathbf{S}. \tag{1.47}$$

In fact, for arbitrary $\mathbf{u}, \mathbf{v} \in V$, we have

$$(\mathbf{S}\mathbf{A})^T\mathbf{u} \cdot \mathbf{v} = \mathbf{u} \cdot (\mathbf{S}\mathbf{A})\mathbf{v} = \mathbf{u} \cdot \mathbf{S}(\mathbf{A}\mathbf{v}) = \mathbf{S}^T\mathbf{u} \cdot \mathbf{A}\mathbf{v} = \mathbf{A}^T\mathbf{S}^T\mathbf{u} \cdot \mathbf{v}, \tag{1.48}$$

and we arrive at (1.46). Analogously

$$(\mathbf{S}^T)^T\mathbf{u} \cdot \mathbf{v} = \mathbf{u} \cdot (\mathbf{S}^T)\mathbf{v} = \mathbf{S}\mathbf{u} \cdot \mathbf{v}, \tag{1.49}$$

and then (1.197) is verified.

A tensor $\mathbf{S}$ is called symmetric if

$$\mathbf{S}\mathbf{u} \cdot \mathbf{v} = \mathbf{u} \cdot \mathbf{S}\mathbf{v} \qquad \forall \mathbf{u}, \mathbf{v} \in V, \tag{1.50}$$

and in such a case we conclude that $\mathbf{S} = \mathbf{S}^T$. A tensor is called skew-symmetric if

$$\mathbf{S}\mathbf{u} \cdot \mathbf{v} = -(\mathbf{u} \cdot \mathbf{S}\mathbf{v}) \qquad \forall \mathbf{u}, \mathbf{v} \in V, \tag{1.51}$$

which implies $\mathbf{S} = -\mathbf{S}^T$.

The set of all symmetric tensors will be denoted by *Sym* and the set of all skew-symmetric tensors will be denoted by *Skw*. In particular, the null tensor $\mathbf{O}$ is symmetric and skew-symmetric.

Any tensor $\mathbf{S} \in Lin$ can be univocally represented by the addition of a symmetric tensor (called $\mathbf{S}^s$) and a skew-symmetric tensor (called $\mathbf{S}^a$), that is,

$$\mathbf{S} = \mathbf{S}^s + \mathbf{S}^a, \tag{1.52}$$

where

$$\mathbf{S}^s = \frac{1}{2}(\mathbf{S} + \mathbf{S}^T), \tag{1.53}$$

$$\mathbf{S}^a = \frac{1}{2}(\mathbf{S} - \mathbf{S}^T), \tag{1.54}$$

are, respectively, called the symmetric component and the skew-symmetric component of $\mathbf{S}$.

Since the transpose operation yields a unique tensor transpose, it follows that the linear combination of a symmetric (skew-symmetric) tensor results in a symmetric (skew-symmetric) tensor. Then, *Sym* and *Skw* are vector subspaces of *Lin*. Moreover, from the uniqueness of the decomposition into symmetric and skew-symmetric components, it is concluded that *Lin* can be written as the direct sum of these two subspaces

$$Lin = Sym \oplus Skw. \tag{1.55}$$

Consider an arbitrary $\mathbf{W} \in Skw$ and an arbitrary $\mathbf{T} \in Lin$. Then, it is possible to show that for any $\mathbf{u} \in V$ the following holds

$$\mathbf{u} \cdot \mathbf{W}\mathbf{u} = 0, \tag{1.56}$$

$$\mathbf{u} \cdot \mathbf{T}\mathbf{u} = \mathbf{u} \cdot \mathbf{T}^s\mathbf{u}. \tag{1.57}$$

In fact, for arbitrary $\mathbf{u} \in V$ we have

$$\mathbf{u} \cdot \mathbf{W}\mathbf{u} = \mathbf{W}^T\mathbf{u} \cdot \mathbf{u} = -\mathbf{W}\mathbf{u} \cdot \mathbf{u}, \tag{1.58}$$

from where (1.56) is proved. Similarly, and using (1.56), we have

$$\mathbf{u} \cdot \mathbf{T}\mathbf{u} = \mathbf{u} \cdot (\mathbf{T}^s + \mathbf{T}^a)\mathbf{u} = \mathbf{u} \cdot \mathbf{T}^s\mathbf{u} + \mathbf{u} \cdot \mathbf{T}^a\mathbf{u} = \mathbf{u} \cdot \mathbf{T}^s\mathbf{u}, \tag{1.59}$$

and (1.57) follows.

The tensor product between two vectors $\mathbf{a}, \mathbf{b} \in V$ is the second-order tensor $\mathbf{a} \otimes \mathbf{b}$ that transforms any vector $\mathbf{v} \in V$ into vector $(\mathbf{b} \cdot \mathbf{v})\mathbf{a}$, that is,

$$(\mathbf{a} \otimes \mathbf{b})\mathbf{v} = (\mathbf{b} \cdot \mathbf{v})\mathbf{a} \qquad \forall \mathbf{v} \in V. \tag{1.60}$$

From the previous definition, and given arbitrary $\mathbf{a}, \mathbf{b}, \mathbf{c}, \mathbf{d} \in V$, we obtain the following results

$$(\mathbf{a} \otimes \mathbf{b})^T = \mathbf{b} \otimes \mathbf{a}, \tag{1.61}$$

$$(\mathbf{a} \otimes \mathbf{b})(\mathbf{c} \otimes \mathbf{d}) = (\mathbf{b} \cdot \mathbf{c})(\mathbf{a} \otimes \mathbf{d}). \tag{1.62}$$

In addition, given $\mathbf{T} \in Lin$, it can be shown that

$$\mathbf{T}(\mathbf{a} \otimes \mathbf{b}) = (\mathbf{T}\mathbf{a}) \otimes \mathbf{b}, \tag{1.63}$$

$$(\mathbf{a} \otimes \mathbf{b})\mathbf{T} = \mathbf{a} \otimes (\mathbf{T}^T\mathbf{b}). \tag{1.64}$$

Indeed, consider arbitrary $\mathbf{u}, \mathbf{v} \in V$, using definition (1.60) forward and backward, we have

$$(\mathbf{a} \otimes \mathbf{b})^T\mathbf{u} \cdot \mathbf{v} = \mathbf{u} \cdot (\mathbf{a} \otimes \mathbf{b})\mathbf{v} = \mathbf{u} \cdot (\mathbf{b} \cdot \mathbf{v})\mathbf{a} = (\mathbf{u} \cdot \mathbf{a})(\mathbf{b} \cdot \mathbf{v}) = (\mathbf{b} \otimes \mathbf{a})\mathbf{u} \cdot \mathbf{v}, \tag{1.65}$$

and then (1.61) holds. Now, observe that

$$\begin{aligned}(\mathbf{a} \otimes \mathbf{b})(\mathbf{c} \otimes \mathbf{d})\mathbf{u} &= (\mathbf{a} \otimes \mathbf{b})(\mathbf{d} \cdot \mathbf{u})\mathbf{c} = (\mathbf{d} \cdot \mathbf{u})(\mathbf{b} \cdot \mathbf{c})\mathbf{a} \\ &= (\mathbf{b} \cdot \mathbf{c})(\mathbf{d} \cdot \mathbf{u})\mathbf{a} = (\mathbf{b} \cdot \mathbf{c})(\mathbf{a} \otimes \mathbf{d})\mathbf{u},\end{aligned} \tag{1.66}$$

and we arrive at (1.62). Similarly, we have

$$\mathbf{T}(\mathbf{a} \otimes \mathbf{b})\mathbf{u} = (\mathbf{b} \cdot \mathbf{u})\mathbf{T}\mathbf{a} = ((\mathbf{T}\mathbf{a}) \otimes \mathbf{b})\mathbf{u}, \tag{1.67}$$

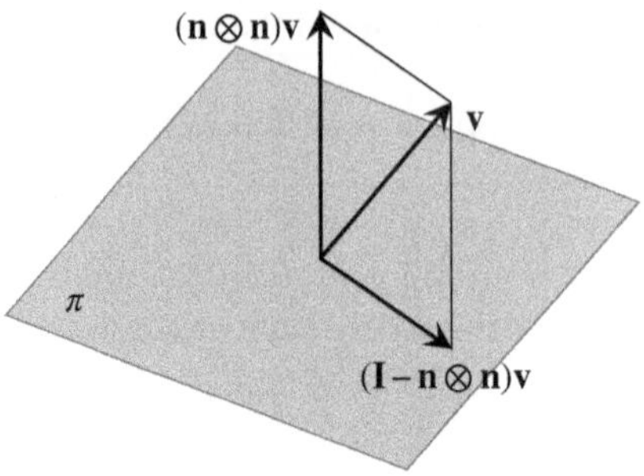

Figure 1.1 Geometric concept of orthogonal projection over the plane π whose normal vector is **n**.

which yields (1.63). Lastly, note that

$$(\mathbf{a} \otimes \mathbf{b})\mathbf{T}\mathbf{u} = (\mathbf{b} \cdot \mathbf{T}\mathbf{u})\mathbf{a} = (\mathbf{T}^T\mathbf{b} \cdot \mathbf{u})\mathbf{a} = (\mathbf{a} \otimes (\mathbf{T}^T\mathbf{b}))\mathbf{u}, \tag{1.68}$$

and then (1.64) is proved.

Analogously to the definition of the inner product in V, it is also possible to define the inner product in *Lin*. Consider two elements $\mathbf{T}, \mathbf{S} \in Lin$ which can be written in the forms $\mathbf{T} = \mathbf{t}_1 \otimes \mathbf{t}_2$ and $\mathbf{S} = \mathbf{s}_1 \otimes \mathbf{s}_2$, respectively, with $\mathbf{t}_1, \mathbf{t}_2, \mathbf{s}_1, \mathbf{s}_2 \in V$. Then, we define the inner product $\mathbf{T} \cdot \mathbf{S}$ in *Lin* as

$$\mathbf{T} \cdot \mathbf{S} = (\mathbf{t}_1 \otimes \mathbf{t}_2) \cdot (\mathbf{s}_1 \otimes \mathbf{s}_2) = (\mathbf{t}_1 \cdot \mathbf{s}_1)(\mathbf{t}_2 \cdot \mathbf{s}_2). \tag{1.69}$$

With definition (1.69), it is straightforward to prove that, for arbitrary $\mathbf{T} \in Lin$ and $\mathbf{u}, \mathbf{v} \in V$, the following result holds

$$\mathbf{T} \cdot (\mathbf{u} \otimes \mathbf{v}) = \mathbf{u} \cdot \mathbf{T}\mathbf{v}. \tag{1.70}$$

In fact, putting $\mathbf{T} = \mathbf{t}_1 \otimes \mathbf{t}_2$, and making use of definitions (1.69) and (1.60) we obtain

$$\begin{aligned}\mathbf{T} \cdot (\mathbf{u} \otimes \mathbf{v}) &= (\mathbf{t}_1 \otimes \mathbf{t}_2) \cdot (\mathbf{u} \otimes \mathbf{v}) = (\mathbf{t}_1 \cdot \mathbf{u})(\mathbf{t}_2 \cdot \mathbf{v}) \\ &= \mathbf{u} \cdot [(\mathbf{t}_2 \cdot \mathbf{v})\mathbf{t}_1] = \mathbf{u} \cdot [(\mathbf{t}_1 \otimes \mathbf{t}_2)\mathbf{v}] = \mathbf{u} \cdot \mathbf{T}\mathbf{v}. \end{aligned} \tag{1.71}$$

Let us denote **n** the unit normal vector to the plane π (see Figure 1.1). The tensor $\mathbf{n} \otimes \mathbf{n}$ applied over any vector $\mathbf{v} \in V$ gives

$$(\mathbf{n} \otimes \mathbf{n})\mathbf{v} = (\mathbf{n} \cdot \mathbf{v})\mathbf{n}, \tag{1.72}$$

which is the orthogonal projection of $\mathbf{v} \in V$ over the direction **n**. In turn, the second-order tensor $\mathbf{P} = \mathbf{I} - \mathbf{n} \otimes \mathbf{n}$ applied over any vector $\mathbf{v} \in V$ yields

$$\mathbf{P}\mathbf{v} = (\mathbf{I} - \mathbf{n} \otimes \mathbf{n})\mathbf{v} = \mathbf{v} - (\mathbf{n} \cdot \mathbf{v})\mathbf{n}, \tag{1.73}$$

which is the orthogonal projection of **v** over the plane π.

It can be appreciated that tensor **P** is symmetric and also verifies $\mathbf{P}^2 = \mathbf{P}$. In fact

$$\mathbf{P}^T = (\mathbf{I} - \mathbf{n} \otimes \mathbf{n})^T = \mathbf{I}^T - (\mathbf{n} \otimes \mathbf{n})^T = \mathbf{I} - \mathbf{n} \otimes \mathbf{n} = \mathbf{P}, \tag{1.74}$$

and, for arbitrary $\mathbf{v} \in V$, it is

$$\mathbf{P}^2\mathbf{v} = \mathbf{P}(\mathbf{P}\mathbf{v}) = \mathbf{P}[\mathbf{v} - (\mathbf{n} \cdot \mathbf{v})\mathbf{n}] = \mathbf{P}\mathbf{v} - (\mathbf{n} \cdot \mathbf{v})\mathbf{P}\mathbf{n} = \mathbf{P}\mathbf{v}, \tag{1.75}$$

and the previous statements hold.

Tensors satisfying these two properties, that is, $\mathbf{P} \in Sym$ and $\mathbf{P}^2 = \mathbf{P}$, are called orthogonal projection tensors. Examples of this kind of tensor are

$$\mathbf{I}, \qquad \mathbf{I} - \mathbf{n} \otimes \mathbf{n}, \qquad \mathbf{n} \otimes \mathbf{n}. \tag{1.76}$$

It is possible to show that

$$\dim(Lin) = \dim V \times \dim V = 9, \tag{1.77}$$

and if we take a basis $\{\mathbf{e}_i\}$ for V, the sets

$$\{(\mathbf{e}_i \otimes \mathbf{e}_j); i,j = 1,2,3\}, \tag{1.78}$$

$$\{(\mathbf{e}_i \otimes \mathbf{e}^j); i,j = 1,2,3\}, \tag{1.79}$$

$$\{(\mathbf{e}^i \otimes \mathbf{e}_j); i,j = 1,2,3\}, \tag{1.80}$$

$$\{(\mathbf{e}^i \otimes \mathbf{e}^j); i,j = 1,2,3\}, \tag{1.81}$$

are different possible bases for Lin. In this way, any tensor $\mathbf{T} \in Lin$ can be expressed by a unique linear combination of the element of the chosen basis. The components of tensor $\mathbf{T}$ in the chosen basis are defined in an analogous manner to that for vectors. Hence, we have

$$T_{ij} = \mathbf{T} \cdot (\mathbf{e}_i \otimes \mathbf{e}_j) = \mathbf{e}_i \cdot \mathbf{T}\mathbf{e}_j, \tag{1.82}$$

$$T^{ij} = \mathbf{T} \cdot (\mathbf{e}^i \otimes \mathbf{e}^j) = \mathbf{e}^i \cdot \mathbf{T}\mathbf{e}^j, \tag{1.83}$$

$$T^i_{.j} = \mathbf{T} \cdot (\mathbf{e}^i \otimes \mathbf{e}_j) = \mathbf{e}^i \cdot \mathbf{T}\mathbf{e}_j, \tag{1.84}$$

$$T_i^{.j} = \mathbf{T} \cdot (\mathbf{e}_i \otimes \mathbf{e}^j) = \mathbf{e}_i \cdot \mathbf{T}\mathbf{e}^j, \tag{1.85}$$

where the inner product between elements of Lin is defined in (1.69). Also, this inner product can be introduced in terms of the trace operation, as shown below. This way, the representation of $\mathbf{T} \in Lin$ in terms of these components is given by

$$\mathbf{T} = T_{ij}(\mathbf{e}^i \otimes \mathbf{e}^j) = T^{ij}(\mathbf{e}_i \otimes \mathbf{e}_j) = T^i_{.j}(\mathbf{e}_i \otimes \mathbf{e}^j) = T_i^{.j}(\mathbf{e}^i \otimes \mathbf{e}_j), \tag{1.86}$$

where the implicit summation of repeated indices is considered. Coefficients T_{ij}, T^{ij}, $T^i_{.j}$ and $T_i^{.j}$ are, respectively, covariant, contravariant, and mixed components of tensor $\mathbf{T}$.

In particular, the identity tensor $\mathbf{I}$ is

$$\mathbf{I} = \mathbf{e}_i \otimes \mathbf{e}^i, \tag{1.87}$$

again with implicit summation over index i. In fact, given arbitrary $\mathbf{u} = u^j\mathbf{e}_j = u_j\mathbf{e}^j \in V$, we have

$$\mathbf{I}\mathbf{u} = (\mathbf{e}_i \otimes \mathbf{e}^i)u^j\mathbf{e}_j = u^j\mathbf{e}_i\delta_{ij} = u^i\mathbf{e}_i = \mathbf{u}, \tag{1.88}$$

or equivalently,

$$\mathbf{I}\mathbf{u} = (\mathbf{e}_i \otimes \mathbf{e}^i)u_j\mathbf{e}^j = u_j g^{ij}\mathbf{e}_i = u_j\mathbf{e}^j = \mathbf{u}. \tag{1.89}$$

For a Cartesian basis $\{\mathbf{e}_i\}$ for V, there is no difference between components of $\mathbf{T}$. In this case, we simply refer to the Cartesian components of the tensor.

The advantage of employing compact notation is again evident when comparing to indicial notation. A tensor is a concept (linear transformation in V) which does not depend on the basis chosen for V. The same is valid for the trace operation and inner product between tensors.

To further illustrate the conceptual aspects highlighted by the intrinsic notation, let us introduce some of the definitions already presented in terms of both notations.

- Tensor. Compact notation: $\mathbf{T} \in Lin$. Indicial notation

$$T_{ij} = T^{kl} g_{ki} g_{lj} = T^{k}_{\cdot j} g_{ki} = T_{i}^{\cdot k} g_{kj}, \tag{1.90}$$

$$T^{ij} = T_{kl} g^{ki} g^{lj} = T^{i}_{\cdot k} g^{kj} = T_{k}^{\cdot j} g^{ki}, \tag{1.91}$$

$$T^{i}_{\cdot j} = T_{kj} g^{ki} = T^{ik} g_{kj} = T_{k}^{\cdot l} g^{ki} g_{lj}, \tag{1.92}$$

$$T_{i}^{\cdot j} = T_{ik} g^{kj} = T^{kj} g_{ki} = T^{k}_{\cdot l} g_{ki} g_{lj}. \tag{1.93}$$

- Application of a tensor over a vector. Compact notation: $\mathbf{u} = \mathbf{T}\mathbf{v}$, $\mathbf{u}, \mathbf{v} \in V$ and $\mathbf{T} \in Lin$. Indicial notation (just some of the possible expressions)

$$u_i = T_{ij} v^j = T_{ij} g^{jk} v_k = T_{i}^{\cdot j} v_j = T_{i}^{\cdot j} g_{jk} v^k, \tag{1.94}$$

$$u^i = T^{ij} v_j = T^{ij} g_{jk} v^k = T^{i}_{\cdot j} v^j = T_{i}^{\cdot j} g^{jk} v_k. \tag{1.95}$$

- Composition of tensors. Compact notation: $\mathbf{T} = \mathbf{A}\mathbf{B}$, $\mathbf{T}, \mathbf{A}, \mathbf{B} \in Lin$. Indicial notation

$$T_{ij} = A_{ik} B^{k}_{\cdot j} = A_{i}^{\cdot k} B_{kj}, \tag{1.96}$$

$$T^{ij} = A^{ik} B_{k}^{\cdot j} = A^{i}_{\cdot k} B^{kj}, \tag{1.97}$$

$$T^{i}_{\cdot j} = A^{i}_{\cdot k} B^{k}_{\cdot j} = A^{ik} B_{kj}, \tag{1.98}$$

$$T_{i}^{\cdot j} = A_{i}^{\cdot k} B_{k}^{\cdot j} = A_{ik} B^{kj}, \tag{1.99}$$

  and variants including the tensor g^{ij} or g_{ij}.

- Tensor product between vectors. Compact notation: $\mathbf{u} \otimes \mathbf{v}$, $\mathbf{u}, \mathbf{v} \in V$. Indicial notation

$$(\mathbf{u} \otimes \mathbf{v})_{ij} = u_i v_j, \tag{1.100}$$

$$(\mathbf{u} \otimes \mathbf{v})^{ij} = u^i v^j, \tag{1.101}$$

$$(\mathbf{u} \otimes \mathbf{v})^{i}_{\cdot j} = u^i v_j, \tag{1.102}$$

$$(\mathbf{u} \otimes \mathbf{v})_{i}^{\cdot j} = u_i v^j. \tag{1.103}$$

- Symmetric component of a tensor. Compact notation: $\mathbf{S} \in Sym \Leftrightarrow \mathbf{S} = \mathbf{S}^T$, and equivalently $\mathbf{S}^a = \mathbf{O}$. Indicial notation

$$S_{ij} = S_{ji} \quad \Longrightarrow \quad [S_{ij}] = [S_{ij}]^T, \tag{1.104}$$

$$S^{ij} = S^{ji} \quad \Longrightarrow \quad [S^{ij}] = [S^{ij}]^T, \tag{1.105}$$

$$S^{i}_{\cdot j} = S_{j}^{\cdot i}. \tag{1.106}$$

  Note that for a symmetric tensor the matrix of covariant components is also symmetric, and the same holds for contravariant components. In contrast, the matrix representation of a symmetric tensor in mixed components is not symmetric in general. Indeed, since $S^{i}_{\cdot j} = S_{j}^{\cdot i}$, we conclude that the same components are symmetrically placed in the two (different) matrix representations of the tensor.

- Skew-symmetric component of a tensor. Compact notation: $\mathbf{W} \in Skw \Leftrightarrow \mathbf{W} = -\mathbf{W}^T$, and then $\mathbf{W}^s = \mathbf{O}$. Indicial notation

$$W_{ij} = -W_{ji} \implies [W_{ij}] = -[W_{ij}]^T \implies W_{ii} = 0, \tag{1.107}$$

$$W^{ij} = -W^{ji} \implies [W^{ij}] = -[W^{ij}]^T \implies W^{ii} = 0, \tag{1.108}$$

$$W^i_{.j} = -W_j^{.i}. \tag{1.109}$$

That is, for a skew-symmetric tensor the matrix representations in covariant and contravariant components are skew-symmetric, while the matrix representations in mixed components are not skew-symmetric. As before, the expression $W^i_{.j} = -W_j^{.i}$ indicates that these coefficients are symmetrically placed, but in the two different matrix representations in mixed components.

The trace operation of a second-order tensor $\mathbf{T} \in Lin$ is a linear functional which associates to each tensor $\mathbf{T} \in Lin$ a real number denoted by $\mathrm{tr}\mathbf{T}$, that is,

$$\begin{aligned} \mathrm{tr} : Lin &\to \mathbb{R}, \\ \mathbf{T} &\mapsto \mathrm{tr}\mathbf{T}, \end{aligned} \tag{1.110}$$

with the property

$$\mathrm{tr}(\mathbf{a} \otimes \mathbf{b}) = \mathbf{a} \cdot \mathbf{b} \qquad \forall \mathbf{a}, \mathbf{b} \in V. \tag{1.111}$$

From the linearity of the trace operation it follows that

$$\begin{aligned} \mathrm{tr}\mathbf{T} &= T^{ij}\mathrm{tr}(\mathbf{e}_i \otimes \mathbf{e}_j) = T_{ij}\mathrm{tr}(\mathbf{e}^i \otimes \mathbf{e}^j) = T^i_{.j}\mathrm{tr}(\mathbf{e}_i \otimes \mathbf{e}^j) = T_i^{.j}\mathrm{tr}(\mathbf{e}^i \otimes \mathbf{e}_j) \\ &= T^{ij}g_{ij} = T_{ij}g^{ij} = T^i_{.i} = T_i^{.i}. \end{aligned} \tag{1.112}$$

These expressions allow us to evaluate $\mathrm{tr}\mathbf{T}$ in terms of the components of the tensor. Evidently, the result is independent of the adopted basis.

As the transpose operation is a linear operation, we have

$$\mathrm{tr}\mathbf{T} = \mathrm{tr}\mathbf{T}^T, \tag{1.113}$$

$$\mathrm{tr}(\mathbf{AB}) = \mathrm{tr}(\mathbf{BA}), \tag{1.114}$$

$$\mathrm{tr}\mathbf{I} = 3. \tag{1.115}$$

Indeed, putting $\mathbf{T} = \mathbf{t}_1 \otimes \mathbf{t}_2$, $\mathbf{A} = \mathbf{a}_1 \otimes \mathbf{a}_2$, and $\mathbf{B} = \mathbf{b}_1 \otimes \mathbf{b}_2$, and recalling that $\mathbf{I} = \mathbf{e}_i \otimes \mathbf{e}^i$, it is

$$\mathrm{tr}\mathbf{T} = \mathrm{tr}(\mathbf{t}_1 \otimes \mathbf{t}_2) = \mathbf{t}_1 \cdot \mathbf{t}_2 = \mathbf{t}_2 \cdot \mathbf{t}_1 = \mathrm{tr}(\mathbf{t}_2 \otimes \mathbf{t}_1) = \mathrm{tr}\mathbf{T}^T, \tag{1.116}$$

and then (1.113) follows. Also, using (1.62) and (1.113) yields

$$\begin{aligned} \mathrm{tr}(\mathbf{AB}) &= \mathrm{tr}[(\mathbf{a}_1 \otimes \mathbf{a}_2)(\mathbf{b}_1 \otimes \mathbf{b}_2)] = \mathrm{tr}[(\mathbf{a}_2 \cdot \mathbf{b}_1)(\mathbf{a}_1 \otimes \mathbf{b}_2)] = (\mathbf{a}_2 \cdot \mathbf{b}_1)(\mathbf{a}_1 \cdot \mathbf{b}_2) \\ &= \mathrm{tr}[(\mathbf{a}_1 \cdot \mathbf{b}_2)(\mathbf{a}_2 \otimes \mathbf{b}_1)] = \mathrm{tr}[(\mathbf{a}_1 \cdot \mathbf{b}_2)(\mathbf{b}_1 \otimes \mathbf{a}_2)] \\ &= \mathrm{tr}[(\mathbf{b}_1 \otimes \mathbf{b}_2)(\mathbf{a}_1 \otimes \mathbf{a}_2)] = \mathrm{tr}(\mathbf{BA}), \end{aligned} \tag{1.117}$$

and (1.114) is proved. Finally, with definition (1.87), we have

$$\mathrm{tr}\mathbf{I} = \mathrm{tr}(\mathbf{e}_i \otimes \mathbf{e}^i) = \mathbf{e}_i \cdot \mathbf{e}^i = \delta_{ii} = 3, \tag{1.118}$$

thus proving (1.115).

Notice that using the trace operation makes it possible to define an inner product in *Lin*

$$\mathbf{S} \cdot \mathbf{T} = \mathrm{tr}(\mathbf{S}^T\mathbf{T}) \qquad \forall \mathbf{S}, \mathbf{T} \in Lin. \tag{1.119}$$

With this definition, the following properties are satisfied

$$\mathbf{S} \cdot \mathbf{T} = \mathbf{T} \cdot \mathbf{S}, \tag{1.120}$$

$$\mathbf{S} \cdot \mathbf{S} \geq 0 \qquad \forall \mathbf{S} \in Lin, \tag{1.121}$$

$$\mathbf{S} \cdot \mathbf{S} = 0 \Leftrightarrow \mathbf{S} = \mathbf{O}, \tag{1.122}$$

and the inner product, in turn, induces a norm in *Lin*

$$\|\mathbf{S}\| = \sqrt{\mathbf{S} \cdot \mathbf{S}} = \sqrt{\mathrm{tr}(\mathbf{S}^T\mathbf{S})}. \tag{1.123}$$

Expressions of the inner product in terms of some of the different tensor components are

$$\mathbf{A} \cdot \mathbf{T} = A_{ij}T^{ij} = A^{ij}T_{ij} = A_i^{\ j}T^i_{\ .j} = A^i_{\ .j}T_i^{\ .j}, \tag{1.124}$$

and variants also involving g^{ij} or g_{ij}.

Now, consider arbitrary $\mathbf{a}, \mathbf{b}, \mathbf{c}, \mathbf{d} \in V$, $\mathbf{T}, \mathbf{A}, \mathbf{B} \in Lin$, $\mathbf{S} \in Sym$ and $\mathbf{W} \in Skw$. Then, some properties of the inner product in *Lin* are

$$\mathbf{I} \cdot \mathbf{T} = \mathrm{tr}\mathbf{T}, \tag{1.125}$$

$$\mathbf{A} \cdot \mathbf{B}\mathbf{T} = \mathbf{B}^T\mathbf{A} \cdot \mathbf{T}, \tag{1.126}$$

$$\mathbf{T} \cdot (\mathbf{u} \otimes \mathbf{v}) = \mathbf{u} \cdot \mathbf{T}\mathbf{v}, \tag{1.127}$$

$$(\mathbf{a} \otimes \mathbf{b}) \cdot (\mathbf{c} \otimes \mathbf{d}) = (\mathbf{a} \cdot \mathbf{c})(\mathbf{b} \cdot \mathbf{d}), \tag{1.128}$$

$$\mathbf{S} \cdot \mathbf{T} = \mathbf{S} \cdot \mathbf{T}^s, \tag{1.129}$$

$$\mathbf{W} \cdot \mathbf{T} = \mathbf{W} \cdot \mathbf{T}^a, \tag{1.130}$$

$$\mathbf{S} \cdot \mathbf{W} = 0, \tag{1.131}$$

$$\mathbf{T} \cdot \mathbf{A} = 0 \implies \mathbf{T} = \mathbf{O}, \tag{1.132}$$

$$\mathbf{T} \cdot \mathbf{S} = 0 \implies \mathbf{T} \in Skw, \tag{1.133}$$

$$\mathbf{T} \cdot \mathbf{W} = 0 \implies \mathbf{T} \in Sym, \tag{1.134}$$

$$\mathbf{A} \cdot \mathbf{B} = \mathbf{A}^s \cdot \mathbf{B}^s + \mathbf{A}^a \cdot \mathbf{B}^a. \tag{1.135}$$

In fact, from definition (1.119), expression (1.125) follows directly because

$$\mathbf{I} \cdot \mathbf{T} = \mathrm{tr}(\mathbf{I}^T\mathbf{T}) = \mathrm{tr}\mathbf{T}. \tag{1.136}$$

Identity (1.126) is also proved directly, since

$$\mathbf{A} \cdot \mathbf{B}\mathbf{T} = \mathrm{tr}(\mathbf{A}^T\mathbf{B}\mathbf{T}) = \mathrm{tr}[(\mathbf{B}^T\mathbf{A})^T\mathbf{T}] = \mathbf{B}^T\mathbf{A} \cdot \mathbf{T}. \tag{1.137}$$

Exercise 1.2 Prove that identities (1.127)–(1.135) hold using the properties of the trace operation.

The determinant of a tensor $\mathbf{A} \in Lin$ can be defined as

$$\det \mathbf{A} = \frac{1}{6}\left[2\mathrm{tr}(\mathbf{A}^3) + (\mathrm{tr}\mathbf{A})^3 - 3(\mathrm{tr}\mathbf{A})\mathrm{tr}(\mathbf{A}^2)\right]. \tag{1.138}$$

For arbitrary $\mathbf{A}, \mathbf{T} \in Lin$, and $\alpha \in \mathbb{R}$, the determinant operation satisfies

$$\det(\mathbf{AT}) = \det(\mathbf{TA}), \tag{1.139}$$

$$\det(\mathbf{AT}) = (\det \mathbf{A})(\det \mathbf{T}), \tag{1.140}$$

$$\det(\alpha\mathbf{T}) = \alpha^3 \det \mathbf{T}, \tag{1.141}$$

$$\det \mathbf{I} = 1. \tag{1.142}$$

The inverse of a tensor $\mathbf{T}$ is also a tensor $\mathbf{T}^{-1}$ with the property

$$\mathbf{T}^{-1}\mathbf{T} = \mathbf{I}. \tag{1.143}$$

A tensor $\mathbf{T}$ is called invertible if $\det \mathbf{T} \neq 0$, and if $\mathbf{T}$ is invertible, its transpose tensor $\mathbf{T}^T$ is also invertible, verifying

$$(\mathbf{T}^T)^{-1} = (\mathbf{T}^{-1})^T = \mathbf{T}^{-T}. \tag{1.144}$$

Tensor $\mathbf{Q}$ is called an orthogonal tensor if it preserves the inner product between vectors, that is,

$$\mathbf{Qu} \cdot \mathbf{Qv} = \mathbf{u} \cdot \mathbf{v} \qquad \forall \mathbf{u}, \mathbf{v} \in V. \tag{1.145}$$

The necessary and sufficient condition for a tensor $\mathbf{Q}$ to be orthogonal is

$$\mathbf{QQ}^T = \mathbf{Q}^T\mathbf{Q} = \mathbf{I}, \tag{1.146}$$

or, equivalently,

$$\mathbf{Q}^T = \mathbf{Q}^{-1}. \tag{1.147}$$

The set of all orthogonal tensors is denoted by *Orth*.

Exercise 1.3 Show that a tensor is orthogonal if and only if $\mathbf{QQ}^T = \mathbf{Q}^T\mathbf{Q} = \mathbf{I}$.

Any orthogonal tensor with positive determinant is called a rotation tensor. In particular, the set of all rotation tensors is denoted by *Rot*.

A tensor $\mathbf{T}$ is called positive definite if

$$\mathbf{v} \cdot \mathbf{Tv} > 0 \qquad \forall \mathbf{v} \in V \qquad \mathbf{v} \neq \mathbf{0}, \tag{1.148}$$

$$\mathbf{v} \cdot \mathbf{Tv} = 0 \Leftrightarrow \mathbf{v} = \mathbf{0}. \tag{1.149}$$

Given a basis $\{\mathbf{e}_k\}$, it is said that all the bases $\{\overline{\mathbf{e}}_k\}$ have the same orientation as $\{\mathbf{e}_k\}$ if they can be obtained from a rotation of the latter. That means

$$\overline{\mathbf{e}}_k = \mathbf{Re}_k \qquad k = 1, 2, 3. \tag{1.150}$$

Since any orthogonal tensor $\mathbf{Q}$ is a rotation $\mathbf{R}$, or it is a rotation multiplied by -1, there exist two classes of bases each associated with an orientation. Hereafter we will assume that one of these orientations has been chosen.

The cross product $\mathbf{u} \times \mathbf{v}$ between vectors $\mathbf{u}, \mathbf{v} \in V$, whose angle between them is θ, is another vector $\mathbf{w}$ such that

$$\mathbf{w} \perp \text{ to the plane defined by } \mathbf{u} \text{ and } \mathbf{v}, \tag{1.151}$$

$$||\mathbf{w}|| = ||\mathbf{u}|| ||\mathbf{v}|| \sin\theta, \tag{1.152}$$

$$\{\mathbf{u}, \mathbf{v}, \mathbf{w}\} \text{ have the same orientation as the adopted basis.} \tag{1.153}$$

Then, it follows that

$$\mathbf{u} \times \mathbf{v} = -(\mathbf{v} \times \mathbf{u}), \tag{1.154}$$

$$\mathbf{u} \times \mathbf{u} = \mathbf{0}, \tag{1.155}$$

$$\mathbf{u} \cdot \mathbf{v} \times \mathbf{w} = \mathbf{w} \cdot \mathbf{u} \times \mathbf{v} = \mathbf{v} \cdot \mathbf{w} \times \mathbf{u}. \tag{1.156}$$

Given three arbitrary elements $\mathbf{u}, \mathbf{v}, \mathbf{w} \in V$, the cross product satisfies

$$\mathbf{u} \times (\mathbf{v} \times \mathbf{w}) = (\mathbf{u} \cdot \mathbf{w})\mathbf{v} - (\mathbf{u} \cdot \mathbf{v})\mathbf{w}, \tag{1.157}$$

or equivalently,

$$\mathbf{u} \times (\mathbf{v} \times \mathbf{w}) = (\mathbf{v} \otimes \mathbf{w})^a \mathbf{u}. \tag{1.158}$$

Exercise 1.4 Prove that (1.157) holds.

When $\mathbf{u}, \mathbf{v}, \mathbf{w}$ are linearly independent, the value of the product $\mathbf{u} \cdot (\mathbf{v} \times \mathbf{w})$ represents the volume of the parallelepiped $\mathscr{P}$ determined by the vectors $\mathbf{u}, \mathbf{v}, \mathbf{w}$. Then, given a tensor $\mathbf{T}$, it is not difficult to verify that

$$\det \mathbf{T} = \frac{\mathbf{T}\mathbf{u} \cdot \mathbf{T}\mathbf{v} \times \mathbf{T}\mathbf{w}}{\mathbf{u} \cdot \mathbf{v} \times \mathbf{w}}, \tag{1.159}$$

from where it follows that

$$\det \mathbf{T} = \frac{\mathrm{vol}(\mathbf{T}(\mathscr{P}))}{\mathrm{vol}(\mathscr{P})}, \tag{1.160}$$

which provides a geometric interpretation of the determinant of a tensor $\mathbf{T}$, and where $\mathbf{T}(\mathscr{P})$ is the image of the parallelepiped $\mathscr{P}$ under the linear transformation (tensor) $\mathbf{T}$ and $\mathrm{vol}(\cdot)$ stands for the volume.

Given a second-order tensor field $\mathbf{T} \in \mathit{Lin}$, we introduce the principal invariants of $\mathbf{T}$ as follows

$$I_1(\mathbf{T}) = \mathrm{tr}\mathbf{T}, \tag{1.161}$$

$$I_2(\mathbf{T}) = \frac{1}{2}(\mathrm{tr}(\mathbf{T}^2) - (\mathrm{tr}\mathbf{T})^2), \tag{1.162}$$

$$I_3(\mathbf{T}) = \det \mathbf{T}. \tag{1.163}$$

The invariance property is established by the fact that

$$I_i(\mathbf{T}) = I_i(\mathbf{Q}\mathbf{T}\mathbf{Q}^T) \qquad \forall \mathbf{Q} \in \mathit{Orth} \qquad i = 1, 2, 3. \tag{1.164}$$

Then, any tensor $\mathbf{T}$ admits the representation established by the Cayley–Hamilton theorem [132]

$$\det(\alpha\mathbf{I} - \mathbf{T}) = \alpha^3 - \alpha^2 I_1(\mathbf{T}) - \alpha I_2(\mathbf{T}) - I_3(\mathbf{T}). \tag{1.165}$$

By putting

$$p(\alpha) = \det(\alpha \mathbf{I} - \mathbf{T}), \tag{1.166}$$

it can be shown that all second-order tensors satisfy the following characteristic equation

$$p(\mathbf{T}) = \mathbf{O}. \tag{1.167}$$

Therefore, we have

$$\mathbf{T}^3 - I_1(\mathbf{T})\mathbf{T}^2 - I_2(\mathbf{T})\mathbf{T} - I_3(\mathbf{T})\mathbf{I} = \mathbf{O}, \tag{1.168}$$

and taking the trace of this equation, and considering the definition of the invariants (1.161), (1.162) and (1.163), yields

$$\mathrm{tr}(\mathbf{T}^3) - (\mathrm{tr}\mathbf{T})\mathrm{tr}(\mathbf{T}^2) + \frac{1}{2}((\mathrm{tr}\mathbf{T})^3 - (\mathrm{tr}\mathbf{T})\mathrm{tr}(\mathbf{T}^2)) = 3\det \mathbf{T}, \tag{1.169}$$

or equivalently,

$$\det \mathbf{T} = \frac{1}{3}\mathrm{tr}(\mathbf{T}^3) + \frac{1}{6}(\mathrm{tr}\mathbf{T})^3 - \frac{1}{2}(\mathrm{tr}\mathbf{T})\mathrm{tr}(\mathbf{T}^2), \tag{1.170}$$

which is exactly (1.138).

Let $\mathbf{T} = T_{ij}(\mathbf{e}_i \otimes \mathbf{e}_j)$ be a second-order tensor given in any orthonormal basis $\{\mathbf{e}_i\}$ for V, then it results in

$$\det \mathbf{T} = \varepsilon_{ijk} T_{i1} T_{j2} T_{k3} = \varepsilon_{ijk} T_{1i} T_{2j} T_{3k}, \tag{1.171}$$

where ε_{ijk} is the permutation symbol[6] defined by

$$\varepsilon_{ijk} = \begin{cases} 1 & \text{if } (i,j,k) \text{ is an even permutation of } (1,2,3), \\ -1 & \text{if } (i,j,k) \text{ is an odd permutation of } (1,2,3), \\ 0 & \text{otherwise.} \end{cases} \tag{1.172}$$

Finally, given a tensor $\mathbf{W} \in Skw$, there exists a unique vector $\mathbf{w} \in V$ such that

$$\mathbf{W}\mathbf{v} = \mathbf{w} \times \mathbf{v} \qquad \forall \mathbf{v} \in V. \tag{1.173}$$

Vector $\mathbf{w}$ associated with the skew-symmetric tensor $\mathbf{W}$ is called the axial vector of $\mathbf{W}$. From this definition, it follows that the kernel of $\mathbf{W} \in Skw$, denoted by $\mathcal{N}(\mathbf{W})$, is the one-dimensional subspace of V spanned by the axial vector $\mathbf{w}$ of $\mathbf{W}$, that is,

$$\mathcal{N}(\mathbf{W}) = \{\mathbf{v} \in V;\ \mathbf{W}\mathbf{v} = \mathbf{0}\} = \{\mathbf{v} \in V;\ \mathbf{v} = \lambda \mathbf{w},\ \forall \lambda \in \mathbb{R}\}. \tag{1.174}$$

1.3 Third-Order Tensors

In this section we will extend the concept of linear transformations to embrace linear operators between V and *Lin*, and vice versa. This context gives rise to the notion of third-order tensors.

---

6 Actually, this is a very peculiar third-order tensor, also known as the Levi–Civita symbol.

In short, third-order tensors are linear applications of the form

$$\begin{aligned}\mathsf{S} : \mathit{Lin} &\rightarrow V,\\ \mathbf{T} &\mapsto \mathsf{S}\mathbf{T},\end{aligned} \tag{1.175}$$

as well as

$$\begin{aligned}\mathsf{S} : V &\rightarrow \mathit{Lin},\\ \mathbf{a} &\mapsto \mathsf{S}\mathbf{a},\end{aligned} \tag{1.176}$$

which implies that these entities can be applied to elements (vectors) of V, and also to elements (second-order tensors) of Lin. The set, actually vector space when endowed with the standard operations, of all third-order tensors will be called Lin. The null element in Lin is denoted by $\mathsf{0}$, and is such that

$$\mathsf{0}\mathbf{a} = \mathbf{O} \quad \forall \mathbf{a} \in V, \tag{1.177}$$

$$\mathsf{0}\mathbf{T} = \mathbf{0} \quad \forall \mathbf{T} \in \mathit{Lin}. \tag{1.178}$$

However, applications (1.175) and (1.176) are not fully characterized yet. Consider $\mathbf{a} \in V, \mathbf{T} = \mathbf{t}_1 \otimes \mathbf{t}_2 \in \mathit{Lin}$ and $\mathsf{S} = \mathbf{s}_1 \otimes \mathbf{s}_2 \otimes \mathbf{s}_3 \in \mathsf{Lin}$, with $\mathbf{t}_1, \mathbf{t}_2, \mathbf{s}_1, \mathbf{s}_2, \mathbf{s}_3 \in V$. Hence, we define application (1.175) as follows

$$\mathsf{S}\mathbf{T} = (\mathbf{s}_1 \otimes \mathbf{s}_2 \otimes \mathbf{s}_3)(\mathbf{t}_1 \otimes \mathbf{t}_2) = (\mathbf{s}_2 \cdot \mathbf{t}_1)(\mathbf{s}_3 \cdot \mathbf{t}_2)\mathbf{s}_1, \tag{1.179}$$

and, analogously, application (1.176) is defined by

$$\mathsf{S}\mathbf{a} = (\mathbf{s}_1 \otimes \mathbf{s}_2 \otimes \mathbf{s}_3)\mathbf{a} = (\mathbf{s}_3 \cdot \mathbf{a})(\mathbf{s}_1 \otimes \mathbf{s}_2). \tag{1.180}$$

In the space Lin the transpose operation of a second-order tensor was defined as $\mathbf{a} \cdot \mathbf{T}\mathbf{b} = \mathbf{T}^T\mathbf{a} \cdot \mathbf{b}$. In the same manner, we can define different transpose operations for elements in Lin. We then define the transpose operations[7]

$$\mathbf{a} \cdot \mathsf{S}\mathbf{T} = \mathsf{S}^T\mathbf{a} \cdot \mathbf{T}, \tag{1.181}$$

$$\mathsf{S}\mathbf{a} \cdot \mathbf{T} = \mathbf{a} \cdot \mathsf{S}^{\frac{1}{T}}\mathbf{T}. \tag{1.182}$$

Let us show that these third-order tensors are unique, and also that $(\mathsf{S}^{\frac{1}{T}})^T = \mathsf{S}$ holds. To do this, assume that $\mathsf{S}_1^T \neq \mathsf{S}_2^T$, then

$$\begin{aligned}\mathbf{a} \cdot \mathsf{S}\mathbf{T} &= \mathsf{S}_1^T\mathbf{a} \cdot \mathbf{T} \qquad && \forall \mathbf{a} \in V, \ \forall \mathbf{T} \in \mathit{Lin},\\ \mathbf{a} \cdot \mathsf{S}\mathbf{T} &= \mathsf{S}_2^T\mathbf{a} \cdot \mathbf{T} \qquad && \forall \mathbf{a} \in V, \ \forall \mathbf{T} \in \mathit{Lin}.\end{aligned}$$

Subtracting both expressions above, we obtain

$$(\mathsf{S}_1^T - \mathsf{S}_2^T)\mathbf{a} \cdot \mathbf{T} = 0 \qquad \forall \mathbf{a} \in V, \ \forall \mathbf{T} \in \mathit{Lin}, \tag{1.183}$$

---

7 An important point to stress here is that the definition of the transpose operation is not unique in the case of third-order tensors. As a matter of fact, the transpose operation could equivalently be defined as

$$\mathsf{S}\mathbf{a} \cdot \mathbf{T} = \mathbf{a} \cdot \mathsf{S}^T\mathbf{T},$$

$$\mathbf{a} \cdot \mathsf{S}\mathbf{T} = \mathsf{S}^{\frac{1}{T}}\mathbf{a} \cdot \mathbf{T}.$$

Comparing the definitions, we see that the operation $(\cdot)^T$ defined here is equivalent to the operation $(\cdot)^{\frac{1}{T}}$ defined in the main text (also equivalent to $((\cdot)^T)^T$). In any case, the subsequent mathematical developments have to consistently follow the chosen definition of this operation.

which implies

$$(\mathbf{S}_1^T - \mathbf{S}_2^T)\mathbf{a} = \mathbf{O} \qquad \forall \mathbf{a} \in V, \tag{1.184}$$

and, from (1.177), we finally conclude that

$$\mathbf{S}_1^T = \mathbf{S}_2^T, \tag{1.185}$$

leading to a contradiction.

Analogously, assuming that there exist $\mathbf{S}_1^{\frac{1}{T}} \neq \mathbf{S}_2^{\frac{1}{T}}$, a similar argument yields

$$\mathbf{a} \cdot (\mathbf{S}_1^{\frac{1}{T}} - \mathbf{S}_2^{\frac{1}{T}})\mathbf{T} = 0 \qquad \forall \mathbf{a} \in V,\ \forall \mathbf{T} \in Lin, \tag{1.186}$$

from which we obtain

$$(\mathbf{S}_1^{\frac{1}{T}} - \mathbf{S}_2^{\frac{1}{T}})\mathbf{T} = \mathbf{0} \qquad \forall \mathbf{T} \in Lin, \tag{1.187}$$

and, using (1.178), we obtain a contradiction, from which it follows that

$$\mathbf{S}_1^{\frac{1}{T}} = \mathbf{S}_2^{\frac{1}{T}}. \tag{1.188}$$

Finally, from the transpose definition we have

$$\mathbf{S}\mathbf{a} \cdot \mathbf{T} = \mathbf{a} \cdot \mathbf{S}^{\frac{1}{T}}\mathbf{T} = (\mathbf{S}^{\frac{1}{T}})^T\mathbf{a} \cdot \mathbf{T}, \tag{1.189}$$

and then

$$(\mathbf{S} - (\mathbf{S}^{\frac{1}{T}})^T)\mathbf{a} \cdot \mathbf{T} = 0 \qquad \forall \mathbf{a} \in V,\ \forall \mathbf{T} \in Lin, \tag{1.190}$$

that is,

$$(\mathbf{S} - (\mathbf{S}^{\frac{1}{T}})^T)\mathbf{a} = \mathbf{O} \qquad \forall \mathbf{a} \in V, \tag{1.191}$$

and, similarly, from (1.177), we conclude that

$$\mathbf{S} = (\mathbf{S}^{\frac{1}{T}})^T. \tag{1.192}$$

Let us see now that, if a third-order tensor is of the form $\mathbf{S} = \mathbf{s}_1 \otimes \mathbf{s}_2 \otimes \mathbf{s}_3$, the following is verified

$$(\mathbf{s}_1 \otimes \mathbf{s}_2 \otimes \mathbf{s}_3)^T = \mathbf{s}_2 \otimes \mathbf{s}_3 \otimes \mathbf{s}_1. \tag{1.193}$$

To obtain this result, consider definitions (1.179) and (1.180). Then, for arbitrary $\mathbf{a} \in V$ and $\mathbf{T} = \mathbf{t}_1 \otimes \mathbf{t}_2 \in Lin$, we have

$$\begin{aligned}\mathbf{a} \cdot \mathbf{S}\mathbf{T} &= \mathbf{a} \cdot (\mathbf{s}_1 \otimes \mathbf{s}_2 \otimes \mathbf{s}_3)(\mathbf{t}_1 \otimes \mathbf{t}_2) = (\mathbf{s}_2 \cdot \mathbf{t}_1)(\mathbf{s}_3 \cdot \mathbf{t}_2)(\mathbf{a} \cdot \mathbf{s}_1) \\ &= (\mathbf{s}_2 \otimes \mathbf{s}_3 \otimes \mathbf{s}_1)\mathbf{a} \cdot (\mathbf{t}_1 \otimes \mathbf{t}_2) = (\mathbf{s}_1 \otimes \mathbf{s}_2 \otimes \mathbf{s}_3)^T\mathbf{a} \cdot (\mathbf{t}_1 \otimes \mathbf{t}_2) = \mathbf{S}^T\mathbf{a} \cdot \mathbf{T},\end{aligned} \tag{1.194}$$

and (1.193) follows.

In the same manner, it can be proved that

$$(\mathbf{s}_1 \otimes \mathbf{s}_2 \otimes \mathbf{s}_3)^{\frac{1}{T}} = \mathbf{s}_3 \otimes \mathbf{s}_1 \otimes \mathbf{s}_2. \tag{1.195}$$

In fact, with (1.179) and (1.180) we get

$$\begin{aligned}\mathbf{S}\mathbf{a} \cdot \mathbf{T} &= (\mathbf{s}_1 \otimes \mathbf{s}_2 \otimes \mathbf{s}_3)\mathbf{a} \cdot (\mathbf{t}_1 \otimes \mathbf{t}_2) = (\mathbf{s}_3 \cdot \mathbf{a})(\mathbf{s}_1 \cdot \mathbf{t}_1)(\mathbf{s}_2 \cdot \mathbf{t}_2) \\ &= \mathbf{a} \cdot (\mathbf{s}_3 \otimes \mathbf{s}_1 \otimes \mathbf{s}_2)(\mathbf{t}_1 \otimes \mathbf{t}_2) = \mathbf{a} \cdot (\mathbf{s}_1 \otimes \mathbf{s}_2 \otimes \mathbf{s}_3)^{\frac{1}{T}}(\mathbf{t}_1 \otimes \mathbf{t}_2),\end{aligned} \tag{1.196}$$

proving then (1.195).

With the previous results, we can directly obtain the following relations

$$(\mathsf{S}^T)^T = \mathsf{S}^{\frac{1}{T}}, \tag{1.197}$$

$$((\mathsf{S}^T)^T)^T = \mathsf{S}. \tag{1.198}$$

Exercise 1.5 Prove that (1.197) and (1.198) are verified.

Let $\mathbf{a} \in V$ and $\mathbf{T} = \mathbf{t}_1 \otimes \mathbf{t}_2 \in Lin$ be arbitrary, then we can build an element of Lin using the tensor product as follows

$$\mathbf{a} \otimes \mathbf{T} = \mathbf{a} \otimes \mathbf{t}_1 \otimes \mathbf{t}_2, \tag{1.199}$$

$$\mathbf{T} \otimes \mathbf{a} = \mathbf{t}_1 \otimes \mathbf{t}_2 \otimes \mathbf{a}. \tag{1.200}$$

In this way, considering $\mathbf{u}, \mathbf{v} \in V, \mathbf{T}, \mathbf{S} \in Lin$ and $\mathsf{S} \in$ Lin, it is quite straightforward to show that

$$(\mathbf{u} \otimes \mathbf{T})^T = \mathbf{T} \otimes \mathbf{u}, \tag{1.201}$$

$$(\mathbf{T} \otimes \mathbf{u})^{\frac{1}{T}} = \mathbf{u} \otimes \mathbf{T}, \tag{1.202}$$

$$(\mathbf{u} \otimes \mathbf{T})\mathbf{S} = (\mathbf{T} \cdot \mathbf{S})\mathbf{u}, \tag{1.203}$$

$$(\mathbf{u} \otimes \mathbf{T})\mathbf{v} = \mathbf{u} \otimes (\mathbf{T}\mathbf{v}), \tag{1.204}$$

$$(\mathbf{T} \otimes \mathbf{u})\mathbf{S} = \mathbf{T}\mathbf{S}\mathbf{u}, \tag{1.205}$$

$$(\mathbf{T}^T \otimes \mathbf{u})^T\mathbf{S}^T = \mathbf{T}\mathbf{S}\mathbf{u}, \tag{1.206}$$

$$(\mathbf{T} \otimes \mathbf{u})\mathbf{v} = (\mathbf{u} \cdot \mathbf{v})\mathbf{T}, \tag{1.207}$$

$$(\mathbf{T} \otimes \mathbf{v})^T\mathbf{u} = (\mathbf{T}^T\mathbf{u}) \otimes \mathbf{v}, \tag{1.208}$$

$$(\mathbf{T} \otimes \mathbf{v})^T\mathbf{S} = \mathbf{T}^T\mathbf{S}^T\mathbf{v}, \tag{1.209}$$

$$(\mathsf{S}^T\mathbf{u})\mathbf{v} = (\mathsf{S}\mathbf{v})^T\mathbf{u}, \tag{1.210}$$

$$\mathsf{S}(\mathbf{v} \otimes \mathbf{u}) = (\mathsf{S}\mathbf{u})\mathbf{v}. \tag{1.211}$$

Exercise 1.6 Prove that tensor identities (1.201)–(1.211) hold.

The vector space Lin can also be endowed with an inner product structure. Consider the elements $\mathsf{T}, \mathsf{S} \in$ Lin, with $\mathsf{T} = \mathbf{t}_1 \otimes \mathbf{t}_2 \otimes \mathbf{t}_3$ and $\mathsf{S} = \mathbf{s}_1 \otimes \mathbf{s}_2 \otimes \mathbf{s}_3$, $\mathbf{t}_1, \mathbf{t}_2, \mathbf{t}_3, \mathbf{s}_1, \mathbf{s}_2, \mathbf{s}_3 \in V$. Then, we define the inner product $\mathsf{T} \cdot \mathsf{S}$ as follows

$$\mathsf{T} \cdot \mathsf{S} = (\mathbf{t}_1 \otimes \mathbf{t}_2 \otimes \mathbf{t}_3) \cdot (\mathbf{s}_1 \otimes \mathbf{s}_2 \otimes \mathbf{s}_3) = (\mathbf{t}_1 \cdot \mathbf{s}_1)(\mathbf{t}_2 \cdot \mathbf{s}_2)(\mathbf{t}_3 \cdot \mathbf{s}_3). \tag{1.212}$$

So, for $\mathbf{S}, \mathbf{T} \in Lin$ (second-order tensors), it is possible to show that

$$(\mathbf{u} \otimes \mathbf{S}) \cdot (\mathbf{v} \otimes \mathbf{T}) = (\mathbf{u} \cdot \mathbf{v})(\mathbf{S} \cdot \mathbf{T}), \tag{1.213}$$

$$(\mathbf{u} \otimes \mathbf{S}) \cdot (\mathbf{T} \otimes \mathbf{v}) = \mathbf{T}\mathbf{S}\mathbf{v} \cdot \mathbf{u}, \tag{1.214}$$

$$(\mathbf{u} \otimes \mathbf{S}) \cdot (\mathbf{T} \otimes \mathbf{v})^T = \mathbf{S}\mathbf{T}\mathbf{u} \cdot \mathbf{v}. \tag{1.215}$$

Exercise 1.7 Prove that tensor identities (1.213)–(1.215) hold.

Another transpose operation that can be introduced in Lin is the following. Consider an arbitrary $\mathsf{S} = \mathbf{s}_1 \otimes \mathbf{s}_2 \otimes \mathbf{s}_3$, then we define

$$\mathsf{S}^{\mathrm{t}} = (\mathbf{s}_1 \otimes \mathbf{s}_2 \otimes \mathbf{s}_3)^{\mathrm{t}} = \mathbf{s}_1 \otimes \mathbf{s}_3 \otimes \mathbf{s}_2. \tag{1.216}$$

With this operation, for an arbitrary $\mathbf{a} \in V$, the following holds

$$\mathsf{S}^{\mathrm{t}}\mathbf{a} = (\mathbf{s}_1 \otimes \mathbf{s}_2 \otimes \mathbf{s}_3)^{\mathrm{t}}\mathbf{a} = (\mathbf{a} \cdot \mathbf{s}_2)(\mathbf{s}_1 \otimes \mathbf{s}_3), \tag{1.217}$$

and also

$$(\mathsf{S}^{\mathrm{t}})^T = (\mathsf{S}^{\frac{1}{T}})^{\mathrm{t}}. \tag{1.218}$$

Also, consider two arbitrary vectors $\mathbf{u}, \mathbf{v} \in V$, then the next result holds

$$(\mathsf{S}^{\mathrm{t}}\mathbf{u})\mathbf{v} = (\mathsf{S}\mathbf{v})\mathbf{u}. \tag{1.219}$$

Exercise 1.8 Prove that (1.219) holds.

Moreover, and similarly to the composition of second-order tensors (composition of linear transformations), it is possible to define the composition of third-order tensors. Consider arbitrary $\mathsf{S} = \mathbf{s}_1 \otimes \mathbf{s}_2 \otimes \mathbf{s}_3 \in$ Lin and $\mathbf{T} = \mathbf{t}_1 \otimes \mathbf{t}_2 \in Lin$. Then we can define the composition of S with $\mathbf{T}$ yielding an element in Lin. This can be fully characterized as follows

$$\mathsf{S} \circ \mathbf{T} = (\mathbf{s}_1 \otimes \mathbf{s}_2 \otimes \mathbf{s}_3) \circ (\mathbf{t}_1 \otimes \mathbf{t}_2) = (\mathbf{s}_3 \cdot \mathbf{t}_1)(\mathbf{s}_1 \otimes \mathbf{s}_2 \otimes \mathbf{t}_2). \tag{1.220}$$

Clearly, this operation gives $\mathsf{S} \circ \mathbf{T} \in$ Lin. Differently, for arbitrary $\mathsf{S}, \mathsf{R} \in$ Lin, we can proceed as

$$\mathsf{S} \circ \mathsf{R} = (\mathbf{s}_1 \otimes \mathbf{s}_2 \otimes \mathbf{s}_3) \circ (\mathbf{r}_1 \otimes \mathbf{r}_2 \otimes \mathbf{r}_3) = (\mathbf{s}_2 \cdot \mathbf{r}_1)(\mathbf{s}_3 \cdot \mathbf{r}_2)(\mathbf{s}_1 \otimes \mathbf{r}_3), \tag{1.221}$$

noting that this case results in $\mathsf{S} \circ \mathsf{R} \in Lin$.

With these definitions we obtain, for example,

$$(\mathbf{u} \otimes \mathbf{T}) \circ (\mathbf{S} \otimes \mathbf{v}) = (\mathbf{T} \cdot \mathbf{S})(\mathbf{u} \otimes \mathbf{v}). \tag{1.222}$$

This introduction should help the reader to become familiar with the mathematical manipulation of vector and tensor algebra, as well as illustrating the introduction of definitions and their consequences, always working with intrinsic notations, that is, independently from the coordinate system.

In effect, it is possible to extend all the operations with vectors and second-order tensors to second- and third-order tensors. An example is the definition of the cross product between a second-order tensor $\mathbf{S} = \mathbf{s}_1 \otimes \mathbf{s}_2$ and a vector $\mathbf{a}$, or between two second-order tensors $\mathbf{S} = \mathbf{s}_1 \otimes \mathbf{s}_2$ and $\mathbf{T} = \mathbf{t}_1 \otimes \mathbf{t}_2$, for instance

$$\mathbf{S} \times \mathbf{a} = (\mathbf{s}_1 \otimes \mathbf{s}_2) \times \mathbf{a} = \mathbf{s}_1 \otimes (\mathbf{s}_2 \times \mathbf{a}), \tag{1.223}$$

$$\mathbf{a} \times \mathbf{S} = \mathbf{a} \times (\mathbf{s}_1 \otimes \mathbf{s}_2) = (\mathbf{a} \times \mathbf{s}_1) \otimes \mathbf{s}_2, \tag{1.224}$$

$$\mathbf{S} \times \mathbf{T} = (\mathbf{s}_1 \otimes \mathbf{s}_2) \times (\mathbf{t}_1 \otimes \mathbf{t}_2) = (\mathbf{s}_1 \times \mathbf{t}_1) \times (\mathbf{s}_2 \times \mathbf{t}_2). \tag{1.225}$$

With these definitions we can obtain, for example, the following result

$$\mathbf{S} \times \mathbf{a} = -[\mathbf{a} \times \mathbf{S}^T]^T. \tag{1.226}$$

Furthermore, we can define operations for a second-order tensor with the structure $\mathbf{S} = \mathbf{s}_1 \otimes \mathbf{s}_2$ such as the following

$$\mathbf{S}_\times = (\mathbf{s}_1 \otimes \mathbf{s}_2)_\times = \mathbf{s}_1 \times \mathbf{s}_2, \tag{1.227}$$

which is called the Gibbs product, and this verifies

$$(\mathbf{S}^T)_\times = -\mathbf{S}_\times. \tag{1.228}$$

In addition, given the axial vector $\mathbf{a}$ of an arbitrary $\mathbf{A} \in Skw$, and letting $\mathbf{B}$ be arbitrary, the following result holds

$$\mathbf{A} \cdot \mathbf{B} = -\mathbf{a} \cdot \mathbf{B}_\times. \tag{1.229}$$

Exercise 1.9 Prove that (1.229) holds.

The treatment of higher-order tensors, fourth-order tensors for example, is entirely analogous. Nevertheless, as we increase the order of the tensor it is evident that the number of possible operations and number of alternative equivalent definitions grows. The simplest examples are the different transpose operations that can be introduced. For instance, consider the fourth-order tensor $\mathbb{D}$

$$\mathbb{D} = \mathbf{d}_1 \otimes \mathbf{d}_2 \otimes \mathbf{d}_3 \otimes \mathbf{d}_4. \tag{1.230}$$

The following transpose operations can be defined

$$\mathbb{D}^T = (\mathbf{d}_1 \otimes \mathbf{d}_2 \otimes \mathbf{d}_3 \otimes \mathbf{d}_4)^T = \mathbf{d}_3 \otimes \mathbf{d}_4 \otimes \mathbf{d}_1 \otimes \mathbf{d}_2, \tag{1.231}$$

$$\mathbb{D}^{\mathcal{T}} = (\mathbf{d}_1 \otimes \mathbf{d}_2 \otimes \mathbf{d}_3 \otimes \mathbf{d}_4)^{\mathcal{T}} = \mathbf{d}_2 \otimes \mathbf{d}_1 \otimes \mathbf{d}_3 \otimes \mathbf{d}_4, \tag{1.232}$$

$$\mathbb{D}^{\mathfrak{T}} = (\mathbf{d}_1 \otimes \mathbf{d}_2 \otimes \mathbf{d}_3 \otimes \mathbf{d}_4)^{\mathfrak{T}} = \mathbf{d}_1 \otimes \mathbf{d}_2 \otimes \mathbf{d}_4 \otimes \mathbf{d}_3, \tag{1.233}$$

$$\mathbb{D}^{\mathrm{T}} = (\mathbf{d}_1 \otimes \mathbf{d}_2 \otimes \mathbf{d}_3 \otimes \mathbf{d}_4)^{\mathrm{T}} = \mathbf{d}_4 \otimes \mathbf{d}_2 \otimes \mathbf{d}_3 \otimes \mathbf{d}_1, \tag{1.234}$$

among others.

1.4 Complementary Reading

For the reader interested in deepening their knowledge of the topics briefly addressed in this chapter, we recommend reading the appendices, particularly Appendix A, where the reader will find concepts and additional information that will ease the reading of the following works: [23], [47], [84], [93–95], [106], [132], [135], [171], [265], [299] and [300].

2

Vector and Tensor Analysis

2.1 Differentiation

In this chapter we will introduce a notation for the concept of differentiation abstract enough to include in the definition functions f of the kind

$$D(f) \subset \mathbb{R}, \mathscr{E}, V, Lin, \mathsf{Lin}, \tag{2.1}$$

$$R(f) \subset \mathbb{R}, \mathscr{E}, V, Lin, \mathsf{Lin}, \tag{2.2}$$

where $D(f)$ and $R(f)$ are, respectively, the domain of f and the image under the transformation f, also called the codomain of f.

Let $\mathscr{X}$ and $\mathscr{Y}$ be two normed vector spaces. For the purpose of the present chapter, we will consider that these spaces are finite-dimensional. Suppose $f : \mathscr{X} \to \mathscr{Y}$ is defined in a neighborhood of the origin (null element) of $\mathscr{X}$. Then, $f(u)$ approaches the origin faster than u if

$$\lim_{\substack{u\to 0\\ u\neq 0}} \frac{\|f(u)\|_{\mathscr{Y}}}{\|u\|_{\mathscr{X}}} = 0. \tag{2.3}$$

This will be represented by

$$f(u) = o(u), \qquad u \to 0, \tag{2.4}$$

or, equivalently, by

$$f(u) = o(u). \tag{2.5}$$

Similarly, $f(u) = g(u) + o(u)$ stands for

$$f(u) - g(u) = o(u). \tag{2.6}$$

Note that this last definition has a clear meaning even when $R(f)$ and $R(g)$ are subsets of $\mathscr{E}$ because in such cases $f(u) - g(u) \in V$.

Example 2.1 Let $\phi : \mathbb{R} \to \mathbb{R}$ be such that for each $t \in \mathbb{R}$ we have $\phi(t) = t^\alpha$. Then, $\phi(t) = o(t)$ if $\alpha > 1$.

Consider a function g defined in an open set $\mathscr{J}$ of $\mathbb{R}$, whose image is in the set $\mathbb{R}$ (g is a real-valued function), or in V (g is a vector-valued function), or in Lin (g is a second-order tensor-valued function), or in Lin (g is a third-order tensor-valued

Introduction to the Variational Formulation in Mechanics: Fundamentals and Applications, First Edition.
Edgardo O. Taroco, Pablo J. Blanco and Raúl A. Feijóo.

function), or in $\mathscr{E}$ (g is said to be point-valued function). The derivative of g at t, $\dot{g}(t)$, if exists, it is defined by

$$\dot{g}(t) = \frac{d}{dt}g(t) = \lim_{\alpha \to 0} \frac{1}{\alpha}[g(t+\alpha) - g(t)]. \tag{2.7}$$

Observe that if $R(g) \subset \mathscr{E}$ then $(g(t+\alpha) - g(t)) \in V$, and then the derivative is a vector field. Likewise, the derivative of a real-valued function is a real number, the derivative of a vector-valued function will be a vector field, and that of a tensor-valued function will be a tensor field.

Function g is said to be differenciable at $t \in \mathscr{I}$ if there exists the derivative at t. Thus, from the definition of the derivative we have that

$$\lim_{\alpha \to 0} \frac{1}{\alpha}[g(t+\alpha) - g(t) - \alpha\dot{g}(t)] = 0, \tag{2.8}$$

or in an equivalent manner

$$g(t+\alpha) - g(t) - \alpha\dot{g}(t) = o(\alpha), \tag{2.9}$$

from where

$$g(t+\alpha) - g(t) = \alpha\dot{g}(t) + o(\alpha). \tag{2.10}$$

Clearly, $\alpha\dot{g}(t)$ is linear with respect to α, then, $g(t+\alpha) - g(t)$ can be written as the sum of a linear term in α and a term that approaches zero faster than α. This is a rather useful definition for a derivative. Therefore, we define the derivative of g at t as the linear transformation in $\mathbb{R}$ that approaches $g(t+\alpha) - g(t)$ for small values of α.

Hence, consider two finite-dimensional normed vector spaces $\mathscr{X}$ and $\mathscr{Y}$, $\mathscr{A}$ is an open set of $\mathscr{X}$ and let $g : \mathscr{A} \to \mathscr{Y}$ be a function. Function g is said to be differentiable at $x \in \mathscr{A}$ if there exists the linear transformation

$$\begin{aligned} \mathscr{D}g(x) : \mathscr{X} &\to \mathscr{Y}, \\ u &\mapsto \mathscr{D}g(x)[u], \end{aligned} \tag{2.11}$$

such that

$$g(x+u) = g(x) + \mathscr{D}g(x)[u] + o(u), \qquad u \to 0. \tag{2.12}$$

The linear transformation $\mathscr{D}g(x)$ is then called the derivative of g at x. Given that in a finite-dimensional normed space all norms are equivalent, $\mathscr{D}g(x)$ is independent from the chosen norms for $\mathscr{X}$ and $\mathscr{Y}$. In turn, if there exists $\mathscr{D}g(x)$, it is unique. In fact, for each $u \in \mathscr{X}$ it is

$$\mathscr{D}g(x)[u] = \lim_{\alpha \to 0} \frac{1}{\alpha}[g(x+\alpha u) - g(x)] = \left.\frac{d}{d\alpha}g(x+\alpha u)\right|_{\alpha=0}. \tag{2.13}$$

If g is differentiable for all $x \in \mathscr{A}$, $\mathscr{D}g$ represents the transformation which, for each $x \in \mathscr{A}$, gives $\mathscr{D}g(x) \in Lin(\mathscr{X}, \mathscr{Y})$, where $Lin(\mathscr{X}, \mathscr{Y})$ is the space whose elements are the linear transformations from $\mathscr{X}$ into $\mathscr{Y}$. This space is also finite-dimensional, in fact, $\dim(Lin(\mathscr{X}, \mathscr{Y})) = (\dim \mathscr{X})(\dim \mathscr{Y})$, and then we can define a norm. This fact allows us to have notions of continuity and differentiability for $\mathscr{D}g$. In particular, g is said to be regular, or of class C^1 in $\mathscr{A}$ if g is differentiable for all points $x \in \mathscr{A}$ and if $\mathscr{D}g$ is continuous. Also, g is said to be of class C^2 in $\mathscr{A}$ if g is of class C^1 and $\mathscr{D}g$ is also regular, and so on.

Let us consider some examples which will be useful for the forthcoming developments.

- Consider

$$g : V \to \mathbb{R}, \qquad \mathbf{v} \mapsto g(\mathbf{v}) = \mathbf{v} \cdot \mathbf{v}. \tag{2.14}$$

  Then

$$g(\mathbf{v} + \mathbf{u}) = (\mathbf{v} + \mathbf{u}) \cdot (\mathbf{v} + \mathbf{u}) = \mathbf{v} \cdot \mathbf{v} + 2\mathbf{v} \cdot \mathbf{u} + \mathbf{u} \cdot \mathbf{u} = g(\mathbf{v}) + 2\mathbf{v} \cdot \mathbf{u} + o(\mathbf{u}). \tag{2.15}$$

  From where

$$\mathscr{D}g(\mathbf{v})[\mathbf{u}] = 2\mathbf{v} \cdot \mathbf{u}. \tag{2.16}$$

- Consider

$$g : Lin \to Lin, \qquad \mathbf{A} \mapsto g(\mathbf{A}) = \mathbf{A}^2 = \mathbf{A}\mathbf{A}. \tag{2.17}$$

  Then

$$g(\mathbf{A} + \mathbf{U}) = (\mathbf{A} + \mathbf{U})^2 = \mathbf{A}^2 + (\mathbf{A}\mathbf{U} + \mathbf{U}\mathbf{A}) + o(\mathbf{U}). \tag{2.18}$$

  From where

$$\mathscr{D}g(\mathbf{A})[\mathbf{U}] = \mathbf{A}\mathbf{U} + \mathbf{U}\mathbf{A}. \tag{2.19}$$

- Consider

$$g : \mathscr{X} \to \mathscr{Y} \text{ linear}, \qquad x \mapsto g(x). \tag{2.20}$$

  Then

$$g(x + u) = g(x) + g(u). \tag{2.21}$$

  From where

$$\mathscr{D}g(x)[u] = g(u). \tag{2.22}$$

Frequently, it will be necessary to calculate the derivative of a product $\pi(f, g)$ between two functions f and g. In particular, we have already presented some product operations, for example:

- Product of a real number α by a vector $\mathbf{v} \in V$. In this case we have $\pi(\alpha, \mathbf{v}) = \alpha\mathbf{v}$.
- Inner product between two vectors in V. In this case $\pi(\mathbf{u}, \mathbf{v}) = \mathbf{u} \cdot \mathbf{v}$.
- Inner product between two tensors, then $\pi(\mathbf{A}, \mathbf{T}) = \mathbf{A} \cdot \mathbf{T}$.
- Tensor product between vectors, then $\pi(\mathbf{u}, \mathbf{v}) = \mathbf{u} \otimes \mathbf{v}$.
- Application product of a second-order tensor $\mathbf{T}$ over a vector $\mathbf{v}$. In this case it is $\pi(\mathbf{T}, \mathbf{v}) = \mathbf{T}\mathbf{v}$.

All these operations have a common feature: they are all linear with respect to each argument. In this case the operation is said to be bilinear. Hence, and seeking a general rule, let us consider the following abstract product operation

$$\pi : \mathscr{F} \times \mathscr{G} \to \mathscr{Y} \quad \text{bilinear}, \qquad (f, g) \mapsto \pi(f, g), \tag{2.23}$$

where $\mathscr{F}$, $\mathscr{G}$ and $\mathscr{Y}$ are finite-dimensional normed vector spaces. Consider now the functions

$$f : \mathscr{A} \subset \mathscr{X} \to \mathscr{F}, \qquad g : \mathscr{A} \subset \mathscr{X} \to \mathscr{G}, \tag{2.24}$$

where $\mathscr{X}$ is a finite-dimensional vector space and $\mathscr{A}$ is an open subset of $\mathscr{X}$. With these elements, the product $h = \pi(f, g)$ is the function

$$\begin{aligned} h &: \mathscr{A} \to \mathscr{Y}, \\ &x \mapsto h(x) = \pi(f(x), g(x)). \end{aligned} \tag{2.25}$$

Therefore, we have the following differentiation rule.

Product Rule. Let f and g be differentiable functions at $x \in \mathscr{A}$. Then, its product $h = \pi(f, g)$ is also differentiable at x, and it is such that

$$\mathscr{D}h(x)[u] = \pi(\mathscr{D}f(x)[u], g(x)) + \pi(f(x), \mathscr{D}g(x)[u]). \tag{2.26}$$

In particular, when $\mathscr{A}$ is an open subset of $\mathbb{R}$, the product rule and the fact that π is bilinear lead to

$$\dot{h}(t) = \pi(\dot{f}(t), g(t)) + \pi(f(t), \dot{g}(t)). \tag{2.27}$$

Making use of this result, we obtain the following proposition. Let $\phi, \mathbf{v}, \mathbf{w}, \mathbf{S}$ and $\mathbf{T}$ smooth functions in an open subset of $\mathbb{R}$ such that the image of ϕ is $\mathbb{R}$, that of $\mathbf{v}, \mathbf{w}$ is V, and for $\mathbf{S}, \mathbf{T}$ it is *Lin*. Then, it is straightforward to prove that

$$\overline{(\phi\mathbf{w})}^{\cdot} = \dot{\phi}\mathbf{v} + \phi\dot{\mathbf{v}}, \tag{2.28}$$

$$\overline{(\mathbf{v} \cdot \mathbf{w})}^{\cdot} = \dot{\mathbf{v}} \cdot \mathbf{w} + \mathbf{v} \cdot \dot{\mathbf{w}}, \tag{2.29}$$

$$\overline{(\mathbf{T}\mathbf{S})}^{\cdot} = \dot{\mathbf{T}}\mathbf{S} + \mathbf{T}\dot{\mathbf{S}}, \tag{2.30}$$

$$\overline{(\mathbf{T} \cdot \mathbf{S})}^{\cdot} = \dot{\mathbf{T}} \cdot \mathbf{S} + \mathbf{T} \cdot \dot{\mathbf{S}}, \tag{2.31}$$

$$\overline{(\mathbf{T}\mathbf{v})}^{\cdot} = \dot{\mathbf{T}}\mathbf{v} + \mathbf{T}\dot{\mathbf{v}}. \tag{2.32}$$

Exercise 2.1 Prove identities (2.28)–(2.32).

Another result frequently used in practice is the chain rule. Let $\mathscr{X}$, $\mathscr{Y}$ and $\mathscr{F}$ be finite-dimensional vector spaces (here we include the Euclidean point space $\mathscr{E}$). $\mathscr{A}$ and $\mathscr{C}$ are open subsets of $\mathscr{X}$ and $\mathscr{Y}$, respectively. Let f and g be functions such that

$$g : \mathscr{A} \to \mathscr{Y}, \qquad f : \mathscr{C} \to \mathscr{F}, \tag{2.33}$$

and where $R(g) \subset \mathscr{C}$. Then, we have the following differentiation rule for the composition of functions.

Chain Rule. Let g and f be differentiable functions at $x \in \mathscr{A}$ and $y = g(x)$, correspondingly. Then, the composition

$$h = f \circ g : \mathscr{A} \to \mathscr{F}, \tag{2.34}$$

defined by

$$h(x) = f(g(x)), \tag{2.35}$$

is differentiable at x, and the derivative is given by

$$\mathcal{D}h(x) = \mathcal{D}f(y) \circ \mathcal{D}g(x), \tag{2.36}$$

and its differential at x in the direction of $u \in \mathcal{X}$ is

$$\mathcal{D}h(x)[u] = \mathcal{D}f(g(x))[\mathcal{D}g(x)[u]]. \tag{2.37}$$

In particular, when $\mathcal{X} \equiv \mathbb{R}$, g and therefore h are real-valued functions. Putting t instead of x in the previous expression we have

$$\mathcal{D}h(t)[\alpha] = \mathcal{D}f(g(t))[\mathcal{D}g(t)[\alpha]], \tag{2.38}$$

which leads us to

$$\frac{d}{dt}h(t) = \frac{d}{dt}f(g(t)) = \mathcal{D}f(g(t))\left[\frac{d}{dt}g(t)\right]. \tag{2.39}$$

With this result we can prove that, for instance, the following holds

$$\dot{\overline{(\mathbf{S}^T)}} = (\dot{\mathbf{S}})^T. \tag{2.40}$$

Exercise 2.2 Prove identity (2.40).

A particularly useful result is to calculate the derivative of the determinant of a second-order tensor. In order to do this we recall (1.165) for a second-order tensor $\mathbf{T}$, and with $\alpha = -1$, which leads to

$$\det(\mathbf{I} + \mathbf{T}) = 1 + \mathrm{tr}\mathbf{T} + \frac{1}{2}((\mathrm{tr}\mathbf{T})^2 + \mathrm{tr}(\mathbf{T}^2)) + \det\mathbf{T}, \tag{2.41}$$

and hence

$$\det(\mathbf{I} + \mathbf{T}) = 1 + \mathrm{tr}\mathbf{T} + o(\mathbf{T}). \tag{2.42}$$

Consider now the following

$$\det(\mathbf{A} + \mathbf{B}) = \det((\mathbf{I} + \mathbf{B}\mathbf{A}^{-1})\mathbf{A}) = (\det\mathbf{A})\det(\mathbf{I} + \mathbf{B}\mathbf{A}^{-1}). \tag{2.43}$$

Using now (2.42) into (2.43) we obtain

$$\begin{aligned}\det(\mathbf{A} + \mathbf{B}) &= \det\mathbf{A}(1 + \mathrm{tr}(\mathbf{B}\mathbf{A}^{-1}) + o(\mathbf{B}\mathbf{A}^{-1}))\\ &= \det\mathbf{A} + (\det\mathbf{A})\mathrm{tr}(\mathbf{B}\mathbf{A}^{-1}) + (\det\mathbf{A})o(\mathbf{B}\mathbf{A}^{-1}).\end{aligned} \tag{2.44}$$

Clearly, from definition (2.12), we obtain

$$\mathcal{D}(\det\mathbf{A})[\mathbf{B}] = (\det\mathbf{A})\mathrm{tr}(\mathbf{B}\mathbf{A}^{-1}). \tag{2.45}$$

From this result, it is straightforward to see that

$$\dot{\overline{(\det\mathbf{A})}} = (\det\mathbf{A})\mathrm{tr}(\dot{\mathbf{A}}\mathbf{A}^{-1}). \tag{2.46}$$

2.2 Gradient

In this section the definition given in (2.11) will be made specific for functions defined in the open set Ω of the Euclidean space $\mathscr{E}$ where an origin O will be adopted, allowing the biunivocal correspondence between points $X \in \mathscr{E}$ and vectors $\mathbf{x} = X - O \in V$. Function ϕ in Ω is said to be real-valued, vector-valued, tensor-valued (of any order), or point-valued if the elements in the image are, for each $\mathbf{x} \in \Omega$, real numbers, vectors, tensors (of any order), or points from $\mathscr{E}$ (equivalently vectors from V). For such functions ϕ (assumed to be differentiable) we have seen that $\mathscr{D}\phi(x) \in Lin(\mathscr{V}, \mathscr{Y})$, where $\mathscr{V}$ is the normed space associated with V, and where the normed space $\mathscr{Y}$ will denote the normed spaces associated with $\mathbb{R}$, V, Lin, Lin, accordingly.

This linear application $\mathscr{D}\phi(\mathbf{x})$ will be called the *gradient* (and will be denoted by $\nabla\phi(\mathbf{x})$ or simply by $\nabla\phi$) of the field ϕ which, as seen in (2.12), is such that it approaches the value of ϕ at $\mathbf{x} + \mathbf{u} \in \Omega$ with an error of order $o(\mathbf{u})$, in the following sense

$$\phi(\mathbf{x} + \mathbf{u}) = \phi(\mathbf{x}) + \nabla\phi(\mathbf{x})[\mathbf{u}] + o(\mathbf{u}). \tag{2.47}$$

If we adopt an orthonormal basis for V, say $\{\mathbf{e}_i\}$, the previous expression is equivalent to

$$\nabla\phi(\mathbf{x})[\mathbf{e}_k] = \lim_{\alpha\to 0}\frac{1}{\alpha}[\phi(\mathbf{x} + \alpha\mathbf{e}_k) - \phi(\mathbf{x})]. \tag{2.48}$$

As can be easily seen, the limit is simply the partial derivative of ϕ with respect to x_k, at point $\mathbf{x}$, that is

$$\nabla\phi(\mathbf{x})[\mathbf{e}_k] = \frac{\partial\phi(\mathbf{x})}{\partial x_k}. \tag{2.49}$$

Observe that, for a field in a finite-dimensional space of dimension $\dim\mathscr{Y}$, the gradient is a transformation which belongs to a space of dimension $(\dim\mathscr{Y})(\dim V)$. Result (2.49) allows us to easily define the representation of real-valued functions, vector-valued functions, and tensor-valued functions, both in compact (intrinsic) notation as well as in indicial notation, as shown in the following.

- Let ϕ be a real-valued function (also called scalar field) φ, then

$$\nabla\varphi = \frac{\partial\varphi}{\partial x_k}\mathbf{e}_k. \tag{2.50}$$

- Let ϕ be a vector-valued function (also called a vector field) $\mathbf{u}$, then

$$\nabla\mathbf{u} = \frac{\partial\mathbf{u}}{\partial x_k}\otimes\mathbf{e}_k = \frac{\partial u_j}{\partial x_k}(\mathbf{e}_j\otimes\mathbf{e}_k). \tag{2.51}$$

- Let ϕ be a second-order tensor-valued function (also called a second-order tensor field) $\mathbf{S}$, then

$$\nabla\mathbf{S} = \frac{\partial\mathbf{S}}{\partial x_k}\otimes\mathbf{e}_k = \frac{\partial S_{ij}}{\partial x_k}(\mathbf{e}_i\otimes\mathbf{e}_j\otimes\mathbf{e}_k). \tag{2.52}$$

- Let ϕ be a third-order tensor-valued function (also called a third-order tensor field) S, then

$$\nabla\mathsf{S} = \frac{\partial\mathsf{S}}{\partial x_k}\otimes\mathbf{e}_k = \frac{\partial S_{ijr}}{\partial x_k}(\mathbf{e}_i\otimes\mathbf{e}_j\otimes\mathbf{e}_r\otimes\mathbf{e}_k). \tag{2.53}$$

- Let ϕ be a fourth-order tensor-valued function (also called a fourth-order tensor field) $\mathbb{S}$, then

$$\nabla \mathbb{S} = \frac{\partial \mathbb{S}}{\partial x_k} \otimes \mathbf{e}_k = \frac{\partial S_{ijmn}}{\partial x_k}(\mathbf{e}_i \otimes \mathbf{e}_j \otimes \mathbf{e}_m \otimes \mathbf{e}_n \otimes \mathbf{e}_k). \tag{2.54}$$

In intrinsic notation, the gradient of a vector field $\mathbf{u}$ can be defined as the second-order tensor $\nabla \mathbf{u}$ that satisfies

$$(\nabla \mathbf{u})^T \mathbf{a} = \nabla(\mathbf{u} \cdot \mathbf{a}), \tag{2.55}$$

for any arbitrary constant vector $\mathbf{a} \in V$.

Analogously, the gradient of a second-order tensor field $\mathbf{S}$ can be defined as the third-order tensor $\nabla \mathbf{S}$ that satisfies

$$(\nabla \mathbf{S})^T \mathbf{a} = \nabla(\mathbf{S}^T \mathbf{a}), \tag{2.56}$$

for any arbitrary constant vector $\mathbf{a} \in V$.

Alternatively, we can define the gradient of the tensor field $\mathbf{S}$ as the third-order tensor $\nabla \mathbf{S}$ such that, for any arbitrary constant $\mathbf{a} \in V$ verifies

$$(\nabla \mathbf{S})^{\mathrm{t}} \mathbf{a} = \nabla(\mathbf{S}\mathbf{a}). \tag{2.57}$$

Now, consider a vector field of the form $\phi \mathbf{u}$, with constant and arbitrary $\mathbf{u} \in V$. Then, the following holds

$$\nabla(\phi \mathbf{u}) = \mathbf{u} \otimes \nabla \phi. \tag{2.58}$$

In fact, considering an arbitrary constant $\mathbf{a} \in V$, from (2.55) note that

$$[\mathbf{u} \otimes (\nabla \phi)]^T \mathbf{a} = [(\nabla \phi) \otimes \mathbf{u}]\mathbf{a} = (\nabla \phi)\mathbf{u} \cdot \mathbf{a} = \nabla(\phi \mathbf{u} \cdot \mathbf{a}) = [\nabla(\phi \mathbf{u})]^T \mathbf{a}. \tag{2.59}$$

Then, expression (2.58) holds.

Consider now that we have a second-order tensor with the structure $\mathbf{S} = \mathbf{u} \otimes \mathbf{v}$. Then, the following result can be proved

$$\nabla(\mathbf{u} \otimes \mathbf{v}) = (\mathbf{u} \otimes \nabla \mathbf{v}) + ((\nabla \mathbf{u})^T \otimes \mathbf{v})^T. \tag{2.60}$$

Indeed, from definition (2.56), notice that

$$\begin{aligned} [\nabla(\mathbf{u} \otimes \mathbf{v})]^T \mathbf{a} &= \nabla((\mathbf{v} \otimes \mathbf{u})\mathbf{a}) = \nabla((\mathbf{a} \cdot \mathbf{u})\mathbf{v}) = (\mathbf{a} \cdot \mathbf{u})\nabla \mathbf{v} + \mathbf{v} \otimes \nabla(\mathbf{a} \cdot \mathbf{u}) \\ &= (\nabla \mathbf{v} \otimes \mathbf{u})\mathbf{a} + \mathbf{v} \otimes ((\nabla \mathbf{u})^T \mathbf{a}) = [(\nabla \mathbf{v} \otimes \mathbf{u}) + (\mathbf{v} \otimes (\nabla \mathbf{u})^T)]\mathbf{a}. \end{aligned} \tag{2.61}$$

Then, using (1.195), this implies that

$$\begin{aligned} \nabla(\mathbf{u} \otimes \mathbf{v}) &= ((\nabla(\mathbf{u} \otimes \mathbf{v}))^T)^{\frac{1}{T}} = (\nabla \mathbf{v} \otimes \mathbf{u})^{\frac{1}{T}} + (\mathbf{v} \otimes (\nabla \mathbf{u})^T)^{\frac{1}{T}} \\ &= (\mathbf{u} \otimes \nabla \mathbf{v}) + ((\nabla \mathbf{u})^T \otimes \mathbf{v})^T, \end{aligned} \tag{2.62}$$

then identity (2.60) is verified.

As a consequence, we can define the second gradient of a vector field $\mathbf{u}$ as the third-order tensor $\nabla\nabla \mathbf{u}$ which satisfies the following

$$(\nabla\nabla \mathbf{u})^T \mathbf{a} = \nabla((\nabla \mathbf{u})^T \mathbf{a}) = \nabla(\nabla(\mathbf{u} \cdot \mathbf{a})), \tag{2.63}$$

for any arbitrary constant $\mathbf{a} \in V$. In a completely equivalent manner, $\nabla\nabla\mathbf{u}$ can be defined as

$$(\nabla\nabla\mathbf{u})\mathbf{a} = \nabla((\nabla\mathbf{u})\mathbf{a}). \tag{2.64}$$

Several results can be proved involving different types of functions. In effect, for φ, $\mathbf{u}$ ($\mathbf{v}$), $\mathbf{S}$ ($\mathbf{T}$) and S scalar, vector, tensor (second and third order), respectively, the following results hold

$$\nabla(\mathbf{S}^T\mathbf{u}) = (\nabla\mathbf{S})^T\mathbf{u} + \mathbf{S}^T\nabla\mathbf{u}, \tag{2.65}$$

$$\nabla(\mathsf{S}\mathbf{u}) = (\nabla\mathsf{S})^{\mathrm{t}}\mathbf{u} + \mathsf{S}\nabla\mathbf{u}, \tag{2.66}$$

$$\nabla(\varphi\mathbf{u}) = \varphi\nabla\mathbf{u} + \mathbf{u} \otimes \nabla\varphi, \tag{2.67}$$

$$\nabla(\mathbf{v} \cdot \mathbf{u}) = (\nabla\mathbf{v})^T\mathbf{u} + (\nabla\mathbf{u})^T\mathbf{v}, \tag{2.68}$$

$$\nabla(\varphi\mathbf{S}) = \varphi(\nabla\mathbf{S}) + \mathbf{S} \otimes \nabla\varphi, \tag{2.69}$$

$$\nabla(\mathbf{S} \cdot \mathbf{T}) = (\nabla\mathbf{S})^{\frac{1}{T}}\mathbf{T} + (\nabla\mathbf{T})^{\frac{1}{T}}\mathbf{S}, \tag{2.70}$$

$$(\nabla(\nabla\mathbf{u}))^{\frac{1}{T}} = ((\nabla\nabla\mathbf{u})^T)^{\mathrm{t}}, \tag{2.71}$$

$$(\nabla(\nabla\mathbf{u}))^{\frac{1}{T}} = \nabla((\nabla\mathbf{u})^T). \tag{2.72}$$

Exercise 2.3 Prove the validity of expressions (2.65)–(2.72).

2.3 Divergence

With the definition of the gradient of a vector field, as well as the gradient of a tensor field (of any order), we can introduce a new linear operator called divergence of ϕ and denoted by div ϕ. This operator can be represented by a (linear) operation involving the gradient $\nabla\phi$ and the second-order identity tensor $\mathbf{I}$, which takes the gradient $\nabla\phi$, which belongs to a space of dimension (dim $\mathscr{Y}$)(dim V), into a space of dimension dim $\mathscr{Y}$. We can reinterpret the divergence operation as opposed to the gradient in this specific sense. Then, we have

$$\operatorname{div}\phi = c(\nabla\phi, \mathbf{I}), \tag{2.73}$$

which corresponds to the following specific definitions.

- Let ϕ be a vector field $\mathbf{u}$, then

  $$\operatorname{div}\mathbf{u} = \nabla\mathbf{u} \cdot \mathbf{I}, \tag{2.74}$$

  which in Cartesian coordinates is

  $$\operatorname{div}\mathbf{u} = \frac{\partial\mathbf{u}}{\partial x_k} \cdot \mathbf{e}_k = \frac{\partial u_k}{\partial x_k}. \tag{2.75}$$

- Let ϕ be a second-order tensor field $\mathbf{S}$, then

  $$\operatorname{div}\mathbf{S} = (\nabla\mathbf{S})\mathbf{I}, \tag{2.76}$$

  which in Cartesian coordinates is

  $$\operatorname{div}\mathbf{S} = \frac{\partial\mathbf{S}}{\partial x_k}\mathbf{e}_k = \frac{\partial S_{jk}}{\partial x_k}\mathbf{e}_j. \tag{2.77}$$

- Let ϕ be a third-order tensor field $\mathbf{S}$, then

$$\text{div}\,\mathbf{S} = (\nabla\mathbf{S})\mathbf{I}, \tag{2.78}$$

which in Cartesian coordinates is

$$\text{div}\,\mathbf{S} = \frac{\partial \mathbf{S}}{\partial x_k}\mathbf{e}_k = \frac{\partial S_{ijk}}{\partial x_k}(\mathbf{e}_i \otimes \mathbf{e}_j). \tag{2.79}$$

- Let ϕ be a fourth-order tensor field $\mathbb{S}$, then

$$\text{div}\,\mathbb{S} = (\nabla\mathbb{S})\mathbf{I}, \tag{2.80}$$

which in Cartesian coordinates is

$$\text{div}\,\mathbb{S} = \frac{\partial \mathbb{S}}{\partial x_k}\mathbf{e}_k = \frac{\partial S_{ijrk}}{\partial x_k}(\mathbf{e}_i \otimes \mathbf{e}_j \otimes \mathbf{e}_r). \tag{2.81}$$

Alternatively, but in a completely analogous way, we can define the divergence of a second-order tensor field $\mathbf{S}$ as the unique vector field div $\mathbf{S}$ with the property

$$(\text{div}\,\mathbf{S}) \cdot \mathbf{a} = \text{div}\,(\mathbf{S}^T\mathbf{a}), \tag{2.82}$$

for any arbitrary constant $\mathbf{a} \in V$.

Similarly, for a third-order tensor $\mathbf{S}$ we can define

$$(\text{div}\,\mathbf{S}) \cdot \mathbf{A} = \text{div}\,(\mathbf{S}^{\frac{1}{T}}\mathbf{A}), \tag{2.83}$$

for any arbitrary constant $\mathbf{A} \in Lin$.

Now, it is possible to show that

$$\text{div}\,(\mathbf{ST}) = (\nabla\mathbf{S})\mathbf{T} + \mathbf{S}\text{div}\,\mathbf{T}. \tag{2.84}$$

In fact, using definitions (2.74) and (2.82) we have

$$\begin{aligned} \text{div}\,(\mathbf{ST}) \cdot \mathbf{a} &= \text{div}\,((\mathbf{ST})^T\mathbf{a}) = \text{div}\,(\mathbf{T}^T\mathbf{S}^T\mathbf{a}) = \mathbf{I} \cdot [\mathbf{T}^T\nabla(\mathbf{S}^T\mathbf{a})] + \mathbf{S}^T\mathbf{a} \cdot \text{div}\,\mathbf{T} \\ &= \mathbf{T} \cdot \nabla(\mathbf{S}^T\mathbf{a}) + \mathbf{a} \cdot \mathbf{S}\,\text{div}\,\mathbf{T} = (\nabla\mathbf{S})\mathbf{T} \cdot \mathbf{a} + \mathbf{a} \cdot \mathbf{S}\,\text{div}\,\mathbf{T} \\ &= [(\nabla\mathbf{S})\mathbf{T} + \mathbf{S}\,\text{div}\,\mathbf{T}] \cdot \mathbf{a}, \end{aligned} \tag{2.85}$$

and identity (2.84) follows.

Consider now the identity tensor $\mathbf{T} = \mathbf{I}$, then it follows directly that

$$\text{div}\,\mathbf{S} = (\nabla\mathbf{S})\mathbf{I}, \tag{2.86}$$

which establishes the connection with definition (2.76).

Consider now a second-order tensor with the structure $\mathbf{S} = \mathbf{u} \otimes \mathbf{v}$, then from the previous results it follows that

$$\text{div}\,(\mathbf{u} \otimes \mathbf{v}) = \mathbf{u}\,\text{div}\,\mathbf{v} + (\nabla\mathbf{u})\mathbf{v}. \tag{2.87}$$

In fact, from definition (2.76) and using expression (2.60) yields

$$\begin{aligned} \text{div}\,(\mathbf{u} \otimes \mathbf{v}) &= (\nabla(\mathbf{u} \otimes \mathbf{v}))\mathbf{I} = (\mathbf{u} \otimes \nabla\mathbf{v})\mathbf{I} + ((\nabla\mathbf{u})^T \otimes \mathbf{v})^T\mathbf{I} \\ &= \mathbf{u}(\nabla\mathbf{v} \cdot \mathbf{I}) + (\nabla\mathbf{u}\;\mathbf{I})\mathbf{v} = \mathbf{u}\,\text{div}\,\mathbf{v} + (\nabla\mathbf{u})\mathbf{v}, \end{aligned} \tag{2.88}$$

and (2.87) follows.

Additional results can be proved in a similar manner. For example, for φ, $\mathbf{u}$, $\mathbf{S}$ ($\mathbf{T}$) and S, scalar, vector and tensor fields (second and third order), respectively, the following results can be verified

$$\operatorname{div}\,(\mathbf{S}\overset{1}{\bar{T}}\mathsf{T}) = \mathbf{S}\cdot\nabla\mathsf{T} + \mathsf{T}\cdot\operatorname{div}\mathbf{S}, \tag{2.89}$$

$$\operatorname{div}\,(\mathbf{S}^T\mathbf{u}) = \mathbf{S}\cdot\nabla\mathbf{u} + \operatorname{div}\mathbf{S}\cdot\mathbf{u}, \tag{2.90}$$

$$\operatorname{div}\,(\mathsf{S}^T\mathbf{u}) = (\operatorname{div}\mathsf{S})^T\mathbf{u} + \mathsf{S}^T(\nabla\mathbf{u})^T, \tag{2.91}$$

$$\operatorname{div}\,(\varphi\mathbf{u}) = \varphi\,\operatorname{div}\mathbf{u} + \nabla\varphi\cdot\mathbf{u}, \tag{2.92}$$

$$\operatorname{div}\,(\varphi\mathbf{S}) = \varphi\,\operatorname{div}\mathbf{S} + \mathbf{S}\nabla\varphi, \tag{2.93}$$

$$\operatorname{div}\,(\varphi\mathsf{S}) = \varphi\,\operatorname{div}\mathsf{S} + \mathsf{S}\nabla\varphi, \tag{2.94}$$

$$\operatorname{div}\,(\mathbf{S}\otimes\mathbf{u}) = (\operatorname{div}\mathbf{u})\mathbf{S} + (\nabla\mathbf{S})\mathbf{u}, \tag{2.95}$$

$$\operatorname{div}\,(\mathbf{u}\otimes\mathbf{S}) = \mathbf{u}\otimes\operatorname{div}\mathbf{S} + (\nabla\mathbf{u})\mathbf{S}^T. \tag{2.96}$$

Exercise 2.4 Prove that identities (2.89)–(2.96) hold.

An interesting example of the potential of combining compact notation and third-order tensors is offered by the following identity for a vector-valued function $\mathbf{u}$

$$\nabla(\operatorname{div}\mathbf{u}) = \operatorname{div}\,(\nabla\mathbf{u})^T. \tag{2.97}$$

On one side, with $\mathbf{A} \in \mathit{Lin}$ being a constant second-order tensor, we know, from (2.70), that the following is true

$$\nabla(\mathbf{S}\cdot\mathbf{A}) = (\nabla\mathbf{S})\overset{1}{\bar{T}}\mathbf{A}. \tag{2.98}$$

Then, putting the tensor $\nabla\mathbf{u}$ into definition (2.76), and taking into account (2.72), we obtain

$$\nabla(\operatorname{div}\mathbf{u}) = \nabla(\nabla\mathbf{u}\cdot\mathbf{I}) = (\nabla(\nabla\mathbf{u}))\overset{1}{\bar{T}}\mathbf{I} = (\nabla(\nabla\mathbf{u})^T)\mathbf{I} = \operatorname{div}\,(\nabla\mathbf{u})^T. \tag{2.99}$$

2.4 Curl

In addition, we can define in intrinsic notation the curl operation. Given the reduced applicability of this kind of operation in the forthcoming sections, we limit this to introducing the curl operation for vector-valued functions and for second-order tensor-valued functions.

Let $\mathbf{u}$ be a vector field. We define the curl of $\mathbf{u}$, which is also a vector field, denoted by $\operatorname{curl}\mathbf{u}$, through the divergence operation as follows

$$(\operatorname{curl}\mathbf{u})\cdot\mathbf{a} = \operatorname{div}\,(\mathbf{u}\times\mathbf{a}), \tag{2.100}$$

for any arbitrary constant $\mathbf{a} \in V$.

In turn, the curl of a second-order tensor field $\mathbf{S}$, which yields another second-order tensor field denoted by $\operatorname{curl}\mathbf{S}$, is defined as follows

$$(\operatorname{curl}\mathbf{S})\mathbf{a} = \operatorname{curl}(\mathbf{S}\mathbf{a}), \tag{2.101}$$

for any arbitrary constant $\mathbf{a} \in V$.

In Cartesian coordinates, previous definitions lead us to the classical definitions.

- Let $\mathbf{u} \in V$ be a vector field, then

$$\text{curl}\mathbf{u} = \epsilon_{ijk}\frac{\partial u_k}{\partial x_j}\mathbf{e}_i. \tag{2.102}$$

- Let $\mathbf{S} \in Lin$ be a second-order tensor field, then

$$\text{curl}\mathbf{S} = \epsilon_{ijk}\frac{\partial S_{kl}}{\partial x_j}(\mathbf{e}_i \otimes \mathbf{e}_l). \tag{2.103}$$

From this, it follows that for a second-order tensor with the structure $\mathbf{S} = \mathbf{u} \otimes \mathbf{v}$, the curl is given by

$$\text{curl}(\mathbf{u} \otimes \mathbf{v}) = (\text{curl}\mathbf{u}) \otimes \mathbf{v} + [(\nabla\mathbf{v}) \times \mathbf{u}]^T. \tag{2.104}$$

Exercise 2.5 Prove that (2.104) holds.

Let us now calculate the axial vector, say $\mathbf{w}$, of the skew-symmetric component of the gradient of a vector field $\mathbf{u}$, that is, of the tensor $(\nabla\mathbf{u})^a$. From the definition (1.173), and considering an arbitrary constant vector $\mathbf{a} \in V$, we are looking for the vector $\mathbf{w}$ such that

$$\mathbf{w} \times \mathbf{a} = (\nabla\mathbf{u} - (\nabla\mathbf{u})^T)\mathbf{a} = (\nabla\mathbf{u})^a\mathbf{a}. \tag{2.105}$$

Consider another arbitrary vector $\mathbf{b} \in V$. We then can write

$$\mathbf{w} \cdot (\mathbf{a} \times \mathbf{b}) = (\mathbf{w} \times \mathbf{a}) \cdot \mathbf{b} = [(\nabla\mathbf{u})^a\mathbf{a}] \cdot \mathbf{b} = (\nabla\mathbf{u})^a \cdot (\mathbf{b} \otimes \mathbf{a}). \tag{2.106}$$

By using (2.90), also exploiting the relation (1.157), and finally using the definition (2.100), we get

$$\begin{aligned}(\nabla\mathbf{u})^a \cdot (\mathbf{b} \otimes \mathbf{a}) &= \text{div}\,[(\mathbf{b} \otimes \mathbf{a})^T\mathbf{u} - (\mathbf{a} \otimes \mathbf{b})^T\mathbf{u}] = \text{div}\,[\mathbf{a}(\mathbf{b} \cdot \mathbf{u}) - \mathbf{b}(\mathbf{a} \cdot \mathbf{u})] \\ &= \text{div}\,[\mathbf{u} \times (\mathbf{a} \times \mathbf{b})] = \text{curl}\mathbf{u} \cdot (\mathbf{a} \times \mathbf{b}).\end{aligned} \tag{2.107}$$

Comparing (2.106) with (2.107) we obtain

$$\mathbf{w} \cdot (\mathbf{a} \times \mathbf{b}) = \text{curl}\mathbf{u} \cdot (\mathbf{a} \times \mathbf{b}), \tag{2.108}$$

and since $\mathbf{a}, \mathbf{b} \in V$ are arbitrary, the cross product $\mathbf{c} = \mathbf{a} \times \mathbf{b} \in V$ is also arbitrary, and therefore we obtain

$$\mathbf{w} = \text{curl}\mathbf{u}. \tag{2.109}$$

This shows that the axial vector of the skew-symmetric component of the gradient of a vector field is the curl of such field, that is,

$$\text{curl}\mathbf{u} \times \mathbf{a} = (\nabla\mathbf{u} - (\nabla\mathbf{u})^T)\mathbf{a}, \tag{2.110}$$

for any vector $\mathbf{a} \in V$.

2.5 Laplacian

In previous sections we defined the gradient and divergence operations of a field ϕ. If we now admit that the field is of class C^2, we can introduce another operation called Laplacian, which will be denoted by $\Delta\phi$, as follows

$$\Delta\phi = \text{div}\,\nabla\phi = c(\nabla(\nabla\phi), \mathbf{I}). \tag{2.111}$$

Recalling the definitions of gradient and divergence, it is easy to notice that the Laplacian yields a field of the same nature of ϕ (less regular, however). Thus, in intrinsic notation, and in the particular case of Cartesian coordinates, we have the following definitions

- Let ϕ be a scalar field φ, then

$$\Delta\varphi = \nabla(\nabla\varphi)\cdot\mathbf{I} = \frac{\partial^2\varphi}{\partial x_k \partial x_k}. \tag{2.112}$$

- Let ϕ be a vector field $\mathbf{u}$, then

$$\Delta\mathbf{u} = \nabla(\nabla\mathbf{u})\mathbf{I} = \frac{\partial^2\mathbf{u}}{\partial x_k \partial x_k} = \frac{\partial^2 u_i}{\partial x_k \partial x_k}\mathbf{e}_i. \tag{2.113}$$

- Let ϕ be a second-order tensor field $\mathbf{S}$, then

$$\Delta\mathbf{S} = \nabla(\nabla\mathbf{S})\mathbf{I} = \frac{\partial^2\mathbf{S}}{\partial x_k \partial x_k} = \frac{\partial^2 S_{ij}}{\partial x_k \partial x_k}(\mathbf{e}_i \otimes \mathbf{e}_j). \tag{2.114}$$

- Let ϕ be a third-order tensor field S, then

$$\Delta\mathsf{S} = \nabla(\nabla\mathsf{S})\mathbf{I} = \frac{\partial^2\mathsf{S}}{\partial x_m \partial x_m} = \frac{\partial^2 \mathsf{S}_{ijk}}{\partial x_m \partial x_m}(\mathbf{e}_i \otimes \mathbf{e}_j \otimes \mathbf{e}_k). \tag{2.115}$$

Examining the previous expressions, we note that the Laplacian of a field ϕ is another field of the same kind, whose components are given by the Laplacian of the corresponding components of ϕ.

In order to illustrate the use of components, let us calculate the Cartesian components of the Laplacian of the second-order tensor field $\mathbf{S}$ shown in (2.114). Hence, the components of $\Delta\mathbf{S}$ are

$$\begin{aligned}
(\Delta\mathbf{S})_{ij} = ([\nabla(\nabla\mathbf{S})]\mathbf{I})_{ij} &= \left[\frac{\partial}{\partial x_l}\left[\frac{\partial}{\partial x_k} S_{ij}(\mathbf{e}_i \otimes \mathbf{e}_j) \otimes \mathbf{e}_k\right] \otimes \mathbf{e}_l\right] \delta_{rs}(\mathbf{e}_r \otimes \mathbf{e}_s) \\
&= \frac{\partial^2 S_{ij}}{\partial x_k \partial x_l}(\mathbf{e}_i \otimes \mathbf{e}_j \otimes \mathbf{e}_k \otimes \mathbf{e}_l)\delta_{rs}(\mathbf{e}_r \otimes \mathbf{e}_s) \\
&= \frac{\partial^2 S_{ij}}{\partial x_k \partial x_l}(\mathbf{e}_i \otimes \mathbf{e}_j)\delta_{kl} = \frac{\partial^2 S_{ij}}{\partial x_k \partial x_k}(\mathbf{e}_i \otimes \mathbf{e}_j).
\end{aligned} \tag{2.116}$$

Before ending this section, observe that with the available results we can directly show that, as $\mathbf{u}$ is a vector field, the following relation holds

$$\text{div}\,[\nabla\mathbf{u} \pm (\nabla\mathbf{u})^T] = \Delta\mathbf{u} \pm \nabla(\text{div}\,\mathbf{u}). \tag{2.117}$$

Exercise 2.6 Prove that (2.117) is valid.

A useful identity encountered in the field of mechanics is stated next. Consider a vector field $\mathbf{u}$. It holds that

$$\Delta \mathbf{u} = -\text{curlcurl}\mathbf{u} + \nabla(\text{div }\mathbf{u}). \tag{2.118}$$

Let us show that (2.118) holds. Consider a constant arbitrary vector $\mathbf{a} \in V$, then, from the definition of the divergence of a tensor field and the fact that the tensor under consideration is skew-symmetric,

$$-\text{div }[(\nabla\mathbf{u} - (\nabla\mathbf{u})^T)\mathbf{a}] = [\text{div }(\nabla\mathbf{u} - (\nabla\mathbf{u})^T)] \cdot \mathbf{a}, \tag{2.119}$$

and now using (2.117) renders

$$[\text{div }(\nabla\mathbf{u} - (\nabla\mathbf{u})^T)] \cdot \mathbf{a} = [\Delta\mathbf{u} - \nabla(\text{div }\mathbf{u})] \cdot \mathbf{a}. \tag{2.120}$$

In turn, from the definition of the axial vector of a skew-symmetric tensor, the fact that the axial vector of the $\nabla\mathbf{u}$ tensor field is curl$\mathbf{u}$, and from the definition of curl operation given in (2.100), we obtain

$$-\text{div }[(\nabla\mathbf{u} - (\nabla\mathbf{u})^T)\mathbf{a}] = -\text{div }(\text{curl}\mathbf{u} \times \mathbf{a}) = -\text{curlcurl}\mathbf{u} \cdot \mathbf{a}. \tag{2.121}$$

Comparing (2.120) and (2.121) the result (2.118) follows directly.

2.6 Integration

This section is devoted to presenting some results and relations involving integral operations, which will be useful in forthcoming sections.

Let us recall the *divergence theorem*. Let Ω be a bounded region with smooth boundary.[1] In turn, let us consider the smooth fields $\phi : \Omega \to \mathbb{R}$ and $\mathbf{u} : \Omega \to V$. Then, the following relations hold

$$\int_{\partial\Omega} \phi\mathbf{n}\, d\partial\Omega = \int_{\Omega} \nabla\phi\, d\Omega, \tag{2.122}$$

$$\int_{\partial\Omega} \mathbf{u} \cdot \mathbf{n}\, d\partial\Omega = \int_{\Omega} \text{div }\mathbf{u}\, d\Omega, \tag{2.123}$$

where $\mathbf{n}$ is the outward unit normal vector to $\partial\Omega$.

Let us see that analogous results are valid when considering smooth tensor fields. In particular, for fields $\mathbf{S} : \Omega \to Lin$ and $\mathsf{S} : \Omega \to \mathsf{Lin}$ it is verified that

$$\int_{\partial\Omega} \mathbf{S}\mathbf{n}\, d\partial\Omega = \int_{\Omega} \text{div }\mathbf{S}\, d\Omega, \tag{2.124}$$

$$\int_{\partial\Omega} \mathsf{S}\mathbf{n}\, d\partial\Omega = \int_{\Omega} \text{div }\mathsf{S}\, d\Omega. \tag{2.125}$$

In fact, from definition (2.82) for an arbitrary constant vector field $\mathbf{a} \in V$, we have

$$\left[\int_{\Omega} \text{div }\mathbf{S}\, d\Omega\right] \cdot \mathbf{a} = \int_{\Omega} \text{div }(\mathbf{S}^T\mathbf{a})\, d\Omega = \int_{\partial\Omega} \mathbf{S}^T\mathbf{a} \cdot \mathbf{n}\, d\partial\Omega = \left[\int_{\partial\Omega} \mathbf{S}\mathbf{n}\, d\partial\Omega\right] \cdot \mathbf{a}, \tag{2.126}$$

and then we arrive at (2.124).

1 Roughly speaking, a smooth region is an open region Ω with boundary $\partial\Omega$ piece-wise smooth, that is, the normal vector is uniquely defined almost everywhere in $\partial\Omega$, except in a finite number of points (2D) or curves (3D).

Analogously, from definition (2.83), and given an arbitrary constant tensor field $\mathbf{A} \in$ *Lin*,

$$\left[\int_\Omega \operatorname{div} \mathsf{S}\ d\Omega\right] \cdot \mathbf{A} = \int_\Omega \operatorname{div} (\mathsf{S}^{\frac{1}{T}}\mathbf{A})\ d\Omega = \int_{\partial\Omega} \mathsf{S}^{\frac{1}{T}}\mathbf{A} \cdot \mathbf{n}\ d\partial\Omega = \left[\int_{\partial\Omega} \mathsf{S}\mathbf{n}\ d\partial\Omega\right] \cdot \mathbf{A}, \tag{2.127}$$

and so (2.125) is demonstrated.

From the result given by (2.90), and keeping in mind relation (2.123), it is easy to see that

$$\int_{\partial\Omega} \mathbf{S}\mathbf{n} \cdot \mathbf{v}\ d\partial\Omega = \int_\Omega (\mathbf{S} \cdot \nabla\mathbf{v} + \operatorname{div} \mathbf{S} \cdot \mathbf{v})\ d\Omega. \tag{2.128}$$

This expression will be employed repeatedly through this work.

Let us exemplify the applicability of these results in combination with the use of third-order tensors by verifying the following identity

$$\int_\Omega [\mathbf{u} \otimes \operatorname{div} \mathbf{S} + (\nabla\mathbf{u})\mathbf{S}^T]\ d\Omega = \int_{\partial\Omega} \mathbf{u} \otimes (\mathbf{S}\mathbf{n})\ d\partial\Omega. \tag{2.129}$$

In fact, putting $\mathsf{S} = \mathbf{u} \otimes \mathbf{S}$, from (2.96) we have that

$$\int_\Omega \operatorname{div} \mathsf{S}\ d\Omega = \int_\Omega \operatorname{div} (\mathbf{u} \otimes \mathbf{S})\ d\Omega = \int_\Omega [\mathbf{u} \otimes \operatorname{div} \mathbf{S} + (\nabla\mathbf{u})\mathbf{S}^T]\ d\Omega, \tag{2.130}$$

while, from (1.204), we also have

$$\int_{\partial\Omega} \mathsf{S}\mathbf{n}\ d\partial\Omega = \int_{\partial\Omega} (\mathbf{u} \otimes \mathbf{S})\mathbf{n}\ d\partial\Omega = \int_{\partial\Omega} \mathbf{u} \otimes (\mathbf{S}\mathbf{n})\ d\partial\Omega. \tag{2.131}$$

Thus, using expression (2.125), identity (2.129) is verified.

In the same way, the following result also holds

$$\int_\Omega [(\operatorname{div} \mathbf{u})\mathbf{S} + (\nabla\mathbf{S})\mathbf{u}]\ d\Omega = \int_{\partial\Omega} (\mathbf{u} \cdot \mathbf{n})\mathbf{S}\ d\partial\Omega. \tag{2.132}$$

In fact, considering $\mathsf{S} = \mathbf{S} \otimes \mathbf{u}$, from (2.95) we have that

$$\int_\Omega \operatorname{div} \mathsf{S}\ d\Omega = \int_\Omega \operatorname{div} (\mathbf{S} \otimes \mathbf{u})\ d\Omega = \int_\Omega [(\operatorname{div} \mathbf{u})\mathbf{S} + (\nabla\mathbf{S})\mathbf{u}]\ d\Omega, \tag{2.133}$$

while, from (1.207)

$$\int_{\partial\Omega} \mathsf{S}\mathbf{n}\ d\partial\Omega = \int_{\partial\Omega} (\mathbf{S} \otimes \mathbf{u})\mathbf{n}\ d\partial\Omega = \int_{\partial\Omega} (\mathbf{u} \cdot \mathbf{n})\mathbf{S}\ d\partial\Omega, \tag{2.134}$$

from what we proved in relation (2.132).

Let us demonstrate that the following relation holds

$$\int_\Omega \nabla\mathbf{S}\ d\Omega = \int_{\partial\Omega} \mathbf{S} \otimes \mathbf{n}\ d\partial\Omega. \tag{2.135}$$

In order to do this, let us take expression (2.122) with $\varphi = \mathbf{u} \cdot \mathbf{a}$, with $\mathbf{a}$ arbitrary and constant. Hence, we write

$$\left[\int_\Omega (\nabla\mathbf{u})^T\ d\Omega\right] \mathbf{a} = \int_\Omega \nabla(\mathbf{u} \cdot \mathbf{a})\ d\Omega = \int_{\partial\Omega} (\mathbf{u} \cdot \mathbf{a})\mathbf{n}\ d\partial\Omega = \left[\int_{\partial\Omega} (\mathbf{n} \otimes \mathbf{u})\ d\partial\Omega\right] \mathbf{a}, \tag{2.136}$$

and therefore we conclude that

$$\int_{\Omega} \nabla \mathbf{u} \; d\Omega = \int_{\partial\Omega} \mathbf{u} \otimes \mathbf{n} \; d\partial\Omega. \tag{2.137}$$

So, from definition (2.56), and using (2.137) and (1.208), leads to

$$\begin{aligned}\left[\int_{\Omega} (\nabla \mathbf{S})^T \; d\Omega\right] \mathbf{a} &= \int_{\Omega} \nabla(\mathbf{S}^T\mathbf{a}) \; d\Omega = \int_{\partial\Omega} (\mathbf{S}^T\mathbf{a}) \otimes \mathbf{n} \; d\partial\Omega \\ &= \int_{\partial\Omega} (\mathbf{S} \otimes \mathbf{n})^T \mathbf{a} \; d\partial\Omega = \left[\int_{\partial\Omega} (\mathbf{S} \otimes \mathbf{n})^T \; d\partial\Omega\right] \mathbf{a},\end{aligned} \tag{2.138}$$

and then (2.135) is demonstrated.

As a last example let us prove that the following result holds

$$\int_{\Omega} [(\Delta\mathbf{S})\mathbf{u} - \mathbf{S}(\Delta\mathbf{u})] \; d\Omega = \int_{\partial\Omega} [((\nabla\mathbf{S})\mathbf{n})\mathbf{u} - \mathbf{S}(\nabla\mathbf{u})\mathbf{n}] \; d\partial\Omega. \tag{2.139}$$

Recall first that for scalar fields φ, ϕ smooth enough, the following is true

$$\int_{\Omega} (\varphi\Delta\phi - \phi\Delta\varphi) \; d\Omega = \int_{\partial\Omega} (\varphi\nabla\phi - \phi\nabla\varphi) \cdot \mathbf{n} \; d\partial\Omega. \tag{2.140}$$

Assume now that we have a vector field $\mathbf{u} \in V$ and a second-order tensor field $\mathbf{S} \in Lin$. From (2.84), and putting $\mathbf{T} = \nabla\mathbf{u}$, yields

$$\text{div}\,(\mathbf{S}\nabla\mathbf{u}) = \mathbf{S}\text{div}\,(\nabla\mathbf{u}) + (\nabla\mathbf{S})(\nabla\mathbf{u}), \tag{2.141}$$

and therefore, since $\text{div}\,(\nabla\mathbf{u}) = \Delta\mathbf{u}$,

$$\mathbf{S}(\Delta\mathbf{u}) = \text{div}\,(\mathbf{S}\nabla\mathbf{u}) - (\nabla\mathbf{S})(\nabla\mathbf{u}). \tag{2.142}$$

Let $\mathbf{a} \in V$ be an arbitrary constant vector. Taking into account the definition (2.82) of the divergence of a second-order tensor for the tensor $(\nabla\mathbf{S})^{\frac{1}{T}}\mathbf{u}$, and making use of (1.211), (2.65) and (2.60), it follows that

$$\begin{aligned}\left[\text{div}\left[((\nabla\mathbf{S})^{\frac{1}{T}}\mathbf{u})^T\right]\right] \cdot \mathbf{a} &= \text{div}\left[((\nabla\mathbf{S})^{\frac{1}{T}}\mathbf{u})\mathbf{a}\right] = \text{div}\left[(\nabla\mathbf{S})^{\frac{1}{T}}(\mathbf{a} \otimes \mathbf{u})\right] \\ &= (\nabla\mathbf{S}) \cdot \nabla(\mathbf{a} \otimes \mathbf{u}) + (\mathbf{a} \otimes \mathbf{u}) \cdot \text{div}\,(\nabla\mathbf{S}) \\ &= (\nabla\mathbf{S}) \cdot (\mathbf{a} \otimes \nabla\mathbf{u}) + (\mathbf{a} \otimes \mathbf{u}) \cdot \Delta\mathbf{S} \\ &= [(\nabla\mathbf{S})(\nabla\mathbf{u}) + (\Delta\mathbf{S})\mathbf{u}] \cdot \mathbf{a},\end{aligned} \tag{2.143}$$

that is,

$$\text{div}\left[((\nabla\mathbf{S})^{\frac{1}{T}}\mathbf{u})^T\right] = (\nabla\mathbf{S})(\nabla\mathbf{u}) + (\Delta\mathbf{S})\mathbf{u}, \tag{2.144}$$

and therefore

$$(\Delta\mathbf{S})\mathbf{u} = \text{div}\left[((\nabla\mathbf{S})^{\frac{1}{T}}\mathbf{u})^T\right] - (\nabla\mathbf{S})(\nabla\mathbf{u}). \tag{2.145}$$

Subtracting (2.142) from (2.145) yields

$$(\Delta\mathbf{S})\mathbf{u} - \mathbf{S}(\Delta\mathbf{u}) = \text{div}\left[((\nabla\mathbf{S})^{\frac{1}{T}}\mathbf{u})^T\right] - \text{div}\,(\mathbf{S}\nabla\mathbf{u}). \tag{2.146}$$

Integrating both sides of this expression in the domain Ω, using the divergence theorem (2.124), and employing expression (1.210), leads us to

$$\int_{\Omega} [(\Delta\mathbf{S})\mathbf{u} - \mathbf{S}(\Delta\mathbf{u})]\ d\Omega = \int_{\Omega} \left[\operatorname{div}\left[((\nabla\mathbf{S})^{\frac{1}{T}}\mathbf{u})^T\right] - \operatorname{div}(\mathbf{S}\nabla\mathbf{u})\right] d\Omega$$
$$= \int_{\partial\Omega} \left[((\nabla\mathbf{S})^{\frac{1}{T}}\mathbf{u})^T - (\mathbf{S}\nabla\mathbf{u})\right]\mathbf{n}\ d\partial\Omega = \int_{\partial\Omega} [((\nabla\mathbf{S})\mathbf{n})\mathbf{u} - (\mathbf{S}\nabla\mathbf{u})\mathbf{n}]\ d\partial\Omega, \tag{2.147}$$

and the result is thus demonstrated.

2.7 Coordinates

Let us see in more detail some aspects related to bases and coordinates.

Cartesian space $\mathbb{R}^n$ is the n-dimensional vector space whose elements are ordered lists of n real numbers $\{x^1, x^2, \dots, x^n\}$, and in which addition and multiplication by a real number are defined in the usual sense.

A coordinate system in an open subset $\mathscr{X}$ of the n-dimensional Euclidean space is a bijective function from $\mathscr{X}$ into $\mathbb{R}^n$ such that its gradient is invertible and the second gradient is continuous. That is, the transformation is of class C^2. Thus, if we denote by $\overline{\mathbf{x}}$[2] this transformation, we have

$$\begin{aligned} &\overline{\mathbf{x}} : \mathscr{X} \to \mathbb{R}^n, \\ &\quad \mathbf{x} \mapsto \overline{\mathbf{x}} = \{\overline{x}^1(\mathbf{x}), \dots, \overline{x}^n(\mathbf{x})\} = \{x^1, \dots, x^n\}, \end{aligned} \tag{2.148}$$

where $\overline{x}^k$ is a scalar field with the same regularity of $\overline{\mathbf{x}}$. For a given point $\mathbf{x}$, the number $x^k = \overline{x}^k(\mathbf{x})$ represents the kth coordinate of point $\mathbf{x}$ in the coordinate system $\overline{\mathbf{x}}$.

Denoting by $\underline{\mathbf{x}}$ the inverse of the transformation $\overline{\mathbf{x}}$, we have

$$\overline{x}^k(\underline{\mathbf{x}}(x^1, \dots, x^n)) = x^k \qquad k = 1, \dots, n. \tag{2.149}$$

By virtue of the regularity assumed for the transformation $\overline{\mathbf{x}}$, at each point $\mathbf{x}$ it is possible to define the following vectors

$$\mathbf{e}^k = \nabla\overline{x}^k(\mathbf{x}) \qquad\qquad k \in 1, \dots, n, \tag{2.150}$$

$$\mathbf{e}_k = \left.\frac{\partial}{\partial x^k}\underline{\mathbf{x}}(x^1, \dots, x^n)\right|_{x^j = \overline{x}^j(\mathbf{x})\text{fixed } \forall\ j,\ j\neq k} \qquad k \in 1, \dots, n. \tag{2.151}$$

In particular, the vector field $\mathbf{e}^k(\mathbf{x})$ is normal to the coordinate surface $x^k = C$ (constant) which goes through point $\mathbf{x}$. In turn, the vector field $\mathbf{e}_k(\mathbf{x})$ is tangent to the kth coordinate curve that goes through $\mathbf{x}$. This coordinate curve is the locus of points close to point $\mathbf{x}$ for which all its coordinates, except coordinate x^k, have the same value as those from point $\mathbf{x}$ (see Figure 2.1).

From the previous considerations, it turns out that $\{\mathbf{e}^1(\mathbf{x}), \dots, \mathbf{e}^n(\mathbf{x})\}$ and also $\{\mathbf{e}_1(\mathbf{x}), \dots, \mathbf{e}_n(\mathbf{x})\}$ are bases of the vector space V which is associated with the n-dimensional Euclidean space. In particular, the basis $\{\mathbf{e}_1(\mathbf{x}), \dots, \mathbf{e}_n(\mathbf{x})\}$ is called the

2 Note that here we employ bold format for points because of the already noted one-to-one correspondence between the space $\mathscr{E}$ and V.

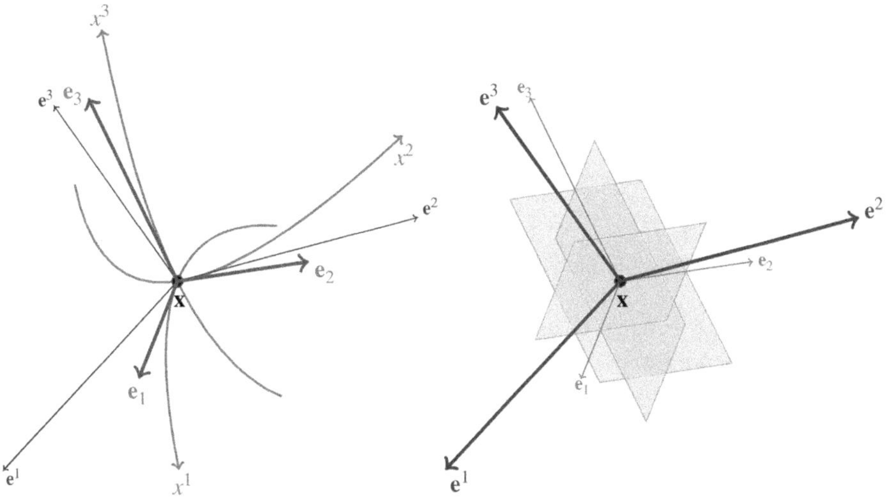

Figure 2.1 Basic elements of covariant and contravariant coordinate systems.

natural basis at point $\mathbf{x}$ of the coordinate system $\overline{\mathbf{x}}$. In turn, the basis $\{\mathbf{e}^1(\mathbf{x}), \ldots, \mathbf{e}^n(\mathbf{x})\}$ is called the dual basis (also the reciprocal basis). Note that, when the point $\mathbf{x}$ is changed within the open set $\mathscr{X}$ (domain of definition of the transformation $\overline{\mathbf{x}}$), we obtain fields of dual and natural bases. In general, these bases need not be orthonormal. If the coordinate surfaces are orthogonal, then the coordinate curves are orthogonal to these surfaces. In such cases, $\mathbf{e}^k$ and $\mathbf{e}_k$ are parallel, but not identical in general (this is the case, for example, for cylindrical and spherical coordinates). In the case in which the natural basis is constant, the coordinates are called Cartesian coordinates and if, in addition, the elements of the natural basis are orthonormal, these are called rectangular Cartesian coordinates.

In the field of mechanics, the Euclidean space is three-dimensional. In this case, we can define the volume, denoted by V, of the parallelepiped determined by the basis $\{\mathbf{e}_1, \mathbf{e}_2, \mathbf{e}_3\}$ as

$$\mathsf{V} = \mathbf{e}_1 \cdot (\mathbf{e}_2 \times \mathbf{e}_3), \tag{2.152}$$

and similarly,

$$\mathsf{V}' = \mathbf{e}^1 \cdot (\mathbf{e}^2 \times \mathbf{e}^3). \tag{2.153}$$

Then, the natural and dual bases are related through the following relations

$$\mathbf{e}^1 = \frac{1}{\mathsf{V}}(\mathbf{e}_2 \times \mathbf{e}_3), \tag{2.154}$$

$$\mathbf{e}^2 = \frac{1}{\mathsf{V}}(\mathbf{e}_3 \times \mathbf{e}_1), \tag{2.155}$$

$$\mathbf{e}^3 = \frac{1}{\mathsf{V}}(\mathbf{e}_1 \times \mathbf{e}_2). \tag{2.156}$$

Equivalently, we have

$$\mathbf{e}_1 = \frac{1}{\mathsf{V}'}(\mathbf{e}^2 \times \mathbf{e}^3), \tag{2.157}$$

$$\mathbf{e}_2 = \frac{1}{V'}(\mathbf{e}^3 \times \mathbf{e}^1), \tag{2.158}$$

$$\mathbf{e}_3 = \frac{1}{V'}(\mathbf{e}^1 \times \mathbf{e}^2), \tag{2.159}$$

where it follows that

$$V = \frac{1}{V'}. \tag{2.160}$$

Exercise 2.7 Prove that (2.160) is verified.

In some problems, it is usual to use cylindrical coordinates. In such cases, for each point $\mathbf{x} \in \mathscr{E}$ we attach a triplet of real numbers $\overline{\mathbf{x}} = \{x^1, x^2, x^3\} = \{r, \theta, \xi\}$ which represent, respectively, the distance of point $\mathbf{x}$ to a given line called the axis of the cylindrical coordinate system, the angle between a plane containing the axis (which is considered as the origin plane to measure angles) and the plane that goes through the axis and the point $\mathbf{x}$, and, finally, the distance from $\mathbf{x}$ to a plane orthogonal to the axis, taken as the origin to measure this distance (see Figure 2.2).

Indeed, for the cylindrical coordinate system we have $\overline{\mathbf{x}} = \{\overline{x}^1, \overline{x}^2, \overline{x}^3\} = \{r, \theta, \xi\}$ and $\underline{\mathbf{x}} = \{\underline{x}^1, \underline{x}^2, \underline{x}^3\} = \{x, y, z\}$ with

$$\overline{x}^1 = r = \sqrt{x^2 + y^2}, \qquad \overline{x}^2 = \theta = \arctan\frac{y}{x}, \qquad \overline{x}^3 = \xi = z, \tag{2.161}$$

and the inverse relations

$$\underline{x}^1 = x = r\cos\theta, \qquad \underline{x}^2 = y = r\sin\theta, \qquad \underline{x}^3 = z = \xi. \tag{2.162}$$

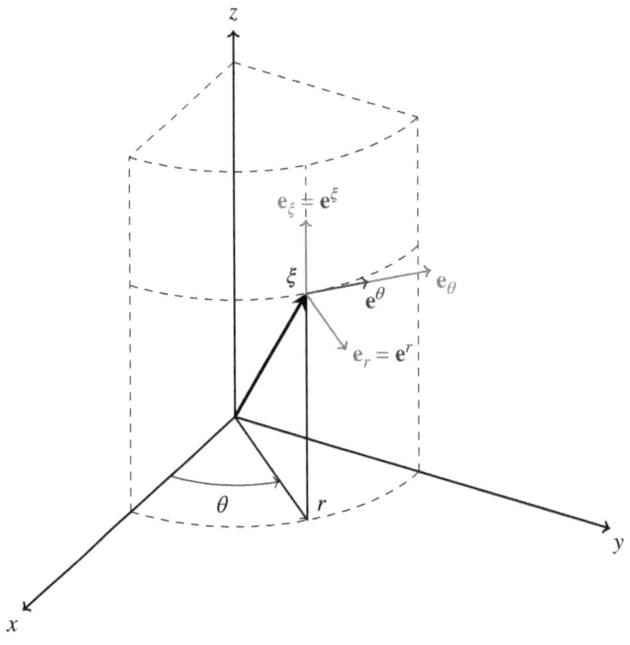

Figure 2.2 Elements of cylindrical coordinates.

Then it is easy to see that

$$\mathbf{e}^r = \nabla \overline{x}^1 = \nabla r = \begin{pmatrix} \frac{x}{\sqrt{x^2+y^2}} \\ \frac{y}{\sqrt{x^2+y^2}} \\ 0 \end{pmatrix}, \quad \mathbf{e}^\theta = \nabla \overline{x}^2 = \nabla \theta = \begin{pmatrix} \frac{-y}{x^2+y^2} \\ \frac{x}{x^2+y^2} \\ 0 \end{pmatrix},$$

$$\mathbf{e}^\xi = \nabla \overline{x}^3 = \nabla \xi = \begin{pmatrix} 0 \\ 0 \\ 1 \end{pmatrix}, \tag{2.163}$$

and also

$$\mathbf{e}_r = \frac{\partial \underline{\mathbf{x}}}{\partial r} = \begin{pmatrix} \cos\theta \\ \sin\theta \\ 0 \end{pmatrix}, \quad \mathbf{e}_\theta = \frac{\partial \underline{\mathbf{x}}}{\partial \theta} = \begin{pmatrix} -r\sin\theta \\ r\cos\theta \\ 0 \end{pmatrix}, \quad \mathbf{e}_\xi = \frac{\partial \underline{\mathbf{x}}}{\partial \xi} = \begin{pmatrix} 0 \\ 0 \\ 1 \end{pmatrix}, \tag{2.164}$$

from which it follows that

$$\mathbf{e}_r = \mathbf{e}^r, \qquad \mathbf{e}_\theta = r^2 \mathbf{e}^\theta, \qquad \mathbf{e}_\xi = \mathbf{e}^\xi. \tag{2.165}$$

Exercise 2.8 Prove the validity of relations (2.165).

Another frequently used coordinate system is the spherical system. In this case, to each point $\mathbf{x}$ we associate three real numbers $\{r, \theta, \phi\}$ whose meaning is clear from Figure 2.3, where r is the radial distance from the origin, ϕ is the azimuthal angle, and θ is the polar angle. Similarly to the previous case, we have now $\overline{\mathbf{x}} = \{\overline{x}^1, \overline{x}^2, \overline{x}^3\} = \{r, \theta, \phi\}$ and $\underline{\mathbf{x}} = \{\underline{x}^1, \underline{x}^2, \underline{x}^3\} = \{x, y, z\}$ with

$$\overline{x}^1 = r = \sqrt{x^2+y^2+z^2}, \qquad \overline{x}^2 = \theta = \arctan\frac{\sqrt{x^2+y^2}}{z},$$

$$\overline{x}^3 = \phi = \arctan\frac{y}{x}, \tag{2.166}$$

and the corresponding inverse relations

$$\underline{x}^1 = x = r\sin\theta\cos\phi, \qquad \underline{x}^2 = y = r\sin\theta\sin\phi, \qquad \underline{x}^3 = z = r\cos\theta. \tag{2.167}$$

In this case this gives

$$\mathbf{e}^r = \nabla \overline{x}^1 = \nabla r = \begin{pmatrix} \frac{x}{\sqrt{x^2+y^2+z^2}} \\ \frac{y}{\sqrt{x^2+y^2+z^2}} \\ \frac{z}{\sqrt{x^2+y^2+z^2}} \end{pmatrix},$$

$$\mathbf{e}^\theta = \nabla \overline{x}^2 = \nabla \theta = \begin{pmatrix} \frac{xz}{(x^2+y^2+z^2)\sqrt{x^2+y^2}} \\ \frac{yz}{(x^2+y^2+z^2)\sqrt{x^2+y^2}} \\ \frac{-\sqrt{x^2+y^2}}{x^2+y^2+z^2} \end{pmatrix}, \tag{2.168}$$

$$\mathbf{e}^\phi = \nabla \overline{x}^3 = \nabla \phi = \begin{pmatrix} \frac{-y}{x^2+y^2} \\ \frac{x}{x^2+y^2} \\ 0 \end{pmatrix},$$

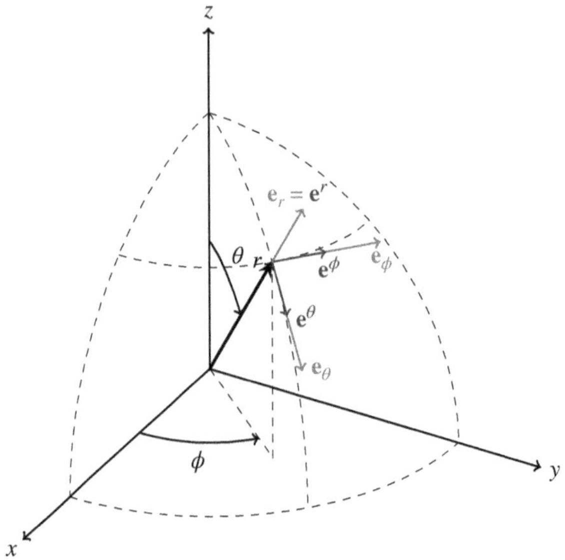

Figure 2.3 Elements of spherical coordinates.

and

$$\mathbf{e}_r = \frac{\partial \underline{\underline{\mathbf{x}}}}{\partial r} = \begin{pmatrix} \sin\theta\cos\phi \\ \sin\theta\sin\phi \\ \cos\theta \end{pmatrix}, \quad \mathbf{e}_\theta = \frac{\partial \underline{\underline{\mathbf{x}}}}{\partial \theta} = \begin{pmatrix} r\cos\theta\cos\phi \\ r\cos\theta\sin\phi \\ -r\sin\theta \end{pmatrix}, \quad \mathbf{e}_\phi = \frac{\partial \underline{\underline{\mathbf{x}}}}{\partial \phi} = \begin{pmatrix} -r\sin\theta\sin\phi \\ r\sin\theta\cos\phi \\ 0 \end{pmatrix}. \tag{2.169}$$

Therefore, in this case the following identities hold

$$\mathbf{e}_r = \mathbf{e}^r, \qquad \mathbf{e}_\theta = r^2\mathbf{e}^\theta, \qquad \mathbf{e}_\phi = r^2\sin^2\theta\mathbf{e}^\phi. \tag{2.170}$$

Exercise 2.9 Demonstrate that the relations in (2.170) are valid.

Now, let $\mathbf{v}$ be a vector field defined in $\mathscr{X}$. Then, for each point $\mathbf{x}$, $\mathbf{v}(\mathbf{x})$ is a vector which can be expressed in terms of any basis, that is, the natural or the dual basis. This means

$$\mathbf{v}(\mathbf{x}) = v^i(\mathbf{x})\mathbf{e}_i(\mathbf{x}) = v_i(\mathbf{x})\mathbf{e}^i(\mathbf{x}). \tag{2.171}$$

Fields $v^1, \dots, v^n$ are the contravariant components of $\mathbf{v}$ relative to the coordinate system $\overline{\mathbf{x}}$, which is when the natural basis has been adopted. In the same way, fields $v_1, \dots, v_n$ are the covariant components of $\mathbf{v}$ relative to the coordinate system $\overline{\mathbf{x}}$ when the dual basis has been adopted. The same procedure can be followed with tensor fields of any order (see also Chapter 1).

However, as it is usual in physics, vector and tensor fields possess their own dimensional units (length, surface, force, etc.) with a precise physical meaning. For instance, the velocity vector field of a system of particles has the meaning of the time rate of the space length traveled by particles. The components of this field in the different coordinate systems presented before possess different physical meaning, and this is because of the different physical meaning owned by the elements that form the different bases. For instance, for a cylindrical coordinate system, $\mathbf{e}^r$ has no dimensional units, while $\mathbf{e}^\theta$ has the dimension of the inverse of length.

Thus, in mechanics it can be useful to supply all components of a given entity (a vector or a tensor) with the same physical identity of the element they represent. This motivated the development of the so-called physical components.

For an orthogonal coordinate system, these components are precisely defined as being the components with respect to the basis

$$\mathbf{i}_{\langle k\rangle} = \frac{\mathbf{e}_k}{\|\mathbf{e}_k\|} = \frac{\mathbf{e}^k}{\|\mathbf{e}^k\|} \qquad k = 1, \dots, n. \tag{2.172}$$

This new orthonormal basis is defined by a set of vectors $\{\mathbf{i}_{\langle 1\rangle}, \dots, \mathbf{i}_{\langle n\rangle}\}$ which are always tangent to the coordinate curves and orthogonal to the coordinate surfaces. Hence, we have

$$\mathbf{v}(\mathbf{x}) = v^k(\mathbf{x})\mathbf{e}_k(\mathbf{x}) = v_k(\mathbf{x})\mathbf{e}^k(\mathbf{x}) = v_{\langle k\rangle}(\mathbf{x})\mathbf{i}_{\langle k\rangle}(\mathbf{x}), \tag{2.173}$$

where $v_{\langle k\rangle}(\mathbf{x})$ is the kth physical component of $\mathbf{v}$ at point $\mathbf{x}$. In turn, from the previous expression we see that

$$v_{\langle k\rangle}(\mathbf{x}) = v^k(\mathbf{x})\|\mathbf{e}_k(\mathbf{x})\| = v_k(\mathbf{x})\|\mathbf{e}^k(\mathbf{x})\|. \tag{2.174}$$

Analogous definitions and rules can be obtained for second-order tensor fields. In effect

$$\mathbf{T} = T^{ij}(\mathbf{e}_i \otimes \mathbf{e}_j) = T_{ij}(\mathbf{e}^i \otimes \mathbf{e}^j) = T^i_{\ .j}(\mathbf{e}_i \otimes \mathbf{e}^j) = T_{\langle ij\rangle}(\mathbf{i}_{\langle i\rangle} \otimes \mathbf{i}_{\langle j\rangle}), \tag{2.175}$$

at each point $\mathbf{x}$. From this last expression, the following relations among the different components of a second-order tensor hold

$$T_{\langle ij\rangle} = T^{ij}\|\mathbf{e}_i\|\|\mathbf{e}_j\| = T_{ij}\|\mathbf{e}^i\|\|\mathbf{e}^j\| = T^i_{\ .j}\|\mathbf{e}_i\|\|\mathbf{e}^j\| = T_i^{\ .j}\|\mathbf{e}^i\|\|\mathbf{e}_j\|. \tag{2.176}$$

Making use of the metric tensor g_{ij}, or g^{ij}, the physical components of the vector and tensor fields are written as

$$v_{\langle k\rangle} = v^k\sqrt{g_{kk}} = v_k\sqrt{g^{kk}}, \tag{2.177}$$

$$T_{\langle ij\rangle} = T^{ij}\sqrt{g_{ii}}\sqrt{g_{jj}} = T_{ij}\sqrt{g^{ii}}\sqrt{g^{jj}} = T^i_{\ .j}\sqrt{g_{ii}}\sqrt{g^{jj}} = T_i^{\ .j}\sqrt{g^{ii}}\sqrt{g_{jj}}. \tag{2.178}$$

Exercise 2.10 Prove that relations (2.177) and (2.178) are verified.

Exercise 2.11 Consider $\mathbf{u} \in V$ and $\mathbf{T} \in Lin$. Write the covariant, contravariant, and physical components of these fields for the cylindrical and spherical coordinate systems.

Moreover, from the definition of the coordinate system, the natural basis (also called the intrinsic basis), is a differentiable vector field. Therefore, there exists the gradient associated with each vector basis (understood as a vector field), which is a continuous second-order tensor field denoted by

$$\Gamma_{(k)} = \nabla\mathbf{e}_k \qquad k = 1, \dots, n. \tag{2.179}$$

In particular, the mixed components $\Gamma_{(k)\ .j}^{\ \ i}$ of these gradients are called Christoffel symbols of the coordinate system, that is,

$$\Gamma_{(k)\ .j}^{\ \ i} = \mathbf{e}^i \cdot \Gamma_{(k)}\mathbf{e}_j. \tag{2.180}$$

For this object, it is possible to show that

$$\Gamma_{(k).j}^{\ \ i} = \Gamma_{(j).k}^{\ \ i}, \tag{2.181}$$

and from the very definition of the gradient operation the following holds

$$\frac{\partial \mathbf{e}_i}{\partial x^j} = \Gamma_{(i).j}^{\ \ k}\mathbf{e}_k. \tag{2.182}$$

Employing the metric tensor associated with the coordinate system, the Christoffel symbols can be calculated through

$$\Gamma_{(i).j}^{\ \ k} = \frac{1}{2}g^{kr}(g_{ri,j} + g_{rj,i} - g_{ij,r}), \tag{2.183}$$

where we denote

$$g_{ij,r} = \frac{\partial g_{ij}}{\partial x^r}. \tag{2.184}$$

Furthermore, it can be verified that the Christoffel symbols of a coordinate system are equal to zero if and only if the natural basis is constant. These coordinates are said to be Cartesian (but not necessarily orthogonal).

Now let $\mathbf{u}$ be a differentiable vector field. The gradient, $\nabla\mathbf{u}$, results in a second-order tensor field. The different components of $\nabla\mathbf{u}$ with respect to the natural basis are called covariant derivatives. From this, we notice that there exist four kinds of covariant derivatives

$$u^i|_j = \mathbf{e}^i \cdot (\nabla\mathbf{u})\mathbf{e}_j, \tag{2.185}$$

$$u_i|_j = \mathbf{e}_i \cdot (\nabla\mathbf{u})\mathbf{e}_j, \tag{2.186}$$

$$u_i|^j = \mathbf{e}_i \cdot (\nabla\mathbf{u})\mathbf{e}^j, \tag{2.187}$$

$$u^i|^j = \mathbf{e}^i \cdot (\nabla\mathbf{u})\mathbf{e}^j. \tag{2.188}$$

In order to compute these components, consider expression (2.67), which in this case is

$$\nabla\mathbf{u} = \nabla(u^k\mathbf{e}_k) = \mathbf{e}_k \otimes \nabla u^k + u^k\nabla\mathbf{e}_k. \tag{2.189}$$

Hence, for instance, we obtain

$$\begin{aligned} u^i|_j &= \mathbf{e}^i \cdot (\mathbf{e}_k \otimes \nabla u^k + u^k\nabla\mathbf{e}_k)\mathbf{e}_j = (\nabla u^k \cdot \mathbf{e}_j)(\mathbf{e}^i \cdot \mathbf{e}_k) + u^k\mathbf{e}^i \cdot (\nabla\mathbf{e}_k)\mathbf{e}_j \\ &= \nabla u^i \cdot \mathbf{e}_j + u^k\Gamma_{(k).j}^{\ \ i} = \frac{\partial u^i}{\partial x^j} + u^k\Gamma_{(k).j}^{\ \ i}. \end{aligned} \tag{2.190}$$

Analogously, we obtain

$$u_i|_j = \frac{\partial u_i}{\partial x^j} - u_k\Gamma_{(i).j}^{\ \ k}. \tag{2.191}$$

It can be appreciated that the covariant derivatives are the same as the partial derivatives for all vector fields $\mathbf{u}$ if and only if the coordinate system is Cartesian. An equivalent result can be proven for tensor fields of any order.

Following the same ideas, we can express the divergence of a second-order tensor field. For example, the kth covariant component of the vector div $\mathbf{L}$, where $\mathbf{L}$ is a smooth second-order tensor field, can be put in terms of the covariant derivatives and Christoffel symbols, which yields

$$(\operatorname{div}\mathbf{L})^i = L^{ij}|_j = \frac{1}{\sqrt{g}}\frac{\partial}{\partial x^j}(\sqrt{g}L^{ij}) + L^{kj}\Gamma_{(k).j}^{\ \ \ i}, \tag{2.192}$$

where g denotes the determinant of the metric tensor g_{ij}.

Exercise 2.12 Prove that expression (2.192) is verified.

2.8 Complementary Reading

For the reader interested in increasing their knowledge of the topics briefly addressed in this chapter we recommend the following complementary material: [23], [47], [59], [60], [61], [84], [106], [132], [135], [161], [171], [174], [183], [209], [265], [299] and [300].

Part II

Variational Formulations in Mechanics

The goal of the second part of this book is to introduce the reader to the foundations and developments of variational formulations in the field of mechanics. To do this, Chapter 3 is devoted to the basic underlying ingredients that play a major role in variational formalism. This way of approaching the modeling of a physical phenomenon is called the *Method of Virtual Power*. From the basic kinematical setting, going through the mathematical duality and reaching the so-called *Principle of Virtual Power*, this variational structure provides a clear and elegant framework to unambiguously understand the interplay between model kinematical hypotheses and the way they shape the realm of forces and stresses. This establishes direct consequences in terms of governing equations, boundary conditions, jump conditions, and constitutive modeling, among others.

The modeling of the mechanics of deformable bodies in the most general scenario is addressed in Chapter 3. In addition, classical kinematical constraints considered over the bodies are discussed in detail, as well as issues related to different observational descriptions of the phenomena, namely Eulerian and Lagrangian descriptions, and their application in different specific problems. Chapters 4, 5 and 6 present the application of variational formulations in classical problems of mechanics in the infinitesimal strain regime, including hyperelastic behavior and its different formulations, as well as the modeling of plasticity and creep phenomena.

3

Method of Virtual Power

3.1 Introduction

Variational principles and methods have had a pivotal role in establishing a modern point of view for both theoretical and applied mechanics. As we will see in this chapter, a variational formalism for the basic laws that rule the behavior of continuum media is, in essence, the most natural and rigorous foundational framework.

The use of the variational formulation allows us to reduce to a single integral expression all the ingredients that partake in the mathematical formulation of the physical problem under analysis, namely, equilibrium equations, constitutive equations (material behavior), boundary conditions, initial conditions, and jump conditions, among others.

The local form[1] of the governing equations which rule the motion of a body spring as a mere consequence of the variational formulation. In turn, and this is extremely important from the perspective of applications in mechanics, the variational formulation naturally induces the most appropriate and mathematically sound structure in which numerical methods to obtain approximate solutions can be framed. These methods, called *variational methods*, allow approximate solutions to be obtained using the same unified mathematical context as the continuum problem, and facilitating the computational implementation regardless of the complexity of the mechanical problem.

Another fundamental aspect of the variational formulation in mechanics, when compared to the *classical* (local) form, is that it is capable of reconciling, within a unified fabric, different mechanical problems that are not apparently related to each other. This power of synthesis permits the basic hypotheses and essential aspects of the model which is being studied to be discriminated. Finally, the variational formulation provides, through the mathematical framework of functional analysis, the elements required for the study of the existence and uniqueness of solutions, as well as, in the numerical realm, *a priori* and *a posteriori* estimates of the error in approximate solutions.

In the light of these observations, we can conclude (see also [167]) that the variational formulation in mechanics has embraced a paramount role in the fabrication of mechanical models, becoming the most *natural* way of approaching their formulation.

1 This refers to the sense of the point-wise mathematical relation occurring in a continuum system.

Introduction to the Variational Formulation in Mechanics: Fundamentals and Applications, First Edition.
Edgardo O. Taroco, Pablo J. Blanco and Raúl A. Feijóo.

Starting with the Principle of Virtual Power,[2] admitted as a Fundamental Principle, in combination with the kinematics of continuum media and with the notion of mathematical duality between motion and forces, in this chapter it will be seen how the concepts of external forces and internal stresses a given body can be subjected to, and the concepts of mechanical equilibrium and compatibility in its local forms, emerge as consequences of this theoretical tapestry.

For instance, in the *classic* formulation of mechanics the existence of *external forces* represented by vector fields is accepted *a priori*. Then, it is referred to *forces per unit volume* and *forces per unit surface* as already acknowledged, and sometimes far-fetched, entities.

Nevertheless, there exists a second approach, that is the variational overture, to characterize the external forces that act over the given body. In this alternative framework, the external forces are reworked through the concept of duality between such forces and the motion actions performed over the body, and therefore this duality characterizes the power exerted in their execution. As will be seen in the forthcoming sections, this way to define the concept of force is more natural than the classical notion of force introduced by Newton because it makes use of a strong connection with the mechanical reality we live in: the existence of these forces is only revealed once motion actions are performed over the body.

In this sense, we recommend reading Chapters 3 and 4 in the book by C. Lanczos ([167]), the books by Duvaut and Lions ([76]), by Ekeland and Temam ([78]), by P. Germain ([117]), by Y. C. Fung ([111]), by M. Fremond ([109]), by J. T. Oden and coworkers ([220], [225–227], [230]), by S. G. Mikhlin ([198]), by P. D. Panagiotopoulos ([236]), by K. Rektorys ([250]), the works by P. Germain ([114–116], [118]), by M. A. Maugin ([189]), by G. Romano and coworkers ([253–262]), and the works [93–95], and [286].

3.2 Kinematics

3.2.1 Body and Deformations

The most singular physical characteristic of bodies is that they occupy regions of the Euclidean space $\mathscr{E}$. Even if a body can occupy several regions of the space at different moments, none of them is intrinsically associated with the body. However, we can always select any of these regions, which will be denoted by $\mathscr{B}$, and establish a bijective correspondence between each of the particles that compose the body and the point X occupied by the particle in the region $\mathscr{B}$.

Once this identification has been made, the body comes to be formally a region, denoted by $\mathscr{B}$, of $\mathscr{E}$. This region $\mathscr{B}$ is called the *reference configuration*. Notice that, by virtue of its own definition, innumerous reference configurations could be adopted for the same body. This reference configuration is, in general, selected in order to facilitate further analyses. Hence, for instance, if one wants to study a plate with initial deformations, it would be good practice to choose as the reference configuration the initial deformation-free (plane) configuration. This example shows us another

2 Initiated by Aristóteles around 300 years BC, formulated for the first time in the celebrated works by J. Bernoulli, and definitely established in the D'Alembert contributions.

characteristic of the reference configuration: such a configuration is not necessarily occupied by the body during its real motion.

Given the reference configuration $\mathscr{B}$, each point $X \in \mathscr{B}$ is also called a material point, and subdomains $\mathscr{P}$ from $\mathscr{B}$ are called parts of the body $\mathscr{B}$. Because of certain actions, whose origin will be discussed in due course, the body $\mathscr{B}$ can occupy different regions (also called configurations). Thus, the introduction of a real parameter, say $t \in [t_0, t_f]$, is mandatory in order to identify the region $\mathscr{B}_t$ occupied by the body, which corresponds to the configuration at instant t. It is worth highlighting that t is not necessarily related to physical time. In fact, in problems which are time-independent, parameter t will simply establish the order of precedence for these configurations.

In order to make the presentation simple, let us assume that we have adopted a point $O \in \mathscr{E}$, called the origin in the Euclidean space. In this manner, we can automatically identify any point $X \in \mathscr{E}$, with the corresponding vector defined by $\mathbf{X} = X - O$.

One of the most remarkable aspects of mechanics is the study of the deformation process a body undergoes. Physically, we say that a body deforms when it goes from a given configuration $\mathscr{B}$ to the so-called actual (or current, or also deformed) configuration $\mathscr{B}_t$. The deformation is thus fully characterized through the application

$$\begin{aligned} \mathscr{X}_t : \mathscr{B} &\rightarrow \mathscr{B}_t, \\ \mathbf{X} &\mapsto \mathbf{x} = \mathscr{X}_t(\mathbf{X}). \end{aligned} \tag{3.1}$$

This application has to meet some requirements for it to be acknowledged as the representation of a deformation process. It is mandatory to require that there exists no material interpenetration. In mathematical terms, this means that the mapping $\mathscr{X}_t$ is bijective. In other words, for each point $\mathbf{X}$ in $\mathscr{B}$ there will be a single point $\mathbf{x}$ in $\mathscr{B}_t$, and vice versa. In turn, since $\mathscr{X}_t$ is an application that maps points from $\mathscr{E}$ into points from $\mathscr{E}$, its tangent operator, that is, the gradient, denoted by $\nabla\mathscr{X}_t$, is a second-order tensor field. Recalling that $\det \nabla\mathscr{X}_t$ is related to the change of volume of a given geometrical object (see expression (1.160) and also [132]), it is reasonable to require that

$$\det \nabla\mathscr{X}_t > 0 \qquad \forall t \in [t_0, t_f], \tag{3.2}$$

since, otherwise, there is the possibility of mapping a piece of continuum into a point, which is physically inconsistent.

Hence, we call the deformation of a body, when changing from configuration $\mathscr{B}$ to $\mathscr{B}_t$, the application (3.1), which is bijective and satisfies

$$\begin{aligned} &\det \nabla\mathscr{X}_t(\mathbf{X}) > 0 \qquad \forall \mathbf{X} \in \mathscr{B}, \\ &\mathscr{X}_t(\mathscr{B}) = \mathscr{B}_t, \\ &\mathscr{X}_t(\partial\mathscr{B}) = \partial\mathscr{B}_t, \end{aligned} \tag{3.3}$$

where $\partial\mathscr{B}$ and $\partial\mathscr{B}_t$ represent the boundaries of the configurations $\mathscr{B}$ and $\mathscr{B}_t$, respectively.

This deformation mapping can be materialized through a vector field defined between the positions in space occupied by all the particles in the body $\mathscr{B}$ before and after the deformation, as shown in Figure 3.1.

Putting $\mathbf{x} = \mathscr{X}_t(\mathbf{X})$, we can then write

$$\mathbf{U}_t = \mathbf{U}_t(\mathbf{X}) = \mathbf{x} - \mathbf{X} = \mathscr{X}_t(\mathbf{X}) - \mathbf{X}, \tag{3.4}$$

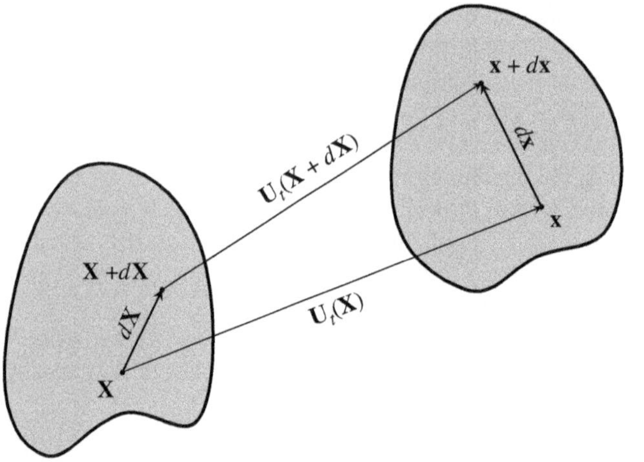

Figure 3.1 Mapping $\mathscr{X}_t$ characterizing the deformation of the body from configuration $\mathscr{B}$ to configuration $\mathscr{B}_t$.

that is,

$$\mathbf{x} = \mathbf{X} + \mathbf{U}_t(\mathbf{X}), \tag{3.5}$$

where we have introduced the vector field $\mathbf{U}_t$, which stands for the so-called *displacement field* relative to configuration $\mathscr{B}$. Evidently, $\mathbf{U}_t$ has to satisfy certain restrictions so that expressions (3.3) are verified.

Now, we define the second-order tensor field, denoted by $\mathbf{F}_t$, given by

$$\mathbf{F}_t = \nabla \mathscr{X}_t = \nabla \mathbf{X} + \nabla \mathbf{U}_t = \mathbf{I} + \nabla \mathbf{U}_t, \tag{3.6}$$

where $\mathbf{I}$ is the identity second-order tensor field. Tensor $\mathbf{F}_t$ represents the gradient, with respect to the material point $\mathbf{X}$, of the deformation mapping. Hereafter and for simplicity, unless stated otherwise, script t is dropped in the notation for the displacement field $\mathbf{U}_t$ (now simply $\mathbf{U}$) and in the deformation gradient $\mathbf{F}_t$ (now simply $\mathbf{F}$).

A deformation $\mathscr{X}$ is said to be homogeneous if its gradient is uniform (i.e. it does not depend on $\mathbf{X}$). Thus, any homogeneous deformation admits the following representation

$$\mathscr{X}(\mathbf{X}) = \mathscr{X}(\mathbf{X}_o) + \mathbf{F}(\mathbf{X} - \mathbf{X}_o) \qquad \forall \mathbf{X}, \mathbf{X}_o \in \mathscr{B}, \tag{3.7}$$

where $\mathbf{F} = \nabla \mathscr{X}$. Two classes of homogeneous deformations will be useful for forthcoming developments. First, it will be said that $\mathscr{X}$ is a translation if

$$\mathscr{X}(\mathbf{X}) = \mathbf{X} + \mathbf{U}, \tag{3.8}$$

where $\mathbf{U}$ is a uniform vector field which characterizes the translation. Second, it will be said that $\mathscr{X}$ is a rotation around a fixed point $\mathbf{X}_o$ if

$$\mathscr{X}(\mathbf{X}) = \mathbf{X}_o + \mathbf{R}(\mathbf{X} - \mathbf{X}_o), \tag{3.9}$$

where $\mathbf{R}$ is a uniform second-order skew-symmetric tensor field, which means that $\mathbf{R} \in$ *Rot*.

Let $\mathbf{X}_o \in \mathscr{B}$ be an arbitrary point, and denote by $\mathscr{S}(\mathbf{X}_o)$ a sufficiently small neighborhood of $\mathbf{X}_o$. Consider now that the mapping $\mathscr{X}$ is not homogeneous, then we can expand $\mathscr{X}$ as follows

$$\mathscr{X}(\mathbf{X}) = \mathscr{X}(\mathbf{X}_o) + \mathbf{F}(\mathbf{X}_o)(\mathbf{X} - \mathbf{X}_o) + o(\mathbf{X} - \mathbf{X}_o) \qquad \forall \mathbf{X} \in \mathscr{S}(\mathbf{X}_o). \tag{3.10}$$

From this, we conclude that, in the neighborhood of point $\mathbf{X}_o$, every deformation is, up to an error of order $o(\mathbf{X} - \mathbf{X}_o)$, a homogeneous deformation. Hereafter we remain in the context provided by (3.10), known as zero-order continua (also referred to as the classical theory of continuum mechanics). Higher-order theories are possible by incorporating into the developments subsequent terms arising from (3.10).

The next step consists of searching for a measure to further characterize the deformation. To do this, consider $\mathbf{X} = \mathbf{X}_o + d\mathbf{X}$ (see also Figure 3.1), which replaced in (3.10) leads to

$$d\mathbf{x} = \nabla\mathscr{X}(\mathbf{X}_o)d\mathbf{X} = \mathbf{F}d\mathbf{X}, \tag{3.11}$$

and therefore

$$d\mathbf{x} \cdot d\mathbf{x} = \mathbf{F}d\mathbf{X} \cdot \mathbf{F}d\mathbf{X} = \mathbf{F}^T\mathbf{F}d\mathbf{X} \cdot d\mathbf{X}. \tag{3.12}$$

Hence, a measure of the deformation of a given *fiber*, denoted by $d\mathbf{X}$, when the fiber goes to the actual configuration $d\mathbf{x}$ is given by

$$d\mathbf{x} \cdot d\mathbf{x} - d\mathbf{X} \cdot d\mathbf{X} = (\mathbf{F}^T\mathbf{F} - \mathbf{I})d\mathbf{X} \cdot d\mathbf{X} = 2\mathbf{E}d\mathbf{X} \cdot d\mathbf{X}, \tag{3.13}$$

where

$$\mathbf{E} = \frac{1}{2}(\mathbf{F}^T\mathbf{F} - \mathbf{I}) = \frac{1}{2}(\nabla\mathbf{U} + (\nabla\mathbf{U})^T + (\nabla\mathbf{U})^T\nabla\mathbf{U}), \tag{3.14}$$

is designated as the Green deformation tensor.[3]

So far, we have adopted a frame of reference to describe the deformation known as the Lagrangian description, in which the functional dependence of the fields is expressed in terms of the material point $\mathbf{X}$ along the whole deformation process, that is, regardless of the parameter t. Given the properties specified to the deformation mapping $\mathscr{X}_t$, it follows that there exists the inverse mapping, denoted by $\mathscr{X}_t^{-1}$, which is also sufficiently smooth. Then, our line of thought can be inverted, expressing the physical fields in terms of the point $\mathbf{x}$, related to $\mathbf{X}$ through such inverse mapping, that is $\mathbf{X} = \mathscr{X}_t^{-1}(\mathbf{x})$.

It is important to notice here that, as the parameter t runs, point $\mathbf{x}$ is the position occupied by different material points $\mathbf{X}$. This kind of description is termed the Eulerian description or spatial description. Hence, instead of placing the analysis on top of material points $\mathbf{X}$ from the material configuration $\mathscr{B}$, the analysis is placed in the corresponding spatial points $\mathbf{x}$ (under the transformation $\mathscr{X}_t$) from configuration $\mathscr{B}_t$, and we obtain the spatial description, in contrast to the material description (also reference description) adopted before.

Introducing this description, and following a line of thought analogous to that already employed with the Lagrangian description, we have

$$\mathbf{X} = \mathbf{x} - \mathbf{u}(\mathbf{x}), \tag{3.15}$$

3 Some authors also call it the Lagrange deformation tensor even if, according to Truesdell (see [299]), tensor $\mathbf{E}$ was firstly introduced by Green in 1841 and by St. Venant in 1844.

where the spatial description of the displacement field $\mathbf{u}$[4] has been introduced. Then we obtain

$$d\mathbf{X} = \mathbf{F}^{-1} d\mathbf{x}, \tag{3.16}$$

with

$$(\mathbf{F}^{-1})_s = \text{grad}\, \mathcal{X}^{-1} = \text{grad}\, \mathbf{x} - \text{grad}\, \mathbf{u} = \mathbf{I} - \text{grad}\, \mathbf{u}, \tag{3.17}$$

where grad $(\cdot)$ denotes the gradient with respect to the spatial variable $\mathbf{x}$ and notation $(\cdot)_s$ stands for the spatial description of the field $(\cdot)$. Hence, $\mathbf{F}(\mathbf{X})$ is a material field, and so its inverse $\mathbf{F}^{-1}(\mathbf{X})$. Then, this gives $(\mathbf{F}^{-1})_s(\mathbf{x}) = \mathbf{F}^{-1}(\mathcal{X}_t^{-1}(\mathbf{x}))$ (see also Section 3.2.2). From previous expressions we have

$$d\mathbf{x} \cdot d\mathbf{x} - d\mathbf{X} \cdot d\mathbf{X} = (\mathbf{I} - \mathbf{F}^{-T}\mathbf{F}^{-1}) d\mathbf{x} \cdot d\mathbf{x} = 2\mathbf{A} d\mathbf{x} \cdot d\mathbf{x}, \tag{3.18}$$

where $\mathbf{A}$ is known as the Almansi deformation tensor, and is given by

$$\mathbf{A} = \frac{1}{2}((\text{grad}\, \mathbf{u})^T + \text{grad}\, \mathbf{u} - (\text{grad}\, \mathbf{u})^T \text{grad}\, \mathbf{u}), \tag{3.19}$$

which is the spatial description of the measure of deformation.

Moreover, assuming now that the displacements and the corresponding gradients are such that

$$\|\mathbf{u}\| < \epsilon, \quad \|\nabla \mathbf{u}\| < \epsilon \quad \text{and} \quad \|\text{grad}\, \mathbf{u}\| < \epsilon, \tag{3.20}$$

where $\epsilon > 0$ is sufficiently small, the high-order terms arising in (3.14) and (3.19), which are given by $(\nabla \mathbf{U})^T \nabla \mathbf{U}$ or by $(\text{grad}\, \mathbf{u})^T \text{grad}\, \mathbf{u}$, accordingly, can be neglected in comparison with the linear terms $\nabla \mathbf{U}$ and grad $\mathbf{u}$, respectively, and, furthermore, it is possible to assume that the displacement fields under both material and spatial descriptions coincide. As a consequence, and using expressions (3.13) and (3.18), it is verified that

$$\mathbf{E} d\mathbf{X} \cdot d\mathbf{X} = \mathbf{A} d\mathbf{x} \cdot d\mathbf{x} = (\mathbf{A}_m \mathbf{F} d\mathbf{X}) \cdot (\mathbf{F} d\mathbf{X}) = \mathbf{F}^T \mathbf{A}_m \mathbf{F} d\mathbf{X} \cdot d\mathbf{X}, \tag{3.21}$$

where $(\cdot)_m$ indicates the material description of the spatial field $(\cdot)$. In this case, the material description of the spatial field $\mathbf{A}(\mathbf{x})$ is $\mathbf{A}_m(\mathbf{X}) = \mathbf{A}(\mathcal{X}_t(\mathbf{X}))$ (see also Section 3.2.2). Now, observe that

$$\begin{aligned} \mathbf{E} = \mathbf{F}^T \mathbf{A}_m \mathbf{F} &= (\mathbf{I} + (\nabla \mathbf{U})^T) \mathbf{A}_m (\mathbf{I} + \nabla \mathbf{U}) \\ &= \mathbf{A}_m + (\nabla \mathbf{U})^T \mathbf{A}_m + \mathbf{A}_m \nabla \mathbf{U} + (\nabla \mathbf{U})^T \mathbf{A}_m \nabla \mathbf{U} = \mathbf{A}_m + o(\mathbf{A}_m). \end{aligned} \tag{3.22}$$

This result shows us that, under the hypotheses of small displacements and small gradients, the Green and the Almansi tensor fields differ only if high-order terms are taken into consideration. Neglecting these terms, we reach the conclusion that $\nabla(\cdot) = \text{grad}\,(\cdot)$, that is, material and spatial gradients agree, and therefore

$$\mathbf{E} = \mathbf{A} = \varepsilon = \frac{1}{2}(\nabla \mathbf{u} + (\nabla \mathbf{u})^T), \tag{3.23}$$

which is known as the infinitesimal deformation tensor field.

---

4 Note here that the relation $\mathbf{u}(\mathbf{x})|_{\mathbf{x}=\mathcal{X}(\mathbf{X})} = \mathbf{u}(\mathcal{X}(\mathbf{X})) = \mathbf{U}(\mathbf{X})$ holds.

An infinitesimal deformation is said to be rigid if the deformation measure given by the tensor field ε is nullified for all points of the body. Thus, for a rigid deformation the following is verified

$$\nabla \mathbf{u} = -(\nabla \mathbf{u})^T. \tag{3.24}$$

That is, the gradient of the displacement field corresponding to an infinitesimal rigid deformation is a skew-symmetric second-order tensor field.

This allows us to introduce the following definition: an infinitesimal displacement field is called rigid if its gradient is uniform and skew-symmetric or, equivalently, if it admits the following representation

$$\mathbf{u}(\mathbf{X}) = \mathbf{u}(\mathbf{X}_o) + \mathbf{W}(\mathbf{X} - \mathbf{X}_o) \qquad \forall \mathbf{X}, \mathbf{X}_o \in \mathscr{B}, \tag{3.25}$$

where $\mathbf{W}$ is a skew-symmetric uniform tensor. In turn, introducing the axial vector of $\mathbf{W}$, here denoted by the uniform vector field $\mathbf{w}$ (see expression (1.173)), the previous statement is equivalent to

$$\mathbf{u}(\mathbf{X}) = \mathbf{u}(\mathbf{X}_o) + \mathbf{w} \times (\mathbf{X} - \mathbf{X}_o) \qquad \forall \mathbf{X}, \mathbf{X}_o \in \mathscr{B}. \tag{3.26}$$

Noting that

$$\begin{aligned}(\mathbf{u}(\mathbf{X}) - \mathbf{u}(\mathbf{X}_o)) \cdot (\mathbf{X} - \mathbf{X}_o) &= \mathbf{W}(\mathbf{X} - \mathbf{X}_o) \cdot (\mathbf{X} - \mathbf{X}_o) \\ &= -\mathbf{W}(\mathbf{X} - \mathbf{X}_o) \cdot (\mathbf{X} - \mathbf{X}_o) = 0,\end{aligned} \tag{3.27}$$

since $\mathbf{W}$ is skew-symmetric, this gives

$$\mathbf{u}(\mathbf{X}) - \mathbf{u}(\mathbf{X}_o) \perp \mathbf{X} - \mathbf{X}_o. \tag{3.28}$$

In other words, for any infinitesimal rigid deformation, the displacement of point $\mathbf{X}$, relative to $\mathbf{X}_o$, is orthogonal to the vector $\mathbf{X} - \mathbf{X}_o$.

3.2.2 Motion: Deformation Rate

We define the motion of a body $\mathscr{B}$ as the uniparametric family of applications

$$\begin{aligned}\mathscr{X} : \mathscr{B} \times [t_0, t_f] &\to \mathscr{E}, \\ (\mathbf{X}, t) &\mapsto \mathbf{x} = \mathscr{X}(\mathbf{X}, t),\end{aligned} \tag{3.29}$$

where $\mathscr{X}$ is a deformation.[5] Then, the point

$$\mathbf{x} = \mathscr{X}(\mathbf{X}, t), \tag{3.30}$$

is the place occupied by particle $\mathbf{X}$ at time t. Then, the region occupied by the body at time t is given by

$$\mathscr{B}_t = \mathscr{X}(\mathscr{B}, t). \tag{3.31}$$

As discussed before, in some cases it is convenient to work with variables $(\mathbf{x}, t)$ instead of $(\mathbf{X}, t)$. This is formalized by introducing the concept of trajectory

$$\mathscr{T} = \{(\mathbf{x}, t); \mathbf{x} \in \mathscr{B}_t, t \in [t_0, t_f]\}. \tag{3.32}$$

5 Note the notational equivalence $\mathscr{X}_t(\mathbf{X}) = \mathscr{X}(\mathbf{X}, t)$, see (3.1).

Since for each t the application $\mathcal{X}(\cdot, t) : \mathcal{B} \to \mathcal{B}_t$ is bijective, there exists the inverse mapping $\mathcal{X}_t^{-1}$

$$\begin{aligned} \mathcal{X}_t^{-1} : \mathcal{B}_t &\to \mathcal{B}, \\ \mathbf{x} &\mapsto \mathcal{X}^{-1}(\mathbf{x}, t) = \mathbf{X}, \end{aligned} \tag{3.33}$$

which permits the following identities to be obtained

$$\mathcal{X}(\mathcal{X}^{-1}(\mathbf{x}, t), t) = \mathbf{x}, \tag{3.34}$$

$$\mathcal{X}^{-1}(\mathcal{X}(\mathbf{X}, t), t) = \mathbf{X}. \tag{3.35}$$

Exploiting these applications, we can write any field either in terms of $(\mathbf{X}, t)$, and in such a case we obtain the material description of the field, or in terms of $(\mathbf{x}, t)$, and in such a case we have the spatial description of the field. Let, for example, the material field be

$$\phi : \mathcal{B} \times [t_0, t_f] \mapsto \phi(\mathbf{X}, t), \tag{3.36}$$

then the spatial description of ϕ, denoted shortly by ϕ_s, is

$$\phi_s(\mathbf{x}, t) = \phi(\mathcal{X}^{-1}(\mathbf{x}, t), t). \tag{3.37}$$

Likewise, the material description of the spatial field $\varphi = \varphi(\mathbf{x}, t)$ is given by

$$\varphi_m(\mathbf{X}, t) = \varphi(\mathcal{X}(\mathbf{X}, t), t). \tag{3.38}$$

Evidently, from the previous relations it follows that

$$(\phi_s)_m = \phi, \tag{3.39}$$

$$(\varphi_m)_s = \varphi. \tag{3.40}$$

In this manner, it will be necessary to distinguish derivatives with respect to material variables from the derivatives with respect to spatial variables. In fact, we may have the following operations.

- For the material field ϕ, the expressions

$$\dot{\phi}(\mathbf{X}, t) = \frac{\partial \phi}{\partial t}(\mathbf{X}, t)\bigg|_{\mathbf{X}\ \text{fixed}}, \tag{3.41}$$

$$\nabla \phi(\mathbf{X}, t) = \nabla_{\mathbf{X}} \phi(\mathbf{X}, t)\big|_{t\ \text{fixed}}, \tag{3.42}$$

  represent, respectively, the derivatives with respect to t keeping the material point $\mathbf{X}$ fixed, and the gradient with respect to $\mathbf{X}$ for a fixed t. These derivatives are known, respectively, as the material rate (or material time derivative, or total time derivative) and as the material gradient of ϕ.
- Given the spatial field φ, the expressions

$$\varphi'(\mathbf{x}, t) = \frac{\partial \varphi}{\partial t}(\mathbf{x}, t)\bigg|_{\mathbf{x}\ \text{fixed}}, \tag{3.43}$$

$$\text{grad}\, \varphi(\mathbf{x}, t) = \nabla_{\mathbf{x}} \varphi(\mathbf{x}, t)\big|_{t\ \text{fixed}}, \tag{3.44}$$

  are called, respectively, the spatial rate (or spatial time derivative) and the *spatial gradient* of φ.

Let us see some specific examples which will be useful in forthcoming developments. Consider the motion (3.29), then

$$\dot{\mathbf{x}} = \frac{\partial \mathscr{X}}{\partial t}(\mathbf{X}, t)\bigg|_{\mathbf{X}\ \text{fixed}}, \tag{3.45}$$

is the material description of the velocity of particle **X**, while

$$\ddot{\mathbf{x}} = \frac{\partial^2 \mathscr{X}}{\partial t^2}(\mathbf{X}, t)\bigg|_{\mathbf{X}\ \text{fixed}}, \tag{3.46}$$

is the material description of the acceleration of the same particle.

Making use of the inverse $\mathscr{X}^{-1}$, we can obtain the spatial description of the velocity, which will be denoted by $\mathbf{v}(\mathbf{x}, t)$. In fact

$$\mathbf{v}(\mathbf{x}, t) = \frac{\partial \mathscr{X}}{\partial t}(\mathscr{X}^{-1}(\mathbf{x}, t), t). \tag{3.47}$$

Observe that $\mathbf{v}(\mathbf{x}, t)$ is the velocity of the material point **X**, which at time t is positioned at **x** in the Euclidean space.

Analogously, we can calculate the time derivative of the inverse mapping (3.33), defined by

$$\mathbf{X}' = \frac{\partial \mathscr{X}^{-1}}{\partial t}(\mathbf{x}, t)\bigg|_{\mathbf{x}=\mathscr{X}(\mathbf{X},t)\ \text{fixed}}, \tag{3.48}$$

which is termed the inverse velocity [133, 188], and for which the expression of its material description is

$$\mathbf{V}(\mathbf{X}, t) = \frac{\partial \mathscr{X}^{-1}}{\partial t}(\mathscr{X}(\mathbf{X}, t), t). \tag{3.49}$$

This inverse velocity is understood as the time rate of change of the velocity of different particles which go through the same spatial location **x** at time t.

Note that from (3.45) and (3.48), and using (3.34), gives

$$\begin{aligned}
\mathbf{0} = \frac{\partial \mathbf{x}}{\partial t}\bigg|_{\mathbf{x}\ \text{fixed}} &= \frac{\partial}{\partial t}\mathscr{X}(\mathscr{X}^{-1}(\mathbf{x}, t), t)\bigg|_{\mathbf{x}\ \text{fixed}} \\
&= \frac{\partial \mathscr{X}}{\partial t}(\mathbf{X}, t)\bigg|_{\mathbf{X}\ \text{fixed}} + \nabla \mathscr{X}(\mathbf{X}, t)\frac{\partial \mathscr{X}^{-1}}{\partial t}(\mathbf{x}, t)\bigg|_{\mathbf{x}=\mathscr{X}(\mathbf{X},t)\ \text{fixed}} \\
&= \dot{\mathbf{x}}(\mathbf{X}, t) + \mathbf{F}(\mathbf{X}, t)\mathbf{X}'(\mathscr{X}(\mathbf{X}, t), t).
\end{aligned} \tag{3.50}$$

This implies the following relation between the material description of the real velocity of particles and the inverse velocity

$$\mathbf{v}_m = -\mathbf{F}\mathbf{V}. \tag{3.51}$$

In addition, in many applications it is necessary to calculate the material time derivative (total time derivative) of a spatial field $\varphi = \varphi(\mathbf{x}, t)$. As the name indicates, we need to look for the time rate of change of the quantity φ keeping the material point **X** constant. To perform this calculation, first we obtain the material description of the spatial field φ, and then we calculate the material rate and, finally, the result is expressed in its spatial description. This amounts to performing the following steps

$$\dot{\varphi} = \left(\dot{\overline{(\varphi_m)}}\right)_s, \tag{3.52}$$

or, in extended form,

$$\dot{\varphi}(\mathbf{x}, t) = \frac{\partial \varphi}{\partial t}(\mathscr{X}(\mathbf{X}, t), t)\bigg|_{\mathbf{X}=\mathscr{X}^{-1}(\mathbf{x},t)}. \tag{3.53}$$

It is possible to demonstrate (see [132, page 62]) that the material rate is inter-changeable with the material and spatial transformations, that is,

$$(\dot{\phi})_s = \dot{\overline{(\phi_s)}} = \dot{\phi}_s, \tag{3.54}$$

$$(\dot{\varphi})_m = \dot{\overline{(\varphi_m)}} = \dot{\varphi}_m. \tag{3.55}$$

In particular, assuming the spatial field φ to be the spatial description of the velocity field $\mathbf{v}$ we have, as a consequence of the previous result, that

$$(\dot{\mathbf{v}})_m = \dot{\overline{(\mathbf{v}_m)}} = \ddot{\mathbf{x}}, \tag{3.56}$$

where $\dot{\mathbf{v}}$ is the spatial description of the acceleration.

Using the chain rule, and with ϕ, $\mathbf{u}$ and $\mathbf{T}$ being sufficiently smooth scalar, vector, and second-order tensor fields, respectively, the following relations are verified

$$\dot{\phi} = \phi' + \text{grad}\, \phi \cdot \mathbf{v}, \tag{3.57}$$

$$\dot{\mathbf{u}} = \mathbf{u}' + (\text{grad}\, \mathbf{u})\mathbf{v}, \tag{3.58}$$

$$\dot{\mathbf{T}} = \mathbf{T}' + (\text{grad}\, \mathbf{T})\mathbf{v}. \tag{3.59}$$

Indeed, using the chain rule we obtain

$$\begin{aligned}
\dot{\phi}(\mathbf{x}, t) &= \left(\frac{\partial \phi}{\partial t}(\mathscr{X}(\mathbf{X}, t), t)\bigg|_{\mathbf{X}\ \text{fixed}} \right)_s \\
&= \left(\frac{\partial \phi}{\partial t}(\mathscr{X}(\mathbf{X}, t), t)\bigg|_{\mathbf{x}=\mathscr{X}(\mathbf{X},t)\ \text{fixed}} + \text{grad}\, \phi(\mathscr{X}(\mathbf{X}, t), t) \cdot \frac{\partial \mathscr{X}}{\partial t}(\mathbf{X}, t)\bigg|_{\mathbf{X}\ \text{fixed}} \right)\bigg|_{\mathbf{X}=\mathscr{X}^{-1}(\mathbf{x},t)} \\
&= \phi'(\mathbf{x}, t) + \text{grad}\, \phi(\mathbf{x}, t) \cdot \mathbf{v}(\mathbf{x}, t),
\end{aligned} \tag{3.60}$$

and therefore (3.57) is proved. Now, let $\mathbf{a}$ be an arbitrary constant vector field. Putting $\phi = \mathbf{u} \cdot \mathbf{a}$ into (3.57) we obtain

$$\begin{aligned}
\dot{\mathbf{u}} \cdot \mathbf{a} &= \dot{\overline{(\mathbf{u} \cdot \mathbf{a})}} = (\mathbf{u} \cdot \mathbf{a})' + [\text{grad}\, (\mathbf{u} \cdot \mathbf{a})] \cdot \mathbf{v} \\
&= \mathbf{u}' \cdot \mathbf{a} + (\text{grad}\, \mathbf{u})^T \mathbf{a} \cdot \mathbf{v} = [\mathbf{u}' + (\text{grad}\, \mathbf{u})\mathbf{v}] \cdot \mathbf{a},
\end{aligned} \tag{3.61}$$

and so (3.58) is verified. In a similar manner, considering $\mathbf{u} = \mathbf{T}\mathbf{a}$ in (3.58), and using the results derived in Chapters 1 and 2, yields

$$\begin{aligned}
\dot{\mathbf{T}}\mathbf{a} &= \dot{\overline{(\mathbf{T}\mathbf{a})}} = (\mathbf{T}\mathbf{a})' + [\text{grad}\, (\mathbf{T}\mathbf{a})]\mathbf{v} \\
&= \mathbf{T}'\mathbf{a} + [(\text{grad}\, \mathbf{T})^t \mathbf{a}]\mathbf{v} = [\mathbf{u}' + (\text{grad}\, \mathbf{T})\mathbf{v}]\mathbf{a},
\end{aligned} \tag{3.62}$$

from which (3.59) follows directly.

For the particular case when we consider the vector field $\mathbf{u}$ to be the velocity field $\mathbf{v}$ we reach

$$\dot{\mathbf{v}} = \mathbf{v}' + (\text{grad}\, \mathbf{v})\mathbf{v}. \tag{3.63}$$

Importantly, we note that by using these previous expressions it is possible to relate the material and spatial rates for the physical fields.

A similar relation can be established between the material and spatial gradients. In effect, let φ and $\mathbf{u}$ be a scalar and a vector field, respectively, in their spatial descriptions. Then, we can write

$$\nabla\varphi_m = \mathbf{F}^T(\text{grad}\,\varphi)_m, \tag{3.64}$$

$$\nabla\mathbf{u}_m = (\text{grad}\,\mathbf{u})_m\mathbf{F}, \tag{3.65}$$

where $\mathbf{F}$ is the deformation gradient tensor. These results can be proved straightforwardly through the chain rule in the following expressions

$$\varphi_m(\mathbf{X},t) = (\varphi(\mathbf{x},t))_m = \varphi(\mathcal{X}(\mathbf{X},t),t), \tag{3.66}$$

$$\mathbf{u}_m(\mathbf{X},t) = (\mathbf{u}(\mathbf{x},t))_m = \mathbf{u}(\mathcal{X}(\mathbf{X},t),t), \tag{3.67}$$

then

$$\nabla\varphi_m = (\nabla\mathcal{X})^T(\text{grad}\,\varphi)_m = \mathbf{F}^T(\text{grad}\,\varphi)_m, \tag{3.68}$$

$$\nabla\mathbf{u}_m = (\text{grad}\,\mathbf{u})_m\nabla\mathcal{X} = (\text{grad}\,\mathbf{u})_m\mathbf{F}. \tag{3.69}$$

An alternative (more operational) strategy to prove these results is shown next. By definition we have

$$d\varphi = (\text{grad}\,\varphi)\cdot d\mathbf{x}, \tag{3.70}$$

$$d\mathbf{u} = (\text{grad}\,\mathbf{u})d\mathbf{x}, \tag{3.71}$$

however, we know that $d\mathbf{x} = \mathbf{F}d\mathbf{X}$, and so we write

$$(d\varphi)_m = (\text{grad}\,\varphi)_m\cdot\mathbf{F}d\mathbf{X} = \mathbf{F}^T(\text{grad}\,\varphi)_m\cdot d\mathbf{X}, \tag{3.72}$$

$$(d\mathbf{u})_m = (\text{grad}\,\mathbf{u})_m\mathbf{F}d\mathbf{X}. \tag{3.73}$$

In turn, this is

$$(d\varphi)_m = \nabla\varphi_m\cdot d\mathbf{X}, \tag{3.74}$$

$$(d\mathbf{u})_m = \nabla\mathbf{u}_m d\mathbf{X}, \tag{3.75}$$

and comparing (3.72) with (3.74), and (3.73) with (3.75) the results follow.

As already discussed, the rigid motion is a characteristic mapping which appears several times in continuum mechanics. A motion is said to be rigid, for all time t, if

$$\frac{\partial}{\partial t}\|\mathcal{X}(\mathbf{X},t) - \mathcal{X}(\mathbf{X}_o,t)\| = \mathbf{0} \qquad \forall\mathbf{X},\mathbf{X}_o\in\mathscr{B}. \tag{3.76}$$

In other words, the motion $\mathcal{X}$ is rigid if the distance between any two material points of the body remains unchanged in time.

Let $\mathcal{X}$ be a motion and $\mathbf{v}$ the corresponding velocity field, that is,

$$\begin{aligned} &\mathbf{x} = \mathcal{X}(\mathbf{X},t), \qquad \mathbf{X}\in\mathscr{B},\ t\in\mathbb{R}, \\ &\mathbf{v}(\mathbf{x},t) = \frac{\partial\mathcal{X}}{\partial t}(\mathbf{X},t)\bigg|_{\mathbf{X}=\mathcal{X}^{-1}(\mathbf{x},t)}, \end{aligned} \tag{3.77}$$

then the following propositions are equivalent (see [132, page 70] for these proofs):

- $\mathscr{X}$ is rigid
- for each t, $\mathbf{v}(\cdot, t)$ has the structure of an infinitesimal rigid displacement in $\mathscr{B}_t$, that is, $\mathbf{v}$ admits the representation

$$\mathbf{v}(\mathbf{x}, t) = \mathbf{v}(\mathbf{y}, t) + \mathbf{W}(t)(\mathbf{x} - \mathbf{y}), \tag{3.78}$$

  for all $\mathbf{x}, \mathbf{y} \in \mathscr{B}_t$, and where $\mathbf{W}(t)$ is a skew-symmetric second-order tensor not depending on $\mathbf{x}$
- the spatial field $\mathbf{L} = \operatorname{grad} \mathbf{v}$ is skew-symmetric for all $(\mathbf{x}, t) \in \mathscr{T}$.

If the velocity field is not rigid, and for any point $\mathbf{x} \in \mathscr{B}_t$ close enough to another point $\mathbf{y} \in \mathscr{B}_t$, we have

$$\mathbf{v}(\mathbf{x}, t) = \mathbf{v}(\mathbf{y}, t) + \operatorname{grad} \mathbf{v}(\mathbf{y}, t)(\mathbf{x} - \mathbf{y}) + o(\mathbf{x} - \mathbf{y}). \tag{3.79}$$

Denoting by $\mathbf{L} = \operatorname{grad} \mathbf{v}(\mathbf{y}, t)$, and recalling that we can always decompose a second-order tensor field into symmetric and skew-symmetric components, which are given by

$$\mathbf{D} = \frac{1}{2}(\mathbf{L} + \mathbf{L}^T) = \frac{1}{2}(\operatorname{grad} \mathbf{v} + (\operatorname{grad} \mathbf{v})^T) = (\operatorname{grad} \mathbf{v})^s, \tag{3.80}$$

$$\mathbf{W} = \frac{1}{2}(\mathbf{L} - \mathbf{L}^T) = \frac{1}{2}(\operatorname{grad} \mathbf{v} - (\operatorname{grad} \mathbf{v})^T) = (\operatorname{grad} \mathbf{v})^a, \tag{3.81}$$

we have

$$\mathbf{L} = \mathbf{D} + \mathbf{W}. \tag{3.82}$$

Hence, expression (3.79) can be rewritten as

$$\mathbf{v}(\mathbf{x}, t) = \mathbf{v}(\mathbf{y}, t) + \mathbf{W}(\mathbf{y}, t)(\mathbf{x} - \mathbf{y}) + \mathbf{D}(\mathbf{y}, t)(\mathbf{x} - \mathbf{y}) + o(\mathbf{x} - \mathbf{y}). \tag{3.83}$$

Therefore, in the neighborhood of a point $\mathbf{y}$, and within an error of the order of $o(\mathbf{x} - \mathbf{y})$, the velocity field $\mathbf{v}$ is the sum of a rigid velocity field and a velocity field of the form $\mathbf{D}(\mathbf{y}, t)(\mathbf{x} - \mathbf{y})$. Furthermore, it is rather direct to show that the second-order tensor $\mathbf{D}$ (the symmetric component of the velocity gradient spatial tensor field) is associated with the rate of change of the squared length of an infinitesimal fiber, in the actual configuration $\mathscr{B}_t$, at point $\mathbf{y}$ and at time t. This is why the spatial tensor field $\mathbf{D}$ has been termed the deformation rate. In order to see this, consider first

$$\dot{\overline{(d\mathbf{x} \cdot d\mathbf{x})}} = 2d\mathbf{x} \cdot \dot{\overline{(d\mathbf{x})}} = 2d\mathbf{x} \cdot (\dot{\mathbf{F}} d\mathbf{X})_s. \tag{3.84}$$

Second, using (3.65) we have

$$\dot{\mathbf{F}} = \frac{\partial \nabla \mathscr{X}}{\partial t}(\mathbf{X}, t) = \nabla \frac{\partial \mathscr{X}}{\partial t}(\mathbf{X}, t) = \nabla \dot{\mathbf{x}} = \nabla \mathbf{v}_m = (\operatorname{grad} \mathbf{v})_m \mathbf{F}, \tag{3.85}$$

that is,

$$\dot{\mathbf{F}} = \mathbf{L}_m \mathbf{F}, \tag{3.86}$$

which, when replaced in (3.84), leads to

$$\dot{\overline{(d\mathbf{x} \cdot d\mathbf{x})}} = 2d\mathbf{x} \cdot (\mathbf{L}_m \mathbf{F} d\mathbf{X})_s = 2d\mathbf{x} \cdot \mathbf{L} d\mathbf{x} = 2d\mathbf{x} \cdot (\mathbf{D} + \mathbf{W}) d\mathbf{x} = 2\mathbf{D} d\mathbf{x} \cdot d\mathbf{x}, \tag{3.87}$$

and the statement is thus demonstrated.

Another class of motion to be used in forthcoming developments is the isochoric motion, that is, the motion which preserves the volume. Let $\mathscr{X}$ be the relative motion to the reference configuration $\mathscr{B}$. For any given part $\mathscr{P}$ of $\mathscr{B}$ we have that

$$\mathscr{P}_t = \mathscr{X}(\mathscr{P}, t), \tag{3.88}$$

is the region which will be occupied by part $\mathscr{P}$ at time t. The volume of this part is given by

$$\text{vol}(\mathscr{P}_t) = \int_{\mathscr{P}_t} dV. \tag{3.89}$$

Recalling the meaning of $\det \mathbf{F} = \det \nabla \mathscr{X}$, this calculation can be carried out in the reference configuration,[6] so we change variables and write

$$\text{vol}\,(\mathscr{P}_t) = \int_{\mathscr{P}} \det \mathbf{F}\, dV. \tag{3.90}$$

By definition, a motion is said to be isochoric if, for every part $\mathscr{P}_t$ of $\mathscr{B}_t$ and for all time t, the volume remains invariant, that is,

$$\overline{(\text{vol}(\mathscr{P}_t))}^{\cdot} = 0. \tag{3.91}$$

Thus

$$\overline{(\text{vol}(\mathscr{P}_t))}^{\cdot} = \overline{\left(\int_{\mathscr{P}} \det \mathbf{F}\, dV\right)}^{\cdot} = \int_{\mathscr{P}} \overline{(\det \mathbf{F})}^{\cdot}\, dV. \tag{3.92}$$

Considering the derivative of the determinant given by (2.46), and using (3.86), yields

$$\begin{aligned}\overline{(\det \mathbf{F})}^{\cdot} &= (\det \mathbf{F})\text{tr}\,(\dot{\mathbf{F}}\mathbf{F}^{-1}) = (\det \mathbf{F})\text{tr}\,\mathbf{L}_m \\ &= (\det \mathbf{F})(\text{tr}\,(\text{grad}\,\mathbf{v}))_m = (\det \mathbf{F})(\text{div}\,\mathbf{v})_m,\end{aligned} \tag{3.93}$$

and hence,

$$\overline{(\text{vol}(\mathscr{P}_t))}^{\cdot} = \int_{\mathscr{P}} (\text{div}\,\mathbf{v})_m \det \mathbf{F}\, dV = \int_{\mathscr{P}_t} \text{div}\,\mathbf{v}\, dV. \tag{3.94}$$

Moreover, since for an isochoric motion the previous expression must be nullified for every part $\mathscr{P}_t$, locally we obtain

$$\text{div}\,\mathbf{v} = \text{tr}\,\mathbf{L} = \text{tr}\,\mathbf{D} = 0. \tag{3.95}$$

3.2.3 Motion Actions: Kinematical Constraints

Previous sections were committed to the presentation of the basic concepts related to the kinematical description of deformable bodies. We have seen that all possible configurations $\mathscr{B}_t$ a body can reach in $\mathscr{E}$ can be in one-to-one correspondence with the displacement fields with respect to a given reference configuration $\mathscr{B}$.

In other words, given the configuration $\mathscr{B}_t$ and $t \in [t_0, t_f]$, such a configuration can be obtained from the displacement field $\mathbf{U}_t$ defined in $\mathscr{B}$. Therefore, it is absolutely

6 See [132, page 51]. For a rigorous proof of the first expression of equation (14) in this work, see [60, pages 247–254].

equivalent either to make reference to configuration $\mathscr{B}_t$ or to refer to the configuration associated with $\mathbf{U}_t$, provided the reference configuration $\mathscr{B}$ is known. Also, this is equivalent to referring to the spatial description of the displacement field $\mathbf{U}_t$, which is denoted by $\mathbf{u}_t$, and that is defined in $\mathscr{B}_t$.

The set of all possible configurations the body can take define a vector space that we call $\mathscr{U}$. This space is equipped with a topology suitable for the mechanical problem under study. In this sense, it is important to highlight that this part of the present document lies fundamentally in the algebraic realm, which allows us to recurrently resort to a geometric treatment of the concepts and notions which will allow us to clearly expose the mechanical foundations without obscuring the presentation with excessively intricate technical aspects.

Consider the body in the configuration $\mathbf{u}_t \in \mathscr{U}$. Then, an arbitrary motion from this configuration $\mathbf{u}_t$ is characterized by an uniparametric family of possible configurations $\mathbf{u}_t(\mathbf{x}, \tau)$, $\tau \in [t, t_f]$, such that $\mathbf{u}_t(\mathbf{x}, \tau)$ at $\tau = t$ coincides with $\mathbf{u}_t$.

For each possible motion from configuration $\mathbf{u}_t$ we can attach, at time $\tau = t$, a velocity field $\mathbf{v}$ (in its spatial description) which is called motion action, or simply action, from $\mathbf{u}_t$. This motion action is given by

$$\mathbf{v}(\mathbf{x}, t) = \left.\frac{\partial \mathbf{U}_t}{\partial \tau}(\mathbf{X}, \tau)\right|_{\substack{\tau = t \\ \mathbf{X} = \mathscr{X}^{-1}(\mathbf{x}, t)}} . \tag{3.96}$$

The set of all possible motion actions from configuration $\mathbf{u}_t \in \mathscr{U}$ defines the vector space $\mathscr{V}$ formed by all velocity fields that can be prescribed to the body departing from $\mathbf{u}_t$. Evidently, the real velocity field featured by the body at time t is an element of $\mathscr{V}$, while the rest of the elements of $\mathscr{V}$ are called to be virtual velocities.

Overall, the motion of a body will have to satisfy certain kinematical constraints (see, for instance, Figure 3.2). All possible configurations that satisfy these constraints are called admissible configurations. The subset of $\mathscr{U}$ formed by all the admissible configurations is designated by

$$\text{Kin}_u = \{\mathbf{u} \in \mathscr{U};\ \mathbf{u} \text{ is a kinematically admissible configuration}\}. \tag{3.97}$$

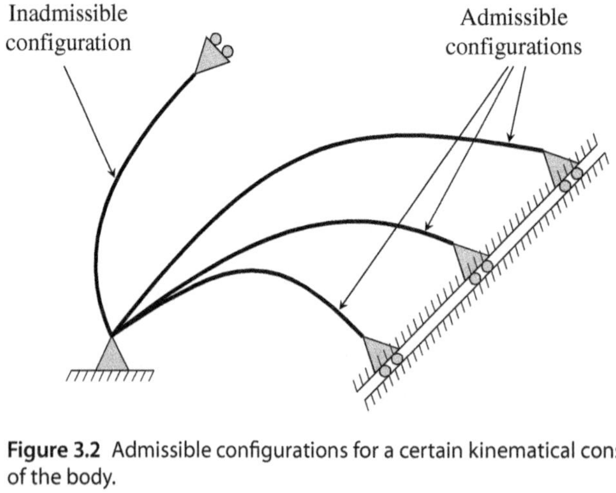

Figure 3.2 Admissible configurations for a certain kinematical constraint prescribed over the motion of the body.

Every motion from the admissible configuration $\mathbf{u}_t \in \text{Kin}_u$, that is to say, every uniparametric family $\mathbf{u}_t(\mathbf{x}, \tau)$, $\tau \in [t, t_f]$ such that for all $\tau \in [t, t_f]$ results $\mathbf{u}_t(\mathbf{x}, \tau) \in \text{Kin}_u$, is said to be an admissible motion. In other words, a motion from an admissible configuration is said to be admissible if all the configurations reached by that motion are also admissible.

For each admissible motion starting from configuration $\mathbf{u}_t \in \text{Kin}_u$ there is a motion action $\mathbf{v} \in \mathscr{V}$ called kinematically admissible motion action, and the set of all kinematically admissible motion actions $\text{Kin}_v \subset \mathscr{V}$ is

$$\text{Kin}_v = \{\mathbf{v} \in \mathscr{V} \,;\ \mathbf{v} \text{ kinematically admissible}\}. \tag{3.98}$$

From the very definition of Kin_v, it is clear that this set depends upon the configuration $\mathbf{u}_t$. That is, given two different admissible configurations, the sets Kin_v associated with each configuration are not necessarily the same.

In particular, the configuration $\mathbf{u}_t \in \text{Kin}_u$ is said to have kinematical constraints if $\text{Kin}_v \subseteq \mathscr{V}$, that is, Kin_v is a proper subset of $\mathscr{V}$ (Kin_v is not $\mathscr{V}$). Consider the example shown in Figure 3.2. For any admissible configuration, the admissible motion actions have to be such that the velocity is zero at the pinned support, while it has to be parallel to the inclined plane in the roller support. In this example, all admissible configurations feature kinematical constraints.

Observational data suggest that there exist different kinds of kinematical constraints. Here, we limit the presentation to the following classes[7]

- without constraints
- with frictionless bilateral constraints
- with nonadhesive frictionless unilateral constraints.

When the body is not subjected to any kind of kinematical constraints, it is said that the body is free, and in such a case it follows that

$$\text{no constraints} \quad \rightarrow \quad \text{Kin}_v = \mathscr{V}. \tag{3.99}$$

Frictionless bilateral constraints are those constraints for which if the motion is constrained in one direction, then it is also constrained in the opposite direction (see Figure 3.3). Moreover, in the directions in which motion is allowed, it is performed without facing any kind of resistance (this concept will be formalized in forthcoming sections). For this kind of constraint, it is easy to show that

$$\text{bilateral constraints} \quad \rightarrow \quad \text{Kin}_v = \bar{\mathbf{v}} + \text{Var}_v, \tag{3.100}$$

where

- $\bar{\mathbf{v}} \in \text{Kin}_v$, is an arbitrary motion action which satisfies the prescribed kinematical constraints
- $\text{Var}_v = \{\mathbf{v} \in \mathscr{V} \,;\ \mathbf{v} = \mathbf{0} \text{ over points where motion actions are constrained}\}$.

The previous definition tells us that if $\mathbf{v} \in \text{Var}_v$, it is also verified that $\alpha\mathbf{v} \in \text{Var}_v$. So we note that Var_v is a vector subspace of $\mathscr{V}$ and Kin_v is, therefore, a translation of this vector subspace (see Appendix A.11). Equivalently, we refer to Kin_v as a linear manifold.

---

7 Other kinds of kinematical constraints can be encountered in mechanics. This is, for example, the case for isochoric motions. Such specific problems will be addressed at the end of this chapter.

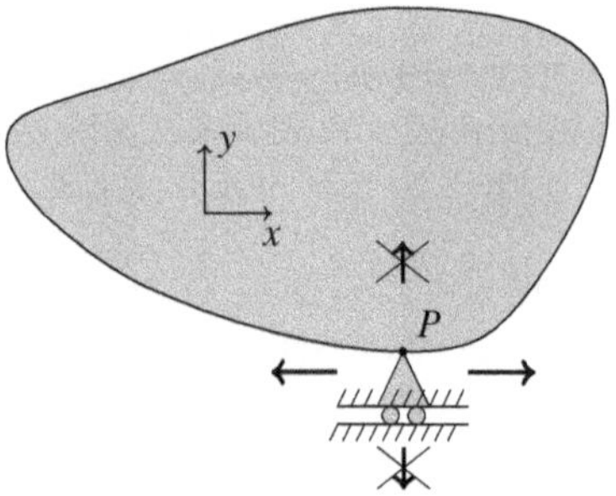

Figure 3.3 Example of a bilateral constraint. Motion at point *P* is allowed in both horizontal directions (arrows), while vertical motion is constrained in both directions (crossed-out arrows).

Also, it can be appreciated that in the case of homogeneous bilateral constraints we can always take $\bar{\mathbf{v}} = \mathbf{0}$, from which it follows that $\mathrm{Kin}_v = \mathrm{Var}_v$. In the most general case, for non-homogeneous bilateral constraints,

$$\mathrm{Var}_v = \mathrm{Kin}_v - \bar{\mathbf{v}}. \tag{3.101}$$

That is to say, every motion action belonging to Var_v can be described as the difference between kinematically admissible motion actions, regardless of the fact that constraints are homogeneous or non-homogeneous. In the literature, the motion actions $\mathbf{v} \in \mathrm{Var}_v$ are called kinematically admissible virtual motion actions.

Non-adhesive frictionless unilateral constraints are characterized by the fact that if the motion action is constrained in a given direction, it is not constrained in the opposite direction (see Figure 3.4). In the directions in which motion is allowed, it occurs without any kind of resistance. Moreover, for this kind of constraint, again it is verified that

$$\text{unilateral constraints} \quad \rightarrow \quad \mathrm{Kin}_v = \bar{\mathbf{v}} + \mathrm{Var}_v, \tag{3.102}$$

where now Kin_v is the translation, dictated by $\bar{\mathbf{v}}$, of the convex cone whose apex is at the origin, denoted by Var_v (see Appendix A.12). It is important to highlight here that this kind of constraint introduces a non-linearity in the definition of the kinematics. As a matter of fact, depending on the configuration $\mathbf{u}_t$ occupied by the body, a given unilateral constraint may or may not be active for such a configuration.[8] For the example in Figure 3.4, the set Var_v is the cone given by

$$\mathrm{Var}_v = \{\mathbf{v} \in V;\ (\mathbf{v} \cdot \mathbf{e}_y)(P) \geq 0\}, \tag{3.103}$$

where $(\mathbf{v} \cdot \mathbf{e}_y)(P)$ represents the component of the velocity field $\mathbf{v}$ in the direction $\mathbf{e}_y$, evaluated at point *P*. In this simple example it is possible to illustrate the main feature of set Var_v. In effect, if $\mathbf{v} \in \mathrm{Var}_v$, then $\lambda\mathbf{v}$, with $\lambda \in \mathbb{R}^+$, also belongs to Var_v. Hence, it is not difficult to show that (3.101) is also verified for this kind of constraint. Furthermore, this example also helps to conclude that the unilateral constraint is a more general constraint than the bilateral one, the latter being a particular case of the former.

In this way, regardless the kind of kinematical constraint prescribed over the body, it is $\mathrm{Kin}_v = \bar{\mathbf{v}} + \mathrm{Var}_v$, where $\bar{\mathbf{v}} \in \mathrm{Kin}_v$ is an arbitrary motion action compatible with the kinematical constraints. In addition, depending on the type of kinematical constraint,

8 In most undergraduate and graduate courses, mechanics is introduced considering exclusively bilateral constraints, while unilateral constraints are completely overlooked. In this sense, is a constraint less frequently encountered in such courses. In contrast, this kind of constraint does happen in almost every mechanical component. Ignoring this kind of constraint makes the student prone to extend to the unilateral case, the mechanical models valid for bilateral constraints, and, consequently, increases the susceptibility to errors in the analysis.

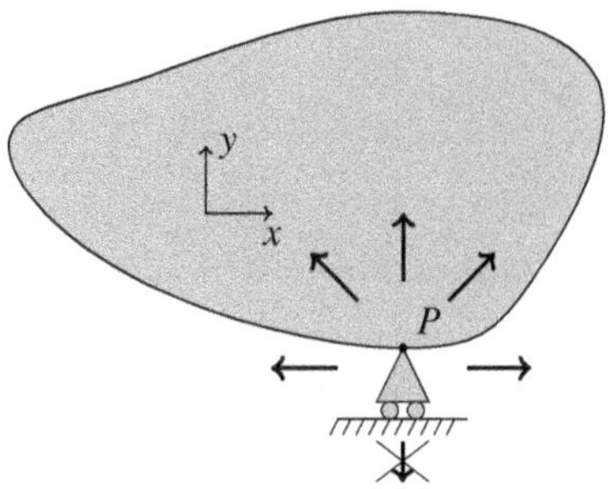

Figure 3.4 Example of a unilateral constraint. Motion at point P is not allowed in the vertical downward direction (crossed-out arrow), while it is allowed otherwise (arrows).

Var_v can be the very vector space $\mathscr{V}$ (no constraints), a vector subspace of $\mathscr{V}$ (bilateral constraints), or a convex cone with the apex in the origin (unilateral constraints).

Going ahead with the presentation, we have previously seen that exploiting the data intrinsic to the velocity field $\mathbf{v}$, we can define the deformation rate field as in (3.80). We can, therefore, introduce the vector space $\mathscr{W}$, whose elements are all the second-order symmetric tensor fields, defined in the actual (current) configuration. From this definition, it is easy to note that not every $\mathbf{D} \in \mathscr{W}$ emerges from a motion action $\mathbf{v} \in \mathscr{V}$. Indeed, given $\mathbf{D}$, there must be a field $\mathbf{v}$ such that

$$\mathbf{D} = (\text{grad}\, \mathbf{v})^s = \mathscr{D}\mathbf{v}, \tag{3.104}$$

where we have introduced the operator $\mathscr{D}$ not only to simplify the notation, but, fundamentally, to highlight the most relevant aspects: first we define the primal (or primary) variables that are of importance in the description of a given problem, and then we define the strategy to approximate these fields in the vicinity of a given point by means of a certain linear operator, denoted here by $\mathscr{D}$.

In particular, if, for a given $\mathbf{D} \in \mathscr{W}$, there exists the field $\mathbf{v} \in \text{Kin}_v$ such that (3.104) is verified, then $\mathbf{D}$ is said to be a kinematically admissible compatible deformation rate. In addition, the set of all possible rigid motion actions $\mathbf{v} \in \mathscr{V}$ constitutes a vector subspace, denoted by $\mathscr{N}(\mathscr{D})$, of $\mathscr{V}$, which is called the null space (or also kernel) of the deformation rate operator, and whose definition is

$$\mathscr{N}(\mathscr{D}) = \{\mathbf{v} \in \mathscr{V};\ \mathscr{D}\mathbf{v} = \mathbf{O}\ \ \forall \mathbf{x} \in \mathscr{B}_t\}. \tag{3.105}$$

Let us briefly recapitulate the main concepts introduced in this section. The study of the kinematics of deformable bodies, within a classical continuum theory, and using a description in its current configuration $\mathscr{B}_t$, requires the introduction of the following components:

- the vector space $\mathscr{V}$ of possible motion actions
- the deformation rate linear operator $\mathscr{D}(\cdot) = (\text{grad}\,(\cdot))^s$
- the vector space $\mathscr{W}$ of the deformation rates
- the vector subspace $\mathscr{N}(\mathscr{D})$ of rigid motion actions
- the subset Kin_v of $\mathscr{V}$ of kinematically admissible motion actions, that is, actions compatible with the kinematical constraints; this leads to three possibilities:
  - the whole space $\mathscr{V}$ (unconstrained body)
  - the translation of the vector subspace Var_v (bilateral constraints)
  - the translation of the convex cone with apex in the origin, also called Var_v (unilateral constraints).

These elements of the kinematical model are illustrated in Figure 3.5.

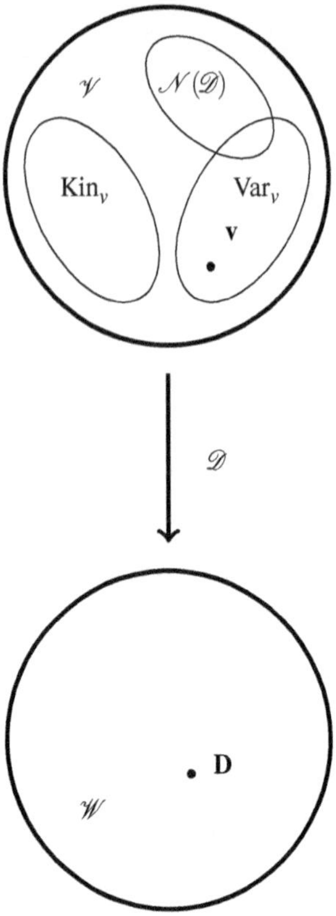

Figure 3.5 Vector spaces and basic components introduced to describe the kinematics of deformable bodies.

3.3 Duality and Virtual Power

Another fundamental concept in the field of mechanics is that of the "effort" needed to accomplish some "action" over a body $\mathscr{B}$ in its current configuration $\mathscr{B}_t$. The typical approach to characterize this effort is through a vector field termed "force". In this manner, this concept rises as an *a priori* entity, fully independent of the kinematics chosen to model the observed phenomena.

Notwithstanding the ingrained idea of forces and the success attained by this conceptual framework, which would coerce the reader to viscerally embrace these entities towards the construction of mechanical models, in what follows a second line of reasoning will be followed. The alternative theoretical edifice proposed here, although apparently more abstract at first sight, comes to be seen as a more suitable tool to realize the concrete experience faced in our routine interaction with reality. The hallmark feature that differentiates such a framework from the classical approach is that the efforts (both external and internal), also called forces, experienced by a system are malleable constructs fully determined by the kinematical model under consideration. That is to say, in this setting, the forces emerge naturally as entities in mathematical duality to given motion actions. The mathematical duality employed in this journey is defined through the primary concept of virtual power or virtual work. This dual incarnation is in absolute contrast to the classical *a priori* idealization of these forces.

Contrary to what could be believed, this schematization is as old as the very field of mechanics. At the same time, from the first attempts to equip mechanics with an increasingly precise mathematical structure, the concept of power emerged as an essential pillar. In this direction, the pioneer works by J. Bernoulli (1717), definitively consolidated by D'Alembert (1743), have to be mentioned. Moreover, such a strategy for the definition of forces is more natural than the first one because it is capable of expressing rather common physical experiences. Indeed, it is possible to realize that the introduction of motion actions to reveal external and internal forces in a body and at a given configuration bears an unquestionable physical meaning. Furthermore, we can assert that this is inherent to the human condition, as will be clear in the examples cited in the forthcoming sections.

3.3.1 Motion Actions and Forces

Consider the problem of establishing which of two objects is heavier. Moreover, we present this question to a child. The normal procedure through which the child will provide an answer is based on extremely intuitive reasoning. In fact, the child will prescribe motion actions to these objects, and will precisely come up with the answer to the question.

In fact, if we want to know the weight of a given object, for example our luggage, which is lying on the floor, we lightly lift it, assessing the weight through the power (or work) required to execute the motion action (the small lifting). In other words, what we naturally try to do is to introduce a virtual motion action to move the luggage away from its current configuration (lying on the floor).

In order to formalize these ideas, consider the body occupying configuration $\mathscr{B}_t$ and without constraints. In this case it is $\text{Kin}_u \equiv \mathscr{U}$, and the set of all kinematically admissible motion actions Kin_v is such that $\text{Kin}_v \equiv \mathscr{V}$, where $\mathscr{V}$ is the (real) vector space of all possible motion actions. The system of external forces, denoted by f, which at time t are applied over the body, is characterized by a linear continuous functional in $\mathscr{V}$,[9] whose value in $\mathbb{R}$,[10] for each motion action $\mathbf{v} \in \mathscr{V}$, is called the external power expended by the system of forces f for such motion action $\mathbf{v}$, which is denoted by $P_e = \langle f, \mathbf{v} \rangle = f(\mathbf{v})$. The set of all systems of forces f, that is, the set of all linear and continuous functionals in $\mathscr{V}$, defines the vector space $\mathscr{V}'$, called the space of external forces. It is easy to see that, from a mathematical perspective, this space is the dual space of $\mathscr{V}$. For more details see Appendix A.16.

9 If $\mathscr{V}$ is not finite-dimensional, for the continuity of the functional we will have to define a proper topology in $\mathscr{V}$. This topology is closely related to the mechanical problem under consideration. Seeking to keep the presentation of this chapter as simple as possible, this last property will be ignored, since it will only be necessary in some specific and more technical than conceptual aspects. More details on this can be found in [220] and in Appendices A.14 and A.16.

10 For the purpose and scope of this document, it is sufficient to consider the hypothesis of linear and continuous functionals whose number field is the set of real numbers $\mathbb{R}$. This is motivated by the fact that the elements in $\mathscr{V}$ are real-valued vector fields. However, the same ideas and concepts can directly be extended when $\mathscr{V}$ is formed by vectors in the field of complex numbers $\mathbb{C}$. In such a situation, it is enough to assume that forces are modeled through linear and continuous functionals over $\mathbb{C}$. Hereafter, it will always be considered that these cost functionals are defined over the set $\mathbb{R}$.

We see then that two aspects emerge as fundamental issues

- In a given mechanical problem, the space of motion actions $\mathscr{V}$ is the first aspect to be defined, together with the associated notion of rigid motions.
- The mathematical duality between the previously defined space $\mathscr{V}$ and the force space $\mathscr{V}'$ is the second hypothesis. In this way, for a given kinematical model characterized by the space $\mathscr{V}$, the system of forces compatible with this model is automatically shaped by the duality, through the bilinear form

$$\begin{aligned} \langle \cdot , \cdot \rangle : \mathscr{V}' \times \mathscr{V} &\to \mathbb{R}, \\ (f, \mathbf{v}) &\mapsto P_e = \langle f, \mathbf{v} \rangle. \end{aligned} \tag{3.106}$$

This also shows us that, the richer the space $\mathscr{V}$, the wider the scope reached by the definition of motion actions and the more refined the characterization of forces compatible with the model (see [115, 116, 117] and Appendix B).

The applicability of these concepts in basic examples is illustrated next.

Example 3.1 Consider that the body is a single particle P free to move in $\mathscr{E}$. Then $\text{Kin}_v = \mathscr{V} \equiv V$, where V is the three-dimensional vector space associated with the Euclidean space $\mathscr{E}$. A linear form in V, which could characterize the power expended for a given action, can be represented as an inner product of vectors

$$P_e = \langle f, \mathbf{v} \rangle = \mathbf{F} \cdot \mathbf{v} \qquad \mathbf{F} \in V', \ \mathbf{v} \in V, \tag{3.107}$$

where $V = \mathbb{R}^3$ and therefore $V' = \mathbb{R}^3$. From this, we see that a force f that can be applied over the particle P, compatible with the adopted kinematics, can be represented by a vector acting on P, and whose direction and intensity are given by $\mathbf{F}$. Hence, the duality provides the classical notion of force.

Example 3.2 Consider now the case of a rigid body, which in the configuration at time t is free of constraints. Then the admissible motion actions will be those which ensure the body remains rigid. The motion actions at time t will only be rigid, that is, only rigid velocity fields are admissible. In this manner, an arbitrary element of $\mathscr{V}$ is given by

$$\mathbf{v}(\mathbf{x}) = \mathbf{v}_O + \mathbf{W}(\mathbf{x} - O) = \mathbf{v}_O + \mathbf{w} \times (\mathbf{x} - O), \tag{3.108}$$

where $\mathbf{v}(\mathbf{x})$ is the velocity of point $\mathbf{x} \in \mathscr{B}_t$, $\mathbf{v}_O$ is the velocity of point O, being $O \in \mathscr{B}_t$ arbitrary, $\mathbf{W}$ is a uniform skew-symmetric tensor, and $\mathbf{w}$ its axial vector, which characterizes the angular rotation around the axis defined by $\mathbf{w}$, which goes through O. From the previous considerations, it follows that the space of possible (and also admissible) motion actions $\mathscr{V}$ is a vector space of dimension six, three associated with $\mathbf{v}_O$ and three associated with $\mathbf{w}$, that is,

$$\mathscr{V}' = V \times V. \tag{3.109}$$

Again, the linear forms in this six-dimensional space can be represented through a proper inner product, appearing in this way as the pair of vectors $(\mathbf{F}_O, \mathbf{m}_O) \in V' \times V'$ which represent, respectively, the resultant force of the system of forces applied over

$\mathscr{B}_t$ and the resultant moment exerted by this system at point O. In fact, due to the linearity in the representation, the power expended by the pair force-velocity is

$$
\begin{aligned}
P_e = \langle f, \mathbf{v} \rangle &= \langle f, \mathbf{v}_O + \mathbf{w} \times (\mathbf{x} - O) \rangle = \langle f, \mathbf{v}_O \rangle + \langle f, \mathbf{w} \times (\mathbf{x} - O) \rangle \\
&= \mathbf{F}_O \cdot \mathbf{v}_O + \mathbf{F}_O \cdot \mathbf{w} \times (\mathbf{x} - O) = \mathbf{F}_O \cdot \mathbf{v}_O + (\mathbf{x} - O) \times \mathbf{F}_O \cdot \mathbf{w} \\
&= \mathbf{F}_O \cdot \mathbf{v}_O + \mathbf{m}_O \cdot \mathbf{w}. \qquad (3.110)
\end{aligned}
$$

Moreover, since the representation of the rigid velocity field is independent from point O, we can take another point of $\mathscr{B}_t$ to describe these motions. Let P be this alternative point, then

$$\mathbf{v}(\mathbf{x}) = \mathbf{v}_P + \mathbf{W}(\mathbf{x} - P) = \mathbf{v}_P + \mathbf{w} \times (\mathbf{x} - P), \qquad (3.111)$$

where

$$\mathbf{v}_P = \mathbf{v}_O + \mathbf{w} \times (P - O), \qquad (3.112)$$

and the power expended by the same force is given by

$$P_e = \langle f, \mathbf{v} \rangle = \mathbf{F}_P \cdot \mathbf{v}_P + \mathbf{m}_P \cdot \mathbf{w}. \qquad (3.113)$$

Subtracting equation (3.113) from (3.110) yields

$$0 = (\mathbf{F}_O - \mathbf{F}_P) \cdot \mathbf{v}_O + \left(\mathbf{m}_O - [\mathbf{m}_P + (P - O) \times \mathbf{F}_P] \right) \cdot \mathbf{w}. \qquad (3.114)$$

Since $\mathbf{v}_O$ and $\mathbf{w}$ are arbitrary, we conclude that

$$\mathbf{F}_O = \mathbf{F}_P = \mathbf{F}, \qquad (3.115)$$

$$\mathbf{m}_O = \mathbf{m}_P + (P - O) \times \mathbf{F}. \qquad (3.116)$$

This allows us to recover the classical results of the mechanics of rigid bodies. Forces f are, therefore, characterized by a vector $\mathbf{F}$, called the resultant force, and by a vector $\mathbf{m}_O$, called the resultant moment. As is evident, $\mathbf{F}$ is independent from the point O chosen for the description of rigid motion actions, while the vector $\mathbf{m}_O$ depends on this choice.

The previous examples demonstrate that the strategy based on the consideration (actually a hypothesis) of employing linear and continuous functionals defined over the real numbers provides the reader with a great power of synthesis melded with a strong mechanical flavor. In a few lines, we have unveiled the reasons behind the representation of forces through vectors, as well as the fact that the moment depends on a point. In classic textbooks, in contrast, this definitely takes more space and time to be conveyed to the reader.

3.3.2 Deformation Actions and Internal Stresses

Another fundamental concept in the realm of continuum mechanics is that of internal forces or internal stresses. To illustrate this concept, let us resort to a simple example. Consider the fan belt of an engine. If one wants to test if the fan belt is adjusted properly, we make an attempt to displace the belt from its current configuration. Then, if our motion action is able to deflect the belt, the quantity of power put into play in the process will inform us about the value of the belt tension. In this manner, the nature of the internal force is revealed once we perform a non-rigid motion action over the belt (a deflection), through the resistance opposed by the system to the belt deflection.

The previous example leads us to adopt the following hypotheses

- The internal stresses are characterized by a linear and continuous functional, whose arguments are the motion actions and the first spatial gradient. The value of this functional, denoted by $p_i(\mathbf{v}, \text{grad}\,\mathbf{v})$, for each pair $(\mathbf{v}, \text{grad}\,\mathbf{v})$ is called the internal power per unit volume. By definition, $p_i(\mathbf{v}, \text{grad}\,\mathbf{v})$ is a scalar field defined in the current configuration occupied by the body $\mathscr{B}_t$. The integral of this scalar field over the entire volume defined by $\mathscr{B}_t$ is termed the total internal power, or simply the internal power, which in short is denoted by P_i.
- In the realm of continuum mechanics, the internal power P_i can be typified through the expression

$$P_i = \int_{\mathscr{B}_t} p_i \, dV = -\int_{\mathscr{B}_t} (\mathbf{f} \cdot \mathbf{v} + \mathbf{T} \cdot \text{grad}\,\mathbf{v}) \, dV, \tag{3.117}$$

where the negative sign is taken for convenience, as will become clear in forthcoming developments.
- From the observation that any rigid body motion action is unable to assess the internal structure of the forces taking place in the body, it is reasonable to assume that

$$P_i = 0 \qquad \forall \mathbf{v} \in \mathcal{N}(\mathscr{D}). \tag{3.118}$$

Let us now inspect expression (3.117) for different motion actions $\mathbf{v}$.
- Let $\mathbf{v}$ be a translation, then $\mathbf{v} = \mathbf{c}$, $\forall \mathbf{x} \in \mathscr{B}_t$ ($\mathbf{c}$ is a constant vector in $\mathbb{R}^3$), and we have that $\text{grad}\,\mathbf{v} = \mathbf{O}$ implies $\mathbf{D} = \mathbf{W} = \mathbf{O}$, then

$$P_i = -\int_{\mathscr{B}_t} \mathbf{f} \cdot \mathbf{v} \, dV = 0, \tag{3.119}$$

for any translational motion $\mathbf{v}$, which in turn leads to

$$\left(\int_{\mathscr{B}_t} \mathbf{f} \, dV \right) \cdot \mathbf{c} = 0 \qquad \forall \mathbf{c} \in \mathbb{R}^3, \tag{3.120}$$

and with this we have

$$\int_{\mathscr{B}_t} \mathbf{f} \, dV = \mathbf{0}. \tag{3.121}$$

Since this result must hold for every part $\mathscr{P}_t$ of the body $\mathscr{B}_t$, then we conclude that

$$\mathbf{f} = \mathbf{0} \qquad \forall \mathbf{x} \in \mathscr{B}_t. \tag{3.122}$$

- Let $\mathbf{v}$ be a rigid motion, then $\mathbf{v} = \mathbf{v}_O + \mathbf{W}(\mathbf{x} - O)$ where $\mathbf{v}_O$ is a constant vector and $\mathbf{W}$ is a constant skew-symmetric tensor. Thus, we have

$$\text{grad}\,\mathbf{v} = \mathbf{W}. \tag{3.123}$$

From this result, and using (3.122) we reach

$$P_i = -\int_{\mathscr{B}_t} \mathbf{T} \cdot \mathbf{W} \, dV = 0, \tag{3.124}$$

for all $\mathbf{W}$ constant and skew-symmetric. This allows us to conclude that

$$\left(\int_{\mathscr{B}_t} \mathbf{T}\, dV\right) \in Sym, \tag{3.125}$$

and because this must be valid for every part $\mathscr{P}_t$ of $\mathscr{B}_t$ we have

$$\mathbf{T} \in Sym \qquad \forall \mathbf{x} \in \mathscr{B}_t. \tag{3.126}$$

In this manner, the characterization of the internal power expended through a given motion action $\mathbf{v}$ is given by the cost functional

$$P_i = \int_{\mathscr{B}_t} p_i\, dV = -\int_{\mathscr{B}_t} \mathbf{T} \cdot (\text{grad}\, \mathbf{v})^s\, dV = -\int_{\mathscr{B}_t} \mathbf{T} \cdot \mathbf{D}\, dV = -(\mathbf{T}, \mathbf{D}), \tag{3.127}$$

which means that the internal stress state of the body is fully characterized by a linear and continuous functional defined over the space of deformation rate actions $\mathscr{W}$. The set of all these functionals is the space of internal stresses $\mathscr{W}'$, dual to $\mathscr{W}$ in the above sense, and whose elements are shaped as symmetric second-order tensors $\mathbf{T}$. These tensors are acknowledged in the literature as Cauchy stress tensors.

As an outcome of the hypotheses considered above, we can see that the internal power P_i is invariant to rigid motions actions. This implies that the quantity P_i is objective, and by virtue of the objectivity of the tensor field $\mathbf{D}$ it follows that the internal stress $\mathbf{T}$ is also objective.

3.3.3 Mechanical Models and the Equilibrium Operator

We have seen in the previous sections that the theoretical fabric underlying the construction of mechanical models rests on the following procedure

1) Consider the body $\mathscr{B}$ occupying at time t the current configuration $\mathscr{B}_t$ (or equivalently $\mathbf{u}_t$) in $\mathscr{E}$. For simplicity, it will also be assumed that the region $\mathscr{B}_t$ is connected, open and limited by a sufficiently smooth boundary $\partial\mathscr{B}_t$.[11]
2) For the configuration $\mathscr{B}_t$, the vector space $\mathscr{V}$ is defined whose elements are called motion actions. The construction of $\mathscr{V}$ depends upon the kinematical hypotheses of the model.[12]
3) Having defined $\mathscr{V}$, the deformation rate linear operator typifies the space of deformation rate actions $\mathscr{W}$.
4) From the deformation rate operator $\mathscr{D}$, which in the realm of three-dimensional continuum mechanics is $\mathscr{D}(\cdot) = (\text{grad}\,(\cdot))^s$, the space of rigid motion actions $\mathscr{N}(\mathscr{D})$ is characterized.

11 By smooth we mean a Lipschitz boundary (see [250, page 323]).

12 For example, when working with rigid bodies, $\mathscr{V}$ will be formed by fields $\mathbf{v}$ which are rigid motions. In the case of bending of beams using, for instance, Bernoulli hypotheses (see Chapter 7), fields $\mathbf{v}$ must be such that transversal sections to the axis of the beam remain plane and orthogonal to the axis. In the case of torsion of beams (see Chapter 8), fields $\mathbf{v}$ must ensure rigid rotation of the cross-sections around the beam axis. In the case of modeling shells under Kirchhoff–Love hypotheses (see Chapter 9), fields $\mathbf{v}$ have to be such that the normal vector to the mean surface remains normal during the motion action. In the case of incompressible media (see Chapter 11), fields $\mathbf{v}$ must be such that the motion action is isochoric. Evidently, each mechanical model requires the motion actions $\mathbf{v} \in \mathscr{V}$ to have different regularity, which defines the topology of space $\mathscr{V}$, such that the mathematical operations to be performed are well-posed. These considerations will be analyzed in detail in forthcoming chapters.

5) Through the consideration of kinematical constraints to the body motion, the set Kin_v and the space Var_v are defined.
6) The previous basic ingredients allow the following hypotheses to be introduced
   - External forces acting f over the body at configuration $\mathscr{B}_t$ are characterized by a linear and continuous functional defined in $\mathscr{V}$, that is, $f \in \mathscr{V}'$, being $\mathscr{V}'$ the dual space of $\mathscr{V}$, and its value for a given motion action $\mathbf{v} \in \mathscr{V}$ is the external virtual power

$$P_e = \langle f, \mathbf{v} \rangle, \tag{3.128}$$

   where $\langle \cdot, \cdot \rangle : \mathscr{V}' \times \mathscr{V} \to \mathbb{R}$ is a bilinear form (not specified yet), which provides the duality sense between spaces $\mathscr{V}'$ and $\mathscr{V}$.
   - Internal stresses $\mathbf{T}$ experienced by the body at configuration $\mathscr{B}_t$ are characterized by a linear and continuous functional defined in $\mathscr{W}$, that is, $\mathbf{T} \in \mathscr{W}'$, being $\mathscr{W}'$ the dual space to $\mathscr{W}$, and its value for a given deformation rate action $\mathbf{D} \in \mathscr{W}$ is the internal virtual power

$$P_i = -\int_{\mathscr{B}_t} \mathbf{T} \cdot \mathbf{D} \, dV = -(\mathbf{T}, \mathbf{D}). \tag{3.129}$$

   It is important to note here that a specific form for the duality product $(\mathbf{T}, \mathbf{D})$ has been adopted consistently with the mechanics of deformable bodies in the classical continuum mechanics framework.
   - The fact that $P_i = 0$ must be ensured for all rigid motion actions.

First, it is important to note here that expressions (3.128) and (3.129) arise as a consequence of the continuity assumed in the model. Mathematically, these expressions are a consequence of the Riesz representation theorem, which establishes the connection between a given space and its dual (see [250, page 111]). In a nutshell, and for the case of Hilbert spaces, this theorem tells us that every linear and continuous functional can be expressed through an inner product.

We have seen that spaces $\mathscr{V}'$ and $\mathscr{W}'$ are the (topological) dual spaces to $\mathscr{V}$ and $\mathscr{W}$, respectively, and the forms $\langle \cdot, \cdot \rangle$ and $(\cdot, \cdot)$ represent, correspondingly, duality pairings in $\mathscr{V}' \times \mathscr{V}$ and $\mathscr{W}' \times \mathscr{W}$. Hence, we can introduce a further operator

$$\begin{aligned} \mathscr{D}^* : \mathscr{W}' &\to \mathscr{V}', \\ \mathbf{T} &\mapsto f = \mathscr{D}^*\mathbf{T}, \end{aligned} \tag{3.130}$$

called the adjoint operator (see [19], [166] and [178]) or the transpose operator (see [220] and [230]). Here, we will interchangeably call it the adjoint operator or the equilibrium operator since, as will become clear later, this illustrates the mechanical meaning of this entity. This adjoint operator is defined as

$$(\mathbf{T}, \mathscr{D}\mathbf{v}) = \langle \mathscr{D}^*\mathbf{T}, \mathbf{v} \rangle \qquad \mathbf{v} \in \mathscr{V}. \tag{3.131}$$

As a second and fundamental remark, observe that, from the expression of the internal power P_i, given by the inner product $(\cdot, \cdot)$ from (3.129), the specific structure for the external forces f is automatically established, through the consistent characterization of the duality product $\langle \cdot, \cdot \rangle$ in $\mathscr{V}$. Moreover, this puts in evidence the great difference between the present approach and the classical one in which the forces compatible with a given kinematical model are defined *a priori* and in an *ad hoc* manner. This will be

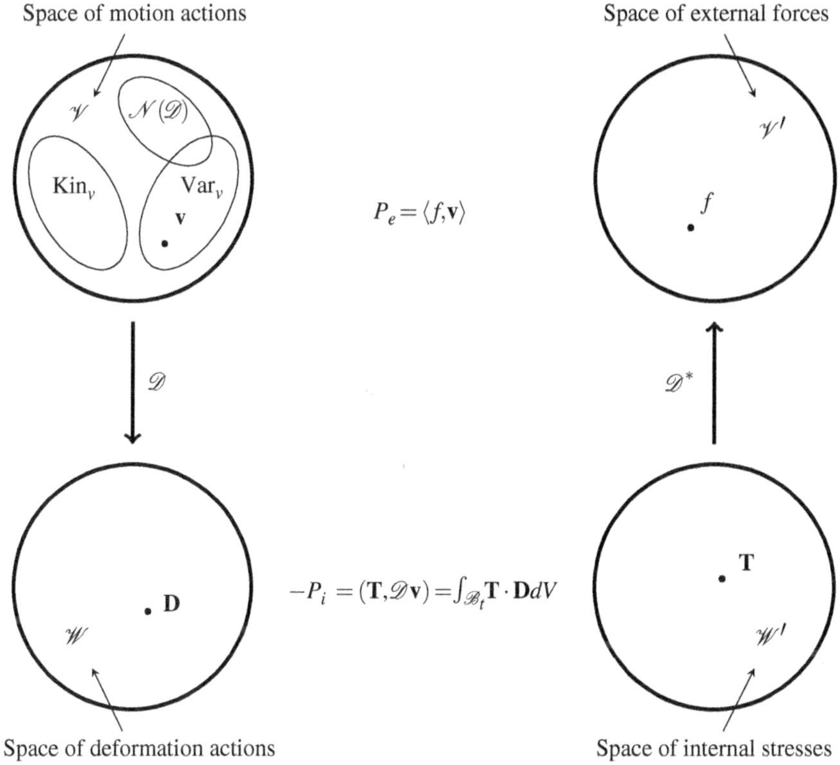

Figure 3.6 The main components in the unified modeling framework.

analyzed for the case of three-dimensional deformable bodies in the typical continuum mechanics framework in this chapter. The application of the same concepts in problems of beam bending and torsion, shells and plates, heat conduction, fluid flow and high order continua will be addressed in due course in specific chapters. The basic procedure to unveil the structure of the external force f, in the context of continuum mechanics, consists of integrating by parts.

All the ingredients that play a role in this unified modeling framework are exposed in Figure 3.6, which complements Figure 3.5 by adding the dual characterization of external and internal forces.

Exercise 3.1 For each of the following inner products, which define a certain measure of internal power, find the equilibrium operator $\mathscr{D}^*$. Disregard the existence of kinematical constraints.

i) v is a sufficiently smooth scalar field, $\mathscr{D}(\cdot) = \frac{d(\cdot)}{dx}$, t is a scalar field

$$P_i = -\left(t, \frac{dv}{dx}\right) = -\int_0^L t\frac{dv}{dx}\, dx. \tag{3.132}$$

ii) v is a sufficiently smooth scalar field, $\mathscr{D}(\cdot) = \frac{d^2(\cdot)}{dx^2}$, p is a scalar field

$$P_i = -\left(p, \frac{d^2 v}{dx^2}\right) = -\int_0^L p \frac{d^2 v}{dx^2}\, dx. \tag{3.133}$$

iii) $\mathbf{v}$ is a sufficiently smooth vector field, $\mathscr{D}(\cdot) = (\text{grad}\,(\cdot))^s$, $\mathbf{T}$ is a second-order tensor field

$$P_i = -(\mathbf{T}, \text{grad}\,\mathbf{v}) = -\int_{\mathscr{B}_t} \mathbf{T} \cdot (\text{grad}\,\mathbf{v})^s\, dV. \tag{3.134}$$

iv) $\mathbf{v}$ is a sufficiently smooth vector field, $\mathscr{D}(\cdot) = \text{grad}\,(\text{grad}\,(\cdot))$, M is a third-order tensor field

$$P_i = -(\mathsf{M}, \text{grad}\,(\text{grad}\,\mathbf{v})) = -\int_{\mathscr{B}_t} \mathsf{M} \cdot \text{grad}\,(\text{grad}\,\mathbf{v})\, dV. \tag{3.135}$$

v) $\mathbf{v}$ is a sufficiently smooth vector field, $\mathscr{D}(\cdot) = \text{curl}(\cdot)$, $\mathbf{s}$ is a vector field

$$P_i = (\mathbf{s}, \text{curl}\,\mathbf{v}) = \int_{\mathscr{B}_t} \mathbf{s} \cdot \text{curl}\,\mathbf{v}\, dV. \tag{3.136}$$

The other principle on top of which we build the mechanics is the Principle of Virtual Power (PVP). This principle establishes the conditions in which a sense of mechanical equilibrium takes place as a whole. Attached to this is the fact that the PVP also provides a definite characterization of the system of external forces f which are compatible with the kinematical model. Remarkably, the very same PVP univocally leads to the typification of equilibrium equations in differential form, that is, the local or point-wise relations the fields must satisfy.

The path to follow in the next section is of increasingly complex scenarios, starting with unconstrained bodies, analyzing bodies with bilateral constraints, and finally reaching the case of bodies with unilateral constraints.

3.4 Bodies without Constraints

Let us consider that, at time t, a given body $\mathscr{B}$ is at configuration $\mathscr{B}_t$ free of constraints. In this case we saw that $\text{Kin}_u = \mathscr{U}$ and $\text{Kin}_v = \mathscr{V}$, that is, the vector space of all possible configurations coincides with that of admissible configurations, and the vector space of all possible motion actions is the same than the space of admissible ones. As can be appreciated, these motion actions are executed over $\mathbf{u}_t$[13] and tend to take the body away from its natural state. What is really fundamental is that these motion actions are not necessarily experienced by the body during the real motion at time t. This is why we designate these actions as virtual motion actions, or variations of the real motion actions. Finally, recall that the space of all virtual motion actions is denoted by Var_v. We next enunciate the PVP for the case of unconstrained bodies.

13 As already said, we talk about configuration $\mathscr{B}_t$ and the corresponding spatial description of the displacement field $\mathbf{u}_t$ from the reference configuration $\mathscr{B}$ interchangeably.

3.4.1 Principle of Virtual Power

For any Galilean frame of reference[14] and for every time t, the body $\mathscr{B}$ is at (static) equilibrium in the configuration free of constraints $\mathscr{B}_t$ under the action (not to be confused with motion actions) of a system of forces $f \in \mathscr{V}'$ if

$$P_e = \langle f, \hat{\mathbf{v}} \rangle = 0 \qquad \forall \hat{\mathbf{v}} \in \mathrm{Var}_v \cap \mathscr{N}(\mathscr{D}) = \mathscr{N}(\mathscr{D}), \tag{3.137}$$

that is, if the external virtual power exerted by the forces that act over the body in configuration $\mathscr{B}_t$ is null for all rigid virtual motion actions, and, in addition, the following is verified

$$P_i + P_e = -(\mathbf{T}, \mathscr{D}\hat{\mathbf{v}}) + \langle f, \hat{\mathbf{v}} \rangle = 0 \qquad \forall \hat{\mathbf{v}} \in \mathrm{Var}_v, \tag{3.138}$$

that is, if the sum of the internal virtual power and the external virtual power is nullified for all kinematically admissible virtual motion actions. In this manner, we postulate the concept of mechanical equilibrium through the variational equations (3.137) and (3.138).

If we want to extend this principle to encompass the concept of dynamic equilibrium, we have to add the power corresponding to the inertial forces. In other words, the external forces f applied over the body are now combined with the inertial force given by $\rho\dot{\mathbf{v}}$.[15] Hence, we have

$$P_e = \langle f^*, \hat{\mathbf{v}} \rangle \qquad f^* = f - \rho\dot{\mathbf{v}}. \tag{3.139}$$

It is important to highlight here that, in the dynamic case, the PVP is nothing but D'Alembert's principle (see [167]).

Going back to the PVP, observe that the first statement allows us to characterize the forces $f \in \mathscr{V}'$ compatible with the model. Indeed, for the present case it is

$$\mathscr{N}(\mathscr{D}) = \{\mathbf{v} \in \mathscr{V}\,;\; \mathscr{D}\mathbf{v} = \mathbf{O}\}, \tag{3.140}$$

which implies

$$\mathscr{N}(\mathscr{D}) = \{\mathbf{v} \in \mathscr{V}\,;\; \mathbf{v}(\mathbf{x}) = \mathbf{v}_O + \mathbf{w} \times (\mathbf{x} - O)\}, \tag{3.141}$$

where O is an arbitrary point and $\mathbf{v}_O \in \mathscr{V}$ and $\mathbf{w} \in \mathscr{V}$ are also arbitrary elements. Then, it turns out to be $\dim \mathscr{N}(\mathscr{D}) = 6$, and for all $\hat{\mathbf{v}} \in \mathscr{N}(\mathscr{D})$ yields

$$\begin{aligned} P_e = \langle f, \hat{\mathbf{v}} \rangle &= \langle f, \hat{\mathbf{v}}_O + \hat{\mathbf{w}} \times (\mathbf{x} - O) \rangle = \langle f, \hat{\mathbf{v}}_O \rangle + \langle f, \hat{\mathbf{w}} \times (\mathbf{x} - O) \rangle \\ &= \mathbf{R}_f \cdot \hat{\mathbf{v}}_O + \mathbf{R}_f \cdot \hat{\mathbf{w}} \times (\mathbf{x} - O) = \mathbf{R}_f \cdot \hat{\mathbf{v}}_O + (\mathbf{x} - O) \times \mathbf{R}_f \cdot \hat{\mathbf{w}} \\ &= \mathbf{R}_f \cdot \hat{\mathbf{v}}_O + \mathbf{m}_O \cdot \hat{\mathbf{w}} = 0 \qquad \forall (\hat{\mathbf{v}}_O, \hat{\mathbf{w}}) \in \mathscr{V} \times \mathscr{V}, \end{aligned} \tag{3.142}$$

which implies

$$\begin{aligned} \mathbf{R}_f &= \mathbf{0}, \\ \mathbf{m}_O &= \mathbf{0}. \end{aligned} \tag{3.143}$$

This result reveals that the external forces which are compatible with the model have to be such that the resultant force $\mathbf{R}_f$ is null, and the resultant of moments $\mathbf{m}_O$ with respect to an arbitrary point O is also null. If the external force f is known, such

14 Here, the Galilean frame of reference stands for the absolute frame of reference.

15 Notice that $\dot{\mathbf{v}}$ is the spatial description of the acceleration determined by the real motion.

datum will be compatible with the mechanical model if such conditions are satisfied. Also, note that the set of external forces $f \in V'$ which satisfy the conditions established by the PVP is not empty, since $f = 0$ is an element in this set. In mechanical terms, this tells us that the null force is compatible with the equilibrium of the body $\mathscr{B}$ at configuration $\mathscr{B}_t$.

The second statement of the PVP allows us to extend the definition of the mechanical equilibrium to motion actions which are not necessarily rigid. As a matter of fact, the second statement embraces the first statement as a particular case since, by hypothesis, we have admitted that it is $P_i = 0$ for all $\mathbf{v}$ rigid. In turn, the utilization of the second part of the PVP brings an additional difficulty, since it establishes a relation, in terms of mechanical equilibrium, between the internal stress $\mathbf{T} \in \mathscr{W}'$ and the external force $f \in \mathscr{V}'$ compatible with the equilibrium (i.e. $P_e = 0$, $\forall \mathbf{v} \in \mathscr{N}(\mathscr{D})$). Therefore, for the second part of the PVP to make sense, it is necessary to show that there exists $\mathbf{T} \in \mathscr{W}'$ such that (3.138) is satisfied for a given $f \in \mathscr{V}'$ at equilibrium. The following theorem provides this important result.

Theorem 3.1 Representation Theorem. Let $\mathbf{u}_t \in \text{Kin}_u$ be an equilibrium configuration under the action of the system of external forces $f \in \mathscr{V}'$. Then, there exists $\mathbf{T} \in \mathscr{W}'$ such that

$$f = \mathscr{D}^*\mathbf{T}. \tag{3.144}$$

Proof. By hypothesis we know that $f \in \mathscr{V}'$ is such that $\langle f, \hat{\mathbf{v}} \rangle = 0$, $\forall \hat{\mathbf{v}} \in \mathscr{N}(\mathscr{D})$, which implies that $f \perp \mathscr{N}(\mathscr{D})$, that is, f is orthogonal to $\mathscr{N}(\mathscr{D})$, or equivalently $f \in \mathscr{N}(\mathscr{D})^\perp$. We then say that f belongs to the orthogonal complement of $\mathscr{N}(\mathscr{D})$.

Moreover, we know that $\mathbf{v} \in \mathscr{N}(\mathscr{D})$ implies $\mathscr{D}\mathbf{v} = \mathbf{O}$, $\forall \mathbf{v} \in \mathscr{N}(\mathscr{D})$, which is true if and only if $(\mathbf{T}, \mathscr{D}\mathbf{v}) = \langle \mathscr{D}^*\mathbf{T}, \mathbf{v} \rangle = 0$, $\forall \mathbf{T} \in \mathscr{W}'$, which is equivalent to $\mathbf{v} \in R(\mathscr{D}^*)^\perp$. We thus conclude that

$$\mathscr{N}(\mathscr{D}) = R(\mathscr{D}^*)^\perp, \tag{3.145}$$

from which (since $R(\mathscr{D})$ is closed in $\mathscr{W}$) we have

$$\mathscr{N}(\mathscr{D})^\perp = (R(\mathscr{D}^*)^\perp)^\perp = R(\mathscr{D}^*). \tag{3.146}$$

This result is a generalization of the notion of orthogonality in finite-dimensional spaces. Figure 3.7 features $R(\mathscr{D}^*)$ represented by the two-dimensional subspace of hypothetical three-dimensional space $\mathscr{V}'$. From this, we obtain that if f is at equilibrium in $\mathbf{u}_t \in \text{Kin}_u$, then

$$f \in \mathscr{N}(\mathscr{D})^\perp = R(\mathscr{D}^*) \Rightarrow \exists \mathbf{T} \in \mathscr{W}' \text{such that } \mathscr{D}^*\mathbf{T} = f, \tag{3.147}$$

and the result follows. ■

From Theorem 3.1, and for every $\hat{\mathbf{v}} \in \text{Var}_v$, we have

$$\langle f, \hat{\mathbf{v}} \rangle = \langle \mathscr{D}^*\mathbf{T}, \hat{\mathbf{v}} \rangle = (\mathbf{T}, \mathscr{D}\hat{\mathbf{v}}), \tag{3.148}$$

which is the same expression as (3.138). In other words, the second part of the PVP results a mere consequence of the first one.

This expression (3.148) forges the representation of the external force f, or, equivalently, shapes the expression of the duality pairing $\langle \cdot, \cdot \rangle$ between $\mathscr{V}'$ and $\mathscr{V}$. From an

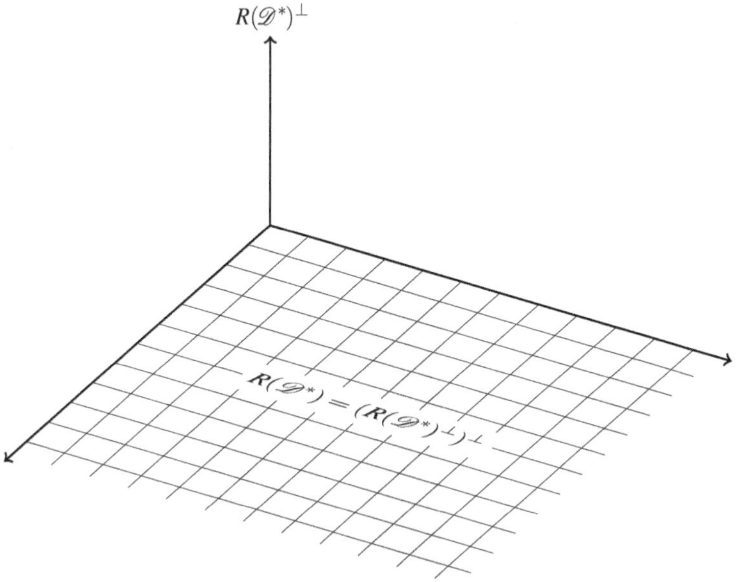

Figure 3.7 Illustration of the orthogonality concept in finite-dimensional spaces.

operational point of view, and in the context of continuum mechanics, this amounts to determining the adjoint operator $\mathscr{D}^*$ through the process of integrating by parts the expression of the internal power $P_i = -(\mathbf{T}, \mathscr{D}\hat{\mathbf{v}})$, recalling that $\hat{\mathbf{v}} \in \mathrm{Var}_v$.

Additional elements come into play from the PVP. For example, we can define the set

$$\mathrm{Est}_T = \{\mathbf{T} \in \mathscr{W}'; -(\mathbf{T}, \mathscr{D}\hat{\mathbf{v}}) + \langle f, \hat{\mathbf{v}}\rangle = 0 \quad \forall \hat{\mathbf{v}} \in \mathrm{Var}_v\}, \tag{3.149}$$

which is the set of all internal stresses equilibrated with the system of external forces $f \in \mathscr{V}'$ (from Theorem 3.1 we see that Est_T is not empty), and also the set

$$\mathrm{Var}_T = \{\hat{\mathbf{T}} \in \mathscr{W}'; \ (\hat{\mathbf{T}}, \mathscr{D}\hat{\mathbf{v}}) = 0 \quad \forall \hat{\mathbf{v}} \in \mathrm{Var}_v\}, \tag{3.150}$$

which is the vector subspace of $\mathscr{W}'$ composed by all internal stresses $\hat{\mathbf{T}}$ at equilibrium with $f = 0$, from which any $\hat{\mathbf{T}} \in \mathrm{Var}_T$ is referred to as a self-equilibrated internal stress.

A useful result that will be employed in forthcoming developments is the following.

Proposition 3.1 The orthogonal complement of the kernel space of the adjoint operator, denoted by $\mathscr{N}(\mathscr{D}^*)^\perp$, is such that

$$\mathscr{N}(\mathscr{D}^*)^\perp = R(\mathscr{D}). \tag{3.151}$$

Proof. From definition (3.150), we have that $\hat{\mathbf{T}} \in \mathrm{Var}_T$ entails $\mathscr{D}^*\hat{\mathbf{T}} = \mathbf{0}$. This means that $\hat{\mathbf{T}} \in \mathscr{N}(\mathscr{D}^*)$ and therefore $\mathrm{Var}_T = \mathscr{N}(\mathscr{D}^*)$. Now, consider $R(\mathscr{D}) = \{\mathbf{D} \in \mathscr{W}; \ \exists \mathbf{v} \in \mathscr{V}$ such that $\mathscr{D}\mathbf{v} = \mathbf{D}\}$, then $\hat{\mathbf{T}} \in \mathrm{Var}_T$ is equivalent to $(\hat{\mathbf{T}}, \mathscr{D}\hat{\mathbf{v}}) = 0 \ \ \forall \hat{\mathbf{v}} \in \mathrm{Var}_v$, which holds if and only if $\hat{\mathbf{T}} \in \mathscr{D}(\mathrm{Var}_v)^\perp = R(\mathscr{D})^\perp$. With this, we have $\mathrm{Var}_T = R(\mathscr{D})^\perp$. Hence,

$$\mathrm{Var}_T = \mathscr{N}(\mathscr{D}^*) = R(\mathscr{D})^\perp. \tag{3.152}$$

Now, recalling that $R(\mathscr{D})$ is a closed subspace of $\mathscr{W}$ we have

$$\mathrm{Var}_T^\perp = \mathscr{N}(\mathscr{D}^*)^\perp = R(\mathscr{D}), \tag{3.153}$$

and (3.151) is demonstrated. ■

Furthermore, we can also note that

$$\mathrm{Est}_T = \mathbf{T}_o + \mathrm{Var}_T, \tag{3.154}$$

where $\mathbf{T}_o \in \mathrm{Est}_T$ is arbitrary, that is, Est_T is the translation of the vector subspace Var_T whose elements are all the self-equilibrated internal stresses. Therefore, we also refer to Est_T as a linear manifold. In addition, it is easy to see that any element in Est_T can be represented by the sum of the element $\mathbf{T}_o$ and an element in Var_T[16] and vice versa, any element in Var_T can be written as the difference of elements in Est_T. Finally, it is important to underline that Theorem 3.1 ensures, for any $f \in \mathscr{V}'$ at equilibrium, the existence of the field $\mathbf{T} \in \mathscr{W}'$ such that (3.144) is true. However, the uniqueness of $\mathbf{T}$ is not guaranteed. In fact, if $\mathbf{T} \in \mathscr{W}'$ is associated with $f \in \mathscr{V}'$ through (3.144), then so is the element $\mathbf{T} + \hat{\mathbf{T}}$, for all $\hat{\mathbf{T}} \in \mathrm{Var}_T$. That is, the internal stress that equilibrates the force f is entirely characterized up to a self-equilibrated internal stress.

Two fundamental aspects of the PVP and of the Representation Theorem 3.1 is that they provide not only a representation for the system of external forces f, which has not been specified so far, but also the local form of the equilibrium equations. Indeed, if $f \in \mathscr{V}'$ is at equilibrium, the variational equation that governs the mechanical problem reads

$$-(\mathbf{T}, \mathscr{D}\hat{\mathbf{v}}) + \langle f, \hat{\mathbf{v}}\rangle = 0 \qquad \forall \hat{\mathbf{v}} \in \mathrm{Var}_v, \tag{3.155}$$

and because the body is free of constraints

$$\langle f, \hat{\mathbf{v}}\rangle = (\mathbf{T}, \mathscr{D}\hat{\mathbf{v}}) = \langle \mathscr{D}^*\mathbf{T}, \hat{\mathbf{v}}\rangle \qquad \forall \hat{\mathbf{v}} \in \mathrm{Var}_v = \mathscr{V}, \tag{3.156}$$

which gives us the representation for f. Let us discuss in more detail the case of a three-dimensional body. Supposing that all the fields are smooth enough so that we can operate mathematically, and knowing that $\mathbf{T} \in Sym$, this gives

$$\begin{aligned}\langle f, \hat{\mathbf{v}}\rangle = (\mathbf{T}, \mathscr{D}\hat{\mathbf{v}}) &= \int_{\mathscr{B}_t} \mathbf{T}\cdot(\operatorname{grad}\hat{\mathbf{v}})^s\, dV = \int_{\mathscr{B}_t} \mathbf{T}\cdot\operatorname{grad}\hat{\mathbf{v}}\, dV \\ &= \int_{\mathscr{B}_t} [\operatorname{div}(\mathbf{T}\hat{\mathbf{v}}) - \operatorname{div}\mathbf{T}\cdot\hat{\mathbf{v}}]\, dV = -\int_{\mathscr{B}_t} \operatorname{div}\mathbf{T}\cdot\hat{\mathbf{v}}\, dV \\ &\quad + \int_{\partial\mathscr{B}_t} \mathbf{T}\mathbf{n}\cdot\hat{\mathbf{v}}\, dS = \langle \mathscr{D}^*\mathbf{T}, \hat{\mathbf{v}}\rangle,\end{aligned} \tag{3.157}$$

where $\mathbf{n}$ is the unit normal vector pointing outward from $\partial\mathscr{B}_t$. In this manner, we characterize the equilibrium operator, which is

$$\mathscr{D}^*(\cdot) = \begin{cases} -\operatorname{div}(\cdot) & \text{in } \mathscr{B}_t, \\ (\cdot)\mathbf{n} & \text{on } \partial\mathscr{B}_t. \end{cases} \tag{3.158}$$

From this, we observe that the external force $f \in \mathscr{V}'$ compatible with the kinematical model here considered is characterized by a body force density $\mathbf{b} = -\operatorname{div}\mathbf{T}$ defined per unit volume in $\mathscr{B}_t$ and by a surface force density $\mathbf{a} = \mathbf{T}\mathbf{n}$ defined per unit surface on $\partial\mathscr{B}_t$.

---

16 This property of the elements in Var_T leads them to be known in the literature as admissible variations of elements in Est_T.

In conclusion, from the PVP and from the hypothesis that states that the internal power reads $P_i = -(\mathbf{T}, \mathbf{D}) = \int_{\mathscr{B}_t} \mathbf{T} \cdot \mathbf{D}\, dV$, we forged the linear functional $f \in \mathscr{V}\,'$ to be

$$\langle f, \hat{\mathbf{v}} \rangle = \int_{\mathscr{B}_t} \mathbf{b} \cdot \hat{\mathbf{v}}\, dV + \int_{\partial\mathscr{B}_t} \mathbf{a} \cdot \hat{\mathbf{v}}\, dS. \tag{3.159}$$

From the characterization given by (3.159), and for the model treated in this section, the PVP takes the following variational form

$$-\int_{\mathscr{B}_t} \mathbf{T} \cdot (\operatorname{grad} \mathbf{v})^s\, dV + \int_{\mathscr{B}_t} \mathbf{b} \cdot \hat{\mathbf{v}}\, dV + \int_{\partial\mathscr{B}_t} \mathbf{a} \cdot \hat{\mathbf{v}}\, dS = 0 \qquad \forall \hat{\mathbf{v}} \in \mathrm{Var}_v, \tag{3.160}$$

which, under proper regularity assumptions, yields

$$\begin{aligned} \int_{\mathscr{B}_t} \mathbf{b} \cdot \hat{\mathbf{v}}\, dV + \int_{\partial\mathscr{B}_t} \mathbf{a} \cdot \hat{\mathbf{v}}\, dS &= \int_{\mathscr{B}_t} \mathbf{T} \cdot (\operatorname{grad} \hat{\mathbf{v}})^s\, dV \\ &= \int_{\mathscr{B}_t} \mathbf{T} \cdot \operatorname{grad} \hat{\mathbf{v}}\, dV = \int_{\mathscr{B}_t} [\operatorname{div}(\mathbf{T}\hat{\mathbf{v}}) - \operatorname{div} \mathbf{T} \cdot \hat{\mathbf{v}}]\, dV \\ &= \int_{\mathscr{B}_t} - \operatorname{div} \mathbf{T} \cdot \hat{\mathbf{v}}\, dV + \int_{\partial\mathscr{B}_t} \mathbf{T}\mathbf{n} \cdot \hat{\mathbf{v}}\, dS \qquad \forall \hat{\mathbf{v}} \in \mathrm{Var}_v. \end{aligned} \tag{3.161}$$

Rearranging terms, this leads to

$$\int_{\mathscr{B}_t} (\operatorname{div} \mathbf{T} + \mathbf{b}) \cdot \hat{\mathbf{v}}\, dV - \int_{\partial\mathscr{B}_t} (\mathbf{T}\mathbf{n} - \mathbf{a}) \cdot \hat{\mathbf{v}}\, dS = 0 \qquad \forall \hat{\mathbf{v}} \in \mathrm{Var}_v = \mathscr{V}. \tag{3.162}$$

We then arrive at the following equations

$$\operatorname{div} \mathbf{T} + \mathbf{b} = \mathbf{0} \quad \text{in } \mathscr{B}_t, \tag{3.163}$$

$$\mathbf{T}\mathbf{n} = \mathbf{a} \quad \text{on } \partial\mathscr{B}_t, \tag{3.164}$$

which are known as the local equilibrium equations, or the Euler–Lagrange equations within the terminology of the calculus of variations (see [112, 265, 305]).

Remarkably, equations (3.163) and (3.164) could be obtained only after introducing the additional hypothesis that all fields involved in the operations were sufficiently smooth. When such a regularity assumption is not verified, the integration by parts cannot be carried out, and the Euler–Lagrange equations have to be understood in a generalized sense (distributional sense). Nevertheless, even in such scenario, the PVP, that is, the variational equation (3.160), maintains its validity. This lack of regularity of the fields gives rise to jump equations, which can be easily found following the same procedure as before and ensuring that the regularity of field $\mathbf{T}$ is enough so that integration by parts is allowed. In fact, if we admit that over the surface S (an inner surface to the body) we have that the field $\mathbf{T}$ is discontinuous, when integrating by parts a further term of the kind $\int_S [[\mathbf{T}]]\mathbf{n} \cdot \hat{\mathbf{v}} dS$ will emerge. Hence, in this case of lack of regularity, the external force f could include a further representation given by the force per unit surface $\mathbf{a}_S$ defined over the discontinuity surface S. Notice, then, that once the equilibrium is satisfied, the surface force $\mathbf{a}$ is related to the internal stress $\mathbf{T}$ through

$$\mathbf{T}\mathbf{n} = \mathbf{a} \qquad \text{on } \partial\mathscr{B}_t, \tag{3.165}$$

and, if there is a discontinuity over S, through

$$[[\mathbf{T}]]\mathbf{n} = \mathbf{a}_S \qquad \text{on } S. \tag{3.166}$$

This result is nothing but the Cauchy theorem which, in this manner, can be obtained as a simple consequence of the PVP (see [134]).

Importantly, given a system of forces $f \in \mathscr{V}'$, the PVP allows us to determine, first, if it is at equilibrium and, second, the internal stresses $\mathbf{T} \in \mathscr{W}'$ for which the equilibrium with f is verified. Contrariwise, the PVP allows, for a given $\mathbf{T} \in \mathrm{Est}_T$ to determine $f \in \mathscr{V}'$ which satisfies the equilibrium. Fundamentally, the PVP allows us to discern which types of external forces f are compatible with the kinematical model.

3.4.2 Principle of Complementary Virtual Power

In this section we will see the conditions that must be satisfied for a given deformation rate action $\mathbf{D} \in \mathscr{W}$ to be compatible, that is, that $\exists \mathbf{v} \in \mathscr{V}$ such that $\mathbf{D} = \mathscr{D}\mathbf{v}$. Let us present an important result which establishes the Principle of Complementary Virtual Power (PCVP), which provides a characterization of these compatible deformation rate actions.

Proposition 3.2 Principle of Complementary Virtual Power. The second-order tensor field $\mathbf{D}$, known as deformation rate action, defined in the configuration $\mathbf{u}_t \in \mathrm{Kin}_u$ which is at equilibrium, is compatible if and only if

$$(\hat{\mathbf{T}}, \mathbf{D}) = 0 \qquad \forall \hat{\mathbf{T}} \in \mathrm{Var}_T = \mathscr{N}(\mathscr{D}^*). \tag{3.167}$$

Proof. Indeed, if $\mathbf{D}$ is compatible there exists $\mathbf{v} \in \mathscr{V}$ such that $\mathbf{D} = \mathscr{D}\mathbf{v}$. Then, for all $\hat{\mathbf{T}} \in \mathrm{Var}_T = \mathscr{N}(\mathscr{D}^*)$ results

$$(\hat{\mathbf{T}}, \mathbf{D}) = (\hat{\mathbf{T}}, \mathscr{D}\mathbf{v}) = \langle \mathscr{D}^*\hat{\mathbf{T}}, \mathbf{v}\rangle = \langle \mathbf{0}, \mathbf{v}\rangle = 0. \tag{3.168}$$

Now, suppose that

$$(\hat{\mathbf{T}}, \mathbf{D}) = 0 \qquad \forall \hat{\mathbf{T}} \in \mathrm{Var}_T = \mathscr{N}(\mathscr{D}^*), \tag{3.169}$$

then

$$\mathbf{D} \in \mathscr{N}(\mathscr{D}^*)^{\perp}. \tag{3.170}$$

From relation (3.151) we have

$$\mathbf{D} \in \mathscr{N}(\mathscr{D}^*)^{\perp} = R(\mathscr{D}) \quad \Rightarrow \quad \exists \mathbf{v} \in \mathscr{V} \text{ such that } \mathbf{D} = \mathscr{D}\mathbf{v}, \tag{3.171}$$

and therefore the PCVP follows. ■

Essentially, the PCVP allows us to determine whether or not $\mathbf{D}$ is a compatible deformation rate action. The existence of $\mathbf{v}$ is ensured by this principle, but not the uniqueness. In fact, if $\mathbf{v}$ is the solution of the variational equation (3.167), which means that the PCVP is satisfied, then the motion action

$$\mathbf{v}_1 = \mathbf{v} + \mathbf{w} \qquad \mathbf{w} \in \mathscr{N}(\mathscr{D}), \tag{3.172}$$

is also a solution of (3.167). In other words, the solution is determined up to a rigid motion action. The interested reader can find more information on this topic in [220, Theorem 5.10.9, page 314].

3.5 Bodies with Bilateral Constraints

This section is devoted to the case in which the body, in its configuration $\mathscr{B}_t$ (equivalently $\mathbf{u}_t \in \text{Kin}_u$), experiences bilateral constraints in the entire boundary $\partial\mathscr{B}_t$ of $\mathscr{B}_t$.[17] Calling $\bar{\mathbf{v}}$ the velocity field prescribed on $\partial\mathscr{B}_t$, we have

$$\text{Kin}_v = \{\mathbf{v} \in \mathscr{V} \,;\, \mathbf{v}|_{\partial\mathscr{B}_t} = \bar{\mathbf{v}}\}. \tag{3.173}$$

Then, Var_v is the vector subspace of $\mathscr{V}$ given by

$$\text{Var}_v = \{\hat{\mathbf{v}} \in \mathscr{V} \,;\, \hat{\mathbf{v}}|_{\partial\mathscr{B}_t} = \mathbf{0}\}. \tag{3.174}$$

As we have seen, given an arbitrary $\mathbf{w} \in \text{Kin}_v$, we get

$$\text{Kin}_v = \mathbf{w} + \text{Var}_v, \tag{3.175}$$

that is, Kin_v is the linear manifold which results from the translation of the vector subspace Var_v of $\mathscr{V}$.

3.5.1 Principle of Virtual Power

For any Galilean frame of reference, and for every time t, the body $\mathscr{B}$ is at (static) equilibrium in the configuration $\mathscr{B}_t$ with bilateral kinematical constraints, under the action of a system of forces $f \in \mathscr{V}'$ if

$$P_e = \langle f, \hat{\mathbf{v}} \rangle = 0 \qquad \forall \hat{\mathbf{v}} \in \text{Var}_v \cap \mathscr{N}(\mathscr{D}), \tag{3.176}$$

and

$$P_i + P_e = -(\mathbf{T}, \mathscr{D}\hat{\mathbf{v}}) + \langle f, \hat{\mathbf{v}} \rangle = 0 \qquad \forall \hat{\mathbf{v}} \in \text{Var}_v. \tag{3.177}$$

Variational equations (3.176) and (3.177) fully define the concept of mechanical equilibrium.

As for the case of bodies without constraints, the first part of the PVP characterizes the external forces f which are compatible with the kinematical model. In particular, if

$$\text{Var}_v \cap \mathscr{N}(\mathscr{D}) = \{\mathbf{0}\}, \tag{3.178}$$

that is, if the kinematical constraints are such that rigid motions are not allowed, we have that any $f \in \mathscr{V}'$ is applicable over the body, since

$$P_e = \langle f, \hat{\mathbf{v}} \rangle = \langle f, \mathbf{0} \rangle = 0 \qquad \forall f \in \mathscr{V}'. \tag{3.179}$$

Consider that all the kinematical constraints are not enough to preclude all rigid motion actions. Then

$$\text{Var}_v \cap \mathscr{N}(\mathscr{D}) = \mathscr{V}_\pi, \tag{3.180}$$

from where applying the first part of the PVP leads us to the fact that $f \in \mathscr{V}'$ has to be orthogonal to the subspace of allowed rigid motions $\mathscr{V}_\pi$, that is

$$f \in \text{Var}_v^\perp + \mathscr{N}(\mathscr{D})^\perp = (\mathscr{V}_\pi)^\perp. \tag{3.181}$$

---

17 This hypothesis is made here without loss of generality. In fact, we could have considered bilateral constraints only over a portion of the boundary, say $\partial\mathscr{B}_{tv} \subset \partial\mathscr{B}_t$.

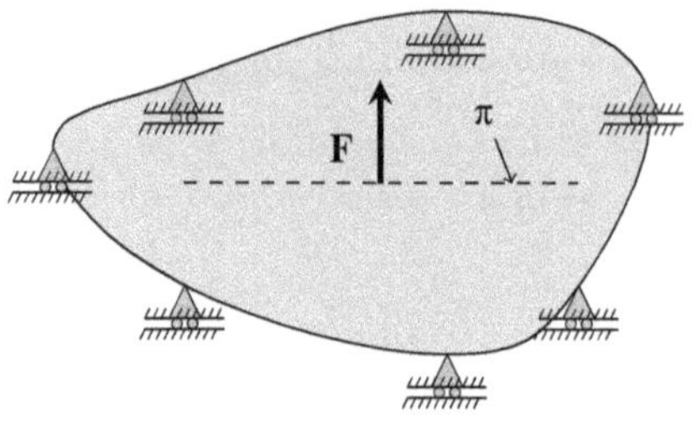

Figure 3.8 Body with bilateral kinematical constraints and resultant force of a compatible system of external forces.

For example, consider the case of Figure 3.8. Here, $\mathscr{V}_\pi$ is the subspace composed by vectors parallel to the plane π. Then, from the PVP we have

$$P_e = \langle f, \hat{\mathbf{v}} \rangle = \mathbf{F} \cdot \hat{\mathbf{v}} = 0 \qquad \forall \hat{\mathbf{v}} \in \mathscr{V}_\pi, \tag{3.182}$$

which shows us that the resultant force $\mathbf{F}$ of the applied system of forces f must be orthogonal to the plane π in order to have static equilibrium.

The previous example shows us the way in which, for a given kinematics established in the body, we can determine the set $\text{Var}_v \cap \mathscr{N}(\mathscr{D})$ and how, from this set and from the PVP, it is possible to completely characterize the external force f compatible with the equilibrium.

Similarly to the case analyzed in the previous section, the second part of the PVP permits the internal stress $\mathbf{T} \in \mathscr{W}'$ that equilibrates the force $f \in \mathscr{V}'$ to be determined. The next theorem is about the existence of this internal stress.

Theorem 3.2 Representation Theorem. Let $\mathbf{u}_t \in \text{Kin}_u$ be a configuration with bilateral kinematical constraints and at equilibrium under the action of the external force $f \in \mathscr{V}'$. Then, there exist $\mathbf{T} \in \mathscr{W}'$ and $r \in \mathscr{V}'$ such that

$$f = \mathscr{D}^*\mathbf{T} - r, \tag{3.183}$$

where r is called the reactive force to the prescribed constraint, or simply the reaction, and it is such that

$$\langle r, \hat{\mathbf{v}} \rangle = 0 \qquad \forall \hat{\mathbf{v}} \in \text{Var}_v, \tag{3.184}$$

or equivalently, $r \in \text{Var}_v^\perp$. Furthermore, $\mathbf{T}$ and r are said to equilibrate f.

Proof. By hypothesis, we have

$$f \in (\text{Var}_v \cap \mathscr{N}(\mathscr{D}))^\perp \quad \Leftrightarrow \quad f \in \text{Var}_v^\perp + \mathscr{N}(\mathscr{D})^\perp. \tag{3.185}$$

Figure 3.9 geometrically illustrates this result in the case of a finite-dimensional space. Using (3.185) and relation (3.146), which also holds in this case, leads to

$$f \in \text{Var}_v^\perp + R(\mathscr{D}^*). \tag{3.186}$$

This implies that there exist $-r \in \text{Var}_v^\perp$ and $\mathbf{T} \in \mathscr{W}'$ such that

$$f = \mathscr{D}^*\mathbf{T} - r, \tag{3.187}$$

or equivalently,

$$\mathscr{D}^*\mathbf{T} = f + r. \tag{3.188}$$

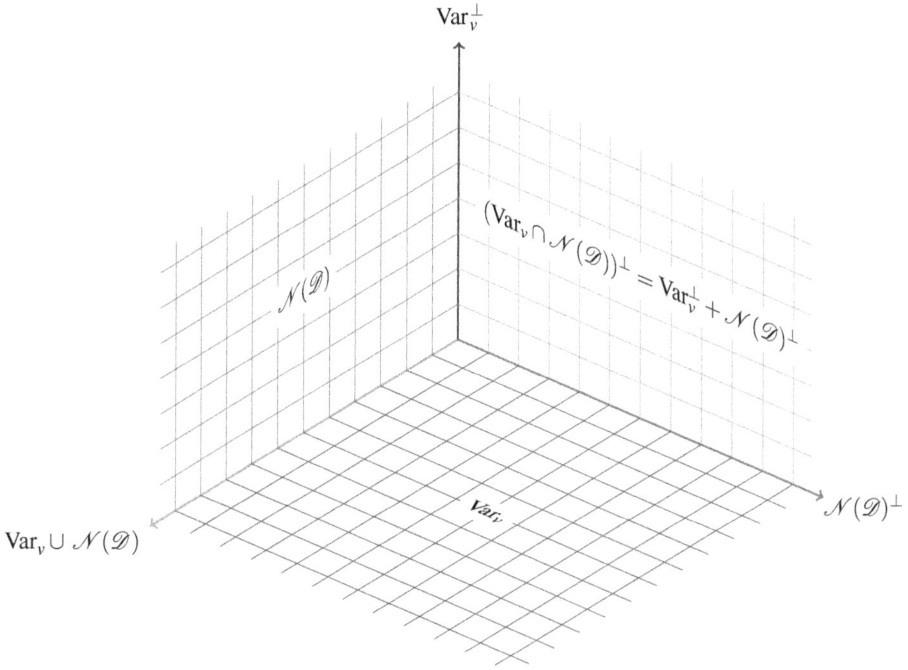

Figure 3.9 Illustration of the orthogonality with respect to the direct sum space in finite-dimensional spaces: $(\mathrm{Var}_v \cap \mathcal{N}(\mathcal{D}))^\perp = \mathrm{Var}_v^\perp + \mathcal{N}(\mathcal{D})^\perp$.

In turn, since $-r \in \mathrm{Var}_v^\perp$, it is

$$\langle -r, \hat{\mathbf{v}} \rangle = 0 \Leftrightarrow \langle r, \hat{\mathbf{v}} \rangle = 0 \qquad \forall \hat{\mathbf{v}} \in \mathrm{Var}_v, \tag{3.189}$$

and the result follows. ■

As a consequence of Theorem 3.2 we have that if $\mathbf{u}_t \in \mathrm{Kin}_u$, with bilateral constraints, is at equilibrium under the action of the force $f \in \mathscr{V}'$ then

$$\langle f, \hat{\mathbf{v}} \rangle = \langle \mathcal{D}^*\mathbf{T} - r, \hat{\mathbf{v}} \rangle = \langle \mathcal{D}^*\mathbf{T}, \hat{\mathbf{v}} \rangle = (\mathbf{T}, \mathcal{D}\mathbf{v}) \qquad \forall \hat{\mathbf{v}} \in \mathrm{Var}_v, \tag{3.190}$$

and, therefore, we arrive at the second part of the PVP, which allows us to determine the internal stress $\mathbf{T}$ which equilibrates the force f, or, reciprocally, given $\mathbf{T}$ we can find the external force such that the equilibrium is reached. Importantly, the possibility of calculating $\mathbf{T}$ as dictated by the second part of the PVP does not depend upon the reactive force r because this reaction is orthogonal to motion actions in Var_v. In other words, the power expended by r, for all $\hat{\mathbf{v}} \in \mathrm{Var}_v$, is null.

A direct effect of the orthogonality between the reaction r and motion actions $\hat{\mathbf{v}} \in \mathrm{Var}_v$ is that r cannot be calculated by assessing the power expended. To calculate these reactive forces, we resort again to Theorem 3.2, since it states that

$$r = \mathcal{D}^*\mathbf{T} - f. \tag{3.191}$$

Exploiting the PVP, expression (3.191) implies

$$\langle r, \hat{\mathbf{v}} \rangle = \langle \mathcal{D}^*\mathbf{T} - f, \hat{\mathbf{v}} \rangle = (\mathbf{T}, \mathcal{D}\hat{\mathbf{v}}) - \langle f, \hat{\mathbf{v}} \rangle \qquad \forall \hat{\mathbf{v}} \in \mathscr{V}. \tag{3.192}$$

In mechanical terms, the previous procedure amounted to replacing the kinematical constraint by the corresponding dual entity, which is the reactive force r, yielding a configuration of the body without constraints. Thus, once $\mathbf{T}$ has been obtained as a function of f using the PVP, the body is released from all kinematical constraints prescribed in Var_v. With this, again, we make use of the PVP, but now in the scenario of a body free of constraints which is subjected to a system of forces given by $r - \mathcal{D}^*\mathbf{T} + f$. In this manner we are able to calculate r.[18]

In an entirely similar manner to that of Section 3.4, given $f \in \mathcal{V}'$ compatible with the equilibrium, we can define the set

$$\mathrm{Est}_T = \{\mathbf{T} \in \mathcal{W}'; -(\mathbf{T}, \mathcal{D}\hat{\mathbf{v}}) + \langle f, \hat{\mathbf{v}}\rangle = 0 \quad \forall \hat{\mathbf{v}} \in \mathrm{Var}_v\}, \tag{3.193}$$

containing all the internal stresses $\mathbf{T}$ at equilibrium with the external force f (from Theorem 3.2 this set is not empty). We also can define

$$\mathrm{Var}_T = \{\hat{\mathbf{T}} \in \mathcal{W}'; \ (\hat{\mathbf{T}}, \mathcal{D}\hat{\mathbf{v}}) = 0 \quad \forall \hat{\mathbf{v}} \in \mathrm{Var}_v\} = (\mathcal{D}(\mathrm{Var}_v))^{\perp}, \tag{3.194}$$

as the vector subspace consisting of all self-equilibrated internal stresses (equilibrated with $f = 0$). Definitions (3.193) and (3.194) imply that

$$\mathrm{Est}_T = \mathbf{T}_o + \mathrm{Var}_T, \tag{3.195}$$

where $\mathbf{T}_o \in \mathrm{Est}_T$ is arbitrary. Once again, we have that Est_T is the linear manifold which results from the translation of a vector subspace of $\mathcal{W}'$.

Let us now consider the specific case of a three-dimensional body with bilateral constraints. Since $f \in \mathcal{V}'$ is compatible with the equilibrium, then the internal stress $\mathbf{T}$ verifies the variational equation

$$(\mathbf{T}, \mathcal{D}\hat{\mathbf{v}}) = \langle f, \hat{\mathbf{v}}\rangle \qquad \forall \hat{\mathbf{v}} \in \mathrm{Var}_v. \tag{3.196}$$

In turn, assuming that $\mathbf{T}$ is regular enough to operate mathematically, we go ahead as follows

$$\begin{aligned}\int_{\mathcal{B}_t} \mathbf{T} \cdot (\mathrm{grad}\,\hat{\mathbf{v}})^s \, dV &= \int_{\mathcal{B}_t} \mathbf{T} \cdot \mathrm{grad}\,\hat{\mathbf{v}} \, dV = \int_{\mathcal{B}_t} [\mathrm{div}\,(\mathbf{T}\hat{\mathbf{v}}) - \mathrm{div}\,\mathbf{T} \cdot \hat{\mathbf{v}}] \, dV \\ &= \int_{\partial\mathcal{B}_t} \mathbf{T}\mathbf{n} \cdot \hat{\mathbf{v}} \, dV - \int_{\mathcal{B}_t} \mathrm{div}\,\mathbf{T} \cdot \hat{\mathbf{v}} \, dV = -\int_{\mathcal{B}_t} \mathrm{div}\,\mathbf{T} \cdot \hat{\mathbf{v}} \, dV,\end{aligned} \tag{3.197}$$

where the fact that $\hat{\mathbf{v}} = 0$ all over $\partial\mathcal{B}_t$ has been used. Replacing (3.197) into (3.196) results in the representation for $f \in \mathcal{V}'$. In the present case, the force f compatible with

18 As examples of reactive forces we can cite the reaction to the kinematical constraint associated with the incompressible behavior of some media ($\mathrm{div}\,\mathbf{v} = 0$, see (3.95)). In this case, the reactive force r is given by a scalar field, say p, which corresponds to the hydrostatic pressure which cannot be determined during the calculation of the internal stress $\mathbf{T}$. Once $\mathbf{T}$ is known, such that equilibrates f, it is possible to know p by releasing the body from the incompressibility constraint. This amounts to applying over the body motion actions which do not satisfy the incompressibility constraint. In this way, $\mathbf{T}, f$ and p expend power and, at the same time, satisfy the PVP, yielding p. Another typical example can be found in the theory of plates and shells. In this field, a common hypothesis (therefore a kinematical constraint) is that normal fibers to the mean surface must not be deformed. Then, the reactive force which guarantees the rigidity of these fibers remains undetermined (null power expended) for all kinematically admissible motion actions.

the kinematical constraint considered over the entire boundary is given by a force per unit volume, say **b**, defined in $\mathscr{B}_t$. Hence

$$\langle f, \hat{\mathbf{v}} \rangle = \int_{\mathscr{B}_t} \mathbf{b} \cdot \hat{\mathbf{v}} dV \qquad \hat{\mathbf{v}} \in \mathrm{Var}_v. \tag{3.198}$$

As it is evident, we are not allowed to prescribe forces over the boundary $\partial\mathscr{B}_t$.

In this way, the variational equation (3.196) that expresses the PVP takes the form

$$\int_{\mathscr{B}_t} \mathbf{T} \cdot (\mathrm{grad}\,\hat{\mathbf{v}})^s \, dV = \int_{\mathscr{B}_t} \mathbf{b} \cdot \hat{\mathbf{v}} \, dV \qquad \forall \hat{\mathbf{v}} \in \mathrm{Var}_v. \tag{3.199}$$

The local form of the equilibrium, which in abstract form is given by $\mathscr{D}^*\mathbf{T} = f + r$ (see (3.191)), is easily found in explicit form through the same integration by parts procedure,

$$-\int_{\mathscr{B}_t} \mathrm{div}\,\mathbf{T} \cdot \hat{\mathbf{v}} dV = \int_{\mathscr{B}_t} \mathbf{b} \cdot \hat{\mathbf{v}} dV \qquad \forall \hat{\mathbf{v}} \in \mathrm{Var}_v, \tag{3.200}$$

which yields

$$\mathrm{div}\,\mathbf{T} + \mathbf{b} = \mathbf{0} \qquad \text{in } \mathscr{B}_t. \tag{3.201}$$

The reactive force r is found from the PVP in the case where the body is free of constraints. Let us then release the constraints present in Var_v as follows

$$\langle r, \hat{\mathbf{v}} \rangle = \int_{\mathscr{B}_t} \mathbf{T} \cdot (\mathrm{grad}\,\hat{\mathbf{v}}) \, dV - \int_{\mathscr{B}_t} \mathbf{b} \cdot \hat{\mathbf{v}} \, dV \qquad \forall \hat{\mathbf{v}} \in \mathscr{V}, \tag{3.202}$$

from where

$$\begin{aligned} \langle r, \hat{\mathbf{v}} \rangle &= \int_{\partial\mathscr{B}_t} \mathbf{Tn} \cdot \hat{\mathbf{v}} \, dS - \int_{\mathscr{B}_t} (\mathrm{div}\,\mathbf{T} + \mathbf{b}) \cdot \hat{\mathbf{v}} \, dV \\ &= \int_{\partial\mathscr{B}_t} \mathbf{Tn} \cdot \hat{\mathbf{v}} \, dS \qquad \forall \hat{\mathbf{v}} \in \mathscr{V}. \end{aligned} \tag{3.203}$$

Hence, we arrive at

$$\int_{\partial\mathscr{B}_t} (\mathbf{r} - \mathbf{Tn}) \cdot \hat{\mathbf{v}} \, dS = 0 \qquad \forall \hat{\mathbf{v}} \in \mathscr{V}, \tag{3.204}$$

which, in local form, reads as

$$\mathbf{r} = \mathbf{Tn} \qquad \text{on } \partial\mathscr{B}_t. \tag{3.205}$$

So far in this section, the analysis has focused on bodies whose boundary was entirely constrained. Consider now the situation in which just a portion of the boundary $\partial\mathscr{B}_t$ of the body, called $\partial\mathscr{B}_{tv}$, is prescribed with kinematical constraints. In such a case, we call $\partial\mathscr{B}_{ta}$ the complementary part of the boundary, in such a way that $\partial\mathscr{B}_t = \partial\mathscr{B}_{tv} \cup \partial\mathscr{B}_{ta}$ and $\partial\mathscr{B}_{tv} \cap \partial\mathscr{B}_{ta} = \emptyset$ are verified. Consider, for example, that $\mathbf{v} = \overline{\mathbf{v}}$ in $\partial\mathscr{B}_{tv}$. In this case, we have

$$\mathrm{Kin}_v = \{\mathbf{v} \in \mathscr{V}\,;\ \mathbf{v}|_{\partial\mathscr{B}_{tv}} = \overline{\mathbf{v}}\}, \tag{3.206}$$

$$\mathrm{Var}_v = \{\mathbf{v} \in \mathscr{V}\,;\ \mathbf{v}|_{\partial\mathscr{B}_{tv}} = \mathbf{0}\}. \tag{3.207}$$

An entirely similar analysis to that carried out before leads us to the following characterization of the external forces compatible with the kinematical model

$$P_e = \langle f, \hat{\mathbf{v}} \rangle = \int_{\mathcal{B}_t} \mathbf{b} \cdot \hat{\mathbf{v}}\, dV + \int_{\partial\mathcal{B}_{ta}} \mathbf{a} \cdot \hat{\mathbf{v}}\, dS \qquad \hat{\mathbf{v}} \in \mathrm{Var}_v. \tag{3.208}$$

In this case, f is characterized by a force per unit volume defined in $\mathcal{B}_t$, here denoted by $\mathbf{b}$, and by a force per unit surface defined in $\partial\mathcal{B}_{ta}$, here denoted by $\mathbf{a}$.

It is worthwhile keeping in mind that, when referring to $\partial\mathcal{B}_{tv}$ as the portion of the boundary with kinematical constraints, over the same portion of the boundary some components of the motion action can be prescribed and some others free.

Therefore, in this more general case the PVP gives, in extended form, the following governing variational equations

$$\int_{\mathcal{B}_t} \mathbf{b} \cdot \hat{\mathbf{v}}\, dV + \int_{\partial\mathcal{B}_{ta}} \mathbf{a} \cdot \hat{\mathbf{v}}\, dS = 0 \qquad \forall \hat{\mathbf{v}} \in \mathrm{Var}_v \cap \mathcal{N}(\mathcal{D}), \tag{3.209}$$

and

$$\int_{\mathcal{B}_t} \mathbf{T} \cdot (\mathrm{grad}\, \hat{\mathbf{v}})^s\, dV = \int_{\mathcal{B}_t} \mathbf{b} \cdot \hat{\mathbf{v}}\, dV + \int_{\partial\mathcal{B}_{ta}} \mathbf{a} \cdot \hat{\mathbf{v}}\, dS \qquad \forall \hat{\mathbf{v}} \in \mathrm{Var}_v. \tag{3.210}$$

Finally, the local form of the equilibrium given in abstract form by $\mathcal{D}^*\mathbf{T} = f + r$, for the present case it turns out to be

$$\mathrm{div}\, \mathbf{T} + \mathbf{b} = \mathbf{0} \quad \text{in } \mathcal{B}_t, \tag{3.211}$$

$$\mathbf{Tn} = \mathbf{a} \quad \text{on } \partial\mathcal{B}_{ta}, \tag{3.212}$$

$$\mathbf{Tn} = \mathbf{r} \quad \text{on } \partial\mathcal{B}_{tv}. \tag{3.213}$$

In this case of bilateral constraints, the same discussion presented at the end of Section 3.4 about the regularity of the fields holds. That is, the variational equation (3.199) behind the PVP remains valid even when the regularity of the fields is not sufficient to perform integration by parts. In these cases where fields can become discontinuous, the PVP implicitly provides the characterization of jump conditions over the surfaces where the discontinuity occurs, analogously to expression (3.166).

3.5.2 Principle of Complementary Virtual Power

The problem of compatibility is regarded as a dual problem to that of the mechanical equilibrium. This amounts to formulating the PCVP for the present case. Let us then analyze the compatibility of deformation rate actions in the case of bodies with bilateral constraints.

We say that the deformation rate action $\mathbf{D}$, defined in the current configuration $\mathbf{u}_t \in \mathrm{Kin}_u$, which is at equilibrium, is compatible if and only if

$$(\hat{\mathbf{T}}, \mathbf{D}) = (\hat{\mathbf{T}}, \mathcal{D}\mathbf{w}) \qquad \forall \hat{\mathbf{T}} \in \mathrm{Var}_T, \tag{3.214}$$

for an arbitrary element $\mathbf{w} \in \mathrm{Kin}_v$.

Before going deeper into the analysis, let us write (3.214) in its extended form

$$\int_{\mathcal{B}_t} \hat{\mathbf{T}} \cdot \mathbf{D}\, dV = \int_{\mathcal{B}_t} \hat{\mathbf{T}} \cdot (\text{grad}\, \mathbf{w})^s\, dV = \int_{\mathcal{B}_t} [\text{div}\,(\hat{\mathbf{T}}\mathbf{w}) - \text{div}\, \hat{\mathbf{T}} \cdot \mathbf{w}]\, dV$$
$$= \int_{\partial\mathcal{B}_t} \hat{\mathbf{T}}\mathbf{n} \cdot \mathbf{w}\, dS = \int_{\partial\mathcal{B}_t} \hat{\mathbf{T}}\mathbf{n} \cdot \bar{\mathbf{v}}\, dS, \tag{3.215}$$

where we used $\text{div}\, \hat{\mathbf{T}} = \mathbf{0}$ by virtue of being $\hat{\mathbf{T}} \in \text{Var}_T$ and $\mathbf{w}$ is constrained all over $\partial\mathcal{B}_t$ to be equal to $\bar{\mathbf{v}}$.

Proposition 3.3 Principle of Complementary Virtual Power. The deformation rate action $\mathbf{D}$, defined in the current configuration $\mathbf{u}_t \in \text{Kin}_u$ which is at equilibrium, is compatible if and only if

$$(\hat{\mathbf{T}} \cdot \mathbf{D}) = \int_{\partial\mathcal{B}_t} \hat{\mathbf{T}}\mathbf{n} \cdot \bar{\mathbf{v}}\, dS \qquad \forall \hat{\mathbf{T}} \in \text{Var}_T. \tag{3.216}$$

Proof. If $\mathbf{D}$ is compatible, there exists $\mathbf{v} \in \text{Kin}_v$ such that $\mathcal{D}\mathbf{v} = \mathbf{D}$. Recalling that Kin_v is a translation of Var_v, that is $\text{Kin}_v = \mathbf{w} + \text{Var}_v$ with $\mathbf{w} \in \text{Kin}_v$ arbitrary, we have that $\mathbf{v} - \mathbf{w} \in \text{Var}_v$. Hence, this yields

$$(\hat{\mathbf{T}}, \mathcal{D}(\mathbf{v} - \mathbf{w})) = 0 \qquad \forall \hat{\mathbf{T}} \in \text{Var}_T, \tag{3.217}$$

which holds from the very definition of Var_T. This shows us that the necessary condition is true.

To demonstrate sufficiency let us suppose that

$$(\hat{\mathbf{T}}, \mathbf{D}) = (\hat{\mathbf{T}}, \mathcal{D}\mathbf{w}) \qquad \forall \hat{\mathbf{T}} \in \text{Var}_T, \tag{3.218}$$

where $\mathbf{w} \in \text{Kin}_v$. Then

$$(\hat{\mathbf{T}}, \mathbf{D} - \mathcal{D}\mathbf{w}) = 0 \qquad \forall \hat{\mathbf{T}} \in \text{Var}_T. \tag{3.219}$$

From (3.194), we conclude that

$$\mathbf{D} - \mathcal{D}\mathbf{w} \in \text{Var}_T^{\perp} = \mathcal{D}(\text{Var}_v). \tag{3.220}$$

Thus

$$\mathbf{D} \in \mathcal{D}\mathbf{w} + \mathcal{D}(\text{Var}_v) = \mathcal{D}(\mathbf{w} + \text{Var}_v) = \mathcal{D}(\text{Kin}_v), \tag{3.221}$$

that is, $\exists \mathbf{v} \in \text{Kin}_v$ such that $\mathcal{D}\mathbf{v} = \mathbf{D}$, and the PCVP holds. ■

3.6 Bodies with Unilateral Constraints

In this section we address the situation in which the body is at configuration $\mathbf{u}_t \in \text{Kin}_u$ with kinematical constraints of frictionless unilateral kind. In all that follows, the presentation is limited to the case in which the set of all admissible motion actions Kin_v that can be considered over the body at configuration $\mathcal{B}_t$ is a convex cone with apex at the origin, or a translation of the cone. That is

$$\text{Kin}_v = \mathbf{w} + \text{Var}_v, \tag{3.222}$$

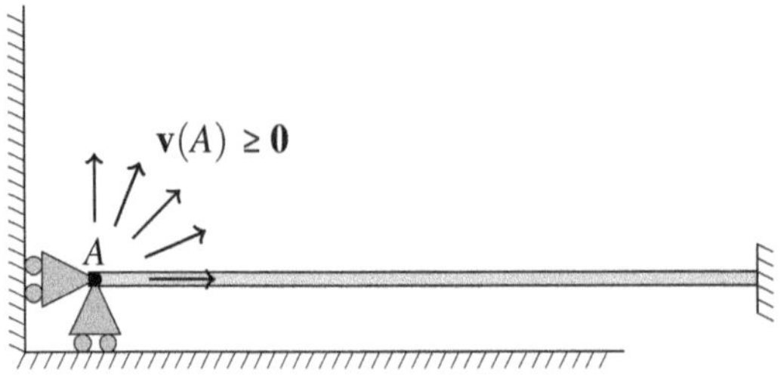

Figure 3.10 Beam with unilateral constraint at point A.

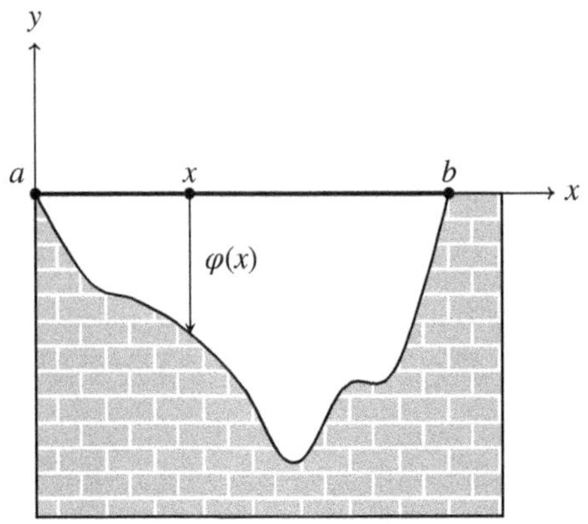

Figure 3.11 Cord with unilateral constraint over the entire domain given by the rigid obstacle.

where Var_v is a convex cone whose apex is at the origin and $\mathbf{w}$ is the translation from the origin.

There exists a large variety of mechanical problems in which this kind of constraint is manifested. Let us present two examples.

Consider the scheme shown in Figure 3.10, where we have a beam with a fixed support over the right extreme and an unilateral support over the left point. It can be appreciated that the set of admissible motion actions is such that at point A the inequality $\mathbf{v}(A) \geq \mathbf{0}$ (understood in a component-wise sense) must be satisfied. Then, we have

$$\mathrm{Kin}_v = \{\mathbf{v} \in \mathscr{V}\ ;\ \mathbf{v}(A) \geq \mathbf{0}\}. \tag{3.223}$$

Consider the body shown in Figure 3.11. The body in this example is a cord whose reference configuration is given by the horizontal line $\overline{ab}$. Let us also suppose that the displacements of particles in the cord, denoted by P, can only be given in the vertical direction, that is along the y axis. These vertical displacements are denoted by the vector $\mathbf{u}_t$ (which, again, has only the vertical component different from zero). Due to the presence of the rigid obstacle defined by the curve $\varphi(x)$, with $x \in [a, b]$, it turns out that $\mathbf{u}_t$ must satisfy

$$\mathbf{u}_t \cdot \mathbf{e}_y \geq \varphi(x) \qquad \forall x \in [a, b] \text{ and } \forall t \in [t_0, t_f], \tag{3.224}$$

and also

$$\mathbf{u}_t(a) = \mathbf{u}_t(b) = \mathbf{0} \qquad \forall t \in [t_0, t_f]. \tag{3.225}$$

Then, calling

$$\mathrm{Var}_u = \{\hat{\mathbf{u}} \in \mathscr{U};\ \hat{\mathbf{u}} \cdot \mathbf{e}_y \geq 0\ \ \forall x \in [a, b] \text{ and } \hat{\mathbf{u}}(a) = \hat{\mathbf{u}}(b) = \mathbf{0}\}, \tag{3.226}$$

it is easy to see that the set of all admissible configurations is given by

$$\text{Kin}_u = \varphi + \text{Var}_u, \tag{3.227}$$

which, thus, results in the translation φ of the convex cone Var_u whose apex is at the origin. In this example, note that Kin_v depends on the configuration $\mathscr{B}_t$. In effect, if $\mathbf{u}_t \in \text{Kin}_u$ is such that no particle of the cord is lying over the obstacle, then all vertical motion actions $\mathbf{v}$ satisfying $\mathbf{v}(a) = \mathbf{v}(b) = \mathbf{0}$ are kinematically admissible motion actions. Then, Kin_v for this configuration $\mathscr{B}_t$ is a vector subspace of the space of all possible motion actions $\mathscr{V}$. In contrast, this does not happen if, at configuration $\mathscr{B}_t$, some portion of the cord is lying over the obstacle. In such case, the set of admissible motion actions Kin_v is given by fields $\mathbf{v}$ such that:

- $\mathbf{v}(a) = \mathbf{v}(b) = \mathbf{0}$
- $\mathbf{v} \cdot \mathbf{e}_y \geq 0$ for every point of the cord lying on the obstacle
- $\mathbf{v}$ is free of constraints for the rest of the points in the cord.

The previous exposition shows us that Kin_v, for the configuration $\mathscr{B}_t$, is the very space Var_v defined by the motion actions $\mathbf{v}$ characterized above. It is rather direct to verify that if $\mathbf{v} \in \text{Kin}_v$, then $\alpha\mathbf{v} \in \text{Kin}_v$, $\forall \alpha \in \mathbb{R}^+$. Hence, it becomes clear that Kin_v is, in this case, a convex cone with apex at the origin.

It is interesting, and also important, to highlight once again that the sets of admissible motion actions Var_v analyzed in Sections 3.4 and 3.5 are particular cases of the situation exposed in this section. The reader should note that every space, and also subspace, is a convex cone with apex at the origin, while the converse is not true. Therefore, unilateral constraints are nothing but a generalization of the kinematical constraints studied so far.

Before going ahead with the exposition, let us explore whether the concept of mechanical equilibrium established in previous sections remains unchanged in this kinematical framework or not. In order to provide a definite answer to this issue let us consider a rigid ball lying over a table, as shown in Figure 3.12. In this case, the admissible motion actions are those rigid actions that are parallel to the plane defined by the table, as well as those that tend to lift the ball from the table. In this example, it is easy to see that $\text{Kin}_v = \text{Var}_v$ holds. In particular, for all rigid motion actions over the plane of the table we have that the associated external power expended is null because of the orthogonality between the external force and the motion actions over the plane. In turn, for the admissible motion actions that have the tendency to lift the ball from the table, we will have to expend our own power to carry them out. We can express this fact in a mathematical fashion as

$$P_e = \langle f, \mathbf{v} \rangle \leq 0 \qquad \forall \mathbf{v} \in \text{Var}_v. \tag{3.228}$$

This shows us that, for unilateral constraints, the definition of the mechanical equilibrium must be consistent with this simple observation of reality.

3.6.1 Principle of Virtual Power

For any Galilean frame of reference, and for every time t, the body $\mathscr{B}$ is at (static) equilibrium in the configuration $\mathscr{B}_t$ with unilateral kinematical constraints, subjected to the system of forces $f \in \mathscr{V}'$ if

$$P_e = \langle f, \hat{\mathbf{v}} \rangle \leq 0 \qquad \forall \hat{\mathbf{v}} \in \text{Var}_v \cap \mathscr{N}(\mathscr{D}), \tag{3.229}$$

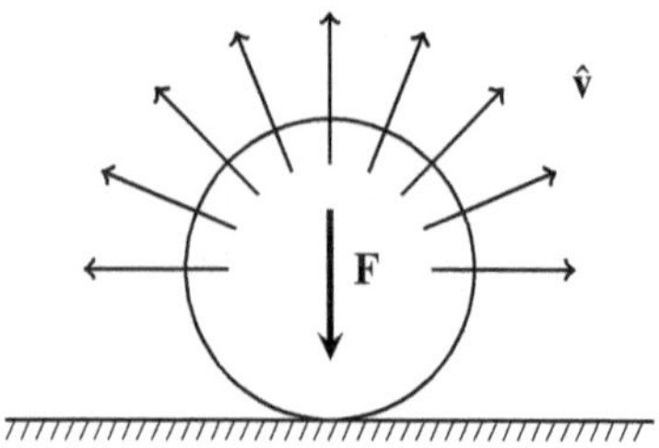

Figure 3.12 Unilateral constraint for the problem of a ball lying on a table.

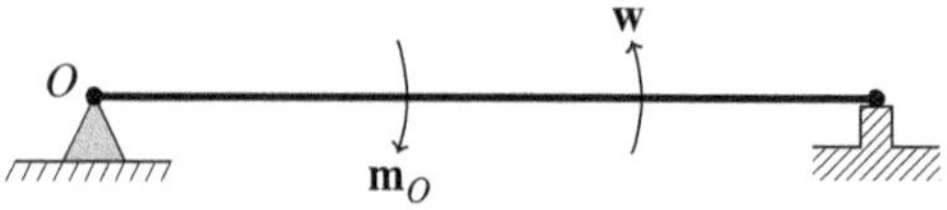

Figure 3.13 Beam with unilateral kinematical constraint.

and, also, if

$$P_i + P_e = -(\mathbf{T}, \mathcal{D}\hat{\mathbf{v}}) + \langle f, \hat{\mathbf{v}} \rangle \leq 0 \qquad \forall \hat{\mathbf{v}} \in \mathrm{Var}_v. \tag{3.230}$$

In this case, and differently from the kinematical constraints studied in previous sections, the concept of mechanical equilibrium as dictated by the PVP is governed by variational inequalities, given by expressions (3.229) and (3.230).

As seen before, the first part of the PVP gives the structure of those systems of forces $f \in \mathcal{V}'$ which are compatible with the kinematical model. For example, consider the beam illustrated in Figure 3.13. The right extremal point is a unilateral support, so it is free to move horizontally as well as in the upward direction, but not downward. Over the left point a pinned support is responsible for the bilateral kinematical constraint. Thus, in such a case it is

$$\mathrm{Var}_v \cap \mathcal{N}(\mathcal{D}) = Rot^-, \tag{3.231}$$

where Rot^- is the set of all counterclockwise rotation actions around point O. As a consequence, a force $f \in \mathcal{V}'$ is compatible with the model provided that

$$\begin{aligned} P_e = \langle f, \hat{\mathbf{v}} \rangle = \langle f, \mathbf{w} \times (\mathbf{x} - O) \rangle = \mathbf{F} \cdot \mathbf{w} \times (\mathbf{x} - O) = (\mathbf{x} - O) \times \mathbf{F} \cdot \mathbf{w} \\ = \mathbf{m}_O \cdot \mathbf{w} \leq 0 \qquad \forall \mathbf{w} \text{ counterclockwise.} \end{aligned} \tag{3.232}$$

This result informs us that the resultant moment of applied forces f with respect to point O, denoted by $\mathbf{m}_O$, must oppose that of $\mathbf{w}$, that is, the resultant moment $\mathbf{m}_O$ has to have clockwise direction. Hence, the system of forces f has to satisfy this condition, otherwise it is incompatible with the static equilibrium of the body.

As we already know, the second part of the PVP provides a mean to characterize the internal stress $\mathbf{T}$ which is at equilibrium with the applied force f. Next, we show a result of the existence of this internal stress $\mathbf{T} \in \mathcal{W}'$ satisfying the PVP, for any given external force $f \in \mathcal{V}'$.

Theorem 3.3 Representation Theorem. Let $\mathbf{u}_t \in \mathrm{Kin}_u$ be a configuration of the body featuring unilateral (frictionless) kinematical constraints, and at equilibrium with the external force $f \in \mathscr{V}'$. Then, there exist $\mathbf{T} \in \mathscr{W}'$ and $r \in \mathscr{V}'$ such that

$$f = \mathscr{D}^*\mathbf{T} - r, \tag{3.233}$$

where r is the reaction to the kinematical constraint, which satisfies

$$\langle r, \hat{\mathbf{v}} \rangle \geq 0 \qquad \forall \hat{\mathbf{v}} \in \mathrm{Var}_v, \tag{3.234}$$

that is, $r \in \mathrm{Var}_v^+$. Moreover, $\mathbf{T}$ and r equilibrate the external force f.

Proof. By hypothesis, we have

$$\langle f, \hat{\mathbf{v}} \rangle \leq 0 \qquad \forall \hat{\mathbf{v}} \in \mathrm{Var}_v \cap \mathscr{N}(\mathscr{D}), \tag{3.235}$$

which means

$$f \in (\mathrm{Var}_v \cap \mathscr{N}(\mathscr{D}))^- = \mathrm{Var}_v^- + \mathscr{N}(\mathscr{D})^-. \tag{3.236}$$

However, $\mathscr{N}(\mathscr{D})$ is a subspace of $\mathscr{V}$, therefore

$$\mathscr{N}(\mathscr{D})^- = \mathscr{N}(\mathscr{D})^+ = \mathscr{N}(\mathscr{D})^\perp. \tag{3.237}$$

Remembering expression (3.146), we are led to

$$f \in \mathrm{Var}_v^- + R(\mathscr{D}^*). \tag{3.238}$$

This expression signifies that there exist $\mathbf{T} \in \mathscr{W}'$ and $-r \in \mathrm{Var}_v^-$ (that is, $\langle -r, \hat{\mathbf{v}} \rangle \leq 0$, $\forall \hat{\mathbf{v}} \in \mathrm{Var}_v$), such that

$$f = \mathscr{D}^*\mathbf{T} - r, \tag{3.239}$$

or, equivalently,

$$\mathscr{D}^*\mathbf{T} = f + r. \tag{3.240}$$

What is more, and given that $-r \in \mathrm{Var}_v^-$, it gives

$$\langle -r, \hat{\mathbf{v}} \rangle \leq 0 \Leftrightarrow \langle r, \hat{\mathbf{v}} \rangle \geq 0 \qquad \forall \hat{\mathbf{v}} \in \mathrm{Var}_v, \tag{3.241}$$

and the result is proved. ∎

Theorem 3.3 establishes that the reactive forces belong to the convex cone whose apex is at the origin and which is positive conjugate to the cone of all kinematically admissible virtual actions Var_v.

With the result given by this theorem, observe that now we have

$$\langle f, \hat{\mathbf{v}} \rangle = \langle \mathscr{D}^*\mathbf{T} - r, \hat{\mathbf{v}} \rangle = \langle \mathscr{D}^*\mathbf{T}, \hat{\mathbf{v}} \rangle - \langle r, \hat{\mathbf{v}} \rangle = (\mathbf{T}, \mathscr{D}\hat{\mathbf{v}}) - \langle r, \hat{\mathbf{v}} \rangle \qquad \forall \hat{\mathbf{v}} \in \mathrm{Var}_v, \tag{3.242}$$

and recalling that $r \in \mathrm{Var}_v^+$, finally we obtain

$$\langle f, \hat{\mathbf{v}} \rangle \leq (\mathbf{T}, \mathscr{D}\hat{\mathbf{v}}) \quad \forall \hat{\mathbf{v}} \in \mathrm{Var}_v, \tag{3.243}$$

which is the second part of the PVP.

Let us underline some important aspects of the present variational structure. Similarly to the case of bodies without constraints and bodies with bilateral constraints, reactive forces play no role in the definition of mechanical equilibrium (the first part of the PVP). This is an additional asset because in general these reactions are unknown. In the present framework, reactions are naturally purged from the formulation as a consequence of the relation with motion actions in Var_v. For example, for free bodies (or with bilateral constraints) it is $r \in \mathrm{Var}_v^{\perp}$, while in the case of unilateral constraints it is $r \in \mathrm{Var}_v^{+}$. It is noteworthy that in the three cases we have

$$\langle f, \hat{\mathbf{v}} \rangle + \langle r, \hat{\mathbf{v}} \rangle = 0 \qquad \forall \hat{\mathbf{v}} \in \mathrm{Var}_v \cap \mathcal{N}(\mathcal{D}). \tag{3.244}$$

For free bodies (or with bilateral constraints), the second term of the left-hand side in (3.244) is null because $r \in \mathrm{Var}_v^{\perp}$ (particularly, for free bodies $\mathrm{Var}_v = \mathcal{V}$, then $r = 0$!). For bodies with unilateral constraints such a term is positive ($\langle r, \hat{\mathbf{v}} \rangle \geq 0$) because $r \in \mathrm{Var}_v^{+}$. This makes it possible to dismiss r from (3.244), yielding $\langle f, \hat{\mathbf{v}} \rangle \leq 0, \forall \hat{\mathbf{v}} \in \mathrm{Var}_v \cap \mathcal{N}(\mathcal{D})$. However, when we do this, we pay the price of dealing with a variational inequality. With this, the (once linear) problem of determining the internal stress $\mathbf{T}$ caused by a given force f, the latter compatible with the equilibrium, turns, in the case of unilateral constraints, into a nonlinear problem.

As before, for $f \in \mathcal{V}'$ at equilibrium, we now introduce the definitions

$$\mathrm{Est}_T = \{\mathbf{T} \in \mathcal{W}; -(\mathbf{T}, \mathcal{D}\hat{\mathbf{v}}) + \langle f, \hat{\mathbf{v}} \rangle \leq 0 \quad \forall \hat{\mathbf{v}} \in \mathrm{Var}_v\}, \tag{3.245}$$

and

$$\mathrm{Var}_T = \{\hat{\mathbf{T}} \in \mathcal{W}'; -(\hat{\mathbf{T}}, \mathcal{D}\hat{\mathbf{v}}) \leq 0 \quad \forall \hat{\mathbf{v}} \in \mathrm{Var}_v\} = (\mathcal{D}(\mathrm{Var}_v))^{+}. \tag{3.246}$$

Here, Var_T is the set of all internal stresses at equilibrium with the null force. Therefore, they are acknowledged as self-equilibrated stresses. As can be seen from the definition (3.246), Var_T is the positive conjugate cone to the convex cone $\mathcal{D}(\mathrm{Var}_v)$ whose vertex is at the origin. As in the previous sections, the following holds

$$\mathrm{Est}_T = \mathbf{T}_o + \mathrm{Var}_T, \tag{3.247}$$

where $\mathbf{T}_o$ is an arbitrary element of Est_T. Hence, Est_T is the translation of the convex cone Var_T.

Before ending this section, Let us write down the variational inequality of the PVP (3.230) for the case of a three-dimensional body

$$\int_{\partial\mathcal{B}_t} \mathbf{T} \cdot (\mathrm{grad}\,\hat{\mathbf{v}})^s \, dV \geq \int_{\mathcal{B}_t} \mathbf{b} \cdot \hat{\mathbf{v}} \, dV + \int_{\partial\mathcal{B}_{ta}} \mathbf{a} \cdot \hat{\mathbf{v}} \, dS \qquad \forall \hat{\mathbf{v}} \in \mathrm{Var}_v, \tag{3.248}$$

where $\partial\mathcal{B}_{ta}$ is the part of the boundary $\partial\mathcal{B}_t$ which is free of kinematical constraints of any kind. The complementary part of the boundary $\partial\mathcal{B}_t \setminus \partial\mathcal{B}_{ta}$ is somehow prescribed with unilateral and, also possibly bilateral, kinematical constraints. Also, as before, $\mathbf{b}$ is a force per unit volume defined all over $\mathcal{B}_t$ and $\mathbf{a}$ is a force per unit surface defined on $\partial\mathcal{B}_{ta}$.

3.6.2 Principle of Complementary Virtual Power

Let us inspect now the problem dual to the mechanical equilibrium problem, that is, the compatibility problem for bodies with unilateral constraints. The PCVP, as before, characterizes the compatibility of deformation rate actions $\mathbf{D}$, that is the fact that there exists $\mathbf{v}$ such that $\mathbf{D} = \mathcal{D}\mathbf{v}$.

Proposition 3.4 It is said that the deformation rate action $\mathbf{D}$, defined in $\mathbf{u}_t \in \mathrm{Kin}_u$, with unilateral kinematical constraints, and at equilibrium when subjected to the force $f \in \mathscr{V}'$, is compatible if and only if

$$(\hat{\mathbf{T}}, \mathbf{D}) \geq (\hat{\mathbf{T}}, \mathscr{D}\mathbf{w}) \qquad \forall \hat{\mathbf{T}} \in \mathrm{Var}_T, \tag{3.249}$$

for an arbitrary $\mathbf{w} \in \mathrm{Kin}_v$.

Proof. If $\mathbf{D}$ is compatible, then there exists $\mathbf{v} \in \mathrm{Kin}_v$ such that $\mathbf{D} = \mathscr{D}\mathbf{v}$. In addition, since $\mathbf{v} \in \mathrm{Kin}_v = \mathbf{w} + \mathrm{Var}_v$, we have $\mathbf{v} - \mathbf{w} \in \mathrm{Var}_v$ and, thus, we can write

$$(\hat{\mathbf{T}}, \mathscr{D}(\mathbf{v} - \mathbf{w})) \geq 0 \qquad \forall \hat{\mathbf{T}} \in \mathrm{Var}_T, \tag{3.250}$$

that is

$$(\hat{\mathbf{T}}, \mathscr{D}\mathbf{v}) \geq (\hat{\mathbf{T}}, \mathscr{D}\mathbf{w}) \qquad \forall \hat{\mathbf{T}} \in \mathrm{Var}_T, \tag{3.251}$$

and so the necessary condition is proved.

Assume now that $\mathbf{D}$ satisfies the PCVP, then

$$(\hat{\mathbf{T}}, \mathbf{D} - \mathscr{D}\mathbf{w}) \geq 0 \qquad \forall \hat{\mathbf{T}} \in \mathrm{Var}_T, \tag{3.252}$$

which means that

$$\mathbf{D} - \mathscr{D}\mathbf{w} \in \mathrm{Var}_T^+, \tag{3.253}$$

and therefore

$$\mathbf{D} \in \mathscr{D}\mathbf{w} + \mathrm{Var}_T^+. \tag{3.254}$$

From (3.246), we know that $\mathrm{Var}_T = (\mathscr{D}(\mathrm{Var}_v))^+$ and, recalling that $\mathscr{D}(\mathrm{Var}_v)$ is a closed convex cone with apex at the origin, we finally get

$$\mathbf{D} \in \mathscr{D}\mathbf{w} + ((\mathscr{D}(\mathrm{Var}_v))^+)^+ = \mathscr{D}\mathbf{w} + \mathscr{D}(\mathrm{Var}_v) = \mathscr{D}(\mathrm{Kin}_v), \tag{3.255}$$

implying that there exists $\mathbf{v} \in \mathrm{Kin}_v$ such that $\mathbf{D} = \mathscr{D}\mathbf{v}$, and so the sufficient condition is proved. Therefore, the characterization of compatible deformation rate actions given by the PCVP holds. ■

Notably, the concepts presented for the case of unilateral kinematical constraints are absolutely general, in the sense that they also hold for bodies with bilateral kinematical constraints as well as for bodies without constraints. These latter cases are particular situations of the unilateral one. Mathematically, this generalization occurs because the admissible motion actions characterized by Var_v belong to a convex cone (with apex at the origin) in the case of unilateral constraints. In the other cases (bilateral and no constraints) Var_v was, simply, a particular case of such a convex cone. As a matter of fact, we have seen that the mechanical equilibrium for bodies with unilateral constraints is formulated by the expression of the PVP given by the following variational inequality

$$-(\mathbf{T}, \mathscr{D}\hat{\mathbf{v}}) + \langle f, \hat{\mathbf{v}} \rangle \leq 0 \qquad \forall \hat{\mathbf{v}} \in \mathrm{Var}_v. \tag{3.256}$$

Now, given Var_v is a vector subspace of $\mathscr{V}$ (e.g. in bodies with bilateral constraints), if $\hat{\mathbf{v}} \in \mathrm{Var}_v$, then $\alpha\hat{\mathbf{v}} \in \mathrm{Var}_v$ for all $\alpha \in \mathbb{R}$. Thus

$$-(\mathbf{T}, \mathscr{D}(\alpha\hat{\mathbf{v}})) + \langle f, \alpha\hat{\mathbf{v}} \rangle \leq 0 \qquad \forall \hat{\mathbf{v}} \in \mathrm{Var}_v, \ \forall \alpha \in \mathbb{R}. \tag{3.257}$$

From the linearity of the functionals it follows that

$$\alpha[-(\mathbf{T}, \mathcal{D}\hat{\mathbf{v}}) + \langle f, \hat{\mathbf{v}}\rangle] \leq 0 \qquad \forall \hat{\mathbf{v}} \in \mathrm{Var}_v,\ \forall \alpha \in \mathbb{R}. \tag{3.258}$$

Since α can take arbitrary positive and/or negative values in $\mathbb{R}$, the only possibility for (3.258) to hold is that

$$-(\mathbf{T}, \mathcal{D}\hat{\mathbf{v}}) + \langle f, \hat{\mathbf{v}}\rangle = 0 \qquad \forall \hat{\mathbf{v}} \in \mathrm{Var}_v, \tag{3.259}$$

and the expression of the PVP for bodies with bilateral constraints is recovered.

3.7 Lagrangian Description of the Principle of Virtual Power

At this point, it is worthwhile recalling that the exposition of the PVP, as well as that of its dual problem, the PCVP, is carried out for a body occupying its actual configuration in space. In the mechanics realm, this amounts to considering the spatial, also known as Eulerian, description of the fields involved in the PVP (or PCVP). This entails that all variables, namely motion actions, deformation rate actions, external forces, and internal stresses, are fields defined in the current spatial configuration $\mathscr{B}_t$. In this way, for example, in the case of bilateral constraints, the mechanical equilibrium was mathematically expressed through the variational equation

$$P_i + P_e = -(\mathbf{T}, \mathcal{D}\hat{\mathbf{v}}) + \langle f, \hat{\mathbf{v}}\rangle = 0 \qquad \forall \hat{\mathbf{v}} \in \mathrm{Var}_v. \tag{3.260}$$

For a three-dimensional body, the internal power functional, P_i, reads

$$-P_i = (\mathbf{T}, \mathcal{D}\hat{\mathbf{v}}) = \int_{\mathscr{B}_t} \mathbf{T} \cdot (\operatorname{grad} \hat{\mathbf{v}})^s \, dV. \tag{3.261}$$

Expression (3.261) is the way to calculate the internal power (whose value is independent of the configuration of the body) by using quantities defined in the spatial configuration of the body.

For certain problems, performing these calculations in this spatial description poses no difficulties, since the region $\mathscr{B}_t$ is already known. In other problems, like in the general case of deformable bodies (with kinematics described as in Section 3.2), the very definition of $\mathscr{B}_t$ is unknown. Notwithstanding this, in such cases we can always resort to some known configuration that we call now reference configuration, denote by $\mathscr{B}_r$. In many problems, this reference configuration is the already introduced material configuration $\mathscr{B}$. So, in those cases, it is fundamental to determine the material, also known Lagrangian, description of the PVP (or PCVP).

To achieve the goal stated above, we proceed with a change of coordinates and write

$$-P_i = (\mathbf{T}, \mathcal{D}\hat{\mathbf{v}}) = \int_{\mathscr{B}_t} \mathbf{T} \cdot (\operatorname{grad} \hat{\mathbf{v}})^s \, dV = \int_{\mathscr{B}} \mathbf{T}_m \cdot ((\operatorname{grad} \hat{\mathbf{v}})^s)_m \det \mathbf{F} \, dV. \tag{3.262}$$

Recall (3.29) and (3.38), and so we have $\mathcal{X}_t : \mathscr{B} \to \mathscr{B}_t$ and $\mathbf{T}_m(\mathbf{X}) = \mathbf{T}(\mathcal{X}_t(\mathbf{X}))$. Let us define the displacement field in the material configuration as $\mathbf{U}(\mathbf{X})$ such that $\mathbf{x} = \mathbf{X} + \mathbf{U}(\mathbf{X})$ (see (3.5)). Moreover, from the transformation of the differential element we obtained $\det \mathbf{F} = \det \nabla \mathcal{X}_t$. In turn, from (3.69) we know that $(\operatorname{grad} \mathbf{v})_m = \nabla \mathbf{v}_m \mathbf{F}^{-1}$.

Replacing these results into (3.262), and noting that $\mathbf{T}$ is a symmetric second-order tensor, gives

$$
\begin{aligned}
-P_i &= \int_{\mathscr{B}} \mathbf{T}_m \cdot ((\operatorname{grad} \hat{\mathbf{v}})^s)_m \det \mathbf{F} dV = \int_{\mathscr{B}} \mathbf{T}_m \cdot (\nabla \hat{\mathbf{v}}_m \mathbf{F}^{-1})^s \det \mathbf{F} dV = \\
&= \int_{\mathscr{B}} \mathbf{T}_m \cdot \nabla \hat{\mathbf{v}}_m \mathbf{F}^{-1} \det \mathbf{F} dV = \int_{\mathscr{B}} (\det \mathbf{F} \mathbf{T}_m \mathbf{F}^{-T}) \cdot \nabla \hat{\mathbf{v}}_m dV \\
&= \int_{\mathscr{B}} \tilde{\mathbf{T}} \cdot \nabla \hat{\mathbf{v}}_m dV. \qquad (3.263)
\end{aligned}
$$

The field $\tilde{\mathbf{T}}$ is known in the literature as the first Piola–Kirchhoff stress tensor, and is defined by

$$
\tilde{\mathbf{T}} = \det \mathbf{F} \mathbf{T}_m \mathbf{F}^{-T}. \qquad (3.264)
$$

As can be appreciated, $\tilde{\mathbf{T}}$ is not symmetric and is power-conjugate to $\nabla \hat{\mathbf{v}}_m$. In this fashion, we have arrived at a material description for the calculation of the internal power P_i.

Let us specify kinematical constraints of bilateral type over $\partial \mathscr{B}_{tv}$, which, as we already know (see (3.208)), allows us to specify a force per unit surface over the complementary boundary $\partial \mathscr{B}_{ta}$, whose representation in the material configuration is denoted by $\partial \mathscr{B}_a$, featuring an outward normal unit vector $\mathbf{N}$. Then, the material description to calculate the external power P_e is reached as shown,

$$
\begin{aligned}
\langle f, \hat{\mathbf{v}} \rangle &= \int_{\mathscr{B}_t} \mathbf{b} \cdot \hat{\mathbf{v}} \, dV + \int_{\partial \mathscr{B}_{ta}} \mathbf{a} \cdot \hat{\mathbf{v}} \, dS \\
&= \int_{\mathscr{B}} \det \mathbf{F} \mathbf{b}_m \cdot \hat{\mathbf{v}}_m \, dV + \int_{\partial \mathscr{B}_a} \mathbf{a}_m \|\mathbf{F}^{-1} \mathbf{N}\| \det \mathbf{F} \cdot \hat{\mathbf{v}}_m \, dS \\
&= \int_{\mathscr{B}} \mathbf{b}^* \cdot \hat{\mathbf{v}}_m \, dV + \int_{\partial \mathscr{B}_a} \mathbf{a}^* \cdot \hat{\mathbf{v}}_m \, dS, \qquad (3.265)
\end{aligned}
$$

where we have used the transformation of the element of surface to the material configuration, and we have introduced the following transformed external forces

$$
\mathbf{b}^* = \det \mathbf{F} \mathbf{b}_m, \qquad (3.266)
$$

$$
\mathbf{a}^* = \mathbf{a}_m \|\mathbf{F}^{-1} \mathbf{N}\| \det \mathbf{F}. \qquad (3.267)
$$

Combining (3.263) and (3.265), we arrive at the variational equation underlying the material, or Lagrangian, description of the PVP, which reads

$$
\int_{\mathscr{B}} \mathbf{b}^* \cdot \hat{\mathbf{v}}_m \, dV + \int_{\partial \mathscr{B}_a} \mathbf{a}^* \cdot \hat{\mathbf{v}}_m \, dS = 0 \qquad \forall \hat{\mathbf{v}} \in (\mathrm{Var}_v \cap \mathscr{N}(\mathscr{D}))_m, \qquad (3.268)
$$

and

$$
\int_{\mathscr{B}} \tilde{\mathbf{T}} \cdot \nabla \hat{\mathbf{v}}_m dV = \int_{\mathscr{B}} \mathbf{b}^* \cdot \hat{\mathbf{v}}_m dV + \int_{\partial \mathscr{B}_a} \mathbf{a}^* \cdot \hat{\mathbf{v}}_m dS \qquad \forall \hat{\mathbf{v}}_m \in (\mathrm{Var}_v)_m, \qquad (3.269)
$$

where $(\mathrm{Var}_v)_m$ groups all the kinematically admissible motion actions described in the material configuration.

The main conceptual drawback in working with expression (3.269) is that the tensor field $\tilde{\mathbf{T}}$ is not necessarily symmetric. In order to recover a variational formulation of the

PVP incorporating a symmetric measure of the internal stress in the material configuration we proceed with

$$\begin{aligned}
-P_i &= \int_{\mathscr{B}} \tilde{\mathbf{T}} \cdot \nabla\hat{\mathbf{v}}_m dV = \int_{\mathscr{B}} (\det \mathbf{F}\mathbf{T}_m\mathbf{F}^{-T}) \cdot \nabla\hat{\mathbf{v}}_m dV \\
&= \int_{\mathscr{B}} (\det \mathbf{F}(\mathbf{F}\mathbf{F}^{-1})\mathbf{T}_m\mathbf{F}^{-T}) \cdot \nabla\hat{\mathbf{v}}_m dV = \int_{\mathscr{B}} (\det \mathbf{F}\mathbf{F}^{-1}\mathbf{T}_m\mathbf{F}^{-T}) \cdot \mathbf{F}^T\nabla\hat{\mathbf{v}}_m dV \\
&= \int_{\mathscr{B}} \overline{\mathbf{T}} \cdot \mathbf{F}^T\nabla\hat{\mathbf{v}}_m dV = \int_{\mathscr{B}} \overline{\mathbf{T}} \cdot (\mathbf{F}^T\nabla\hat{\mathbf{v}}_m)^s dV,
\end{aligned} \tag{3.270}$$

where

$$\overline{\mathbf{T}} = \det \mathbf{F}\mathbf{F}^{-1}\mathbf{T}_m\mathbf{F}^{-T}, \tag{3.271}$$

is a symmetric second-order tensor field defined in $\mathscr{B}$, which is known in the literature as second Piola–Kirchhoff stress tensor. Furthermore, this stress measure is power-conjugate to the kinematical entity $(\mathbf{F}^T\nabla\hat{\mathbf{v}}_m)^s$.

In this manner, we have reshaped the material description of the PVP given by (3.269), resulting in

$$\int_{\mathscr{B}} \overline{\mathbf{T}} \cdot (\mathbf{F}^T\nabla\hat{\mathbf{v}}_m)^s dV = \int_{\mathscr{B}} \mathbf{b}^* \cdot \hat{\mathbf{v}}_m dV + \int_{\partial\mathscr{B}_a} \mathbf{a}^* \cdot \hat{\mathbf{v}}_m dS \quad \forall\hat{\mathbf{v}}_m \in (\mathrm{Var}_v)_m. \tag{3.272}$$

Let us uncover the physical meaning of the second-order tensor $(\mathbf{F}^T\nabla\hat{\mathbf{v}}_m)^s$. To do this, recall the Green deformation tensor $\mathbf{E}(\mathbf{U})$, whose definition is in (3.14). Noticing that the field $\hat{\mathbf{v}}_m$ defines for each point $\mathbf{X} \in \mathscr{B}$ a motion action, we now explore the rate of the Green deformation tensor at $\mathbf{U}$ in the direction of $\hat{\mathbf{v}}_m$. Hence, for $\alpha \in \mathbb{R}$ small enough we have

$$\mathbf{E}(\mathbf{U} + \alpha\hat{\mathbf{v}}_m) = \mathbf{E}(\mathbf{U}) + \alpha\mathscr{D}(\mathbf{E}(\mathbf{U}))[\hat{\mathbf{v}}_m] + o(\alpha\hat{\mathbf{v}}_m), \tag{3.273}$$

where the notation $\mathscr{D}(\mathbf{E}(\mathbf{U}))[\hat{\mathbf{v}}_m]$ characterizes the Gâteaux derivative of $\mathbf{E}$ at $\mathbf{U}$ and in the direction established by $\hat{\mathbf{v}}_m$, that is

$$\mathscr{D}(\mathbf{E}(\mathbf{U}))[\hat{\mathbf{v}}_m] = \left.\frac{d}{d\alpha}\mathbf{E}(\mathbf{U} + \alpha\hat{\mathbf{v}}_m)\right|_{\alpha=0}, \tag{3.274}$$

which is nothing but a specific case of (2.12). We now introduce the perturbed element $\mathbf{U} + \alpha\hat{\mathbf{v}}_m$ into definition (3.14) and operate

$$\begin{aligned}
\mathbf{E}(\mathbf{U} + \alpha\hat{\mathbf{v}}_m) &= \frac{1}{2}\left[\nabla(\mathbf{U} + \alpha\hat{\mathbf{v}}_m) + (\nabla(\mathbf{U} + \alpha\hat{\mathbf{v}}_m))^T + (\nabla(\mathbf{U} + \alpha\hat{\mathbf{v}}_m))^T\nabla(\mathbf{U} + \alpha\hat{\mathbf{v}}_m)\right] \\
&= \mathbf{E}(\mathbf{U}) + \frac{\alpha}{2}(\nabla\hat{\mathbf{v}}_m + (\nabla\hat{\mathbf{v}}_m)^T + (\nabla\hat{\mathbf{v}}_m)^T\nabla\mathbf{U} + (\nabla\mathbf{U})^T\nabla\hat{\mathbf{v}}_m) \\
&\quad + \frac{\alpha^2}{2}(\nabla\hat{\mathbf{v}}_m)^T\nabla\hat{\mathbf{v}}_m.
\end{aligned} \tag{3.275}$$

Therefore, this gives

$$\left.\frac{d}{d\alpha}\mathbf{E}(\mathbf{U} + \alpha\hat{\mathbf{v}}_m)\right|_{\alpha=0} = \frac{1}{2}(\nabla\hat{\mathbf{v}}_m + (\nabla\hat{\mathbf{v}}_m)^T + (\nabla\hat{\mathbf{v}}_m)^T\nabla\mathbf{U} + (\nabla\mathbf{U})^T\nabla\hat{\mathbf{v}}_m). \tag{3.276}$$

In turn, we have

$$\mathbf{F}^T\nabla\hat{\mathbf{v}}_m = (\mathbf{I} + (\nabla\mathbf{U})^T)\nabla\hat{\mathbf{v}}_m = \nabla\hat{\mathbf{v}}_m + (\nabla\mathbf{U})^T\nabla\hat{\mathbf{v}}_m. \tag{3.277}$$

Comparing (3.276) and (3.277) we conclude that

$$(\mathbf{F}^T\nabla\hat{\mathbf{v}}_m)^s = \frac{d}{d\alpha}\mathbf{E}(\mathbf{U}+\alpha\hat{\mathbf{v}}_m)\Big|_{\alpha=0}. \tag{3.278}$$

The physical meaning is now clear. The symmetric tensor field $(\mathbf{F}^T\nabla\hat{\mathbf{v}}_m)^s$ stands for the rate of the Green deformation measure $\mathbf{E}$ for the configuration $\mathbf{U}$, according to the direction $\hat{\mathbf{v}}_m$.

Some fundamental aspects of the variational formulation underlying the PVP are worth recalling before ending this section.

In the sections preceding the present one, the PVP was expressed in its spatial description $\mathscr{B}_t$ (see (3.210) for the case of bilateral constraints). This spatial description is the natural configuration in which to define the notion of mechanical equilibrium between the external force f and the internal stress $\mathbf{T}$ because it is the configuration truly occupied by the body. That is to say, the equilibrium inherently takes place in the spatial configuration $\mathscr{B}_t$. Although, in practice, there is a variety of problems in which this spatial configuration is known, there are many other situations for which the region $\mathscr{B}_t$ is part of the unknowns in the problem at hand. In spite of this, a common feature in these cases is that another configuration is actually known, which here we have called the material, or reference, configuration. The existence of other configurations of mechanical relevance motivated us to rewrite the PVP in the so-called Lagrangian, or material, description. Nonetheless, the fact that the calculations are being carried out in this reference configuration should not be confused with the actual configuration where equilibrium takes place.

In the spatial description of the PVP, the dependence of the variational formulation with respect to the current configuration $\mathbf{u}_t$ remains implicit in the fact that the fields and the domain of integration are defined at configuration $\mathscr{B}_t$. In contrast, the material description of the PVP explicitly exhibits the dependence on the current configuration through the tensor $\mathbf{F}$ (which is a function of $\mathbf{U}$).

Should the problem require the incorporation of dynamics, it is necessary to add into the spatial description of the PVP (see expression (3.210)) the power expended associated with the inertial forces,

$$P_e^{inerc} = -\int_{\mathscr{B}_t} \rho\dot{\mathbf{v}}\cdot\hat{\mathbf{v}}dV. \tag{3.279}$$

In the material description, this term assumes the form

$$P_e^{inerc} = -\int_{\mathscr{B}} \rho_m\ddot{\mathbf{U}}\cdot\hat{\mathbf{v}}_m \det\mathbf{F}dV = -\int_{\mathscr{B}} \rho^*\ddot{\mathbf{U}}\cdot\hat{\mathbf{v}}_m dV, \tag{3.280}$$

which is a linear functional in $\hat{\mathbf{v}}_m$ (but not in terms of $\mathbf{U}$).

3.8 Configurations with Preload and Residual Stresses

We have already stressed the fact that the variational expression of the PVP holds regardless the constitutive behavior of the matter that composed the continuum body. Actually, the PVP provides a characterization of the internal stress $\mathbf{T}$ which equilibrates the external force f. Even more, we have pointed out that this internal stress is not unique, but it is defined up to a self-equilibrated internal stress. Purposely, the PVP is not able to characterize the deformation of the body in the equilibrium configuration. This characterization is achieved only by introducing the material behavior into the variational equation of the PVP. This is attained by specifying a proper constitutive relation to bridge a suitable deformation measure, and possibly its entire history, with the internal stress $\mathbf{T}$.

In the field of solid mechanics, this is accomplished by experimental tests carried out over material specimens at a virgin state. This virgin state implies that the material is free of deformations and free of initial internal stresses. In other words, and simplifying to the case of materials without memory (independence of deformation history), the integration of thermodynamic principles with adequate experimental testing allows constitutive equations to be produced, for example, of the kind $\overline{\mathbf{T}} = \overline{\mathbf{T}}(\mathbf{E})$. These equations connect the measure of deformation given by $\mathbf{E}$ (equivalently $\mathbf{F}^T\mathbf{F} - \mathbf{I}$) and the second Piola–Kirchhoff stress tensor.

By introducing this constitutive relation into the PVP, the equilibrium problem, which is linear in terms of the stress field $\mathbf{T}$, turns into, generally, a highly nonlinear problem in terms of the displacement field $\mathbf{u}_t$. This variational problem is enunciated as follows: find $\mathbf{u}_t \in \mathrm{Kin}_u$ (spatial configuration of the displacement) such that

$$\int_{\mathscr{B}_t} \left[\frac{1}{\det \mathbf{F}} \mathbf{F}\overline{\mathbf{T}}(\mathbf{E}((\mathbf{u}_t)_m))\mathbf{F}^T\right]_s \cdot (\mathrm{grad}\, \hat{\mathbf{v}})^s dV = \langle f, \hat{\mathbf{v}}\rangle \qquad \forall \hat{\mathbf{v}} \in \mathrm{Var}_v, \tag{3.281}$$

where $[\cdot]_s$ stands for the spatial description of the field. Then, in the case in which the spatial configuration $\mathscr{B}_t$ is known, the solution of equation (3.281) delivers not only the spatial description of the displacement field $\mathbf{u}_t$, but also the deformation state and, as a consequence of the constitutive relation, the internal stress state $\mathbf{T}$. In such a case, the material configuration is known through the transformation $\mathbf{X} = \mathbf{x} - \mathbf{u}_t(\mathbf{x}) \in \mathscr{B}$.

A typical example of this situation is encountered in the planning of structures whose mechanical regime naturally encompasses known load settings. For instance, in the case of bridges, or aircraft wings, the deformed configuration $\mathscr{B}_t$ is a target in the design and can be supposed to be known once a given set of external forces is applied. In such problems, the material configuration $\mathscr{B}$, free of deformations and stresses, is therefore unknown and must be calculated in order to manufacture the structural component. Another, more recent, example is found in the hemodynamics field, where the interaction between the vascular wall and the blood flow poses the problem of computing the stress state across the thickness of these vessels. Medical imaging technologies are capable of delivering *in vivo* configurations which are neither deformation-free nor stress-free. In fact, these configurations are at equilibrium with loads such as those provided by the blood pressure as well as by tethering forces.

The common feature in the examples discussed in the previous paragraph is that the known configuration is the spatial configuration $\mathscr{B}_t$ where the equilibrium occurs. For such a configuration, the external forces f applied are also known, and the problem turns into determining the deformation mapping which led the body to its current configuration $\mathscr{B}_t$ under the action of forces f. This is why these problems are classified as mechanical equilibrium with preload.

In contrast to (3.281), when the spatial configuration $\mathscr{B}_t$ is not known, and, instead, the material configuration $\mathscr{B}$ is at hand, the PVP consists of the following: determine $\mathbf{U}_t \in (\mathrm{Kin}_u)_m$ (material configuration of the displacement) such that

$$\int_{\mathscr{B}} \overline{\mathbf{T}}(\mathbf{E}(\mathbf{U}_t)) \cdot (\mathbf{F}^T \nabla \hat{\mathbf{v}}_m)^s dV = \langle f_m, \hat{\mathbf{v}}_m \rangle \qquad \forall \hat{\mathbf{v}}_m \in (\mathrm{Var}_v)_m, \tag{3.282}$$

where, now, the motion actions are characterized through their corresponding material descriptions (see also expression (3.272)). The solution to this problem yields the material description of the displacement field $\mathbf{U}_t$, the corresponding deformation measure $\mathbf{E}$, the stress state constitutively associated $\overline{\mathbf{T}}$, and, evidently, the configuration $\mathscr{B}_t$ at equilibrium, given by $\mathbf{x} = \mathbf{X} + \mathbf{U}_t(\mathbf{X}) \in \mathscr{B}_t$.

As discussed above, there exist problems in which even the material configuration is a configuration in which the body is not stress free. In these situations the mechanical equilibrium takes place in the presence of the so-called residual stresses. The genesis of these residual stresses may be caused by different phenomena, such as thermal processes, deformation processes, and growth processes, among others. Importantly, such residual stresses may have a remarkable effect on the internal stress, and therefore in the performance of a given structural component in daily operation. This happens in lamination processes, for instance. In another example, residual stresses are purposely incorporated in the mechanical component so that its performance improves when subjected to certain external forces. In other examples, residual stresses originate because of adaptation processes, such as growth in biological tissues or solidification processes during welding.

On the one hand, as already noted, the preload mechanical equilibrium is nothing but the usage of (3.281), for which, instead of $\mathscr{B}_t$, we may know a given reference configuration called $\mathscr{B}_r$, at equilibrium with certain external forces f_r. On the other hand, let us address the mechanical equilibrium admitting the existence of residual stresses. This problem will be tackled first, assuming that we know the spatial configuration $\mathscr{B}_t$ of the body, and, second, when we know the material configuration $\mathscr{B}$ of the body. Because of the nonlinearity of these equations, consistent linearization procedures will be exposed in next section by resorting to the Gâteaux derivative of the corresponding variational equations.

The presentation is limited to the case with a stress field $\mathbf{T}$ at equilibrium, in the spatial configuration $\mathscr{B}_t$, with the system of external forces f characterized by a pressure, denoted by p, applied over the portion of the boundary $\partial\mathscr{B}_{tp} \subset \partial\mathscr{B}_t$, with unit normal vector $\mathbf{n}$. In addition, we will assume that the stress $\mathbf{T}$ is formed by a constitutive component, denoted by $\mathbf{T}_C$, where C stands for the fact that this component is originated constitutively and by a residual component, called $\mathbf{T}_R$, which will be assumed to be given by the known second Piola–Kirchhoff stress tensor $\overline{\mathbf{T}}_R$. Then, the Cauchy stress tensor results

$$\mathbf{T} = \mathbf{T}_R + \mathbf{T}_C = \left[\frac{1}{\det \mathbf{F}} \mathbf{F} \left[\overline{\mathbf{T}}_R + \overline{\mathbf{T}}(\mathbf{E}((\mathbf{u}_t)_m)) \right] \mathbf{F}^T \right]_s. \tag{3.283}$$

It is important to note that, in the case of a nonlinear material behavior, the constitutive equation which allows $\overline{\mathbf{T}}$ to be fully characterized in terms of $\mathbf{E}((\mathbf{u}_t)_m)$ has to take into account the impact of the residual stress state $\overline{\mathbf{T}}_R$.

Hence, applying the PVP we have the following variational statement: find $\mathbf{u}_t \in \mathrm{Kin}_u$ such that

$$\int_{\mathscr{B}_t} \left[\frac{1}{\det \mathbf{F}}\mathbf{F}\left[\overline{\mathbf{T}}_R + \overline{\mathbf{T}}(\mathbf{E}((\mathbf{u}_t)_m))\right]\mathbf{F}^T\right]_s \cdot \mathrm{grad}\,\hat{\mathbf{v}}^s dV = \int_{\partial\mathscr{B}_{tp}} p\mathbf{n}\cdot\hat{\mathbf{v}}dS \quad \forall\hat{\mathbf{v}} \in \mathrm{Var}_v. \tag{3.284}$$

On the other hand, as already shown, the material description of the PVP is given by the following statement: find $\mathbf{U}_t \in (\mathrm{Kin}_u)_m$ such that

$$\int_{\mathscr{B}} \left[\overline{\mathbf{T}}_R + \overline{\mathbf{T}}(\mathbf{E}(\mathbf{U}_t))\right] \cdot (\mathbf{F}^T\nabla\hat{\mathbf{v}}_m)^s dV = \int_{\partial\mathscr{B}_p} p_m \mathbf{F}^{-T}\mathbf{N}\cdot\hat{\mathbf{v}}_m \det\mathbf{F}dS \quad \forall\hat{\mathbf{v}}_m \in (\mathrm{Var}_v)_m, \tag{3.285}$$

where $\mathbf{N}$ is the unit normal vector to $\partial\mathscr{B}_p \subset \partial\mathscr{B}$, where the material description of the pressure p_m is applied.

Before going into the following section in which we will deal with the linearization procedures for such problems, let us recall the relations between the displacement fields $\mathbf{u}_t$ and $\mathbf{U}_t$ which characterize the transformation (deformation process) which take $\mathscr{B}$ into $\mathscr{B}_t$ and vice versa

$$\begin{aligned}
&\mathbf{x} = \mathscr{X}_t(\mathbf{X}) = \mathbf{X} + \mathbf{U}_t(\mathbf{X}), \quad \mathbf{x} \in \mathscr{B}_t,\ \mathbf{X} \in \mathscr{B},\\
&\mathbf{X} = \mathscr{X}_t^{-1}(\mathbf{x}) = \mathbf{x} - \mathbf{u}_t(\mathbf{x}), \quad \mathbf{X} \in \mathscr{B},\ \mathbf{x} \in \mathscr{B}_t,\\
&\mathbf{u}_t(\mathbf{x}) = (\mathbf{U}_t(\mathbf{X}))_s = \mathbf{U}_t(\mathscr{X}_t^{-1}(\mathbf{x})),\\
&\mathbf{F} = \nabla\mathscr{X}_t = \mathbf{I} + \nabla\mathbf{U}_t,\\
&\mathbf{f} = \mathrm{grad}\,\mathscr{X}_t^{-1} = \mathbf{I} - \mathrm{grad}\,\mathbf{u}_t,
\end{aligned} \tag{3.286}$$

where $\nabla(\cdot)$ and $\mathrm{grad}\,(\cdot)$ represent, respectively, the spatial gradients with respect to material and spatial coordinates. Thus, we also have

$$\begin{aligned}
&[\mathbf{F}^{-1}]_s = \mathbf{f},\\
&[\mathbf{f}^{-1}]_m = \mathbf{F}.
\end{aligned} \tag{3.287}$$

3.9 Linearization of the Principle of Virtual Power

As already pointed out, both problems of mechanical equilibrium written through (3.284) and (3.285) are highly nonlinear. Then, as with any nonlinear function, it is interesting to explore its tangent behavior. This is particularly important in the development of numerical methods for the approximate solution of such problems. In fact, methods of Newton type rely on some tangent (or quase-tangent) operator at each step in the iterative algorithm. These tangent operators are obtained through the calculation of the Gâteaux derivative of the equilibrium equation (see Appendix C.4).In the first place, we will see some basic operations which are useful in the development of these calculations for both the material and the spatial descriptions of the equilibrium.

3.9.1 Preliminary Results

Let us consider the displacement field described in the material configuration, denoted by $\mathbf{U}$ ($\mathbf{u}$ is its spatial description),[19] which is perturbed in the direction $\delta\mathbf{U}$ as $\mathbf{U}_\tau = \mathbf{U} + \tau\delta\mathbf{U}$. With this, the deformation gradient, originally given by $\mathbf{F} = \mathbf{I} + \nabla\mathbf{U}$, results in $\mathbf{F}_\tau = \mathbf{I} + \nabla\mathbf{U}_\tau = \mathbf{I} + \nabla(\mathbf{U} + \tau\delta\mathbf{U})$. We are able to carry out the calculation of the Gâteaux derivative of the following operators involving $\mathbf{F}_\tau$. The next results are useful to perform the linearization of the equilibrium equations. Thus, we have

$$\left.\frac{d}{d\tau}\mathbf{F}_\tau\right|_{\tau=0} = \nabla\delta\mathbf{U} = (\text{grad}\,\delta\mathbf{u})_m\mathbf{F}, \tag{3.288}$$

$$\left.\frac{d}{d\tau}\mathbf{F}_\tau^{-1}\right|_{\tau=0} = -\mathbf{F}^{-1}(\nabla\delta\mathbf{U})\mathbf{F}^{-1} = -\mathbf{F}^{-1}(\text{grad}\,\delta\mathbf{u})_m, \tag{3.289}$$

$$\left.\frac{d}{d\tau}\det\mathbf{F}_\tau\right|_{\tau=0} = \det\mathbf{F}(\mathbf{F}^{-T}\cdot\nabla\delta\mathbf{U}) = \det\mathbf{F}(\text{div}\,\delta\mathbf{u})_m, \tag{3.290}$$

$$\left.\frac{d}{d\tau}\mathbf{E}_\tau\right|_{\tau=0} = (\mathbf{F}^T(\nabla\delta\mathbf{U}))^s = \mathbf{F}^T(\text{grad}\,\delta\mathbf{u})_m^s\mathbf{F}. \tag{3.291}$$

Exercise 3.2 Prove that identities (3.288)–(3.291) hold.

It can be appreciated that, in expressions (3.288)–(3.291), the Gâteaux differentials were written following two different methods, namely, in terms of the differential defined in the material configuration, $\delta\mathbf{U}$, and in terms of that defined in the spatial configuration, $\delta\mathbf{u}$. This will be employed in the calculus of the linear equation associated with the mechanical equilibrium (3.285).

Let us consider now the spatial description of the displacement field $\mathbf{u}$, which is perturbed yielding $\mathbf{u}_\tau = \mathbf{u} + \tau\delta\mathbf{u}$. Then, from $[\mathbf{F}^{-1}]_s = \mathbf{I} - \text{grad}\,\mathbf{u}$ it is found that $[\mathbf{F}^{-1}]_{s,\tau} = \mathbf{I} - \text{grad}\,\mathbf{u}_\tau = \mathbf{I} - \text{grad}\,(\mathbf{u} + \tau\delta\mathbf{u})$. We therefore have the following Gâteaux derivatives for operators involving $\mathbf{F}_{s,\tau}$

$$\left.\frac{d}{d\tau}[\mathbf{F}^{-1}]_{s,\tau}\right|_{\tau=0} = -\text{grad}\,\delta\mathbf{u}, \tag{3.292}$$

$$\left.\frac{d}{d\tau}\mathbf{F}_{s,\tau}\right|_{\tau=0} = \mathbf{F}_s(\text{grad}\,\delta\mathbf{u})\mathbf{F}_s, \tag{3.293}$$

$$\left.\frac{d}{d\tau}\det\mathbf{F}_{s,\tau}\right|_{\tau=0} = \det\mathbf{F}_s(\mathbf{F}_s^T\cdot\text{grad}\,\delta\mathbf{u}), \tag{3.294}$$

$$\left.\frac{d}{d\tau}\mathbf{E}_{s,\tau}\right|_{\tau=0} = \mathbf{F}_s^T(\mathbf{F}_s(\text{grad}\,\delta\mathbf{u}))^s\mathbf{F}_s. \tag{3.295}$$

Exercise 3.3 Prove that (3.292)–(3.295) hold.

19 Here we have simplified the notation, putting $\mathbf{U} = \mathbf{U}_t$ for the material description of the displacement and $\mathbf{u} = \mathbf{u}_t$ for the spatial counterpart.

Finally, in the expressions above recall that the notation $(\cdot)^s$ indicates the symmetric part of the second order tensor field $(\cdot)$. Expressions (3.292)–(3.295) will be utilized in the linearization of the mechanical equilibrium equation (3.284).

3.9.2 Known Spatial Configuration

Variational equation (3.284) can be written in compact form as follows: find $\mathbf{u} \in \text{Kin}_u$ such that

$$\langle \mathscr{R}_s(\mathbf{u}), \hat{\mathbf{v}} \rangle_{\mathscr{B}_t} = 0 \qquad \forall \hat{\mathbf{v}} \in \text{Var}_v. \tag{3.296}$$

Consider the Newton–Raphson technique as a mean to linearize equation (3.296): for a point $\mathbf{u}^k \in \text{Kin}_u$, find the increment $\delta\mathbf{u} \in \text{Var}_v$ such that

$$\langle \mathscr{R}_s(\mathbf{u}^k), \hat{\mathbf{v}} \rangle_{\mathscr{B}_t} + \frac{d}{d\tau} \langle \mathscr{R}_s(\mathbf{u}^k + \tau\delta\mathbf{u}), \hat{\mathbf{v}} \rangle_{\mathscr{B}_t} \bigg|_{\tau=0} = 0 \qquad \forall \hat{\mathbf{v}} \in \text{Var}_v. \tag{3.297}$$

As the iteration number is incremented (k grows) it is expected that $\mathbf{u}^k \to \mathbf{u}$ and, also, that $\delta\mathbf{u} \to \mathbf{0}$. Taking into consideration the results obtained in Section 3.9.1, and dropping the iteration index k for ease of notation, it is straightforward to explicitly obtain the terms in (3.297), reaching the following linear variational equation: given $\mathbf{u} \in \text{Kin}_u$, find $\delta\mathbf{u} \in \text{Var}_v$ such that

$$\begin{aligned} &-\int_{\mathscr{B}_t} (\mathbf{F}_s^T \cdot \text{grad}\,\delta\mathbf{u})\mathbf{T} \cdot (\text{grad}\,\hat{\mathbf{v}})^s \, dV + \int_{\mathscr{B}_t} 2(\mathbf{F}_s(\text{grad}\,\delta\mathbf{u})\mathbf{T}) \cdot (\text{grad}\,\hat{\mathbf{v}})^s \, dV \\ &+ \int_{\mathscr{B}_t} \mathbf{D}_s(\mathbf{F}_s \text{grad}\,\delta\mathbf{u})^s \cdot (\text{grad}\,\hat{\mathbf{v}})^s \, dV = -\int_{\mathscr{B}_t} \mathbf{T} \cdot (\text{grad}\,\hat{\mathbf{v}})^s \, dV \\ &+ \int_{\partial\mathscr{B}_{tp}} p\mathbf{n} \cdot \hat{\mathbf{v}} dS \qquad \forall \hat{\mathbf{v}} \in \text{Var}_v, \end{aligned} \tag{3.298}$$

where

$$\mathbf{T} = \left[\frac{1}{\det \mathbf{F}} \mathbf{F} \left(\overline{\mathbf{T}}_R + \overline{\mathbf{T}}(\mathbf{E}(\mathbf{u}_m)) \right) \mathbf{F}^T \right]_s, \tag{3.299}$$

$$\mathbf{D}_s(\mathbf{F}_s \text{grad}\,\delta\mathbf{u})^s = \frac{1}{\det \mathbf{F}_s} \mathbf{F}_s \left[\left(\frac{\partial \overline{\mathbf{T}}}{\partial \mathbf{E}} \right)_s \mathbf{F}_s^T (\mathbf{F}_s \text{grad}\,\delta\mathbf{u})^s \mathbf{F}_s \right] \mathbf{F}_s^T. \tag{3.300}$$

Clearly, $\mathbf{D}_s$ is a fourth-order tensor which maps symmetric second-order tensors into symmetric second-order tensors, and whose components in a Cartesian frame of reference are given by

$$[\mathbf{D}_s]_{ijkl} = \frac{1}{\det \mathbf{F}_s} [\mathbf{F}_s]_{ia} [\mathbf{F}_s]_{jb} [\mathbf{F}_s]_{kc} [\mathbf{F}_s]_{ld} \left[\left(\frac{\partial \overline{\mathbf{T}}}{\partial \mathbf{E}} \right)_s \right]_{abcd}. \tag{3.301}$$

Exercise 3.4 Prove that the linear form (3.298) is verified.

3.9.3 Known Material Configuration

Similarly to the previous section, variational equation (3.285) can be rewritten in compact form as follows: find $\mathbf{U} \in (\text{Kin}_u)_m$ such that

$$\langle \mathscr{R}_m(\mathbf{U}), \hat{\mathbf{v}}_m \rangle_{\mathscr{B}} = 0 \qquad \forall \hat{\mathbf{v}}_m \in (\text{Var}_v)_m. \tag{3.302}$$

Again, a Newton–Raphson linearization procedure applied to (3.302) renders the following linear variational equation: for a point $\mathbf{U}^k \in (\mathrm{Kin}_u)_m$ find the increment $\delta\mathbf{U} \in (\mathrm{Var}_v)_m$ such that

$$\langle \mathscr{R}_m(\mathbf{U}), \hat{\mathbf{v}}_m \rangle_{\mathscr{B}} + \frac{d}{d\tau} \langle \mathscr{R}_m(\mathbf{U}^k + \tau\delta\mathbf{U}), \hat{\mathbf{v}}_m \rangle_{\mathscr{B}} \bigg|_{\tau=0} = 0 \quad \forall \hat{\mathbf{v}}_m \in (\mathrm{Var}_v)_m. \tag{3.303}$$

In this case, as the iteration number grows (k grows) it is expected that $\mathbf{U}^k \to \mathbf{U}$ and, in addition, that $\delta\mathbf{U} \to \mathbf{0}$. Taking into account the preliminary results from Section 3.9.1 it becomes an exercise to calculate the derivative required in (3.303). Nevertheless, instead of using material configuration $\mathscr{B}$, it is usual to write these derivatives in the spatial configuration updated at the configuration found at iteration k, denoted by $\mathscr{B}_t^k$, which is found using the transformation $\mathbf{x}^k = \mathbf{X} + \mathbf{U}^k(\mathbf{X})$. This was the reason for expressing the Gâteaux differentials in the spatial configuration in Section 3.9.1. Hereafter, in order to ease the notation, we will drop superscript k, and the updated configuration $\mathscr{B}_t^k$ becomes now $\mathscr{B}_t$. However, this should not be confused with the actual spatial configuration where the true mechanical equilibrium occurs. In the same manner, all the fields will be described in such updated configuration, and then we will not make differences between denomination $(\cdot)_{s^k}$ and $(\cdot)_s$. Moreover, we will make abuse of notation by writing $\mathrm{Kin}_u = ((\mathrm{Kin}_u)_m)_{s^k}$ and $\mathrm{Var}_u = ((\mathrm{Var}_u)_m)_{s^k}$. Thus, we obtain the following linear variational equation: given $\mathbf{U}_s \in \mathrm{Kin}_u$, find $\delta\mathbf{U}_s \in \mathrm{Var}_v$ such that

$$\begin{aligned} &\int_{\mathscr{B}_t} \mathbf{D}_s(\mathrm{grad}\,\delta\mathbf{U}_s)^s \cdot (\mathrm{grad}\,\hat{\mathbf{v}})^s \, dV + \int_{\mathscr{B}_t} (\mathrm{grad}\,\delta\mathbf{U}_s)\mathbf{T}_s \cdot (\mathrm{grad}\,\hat{\mathbf{v}}) \, dV \\ &\quad + \int_{\partial\mathscr{B}_{tp}} [p((\mathrm{grad}\,\delta\mathbf{U}_s)^T - (\mathrm{div}\,\delta\mathbf{U}_s)\mathbf{I})\mathbf{n} \cdot \hat{\mathbf{v}}] \, dS \\ &\quad = -\int_{\mathscr{B}_t} \mathbf{T}_s \cdot (\mathrm{grad}\,\hat{\mathbf{v}}) \, dV + \int_{\partial\mathscr{B}_{tp}} p\mathbf{n} \cdot \hat{\mathbf{v}} \, dS \qquad \forall \hat{\mathbf{v}} \in \mathrm{Var}_v, \end{aligned} \tag{3.304}$$

where

$$\mathbf{T}_s = \left[\frac{1}{\det \mathbf{F}} \mathbf{F} \left(\overline{\mathbf{T}}_R + \overline{\mathbf{T}}(\mathbf{E}(\mathbf{U})) \right) \mathbf{F}^T \right]_s, \tag{3.305}$$

$$\mathbf{D}_s(\mathrm{grad}\,\delta\mathbf{U}_s)^s = \frac{1}{\det \mathbf{F}_s} \mathbf{F}_s \left[\left(\frac{\partial \overline{\mathbf{T}}}{\partial \mathbf{E}} \right)_s \mathbf{F}_s^T (\mathrm{grad}\,\delta\mathbf{U}_s)^s \mathbf{F}_s \right] \mathbf{F}_s^T. \tag{3.306}$$

Exercise 3.5 Show that the linear equation (3.304) holds.

3.10 Infinitesimal Deformations and Small Displacements

Many practical problems in the field of continuum mechanics can be modeled taking into account hypotheses of small deformations and small displacements. In such cases, all configurations $\mathscr{B}_t$, $t \in [t_0, t_f]$, can be regarded as a single configuration, say $\mathscr{B}$, which the body occupies in the Euclidean space. When this happens, and assuming that the reference configuration coincides with $\mathscr{B}$, we have already seen in (3.23) that

$$\mathbf{F} \equiv \mathbf{I}, \quad \nabla(\cdot) \equiv \mathrm{grad}\,(\cdot), \quad \mathbf{E} \equiv (\nabla\mathbf{u})^s, \tag{3.307}$$

where $\mathbf{u}$ is the displacement field. Furthermore, the equivalence between stress measures also holds (see (3.264) and (3.271))

$$\mathbf{T} \equiv \tilde{\mathbf{T}} \equiv \overline{\mathbf{T}}. \tag{3.308}$$

That is to say, the Cauchy stress tensor can be confounded with both the first and the second Piola–Kirchhoff stress tensors.

Under these hypotheses let us revise the expression for the PVP and for the PVCP for the cases of bilateral and unilateral kinematical constraints studied in this chapter.

3.10.1 Bilateral Constraints

Let us first assume that the set of kinematically admissible configurations Kin_u is invariant through time. In such a case, and for bilateral constraints, Kin_u is a linear manifold associated with the vector subspace Var_u of $\mathscr{U}$. In other words, and for a time increment small enough (infinitesimal) δt, we will always be capable of performing the identification of the subspace of all admissible motion actions, Var_v, with Var_u consisting of all kinematically admissible displacement fields, which is

$$\mathrm{Kin}_u = \mathbf{u}_o + \mathrm{Var}_u, \tag{3.309}$$

with $\mathbf{u}_o \in \mathrm{Kin}_u$ arbitrary. Now, the notion of mechanical equilibrium can be established through the so-called Principle of Virtual Work (PVW), which is nothing but the PVP when we identify any virtual motion action $\hat{\mathbf{v}} \in \mathrm{Var}_v$ with the virtual displacement $\hat{\mathbf{u}} = \hat{\mathbf{v}}\delta t \in \mathrm{Var}_u$.[20] In this manner, the PVW states that the external force f is at (static) equilibrium in configuration $\mathbf{u}$, under the hypotheses of infinitesimal deformations and small displacements if

$$\langle f, \hat{\mathbf{u}} \rangle = 0 \qquad \forall \hat{\mathbf{u}} \in \mathrm{Var}_u \cap \mathcal{N}(\mathcal{D}), \tag{3.310}$$

$$-(\mathbf{T}, \mathcal{D}\hat{\mathbf{u}}) + \langle f, \hat{\mathbf{u}} \rangle = 0 \qquad \forall \hat{\mathbf{u}} \in \mathrm{Var}_u. \tag{3.311}$$

This entails that the virtual work expended by applied forces is null for all kinematically admissible rigid virtual displacements and, moreover, the sum of the internal virtual work and the external virtual work must be zero for all kinematically admissible virtual displacements. Notably, by replacing "power" by "work" and "motion action" by "displacement" we go from the PVP to the PVW.

Exercise 3.6 Following a totally analogous reasoning, formulate the Principle of Complementary Virtual Work for the case of small displacements and infinitesimal deformations.

It should be noted that, under the hypotheses of small displacements and infinitesimal deformations, and with bilateral kinematical constraints, the mechanical equilibrium is a notion which is independent of the configuration. The next section shows that this is not particularly valid for the case of unilateral constraints.

---

20 In the literature, it is common to find a hypothesis of infinitesimal virtual displacements. Such a hypothesis is not required, since the PVP is always linear with respect to virtual motion actions. Hence, the magnitude of the motion action is not relevant, but its direction. In short, the magnitude of virtual displacements plays no role in the PVW.

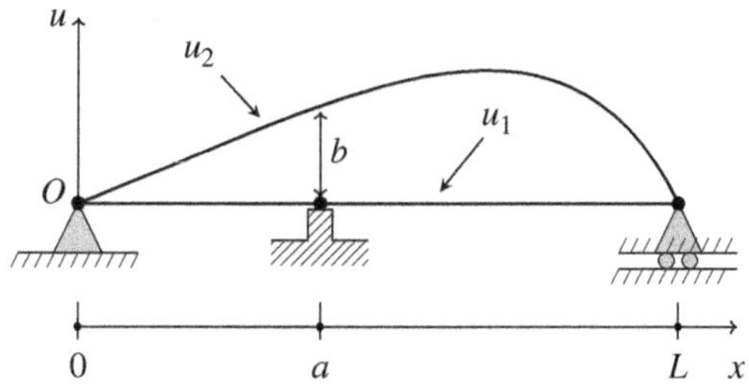

Figure 3.14 Beam with unilateral support at point $x = a$.

3.10.2 Unilateral Constraints

Consider now that the set Kin_u is a conic manifold which depends on the current configuration. We have already seen that

$$\text{Kin}_u(\mathbf{u}) = \mathbf{u} + \text{Var}_u(\mathbf{u}), \tag{3.312}$$

with $\mathbf{u} \in \text{Kin}_u$ arbitrary, and we have explicitly put the dependence of the definition of the cone at the vertex Var_u on the configuration $\mathbf{u}$. In fact, the set

$$\text{Var}_u(\mathbf{u}) = \{\hat{\mathbf{u}} \in \mathscr{U};\ \mathbf{u} + \hat{\mathbf{u}} \in \text{Kin}_u\}, \tag{3.313}$$

is the set of all kinematically admissible virtual displacements we can consider at configuration $\mathbf{u} \in \text{Kin}_u$. Once again, we can identify the convex cone of kinematically admissible motion actions Var_v at configuration $\mathbf{u}$ with the convex cone $\text{Var}_u(\mathbf{u})$ of virtual displacements through $\hat{\mathbf{u}} = \delta t \hat{\mathbf{v}}$. An example is illustrated in Figure 3.14. For this example we have

$$\text{Kin}_u(u_1) = \{u \in \mathscr{U};\ u(0) = u(L) = 0,\ u(a) \geq 0\} = \text{Var}_u(u_1), \tag{3.314}$$

where u_1 is the configuration (straight line) indicated in the figure. Suppose the beam is now in configuration u_2 such that

$$u_2(0) = u_2(L) = 0 \quad \text{and} \quad u_2(a) = b > 0. \tag{3.315}$$

Then, for this configuration we have

$$\text{Kin}_u(u_2) = u_2 + \text{Var}_u(u_2), \tag{3.316}$$

with

$$\text{Var}_u(u_2) = \{u \in \mathscr{U};\ u(0) = u(L) = 0,\ u(a) \geq -b\}. \tag{3.317}$$

This example clearly shows that Var_u depends on the adopted configuration. At this point, it is important to stress that, even if all configurations are regarded as a single configuration by virtue of the hypotheses previously introduced, this does not mean that the same applies to the unilateral constraints. As a matter of fact, for a given configuration the unilateral constraint is active (this happens in configuration u_1), while for another configuration the constraint is not active (this happens in configuration u_2, for which we can consider arbitrarily positive displacements and negative displacements bigger than $-b$).

Thus, the PVW tells us that the external force f is at (static) equilibrium in the configuration $\mathbf{u}$, under the hypotheses of small displacements and infinitesimal deformations, if

$$\langle f, \hat{\mathbf{u}} \rangle \leq 0 \qquad \forall \hat{\mathbf{u}} \in \mathrm{Var}_u(\mathbf{u}) \cap \mathcal{N}(\mathcal{D}), \tag{3.318}$$

$$-(\mathbf{T}, \mathcal{D}\hat{\mathbf{u}}) + \langle f, \hat{\mathbf{u}} \rangle \leq 0 \qquad \forall \hat{\mathbf{u}} \in \mathrm{Var}_u(\mathbf{u}). \tag{3.319}$$

Therefore, we see that even in the case of hypotheses of small displacements and infinitesimal deformations, in the case of unilateral constraints, the concept of mechanical equilibrium depends on the configuration occupied by the body. Even in such a simplified situation, the unilateral constraint introduces an intrinsic nonlinear aspect into the very formulation of the equilibrium.

3.11 Final Remarks

Throughout this chapter, we observed that the variational formulation, as a mathematical ground to express the Principle of Virtual Power, is the best suited formalism to handle problems in the realm of mechanics. This unified variational tapestry is not only mathematically sound, but, and fundamentally, it provides a systematic methodology to elaborate mathematical constructs of reality employing a minimal set of hypotheses. Indeed, from an entirely kinematical perspective, we noticed how the occurrence of dual entities such as forces and stresses are reworked into the variational framework provided by functional analysis, through cost functionals properly defined in functional spaces.

Therefore, if the model predictions are not in agreement with reality, then the description of the kinematics must be revised, seeking to refine the characterization of the kinematics. As a consequence, we have that the concept of forces and stresses is, through duality, inextricably connected to that of admissible motion actions and deformation rate actions. That is, the postulated kinematics will shape the forces and stresses compatible with the model (see [114]–[116] and [189]).

Notice that, along the analysis developed up to the present moment, we have not discussed specifics related to the behavior of the material that composes the system under study. Unmistakably, the response of a given system to external stimuli depends upon the matter composition. This aspect will be addressed in the following chapters. The reader should have noticed that the presentation in this chapter was limited to a single aspect of the mechanics of deformable bodies, which related to the specifics of deformation itself, while other mechanical interactions, such as those originated from thermal coupling or from electromagnetic interactions, have been omitted. Notwithstanding this, the mathematical structure provided by the variational formalism can straightforwardly encompass these phenomena. Such espousal can be made effective by modifying the concept of motion actions, and therefore generalizing the definition of the kinematics defined for our body (see, for example, [189]).

Notably, the variational formulation of a mechanical problem delivers the best-suited groundwork not only to study mathematical aspects, such as the well-posedness of the problem, but also to develop numerical methods towards constructing approximate solutions to these problems in the field of scientific computing. This is another advantage of being embraced by the variational fabric in the modeling of mechanical problems.

3.12 Complementary Reading

For the reader interested in further study of some mathematical and mechanical aspects, we recommend as complementary material the books by P. Chadwic [47], G. Duvaut and J. L. Lions [76], I. Ekeland and R. Temam [78], R. A. Feijóo [84], W. Flugge [106], M. Fremond [109], Y. C. Fung [111], P. Germain [117], and the scientific articles related to the Principle of Virtual Power [114–116] by M. E. Gurtin [131–132], C. Lanczos [167], I-Shih Liu [174], L. E. Malvern [183], G.A. Maugin [189], S. G. Mikhlin [198], J. T. Oden and J. N. Reddy [230], P. D. Panagiotopoulos [236], C. Truesdell [299], C. Truesdell and R. A. Toupin [300], K. Washizu [303], and the works by G. Romano and coworkers [253–260], [93–95], and [286].

4

Hyperelastic Materials at Infinitesimal Strains

4.1 Introduction

Chapter 3 was devoted to the presentation of the foundational aspects of the mechanics regardless of the specifics of the constituent material that forms the bodies.

In this chapter, we will illustrate the application of these principles of the mechanics to the case of bodies undergoing infinitesimal deformations and featuring hyperelastic material behavior. To achieve this goal, we will first discuss the definition and properties of such hyperelastic constitutive equations, and then the attention will be on the variational formulations and extremal principles that can be constructed for such problems.

4.2 Uniaxial Hyperelastic Behavior

When performing uniaxial tests, a linear elastic material is characterized by a relation of the following kind,

$$\sigma = \mathbb{E}\varepsilon, \qquad \mathbb{E} > 0, \tag{4.1}$$

where σ is the stress, ε is the deformation, and $\mathbb{E}$ is the Young's modulus, or the elastic modulus.

In what follows, we will point out some features that, even if they appear to be obvious, will be useful in the forthcoming presentation.

As it is clear from (4.1), $\varepsilon = 0 \Rightarrow \sigma = 0$, that is, we admit that if no initial deformation is present in the body, then no initial stress can develop. Because $\mathbb{E} > 0$, the inverse relation $\varepsilon = \mathbb{E}^{-1}\sigma$ exists. Moreover, we can define the so-called strain energy density function $\pi = \pi(\varepsilon)$ given by

$$\pi(\overline{\varepsilon}) = \int_0^{\overline{\varepsilon}} \sigma(\varepsilon)\, d\varepsilon = \int_0^{\overline{\varepsilon}} \mathbb{E}\varepsilon\, d\varepsilon = \frac{1}{2}\mathbb{E}\overline{\varepsilon}^2 \geq 0. \tag{4.2}$$

Clearly, this strain energy density function is zero if and only if $\overline{\varepsilon} = 0$. In addition, from the definition of function π we get

$$\overline{\sigma} = \left.\frac{\partial \pi}{\partial \varepsilon}\right|_{\overline{\varepsilon}}, \tag{4.3}$$

or, in other words, that the stress $\overline{\sigma}$ associated with the strain $\overline{\varepsilon}$, through the elastic constitutive equation (4.1), can be obtained by calculating the derivative of the strain

Introduction to the Variational Formulation in Mechanics: Fundamentals and Applications, First Edition.
Edgardo O. Taroco, Pablo J. Blanco and Raúl A. Feijóo.

energy density function with respect to the strain and performing the evaluation at point $\overline{\varepsilon}$.

In the current example, function π is a parabolic function with positive second derivative ($\mathbb{E} > 0$) and, therefore, it is a strictly convex function. To put this property in more general terms we write

$$\pi(\varepsilon) = \pi(\varepsilon_0) + \left.\frac{d\pi}{d\varepsilon}\right|_{\varepsilon_0} (\varepsilon - \varepsilon_0) + \frac{1}{2}\left.\frac{d^2\pi}{d\varepsilon^2}\right|_{\varepsilon_0} (\varepsilon - \varepsilon_0)^2$$
$$= \pi(\varepsilon_0) + \mathbb{E}\varepsilon_0(\varepsilon - \varepsilon_0) + \frac{1}{2}\mathbb{E}(\varepsilon - \varepsilon_0)^2 \geq \pi(\varepsilon_0) + \sigma(\varepsilon_0)(\varepsilon - \varepsilon_0), \tag{4.4}$$

where the equality is verified if and only if $\varepsilon = \varepsilon_0$.

Similarly to the definition of the strain energy density function π, we can introduce the function $\pi^* = \pi^*(\sigma)$, called the complementary strain energy density function, defined by

$$\pi^*(\overline{\sigma}) = \int_0^{\overline{\sigma}} \varepsilon(\sigma) d\sigma = \int_0^{\overline{\sigma}} \frac{1}{\mathbb{E}}\sigma\, d\sigma = \frac{1}{2\mathbb{E}}\overline{\sigma}^2 \geq 0, \tag{4.5}$$

and which is zero if and only if $\overline{\sigma} = 0$. From this definition we observe that

$$\overline{\varepsilon} = \left.\frac{d\pi^*}{d\sigma}\right|_{\overline{\sigma}}, \tag{4.6}$$

which implies that the strain $\overline{\varepsilon}$ associated with the stress $\overline{\sigma}$ through the elastic constitutive equation can be obtained through the derivative of the complementary strain energy density function at point $\overline{\sigma}$.

Analogously to (4.4), we also note that we can write

$$\pi^*(\sigma) = \pi^*(\sigma_0) + \left.\frac{d\pi^*}{d\sigma}\right|_{\sigma_0} (\sigma - \sigma_0) + \frac{1}{2}\left.\frac{d^2\pi^*}{d\sigma^2}\right|_{\sigma_0} (\sigma - \sigma_0)^2 \geq \pi^*(\sigma_0) + \varepsilon(\sigma_0)(\sigma - \sigma_0), \tag{4.7}$$

and the equality holds for $\sigma = \sigma_0$. That is to say, the complementary strain energy function is also a strictly convex function.

Previous results are valid even in the case of materials featuring a nonlinear elastic behavior, provided the constitutive equation $\sigma = \sigma(\varepsilon)$ (or its inverse $\varepsilon = \varepsilon(\sigma)$) is a monotonically increasing function, that is

$$(\sigma^B - \sigma^A)(\varepsilon^B - \varepsilon^A) \geq 0 \qquad \forall \varepsilon^A, \varepsilon^B, \tag{4.8}$$

and it is equal to zero if and only if $\varepsilon^A = \varepsilon^B$, where $\sigma^B = \sigma(\varepsilon^B)$ and $\sigma^A = \sigma(\varepsilon^A)$. Particularly, for $\varepsilon^B = \varepsilon^A + d\varepsilon = \varepsilon$, and putting $\sigma = \sigma(\varepsilon)$, we have

$$(\sigma - \sigma^A) d\varepsilon \geq 0, \tag{4.9}$$

and integrating between ε^A and ε yields

$$\int_{\varepsilon^A}^{\varepsilon} (\sigma - \sigma^A) d\varepsilon \geq 0, \tag{4.10}$$

from where

$$\int_{\varepsilon^A}^{\varepsilon} \sigma d\varepsilon - \sigma^A(\varepsilon - \varepsilon^A) \geq 0 \quad \Rightarrow \quad \pi(\varepsilon) - \pi(\varepsilon^A) \geq \sigma^A(\varepsilon - \varepsilon^A), \tag{4.11}$$

where the equality holds if and only if $\varepsilon = \varepsilon^A$.

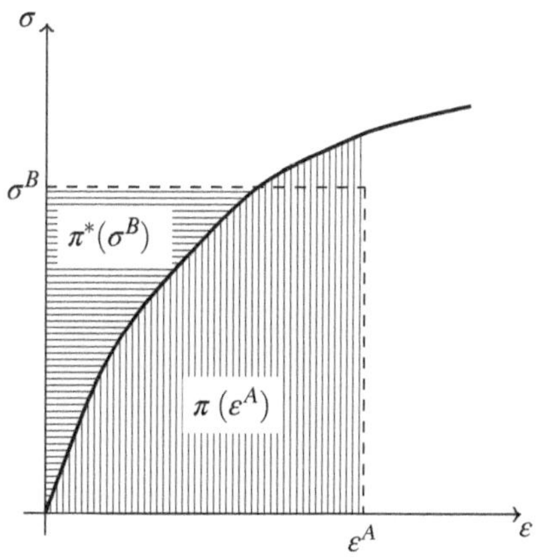

Figure 4.1 Strain energy density function π and complementary strain energy density function π^*.

Similarly, consider that the function $\varepsilon = \varepsilon(\sigma)$ is monotonically increasing, that is

$$(\varepsilon^B - \varepsilon^A)(\sigma^B - \sigma^A) \geq 0 \qquad \forall \sigma^A, \sigma^B, \tag{4.12}$$

and equal to zero if and only if $\sigma^B = \sigma^A$. Again, for $\sigma^B = \sigma = \sigma^A + d\sigma$ we have

$$(\varepsilon - \varepsilon^A) d\sigma \geq 0, \tag{4.13}$$

and integrating between σ^A and σ results in

$$\int_{\sigma^A}^{\sigma} (\varepsilon - \varepsilon^A) d\sigma \geq 0, \tag{4.14}$$

from which

$$\int_{\sigma^A}^{\sigma} \varepsilon d\sigma - \varepsilon^A(\sigma - \sigma^A) \geq 0 \quad \Rightarrow \quad \pi^*(\sigma) - \pi^*(\sigma^A) \geq \varepsilon^A(\sigma - \sigma^A), \tag{4.15}$$

where the equality is verified if and only if $\sigma = \sigma^A$.

Next, let us unveil an additional property implicit in functions π and π^*. Let σ^B and ε^A be a stress and a strain which are not related by the elastic constitutive equation, then it turns out that the following holds

$$\pi(\varepsilon^A) + \pi^*(\sigma^B) \geq \sigma^B \varepsilon^A. \tag{4.16}$$

Figure 4.1 graphically represents the previous expression. It is important to highlight that this result is independent from the linearity of the constitutive equation, while it only depends on its monotonicity.

Inequality (4.16) provides a technique to build π^* from the knowledge of π, and vice versa. This procedure is known as Legendre's transformation, which is also known as a procedure to determine a convex conjugate function to a given function (π^* is dual to π). In effect, suppose that we know $\pi = \pi(\varepsilon)$, then we can define the dual function $\pi^* = \pi^*(\sigma)$ to π such that its value at $\sigma = \sigma^*$ is given by

$$\pi^*(\sigma^*) = \max_{\varepsilon}(\sigma^* \varepsilon - \pi(\varepsilon)). \tag{4.17}$$

For the example of the linear elastic constitutive equation 4.1 we have

$$\pi(\varepsilon) = \frac{1}{2} \mathbb{E} \varepsilon^2. \tag{4.18}$$

Therefore, the maximum in (4.17) is verified at ε^* such that

$$\frac{d}{d\varepsilon}(\sigma^*\varepsilon - \pi(\varepsilon))\bigg|_{\varepsilon^*} = 0, \tag{4.19}$$

which means

$$\sigma^* - \mathbb{E}\varepsilon^* = 0 \quad \Rightarrow \quad \varepsilon^* = \frac{\sigma^*}{\mathbb{E}}. \tag{4.20}$$

Replacing (4.20) into (4.17) we get

$$\pi^*(\sigma^*) = \sigma^*\varepsilon^* - \pi(\varepsilon^*) = \frac{\sigma^{*2}}{\mathbb{E}} - \frac{1}{2}\mathbb{E}\frac{\sigma^{*2}}{\mathbb{E}^2} = \frac{1}{2}\frac{\sigma^{*2}}{\mathbb{E}}, \tag{4.21}$$

which is the same result we arrived at previously through the definition of the strain energy density function.

In a completely analogous procedure, we can define π as the convex conjugate function to π^* through the following problem

$$\pi(\varepsilon^*) = \max_{\sigma}(\sigma\varepsilon^* - \pi^*(\sigma)). \tag{4.22}$$

Following an entirely similar reasoning to that made for π^*, leads us, in the linear case, to

$$\pi(\varepsilon^*) = \frac{1}{2}\mathbb{E}\varepsilon^{*2}. \tag{4.23}$$

We can thus summarize the basic ingredients present in the constitutive equation for uniaxial states. Then, we say that the stress $\overline{\sigma}$ and the strain $\overline{\varepsilon}$ are related through a constitutive equation if any of the following, all equivalent, expressions is verified.

1) The constitutive equation

$$\overline{\sigma} = \sigma(\overline{\varepsilon}), \tag{4.24}$$

or its inverse

$$\overline{\varepsilon} = \varepsilon(\overline{\sigma}), \tag{4.25}$$

is a monotonically increasing function.

2) There exists a strictly convex function, called the strain energy density function π, such that

$$\overline{\sigma} = \frac{d\pi}{d\varepsilon}\bigg|_{\overline{\varepsilon}}, \tag{4.26}$$

or there exists the complementary strain energy density function π^*, such that

$$\overline{\varepsilon} = \frac{d\pi^*}{d\sigma}\bigg|_{\overline{\sigma}}. \tag{4.27}$$

3) Given ε and σ, the following holds

$$\pi(\varepsilon) + \pi^*(\sigma) \geq \sigma\varepsilon, \tag{4.28}$$

where the equality is verified if and only if σ, ε are related through the constitutive equation.

4.3 Three-Dimensional Hyperelastic Constitutive Laws

The generalization of the uniaxial constitutive equation to the three-dimensional case, where different strains and stresses are interwoven, is developed here. We have seen that, even in the uniaxial linear case where $\sigma = \mathbb{E}\varepsilon$, the constitutive equation entails a transformation between the space of strains and its topological dual, the space of stresses.

Thus, the natural generalization of the linear constitutive equation consists of introducing a linear mapping from the space of strains $\mathscr{W}$ into its dual $\mathscr{W}'$. For the case of three-dimensional bodies we have

$$\mathbf{T} = \mathbb{D}\mathbf{E}, \tag{4.29}$$

where $\mathbb{D}$ is the fourth-order elasticity tensor, and $\mathbf{E}$ and $\mathbf{T}$ are symmetric second-order tensors which represent, respectively, the strain tensor and the Cauchy stress tensor.[1] In what follows, the presentation is limited to the case in which the elasticity tensor $\mathbb{D}$ satisfies the following properties

$$\mathbb{D}\mathbf{A} \cdot \mathbf{B} = \mathbb{D}\mathbf{B} \cdot \mathbf{A} \qquad \forall \mathbf{A}, \mathbf{B} \in \mathscr{W} \quad \Rightarrow \mathbb{D} = \mathbb{D}^T, \tag{4.30}$$

$$\mathbb{D}\mathbf{E} \cdot \mathbf{E} \geq 0 \qquad \forall \mathbf{E} \in \mathscr{W}, \tag{4.31}$$

$$\mathbb{D}\mathbf{E} \cdot \mathbf{E} = 0 \qquad \Leftrightarrow \mathbf{E} = \mathbf{O}. \tag{4.32}$$

If the tensor $\mathbb{D}$ is independent from the material point, we say that the body is homogeneous. In general, for the characterization of $\mathbb{D}$, and due to the symmetries assumed, it will be necessary to have 21 elastic constants (recall that $\mathbb{D}$ maps symmetric second-order tensors into symmetric second-order tensors). In this general case, we say that the material is linear elastic and also anisotropic. In the particular case of an isotropic linear elastic material, tensor $\mathbb{D}$ is characterized by just two elastic constants, namely μ and λ, called Lamé constants, and takes the form

$$\mathbb{D} = 2\mu\mathbb{I} + \lambda(\mathbf{I} \otimes \mathbf{I}), \tag{4.33}$$

where $\mathbb{I}$ and $\mathbf{I}$ are, respectively, the fourth- and second-order identity tensors, and where $\otimes$ represents the tensor product, here between second-order tensors.[2]

With a completely analogous analysis to that of the previous section, it is possible to define the strain deformation density function $\pi = \pi(\mathbf{E})$ as

$$\pi(\overline{\mathbf{E}}) = \int_{\mathbf{O}}^{\overline{\mathbf{E}}} \mathbf{T}(\mathbf{E}) \cdot d\mathbf{E} = \int_{\mathbf{O}}^{\overline{\mathbf{E}}} \mathbb{D}\mathbf{E} \cdot d\mathbf{E} = \frac{1}{2}\mathbb{D}\overline{\mathbf{E}} \cdot \overline{\mathbf{E}}, \tag{4.34}$$

and from the properties of $\mathbb{D}$ it follows that

$$\pi(\mathbf{E}) \geq 0, \tag{4.35}$$

which is equal to zero if and only if $\mathbf{E} = \mathbf{O}$.

---

1 Tensor $\mathbf{E}$ here is not to be confused with the Green–Lagrange tensor defined in Chapter 3. In the present case, it is $\mathbf{E} = (\nabla\mathbf{u})^s$, where $\mathbf{u}$ is the displacement field in the body. Also, tensor $\mathbf{T}$ is the Cauchy stress tensor.

2 In this case, and recalling that seen in Section 1.3, observe that, for two given tensors $\mathbf{A} = \mathbf{a}_1 \otimes \mathbf{a}_2$ and $\mathbf{B} = \mathbf{b}_1 \otimes \mathbf{b}_2$, we have $\mathbf{A} \otimes \mathbf{B} = \mathbf{a}_1 \otimes \mathbf{a}_2 \otimes \mathbf{b}_1 \otimes \mathbf{b}_2$.

In this manner, the stress state $\overline{\mathbf{T}}$[3] associated with $\overline{\mathbf{E}}$ through the elastic constitutive equation is such that

$$\overline{\mathbf{T}} = \left.\frac{\partial \pi}{\partial \mathbf{E}}\right|_{\mathbf{E}=\overline{\mathbf{E}}} = \mathbb{D}\overline{\mathbf{E}}. \tag{4.36}$$

In turn, it is

$$\begin{aligned}\pi(\mathbf{E}) &= \pi(\mathbf{E}_0) + \left.\frac{\partial \pi}{\partial \mathbf{E}}\right|_{\mathbf{E}_0} \cdot (\mathbf{E}-\mathbf{E}_0) + \frac{1}{2}\left.\frac{\partial^2 \pi}{\partial \mathbf{E}^2}\right|_{\mathbf{E}_0} (\mathbf{E}-\mathbf{E}_0)\cdot(\mathbf{E}-\mathbf{E}_0)\\ &= \pi(\mathbf{E}_0) + \mathbb{D}\mathbf{E}_0 \cdot (\mathbf{E}-\mathbf{E}_0) + \frac{1}{2}\mathbb{D}(\mathbf{E}-\mathbf{E}_0)\cdot(\mathbf{E}-\mathbf{E}_0),\end{aligned} \tag{4.37}$$

from where

$$\pi(\mathbf{E}) - \pi(\mathbf{E}_0) \geq \mathbb{D}\mathbf{E}_0 \cdot (\mathbf{E}-\mathbf{E}_0) = \mathbf{T}_0 \cdot (\mathbf{E}-\mathbf{E}_0), \tag{4.38}$$

and where the equality is verified if and only if $\mathbf{E} = \mathbf{E}_0$. Hence, we conclude that π is a strictly convex function over the space of deformations.

The extension of the previous result to the case of nonlinear elasticity is straightforward. To address that situation, we consider, as before, that the function is monotonic, which translates into the following restriction

$$(\mathbf{T}^B - \mathbf{T}^A)\cdot(\mathbf{E}^B - \mathbf{E}^A) \geq 0 \qquad \forall \mathbf{E}^A, \mathbf{E}^B \in \mathscr{W}, \tag{4.39}$$

where the equality holds if and only if $\mathbf{E}^A = \mathbf{E}^B$, and where $\mathbf{T}^B = \mathbf{T}(\mathbf{E}^B)$ and $\mathbf{T}^A = \mathbf{T}(\mathbf{E}^A)$. Since expression (4.39) is valid for all elements in the space of deformations, it is particularly valid for $\mathbf{E} = \mathbf{E}^A + d\mathbf{E}$, and calling $\mathbf{T} = \mathbf{T}(\mathbf{E})$ the stress state associated with $\mathbf{E}$ through the constitutive equation, we have

$$(\mathbf{T} - \mathbf{T}^A)\cdot d\mathbf{E} \geq 0. \tag{4.40}$$

Hence, the increment in the strain energy density function produced in the process that takes us from the strain state $\mathbf{E}^A$ to the state $\mathbf{E}^B$ is

$$\int_{\mathbf{E}^A}^{\mathbf{E}^B} (\mathbf{T} - \mathbf{T}^A)\cdot d\mathbf{E} \geq 0, \tag{4.41}$$

from where

$$\pi(\mathbf{E}^B) - \pi(\mathbf{E}^A) \geq \mathbf{T}^A \cdot (\mathbf{E}^B - \mathbf{E}^A), \tag{4.42}$$

where the equality is verified if and only if $\mathbf{E}^A = \mathbf{E}^B$. We thus obtain, as before, the property of convexity of function π, but, in this case, by just making use of the monotonicity of the constitutive equation given by the hypothesis (4.39).

Importantly, since $\mathbb{D}$ is symmetric and positive definite (note that (4.31) and (4.32) must hold), there exists $\mathbb{D}^{-1}$ which is also symmetric and positive definite. This result allows us to determine $\mathbf{E}$ as a function of $\mathbf{T}$, regarded as the inverse constitutive equation, as

$$\mathbf{E} = \mathbf{E}(\mathbf{T}) = \mathbb{D}^{-1}\mathbf{T}, \tag{4.43}$$

---

3 Here tensor $\overline{\mathbf{T}}$ is not to be confused with the Piola–Kirchhoff stress tensor of the second kind defined in (3.271). In this chapter, $\overline{\mathbf{T}}$ refers to a Cauchy stress state defined in the unique configuration of interest considered for the body $\mathscr{B}$.

from which we obtain

$$\pi^*(\overline{\mathbf{T}}) = \int_{\mathbf{O}}^{\overline{\mathbf{T}}} \mathbf{E}(\mathbf{T}) \cdot d\mathbf{T} = \int_{\mathbf{O}}^{\overline{\mathbf{T}}} \mathbb{D}^{-1}\mathbf{T} \cdot d\mathbf{T} = \frac{1}{2}\mathbb{D}^{-1}\overline{\mathbf{T}} \cdot \overline{\mathbf{T}} \geq 0, \tag{4.44}$$

where the equality holds for $\overline{\mathbf{T}} = \mathbf{O}$. In addition

$$\overline{\mathbf{E}} = \left.\frac{\partial \pi^*}{\partial \mathbf{T}}\right|_{\mathbf{T}=\overline{\mathbf{T}}}, \tag{4.45}$$

and

$$\pi^*(\mathbf{T}) - \pi^*(\mathbf{T}_0) \geq \mathbf{E}_0 \cdot (\mathbf{T} - \mathbf{T}_0), \tag{4.46}$$

where $\mathbf{E}_0 = \mathbf{E}(\mathbf{T}_0)$, and the equality holds for $\mathbf{T} = \mathbf{T}_0$. In other words, the complementary strain energy density function is a strictly convex function over the space of stresses $\mathscr{W}'$. Again, this result can be obtained from the fact that the inverse constitutive function is monotonic.

Finally, from the properties of π and π^* we attain the following result

$$\pi(\mathbf{E}) + \pi^*(\mathbf{T}) \geq \mathbf{E} \cdot \mathbf{T}, \tag{4.47}$$

where the equality is valid when $\mathbf{T}$ and $\mathbf{E}$ are related through the constitutive equation.

In a nutshell, the characterization of a hyperelastic material can be performed in three different, but all equivalent, manners.

1) The constitutive equation is characterized by a monotonically increasing function

$$\mathbf{T} = \mathbf{T}(\mathbf{E}), \tag{4.48}$$

or by its inverse

$$\mathbf{E} = \mathbf{E}(\mathbf{T}). \tag{4.49}$$

2) There exists the strain energy density function, which is a strictly convex function, denoted by π, such that

$$\overline{\mathbf{T}} = \left.\frac{\partial \pi}{\partial \mathbf{E}}\right|_{\overline{\mathbf{E}}}, \tag{4.50}$$

or, equivalently, there exists the convex conjugate function, the complementary strain energy density function, denoted by π^*, such that

$$\overline{\mathbf{E}} = \left.\frac{\partial \pi^*}{\partial \mathbf{T}}\right|_{\overline{\mathbf{T}}}. \tag{4.51}$$

3) Given $\mathbf{E} \in \mathscr{W}$ and $\mathbf{T} \in \mathscr{W}'$ we have

$$\pi(\mathbf{E}) + \pi^*(\mathbf{T}) \geq \mathbf{E} \cdot \mathbf{T}, \tag{4.52}$$

where the equality holds if and only if $\mathbf{E}$ and $\mathbf{T}$ are related through the constitutive equation of the material.

For the case of an isotropic elastic material, the previous statements reduce to

$$\mathbf{T} = \mathbb{D}\mathbf{E} = 2\mu\mathbf{E} + \lambda(\mathrm{tr}\mathbf{E})\mathbf{I}, \tag{4.53}$$

$$\mathbf{E} = \mathbb{D}^{-1}\mathbf{T} = \frac{1}{2\mu}\mathbf{T} - \frac{\lambda}{2\mu(2\mu + 3\mu)}(\mathrm{tr}\mathbf{T})\mathbf{I}, \tag{4.54}$$

$$\pi(\mathbf{E}) = \frac{1}{2}\mathbb{D}\mathbf{E}\cdot\mathbf{E} = \frac{1}{2}[2\mu\mathbf{E}\cdot\mathbf{E} + \lambda(\mathrm{tr}\mathbf{E})^2], \tag{4.55}$$

$$\begin{aligned}\pi^*(\mathbf{T}) &= \max_{\mathbf{E}\in\mathscr{W}}(\mathbf{T}\cdot\mathbf{E} - \pi(\mathbf{E})) = \frac{1}{2}\mathbb{D}^{-1}\mathbf{T}\cdot\mathbf{T}\\ &= \frac{1}{2}\left(\frac{1}{2\mu}\mathbf{T}\cdot\mathbf{T} - \frac{\lambda}{2\mu(2\mu+3\mu)}(\mathrm{tr}\mathbf{T})^2\right).\end{aligned} \tag{4.56}$$

4.4 Equilibrium in Bodies without Constraints

The elastostatics problem, considering the hypothesis of infinitesimal strains, consists of the following: determine the fields $\mathbf{u}_0 \in \mathrm{Kin}_u = \mathscr{U}$, $\mathbf{E}_0 \in \mathscr{W}$ and $\mathbf{T}_0 \in \mathscr{W}'$ such that

- $\mathbf{T}_0 \in \mathrm{Est}_T$, that is $\mathbf{T}_0$ is at equilibrium with the applied load f
- $\mathbf{E}_0$ and $\mathbf{T}_0$ are related through the constitutive equation, for example $\mathbf{T}_0 = \mathbb{D}\mathbf{E}_0$, or in the more general case

$$\mathbf{E}_0 = \left.\frac{\partial\pi^*}{\partial\mathbf{T}}\right|_{\mathbf{T}_0}, \qquad \mathbf{T}_0 = \left.\frac{\partial\pi}{\partial\mathbf{E}}\right|_{\mathbf{E}_0}. \tag{4.57}$$

- $\mathbf{E}_0 = \mathscr{D}\mathbf{u}_0 = (\nabla\mathbf{u})^s$, which means that the deformation is compatible.

As seen in Chapter 3, the Principle of Virtual Work (PVW) allows us to characterize the stress state $\mathbf{T}_0$ that, for the case of bodies without constraints, corresponds to

$$(\mathbf{T}_0, \mathscr{D}\hat{\mathbf{u}}) = \langle f, \hat{\mathbf{u}}\rangle \qquad \forall\hat{\mathbf{u}} \in \mathrm{Var}_u = \mathscr{U}, \tag{4.58}$$

noting that $f \in \mathscr{U}'$ must satisfy

$$\langle f, \hat{\mathbf{u}}\rangle = 0 \qquad \forall\hat{\mathbf{u}} \in \mathrm{Var}_u \cap \mathscr{N}(\mathscr{D}) = \mathscr{N}(\mathscr{D}). \tag{4.59}$$

The existence of at least a single stress state $\mathbf{T}_0$ at equilibrium with f is guaranteed by a Representation Theorem (see Theorem 3.1 in Chapter 3). If there exists $\mathbf{T}_0$, then there exists $\mathbf{E}_0 = \mathbb{D}^{-1}\mathbf{T}_0$ (strain associated with $\mathbf{T}_0$ by means of the constitutive equation). The question is now the following: does there exist $\mathbf{u}_0 \in \mathrm{Kin}_u$ such that $\mathbf{E}_0 = \mathscr{D}\mathbf{u}_0$? The answer is positive, and is formalized in the following result.

Theorem 4.1 (Equilibrium in Bodies without Constraints) Given $f \in \mathscr{U}'$ at equilibrium, there exists $\mathbf{u}_0 \in \mathscr{U}$ such that

$$\mathscr{K}\mathbf{u}_0 = f \qquad \mathscr{K} = \mathscr{D}^*\mathbb{D}\mathscr{D}, \tag{4.60}$$

or, equivalently, $f \in R(\mathscr{K})$. Operator $\mathscr{K} : \mathscr{U} \to \mathscr{U}'$ is known as the stiffness operator.

Proof. From the hypothesis of the theorem, $f \in \mathscr{U}'$ is an equilibrated external load. Then, from the PVW we have

$$\langle f, \hat{\mathbf{u}}\rangle = 0 \qquad \forall\hat{\mathbf{u}} \in \mathrm{Var}_u \cap \mathscr{N}(\mathscr{D}) = \mathscr{N}(\mathscr{D}), \tag{4.61}$$

that is

$$f \in \mathscr{N}(\mathscr{D})^{\perp}. \tag{4.62}$$

Now, from the results reported in Appendix C we have

$$f \in R(\mathscr{K}), \tag{4.63}$$

which tells us that there exists $\mathbf{u}_0 \in \mathscr{U} \setminus \mathscr{N}(\mathscr{D})$ (i.e. $\mathbf{u}_0$ is defined up to a rigid motion) such that

$$\mathscr{K}\mathbf{u}_0 = f. \tag{4.64}$$

■

From this existence theorem, for the case of linear materials (the same is straightforwardly obtained for nonlinear materials), we conclude that

$$\langle f, \hat{\mathbf{u}}\rangle = \langle \mathscr{K}\mathbf{u}_0, \hat{\mathbf{u}}\rangle = \langle \mathscr{D}^*\mathbb{D}\mathscr{D}\mathbf{u}_0, \hat{\mathbf{u}}\rangle = (\mathbb{D}\mathscr{D}\mathbf{u}_0, \mathscr{D}\hat{\mathbf{u}}) \qquad \forall \hat{\mathbf{u}} \in \mathscr{U}. \tag{4.65}$$

Then, the elastostatics problem associated with the mechanical equilibrium in the case of bodies without constraints is characterized by the variational expression called the PVW, as we discuss in the next section.

4.4.1 Principle of Virtual Work

The problem of elastostatics, considering the hypothesis of infinitesimal strains and small displacements, and considering a linear hyperelastic material behavior, is characterized by the PVW. As we know, the PVW is mathematically formulated as the variational equation which amounts to finding $\mathbf{u}_0 \in \mathscr{U} \setminus \mathscr{N}(\mathscr{D})$, such that

$$(\mathbb{D}\mathscr{D}\mathbf{u}_0, \mathscr{D}\hat{\mathbf{u}}) = \langle f, \hat{\mathbf{u}}\rangle \qquad \forall \hat{\mathbf{u}} \in \mathrm{Var}_u = \mathscr{U}. \tag{4.66}$$

Noting that $\mathrm{Kin}_u = \mathbf{u}_0 + \mathrm{Var}_u$, we have that $\mathrm{Var}_u = \mathrm{Kin}_u - \mathbf{u}_0$, and therefore the variational equation (4.66) can be rewritten as

$$(\mathbb{D}\mathscr{D}\mathbf{u}_0, \mathscr{D}(\mathbf{u} - \mathbf{u}_0)) = \langle f, (\mathbf{u} - \mathbf{u}_0)\rangle \qquad \forall \mathbf{u} \in \mathrm{Kin}_u = \mathscr{U}. \tag{4.67}$$

In the case of a nonlinear hyperelastic material, the PVW takes the more general form

$$\left(\left.\frac{\partial \pi}{\partial \mathbf{E}}\right|_{\mathbf{E}_0 = \mathscr{D}\mathbf{u}_0}, \mathscr{D}(\mathbf{u} - \mathbf{u}_0) \right) = \langle f, (\mathbf{u} - \mathbf{u}_0)\rangle \qquad \forall \mathbf{u} \in \mathrm{Kin}_u = \mathscr{U}. \tag{4.68}$$

4.4.2 Principle of Minimum Total Potential Energy

Exploiting the hyperelastic properties of the material studied in this chapter, the PVW formulated in the previous section corresponds to the critical point of a certain cost functional. In fact, from the convexity of functional π given by (4.38), for kinematically admissible strains, for linear material behavior we have

$$\pi(\mathscr{D}\mathbf{u}) - \pi(\mathscr{D}\mathbf{u}_0) \geq \mathbb{D}\mathscr{D}\mathbf{u}_0 \cdot \mathscr{D}(\mathbf{u} - \mathbf{u}_0), \tag{4.69}$$

while for the case of nonlinear materials (see (4.46)) we have

$$\pi(\mathscr{D}\mathbf{u}) - \pi(\mathscr{D}\mathbf{u}_0) \geq \left.\frac{\partial \pi}{\partial \mathbf{E}}\right|_{\mathscr{D}\mathbf{u}_0} \cdot \mathscr{D}(\mathbf{u} - \mathbf{u}_0), \tag{4.70}$$

where the equality holds now for all elements of the form $\mathbf{u} - \mathbf{u}_0 \in \mathscr{U} \setminus \mathscr{N}(\mathscr{D})$. Integrating all over $\mathscr{B}$ (region occupied by the body) we obtain

$$\int_{\mathscr{B}} \left(\pi(\mathscr{D}\mathbf{u}) - \pi(\mathscr{D}\mathbf{u}_0)\right) dV \geq \int_{\mathscr{B}} \mathbb{D}\mathscr{D}\mathbf{u}_0 \cdot \mathscr{D}(\mathbf{u} - \mathbf{u}_0) dV \quad \forall \mathbf{u} \in \mathrm{Kin}_u = \mathscr{U}, \tag{4.71}$$

and in the nonlinear case

$$\int_{\mathscr{B}} \left(\pi(\mathscr{D}\mathbf{u}) - \pi(\mathscr{D}\mathbf{u}_0)\right) dV \geq \int_{\mathscr{B}} \left.\frac{\partial \pi}{\partial \mathbf{E}}\right|_{\mathscr{D}\mathbf{u}_0} \cdot \mathscr{D}(\mathbf{u}-\mathbf{u}_0) dV \quad \forall \mathbf{u} \in \mathrm{Kin}_u = \mathscr{U}. \tag{4.72}$$

Replacing (4.71) (or (4.72)) in the PVW (4.67) (or in (4.68)) yields

$$\int_{\mathscr{B}} \left(\pi(\mathscr{D}\mathbf{u}) - \pi(\mathscr{D}\mathbf{u}_0)\right) dV \geq \langle f, (\mathbf{u}-\mathbf{u}_0)\rangle \qquad \forall \mathbf{u} \in \mathrm{Kin}_u = \mathscr{U}. \tag{4.73}$$

We introduce the Total Potential Energy cost functional as

$$\Pi(\mathbf{u}) = \int_{\mathscr{B}} \pi(\mathscr{D}\mathbf{u}) dV - \langle f, \mathbf{u}\rangle, \tag{4.74}$$

then, rearranging terms gives

$$\Pi(\mathbf{u}) = \int_{\mathscr{B}} \pi(\mathscr{D}\mathbf{u}) dV - \langle f, \mathbf{u}\rangle \geq \int_{\mathscr{B}} \pi(\mathscr{D}\mathbf{u}_0) dV - \langle f, \mathbf{u}_0\rangle = \Pi(\mathbf{u}_0) \qquad \forall \mathbf{u} \in \mathrm{Kin}_u, \tag{4.75}$$

where the equality is verified if and only if $\mathbf{u}$ and $\mathbf{u}_0$ are equal up to a rigid body motion. We have thus arrived at a minimum principle, valid for both linear and nonlinear hyperelastic materials. This minimum principle, equivalent to the variational principle of the PVW, is called the Principle of Minimum Total Potential Energy, and, for bodies without constraints, it is enunciated next. The elastostatics problem corresponding to the mechanical equilibrium is equivalent to determining the critical point $\mathbf{u}_0 \in \mathscr{U} \setminus \mathscr{N}(\mathscr{D})$ such that the Total Potential Energy functional is minimized, that is

$$\Pi(\mathbf{u}_0) = \min_{\mathbf{u}\in\mathscr{U}\setminus\mathscr{N}(\mathscr{D})} \Pi(\mathbf{u}) = \min_{\mathbf{u}\in\mathscr{U}\setminus\mathscr{N}(\mathscr{D})} \left[\int_{\mathscr{B}} \pi(\mathscr{D}\mathbf{u}) dV - \langle f, \mathbf{u}\rangle\right]. \tag{4.76}$$

4.4.3 Local Equations and Boundary Conditions

Previous sections addressed the PVW in an abstract form, written in terms of operators between some space and the corresponding dual. Therefore, such a formulation is valid for any continuum composed by solid media featuring hyperelastic behavior, provided the hypotheses of small displacements and infinitesimal strains hold. In this section we frame these results for the case of three-dimensional bodies, with the purpose of unveiling the local form of the equilibrium equations, also called the strong form of the mechanical equilibrium. For this, recall that for a three-dimensional body, according to (3.159), we have

$$\langle f, \mathbf{u}\rangle = \int_{\mathscr{B}} \mathbf{b}\cdot\mathbf{u}\, dV + \int_{\partial\mathscr{B}} \mathbf{a}\cdot\mathbf{u}\, dS, \tag{4.77}$$

where $\mathbf{b}$ and $\mathbf{a}$ are, respectively, the force per unit volume and the force per unit surface defined in $\mathscr{B}$ and $\partial\mathscr{B}$, correspondingly. In turn, for a linear hyperelastic material, it is

$$\pi(\mathscr{D}\mathbf{u}) = \frac{1}{2}\mathbb{D}\mathscr{D}\mathbf{u}\cdot\mathscr{D}\mathbf{u} = \frac{1}{2}\mathbb{D}(\nabla\mathbf{u})^s\cdot(\nabla\mathbf{u})^s, \tag{4.78}$$

and so the problem of finding the critical point of the cost functions (4.76) becomes

$$\Pi(\mathbf{u}_0) = \min_{\mathbf{u}\in\mathscr{U}\setminus\mathscr{N}(\mathscr{D})} \left[\int_{\mathscr{B}} \frac{1}{2}\mathbb{D}(\nabla\mathbf{u})^s\cdot(\nabla\mathbf{u})^s dV - \int_{\mathscr{B}} \mathbf{b}\cdot\mathbf{u}\, dV - \int_{\partial\mathscr{B}} \mathbf{a}\cdot\mathbf{u}\, dS\right]. \tag{4.79}$$

The necessary, and in this case sufficient, condition for $\mathbf{u}_0$ to be a minimum corresponds to

$$\delta\Pi(\mathbf{u}_0, \hat{\mathbf{u}}) = 0 \qquad \forall\hat{\mathbf{u}} \in \mathscr{U}, \tag{4.80}$$

where $\delta\Pi(\mathbf{u}_0, \hat{\mathbf{u}})$ is the first variation of the Total Potential Energy cost functional. In other words, this stands for the Gâteaux derivative of this functional evaluated at $\mathbf{u}_0$ and in the direction of $\hat{\mathbf{u}}$, that is

$$\begin{aligned}\delta\Pi(\mathbf{u}_0, \hat{\mathbf{u}}) &= \left.\frac{d}{d\alpha}\Pi(\mathbf{u}_0 + \alpha\hat{\mathbf{u}})\right|_{\alpha=0} \\ &= \int_{\mathscr{B}} \mathbb{D}(\nabla\mathbf{u}_0)^s \cdot (\nabla\hat{\mathbf{u}})^s \, dV - \int_{\mathscr{B}} \mathbf{b}\cdot\hat{\mathbf{u}}\, dV - \int_{\partial\mathscr{B}} \mathbf{a}\cdot\hat{\mathbf{u}}\, dS = 0 \qquad \forall\hat{\mathbf{u}} \in \mathscr{U},\end{aligned} \tag{4.81}$$

which is nothing but the expression of the PVW for the particular case we are studying.

Assuming the fields are regular enough so we can make use of the tensor identity (2.90), and exploiting the symmetry of tensor $\mathbb{D}(\nabla\mathbf{u}_0)^s$, yields

$$\mathbb{D}(\nabla\mathbf{u}_0)^s \cdot (\nabla\hat{\mathbf{u}})^s = \operatorname{div}[\mathbb{D}(\nabla\mathbf{u}_0)^s\hat{\mathbf{u}}] - \operatorname{div}[\mathbb{D}(\nabla\mathbf{u}_0)^s] \cdot \hat{\mathbf{u}}, \tag{4.82}$$

and then we obtain

$$\begin{aligned}\delta\Pi(\mathbf{u}_0, \hat{\mathbf{u}}) = &-\int_{\mathscr{B}} \left(\operatorname{div}[\mathbb{D}(\nabla\mathbf{u}_0)^s] + \mathbf{b}\right)\cdot\hat{\mathbf{u}}\, dV \\ &+ \int_{\partial\mathscr{B}} \left(\mathbb{D}(\nabla\mathbf{u}_0)^s\mathbf{n} - \mathbf{a}\right)\cdot\hat{\mathbf{u}}\, dS = 0 \qquad \forall\hat{\mathbf{u}} \in \mathscr{U},\end{aligned} \tag{4.83}$$

where $\mathbf{n}$ is the outward normal unit vector defined over $\partial\mathscr{B}$. In this manner, we arrive at the Euler–Lagrange equations associated with the problem of the minimization of the Total Potential Energy functional

$$\begin{cases} -\operatorname{div}[\mathbb{D}(\nabla\mathbf{u}_0)^s] = \mathbf{b} & \text{in } \mathscr{B}, \\ \mathbb{D}(\nabla\mathbf{u}_0)^s\mathbf{n} = \mathbf{a} & \text{on } \partial\mathscr{B}. \end{cases} \tag{4.84}$$

These equations express the strong form[4] in which the mechanical equilibrium is verified within the infinitesimal strain theory for hyperelastic bodies.

Assuming the material is isotropic and homogeneous, the previous equations result in

$$\begin{cases} -\operatorname{div}[2\mu(\nabla\mathbf{u}_0)^s + \lambda\operatorname{tr}(\nabla\mathbf{u}_0)^s\mathbf{I}] = \mathbf{b} & \text{in } \mathscr{B}, \\ 2\mu(\nabla\mathbf{u}_0)^s\mathbf{n} + \lambda\operatorname{tr}(\nabla\mathbf{u}_0)^s\mathbf{n} = \mathbf{a} & \text{on } \partial\mathscr{B}. \end{cases} \tag{4.85}$$

Now, remembering expressions (1.113), (2.93), and (2.117), we obtain, for the case of homogeneous isotropic materials, the Navier equations

$$-\mu(\Delta\mathbf{u}_0 + \nabla(\operatorname{div}\mathbf{u}_0)) - \lambda\nabla(\operatorname{div}\mathbf{u}_0) = \mathbf{b} \qquad \text{in } \mathscr{B}, \tag{4.86}$$

or, equivalently,

$$-\mu\Delta\mathbf{u}_0 - (\lambda + \mu)\nabla(\operatorname{div}\mathbf{u}_0) = \mathbf{b} \qquad \text{in } \mathscr{B}. \tag{4.87}$$

---

4 This is called the strong form in the sense that it is valid for every point in the continuum.

Furthermore, the boundary conditions of mechanical character (force per unit surface is prescribed) give

$$[2\mu(\nabla \mathbf{u}_0)^s + \lambda(\mathrm{div}\mathbf{u}_0)\mathbf{I}]\mathbf{n} = \mathbf{a} \quad \text{on } \partial\mathscr{B}. \tag{4.88}$$

It is worthwhile recalling that the boundary condition (4.88) is not explicitly prescribed in the minimization of the total potential energy functional Π, whose domain of definition is $\mathrm{Kin}_u = \mathscr{U}$. In fact, the only boundary conditions prescribed to $\mathbf{u}$ are of kinematical nature (since $\mathbf{u} \in \mathrm{Kin}_u$). For this reason, the constraints established over $\mathbf{u} \in \mathrm{Kin}_u$ are called principal boundary conditions. It can then be appreciated that the very minimization process is responsible for the selection of the field $\mathbf{u}_0$ which, being a critical point of the functional, also satisfies the mechanical boundary condition (4.88). For this reason, this type of expression is referred to as a natural boundary condition because it is naturally verified in the search for the element that minimizes the cost functional Π.

Finally, Figure 4.2 graphically illustrates the ingredients of the PVW for the case of hyperelastic materials studied here.

4.4.4 Principle of Complementary Virtual Work

According to what we saw in Chapter 3, for the case of bodies without constraints, given the external load $f \in \mathscr{U}'$ at equilibrium, that is

$$\langle f, \hat{\mathbf{u}} \rangle = 0 \qquad \forall \hat{\mathbf{u}} \in \mathrm{Var}_u \cap \mathscr{N}(\mathscr{D}) = \mathscr{N}(\mathscr{D}), \tag{4.89}$$

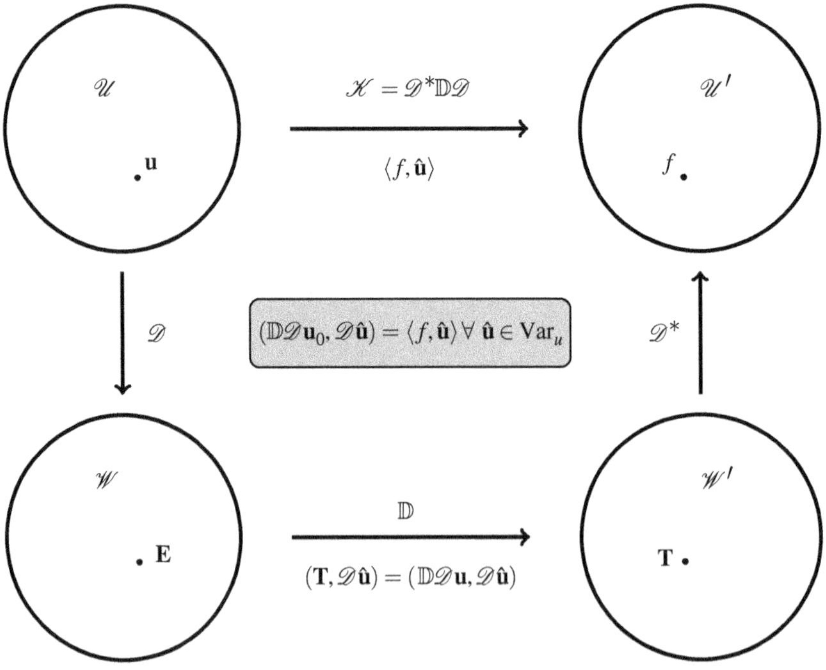

Figure 4.2 Ingredients in the formulation of the PVW for hyperelastic materials.

the linear manifold in the vector space $\mathscr{W}'$ given by all internal stress fields $\mathbf{T} \in \mathscr{W}'$ at equilibrium with f is given by

$$\text{Est}_T = \{\mathbf{T} \in \mathscr{W}'; -(\mathbf{T}, \mathscr{D}\hat{\mathbf{u}}) + \langle f, \hat{\mathbf{u}}\rangle = 0 \quad \forall \hat{\mathbf{u}} \in \text{Var}_u\}, \tag{4.90}$$

and the space of self-equilibrated internal stresses is

$$\text{Var}_T = \{\hat{\mathbf{T}} \in \mathscr{W}'; -(\hat{\mathbf{T}}, \mathscr{D}\hat{\mathbf{u}}) = 0 \quad \forall \hat{\mathbf{u}} \in \text{Var}_u\}, \tag{4.91}$$

from where we know that

$$\text{Est}_T = \mathbf{T}_0 + \text{Var}_T, \tag{4.92}$$

with $\mathbf{T}_0 \in \text{Est}_T$ arbitrary.

Hence, the compatibility of the strain $\mathbf{E}_0 \in \mathscr{W}$ is stated by the variational problem defined by the Principle of Complementary Virtual Power (PCVP) (3.167). In the present case of infinitesimal strains, the PCVP renders the Principle of Complementary Virtual Work (PCVW), which has the form

$$(\hat{\mathbf{T}}, \mathbf{E}_0) = 0 \qquad \forall \hat{\mathbf{T}} \in \text{Var}_T. \tag{4.93}$$

For the case under study, hyperelastic materials feature a relation between the strain $\mathbf{E}_0 \in \mathscr{W}$ and the stress $\mathbf{T}_0 \in \mathscr{W}'$ given by

$$\mathbf{E}_0 = \left.\frac{\partial \pi^*}{\partial \mathbf{T}}\right|_{\mathbf{T}_0}, \tag{4.94}$$

which for linear elastic materials is reduced to $\mathbf{E}_0 = \mathbb{D}^{-1}\mathbf{T}_0$.

Then, the compatibility in the elastostatics problem for bodies without constraints is equivalent to determining the stress field $\mathbf{T}_0 \in \text{Est}_T$ such that

$$\left(\hat{\mathbf{T}}, \left.\frac{\partial \pi^*}{\partial \mathbf{T}}\right|_{\mathbf{T}_0}\right) = 0 \qquad \forall \hat{\mathbf{T}} \in \text{Var}_T. \tag{4.95}$$

Moreover, recalling that $\text{Est}_T = \mathbf{T}_0 + \text{Var}_T$, and so $\text{Var}_T = \text{Est}_T - \mathbf{T}_0$, the PCVW can be rewritten as follows: find $\mathbf{T}_0 \in \text{Est}_T$ such that

$$\left(\mathbf{T} - \mathbf{T}_0, \left.\frac{\partial \pi^*}{\partial \mathbf{T}}\right|_{\mathbf{T}_0}\right) = 0 \qquad \forall \mathbf{T} \in \text{Est}_T, \tag{4.96}$$

and for the linear elastic case it is

$$(\mathbf{T} - \mathbf{T}_0, \mathbb{D}^{-1}\mathbf{T}_0) = 0 \qquad \forall \mathbf{T} \in \text{Est}_T. \tag{4.97}$$

4.4.5 Principle of Minimum Complementary Energy

Now, taking into consideration the convexity of the complementary strain energy density function π^*, which in the general hyperelastic case states that

$$\pi^*(\mathbf{T}) - \pi^*(\mathbf{T}_0) \geq \left.\frac{\partial \pi^*}{\partial \mathbf{T}}\right|_{\mathbf{T}_0} \cdot (\mathbf{T} - \mathbf{T}_0), \tag{4.98}$$

where the equality is verified for $\mathbf{T} = \mathbf{T}_0$, and integrating in the domain occupied by the body, we have

$$\int_{\mathscr{B}} (\pi^*(\mathbf{T}) - \pi^*(\mathbf{T}_0))\, dV \geq \left(\mathbf{T} - \mathbf{T}_0, \left.\frac{\partial \pi^*}{\partial \mathbf{T}}\right|_{\mathbf{T}_0}\right) = 0 \qquad \forall \mathbf{T} \in \text{Est}_T. \tag{4.99}$$

Now we define the Complementary Energy cost functional as follows

$$\Pi^*(\mathbf{T}) = \int_{\mathscr{B}} \pi^*(\mathbf{T})dV. \tag{4.100}$$

Hence, putting (4.99) into the expression of the PCVW given by (4.96) yields

$$\Pi^*(\mathbf{T}) = \int_{\mathscr{B}} \pi^*(\mathbf{T})dV \geq \int_{\mathscr{B}} \pi^*(\mathbf{T}_0)dV = \Pi^*(\mathbf{T}_0) \qquad \forall \mathbf{T} \in \mathrm{Est}_T, \tag{4.101}$$

where, as before, the equality holds for $\mathbf{T} = \mathbf{T}_0$. Therefore, the compatibility problem formulated by the PCVW in the context of the elastostatics of bodies without constraints is equivalent to the problem of finding $\mathbf{T}_0 \in \mathrm{Est}_T$ such that the Complementary Energy functional is minimized, that is

$$\Pi^*(\mathbf{T}_0) = \min_{\mathbf{T}\in\mathrm{Est}_T} \Pi^*(\mathbf{T}) = \min_{\mathbf{T}\in\mathrm{Est}_T} \int_{\mathscr{B}} \pi^*(\mathbf{T})dV. \tag{4.102}$$

This formulation, equivalent to the variational problem behind the PCVW, is called the Principle of Minimum Complementary Energy.

Notice that the existence and uniqueness of the stress $\mathbf{T}_0 \in \mathrm{Est}_T$ is guaranteed by Π^* being a lower semi-continuous strictly convex and coercive[5] functional in the linear manifold Est_T. In fact, the necessary, and in this case sufficient, condition for the critical point to be a minimum is

$$\begin{aligned}\delta\Pi^*(\mathbf{T}_0, \hat{\mathbf{T}}) &= \frac{d}{d\alpha}\Pi^*(\mathbf{T}_0 + \alpha\hat{\mathbf{T}})\Big|_{\alpha=0} \\ &= \int_{\mathscr{B}} \frac{d}{d\alpha}\pi^*(\mathbf{T}_0 + \alpha\hat{\mathbf{T}})\Big|_{\alpha=0} dV = 0 \qquad \forall \hat{\mathbf{T}} \in \mathrm{Var}_T,\end{aligned} \tag{4.103}$$

that is,

$$\delta\Pi^*(\mathbf{T}_0, \hat{\mathbf{T}}) = \int_{\mathscr{B}} \frac{\partial \pi^*}{\partial \mathbf{T}}\Big|_{\mathbf{T}_0} \cdot \hat{\mathbf{T}}\, dV = \left(\hat{\mathbf{T}}, \frac{\partial \pi^*}{\partial \mathbf{T}}\Big|_{\mathbf{T}_0}\right) = 0 \qquad \forall \hat{\mathbf{T}} \in \mathrm{Var}_T, \tag{4.104}$$

which is exactly the variational problem corresponding to the PCVW.

In particular, for the case of a linearly elastic material, the minimization of the Complementary Energy functional consists in finding $\mathbf{T}_0 \in \mathrm{Est}_T$ such that

$$\Pi^*(\mathbf{T}_0) = \min_{\mathbf{T}\in\mathrm{Est}_T} \int_{\mathscr{B}} \frac{1}{2}\mathbb{D}^{-1}\mathbf{T} \cdot \mathbf{T}\, dV. \tag{4.105}$$

4.4.6 Additional Remarks

In this section we have arrived at the following result: the fields $\mathbf{u}_0 \in \mathrm{Kin}_u$, $\mathbf{E}_0$ and $\mathbf{T}_0 \in \mathrm{Est}_T$, which are the solution of the elastostatics problem (remember that $\mathbf{E}_0 = \mathscr{D}\,\mathbf{u}_0$ and $\mathbf{T}_0$ are related through a hyperelastic constitutive equation), are also the solution of the following two problems:

- Equilibrium: Total Potential Energy cost functional Π defined in (4.74)

$$\Pi(\mathbf{u}_0) = \min_{\mathbf{u}\in\mathrm{Kin}_u} \Pi(\mathbf{u}). \tag{4.106}$$

5 Coercive in the sense that $\Pi^*(\mathbf{T}) \to \infty$ para $\|\mathbf{T}\| \to \infty$.

- Compatibility: Complementary Energy cost functional Π^* defined in (4.100)

$$\Pi^*(\mathbf{T}_0) = \min_{\mathbf{T}\in \mathrm{Est}_T} \Pi^*(\mathbf{T}). \tag{4.107}$$

Observe that it is easy to show that $\Pi(\mathbf{u}_0) = -\Pi^*(\mathbf{T}_0)$, and, therefore, the duality between (4.106) and (4.107) is, once again, characterized. As a matter of fact, this result is demonstrated by using (4.47) for the case when $\mathbf{E}_0 = \mathcal{D}\,\mathbf{u}_0$ and $\mathbf{T}_0$ are related through a constitutive function, leading to

$$\pi(\mathbf{E}_0) + \pi^*(\mathbf{T}_0) = \mathbf{E}_0 \cdot \mathbf{T}_0, \tag{4.108}$$

then

$$\begin{aligned}\Pi(\mathbf{u}) \geq \Pi(\mathbf{u}_0) &= \int_{\mathcal{B}} \pi(\mathcal{D}\,\mathbf{u}_0)\, dV - \langle f, \mathbf{u}_0\rangle \\ &= -\int_{\mathcal{B}} \pi^*(\mathbf{T}_0)\, dV + \left[\int_{\mathcal{B}} \mathbf{T}_0 \cdot \mathcal{D}\,\mathbf{u}_0\, dV - \langle f, \mathbf{u}_0\rangle\right].\end{aligned} \tag{4.109}$$

Since $\mathbf{T}_0 \in \mathrm{Est}_T$ and $\mathbf{u}_0 \in \mathrm{Kin}_u = \mathrm{Var}_u$, the term between brackets is the expression of the PVW. As a consequence, it gives

$$\Pi(\mathbf{u}) \geq \Pi(\mathbf{u}_0) = -\Pi^*(\mathbf{T}_0) \geq -\Pi^*(\mathbf{T}) \qquad \forall \mathbf{u} \in \mathrm{Kin}_u, \qquad \forall \mathbf{T} \in \mathrm{Est}_T. \tag{4.110}$$

These inequalities provide a way to quantify upper and lower bounds for the solution which is being sought. This is practically achieved by making use of kinematically admissible displacement fields and stress fields equilibrated with f.

4.5 Equilibrium in Bodies with Bilateral Constraints

Let us, in this section, develop the equivalence between the variational principles corresponding to the PVW and the PCVW and principles of minimum for bodies with prescribed bilateral constraints. In this context, we have seen that

$$\mathrm{Kin}_u = \{\mathbf{u} \in \mathcal{U};\ \mathbf{u}|_{\partial\mathcal{B}_u} = \overline{\mathbf{u}}\}, \tag{4.111}$$

$$\mathrm{Var}_u = \{\hat{\mathbf{u}} \in \mathcal{U};\ \hat{\mathbf{u}}|_{\partial\mathcal{B}_u} = \mathbf{0}\}, \tag{4.112}$$

where Var_u is a subspace of $\mathcal{U}$, Kin_u is a linear manifold (translation of Var_u), and $\partial\mathcal{B}_u$ is the subset of the boundary $\partial\mathcal{B}$ of $\mathcal{B}$ where displacements are prescribed. Recall that, from the PVW, we have that f is at equilibrium if

$$\langle f, \hat{\mathbf{u}}\rangle = 0 \qquad \forall \hat{\mathbf{u}} \in \mathrm{Var}_u \cap \mathcal{N}(\mathcal{D}). \tag{4.113}$$

If f is at equilibrium, there exists at least one stress state $\mathbf{T} \in \mathcal{W}'$ which equilibrates f, defined by

$$-(\mathbf{T}, \mathcal{D}\,\hat{\mathbf{u}}) + \langle f, \hat{\mathbf{u}}\rangle = 0 \qquad \forall \hat{\mathbf{u}} \in \mathrm{Var}_u. \tag{4.114}$$

As in the previous section, the question is whether or not, for f at the equilibrium given, there exists $\mathbf{u}_0 \in \mathrm{Kin}_u$ such that the associated stress state $\mathbf{T}_0$

$$\mathbf{T}_0 = \left.\frac{\partial \pi}{\partial \mathbf{E}}\right|_{\mathbf{E}_0 = \mathcal{D}\,\mathbf{u}_0}, \tag{4.115}$$

equilibrates f, which implies

$$\left(\left. \frac{\partial \pi}{\partial \mathbf{E}} \right|_{\mathbf{E}_0 = \mathscr{D}\mathbf{u}_0}, \mathscr{D}\hat{\mathbf{u}} \right) = \langle f, \hat{\mathbf{u}} \rangle \qquad \forall \hat{\mathbf{u}} \in \mathrm{Var}_u. \tag{4.116}$$

The following result provides the positive answer to this question. This result is demonstrated for linearly elastic materials. The same result holds for general nonlinear hyperelastic behavior.

Theorem 4.2 (Equilibrium in Bodies with Bilateral Constraints) Given $f \in \mathscr{U}'$ at equilibrium, there exists $r \in \mathrm{Var}_u^\perp$ and $\mathbf{u}_0 \in Kin_u$ such that

$$\mathscr{K}\mathbf{u}_0 = f + r, \qquad \mathscr{K} = \mathscr{D}^*\mathbb{D}\mathscr{D}. \tag{4.117}$$

In addition, if the rigid motion has been eliminated, then $\mathbf{u}_0$ is unique.

Proof. By hypothesis we have

$$f \in [\mathrm{Var}_u \cap \mathscr{N}(\mathscr{D})]^\perp. \tag{4.118}$$

Given that $\mathscr{N}(\mathscr{D}) = \mathscr{N}(\mathscr{K})$ and that $\mathrm{Var}_u \cap \mathscr{N}(\mathscr{K}) = \mathrm{Var}_u \cap [\mathscr{K}(\mathrm{Var}_u)]^\perp$ (see the results reported in Appendix C), we have the following

$$\begin{aligned} [\mathrm{Var}_u \cap \mathscr{N}(\mathscr{D})]^\perp &= [\mathrm{Var}_u \cap \mathscr{N}(\mathscr{K})]^\perp \\ &= \left[\mathrm{Var}_u \cap [\mathscr{K}(\mathrm{Var}_u)]^\perp\right]^\perp = \mathrm{Var}_u^\perp \oplus \mathscr{K}(\mathrm{Var}_u). \end{aligned} \tag{4.119}$$

Thus, using the results reported in Appendix C, we can write

$$[\mathrm{Var}_u \cap \mathscr{N}(\mathscr{D})]^\perp = \mathrm{Var}_u^\perp \oplus \mathscr{K}(\mathrm{Var}_u). \tag{4.120}$$

In turn, we know that $[\mathrm{Var}_u \cap \mathscr{N}(\mathscr{D})]^\perp = (\mathrm{Var}_u)^\perp \oplus (\mathscr{N}(\mathscr{D}))^\perp$ (see Appendix A), and also that $(\mathscr{N}(\mathscr{D}))^\perp = R(\mathscr{K})$ (see the results reported in Appendix C). Hence, we have

$$[\mathrm{Var}_u \cap \mathscr{N}(\mathscr{D})]^\perp = \mathrm{Var}_u^\perp \oplus R(\mathscr{K}). \tag{4.121}$$

From (4.120) and (4.121) we arrive at

$$f \in \mathrm{Var}_u^\perp \oplus R(\mathscr{K}) = \mathrm{Var}_u^\perp \oplus \mathscr{K}(\mathrm{Var}_u). \tag{4.122}$$

Let $\mathbf{w} \in \mathrm{Kin}_u$ be arbitrary, then $\mathscr{K}\mathbf{w} \in R(\mathscr{K})$ and the previous expression yields

$$f - \mathscr{K}\mathbf{w} \in \mathrm{Var}_u^\perp \oplus \mathscr{K}(\mathrm{Var}_u), \tag{4.123}$$

and therefore

$$f \in \mathrm{Var}_u^\perp + \mathscr{K}\mathbf{w} + \mathscr{K}(\mathrm{Var}_u) = \mathrm{Var}_u^\perp + \mathscr{K}(\mathbf{w} + \mathrm{Var}_u). \tag{4.124}$$

Remembering that $\mathrm{Kin}_u = \mathbf{w} + \mathrm{Var}_u$ with $\mathbf{w} \in \mathrm{Kin}_u$ arbitrary, results in

$$f \in \mathrm{Var}_u^\perp + \mathscr{K}(\mathrm{Kin}_u). \tag{4.125}$$

This expression tells us that there exists $-r \in \mathrm{Var}_u^\perp$ and $\mathbf{u}_0 \in \mathrm{Kin}_u$ such that

$$f + r = \mathscr{K}\mathbf{u}_0. \tag{4.126}$$

To prove the uniqueness, let $\hat{\mathbf{u}} \in \mathrm{Var}_u$ be an arbitrary element, then

$$\langle f, \hat{\mathbf{u}} \rangle = \langle -r + \mathscr{K}\mathbf{u}_0, \hat{\mathbf{u}} \rangle = \langle \mathscr{K}\mathbf{u}_0, \hat{\mathbf{u}} \rangle = (\mathbb{D}\mathscr{D}\mathbf{u}_0, \mathscr{D}\hat{\mathbf{u}}) \qquad \forall \hat{\mathbf{u}} \in \mathrm{Var}_u. \tag{4.127}$$

Taking $\mathbf{u}_0$ as the element that executes the translation of Var_u, that is $\mathrm{Kin}_u = \mathbf{u}_0 + \mathrm{Var}_u$ and then $\mathrm{Var}_u = \mathrm{Kin}_u - \mathbf{u}_0$, the previous expression can be written as

$$\langle f, \mathbf{u} - \mathbf{u}_0 \rangle = (\mathbb{D}\mathcal{D}\,\mathbf{u}_0, \mathcal{D}(\mathbf{u} - \mathbf{u}_0)) \qquad \forall \mathbf{u} \in \mathrm{Kin}_u. \tag{4.128}$$

If there exists another displacement field $\mathbf{u}_1 \neq \mathbf{u}_0$ satisfying the theorem, then

$$\langle f, \mathbf{u} - \mathbf{u}_1 \rangle = (\mathbb{D}\mathcal{D}\,\mathbf{u}_1, \mathcal{D}(\mathbf{u} - \mathbf{u}_1)) \qquad \forall \mathbf{u} \in \mathrm{Kin}_u. \tag{4.129}$$

Putting, respectively, $\mathbf{u} = \mathbf{u}_1$ in (4.128) and $\mathbf{u} = \mathbf{u}_0$ in (4.129), and adding the equations results in

$$(\mathbb{D}\mathcal{D}(\mathbf{u}_0 - \mathbf{u}_1), \mathcal{D}(\mathbf{u}_0 - \mathbf{u}_1)) = 0. \tag{4.130}$$

From the properties of the elasticity tensor $\mathbb{D}$ we have

$$\mathcal{D}(\mathbf{u}_0 - \mathbf{u}_1) = \mathbf{O} \quad \Rightarrow \quad \mathbf{u}_0 - \mathbf{u}_1 \in \mathcal{N}(\mathcal{D}), \tag{4.131}$$

which implies that $\mathbf{u}_0$ and $\mathbf{u}_1$ are equal up to a rigid body motion. If this motion has been eliminated by the kinematical constraints in $\partial\mathcal{B}_u$, we arrive at the uniqueness of the displacement field $\mathbf{u}_0$. ■

As in the case of bodies without constraints, the previous theorem allows us to conclude that the elastostatics problem associated with the mechanical equilibrium problem for bodies with bilateral constraints corresponds to the variational problem known as the PVW.

4.5.1 Principle of Virtual Work

For bodies with bilateral constraints, the problem of elastostatics, also taking into consideration the hypothesis of infinitesimal strains and small displacements, and for the specific case of hyperelastic material behavior, is characterized by the PVW. The variational formulation for the PVW consists in finding $\mathbf{u}_0 \in \mathrm{Kin}_u$, such that

$$\left(\left.\frac{\partial \pi}{\partial \mathbf{E}}\right|_{\mathcal{D}\mathbf{u}_0}, \mathcal{D}(\mathbf{u} - \mathbf{u}_0) \right) = \langle f, (\mathbf{u} - \mathbf{u}_0) \rangle \qquad \forall \mathbf{u} \in \mathrm{Kin}_u. \tag{4.132}$$

For materials featuring linear elastic behavior, the PVW takes the form

$$(\mathbb{D}\mathcal{D}\,\mathbf{u}_0, \mathcal{D}(\mathbf{u} - \mathbf{u}_0)) = \langle f, (\mathbf{u} - \mathbf{u}_0) \rangle \qquad \forall \mathbf{u} \in \mathrm{Kin}_u. \tag{4.133}$$

4.5.2 Principle of Minimum Total Potential Energy

From the convexity of the strain energy density function (4.46), and in an entirely analogous procedure to that performed in Section 4.4.2, we arrive at the following inequality

$$\int_{\mathcal{B}} \left(\pi(\mathcal{D}\mathbf{u}) - \pi(\mathcal{D}\mathbf{u}_0) \right) dV \geq \int_{\mathcal{B}} \left.\frac{\partial \pi}{\partial \mathbf{E}}\right|_{\mathcal{D}\mathbf{u}_0} \cdot \mathcal{D}(\mathbf{u} - \mathbf{u}_0) dV \qquad \forall \mathbf{u} \in \mathrm{Kin}_u, \tag{4.134}$$

where the equality holds if and only if $\mathbf{u} - \mathbf{u}_0 \in \mathcal{N}(\mathcal{D})$. Replacing (4.134) in the PVW (4.132) we get

$$\int_{\mathcal{B}} \left(\pi(\mathcal{D}\mathbf{u}) - \pi(\mathcal{D}\mathbf{u}_0) \right) dV \geq \langle f, (\mathbf{u} - \mathbf{u}_0) \rangle \qquad \forall \mathbf{u} \in \mathrm{Kin}_u, \tag{4.135}$$

and arranging terms gives

$$\Pi(\mathbf{u}) = \int_{\mathscr{B}} \pi(\mathscr{D}\mathbf{u})dV - \langle f, \mathbf{u}\rangle \geq \int_{\mathscr{B}} \pi(\mathscr{D}\mathbf{u}_0)dV - \langle f, \mathbf{u}_0\rangle = \Pi(\mathbf{u}_0) \qquad \forall \mathbf{u} \in \text{Kin}_u, \tag{4.136}$$

where the equality holds when $\mathbf{u} - \mathbf{u}_0 \in \mathscr{N}(\mathscr{D})$. We thus obtain the Principle of Minimum Total Potential Energy enunciated next. The elastostatics problem corresponding to the mechanical equilibrium is equivalent to determining $\mathbf{u}_0 \in \text{Kin}_u$ such that the Total Potential Energy functional Π is minimized, that is

$$\Pi(\mathbf{u}_0) = \min_{\mathbf{u}\in\text{Kin}_u} \Pi(\mathbf{u}) = \min_{\mathbf{u}\in\text{Kin}_u} \left[\int_{\mathscr{B}} \pi(\mathscr{D}\mathbf{u})dV - \langle f, \mathbf{u}\rangle \right]. \tag{4.137}$$

As before, from the point of view of the minimization problem, we have that Kin_u is a linear manifold and Π is a lower semi-continuous convex and coercive functional. Then, the existence of a minimum is guaranteed, and if $\text{Kin}_u \cap \mathscr{N}(\mathscr{D}) = \{\mathbf{0}\}$, that is, rigid motions are eliminated, we have that $\mathbf{u}_0$ is unique because in such a case Π is strictly convex (see Appendix A.3).

Finally, for the case of a three-dimensional body, the functional Π takes the form

$$\Pi(\mathbf{u}) = \int_{\mathscr{B}} \frac{1}{2}\mathbb{D}(\nabla\mathbf{u})^s \cdot (\nabla\mathbf{u})^s dV - \int_{\mathscr{B}} \mathbf{b}\cdot\mathbf{u}\, dV - \int_{\partial\mathscr{B}_a} \mathbf{a}\cdot\mathbf{u}\, dS, \tag{4.138}$$

where $\mathbf{b}$ and $\mathbf{a}$ are, respectively, the force per unit volume (defined in $\mathscr{B}$) and the force per unit surface over the boundary $\partial\mathscr{B}_a = \partial\mathscr{B} \setminus \partial\mathscr{B}_u$.

4.5.3 Principle of Complementary Virtual Work

Now, we will study the compatibility problem. As we have already discussed, the PCVW allows us to characterize, independently from the material behavior, the compatibility of the strain field $\mathbf{E}_0$. In effect, $\mathbf{E}_0$ is compatible if and only if

$$(\hat{\mathbf{T}}, \mathbf{E}_0) = (\hat{\mathbf{T}}, \mathscr{D}\mathbf{u}_0) \qquad \forall \hat{\mathbf{T}} \in \text{Var}_T, \tag{4.139}$$

where $\mathbf{u}_0 \in \text{Kin}_u$.

Applying this variational principle to the case of hyperelastic materials, we have that the compatibility problem framed in the elastostatics field is equivalent to determining $\mathbf{T}_0 \in \text{Est}_T$ such that the corresponding strain $\mathbf{E}_0$ related via a hyperelastic constitutive equation $\mathbf{E}_0 = \left.\frac{\partial\pi^*}{\partial\mathbf{T}}\right|_{\mathbf{T}_0}$ (in the linear case $\mathbf{E}_0 = \mathbb{D}^{-1}\mathbf{T}_0$) is compatible, that is, there exists $\mathbf{u}_0 \in \text{Kin}_u$ such that $\mathbf{E}_0 = \mathscr{D}\mathbf{u}_0$. Hence, the variational formulation corresponding to the PCVW in elastostatics consists of finding $\mathbf{T}_0 \in \text{Est}_T$ such that

$$\left(\hat{\mathbf{T}}, \left.\frac{\partial\pi^*}{\partial\mathbf{T}}\right|_{\mathbf{T}_0}\right) = (\hat{\mathbf{T}}, \mathscr{D}\mathbf{u}_0) \qquad \forall \hat{\mathbf{T}} \in \text{Var}_T. \tag{4.140}$$

Since Est_T is the translation of Var_T, that is, $\text{Var}_T = \text{Est}_T - \mathbf{T}_0$, expression (4.140) can be rewritten as

$$\left(\mathbf{T} - \mathbf{T}_0, \left.\frac{\partial\pi^*}{\partial\mathbf{T}}\right|_{\mathbf{T}_0}\right) = (\mathbf{T} - \mathbf{T}_0, \mathscr{D}\mathbf{u}_0) \qquad \forall \mathbf{T} \in \text{Est}_T, \tag{4.141}$$

and in the case of linear elastic behavior,

$$(\mathbf{T} - \mathbf{T}_0, \mathbb{D}^{-1}\mathbf{T}_0) = (\mathbf{T} - \mathbf{T}_0, \mathscr{D}\mathbf{u}_0) \qquad \forall \mathbf{T} \in \text{Est}_T. \tag{4.142}$$

4.5.4 Principle of Minimum Complementary Energy

Let us carry out a procedure completely analogous to that of Section 4.4.5, that is, let us make use of the convexity property of the complementary strain energy function and integrate in the domain occupied by the body $\mathscr{B}$ to obtain

$$\int_{\mathscr{B}} \left(\pi^*(\mathbf{T}) - \pi^*(\mathbf{T}_0)\right) dV \geq \left(\mathbf{T} - \mathbf{T}_0, \left.\frac{\partial \pi^*}{\partial \mathbf{T}}\right|_{\mathbf{T}_0} \right), \tag{4.143}$$

where the equality holds for $\mathbf{T} = \mathbf{T}_0$. Now, replacing (4.143) into the PCVW given by (4.141) leads to

$$\int_{\mathscr{B}} \left(\pi^*(\mathbf{T}) - \pi^*(\mathbf{T}_0)\right) dV \geq (\mathbf{T} - \mathbf{T}_0, \mathscr{D}\,\mathbf{u}_0) \qquad \forall \mathbf{T} \in \mathrm{Est}_T. \tag{4.144}$$

By virtue of $\mathbf{T} - \mathbf{T}_0 \in \mathrm{Var}_T$ and $\mathbf{u}_0 \in \mathrm{Kin}_u$, the bilinear form $(\mathbf{T} - \mathbf{T}_0, \mathscr{D}\,\mathbf{u}_0)$ reduces to a linear form in the argument $\mathbf{T} - \mathbf{T}_0$, defined over the part of the boundary $\partial\mathscr{B}_u$ where displacements are kinematically prescribed. In fact, for the case of the three-dimensional body under study, we use the symmetry of the stress tensor field, together with (2.90) and the fact that $\mathbf{T} - \mathbf{T}_0 \in \mathrm{Var}_T$, resulting in

$$\begin{aligned}
(\mathbf{T} - \mathbf{T}_0, \mathscr{D}\,\mathbf{u}_0) &= \int_{\mathscr{B}} (\mathbf{T} - \mathbf{T}_0) \cdot (\nabla \mathbf{u}_0)^s dV = \int_{\mathscr{B}} (\mathbf{T} - \mathbf{T}_0) \cdot \nabla \mathbf{u}_0 \, dV \\
&= \int_{\mathscr{B}} \left(\mathrm{div}[(\mathbf{T} - \mathbf{T}_0)\mathbf{u}_0] - \mathrm{div}(\mathbf{T} - \mathbf{T}_0) \cdot \mathbf{u}_0\right) \, dV \\
&= \int_{\partial\mathscr{B}} (\mathbf{T} - \mathbf{T}_0)\mathbf{n} \cdot \mathbf{u}_0 \, dS = \int_{\partial\mathscr{B}_u} (\mathbf{T} - \mathbf{T}_0)\mathbf{n} \cdot \overline{\mathbf{u}} \, dS.
\end{aligned} \tag{4.145}$$

With the purpose of highlighting this feature, we introduce the following notation

$$(\mathbf{T} - \mathbf{T}_0, \mathscr{D}\,\mathbf{u}_0) = ((\mathbf{T} - \mathbf{T}_0, \overline{\mathbf{u}})). \tag{4.146}$$

Therefore, expression (4.144) turns into

$$\int_{\mathscr{B}} \left(\pi^*(\mathbf{T}) - \pi^*(\mathbf{T}_0)\right) dV \geq ((\mathbf{T} - \mathbf{T}_0, \overline{\mathbf{u}})) \qquad \forall \mathbf{T} \in \mathrm{Est}_T, \tag{4.147}$$

and after arranging terms we obtain

$$\begin{aligned}
\Pi^*(\mathbf{T}) &= \int_{\mathscr{B}} \pi^*(\mathbf{T}) dV - ((\mathbf{T}, \overline{\mathbf{u}})) \\
&\geq \int_{\mathscr{B}} \pi^*(\mathbf{T}_0) dV - ((\mathbf{T}_0, \overline{\mathbf{u}})) = \Pi^*(\mathbf{T}_0) \qquad \forall \mathbf{T} \in \mathrm{Est}_T,
\end{aligned} \tag{4.148}$$

where the equality is verified if and only if $\mathbf{T} = \mathbf{T}_0$. Hence, we have that the PCVW is equivalent to finding the minimum of the Complementary Energy functional Π^*.

In this way, in the context of hyperelastic materials and within the hypothesis of small displacements and infinitesimal strains, we have achieved the characterization of the compatibility problem through the so-called Principle of Minimum Complementary Energy. In elastostatics, this consists of finding $\mathbf{T}_0 \in \mathrm{Est}_T$ such that the Complementary Energy functional is minimized, that is

$$\Pi^*(\mathbf{T}_0) = \min_{\mathbf{T} \in \mathrm{Est}_T} \Pi^*(\mathbf{T}) = \min_{\mathbf{T} \in \mathrm{Est}_T} \left[\int_{\mathscr{B}} \pi^*(\mathbf{T}) dV - ((\mathbf{T}, \overline{\mathbf{u}})) \right]. \tag{4.149}$$

We have also seen that if there exists a solution to this problem, it is unique. The existence is guaranteed because Π^* is a lower semi-continuous strictly convex and coercive[6] functional in the non-empty linear manifold Est_T (see Appendix A.3).

Finally, in a completely analogous fashion to that presented for bodies without constraints in Section 4.4.6, it is possible to show that

$$\Pi(\mathbf{u}) \geq \Pi(\mathbf{u}_0) = -\Pi^*(\mathbf{T}_0) \geq -\Pi^*(\mathbf{T}) \qquad \forall \mathbf{u} \in \mathrm{Kin}_u, \qquad \forall \mathbf{T} \in \mathrm{Est}_T, \tag{4.150}$$

where $\mathbf{u}_0 \in \mathrm{Kin}_u$ and $\mathbf{T}_0 \in \mathrm{Est}_T$ are the solutions to the Total Potential Energy and Complementary Energy minimum problems, respectively.

4.6 Equilibrium in Bodies with Unilateral Constraints

Let us pay attention now to the case of bodies with unilateral constraints, under the assumptions of Chapter 3. Specifically, in Section 3.6 the presentation was limited to the case of Var_u being a convex cone with an apex at the origin. As always, we consider that Kin_u is the translation of Var_u. More general kinematical constraints in which Kin_u is a convex closed cone have been analyzed by P. D. Panagiotopoulos [236], G. Duvaut and J. L. Lions [76], P. H. Brézis [42], I. Ekeland and R. Temam [78], and by G. Fichera [103] and [104].

4.6.1 Principle of Virtual Work

Within the present context, that is, Var_u is a convex cone, given the load at equilibrium $f \in \mathscr{U}'$, we have seen that there exists $\mathbf{T} \in \mathscr{W}'$ such that

$$-(\mathbf{T}, \mathscr{D}\hat{\mathbf{u}}) + \langle f, \hat{\mathbf{u}} \rangle \leq 0 \qquad \forall \hat{\mathbf{u}} \in \mathrm{Var}_u. \tag{4.151}$$

Once again, the elastostatics problem associated with the mechanical equilibrium consists of finding $\mathbf{u}_0 \in \mathrm{Kin}_u$ such that the stress state $\mathbf{T}_0$, associated with this field through a constitutive relation

$$\mathbf{T}_0 = \left.\frac{\partial \pi}{\partial \mathbf{E}}\right|_{\mathscr{D}\mathbf{u}_0}, \tag{4.152}$$

equilibrates the external load f. Thus, the PVW takes the variational form enunciated next. The elastostatics problem associated with the equilibrium in bodies with unilateral constraints consists of determining $\mathbf{u}_0 \in \mathrm{Kin}_u$ such that

$$-\left(\left.\frac{\partial \pi}{\partial \mathbf{E}}\right|_{\mathscr{D}\mathbf{u}_0}, \mathscr{D}\hat{\mathbf{u}} \right) + \langle f, \hat{\mathbf{u}} \rangle \leq 0 \qquad \forall \hat{\mathbf{u}} \in \mathrm{Var}_u(\mathbf{u}_0). \tag{4.153}$$

4.6.2 Principle of Minimum Total Potential Energy

As in previous sections, given that Kin_u is a translation of Var_u, the expression (4.153) can be written as a function only in terms of kinematically admissible fields, that is,

$$\left(\left.\frac{\partial \pi}{\partial \mathbf{E}}\right|_{\mathscr{D}\mathbf{u}_0}, \mathscr{D}(\mathbf{u} - \mathbf{u}_0) \right) - \langle f, \mathbf{u} - \mathbf{u}_0 \rangle \geq 0 \qquad \forall \mathbf{u} \in \mathrm{Kin}_u. \tag{4.154}$$

6 That is, $\Pi^*(\mathbf{T}) \to \infty$ for $\|\mathbf{T}\| \to \infty$.

Note that the first term in (4.154) is nothing but the Gâteaux differential at point $\mathbf{u}_0$ according to the direction given by $\mathbf{u} - \mathbf{u}_0$ of the Total Potential Energy functional

$$\Pi(\mathbf{u}) = \int_{\mathscr{B}} \pi(\mathscr{D}\mathbf{u})dV - \langle f, \mathbf{u}\rangle. \tag{4.155}$$

As a matter of fact, notice that

$$\begin{aligned} \delta\Pi(\mathbf{u}_0, \mathbf{u} - \mathbf{u}_0) &= \left.\frac{d}{d\alpha}\Pi(\mathbf{u}_0 + \alpha(\mathbf{u} - \mathbf{u}_0))\right|_{\alpha=0} \\ &= \int_{\mathscr{B}} \left.\frac{\partial \pi}{\partial \mathbf{E}}\right|_{\mathscr{D}\mathbf{u}_0} \cdot \mathscr{D}(\mathbf{u} - \mathbf{u}_0)dV - \langle f, \mathbf{u} - \mathbf{u}_0\rangle, \end{aligned} \tag{4.156}$$

then, expression (4.154) is equivalent to

$$\delta\Pi(\mathbf{u}_0, \mathbf{u} - \mathbf{u}_0) \geq 0 \qquad \forall \mathbf{u} \in \mathrm{Kin}_u. \tag{4.157}$$

From the properties of the cost functional Π, which is differentiable, lower semi-continuous strictly convex, and coercive (see Appendix A.3), it follows that (4.157) is the necessary and sufficient (see [103]) condition for $\mathbf{u}_0$ to be a minimum of the Total Potential Energy functional in the convex domain (translation of a convex cone) Kin_u. We have thus arrived at the formulation of the Principle of Minimum Total Potential Energy, which states that the elastostatics problem associated with the mechanical equilibrium with unilateral constraints is equivalent to determining $\mathbf{u}_0 \in \mathrm{Kin}_u$ such that the Total Potential Energy is minimized, that is,

$$\Pi(\mathbf{u}_0) = \min_{\mathbf{u}\in\mathrm{Kin}_u} \Pi(\mathbf{u}) = \min_{\mathbf{u}\in\mathrm{Kin}_u} \left[\int_{\mathscr{B}} \pi(\mathscr{D}\mathbf{u})dV - \langle f, \mathbf{u}\rangle\right]. \tag{4.158}$$

4.6.3 Principle of Complementary Virtual Work

As we have seen, for the case of unilateral constraints, the strain $\mathbf{E}$ is compatible if and only if

$$(\hat{\mathbf{T}}, \mathbf{E}) \geq (\hat{\mathbf{T}}, \mathscr{D}\mathbf{w}) \qquad \forall \hat{\mathbf{T}} \in \mathrm{Var}_T. \tag{4.159}$$

With this statement, we straightforwardly formulate the PCVW for bodies with unilateral constraints in elastostatics, which consists of finding $\mathbf{T}_0 \in \mathrm{Est}_T$ such that

$$\left(\hat{\mathbf{T}}, \left.\frac{\partial \pi^*}{\partial \mathbf{T}}\right|_{\mathbf{T}_0}\right) \geq (\hat{\mathbf{T}}, \mathscr{D}\mathbf{u}_0) \qquad \forall \hat{\mathbf{T}} \in \mathrm{Var}_T. \tag{4.160}$$

Recalling that

$$\mathrm{Var}_T = \{\hat{\mathbf{T}} \in \mathscr{W}'; \ (\hat{\mathbf{T}}, \mathscr{D}\hat{\mathbf{u}}) \geq 0 \ \ \forall \hat{\mathbf{u}} \in \mathrm{Var}_u(\mathbf{u}_0)\}, \tag{4.161}$$

$$\mathrm{Est}_T = \{\mathbf{T} \in \mathscr{W}'; -(\mathbf{T}, \mathscr{D}\hat{\mathbf{u}}) + \langle f, \hat{\mathbf{u}}\rangle \leq 0 \ \ \forall \hat{\mathbf{u}} \in \mathrm{Var}_u(\mathbf{u}_0)\}, \tag{4.162}$$

we have $\mathrm{Est}_T = \mathbf{T} + \mathrm{Var}_T$ for $\mathbf{T} \in \mathrm{Est}_T$ arbitrary, and taking into account arbitrary elements $\mathbf{T} \in \mathrm{Est}_T$ and $\hat{\mathbf{T}} \in \mathrm{Var}_T$, then

$$-(\hat{\mathbf{T}}, \mathscr{D}\hat{\mathbf{u}}) \leq 0 \qquad \forall \hat{\mathbf{u}} \in \mathrm{Var}_u(\mathbf{u}_0), \tag{4.163}$$

$$-(\mathbf{T}, \mathscr{D}\hat{\mathbf{u}}) + \langle f, \hat{\mathbf{u}}\rangle \leq 0 \qquad \forall \hat{\mathbf{u}} \in \mathrm{Var}_u(\mathbf{u}_0). \tag{4.164}$$

Adding these two inequalities we have

$$-(\mathbf{T}+\hat{\mathbf{T}},\mathscr{D}\,\hat{\mathbf{u}})+\langle f,\hat{\mathbf{u}}\rangle \le 0 \qquad \forall \hat{\mathbf{u}} \in \mathrm{Var}_u(\mathbf{u}_0), \tag{4.165}$$

from where we conclude that $\mathbf{T}+\hat{\mathbf{T}} \in \mathrm{Est}_T$. Since $\mathbf{T}$ and $\hat{\mathbf{T}}$ are arbitrary, we thus obtain that $\mathrm{Est}_T = \mathbf{T}+\mathrm{Var}_T$ for all $\mathbf{T} \in \mathrm{Est}_T$. This result allows us to rewrite the PCVW in the following manner: determine $\mathbf{T}_0 \in \mathrm{Est}_T$ such that

$$\left(\mathbf{T}-\mathbf{T}_0, \left.\frac{\partial \pi^*}{\partial \mathbf{T}}\right|_{\mathbf{T}_0}\right) \ge (\mathbf{T}-\mathbf{T}_0,\mathscr{D}\,\mathbf{u}_0) \qquad \forall \mathbf{T} \in \mathrm{Est}_T, \tag{4.166}$$

and for the specific case of linear elastic behavior

$$(\mathbf{T}-\mathbf{T}_0,\mathbb{D}^{-1}\mathbf{T}_0) \ge (\mathbf{T}-\mathbf{T}_0,\mathscr{D}\,\mathbf{u}_0) \qquad \forall \mathbf{T} \in \mathrm{Est}_T. \tag{4.167}$$

4.6.4 Principle of Minimum Complementary Energy

Utilizing the convexity of the complementary strain energy density function π^* and integrating in the body $\mathscr{B}$, we have

$$\int_{\mathscr{B}} \left(\pi^*(\mathbf{T})-\pi^*(\mathbf{T}_0)\right) dV \ge \left(\mathbf{T}-\mathbf{T}_0, \left.\frac{\partial \pi^*}{\partial \mathbf{T}}\right|_{\mathbf{T}_0}\right), \tag{4.168}$$

where the equality holds for $\mathbf{T}=\mathbf{T}_0$. Putting (4.168) into (4.166) gives

$$\int_{\mathscr{B}} \left(\pi^*(\mathbf{T})-\pi^*(\mathbf{T}_0)\right) dV \ge (\mathbf{T}-\mathbf{T}_0,\mathscr{D}\,\mathbf{u}_0) \qquad \forall \mathbf{T} \in \mathrm{Est}_T. \tag{4.169}$$

Therefore, because $\mathbf{T}$ and $\mathbf{T}_0$ are stress fields which are equilibrated with the external load f, the bilinear form $(\mathbf{T}-\mathbf{T}_0,\mathscr{D}\,\mathbf{u}_0)$ reduces to a linear form in $\mathbf{T}-\mathbf{T}_0$ over the part of the boundary $\partial\mathscr{B}_u$ where kinematical constraints are being considered (see (4.145)). This is represented through the following notation

$$(\mathbf{T}-\mathbf{T}_0,\mathscr{D}\,\mathbf{u}_0) = ((\mathbf{T}-\mathbf{T}_0,\overline{\mathbf{u}})). \tag{4.170}$$

Putting this into (4.169), and arranging terms we obtain finally

$$\begin{aligned} \Pi^*(\mathbf{T}) &= \int_{\mathscr{B}} \pi^*(\mathbf{T})dV - ((\mathbf{T},\overline{\mathbf{u}})) \\ &\ge \int_{\mathscr{B}} \pi^*(\mathbf{T}_0)dV - ((\mathbf{T}_0,\overline{\mathbf{u}})) = \Pi^*(\mathbf{T}_0) \qquad \forall \mathbf{T} \in \mathrm{Est}_T, \end{aligned} \tag{4.171}$$

where the equality is verified if and only if $\mathbf{T}=\mathbf{T}_0$. Then, we have that the elastostatics problem associated with the compatibility problem is equivalent to the minimization of the Complementary Energy cost functional. This is called the Principle of Minimum Complementary Energy and consists of determining $\mathbf{T}_0 \in \mathrm{Est}_T$ such that

$$\Pi^*(\mathbf{T}_0) = \min_{\mathbf{T}\in\mathrm{Est}_T} \Pi^*(\mathbf{T}) = \min_{\mathbf{T}\in\mathrm{Est}_T} \left[\int_{\mathscr{B}} \pi^*(\mathbf{T})dV - ((\mathbf{T},\overline{\mathbf{u}}))\right]. \tag{4.172}$$

The uniqueness of this field has already been established. The existence of $\mathbf{T}_0$ follows from the properties of Π^*, which is lower semi-continuous strictly convex and coercive, and from the fact that Est_T is a non-empty closed convex set (see [103]).

4.7 Min–Max Principle

4.7.1 Hellinger–Reissner Functional

In previous sections we have seen that, for hyperelastic materials defined through the convex strain energy density function π, or equivalently through the, also convex, complementary strain energy density function π^*, and within the hypothesis of small displacements and infinitesimal strains, the problem of determining the fields $\mathbf{u}_0$, $\mathbf{E}_0$ and $\mathbf{T}_0$ that satisfy

- kinematical admissibility
- compatibility
- constitutive equation
- mechanical equilibrium,

is equivalent to the following problems of minimum: the Principle of Minimum Total Potential Energy

$$\Pi(\mathbf{u}_0) = \min_{\mathbf{u}\in \mathrm{Kin}_u} \Pi(\mathbf{u}) = \min_{\mathbf{u}\in \mathrm{Kin}_u} \left[\int_{\mathscr{B}} \pi(\mathscr{D}\mathbf{u}) dV - \langle f, \mathbf{u} \rangle \right], \tag{4.173}$$

and the Principle of Minimum Complementary Energy

$$\Pi^*(\mathbf{T}_0) = \min_{\mathbf{T}\in \mathrm{Est}_T} \Pi^*(\mathbf{T}) = \min_{\mathbf{T}\in \mathrm{Est}_T} \left[\int_{\mathscr{B}} \pi^*(\mathbf{T}) dV - ((\mathbf{T}, \overline{\mathbf{u}})) \right]. \tag{4.174}$$

Particularly, the solution $\mathbf{u}_0$ of (4.173) and $\mathbf{T}_0$ of (4.174) are such that

$$\Pi(\mathbf{u}_0) = -\Pi^*(\mathbf{T}_0), \tag{4.175}$$

that is to say

$$\min_{\mathbf{u}\in \mathrm{Kin}_u} \Pi(\mathbf{u}) = - \min_{\mathbf{T}\in \mathrm{Est}_T} \Pi^*(\mathbf{T}), \tag{4.176}$$

from where

$$\min_{\mathbf{u}\in \mathrm{Kin}_u} \Pi(\mathbf{u}) = \max_{\mathbf{T}\in \mathrm{Est}_T} [-\Pi^*(\mathbf{T})]. \tag{4.177}$$

The Principle of Minimum Total Potential Energy is defined in terms of a kinematic entity (the displacement field), and so this problem is known as the Primal formulation. In turn, and due to the fact that internal and external stresses were introduced through duality arguments, the Principle of Minimum Complementary Energy is also known as the dual formulation.

Finally, we note that for the different kinds of kinematical constraints (no constraints, bilateral and unilateral constraints), the cost functionals Π and Π^* remain the same, and the sole changes in the corresponding formulations are the sets Kin_u and Est_T. In the most general scenario posed by unilateral constraints, these sets are closed and convex, and for bilateral constraints and for bodies without constraints these sets are linear manifolds.

In the case of unilateral constraints, the solution is characterized by variational inequalities, as stated by (4.153) for the PVW and by (4.160) for the PCVW. For this class of constraints this clearly shows the nonlinear nature of the problem, even for linear elastic materials. In the case of bilateral constraints (and also when the body has no constraints) the minimum is characterized by variational equations, as stated by (4.132) for the PVW and by (4.140) for the PCVW.

One the one hand, the construction of the set Kin_u is, in general, not a difficult task because its elements must be regular enough and must comply with the kinematical constraints prescribed over the motion. On the other hand, the same cannot be said about the set Est_T because in this case its elements are internal stress tensor fields which are at equilibrium with the system of forces f the body is subjected to. Because of this technical difficulty, the question that emerges from these thoughts is whether or not there is an alternative way to characterize the solution of the elastostatics such that the construction of the involved sets is not that arduous. The answer is affirmative, and this section is devoted to developing such an alternative variational strategy.

In order to do this, recall the duality between functions π and π^*. In particular, consider the Legendre's transformation,

$$\pi(\mathscr{D}\mathbf{u}) = \max_{\mathbf{T}\in\mathscr{W}'} [\mathbf{T}\cdot\mathscr{D}\mathbf{u} - \pi^*(\mathbf{T})], \tag{4.178}$$

where the stress field $\mathbf{T}$ is simply limited to being an element in the space $\mathscr{W}'$. Replacing this in the definition of the cost functional Π in (4.173) leads to

$$\begin{aligned}\Pi(\mathbf{u}) &= \int_{\mathscr{B}} \pi(\mathscr{D}\mathbf{u})dV - \langle f, \mathbf{u}\rangle \\ &= \max_{\mathbf{T}\in\mathscr{W}'} \left[-\int_{\mathscr{B}} \pi^*(\mathbf{T})dV + \int_{\mathscr{B}} \mathbf{T}\cdot\mathscr{D}\mathbf{u}\, dV - \langle f, \mathbf{u}\rangle\right].\end{aligned} \tag{4.179}$$

Defining the cost functional

$$\mathscr{F}_{\mathrm{HR}}(\mathbf{u}, \mathbf{T}) = -\int_{\mathscr{B}} \pi^*(\mathbf{T})dV + \int_{\mathscr{B}} \mathbf{T}\cdot\mathscr{D}\mathbf{u}\, dV - \langle f, \mathbf{u}\rangle, \tag{4.180}$$

which is known as the Hellinger–Reissner functional (after these authors proposed it in 1914 and 1950, respectively), the functional Π is written as

$$\Pi(\mathbf{u}) = \max_{\mathbf{T}\in\mathscr{W}'} \mathscr{F}_{\mathrm{HR}}(\mathbf{u}, \mathbf{T}). \tag{4.181}$$

Since this expression is verified for all $\mathbf{u} \in \mathrm{Kin}_u$, it is particularly verified for $\mathbf{u}_0$. Then

$$\Pi(\mathbf{u}_0) \geq \mathscr{F}_{\mathrm{HR}}(\mathbf{u}_0, \mathbf{T}) \qquad \forall \mathbf{T} \in \mathscr{W}', \tag{4.182}$$

and the equality holds only for $\mathbf{T} = \mathbf{T}_0 = \mathbf{T}(\mathscr{D}\mathbf{u}_0)$, that is to say, if $\mathbf{T}$ is associated with $\mathbf{u}_0$ through the constitutive equation. As a consequence, we have that the Principle of Minimum Total Potential Energy given by (4.173) is equivalent to

$$\min_{\mathbf{u}\in\mathrm{Kin}_u} \Pi(\mathbf{u}) = \min_{\mathbf{u}\in\mathrm{Kin}_u} \max_{\mathbf{T}\in\mathscr{W}'} \mathscr{F}_{\mathrm{HR}}(\mathbf{u}, \mathbf{T}), \tag{4.183}$$

from where we have

$$\Pi(\mathbf{u}_0) = \min_{\mathbf{u}\in\mathrm{Kin}_u} \Pi(\mathbf{u}) \leq \mathscr{F}_{\mathrm{HR}}(\mathbf{u}, \mathbf{T}_0) \qquad \forall \mathbf{u} \in \mathrm{Kin}_u. \tag{4.184}$$

From expressions (4.182) and (4.184) we are led to

$$\mathscr{F}_{\mathrm{HR}}(\mathbf{u}_0, \mathbf{T}) \leq \Pi(\mathbf{u}_0) \leq \mathscr{F}_{\mathrm{HR}}(\mathbf{u}, \mathbf{T}_0) \qquad \forall \mathbf{T} \in \mathscr{W}', \qquad \forall \mathbf{u} \in \mathrm{Kin}_u, \tag{4.185}$$

and from this we obtain as the result

$$\max_{\mathbf{T}\in\mathscr{W}'} \min_{\mathbf{u}\in\mathrm{Kin}_u} \mathscr{F}_{\mathrm{HR}}(\mathbf{u}, \mathbf{T}) = \min_{\mathbf{u}\in\mathrm{Kin}_u} \max_{\mathbf{T}\in\mathscr{W}'} \mathscr{F}_{\mathrm{HR}}(\mathbf{u}, \mathbf{T}). \tag{4.186}$$

4.7.2 Hellinger–Reissner Principle

The Hellinger–Reissner Principle established that the displacement and stress fields, $\mathbf{u}_0$ and $\mathbf{T}_0$, which are, respectively, solutions of (4.173) and (4.174), are the saddle point of the Hellinger–Reissner functional introduced in (4.180). Hence, we have arrived at an alternative variational formulation for the elastostatics problem, which consists of finding the fields $\mathbf{u}_0 \in \mathrm{Kin}_u$ and $\mathbf{T}_0 \in \mathscr{W}'$ such that

$$(\mathbf{u}_0, \mathbf{T}_0) = \min_{\mathbf{u}\in\mathrm{Kin}_u} \max_{\mathbf{T}\in\mathscr{W}'} \mathscr{F}_{\mathrm{HR}}(\mathbf{u}, \mathbf{T})$$
$$= \min_{\mathbf{u}\in\mathrm{Kin}_u} \max_{\mathbf{T}\in\mathscr{W}'} \left[-\int_{\mathscr{B}} \pi^*(\mathbf{T}) dV + \int_{\mathscr{B}} \mathbf{T} \cdot \mathscr{D}\mathbf{u}\, dV - \langle f, \mathbf{u} \rangle \right]. \tag{4.187}$$

From a purely mechanical perspective, the formulation of the Min–Max Principle stated through (4.187) resulted from the following reasoning. First, observe that in the Principle of Minimum Total Potential Energy (4.173) the following equations are implicit: the compatibility of strains, that is, given $\mathbf{u}_0$, then

$$\mathbf{E}_0 = \mathscr{D}\mathbf{u}_0, \tag{4.188}$$

and the constitutive equation which states that $\mathbf{E}_0 = \mathscr{D}\mathbf{u}_0$ and $\mathbf{T}_0$ are associated through the constitutive equation of the hyperelastic material under study

$$\mathbf{E}_0 = \left.\frac{\partial \pi^*}{\partial \mathbf{T}}\right|_{\mathbf{T}_0}, \qquad \mathbf{T}_0 = \left.\frac{\partial \pi}{\partial \mathbf{E}}\right|_{\mathbf{E}_0}. \tag{4.189}$$

Second, for constructing the functional $\mathscr{F}_{\mathrm{HR}}$ we have explicitly incorporated this last constitutive equation into the functional Π. This was specifically achieved by forcing the field $\mathbf{T}$, related to the strain $\mathscr{D}\mathbf{u}$ through the constitutive equation as the solution of the problem

$$\max_{\mathbf{T}\in\mathscr{W}'} [\mathbf{T} \cdot \mathscr{D}\mathbf{u} - \pi^*(\mathbf{T})] = \pi(\mathscr{D}\mathbf{u}). \tag{4.190}$$

Third, through the minimum condition with respect to the variable $\mathbf{u} \in \mathrm{Kin}_u$, we seek for the field $\mathbf{T}$ that, in addition to the fact that it is associated with $\mathscr{D}\mathbf{u}$ by the constitutive equation, satisfies the equilibrium.

As a consequence, the equations associated with the Hellinger–Reissner Principle are the following: the mechanical equilibrium and the constitutive equation. In effect, the necessary condition for $\mathbf{T}$ to be a maximum in the vector space $\mathscr{W}'$, is given by

$$\delta\mathscr{F}_{\mathrm{HR}}((\mathbf{u}_0, \mathbf{T}_0), \hat{\mathbf{T}}) = \left.\frac{d}{d\alpha}\mathscr{F}_{\mathrm{HR}}(\mathbf{u}_0, \mathbf{T}_0 + \alpha\hat{\mathbf{T}})\right|_{\alpha=0}$$
$$= \int_{\mathscr{B}} \left(-\left.\frac{\partial \pi^*}{\partial \mathbf{T}}\right|_{\mathbf{T}_0} + \mathscr{D}\mathbf{u}_0 \right) \cdot \hat{\mathbf{T}}\, dV = 0 \qquad \forall \hat{\mathbf{T}} \in \mathscr{W}', \tag{4.191}$$

from where we conclude that

$$\mathbf{E}_0 = \mathscr{D}\mathbf{u}_0 = \left.\frac{\partial \pi^*}{\partial \mathbf{T}}\right|_{\mathbf{T}_0} \qquad \text{in } \mathscr{B}. \tag{4.192}$$

In turn, the condition for $\mathbf{u} \in \mathrm{Kin}_u$ to be a minimum in the non-empty closed convex set Kin_u is given by the variational inequality

$$\begin{aligned}\delta\mathscr{F}_{\mathrm{HR}}((\mathbf{u}_0,\mathbf{T}_0),\hat{\mathbf{u}}) &= \frac{d}{d\alpha}\mathscr{F}_{\mathrm{HR}}(\mathbf{u}_0+\alpha\hat{\mathbf{u}},\mathbf{T}_0)\Big|_{\alpha=0} = \int_{\mathscr{B}}\mathbf{T}\cdot\mathscr{D}\hat{\mathbf{u}}\,dV - \langle f,\hat{\mathbf{u}}\rangle \\ &= (\mathbf{T},\mathscr{D}\hat{\mathbf{u}}) - \langle f,\hat{\mathbf{u}}\rangle \geq 0 \qquad \forall\hat{\mathbf{u}} \in \mathrm{Var}_u(\mathbf{u}_0),\end{aligned} \tag{4.193}$$

which is the PVW that, as we already know, characterizes the mechanical equilibrium between the stress field $\mathbf{T}$ and the force f.

Finally, it is interesting to highlight that by solving the min–max problem we simultaneously obtain the fields $\mathbf{u}_0$ and $\mathbf{T}_0$. This acquires relevance not only in the theoretical realm, but also in practical situations, when approximate solutions are sought via numerical methods.

4.8 Three-Field Functional

In this section we will construct a new cost functional $\mathscr{F}(\mathbf{u},\mathbf{E},\mathbf{T})$, known as the Generalized Functional or Three-Field Functional. For this, we remember the Principle of Minimum Total Potential Energy

$$\Pi(\mathbf{u}_0) = \min_{\mathbf{u}\in\mathrm{Kin}_u}\Pi(\mathbf{u}) = \min_{\mathbf{u}\in\mathrm{Kin}_u}\left[\int_{\mathscr{B}}\pi(\mathscr{D}\mathbf{u})\,dV - \langle f,\mathbf{u}\rangle\right]. \tag{4.194}$$

Now, the Total Potential Energy can be subtly rewritten in the following form

$$\Pi(\mathbf{u}) = \int_{\mathscr{B}}\pi(\mathbf{E})\,dV - \langle f,\mathbf{u}\rangle, \tag{4.195}$$

where $\mathbf{E}$ satisfies the subsidiary condition

$$\mathbf{E} = \mathscr{D}\mathbf{u} \quad \Leftrightarrow \quad \mathbf{E} \in R(\mathscr{D}). \tag{4.196}$$

Exploiting the theoretical formalism provided by the Lagrange multipliers, we can build another functional, say $\mathscr{F}(\mathbf{u},\mathbf{E},\mathbf{T})$, given by

$$\mathscr{F}(\mathbf{u},\mathbf{E},\mathbf{T}) = \int_{\mathscr{B}}\pi(\mathbf{E})\,dV - \langle f,\mathbf{u}\rangle - \int_{\mathscr{B}}\mathbf{T}\cdot(\mathbf{E}-\mathscr{D}\mathbf{u})\,dV. \tag{4.197}$$

To see this, consider first the functional

$$\mathscr{L}(\mathbf{u},\mathbf{E}) = \int_{\mathscr{B}}\pi(\mathbf{E})\,dV - \langle f,\mathbf{u}\rangle, \tag{4.198}$$

defined for $\mathbf{u} \in \mathrm{Kin}_u$ and for $\mathbf{E} \in \mathscr{W}$. From its definition, functional $\mathscr{L}$ is such that

$$\Pi(\mathbf{u}) = \mathscr{L}(\mathbf{u},\mathbf{E}) \qquad \forall\mathbf{E} \in R(\mathscr{D}). \tag{4.199}$$

Consider now the concept of indicator (or characteristic) function, denoted by $\mathscr{I}_{\mathscr{C}}$, of a subset $\mathscr{C} \subset \mathscr{V}$

$$\begin{aligned}&\mathscr{I}_{\mathscr{C}} : \mathscr{V} \to \{0,+\infty\},\\ &v \mapsto \mathscr{I}_{\mathscr{C}}(v) = \begin{cases}0 & \text{if } v \in \mathscr{C},\\ +\infty & \text{if } v \notin \mathscr{C}.\end{cases}\end{aligned} \tag{4.200}$$

Thus, we can appreciate that

$$\Pi(\mathbf{u}) = \min_{\mathbf{E}\in\mathscr{W}} \left[\mathscr{L}(\mathbf{u},\mathbf{E}) + \mathscr{I}_{R(\mathscr{D})}(\mathbf{E})\right]. \tag{4.201}$$

Making use of the duality between internal stresses and strains, we have that the indicator function $\mathscr{I}_{R(\mathscr{D})}$ can be exactly built by maximizing the linear functional

$$\mathscr{I}_{R(\mathscr{D})}(\mathbf{E}) = \max_{\Lambda\in\mathscr{W}'} \left[-\int_{\mathscr{B}} \Lambda\cdot(\mathbf{E}-\mathscr{D}\mathbf{u})\, dV\right], \tag{4.202}$$

where the field $\Lambda \in \mathscr{W}'$ plays the role of Lagrange multiplier associated with the subsidiary condition $\mathbf{E} \in R(\mathscr{D})$. As will be seen next, this multiplier will correspond to the stress field $\mathbf{T} \in \mathscr{W}'$ which, associated with the strain $\mathbf{E} = \mathscr{D}\mathbf{u}$ through a constitutive equation, is at equilibrium with the external force f. Introducing (4.202) into expression (4.201) we have

$$\Pi(\mathbf{u}) = \min_{\mathbf{E}\in\mathscr{W}} \max_{\Lambda\in\mathscr{W}'} \left[\mathscr{L}(\mathbf{u},\mathbf{E}) - \int_{\mathscr{B}} \Lambda\cdot(\mathbf{E}-\mathscr{D}\mathbf{u})\, dV\right]. \tag{4.203}$$

Hence, we finally get

$$\begin{aligned} \min_{\mathbf{u}\in\mathrm{Kin}_u} \Pi(\mathbf{u}) &= \min_{\mathbf{u}\in\mathrm{Kin}_u} \min_{\mathbf{E}\in\mathscr{W}} \max_{\Lambda\in\mathscr{W}'} \left[\mathscr{L}(\mathbf{u},\mathbf{E}) - \int_{\mathscr{B}} \Lambda\cdot(\mathbf{E}-\mathscr{D}\mathbf{u})\, dV\right] \\ &= \min_{\mathbf{u}\in\mathrm{Kin}_u} \min_{\mathbf{E}\in\mathscr{W}} \max_{\Lambda\in\mathscr{W}'} \mathscr{F}(\mathbf{u},\mathbf{E},\Lambda). \end{aligned} \tag{4.204}$$

This result leads us to a Generalized Variational Principle, which consists of minimizing the three-field functional $\mathscr{F}$. In the context of elastostatics, this variational principle is equivalent to finding the triple $\mathbf{u}_0 \in \mathrm{Kin}_u$, $\mathbf{E}_0 \in \mathscr{W}$, and $\mathbf{T}_0 \in \mathscr{W}'$ such that

$$\begin{aligned} (\mathbf{u}_0,\mathbf{E}_0,\mathbf{T}_0) &= \min_{\mathbf{u}\in\mathrm{Kin}_u} \min_{\mathbf{E}\in\mathscr{W}} \max_{\mathbf{T}\in\mathscr{W}'} \mathscr{F}(\mathbf{u},\mathbf{E},\mathbf{T}) \\ &= \min_{\mathbf{u}\in\mathrm{Kin}_u} \min_{\mathbf{E}\in\mathscr{W}} \max_{\mathbf{T}\in\mathscr{W}'} \left[\int_{\mathscr{B}} \pi(\mathbf{E})\, dV - \langle f,\mathbf{u}\rangle - \int_{\mathscr{B}} \mathbf{T}\cdot(\mathbf{E}-\mathscr{D}\mathbf{u})\, dV\right]. \end{aligned} \tag{4.205}$$

As a matter of fact, since the problem is defined in $\mathrm{Kin}_u \times \mathscr{W} \times \mathscr{W}'$, the necessary condition for a triple to be a stationary point leads to

$$\delta\mathscr{F}((\mathbf{u}_0,\mathbf{E}_0,\mathbf{T}_0),\hat{\mathbf{u}}) \ge 0 \qquad \forall\hat{\mathbf{u}} \in \mathrm{Var}_u(\mathbf{u}_0), \tag{4.206}$$

$$\delta\mathscr{F}((\mathbf{u}_0,\mathbf{E}_0,\mathbf{T}_0),\hat{\mathbf{E}}) = 0 \qquad \forall\hat{\mathbf{E}} \in \mathscr{W}, \tag{4.207}$$

$$\delta\mathscr{F}((\mathbf{u}_0,\mathbf{E}_0,\mathbf{T}_0),\hat{\mathbf{T}}) = 0 \qquad \forall\hat{\mathbf{T}} \in \mathscr{W}'. \tag{4.208}$$

In extended form, this reads

$$\begin{aligned} \delta\mathscr{F}((\mathbf{u}_0,\mathbf{E}_0,\mathbf{T}_0),\hat{\mathbf{u}}) &= \left.\frac{d}{d\alpha}\mathscr{F}(\mathbf{u}_0+\alpha\hat{\mathbf{u}},\mathbf{E}_0,\mathbf{T}_0)\right|_{\alpha=0} \\ &= (\mathbf{T}_0,\mathscr{D}\hat{\mathbf{u}}) - \langle f,\hat{\mathbf{u}}\rangle \ge 0 \qquad \forall\hat{\mathbf{u}} \in \mathrm{Var}_u(\mathbf{u}_0), \end{aligned} \tag{4.209}$$

which corresponds to the expression of the PVW, that is $\mathbf{T}_0 \in \mathrm{Est}_T$. In addition

$$\begin{aligned} \delta\mathscr{F}((\mathbf{u}_0,\mathbf{E}_0,\mathbf{T}_0),\hat{\mathbf{E}}) &= \left.\frac{d}{d\alpha}\mathscr{F}(\mathbf{u}_0,\mathbf{E}_0+\alpha\hat{\mathbf{E}},\mathbf{T}_0)\right|_{\alpha=0} \\ &= \int_{\mathscr{B}} \left(\left.\frac{\partial\pi}{\partial\mathbf{E}}\right|_{\mathbf{E}_0} - \mathbf{T}_0\right)\cdot\hat{\mathbf{E}}\, dV = 0 \qquad \forall\hat{\mathbf{E}} \in \mathscr{W}, \end{aligned} \tag{4.210}$$

from where it follows that

$$\mathbf{T}_0 = \left.\frac{\partial \pi}{\partial \mathbf{E}}\right|_{\mathbf{E}_0} \qquad \text{in } \mathscr{B}, \tag{4.211}$$

implying that $\mathbf{E}_0$ and $\mathbf{T}_0$ are related through the constitutive equation. At last, we have

$$\begin{aligned}\delta\mathscr{F}((\mathbf{u}_0, \mathbf{E}_0, \mathbf{T}_0), \hat{\mathbf{T}}) &= \left.\frac{d}{d\alpha}\mathscr{F}(\mathbf{u}_0, \mathbf{E}_0, \mathbf{T}_0 + \alpha\hat{\mathbf{T}})\right|_{\alpha=0} \\ &= \int_{\mathscr{B}} \hat{\mathbf{T}} \cdot (\mathbf{E}_0 - \mathscr{D}\mathbf{u}_0) dV = 0 \qquad \forall \hat{\mathbf{T}} \in \mathscr{W}',\end{aligned} \tag{4.212}$$

which leads us to conclude that

$$\mathbf{E}_0 = \mathscr{D}\mathbf{u}_0 \qquad \text{in } \mathscr{B}, \tag{4.213}$$

that is, the strain field $\mathbf{E}_0$ is compatible with the displacement field $\mathbf{u}_0 \in \mathrm{Kin}_u$.

With this, we observe that the point in the set $\mathrm{Kin}_u \times \mathscr{W} \times \mathscr{W}'$ which makes the generalized functional $\mathscr{F}$ stationary is the solution of the elastostatics problem, and vice versa, the solution of the elastostatics problem is a stationary point of $\mathscr{F}$. Therefore, the equivalence between both problems is demonstrated.

From the point of view of numerical approximation of these variational problems, the appeal of this generalized functional is that its solution simultaneously delivers the triple $(\mathbf{u}_0, \mathbf{E}_0, \mathbf{T}_0)$ which fully characterizes the mechanics in the elastostatics problem. Again, the construction of functional spaces in this formulation is simplified because the only constraints to be satisfied are the kinematical constraints over the portion of the boundary $\partial\mathscr{B}_u$ for fields $\mathbf{u}$. The construction of approximation spaces for the fields $\mathbf{E}$ and $\mathbf{T}$ is straightforward, as these spaces do not involve constraints, either of mechanical character for $\mathbf{T}$ or of kinematical character for $\mathbf{E}$. In any case, the proper selection of these spaces in the finite-dimensional realm demands additional considerations in order to guarantee the correct behavior of the approximate solution (see [49], [78], [151], [225–227], [306], and [307]).

Finally, Figure 4.3 schematically displays the relations established among the problems presented so far.

4.9 Castigliano Theorems

In this section all the developments are limited to the case of bodies with bilateral kinematical constraints, that is, Kin_u is a linear manifold resulting from the translation of Var_u.

4.9.1 First and Second Theorems

Consider that the system of external forces f to which the body is subjected to in the configuration $\mathscr{B}$ is a system characterized by n parameters, that is,

$$f = \sum_{i=1}^{n} Q_i f_i, \tag{4.214}$$

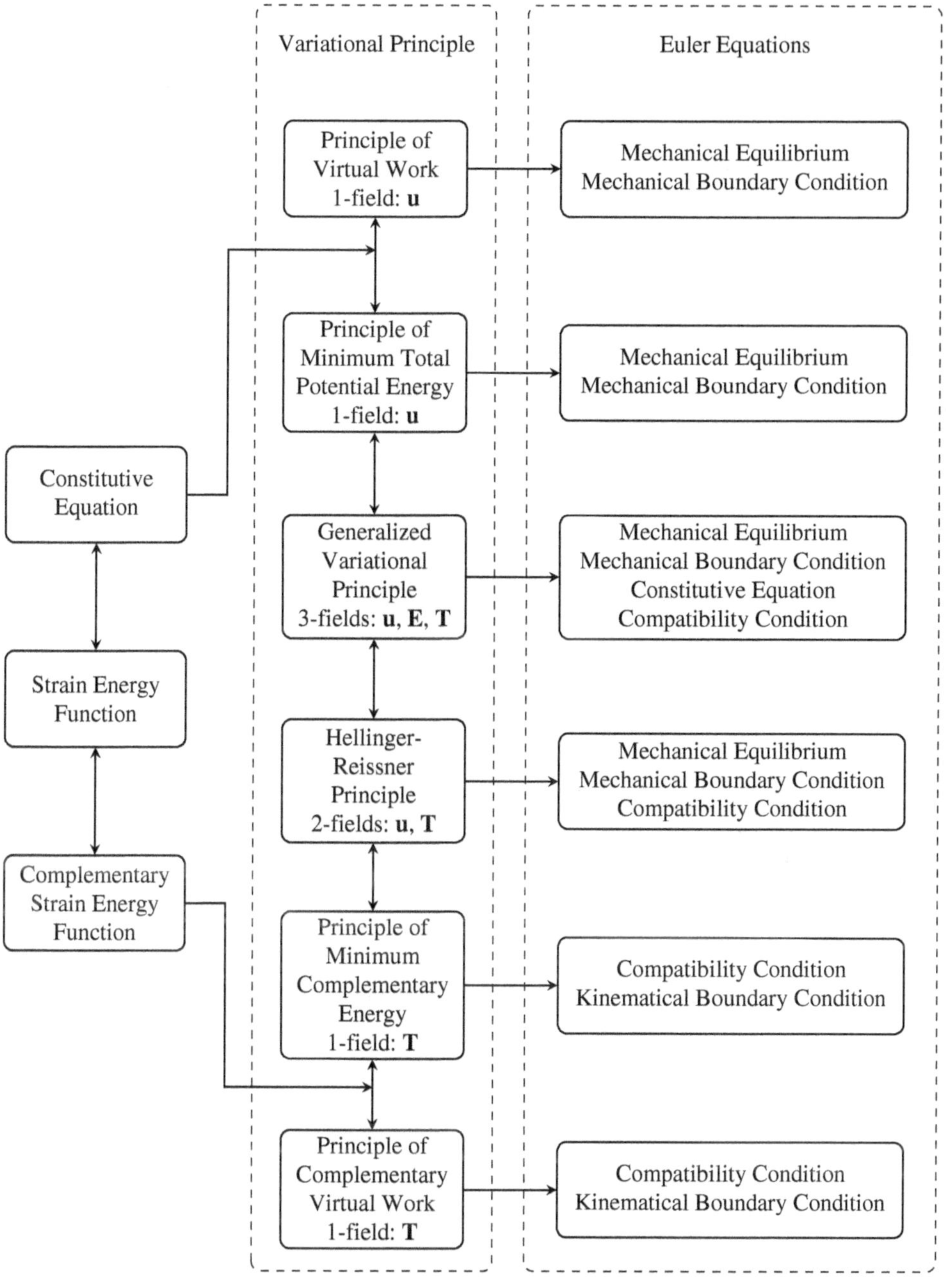

Figure 4.3 Relations among variational principles in the elastostatics problem.

where $f_i \in \mathscr{V}'$ is the ith force mode and $Q_i \in \mathbb{R}$ is the load parameter associated with the ith mode, also called the generalized load parameter. In this case, the Total Potential Energy functional, $\Pi(\mathbf{u})$, takes the form

$$\Pi(\mathbf{u}) = \int_{\mathscr{B}} \pi(\mathbf{u})\, dV - \langle f, \mathbf{u}\rangle = \int_{\mathscr{B}} \pi(\mathbf{u})\, dV - \sum_{i=1}^{n} Q_i \langle f_i, \mathbf{u}\rangle$$
$$= \int_{\mathscr{B}} \pi(\mathbf{u})\, dV - \sum_{i=1}^{n} Q_i \mathsf{q}_i, \tag{4.215}$$

where q_i is called the generalized displacement parameter and is related to $\mathbf{u}$ through

$$\mathsf{q}_i = \langle f_i, \mathbf{u}\rangle \qquad i = 1, \ldots, n. \tag{4.216}$$

By virtue of (4.216), it is possible to write $\mathbf{u}$ as a function of the generalized parameters q_i. Putting $\mathbf{q} = (\mathsf{q}_1, \ldots, \mathsf{q}_n) \in \mathbb{R}^n$, the previous statement implies

$$\mathbf{u} = \mathbf{u}(\mathbf{q}). \tag{4.217}$$

Then, the Principle of Minimum Total Potential Energy gives

$$\min_{\mathbf{u} \in \mathrm{Kin}_u} \Pi(\mathbf{u}) = \min_{\mathsf{q} \in \mathbb{R}^n} \tilde{\Pi}(\mathbf{q}) = \min_{\mathsf{q} \in \mathbb{R}^n} \left[\int_{\mathscr{B}} \pi(\mathbf{u}(\mathbf{q}))\, dV - \sum_{i=1}^{n} Q_i \mathsf{q}_i \right]. \tag{4.218}$$

From the necessary condition for a stationary point of Π, we obtain

$$\frac{\partial}{\partial \mathsf{q}_i} \int_{\mathscr{B}} \pi(\mathbf{u}(\mathbf{q}))\, dV = Q_i \qquad i = 1, \ldots, n. \tag{4.219}$$

Expression (4.219) corresponds to Castigliano's first theorem extended to the case of nonlinear hyperelasticity.

Let us consider now the Principle of Minimum Complementary Energy, which we repeat here for completeness

$$\min_{\mathbf{T} \in \mathrm{Est}_T} \Pi^*(\mathbf{T}) = \min_{\mathbf{T} \in \mathrm{Est}_T} \left[\int_{\mathscr{B}} \pi^*(\mathbf{T})\, dV - ((\mathbf{T}, \overline{\mathbf{u}})) \right]. \tag{4.220}$$

Suppose that the kinematical data over $\partial\mathscr{B}_u$, that is $\overline{\mathbf{u}}$, can be written as

$$\overline{\mathbf{u}} = \sum_{i=1}^{n} \overline{q}_i \mathbf{u}_i, \tag{4.221}$$

where $\overline{q}_i \in \mathbb{R}$, and $\mathbf{u}_i$ is the prescribed displacement mode. Introducing (4.221) into the Complementary Energy functional Π^* yields

$$\Pi^*(\mathbf{T}) = \int_{\mathscr{B}} \pi^*(\mathbf{T})\, dV - \sum_{i=1}^{n} ((\mathbf{T}, \mathbf{u}_i))\overline{q}_i = \int_{\mathscr{B}} \pi^*(\mathbf{T})\, dV - \sum_{i=1}^{n} \mathsf{Q}_i \overline{q}_i, \tag{4.222}$$

where we have defined

$$\mathsf{Q}_i = ((\mathbf{T}, \mathbf{u}_i)) \qquad i = 1, \ldots, n, \tag{4.223}$$

that is, $\mathsf{Q}_i = \mathsf{Q}_i(\mathbf{T})$. Therefore, in general, we can express the stress fields $\mathbf{T} \in \mathrm{Est}_T$ as a function of the parameters Q_i. Putting $\mathbf{Q} = (\mathsf{Q}_1, \ldots, \mathsf{Q}_n) \in \mathbb{R}^n$, this means

$$\mathbf{T} = \mathbf{T}(\mathbf{Q}). \tag{4.224}$$

Then, the Principle of Minimum Complementary Energy turns into

$$\min_{\mathbf{T}\in \mathrm{Est}_T} \Pi^*(\mathbf{T}) = \min_{\mathsf{Q}\in\mathbb{R}^n} \tilde{\Pi}^*(\mathbf{Q}) = \min_{\mathsf{Q}\in\mathbb{R}^n} \left[\int_{\mathscr{B}} \pi^*(\mathbf{T}(\mathbf{Q}))\, dV - \sum_{i=1}^{n} \mathsf{Q}_i \overline{q}_i \right]. \tag{4.225}$$

The necessary condition to achieve a stationary point yields

$$\frac{\partial}{\partial \mathsf{Q}_i} \int_{\mathscr{B}} \pi^*(\mathbf{T}(\mathbf{Q}))\, dV = \overline{q}_i \qquad i = 1, \dots, n. \tag{4.226}$$

Expression (4.226) corresponds to Castigliano's second theorem extended to bodies featuring nonlinear hyperelastic behavior.

4.9.2 Bounds for Displacements and Generalized Loads

In Section 4.5.4 we showed the relation between the Total Potential Energy and the Complementary Energy through inequality (4.150). As $\mathbf{u}_0 \in \mathrm{Kin}_u$ and $\mathbf{T}_0 \in \mathrm{Est}_T$ are the fields that are the solution to the elastostatics problem, this inequality established that

$$\begin{aligned} \Pi(\mathbf{u}) &= \int_{\mathscr{B}} \pi(\mathscr{D}\mathbf{u})dV - \langle f, \mathbf{u}\rangle \geq \Pi(\mathbf{u}_0) = \int_{\mathscr{B}} \pi(\mathscr{D}\mathbf{u}_0)dV - \langle f, \mathbf{u}_0\rangle \\ &= -\Pi^*(\mathbf{T}_0) = -\int_{\mathscr{B}} \pi^*(\mathbf{T}_0)dV + ((\mathbf{T}_0, \overline{\mathbf{u}})) \geq -\int_{\mathscr{B}} \pi^*(\mathbf{T})dV + ((\mathbf{T}, \overline{\mathbf{u}})) \\ &= -\Pi^*(\mathbf{T}) \quad \forall \mathbf{u} \in \mathrm{Kin}_u, \qquad \forall \mathbf{T} \in \mathrm{Est}_T. \end{aligned} \tag{4.227}$$

For simplicity, consider that the kinematical constraints prescribed on $\partial\mathscr{B}_u$ are homogeneous, that is, $\overline{\mathbf{u}} = \mathbf{0}$, and that the external load f is a 1-parameter load, that is, $f = Qf_1$. In this case, (4.227) becomes

$$\begin{aligned} \int_{\mathscr{B}} \pi(\mathscr{D}\mathbf{u})dV - Q\mathsf{q} &\geq \int_{\mathscr{B}} \pi(\mathscr{D}\mathbf{u}_0)dV - Q\mathsf{q}_0 = -\int_{\mathscr{B}} \pi^*(\mathbf{T}_0)dV \\ &\geq -\int_{\mathscr{B}} \pi^*(\mathbf{T})dV \quad \forall \mathbf{u} \in \mathrm{Kin}_u, \qquad \forall \mathbf{T} \in \mathrm{Est}_T, \end{aligned} \tag{4.228}$$

where

$$\mathsf{q} = \langle f_1, \mathbf{u}\rangle \qquad \mathsf{q}_0 = \langle f_1, \mathbf{u}_0\rangle, \tag{4.229}$$

and for which the linear manifold Est_T is

$$\mathrm{Est}_T = \{\mathbf{T} \in \mathscr{W}'; \ (\mathbf{T}, \hat{\mathbf{u}}) = Q\langle f_1, \hat{\mathbf{u}}\rangle \quad \forall \hat{\mathbf{u}} \in \mathrm{Var}_u\}. \tag{4.230}$$

Then, given an arbitrary displacement direction $\mathbf{u}^* \in \mathrm{Kin}_u$, and defining $\mathsf{q}^* = \langle f_1, \mathbf{u}^*\rangle$, we can select the element $\lambda\mathbf{u}^*$ seeking to obtain a sharp lower bound for $\int_{\mathscr{B}} \pi^*(\mathbf{T}_0)dV$ by solving the following problem

$$\begin{aligned} \max_{\lambda\in\mathbb{R}} \left[\lambda Q\mathsf{q}^* - \int_{\mathscr{B}} \pi(\lambda\mathscr{D}\mathbf{u}^*)dV \right] &\leq Q\mathsf{q}_0 - \int_{\mathscr{B}} \pi(\mathscr{D}\mathbf{u}_0)dV \\ = \int_{\mathscr{B}} \pi^*(\mathbf{T}_0)dV &\leq \int_{\mathscr{B}} \pi^*(\mathbf{T})dV \qquad \forall \mathbf{u}^* \in \mathrm{Kin}_u, \qquad \forall \mathbf{T} \in \mathrm{Est}_T. \end{aligned} \tag{4.231}$$

In the case of a linear elastic material, the maximum can be explicitly calculated. In fact, for these materials it is

$$\pi(\lambda\mathbf{E}) = \lambda^2 \pi(\mathbf{E}), \tag{4.232}$$

and back to (4.231) yields

$$\max_{\lambda\in\mathbb{R}} \left[\lambda \mathsf{Q}\mathsf{q}^* - \lambda^2 \int_{\mathscr{B}} \pi(\mathscr{D}\mathbf{u}^*)dV\right] \leq \mathsf{Q}\mathsf{q}_0 - \int_{\mathscr{B}} \pi(\mathscr{D}\mathbf{u}_0)dV$$
$$= \int_{\mathscr{B}} \pi^*(\mathbf{T}_0)dV \leq \int_{\mathscr{B}} \pi^*(\mathbf{T})dV \qquad \forall \mathbf{u}^* \in \mathrm{Kin}_u, \qquad \forall \mathbf{T} \in \mathrm{Est}_T. \tag{4.233}$$

In this manner, the element $\lambda_{\max}$ for which the maximum is attained is characterized by the equation

$$\mathsf{Q}\mathsf{q}^* - 2\lambda_{\max} \int_{\mathscr{B}} \pi(\mathscr{D}\mathbf{u}^*)dV = 0, \tag{4.234}$$

that is,

$$\lambda_{\max} = \frac{\mathsf{Q}\mathsf{q}^*}{2\int_{\mathscr{B}} \pi(\mathscr{D}\mathbf{u}^*)dV}. \tag{4.235}$$

Introducing (4.235) into (4.231) leads to

$$\frac{\lambda_{\max}}{2}\mathsf{Q}\mathsf{q}^* \leq \mathsf{Q}\mathsf{q}_0 - \int_{\mathscr{B}} \pi(\mathscr{D}\mathbf{u}_0)dV = \int_{\mathscr{B}} \pi^*(\mathbf{T}_0)dV$$
$$\leq \int_{\mathscr{B}} \pi^*(\mathbf{T})dV \quad \forall \mathbf{u}^* \in \mathrm{Kin}_u, \qquad \forall \mathbf{T} \in \mathrm{Est}_T. \tag{4.236}$$

In particular, if this result is applied to the solution $\mathbf{u}_0 \in \mathrm{Kin}_u$, and recalling that the material is linear, we have

$$\lambda_{\max} = \frac{\mathsf{Q}\mathsf{q}_0}{2\int_{\mathscr{B}} \pi(\mathscr{D}\mathbf{u}_0)dV} = 1, \tag{4.237}$$

and then

$$\frac{1}{2}\mathsf{Q}\mathsf{q}_0 = \int_{\mathscr{B}} \pi^*(\mathbf{T}_0)dV \leq \int_{\mathscr{B}} \pi^*(\mathbf{T})dV \qquad \forall \mathbf{T} \in \mathrm{Est}_T, \tag{4.238}$$

from where

$$\mathsf{q}_0 \leq \frac{2}{\mathsf{Q}} \int_{\mathscr{B}} \pi^*(\mathbf{T})dV \qquad \forall \mathbf{T} \in \mathrm{Est}_T. \tag{4.239}$$

With this procedure, we have obtained an upper bound for the generalized displacement q_0 (a real number) of the body, which is associated with the generalized load Q solely in terms of stress fields which are at equilibrium with such generalized load Q.

Example 4.1 This example illustrates an application of (4.239) for the case of a linear elastic beam under hypotheses established by Bernoulli (see Chapter 7). Thus, consider the cantilevered beam as seen in Figure 4.4. A single load denoted by Q is applied over the left extreme.

For this example we have

$$\mathrm{Est}_T = \{M;\ M = Qx\}, \tag{4.240}$$

$$\pi^*(\mathbf{T}) = \frac{1}{2}\frac{M^2}{\mathbb{E}I} = \frac{1}{2}\frac{Q^2x^2}{\mathbb{E}I}, \tag{4.241}$$

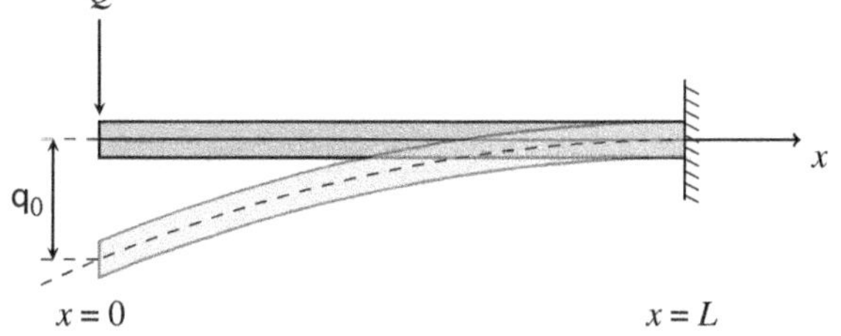

Figure 4.4 Upper bounds for the displacement of a cantilevered linear elastic beam within the theory of Bernoulli.

where M is the bending moment, Q is the applied load at the point where we want to find the upper bound for the displacement (the left boundary in the case of the figure), $\mathbb{E}$ is the Young elastic modulus, and I is the moment of inertia of the beam cross-section. We are considering here that $\mathbb{E}$ and I are constant along the beam. Then

$$\mathsf{q}_0 \leq \frac{2}{Q}\int_0^L \frac{1}{2}\frac{Q^2x^2}{\mathbb{E}I}dx = \frac{QL^3}{3\mathbb{E}I}, \tag{4.242}$$

and we obtain the upper bound, which in this particular case coincides with exact value of the displacement. Remarkably, this result was achieved only by knowing the set of internal stress fields which are at equilibrium with the applied load. In turn, if we apply a unitary generalized load ($Q = 1$), an upper bound for the displacement is given by the ratio $\frac{L^3}{3\mathbb{E}I}$.

Now we will proceed to obtain an upper bound for the loads. For this, recall the inequality established by (4.227), valid for all hyperelastic materials. Assuming $f = 0$, leads to

$$\begin{aligned} -\Pi^*(\mathbf{T}) = ((\mathbf{T},\overline{\mathbf{u}})) - \int_{\mathscr{B}} \pi^*(\mathbf{T})dV \leq ((\mathbf{T}_0,\overline{\mathbf{u}})) - \int_{\mathscr{B}} \pi^*(\mathbf{T}_0)dV \\ = \int_{\mathscr{B}} \pi(\mathscr{D}\mathbf{u}_0)dV \leq \int_{\mathscr{B}} \pi(\mathscr{D}\mathbf{u})dV \qquad \forall \mathbf{u} \in \mathrm{Kin}_u, \qquad \forall \mathbf{T} \in \mathrm{Est}_T. \end{aligned} \tag{4.243}$$

Consider the arbitrary stress element given by the direction $\mathbf{T}^* \in \mathrm{Est}_T$. Then we can pick the stress $\lambda\mathbf{T}^*$ which renders a sharp lower bound for $\int_{\mathscr{B}} \pi(\mathscr{D}\mathbf{u}_0)dV$. For the case of linear elastic materials, where we can use an analogous argument to (4.232), this implies

$$\begin{aligned} \max_{\lambda\in\mathbb{R}} \left[\lambda((\mathbf{T}^*,\overline{\mathbf{u}})) - \lambda^2 \int_{\mathscr{B}} \pi^*(\mathbf{T}^*)dV\right] \leq ((\mathbf{T}_0,\overline{\mathbf{u}})) - \int_{\mathscr{B}} \pi^*(\mathbf{T}_0)dV \\ = \int_{\mathscr{B}} \pi(\mathscr{D}\mathbf{u}_0)dV \leq \int_{\mathscr{B}} \pi(\mathscr{D}\mathbf{u})dV \qquad \forall \mathbf{u} \in \mathrm{Kin}_u, \qquad \forall \mathbf{T}^* \in \mathrm{Est}_T. \end{aligned} \tag{4.244}$$

Hence, we obtain

$$\lambda_{\max} = \frac{((\mathbf{T}^*,\overline{\mathbf{u}}))}{2\int_{\mathscr{B}} \pi^*(\mathbf{T}^*)dV}. \tag{4.245}$$

Putting (4.245) into (4.244) gives

$$\begin{aligned} \lambda_{\max} \left[((\mathbf{T}^*,\overline{\mathbf{u}})) - \lambda_{\max} \int_{\mathscr{B}} \pi^*(\mathbf{T}^*)dV\right] \leq ((\mathbf{T}_0,\overline{\mathbf{u}})) - \int_{\mathscr{B}} \pi^*(\mathbf{T}_0)dV \\ = \int_{\mathscr{B}} \pi(\mathscr{D}\mathbf{u}_0)dV \leq \int_{\mathscr{B}} \pi(\mathscr{D}\mathbf{u})dV \qquad \forall \mathbf{u} \in \mathrm{Kin}_u, \qquad \forall \mathbf{T}^* \in \mathrm{Est}_T, \end{aligned} \tag{4.246}$$

from where

$$\lambda_{\max}\frac{1}{2}((\mathbf{T}^*,\overline{\mathbf{u}})) \leq \int_{\mathscr{B}} \pi(\mathscr{D}\,\mathbf{u})dV. \tag{4.247}$$

In particular, for the direction which is the solution of the elastostatics problem, $\mathbf{T}_0$ gives $\lambda_{\max} = 1$, and then

$$\frac{1}{2}((\mathbf{T}_0,\overline{\mathbf{u}})) \leq \int_{\mathscr{B}} \pi(\mathscr{D}\,\mathbf{u})dV \qquad \forall \mathbf{u} \in \mathrm{Kin}_u. \tag{4.248}$$

If we admit now that

$$\overline{\mathbf{u}} = \overline{q}\mathbf{u}_1, \tag{4.249}$$

that is, the prescribed displacement depends only on one generalized displacement parameter, and defining

$$\mathsf{Q}_0 = ((\mathbf{T}_0,\mathbf{u}_1)), \tag{4.250}$$

then we have

$$\frac{1}{2}\mathsf{Q}_0\overline{q} \leq \int_{\mathscr{B}} \pi(\mathscr{D}\,\mathbf{u})dV \qquad \forall \mathbf{u} \in \mathrm{Kin}_u, \tag{4.251}$$

and finally

$$\mathsf{Q}_0 \leq \frac{2}{\overline{q}}\int_{\mathscr{B}} \pi(\mathscr{D}\,\mathbf{u})dV \qquad \forall \mathbf{u} \in \mathrm{Kin}_u. \tag{4.252}$$

This results tells us that by making use of kinematically admissible displacement fields, we can obtain an upper bound for the generalized load which would be required to be applied to the body to produce a generalized displacement $\overline{q}$.

So far, the analysis to obtain upper bounds for loads and displacements has been limited to the very same points where the load is applied, or where the displacement is prescribed. Let us extend the analysis to obtain bounds for the displacement field at any point in the body. In order do this, consider the body $\mathscr{B}$ subjected to the load $f \in \mathscr{V}'$, and to homogeneous kinematical constraints on $\partial\mathscr{B}_u$, that is, $\overline{\mathbf{u}} = \mathbf{0}$. Field $\mathbf{u}_0$ denotes the solution of the corresponding elastostatics problem. Let $\mathbf{T}^* \in \mathscr{W}'$ be an arbitrary internal stress field equilibrated with the load f^*. From (4.47), we have

$$\pi(\mathscr{D}\,\mathbf{u}_0) + \pi^*(\mathbf{T}^*) \geq \mathbf{T}^* \cdot \mathscr{D}\,\mathbf{u}_0, \tag{4.253}$$

and then

$$\pi^*(\mathbf{T}^*) \geq \mathbf{T}^* \cdot \mathscr{D}\,\mathbf{u}_0 - \pi(\mathscr{D}\,\mathbf{u}_0). \tag{4.254}$$

Integrating all over the domain occupied by the body $\mathscr{B}$ and since $\mathbf{T}^*$ is equilibrated with f^* gives

$$\int_{\mathscr{B}} \pi^*(\mathbf{T}^*)dV \geq \int_{\mathscr{B}} \mathbf{T}^* \cdot \mathscr{D}\,\mathbf{u}_0 dV - \int_{\mathscr{B}} \pi(\mathscr{D}\,\mathbf{u}_0)dV = \langle f^*,\mathbf{u}_0\rangle - \int_{\mathscr{B}} \pi(\mathscr{D}\,\mathbf{u}_0)dV. \tag{4.255}$$

For a linear elastic material, we have

$$\pi(\mathscr{D}\,\mathbf{u}_0) = \frac{1}{2}\mathbf{T}_0 \cdot \mathscr{D}\,\mathbf{u}_0, \quad \mathbf{T}_0 = \mathbb{D}\mathscr{D}\,\mathbf{u}_0, \tag{4.256}$$

where $\mathbf{T}_0$ is a stress field equilibrated with the external load f, then

$$\int_{\mathscr{B}} \pi^*(\mathbf{T}^*)dV \geq \langle f^*, \mathbf{u}_0 \rangle - \frac{1}{2}\langle f, \mathbf{u}_0 \rangle = \left\langle f^* - \frac{1}{2}f, \mathbf{u}_0 \right\rangle. \tag{4.257}$$

For example, if we want to provide a bound for $\mathbf{u}_0$ at an arbitrary point, say $\overline{\mathbf{x}} \in \mathscr{B}$, in the direction of the displacement mode $\mathbf{u}_1$, also possibly arbitrary, it is sufficient to adopt f^* as being

$$f^* = \frac{1}{2}f + \delta_{\overline{\mathbf{x}}} Q^* \mathbf{u}_1, \tag{4.258}$$

where $\delta_{\overline{\mathbf{x}}}$ is the Dirac delta function defined at $\overline{\mathbf{x}}$, and Q^* is an arbitrary real number. In fact, replacing f^* given by (4.258) in (4.257) gives

$$\int_{\mathscr{B}} \pi^*(\mathbf{T}^*)dV \geq \langle \delta_{\overline{\mathbf{x}}} Q^* \mathbf{u}_1, \mathbf{u}_0 \rangle = Q^*(\mathbf{u}_0 \cdot \mathbf{u}_1)_{\overline{\mathbf{x}}}, \tag{4.259}$$

from where

$$(\mathbf{u}_0 \cdot \mathbf{u}_1)_{\overline{\mathbf{x}}} \leq \frac{1}{Q^*} \int_{\mathscr{B}} \pi^*(\mathbf{T}^*)dV. \tag{4.260}$$

It is important to mention again that the field $\mathbf{T}^*$ is equilibrated with the load f^* that was chosen. Since $\mathbf{T}^*$ will depend on Q^* we will be able to select Q^* so as to minimize the right hand side of (4.260). Hence, the optimal upper bound corresponds to

$$(\mathbf{u}_0 \cdot \mathbf{u}_1)_{\overline{\mathbf{x}}} \leq \min_{Q^* \in \mathbb{R}} \left[\frac{1}{Q^*} \int_{\mathscr{B}} \pi^*(\mathbf{T}^*)dV \right]. \tag{4.261}$$

Let us illustrate the usage of these results through an example.

Example 4.2 Consider the beam component displayed in Figure 4.5, of length L, with homogeneous linear elastic behavior characterized by $\mathbb{E}$ and with a constant moment of inertia I. A distributed load per unit length equal to p is uniformly applied over the beam.

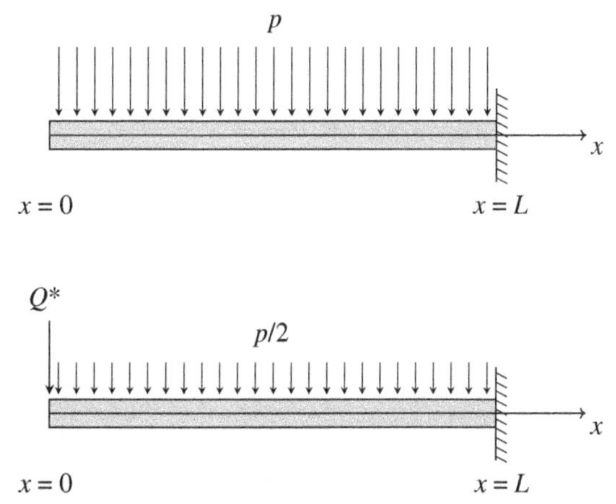

Figure 4.5 Upper bounds for the displacement of a cantilevered linear elastic beam within the theory of Bernoulli, with distributed load.

Then, we have:

$$f \to p \text{ uniformly distributed load,} \tag{4.262}$$

$$f^* \to \frac{1}{2}p + Q^*, \text{where } Q^* \text{ is a load entirely concentrated at } \overline{x} = 0. \tag{4.263}$$

The stress state $\mathbf{T}^*$ is characterized by the bending moment M^*, which must be equilibrated with the load f^*. As a consequence, we have

$$M^* = \frac{1}{2}p\frac{x^2}{2} + Q^*x = \frac{1}{4}px^2 + Q^*x, \tag{4.264}$$

and thus

$$\begin{aligned}\int_{\mathscr{B}} \pi^*(\mathbf{T}^*)dV &= \int_0^L \frac{1}{2}\frac{M^{*2}}{\mathbb{E}I}dx = \int_0^L \frac{1}{2\mathbb{E}I}\left(\frac{1}{4}px^2 + Q^*x\right)^2 dx \\ &= \frac{1}{2\mathbb{E}I}\left(\frac{1}{80}p^2L^5 + \frac{1}{8}pQ^*L^4 + \frac{1}{3}Q^{*2}L^3\right),\end{aligned} \tag{4.265}$$

from where

$$q_0 \le \min_{Q^* \in \mathbb{R}} \left[\frac{1}{2\mathbb{E}I}\left(\frac{1}{80}\frac{p^2L^5}{Q^*} + \frac{1}{8}pL^4 + \frac{1}{3}Q^*L^3\right)\right]. \tag{4.266}$$

The condition for a stationary point of this functional gives

$$\frac{d}{dQ^*}\left(\frac{1}{80}\frac{p^2L^5}{Q^*} + \frac{1}{8}pL^4 + \frac{1}{3}Q^*L^3\right) = 0, \tag{4.267}$$

leading to

$$Q^* = \sqrt{\frac{3}{80}}pL, \tag{4.268}$$

which provides the optimal bound

$$\mathsf{q}_0 \le \left(\frac{1}{\sqrt{240}} + \frac{1}{16}\right)\frac{pL^4}{\mathbb{E}I} \approx 0.127\frac{pL^4}{\mathbb{E}I}. \tag{4.269}$$

Noting that the exact displacement is $\mathsf{q}_0 = 0.125\frac{qL^4}{EI}$, we have that the bound for the displacement is quite sharp, with an error of 1.6%.

4.10 Elastodynamics Problem

Previous sections concentrated on the elastostatics problem. However, in Chapter 3 we saw that the Principle of Virtual Power provides a theoretical framework which is also valid when the equilibrium is affected by dynamical phenomena. In such a scenario it was necessary to include the power expended related to the forces caused by an accelerated frame of reference within the external power functional P_e. This power is expressed as $-\rho\dot{\mathbf{v}}$, where $\dot{\mathbf{v}}$ is the spatial description of the acceleration field.

To simplify things, we suppose that the set Kin_u is independent from time, and remains convex. Then, within the hypotheses of small displacements and infinitesimal strains, the elastodynamics problem consists of the following: for each time instant $t \in [0, t_f)$, determine the fields $\mathbf{u}_0 \in \text{Kin}_u$, $\mathbf{E}_0 \in \mathscr{W}$ and $\mathbf{T}_0 \in \mathscr{W}'$, such that:

- the field $\mathbf{T}_0$ is at (dynamic) equilibrium with the load $f_t \in \mathscr{W}'$, where f_t represents the load at time t
- the fields $\mathbf{E}_0$ and $\mathbf{T}_0$ are related through the hyperelastic constitutive equation of the material under study
- the field $\mathbf{E}_0 \in \mathscr{W}$ is compatible, that is $\mathbf{E}_0 = \mathscr{D}\,\mathbf{u}_0$
- the field $\mathbf{u}_0$ satisfies the initial conditions

$$\mathbf{u}_0(\mathbf{x}, 0) = \tilde{\mathbf{u}}_0(\mathbf{x}) \qquad \forall \mathbf{x} \in \mathscr{B}, \tag{4.270}$$

$$\dot{\mathbf{u}}_0(\mathbf{x}, 0) = \tilde{\dot{\mathbf{u}}}_0(\mathbf{x}) = \tilde{\mathbf{v}}_0(\mathbf{x}) \qquad \forall \mathbf{x} \in \mathscr{B}. \tag{4.271}$$

Equivalently to the elastostatics problem, we can formulate the elastodynamics problem as a variational problem written only in terms of the displacement field (primal variational formulation), as a function of stress fields dynamically equilibrated with the load f_t (dual variational formulation), in terms of displacement and stress fields (mixed variational formulations) or as a function of the three fields, namely displacements, strains, and stresses (generalized variational formulations).

The current presentation will be limited to primal variational formulations, that is, formulations fully characterized by the displacement field. In such cases, the elastodynamics problem consists of finding, for each time instant $t \in [0, t_f)$, the displacement field $\mathbf{u}_0 \in \mathrm{Kin}_u$ such that it satisfies the dynamic equilibrium

$$\left(\left.\frac{\partial \pi}{\partial \mathbf{E}}\right|_{\mathscr{D}\mathbf{u}_0}, \mathscr{D}\,\hat{\mathbf{u}} \right) \geq \langle f_t - \rho \ddot{\mathbf{u}}_0, \hat{\mathbf{u}} \rangle \quad \forall \hat{\mathbf{u}} \in \mathrm{Var}_u(\mathbf{u}_0(\cdot, t)), \tag{4.272}$$

and satisfies the initial conditions

$$\begin{cases} \mathbf{u}_0(\mathbf{x}, 0) = \tilde{\mathbf{u}}_0(\mathbf{x}) & \forall \mathbf{x} \in \mathscr{B}, \\ \dot{\mathbf{u}}_0(\mathbf{x}, 0) = \tilde{\dot{\mathbf{u}}}_0(\mathbf{x}) = \tilde{\mathbf{v}}_0(\mathbf{x}) & \forall \mathbf{x} \in \mathscr{B}. \end{cases} \tag{4.273}$$

Because of the kinematical constraints which are implicitly defined in the time-independent set Kin_u, the initial conditions must be compatible with such kinematical constraints.

If we consider bodies with bilateral kinematical constraints, the set Var_u becomes a vector subspace of $\mathscr{U}$, and the variational inequality (4.272) turns into a variational equation. In this particular case, and assuming the existence of a solution, the uniqueness follows easily. In fact, remember that Var_u was considered to be the same as Var_v, being capable of representing the set of all kinematically admissible virtual motion actions. Therefore, introducing the set Kin_v of kinematically admissible velocity fields, we have that the characterization of the mechanical equilibrium can be reframed as follows

$$\left(\left.\frac{\partial \pi}{\partial \mathbf{E}}\right|_{\mathscr{D}\mathbf{u}_0}, \mathscr{D}(\dot{\mathbf{u}} - \dot{\mathbf{u}}_0) \right) + \langle \rho \ddot{\mathbf{u}}_0, \dot{\mathbf{u}} - \dot{\mathbf{u}}_0 \rangle = \langle f_t, \dot{\mathbf{u}} - \dot{\mathbf{u}}_0 \rangle \qquad \forall \dot{\mathbf{u}} \in \mathrm{Kin}_v. \tag{4.274}$$

Then, if there exist two different solutions, say $\mathbf{u}_0$ and $\mathbf{u}_1$, where $\mathbf{u}_1$ is also a solution of (4.274) (putting $\mathbf{u}_1$ instead of $\mathbf{u}_0$), we have

$$\left(\left.\frac{\partial \pi}{\partial \mathbf{E}}\right|_{\mathscr{D}\mathbf{u}_0} - \left.\frac{\partial \pi}{\partial \mathbf{E}}\right|_{\mathscr{D}\mathbf{u}_1}, \mathscr{D}(\dot{\mathbf{u}}_1 - \dot{\mathbf{u}}_0) \right) + \langle \rho(\ddot{\mathbf{u}}_0 - \ddot{\mathbf{u}}_1), \dot{\mathbf{u}}_1 - \dot{\mathbf{u}}_0 \rangle = 0. \tag{4.275}$$

For a linear elastic material, this expression becomes

$$(\mathbb{D}\mathscr{D}(\mathbf{u}_1 - \mathbf{u}_0), \mathscr{D}(\dot{\mathbf{u}}_1 - \dot{\mathbf{u}}_0)) + \langle \rho(\ddot{\mathbf{u}}_1 - \ddot{\mathbf{u}}_0), \dot{\mathbf{u}}_1 - \dot{\mathbf{u}}_0 \rangle = 0. \tag{4.276}$$

Using the symmetry of the elastic tensor $\mathbb{D}$, we reach the following identity

$$\frac{1}{2}\frac{d}{dt}\left[(\mathbb{D}\mathcal{D}(\mathbf{u}_1-\mathbf{u}_0),\mathcal{D}(\mathbf{u}_1-\mathbf{u}_0))+\langle\rho(\dot{\mathbf{u}}_1-\dot{\mathbf{u}}_0),\dot{\mathbf{u}}_1-\dot{\mathbf{u}}_0\rangle\right]=0. \tag{4.277}$$

Now, integrating between the initial time $t=0$, and a generic time instant t, and given that $\mathbf{u}_0$ and $\mathbf{u}_1$ both satisfy the same initial conditions, and exploiting the positivity of the tensor $\mathbb{D}$ and of ρ, it follows that

$$\mathcal{D}(\mathbf{u}_1-\mathbf{u}_0)=\mathbf{O} \qquad \forall t\in[0,t_f) \qquad \forall \mathbf{x}\in\mathcal{B}, \tag{4.278}$$

$$\dot{\mathbf{u}}_1-\dot{\mathbf{u}}_0=\mathbf{0} \qquad \forall t\in[0,t_f) \qquad \forall \mathbf{x}\in\mathcal{B}. \tag{4.279}$$

Finally, we conclude that

$$\mathbf{u}_1=\mathbf{u}_0, \tag{4.280}$$

and the uniqueness in the elastodynamics problem is demonstrated. The same result trivially follows for nonlinear hyperelastic materials by making use of the fact that the constitutive relation is monotonically increasing and from the positivity of ρ.

Next, we will show that the combination of the PVW with some properties of the strain energy function and of the mass density allows us to obtain an alternative variational formulation to characterize the dynamic equilibrium. Although the forthcoming developments are elaborated within the context of infinitesimal strains and time-independent kinematical constraints, the obtained results can be extended to more complex scenarios, as seen, for example, in the works by P. D. Panagiotopoulos ([236]), G. Fichera ([103, 104]), G. Duvaut and J. L. Lions ([76]), and I. Ekeland and R. Temam ([78]).

Observe that the PVW modified to account for the power expended by inertia forces results in the variational equation (4.274), which must hold for all time instants $t\in[0,t_f)$. Therefore, integrating along that time interval gives

$$\int_0^{t_f}\left[\left(\left.\frac{\partial\pi}{\partial\mathbf{E}}\right|_{\mathcal{D}\mathbf{u}_0},\mathcal{D}(\dot{\mathbf{u}}-\dot{\mathbf{u}}_0)\right)+\langle\rho\ddot{\mathbf{u}}_0,\dot{\mathbf{u}}-\dot{\mathbf{u}}_0\rangle-\langle f_t,\dot{\mathbf{u}}-\dot{\mathbf{u}}_0\rangle\right]dt=0 \quad \forall\dot{\mathbf{u}}\in\mathrm{Kin}_v, \tag{4.281}$$

and from the properties of Kin_v and of Var_u, the previous expression can be rewritten as

$$\int_0^{t_f}\left[\left(\left.\frac{\partial\pi}{\partial\mathbf{E}}\right|_{\mathcal{D}\mathbf{u}_0},\mathcal{D}\hat{\mathbf{u}}\right)+\langle\rho\ddot{\mathbf{u}}_0,\hat{\mathbf{u}}\rangle-\langle f_t,\hat{\mathbf{u}}\rangle\right]dt=0 \qquad \forall\hat{\mathbf{u}}\in\mathrm{Var}_u. \tag{4.282}$$

From the definition of Var_u, we have

$$\hat{\mathbf{u}}(\mathbf{x},0)=\hat{\mathbf{u}}(\mathbf{x},t_f)=\mathbf{0} \qquad \forall\mathbf{x}\in\mathcal{B}. \tag{4.283}$$

Hence, assuming that ρ does not change in time, the second term in (4.282) can be recast as

$$\begin{aligned}
\int_0^{t_t} \langle \rho\ddot{\mathbf{u}}_0, \hat{\mathbf{u}}\rangle\, dt &= \int_0^{t_t} \int_{\mathscr{B}} \rho\ddot{\mathbf{u}}_0 \cdot \hat{\mathbf{u}}\, dV\, dt = \int_{\mathscr{B}} \int_0^{t_t} \rho\ddot{\mathbf{u}}_0 \cdot \hat{\mathbf{u}}\, dt\, dV \\
&= \int_{\mathscr{B}} \left\{ \rho[\dot{\mathbf{u}}_0 \cdot \hat{\mathbf{u}}]\Big|_0^{t_f} - \int_0^{t_f} \rho\dot{\mathbf{u}}_0 \cdot \hat{\dot{\mathbf{u}}}\, dt \right\} dV = -\int_{\mathscr{B}} \int_0^{t_f} \rho\dot{\mathbf{u}}_0 \cdot \hat{\dot{\mathbf{u}}}\, dt\, dV \\
&= -\int_0^{t_f} \int_{\mathscr{B}} \rho\dot{\mathbf{u}}_0 \cdot \hat{\dot{\mathbf{u}}}\, dV\, dt = -\frac{d}{d\alpha} \int_0^{t_f} \int_{\mathscr{B}} \pi_{\text{kin}}(\dot{\mathbf{u}}_0 + \alpha\hat{\dot{\mathbf{u}}})\, dV\, dt\Big|_{\alpha=0} \\
&= -\frac{d}{d\alpha} \int_0^{t_f} \Pi_{\text{kin}}(\dot{\mathbf{u}}_0 + \alpha\hat{\dot{\mathbf{u}}})\, dt\Big|_{\alpha=0},
\end{aligned} \tag{4.284}$$

where $\pi_{\text{kin}}(\dot{\mathbf{u}})$ and $\Pi_{\text{kin}}(\dot{\mathbf{u}})$ are, respectively, the kinetic energy density and the total kinetic energy of the body

$$\pi_{kin}(\dot{\mathbf{u}}) = \frac{1}{2}\rho\dot{\mathbf{u}} \cdot \dot{\mathbf{u}}, \tag{4.285}$$

$$\Pi_{kin}(\dot{\mathbf{u}}) = \int_{\mathscr{B}} \pi_{kin}(\dot{\mathbf{u}})\, dV. \tag{4.286}$$

Similarly, the other two terms in (4.282) lead to

$$\int_0^{t_f} \int_{\mathscr{B}} \frac{\partial \pi}{\partial \mathbf{E}}\Big|_{\mathscr{D}\mathbf{u}_0} \cdot \mathscr{D}\hat{\mathbf{u}}\, dV\, dt = \frac{d}{d\alpha} \int_0^{t_f} \int_{\mathscr{B}} \pi(\mathscr{D}(\mathbf{u}_0 + \alpha\hat{\mathbf{u}}))\, dV\, dt\Big|_{\alpha=0}, \tag{4.287}$$

$$\int_0^{t_f} \langle f, \hat{\mathbf{u}}\rangle\, dt = \frac{d}{d\alpha} \int_0^{t_f} \langle f, \mathbf{u}_0 + \alpha\hat{\mathbf{u}}\rangle\, dt\Big|_{\alpha=0}. \tag{4.288}$$

Thus, we recover an expression in terms of the total potential energy functional as

$$\int_0^{t_f} \left[\int_{\mathscr{B}} \frac{\partial \pi}{\partial \mathbf{E}}\Big|_{\mathscr{D}\mathbf{u}} \cdot \mathscr{D}\hat{\mathbf{u}}\, dV - \langle f, \hat{\mathbf{u}}\rangle \right] dt = \frac{d}{d\alpha} \int_0^{t_f} \Pi(\mathbf{u} + \alpha\hat{\mathbf{u}})\, dt\Big|_{\alpha=0}. \tag{4.289}$$

Replacing (4.286) and (4.289) in expression (4.282) yields

$$\begin{aligned}
&\frac{d}{d\alpha} \int_0^{t_f} \left(\Pi_{kin}(\dot{\mathbf{u}}_0 + \alpha\hat{\dot{\mathbf{u}}}) - \Pi(\mathbf{u}_0 + \alpha\hat{\mathbf{u}}) \right) dt\Big|_{\alpha=0} \\
&= \frac{d}{d\alpha} \int_0^{t_f} \mathscr{L}(\mathbf{u}_0 + \alpha\hat{\mathbf{u}})\, dt\Big|_{\alpha=0} = 0 \qquad \forall \hat{\mathbf{u}} \in \text{Var}_u,
\end{aligned} \tag{4.290}$$

where we have defined $\mathscr{L}(\mathbf{u}_0)$, which is known as the Lagrangian for the system under study. This result shows us that, departing from the PVW, and for the case of hyperelastic materials, equation (4.290) (in general for systems which, in some form, store total potential energy), it is possible to arrive to an alternative variational formulation known as Principle of Least Action or Hamilton's Principle. In the present context, this variational principle characterizes the elastodynamics problem, and states that, from all possible admissible configurations a hyperelastic body can assume from its initial admissible configuration to the admissible configuration at the final time instant t_f, the solution is the sequence of configurations, or the path, which is a stationary point of the associated Lagrangian cost functional. That is, the variational problem consists of $\forall t \in [0, t_f)$, finding $\mathbf{u}_0 \in \mathrm{Kin}_u$, such that

$$\frac{d}{d\alpha} \int_0^{t_f} \mathscr{L}(\mathbf{u}_0 + \alpha\hat{\mathbf{u}})\, dt \Big|_{\alpha=0} = 0 \qquad \forall \hat{\mathbf{u}} \in \mathrm{Var}_u, \tag{4.291}$$

where

$$\mathscr{L}(\mathbf{u}) = \Pi_{\mathrm{kin}}(\dot{\mathbf{u}}) - \Pi(\mathbf{u}) = \int_{\mathscr{B}} \left(\frac{1}{2}\rho\dot{\mathbf{u}}\cdot\dot{\mathbf{u}} - \pi(\mathbf{u})\right) dV + \langle f, \mathbf{u}\rangle. \tag{4.292}$$

The reader interested in pursuing further knowledge about the existence, uniqueness, and variational formulations in problems involving dynamics is directed to the works by G. Fichera ([103, 104]), P. H. Brézis ([42]), G. Duvaut and J. L. Lions ([76]), and I. Ekeland and R. Temam ([78]).

4.11 Approximate Solution to Variational Problems

In this section we will venture into the domain of numerical algorithms to find approximate solutions in the field of mechanics, illustrating the way in which the variational formulation stands out as a simple and natural conceptual schema also in the approximate realm.

Let us first work with variational formulations for bodies with bilateral kinematical constraints, and then extend some analysis to bodies with unilateral constraints.

4.11.1 Elastostatics Problem

As we have seen in previous sections, the elastostatics problem is equivalent to the following variational problems

$$\text{Principle of Minimum Total Potential Energy} \qquad \min_{\mathbf{u}\in\mathrm{Kin}_u} \Pi(\mathbf{u}), \tag{4.293}$$

$$\text{Principle of Minimum Complementary Energy} \qquad \min_{\mathbf{T}\in\mathrm{Est}_T} \Pi^*(\mathbf{T}). \tag{4.294}$$

As we can appreciate, these two problems are defined in the sets Kin_u and Est_T which, for bilateral kinematical constraints, are translations of the infinite-dimensional vector subspaces Var_u and Var_T, that is, $\mathrm{Kin}_u = \overline{\mathbf{u}} + \mathrm{Var}_u$, $\overline{\mathbf{u}} \in \mathrm{Kin}_u$ arbitrary, and $\mathrm{Est}_T = \overline{\mathbf{T}} + \mathrm{Var}_T$, $\overline{\mathbf{T}} \in \mathrm{Kin}_T$ arbitrary.

To obtain approximate solutions to these problems, we have to proceed to cast these variational problems in finite-dimensional spaces. In order to do this, let us write $\{\phi_i\}_{i=1}^{\infty}$ as a complete set of coordinate functions, or basis functions, for the space Var_u (in the case of a three-dimensional body each function ϕ_i is a vector field). This amounts to say that, for a given $\epsilon > 0$ arbitrarily small, it will always be possible to find an integer $N = N(\epsilon)$ such that for all $n > N$ it is possible to define $a_i \in \mathbb{R}$, $i = 1, \dots, n$, for which the following holds

$$\left\| \mathbf{u} - \sum_{i=1}^{n} a_i\phi_i \right\|_{\mathscr{U}} < \epsilon, \tag{4.295}$$

where $\|\cdot\|_{\mathscr{U}}$ is a proper norm adopted in $\mathscr{U}$. Based on this, and for a finite n, we can define the finite-dimensional space

$$\text{Var}_u^n = \text{span}\{\phi_i\}_{i=1}^n, \tag{4.296}$$

which verifies $\text{Var}_u^n \subset \text{Var}_u$. Since Kin_u is a translation of Var_u given by $\bar{\mathbf{u}}$ ($\bar{\mathbf{u}} \in \text{Kin}_u$ arbitrary), it is possible to define in Kin_u the subset $\text{Kin}_u^n \subset \text{Kin}_u$ given by

$$\text{Kin}_u^n = \bar{\mathbf{u}} + \text{Var}_u^n = \bar{\mathbf{u}} + \text{span}\{\phi_i\}_{i=1}^n. \tag{4.297}$$

Similarly, we can perform these steps to rework the space Var_T and the set Est_T as finite-dimensional sets. Then, calling $\{\psi_i\}_{i=1}^{\infty}$ the complete set of coordinate functions (for three-dimensional bodies each function ψ_i is a second-order tensor field) in the space Var_T we have, for each finite r, that

$$\text{Var}_T^r = \text{span}\{\psi_i\}_{i=1}^r \subset \text{Var}_T, \tag{4.298}$$

$$\text{Est}_T^r = \bar{\mathbf{T}} + \text{Var}_T^r = \bar{\mathbf{T}} + \text{span}\{\psi_i\}_{i=1}^r \subset \text{Est}_T. \tag{4.299}$$

With these new sets, we can frame the variational problems in the finite-dimensional realm posed by Kin_u^n and Est_T^r accordingly. As it will be seen, this will lead us to the problem of minimizing a function in $\mathbb{R}^n$ and $\mathbb{R}^r$, respectively. The numerical algorithms for the solution of these finite-dimensional problems can be grouped into two categories: direct methods formed by algorithms, which, in general, make the attempt to minimize the objective function at each stage, and indirect methods, corresponding to algorithms that seek the solution to those algebraic equations which characterize the stationary point.

The Principle of Minimum Total Potential Energy and the Principle of Minimum Complementary Energy in the finite-dimensional domain are

$$\min_{\mathbf{u}\in\text{Kin}_u^n} \Pi(\mathbf{u}), \tag{4.300}$$

$$\min_{\mathbf{T}\in\text{Est}_T^r} \Pi^*(\mathbf{T}). \tag{4.301}$$

Given that $\text{Kin}_u^n = \bar{\mathbf{u}} + \text{Var}_u^n$ and $\text{Est}_T^r = \bar{\mathbf{T}} + \text{Var}_T^r$, we have

$$\mathbf{u} \in \text{Kin}_u^n \quad \Rightarrow \quad \mathbf{u} = \bar{\mathbf{u}} + \sum_{i=1}^n a_i\phi_i, \tag{4.302}$$

$$\mathbf{T} \in \text{Est}_T^r \quad \Rightarrow \quad \mathbf{T} = \bar{\mathbf{T}} + \sum_{i=1}^r b_i\psi_i, \tag{4.303}$$

where, by changing $a_i \in \mathbb{R}$, $i = 1, \dots, n$ and $b_i \in \mathbb{R}$, $i = 1, \dots, r$, we are able to represent any element in Kin_u^n and Est_T^r, respectively.

Based on this, the Total Potential Energy functional $\Pi(\mathbf{u})$ in Kin_u^n takes the form

$$\Pi\left(\bar{\mathbf{u}} + \sum_{i=1}^n a_i\phi_i\right) = \int_{\mathscr{B}} \pi\left(\bar{\mathbf{u}} + \sum_{i=1}^n a_i\phi_i\right) dV - \langle f, \bar{\mathbf{u}}\rangle - \sum_{i=1}^n a_i\langle f, \phi_i\rangle$$
$$= \tilde{\Pi}(a_1, \dots, a_n) = \tilde{\Pi}(\mathbf{a}), \tag{4.304}$$

where $\tilde{\Pi}$ is now a function whose domain is in $\mathbb{R}^n$. Hence, the following equivalence holds for the Principle of Minimum Total Potential Energy

$$\min_{\mathbf{u}\in\text{Kin}_u^n} \Pi(\mathbf{u}) \quad \Leftrightarrow \quad \min_{\mathbf{a}\in\mathbb{R}^n} \tilde{\Pi}(\mathbf{a}). \tag{4.305}$$

In other words, when moving to the finite-dimensional setting we have transformed the original problem of minimizing the cost functional Π in a infinite-dimensional function space into the problem of minimizing the function $\tilde{\Pi}$ defined in $\mathbb{R}^n$.

Similarly, defining $\mathbf{b} = (b_1, \dots, b_r)$, the Principle of Minimum Complementary Energy becomes

$$\min_{\mathbf{T}\in \mathrm{Est}_T^r} \Pi^*(\mathbf{T}) \quad \Leftrightarrow \quad \min_{\mathbf{b}\in\mathbb{R}^r} \tilde{\Pi}^*(\mathbf{b}), \tag{4.306}$$

where

$$\tilde{\Pi}^*(\mathbf{b}) = \int_{\mathscr{B}} \pi^*\left(\overline{\mathbf{T}} + \sum_{i=1}^{r} b_i \psi_i\right) dV - \left(\left(\overline{\mathbf{T}} + \sum_{i=1}^{r} b_i \psi_i, \overline{\mathbf{u}}\right)\right). \tag{4.307}$$

Consider now a linear elastic material behavior. From (4.304) we have

$$\begin{aligned} \tilde{\Pi}(\mathbf{a}) = &\frac{1}{2}\int_{\mathscr{B}} \mathbb{D}\left(\nabla\left(\overline{\mathbf{u}} + \sum_{i=1}^{n} a_i\phi_i\right)\right)^s \cdot \left(\nabla\left(\overline{\mathbf{u}} + \sum_{j=1}^{n} a_j\phi_j\right)\right)^s dV \\ &- \langle f, \overline{\mathbf{u}}\rangle - \sum_{i=1}^{n} a_i \langle f, \phi_i\rangle. \end{aligned} \tag{4.308}$$

From the symmetry and positivity of $\mathbb{D}$, it follows that the necessary and sufficient condition for a minimum of $\tilde{\Pi}$ is characterized by

$$\frac{\partial \tilde{\Pi}}{\partial a_i} = 0 \qquad i = 1, \dots, n, \tag{4.309}$$

and in explicit form

$$\sum_{j=1}^{n} \left[\int_{\mathscr{B}} \mathbb{D}(\nabla\phi_j)^s \cdot (\nabla\phi_i)^s dV\right] a_j = \langle f, \phi_i\rangle - \int_{\mathscr{B}} \mathbb{D}(\nabla\overline{\mathbf{u}})^s \cdot (\nabla\phi_i)^s dV \quad i = 1, \dots, n. \tag{4.310}$$

As expected, expression (4.310) can equivalently be derived directly from the PVW by considering a finite-dimensional counterpart of the space of admissible variations Var_u, say Var_u^n.

As a result, the minimization problem stated by (4.305) in the finite-dimensional setting is equivalent to solving the following system of linear equations

$$\mathbf{K}\mathbf{a} = \mathbf{f}, \tag{4.311}$$

where $\mathbf{K}$ is the coefficient matrix, which is symmetric and positive definite (by virtue of the properties of $\mathbb{D}$ and because the rigid motion was eliminated). Thus, a generic element $[\mathbf{K}]_{ij} = K_{ij}$ is given by

$$K_{ij} = \int_{\mathscr{B}} \mathbb{D}(\nabla\phi_j)^s \cdot (\nabla\phi_i)^s dV, \tag{4.312}$$

$\mathbf{f} \in \mathbb{R}^n$ is the right-hand side vector, which depends on the load applied f and the boundary conditions over $\partial\mathscr{B}_u$. Then, a generic element $[\mathbf{f}]_i = f_i$ is given by

$$f_i = \langle f, \phi_i\rangle - \int_{\mathscr{B}} \mathbb{D}(\nabla\overline{\mathbf{u}})^s \cdot (\nabla\phi_i)^s dV. \tag{4.313}$$

Moreover, from the properties of matrix $\mathbf{K}$ it follows that there exists the inverse matrix $\mathbf{K}^{-1}$, and then the coefficients of the approximate solution will be given by $\mathbf{a} = \mathbf{K}^{-1}\mathbf{f}$.

Here, it is clearly seen that if the basis functions ϕ_i are picked in such a way that they are orthogonal in the following sense

$$K_{ij} \begin{cases} = 0 & \text{if } i \neq j, \\ \neq 0 & \text{if } i = j, \end{cases} \tag{4.314}$$

then the coefficient matrix in the system of linear equations (4.311) is diagonal. The construction of such functions is in many cases an arduous task, possibly impossible to be implemented in practice. However, a strategy aimed at generating a coefficient matrix $\mathbf{K}$ with the largest number of zero entries consists of choosing basis functions ϕ_i of compact support, that is, ϕ_i is different from zero in just a compact region of $\mathscr{B}$, say $\mathscr{B}_i$, while it is exactly zero in the complementary region $\mathscr{B} \setminus \mathscr{B}_i$. Using this approach, we will have

$$K_{ij} = \int_{\mathscr{B}} \mathbb{D}(\nabla\phi_i)^s \cdot (\nabla\phi_j)^s dV \begin{cases} = 0 & \text{if } \mathscr{B}_i \cap \mathscr{B}_j = \emptyset, \\ \neq 0 & \text{otherwise.} \end{cases} \tag{4.315}$$

This fundamental aspect, jointly with the need for facilitating the construction of arbitrary elements $\bar{\mathbf{u}}$, has made of the finite element method (FEM) one of the most attractive methodologies towards the construction of basis functions [45, 151, 225–227, 294, 306].

For the case of nonlinear hyperelastic materials, the necessary and sufficient conditions that characterize a minimum of the total potential energy functional is given by

$$\int_{\mathscr{B}} \left.\frac{\partial\pi}{\partial\mathbf{E}}\right|_{\left(\nabla\left(\bar{\mathbf{u}}+\sum_{j=1}^{n} a_j\phi_j\right)\right)^s} \cdot (\nabla\phi_i)^s dV = \langle f, \phi_i \rangle \qquad i = 1, \dots, n, \tag{4.316}$$

which corresponds to a system of nonlinear equations. Again, this result can be easily obtained from the PVW by considering a finite-dimensional space $\mathrm{Var}_u^n \in \mathrm{Var}_u$.

On the one hand, the Principle of Minimum Total Potential Energy given by (4.305) corresponds, in the finite-dimensional setting, to a classical problem of mathematical programming which amounts, for the case of linear elastic materials, to the unconstrained minimization of a quadratic function in $\mathbb{R}^n$. For nonlinear materials the mathematical programming problem is transformed into the unconstrained minimization of a strictly convex function (recall the convexity of the strain energy function and consider that rigid motions have been removed). For these two problems (linear and nonlinear material behavior), the methods constructed to find the minimum of the functions in $\mathbb{R}^n$ are called direct methods. On the other hand, a further finite-dimensional characterization of the minimum of the total potential energy is given by the system of linear equations given by (4.311) for linear elastic materials, and by (4.316) for nonlinear hyperelastic materials. In this case, the methods employed for the solution of the problem aim at solving this system of equations.

An entirely analogous reasoning allows us to conclude that the Principle of Minimum Complementary Energy given by (4.306) corresponds, in the finite-dimensional setting, to the classical problem of mathematical programming which amounts to the unconstrained minimization of a strictly convex function in $\mathbb{R}^r$ (recall the convexity of the complementary strain energy function π^*) when the material behavior is nonlinear, while it is the unconstrained minimization of a quadratic function in $\mathbb{R}^r$ in the case of linear elastic materials. In the latter case, the functional becomes

$$\tilde{\Pi}^*(\mathbf{b}) = \frac{1}{2}\int_{\mathscr{B}} \mathbb{D}^{-1}\left(\overline{\mathbf{T}} + \sum_{i=1}^{r} b_i\psi_i\right)\cdot\left(\overline{\mathbf{T}} + \sum_{j=1}^{r} b_j\psi_j\right)dV - \left(\left(\overline{\mathbf{T}} + \sum_{i=1}^{r} b_i\psi_i, \overline{\mathbf{u}}\right)\right). \tag{4.317}$$

In the nonlinear case, the characterization of the minimum is given by the system of nonlinear equations

$$\frac{\partial\tilde{\Pi}^*}{\partial b_i} = 0 \qquad i = 1, \dots, r, \tag{4.318}$$

or explicitly

$$\int_{\mathscr{B}} \left.\frac{\partial\pi^*}{\partial\mathbf{T}}\right|_{\overline{\mathbf{T}}+\sum_{j=1}^{r} b_j\psi_j} \cdot \psi_i dV = ((\psi_i, \overline{\mathbf{u}})) \qquad i = 1, \dots, r, \tag{4.319}$$

and when the material is linear elastic it is characterized by the following system of linear equations

$$\sum_{j=1}^{r}\left[\int_{\mathscr{B}} \mathbb{D}^{-1}\psi_j \cdot \psi_i dV\right] b_j = ((\psi_i, \overline{\mathbf{u}})) - \int_{\mathscr{B}} \mathbb{D}^{-1}\overline{\mathbf{T}} \cdot \psi_i dV \qquad i = 1, \dots, r, \tag{4.320}$$

where the coefficient matrix is symmetric and positive definite. In compact algebraic form, this expression results in

$$\mathbf{Mb} = \mathbf{g}, \tag{4.321}$$

where the entries of the coefficient matrix $[\mathbf{M}]_{ij} = M_{ij}$ are

$$M_{ij} = \int_{\mathscr{B}} \mathbb{D}^{-1}\psi_j \cdot \psi_i dV, \tag{4.322}$$

and those of the right-hand side vector $[\mathbf{g}]_i = g_i$ are

$$g_i = ((\psi_i, \overline{\mathbf{u}})) - \int_{\mathscr{B}} \mathbb{D}^{-1}\overline{\mathbf{T}} \cdot \psi_i dV. \tag{4.323}$$

It is important to observe that the approximation of the primal and dual variational principles reduces, in the case of linear elastic materials, to the solution of corresponding systems of linear equations. For the construction of these systems of linear equations it is required to specify the basis functions (ϕ_i or ψ_i, respectively), which, through a linear combination, render the admissible approximate fields. In the jargon of the FEM, this amounts to defining the type of finite element to be employed. Then, for the numerical calculation of the entries of the coefficient matrices it is required to integrate positive definite symmetric bilinear forms that render the matrix entries (4.312) and (4.322), respectively. In turn, for the numerical determination of the right-hand sides, integration of the linear forms in (4.313) and (4.323) is mandatory. These integrals can be solved either analytically or numerically.

So the determination of the approximate solution to the problems of minimum implies the characterization of the coordinate functions $\{\phi_i\}_{i=1}^{\infty}$ and $\{\psi_i\}_{i=1}^{\infty}$. It is also worth noting that the choice of the arbitrary fields $\overline{\mathbf{u}}$ and $\overline{\mathbf{T}}$ (related to the kinematical constraints over $\partial\mathscr{B}_u$ and to the self-equilibrium constraints over $\mathscr{B}$ and $\partial\mathscr{B}_a$, respectively) can be troublesome. Having said this, in the context of the FEM, we notice that:

- the FEM is nothing but a rather convenient systematic manner to build the basis functions ϕ_i, $i = 1, \dots, n$, and ψ_i, $i = 1, \dots, r$
- the determination of the arbitrary elements $\overline{\mathbf{u}}$ or $\overline{\mathbf{T}}$ is a simple and quite straightforward task using the very same basis functions.

Some questions can now be posed regarding the basis functions. For example, which are the general features that these coordinate functions must have? The answer to this question implicitly lies in the very variational formulation. Indeed, consider the Principle of Minimum Total Potential Energy and assume, for simplicity, that $\overline{\mathbf{u}} = \mathbf{0}$ over $\partial\mathscr{B}_u$. In this case $\text{Kin}_u \equiv \text{Var}_u$ is a vector subspace of $\mathscr{U}$, and from the properties of $\mathbb{D}$ we have that the bilinear form

$$a(\mathbf{u}, \mathbf{v}) = \int_{\mathscr{B}} \mathbb{D}(\nabla\mathbf{u})^s \cdot (\nabla\mathbf{v})^s dV \qquad \mathbf{u}, \mathbf{v} \in \text{Kin}_u, \tag{4.324}$$

satisfies all the properties to be regarded as an inner product in $\mathscr{U}$. This inner product induces the so-called energy norm

$$||\mathbf{u}||_E = \sqrt{a(\mathbf{u}, \mathbf{u})} \qquad \mathbf{u} \in \text{Kin}_u. \tag{4.325}$$

The closure of Kin_u with respect to this energy norm gives rise to a Hilbert space, denoted by H, called the energy space, which is composed of all functions (actually classes of functions) which are square integrable in $\mathscr{B}$, and whose first gradients are also square integrable functions in $\mathscr{B}$, and such that over $\partial\mathscr{B}_u$ they are null.[7]

Under similar assumptions to turn Est_T into a vector space and have $\text{Est}_T \equiv \text{Var}_T$, the Principle of Minimum Complementary Energy, the bilinear form becomes

$$c(\mathbf{T}, \mathbf{S}) = \int_{\mathscr{B}} \mathbf{T} \cdot \mathbb{D}^{-1}\mathbf{S}dV \qquad \mathbf{T}, \mathbf{S} \in \text{Est}_T, \tag{4.326}$$

which induces the norm

$$||\mathbf{T}|| = \sqrt{c(\mathbf{T}, \mathbf{T})} \qquad \mathbf{T} \in \text{Est}_T. \tag{4.327}$$

Then, the coordinate functions ψ_i must only be square integrable in $\mathscr{B}$ as dictated by this bilinear form.

A fundamental difference exists between the primal and the dual variational principles which, from the numerical standpoint, turns the primal formulation into a much more appealing strategy than the dual formulation for solving the elastostatics problem. In fact, the Principle of Minimum Total Potential Energy (displacement-based model) requires the functions ϕ_i to be regular enough in the sense established above, and zero over the portion of the boundary $\partial\mathscr{B}_u$ (we are assuming homogeneous kinematical constraints there). Making use of the FEM, the construction of such basis functions is a simple task, even in the case of non-homogeneous kinematical constraints. In contrast,

7 In the case of structural components and high-order continua, some models require that admissible functions feature higher-order derivatives that must be square integrable functions. This is directly related to the fact that in these situations the operator $\mathscr{D}$ contains higher-order derivatives. These specific cases will be addressed in Chapter 7 for variational theories of beam components, in Chapter 9 for variational theories of plates and shells, and finally in Chapter 12 for variational formulations of higher-order continuum media.

the Principle of Minimum Complementary Energy (stress-based model) requires the basis functions ψ_i to be self-equilibrated internal stresses. It is at this point that the complexity of the dual formulation arises. In fact, the construction of these self-equilibrated functions is far from simple and represents an unavoidable obstacle that makes this formulation practically intricate, if not impossible in many cases, precluding it from being as popular as the primal formulation.

4.11.2 Hellinger–Reissner Principle

With an analogous reasoning to that of the previous section, in this section we tackle the approximation of the variational Hellinger–Reissner principle. Thus, the approach consists of formulating this problem in finite-dimensional spaces. We then call $\{\phi_i\}_{i=1}^{\infty}$ to the complete set of coordinate functions in Var_u and $\{\psi_i\}_{i=1}^{\infty}$ to the complete set of basis functions in $\mathscr{W}'$. In this way, we have

$$\mathbf{u} \in \text{Kin}_u^n \quad \Leftrightarrow \quad \mathbf{u} = \bar{\mathbf{u}} + \sum_{i=1}^{n} a_i \phi_i \quad \bar{\mathbf{u}} \in \text{Kin}_u \text{ arbitrary}, \tag{4.328}$$

$$\mathbf{T} \in \mathscr{W}'^r \quad \Leftrightarrow \quad \mathbf{T} = \sum_{i=1}^{r} c_i \psi_i. \tag{4.329}$$

From this, the Hellinger–Reissner variational principle consists of finding $\mathbf{u}_0 \in \text{Kin}_u^n$ and $\mathbf{T}_0 \in \mathscr{W}'^r$ such that they are the stationary points of the following functional

$$\begin{aligned}(\mathbf{u}_0, \mathbf{T}_0) &= \min_{\mathbf{u}\in\text{Kin}_u^n} \max_{\mathbf{T}\in\mathscr{W}'^r} \mathscr{F}_{\text{HR}}(\mathbf{u}, \mathbf{T}) \\ &= \min_{\mathbf{u}\in\text{Kin}_u^n} \max_{\mathbf{T}\in\mathscr{W}'^r} \left[-\int_{\mathscr{B}} \pi^*(\mathbf{T}) dV + \int_{\mathscr{B}} \mathbf{T} \cdot (\nabla \mathbf{u})^s dV - \langle f, \mathbf{u} \rangle \right]. \end{aligned} \tag{4.330}$$

This is equivalent to finding the real numbers a_i, $i = 1, \dots, n$, and c_j, $j = 1, \dots, r$ such that the following function is stationary

$$\begin{aligned}\tilde{\mathscr{F}}_{\text{HR}}(\mathbf{a}, \mathbf{c}) &= \tilde{\mathscr{F}}_{\text{HR}}(a_1, \dots, a_n, c_1, \dots, c_r) = -\int_{\mathscr{B}} \pi^* \left(\sum_{j=1}^{r} c_j \psi_j \right) dV \\ &+ \int_{\mathscr{B}} \left(\sum_{j=1}^{r} c_j \psi_j \right) \cdot \left(\nabla \left(\bar{\mathbf{u}} + \sum_{i=1}^{n} a_i \phi_i \right) \right)^s dV - \left\langle f, \left(\bar{\mathbf{u}} + \sum_{i=1}^{n} a_i \phi_i \right) \right\rangle. \end{aligned} \tag{4.331}$$

If the material behavior is linear elastic, (4.331) becomes

$$\begin{aligned}\tilde{\mathscr{F}}_{\text{HR}}(\mathbf{a}, \mathbf{c}) &= -\frac{1}{2} \sum_{j=1}^{r} \sum_{s=1}^{r} c_j c_s \int_{\mathscr{B}} \mathbb{D}^{-1} \psi_j \cdot \psi_s dV + \sum_{j=1}^{r} \sum_{i=1}^{n} c_j a_i \int_{\mathscr{B}} \psi_j \cdot (\nabla \phi_i)^s dV \\ &+ \sum_{j=1}^{r} c_j \int_{\mathscr{B}} \psi_j \cdot (\nabla \bar{\mathbf{u}})^s dV - \sum_{i=1}^{n} a_i \langle f, \phi_i \rangle - \langle f, \bar{\mathbf{u}} \rangle. \end{aligned} \tag{4.332}$$

As before, the methods to solve this problem can be categorized into two groups: direct methods aimed at finding the solution of the min-max problem, that is, the saddle-point problem, and indirect methods aimed at solving the associated system of linear (or nonlinear depending on the material behavior) equations. As for the latter case, the min–max problem is, for linear elastic materials, characterized by the following system of linear equations

$$\frac{\partial \tilde{\mathscr{F}}_{HR}}{\partial a_i} = 0 \qquad i = 1, \dots, n, \tag{4.333}$$

$$\frac{\partial \tilde{\mathscr{F}}_{HR}}{\partial c_j} = 0 \qquad j = 1, \dots, r, \tag{4.334}$$

or, in explicit form,

$$\sum_{j=1}^{r} \left[\int_{\mathscr{B}} \psi_j \cdot (\nabla \phi_i)^s dV \right] c_j = \langle f, \phi_i \rangle \qquad i = 1, \dots, n, \tag{4.335}$$

and

$$-\sum_{s=1}^{r} \left[\int_{\mathscr{B}} \mathbb{D}^{-1} \psi_s \cdot \psi_j dV \right] c_s + \sum_{i=1}^{n} \left[\int_{\mathscr{B}} \psi_j \cdot (\nabla \phi_i)^s dV \right] a_i$$
$$= -\int_{\mathscr{B}} \psi_j \cdot (\nabla \overline{\mathbf{u}})^s dV \quad j = 1, \dots, r. \tag{4.336}$$

In compact form, the previous system of equations can be written as

$$\mathbf{K}_{HR}\mathbf{d} = \mathbf{f}_{HR}, \tag{4.337}$$

where

$$\mathbf{K}_{HR} = \begin{pmatrix} \mathbf{O} & \mathbf{A}^T \\ \mathbf{A} & \mathbf{C} \end{pmatrix} \in \mathbb{R}^{(n+r)\times(n+r)}, \tag{4.338}$$

$$\mathbf{d} = \begin{pmatrix} \mathbf{a} \\ \mathbf{c} \end{pmatrix} \in \mathbb{R}^{(n+r)}, \tag{4.339}$$

$$\mathbf{f}_{HR} = \begin{pmatrix} \mathbf{f}_1 \\ \mathbf{f}_2 \end{pmatrix} \in \mathbb{R}^{(n+r)}, \tag{4.340}$$

where the block matrix $[\mathbf{A}]_{ji} = A_{ji}$ and the symmetric matrix $[\mathbf{C}]_{js} = C_{js}$ are

$$A_{ji} = \int_{\mathscr{B}} \psi_j \cdot (\nabla \phi_i)^s dV, \tag{4.341}$$

$$C_{js} = -\int_{\mathscr{B}} \psi_j \cdot \mathbb{D}^{-1} \psi_s dV, \tag{4.342}$$

and the vectors on the right-hand side, $[\mathbf{f}_1]_i = f_{1i}$ and $[\mathbf{f}_2]_j = f_{2j}$, are given by

$$f_{1i} = \langle f, \phi_i \rangle, \tag{4.343}$$

$$f_{2j} = -\int_{\mathscr{B}} \psi_j \cdot (\nabla \overline{\mathbf{u}})^s dV. \tag{4.344}$$

Observe that $\mathbf{K}_{HR}$ is symmetric, then using the fact that block $\mathbf{C}$ is symmetric and positive definite, we write the system of equations (4.337) as

$$\mathbf{A}^T \mathbf{c} = \mathbf{f}_1, \tag{4.345}$$

$$\mathbf{A}\mathbf{a} + \mathbf{C}\mathbf{c} = \mathbf{f}_2, \tag{4.346}$$

and, therefore, from (4.346) we get

$$\mathbf{C}\mathbf{c} = \mathbf{f}_2 - \mathbf{A}\mathbf{a}. \tag{4.347}$$

Since $\mathbf{C}$ is positive definite, its inverse exists, so we can find $\mathbf{c}$ and replace it into (4.345), arriving at

$$\mathbf{A}^T\mathbf{C}^{-1}\mathbf{A}\mathbf{a} = \mathbf{f}_1 - \mathbf{A}^T\mathbf{C}^{-1}\mathbf{f}_2. \tag{4.348}$$

Hence, the solution $(\mathbf{a}, \mathbf{c})$ of the min–max problem is characterized, for linear elastic materials, by the solution $\mathbf{a}$ of equation (4.348), which is then replaced into (4.347) to yield $\mathbf{c}$. Given that there exists $\mathbf{C}^{-1}$, the matrix $\mathbf{A}^T\mathbf{C}^{-1}\mathbf{A}$ will be invertible if the rank of matrix $\mathbf{A}$ is r satisfying $r \leq n$. This shows us that the construction of spaces Var_u^n and $\mathscr{W}'^r$ must be executed in such a way that this property of block $\mathbf{A}$ is satisfied.

4.11.3 Generalized Variational Principle

The Generalized Variational Principle seeks for the element $(\mathbf{u}_0, \mathbf{E}_0, \mathbf{T}_0)$ in the set $\mathrm{Kin}_u \times \mathscr{W} \times \mathscr{W}'$ such that minimizes (4.205). As before, the construction of approximate solutions is carried out within finite-dimensional sets. Let us then call $\{\phi_i\}_{i=1}^{\infty}$ the complete set of basis functions for Var_u, $\{\Xi_i\}_{i=1}^{\infty}$ is the complete set of coordinate functions for $\mathscr{W}$, and $\{\psi_i\}_{i=1}^{\infty}$ is the set of coordinate functions for $\mathscr{W}'$. Therefore, we write

$$\mathbf{u} \in \mathrm{Kin}_u^n \quad \Rightarrow \quad \mathbf{u} = \bar{\mathbf{u}} + \sum_{i=1}^{n} a_i\phi_i, \tag{4.349}$$

$$\mathbf{E} \in \mathscr{W}^m \quad \Rightarrow \quad \mathbf{E} = \sum_{i=1}^{m} b_i\Xi_i, \tag{4.350}$$

$$\mathbf{T} \in \mathscr{W}'^r \quad \Rightarrow \quad \mathbf{T} = \sum_{i=1}^{r} c_i\psi_i. \tag{4.351}$$

Then, the three-field variational principle redefined in the finite-dimensional setting takes the following form: find $\mathbf{u}_0 \in \mathrm{Kin}_u^n$, $\mathbf{E}_0 \in \mathscr{W}^m$ and $\mathbf{T}_0 \in \mathscr{W}'^r$ such that

$$\begin{aligned}(\mathbf{u}_0, \mathbf{E}_0, \mathbf{T}_0) &= \min_{\mathbf{u}\in\mathrm{Kin}_u^n} \min_{\mathbf{E}\in\mathscr{W}^m} \max_{\mathbf{T}\in\mathscr{W}'^r} \mathscr{F}(\mathbf{u}, \mathbf{E}, \mathbf{T}) \\ &= \min_{\mathbf{u}\in\mathrm{Kin}_u^n} \min_{\mathbf{E}\in\mathscr{W}^m} \max_{\mathbf{T}\in\mathscr{W}'^r} \left[\int_{\mathscr{B}} \pi(\mathbf{E})dV - \langle f, \mathbf{u}\rangle - \int_{\mathscr{B}} \mathbf{T}\cdot(\mathbf{E} - \nabla\mathbf{u}^s)dV\right].\end{aligned} \tag{4.352}$$

This amounts to finding the real numbers a_i, $i = 1, \dots, n$, b_k, $k = 1, \dots, m$ and c_j, $j = 1, \dots, r$ for which the function

$$\begin{aligned}\tilde{\mathscr{F}}(\mathbf{a}, \mathbf{b}, \mathbf{c}) &= \tilde{\mathscr{F}}(a_1, \dots, a_n, b_1, \dots, b_m, c_1, \dots, c_r) \\ &= \int_{\mathscr{B}} \pi\left(\sum_{k=1}^{m} b_k\Xi_k\right)dV - \left\langle f, \left(\bar{\mathbf{u}} + \sum_{i=1}^{n} a_i\phi_i\right)\right\rangle \\ &\quad - \int_{\mathscr{B}} \left(\sum_{j=1}^{r} c_j\psi_j\right)\cdot\left[\left(\sum_{k=1}^{m} b_k\Xi_k\right) - \left(\nabla\left(\bar{\mathbf{u}} + \sum_{i=1}^{n} a_i\phi_i\right)\right)^s\right]dV,\end{aligned} \tag{4.353}$$

attains a stationary point.

We have discussed that numerical algorithms to address this problem can be classified into direct methods aimed at finding the stationary point of function $\tilde{\mathscr{F}}$ or indirect methods whose goal is to solve the associated system of algebraic equations. In the latter case, these equations are

$$\frac{\partial \tilde{\mathscr{F}}}{\partial a_i} = 0 \qquad i = 1, \dots, n, \tag{4.354}$$

$$\frac{\partial \tilde{\mathscr{F}}}{\partial b_k} = 0 \qquad k = 1, \dots, m, \tag{4.355}$$

$$\frac{\partial \tilde{\mathscr{F}}}{\partial c_j} = 0 \qquad j = 1, \dots, r. \tag{4.356}$$

Explicitly, these equations correspond to

$$\sum_{j=1}^{r} \left[\int_{\mathscr{B}} \psi_j \cdot (\nabla \phi_i)^s dV \right] c_j = \langle f, \phi_i \rangle \qquad i = 1, \dots, n, \tag{4.357}$$

$$\int_{\mathscr{B}} \left. \frac{\partial \pi}{\partial \mathbf{E}} \right|_{\sum_{t=1}^{m} b_t \Xi_t} \cdot \Xi_k dV - \sum_{j=1}^{r} \left[\int_{\mathscr{B}} \psi_j \cdot \Xi_k dV \right] c_j = 0 \qquad k = 1, \dots, m, \tag{4.358}$$

$$- \sum_{k=1}^{m} \left[\int_{\mathscr{B}} \psi_j \cdot \Xi_k dV \right] b_k + \sum_{i=1}^{n} \left[\int_{\mathscr{B}} \psi_j \cdot (\nabla \phi_i)^s dV \right] a_i$$

$$= - \int_{\mathscr{B}} \psi_j \cdot (\nabla \overline{\mathbf{u}})^s dV \quad j = 1, \dots, r, \tag{4.359}$$

while for a linear elastic material we have to replace (4.358) by

$$\sum_{t=1}^{m} \left[\int_{\mathscr{B}} \mathbb{D} \Xi_t \cdot \Xi_k dV \right] b_t - \sum_{j=1}^{r} \left[\int_{\mathscr{B}} \psi_j \cdot \Xi_k dV \right] c_j = 0 \qquad k = 1, \dots, m. \tag{4.360}$$

In matrix form, this system of equations has the structure

$$\mathbf{K}_G \mathbf{e} = \mathbf{f}_G, \tag{4.361}$$

where

$$\mathbf{K}_G = \begin{pmatrix} \mathbf{O} & \mathbf{O} & \mathbf{A}^T \\ \mathbf{O} & \mathbf{B} & -\mathbf{H} \\ \mathbf{A} & -\mathbf{H}^T & \mathbf{O} \end{pmatrix} \in \mathbb{R}^{(n+m+r)\times(n+m+r)}, \tag{4.362}$$

$$\mathbf{e} = \begin{pmatrix} \mathbf{a} \\ \mathbf{b} \\ \mathbf{c} \end{pmatrix} \in \mathbb{R}^{(n+m+r)}, \tag{4.363}$$

$$\mathbf{f}_G = \begin{pmatrix} \mathbf{f}_1 \\ \mathbf{0} \\ \mathbf{f}_2 \end{pmatrix} \in \mathbb{R}^{(n+m+r)}, \tag{4.364}$$

where the block matrices $[\mathbf{A}]_{ji} = A_{ji}$, $[\mathbf{B}]_{kt} = B_{kt}$, and $[\mathbf{H}]_{kj} = H_{kj}$ are

$$A_{ji} = \int_{\mathscr{B}} \psi_j \cdot (\nabla \phi_i)^s dV, \tag{4.365}$$

$$B_{kt} = \int_{\mathscr{B}} \mathbb{D}\Xi_t \cdot \Xi_k dV, \tag{4.366}$$

$$H_{kj} = \int_{\mathscr{B}} \psi_j \cdot \Xi_k dV, \tag{4.367}$$

and where the right-hand side vectors $[\mathbf{f}_1]_i = f_{1i}$ and $[\mathbf{f}_2]_j = f_{2j}$ are defined by

$$f_{1i} = \langle f, \phi_i \rangle, \tag{4.368}$$

$$f_{2j} = -\int_{\mathscr{B}} \psi_j \cdot (\nabla \overline{\mathbf{u}})^s dV. \tag{4.369}$$

The system of linear equations (4.361) can be written in expanded form

$$\mathbf{A}^T \mathbf{c} = \mathbf{f}_1, \tag{4.370}$$

$$\mathbf{Bb} - \mathbf{Hc} = \mathbf{0}, \tag{4.371}$$

$$\mathbf{Aa} - \mathbf{H}^T \mathbf{b} = \mathbf{f}_2. \tag{4.372}$$

Since $\mathbf{B}$ is symmetric and positive definite, from (4.371) we have

$$\mathbf{b} = \mathbf{B}^{-1}\mathbf{Hc}, \tag{4.373}$$

and putting this into (4.372),

$$-(\mathbf{H}^T\mathbf{B}^{-1}\mathbf{H})\mathbf{c} + \mathbf{Aa} = \mathbf{f}_2. \tag{4.374}$$

If the rank of the block matrix $\mathbf{H}$ is r with $r \leq m$, we have that there exists the inverse of the matrix $\mathbf{H}^T\mathbf{B}^{-1}\mathbf{H}$, therefore

$$\mathbf{c} = -(\mathbf{H}^T\mathbf{B}^{-1}\mathbf{H})^{-1}(\mathbf{f}_2 - \mathbf{Aa}). \tag{4.375}$$

Finally, putting this result into (4.370) yields

$$[\mathbf{A}^T(\mathbf{H}^T\mathbf{B}^{-1}\mathbf{H})^{-1}\mathbf{A}]\mathbf{a} = \mathbf{A}^T(\mathbf{H}^T\mathbf{B}^{-1}\mathbf{H})^{-1}\mathbf{f}_2 + \mathbf{f}_1. \tag{4.376}$$

Again, if the rank of the block matrix $\mathbf{A}$ is n satisfying $n \leq r$, expressions (4.376), (4.375), and (4.373) give the solution $(\mathbf{a}, \mathbf{b}, \mathbf{c})$ of the system of equations we are looking for.

4.11.4 Contact Problems in Elastostatics

According to previous sections, the elastostatics problem involving unilateral kinematical constraints consists of the following variational problem: find $\mathbf{u}_0 \in \text{Kin}_u$ such that

$$\Pi(\mathbf{u}_0) = \min_{\mathbf{u}\in\text{Kin}_u} \Pi(\mathbf{u}) = \min_{\mathbf{u}\in\text{Kin}_u} \left[\int_{\mathscr{B}} \pi((\nabla\mathbf{u})^s) - \langle f, \mathbf{u} \rangle \right], \tag{4.377}$$

where Kin_u is the translation of a convex cone.

In order to obtain an approximate solution for problem (4.377) we frame the variational equation in the finite-dimensional counterpart of the convex cone Kin_u.

Denoting $\{\phi_i\}_{i=1}^{\infty}$ to the complete set of coordinate functions for the space of all possible displacements $\mathscr{U}$, the set Kin_u^n can be defined as

$$\text{Kin}_u^n = \left\{ \mathbf{u} \in \mathscr{U};\ \mathbf{u} = \sum_{i=1}^{n} a_i\phi_i, \mathbf{a} = (a_1, \ldots, a_n) \in \mathcal{M} \subset \mathbb{R}^n \right\}, \tag{4.378}$$

where $\mathcal{M}$ is a convex set in $\mathbb{R}^n$.

Therefore, the Principle of Minimum Total Potential Energy for bodies featuring unilateral constraints is reduced to the problem of mathematical programming which consists of finding $\mathbf{a}_0 \in \mathcal{M}$ such that

$$\tilde{\Pi}(\mathbf{a}_0) = \min_{\mathbf{a}\in\mathcal{M}} \tilde{\Pi}(\mathbf{a}) = \min_{\mathbf{a}\in\mathcal{M}} \left[\int_{\mathcal{B}} \pi \left(\left(\nabla \left(\sum_{i=1}^{n} a_i \phi_i \right) \right)^s \right) dV - \left\langle f, \sum_{i=1}^{n} a_i \phi_i \right\rangle \right]. \tag{4.379}$$

In general, to make the nonlinear problem presented above numerically tractable the convex $\mathcal{M}$ is approximated, leading to a new convex set, say $\tilde{\mathcal{M}}$. In the context of the FEM, a strategy to build $\tilde{\mathcal{M}}$ is to establish that the unilateral constraints are satisfied only at certain points in the domain of the body, for example at nodal points, or at points employed in the integration of the expressions that play a role in the variational principle (see [22, 86, 101]).

The outcome of the approximation that leads us to $\tilde{\mathcal{M}}$ is that the unilateral constraints can be described by a set of m inequalities of the type

$$\mathbf{Aa} \leq \mathbf{c}, \tag{4.380}$$

where $\mathbf{A} \in \mathbb{R}^{m\times n}$, and whose rank is m.

For example, for the case of unilateral discrete supports, the unilateral constraints read [86]

$$\mathbf{d} \leq \mathbf{a} \leq \mathbf{b}, \tag{4.381}$$

where the inequalities must be understood component-wise.

Thus, the variational principle (4.377), with unilateral constraints of the type (4.380), takes the form: find $\mathbf{a}_0 \in \mathbb{R}^n$ such that

$$\Pi(\mathbf{a}_0) = \min_{\substack{\mathbf{a}\in\mathbb{R}^n \\ \mathbf{Aa}\leq\mathbf{c}}} \Pi(\mathbf{a}) = \min_{\substack{\mathbf{a}\in\mathbb{R}^n \\ \mathbf{Aa}\leq\mathbf{c}}} \left[\int_{\mathcal{B}} \pi \left(\left(\nabla \left(\sum_{i=1}^{n} a_i \phi_i \right) \right)^s \right) dV - \left\langle f, \sum_{i=1}^{n} a_i \phi_i \right\rangle \right]. \tag{4.382}$$

For the particular case of linear elastic materials, this is reduced to the classical quadratic programming problem of finding $\mathbf{a}_0 \in \mathbb{R}^n$ such that

$$\Pi(\mathbf{a}_0) = \min_{\substack{\mathbf{a}\in\mathbb{R}^n \\ \mathbf{Aa}\leq\mathbf{c}}} \Pi(\mathbf{a}) = \min_{\substack{\mathbf{a}\in\mathbb{R}^n \\ \mathbf{Aa}\leq\mathbf{c}}} \left[\frac{1}{2} \mathbf{a} \cdot \mathbf{Ka} - \mathbf{f} \cdot \mathbf{a} \right], \tag{4.383}$$

where $\mathbf{K} \in \mathbb{R}^{n\times n}$ is a symmetric positive definite matrix whose entries $[\mathbf{K}]_{ij} = K_{ij}$ are given by

$$K_{ij} = \int_{\mathcal{B}} \mathbb{D}(\nabla \phi_j)^s \cdot (\nabla \phi_i)^s dV, \tag{4.384}$$

and the vector $\mathbf{f} \in \mathbb{R}^n$ has components $[\mathbf{f}]_i = f_i$ defined by

$$f_i = \langle f, \phi_i \rangle. \tag{4.385}$$

The set of constraints over the elements $\mathbf{a} \in \mathbb{R}^n$ in the variational problem (4.383) can be dealt with by introducing the corresponding Lagrange multipliers, denoted by $\lambda \in \mathbb{R}^m$ such that $\lambda \geq \mathbf{0}$. In fact, noticing that

$$\max_{\lambda \geq \mathbf{0}} (\mathbf{A}\mathbf{a} - \mathbf{c}) \cdot \lambda = \begin{cases} +\infty & \forall \mathbf{a} \text{ such that } \mathbf{A}\mathbf{a} - \mathbf{c} > \mathbf{0}, \\ 0 & \forall \mathbf{a} \text{ such that } \mathbf{A}\mathbf{a} - \mathbf{c} \leq \mathbf{0}, \end{cases} \tag{4.386}$$

problem (4.383) is equivalent to the following min–max (saddle point) problem: find $(\mathbf{a}_0, \lambda_0) \in \mathbb{R}^n \times \mathbb{R}^m$ such that

$$\begin{aligned} \mathscr{L}(\mathbf{a}_0, \lambda_0) &= \min_{\mathbf{a} \in \mathbb{R}^n} \max_{\substack{\lambda \in \mathbb{R}^m \\ \lambda \geq \mathbf{0}}} \mathscr{L}(\mathbf{a}, \lambda) \\ &= \min_{\mathbf{a} \in \mathbb{R}^n} \max_{\substack{\lambda \in \mathbb{R}^m \\ \lambda \geq \mathbf{0}}} \left[\frac{1}{2} \mathbf{a} \cdot \mathbf{K}\mathbf{a} - \mathbf{f} \cdot \mathbf{a} + (\mathbf{A}\mathbf{a} - \mathbf{c}) \cdot \lambda \right]. \end{aligned} \tag{4.387}$$

Regarding the unknown **a**, the minimization problem (4.387) is unconstrained, so the minimization implies searching for the vector $\mathbf{a} \in \mathbb{R}^n$, which is associated with $\lambda \in \mathbb{R}^m$ through the necessary condition

$$\mathbf{K}\mathbf{a} - \mathbf{f} + \mathbf{A}^T \lambda = \mathbf{0}, \tag{4.388}$$

and, because **K** is symmetric and positive definite, we can write

$$\mathbf{a} = \mathbf{K}^{-1}(\mathbf{f} - \mathbf{A}^T \lambda), \tag{4.389}$$

which placed into (4.387) leads to a new problem exclusively defined in terms of the Lagrange multiplier, known as the dual problem, which states the following: find λ_0 such that

$$\mathscr{Q}(\lambda_0) = \min_{\substack{\lambda \in \mathbb{R}^m \\ \lambda \geq \mathbf{0}}} \mathscr{Q}(\lambda) = \min_{\substack{\lambda \in \mathbb{R}^m \\ \lambda \geq \mathbf{0}}} \left[\frac{1}{2} \lambda \cdot \mathbf{P}\lambda - \lambda \cdot \mathbf{e} \right], \tag{4.390}$$

where $\mathbf{P} \in \mathbb{R}^{m \times m}$, a symmetric matrix, and $\mathbf{e} \in \mathbb{R}^m$, are given by

$$\mathbf{P} = \mathbf{A}\mathbf{K}^{-1}\mathbf{A}^T, \tag{4.391}$$

$$\mathbf{e} = \mathbf{A}\mathbf{K}^{-1}\mathbf{f} - \mathbf{c}. \tag{4.392}$$

This dual problem is still a quadratic programming problem, but a fundamental difference is that this dual problem is defined in a simpler convex set than the convex $\mathcal{M}$ in problem (4.383). In fact, in the saddle point problem (4.387), as well as in the dual problem (4.390), this kind of simpler constraint $\lambda \geq \mathbf{0}$ characterizes the convex cone at the origin. This allows the characterization of the solution to (4.387), and also to (4.390), to be a linear complementarity problem [58].

In effect, let us analyze the min–max problem (4.387). The solution, denoted by $(\mathbf{a}_0, \lambda_0)$, is characterized by

$$(\mathbf{K}\mathbf{a}_0 - \mathbf{f} + \mathbf{A}^T \lambda_0) \cdot (\mathbf{a} - \mathbf{a}_0) = 0 \qquad \forall \mathbf{a} \in \mathbb{R}^n, \tag{4.393}$$

$$(\mathbf{A}\mathbf{a}_0 - \mathbf{c}) \cdot (\lambda - \lambda_0) \leq 0 \qquad \forall \lambda \in \mathbb{R}^m, \ \lambda \geq \mathbf{0}. \tag{4.394}$$

Expression (4.393) corresponds to the minimum necessary condition with respect to the unconstrained variable **a**. In turn, expression (4.394) corresponds to the maximum necessary condition over the variable λ which is constrained to belonging to the convex cone given by $\lambda \geq \mathbf{0}$. From (4.393) we conclude that

$$\mathbf{K}\mathbf{a}_0 - \mathbf{f} + \mathbf{A}^T\lambda_0 = \mathbf{0}. \tag{4.395}$$

From (4.394) we have that for $\lambda = \mathbf{0}$ this is $(\mathbf{A}\mathbf{a}_0 - \mathbf{c}) \cdot \lambda_0 \geq 0$ and $\lambda = 2\lambda_0$ yields $(\mathbf{A}\mathbf{a}_0 - \mathbf{c}) \cdot \lambda_0 \leq 0$. From these two inequalities, we conclude that it is necessarily $(\mathbf{A}\mathbf{a}_0 - \mathbf{c}) \cdot \lambda_0 = 0$. Hence, $(\mathbf{A}\mathbf{a}_0 - \mathbf{c}) \cdot \lambda \leq 0$, $\forall \lambda \geq \mathbf{0}$, and we obtain

$$\mathbf{A}\mathbf{a}_0 - \mathbf{c} \leq \mathbf{0}. \tag{4.396}$$

Therefore, we see that the solution $(\mathbf{a}_0, \lambda_0)$ to problem (4.387) is the solution of the following linear complementarity problem

$$\begin{aligned}
&\mathbf{K}\mathbf{a} + \mathbf{A}^T\lambda = \mathbf{f},\\
&\mathbf{A}\mathbf{a} - \mathbf{c} + \mathbf{z} = \mathbf{0},\\
&\mathbf{z} \cdot \lambda = 0,\\
&\mathbf{z} \geq \mathbf{0},\\
&\lambda \geq \mathbf{0},
\end{aligned} \tag{4.397}$$

where the variable $\mathbf{z} \in \mathbb{R}^m$ is called the slack variable and is introduced to transform inequality $\mathbf{A}\mathbf{a} - \mathbf{c} \leq \mathbf{0}$ into an equality.

An analogous reasoning shows us that the solution λ_0 to problem (4.390) is characterized by

$$(\mathbf{P}\lambda_0 - \mathbf{e}) \cdot (\lambda - \lambda_0) \geq 0 \qquad \forall \lambda \geq \mathbf{0}, \tag{4.398}$$

from where

$$(\mathbf{P}\lambda_0 - \mathbf{e}) \cdot \lambda_0 = 0, \tag{4.399}$$

$$(\mathbf{P}\lambda_0 - \mathbf{e}) \geq \mathbf{0}, \tag{4.400}$$

and then we conclude that the solution λ_0 of problem (4.390) is also solution of the following complementarity linear problem

$$\begin{aligned}
&\mathbf{P}\lambda - \mathbf{e} - \mathbf{z} = \mathbf{0},\\
&\mathbf{z} \cdot \lambda = 0,\\
&\mathbf{z} \geq \mathbf{0},\\
&\lambda \geq \mathbf{0}.
\end{aligned} \tag{4.401}$$

In this manner, by solving (4.390) we can resort, for example, to two algorithms. The first is the Gauss–Seidel method with relaxation and projection, and the second is known as the Lemke method, which after a finite number of steps provides the solution to the associated linear complementarity problem [24].

Let us briefly describe the Gauss–Seidel algorithm.

1) Take an admissible λ^0, that is $\lambda^0 \geq \mathbf{0}$.
2) Take $w \in (0, 2)$.

3) For $k = 0, 1, 2, \ldots$, and for all $i = 1, \ldots, m$ compute

$$\lambda_i^* = \frac{e_i - \sum_{j=1}^{i-1} P_{ij}\lambda_j^{k+1} - \sum_{j=i+1}^{n} P_{ij}\lambda_j^k}{P_{ii}}, \tag{4.402}$$

$$\lambda_i^{k+1} = \Gamma((1-w)\lambda_i^k + w\lambda_i^*), \tag{4.403}$$

until

$$\frac{\|\lambda^{k+1} - \lambda^k\|}{\|\lambda^k\|} \le \epsilon, \tag{4.404}$$

where Γ is the projection operator in $[0, +\infty)$ given by

$$\Gamma(u) = \max\{0, u\}, \tag{4.405}$$

where ϵ is a sufficiently small positive number that defines the convergence tolerance, and $\| \cdot \|$ is the adopted norm in $\mathbb{R}^n$.

4.12 Complementary Reading

To increase knowledge of the topics addressed in this chapter, the interested reader can resort to the following works:

1) **Books**: N. I. Akhiezer [1], W. F. Ames [2, 3], D. G. Ashwell and R. H. Gallagher [17], J. P. Aubin [18], M. S. Bazaraa and C. M. Shetty [24], G. A. Bliss [36], O. Bolza [37], C. A. Brebbia and H. Tottenham [41], B. Carnahan, H. A. Luther and J. Wilkes [46], E. W. Cheney [48], P. G. Ciarlet [49, 50], L. Collatz [52], B. Courant [60], B. Courant and D. Hilbert [61], G. Dahlquist and A. Björck [64], P. J. Davis [65], J. W. Detteman [71], G. Duvaut and J. L. Lions [76], C. Dym and I. H. Shames [77], L. Elsgoltz [79], I. Ekeland and R. Temam [78], Fadeev Faddeeva [82], R. A. Feijóo [84], R. A. Feijóo and E. Taroco [92–95], G. Fichera [103, 104], B. A. Finlayson [105], G. E. Forsythe and W. R.Wason [108], G. E. Forsythe and C. Moler [107], M. Fremond [109], Y. C. Fung [111], I. M. Gelfand and S. V. Fomin [112], P. Germain [117, 118], R. Glowinsk J. L. Lions and R. Tremoliers [122], M. E. Gurtin [131, 132], E. Isaacson and H. B. Keller [149], C. Johnson [151], A. N. Kolmogorov and S. V. Fomin [161], E. Kreyszig [166], C. Lanczos [167], H. Leipholz [172], D. G. Luenberger [178], G. I. Marchuk [185], H. D. Meyer [192], S. G. Mikhlin [198, 199, 200, 201, 202], S. G. Mikhlin and K. L. Smolitskiy [203], M. Morse [204], A. W. Naylor and G. R. Sell [209], J. Necas [210], J. T. Oden [217, 218, 220–222], J. T. Oden and G. F. Carey [225, 226], J. T. Oden, G. F. Carey and E. B. Becker [227], J. T. Oden and J. N. Reddy [229, 230], P. M. Prenter [247], J. S. Prezemieniecki [248], K. Rektorys [250], R. D. Richtmyer and K. W. Morton [251], H. Sagan [264, 265], G. Strang and G. Fix [279], E. Taroco and R. A. Feijóo [283, 286], T. R. Tauchert [288], A. E. Taylor [289], S. Timohenko and J. N. Goodier [295], M. N. Vainberg [301], K. Washizu [303], R. Weinstock [305], O. C. Zienkiewicz and R. L. Taylor [306, 307],
2) **Scientific articles**: J. H. Argyris [6–11], J. H. Argyris and S. Kelsey [13], J. H. Argyris and D. W. Scharpf [14, 15], J. H. Argyris, I. Fried and D. W. Scharpf [12], H. J. C. Barbosa and R. A. Feijóo [22], L. Bevilacqua, R. A. Feijóo and L. Rojas [30, 31], J. H.

Bramble and S. R. Hilbert [40], P. H. Brézis [42], B. Courant [59], Z. Da Fonseca [63], G. del Piero [239], G. del Piero and P. Podio-Guidugli [240], De Veubeke [66–69], R. A. Feijóo [83, 85], R. A. Feijóo and H. J. C. Barbosa [86], R. A. Feijóo and L. Bevilacqua [87], R. A. Feijóo, L. Rojas and L. Bevilacqua [88, 89], R. A. Feijóo, L. Rojas, E. Taroco and L. Bevilacqua [90], R. A. Feijóo and E. Taroco [91], M. Geradin [113], P. Germain [114–116], C. Gonzales Guirnaldes [124], J. T. Oden [214–216, 219], J. T. Oden and G. Aguirre-Ramirez [223], J. T. Oden and W. H. Armstrong [224], J. T. Oden and B. E. Kelley [228], J. T. Oden and D. Somogyi [231], G. Romano [253–256], G. Romano and G. Alfano [257], G. Romano and M. Romano [258], G. Romano, L. Rosati and F. Marotti de Sciarra [259–261], G. Romano, L. Rosati, F. Marotti de Sciarra and P. Bisegna [262], E. Taroco and R. A. Feijóo [282, 284], E. Taroco, R. A. Feijóo and L. C. Martins [287], J. M. Thomas [293], O. C. Zienkiewicz and J. Z. Zhu [308], N. Zouain, R. A. Feijóo and E. Taroco [312].

5

Materials Exhibiting Creep

5.1 Introduction

The existence of components in engines and mechanisms subjected to high temperatures is commonplace. Therefore, understanding the mechanical properties of constitutive materials is fundamental in order to predict the behavior of the component during its useful life.

When materials are subjected to high temperatures, the strains that take place over long periods of time, phenomenon known as creep, become relevant. Hence, it is necessary to determine the mechanical models which are capable of representing the behavior of such materials to estimate the strains and stresses developed in the components.

5.2 Phenomenological Aspects of Creep in Metals

Although L. T. Vicat carried out systematic observations of slow strains in metallic bars in the 19th century [302], in general it is widely acknowledged that the phenomenological theory of fluency (also known as creep), started at the beginning of the 20th century with the contributions of the English physicist E. N. Andrade [4, 5], who introduced the terminology used these days in the description of the different phases of the creep phenomenon.

A typical creep experiment aimed at establishing a relation between the uniaxial deformation ε and the stress σ, through time t, at a given temperature consists of placing the body to be tested in an oven at uniform and controlled temperature and subjecting it to a tensile test. Note that, for a constant tensile load, when the body increases its length the transversal area is diminished, the stress has the tendency to increase along time. This aspect was carefully considered even in the early stages when performing experimentation in the creep domain, and ingenious mechanisms were elaborated to guarantee the constancy of the tensile load [4, 5]. The results of such experiments are put in the form of ε (accumulated strain) versus t (time) plots, as shown in Figure 5.1. The instantaneous strain that occurs when the load is applied is designated ε_0. This instantaneous strain includes elastic and inelastic strains (e.g. plastic strains), which do not depend on time, but on the level of load applied, that is $\varepsilon_0 = \varepsilon_0(P)$, where P is the applied load.

From the observation of the creep curves, we can distinguish, more obviously for the load P_2, three regions in which the material exhibits different behavior. At the beginning, the segment of the curve marked with I is characterized by a decreasing strain rate $\dot{\varepsilon}$. This

Introduction to the Variational Formulation in Mechanics: Fundamentals and Applications, First Edition.
Edgardo O. Taroco, Pablo J. Blanco and Raúl A. Feijóo.

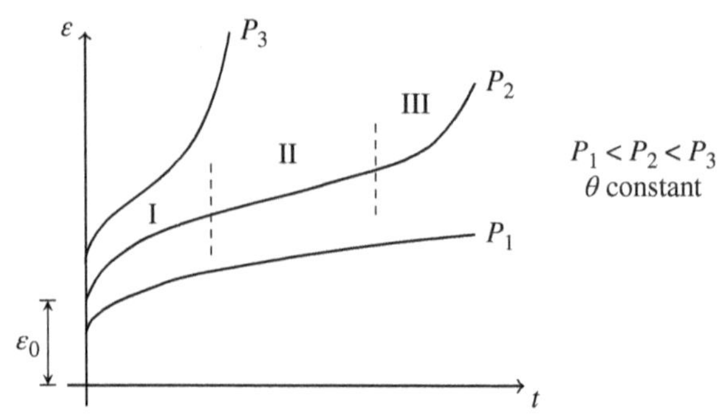

Figure 5.1 Characteristic behavior of a material experiencing creep when a constant load P is applied.

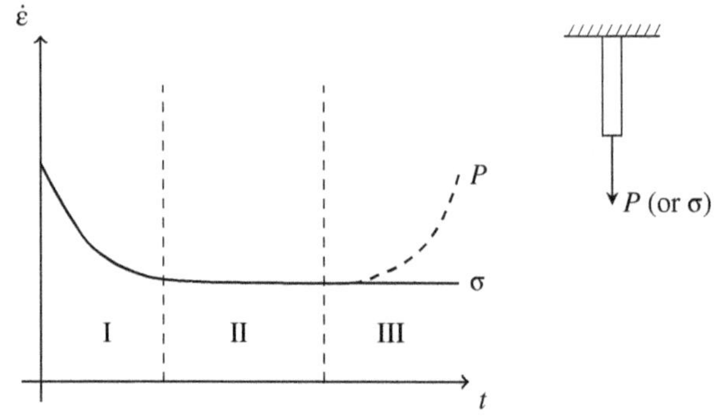

Figure 5.2 Creep strain rate when applying constant load P or constant stress σ.

stage is known as primary creep. In the region denoted by II, the strain rate is approximately constant, being the minimum value achieved during primary creep. This stage is called secondary creep or steady creep. Finally, region III is characterized by an increasing strain rate, leading to material failure. This third region is known as tertiary creep.

It can be appreciated that during the primary and secondary creep stages the stress remains almost constant. This does not happen in tertiary creep, where the changes in the cross-sectional area of the component become noticeable. Moreover, the nucleation of internal voids in the bulk of the component substantially modifies the stress distribution along the transversal area. Because of this intrinsic complexity in the material behavior, great care is required when modeling this tertiary region.

Figure 5.2 displays the strain rate $\dot{\varepsilon}$ versus time t, and the three regions can be more clearly distinguished.

The identification of the regions in the creep curves is made using conventional considerations. For example, the linear part corresponding to the steady regime (secondary creep) is not always present. For low levels of applied load, the creep strain rate reaches almost no minimum point during the whole experiment (see Figure 5.1, curve for load P_1). In contrast, for higher levels of applied load, the steady part is reduced to a single inflection point.

The information that can be extracted from this kind of creep experiment is usually represented in different ways according to the specific interest in the relations among the variables which play a role in the problem. Hence, to highlight the influence of time in the creep strain, it is convenient to turn the creep curves, as represented in Figure 5.1, into a

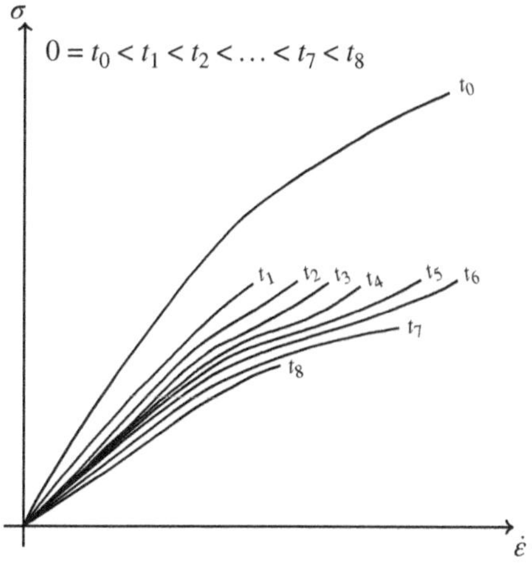

Figure 5.3 Qualitative isochronous creep curves for a chromium-nickel-molybdenum steel.

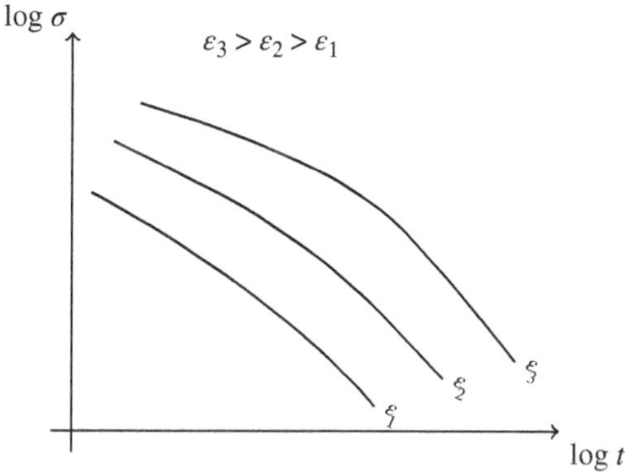

Figure 5.4 Same strain creep curves in a stress–time log–log plot.

new system of coordinates $\dot{\varepsilon}$ versus σ, with the time t being a constant parameter for each curve. This kind of curve is called an isochronous creep curve. Figure 5.3 qualitatively displays a series of isochronous curves, obtained from experiments lasting 100 000 hours for a chromium-nickel-molybdenum steel by E. L. Robinson (see [252]).

Another kind of representation is given by stress σ versus time t curves on a log–log scale. These are known as curves of equal strain, as seen in Figure 5.4.

A useful characterization is also provided by the curve representing the behavior of the minimum creep strain rate, denoted by $\dot{\varepsilon}_{\text{min}}$, as a function of the stress σ. This curve is usually reported in log scale, as illustrated in Figure 5.5. Clearly, for low levels of stress this relation is almost linear, while it becomes a function of a power of σ when the stress is high.

Another creep curve relates the level of stress the component is subjected to and the time elapsed until rupture of the component occurs. This curve is also plotted in log scale, as shown in Figure 5.6. This kind of creep curve demonstrates a transition in the material behavior according to the level of stress. The first part of the curve is usually associated with ductile fracture, while the second part is associated with brittle fracture.

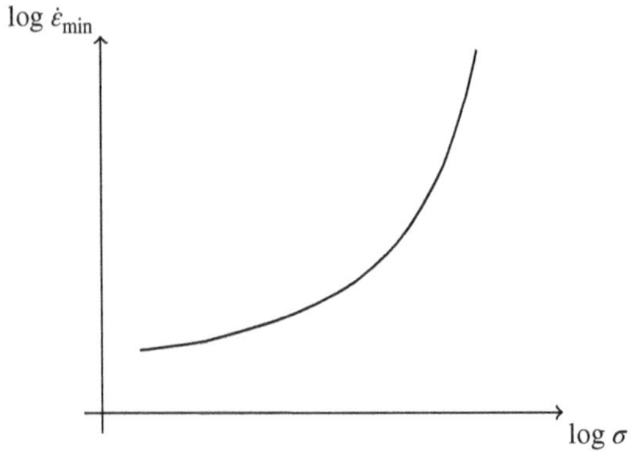

Figure 5.5 Minimum creep strain rate $\dot{\varepsilon}_{\min}$ as a function of the stress level σ.

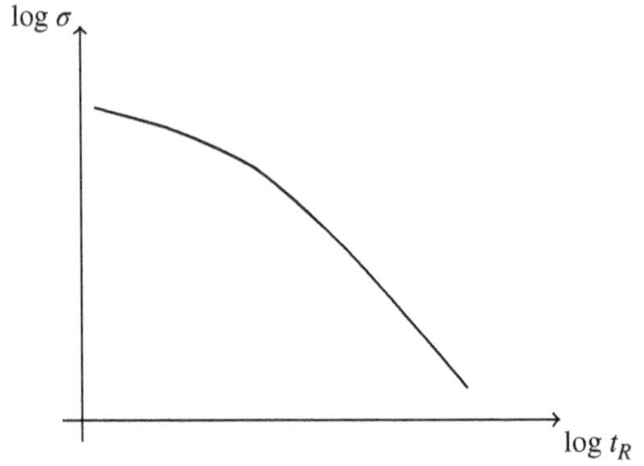

Figure 5.6 Typical creep curve depicting the time elapsed until rupture.

5.3 Influence of Temperature

In the previous section the creep phenomenon was described assuming the temperature in the component was maintained constant during the test. As a general rule, we can say that the higher temperature, the higher the creep strain rate.

Nowadays, it is widely known that creep phenomena are governed by different thermally activated mechanisms. Hence, the different creep theories assume that inelastic strains are associated with the motion of certain structural elements at the crystal scale induced by thermal action.

For a given time instant t, each of these elements owns a certain energy, say u, associated with the mechanism under consideration. When this energy u reaches a critical value u_o, called the activation energy, in a structural element, the element is displaced, inducing a strain.

Assuming that the energy distribution among the elements follows the Maxwell law, it is possible to show that the number of potentially activated elements at that time is given by

$$\exp\left(-\frac{u_o}{R\theta}\right), \tag{5.1}$$

where R is the gas constant and θ is the absolute temperature.

Focusing on the relation between element activation and the appearance of an associated strain, it is natural to accept that the corresponding creep strain rate related to

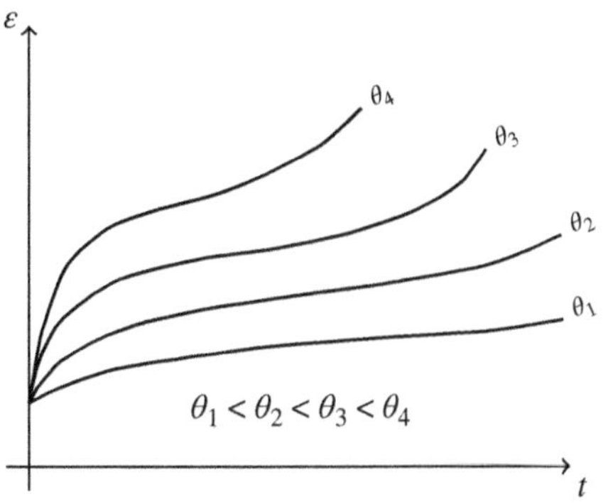

Figure 5.7 Influence of temperature θ on the total strain ε.

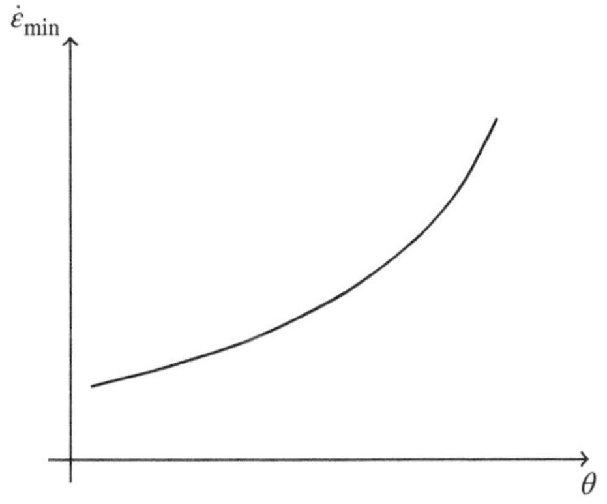

Figure 5.8 Influence of temperature θ on the minimum strain rate $\dot{\varepsilon}_{min}$.

a given mechanism is proportional to the number of activated elements. Following this reasoning, we reach an exponential function, known as the Arrhenius law. Figures 5.7 and 5.8 illustrate the influence of the temperature in the total strain ε and in the minimum strain rate $\dot{\varepsilon}_{min}$.

Several authors (see the works by M. F. Ashby, [16], J. E. Dorn, [74, 75], J. J. Frost et al. [110], A. K. Mukherjee et al. [205], and H. Oikawa [234], among others) have investigated a series of mechanisms which are able to drive material creep. Of these mechanisms, the most relevant are dislocation creep and diffusion creep. In the former case, the defects in the crystal structure (or dislocations) can move by virtue of thermal activation, as well as by the existence of high stresses. At low temperatures and/or stresses this motion diminishes. Nevertheless, creep phenomena may continue to unfold due to the influence of the latter mechanism, that is, diffusion creep. While dislocation creep is highly nonlinear dependent on the level of stress in the material, diffusion creep is approximately linear with respect to the stress.

Figure 5.9 represents a map characterizing deformation mechanisms, which is qualitatively similar to that reported in [110] for iron. The mechanisms which drive creep phenomena represented in Figure 5.9 are

- dislocation glide (plastic deformation)
- dislocation creep
- Coble diffusion creep
- Nabarro–Herring diffusion creep.

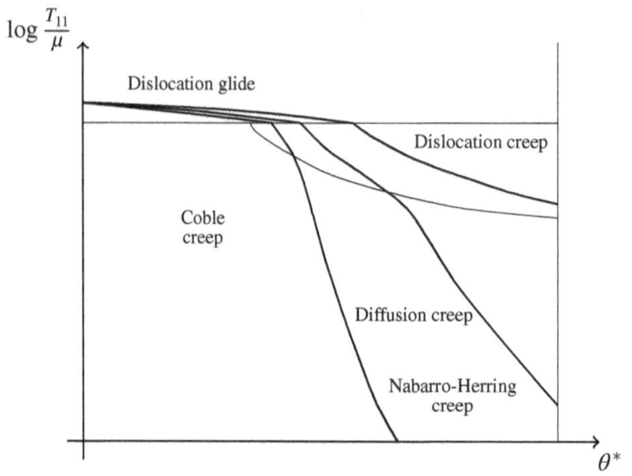

Figure 5.9 Deformation mechanisms map qualitatively similar to that reported in [110] for an AISI316 steel. Stress T_{11} is normalized by the shear modulus μ, which is then plotted against a normalized temperature θ^*.

The constitutive equations for such mechanisms are summarized in Table 5.1, where the so-called self-diffusion coefficient D_v and the grain-boundary diffusion coefficient D_{gb} are governed by an expression including the factor

$$\exp\left(-\frac{\cdot}{R\theta}\right). \tag{5.2}$$

This implies that the influence of the temperature in the material behavior is explicitly introduced through a factor that is equivalent to that presented in (5.1).

When the qualitative behavior exemplified in Figure 5.9 is quantitatively formalized through experiments of specific materials, in creep experiments that last thousands of hours (in the scale of years), the dominant mechanism is that of dislocation creep, whose model, as reported in Table 5.1, is governed by the Norton law [53, 54, 212]. It is interesting to point out, however, that in industrial applications, where the component useful life reaches hundreds of thousands of hours, the dominant creep phenomenon is either that of Coble diffusion creep or Nabarro–Herring diffusion creep. The constitutive equations for these phenomena presented in Table 5.1 suggest a viscous material behavior of Newtonian type, quite different to that proposed by the Norton law.

5.4 Recovery, Relaxation, Cyclic Loading, and Fatigue

Previous sections were devoted to the discussion of some phenomenological aspects of creep phenomena for the particular case of applied load (or stress) under isothermal conditions. In the literature, these characteristics have been reported through simple creep testing. In general, experience tells us that the real-life conditions to which mechanical components are subjected differ markedly, both in the level of stress as well as in the exposure to temperature variations, from those idealizations made in controlled laboratory conditions. For example, finding a mechanical component subjected

Table 5.1 Constitutive equations from [110] for the deformation mechanisms map in Figure 5.9.

Dislocation glide	
$D_{11} = A\exp\left[-ba\frac{\left(\frac{\mu b}{c} - T_{11}\right)}{R\theta}\right]$	if $\frac{T_{11}}{\mu} \geq \frac{T^0_{11}}{\mu}$
$D_{11} = 0$	if $\frac{T_{11}}{\mu} < \frac{T^0_{11}}{\mu}$
Dislocation creep	
$D_{11} = A\mu b\frac{D_v}{R\theta}\left(\frac{T_{11}}{\mu}\right)^n$	
Diffusion creep	
$D_{11} = A\mu b\frac{D_v}{R\theta}\left(\frac{b}{d}\right)^2\frac{T_{11}}{\mu}$	Nabarro–Herring
$D_{11} = A\mu b\frac{D_{gb}}{R\theta}\left(\frac{b}{d}\right)^3\frac{T_{11}}{\mu}$	Coble

where
a: activation area
b: Burger vector
c: obstacle spacing
d: grain size
R: Boltzmann constant
θ: temperature
μ: shear modulus
D_{11}: creep strain rate
T_{11}: stress
T^0_{11}: yield stress
D_v: self-diffusion coefficient
D_{gb}: grain-boundary diffusion coefficient
A: constant

to time-dependent loads as well as thermal processes is regular practice. Examples of this are

- pressure vessels, for which the load (pressure) is kept constant, while the fluid temperature in the vessel changes according to a given process
- gas turbine disks, for which centrifugal forces and temperature vary according to the power expended by the turbine, which is a function of time.

These scenarios pose the problem of understanding creep mechanisms and material models in more general conditions.

Thus, if in a simple creep experiment we fully unload the material specimen, we note an instantaneous (elastic) response followed by a response deferred in time, resulting, at the end of the experiment, in a permanent strain, as seen in Figure 5.10.

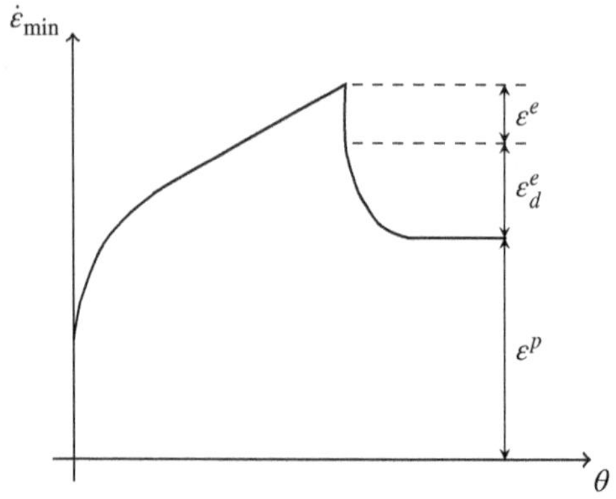

Figure 5.10 Strain recovery from a creep experiment. ε^e, elastic recovery; ε^e_d, creep strain recovery; ε^p, permanent strain.

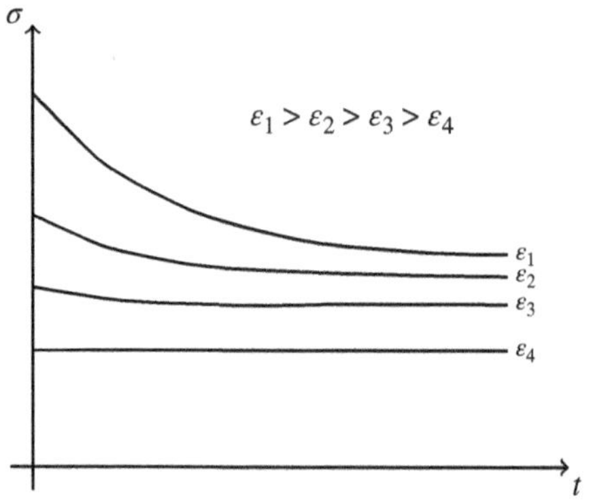

Figure 5.11 Stress relaxation phenomenon at constant temperature.

Another interesting phenomenon is the stress-relaxation. This phenomenon occurs when the specimen is subjected to a creep test, and from a time instant onwards we fix the loading conditions in order to maintain a constant strain. From that instant, it is observed that the stress magnitude falls off in time, giving rise to what has been called stress-relaxation. Figure 5.11 qualitatively illustrates this observation for different levels of initial stress. Notice that there is a threshold below which no relaxation is appreciated whatsoever.

A further fundamental aspect worth mentioning is related to the material response when cyclic loads are applied to the component. Such a response, as expected, depends upon the temperature. Figure 5.12 presents the amplitude of the strain as a function of the number of cycles until the rupture point is reached. This rupture is called fatigue rupture. Many metals feature a level of strain amplitude under which the material can bear an "infinite" number of cycles without reaching the rupture point.

Nonetheless, if the cyclic load occurs at high temperature, a strong interaction between mechanisms related to material degradation caused by the very cyclic nature of the load, and the creep and relaxation mechanisms is observed.

Overall, in an experiment in which the strain rate $\dot{\varepsilon}$ is controlled in such a way that the specimen is subjected to continuous strain cycles, as seen, for example, in Figure 5.13(a), the fatigue curves take the form indicated in Figure 5.13(c), and the stress response consists of a curve similar to that in Figure 5.13(b). By analyzing this material response, we can appreciate that the smaller the strain rate, the smaller the number of cycles required

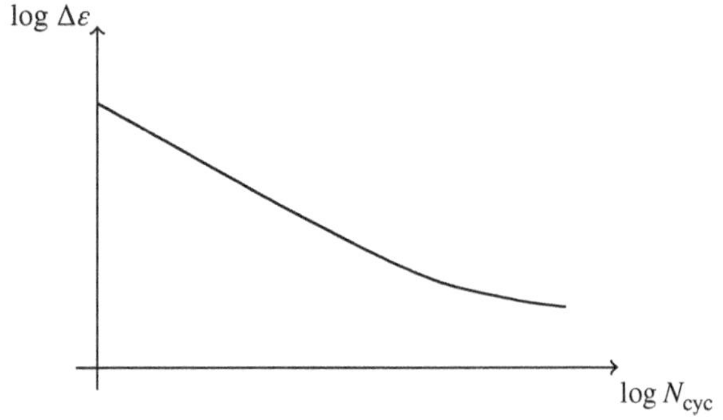

Figure 5.12 Creep rupture curve, caused by fatigue, displaying the strain amplitude $\Delta\varepsilon$ versus the number of cycles N_{cyc} until rupture, at constant temperature.

to reach the rupture point. The explanation for this result arrives from the interaction between mechanisms, namely, fatigue caused by the cyclic load and creep degradation. In fact, the smaller the strain rate, the more the time during which spikes of stress are affecting the specimen, which leads to an increase in damage by creep.

If the experimental setup is such that relaxation phenomena are allowed, it is observed that the larger the relaxation time, the smaller the number of cycles to reach the rupture, as shown in Figure 5.14. Moreover, a saturation phenomenon with respect to the time interval in which the relaxation takes place is also observed.

When the experiment is executed such that the stress is maintained constant along a period of time, say t_h, the behavior is similar to that shown in Figure 5.13. This can be observed in Figure 5.15. As this period of time t_h is increased, the number of cycles to reach the rupture is reduced, but the saturation observed in Figure 5.14 is not found.

It is also important to underline that, in the previous analyses, we limit ourselves to studying the material response when the component is subjected to extremely simple load-temperature protocols. In real-life conditions, such as those existing in the chemical, naval, mechanical, and nuclear industries, among others, the load-temperature conditions are much more complicated than the idealizations just analyzed. In addition, as already pointed out, creep can be extended for long periods of time, for which extrapolation techniques are employed, but with scarce experimental validation. The discussion above should serve to illustrate the complexity underlying the modeling of creep phenomena.

5.5 Uniaxial Constitutive Equations

For the mechanical analysis of structural components exhibiting creep along their useful life it will be necessary to build constitutive equations able to model such phenomena.

For moderate stress levels and loads acting for long periods of time, steady-state creep phenomena are physically predominant over phenomenology related to unsteadiness. This has motivated the development of most constitutive equations, which in general take the form

$$\dot{\varepsilon}_c = f_1(\sigma), \tag{5.3}$$

that is, the creep strain rate at constant temperature in the secondary creep phase depends on the stress.

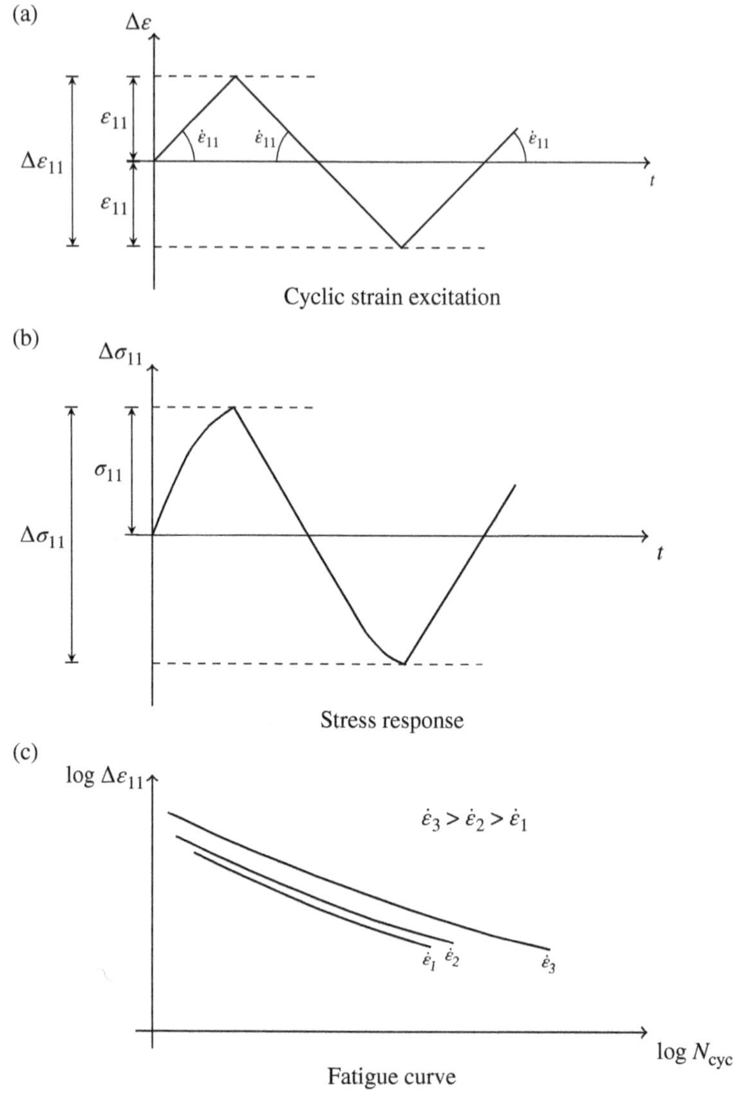

Figure 5.13 Creep rupture curve caused by cyclic strain.

One of the first authors to propose a constitutive expression for the steady-state component was Norton in [212] who, like Bailey in [20, 21], established the following power law

$$\dot{\varepsilon}_c = A\sigma^n, \tag{5.4}$$

where A and n are material parameters. For example, for some steels n can be in the range [3, 8], which gives an idea of the strong nonlinearity of equations of the kind of (5.4).

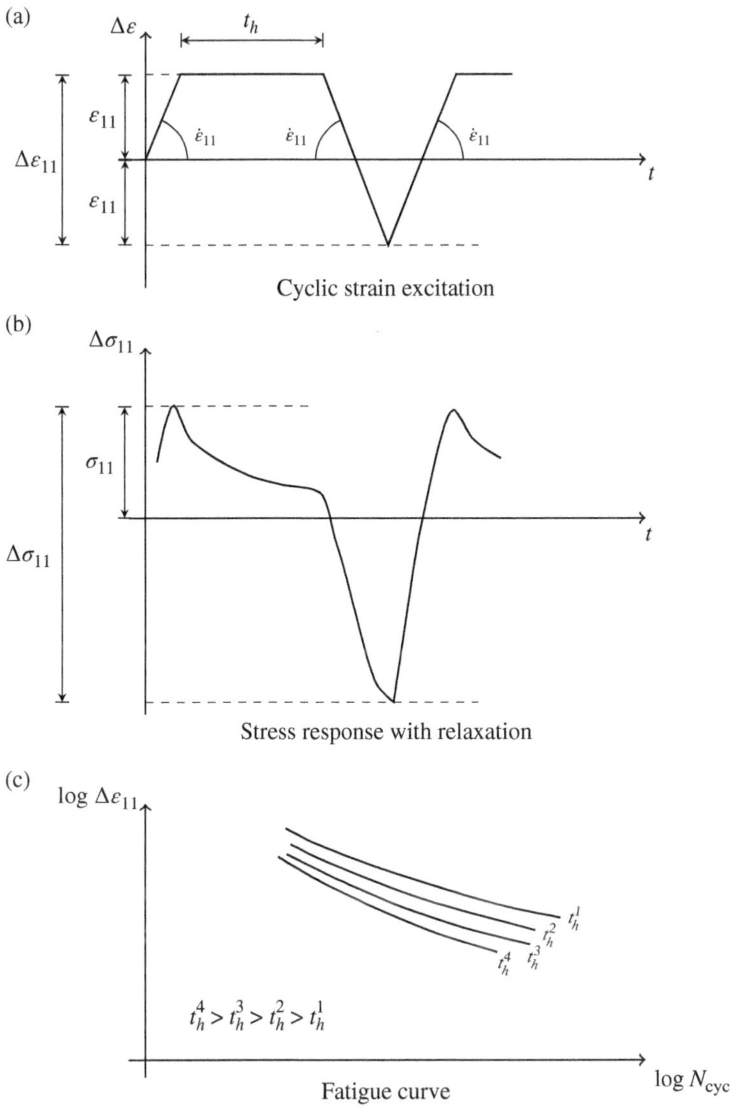

Figure 5.14 Creep rupture curve caused by cyclic strain with stress relaxation.

Before going ahead, it is important to highlight that, for low levels of stress, the predominant creep mechanism is that of diffusion creep (see Figure 5.9). This mechanism is characterized by an expression of the form

$$\dot{\varepsilon}_c = A\sigma. \tag{5.5}$$

In turn, for high stress levels, the dominant mechanism is that of dislocation creep, characterized by expressions of the form of (5.4) with $n > 1$.

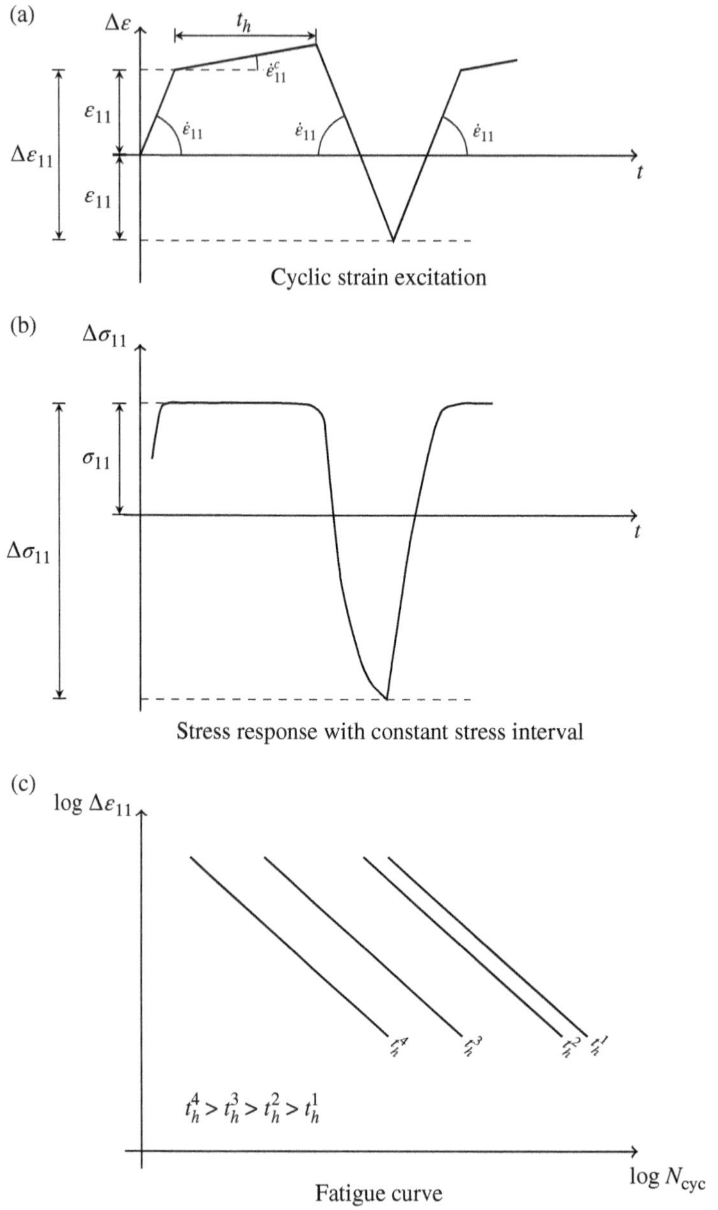

Figure 5.15 Creep rupture curve caused by cyclic strain with constant stress period.

In particular, material constants A and n are determined through the curves $\log \dot{\varepsilon}_{min}$ versus $\log \sigma$ (see Figure 5.5) obtained in creep experiments. Then, this gives

$$\log \dot{\varepsilon}_c = \log A + n \log \sigma. \tag{5.6}$$

In this manner, the inclination of the straight line obtained in the experiments yields n, and the intersection of this line with the ordinate axis yields A.

In general, it is convenient to work with dimensionless constitutive equations. Thus, proper reference values $\dot{\varepsilon}_c^*$ and σ^* are chosen to characterize the creep strain rate and the stress, respectively. Hence, we have the dimensionless equation

$$\frac{\dot{\varepsilon}_c}{\dot{\varepsilon}_c^*} = \left(\frac{\sigma}{\sigma^*}\right)^n. \tag{5.7}$$

Trivially, for $\sigma = \sigma^*$ this is $\dot{\varepsilon}_c = \dot{\varepsilon}_c^*$, and so σ^* is called reference stress, which corresponds to the creep strain rate $\dot{\varepsilon}_c^*$, which usually is taken to be $\dot{\varepsilon}_c^* = 10^{-9} seg^{-1}$ [147].

However, for even higher stress levels it has been shown that the strain rate is actually larger than the one predicted by the power law. Motivated by this, exponential expressions of the following form have been proposed [232, 233]

$$\frac{\dot{\varepsilon}_c}{\dot{\varepsilon}_c^*} = \exp\left(\frac{\sigma}{\sigma^*}\right). \tag{5.8}$$

One of the disadvantages of this expression is that the strain rate is not zeroed for $\sigma = 0$. A corrected expression was then proposed by Nadai in [206], which takes the form

$$\frac{\dot{\varepsilon}_c}{\dot{\varepsilon}_c^*} = \sinh\left(\frac{\sigma}{\sigma^*}\right). \tag{5.9}$$

For values $\sigma \gg \sigma^*$, the exponential law practically coincides with the hyperbolic sine law, while for low stress levels the hyperbolic sine can be approximated by

$$\dot{\varepsilon}_c = \frac{\dot{\varepsilon}_c^*}{\sigma^*}\sigma, \tag{5.10}$$

which implies that a linear relation with σ is obtained, in correspondence with the diffusion creep phenomenon.

Through an exhaustive program of creep experiments, in [191] it was concluded that the hyperbolic sine law better predicts the creep phenomenon than the Norton law. Nevertheless, the Norton power law is more attractive from the computational point of view. What is more, for $n = 1$ the Norton law reduces to the elasticity equation, while for $n \to \infty$ the Norton equation tends to that of an ideally plastic material.

Alternative expressions have been proposed to correct the deficiency of the exponential law pointed out above. For instance, in [237] the following law was postulated

$$\frac{\dot{\varepsilon}_c}{\dot{\varepsilon}_c^*} = \exp\left(\frac{\sigma}{\sigma^*}\right) - 1. \tag{5.11}$$

All the expressions presented above have a range of applicability limited to the context of steady-state creep. Innumerous contributions have made attempts to model the primary and secondary creep stages through a single unified law.

Thus, in [4, 5] the creep experiments at constant stress yielded the so-called Andrade expression

$$\ell = \ell_0(1 + \beta t^{1/3})e^{kt}, \tag{5.12}$$

establishing a relation between the length ℓ of the specimen for each time instant t and its initial length ℓ_0. Since these works were in the context of finite strains and constant stress levels, we can write

$$\dot{\varepsilon}_c = \frac{d\ell/dt}{\ell} = \frac{\beta}{3}\frac{1}{(t^{2/3} + \beta t)} + k. \tag{5.13}$$

Clearly, as t grows, the creep strain rate tends to a constant value k (k-creep in the terminology of [4, 5]). In addition, and assuming β is small enough, we can write

$$\varepsilon_c = \int_{\ell_0}^{\ell} \frac{d\ell}{\ell} = kt + \beta t^{1/3}. \tag{5.14}$$

From this identity we notice that the creep strain given by the Andrade formula consists of the superposition of two creep modes, namely β-creep, occurring at a decreasing rate, and k-creep, taking place at constant rate.

In general, it is possible to approximate the creep strain through an expression of the kind

$$\varepsilon_c = g_1(\sigma)t^m + g_2(\sigma)t, \tag{5.15}$$

where m is approximately 1/3. If the analysis focuses on creep phenomena happening during short periods of time, it will be sufficient to work with the first term from the right-hand side in (5.15).

A similar expression to (5.15) was proposed in [186], in which the total creep strain is the sum of a part which linearly grows with t and another part which exponentially decreases with t

$$\varepsilon_c = k(1 - e^{-\alpha t}) + \beta t. \tag{5.16}$$

Moreover, if we want to include the effect of temperature we can resort to the Arrhenius law, as seen in (5.1).

As can be appreciated, most of the contributions addressing creep phenomena at constant stress and temperature fall into a constitutive equation of the form

$$\varepsilon_c = f(\sigma, t, \theta), \tag{5.17}$$

which in general is postulated to be separable with respect to the variables, that is,

$$\varepsilon_c = f_1(\sigma) f_2(t) f_3(\theta). \tag{5.18}$$

Next, we present some examples of the different functional components in (5.18).

- Stress function f_1 (A, β and n are material constants)
  - $f_1(\sigma) = A\sigma^n$ (Norton–Bailey)
  - $f_1(\sigma) = A \sinh(\beta\sigma)$ (Prandtl–Nadai)
  - $f_1(\sigma) = A \exp(\beta\sigma)$ (Dorn–Ludwik)
  - $f_1(\sigma) = A(\sinh(\beta\sigma))^n$ (Garófalo)
  - $f_1(\sigma) = A(\exp(\beta\sigma) - 1)$ (Soderberg)
- Time function f_2 (β, k, α, b and m are material constants)
  - $f_2(t) = t$ (secondary creep)
  - $f_2(t) = bt^m$ (Bailey)
  - $f_2(t) = (1 + bt^{1/3})e^{kt}$ (Andrade)
  - $f_2(t) = k(1 - e^{-\alpha t}) + \beta t$ (Pao–Marin)
- Temperature function f_3
  - $f_3(\theta) = A \ \exp\left(-\frac{u_0}{R\theta}\right)$

The discussion developed so far has addressed the problem of creep under constancy of stress and temperature. For the analysis of real-life problems, the framework presented above must be extended to embrace variable stress and temperature conditions. Two approaches can be considered to effectively account for this more general setting.

1) In the first approach it is postulated that the material response depends solely on the current state of the body.
2) In the second approach, the material response is assumed to explicitly depend on the history of the whole mechanical regime, which means dependence on the entire loading process, including history of strains and temperatures the body has experienced up to the current time instant.

While the first approach leads us to state-like constitutive equations, the second approach results in the so-called materials with memory.

On the one hand, even if experimental reality points toward the direction of materials with memory, it is also true that little experimental information is known. This affects the practical applicability of constitutive equations of this kind. On the other hand, state equations have been extensively applied, and experimental data is easily available. At the same time, such equations can easily be manipulated from the analytical and numerical points of view. These advantages allow state equations to be verified (and trusted) and easily introduced in existing computational programs.

In order to construct these state equations, let us first rewrite the constitutive equation $\varepsilon_c = f_1(\sigma) f_2(t) f_3(\theta)$ in terms of $\dot{\varepsilon}_c$. Thus

$$\dot{\varepsilon}_c = f_1(\sigma)\,\frac{df_2(t)}{dt}\,f_3(\theta), \tag{5.19}$$

where let us recall that we are assuming σ and θ remain constant.

The so-called time hardening theory consists of accepting that, even for variable stress and temperature, the creep strain rate is given by an expression similar to

$$\dot{\varepsilon}_c = g_1(\sigma) g_2(t) g_3(\theta). \tag{5.20}$$

The material response that results from this theory, for a loading protocol as illustrated in Figure 5.16 (bottom panel), is represented in the same figure (top panel). As can be seen, for the stress level σ_1 the response follows the corresponding path. When, at time instant t_1 the stress jumps from σ_1 to σ_2, the strain rate corresponds to that of a material which at time t_1 is subjected to σ_2 (dashed curve). For this reason, this is called the time hardening phenomenon, since the material response explicitly depends on the elapsed time.

Going back to the expression (5.19) of $\dot{\varepsilon}_c$ deduced for constant stress and temperature, and recalling (5.18), it is possible, by using these two functional forms, to remove the time variable t, leading to

$$\dot{\varepsilon}_c = h_1(\sigma) h_2(\varepsilon_c) h_3(\theta). \tag{5.21}$$

In this manner, we have reached an expression which relates the strain rate $\dot{\varepsilon}_c$ with the current stress state, the accumulated creep strain (which somehow characterizes the history of the entire process), and the temperature.

Again, the strain hardening theory departs from the hypothesis that expression (5.21) holds even in the case of constant stress and temperature. Figure 5.16 also illustrates the case of variable stress assuming the strain hardening theory is valid.

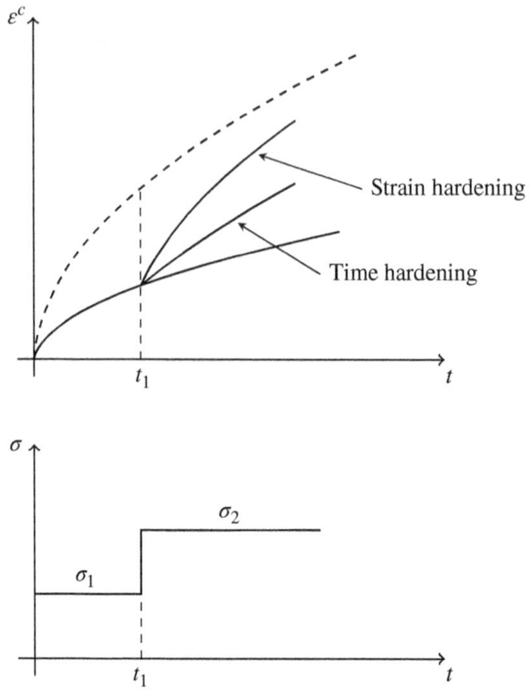

Figure 5.16 Strain and time hardening.

Given the premises behind expressions (5.19) and (5.20), we conclude that both coincide if the stress is maintained constant. The question is to what extent are these theories truly predictive in the case of experiments with variable load programs. Experiments show that such phenomenology is better represented by the strain hardening theory when compared to the time hardening theory. It is worth mentioning that the latter theory has been widely used because of its convenience from the computational perspective.

In effect, admitting that in this theory the following is verified

$$\dot{\varepsilon}_c = \Phi(\sigma, \theta)\,\frac{d\psi(t)}{dt}, \tag{5.22}$$

and assuming ψ to be monotonically increasing in $\mathbb{R}^+$ (ψ is monotonically increasing if for $t_2 > t_1$, gives $\psi(t_2) > \psi(t_1)$), so we can introduce a new variable $\zeta = \psi(t)$, and therefore $d\zeta = \frac{d\psi(t)}{dt}dt$, which, replaced in (5.22), renders

$$\dot{\varepsilon}_c = \Phi(\sigma, \theta)\,\frac{d\zeta}{dt}. \tag{5.23}$$

From this, we can then write

$$\frac{d\varepsilon_c}{d\zeta} = \Phi(\sigma, \theta). \tag{5.24}$$

The left-hand side in the above identity is the creep strain rate with respect to the new variable ζ, and the right-hand side is, as we have seen, a constitutive equation responsible for representing secondary creep. Using this strategy, the steady-state creep analysis can, through such a simple change of scale, be framed as a time hardening case.

Importantly, strain and time hardening theories are not good candidates to account for alternating loading-unloading processes, in which recovery and relaxation phenomena need to be taken into consideration. As a consequence, we need more comprehensive constitutive equations that enable modeling in such mechanical regimes.

Several possibilities have been proposed (see, for example, [38, 39]) by extending the ideas underlying the previously discussed constitutive equations. Basically, these extended equations introduce additional variables, called internal variables, which somehow characterize the dissipative mechanisms responsible for the creep (see also [119]).

Following [39], let us develop a creep constitutive model with two internal variables α and R (more general inelastic processes than creep can be included in this kind of model). Then, we have a state equation

$$\dot{\varepsilon}_c = f(\sigma, \alpha, R), \tag{5.25}$$

and equations (also referred to as state equations) which govern the evolution of these internal variables

$$\dot{\alpha} = g_1(\sigma, \alpha, R)\dot{\varepsilon}_c - g_2(\sigma, \alpha, R)\alpha, \tag{5.26}$$

$$\dot{R} = h_1(\sigma, \alpha, R)|\dot{\varepsilon}_c| - h_2(\sigma, \alpha, R)(R - R_0). \tag{5.27}$$

In the equations above, the existence of two mechanisms is considered: a hardening mechanism (first term) and a softening one (second term).

Equations (5.26) and (5.27) must be complemented with proper initial conditions, for example

$$\alpha = 0, \quad R = R_0, \quad \varepsilon_c = 0 \quad \text{at} \quad t = 0. \tag{5.28}$$

Functions g_1, g_2, h_1, and h_2 take different forms for each theory [39]. Let us see two examples.

The Boiley–Orowan constitutive equation assumes $\alpha = 0$ and also

$$\dot{\varepsilon}_c = \text{sgn}(\sigma)f(|\sigma|, R), \tag{5.29}$$

$$\dot{R} = h_1(R)|\dot{\varepsilon}_c| - h_2(R), \tag{5.30}$$

where

$$f \geq 0 \qquad \text{if } |\sigma| \geq R, \tag{5.31}$$

$$f = 0 \qquad \text{if } |\sigma| < R, \tag{5.32}$$

$$h_2(R) = k_1|R|^{n-m}, \tag{5.33}$$

$$h_1(R) = \frac{1}{k_2}|R|^{-m}. \tag{5.34}$$

This functional form enables us to model Norton-like secondary creep phenomena. In fact, making $\dot{R} = 0$ we have

$$|\dot{\varepsilon}_c| = \frac{h_2(R)}{h_1(R)} = f(|\sigma|, R) = k_1 k_2 |R|^n, \tag{5.35}$$

which is exactly a Norton law.

In this model, parameters k_1, k_2, n, and m are determined through experiments.

The Robinson constitutive equation assumes $R = R_0$ for all time instants, yielding

$$\dot{\varepsilon}_c = f(\sigma, \alpha) = f(\sigma - \alpha), \tag{5.36}$$

$$\dot{\alpha} = g_1(\alpha)\dot{\varepsilon}_c - g_2(\alpha)\alpha, \tag{5.37}$$

where

$$f = \begin{cases} \mu F^{(n-1)/2}\,(\sigma - \alpha), & \text{for } F > 0 \text{ and } \sigma(\sigma - \alpha) > 0, \\ 0, & \text{otherwise,} \end{cases} \tag{5.38}$$

with

$$F = \frac{(\sigma - \alpha)^2}{R_0^2} - 1, \tag{5.39}$$

and

$$g_1(\alpha) = \begin{cases} c_1(\frac{\alpha}{R_0})^q & \text{for } \alpha > \alpha_0 \text{ and } \sigma\alpha \leq 0, \\ c_1(\frac{\alpha_0}{R_0})^q & \text{otherwise,} \end{cases} \tag{5.40}$$

$$g_2(\alpha) = \begin{cases} c_2(\frac{\alpha}{R_0})^{m-q-1} & \text{for } \alpha > \alpha_0 \text{ and } \sigma\alpha \leq 0, \\ c_2(\frac{\alpha_0}{R_0})^{m-q-1} & \text{otherwise,} \end{cases} \tag{5.41}$$

where, now, we have eight material constants to be determined: μ, n, R_0, α_0, c_1, c_2, m, and q.

It is evident that as long as the complexity of the constitutive equation is increased, the number of parameters to be defined grows. Experiments to accomplish such a task are, in turn, far more complicated. This, added to variable temperature experiments, makes the definition of a material response during creep an arduous task.

Finally, it is important to highlight that no models for tertiary creep and the subsequent rupture have been discussed, even though this problem is of high practical relevance. However, the mechanical formulations for unsteady creep to be written in the variational formalism along this chapter can straightforwardly be extended to incorporate the tertiary creep phase.

5.6 Three-Dimensional Constitutive Equations

This section addresses the generalization of the uniaxial constitutive equations for the case of three-dimensional stress states. As for theories of plasticity (see Chapter 6), the creep theory for arbitrarily complex multiaxial stress states is developed supported by a set of considerations with strong experimental foundations. Let us present these considerations next.

In the first place, any constitutive equation for multiaxial stress states must be capable of being reduced to a uniaxial counterpart.

A multiaxial constitutive equation has to take into account the fact that primary and secondary creep develop at constant volume. For three-dimensional continua, this entails that the creep strain rate must satisfy

$$\text{tr}\mathbf{D} = 0, \tag{5.42}$$

or the equivalent condition

$$\text{div}\,\mathbf{v} = 0, \tag{5.43}$$

where $\mathbf{D}$ is the strain rate tensor and $\mathbf{v}$ is the velocity field, which are related through

$$\mathbf{D} = \frac{1}{2}(\nabla\mathbf{v} + (\nabla\mathbf{v})^T). \tag{5.44}$$

Furthermore, the constitutive equation must be independent from the hydrostatic component of the stress state. Indeed, since we can write

$$\mathbf{T} = \frac{1}{3}(\text{tr}\mathbf{T})\mathbf{I} + \mathbf{S}, \tag{5.45}$$

where $\mathbf{I}$ is the identity tensor and $\mathbf{S}$ is the deviatoric component of the stress tensor $\mathbf{T}$, we have that the creep strain rate $\mathbf{D}$ will be a function only of $\mathbf{S}$.

The final consideration is that, for an isotropic material, the principal directions of the strain rate tensor $\mathbf{D}$ coincide with those of the stress tensor $\mathbf{T}$. In fact, consider an eigenvector $\mathbf{a}$ of the tensor $\mathbf{T}$ and λ the corresponding eigenvalue, that is

$$\mathbf{T}\mathbf{a} = \lambda\mathbf{a}, \tag{5.46}$$

then, from (5.45) we deduce

$$\mathbf{S}\mathbf{a} = \left(\mathbf{T} - \frac{1}{3}(\text{tr}\mathbf{T})\mathbf{I}\right)\mathbf{a} = \mathbf{T}\mathbf{a} - \frac{1}{3}(\text{tr}\mathbf{T})\mathbf{a} = \left(\lambda - \frac{1}{3}\text{tr}\mathbf{T}\right)\mathbf{a} = \gamma\mathbf{a}, \tag{5.47}$$

which means that $\mathbf{a}$ is also eigenvector of the deviatoric tensor $\mathbf{S}$, whose eigenvalue is γ. In other words, the creep strain rate $\mathbf{D}$ must be aligned with $\mathbf{S}$.

We will show that these considerations are all embedded in the most famous constitutive equations for creep.

In 1930, R. W. Bailey experimentally observed and reported in [20, 21] that, in metals, creep strains developed keeping the volume constant, not being affected by the level of hydrostatic pressure. From these observations, and assuming an isotropic behavior, F. K. Odqvist, in [232, 233], postulated in 1934 the first constitutive law for secondary creep considering the three-dimensionality of the stress state

$$\mathbf{D} = f(\mathbf{S})\mathbf{S} = \frac{3}{2}kT_e^{n-1}\mathbf{S}, \tag{5.48}$$

where k and n are material constants and T_e is the effective stress given by

$$T_e = \sqrt{\frac{3}{2}\mathbf{S}\cdot\mathbf{S}}. \tag{5.49}$$

Let us see that Odqvist law satisfies all the considerations described above. First, note that the hydrostatic component of the stress brings no influence in the resulting strain rate. That is, two stress states equal up to a hydrostatic pressure yield the same strain rate $\mathbf{D}$. Indeed, let $\mathbf{T}_1$ and $\mathbf{T}_2$ be these two stress states such that

$$\mathbf{T}_2 - \mathbf{T}_1 = p\mathbf{I}, \tag{5.50}$$

then

$$\mathbf{T}_2 = p\mathbf{I} + \mathbf{T}_1, \tag{5.51}$$

and

$$\mathrm{tr}\mathbf{T}_2 = 3p + \mathrm{tr}\mathbf{T}_1. \tag{5.52}$$

The deviatoric component of $\mathbf{T}_2$, say $\mathbf{S}_2$, results in

$$\mathbf{S}_2 = \mathbf{T}_2 - \frac{1}{3}(\mathrm{tr}\mathbf{T}_2)\mathbf{I} = p\mathbf{I} + \mathbf{T}_1 - p\mathbf{I} - \frac{1}{3}(\mathrm{tr}\mathbf{T}_1)\mathbf{I} = \mathbf{T}_1 - \frac{1}{3}(\mathrm{tr}\mathbf{T}_1)\mathbf{I} = \mathbf{S}_1, \tag{5.53}$$

from which we conclude that

$$\mathbf{D}(\mathbf{T}_1) = \mathbf{D}(\mathbf{T}_2). \tag{5.54}$$

In addition, the strain rate obtained from the constitutive law $\mathbf{D} = f(\mathbf{S})\mathbf{S}$ corresponds to an isochoric velocity field because it verifies

$$\mathrm{tr}\mathbf{D} = f(\mathbf{S})\mathrm{tr}\mathbf{S} = 0. \tag{5.55}$$

Now, the effective stress T_e defined in (5.49) and the strain rate D_e given by

$$D_e = \sqrt{\frac{2}{3}\mathbf{D} \cdot \mathbf{D}}, \tag{5.56}$$

verify a similar relation to the Norton law, that is,

$$D_e = kT_e^n. \tag{5.57}$$

As a matter of fact, from (5.48) it follows that

$$\mathbf{D} \cdot \mathbf{D} = \frac{9}{4}k^2 T_e^{2n-2}\mathbf{S} \cdot \mathbf{S} = \frac{3}{2}k^2 T_e^{2n}, \tag{5.58}$$

from where

$$D_e = \sqrt{\frac{2}{3}\mathbf{D} \cdot \mathbf{D}} = kT_e^n. \tag{5.59}$$

Let us show that for the particular uniaxial stress state, the constitutive law (5.48) reduces to the Norton law. Let $\{\mathbf{e}_1, \mathbf{e}_2, \mathbf{e}_3\}$ be a unitary orthogonal triple of vectors. For uniaxial stress in the direction of $\mathbf{e}_1$, we get

$$\mathbf{T} = T_{11}(\mathbf{e}_1 \otimes \mathbf{e}_1), \tag{5.60}$$

$$\mathrm{tr}\mathbf{T} = T_{11}\mathrm{tr}(\mathbf{e}_1 \otimes \mathbf{e}_1) = T_{11}\mathbf{e}_1 \cdot \mathbf{e}_1 = T_{11}, \tag{5.61}$$

then

$$\mathbf{S} = \mathbf{T} - \frac{1}{3}(\mathrm{tr}\mathbf{T})\mathbf{I} = T_{11}(\mathbf{e}_1 \otimes \mathbf{e}_1) - \frac{1}{3}T_{11}\mathbf{I}, \tag{5.62}$$

$$T_e^2 = \frac{3}{2}\mathbf{S} \cdot \mathbf{S} = \frac{3}{2}\left(T_{11}^2 - \frac{2}{3}T_{11}^2 + \frac{1}{3}T_{11}^2\right) = T_{11}^2, \tag{5.63}$$

and therefore

$$D_{11} = \frac{3}{2}kT_{11}^{n-1}\left(T_{11} - \frac{1}{3}T_{11}\right) = kT_{11}^n. \tag{5.64}$$

Interestingly, note that this constitutive law can be associated with a potential function, denoted by $\phi = \phi(\mathbf{T})$, such that

$$\mathbf{D} = \phi_{\mathrm{T}} = \frac{\partial \phi}{\partial \mathbf{T}}. \tag{5.65}$$

This potential function is specifically given by

$$\phi = \frac{k}{n+1} T_e^{n+1} = \frac{k}{n+1}\left(\frac{3}{2}\mathbf{S}\cdot\mathbf{S}\right)^{(n+1)/2}, \tag{5.66}$$

and depends explicitly on **S** which, in turn, is a function of **T**. So, ϕ is an implicit function of **T**. Using the chain rule, and introducing the fourth-order tensor $\mathbf{S}_\mathbf{T} = \frac{\partial \mathbf{S}}{\partial \mathbf{T}}$, we have

$$\phi_\mathbf{T} = (\mathbf{S}_\mathbf{T})^T \phi_\mathbf{S}. \tag{5.67}$$

Putting

$$\mathbf{S} = \mathbf{T} - \frac{1}{3}(\mathrm{tr}\mathbf{T})\mathbf{I} = \mathbf{T} - \frac{1}{3}(\mathbf{I}\otimes\mathbf{I})\mathbf{T} = \left(\mathbb{I} - \frac{1}{3}\mathbf{I}\otimes\mathbf{I}\right)\mathbf{T} = \mathbb{C}\mathbf{T}, \tag{5.68}$$

we can appreciate that the fourth-order tensor $\mathbb{C}$ is symmetric in the sense given by (1.231), that is, $\mathbb{C} = \mathbb{C}^T$. Hence

$$\mathbf{S}_\mathbf{T} = \mathbb{C} = \mathbb{C}^T, \tag{5.69}$$

and replacing this result into (5.67), and noting that $\mathbb{C}\mathbf{S} = \mathbf{S}$, we obtain

$$\phi_\mathbf{T} = \mathbb{C}\phi_\mathbf{S} = \mathbb{C}\left(kT_e^n \frac{3}{2}\frac{\mathbf{S}}{T_e}\right) = \frac{3}{2}kT_e^{n-1}\mathbb{C}\mathbf{S} = \frac{3}{2}kT_e^{n-1}\mathbf{S} = \phi_\mathbf{S}. \tag{5.70}$$

When it is possible to construct the potential function ϕ, leading up to the strain rate constitutive law, the constitutive law is called an associative law.

Another mathematically and physically important aspect of the potential function ϕ is the convexity. Given that ϕ is regular, the convexity can be expressed as

$$\phi(\mathbf{T}^*) - \phi(\mathbf{T}_0) \geq \phi_\mathbf{T}|_{\mathbf{T}_0} \cdot (\mathbf{T}^* - \mathbf{T}_0) \qquad \mathbf{T}^* \in \mathscr{W}'. \tag{5.71}$$

In fact

$$\phi(\mathbf{T}^*) = \phi(\mathbf{T}) + \phi_\mathbf{T}|_{\mathbf{T}_0} \cdot (\mathbf{T}^* - \mathbf{T}_0) + \frac{1}{2}\phi_{\mathbf{TT}}\bigg|_{\mathbf{T}_0} (\mathbf{T}^* - \mathbf{T}_0)\cdot(\mathbf{T}^* - \mathbf{T}_0) + \dots. \tag{5.72}$$

If $\phi_{\mathbf{TT}}$ is a positive definite fourth-order tensor, implying that

$$\phi_{\mathbf{TT}}|_{\mathbf{T}_0}(\mathbf{T}^* - \mathbf{T}_0)\cdot(\mathbf{T}^* - \mathbf{T}_0) \geq 0, \tag{5.73}$$

and equal to zero for $\mathbf{T}^* = \mathbf{T}_0$, then ϕ is a convex potential function. For the Odqvist law in (5.48), this gives

$$\phi_{\mathbf{TT}}(\mathbf{T}) = \mathbb{C}\phi_{\mathbf{SS}} = \mathbb{C}\left[\frac{3}{2}kT_e^{n-1}\mathbb{I} + \frac{9}{4}k(n-1)T_e^{n-3}\mathbf{S}\otimes\mathbf{S}\right]. \tag{5.74}$$

When the stress state is different from zero we get

$$\phi_{\mathbf{TT}}(\mathbf{T}) = \alpha\mathbb{C} + \beta\mathbf{S}\otimes\mathbf{S} \qquad \alpha, \beta > 0. \tag{5.75}$$

Thus, for an arbitrary $\mathbf{T}^*$, it follows that

$$\begin{aligned}\phi_{\mathbf{TT}}(\mathbf{T})(\mathbf{T}^* - \mathbf{T})\cdot(\mathbf{T}^* - \mathbf{T}) &= [\alpha\mathbb{C}(\mathbf{T}^* - \mathbf{T}) + \beta(\mathbf{S}\cdot(\mathbf{T}^* - \mathbf{T}))\mathbf{S}]\cdot(\mathbf{T}^* - \mathbf{T})\\ &= \alpha(\mathbf{T}^* - \mathbf{T})^D\cdot(\mathbf{T}^* - \mathbf{T}) + \beta[\mathbf{S}\cdot(\mathbf{T}^* - \mathbf{T})]^2,\end{aligned} \tag{5.76}$$

where $(\cdot)^D$ denotes the deviatoric component of $(\cdot)$. Then

$$\begin{aligned}\phi_{\mathbf{TT}}(\mathbf{T})(\mathbf{T}^* - \mathbf{T})\cdot(\mathbf{T}^* - \mathbf{T}) &= \alpha(\mathbf{T}^* - \mathbf{T})^D\cdot(\mathbf{T}^* - \mathbf{T})^D + \beta[\mathbf{S}\cdot(\mathbf{T}^* - \mathbf{T})^D]^2\\ &\geq \alpha(\mathbf{T}^* - \mathbf{T})^D\cdot(\mathbf{T}^* - \mathbf{T})^D \geq 0,\end{aligned} \tag{5.77}$$

and equal to zero if and only if $(\mathbf{T}^* - \mathbf{T})^D = \mathbf{O}$. We have thus proved the convexity of ϕ given by (5.66).

Moreover, putting $\mathbf{D}$ as the strain rate associated with the stress state $\mathbf{T}$, for a stress of the form

$$\mathbf{T}_1 = \lambda \mathbf{T}, \tag{5.78}$$

corresponds to a strain rate

$$\mathbf{D}_1 = \frac{3}{2}k\left(\frac{3}{2}\lambda\mathbf{S} \cdot \lambda\mathbf{S}\right)^{(n-1)/2} \lambda\mathbf{S} = \lambda^n \frac{3}{2}k\left(\frac{3}{2}\mathbf{S} \cdot \mathbf{S}\right)^{n-1} \mathbf{S} = \lambda^n \mathbf{D}, \tag{5.79}$$

that is, the strain rate $\mathbf{D}_1$, associated with the stress $\mathbf{T}_1$, is λ^n times the strain rate $\mathbf{D}$ associated with $\mathbf{T}$.

The previously obtained results allow us to conclude that, in the stress space, the region characterized by $\mathbf{T}$ satisfying

$$\phi(\mathbf{T}) \leq K, \qquad K \text{ constant}, \tag{5.80}$$

is convex, and for different values of K these regions are similar and concentric. Also, any strain rate is normal to the surface $\phi = K$ at the corresponding stress state and, what is more, strain rates correspondingly associated with proportionally related stress states are parallel, as illustrated in Figure 5.17.

The Odqvist constitutive law (5.48) enables the determination of $\mathbf{D}$ from the stress state. The inverse constitutive equation can be found readily. In order to do this, recall that $\mathbf{D}$ satisfies tr$\mathbf{D} = 0$. Then, the inverse constitutive equation will be defined for all strain rates satisfying the fact that the motion underlying the strain rate is isochoric.

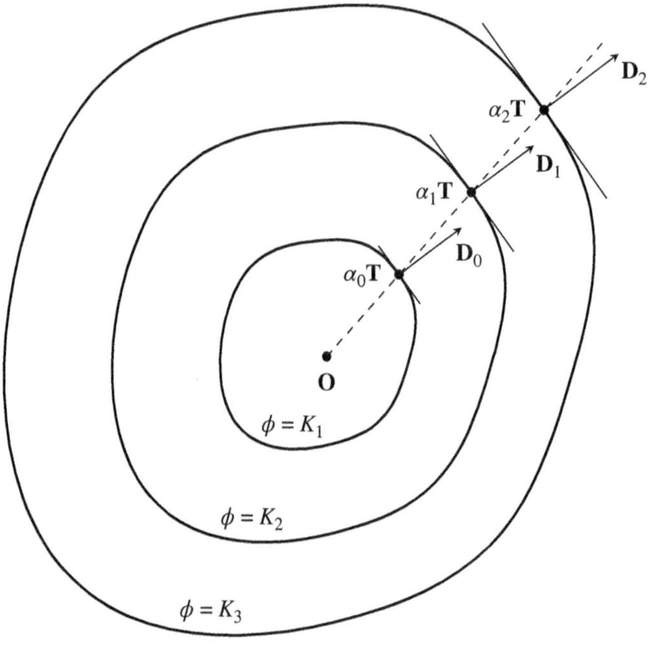

Figure 5.17 Constitutive potential function for isotropic secondary creep.

Inverting (5.57), we have

$$T_e = \left(\frac{D_e}{k}\right)^{1/n}, \tag{5.81}$$

and replacing it into the Odqvist law (5.48) leads us to

$$\mathbf{D} = \frac{3}{2}k\left(\frac{D_e}{k}\right)^{(n-1)/n}\mathbf{S} = \frac{3}{2}k^{1/n}D_e^{1-1/n}\mathbf{S}, \tag{5.82}$$

from where

$$\mathbf{S} = \frac{2}{3}k^{-1/n}D_e^{(1/n)-1}\mathbf{D}, \tag{5.83}$$

which is the inverse constitutive equation we were looking for.

As before, there exists a potential function

$$\psi(\mathbf{D}) = \frac{n}{n+1}\frac{D_e^{(1+n)/n}}{k^{1/n}}, \tag{5.84}$$

such that

$$\psi_{\mathbf{D}} = \frac{1}{k^{1/n}}D_e^{[(1+n)/n]-1}\frac{1}{2}\frac{4}{3}\frac{\mathbf{D}}{D_e} = \frac{2}{3}\frac{1}{k^{1/n}}D_e^{(1-n)/n}\mathbf{D} = \mathbf{S}. \tag{5.85}$$

In turn, ψ is convex and features similar properties to potential function ϕ defined in (5.66). Note that ψ is defined in a space of traceless strain rate tensors.

Furthermore, for a given material point in the body we have a certain stress state $\mathbf{T}$, which has associated a strain rate $\mathbf{D} = f(\mathbf{S})\mathbf{S}$, where $\mathbf{S}$ is the deviatoric part of $\mathbf{T}$. Then, since $\mathrm{tr}\mathbf{D} = 0$, the dissipation power will be

$$\mathbf{T}\cdot\mathbf{D} = \mathbf{S}\cdot\mathbf{D} = \mathbf{S}\cdot f(\mathbf{S})\mathbf{S} = \mathbf{S}\cdot\left(\frac{3}{2}kT_e^{n-1}\mathbf{S}\right) = kT_e^{n+1} = (n+1)\phi(\mathbf{T}). \tag{5.86}$$

This implies that the surface ϕ = constant in the space of stresses physically represents a surface in which the dissipation is constant.

Likewise

$$\mathbf{S}\cdot\mathbf{D} = \frac{2}{3}\frac{1}{k^{1/n}}D_e^{(1-n)/n}\mathbf{D}\cdot\mathbf{D} = \frac{1}{k^{1/n}}D_e^{(1+n)/n} = \frac{1+n}{n}\psi(\mathbf{D}), \tag{5.87}$$

and so the surface ψ = constant represents, in the space of strain rates, a constant dissipation power.

Moreover, for the same dissipation power, from (5.86) and (5.87) it follows that

$$n\phi(\mathbf{T}) = \psi(\mathbf{D}), \tag{5.88}$$

or, equivalently, if $\mathbf{T}$ and $\mathbf{D}$ are related by the Odqvist constitutive law, then

$$\psi(\mathbf{D}) + \phi(\mathbf{T}) = \mathbf{T}\cdot\mathbf{D}. \tag{5.89}$$

The similarity of this property with that satisfied by the hyperelastic materials studied in Chapter 4 is readily identified.

Finally, and following [44], let us consider, in the space of stresses, the surface for which the dissipation energy is an arbitrary constant K_1, then

$$(n+1)\phi_n(T) = K_1, \tag{5.90}$$

where $\phi_n(\mathbf{T})$ denotes a material with exponent n. For a different material, with exponent $r < n$, the same dissipation energy corresponds to

$$(r + 1)\phi_r(\mathbf{T}) = K_1. \tag{5.91}$$

Hence

$$(n + 1)\phi_n(\mathbf{T}) = (r + 1)\phi_r(\mathbf{T}), \tag{5.92}$$

and, because $n + 1 > r + 1$, we obtain

$$\phi_n(\mathbf{T}) < \phi_r(\mathbf{T}). \tag{5.93}$$

That is, the constant dissipation energy surfaces for larger values of n are enclosed by those with smaller values of the exponent. In particular, all these surfaces will be contained in the region limited by the surface $n = 1$ (corresponding to an incompressible elastic material) and by the surface $n = \infty$ (corresponding to a rigid ideal plastic material). Reciprocally, in terms of the space of strain rates, we have

$$\frac{1+n}{n}\psi_n(\mathbf{D}) = \frac{1+r}{r}\psi_r(\mathbf{D}), \tag{5.94}$$

and for $n > r$ this results in

$$1 + \frac{1}{n} < 1 + \frac{1}{r}, \tag{5.95}$$

therefore

$$\psi_n(\mathbf{D}) > \psi_r(\mathbf{D}). \tag{5.96}$$

This means that surfaces for larger values of n wrap those for lower values.

This kind of result is used in the development of collapse loads by E. Taroco et al. [282] and R. A. Feijóo et al. [99]. Also, these expressions form the basis for the determination of generalized constitutive laws (see, for example, V. I. Rozenblium [263]).

An alternative constitutive equation is due to Tresca [284], which is also defined through convex potential functions, $\phi(\mathbf{T})$ and $\psi(\mathbf{D})$, and which, notwithstanding featuring vertices, possess the same properties than the Odqvist law.

5.7 Generalization of the Constitutive Law

Making use of the basic considerations discussed at the beginning of Section 5.6, limiting the analysis to the case of isotropic material behavior and assuming identical material behavior for compression and traction, it is possible to formulate a general expression which meets all the premises to become a constitutive law (see the work by W. Prager [245]). Such an expression takes the form

$$\mathbf{D} = \alpha_1\mathbf{S} + \alpha_3\mathbf{S}^3 + \alpha_5\mathbf{S}^5 + \dots, \tag{5.97}$$

where the strain rate is given by a polynomial in powers of the deviatoric stress $\mathbf{S}$ associated with $\mathbf{T}$.

However, the powers of $\mathbf{S}$ are not independent. In fact, the Cayley–Hamilton theorem, through (1.168), states that every second-order traceless tensor satisfies

$$\mathbf{S}^3 - J_2\mathbf{S} - J_3\mathbf{I} = \mathbf{O} \quad \rightarrow \quad \mathbf{S}^3 = J_2\mathbf{S} + J_3\mathbf{I}, \tag{5.98}$$

where the invariants of $\mathbf{S}$, J_2 and J_3, result (see (1.161)–(1.163) and (1.170))

$$J_2 = \frac{1}{2}\mathrm{tr}(\mathbf{S}^2) = \frac{1}{2}\mathbf{S} \cdot \mathbf{S}, \tag{5.99}$$

$$J_3 = \det \mathbf{S} = \frac{1}{3}\mathrm{tr}(\mathbf{S}^3) = \frac{1}{3}\mathbf{S}^3 \cdot \mathbf{I}, \tag{5.100}$$

where we have used $J_1 = \mathrm{tr}\mathbf{S} = 0$ because $\mathbf{S}$ is a deviatoric tensor.

With this, it is not difficult to write $\mathbf{S}^{2n+1}$, $n = 1, 2, \dots$ in terms of $\mathbf{I}$, $\mathbf{S}$ and $\mathbf{S}^2$, in particular

$$\begin{aligned} \mathbf{S}^5 &= J_3\mathbf{S}^2 + J_2^2\mathbf{S} + J_2J_3\mathbf{I}, \\ \mathbf{S}^7 &= 2J_2J_3\mathbf{S}^2 + (J_3^2 + J_2^3)\mathbf{S} + J_2^2J_3\mathbf{I}, \\ &\vdots \quad \vdots \end{aligned} \tag{5.101}$$

Introducing these results into (5.97), gives

$$\mathbf{D} = p_1\mathbf{I} + p_2\mathbf{S} + p_3\mathbf{S}^2, \tag{5.102}$$

where p_i, $i = 1, 2, 3$, are polynomials of J_2 and J_3.

The condition of isochoric deformation is achieved by

$$\mathrm{tr}\mathbf{D} = 3p_1 + p_3\mathrm{tr}(\mathbf{S}^2) = 3p_1 + 2p_3J_2 = 0, \tag{5.103}$$

from where

$$p_1 = -\frac{2}{3}p_3J_2, \tag{5.104}$$

and, putting this into (5.102), this gives

$$\mathbf{D} = p_2\mathbf{S} + p_3\left[\mathbf{S}^2 - \frac{2}{3}J_2\mathbf{I}\right]. \tag{5.105}$$

In this way, we have a general constitutive law for isotropic secondary creep for multiple stress states. Setting different expressions for p_2 and p_3, we recover the well-known constitutive equations. For example, by setting

$$p_3 = 0, \tag{5.106}$$

$$p_2 = \frac{3}{2}k(3J_2)^{(n-1)/2}, \tag{5.107}$$

the Odqvist law (5.48) is obtained.

Knowing that $\mathbf{S} = \mathbb{C}\mathbf{T}$, with $\mathbb{C} = \mathbb{I} - \frac{1}{3}\mathbf{I} \otimes \mathbf{I}$, and taking into consideration (5.99) and (5.100), yields

$$\frac{\partial J_2}{\partial \mathbf{T}} = \mathbb{C}\frac{\partial J_2}{\partial \mathbf{S}} = \mathbb{C}\mathbf{S} = \mathbf{S}, \tag{5.108}$$

$$\frac{\partial J_3}{\partial \mathbf{T}} = \mathbb{C}\mathbf{S}^2 = \mathbf{S}^2 - \frac{1}{3}[\mathrm{tr}(\mathbf{S}^2)]\mathbf{I} = \mathbf{S}^2 - \frac{2}{3}J_2\mathbf{I}. \tag{5.109}$$

Thus, the generalized constitutive equation for $\mathbf{D}$ gives

$$\mathbf{D} = p_2\frac{\partial J_2}{\partial \mathbf{T}} + p_3\frac{\partial J_3}{\partial \mathbf{T}}. \tag{5.110}$$

With this, if there exists a potential

$$\phi = \phi(J_2, J_3), \tag{5.111}$$

such that

$$p_2 = \frac{\partial \phi}{\partial J_2} \qquad p_3 = \frac{\partial \phi}{\partial J_3}, \tag{5.112}$$

the strain rate is written as

$$\mathbf{D} = \frac{\partial \phi}{\partial \mathbf{T}}. \tag{5.113}$$

An alternative manner to establish a general constitutive law for secondary creep is directly from the definition of a potential function, say ϕ, from which the strain rate is to be derived. To guarantee isotropy, this function has to be of the form

$$\phi = \phi(I_1, I_2, I_3), \tag{5.114}$$

where I_i, $i = 1, 2, 3$, are the invariants associated with the stress state $\mathbf{T}$, whose definitions are in (1.161), (1.162), and (1.163), respectively. With these definitions, it is easy to show that

$$I_2 = J_2 - \frac{1}{3} I_1^2, \tag{5.115}$$

$$I_3 = J_3 + \frac{1}{27} I_1^3 - \frac{1}{3} I_1 J_2, \tag{5.116}$$

where J_2 and J_3 are the invariants of the deviatoric stress $\mathbf{S}$ defined in (5.99) and (5.100), respectively. Then, putting this into the expression of ϕ we have

$$\phi = \phi(I_1, J_2, J_3). \tag{5.117}$$

Noting that

$$\mathbf{D} = \frac{\partial \phi}{\partial \mathbf{T}} = \frac{\partial \phi}{\partial I_1}\frac{\partial I_1}{\partial \mathbf{T}} + \frac{\partial \phi}{\partial J_2}\frac{\partial J_2}{\partial \mathbf{T}} + \frac{\partial \phi}{\partial J_3}\frac{\partial J_3}{\partial \mathbf{T}}, \tag{5.118}$$

considering $\operatorname{tr}\mathbf{D} = 0$, taking into account that $\frac{\partial I_1}{\partial \mathbf{T}} = \mathbf{I}$, and utilizing expressions (5.108) and (5.109), results in

$$\begin{aligned} \operatorname{tr}\mathbf{D} &= \frac{\partial \phi}{\partial I_1}\operatorname{tr}\left(\frac{\partial I_1}{\partial \mathbf{T}}\right) + \frac{\partial \phi}{\partial J_2}\operatorname{tr}\left(\frac{\partial J_2}{\partial \mathbf{T}}\right) + \frac{\partial \phi}{\partial J_3}\operatorname{tr}\left(\frac{\partial J_3}{\partial \mathbf{T}}\right) \\ &= \frac{\partial \phi}{\partial I_1}3 + \frac{\partial \phi}{\partial J_2}\operatorname{tr}\mathbf{S} + \frac{\partial \phi}{\partial J_3}\operatorname{tr}\left(\mathbf{S}^2 - \frac{2}{3}J_2\mathbf{I}\right) = 3\frac{\partial \phi}{\partial I_1} = 0, \end{aligned} \tag{5.119}$$

which must hold for every stress state $\mathbf{T}$. This result indicates that ϕ can no longer be a function of I_1, and so we reach the final functional form for the potential function ϕ

$$\phi = \phi(J_2, J_3), \tag{5.120}$$

which, in addition, must be convex, that is

$$\phi(\mathbf{T}^*) - \phi(\mathbf{T}_0) \geq \phi_{\mathbf{T}}\big|_{\mathbf{T}_0} \cdot (\mathbf{T}^* - \mathbf{T}_0) \qquad \mathbf{T}^* \in \mathscr{W}'. \tag{5.121}$$

If there exists this potential function, it will always be possible to construct a dual function, say $\psi = \psi(\mathbf{D})$, such that

$$\psi(\mathbf{D}) = \mathbf{T} \cdot \mathbf{D} - \phi(\mathbf{T}), \tag{5.122}$$

or, equivalently,

$$\psi(\mathbf{D}) + \phi(\mathbf{T}) = \mathbf{T} \cdot \mathbf{D}. \tag{5.123}$$

Keeping in mind that in the field of creep the product $\mathbf{T} \cdot \mathbf{D}$ stands for the dissipation power, potentials ψ and ϕ are called dissipation potentials.

From the very definition of ψ, it follows that

$$\mathbf{T} = \psi_{\mathbf{D}}, \tag{5.124}$$

and

$$\psi(\mathbf{D}^*) - \psi(\mathbf{D}_0) \geq \psi_{\mathbf{D}}\big|_{\mathbf{D}_0} \cdot (\mathbf{D}^* - \mathbf{D}_0) \qquad \mathbf{D}^* \in \mathscr{W}\,. \tag{5.125}$$

Denoting by $\mathbf{T}$ and $\mathbf{D}$ the stress state and the corresponding strain rate related through the creep constitutive equation, expression (5.125) can be rewritten as

$$\psi(\mathbf{D}^*) - \psi(\mathbf{D}) \geq \mathbf{T} \cdot (\mathbf{D}^* - \mathbf{D}), \tag{5.126}$$

and, with (5.123), the inequality (5.126) takes the form

$$\psi(\mathbf{D}^*) - \psi(\mathbf{D}) \geq \mathbf{T} \cdot \mathbf{D}^* - \mathbf{T} \cdot \mathbf{D} = \mathbf{T} \cdot \mathbf{D}^* - \psi(\mathbf{D}) - \phi(\mathbf{T}), \tag{5.127}$$

yielding

$$\psi(\mathbf{D}^*) + \phi(\mathbf{T}) \geq \mathbf{T} \cdot \mathbf{D}^*, \tag{5.128}$$

where the equality is verified if and only if $\mathbf{T}$ and $\mathbf{D}^*$ are related via the constitutive equation.

In this manner, admitting the existence of the dissipation potentials, the constitutive law for secondary creep can be shaped in the following ways

$$\psi(\mathbf{D}) + \phi(\mathbf{T}) = \mathbf{T} \cdot \mathbf{D}, \tag{5.129}$$

$$\mathbf{D} = \left.\frac{\partial \phi}{\partial \mathbf{T}}\right|_{\mathbf{T}}, \tag{5.130}$$

$$\mathbf{T} = \left.\frac{\partial \psi}{\partial \mathbf{D}}\right|_{\mathbf{D}}. \tag{5.131}$$

In the process of establishing this constitutive law for secondary creep, a series of hypotheses have been considered. These considerations provide a similar constitutive structure to that encountered in Chapter 4 when dealing with hyperelastic materials. In the following sections we will show that this will enable the modeling of mechanical problems with materials exhibiting secondary creep by means of minimum problems that are entirely similar to those formulated in Chapter 4.

Finally, as anticipated at the beginning of this chapter, we will not address material behavior featuring time or strain hardening. Also, variable loading processes will not be addressed. The reader will find these topics addressed in the book by Y. N. Rabotnov [249].

5.8 Constitutive Equations for Structural Components

When modeling a mechanical problem, the introduction of kinematical constraints leads, as in the case of beams (see Chapter 7), and plates and shells (see Chapter 9), to the appearance of generalized stresses in the corresponding Principles of Virtual Power (PVPs). For example, in the case of plates and shells we refer to the strains over the middle surface, and the stresses associated with these strains, namely the membrane stress tensor $\mathbf{N}_t$ and the membrane moment tensor $\mathbf{M}_t$.

The kinematics of these structural components as well as the formulation of the PVP in each case will be carefully considered in Chapter 7 for the case of beams and in Chapter 9 for plates and shells. The reader who is unfamiliar with structural components can jump to these chapters before going ahead with the present section, where the details of the different kinematical theories are presented and discussed, as well as the consequences in terms of the generalized stresses. This section only presents fundamental definitions required in the treatment of the constitutive modeling.

When the material is purely elastic, the construction of these constitutive equations for the generalized stresses $\mathbf{N}_t$ and $\mathbf{M}_t$ presents no difficulty whatsoever. However, the same does not happen when the material features an inelastic response.

The use of plasticity models in generalized variables can be found in the works by W. Prager [246], P. G. Hodge [145, 146], and M. A. Save and C. E. Massonnet [272, 273]. With respect to creep phenomena, the first attempts to deal with generalized stresses emerged from the contributions of L. M. Kachanov [152], C. R. Calladine and D. C. Drucker [44], Y. N. Rabotnov [249], and W. Olszak and A. Sawczuk [235].

Next, we will analyze some constitutive equations for the study of secondary creep making use of generalized stresses for a beam model.

5.8.1 Bending of Beams

Let us consider a beam whose cross-sectional area features two axes of symmetry and which is subjected to the action of loads acting over one of the symmetry planes, for example the plane $\overline{xOz}$, where Ox is the longitudinal axis of the beam (assumed to be straight) and Oz is the vertical axis. The height of the beam is $2h$ and the width b is a function of z/h. The kinematical hypotheses that we will consider correspond to the Bernoulli hypotheses, which establish that (i) transversal sections remain normal to the mid plane after the deformation and (ii) normal fibers (in the Oz direction) do not modify its length during the deformation.

These two hypotheses constrain all kinematically admissible velocity fields to be of the form

$$\mathbf{v} = w\mathbf{n} - z\frac{dw}{dx}\mathbf{e}_x, \tag{5.132}$$

where $w = w(x)$ is the velocity in the direction normal to the mid plane of the beam, whose coordinate is x, and $\mathbf{n}$ is the unit normal vector to such plane, which is considered constant because the beam is assumed to be straight, and $\mathbf{e}_x$ is the unit vector in the positive direction of this axis.

Following the theoretical developments presented in Chapter 7 (see also [286]), the generalized strain measure is given by

$$\chi = -\frac{d^2w}{dx^2}, \tag{5.133}$$

which, by duality, has a generalized stress denoted by M (bending moment) given by

$$M = \int_{-h}^{h} \sigma z b\left(\frac{z}{h}\right) dz, \tag{5.134}$$

where σ is the stress in the direction of the beam axis.

Now, the uniaxial constitutive relation between σ and the uniaxial strain rate $\varepsilon = z\chi$ is assumed to be known. For example, consider a law of the Norton type

$$\sigma = s(\varepsilon) = \sigma_n \left(\frac{\varepsilon}{\varepsilon_n} \right)^{1/n}, \tag{5.135}$$

where n, σ_n, and ε_n are parameters that depend on the specific material. Replacing (5.135) into (5.134) results in

$$M = \int_{-h}^{h} s(\varepsilon) z b \left(\frac{z}{h} \right) dz. \tag{5.136}$$

Given the symmetry of the beam, and performing a change of variables from z to ε, with $\varepsilon' = h\chi$, yields

$$M = 2 \int_0^h s(\varepsilon) z b \left(\frac{z}{h} \right) dz = \frac{2}{\chi^2} \int_0^{\varepsilon'} s(\varepsilon) \varepsilon b \left(\frac{\varepsilon}{\varepsilon'} \right) d\varepsilon = \frac{2h^2}{\varepsilon'^2} \int_0^{\varepsilon'} s(\varepsilon) \varepsilon b \left(\frac{\varepsilon}{\varepsilon'} \right) d\varepsilon. \tag{5.137}$$

If we denote b_0 to the value of b at $z = 0$ and if σ^* is an arbitrary stress-like constant, we can introduce

$$m(\varepsilon') = m(h\chi) = \frac{2}{\sigma^* b_0 \varepsilon'^2} \int_0^{\varepsilon'} s(\varepsilon) \varepsilon b \left(\frac{\varepsilon}{\varepsilon'} \right) d\varepsilon, \tag{5.138}$$

and we write

$$M = \sigma^* b_0 h^2 m(h\chi). \tag{5.139}$$

This expression relates the generalized strain rate χ (curvature change rate) to the generalized stress M.

For the particular case of a beam with rectangular cross-section, and for the Norton law, the constitutive equation becomes

$$m(h\chi) = \frac{2}{\sigma^* b_0 \varepsilon'^2} \int_0^{\varepsilon'} \sigma_n \left(\frac{\varepsilon}{\varepsilon_n} \right)^{1/n} \varepsilon b_0 d\varepsilon = \frac{2n}{2n+1} \frac{\sigma_n}{\sigma^*} \left(\frac{\varepsilon'}{\varepsilon_n} \right)^{1/n}, \tag{5.140}$$

and since σ^* is arbitrary, we can take

$$\sigma^* = \sigma^n, \tag{5.141}$$

then

$$m(h\chi) = \frac{2n}{2n+1} \left(\frac{h\chi}{\varepsilon_n} \right)^{1/n}. \tag{5.142}$$

Putting this result into (5.139) we finally reach the following relation

$$M = \sigma_n b_0 h^2 \frac{2n}{2n+1} \left(\frac{h\chi}{\varepsilon_n} \right)^{1/n}, \tag{5.143}$$

whose inverse leads to

$$\chi = \frac{\varepsilon_n}{h} \left(\frac{2n+1}{2n} \frac{1}{\sigma_n b_0 h^2} M \right)^n. \tag{5.144}$$

Calling

$$\chi_n = \frac{\varepsilon_n}{h} \qquad M_n = \frac{2n}{2n+1}\,\sigma_n b_0 h^2, \tag{5.145}$$

results in

$$\chi = \chi_n \left(\frac{M}{M_n}\right)^n. \tag{5.146}$$

Therefore, and based on the previous results, it is straightforward to infer the following creep potential functions

$$\phi(M) = \frac{1}{n+1}\chi_n M_n \left(\frac{M}{M_n}\right)^{n+1}, \tag{5.147}$$

$$\psi(\chi) = \frac{n}{n+1}\chi_n M_n \left(\frac{\chi}{\chi_n}\right)^{(n+1)/n}, \tag{5.148}$$

which, as expected, satisfy the following relation

$$\phi(M) + \psi(\chi) = M\chi. \tag{5.149}$$

Another strategy to present these equations consists of working with the Norton law in dimensional form

$$\varepsilon = B\sigma^n \qquad \sigma = \frac{1}{B^{1/n}}\varepsilon^{1/n}. \tag{5.150}$$

This equation corresponds to the uniaxial behavior for extension or compression. To account for this behavior in a single expression we make use of the sign function

$$\mathrm{sgn}(\varepsilon) = \frac{\varepsilon}{|\varepsilon|}, \tag{5.151}$$

where $|\varepsilon|$ is the absolute value of ε. With this function, and considering that B is a positive constant, the equations take the form

$$\sigma = \frac{1}{B^{1/n}}|\varepsilon|^{1/n}\mathrm{sgn}(\varepsilon), \tag{5.152}$$

$$\varepsilon = B|\sigma|^n\mathrm{sgn}(\sigma). \tag{5.153}$$

In this fashion, and calling A the transversal area of the beam, the expression for the generalized stress results in

$$\begin{aligned} M &= \int_A \sigma z dA = \int_A \frac{1}{B^{1/n}}|z\chi|^{1/n}\mathrm{sgn}(\chi)\mathrm{sgn}(z) z dA \\ &= \frac{1}{B^{1/n}}|\chi|^{1/n}\mathrm{sgn}(\chi)\int_A |z|^{1+1/n} dA. \end{aligned} \tag{5.154}$$

If we define

$$I_n = \int_A |z|^{1+1/n} dA, \tag{5.155}$$

then this gives

$$M = \frac{I_n}{B^{1/n}}|\chi|^{1/n}\mathrm{sgn}(\chi), \tag{5.156}$$

or its inverse,

$$\chi = \frac{B}{I_n^n} |M|^n \text{sgn}(M). \tag{5.157}$$

In this case, the corresponding potential functions are now given by [39]

$$\phi(M) = \frac{1}{n+1} \frac{B}{I_n^n} (M^2)^{(n+1)/2}, \tag{5.158}$$

$$\psi(\chi) = \frac{n}{n+1} \frac{I_n}{B^{1/n}} (\chi^2)^{(n+1)/2}. \tag{5.159}$$

5.8.2 Bending, Extension, and Compression of Beams

If the kinematical description in the problem seen in the previous section is enhanced by adding the strain in the mid plane, the velocity field becomes

$$\mathbf{v} = w\mathbf{n} + \left(v - z\frac{dw}{dx} \right) \mathbf{e}_x, \tag{5.160}$$

where $v = v(x)$ is the velocity field of the mid axis in the direction of x. With this velocity field, the strain rate is now

$$\varepsilon = \frac{dv}{dx} - z\frac{d^2w}{dx^2}. \tag{5.161}$$

Hence, the generalized strain rates are

$$\varepsilon_o = \frac{dv}{dx}, \tag{5.162}$$

$$\chi = -\frac{d^2w}{dx^2}, \tag{5.163}$$

and the dissipation power per unit length of the beam will be $N\varepsilon_o + M\chi$, where N and M are the generalized stresses associated, respectively, with ε_o and χ. Since the beam width is $b(z/h)$, the stresses N and M are given by

$$N = \int_{-h}^{h} \sigma b dz, \tag{5.164}$$

$$M = \int_{-h}^{h} \sigma z b dz. \tag{5.165}$$

Once again, consider a Norton-type material

$$\sigma = \sigma_n \left(\frac{\varepsilon_o + z\chi}{\varepsilon_n} \right)^{1/n}. \tag{5.166}$$

Replacing this into (5.164), and assuming that the width b is uniform, we have

$$N = \int_{-h}^{h} \sigma_n \left(\frac{\varepsilon_o + z\chi}{\varepsilon_n} \right)^{1/n} b dz = \frac{\sigma_n b}{\varepsilon_n^{1/n}} \frac{n}{(n+1)} \frac{1}{\chi} \left[(\varepsilon_o + h\chi)^{1+1/n} - (\varepsilon_o - h\chi)^{1+1/n} \right], \tag{5.167}$$

or, in dimensionless form,

$$\frac{N}{\sigma_n bh} = \frac{1}{\varepsilon_n^{1/n}} \frac{1}{h\chi} \frac{n}{(n+1)} \left[(\varepsilon_o + h\chi)^{1+1/n} - (\varepsilon_o - h\chi)^{1+1/n} \right]. \tag{5.168}$$

Similarly, from (5.165) we have

$$\begin{aligned} M &= \int_{-h}^{h} \sigma_n \left(\frac{\varepsilon_o + z\chi}{\varepsilon_n} \right)^{1/n} zbdz \\ &= \frac{\sigma_n bh^2}{\varepsilon_n^{1/n}} \frac{1}{(h\chi)^2} \frac{n}{2n+1} \left[(\varepsilon_o + h\chi)^{2+1/n} - (\varepsilon_o - h\chi)^{2+1/n} \right] - \frac{\varepsilon_o}{\chi} N, \end{aligned} \tag{5.169}$$

and in dimensionless form

$$\frac{M}{\sigma_n bh^2} = \frac{1}{\varepsilon_n^{1/n}} \frac{1}{(h\chi)^2} \frac{n}{2n+1} \left[(\varepsilon_o + h\chi)^{2+1/n} - (\varepsilon_o - h\chi)^{2+1/n} \right] - \frac{\varepsilon_o}{\chi} \frac{N}{\sigma_n bh^2}. \tag{5.170}$$

We have produced expressions (5.168) and (5.170) which relate N and M with ε_o and χ. These expressions are such that it is not possible to explicitly give either ε_o or χ as functions of N and M. There are several ways to circumvent this technical obstacle. One strategy consists of building an approximate potential function of the following kind,

$$\phi = \phi(R), \tag{5.171}$$

where R, which is a function of N and M, is chosen in such a way that, when $N = 0$ it is $R = M$. Next, we will show how to build such function.

By the definition of the creep potential we have

$$\chi = \frac{\partial \phi}{\partial R} \frac{\partial R}{\partial M}. \tag{5.172}$$

For $N = 0$, from previous section we have seen that

$$\chi = \chi_n \left(\frac{M}{M_n} \right)^n. \tag{5.173}$$

Since for this case we have $M = R$, then we get

$$\phi(R) = \chi_n M_n \frac{1}{n+1} \left(\frac{R}{M_n} \right)^{n+1}. \tag{5.174}$$

In addition, we have seen that, in the space of generalized stresses, the dissipation curves associated with a certain normalization and corresponding to the different values of n are bounded by the curves corresponding to $n = 1$ (stress distribution in elastic incompressible materials) and $n = \infty$ (material perfectly plastic). Thus, by denoting σ_Y the yield stress in the (perfectly plastic) material, the relation between N and M which completely exhausts the resistant capacity of a rectangular cross-section corresponds to

$$N = b(2h - 2z)\sigma_Y, \tag{5.175}$$

$$M = bz(2h - z)\sigma_Y, \tag{5.176}$$

from where

$$\overline{N} = \frac{N}{2bh\sigma_Y} = \frac{1}{h}(h - z), \tag{5.177}$$

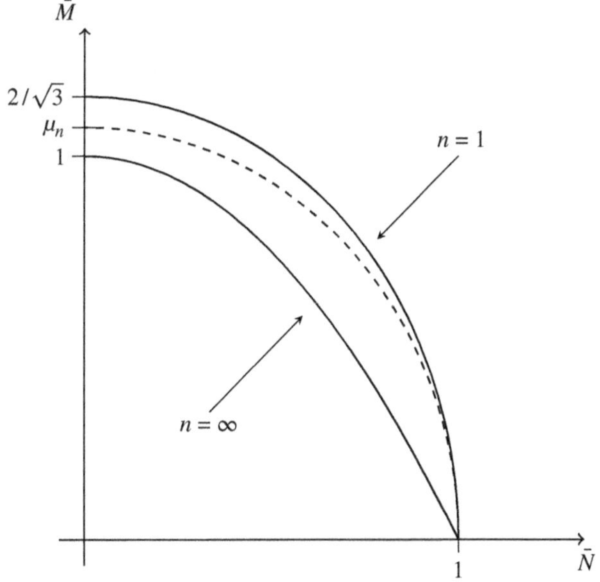

Figure 5.18 Dissipation potential for a beam under bending and extension/compression conditions.

$$\overline{M} = \frac{M}{bh^2\sigma_Y} = \frac{z}{h^2}(2h - z), \tag{5.178}$$

resulting in

$$\overline{N}^2 + \overline{M} = 1. \tag{5.179}$$

In the space of generalized stresses, the previous expression corresponds to a parabola, as observed qualitatively in Figure 5.18.

Moreover, for the case $n = 1$, corresponding to an elastic stress distribution, we have

$$\sigma = \frac{N}{2bh} + \frac{3}{2}\frac{Mz}{bh^3}, \tag{5.180}$$

or

$$\frac{\sigma}{\sigma_n} = \overline{\sigma} = \overline{N} + \frac{3}{2}\overline{M}\frac{z}{h}, \tag{5.181}$$

and adopting the criterion of unitary potential per unit volume as normalization for the dissipation potential, that is

$$b\int_{-h}^{h} \overline{\sigma}^2 dz = 2bh, \tag{5.182}$$

we finally have

$$\overline{N}^2 + \frac{3}{4}\overline{M}^2 = 1, \tag{5.183}$$

which corresponds to an ellipse in the space of generalized stresses, as illustrated in Figure 5.18.

It is now evident that all constant dissipation curves go through the point $(\overline{N}, \overline{M}) = (1, 0)$. To determine the intersection of these curves with the axis $\overline{M}$, it will be sufficient to analyze the dissipated power in the case of pure bending ($\overline{N} = 0$). In order to do this, let us recall that

$$\chi = \chi_n \left(\frac{M}{M_n} \right)^n, \tag{5.184}$$

and then the dissipated power is

$$M\chi = M\chi_n \left(\frac{M}{M_n} \right)^n = \chi_n M_n \left(\frac{M}{M_n} \right)^{n+1}, \tag{5.185}$$

and from the normalization condition given by

$$M\chi = 2bh\sigma_n \varepsilon_n, \tag{5.186}$$

we have

$$\chi_n M_n \left(\frac{M}{M_n} \right)^{n+1} = 2hb\sigma_n \varepsilon_n, \tag{5.187}$$

from where

$$M = \left(\frac{n}{2n+1} \right)^{n/(n+1)} \sigma_n h^2 b, \tag{5.188}$$

and therefore

$$\overline{M} = \frac{M}{\sigma_n h^2 b} = \left(\frac{n}{2n+1} \right)^{n/(n+1)} = \mu_n. \tag{5.189}$$

This expression defines the intersection point we were looking for (see Figure 5.18).

Thus, notice that all the curves are bounded between the curve given by $n = 1$, intersecting the axes at points $(0, 2/\sqrt{3})$ and $(1, 0)$, and the curve for $n = \infty$, intersecting the axes at $(0, 1)$ and $(1, 0)$ (always considering the first quadrant). For any other value of n, the intersection occurs at $(0, \mu_n)$ and $(1, 0)$. Then, the approximate potential function is calculated assuming that this curve is an ellipse which goes through these points, that is

$$\overline{N}^2 + \left(\frac{\overline{M}}{\sqrt{\mu_n}} \right)^2 = 1. \tag{5.190}$$

Supported by these results, the function R is finally found,

$$R^2 = M^2 + \mu_n^2 h^2 N^2, \tag{5.191}$$

and the creep potential results,

$$\phi(R) = \chi_n M_n \frac{1}{(n+1)} \left(\frac{R^2}{M_n^2} \right)^{(n+1)/2}, \tag{5.192}$$

from where

$$\varepsilon_o = \frac{\partial \phi}{\partial R} \frac{\partial R}{\partial N} = \chi_n \left(\frac{R}{M_n} \right)^n (\mu_n h)^2 \frac{N}{R}, \tag{5.193}$$

$$\chi = \frac{\partial \phi}{\partial R} \frac{\partial R}{\partial M} = \chi_n \left(\frac{R}{M_n} \right)^n \frac{M}{R}. \tag{5.194}$$

5.9 Equilibrium Problem for Steady-State Creep

This and forthcoming sections aim to present different methods frequently used in the analysis of structures exposed to creep phenomena. In the spirit of the present notes, the presentation will pursue a variational formalism, which provides a suitable framework for the modeling of physical systems and in particular mechanical models. As discussed already in Chapter 4, the variational formulation intrinsically carries a numerical algorithm for the computation of approximate solutions. The reader interested in methods of analysis other than those of variational flavor can study the references cited in the Complementary Reading section at the end of this chapter.

Our study is focused on the secondary creep phenomena. As in previous chapters, the concept of mechanical equilibrium is defined using the PVP. Assuming that there exist convex creep potentials, minimum principles will be constructed that are entirely similar to those encountered in the domain of elasticity (see Chapter 4).

From the beginning of this chapter, we have seen that in materials subjected to conditions of constant stress and temperature it is possible to observe a plateau for the strain rate. This plateau has been acknowledged to be secondary, or steady-state, creep.

For long periods of time, and neglecting the region known as primary creep, it is reasonable to introduce an approximation for the behavior of the strain at time t as

$$E_{11} = E_{11}^0 + D_{11}^c t, \tag{5.195}$$

where E_{11}^0 is the instantaneous elastic, or elasto-plastic, strain and D_{11}^c is the minimum creep strain rate, corresponding to the steady-state creep region. This approximation is displayed in Figure 5.19.

In many problems relevant to the industry, the strain E_{11}^0 is not relevant and so, for large periods of time, the strain driven by the strain rate D_{11}^c is of primary concern. Thus, the problem of determining the stress state and creep strain rate in a body subjected to constant loads and temperature emerges as a fundamental issue in the domain of mechanics.

5.9.1 Mechanical Equilibrium

Let us consider a body occupying a region of the three-dimensional space $\mathscr{B}$ subjected to the action of a system of external loads $(\mathbf{b}, \mathbf{a})$, which are constant in time. More specifically, $\mathbf{b} = \mathbf{b}(\mathbf{x})$, $\mathbf{x} \in \mathscr{B}$, is a force per unit volume and $\mathbf{a} = \mathbf{a}(\mathbf{x})$, $\mathbf{x} \in \partial\mathscr{B}_a$, is a force per unit surface defined over $\partial\mathscr{B}_a \subset \partial\mathscr{B}$.

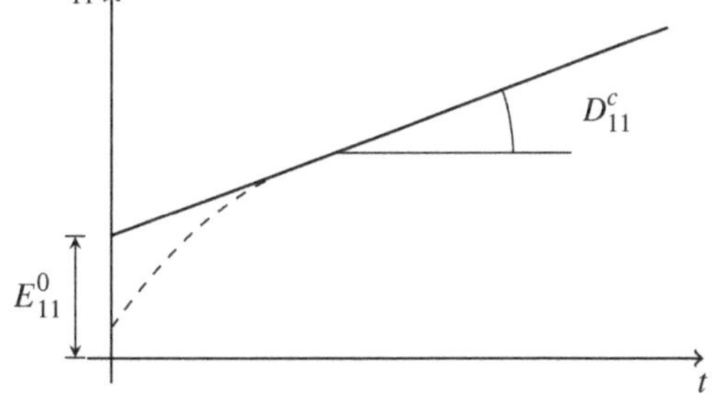

Figure 5.19 Approximation of the strain behavior along time during the secondary creep region.

The body is kinematically constrained by a set of supports. The presentation is limited to the case of bilateral kinematical constraints. This implies that the velocity field, denoted by $\mathbf{v}$, is prescribed at the location of the supports. This means that $\mathbf{v} = \overline{\mathbf{v}}(\mathbf{x})$, $\mathbf{x} \in \partial\mathscr{B}_v$, where $\partial\mathscr{B}_v \subset \partial\mathscr{B}$ is the part of the boundary where these supports are located. In particular, we will assume that $\partial\mathscr{B} = \partial\mathscr{B}_v \cup \partial\mathscr{B}_a$, $\partial\mathscr{B}_v \cap \partial\mathscr{B}_a = \emptyset$.

Furthermore, we admit that the hypothesis of infinitesimal strains holds, and so the actual configurations of the body can be identified with the material (or reference) configuration. Then, if $\mathbf{u} = \mathbf{u}(\mathbf{x}, t)$ is the displacement field experienced by the body when it is subjected to the loads $(\mathbf{b}, \mathbf{a})$, this gives

$$\mathbf{v} = \frac{\partial \mathbf{u}}{\partial t}(\mathbf{x}, t) = \dot{\mathbf{u}}, \tag{5.196}$$

and so the kinematical constraints can be written as

$$\dot{\mathbf{u}} = \overline{\dot{\mathbf{u}}} = \overline{\mathbf{v}} \qquad \text{on } \partial\mathscr{B}_v. \tag{5.197}$$

Hence, the secondary creep mechanical problem consists of determining $\dot{\mathbf{u}}$ in a proper set of kinematically admissible velocity fields Kin_v, the stress tensor field $\mathbf{T}$, and the strain rate tensor field $\mathbf{D}$, such that the variational formulation established by the Principle of Virtual Power and which stands for the definition of the mechanical equilibrium problem (see Chapter 3) is satisfied,

$$-(\mathbf{T}(\mathscr{D}\dot{\mathbf{u}}), \mathscr{D}\hat{\mathbf{v}}) + \langle f, \hat{\mathbf{v}} \rangle = 0 \qquad \forall \hat{\mathbf{v}} \in \text{Var}_v = \text{Var}_{\dot{u}}, \tag{5.198}$$

where the strain rate and the velocity field are related through

$$\mathbf{D} = \mathscr{D}\dot{\mathbf{u}} = (\nabla\dot{\mathbf{u}})^s = \frac{1}{2}(\nabla\dot{\mathbf{u}} + (\nabla\dot{\mathbf{u}})^T), \tag{5.199}$$

or in Cartesian coordinates

$$D_{ij} = \frac{1}{2}\left(\frac{\partial \dot{u}_i}{\partial x_j} + \frac{\partial \dot{u}_j}{\partial x_i}\right). \tag{5.200}$$

The constitutive equation for the deviatoric part of the stress tensor is

$$\mathbf{S} = g(\mathbf{D})\mathbf{D}, \tag{5.201}$$

or in Cartesian coordinates

$$S_{ij} = g(\mathbf{D})D_{ij}. \tag{5.202}$$

It must be remembered that the velocity field satisfies the incompressibility constraint

$$\text{tr}\mathbf{D} = \text{div}\,\dot{\mathbf{u}} = 0, \tag{5.203}$$

or in Cartesian coordinates

$$\frac{\partial \dot{u}_1}{\partial x_1} + \frac{\partial \dot{u}_2}{\partial x_2} + \frac{\partial \dot{u}_3}{\partial x_3} = 0. \tag{5.204}$$

Finally, note that the essential boundary conditions (5.197) are in the very definition of Kin_v and that, for the present case of a three-dimensional body, the system of external loads is shaped as

$$\langle f, \hat{\mathbf{v}} \rangle = \int_{\mathscr{B}} \mathbf{b} \cdot \hat{\mathbf{v}} dV + \int_{\partial\mathscr{B}_a} \mathbf{a} \cdot \hat{\mathbf{v}} dS. \tag{5.205}$$

Notice that the steady-state creep problem is nonlinear by virtue of the nonlinear constitutive equation (5.202). Moreover, it is important to keep in mind that the kinematical constraint of isochoric motion (5.203) must be verified by all kinematically admissible velocity fields. This implies that (5.203) is to be considered in the very definition of the set Kin_v, together with the boundary constraints (5.197) over $\partial\mathscr{B}_v$.

Also worthy of observation is the fact that we have considered the full stress tensor $\mathbf{T}$ in (5.198), notwithstanding that only the deviatoric component $\mathbf{S}$ is responsible for the internal virtual power, once the space Var_v is formed by divergence-free velocity fields, which nullify the contribution to the power exerted by any hydrostatic component, say $p\mathbf{I}$, of the stress tensor $\mathbf{T}$.

5.9.2 Variational Formulation

Let us now take the final step to write the creep problem through the variational framework provided by the PVP.

To do this, first we define the linear manifold of kinematically admissible velocity fields, that is, all the vector fields satisfying the boundary constraints (5.197) and the isochoric motion constraint (5.203). This set, denoted by Kin_v, is then defined by

$$\text{Kin}_v = \{\mathbf{v} \in \mathscr{V}\,;\ \text{div}\,\mathbf{v} = 0,\ \mathbf{v}|_{\partial\mathscr{B}_v} = \overline{\mathbf{v}}\}, \tag{5.206}$$

where $\mathscr{V}$ is a space of sufficiently regular vector functions such that the mathematical operations to be performed make complete sense. Associated with this linear manifold, we have the vector subspace Var_v defined by

$$\text{Var}_v = \{\hat{\mathbf{v}} \in \mathscr{V}\,;\ \text{div}\,\hat{\mathbf{v}} = 0,\ \hat{\mathbf{v}}|_{\partial\mathscr{B}_v} = \mathbf{0}\}, \tag{5.207}$$

also called the space of virtual velocity fields.

Hence, making use of the PVP, the variational problem (compare with (3.210)) associated with the steady-state creep consists of determining $\mathbf{v} \in \text{Kin}_v$ such that

$$\int_{\mathscr{B}} \mathbf{T} \cdot \mathscr{D}\hat{\mathbf{v}}dV = \int_{\mathscr{B}} \mathbf{b} \cdot \hat{\mathbf{v}}dV + \int_{\partial\mathscr{B}_a} \mathbf{a} \cdot \hat{\mathbf{v}}dS \qquad \forall\hat{\mathbf{v}} \in \text{Var}_v, \tag{5.208}$$

where the stress tensor $\mathbf{T}$ is associated with the velocity field $\mathbf{v}$ through the secondary creep constitutive equation, plus the hydrostatic contribution, that is

$$\mathbf{T} = g(\mathscr{D}\mathbf{v})\mathscr{D}\mathbf{v} + p\mathbf{I} = \mathbf{S} + p\mathbf{I}, \tag{5.209}$$

and also

$$\hat{\mathbf{D}} = \mathscr{D}\hat{\mathbf{v}} = (\nabla\hat{\mathbf{v}})^s. \tag{5.210}$$

Introducing the structure of the stress tensor $\mathbf{T}$ given by (5.209) into (5.208), and noting that

$$\mathbf{I} \cdot \hat{\mathbf{D}} = \text{tr}\hat{\mathbf{D}} = \text{div}\hat{\mathbf{v}} = 0, \tag{5.211}$$

by virtue of $\hat{\mathbf{v}} \in \text{Var}_v$, we finally have that the component $p\mathbf{I}$ appearing in (5.209) is irrelevant in the variational problem, which turns to be rewritten as find $\mathbf{v} \in \text{Kin}_v$ such that

$$\int_{\mathscr{B}} g((\nabla\mathbf{v})^s)(\nabla\mathbf{v})^s \cdot (\nabla\hat{\mathbf{v}})^s dV = \int_{\mathscr{B}} \mathbf{b} \cdot \hat{\mathbf{v}}dV + \int_{\partial\mathscr{B}_a} \mathbf{a} \cdot \hat{\mathbf{v}}dS \qquad \forall\hat{\mathbf{v}} \in \text{Var}_v. \tag{5.212}$$

It is now clear that the characterization of the mechanical equilibrium only requires the deviatoric component of the stress tensor $\mathbf{T}$ to be defined in (5.209), that is $\mathbf{S}$.

Similar to the domain of hyperelastic materials, the PVP implicitly carries the strong form of the equilibrium in the form of the Euler–Lagrange equations associated with the variational formulation. This amounts to unveiling the explicit form of the dual (equilibrium) operator $\mathscr{D}^*$. This is achieved by recalling the tensor identity (2.90), which, noting that $\mathbf{S} = \mathbf{S}^T$, after integration yields the divergence theorem,

$$\int_{\mathscr{B}} \mathbf{S} \cdot (\nabla \mathbf{v})^s dV = \int_{\mathscr{B}} \mathbf{S} \cdot \nabla \mathbf{v} dV = \int_{\mathscr{B}} \text{div}\,(\mathbf{S}\mathbf{v}) dV - \int_{\mathscr{B}} \text{div}\,\mathbf{S} \cdot \mathbf{v} dV$$
$$= \int_{\partial\mathscr{B}} \mathbf{S}\mathbf{n} \cdot \mathbf{v} dS - \int_{\mathscr{B}} \text{div}\,\mathbf{S} \cdot \mathbf{v} dV. \tag{5.213}$$

Replacing this result into (5.208), and using the fact that $\hat{\mathbf{v}} = \mathbf{0}$ over $\partial\mathscr{B}_v$, leads to

$$-\int_{\mathscr{B}} (\text{div}\,\mathbf{S} + \mathbf{b}) \cdot \hat{\mathbf{v}} dV + \int_{\partial\mathscr{B}_a} (\mathbf{S}\mathbf{n} - \mathbf{a}) \cdot \hat{\mathbf{v}} dS = 0 \qquad \forall \hat{\mathbf{v}} \in \text{Var}_v. \tag{5.214}$$

Observe that the elements in Var_v are divergence-free. Hence, using the same variational arguments as in previous chapters, from (5.214) we conclude that

$$\begin{cases} \text{div}\,\mathbf{S} + \mathbf{b} = -\nabla\phi & \text{in } \mathscr{B}, \\ \mathbf{S}\mathbf{n} = -\phi\mathbf{n} + \mathbf{a} & \text{on } \partial\mathscr{B}_a, \end{cases} \tag{5.215}$$

where $-\phi$ is a scalar function defined in $\mathscr{B}$, which is a reaction to the kinematical constraint imposed by the isochoric assumption. This result follows from the orthogonality between the space of gradient functions and the space of divergence-free vector fields (see [290] and also Chapter 11).

Since we can write $\nabla\phi = \text{div}\,(\phi\mathbf{I})$, comparing to (5.209), we can perform the identification of ϕ like the pressure field in the continuum, $\phi = p$, which emerges as a consequence of the divergence constraint. Then, from (5.215), together with the kinematical constraints and the constitutive equation we arrive at the secondary creep problem written in strong form, also known as local form

$$\begin{cases} \text{div}\,\mathbf{T} + \mathbf{b} = \mathbf{0} & \text{in } \mathscr{B}, \\ \mathbf{T}\mathbf{n} = \mathbf{a} & \text{on } \partial\mathscr{B}_a, \\ \mathbf{v} = \overline{\mathbf{v}} & \text{on } \partial\mathscr{B}_v, \\ \text{div}\,\mathbf{v} = 0 & \text{in } \mathscr{B}, \\ g((\nabla\mathbf{v})^s)(\nabla\mathbf{v})^s + p\mathbf{I} = \mathbf{T} & \text{in } \mathscr{B}. \end{cases} \tag{5.216}$$

We will now see that, under certain assumptions for function $g(\mathbf{D})$, the variational problem described in (5.212) admits a unique solution for the velocity field.

Consider that $g(\mathbf{D})$ satisfies the following properties: g is a continuous function, also $g = g(\mathbf{D}) \geq 0$ and it is equal to zero if and only if $\mathbf{D} = \mathbf{O}$, and, finally, given arbitrary $\mathbf{D}_1$ and $\mathbf{D}_2$, the following holds

$$(g_2 - g_1)(\mathbf{D}_2 \cdot \mathbf{D}_2 - \mathbf{D}_1 \cdot \mathbf{D}_1) \geq 0, \tag{5.217}$$

where $g_i = g(\mathbf{D}_i)$, $i = 1, 2$.

Provided these properties are satisfied, it is not difficult to prove the following result.

Let $\mathbf{v}_1, \mathbf{v}_2 \in \text{Kin}_v$ be arbitrary velocity fields. Assuming that $g = g(\mathbf{D})$, with $\mathbf{D} = (\nabla \mathbf{v})^s$, satisfies the properties presented before, then

$$(g_2\mathbf{D}_2 - g_1\mathbf{D}_1) \cdot (\mathbf{D}_2 - \mathbf{D}_1) \geq 0, \tag{5.218}$$

and equal to zero if and only if $\mathbf{v}_2 = \mathbf{v}_1$. In turn, recalling that $\mathbf{T}_i = g_i\mathbf{D}_i + p_i\mathbf{I}$, $i = 1, 2$, (5.218) can be rewritten as

$$(\mathbf{T}_2 - \mathbf{T}_1) \cdot (\mathbf{D}_2 - \mathbf{D}_1) \geq 0, \tag{5.219}$$

where the equality holds for $\mathbf{v}_2 = \mathbf{v}_1$.

As a matter of fact, note that

$$\begin{aligned}(g_2\mathbf{D}_2 - g_1\mathbf{D}_1) \cdot (\mathbf{D}_2 - \mathbf{D}_1) &= \frac{1}{2}(g_2 - g_1)(\mathbf{D}_2 \cdot \mathbf{D}_2 - \mathbf{D}_1 \cdot \mathbf{D}_1) \\ &+ \frac{1}{2}(g_2 + g_1)(\mathbf{D}_2 - \mathbf{D}_1) \cdot (\mathbf{D}_2 - \mathbf{D}_1).\end{aligned} \tag{5.220}$$

Clearly, the first term in the right-hand side is positive or zero by virtue of (5.217). In addition, the second term is strictly positive for all $\mathbf{D}_2 - \mathbf{D}_1 \neq \mathbf{O}$ because of the positivity of function g and due to the property of the inner product. Then

$$(g_2\mathbf{D}_2 - g_1\mathbf{D}_1) \cdot (\mathbf{D}_2 - \mathbf{D}_1) \geq 0, \tag{5.221}$$

with the equality holding for $\mathbf{D}_2 = \mathbf{D}_1$.

Now, if $\mathbf{D}_2 = \mathbf{D}_1$, then $\mathbf{D}_2 - \mathbf{D}_1 = \mathbf{O}$ and since $\mathbf{v}_2, \mathbf{v}_1 \in \text{Kin}_v$, we have $\mathbf{v}_2 - \mathbf{v}_1 = \mathbf{0}$ on $\partial\mathcal{B}_v$. Therefore, we have the following boundary value problem

$$\nabla(\mathbf{v}_2 - \mathbf{v}_1) = \mathbf{O} \quad \text{in } \mathcal{B}, \tag{5.222}$$

$$\mathbf{v}_2 - \mathbf{v}_1 = \mathbf{0} \quad \text{on } \partial\mathcal{B}_v, \tag{5.223}$$

whose unique solution is $\mathbf{v}_2 - \mathbf{v}_1 = \mathbf{0}$, that is, $\mathbf{v}_2 = \mathbf{v}_1$. Hence, we finally obtain

$$(g_2\mathbf{D}_2 - g_1\mathbf{D}_1) \cdot (\mathbf{D}_2 - \mathbf{D}_1) \geq 0, \tag{5.224}$$

and the equality is verified if and only if $\mathbf{v}_1 = \mathbf{v}_2$, and the result (5.218) follows.

Now it is easy to demonstrate that the secondary creep variational problem has a unique velocity field as solution, provided it exists. The proof is by contradiction, consider non-trivial velocity fields $\mathbf{v}_2$ and $\mathbf{v}_1$, with $\mathbf{v}_1 \neq \mathbf{v}_2$, which are both solutions of the PVP (5.212), which can be slightly rewritten as follows

$$\int_{\mathcal{B}} g(\mathbf{D}_1)\mathbf{D}_1 \cdot (\mathbf{D}^* - \mathbf{D}_1)dV = \int_{\mathcal{B}} \mathbf{b} \cdot (\mathbf{v}^* - \mathbf{v}_1)dV + \int_{\partial\mathcal{B}_a} \mathbf{a} \cdot (\mathbf{v}^* - \mathbf{v}_1)dS \quad \forall \mathbf{v}^* \in \text{Kin}_v, \tag{5.225}$$

and also

$$\int_{\mathcal{B}} g(\mathbf{D}_2)\mathbf{D}_2 \cdot (\mathbf{D}^* - \mathbf{D}_2)dV = \int_{\mathcal{B}} \mathbf{b} \cdot (\mathbf{v}^* - \mathbf{v}_2)dV + \int_{\partial\mathcal{B}_a} \mathbf{a} \cdot (\mathbf{v}^* - \mathbf{v}_2)dS \quad \forall \mathbf{v}^* \in \text{Kin}_v, \tag{5.226}$$

where $\mathbf{D}_i = (\nabla \mathbf{v}_i)^s$, $i = 1, 2$, $\mathbf{D}^* = (\nabla \mathbf{v}^*)^s$.

Since $\mathbf{v}^*$ is arbitrary, these expressions also hold for a particular $\mathbf{v}^*$, for instance, take $\mathbf{v}^* = \mathbf{v}_2$ and $\mathbf{v}^* = \mathbf{v}_1$ in (5.225) and (5.226), respectively. Thus

$$\int_{\mathcal{B}} g(\mathbf{D}_1)\mathbf{D}_1 \cdot (\mathbf{D}_2 - \mathbf{D}_1)dV = \int_{\mathcal{B}} \mathbf{b} \cdot (\mathbf{v}_2 - \mathbf{v}_1)dV + \int_{\partial\mathcal{B}_a} \mathbf{a} \cdot (\mathbf{v}_2 - \mathbf{v}_1)dS, \tag{5.227}$$

$$\int_{\mathscr{B}} g(\mathbf{D}_2)\mathbf{D}_2 \cdot (\mathbf{D}_1 - \mathbf{D}_2) dV = \int_{\mathscr{B}} \mathbf{b} \cdot (\mathbf{v}_1 - \mathbf{v}_2) dV + \int_{\partial\mathscr{B}_a} \mathbf{a} \cdot (\mathbf{v}_1 - \mathbf{v}_2) dS. \tag{5.228}$$

By adding these two equations we obtain

$$\int_{\mathscr{B}} (g(\mathbf{D}_1)\mathbf{D}_1 - g(\mathbf{D}_2)\mathbf{D}_2) \cdot (\mathbf{D}_2 - \mathbf{D}_1) dV = 0, \tag{5.229}$$

or, equivalently,

$$\int_{\mathscr{B}} (g(\mathbf{D}_2)\mathbf{D}_2 - g(\mathbf{D}_1)\mathbf{D}_1) \cdot (\mathbf{D}_2 - \mathbf{D}_1) dV = 0. \tag{5.230}$$

The result (5.218) tells us that the integrand is strictly positive provided $\mathbf{v}_2 \neq \mathbf{v}_1$. We then reach a contradiction which was originated by the assumption that solutions $\mathbf{v}_2$ and $\mathbf{v}_1$ were different. Therefore, we conclude that, if the solution to the PVP (5.212) exists, it is unique.

With the uniqueness of the velocity field, it follows that the strain rate tensor $\mathbf{D} = (\nabla\mathbf{v})^s$ is also unique and, through the constitutive equation, it follows that the deviatoric component of $\mathbf{T}$, given by $\mathbf{S} = g(\mathbf{D})\mathbf{D}$, is also unique. To demonstrate the uniqueness of the complete stress tensor we have to prove that the hydrostatic component, the pressure p (recall that $\mathbf{T} = \mathbf{S} + p\mathbf{I}$), is also unique.

The proof is carried out using the PVP, which will supply the correct expression involving the pressure field p which is dual to the kinematical constraint associated with the incompressibility. This is achieved by removing this constraint from the space of kinematically admissible variations. So, the divergence-free constraint is not present any more within the space Var_v, denoted now by Var_v^0. Then, from the PVP (5.212) we have that, if there exist two pressure fields $p_1 \neq p_2$, they must be such that

$$-\int_{\mathscr{B}} \nabla(p_2 - p_1) \cdot \hat{\mathbf{v}} dV + \int_{\partial\mathscr{B}_a} (p_2 - p_1)\hat{\mathbf{v}} \cdot \mathbf{n} dS = 0 \qquad \forall \hat{\mathbf{v}} \in \mathrm{Var}_v^0, \tag{5.231}$$

where

$$\mathrm{Var}_v^0 = \{\hat{\mathbf{v}} \in \mathscr{V}\ ;\ \hat{\mathbf{v}}|_{\partial\mathscr{B}_v} = \mathbf{0}\}. \tag{5.232}$$

From the variational equation (5.231) we then have

$$p_2 = p_1, \tag{5.233}$$

and so the uniqueness of the full stress tensor $\mathbf{T}$ follows.

An alternative proof can be constructed using the strong form of the mechanical equilibrium. In effect, replacing $\mathbf{T} = \mathbf{S} + p\mathbf{I}$ in the strong form of the equilibrium (5.215), and since $\mathbf{S}$ is now known, it follows that

$$\nabla p = -(\mathbf{b} + \operatorname{div} \mathbf{S}) \quad \text{in } \mathscr{B}, \tag{5.234}$$

with the boundary condition

$$p = (\mathbf{a} - \mathbf{S}\mathbf{n}) \cdot \mathbf{n} \quad \text{on } \partial\mathscr{B}_a. \tag{5.235}$$

Again, this problem admits a unique solution for p. In effect, if p_1 are p_2 are both solutions, the difference satisfies

$$\nabla(p_1 - p_2) = \mathbf{0} \quad \text{in } \mathscr{B}, \tag{5.236}$$

$$p_1 - p_2 = 0 \quad \text{on } \partial\mathscr{B}_a. \tag{5.237}$$

The solution to this problem is $p_1 = p_2$, and so the uniqueness of the pressure field follows.

We have arrived here at an important result. If the PVP admits a solution, then the velocity field is unique, and so it is the complete stress field, the component constitutively associated as well as the hydrostatic component reactive to the incompressibility constraint.

It is important to highlight that the uniqueness of the entire stress tensor is guaranteed provided there is a part of the boundary, say $\partial\mathscr{B}_a$, with no kinematical constraints. If the steady-state creep variational problem is such that $\partial\mathscr{B} \equiv \partial\mathscr{B}_v$, that is, there are only kinematical constraints over the whole boundary of the body, the stress tensor $\mathbf{T}$ remains uniquely determined up to a hydrostatic component. That is, if $\mathbf{T}^*$ is a solution of the PVP, then $\mathbf{T}^* + \tilde{p}\mathbf{I}$ is also a solution.

5.9.3 Variational Principles of Minimum

Next, let us consider that the secondary creep constitutive equation can be obtained from a dissipation potential function $\psi = \psi(\mathbf{D})$, that is,

$$\mathbf{T} = \psi_{\mathbf{D}} = \frac{\partial \psi}{\partial \mathbf{D}}. \tag{5.238}$$

From the result (5.218) seen in the previous section we also obtained

$$(\mathbf{T}_2 - \mathbf{T}_1) \cdot (\mathbf{D}_2 - \mathbf{D}_1) \geq 0, \tag{5.239}$$

where the pairs $(\mathbf{T}_2, \mathbf{D}_2)$ and $(\mathbf{T}_1, \mathbf{D}_1)$ are associated through the creep constitutive equation. By taking $\mathbf{D}_2 = \mathbf{D}_1 + d\mathbf{D}$ and $\mathbf{T}_2 = \mathbf{T}$ results in

$$(\mathbf{T} - \mathbf{T}_1) \cdot d\mathbf{D} \geq 0. \tag{5.240}$$

Performing the whole process that takes the body from the strain rate state $\mathbf{D}_1$ to the strain rate state $\mathbf{D}_2$ yields

$$\int_{\mathbf{D}_1}^{\mathbf{D}_2} (\mathbf{T} - \mathbf{T}_1) \cdot d\mathbf{D} = \int_{\mathbf{D}_1}^{\mathbf{D}_2} \mathbf{T} \cdot d\mathbf{D} - \mathbf{T}_1 \cdot (\mathbf{D}_2 - \mathbf{D}_1) \geq 0. \tag{5.241}$$

Introducing the potential function ψ such that (5.238) holds, the previous expression takes the form

$$\int_{\mathbf{D}_1}^{\mathbf{D}_2} \psi_{\mathbf{D}} \cdot d\mathbf{D} - \mathbf{T}_1 \cdot (\mathbf{D}_2 - \mathbf{D}_1) \geq 0, \tag{5.242}$$

or, equivalently,

$$\int_{\mathbf{D}_1}^{\mathbf{D}_2} d\psi - \mathbf{T}_1 \cdot (\mathbf{D}_2 - \mathbf{D}_1) \geq 0, \tag{5.243}$$

which finally leads to

$$\psi(\mathbf{D}_2) - \psi(\mathbf{D}_1) \geq \psi_{\mathbf{D}}(\mathbf{D}_1) \cdot (\mathbf{D}_2 - \mathbf{D}_1). \tag{5.244}$$

This last expression states that the potential function is convex. The convexity property is, in this case, a consequence of the property of function $g(\mathbf{D})$ established in (5.218).

Now, replacing the inequality (5.244) in the PVP (5.208), and writing $\hat{\mathbf{v}} = \mathbf{v}^* - \mathbf{v}$, with $\mathbf{v}^* \in \text{Kin}_v$, gives

$$\int_{\mathscr{B}} (\psi(\mathbf{D}^*) - \psi(\mathbf{D}))\, dV \geq \int_{\mathscr{B}} \mathbf{b} \cdot (\mathbf{v}^* - \mathbf{v}) dV + \int_{\partial \mathscr{B}_a} \mathbf{a} \cdot (\mathbf{v}^* - \mathbf{v}) dS \quad \forall \mathbf{v}^* \in \text{Kin}_v, \tag{5.245}$$

and rearranging terms gives

$$\begin{aligned} \Pi(\mathbf{v}^*) = \int_{\mathscr{B}} \psi(\mathbf{D}^*) dV - \int_{\mathscr{B}} \mathbf{b} \cdot \mathbf{v}^* dV - \int_{\partial \mathscr{B}_a} \mathbf{a} \cdot \mathbf{v}^* dS \geq \\ \int_{\mathscr{B}} \psi(\mathbf{D}) dV - \int_{\mathscr{B}} \mathbf{b} \cdot \mathbf{v} dV - \int_{\partial \mathscr{B}_a} \mathbf{a} \cdot \mathbf{v} dS = \Pi(\mathbf{v}) \qquad \forall \mathbf{v}^* \in \text{Kin}_v, \end{aligned} \tag{5.246}$$

where the equality holds for $\mathbf{v}^* = \mathbf{v}$.

In this manner, we have arrived at the secondary creep problem written as a minimization problem. That is, the secondary creep problem corresponding to the mechanical equilibrium is equivalent to determining $\mathbf{v}_0 \in \text{Kin}_v$ such that the functional Π is minimized, that is

$$\begin{aligned} \Pi(\mathbf{v}_0) &= \min_{\mathbf{v} \in \text{Kin}_v} \Pi(\mathbf{v}) \\ &= \min_{\mathbf{v} \in \text{Kin}_v} \left[\int_{\mathscr{B}} \psi((\nabla \mathbf{v})^s) dV - \int_{\mathscr{B}} \mathbf{b} \cdot \mathbf{v} dV - \int_{\partial \mathscr{B}_a} \mathbf{a} \cdot \mathbf{v} dS \right]. \end{aligned} \tag{5.247}$$

Up to this point, we have presented the secondary creep problem within a formulation which describes the creep velocity field. Next, we will address the dual variational formulation, in which the principal variable of interest is the stress state.

In order to construct this variational formulation let us define, analogously to what was done in Chapter 3 (and also in Chapter 4), the set of all stress fields which are statically equilibrated with the system of external loads f characterized in the present context by the forces per unit volume $\mathbf{b}$ and per unit area $\mathbf{a}$, that is

$$\text{Est}_T = \{\mathbf{T} \in \mathscr{W}'; -(\mathbf{T}, \mathscr{D}\hat{\mathbf{v}}) + \langle f, \hat{\mathbf{v}} \rangle = 0 \quad \forall \hat{\mathbf{v}} \in \text{Var}_v\}. \tag{5.248}$$

This set can be expressed in terms of the strong form of the mechanical equilibrium as follows

$$\text{Est}_T = \{\mathbf{T} \in \mathscr{W}'; \ \text{div}\, \mathbf{T} + \mathbf{b} = \mathbf{0} \quad \text{in } \mathscr{B}, \ \mathbf{T}\mathbf{n} = \mathbf{a} \quad \text{on } \partial \mathscr{B}_a\}. \tag{5.249}$$

Conceptually, Est_T contains all the stress fields which satisfy the mechanical equilibrium for a given system of external loads. From this set, we directly have the space of stress fields which are self-equilibrated

$$\text{Var}_T = \{\hat{\mathbf{T}} \in \mathscr{W}'; -(\hat{\mathbf{T}}, \mathscr{D}\hat{\mathbf{v}}) = 0 \quad \forall \hat{\mathbf{v}} \in \text{Var}_v\}, \tag{5.250}$$

or in the strong form

$$\text{Var}_T = \{\hat{\mathbf{T}} \in \mathscr{W}'; \ \text{div}\hat{\mathbf{T}} = \mathbf{0} \quad \text{in } \mathscr{B}, \ \hat{\mathbf{T}}\mathbf{n} = \mathbf{0} \quad \text{on } \partial \mathscr{B}_a\}, \tag{5.251}$$

which means that all elements in Var_T are at equilibrium with the homogeneous system of forces ($\mathbf{b} = \mathbf{0}$ and $\mathbf{a} = \mathbf{0}$). Furthermore, from the linearity of the equilibrium equations it follows that, given arbitrary $\mathbf{T}_1, \mathbf{T}_2 \in \text{Est}_T$, the difference is a self-equilibrated stress field, that is, $\mathbf{T}_1 - \mathbf{T}_2 \in \text{Var}_T$.

From the definition of the secondary creep problem, we see that the stress field satisfying the equilibrium must be given by a field $\mathbf{T} \in \mathrm{Est}_T$ such that the strain rate associated with this field through the constitutive equation is compatible, that is, such that the strain rate can be written as the symmetric gradient of a kinematically admissible velocity field. Thus, the problem consists of determining the stress field meeting all these conditions. The dual variational formulation is basically the Principle of Complementary Virtual Power (PCVP) applied to the case of materials exhibiting creep phenomena (see Chapter 3). This variational equation allows us to characterize the stress field by finding $\mathbf{T} \in \mathrm{Est}_T$ such that

$$\int_{\mathscr{B}} \mathbf{D} \cdot (\mathbf{T}^* - \mathbf{T}) dV = \int_{\partial\mathscr{B}_v} (\mathbf{T}^* - \mathbf{T})\mathbf{n} \cdot \bar{\mathbf{v}} dS \qquad \forall \mathbf{T}^* \in \mathrm{Est}_T, \tag{5.252}$$

or, equivalently,

$$\int_{\mathscr{B}} \mathbf{D} \cdot \hat{\mathbf{T}} dV = \int_{\partial\mathscr{B}_v} \hat{\mathbf{T}}\mathbf{n} \cdot \bar{\mathbf{v}} dS \qquad \forall \hat{\mathbf{T}} \in \mathrm{Var}_T, \tag{5.253}$$

where $\mathbf{D} = h(\mathbf{T})$, implying that the strain rate is associated with $\mathbf{T}$ through the (inverse) constitutive equation.

As before, we now admit the existence of the so-called complementary potential function $\phi = \phi(\mathbf{T})$ such that

$$\mathbf{D} = \phi_{\mathbf{T}} = \frac{\partial \phi}{\partial \mathbf{T}}. \tag{5.254}$$

Recalling the property of function g given by (5.239) (see also (5.218)), if we take $\mathbf{T}_2 = \mathbf{T}_1 + d\mathbf{T}$ and $\mathbf{D}_2 = \mathbf{D}$ we obtain

$$d\mathbf{T} \cdot (\mathbf{D} - \mathbf{D}_1) \geq 0, \tag{5.255}$$

and for every process that goes from the stress state $\mathbf{T}_1$ to the state $\mathbf{T}_2$ we get

$$\int_{\mathbf{T}_1}^{\mathbf{T}_2} d\mathbf{T} \cdot (\mathbf{D} - \mathbf{D}_1) \geq 0, \tag{5.256}$$

which yields

$$\int_{\mathbf{T}_1}^{\mathbf{T}_2} \mathbf{D} \cdot d\mathbf{T} - \mathbf{D}_1 \cdot (\mathbf{T}_2 - \mathbf{T}_1) \geq 0, \tag{5.257}$$

and equivalently

$$\int_{\mathbf{T}_1}^{\mathbf{T}_2} \phi_{\mathbf{T}} \cdot d\mathbf{T} - \phi_{\mathbf{T}}(\mathbf{T}_1) \cdot (\mathbf{T}_2 - \mathbf{T}_1) \geq 0, \tag{5.258}$$

from where

$$\phi(\mathbf{T}_2) - \phi(\mathbf{T}_1) \geq \phi_{\mathbf{T}}(\mathbf{T}_1) \cdot (\mathbf{T}_2 - \mathbf{T}_1), \tag{5.259}$$

implying that the complementary function ϕ is also a convex potential.

Introducing this result into the variational equation (5.252), after rearranging terms, we obtain

$$\begin{aligned}\Pi^*(\mathbf{T}^*) &= \int_{\mathscr{B}} \phi(\mathbf{T}^*)dV - \int_{\partial\mathscr{B}_v} \mathbf{T}^*\mathbf{n}\cdot\bar{\mathbf{v}}dS \\ &\geq \int_{\mathscr{B}} \phi(\mathbf{T})dV - \int_{\partial\mathscr{B}_v} \mathbf{T}\mathbf{n}\cdot\bar{\mathbf{v}}dS = \Pi^*(\mathbf{T}) \qquad \forall \mathbf{T}^* \in \mathrm{Est}_T. \end{aligned} \tag{5.260}$$

In this manner, the secondary creep problem is equivalent to finding $\mathbf{T}_0 \in \mathrm{Est}_T$ such that the functional Π^* is minimized, that is,

$$\Pi^*(\mathbf{T}_0) = \min_{\mathbf{T}\in\mathrm{Est}_T} \Pi^*(\mathbf{T}) = \min_{\mathbf{T}\in\mathrm{Est}_T} \left[\int_{\mathscr{B}} \phi(\mathbf{T})dV - \int_{\partial\mathscr{B}_v} \mathbf{T}\mathbf{n}\cdot\bar{\mathbf{v}}dS \right]. \tag{5.261}$$

As in Chapter 4 (see (4.110)), the functionals Π and Π^* present in (5.247) and (5.261) can be related through a single inequality. In fact, let $\mathbf{v}_0$, $\mathbf{D}_0$ and $\mathbf{T}_0$ be the fields which are the solution of the secondary creep problem, and therefore solutions for the minimum problems. Also, consider a kinematically admissible velocity field $\mathbf{v} \in \mathrm{Kin}_v$ and the associated strain rate $\mathbf{D} = (\nabla\mathbf{v})^s$. Finally, let $\mathbf{T} \in \mathrm{Est}_T$ be a statically equilibrated stress field.

Hence, recalling that

$$\psi(\mathbf{D}_0) + \phi(\mathbf{T}_0) = \mathbf{T}_0 \cdot \mathbf{D}_0, \tag{5.262}$$

for $\mathbf{T}_0$ and $\mathbf{D}_0$ constitutively related, we can write

$$\begin{aligned}\int_{\mathscr{B}} \psi(\mathbf{D}_0)dV &= \int_{\mathscr{B}} \mathbf{T}_0\cdot\mathbf{D}_0 dV - \int_{\mathscr{B}} \phi(\mathbf{T}_0)dV \\ &= \int_{\partial\mathscr{B}_a} \mathbf{a}\cdot\mathbf{v}dS + \int_{\partial\mathscr{B}_v} \mathbf{T}_0\mathbf{n}\cdot\mathbf{v}_0 dS + \int_{\mathscr{B}} \mathbf{b}\cdot\mathbf{v}dV - \int_{\mathscr{B}} \phi(\mathbf{T})dV, \end{aligned} \tag{5.263}$$

and, exploiting the minimum problem (5.247), this yields

$$\begin{aligned}\Pi(\mathbf{v}) \geq \Pi(\mathbf{v}_0) &= \int_{\mathscr{B}} \psi(\mathbf{D}_0)dV - \int_{\partial\mathscr{B}_a} \mathbf{a}\cdot\mathbf{v}dS - \int_{\mathscr{B}} \mathbf{b}\cdot\mathbf{v}dV \\ &= -\int_{\mathscr{B}} \phi(\mathbf{T}_0)dV + \int_{\partial\mathscr{B}_v} \mathbf{T}_0\mathbf{n}\cdot\bar{\mathbf{v}}dS \\ &= -\Pi^*(\mathbf{T}_0) \geq -\Pi^*(\mathbf{T}) \qquad \forall\mathbf{v}\in\mathrm{Kin}_v \qquad \forall\mathbf{T}\in\mathrm{Est}_T. \end{aligned} \tag{5.264}$$

Hence, from known quantities we can provide upper and lower bounds for the quantity $\Pi(\mathbf{v}_0) = -\Pi^*(\mathbf{T}_0)$.

For certain systems of external loads and kinematical boundary constraints, the previous result enables us to bound the dissipation potential $\int_{\mathscr{B}} \psi(\mathbf{D})dV$ and the complementary dissipation potential $\int_{\mathscr{B}} \phi(\mathbf{T})dV$. Let us explore two examples.

Consider a first scenario for which $\mathbf{b} = \mathbf{0}$ in $\mathscr{B}$, and also that $\bar{\mathbf{v}} = \mathbf{0}$ on $\partial\mathscr{B}_v$. Then (5.264) becomes

$$\begin{aligned} -\int_{\mathscr{B}} \psi(\mathbf{D})dV + \int_{\partial\mathscr{B}_a} \mathbf{a}\cdot\mathbf{v}dS &\leq \int_{\mathscr{B}} \phi(\mathbf{T}_0)dV \\ &\leq \int_{\mathscr{B}} \phi(\mathbf{T})dV \qquad \forall\mathbf{v}\in\mathrm{Kin}_v, \qquad \forall\mathbf{T}\in\mathrm{Est}_T. \end{aligned} \tag{5.265}$$

Since $\mathbf{v}$ is satisfies homogeneous boundary constraints on $\partial\mathscr{B}_v$, then the field $\lambda\mathbf{v}$, with $\lambda \in \mathbb{R}$, also satisfies the kinematical constraint regardless the value of λ (note that here $\text{Kin}_v \equiv \text{Var}_v$).

Thus, for a given element $\mathbf{v}$ (a direction for the motion action), we can obtain an optimal lower bound for the complementary dissipation potential through the following unconstrained maximization problem for the variable λ

$$\max_{\lambda\in\mathbb{R}} \left[\lambda \int_{\partial\mathscr{B}_a} \mathbf{a}\cdot\mathbf{v}dS - \int_{\mathscr{B}} \psi(\lambda\mathbf{D})dV\right] \leq \int_{\mathscr{B}} \phi(\mathbf{T}_0)dV. \tag{5.266}$$

In the second scenario, consider that $\mathbf{b} = \mathbf{0}$ in $\mathscr{B}$ and $\mathbf{a} = \mathbf{0}$ on $\partial\mathscr{B}_a$. For this case, (5.264) leads to

$$-\int_{\mathscr{B}} \phi(\mathbf{T})dV + \int_{\partial\mathscr{B}_v} \mathbf{Tn}\cdot\overline{\mathbf{v}}dS \leq \int_{\mathscr{B}} \psi(\mathbf{D}_0)dV$$
$$\leq \int_{\mathscr{B}} \psi(\mathbf{D})dV \qquad \forall\mathbf{v}\in\text{Kin}_v, \qquad \forall\mathbf{T}\in\text{Est}_T. \tag{5.267}$$

Using similar arguments, for $\mathbf{T} \in \text{Est}_T$ (in this case $\text{Est}_T \equiv \text{Var}_T$), it turns out that $\lambda\mathbf{T} \in \text{Est}_T$, for all $\lambda \in \mathbb{R}$. Then, having defined an element $\mathbf{T}$ (a stress direction), we can provide for such direction an optimal lower bound for the dissipation potential following the unconstrained maximization process in the variable λ

$$\max_{\lambda\in\mathbb{R}} \left[\lambda \int_{\partial\mathscr{B}_v} \mathbf{Tn}\cdot\overline{\mathbf{v}}dS - \int_{\mathscr{B}} \phi(\lambda\mathbf{T})dV\right] \leq \int_{\mathscr{B}} \psi(\mathbf{D}_0)dV. \tag{5.268}$$

5.10 Castigliano Theorems

5.10.1 First and Second Theorems

In this section we will assume that the force per unit volume $\mathbf{b}$ is zero, and that the force per unit area $\mathbf{a}$ can be written as a uniparametric function. This implies that there exists a parameter, say Q, called the generalized load parameter, and a vector field $\mathbf{l}_a$, which characterizes the load pattern through

$$\mathbf{a} = Q\mathbf{l}_a \quad \text{on } \partial\mathscr{B}_a. \tag{5.269}$$

Introducing this load into the PVP (5.212) we have

$$\int_{\mathscr{B}} g((\nabla\mathbf{v})^s)(\nabla\mathbf{v})^s\cdot(\nabla\hat{\mathbf{v}})^s dV = Q\int_{\partial\mathscr{B}_a} \mathbf{l}_a\cdot\hat{\mathbf{v}}dS = Q\mathsf{q}(\hat{\mathbf{v}}) \qquad \forall\hat{\mathbf{v}}\in\text{Var}_v, \tag{5.270}$$

where q is called the generalized velocity.

Because (5.270) holds for all $\hat{\mathbf{v}} \in \text{Var}_v$, in particular it is verified for $\hat{\mathbf{v}}_1$ such that

$$\mathsf{q}(\hat{\mathbf{v}}_1) = \int_{\partial\mathscr{B}_a} \mathbf{l}_a\cdot\hat{\mathbf{v}}_1 dS = 1, \tag{5.271}$$

which leads us to

$$Q = \int_{\mathscr{B}} g((\nabla\mathbf{v})^s)(\nabla\mathbf{v})^s\cdot(\nabla\hat{\mathbf{v}}_1)^s dV. \tag{5.272}$$

In this way, we can see that given a velocity field $\mathbf{v}$ and the specific virtual velocity field $\hat{\mathbf{v}}_1$ (associated with a unitary generalized velocity), we can calculate the generalized load Q which is responsible for it. This is known in the literature as the fictitious velocity method.

In an entirely similar manner, it is possible to start from the Principle of Complementary Virtual Power and arrive at an expression that enables us to calculate a generalized velocity, as we will see next.

Suppose that the kinematical constraint is written as a uniparametric vector function

$$\overline{\mathbf{v}} = \overline{q}\mathbf{l}_v, \tag{5.273}$$

where $\overline{q}$ is the generalized velocity parameter and $\mathbf{l}_v$ is a vector field. Thus, putting this expression into (5.253), we obtain

$$\int_{\mathscr{B}} \mathbf{D}(\mathbf{T}) \cdot \hat{\mathbf{T}} dV = \mathsf{Q}(\hat{\mathbf{T}})\overline{q} \qquad \forall \hat{\mathbf{T}} \in \mathrm{Var}_T, \tag{5.274}$$

where

$$\mathsf{Q}(\hat{\mathbf{T}}) = \int_{\partial \mathscr{B}_v} \hat{\mathbf{T}}\mathbf{n} \cdot \mathbf{l}_v dS. \tag{5.275}$$

Then, picking a self-equilibrated stress field $\hat{\mathbf{T}}_1$ such that

$$\mathsf{Q}(\hat{\mathbf{T}}_1) = \int_{\partial \mathscr{B}_v} \hat{\mathbf{T}}_1\mathbf{n} \cdot \mathbf{l}_v dS = 1, \tag{5.276}$$

yields

$$\overline{q} = \int_{\mathscr{B}} \mathbf{D}(\mathbf{T}) \cdot \hat{\mathbf{T}}_1 dV. \tag{5.277}$$

This formula allows us to calculate the generalized velocity field $\overline{q}$ in the direction $\mathbf{l}_v$ produced by the stress field $\mathbf{T}$ through the knowledge of a self-equilibrated stress field conveniently chosen. This procedure is referred to in the literature as the fictitious force method.

Let us now admit the existence of the dissipation potentials, and the variational formulation corresponding to the minimum corresponding problems will lead us to the same results found above. Hence, for the generalized load Q, the minimum problem (5.247) turns into the problem of determining $\mathbf{v} \in \mathrm{Kin}_v$ such that

$$\Pi(\mathbf{v}) = \min_{\mathbf{v}^* \in \mathrm{Kin}_v} \left[\int_{\mathscr{B}} \psi((\nabla \mathbf{v}^*)^s) dV - Q\mathsf{q}(\mathbf{v}^*) \right], \tag{5.278}$$

where

$$\mathsf{q}(\mathbf{v}^*) = \int_{\partial \mathscr{B}_a} \mathbf{l}_a \cdot \mathbf{v}^* dS, \tag{5.279}$$

is the generalized velocity associated with the load Q.

Putting $(\nabla \mathbf{v}^*)^s = \mathbf{D}^*(\mathsf{q})$ we can then rewrite the problem in terms of q as follows,

$$\min_{\mathbf{v}^* \in \mathrm{Kin}_v} \Pi(\mathbf{v}^*) = \min_{\mathsf{q} \in \mathbb{R}} \left[\int_{\mathscr{B}} \psi(\mathbf{D}^*(\mathsf{q})) dV - Q\mathsf{q} \right], \tag{5.280}$$

and the necessary condition for a critical point results,

$$\frac{\partial}{\partial \mathsf{q}} \int_{\mathscr{B}} \psi(\mathbf{D}^*(\mathsf{q}))dV = Q, \tag{5.281}$$

which is nothing but the expression of the Castigliano's first theorem for the creep problem. In short, this theorem states that, if the potential $\int_{\mathscr{B}} \psi dV$ is written as a function of the generalized velocity q, the derivative with respect to this velocity is equal to the associated generalized load.

Consider now the minimum problem (5.261). Then, in such a case the problem can be changed to determining $\mathbf{T} \in \mathrm{Est}_T$ such that

$$\Pi^*(\mathbf{T}) = \min_{\mathbf{T}^* \in \mathrm{Est}_T} \left[\int_{\mathscr{B}} \phi(\mathbf{T}^*)dV - \mathsf{Q}(\mathbf{T}^*)\overline{q} \right], \tag{5.282}$$

where now

$$\mathsf{Q}(\mathbf{T}^*) = \int_{\partial \mathscr{B}_v} \mathbf{T}^* \mathbf{n} \cdot \mathbf{l}_v dS. \tag{5.283}$$

By admitting that we can put $\mathbf{T}^*(\mathsf{Q})$, we have

$$\min_{\mathbf{T}^* \in \mathrm{Est}_T} \Pi^*(\mathbf{T}^*) = \min_{\mathsf{Q} \in \mathbb{R}} \left[\int_{\mathscr{B}} \phi(\mathbf{T}^*(\mathsf{Q}))dV - \mathsf{Q}\overline{q} \right], \tag{5.284}$$

and the condition for a minimum becomes

$$\frac{\partial}{\partial \mathsf{Q}} \int_{\mathscr{B}} \phi(\mathbf{T}^*(\mathsf{Q}))dV = \overline{q}, \tag{5.285}$$

which is the Castigliano's second theorem, which states that, if the complementary potential $\int_{\mathscr{B}} \phi dV$ is written in terms of the generalized load Q, the derivative with respect to this load equals the associated generalized velocity.

The techniques presented in this section to compute generalized loads Q or generalized velocities q are completely equivalent, and the usage depends on the familiarity of the practitioner with either one or the other method.

5.10.2 Bounds for Velocities and Generalized Loads

Some of the results found in previous sections will be particularized to the case of Norton constitutive laws. We have seen that the constitutive equations, through the dissipation potentials (5.84) and (5.66), acquire the following forms

$$\psi(\mathbf{D}) = \frac{n}{n+1}\left(\frac{1}{k}\right)^{\frac{1}{n}}\left(\frac{2}{3}\mathbf{D}\cdot\mathbf{D}\right)^{\frac{n+1}{2n}} = \frac{n}{n+1}\left(\frac{1}{k}\right)^{\frac{1}{n}}(D_e)^{\frac{n+1}{n}}, \tag{5.286}$$

$$\phi(\mathbf{T}) = \frac{k}{n+1}\left(\frac{3}{2}\mathbb{C}\mathbf{T}\cdot\mathbb{C}\mathbf{T}\right)^{\frac{n+1}{2}} = \frac{k}{n+1}T_e^{n+1}. \tag{5.287}$$

Also, recalling (5.57) and (5.88), and since (5.89) holds, it follows that

$$\phi(\mathbf{T}) = \frac{1}{n+1}\mathbf{T}\cdot\mathbf{D}, \qquad \psi(\mathbf{D}) = \frac{n}{n+1}\mathbf{T}\cdot\mathbf{D}. \tag{5.288}$$

In turn, the tensor $\mathbb{C}$ introduced in (5.68) is such that

$$\mathbb{C}\mathbf{T} = \mathbf{T} - \frac{1}{3}(\mathrm{tr}\mathbf{T})\mathbf{I} = \mathbf{S}. \tag{5.289}$$

Therefore, it follows that ψ and ϕ are homogeneous functions of degree $(n+1)/n$ and $n+1$, respectively, which implies that

$$\psi(\lambda \mathbf{D}) = \lambda^{(n+1)/n}\psi(\mathbf{D}), \tag{5.290}$$

$$\phi(\lambda \mathbf{T}) = \lambda^{n+1}\phi(\mathbf{T}). \tag{5.291}$$

The previous observation is important because it enables us to derive sharp estimates. In particular we have the two situations addressed next.

Consider that the force per unit volume is $\mathbf{b} = \mathbf{0}$, and also consider a kinematical constraint $\overline{\mathbf{v}} = \mathbf{0}$. So, given the motion direction $\mathbf{v}$, the lower bound for the cost functional $\int_{\mathscr{B}} \phi(\mathbf{T}_0)dV$ is given through the solution of problem (5.266). Hence, from (5.290) we obtain

$$\max_{\lambda \in \mathbb{R}} \left[\lambda \int_{\partial \mathscr{B}_a} \mathbf{a} \cdot \mathbf{v} dS - \lambda^{(n+1)/n} \int_{\mathscr{B}} \psi(\mathbf{D}) dV \right] \leq \int_{\mathscr{B}} \phi(\mathbf{T}_0) dV \leq \int_{\mathscr{B}} \phi(\mathbf{T}) dV. \tag{5.292}$$

Straightforward calculations lead us to the solution $\lambda_{\max}$ given by

$$\lambda_{\max} = \left(\frac{n}{n+1} \frac{\int_{\partial \mathscr{B}_a} \mathbf{a} \cdot \mathbf{v} dS}{\int_{\mathscr{B}} \psi(\mathbf{D}) dV} \right)^n. \tag{5.293}$$

Therefore, the optimal lower bound follows

$$\lambda_{\max} \int_{\partial \mathscr{B}_a} \mathbf{a} \cdot \mathbf{v} dS - \lambda_{\max}^{(n+1)/n} \int_{\mathscr{B}} \psi(\mathbf{D}) dV \leq \int_{\mathscr{B}} \phi(\mathbf{T}_0) dV \leq \int_{\mathscr{B}} \phi(\mathbf{T}) dV, \tag{5.294}$$

and then gives

$$\lambda_{\max} \int_{\partial \mathscr{B}_a} \mathbf{a} \cdot \mathbf{v} dS \leq (n+1) \int_{\mathscr{B}} \phi(\mathbf{T}_0) dV \leq (n+1) \int_{\mathscr{B}} \phi(\mathbf{T}) dV. \tag{5.295}$$

In the case that $\mathbf{v} = \mathbf{v}_0$ is the solution of the creep problem, we have $\lambda_{\max} = 1$ and so

$$\int_{\partial \mathscr{B}_a} \mathbf{a} \cdot \mathbf{v}_0 dS \leq (n+1) \int_{\mathscr{B}} \phi(\mathbf{T}) dV. \tag{5.296}$$

Assuming that the force per unit area $\mathbf{a}$ applied is associated with the generalized load Q, we finally get

$$\mathsf{Q}\mathsf{q}(\mathbf{v}_0) \leq (n+1) \int_{\mathscr{B}} \phi(\mathbf{T}) dV, \tag{5.297}$$

and so

$$\mathsf{q}(\mathbf{v}_0) \leq \frac{1}{\mathsf{Q}}(n+1) \int_{\mathscr{B}} \phi(\mathbf{T}) dV, \tag{5.298}$$

delivering an upper bound for the generalized velocity associated with Q just by providing a stress state $\mathbf{T}$ equilibrated with the generalized load Q.

Consider now the case where the loads are $\mathbf{b} = \mathbf{0}$ and $\mathbf{a} = \mathbf{0}$. In a similar manner, it is possible to obtain an upper bound for the generalized load. In fact, for the chosen constitutive equation we have that the combination of problem (5.268) with the property

(5.291) gives the optimal lower bound for the cost functional $\int_{\mathscr{B}} \psi(\mathbf{D}_0)dV$ in the direction of $\mathbf{T} \in \mathrm{Est}_T$ as

$$\max_{\lambda \in \mathbb{R}} \left[\lambda \int_{\partial \mathscr{B}_v} \mathbf{Tn} \cdot \overline{\mathbf{v}} dS - \lambda^{n+1} \int_{\mathscr{B}} \phi(\mathbf{T}) dV \right] \le \int_{\mathscr{B}} \psi(\mathbf{D}_0) dV \le \int_{\mathscr{B}} \psi(\mathbf{D}) dV, \quad (5.299)$$

and the solution of the maximization problem, denoted by $\lambda_{\max}$, is given by

$$\lambda_{\max} = \left(\frac{1}{n+1} \frac{\int_{\partial \mathscr{B}_v} \mathbf{Tn} \cdot \overline{\mathbf{v}} dS}{\int_{\mathscr{B}} \phi(\mathbf{T}) dV} \right)^{1/n}. \quad (5.300)$$

Then

$$\lambda_{\max} \int_{\partial \mathscr{B}_v} \mathbf{Tn} \cdot \overline{\mathbf{v}} dS - \lambda_{\max}^{n+1} \int_{\mathscr{B}} \phi(\mathbf{T}) dV \le \int_{\mathscr{B}} \psi(\mathbf{D}_0) dV \le \int_{\mathscr{B}} \psi(\mathbf{D}) dV, \quad (5.301)$$

and

$$\lambda_{\max} \int_{\partial \mathscr{B}_v} \mathbf{Tn} \cdot \overline{\mathbf{v}} dS \le \frac{n+1}{n} \int_{\mathscr{B}} \psi(\mathbf{D}_0) dV \le \frac{n+1}{n} \int_{\mathscr{B}} \psi(\mathbf{D}) dV. \quad (5.302)$$

In addition, if $\mathbf{T} = \mathbf{T}_0$ is the solution of the creep problem, it is $\lambda_{\max} = 1$, and therefore

$$\int_{\partial \mathscr{B}_v} \mathbf{T}_0 \mathbf{n} \cdot \overline{\mathbf{v}} dS \le \frac{n+1}{n} \int_{\mathscr{B}} \psi(\mathbf{D}) dV. \quad (5.303)$$

Assuming that $\overline{\mathbf{v}}$ can be written by means of the generalized velocity q, we finally have

$$\mathrm{Q}(\mathbf{T}_0) q \le \frac{n+1}{n} \int_{\mathscr{B}} \psi(\mathbf{D}) dV, \quad (5.304)$$

and so

$$\mathrm{Q}(\mathbf{T}_0) \le \frac{1}{q} \frac{n+1}{n} \int_{\mathscr{B}} \psi(\mathbf{D}) dV, \quad (5.305)$$

is the upper bound for the generalized load associated with the prescribed generalized velocity q, and where $\mathbf{v} \in \mathrm{Kin}_v$ is compatible with the generalized velocity q.

The estimates derived so far are valid for the velocities and generalized loads. Now, we will present the so-called Martin estimates (see [187] and [39]), which provide bounds for the generalized velocity at any point in the structure.

In order to do this, let us remember the following result, which is a consequence of the convexity of the creep dissipation potentials

$$\psi(\mathbf{D}_1) + \phi(\mathbf{T}_2) \ge \mathbf{D}_1 \cdot \mathbf{T}_2, \quad (5.306)$$

where the equality holds for $\mathbf{D}_1$ and $\mathbf{T}_2$ related through the constitutive law.

Calling $\mathbf{T}_1$ and $\mathbf{D}_2$ the stress state and the strain rate associated, respectively, via the constitutive law, with $\mathbf{D}_1$ and $\mathbf{T}_2$, using (5.288) we have

$$\psi(\mathbf{D}_1) = \frac{n}{n+1} \mathbf{T}_1 \cdot \mathbf{D}_1, \quad (5.307)$$

$$\phi(\mathbf{T}_2) = \frac{1}{n+1} \mathbf{T}_2 \cdot \mathbf{D}_2, \quad (5.308)$$

which placed in (5.306) lead to

$$\frac{n}{n+1}\mathbf{T}_1 \cdot \mathbf{D}_1 + \frac{1}{n+1}\mathbf{T}_2 \cdot \mathbf{D}_2 \geq \mathbf{D}_1 \cdot \mathbf{T}_2. \tag{5.309}$$

Putting $\mathbf{T}_1$ and $\mathbf{D}_1$ as the solution of the creep problem, that is $\mathbf{T}_0$ and $\mathbf{D}_0$, for boundary conditions $\mathbf{a}$ over $\partial\mathscr{B}_a$ and $\overline{\mathbf{v}} = \mathbf{0}$ on $\partial\mathscr{B}_v$, and putting $\mathbf{T}_2$ and $\mathbf{D}_2$ as the solution $\mathbf{T}^*$ and $\mathbf{D}^*$ for a, say arbitrary, load $\mathbf{a}^*$ and assuming $\mathbf{b} = \mathbf{0}$ in $\mathscr{B}$, we obtain

$$\begin{aligned}\frac{1}{n+1}\int_{\mathscr{B}} \mathbf{T}^* \cdot \mathbf{D}^* dV &\geq \int_{\mathscr{B}} \mathbf{T}^* \cdot \mathbf{D}_0 dV - \frac{n}{n+1}\int_{\mathscr{B}} \mathbf{T}_0 \cdot \mathbf{D}_0 dV \\ &= \int_{\partial\mathscr{B}_a} \left(\mathbf{a}^* - \frac{n}{n+1}\mathbf{a}\right) \cdot \mathbf{v}_0 dV. \end{aligned} \tag{5.310}$$

This result makes possible the construction of an estimate for the velocity at any point $\mathbf{x}$ in the component. Indeed, choosing $\mathbf{a}^*$ as

$$\mathbf{a}^* = \frac{n}{n+1}\mathbf{a} + Q^*\mathbf{l}_{a_x}, \tag{5.311}$$

where $Q^*\mathbf{l}_{a_x}$ is a load concentrated at point $\mathbf{x}$ in the direction in which the generalized velocity is to be bounded, the inequality (5.310) yields

$$\frac{1}{n+1}\int_{\mathscr{B}} \mathbf{T}^* \cdot \mathbf{D}^* dV \geq \int_{\partial\mathscr{B}_a} Q^*\mathbf{l}_{a_x} \cdot \mathbf{v}_0 dS = Q^*\mathsf{q}(\mathbf{v}_0), \tag{5.312}$$

therefore

$$\mathsf{q}(\mathbf{v}_0) \leq \frac{1}{Q^*}\frac{1}{n+1}\int_{\mathscr{B}} \mathbf{T}^* \cdot \mathbf{D}^* dV. \tag{5.313}$$

Since the fields $\mathbf{T}^*$ and $\mathbf{D}^*$ depend on Q^*, we can optimize the upper bound, getting

$$\mathsf{q}(\mathbf{v}_0) \leq \min_{Q^* \in \mathbb{R}} \frac{1}{Q^*}\frac{1}{n+1}\int_{\mathscr{B}} \mathbf{T}^* \cdot \mathbf{D}^* dV. \tag{5.314}$$

5.11 Examples of Application

5.11.1 Disk Rotating with Constant Angular Velocity

In this example, the variational formulation for the creep problem is applied to the case of a ring rotating at a constant velocity w and with no loads exerted over its boundary, that is,

$$T_r(r_i) = 0, \tag{5.315}$$

$$T_r(r_e) = 0, \tag{5.316}$$

where r_i and r_e are, respectively, the inner and outer radii which characterize the geometry of the ring.

In cylindrical coordinates, the kinematics is fully defined through the velocities $\dot{u}_r$, $\dot{u}_\theta$, and $\dot{u}_z$, and, given the symmetry in the problem, we get $\dot{u}_\theta = 0$. Hence, the strain rate is characterized by

$$D_r = \frac{d\dot{u}_r}{dr}, \quad D_\theta = \frac{\dot{u}_r}{r}, \quad D_z = \frac{d\dot{u}_z}{dz}, \tag{5.317}$$

which, using duality arguments, yield power-conjugate stresses T_r, T_θ, and T_z. Let us also consider that the body is in a plane stress state, which implies that $T_z = 0$, and let us suppose that the material behavior is well represented by a Norton-type constitutive law (a Tresca–Norton law). This constitutive law is expressed by a piece-wise smooth potential functional. For the problem under analysis, it is not necessary to know the whole potential function, given that only a part of the function is required. In fact, for this example the stress distribution verifies $T_\theta > T_r \geq T_z = 0$ for $r \in [r_i, r_e]$. Based on this, the largest difference between the principal stresses is given by $T_\theta - T_z = T_\theta$. Therefore, the (smooth) part of the Tresca–Norton constitutive equation of interest for the present problem is

$$\phi(\mathbf{T}) = \frac{B}{n+1}(T_\theta - T_z)^{n+1} = \frac{B}{n+1}T_\theta^{n+1}, \tag{5.318}$$

which yields

$$D_\theta = \phi_{T_\theta} = BT_\theta^n, \tag{5.319}$$

$$D_r = \phi_{T_r} = 0, \tag{5.320}$$

$$D_z = \phi_{T_z} = -BT_\theta^n, \tag{5.321}$$

where B and n are material constants and, as expected, the incompressibility constraint is satisfied,

$$\mathrm{tr}\mathbf{D} = D_r + D_\theta + D_z = 0. \tag{5.322}$$

Since we are interested in utilizing the variational principle of minimum in terms of the velocity field (see (5.247)), we have to obtain $\psi(\mathbf{D})$. This is achieved through (5.319), so we get

$$\psi(\mathbf{D}) = \mathbf{T} \cdot \mathbf{D} - \phi(\mathbf{T}) = T_\theta D_\theta - \frac{B}{n+1}T_\theta^{n+1} = \frac{1}{B^{1/n}}\frac{n}{n+1}D_\theta^{1+1/n}. \tag{5.323}$$

Notice that, from the kinematics which is considered for the body, all motion actions are such that

$$D_r = \frac{d\dot{u}_r}{dr} = 0. \tag{5.324}$$

Thus, the internal stress which emerges as a reaction to this kinematical constraint cannot be evaluated through the PVP because the power exerted in duality with D_r will always be null. Hence T_r can only be assessed a posteriori, using the equilibrium concept once the solution of the PVP is known for the corresponding space of kinematically admissible motion actions. From this solution, the remaining stress state can be obtained from the constitutive equation, particularly

$$T_z = 0 \qquad T_\theta = \frac{1}{B^{1/n}}D_\theta^{1/n}. \tag{5.325}$$

With all the elements presented up to here, let us formulate the variational equation that enables the modeling of the behavior of a rotating ring at constant velocity w. This variational equation reads: determine $\dot{u}_r \in \mathrm{Kin}_v$ that minimizes the cost functional

$$\Pi(u_r) = \int_0^{2\pi}\int_{r_i}^{r_e} \frac{n}{n+1}\frac{1}{B^{1/n}}D_\theta^{1+1/n} r\,dr\,d\theta - \int_0^{2\pi}\int_{r_i}^{r_e} \rho w^2 r^2 \dot{u}_r\,dr\,d\theta, \tag{5.326}$$

where $D_\theta = \dot{u}_r/r$ and the set (actually a space) Kin_v is characterized by

$$\text{Kin}_v = \left\{ \dot{u}_r;\ D_r = \frac{d\dot{u}_r}{dr} = 0 \text{ for } r \in (r_i, r_e) \right\}. \tag{5.327}$$

Based on this formulation, we automatically conclude that $\dot{u}_r$ is a constant, that is Kin_v is formed by all constant velocity fields $\dot{u}_r$. In other words, $\text{Kin}_v = \mathbb{R}$, and the variational problem turns into the problem of determining $C_1 \in \mathbb{R}$ such that the following function is minimized

$$\begin{aligned} \Pi(C_1) &= \int_0^{2\pi} \int_{r_i}^{r_e} \frac{n}{n+1} \frac{1}{B^{1/n}} \left(\frac{C_1}{r}\right)^{1+1/n} r\, dr\, d\theta - \int_0^{2\pi} \int_{r_i}^{r_e} \rho w^2 r^2 C_1 dr\, d\theta \\ &= 2\pi \left[\frac{n^2}{n^2-1} \frac{1}{B^{1/n}} C_1^{1+1/n} (r_e^\alpha - r_i^\alpha) - \rho w^2 C_1 (r_e^3 - r_i^3)/3 \right], \end{aligned} \tag{5.328}$$

where $\alpha = 1 - 1/n$. The necessary and sufficient condition that characterizes the minimum is

$$\frac{\partial \Pi}{\partial C_1} = \frac{1}{B^{1/n}} \frac{n}{n-1} C_1^{1/n} (r_e^\alpha - r_i^\alpha) - \rho w^2 \frac{(r_e^3 - r_i^3)}{3} = 0, \tag{5.329}$$

and so we can obtain the solution $\dot{u}_r$ of the problem

$$\dot{u}_r = C_1 = B \left[\frac{n-1}{3n} \rho w^2 \frac{(r_e^3 - r_i^3)}{(r_e^\alpha - r_i^\alpha)} \right]^n. \tag{5.330}$$

With $\dot{u}_r$ known, and using the constitutive equation, we obtain

$$D_\theta = \frac{\dot{u}_r}{r} = \frac{C_1}{r}, \tag{5.331}$$

and from the incompressibility constraint it follows that

$$D_z = -D_\theta. \tag{5.332}$$

Hence, the stress state is written as

$$T_\theta = \frac{1}{B^{1/n}} D_\theta^{1/n} = \left(\frac{C_1}{Br}\right)^{1/n} = \frac{1}{r^{1/n}} \left[\frac{n-1}{3n} \rho w^2 \frac{(r_e^3 - r_i^3)}{(r_e^\alpha - r_i^\alpha)} \right]. \tag{5.333}$$

To calculate T_r (variable dual to the kinematical constraint $D_r = 0$) we resort to the PVP (which provides the concept of mechanical equilibrium)

$$\int_0^{2\pi} \int_{r_i}^{r_e} (T_\theta \hat{D}_\theta + T_r \hat{D}_r) r dr d\theta - \int_0^{2\pi} \int_{r_i}^{r_e} \rho w^2 r^2 \hat{u}_r dr d\theta = 0 \qquad \forall \hat{u}_r \in \mathscr{V}, \tag{5.334}$$

where $\mathscr{V}$ is now the space of all virtual motion actions in which the kinematical constraints originally prescribed are removed, and where $\hat{D}_\theta = \frac{\hat{u}_r}{r}$ and $\hat{D}_r = \frac{d\hat{u}_r}{dr}$. In addition, the stress state T_θ is associated with the solution of the original problem, given by (5.333). Since the problem is independent from θ, the previous definition of the equilibrium is equivalent to the following variational equation

$$\int_{r_i}^{r_e} \left(T_\theta \hat{u}_r + r T_r \frac{d\hat{u}_r}{dr} - \rho w^2 r^2 \hat{u}_r \right) dr = 0 \qquad \forall \hat{u}_r \in \mathscr{V}, \tag{5.335}$$

which, in local form, corresponds to

$$T_\theta - \frac{d}{dr}(rT_r) - \rho w^2 r^2 = 0 \quad \text{in } (r_i, r_e), \tag{5.336}$$

$$T_r = 0 \quad \text{at } r = r_i, \tag{5.337}$$

$$T_r = 0 \quad \text{at } r = r_e, \tag{5.338}$$

and where the stress T_θ is given by (5.333). Integrating equation (5.336),

$$T_r = \frac{1}{r}\int T_\theta dr - \rho w^2 \frac{r^3}{3} + \frac{C_2}{r} = \rho w^2 \frac{1}{3}\frac{r_e^3 - r_i^3}{r_e^\alpha - r_i^\alpha} r^{-1/n} - \frac{1}{3}\rho w^2 r^2 + \frac{C_2}{r}, \tag{5.339}$$

and using the boundary conditions (5.337) and (5.338) we arrive at

$$\frac{1}{3}\rho w^2 \left[\frac{r_e^3 - r_i^3}{r_e^\alpha - r_i^\alpha} r_i^{-1/n} - r_i^2\right] + \frac{C_2}{r_i} = 0, \tag{5.340}$$

$$\frac{1}{3}\rho w^2 \left[\frac{r_e^3 - r_i^3}{r_e^\alpha - r_i^\alpha} r_e^{-1/n} - r_e^2\right] + \frac{C_2}{r_e} = 0. \tag{5.341}$$

Multiplying the first expression by r_i and the second by r_e and adding both equations we get

$$\frac{1}{3}\rho w^2 \left[\frac{r_e^3 - r_i^3}{r_e^\alpha - r_i^\alpha}(r_i^\alpha + r_e^\alpha) - (r_i^3 + r_e^3)\right] + 2C_2 = 0, \tag{5.342}$$

from which

$$C_2 = \frac{1}{3}\rho w^2 \frac{r_i^3 r_e^\alpha - r_e^3 r_i^\alpha}{r_e^\alpha - r_i^\alpha}. \tag{5.343}$$

Introducing this result into (5.339) we finally obtain

$$T_r = \frac{1}{6}\frac{\rho w^2}{r}\left[\frac{r_e^3 - r_i^3}{r_e^\alpha - r_i^\alpha}(r^\alpha - r_i^\alpha) - (r^3 - r_i^3)\right], \tag{5.344}$$

and the problem is fully solved. It is interesting to note that the above expression gives for $\alpha = 0$ ($n = 1$) the stress distribution for an elastic material. On the other hand, for $\alpha = 1$ ($n = \infty$) the stress distribution corresponds to a perfectly plastic material governed by the Tresca law.

5.11.2 Cantilevered Beam with Uniform Load

In this section we will make use of the Martin estimate. Consider a beam cantilevered at one extreme, subjected to a uniform load p. The length of the beam is denoted by L and the constitutive equation is of generalized type

$$\chi = B_n M^n, \tag{5.345}$$

where M is the bending moment in the transversal section of the beam and χ is the curvature rate of change defined by

$$\chi = -\frac{d^2 w}{dx^2}, \tag{5.346}$$

with w being the transversal velocity defined over the beam axis.

The solution of this problem using the variational principle of minimum in terms of the velocity field is found in Boyle-Spence [39] and corresponds to

$$w = -B_n(p/2)^n \left(\frac{L^{2n+1}}{2n+1} + \frac{(L-x)^{2n+2}}{(2n+1)(2n+2)} - \frac{L^{2n+2}}{(2n+1)(2n+2)} \right). \tag{5.347}$$

Then, the maximum deflection occurs at $x = L$ and is given by

$$q = -w(L) = B_n(p/2)^n \frac{L^{2n+2}}{2n+2}, \tag{5.348}$$

which, for the case $n = 1$, results in

$$q_1 = \frac{B_n p L^4}{8}. \tag{5.349}$$

Let us now exploit the procedure proposed by Martin to obtain upper bounds for q. For this, from the constitutive equation we have

$$\chi M = B_n M^{n+1} = (1/B_n)^{1/n} \chi^{1+1/n}, \tag{5.350}$$

$$\phi(M) = \frac{1}{n+1} B_n M^{n+1}, \tag{5.351}$$

$$\psi(\chi) = \frac{n}{n+1} B_n M^{n+1} = \frac{n}{n+1} (1/B_n)^{1/n} \chi^{1+1/n}. \tag{5.352}$$

Thus, as we have seen in Section 5.10.2 (see (5.310)), we have

$$\frac{1}{n+1} \int_0^L \chi^* M^* dx \geq \int_0^L M^* \chi dx - \frac{n}{n+1} \int_0^L M \chi dx, \tag{5.353}$$

where χ and M are the solution of the problem, and χ^*, M^* corresponds to the solution of the beam problem but for a different type of load which is to be conveniently selected.

Introducing (5.346) in the right-hand side of (5.353) and integrating by parts we obtain

$$\begin{aligned} \frac{1}{n+1} \int_0^L \chi^* M^* dx \geq & -M^* \left.\frac{dw}{dx}\right|_0^L + \left.\frac{dM^*}{dx} w\right|_0^L + \frac{n}{n+1} M \left.\frac{dw}{dx}\right|_0^L - \left.\frac{dM}{dx} w\right|_0^L \\ & + \int_0^L \left(\frac{n}{n+1} p - p^* \right) dx = \left.\frac{dM^*}{dx} w\right|_L + \int_0^L \left(\frac{n}{n+1} p - p^* \right) dx. \end{aligned} \tag{5.354}$$

Then, in order to bound the displacement w at $x = L$, it will be enough to pick the following kind of load

$$p^* = \begin{cases} \frac{n}{n+1} p & \text{in } x \in (0, L), \\ -Q & \text{at } x = L, \end{cases} \tag{5.355}$$

which leads us to

$$\frac{1}{n+1} \int_0^L B_n M^{*n+1} dx \geq Q w(L), \tag{5.356}$$

or its equivalent expression

$$w(L) \leq \frac{B_n}{n+1} \frac{1}{Q} \int_0^L M^{*n+1} dx = \frac{B_n}{n+1} \frac{1}{Q} \int_0^L \left[Q(L-x) + \frac{n}{n+1} p \frac{(L-x)^2}{2} \right]^{n+1} dx. \tag{5.357}$$

As observed, the bound depends upon the very value of Q. Therefore, the optimal bound for $w(L)$ will be that value of Q which minimizes the rightmost side in the above expression, that is,

$$w(L) \leq \frac{B_n}{n+1} \min_{Q \in \mathbb{R}^+} \frac{1}{Q} \int_0^L \left[Q(L-x) + \frac{n}{n+1} p \frac{(L-x)^2}{2} \right]^{n+1} dx. \tag{5.358}$$

Since the problem becomes nonlinear, we have to make use of numerical techniques to obtain this kind of bounds. In particular, let us see the result for $n = 1$. In such a case it is

$$w(L) \leq \frac{B_n}{2} \min_{Q \in \mathbb{R}^+} \frac{1}{Q} \int_0^L \left[Q(L-x) + \frac{1}{4} p(L-x)^2 \right]^2 dx. \tag{5.359}$$

The value of Q which minimizes the expression is such that

$$\begin{aligned} &-\int_0^L \left[Q(L-x) + \frac{1}{4} p(L-x)^2 \right]^2 dx \\ &\quad + 2Q \int_0^L \left[Q(L-x) + \frac{1}{4} p(L-x)^2 \right] (L-x) dx = 0, \end{aligned} \tag{5.360}$$

yielding

$$\frac{Q}{pL} = 0.19365, \tag{5.361}$$

which put into (5.359) leads us to

$$\begin{aligned} w(L) &\leq \frac{B_n}{2} \left[\frac{QL^3}{3} + \frac{1}{8} pL^4 + \frac{1}{80} \frac{p^2 L^5}{Q} \right] \\ &= \frac{B_n}{2} pL^4 \left[\frac{0.19365}{3} + \frac{1}{8} + \frac{1}{80} \frac{1}{0.19365} \right] = 0.12705 B_n pL^4. \end{aligned} \tag{5.362}$$

Taking this bound as an approximate value, we have

$$w(L) = q_a = 0.12705 B_n pL^4, \tag{5.363}$$

which compared with the exact solution (5.349) provides

$$\frac{q_a}{q_1} = \frac{0.12705}{0.125} = 1.0164. \tag{5.364}$$

This result tells us that the estimation of the velocity field at that point was achieved with an error of 1.64%.

5.12 Approximate Solution to Steady-State Creep Problems

Taking the velocity field as the primary variable, in this section we present two numerical strategies to deliver approximate solutions to the secondary creep problem.

For simplicity, let us consider that the velocity field satisfies $\overline{\mathbf{v}} = \mathbf{0}$ over $\partial \mathcal{B}_v$. With this in mind, and from that studied in Section 5.9.3, the steady-state creep problem, written in terms of the velocity, consists of finding the field $\mathbf{v}_0 \in \text{Kin}_v$ such that the cost functional is minimized

$$\Pi(\mathbf{v}) = \int_{\mathscr{B}} \psi(\mathbf{D})dV - \int_{\mathscr{B}} \mathbf{b} \cdot \mathbf{v}dV - \int_{\partial\mathscr{B}_a} \mathbf{a} \cdot \mathbf{v}dS, \tag{5.365}$$

defined for elements $\mathbf{v} \in \text{Kin}_v$ and where $\mathbf{D} = (\nabla\mathbf{v})^s$. The space Kin_v is

$$\text{Kin}_v = \{\mathbf{v} \in \mathscr{V}\ ;\ \text{div}\,\mathbf{v} = 0,\ \mathbf{v}|_{\partial\mathscr{B}_v} = \mathbf{0}\}. \tag{5.366}$$

Remember that the constraint $\text{div}\,\mathbf{v} = 0$ ensures that all admissible motion actions are isochoric.

The natural approach to address this problem in an approximate manner consists of redefining the minimization problem in a finite-dimensional space $\text{Kin}_v^n \subset \text{Kin}_v$. That is, consider the set of functions $\{\phi_i\}_{i=1}^{\infty}$ dense[1] in Kin_v, called basis functions or coordinate functions. In particular, we have already discussed that these functions can be constructed through finite element methods. Then, calling

$$\text{Kin}_v^n = \text{span}\{\phi_i\}_{i=1}^n, \tag{5.367}$$

the approximate problem now consists of minimizing the cost functional

$$\Pi(\mathbf{v}) = \int_{\mathscr{B}} \psi(\mathbf{D})dV - \int_{\mathscr{B}} \mathbf{b} \cdot \mathbf{v}dV - \int_{\partial\mathscr{B}_a} \mathbf{a} \cdot \mathbf{v}dS, \tag{5.368}$$

in the finite-dimensional space Kin_v^n. In this finite-dimensional framework it gives

$$\mathbf{v} = \sum_{i=1}^{n} c_i\phi_i, \tag{5.369}$$

and putting $\mathbf{c} = (c_1, c_2, \dots, c_n) \in \mathbb{R}^n$, the problem is equivalent to minimizing the function

$$\begin{aligned}\mathscr{F}(\mathbf{c}) = \Pi\left(\sum_{i=1}^{n} c_i\phi_i\right) &= \int_{\mathscr{B}} \psi\left(\sum_{i=1}^{n} c_i(\nabla\phi_i)^s\right)dV \\ &- \sum_{i=1}^{n} c_i\int_{\mathscr{B}} \mathbf{b} \cdot \phi_i dV - \sum_{i=1}^{n} c_i\int_{\partial\mathscr{B}_a} \mathbf{a} \cdot \phi_i dS.\end{aligned} \tag{5.370}$$

With the convexity of potential function ψ, we have that the necessary and sufficient condition is given by

$$\frac{\partial\mathscr{F}(\mathbf{c})}{\partial c_i} = 0 \qquad i = 1, \dots, n, \tag{5.371}$$

resulting in a system of n coupled nonlinear equations. The solution of this problem yields the coefficients which characterize the approximate solution that minimizes the dissipation functional in Kin_v^n.

The algorithm discussed above has two main drawbacks. The first is related to the very definition of the basis functions ϕ_i which must verify the divergence-free constraint. The second is the fact that a system of nonlinear equations needs to be solved. Therefore, computationally speaking, these features imply that any computational code already

1 This implies that given an arbitrary element $\mathbf{v} \in \text{Kin}_v$ and $\varepsilon > 0$, it will always be possible to find $n(\varepsilon)$ such that for all $m \geq n$ we get $\|\mathbf{v} - \sum_{i=1}^{m} a_i\phi_i\| < \varepsilon$, where $\|\cdot\|$ is the norm in Kin_v.

prepared for numerical approximation of mechanical problems will have to undergo substantial specific changes to address the secondary creep problem.

Motivated by the previous intrinsic difficulties of the problem, the question is about the possibility of developing a numerical algorithm which enables the approximation of the steady-state creep problem and that, at the same time, can be easily implemented in available finite element programs dedicated to the linear elastic analysis of structures. There is a possibility for such an algorithm, called the elasto-creep method.

To this end, let us consider the body occupying the region of space $\mathscr{B}$, with boundary $\partial\mathscr{B}$, subjected to the action of the system of loads $(\mathbf{b}, \mathbf{a})$ and equal to $\overline{\mathbf{v}}$ over a part of the boundary $\partial\mathscr{B}_v$. However, now we will consider that the material that constitutes the body can undergo both elastic deformations and creep phenomena (combined elasto-creep phenomena). That is, the material features two simultaneous behaviors, an instantaneous elastic behavior combined with the secondary creep response.

Hence, the elasto-creep problem consists of determining the fields $\dot{\mathbf{u}} = \dot{\mathbf{u}}(\mathbf{x}, t)$, $\dot{\mathbf{E}} = \dot{\mathbf{E}}(\mathbf{x}, t)$, $\dot{\mathbf{E}}^c = \dot{\mathbf{E}}^c(\mathbf{x}, t)$, and $\dot{\mathbf{T}} = \dot{\mathbf{T}}(\mathbf{x}, t)$ defined in $\mathscr{B} \times [0, \infty)$ such that the following conditions are satisfied. First, it must be $\dot{\mathbf{u}} \in \mathrm{Kin}_v$, which, as expected, states that the velocity field must be kinematically admissible. Then, $\dot{\mathbf{E}}$ has to be compatible, that is to say $\dot{\mathbf{E}} = \mathscr{D}\dot{\mathbf{u}} = (\nabla\dot{\mathbf{u}})^s$ for a body in the three-dimensional space. The stress rate $\dot{\mathbf{T}}$ should be related to the elastic strain rate $\dot{\mathbf{E}} - \dot{\mathbf{E}}^c$ through the constitutive equation of a hyperelastic material, which will be assumed to be linear for simplicity. This means $\dot{\mathbf{T}} = \mathbb{D}(\dot{\mathbf{E}} - \dot{\mathbf{E}}^c)$ where, for simplicity again, $\mathbb{D}$ is the elasticity fourth-order tensor of an isotropic material yielding $\mathbb{D} = 2\mu\mathbb{I} + \lambda\mathbf{I} \otimes \mathbf{I}$ ($\mathbb{I}$ is the fourth-order identity tensor, $\mathbf{I}$ is the second-order identity tensor, and λ and μ are the Lamé constants). The creep strain rate $\dot{\mathbf{E}}^c$ is associated with the stress state through a corresponding constitutive equation of the kind $\dot{\mathbf{E}}^c = f(\mathbf{S})\mathbf{S}$, where $\mathbf{S}$ is the deviatoric component of the stress tensor $\mathbf{T}$. Because we are assuming that the loads $(\mathbf{b}, \mathbf{a})$ are constant in time, the stress rate $\dot{\mathbf{T}}$ must be self-equilibrated for all time instants $t \in (0, \infty)$, that is, $\dot{\mathbf{T}} \in \mathrm{Var}_T$, $\forall t \in (0, \infty)$. In extended form, this entails that it must verify the PVP for all $t \in (0, +\infty)$, which in the present case reads

$$\int_{\mathscr{B}} \dot{\mathbf{T}} \cdot (\nabla\hat{\mathbf{u}})^s dV = 0 \qquad \forall\hat{\mathbf{u}} \in \mathrm{Var}_v, \tag{5.372}$$

or in strong (local) form

$$\mathrm{div}\, \dot{\mathbf{T}} = \mathbf{0} \qquad \text{in } \mathscr{B}, \tag{5.373}$$

$$\dot{\mathbf{T}}\mathbf{n} = \mathbf{0} \qquad \text{on } \partial\mathscr{B}_a. \tag{5.374}$$

As observed, the proposed algorithm relies on an evolution problem. Therefore, proper initial conditions are in order to fully characterize the formulation. These conditions must be compatible with the considered instantaneous elastic behavior. That is, the fields $\mathbf{u}_0 = \mathbf{u}(\mathbf{x}, 0)$, $\mathbf{E}_0 = \mathbf{E}(\mathbf{x}, 0)$, $\mathbf{E}_0^c = \mathbf{E}^c(\mathbf{x}, 0)$, and $\mathbf{T}_0 = \mathbf{T}(\mathbf{x}, 0)$ (at time $t = 0$) must comply with the purely elastic problem (instantaneous response). Hence, at time $t = 0$, these fields have to be such that $\mathbf{T}_0$ is at equilibrium with the loads, which entails that it must satisfy the PVP

$$\int_{\mathscr{B}} \mathbf{T}_0 \cdot (\nabla\hat{\mathbf{u}})^s dV = \int_{\mathscr{B}} \mathbf{b} \cdot \hat{\mathbf{u}} dV + \int_{\partial\mathscr{B}_a} \mathbf{a} \cdot \hat{\mathbf{u}} dS \qquad \forall\hat{\mathbf{u}} \in \mathrm{Var}_u \equiv \mathrm{Kin}_u, \tag{5.375}$$

where $\text{Kin}_u = \{\mathbf{u} \in \mathscr{V}\ ;\ \mathbf{u}|_{\partial\mathscr{B}_u} = \mathbf{0}\}$. We have picked homogeneous conditions to simplify the presentation. In local form, this implies

$$\text{div}\ \mathbf{T}_0 = \mathbf{b} \qquad \text{in } \mathscr{B}, \tag{5.376}$$

$$\mathbf{T}_0\mathbf{n} = \mathbf{a} \qquad \text{on } \partial\mathscr{B}_a. \tag{5.377}$$

Also, $\mathbf{T}_0$ has to be related to the initial strain state $\mathbf{E}_0$ through the elastic constitutive equation, that is, $\mathbf{T}_0 = \mathbb{D}\mathbf{E}_0$. In turn, the initial strain $\mathbf{E}_0$ must be compatible with the initial displacement field, denoted by $\mathbf{u}_0$, that is, $\mathbf{E}_0 = (\nabla\mathbf{u}_0)^s$, and this initial displacement $\mathbf{u}_0$ has to be kinematically admissible, which means $\mathbf{u}_0 \in \text{Kin}_u$. Finally, the initial creep strain must be zero $\mathbf{E}_0^c = \mathbf{O}$.

Let us suppose that the solution $\mathbf{T}$, $\dot{\mathbf{E}}$, $\dot{\mathbf{E}}^c$, and $\dot{\mathbf{u}}$ of the previously described problem converges, for $t \to \infty$, to a steady-state solution $\tilde{\mathbf{T}}$, $\tilde{\dot{\mathbf{E}}}$, $\tilde{\dot{\mathbf{E}}}^c$, and $\tilde{\dot{\mathbf{u}}}$. If there exists such solution, it is not difficult to show that it coincides with the solution of the original steady-state creep problem. As a matter of fact, in such case we will have

$$\lim_{t\to\infty} \dot{\mathbf{T}} = \mathbf{O}, \tag{5.378}$$

then

$$\lim_{t\to\infty} \mathbf{T}(\mathbf{x}, t) = \tilde{\mathbf{T}}, \tag{5.379}$$

leading to

$$\lim_{t\to\infty} \dot{\mathbf{E}}^c(\mathbf{x}, t) = f(\tilde{\mathbf{S}})\tilde{\mathbf{S}}, \tag{5.380}$$

where $\tilde{\mathbf{S}}$ is the deviatoric part of $\tilde{\mathbf{T}}$. In addition, from the constitutive relation $\dot{\mathbf{T}} = \mathbb{D}(\dot{\mathbf{E}} - \dot{\mathbf{E}}^c)$, and by using the positivity of the tensor $\mathbb{D}$, and from expression (5.378), we also have

$$\lim_{t\to\infty} (\dot{\mathbf{E}}(\mathbf{x}, t) - \dot{\mathbf{E}}^c(\mathbf{x}, t)) = \mathbf{O}, \tag{5.381}$$

that is

$$\lim_{t\to\infty} \dot{\mathbf{E}}(\mathbf{x}, t) = \lim_{t\to\infty} \dot{\mathbf{E}}^c(\mathbf{x}, t) = f(\tilde{\mathbf{S}})\tilde{\mathbf{S}} = \tilde{\dot{\mathbf{E}}}. \tag{5.382}$$

We thus conclude that the steady-state solution of the elasto-creep problem $\tilde{\mathbf{T}}$, $\tilde{\dot{\mathbf{E}}}$, $\tilde{\dot{\mathbf{u}}}$ satisfies the same equations that govern the steady-state creep. Therefore, with this elasto-creep formulation we obtain, in the limit $t \to \infty$, that the steady-state strain rate $\tilde{\dot{\mathbf{E}}}$ verifies the creep constitutive law, and is then isochoric.

In the approach discussed above, the secondary creep problem was surrogated to a time-dependent linear elastic problem which, at each time t, features known initial strains given by the creep strain rate. This approach to creep problems is appealing mainly because it permits the incompressibility constraint in the definition of the space of kinematically admissible velocity fields Kin_v to be relaxed. Moreover, the resulting formulation at each time step is that of a linear elastic material and, as a consequence, it enables the easy implementation in available finite element programs of linear elasticity. The downside of this formulation is that now we have to deal with an evolution problem, whose steady-state solution is only of interest. Notably, this is counterbalanced by the fact that, in the original creep problem, an iterative process is required to find the solution to the system of nonlinear equations.

Next, we will address in more detail the elasto-creep approach. Let us recall some definitions

$$\text{Kin}_u = \{\mathbf{u} \in \mathscr{V}\ ;\ \mathbf{u}|_{\partial\mathscr{B}_v} = \mathbf{0}\}, \tag{5.383}$$

$$\text{Kin}_v = \{\dot{\mathbf{u}} \in \mathscr{V}\ ;\ \dot{\mathbf{u}}|_{\partial\mathscr{B}_v} = \overline{\mathbf{v}}\}. \tag{5.384}$$

Then, the elasto-creep variational problem consists of finding $\dot{\mathbf{u}} \in \text{Kin}_v$ such that, for all $t \in (0, \infty)$, it is verified that

$$\int_{\mathscr{B}} \dot{\mathbf{T}} \cdot (\nabla\dot{\mathbf{u}}^* - \nabla\dot{\mathbf{u}})^s dV = 0 \qquad \forall \dot{\mathbf{u}}^* \in \text{Kin}_v, \tag{5.385}$$

with the initial conditions $\mathbf{E}_0^c = \mathbf{O}$ and $\mathbf{u}_0 \in \text{Kin}_u$ satisfying

$$\int_{\mathscr{B}} \mathbf{T}_0 \cdot (\nabla\mathbf{u}^* - \nabla\mathbf{u}_0)^s dV = \int_{\mathscr{B}} \mathbf{b} \cdot (\mathbf{u}^* - \mathbf{u}_0) dV + \int_{\partial\mathscr{B}_a} \mathbf{a} \cdot (\mathbf{u}^* - \mathbf{u}_0) dS \ \ \forall \mathbf{u}^* \in \text{Kin}_u, \tag{5.386}$$

and where

$$\dot{\mathbf{T}} = \mathbb{D}((\nabla\dot{\mathbf{u}})^s - \dot{\mathbf{E}}^c), \tag{5.387}$$

$$\dot{\mathbf{E}}^c = f(\mathbf{S})\mathbf{S}, \tag{5.388}$$

$$\mathbf{T}_0 = \mathbb{D}(\nabla\mathbf{u}_0)^s, \tag{5.389}$$

$$\mathbf{S} = \mathbf{T} - \frac{1}{3}(\text{tr}\mathbf{T})\mathbf{I}. \tag{5.390}$$

Introducing these constitutive relations in (5.385) and (5.386), the problem becomes finding $\dot{\mathbf{u}} \in \text{Kin}_v$ such that for all $t \in (0, \infty)$ it is

$$\int_{\mathscr{B}} \mathbb{D}(\nabla\dot{\mathbf{u}})^s \cdot (\nabla\dot{\mathbf{u}}^* - \nabla\dot{\mathbf{u}})^s dV = \int_{\mathscr{B}} \mathbb{D}f(\mathbf{S})\mathbf{S} \cdot (\nabla\dot{\mathbf{u}}^* - \nabla\dot{\mathbf{u}})^s dV \ \ \forall \dot{\mathbf{u}}^* \in \text{Kin}_v, \tag{5.391}$$

with the initial condition $\mathbf{E}_0^c = \mathbf{O}$ and $\mathbf{u}_0 \in \text{Kin}_u$ verifying

$$\int_{\mathscr{B}} \mathbb{D}(\nabla\mathbf{u}_0)^s \cdot (\nabla\mathbf{u}^* - \nabla\mathbf{u}_0)^s dV = \int_{\mathscr{B}} \mathbf{b} \cdot (\mathbf{u}^* - \mathbf{u}_0) dV + \int_{\partial\mathscr{B}_a} \mathbf{a} \cdot (\mathbf{u}^* - \mathbf{u}_0) dS \ \ \forall \mathbf{u}^* \in \text{Kin}_u. \tag{5.392}$$

These two variational equations stand, respectively, for the PVP at time t and for the PVW at the initial time $t = 0$.

The strategy to obtain approximate solutions is based on casting these problems in finite-dimensional sets Kin_v^n and Kin_u^n. In order to do this, consider

$$\dot{\mathbf{u}} \in \text{Kin}_v^n \quad \Rightarrow \quad \dot{\mathbf{u}} = \overline{\mathbf{v}} + \sum_{i=1}^n \dot{u}_i(t)\boldsymbol{\phi}_i(\mathbf{x}), \tag{5.393}$$

$$\mathbf{u}_0 \in \text{Kin}_u^n \quad \Rightarrow \quad \mathbf{u}_0 = \sum_{i=1}^n u_{0i}\boldsymbol{\phi}_i(\mathbf{x}), \tag{5.394}$$

where $\boldsymbol{\phi}_i$, $i = 1, \dots, n$ are vector coordinate functions such that $\boldsymbol{\phi}_i(\mathbf{x}) = \mathbf{0}$ for $\mathbf{x} \in \partial\mathscr{B}_v$, $i = 1, \dots, n$. The unknowns in this problem are $\dot{u}_i(t)$, $t \in (0, \infty)$ and u_{0i}.

As we have already discussed, the coordinate functions can be systematically constructed using the finite element method. This approach relies on the partition of $\mathscr{B}$ into a number of subdomains called finite elements, and on a posterior approximation of the fields $\dot{\mathbf{u}}$ and $\mathbf{u}_0$ within each subdomain. In this case, the numbers $\dot{u}_i$ and u_{0i} have a clear physical meaning, since they represent the components of the velocity $\dot{\mathbf{u}}$ and displacement $\mathbf{u}_0$ at the vertices (nodes) that define the finite elements. Moreover, the global coordinate functions ϕ_i are constructed through the assembly of the contributions provided by the local interpolation functions defined at each finite element.

With these approximations, the elasto-creep problem consists of finding $\dot{\mathbf{u}} \in \mathrm{Kin}_v^n$ such that for all $t \in (0, \infty)$ the following holds

$$\int_{\mathscr{B}} \mathbb{D}(\nabla\dot{\mathbf{u}})^s \cdot (\nabla\dot{\mathbf{u}}^* - \nabla\dot{\mathbf{u}})^s dV = \int_{\mathscr{B}} \mathbb{D}f(\mathbf{S})\mathbf{S} \cdot (\nabla\dot{\mathbf{u}}^* - \nabla\dot{\mathbf{u}})^s dV \quad \forall\dot{\mathbf{u}}^* \in \mathrm{Kin}_v^n, \tag{5.395}$$

with the initial conditions $\mathbf{E}^c = \mathbf{O}$ and $\mathbf{u}_0 \in \mathrm{Kin}_u^n$ satisfying

$$\int_{\Omega} \mathbb{D}(\nabla\mathbf{u}_0)^s \cdot (\nabla\mathbf{u}^* - \nabla\mathbf{u}_0)^s dV = \int_{\mathscr{B}} \mathbf{b} \cdot (\mathbf{u}^* - \mathbf{u}_0) dV + \int_{\partial\mathscr{B}_a} \mathbf{a} \cdot (\mathbf{u}^* - \mathbf{u}_0) dS \quad \forall\mathbf{u}^* \in \mathrm{Kin}_u^n. \tag{5.396}$$

With this approximation, the problem leads us to the following system of ordinary differential equations in the time variable

$$\sum_{j=1}^{n} K_{ij}\dot{u}_j(t) = h_i^c(t) - f_i^2 \qquad i = 1, \dots, n, \tag{5.397}$$

with the initial condition at $t = 0$ given by $\mathbf{E}_0^c = \mathbf{O}$ and

$$\sum_{j=1}^{n} K_{ij}u_{0j} = f_i^1 \qquad i = 1, \dots, n, \tag{5.398}$$

where

$$K_{ij} = \int_{\mathscr{B}} \mathbb{D}(\nabla\phi_j)^s \cdot (\nabla\phi_i)^s dV, \tag{5.399}$$

$$h_i^c(t) = \int_{\mathscr{B}} \mathbb{D}\dot{\mathbf{E}}^c \cdot (\nabla\phi_i)^s dV = \int_{\mathscr{B}} \mathbb{D}f(\mathbf{S}(t))\mathbf{S}(t) \cdot (\nabla\phi_i)^s dV, \tag{5.400}$$

$$f_i^1 = \int_{\mathscr{B}} \mathbf{b} \cdot \phi_i dV + \int_{\partial\mathscr{B}_a} \mathbf{a} \cdot \phi_i dS, \tag{5.401}$$

$$f_i^2 = \int_{\mathscr{B}} \mathbb{D}(\nabla\overline{\mathbf{v}})^s \cdot (\nabla\phi_i)^s dV. \tag{5.402}$$

The numerical integration for the approximate solution of the system of ordinary differential equations can be performed using any quadrature method available. For simplicity, let us present the algorithm based on the Euler method (see [39] for other alternatives). The Euler method consists of the following sequence of computations.

1) For $t_0 = 0$ the system (5.398) allows us to find $\mathbf{u}_0$, and then we compute the initial strain $\mathbf{E}_0 = (\nabla\mathbf{u}_0)^s$ and the stress $\mathbf{T}_0 = \mathbb{D}\mathbf{E}_0$.

2) Given $\mathbf{T}_0$, find

$$\mathbf{S}(0) = \mathbf{T}_0 - \frac{1}{3}(\mathrm{tr}\mathbf{T}_0)\mathbf{I}, \tag{5.403}$$

$$\dot{\mathbf{E}}^c(0) = f(\mathbf{S}(0))\mathbf{S}(0), \tag{5.404}$$

$$h_i^c(0) = \int_{\mathscr{B}} \mathbb{D}\dot{\mathbf{E}}^c(0) \cdot (\nabla \phi_i)^s dV, \quad i = 1, \dots, n. \tag{5.405}$$

3) With vector $\mathbf{h}^c(0) = (h_1^c(0), \dots, h_n^c(0))$, vector $\mathbf{f}^2 = (f_1^2, \dots, f_n^2)$ defined in (5.402), and matrix $[\mathbf{K}]_{ij} = K_{ij}$ in (5.399), it is possible to solve the system of linear equations (5.397). By solving this system we get $\dot{\mathbf{u}}(0)$, and so we compute $\dot{\mathbf{E}}(0) = (\nabla \dot{\mathbf{u}})^s(0)$, and the corresponding stress rate state $\dot{\mathbf{T}}(0)$ as established by the elastic constitutive equation $\dot{\mathbf{T}}(0) = \mathbb{D}(\dot{\mathbf{E}}(0) - \dot{\mathbf{E}}^c(0))$.
4) Hence, the stress state of the body at time instant $t_1 = t_0 + \Delta t = \Delta t$ is

$$\mathbf{T}(t_1) = \mathbf{T}_0 + \Delta t \dot{\mathbf{T}}(0). \tag{5.406}$$

5) If the steady-state solution was achieved then exit, else, put $\mathbf{T}_0 = \mathbf{T}(t_1)$ and go back to Step 2.

In practice, this amounts to execution of a repetition cycle until the norm of the difference of vector $\mathbf{h}^c$, from time instant t_{n+1} to time instant t_n, is sufficiently small, for example

$$\frac{\|\mathbf{h}^c(t_{n+1}) - \mathbf{h}^c(t_n)\|}{\|\mathbf{h}^c(t_n)\|} \leq \varepsilon, \tag{5.407}$$

with $\varepsilon > 0$ small enough.

In turn, the integration step Δt is chosen in order to satisfy the following criteria

$$\Delta t_{n+1} \leq 1.5 \Delta t_n, \tag{5.408}$$

$$\Delta t_{n+1} \leq \min_{\mathbf{x} \in \mathscr{B}} \tau \frac{\|\mathbf{E}(t_n)\|}{\|\dot{\mathbf{E}}(t_n)\|}, \tag{5.409}$$

where $\tau = 0.1$ and the norm is the classical one in the Euclidean space.

5.13 Unsteady Creep Problem

Now we will extend some of the results presented in previous sections to the case of unsteady creep, including diverse types of hardening models, as well as time-dependent loads. Dynamic effects will be disregarded from the analysis. Hence, the constitutive equations considering creep and incorporating time hardening and strain hardening can be, in general, written as

$$\dot{\mathbf{E}}^c = g(T_e, \theta, E_e^c)\mathbf{S} \qquad \text{strain hardening}, \tag{5.410}$$

$$\dot{\mathbf{E}}^c = f(T_e, \theta, t)\mathbf{S} \qquad \text{time hardening}, \tag{5.411}$$

where θ is the temperature, T_e is the effective stress defined in (5.49), and the effective creep strain rate E_e^c is given by (5.56).

With this in mind, the analysis of the unsteady creep problem can be formulated as follows. Consider a body in the region $\mathscr{B}$, with boundary $\partial\mathscr{B}$, subjected to the action of a force per unit volume $\mathbf{b} = \mathbf{b}(\mathbf{x}, t)$ and a force per unit area $\mathbf{a} = \mathbf{a}(\mathbf{x}, t)$, the latter applied over $\partial\mathscr{B}_a \subset \partial\mathscr{B}$. The body is also subjected to a kinematical constraint given by $\overline{\mathbf{u}} = \overline{\mathbf{u}}(\mathbf{x}, t)$ over $\partial\mathscr{B}_v$ (note that, over that boundary, velocity constraints are also being prescribed). Again, the initial conditions must be compatible with the instantaneous elastic response.

Let us employ the PVP to formulate the equilibrium problem at each time instant, and the PVW to formulate the problem of finding the initial conditions. Let us consider the set $\text{Kin}_u = \{\mathbf{u} \in \mathscr{V}\ ;\ \mathbf{u}|_{\partial\mathscr{B}_v} = \overline{\mathbf{u}}\}$ of all kinematically admissible displacement fields, also satisfying the constraint at $t = 0$ over $\partial\mathscr{B}_v$, where displacements along time (and therefore the velocities) are prescribed. The subspace of all virtual displacements $\text{Var}_u = \{\hat{\mathbf{u}} \in \mathscr{V}\ ;\ \hat{\mathbf{u}}|_{\partial\mathscr{B}_v} = \mathbf{0}\}$, which for bilateral kinematical constraints can be expressed, verifies $\mathbf{u}_1, \mathbf{u}_2 \in \text{Kin}_u \Rightarrow \mathbf{u}_1 - \mathbf{u}_2 \in \text{Var}_u$. Now, let $\text{Kin}_v = \{\mathbf{v} \in \mathscr{V}\ ;\ \mathbf{v}|_{\partial\mathscr{B}_v} = \dot{\overline{\mathbf{u}}}\}$ be the set of kinematically admissible motion actions. Finally, let $\text{Var}_v = \{\hat{\mathbf{v}} \in \mathscr{V}\ ;\ \hat{\mathbf{v}}|_{\partial\mathscr{B}_v} = \mathbf{0}\}$ be the space of admissible virtual motion actions. Once more, in this case it verifies $\mathbf{v}_1, \mathbf{v}_2 \in \text{Kin}_v \Rightarrow \mathbf{v}_1 - \mathbf{v}_2 \in \text{Var}_v$. With these ingredients, the variational formulation for the unsteady creep problem consists of determining $\dot{\mathbf{u}} \in \text{Kin}_v$ such that, for each time t, it verifies

$$\int_{\mathscr{B}} \mathbb{D}((\nabla\dot{\mathbf{u}})^s - \dot{\mathbf{E}}^c) \cdot (\nabla\hat{\mathbf{v}})^s dV = \int_{\mathscr{B}} \dot{\mathbf{b}} \cdot \hat{\mathbf{v}} dV + \int_{\partial\mathscr{B}_a} \dot{\mathbf{a}} \cdot \hat{\mathbf{v}} dS \qquad \forall\hat{\mathbf{v}} \in \text{Var}_v, \tag{5.412}$$

where initial conditions satisfy $\mathbf{E}^c_0 = \mathbf{E}^c(\mathbf{x}, 0) = \mathbf{O}$ in $\mathscr{B}$, and the initial displacement $\mathbf{u}_0 = \mathbf{u}(\mathbf{x}, 0) \in \text{Kin}_u$ is such that

$$\int_{\mathscr{B}} \mathbb{D}(\nabla\mathbf{u}_0)^s \cdot (\nabla\hat{\mathbf{u}})^s dV = \int_{\mathscr{B}} \mathbf{b}_0 \cdot \hat{\mathbf{u}} dV + \int_{\partial\mathscr{B}_a} \mathbf{a}_0 \cdot \hat{\mathbf{u}} dS \qquad \forall\hat{\mathbf{u}} \in \text{Var}_u, \tag{5.413}$$

with $\mathbf{b}_0 = \mathbf{b}(\mathbf{x}, 0)$ and $\mathbf{a}_0 = \mathbf{a}(\mathbf{x}, 0)$.

As in previous developments, it is possible to show that the strong form of the governing equations is associated with the previous variational equations (Euler equations). In this manner, the unsteady creep problem consists of determining $\mathbf{u}$, $\mathbf{E}$, $\mathbf{E}^c$, and $\mathbf{T}$ defined in $\mathscr{B} \times [0, t_f)$, where t_f is the final time for the analysis, such that the displacement and velocity fields satisfy the kinematic boundary constraints

$$\mathbf{u}_0 = \mathbf{u}(\mathbf{x}, 0) \in \text{Kin}_u, \tag{5.414}$$

$$\dot{\mathbf{u}}_0 = \dot{\mathbf{u}}(\mathbf{x}, 0) \in \text{Kin}_v. \tag{5.415}$$

The strain rate and the strain are compatible

$$\mathbf{E} = \mathbf{E}(\mathbf{x}, t) = (\nabla\mathbf{u})^s(\mathbf{x}, t) \qquad \text{in } \mathscr{B} \text{ and } \forall t \in [0, t_f), \tag{5.416}$$

$$\dot{\mathbf{E}} = \dot{\mathbf{E}}(\mathbf{x}, t) = (\nabla\dot{\mathbf{u}})^s(\mathbf{x}, t) \qquad \text{in } \mathscr{B} \text{ and } \forall t \in [0, t_f). \tag{5.417}$$

The creep strain rate follows one of the following constitutive equations

$$\dot{\mathbf{E}}^c = \dot{\mathbf{E}}^c(\mathbf{x}, t) = g(T_e, \theta, E^c_e)\mathbf{S} \qquad \text{strain hardening}, \tag{5.418}$$

$$\dot{\mathbf{E}}^c = \dot{\mathbf{E}}^c(\mathbf{x}, t) = f(T_e, \theta, t)\mathbf{S} \qquad \text{time hardening}. \tag{5.419}$$

The stress rate is

$$\dot{\mathbf{T}} = \dot{\mathbf{T}}(\mathbf{x}, t) = \mathbb{D}(\dot{\mathbf{E}} - \dot{\mathbf{E}}^c), \tag{5.420}$$

where T_e and E_e^c represent, respectively, the effective components of the stress and creep strain rate tensors. Furthermore, the stress state satisfies the local form of the equilibrium

$$\text{div}\,\mathbf{T}_0 + \mathbf{b}_0 = \mathbf{0} \qquad \text{in } \mathscr{B} \text{ and } t = 0, \tag{5.421}$$

$$\text{div}\,\dot{\mathbf{T}} + \dot{\mathbf{b}} = \mathbf{0} \qquad \text{in } \mathscr{B} \text{ and } t \in (0, t_f), \tag{5.422}$$

and the natural (mechanical) boundary conditions

$$\mathbf{T}_0\mathbf{n} = \mathbf{a}_0 \qquad \text{in } \partial\mathscr{B}_a \text{ and } t = 0, \tag{5.423}$$

$$\dot{\mathbf{T}}\mathbf{n} = \dot{\mathbf{a}} \qquad \text{in } \partial\mathscr{B}_a \text{ and } t \in (0, t_f). \tag{5.424}$$

5.14 Approximate Solutions to Unsteady Creep Formulations

The utilization of the finite element method in the variational formulation seen in the previous section leads us to the following system of first-order ordinary differential equations,

$$\mathbf{K}\dot{\mathbf{d}} = \dot{\mathbf{f}} + \dot{\mathbf{h}}^c, \tag{5.425}$$

with the initial condition verifying

$$\mathbf{K}\mathbf{d}_0 = \mathbf{f}_0, \tag{5.426}$$

where $\mathbf{K}$ is the global stiffness matrix of the system and $\dot{\mathbf{f}}$ is the load vector at time t, which is associated with the rate of loads $(\dot{\mathbf{b}}, \dot{\mathbf{a}})$ the body is subjected to and with the non-homogeneous essential velocity boundary condition over $\partial\mathscr{B}_v$. Vector $\dot{\mathbf{h}}^c$ is related to the creep strain rate at time t and $\dot{\mathbf{d}}$ is the generalized velocities at nodes for time t. In addition, $\mathbf{f}_0$ is the vector associated with the loads $(\mathbf{b}, \mathbf{a})$ and with the non-homogeneous essential displacement boundary condition over $\partial\mathscr{B}_v$ at time $t = 0$. Finally, $\mathbf{d}_0$ is the vector of generalized displacements at nodes for time $t = 0$.

As before, the approximate solution of the system of ordinary differential equations can be found using methods based on quadrature integration rules, Taylor series, and Runge–Kutta, among others. For the simple case of the Euler method, the algorithm consists of the following steps:

1) For $t_0 = 0$, compute $\mathbf{d}_0$ from $\mathbf{K}\mathbf{d}_0 = \mathbf{f}_0$.
2) Given $\mathbf{d}_0$, compute the displacement field in the whole body $\mathbf{u}_0$, and the associated initial strain $\mathbf{E}_0 = (\nabla\mathbf{u}_0)^s$ and initial stress $\mathbf{T}_0 = \mathbb{D}\mathbf{E}_0$.
3) Compute $\dot{\mathbf{E}}^c(0)$ using the corresponding constitutive law.
4) Compute $\dot{\mathbf{h}}^c(0)$, and from knowing $\dot{\mathbf{b}}(0)$ and $\dot{\mathbf{a}}(0)$, compute $\dot{\mathbf{f}}(0)$.
5) Compute $\dot{\mathbf{d}}(0)$ from $\mathbf{K}\dot{\mathbf{d}}(0) = \dot{\mathbf{f}}(0) + \dot{\mathbf{h}}^c(0)$.
6) Given $\dot{\mathbf{d}}(0)$, compute the velocity $\dot{\mathbf{u}}(\mathbf{x}, 0)$, the strain rate $\dot{\mathbf{E}}(\mathbf{x}, 0) = (\nabla\dot{\mathbf{u}})^s(\mathbf{x}, 0)$, and the stress rate $\dot{\mathbf{T}}(0) = \mathbb{D}(\dot{\mathbf{E}}(\mathbf{x}, 0) - \dot{\mathbf{E}}^c(0))$.
7) Finally, assuming (Euler method) that the rates are constant along each time integration step, one gets

$$t_1 = t_0 + \Delta t, \tag{5.427}$$

$$\mathbf{d}_1 = \mathbf{d}_0 + \Delta t\dot{\mathbf{d}}(0), \tag{5.428}$$

$$\mathbf{E}_1 = \mathbf{E}_0 + \Delta t \dot{\mathbf{E}}(0), \tag{5.429}$$

$$\mathbf{T}_1 = \mathbf{T}_0 + \Delta t \dot{\mathbf{T}}(0), \tag{5.430}$$

$$\mathbf{E}_1^c = \mathbf{E}_0^c + \Delta t \dot{\mathbf{E}}^c(0) = \Delta t \dot{\mathbf{E}}^c(0). \tag{5.431}$$

Hence, at time t_1 the state of the body is fully characterized. Then, steps 3 to 7 are repeated but replacing $t_0 = 0$ by t_1, until the final time t_f is achieved, or until we arrive at a steady-state solution.

5.15 Complementary Reading

The reader interested in further increasing their knowledge in the area covered in this chapter is directed to the following scientific contributions.

1) **Books**: G. Bernasconi and G. Piatti [28], J. T. Boyle and J. Spence [38, 39], P. G. Hodge [145, 146], J. A. H. Hult [147], L. M. Kachanov [152], H. Kraus [165], J. B. Martin [187], A. Nadai [206], F. H. Norton [212], F. K. Odqvist [233], W. Olszak and A. Sawczuk [235], R. K. Penny and D. L. Marriot [237], W. Prager [246], Y. U. N. Robotnov [249], M. A. Save and C. E. Massonnet [272, 273].
2) **Scientific articles**: E. N. Andrade [4, 5], M. F. Ashby [16], R. W. Bailey [20, 21], J. F. Besseling [29], C. R. Calladine and D. C. Drucker [44], publications of the Comitê da Asociação Brasileira de Ciências Mecânicas [53–55], J. E. Dorn [74, 75], R. A. Feijóo and E. Taroco [91, 92, 94, 95], R. A. Feijóo, E. Taroco and J. N. C. Guerreiro [96–98], R. A. Feijóo, E. Taroco and N. Zouain [99], J. J. Frost and M. F. Ashby [110], G. A. Greenbaum and M. F. Rubinstein [127], P. Germain, Q. S. Nguyen and P. Suquet [119], J. N. C. Guerreiro [129], P. Hault [138], J. Marin and Y. N. Pao [186], A. K. Mukherjee, J. E. Bird and J. E. Dorn [205], F. K. Odqvist [232], H. Oikawa [234], W. Prager [245], E. L. Robinson [252], V. I. Rozenblium [263], E. Taroco and R. A. Feijóo [282–284, 286], L. T. Vicat [302].

6

Materials Exhibiting Plasticity

6.1 Introduction

This chapter tackles the mechanics of bodies featuring elasto-plastic behavior. The term plasticity refers to time-independent inelastic processes, excluding viscous phenomena such as creep and relaxation, among others.

The presentation is divided into two parts. First, we describe the mathematical models which characterize the stress–strain constitutive relation for elasto-plastic materials. From that point, the variational principles specified to plasticity are developed. Such variational problems provide the roots for the posterior computational analysis in plasticity, leading to limit analyses and the study of strain and stress evolution.

6.2 Elasto-Plastic Materials

First, let us introduce some characteristic elements from tensile tests in order to identify the physical phenomena which must be embedded in a mathematical model that represents the material elasto-plastic behavior.

Figure 6.1 illustrates typical results obtained from the tensile testing of a metal specimen (think of a bar). The axial load is monotonically increasing, and with a controlled strain rate $\dot{\varepsilon}$ = constant. The different curves in the figure contain a linear part, that is a region of proportionality between the applied load and the strain in the specimen, and a nonlinear region featuring a reduced slope. The slope in this second part is always positive in a strain rate controlled test.

The so-called rupture strains are typically 10 to 100 times the maximum strain reached during the linear part of the curve. The property through which these strains can be considerably large before the rupture point is called ductility.

The change in the σ–ε curve with respect to the strain rate $\dot{\varepsilon}$ is a manifestation of viscous phenomena and is inherently time-dependent. The case of small $\dot{\varepsilon}$, that is, the static loading process, will be utilized to define the stress–strain relation in the theory of time-independent plasticity. Creep and relaxation phenomena are other expressions of the elasto-visco-plastic behavior of real materials, which will not be addressed in the present chapter.

Introduction to the Variational Formulation in Mechanics: Fundamentals and Applications, First Edition.
Edgardo O. Taroco, Pablo J. Blanco and Raúl A. Feijóo.

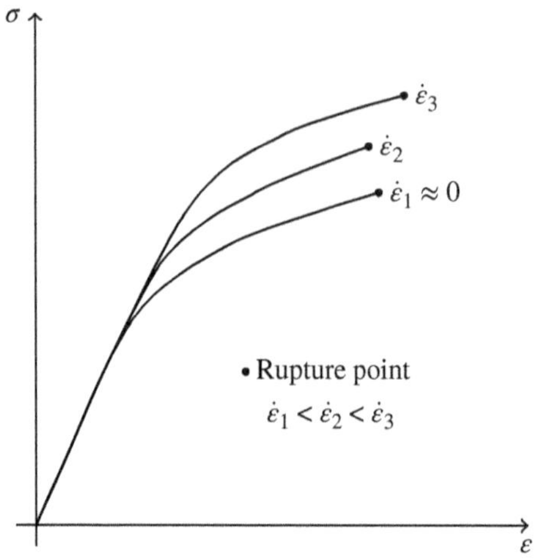

Figure 6.1 Tensile test. Typical curve for the pulling of a metal specimen.

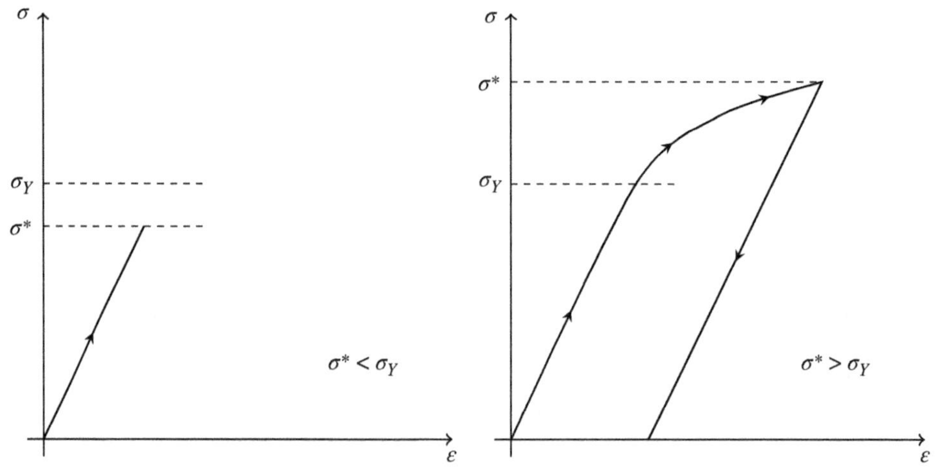

Figure 6.2 Loading/unloading tensile tests.

Consider the tensile test performed at a monotonic load until a certain value σ^* is reached, immediately followed by a monotonic unloading until the load is completely removed. The results will mimic those presented in Figure 6.2, where a dependence on the value chosen for σ^* is observed.

The stress that divides the linear and nonlinear regions is known as the initial yield stress (yield stress in short) or the elastic limit, and is denoted by σ_Y. That is, for any loading/unloading tensile test with $\sigma^* < \sigma_Y$, the stress–strain is reversible.

The plastic behavior is different from the elastic behavior in that it produces permanent (plastic) strains, which are irreversible. This is not a consequence of the nonlinearity in the stress–strain relation. The elastic limit (stress) value and the value of the limit of proportionality are very close for several materials largely used in engineering applications. This has led to confusion about these two different concepts of linearity and

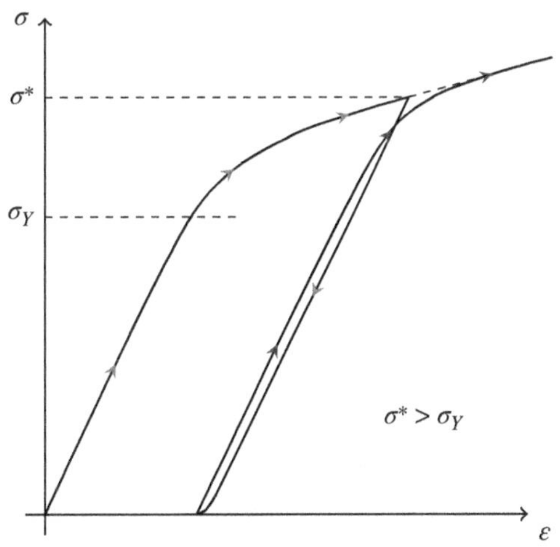

Figure 6.3 Loading/unloading tensile tests followed by a new loading program.

elasticity. The definition followed here considers σ_Y as the minimum value of the stress for which detectable permanent strains are observed in the specimen.

In the loading/unloading tests in which permanent strains occur, plastification takes place only during the stress increase. For decreasing stress there are decrements in the strain which are of reversible (recoverable) nature, and which are practically proportional to the stress decrement, and with the same proportionality value, denoted by $\mathbb{E}$.

Inspection of the material response in a loading/unloading tensile test, when followed by a new loading program, yields a hysteresis together with a new, slightly different, response during the subsequent loading process, as shown in Figure 6.3. These details are disregarded in a simple mathematical model of plastic behavior.

An important observation in relation to this experience is that the final stress achieved during the first loading process, say σ^*, remains as the new elastic limit for the next loading process, after the elastic unloading. Thus, the process of plastic deformation alters the original yield stress, augmenting the range in which elastic behavior is observed during the tensile test. This phenomenon is called strain hardening (also work hardening).

Another illustrative experience to explore the elasto-plastic behavior of some materials consists of a tensile test with plastification followed by an unloading process and compressive solicitation also producing plastification. Figure 6.4 shows that the plastic strain induced during traction causes the elastic resistance during the compression phase to be reduced. That is, the plastification during traction reduces the compressive yield stress, at the same time that the traction yield stress increases. This phenomenon is known as the Bauschinger effect, and implies that, even in an isotropic material in its virgin state, an anisotropic behavior (in the sense $\sigma^o_{Y+} = -\sigma^o_{Y-}$) will emerge as a consequence of the plastification process.

The phenomenological description exercised above shows that the plastic behavior depends upon the loading protocol executed over the body, up to the current mechanical state. In other words, the actual strain in the material is not a function of the current load, but is also a function of the load history to which the material has been exposed.

For example, in Figure 6.5, points 1, 2 and 3 correspond to the same stress. However, at these points different strains are encountered in the body as a result of the different

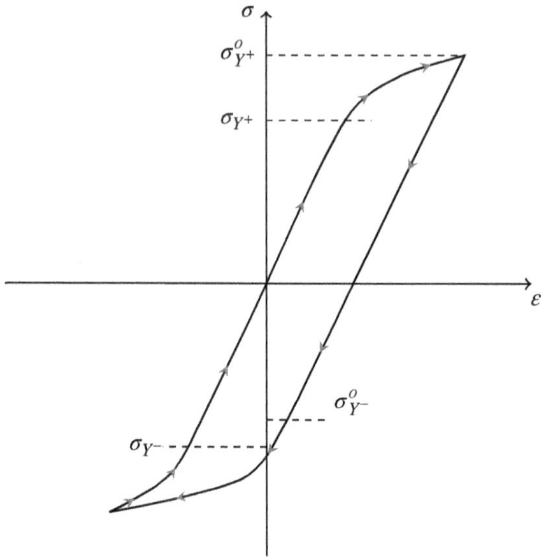

Figure 6.4 Bauschinger effect.

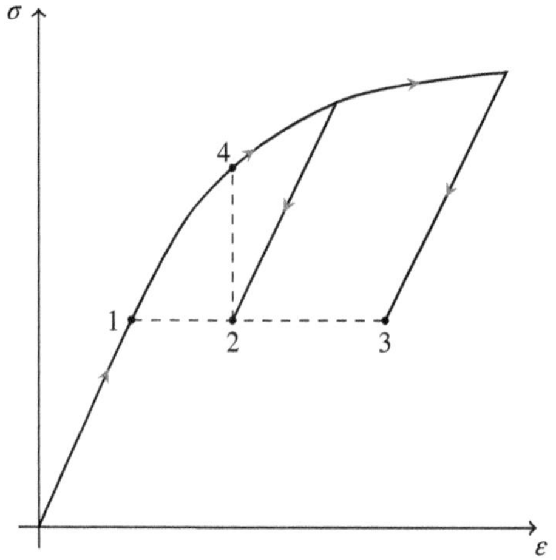

Figure 6.5 Multiple relations between values of stress and strain.

histories of load through which the specimen reached that level of stress. Analogously, points 2 and 4 feature the same strain, but different stress.

In particular, the material is not able to remember the part of the process in which stress and strain variations of elastic nature (recoverable processes) unfold. Then, it is said that the strain is a function of the recorded history, which must be represented by means of the current values of some state parameters such as, for example, the permanent strain, the dissipated plastic power, among others. In this regard, we say that the material behavior depends on the history in the sense that these parameters are known only when the whole loading process is given as a datum.

From a mechanical perspective, neglecting thermodynamic concepts, in this kind of material it is not possible to measure absolute strains, but only strains relative to some reference value. Indeed, imagine an observer receives a specimen of a material which

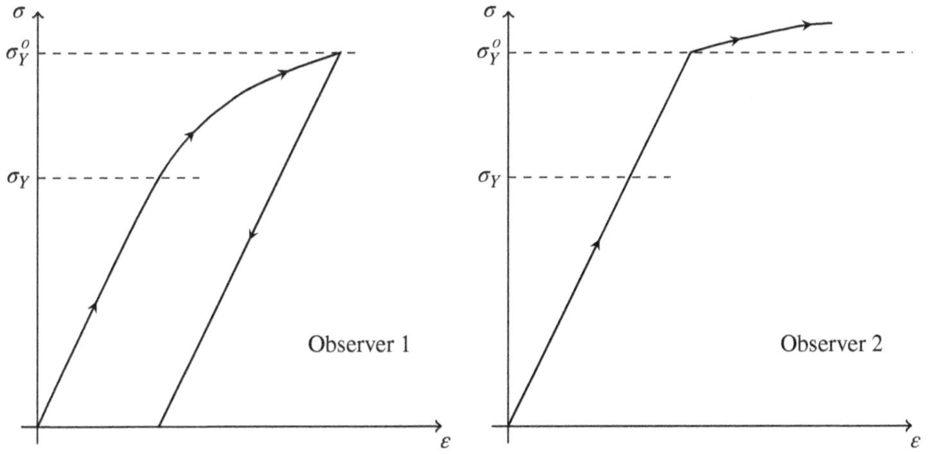

Figure 6.6 Tensile test for a bar which suffered previous plastification processes. Differences of stress–strain relation for two observers.

Figure 6.7 Perfectly plastic material behavior.

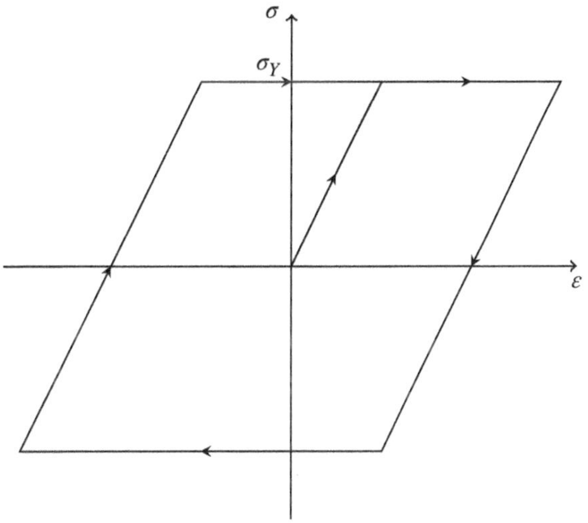

has previously suffered plastic processes (see Figure 6.6). For this observer, the yield stress will be different than the yield stress observed by the person who prepared the specimen from the virgin (load history-free) material. Nevertheless, when the thermodynamic state variables are included in the analysis, it is possible for any observer to define the original virgin state of the material.

Certain alloys present different material behavior to the one just described. Figure 6.7 shows an example in which, even for a constant stress, there is what is known as plastic flow, or plastic yielding. In a way, this mechanical regime is similar to that encountered in fluid flow phenomena. The fundamental difference is that in plastic flow there is no relation between the stress and the strain rate (see Chapter 11). Next, functional relations of the form $\dot{\varepsilon} = \frac{1}{\mu}\sigma$ o $\dot{\gamma} = \frac{1}{\mu}\tau$ will be encountered, which are similar to that of Newtonian fluids. Nevertheless, the parameter μ is not strictly speaking a viscosity because it is a function of the very plastic process. In these materials, the yield stress σ_Y is independent of previously experienced plastification mechanisms, and so does not depend on the recorded history. As a result, these materials feature neither strain hardening nor Bauschinger effects. This particular behavior has been termed perfect plastic model, in contrast to the plastic model with hardening seen before.

The previous expositions lead us to the formalization of elasto-plastic material models using the following hypotheses about material behavior.

- **Time independence**: It is admitted that the resulting strain from a given load (or stress) program is independent from the velocity with which such a program is executed. According to this concept, the purely plastic behavior is time independent because it is independent from the load program. Even so, the plastic behavior depends on the recorded history of the stresses, which is characterized by the values of a set of parameters, denoted by h. A further consequence of this hypothesis is that viscous phenomena as well as creep or relaxation are excluded from the modeling framework. Notwithstanding this hypothesis, we note that time does enter into the equations describing the material model as a parameter that defines the sequence of events. Therefore, the equations are invariant when changes in the time scale are performed. The only flow which is allowed in these conditions is the plastic flow (non-viscous flow), which is characteristic of perfectly plastic materials. In fact, a Newtonian-like viscous relation for a perfectly plastic material

$$\dot{\varepsilon} = \frac{1}{\mu}\sigma, \tag{6.1}$$

  can be integrated if μ is constant, giving

$$\varepsilon(t) = \varepsilon(0) + \frac{1}{\mu}\sigma\, t, \tag{6.2}$$

  determining a relation which is sensitive to the time scale. In the plasticity models, the stress exclusively determines if it is possible for strain rates to take place (independently from the stress level).
- **Unlimited ductility**: The mathematical expressions representing the elasto-plastic material model exclude the possibility of fracturing and ultimately material rupture.
- **Homogeneous temperature**: The body is assumed to be at a uniform temperature, that is, temperature gradients are assumed to be small. In addition, the constitutive relation is insensitive to temperature variations.

Let us now briefly emphasize the main characteristics that differentiate plasticity from elasticity. First, the stress depends on the previous plastic process (and not only on the current level of strain). Second, plastification processes are irreversible, which is manifested in the positive (load with plastification) and negative (elastic unloading) stress infinitesimal variations, which are related to infinitesimal strain variations through different tangent moduli. We thus conclude that a plasticity theory cannot be reduced to a single stress–strain relation as in elasticity, but must be framed in a constitutive formalism which establishes a relation between (pseudo) temporal rates of stress and strain.

6.3 Uniaxial Elasto-Plastic Model

The phenomenological analysis of specimens subjected to pure tensile/compressive loads enables us to introduce in simple terms the equations of an elasto-plastic constitutive model. The elasto-plastic behavior is described, in general, by providing four basic ingredients, namely:

1) a stress–strain elastic relation
2) a yield criterion
3) a hardening law which may modify the yield criterion during the plastic process
4) a plastic flow rule which defines the plastic strain rate for a given stress level satisfying the yield criterion.

Next, each of these ingredients of a mathematical model for plasticity is addressed.

6.3.1 Elastic Relation

To simplify the presentation, let us consider a simple linear elastic[1] relation of the form

$$\varepsilon^e = \frac{1}{\mathbb{E}}\sigma, \tag{6.3}$$

where $\mathbb{E}$ is the elasticity Young modulus of the material. When the material is in its virgin state, and at the beginning of the load process, the material exclusively experiences elastic strains, and so $\varepsilon = \varepsilon^e$. However, whenever plastification occurrs, strains are not only of elastic nature, that is $\varepsilon \neq \varepsilon^e$ and, in this situation, we define the plastic strain as

$$\varepsilon^p = \varepsilon - \varepsilon^e, \tag{6.4}$$

that is

$$\varepsilon^p = \varepsilon - \frac{1}{\mathbb{E}}\sigma. \tag{6.5}$$

This expression states that when the stress is fully removed, that is $\sigma = 0$, the strain coincides with the permanent strain, implying $\varepsilon = \varepsilon^p$, as seen in Figure 6.8. In most of the cases this actually happens, however there are situations in which this statement is not valid.

Now consider the elastic relation for stress and strain variations. From Figure 6.9, the incremental processes of the kind $0 \to 1$, purely elastic, or processes of local unloading, verify the conditions

$$d\varepsilon^e = \frac{1}{\mathbb{E}}d\sigma, \quad d\varepsilon = d\varepsilon^e + d\varepsilon^p, \quad d\varepsilon^p = 0. \tag{6.6}$$

For the plastification processes of the kind $0 \to 2$ (see Figures 6.9 and 6.10), we define in an analogous manner the elastic component of the strain increment by $d\varepsilon^e = \frac{d\sigma}{\mathbb{E}}$, giving

$$d\varepsilon^e = \frac{1}{\mathbb{E}}d\sigma, \quad d\varepsilon = d\varepsilon^e + d\varepsilon^p, \quad d\varepsilon^p \cdot \mathrm{sgn}(\sigma) > 0, \tag{6.7}$$

where $d\sigma$ and $d\varepsilon^e$ equal zero in the case of perfect plasticity, and where $\mathrm{sgn}(\cdot)$ stands for the sign of $(\cdot)$.

---

1 The consideration of a nonlinear model can equally be handled. The main difference is that the tangent modulus is not constant and depends on the strain.

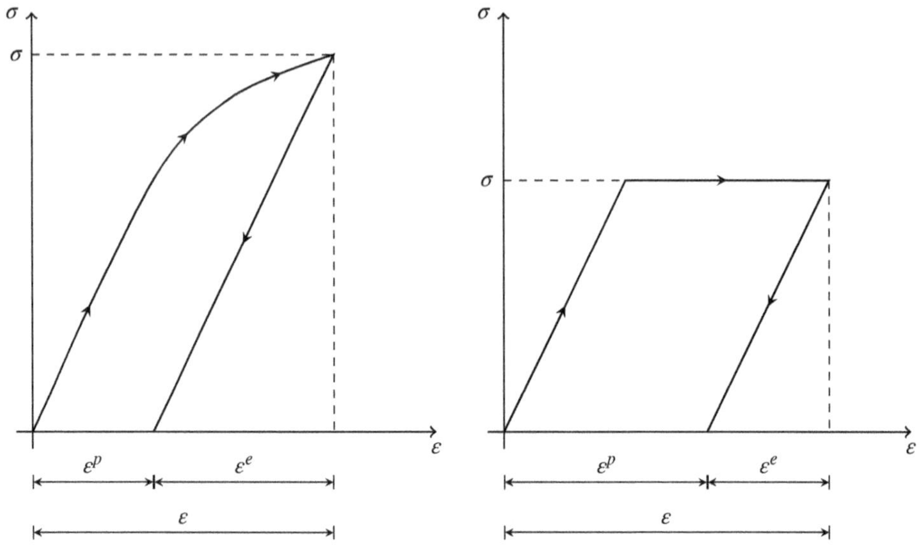

Figure 6.8 Elastic strain and permanent plastic strain. Hardening material (left). Perfectly plastic material (right).

We define the elastic and plastic strain rates by means of the expressions

$$\dot{\varepsilon}^e = \frac{1}{\mathbb{E}}\dot{\sigma}, \quad \dot{\varepsilon}^p = \dot{\varepsilon} - \frac{1}{\mathbb{E}}\dot{\sigma}, \tag{6.8}$$

yielding

$$\dot{\varepsilon} = \dot{\varepsilon}^e + \dot{\varepsilon}^p. \tag{6.9}$$

6.3.2 Yield Criterion

For materials with or without hardening, the plastically admissible stresses, those which can be developed in the material, are necessarily contained in a segment of the stress axis in the diagram σ–ε, as shown in Figure 6.11. Introducing a notation which goes beyond the uniaxial case, we say that there exists an initial set, or initial region, denoted by A^o in the stress space, which defines the admissible stress states for the material. Then, we have

$$\mathsf{A}^o = \{\sigma; -\sigma^o_{Y-} \leq \sigma \leq \sigma^o_{Y+}\}, \tag{6.10}$$

where σ^o_{Y-} and σ^o_{Y+} are, respectively, the yield stresses (or elastic limit) for compression and traction. Notice that we employ a positive parameter σ^o_{Y-} to represent the elastic limit under compressive loads.

For stress states which lie strictly within this admissible region, purely elastic processes can exclusively be developed. For this reason, the interior of region A^o is known as the elastic region. In turn, for stress states lying over the boundary of this admissible domain, plastification (infinitesimal) processes can take place only if they are triggered by an increase in the stress when materials feature hardening or by maintaining a constant stress in perfectly plastic materials. However, for these stress states over the boundary

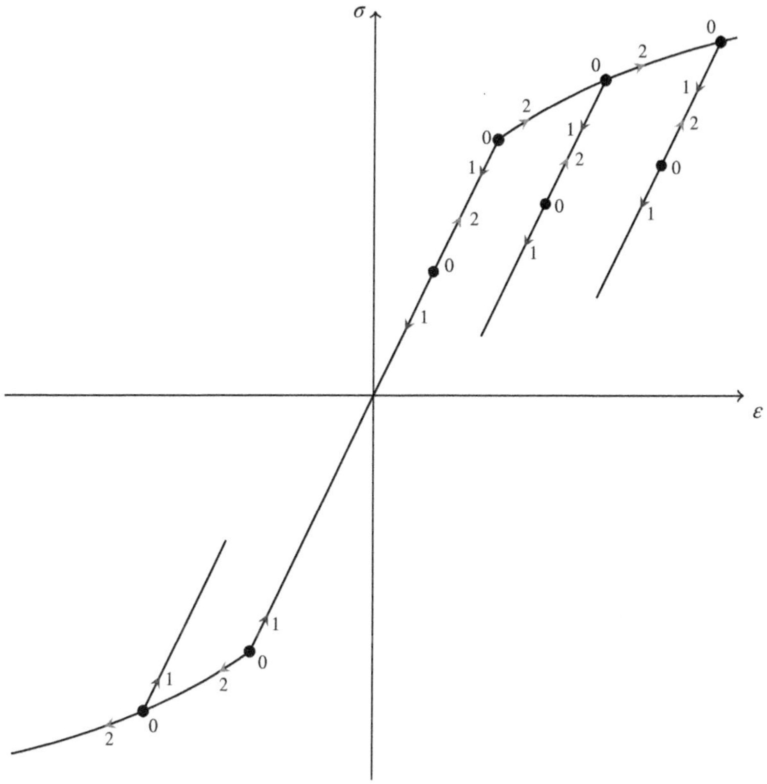

Figure 6.9 Incremental processes.

of A^o, elastic processes can also unfold when diminishing the level of stress (unloading or negative loading). Motivated by this, the boundary of A^o is termed the yield surface. Stresses outside this admissible domain are simply inadmissible in the initial material state and cannot be reached in the case of perfect plasticity.

The admissible domain A^o is conveniently defined through a initial yield function $f^o(\sigma)$ such that the condition $f^o(\sigma) < 0$ characterizes the elastic region and the condition $f^o(\sigma) = 0$ characterizes the yield surface. If the traction and compression plastification modes differ, two plastification functions will be required. In the uniaxial case treated here, these functions are

$$f_1^o(\sigma) = \sigma - \sigma_{Y+}^o, \qquad f_2^o(\sigma) = -\sigma - \sigma_{Y-}^o. \tag{6.11}$$

We can understand the admissible domain A^o as a vector function, whose components are f_1^o and f_2^o, that is,

$$\mathsf{f}^o(\sigma) = (f_1^o(\sigma), f_2^o(\sigma)). \tag{6.12}$$

In this way, the admissible domain is defined by the condition

$$\mathsf{f}^o(\sigma) \leq 0, \tag{6.13}$$

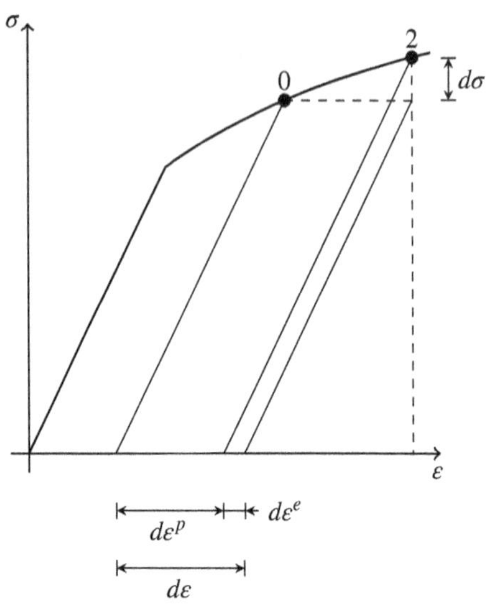

Figure 6.10 Relation between stress and strain increments.

where this condition must be understood in a component-wise sense. Likewise, the elastic region is determined by the condition

$$\mathsf{f}^o(\sigma) < 0, \tag{6.14}$$

and the yield surface is formed by all the stresses which verify one of the conditions

$$f_1^o(\sigma) = 0 \quad \text{and} \quad f_2^o(\sigma) \leq 0, \tag{6.15}$$

or

$$f_1^o(\sigma) \leq 0 \quad \text{and} \quad f_2^o(\sigma) = 0. \tag{6.16}$$

6.3.3 Hardening Law

We have seen that the domain containing plastically admissible stresses is modified along the plastic process and, as a result, the elastic domain and the yield surface are also altered.

A hardening law establishes the way in which the modification of the yield limit is produced. Such a law can be conveniently defined through a yield function of the form $\mathsf{F}(\sigma, \varepsilon^p)$, capable of describing the dependence of the elastic domain $\mathsf{f}(\sigma) < 0$ upon ε^p through (see Figure 6.12)

$$\mathsf{f}(\sigma) = \mathsf{F}(\sigma, \varepsilon^p)|_{\varepsilon^p=\text{cte}}. \tag{6.17}$$

In this manner, we have a single function $\mathsf{F}(\sigma, \varepsilon^p)$ and an infinite number of functions $\mathsf{f}(\sigma)$, one for each value of ε^p. Then, for a given load history that led to a certain accumulated plastic strain $\varepsilon^p = \varepsilon - \frac{\sigma}{\mathbb{E}}$, all the stresses verifying $\mathsf{f}(\sigma) < 0$ are elastic in the sense that purely elastic processes can be launched from this state. Thus, the stress satisfies

$$f_i(\sigma) = 0 \quad \text{and} \quad f_j(\sigma) \leq 0, \quad i \neq j, \quad i,j = 1,2, \tag{6.18}$$

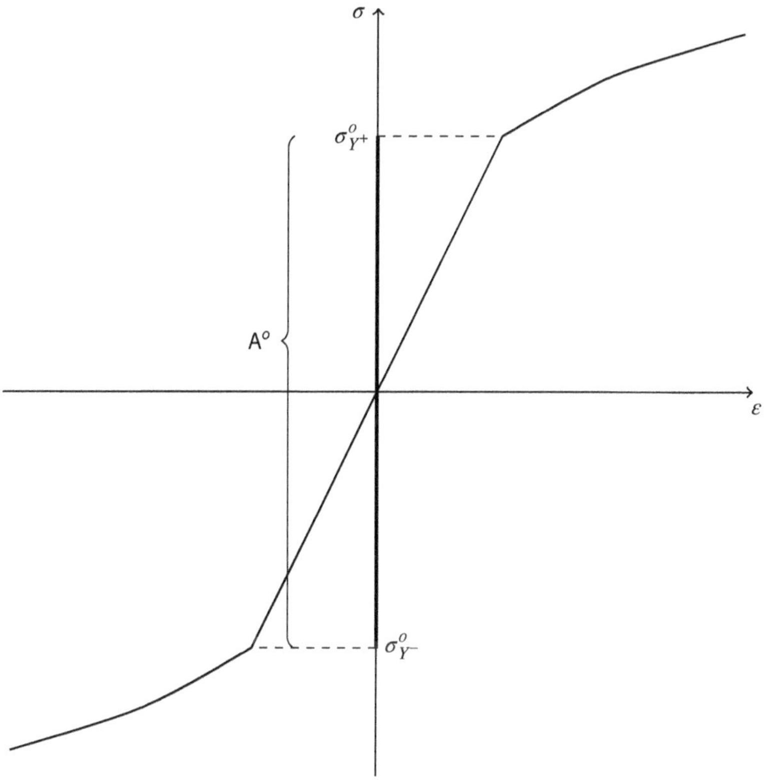

Figure 6.11 Initial region of admissible stresses for the material.

where $f_1(\sigma) = 0$ is the yield function for traction and $f_2(\sigma) = 0$ for compression, then plastic strains can develop in the subsequent loading process. Simultaneously, function f is modified along the plastification process according to $\mathsf{F}(\sigma, \varepsilon^p)$. In particular, the initial yield criterion is related to $\mathsf{F}(\sigma, \varepsilon^p)$ through

$$\mathsf{f}^o(\sigma) = \mathsf{F}(\sigma, \varepsilon^p)|_{\varepsilon^p=0}. \tag{6.19}$$

Differently, a perfectly plastic material possesses a unique yield limit, which is independent from ε^p. Then, for this kind of material, functions F, f and f^o coincide and are independent from ε^p.

In the present uniaxial model, we introduce F as follows

$$\mathsf{F}(\sigma, \varepsilon^p) = (F_1(\sigma, \varepsilon^p), F_2(\sigma, \varepsilon^p)), \tag{6.20}$$

with the traction and compression modes defined by

$$F_1(\sigma, \varepsilon^p) = \sigma - k_1(\varepsilon^p), \tag{6.21}$$

$$F_2(\sigma, \varepsilon^p) = -\sigma - k_2(\varepsilon^p). \tag{6.22}$$

A given traction stress in the yield limit verifies

$$f_1(\sigma) = F_1(\sigma, \varepsilon^p) = 0, \tag{6.23}$$

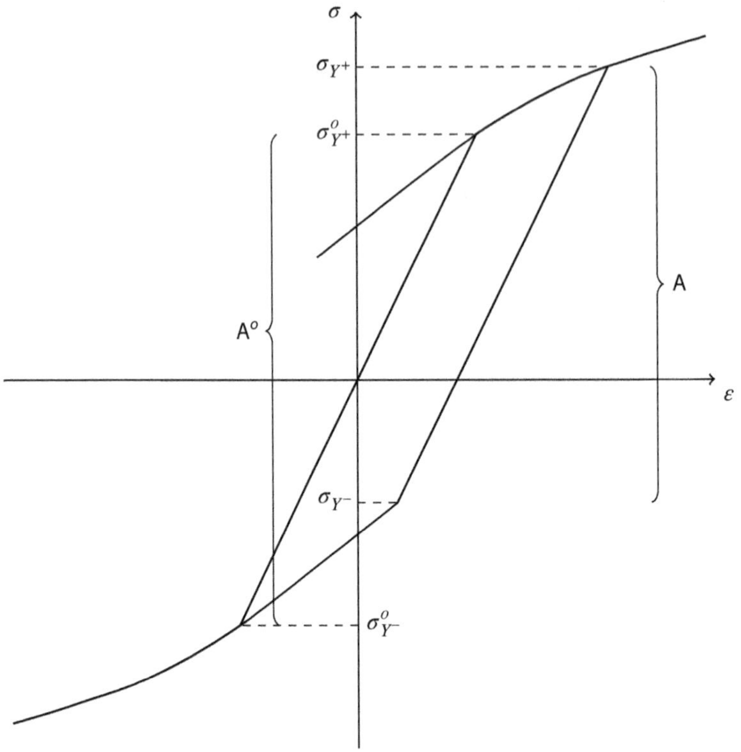

Figure 6.12 Alteration of the admissible domain, from A° to A.

meaning

$$\sigma = k_1(\varepsilon^p). \tag{6.24}$$

Hence, it follows that function $k_1(\varepsilon^p)$ can be established from the diagram with σ_{Y+} as a function of ε obtained in the tensile test, resulting in

$$k_1(\varepsilon^p) = \sigma_{Y+}\left(\varepsilon^p + \frac{1}{\mathbb{E}}\sigma\right), \tag{6.25}$$

$$k_1(0) = \sigma^o_{Y+} = \sigma_{Y+}\left(\frac{1}{\mathbb{E}}\sigma^o_{Y+}\right). \tag{6.26}$$

Analogously for compression

$$k_2(\varepsilon^p) = \sigma_{Y-}\left(\varepsilon^p + \frac{1}{\mathbb{E}}\sigma\right), \tag{6.27}$$

$$k_2(0) = \sigma^o_{Y-} = \sigma_{Y-}\left(\frac{1}{\mathbb{E}}\sigma^o_{Y-}\right). \tag{6.28}$$

6.3.4 Plastic Flow Rule

In keeping with the definition of the elastic strain rate, the elasticity modulus can be interpreted as

$$\mathbb{E} = \frac{d\sigma}{d\varepsilon^e} \quad \Rightarrow \quad \dot{\varepsilon}^e = \frac{1}{\mathbb{E}}\dot{\sigma}. \tag{6.29}$$

Analogously, for a plastic process like that one shown in Figure 6.10, we define the tangent modulus by

$$\mathbb{E}_t = \frac{d\sigma}{d\varepsilon} \quad \Rightarrow \quad \dot{\varepsilon} = \frac{1}{\mathbb{E}_t}\dot{\sigma}, \tag{6.30}$$

and the hardening modulus by

$$\mathbb{E}_p = \frac{d\sigma}{d\varepsilon^p} \quad \Rightarrow \quad \dot{\varepsilon}^p = \frac{1}{\mathbb{E}_p}\dot{\sigma}. \tag{6.31}$$

For standard, or stable, materials, whose stress–strain relation is monotonically increasing, the hardening modulus $\mathbb{E}_p$ is a positive function.

Replacing (6.29), (6.30) and (6.31) into (6.9) gives

$$\frac{1}{\mathbb{E}_t} = \frac{1}{\mathbb{E}} + \frac{1}{\mathbb{E}_p}. \tag{6.32}$$

Evidently, for perfectly plastic materials, $\mathbb{E}_t$ and $\mathbb{E}_p$ are zero and, therefore, this expression makes no sense.

In addition, we have seen that during a traction plastification process (it is similar for compression) the following is verified

$$F(\sigma, \varepsilon^p) = \sigma - k(\varepsilon^p) = 0. \tag{6.33}$$

This relation, derived with respect to ε^p, yields

$$\mathbb{E}_p = \frac{dk}{d\varepsilon^p}. \tag{6.34}$$

The phenomenological description of the elasto-plastic behavior shows that the constitutive law must necessarily be written in incremental form, that is, in terms of rates. Moreover, one also observes that the behavior is intrinsically nonlinear, by virtue of the irreversibility. The constitutive equation must thus be a relation between $\dot{\sigma}$ and $\dot{\varepsilon}$, or just $\dot{\varepsilon}^p$, because the elastic component has already been given by $\dot{\varepsilon}^e = \frac{\dot{\sigma}}{\mathbb{E}}$. Then, the constitutive behavior depends on the current stress level and on the parameters that describe the recorded history which, in the uniaxial case, are characterized by the accumulated plastic strain ($\mathsf{h} \equiv \varepsilon^p$). More precisely, the constitutive relation acquires the form

$$\dot{\varepsilon} = \dot{\varepsilon}(\dot{\sigma}, \sigma, \varepsilon^p) = \frac{1}{\mathbb{E}}\dot{\sigma} + \dot{\varepsilon}^p(\dot{\sigma}, \sigma, \varepsilon^p). \tag{6.35}$$

Next, we explicitly provide this relation assuming, for simplicity, that the material undergoes plastification just under traction loading.

For a material featuring strain hardening, this is

$$\dot{\varepsilon}^p = \begin{cases} 0 & \text{for} \begin{cases} f(\sigma) < 0 & \text{stress in the elastic domain, or} \\ f(\sigma) = 0 & \text{stress in the yield limit, and} \\ \dot{\sigma} \le 0 & \text{unloading local elastic process,} \end{cases} \\ \frac{1}{\mathbb{E}_p}\dot{\sigma} & \text{for} \begin{cases} f(\sigma) = 0 & \text{stress in the yield limit, and} \\ \dot{\sigma} > 0 & \text{loading local process,} \end{cases} \end{cases} \tag{6.36}$$

where f and $\mathbb{E}_p$ depend on the recorded history, for example

$$f(\sigma) = \sigma - k(\varepsilon^p), \tag{6.37}$$

$$\mathbb{E}_p = \frac{dk}{d\varepsilon^p}(\varepsilon^p). \tag{6.38}$$

For a perfectly plastic material, it is

$$\dot{\varepsilon}^p = \begin{cases} 0 & \text{for} \begin{cases} f(\sigma) < 0 & \text{stress in the elastic domain, or} \\ f(\sigma) = 0 & \text{stress in the yield limit, and} \\ \dot{\sigma} < 0 & \text{unloading local elastic process,} \end{cases} \\ \dot{\lambda} \geq 0 & \text{for} \begin{cases} f(\sigma) = 0 & \text{stress in the yield limit, and} \\ \dot{\sigma} = 0 & \text{local flow process.} \end{cases} \end{cases} \tag{6.39}$$

where f is a history-independent function. In turn, the plasticity parameter $\dot{\lambda}$ was introduced in this ideally plastic case. This entity is positive and is unknown, and serves to prescribe the condition that, during the development of the plastic flow under traction loading conditions, only stretching strains can be developed. Evidently, the intensity of such strains is not proportional to the change of stress, since the latter is zero in this case. The plastic strain remains undetermined in this case in the sense that any strain is admissible for the constitutive equation. However, this lack of determination can effectively be removed when the body is at mechanical equilibrium when subjected to certain load conditions and kinematical constraints.

The simple observation of the stress–strain relation for the perfectly plastic material shows us that, even if $\dot{\varepsilon}$ is not determined by $\dot{\sigma}$, the inverse relation is well established. In other words, given $\dot{\varepsilon}$, it is possible to determine $\dot{\sigma}$. As a result, the inverse constitutive equation is free of ambiguities and can be written for materials with or without hardening as

$$\dot{\sigma} = \dot{\sigma}(\dot{\varepsilon}, \sigma, \varepsilon^p) = \mathbb{E}(\dot{\varepsilon} - \dot{\varepsilon}^p(\dot{\varepsilon}, \sigma, \varepsilon^p)), \tag{6.40}$$

where

$$\dot{\varepsilon}^p = \begin{cases} 0 & \text{for} \begin{cases} f(\sigma) < 0 & \text{stress in the elastic domain, or} \\ f(\sigma) = 0 & \text{stress in the yield limit, and} \\ \dot{\varepsilon} \leq 0 & \text{unloading local elastic process,} \end{cases} \\ \dfrac{\mathbb{E}}{\mathbb{E} + \mathbb{E}_p}\dot{\varepsilon} & \text{for} \begin{cases} f(\sigma) = 0 & \text{stress in the yield limit, and} \\ \dot{\varepsilon} > 0 & \text{loading local process.} \end{cases} \end{cases} \tag{6.41}$$

This relation reduces to the perfect plasticity case when $\mathbb{E}_p = 0$ and f is independent from the history. In general, we have

$$f(\sigma) = \sigma - k(\varepsilon^p) \quad \Rightarrow \quad \mathbb{E}_p = \frac{dk}{d\varepsilon^p}, \tag{6.42}$$

which can also be written as

$$\dot{\sigma} = \begin{cases} \mathbb{E}\dot{\varepsilon} & \text{for} \begin{cases} f(\sigma) < 0 & \text{or} \\ f(\sigma) = 0 & \text{and } \dot{\varepsilon} \leq 0, \end{cases} \\ \mathbb{E}_t\dot{\varepsilon} & \text{for} \left\{ f(\sigma) = 0 \right. \quad \text{and } \dot{\varepsilon} > 0, \end{cases} \tag{6.43}$$

with $\mathbb{E}_t = \frac{\mathbb{E}\mathbb{E}_p}{\mathbb{E}+\mathbb{E}_p}$, which becomes identically null in the case of perfect plasticity.

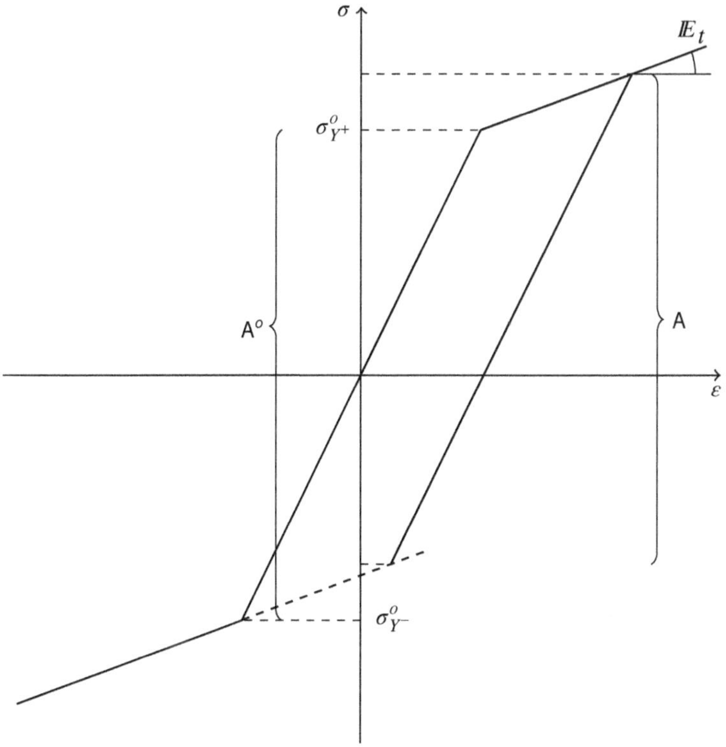

Figure 6.13 Material behavior with linear hardening law.

Let us exemplify the case with linear hardening law. Then, it is

$$k(\varepsilon^p) = a + b\varepsilon^p. \tag{6.44}$$

Coefficients a and b are determined considering the conditions $k(0) = \sigma^o_Y$, implying $a = \sigma^o_Y$, and $\mathbb{E}_p = \frac{dk}{d\varepsilon^p}$ yielding $b = \mathbb{E}_p$. In this way

$$\mathsf{F}(\sigma, \varepsilon^p) = (\sigma - (\sigma^o_{Y+} + \mathbb{E}_{p^+}\varepsilon^p), -\sigma - (\sigma^o_{Y-} - \mathbb{E}_{p^-}\varepsilon^p)). \tag{6.45}$$

Figure 6.13 shows the particular case for $\mathbb{E}_{p^+} = \mathbb{E}_{p^-}$, in which the admissible region only suffers a translation during the plastification process, keeping its original format. This is the well-known kinematical hardening as proposed by Prager. Notice that the Bauschinger effect is automatically incorporated in this model.

6.4 Three-Dimensional Elasto-Plastic Model

Let us generalize now the concepts exposed in the previous section for the case of multiaxial stress. We will develop the theory for the elasto-plastic model in the case of infinitesimal strains. Consider then an infinitesimal volume in the neighborhood of a point in the three-dimensional space which is subjected to a stress state $\mathbf{T}$ and to a given (infinitesimal) strain $\mathbf{E}$.

According to the concepts presented in the previous section, we must introduce in this more general case four basic ingredients for the elasto-plastic model, namely: (i) an elastic relation, (ii) an initial yield criterion, (iii) a hardening law, and (iv) the plastic flow rule.

6.4.1 Elastic Relation

For simplicity, consider a linear elastic relation

$$\mathbf{T} = \mathbb{D}\mathbf{E}^e, \quad \text{or its inverse} \quad \mathbf{E}^e = \mathbb{D}^{-1}\mathbf{T}, \tag{6.46}$$

where $\mathbf{E}^e$ is the elastic component of the strain $\mathbf{E}$, and where $\mathbb{D}$ is the elasticity tensor with the following properties

- $\mathbb{D}$ is symmetric, that is $\mathbb{D} = \mathbb{D}^T$
- $\mathbb{D}$ is positive definite, that is $\mathbb{D}\mathbf{E} \cdot \mathbf{E} > 0$, $\forall \mathbf{E} \in Sym$, $\mathbf{E} \neq \mathbf{O}$ and $\mathbb{D}\mathbf{E} \cdot \mathbf{E} = 0$ if and only if $\mathbf{E} = \mathbf{O}$
- Hence, $\mathbb{D}$ is invertible, that is, there exists $\mathbb{D}^{-1}$ such that $\mathbb{D}^{-1}\mathbb{D} = \mathbb{I}$.

For isotropic materials, this tensor can be explicitly given in terms of only two parameters, for example the Young modulus $\mathbb{E}$ and the Poisson coefficient ν, or, alternatively, as a function of the Lamé parameters $\bar{\lambda}$[2] and μ. In effect, $\mathbb{D}$ as a function of Lamé parameters takes the form

$$\mathbb{D} = 2\mu\mathbb{I} + \bar{\lambda}(\mathbf{I} \otimes \mathbf{I}), \tag{6.47}$$

and in terms of the Young and Poisson parameters reads

$$\mathbb{D} = \frac{\mathbb{E}}{(1+\nu)}\mathbb{I} + \frac{\nu\mathbb{E}}{(1+\nu)(1-2\nu)}(\mathbf{I} \otimes \mathbf{I}). \tag{6.48}$$

These elasticity parameters are related through the expressions

$$\mu = \frac{\mathbb{E}}{2(1+\nu)}, \qquad \bar{\lambda} = \frac{\nu\mathbb{E}}{(1+\nu)(1-2\nu)}, \tag{6.49}$$

$$\mathbb{E} = \frac{\mu(3\bar{\lambda} + 2\mu)}{\bar{\lambda} + \mu}, \qquad \nu = \frac{\bar{\lambda}}{2(\bar{\lambda} + \mu)}. \tag{6.50}$$

In addition, the inverse tensor is

$$\mathbb{D}^{-1} = \frac{1}{2\mu}\mathbb{I} - \frac{\bar{\lambda}}{2\mu(3\bar{\lambda} + 2\mu)}(\mathbf{I} \otimes \mathbf{I}), \tag{6.51}$$

or equivalently

$$\mathbb{D}^{-1} = \frac{1+\nu}{\mathbb{E}}\mathbb{I} - \frac{\nu}{\mathbb{E}}(\mathbf{I} \otimes \mathbf{I}). \tag{6.52}$$

We have thus defined, with these elements, the elastic relation, which is

$$\mathbf{T} = \mathbb{D}\mathbf{E}^e = 2\mu\mathbf{E}^e + \bar{\lambda}(\mathrm{tr}\mathbf{E}^e)\mathbf{I} = \frac{\mathbb{E}}{(1+\nu)}\mathbf{E}^e + \frac{\nu\mathbb{E}}{(1+\nu)(1-2\nu)}(\mathrm{tr}\mathbf{E}^e)\mathbf{I}, \tag{6.53}$$

---

2 The bar was introduced to avoid confusion between this parameter and the plastic parameter λ introduced in the previous section.

while the inverse relation is

$$\mathbf{E}^e = \mathbb{D}^{-1}\mathbf{T} = \frac{1}{2\mu}\mathbf{T} - \frac{\overline{\lambda}}{2\mu(3\overline{\lambda}+2\mu)}(\mathrm{tr}\mathbf{T})\mathbf{I} = \frac{1+\nu}{\mathbb{E}}\mathbf{T} - \frac{\nu}{\mathbb{E}}(\mathrm{tr}\mathbf{T})\mathbf{I}. \tag{6.54}$$

Let us develop an interpretation of these elastic relations which is useful when comparing with the relations found between the plastic strain and the stress. Apply the trace operation in the elastic tensor relations as follows

$$\mathrm{tr}\mathbf{E}^e = \frac{1}{K}T_m, \tag{6.55}$$

where K is the bulk modulus given by

$$K = \overline{\lambda} + \frac{2}{3}\mu = \frac{\mathbb{E}}{3(1-2\nu)}, \tag{6.56}$$

and T_m is the (negative) hydrostatic pressure given by

$$T_m = \frac{1}{3}(\mathrm{tr}\mathbf{T}). \tag{6.57}$$

Recalling now the definition of the deviatoric component of a second-order tensor, say $\mathbf{A}$ ($\mathbf{A}^D = \mathbf{A} - \frac{1}{3}(\mathrm{tr}\mathbf{A})\mathbf{I}$), and calling $\mathbf{S}$ the deviatoric component of $\mathbf{T}$, we have

$$(\mathbf{E}^e)^D = \frac{1}{2\mu}\mathbf{S} = \frac{1+\nu}{\mathbb{E}}\mathbf{S}. \tag{6.58}$$

An energetic interpretation of the elastic relation can be given next. Consider the following notation for the invariants of the stress tensor

$$I_1 = \mathrm{tr}\mathbf{T} = 3T_m, \qquad I_2 = \frac{1}{2}[\mathbf{T}\cdot\mathbf{T} - (\mathrm{tr}\mathbf{T})^2], \qquad I_3 = \det\mathbf{T}, \tag{6.59}$$

where we use $\mathrm{tr}(\mathbf{T}^2) = \mathbf{T}\cdot\mathbf{T}$. Likewise, for the deviatoric component $\mathbf{S} = \mathbf{T} - \frac{1}{3}(\mathrm{tr}\mathbf{T})\mathbf{I}$ it is

$$J_1 = 0, \qquad J_2 = \frac{1}{2}\mathbf{S}\cdot\mathbf{S}, \qquad J_3 = \det\mathbf{S}. \tag{6.60}$$

Analogously for the strain tensor $\mathbf{E}^e$ we have

$$\mathcal{I}_1 = \mathrm{tr}\mathbf{E}^e, \qquad \mathcal{I}_2 = \frac{1}{2}[\mathbf{E}^e\cdot\mathbf{E}^e - (\mathrm{tr}\mathbf{E}^e)^2], \qquad \mathcal{I}_3 = \det\mathbf{E}^e, \tag{6.61}$$

and for its deviatoric component $(\mathbf{E}^e)^D = \mathbf{E}^e - \frac{1}{3}(\mathrm{tr}\mathbf{E}^e)\mathbf{I}$ it gives

$$\mathcal{J}_1 = 0, \qquad \mathcal{J}_2 = \frac{1}{2}(\mathbf{E}^e)^D\cdot(\mathbf{E}^e)^D, \qquad \mathcal{J}_3 = \det(\mathbf{E}^e)^D. \tag{6.62}$$

In the linear elasticity domain, the elastic strain energy and the complementary strain energy per unit volume coincide (see Chapter 4), and verify

$$U = \frac{1}{2}\mathbf{T}\cdot\mathbf{E}^e. \tag{6.63}$$

By replacing in this expression the previous elastic relations we find the energy as a function of the elastic strain

$$U = \frac{K}{2}\mathcal{I}_1^2 + 2\mu\mathcal{J}_2, \tag{6.64}$$

or as a function of the stress

$$U = \frac{1}{18K} I_1^2 + \frac{1}{2\mu} J_2. \tag{6.65}$$

To sum up, a stress state $\mathbf{T}$, characterized by a hydrostatic stress T_m and a deviatoric component $\mathbf{S}$, produces a volume variation proportional to the hydrostatic stress and a distortion (given by $\mathbf{E}^D$) proportional to the deviatoric stress $\mathbf{S}$. In addition, the stored elastic energy caused by an elastic strain is the contribution of a component proportional to the square of the hydrostatic stress (I_1) and a component proportional to the square of the deviatoric stress (J_2).

Finally, taking into account that $\mathbf{E} = \mathbf{E}^e + \mathbf{E}^p$, we obtain

$$\mathbf{E}^p = \mathbf{E} - \mathbf{E}^e = \mathbf{E} - \mathbb{D}^{-1}\mathbf{T}, \tag{6.66}$$

and in terms of the strain rate, with $\mathbf{D} = \dot{\mathbf{E}} = \dot{\mathbf{E}}^e + \dot{\mathbf{E}}^p = \mathbf{D}^e + \mathbf{D}^p$, this gives

$$\mathbf{D}^p = \mathbf{D} - \mathbf{D}^e = \mathbf{D} - \mathbb{D}^{-1}\dot{\mathbf{T}}. \tag{6.67}$$

6.4.2 Yield Criterion and Hardening Law

Let us now admit the existence of a characteristic function of the material that will be termed the yield function, and which has the following form

$$\mathsf{F} = \mathsf{F}(\mathbf{T}, \mathsf{h}), \tag{6.68}$$

whose second argument, h, is a scalar or a vector field representing a set of parameters which enables the reconstruction of the history (h is the recorded history). We saw in the uniaxial case that the accumulated plastic strain could be employed for this purpose.

For a given value of h, the yield function allows us to know the behavior the material must display in a process triggered by a stress state $\mathbf{T}$ verifying the following conditions

- $\mathsf{F}(\mathbf{T}, \mathsf{h}) < 0$, that is, $\mathbf{T}$ belongs to the elastic domain
- $\mathsf{F}(\mathbf{T}, \mathsf{h}) = 0$, that is, $\mathbf{T}$ is in the elastic limit, or equivalently over the yield surface.

By elastic domain we refer to a set in the stress space associated with a given history h, and yield surface means the boundary of this domain. The region for which $\mathsf{F}(\mathbf{T}, \mathsf{h}) > 0$ is inadmissible for the stresses.

Keeping this in mind, we can define a plastically admissible stress if it verifies $\mathsf{F}(\mathbf{T}, \mathsf{h}) \leq 0$. As a consequence, the admissible domain, which depends on the recorded history h, is characterized by the set A in the stress space given by

$$\mathsf{A} = \{\mathbf{T} \in \mathscr{W}';\ \mathsf{F}(\mathbf{T}, \mathsf{h}) \leq 0\}. \tag{6.69}$$

The interior of A is the elastic domain, while its boundary is the yield surface.

Similarly to the uniaxial case, it is also convenient to make use of the functions $\mathsf{f}(\mathbf{T})$ defined for each value of h as follows

$$\mathsf{f}(\mathbf{T}) = \mathsf{F}(\mathbf{T}, \mathsf{h})|_{\bar{\mathsf{h}}}. \tag{6.70}$$

Then, the function f depends on the history and characterizes the material behavior as follows

- $\mathsf{f}(\mathbf{T}) < 0$, elastic domain
- $\mathsf{f}(\mathbf{T}) = 0$, yield surface.

It is worthwhile highlighting that it is possible, for the stress state $\mathbf{T}$, that f features a negative or null value, and there exists $\mathbf{T}^*$ as a result of the process for which f is positive. However, this is only possible if $\mathbf{T}$ precedes $\mathbf{T}^*$, and between these two stress states a plastification takes place, turning f into a new function f^* which verifies $f^*(\mathbf{T}^*) \leq 0$. Notably, function F is always negative or zero for any pair $(\mathbf{T}, h)$ of the loading process.

When a local elastic unloading process is initiated, for a given history characterized by $\overline{h}$, and for a given stress state $\mathbf{T}$ lying over the yield surface, the value of $\overline{h}$ is not modified and, therefore, f remains unchanged. For an infinitesimal elastic unloading process the final state is elastic, and the following holds

$$\begin{cases} f(\mathbf{T}) = 0, & \text{that is } F(\mathbf{T}, h)|_{\overline{h}} = 0, \text{and} \\ f(\mathbf{T} + d\mathbf{T}) < 0, & \text{that is } F(\mathbf{T} + d\mathbf{T}, h)|_{\overline{h}} < 0, \end{cases} \tag{6.71}$$

from where the conditions that identify the local elastic unloading are deduced

$$df < 0 \quad \text{that is} \quad dF|_{\overline{h}} < 0. \tag{6.72}$$

If f is differentiable we have

$$df = \dot{f}dt = f_{\mathbf{T}} \cdot \dot{\mathbf{T}}dt, \tag{6.73}$$

where $f_{\mathbf{T}} = \nabla_{\mathbf{T}} f(\mathbf{T})$, and the local elastic unloading condition results

$$\dot{f} = f_{\mathbf{T}} \cdot \dot{\mathbf{T}} < 0, \tag{6.74}$$

since dt is positive.

Let us explore this condition in terms of the function F. First observe that

$$\dot{F} = F_{\mathbf{T}} \cdot \dot{\mathbf{T}} + F_h \cdot \dot{h} = \dot{f} + F_h \cdot \dot{h}, \tag{6.75}$$

where in the derivation with respect to $\mathbf{T}$ (h) the variable h ($\mathbf{T}$) is kept fixed.

So, in the local elastic unloading there are no changes in the history, that is, $\dot{h} = 0$. Hence, the condition that characterizes this local elastic unloading is given by

$$F_{\mathbf{T}} \cdot \dot{\mathbf{T}} < 0. \tag{6.76}$$

Let us now consider a loading process initiated in a stress state lying over the yield surface, that is

$$f(\mathbf{T}) = 0 \quad \text{or equivalently} \quad F(\mathbf{T}, h) = 0. \tag{6.77}$$

During this process, both the history h and the function f suffer changes. In particular, after an infinitesimal stress variation and history change, the resulting state continues to lie over the yield surface. In other words (see Figure 6.14), one has

$$f^*(\mathbf{T} + d\mathbf{T}) = 0 \quad \text{or equivalently} \quad F(\mathbf{T} + d\mathbf{T}, h + dh) = 0. \tag{6.78}$$

From (6.78) we deduce that the loading process in the yield limit remains characterized by

$$dF = 0. \tag{6.79}$$

In the case where function F is differentiable, this corresponds to

$$\dot{F} = F_{\mathbf{T}} \cdot \dot{\mathbf{T}} + F_h \cdot \dot{h} = 0. \tag{6.80}$$

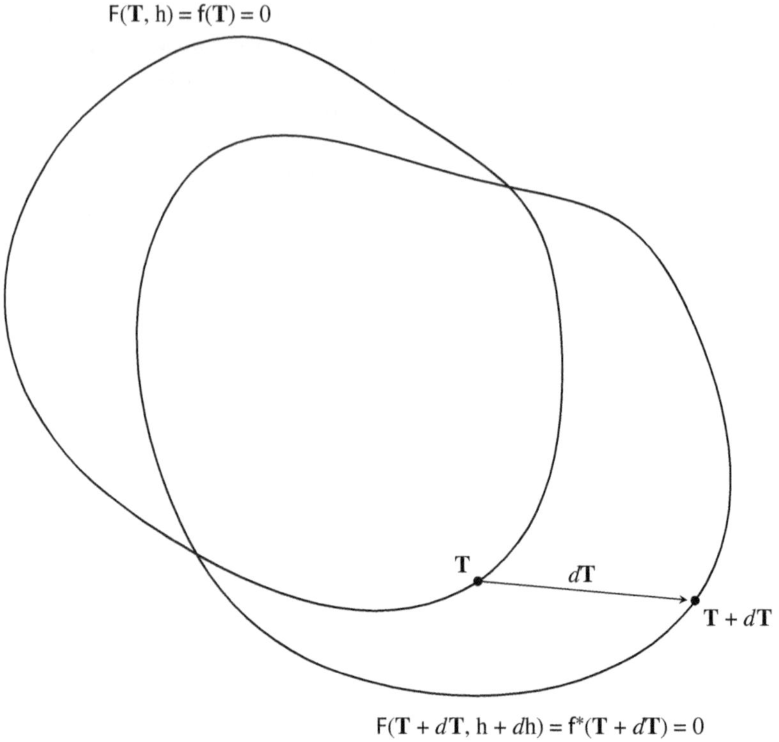

Figure 6.14 Admissible domains in the stress space.

Let us now analyze this condition but written in terms of function $\mathsf{f}(\mathbf{T})$ at the initial state in the loading process. At the initial time it is $\mathsf{f}(\mathbf{T}) = 0$, and the final state is not elastic. Otherwise, the unloading would have been an elastic process. Then

$$\mathsf{f}(\mathbf{T} + d\mathbf{T}) \geq 0, \tag{6.81}$$

and the condition that characterizes this process from an initial state over the yield surface is given by

$$d\mathsf{f} \geq 0 \quad \text{or equivalently} \quad \dot{\mathsf{f}} = \mathsf{f}_{\mathbf{T}} \cdot \dot{\mathbf{T}} \geq 0. \tag{6.82}$$

This leads to two possibilities

- $\dot{\mathsf{f}} > 0$, which characterizes a plastic process driven by effective loading conditions
- $\dot{\mathsf{f}} = 0$, which characterizes a plastic process under neutral loading conditions.

Importantly, observe that while a neutral loading condition cannot exist in uniaxial stress states (only for perfect plasticity), it is commonplace in multiaxial stress states.

In a multiaxial stress state, a perfectly plastic material is characterized by a yield function F, independent from the parameters of the history, h, and that therefore coincides with f (also independent from h). In this material, the only possible plastic process is the neutral process

$$\dot{\mathsf{f}} = \dot{\mathsf{F}} \leq 0 \quad \text{for } \mathbf{T} \text{ such that } \mathsf{F}(\mathbf{T}) = 0. \tag{6.83}$$

Considering that every plastic process is performed verifying $\dot{\mathsf{F}} = \dot{\mathsf{f}} + \mathsf{F}_\mathsf{h} \cdot \dot{\mathsf{h}} = 0$, then

$$\mathsf{F}_\mathsf{h} \cdot \dot{\mathsf{h}} = 0, \tag{6.84}$$

is the necessary condition that must be satisfied when the process is under neutral loading. This condition is trivial in perfect plasticity because $\mathsf{F}_\mathsf{h} = 0$. Nevertheless, if a hardening material undergoes a neutral process without plastic strain, that is, $\mathbf{D}^p = \mathbf{O}$, this process does not alter the history parameter h and, consequently, the condition is verified. This means that

$$\dot{\mathsf{f}} = 0 \quad \Rightarrow \quad \mathbf{D}^p = \mathbf{O}, \tag{6.85}$$

is a condition which ensures that a constitutive relation for a material with strain hardening of the kind $\mathbf{D}^p = \mathbf{D}^p(\dot{\mathbf{T}}, \mathbf{T}, \mathsf{h})$ is consistent with respect to neutral processes.

6.4.3 Potential Plastic Flow

Consider the constitutive relation

$$\mathbf{D}^p = \mathbf{D}^p(\dot{\mathbf{T}}, \mathbf{T}, \mathsf{h}), \tag{6.86}$$

for the situation in which these plastic strain rates can be developed, which means for $\mathbf{T}$ that

- $\mathsf{f}(\mathbf{T}) = 0$ and $\dot{\mathsf{f}}(\mathbf{T}) = \mathsf{f}_\mathbf{T} \cdot \dot{\mathbf{T}} > 0$, for hardening materials
- $\mathsf{f}(\mathbf{T}) = 0$ and $\dot{\mathsf{f}}(\mathbf{T}) = \mathsf{f}_\mathbf{T} \cdot \dot{\mathbf{T}} = 0$, for perfectly plastic materials,

or its equivalent in terms of F

- $\mathsf{F}(\mathbf{T}, \mathsf{h}) = 0$ and $\mathsf{F}_\mathbf{T}(\mathbf{T}, \mathsf{h}) \cdot \dot{\mathbf{T}} > 0$, for hardening materials
- $\mathsf{F}(\mathbf{T}, \mathsf{h}) = 0$ and $\mathsf{F}_\mathbf{T}(\mathbf{T}) \cdot \dot{\mathbf{T}} = 0$, for perfectly plastic materials.

Most models for plasticity hypothesize the existence of a plastic strain rate potential $\mathsf{G}(\mathbf{T}, \mathsf{h})$, or $\mathsf{g}(\mathbf{T}) = \mathsf{G}(\mathbf{T}, \mathsf{h})|_{\bar{\mathsf{h}}}$, whose gradient with respect to $\mathbf{T}$ defines the direction of the plastic strain rate, that is

$$\mathbf{D}^p = \dot{\lambda}\mathsf{G}_\mathbf{T} \quad \text{or} \quad \mathbf{D}^p = \dot{\lambda}\mathsf{g}_\mathbf{T}. \tag{6.87}$$

The plasticity coefficient $\dot{\lambda}$ must be a time rate for the previous expression to be time independent. This plastic multiplier $\dot{\lambda}$ depends upon the process, but not on the velocity with which the process unfolds. In other words, it is not a material-specific parameter, but a process-specific one.

For perfectly plastic materials $\dot{\lambda}$ is undetermined in the context of the constitutive relation, but it is determined when the complete mechanical equilibrium problem is regarded, that is to say, when the body made of elasto-plastic material is subjected to external loads and kinematical constraints, and the mechanical equilibrium is evaluated (without reaching plastic collapse).

For materials featuring strain hardening, we can further restrict the constitutive equation if we want to ensure the validity of the consistency condition for neutral processes discussed in the previous section. Indeed, for this it is sufficient to prescribe that

$$\mathbf{D}^p = \mathbf{O} \quad \text{if} \quad \dot{\mathsf{f}} = 0. \tag{6.88}$$

A simple approach to satisfy this condition is to take $\mathbf{D}^p$ proportional to $\dot{\mathsf{f}}$ and, hence, proportional to $\dot{\mathbf{T}}$. This is achieved in the potential law by taking

$$\dot{\lambda} = \frac{1}{H}\dot{\mathsf{f}}, \tag{6.89}$$

where $H(\mathbf{T}, \mathsf{h})$ is a scalar valued function which defines the strain hardening. In this way, the potential plastic flow rule results in

$$\mathbf{D}^p = \frac{1}{H}\dot{\mathsf{f}}\mathsf{g}_{\mathbf{T}}, \tag{6.90}$$

and recalling that $\dot{\mathsf{f}} = \mathsf{f}_{\mathbf{T}} \cdot \dot{\mathbf{T}}$ we have

$$\mathbf{D}^p = \frac{1}{H}(\mathsf{g}_{\mathbf{T}} \otimes \mathsf{f}_{\mathbf{T}})\dot{\mathbf{T}}. \tag{6.91}$$

The material is then characterized by the yield function F (or f), the plastic potential G (or g), and the hardening function H. In particular, when the plastic potential G coincides with the yield function F the constitutive law is said to be associative, and satisfies the so-called normality law which states that $\mathbf{D}^p = \dot{\lambda}\mathsf{f}_{\mathbf{T}}$. This normality condition states that plastic strain rates are normal to the yield surface $\mathsf{f}(\mathbf{T}) = 0$ at the point determined by the stress $\mathbf{T}$ at the beginning of the process, and which produces that strain rate $\mathbf{D}^p$. This is illustrated in Figure 6.15.

To sum up, the functional form of the potential elasto-plastic constitutive equation must be

$$\mathbf{D} = \mathbf{D}(\dot{\mathbf{T}}, \mathbf{T}, \mathsf{h}) = \mathbb{D}^{-1}\dot{\mathbf{T}} + \dot{\lambda}\mathsf{g}_{\mathbf{T}}, \tag{6.92}$$

where $\dot{\lambda}(\dot{\mathbf{T}}, \mathbf{T}, \mathsf{h})$ is calculated in terms of f and $\dot{\mathsf{f}} = \mathsf{f}_{\mathbf{T}} \cdot \dot{\mathbf{T}}$ by means of the following rules.

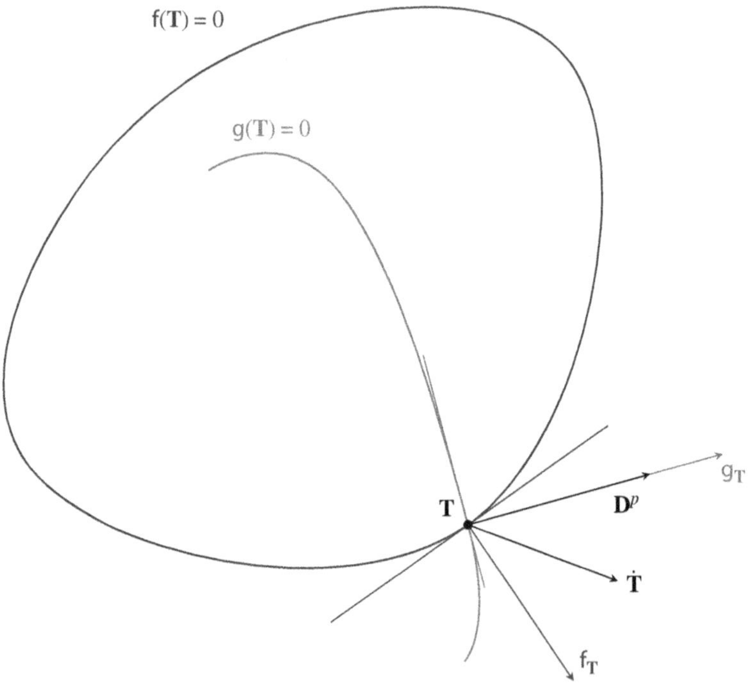

Figure 6.15 Yield function and plastic potential.

For perfectly plastic materials

$$\dot{\lambda}\begin{cases} = 0 & \text{for}\begin{cases} \mathsf{f} < 0 & \text{elastic stress, or} \\ \mathsf{f} = 0 & \text{stress in the yield surface, and} \\ \dot{\mathsf{f}} < 0 & \text{local elastic unloading,} \end{cases} \\ \geq 0 & \text{for}\left\{ \mathsf{f} = 0 \text{ and } \dot{\mathsf{f}} = 0 \right. \quad \text{neutral loading plastic process,} \end{cases} \tag{6.93}$$

where $\dot{\lambda}$ different from zero is indetermined. The situations $\mathsf{f} = 0$ and $\dot{\mathsf{f}} > 0$ are not feasible for perfectly plastic materials.

For strain hardening materials

$$\dot{\lambda} = \begin{cases} 0 & \text{for}\begin{cases} \mathsf{f} < 0 & \text{elastic stress, or} \\ \mathsf{f} = 0 & \text{stress on the yield surface, and} \\ \dot{\mathsf{f}} < 0 & \text{local elastic unloading,} \end{cases} \\ \frac{1}{H}\dot{\mathsf{f}} & \text{for}\begin{cases} \mathsf{f} = 0 & \text{stress on the yield surface, and} \\ \dot{\mathsf{f}} \geq 0 & \text{loading plastic process.} \end{cases} \end{cases} \tag{6.94}$$

It is possible to give a more compact form to these expressions using the Heaviside function

$$\mathscr{H}(c) = \begin{cases} 0 & \text{if } c < 0, \\ 1 & \text{if } c \geq 0, \end{cases} \tag{6.95}$$

and the notation $(\cdot)^+$ implying

$$(c)^+ = \max\{0, c\}.$$

Therefore, the previous constitutive equations are reduced to the following $\mathbf{D} = \mathbb{D}^{-1}\dot{\mathbf{T}} + \mathbf{D}^p$ with $\mathbf{D}^p$ defined according to the type of material. For perfectly plastic materials

$$\mathbf{D}^p = \dot{\lambda}\mathscr{H}(\mathsf{f})\mathscr{H}(\dot{\mathsf{f}})\mathsf{g}_{\mathbf{T}}, \tag{6.96}$$

with $\dot{\lambda} \geq 0$ indetermined. For hardening materials

$$\mathbf{D}^p = \frac{1}{H}\mathscr{H}(\mathsf{f})(\dot{\mathsf{f}})^+\mathsf{g}_{\mathbf{T}}, \tag{6.97}$$

or equivalently

$$\mathbf{D}^p = \dot{\lambda}\mathscr{H}(\mathsf{f})\mathsf{g}_{\mathbf{T}} \quad \text{with} \quad \dot{\lambda} = \frac{(\dot{\mathsf{f}})^+}{H}. \tag{6.98}$$

The construction of the inverse constitutive relation $\dot{\mathbf{T}} = \dot{\mathbf{T}}(\mathbf{D}, \mathbf{T}, \mathsf{h})$ will be addressed after introducing new principles in the theory which, on the one side, justify the potential law and, on the other side, also show that the strain hardening function H and the factor $\dot{\lambda}$ in perfect plasticity are either positive or zero.

Let us now analyze the constraints over the plastic potential g imposed by the additional hypotheses that the material is isotropic and undergoes isochoric strains.

For isotropic materials, the combination of such independence with the directions and the principle of independence with respect to the observer leads us to

$$\mathsf{g}(\mathbf{T}) = \mathsf{g}(\mathbf{R}\mathbf{T}\mathbf{R}^T) \qquad \forall \mathbf{R} \in Rot. \tag{6.99}$$

Recall (see Chapter 1) that Rot is the set of all second-order tensors such that $\mathbf{R}^T = \mathbf{R}$ and $\det \mathbf{R} = 1$. It can be shown that the only functions with this property are those called isotropic functions, which depend solely on the invariants of the tensor argument. Then

$$\mathsf{g} = \mathsf{g}(\mathbf{T}) = \mathsf{g}(I_1, I_2, I_3). \tag{6.100}$$

From (6.59) and (6.60) it is not difficult to see (see equations (5.115) and (5.116) in Chapter 5) that the invariants of $\mathbf{T}$ (I_i, $i = 1, 2, 3$) are related to those of the deviatoric component $\mathbf{S}$ (J_i, $i = 2, 3$) as follows

$$I_2 = J_2 - \frac{1}{3}I_1^2, \tag{6.101}$$

$$I_3 = J_3 + \frac{1}{27}I_1^3 - \frac{1}{3}I_1 J_2, \tag{6.102}$$

and using them in (6.100) yields

$$\mathsf{g} = \mathsf{g}(\mathbf{T}) = \mathsf{g}(I_1, J_2, J_3). \tag{6.103}$$

As a result, we have

$$\mathsf{g}_\mathbf{T} = \mathsf{g}_{I_1}(I_1)_\mathbf{T} + \mathsf{g}_{J_2}(J_2)_\mathbf{T} + \mathsf{g}_{J_3}(J_3)_\mathbf{T}, \tag{6.104}$$

and the derivatives of the invariants result (see (5.108) and (5.109))

$$(I_1)_\mathbf{T} = \mathbf{I}, \tag{6.105}$$

$$(J_2)_\mathbf{T} = \mathbf{S}, \tag{6.106}$$

$$(J_3)_\mathbf{T} = \mathbf{S}\mathbf{S} - \frac{2}{3}J_2\mathbf{I}. \tag{6.107}$$

Introducing these expressions into $\mathbf{D}^p = \dot{\lambda}\mathsf{g}_\mathbf{T}$ and taking the trace we obtain

$$\mathrm{tr}\mathbf{D}^p = 3\dot{\lambda}\mathsf{g}_{I_1}. \tag{6.108}$$

This general result clearly shows us that a potential material displays isochoric motions if and only if its potential does not depend on the invariant I_1 (equivalently from the stress $T_m = \frac{1}{3}I_1$). That means that the function g must be a function of the form

$$\mathsf{g} = \mathsf{g}(J_2, J_3). \tag{6.109}$$

For this kind of material, the constitutive relation becomes

$$\mathbf{D}^p = \frac{1}{H}\mathsf{f}_\mathbf{T} \cdot \dot{\mathbf{T}} \left[\mathsf{g}_{J_2}\mathbf{S} + \mathsf{g}_{J_3}\left(\mathbf{S}\mathbf{S} - \frac{2}{3}J_2\mathbf{I}\right)\right]. \tag{6.110}$$

This class of expression has been extended also to some non-potential materials, replacing g_{J_2} by $\mathsf{g}_2(J_2, J_3)$ and g_{J_3} by $\mathsf{g}_3(J_2, J_3)$, where these functions characterize the behavior of such materials.

6.5 Drucker and Hill Postulates

Examining Figure 6.16, particularly in the case of stable materials, that is materials which feature a one-to-one correspondence between stress and strain for monotonic loading processes, we notice that the stress increments exert positive work, that is, $d\sigma d\varepsilon > 0$, while in so-called unstable materials this is verified $d\sigma d\varepsilon < 0$. This condition generalized for multiaxial stress states can be adopted as an additional postulate in the elasto-plastic theoretical formalism. Even if unstable materials do exist, the theory does not lose generality when these materials are disregarded.

This gives rise to Drucker's postulate (also known as the Drucker–Prager postulate), which establishes the following:

i) During loading processes, the stress increments exert positive work.
ii) For a cycle made up of the application and removal of stress increments, these increments exert either positive or null work in the case where plastic strains occur. For strain hardening materials, this work is null only if the cycle is purely elastic.

It is important to highlight that in the kind of loading cycles addressed here, while the initial and final stresses are equal, the same does not hold for the strains.

Consider now the closed path ABCA displayed in Figure 6.17. In this cycle, the plastic process only takes place in the infinitesimal region corresponding to the segment BC, where the strain increment (notice that it is not the strain rate) has (roughly) the direction given by $\mathbf{D}^p$ in the stress state $\mathbf{T}$. In this cycle ABCA the second part of Drucker's postulate entails the following

$$\oint (\mathbf{T} - \mathbf{T}^*) \cdot d\mathbf{E} > 0, \tag{6.111}$$

for hardening materials. In turn

$$\oint (\mathbf{T} - \mathbf{T}^*) \cdot d\mathbf{E}^e = 0, \tag{6.112}$$

because the integrand is the exact differential of $\frac{1}{2}\mathbb{D}^{-1}(\mathbf{T} - \mathbf{T}^*) \cdot (\mathbf{T} - \mathbf{T}^*)$. As a result of (6.111) and (6.112), we have

$$\oint (\mathbf{T} - \mathbf{T}^*) \cdot d\mathbf{E}^p > 0. \tag{6.113}$$

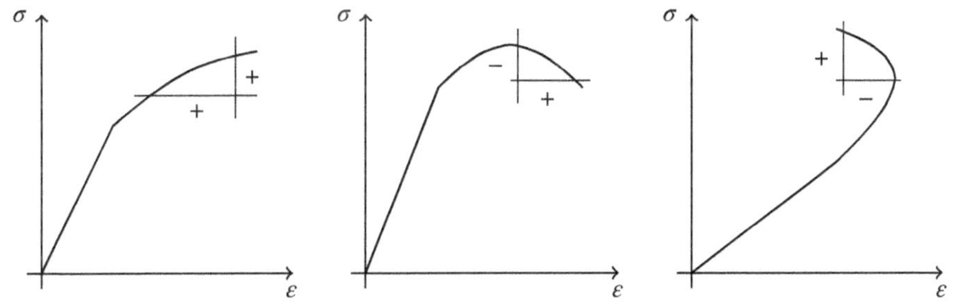

Figure 6.16 Stable and unstable materials.

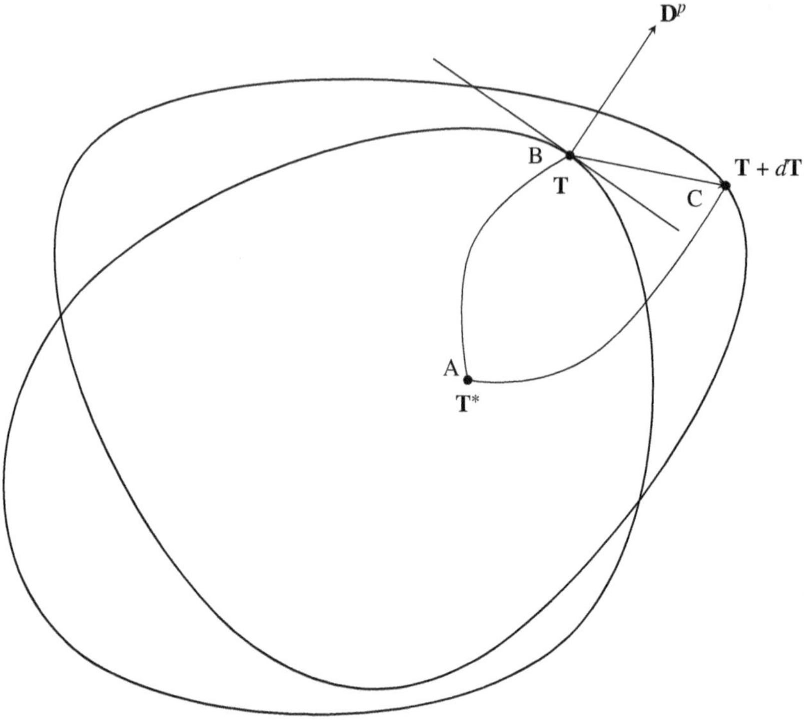

Figure 6.17 Cycle in the stress space.

Nevertheless, the plastic strain only occurs in the BC interval, where $d\mathbf{E}^p = \mathbf{D}^p dt$, then

$$(\mathbf{T} - \mathbf{T}^*) \cdot \mathbf{D}^p > 0. \tag{6.114}$$

This important inequality holds for

- strain hardening materials
- $\mathbf{T}$ and $\mathbf{D}^p \neq \mathbf{O}$ connected by the constitutive relation
- any $\mathbf{T}^*$ which lies within the elastic domain, that is, $f(\mathbf{T}^*) < 0$.

Notice also that for the admissible stress $\mathbf{T}^*$, the inequality (6.114) must also account for the equality.

The inequality (6.114) is known as Principle of Maximal Power (PMP) or Hill's Maximal Principle. This inequality establishes that for a given plastic strain rate $\mathbf{D}^p$, the power dissipated in the real process, $\mathbf{T} \cdot \mathbf{D}^p$, is always larger than all virtual power $\mathbf{T}^* \cdot \mathbf{D}^p$ associated with the real plastic strain rate, and for any virtual stress $\mathbf{T}^*$ within the elastic region (verifying $f(\mathbf{T}^*) < 0$). Let us formalize the PMP next.

The Principle of Maximal Power states that

i) If $\mathbf{T}$ is a stress state over the yield surface, that is, $f(\mathbf{T}) = 0$, for which a (non-null) plastic strain rate occurs $\mathbf{D}^p$, then

$$(\mathbf{T} - \mathbf{T}^*) \cdot \mathbf{D}^p > 0 \qquad \forall \mathbf{T}^* \text{such that } f(\mathbf{T}^*) < 0. \tag{6.115}$$

ii) If $\mathbf{T}$ is a plastically admissible stress state, that is, $f(\mathbf{T}) \leq 0$, for which a (possibly null) plastic strain $\mathbf{D}^p$ occurs, then

$$(\mathbf{T} - \mathbf{T}^*) \cdot \mathbf{D}^p \geq 0 \qquad \forall \mathbf{T}^* \text{such that } f(\mathbf{T}^*) \leq 0. \tag{6.116}$$

The second part of the PMP includes the situation in which $\mathbf{T}$ is within the elastic domain because $\mathbf{D}^p$ is necessarily zero in such a case and, therefore, the inequality is trivially verified. The two statements of the PMP hold for hardening materials, while for perfectly plastic materials only the first statement holds.

Consider again the cycle presented in Figure 6.17. For the loading process given by segment BC, Drucker's postulate (first part), ensures that $d\mathbf{T} \cdot d\mathbf{E} > 0$, then

$$\dot{\mathbf{T}} \cdot \mathbf{D} > 0. \tag{6.117}$$

Considering now the process BCB, the postulate (second part) yields $d\mathbf{T} \cdot d\mathbf{E}^p \geq 0$ because elastic strains do not effectively produce work in a closed cycle, then

$$\dot{\mathbf{T}} \cdot \mathbf{D}^p \geq 0. \tag{6.118}$$

This condition is called the uniqueness condition, and connects real stress rates and plastic strains. It is valid for materials with or without hardening, and for plastic or purely elastic processes.

In particular, for a perfectly plastic material, $\mathbf{T} + \dot{\mathbf{T}}dt$ must be a plastically admissible stress. Hence, by the PMP for the stress state $\mathbf{T}^* = \mathbf{T} + \dot{\mathbf{T}}dt$ we obtain

$$\dot{\mathbf{T}} \cdot \mathbf{D}^p \leq 0, \tag{6.119}$$

which, when combined with (6.118), leads to

$$\dot{\mathbf{T}} \cdot \mathbf{D}^p = 0, \tag{6.120}$$

which is valid for a perfectly plastic material and for any plastic or purely elastic process.

6.6 Convexity, Normality, and Plastic Potential

The PMP proposed by Hill, which in the previous section was derived from Drucker's postulate, poses some constraints for the yield function f which defines the yield limit, as well as for the plastic flow rule which relates the strain rate with the stress and its rate. Next, we describe some consequences of the PMP.

6.6.1 Normality Law and a Rationale for the Potential Law

A clear geometric interpretation can be drawn from the PMP, which establishes that the tensor $\mathbf{T} - \mathbf{T}^*$ forms an acute angle with the tensor $\mathbf{D}^p$. This implies that the admissible region must be fully contained in the half-space whose normal is $\mathbf{D}^p$. If the yield surface is regular at point $\mathbf{T}$, the only way this can happen is that $\mathbf{D}^p$ is normal to the yield surface at $\mathbf{T}$ and, therefore, parallel to $f_{\mathbf{T}}(\mathbf{T})$ and oriented outwards the admissible region. This concept is illustrated in Figure 6.18. That is

$$\mathbf{D}^p = \dot{\lambda} f_{\mathbf{T}}(\mathbf{T}) \qquad \text{with} \qquad \dot{\lambda} \geq 0. \tag{6.121}$$

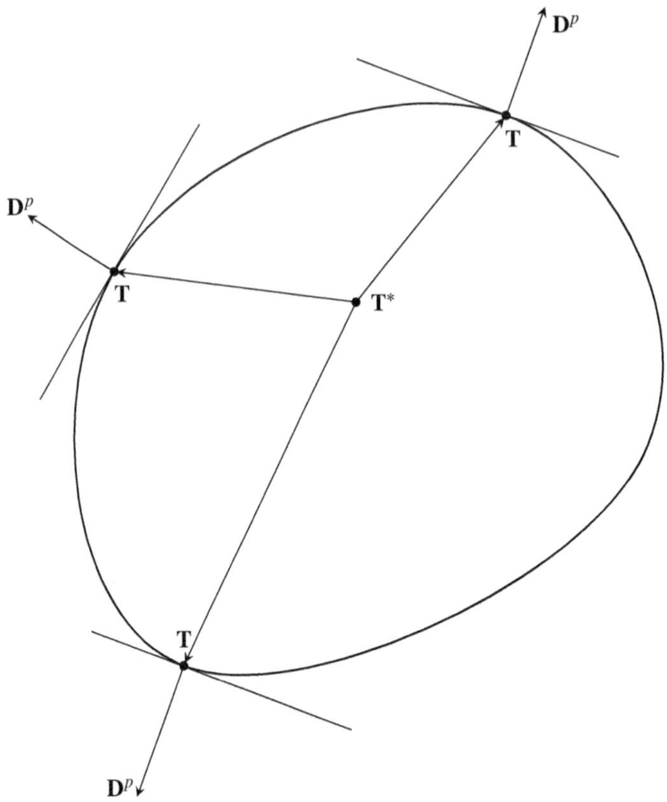

Figure 6.18 Geometric consequences of the PMP for regular points.

In other words, the PMP implies that the plastic flow rule has to be a potential rule (there exists a potential for $\mathbf{D}^p$) and also an associative rule (the potential coincides with f), and that the plastic multiplier $\dot{\lambda}$ must be always positive or null. On the other side, the positivity of $\dot{\lambda}$ allows us to conclude that the hardening function H has also to be positive because $\dot{\lambda} = \frac{1}{H}\dot{\mathsf{f}}$, and $\dot{\mathsf{f}}$ is positive or null in plastic processes.

If the yield surface features singular points in the sense that the tangent plane is not defined (corners), a single yield function f, nondifferentiable at these points, can alternatively be used. Another possibility is to employ a vector-valued yield function whose m components are regular functions which stand for the different yield modes the material can undergo. The admissibility condition for yielding remains unchanged, being $\mathsf{f}(\mathbf{T}) \leq 0$, but now this condition is understood component-wise. Hence, mode ith is active if $[\mathsf{f}(\mathbf{T})]_i = 0$ (hereafter, for simplicity, we use the notation $\mathsf{f}_i(\mathbf{T})$). A single yield mode will be active at regular points over the yield surface, while there will be at least two active yield modes at singular points. At these singular points, the direction of the plastic strain rate $\mathbf{D}^p$ is not univocally determined. Notwithstanding this, the PMP forces this direction to be contained in the positive convex cone defined by the elements which point in the normal direction at

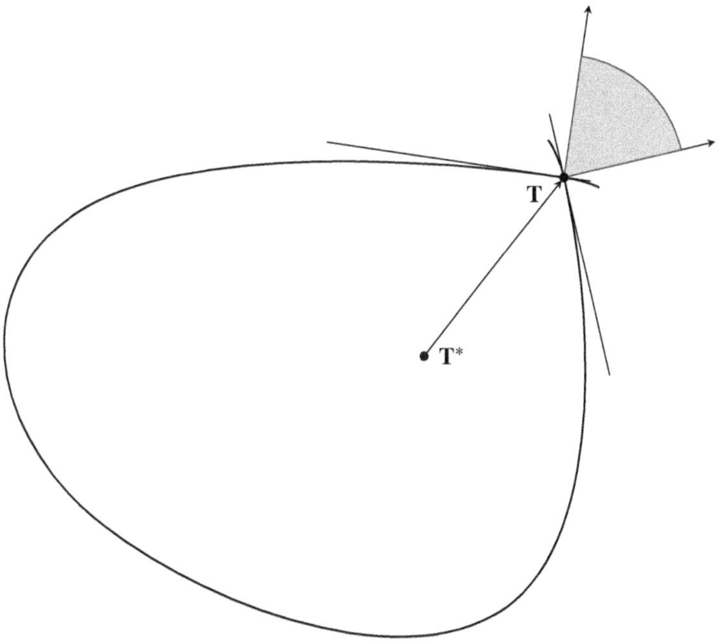

Figure 6.19 Geometric consequences of the PMP for a singular point.

the singular points. This is shown in Figure 6.19, and the condition is analytically written as

$$\mathbf{D}^p = \sum_{i=1}^{m} \dot{\lambda}_i \mathsf{f}_{i\mathbf{T}}(\mathbf{T}), \qquad \dot{\lambda}_i \geq 0, \qquad \dot{\lambda}_i \mathsf{f}_i(\mathbf{T}) = 0 \qquad i = 1, \dots, m. \tag{6.122}$$

This expression tells us that $\mathbf{D}^p$ is a convex linear combination of the gradients of the modes f_i which are active at $\mathbf{T}$ ($\mathsf{f}_i(\mathbf{T}) = 0$) given that inactive modes ($\mathsf{f}_i(\mathbf{T}) < 0$) have null coefficients $\dot{\lambda}_i$.

Expression (6.122) can be rewritten in a more compact form as follows

$$\mathbf{D}^p = \mathsf{f}_{\mathbf{T}}(\mathbf{T})\dot{\lambda}, \qquad \dot{\lambda} \geq 0, \qquad \dot{\lambda} \cdot \mathsf{f}(\mathbf{T}) = 0, \tag{6.123}$$

where now $\dot{\lambda}$ is a vector with m components. Condition $\dot{\lambda} \cdot \mathsf{f}(\mathbf{T}) = 0$, when prescribed in addition to the constraints $\dot{\lambda} \geq 0$ and $\mathsf{f}(\mathbf{T}) \leq 0$, ensures that inactive modes provide no contribution to the characterization of $\mathbf{D}^p$. At the end of this chapter we will discuss an even more compact formulation of this theory using the concept of subdifferentials of a function.

6.6.2 Convexity of the Admissible Region

An important outcome of the normality law imposed by the PMP is the convexity of the admissible region. In effect, from the normality law we can note that for each stress state $\mathbf{T}$ such that $\mathsf{f}(\mathbf{T}) = 0$, the admissible region $\mathsf{f}(\mathbf{T}) \leq 0$ can only be placed either on one side of the tangent plane (in the case of a regular point) or contained within the cone

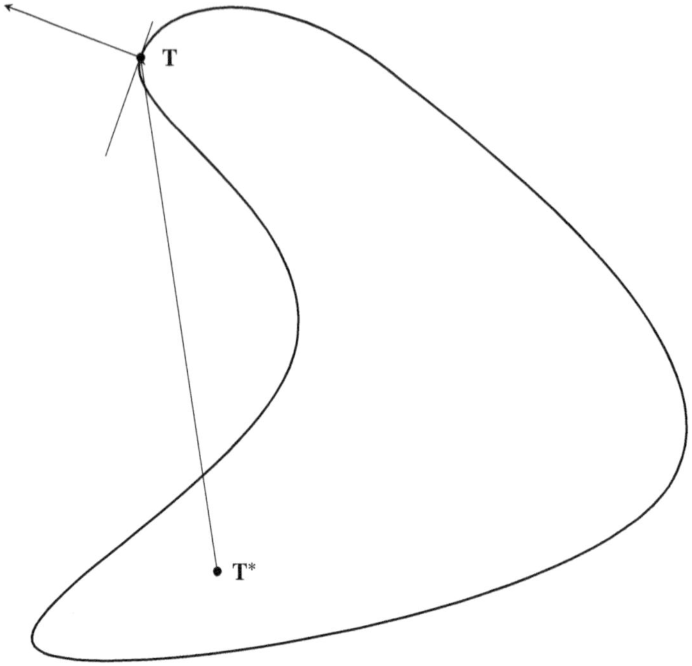

Figure 6.20 Inadmissible non-convex domain for which the PMP is not verified.

formed by the active tangent planes at a singular point, as seen in Figures 6.18 and 6.19, respectively. This can only be satisfied if the admissible region is convex, implying that

$$\mathsf{f}((1-\theta)\mathbf{T}_1 + \theta\mathbf{T}_2) \leq (1-\theta)\mathsf{f}(\mathbf{T}_1) + \theta\mathsf{f}(\mathbf{T}_2) \qquad \forall\theta \in [0,1], \tag{6.124}$$

noting that $\mathsf{f}(\mathbf{T}_1) \leq 0$ and $\mathsf{f}(\mathbf{T}_2) \leq 0$ hold. Therefore, a non-convex domain cannot be an admissible domain (see Figure 6.20).

6.7 Plastic Flow Rule

This section addresses the inverse constitutive relation $\dot{\mathbf{T}} = \dot{\mathbf{T}}(\mathbf{D}, \mathbf{T}, \mathsf{h})$ for the potential law seen in previous sections and, finally, we will wrap up the obtained results. In what follows, and for ease of presentation, we will limit the developments to the case in which the admissible region is fully characterized by a scalar-valued smooth function.

The hardening function was introduced in the model through the relation $\dot{\lambda} = \frac{\dot{\mathsf{f}}}{H}$, which is valid whenever $H \neq 0$. However, we can generalize the validity of the expression

$$\dot{\mathsf{f}} = H\dot{\lambda}, \tag{6.125}$$

for perfect plasticity where $\dot{\mathsf{f}} = 0$ for $\dot{\lambda} > 0$, if we identify a perfectly plastic material with $H = 0$.

Another possible relation for materials with or without hardening is the potential law

$$\mathbf{D} = \mathbb{D}^{-1}\dot{\mathbf{T}} + \dot{\lambda}\mathsf{g}_{\mathbf{T}}. \tag{6.126}$$

Multiplying both sides of this equation by $\mathbb{D}$, and performing the inner product with f_T, we get

$$\dot{\lambda} = \frac{\mathbb{D}\mathbf{D} \cdot \mathsf{f}_\mathrm{T}}{H + \mathbb{D}\mathsf{g}_\mathrm{T} \cdot \mathsf{f}_\mathrm{T}}. \tag{6.127}$$

If the material satisfies the PMP we have $\mathsf{g} = \mathsf{f}$, and the previous expression takes the form

$$\dot{\lambda} = \frac{\mathbb{D}\mathbf{D} \cdot \mathsf{f}_\mathrm{T}}{H + \mathbb{D}\mathsf{f}_\mathrm{T} \cdot \mathsf{f}_\mathrm{T}}. \tag{6.128}$$

Thus, the denominator is positive by virtue of $H \geq 0$, $\mathsf{f}_\mathrm{T} \neq 0$ and $\mathbb{D}\mathsf{f}_\mathrm{T} \cdot \mathsf{f}_\mathrm{T} > 0$ because $\mathbb{D}$ is positive definite. Therefore, expression (6.128) shows us that $\dot{\lambda}$ is defined for all strain rates, both for hardening materials ($H > 0$) and for perfectly plastic materials. As already pointed out, this situation differs from the scenario in which the stress rate was known (instead of the strain rate).

The condition of effective loading, which in terms of $\dot{\mathbf{T}}$ is given by $\mathsf{f}_\mathrm{T} \cdot \dot{\mathbf{T}} > 0$, can be written as a function of $\mathbf{D}$ noting that now $\dot{\lambda}$ is a function of $\mathbf{D}$, as stated by (6.128). Hence, both for perfect plasticity and for hardening materials, $\dot{\lambda}$ is positive if

$$\mathbb{D}\mathbf{D} \cdot \mathsf{f}_\mathrm{T} > 0. \tag{6.129}$$

This expression is the result of the plastic loading condition, written in terms of $\mathbf{D}$ (and $\mathbf{T}$). The elastic unloading remains characterized by

$$\mathbb{D}\mathbf{D} \cdot \mathsf{f}_\mathrm{T} < 0. \tag{6.130}$$

Utilizing expressions (6.126) and (6.128) we finally obtain the constitutive equation

$$\dot{\mathbf{T}} = \mathbb{D}^{ep}\mathbf{D}, \tag{6.131}$$

where

$$\mathbb{D}^{ep} = \mathbb{D} - \frac{c}{H + \mathbb{D}\mathsf{f}_\mathrm{T} \cdot \mathsf{f}_\mathrm{T}} \mathbb{D}\mathsf{f}_\mathrm{T} \otimes \mathbb{D}\mathsf{f}_\mathrm{T}, \tag{6.132}$$

and

$$c = \mathscr{H}(\mathsf{f})\mathscr{H}(\mathbb{D}\mathbf{D} \cdot \mathsf{f}_\mathrm{T}), \tag{6.133}$$

is identically zero for stress states which lie in the elastic domain, as well as for local elastic unloading processes.

The tensor $\mathbb{D}^{ep}$ can only be inverted in the case of hardening materials. In such case it is

$$\mathbf{D} = (\mathbb{D}^{ep})^{-1}\dot{\mathbf{T}}, \tag{6.134}$$

where

$$(\mathbb{D}^{ep})^{-1} = \mathbb{D}^{-1} + \frac{c}{H}\mathsf{f}_\mathrm{T} \otimes \mathsf{f}_\mathrm{T}, \tag{6.135}$$

with c now defined by

$$c = \mathscr{H}(\mathsf{f})\mathscr{H}(\dot{\mathsf{f}}) = \mathscr{H}(\mathsf{f})\mathscr{H}(\mathsf{f}_\mathrm{T} \cdot \dot{\mathbf{T}}). \tag{6.136}$$

Let us summarize next the constitutive equations for elasto-plastic materials which verify the PMP.

- The constitutive equation reads $\dot{\mathbf{T}} = \dot{\mathbf{T}}(\mathbf{D}, \mathbf{T}, \mathsf{h}) = \mathbb{D}(\mathbf{D} - \mathbf{D}^{ep}) = \mathbb{D}(\mathbf{D} - \dot{\lambda}\mathsf{f}_{\mathbf{T}})$ where

$$\dot{\lambda} = \begin{cases} 0 & \text{for} \begin{cases} \mathsf{f} < 0 & \text{elastic stress, or} \\ \mathsf{f} = 0 & \text{stress on the yield surface, and} \\ \mathbb{D}\mathbf{D} \cdot \mathsf{f}_{\mathbf{T}} < 0 & \text{local elastic unloading,} \end{cases} \\ \dfrac{\mathbb{D}\mathbf{D} \cdot \mathsf{f}_{\mathbf{T}}}{H + \mathbb{D}\mathsf{f}_{\mathbf{T}} \cdot \mathsf{f}_{\mathbf{T}}} & \text{for} \begin{cases} \mathsf{f} = 0 & \text{elastic stress,} \\ \mathbb{D}\mathbf{D} \cdot \mathsf{f}_{\mathbf{T}} \geq 0 & \text{loading plastic process,} \end{cases} \end{cases} \tag{6.137}$$

or in compact form

$$\dot{\lambda} = \frac{\mathscr{H}(\mathsf{f})(\mathbb{D}\mathbf{D} \cdot \mathsf{f}_{\mathbf{T}})^{+}}{H + \mathbb{D}\mathsf{f}_{\mathbf{T}} \cdot \mathsf{f}_{\mathbf{T}}}. \tag{6.138}$$

- The inverse constitutive equation reads $\mathbf{D} = \mathbf{D}(\dot{\mathbf{T}}, \mathbf{T}, \mathsf{h}) = \mathbb{D}^{-1}\dot{\mathbf{T}} + \dot{\lambda}\mathsf{f}_{\mathbf{T}}$, where for perfectly plastic materials it is

$$\dot{\lambda} \begin{cases} = 0 & \text{for} \begin{cases} \mathsf{f} < 0 & \text{elastic stress, or} \\ \mathsf{f} = 0 & \text{stress on the yield surface, and} \\ \dot{\mathsf{f}} < 0 & \text{local elastic unloading,} \end{cases} \\ \geq 0 & \text{for} \left\{ \mathsf{f} = 0 \text{ and } \dot{\mathsf{f}} = 0 \right. \quad \text{neutral loading plastic process,} \end{cases} \tag{6.139}$$

with $\dot{\lambda}$ indetermined, and for hardening materials

$$\dot{\lambda} = \begin{cases} 0 & \text{for} \begin{cases} \mathsf{f} < 0 & \text{elastic stress, or} \\ \mathsf{f} = 0 & \text{stress on the yield surface, and} \\ \dot{\mathsf{f}} < 0 & \text{local elastic unloading,} \end{cases} \\ \dfrac{1}{H}\dot{\mathsf{f}} & \text{for} \begin{cases} \mathsf{f} = 0 & \text{stress on the yield surface, and} \\ \dot{\mathsf{f}} \geq 0 & \text{loading plastic process.} \end{cases} \end{cases} \tag{6.140}$$

6.8 Internal Dissipation

Let us now introduce a fundamental ingredient in the theory of plasticity involving the concept of internal plastic dissipation.

The specific power (power per unit volume) of internal plastic dissipation is the scalar-valued field given by

$$\Psi = \mathbf{T} \cdot \mathbf{D}^{p}. \tag{6.141}$$

Considering the hypothesis that the stability postulates are valid, from the PMP we deduce that Ψ is a function, which for hardening materials is defined by

$$\Psi = \Psi(\mathbf{D}^{p}, \mathsf{h}) = \max_{\mathbf{T}^{*} \in \mathsf{A}(\mathsf{h})} \mathbf{T}^{*} \cdot \mathbf{D}^{p} = \max_{\mathbf{T}^{*}} \{\mathbf{T}^{*} \cdot \mathbf{D}^{p},\ \mathsf{F}(\mathbf{T}^{*}, \mathsf{h}) \leq 0\}, \tag{6.142}$$

where $\mathsf{F}(\mathbf{T}^*,\mathsf{h}) \leq 0$ was put in evidence as a subsidiary condition. In turn, for perfectly plastic materials it is

$$\Psi = \Psi(\mathbf{D}^p) = \max_{\mathbf{T}^* \in \mathsf{A}} \mathbf{T}^* \cdot \mathbf{D}^p = \max_{\mathbf{T}^*}\{\mathbf{T}^* \cdot \mathbf{D}^p,\ \mathsf{F}(\mathbf{T}^*) \leq 0\}. \tag{6.143}$$

The dissipation function Ψ is univocally related to the (non-null) strain rate $\mathbf{D}^p$ and depends upon the state stress $\mathbf{T}$ lying over the yield surface ($\mathsf{f}(\mathbf{T}) = 0$) corresponding to $\mathbf{D}^p$ as dictated by the normality law, that is

$$\Psi = \mathbf{T}(\mathbf{D}^p) \cdot \mathbf{D}^p, \tag{6.144}$$

with $\mathbf{T}(\mathbf{D}^p)$ such that $\mathsf{f}(\mathbf{T}) = 0$ and $\mathbf{D}^p = \dot{\lambda}\mathsf{f}_{\mathbf{T}}(\mathbf{T})$. In addition, the uniqueness of the value is obtained even if there exist several states $\mathbf{T}(\mathbf{D}^p)$ associated with the normality law to that strain rate because the product $\mathbf{T}(\mathbf{D}^p) \cdot \mathbf{D}^p$ is the same for all of them. Observe that there is no need for a hypothesis about the differentiability of f to guarantee this property. Figure 6.21 features a schematic representation of this property.

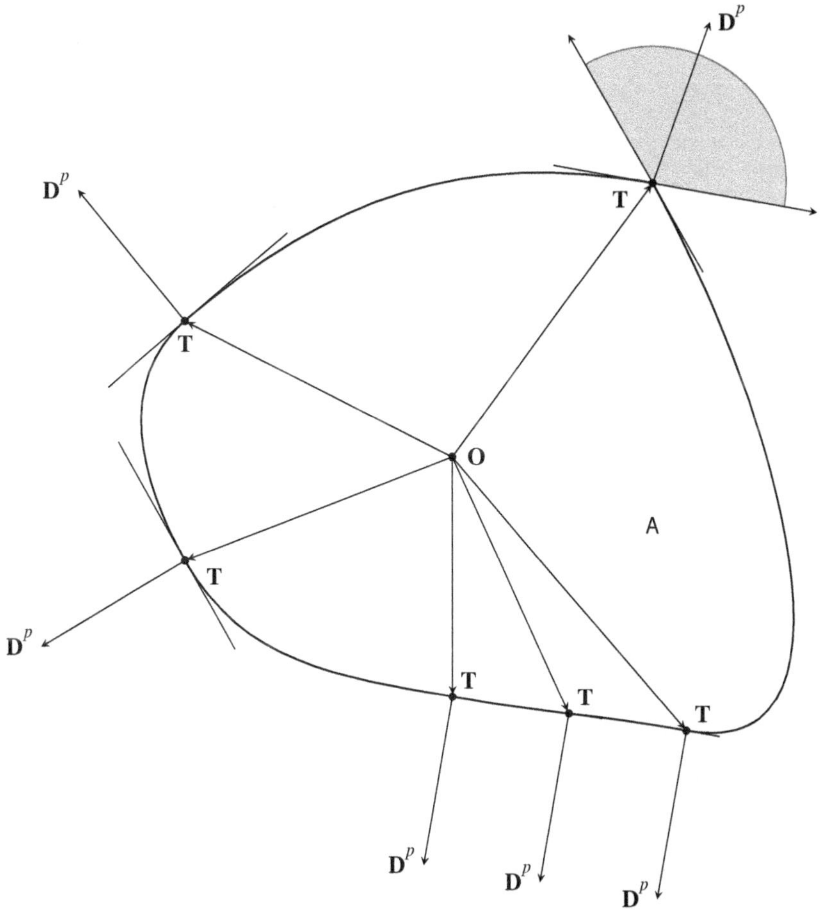

Figure 6.21 Determination of the stress and the internal dissipation as a function of the strain rate.

Depending on the specific form of function f, not every strain rate $\mathbf{D}^p$ can render a plastic strain rate compatible with the normality law. For example, we have seen that if the yield function is independent from the mean stress, then $\mathbf{D}^p$ is traceless. That is to say, every $\mathbf{D}$ whose trace is different from zero is not able to characterize a plastic strain rate. In this case, and considering function Ψ as defined before, but extended to the whole space of strain rates, the maximization can render an infinite value for Ψ. This problem does not exist if the yield region A is bounded in the sense that there exists c such that $\mathbf{T} \cdot \mathbf{T} < c^2$ for all $\mathbf{T} \in \mathsf{A}$.

Moreover, function Ψ is positive definite, and is also homogeneous of degree one in the variable $\mathbf{D}^p$. In fact, the multiplication of $\mathbf{D}^p$ by a number $c > 0$ does not alter the direction and therefore does not change the corresponding stress $\mathbf{T} = \mathbf{T}(\mathbf{D}^p)$ associated by the normality law. In this manner, we have for an arbitrary $c > 0$ that

$$\Psi(c\mathbf{D}^p) = \mathbf{T}(c\mathbf{D}^p) \cdot c\mathbf{D}^p = c\mathbf{T}(\mathbf{D}^p) \cdot \mathbf{D}^p = c\Psi(\mathbf{D}^p). \tag{6.145}$$

On the other side, using the fundamental theorem of homogeneous functions of degree one, yields

$$\Psi(\mathbf{D}^p) = \Psi_{\mathbf{D}^p} \cdot \mathbf{D}^p, \tag{6.146}$$

and, when compared to $\Psi(\mathbf{D}^p) = \mathbf{T}(\mathbf{D}^p) \cdot \mathbf{D}^p$, leads to

$$\mathbf{T} = \mathbf{T}(\mathbf{D}^p) = \Psi_{\mathbf{D}^p} \qquad \Leftrightarrow \qquad \mathsf{f}(\mathbf{T}) = 0 \qquad \text{and} \qquad \mathbf{D}^p = \dot{\lambda}\mathsf{f}_{\mathbf{T}}(\mathbf{T}), \tag{6.147}$$

which, in turn, tells us that Ψ must be a potential function of the stresses defined by the inverse plastic flow rule.

6.9 Common Yield Functions

Let us now address some common yield functions $\mathsf{F}(\mathbf{T}, \mathsf{h})$. In the first place we will consider the yield function $\mathsf{f}(\mathbf{T})$ for a given history of plastic strain and then we will tackle the modification required to account for hardening phenomena.

As we have seen, for isotropic materials the function f must be isotropic and, therefore, a function of the invariants of $\mathbf{T}$, that is

$$\mathsf{f}(\mathbf{T}) = \mathsf{f}(I_1, I_2, I_3) = \mathsf{f}(I_1, J_2, J_3). \tag{6.148}$$

For metals, experience shows that the superposition of a hydrostatic pressure on top of any stress state does not modify the yielding process. As a consequence, for metals it is verified that $\mathsf{f}(\mathbf{T}) = \mathsf{f}(J_2, J_3)$.

Moreover, if the material obeys the stability postulates, its yield limit is also a plastic potential and, as seen, it is $\mathrm{tr}\mathbf{D}^p = 0$ with

$$\mathbf{D}^p = \dot{\lambda}\left[\mathsf{f}_{J_2}\mathbf{S} + \mathsf{f}_{J_3}\left(\mathbf{S}\mathbf{S} - \frac{2}{3}J_2\mathbf{I}\right)\right]. \tag{6.149}$$

The relation $\mathsf{f}(J_2, J_3) = 0$ stands for a straight cylindrical surface with its base in the octahedral plane and whose generatrices are parallel to the direction $\frac{1}{\sqrt{3}}\mathbf{I}$ where the mean stresses are measured ($T_m = \frac{1}{3}\mathrm{tr}\mathbf{T}$).

The base of the cylinder is called the yield locus and can be represented in a plane by taking the principal stresses (T_1, T_2 and T_3) as coordinates. For an isotropic material, this yield locus is given by a symmetric curve with respect to the axis.

6.9.1 The von Mises Criterion

In 1913, von Mises proposed the following yield criterion

$$\mathrm{f}(\mathbf{T}) = \sqrt{J_2} - k \leq 0, \tag{6.150}$$

which can be rewritten as

$$\begin{aligned}\mathrm{f}(\mathbf{T}) &= \sqrt{\frac{1}{2}\mathbf{S}\cdot\mathbf{S}} - k \\ &= \sqrt{\frac{1}{6}[(T_{11}-T_{22})^2 + (T_{11}-T_{33})^2 + (T_{22}-T_{33})^2 + (T_{12}^2 + T_{13}^2 + T_{23}^2)]} - k \leq 0.\end{aligned} \tag{6.151}$$

Parameter k can be interpreted considering that in a pure tensile test, when yielding begins at $T_{11} = \sigma_Y$, we have

$$\mathbf{T} = \begin{pmatrix} T_{11} & 0 & 0 \\ 0 & 0 & 0 \\ 0 & 0 & 0 \end{pmatrix} \quad \Rightarrow \quad \mathrm{f}(\mathbf{T}) = \frac{1}{\sqrt{3}}T_{11} - k = 0 \quad \Rightarrow \quad k = \frac{1}{\sqrt{3}}\sigma_Y, \tag{6.152}$$

while in a pure shear test, when yielding begins at $T_{12} = \tau_{Y\mathrm{Mises}}$, we have

$$\mathbf{T} = \mathbf{S} = \begin{pmatrix} 0 & T_{12} & 0 \\ T_{12} & 0 & 0 \\ 0 & 0 & 0 \end{pmatrix} \quad \Rightarrow \quad \mathrm{f}(\mathbf{T}) = T_{12} - k = 0 \quad \Rightarrow \quad k = \tau_{Y\mathrm{Mises}}. \tag{6.153}$$

A brief exposition can serve as argumentation for this criterion. Let us go through some important aspects related to the Mises criterion.

- Mises considered this criterion as an approximation of the Tresca yield criterion, and this was the original justification. These days, it is widely admitted that the Mises criterion better models the experimental observations for most metals.
- Utilizing the following decomposition for the strain elastic energy

  $$U = \frac{1}{18}I_1^2 + \frac{1}{2\mu}J_2, \tag{6.154}$$

  Hencky proposed in 1924 that the von Mises criterion could be interpreted as a condition which forces the yielding process to be initiated when the distortion energy reaches a critial value

  $$U_{Y\mathrm{distortion}} = \frac{k^2}{2\mu} = \frac{\sigma_Y^2}{6\mu}. \tag{6.155}$$

  Yet this idea had also been proposed by Maxwell in 1856, by Beltrami in 1885, and by Huber in 1904.
- In 1937, Nadai formulated the same yield criterion in the following form: yielding is produced when the octahedral shear stress reaches a critical value. Indeed, the shear stress in this plane is

  $$\tau_{\mathrm{oct}} = \sqrt{\frac{2}{3}J_2}. \tag{6.156}$$

- For metals, in general, we have seen that the criterion should have the functional form $\mathsf{f}(J_2, J_3) = 0$. Noting that J_2 and J_3 are, respectively, of degree 2 and 3 in the stress variable, it turns out that the simplest choice with minimal degree would be to take $\mathsf{f}(J_2) = 0$, which is exactly the von Mises criterion.
- Finally, when probable yield loci in the octahedral plane which meet $\mathsf{f}(J_2, J_3) = 0$ are considered, the simplest equation is a circumference, leading again to the von Mises criterion.

Regarding the plastic flow rule which is employed in conjunction with this criterion, let us first see that

$$\mathsf{f}_{\mathbf{T}} = \frac{1}{2\sqrt{J_2}}(J_2)_{\mathbf{T}} = \frac{1}{2\sqrt{J_2}}\mathbf{S}, \tag{6.157}$$

and recalling that this gradient is used in the yield limit for which $\sqrt{J_2} = k$, gives

$$\mathsf{f}_{\mathbf{T}} = \frac{1}{2k}\mathbf{S}. \tag{6.158}$$

In turn

$$\dot{\mathsf{f}} = \mathsf{f}_{\mathbf{T}} \cdot \dot{\mathbf{T}} = \frac{1}{2\sqrt{J_2}}\mathbf{S} \cdot \dot{\mathbf{T}} = \frac{1}{2\sqrt{J_2}}\mathbf{S} \cdot (\dot{\mathbf{S}} + \dot{T}_m\mathbf{I}) = \frac{1}{2\sqrt{J_2}}\mathbf{S} \cdot \dot{\mathbf{S}}, \tag{6.159}$$

from where $\dot{\mathsf{f}} > 0$ is equivalent to $\mathbf{S} \cdot \dot{\mathbf{S}} > 0$. Finally, putting $\mathsf{f} = 0$ as $\mathbf{S} \cdot \mathbf{S} = 2k^2$, the plastic flow rule emerges as detailed next. For perfectly plastic materials

$$\begin{cases} \mathbf{D}^p = \dfrac{\dot{\lambda}}{2k}\mathbf{S} & \dot{\lambda} \geq 0 \text{ indetermined, } \mathbf{S} \cdot \mathbf{S} = 2k^2 \text{ and } \mathbf{S} \cdot \dot{\mathbf{S}} = 0, \\ \mathbf{D}^p = \mathbf{O} & \text{otherwise.} \end{cases} \tag{6.160}$$

For hardening materials

$$\begin{cases} \mathbf{D}^p = \dfrac{\mathbf{S} \cdot \dot{\mathbf{S}}}{4Hk^2}\mathbf{S} & H = H(\mathbf{T}, \mathsf{h}),\ k = k(\mathsf{h}),\ \mathbf{S} \cdot \mathbf{S} = 2k^2 \text{ and } \mathbf{S} \cdot \dot{\mathbf{S}} \geq 0, \\ \mathbf{D}^p = \mathbf{O} & \text{otherwise.} \end{cases} \tag{6.161}$$

In addition, the dissipation function can be obtained using the expression $\Psi = \mathbf{T}(\mathbf{D}^p) \cdot \mathbf{D}^p$ and this gives perfectly plastic materials

$$\Psi = k\sqrt{2\mathbf{D}^p \cdot \mathbf{D}^p}, \tag{6.162}$$

and for hardening materials

$$\Psi = k(\mathsf{h})\sqrt{2\mathbf{D}^p \cdot \mathbf{D}^p}. \tag{6.163}$$

6.9.2 The Tresca Criterion

In 1864, Tresca proposed the following yield criterion

$$\mathsf{f}(\mathbf{T}) = \max_{i,j=1,2,3} |T_i - T_j| - 2k \leq 0, \tag{6.164}$$

where T_i are the principal stresses. This criterion establishes that the material features a plastic behavior whenever the maximum shear stress reaches a critical value. This function, which is isotropic by definition, can be written in terms of the principal stresses of the deviatoric stress **S** (eigenvalues of **S**) which are given by $S_i = T_i - T_m$. In this manner, the Tresca function can be formulated as follows

$$\mathsf{f}(\mathbf{S}) = \max_{i,j=1,2,3} |S_i - S_j| - 2k \leq 0, \tag{6.165}$$

which clearly indicates that f is independent from T_m (or from I_1). This corresponds to a material that is independent from the mean stress and, therefore, features isochoric plastic strains. For this criterion, the value of k can be interpreted in the pure tensile test, when the material yields for $T_{11} = \sigma_Y$, as follows

$$\mathsf{f} = |T_{11}| - 2k = 0 \qquad \Rightarrow \qquad k = \frac{1}{2}\sigma_Y, \tag{6.166}$$

while in the pure shear stress test, the material yields for $T_{12} = \tau_{Y\text{Tresca}}$, and then

$$\mathsf{f} = 2|T_{12}| - 2k = 0 \qquad \Rightarrow \qquad k = \tau_{Y\text{Tresca}}. \tag{6.167}$$

The yield function which defines the Tresca criterion is nondifferentiable, and can be replaced by the following six-component vector-valued function (yield modes)

$$\mathsf{f}_{ij} = T_i - T_j - 2k \leq 0 \qquad i \neq j \qquad i,j = 1,2,3, \tag{6.168}$$

each of which is a differentiable function. These functions define the six planes which determine a hexagonal straight prismatic surface, whose vertices belong to the Mises circular cylinder. The normality law establishes that, for these cases, the plastic strain is a linear combination of the normals to the active planes of the hexagonal prism at the considered stress. That is, if the (i,j) mode is the only one which is active, and $r \neq i,j$, we have

$$\frac{\partial \mathsf{f}_{ij}}{\partial T_i} = 1, \qquad \frac{\partial \mathsf{f}_{ij}}{\partial T_j} = -1, \qquad \frac{\partial \mathsf{f}_{ij}}{\partial T_r} = 0, \tag{6.169}$$

then, the eigenvalues D_i^p of $\mathbf{D}^p$ are

$$D_i^p = \dot{\lambda}_{ij}, \qquad D_j^p = -\dot{\lambda}_{ij}, \qquad D_r^p = 0. \tag{6.170}$$

If the (i,j) and (i,r) modes are active at the corresponding stress $\mathbf{T}$, we get

$$D_i^p = \dot{\lambda}_{ij} + \dot{\lambda}_{ir}, \qquad D_j^p = -\dot{\lambda}_{ij}, \qquad D_r^p = -\dot{\lambda}_{ir}. \tag{6.171}$$

All these expressions can be written in a unified manner

$$D_i^p = \sum_{\substack{j=1 \\ j\neq i}}^{3} (\dot{\lambda}_{ij} - \dot{\lambda}_{ji}) \qquad i = 1,2,3, \tag{6.172}$$

where for a perfectly plastic material

$$\dot{\lambda}_{ij} \begin{cases} \geq 0 & \text{indetermined for } T_i - T_j = 2k \text{ and } \dot{T}_i = \dot{T}_j, \\ = 0 & \text{otherwise,} \end{cases} \tag{6.173}$$

and for a hardening material

$$\dot{\lambda}_{ij} = \begin{cases} \frac{1}{H}(\dot{T}_i - \dot{T}_j) & \text{for } T_i - T_j = 2k \text{ and } \dot{T}_i - \dot{T}_j \geq 0, \\ 0 & \text{otherwise.} \end{cases} \tag{6.174}$$

Again, the previous results can be cast in a compact form as

$$D_i^p = \sum_{\substack{j=1\\ j\neq i}}^{3} (\dot{\lambda}_{ij} - \dot{\lambda}_{ji})\mathscr{H}(\dot{T}_i - \dot{T}_j)\mathscr{H}(T_i - T_j - 2k) \qquad i = 1, 2, 3, \tag{6.175}$$

for perfect plasticity or

$$D_i^p = \frac{1}{H}\sum_{\substack{j=1\\ j\neq i}}^{3} (T_i - T_j)^+ \, \mathscr{H}(T_i - T_j - 2k) \qquad i = 1, 2, 3, \tag{6.176}$$

for hardening plasticity.

In order to find the dissipation function corresponding to the Tresca criterion it is necessary to address the following linear optimization problem

$$\Psi = \max_{T_1^*, T_2^*, T_3^*} \{(T_1^* D_1^p + T_2^* D_2^p + T_3^* D_3^p),\ T_i^* - T_j^* \leq \sigma_Y,\ i,j = 1, 2, 3,\ i \neq j\}. \tag{6.177}$$

The solution to this problem verifies

$$\Psi = \sigma_Y \max_{i=1,2,3} |D_i^p| = \frac{1}{2}\sigma_Y(|D_1^p| + |D_2^p| + |D_3^p|). \tag{6.178}$$

6.10 Common Hardening Laws

Several measures have been proposed to define parameter h responsible for evaluating the recorded history. Next, we describe some of these alternatives.

- Energetic criterion. In this case, the value of the dissipated plastic work is selected as a hardening parameter, that is

$$\mathsf{h} = \int_0^t \mathbf{T} \cdot \mathbf{D}^p d\tau. \tag{6.179}$$

  In particular, for materials with isochoric plastic strains this gives

$$\mathbf{T} \cdot \mathbf{D}^p = (\mathbf{S} + T_m \mathbf{I}) \cdot \mathbf{D}^p = \mathbf{S} \cdot \mathbf{D}^p. \tag{6.180}$$

  Drucker's postulate ensures that h is positive and monotonically increasing ($\dot{\mathsf{h}} \geq 0$).
- Odqvist's criterion. In this case, the recorded history is characterized by the magnitude of the accumulated plastic strain

$$\mathsf{h} = \int_0^t \sqrt{\frac{1}{2}\mathbf{D}^p \cdot \mathbf{D}^p} d\tau. \tag{6.181}$$

  For isochoric plastic strains $\text{tr}\mathbf{D}^p = 0$ and, consequently, parameter h represents the magnitude of the shear strains.

- Accumulated plastic strain criterion. For infinitesimal strains, the accumulated plastic strain is given by

$$\mathbf{E}^p = \int_0^t \mathbf{D}^p d\tau = \mathbf{E} - \mathbb{D}^{-1}\mathbf{T}. \tag{6.182}$$

  Clearly, this is a tensor-valued parameter suitable to characterize the recorded history because it is altered only when plastic processes take place.
- A useful choice for the parameter h, and with a similar physical meaning to the previous case, is given by

$$\mathsf{h} = \lambda = \int_0^t \dot{\lambda} d\tau, \tag{6.183}$$

  which is a vector-valued parameter with as many components as yield modes present in the definition of the yield criterion for the stress.

Now, once the initial yield criterion $\mathsf{f}^0(\mathbf{T}) \leq 0$ and the strain hardening measure h have been defined, the general yield criterion can be derived from the initial yield criterion by introducing a hardening law. In effect, consider

$$\mathsf{f}^0(\mathbf{T}) = \Psi(\mathbf{T}) - k^0 \quad \text{with} \quad \Psi(\mathbf{O}) = 0, \tag{6.184}$$

then known hardening criteria are discussed next.

- Isotropic hardening. This hardening law was proposed by Taylor and Quinney in 1931 and is defined in the following form

$$\mathsf{F}(\mathbf{T}, \mathsf{h}) = \Psi(\mathbf{T}) - k(\mathsf{h}), \tag{6.185}$$

  where $k(\mathsf{h})$ is a function such that $k(0) = k^0$. This implies that the yield surface uniformly expands according to the value of k.
- Kinematic hardening. This hardening criterion was proposed by Prager in 1958 and establishes that

$$\mathsf{F}(\mathbf{T}, \mathsf{h}) = \Psi(\mathbf{T} - \mathbf{T}^0(\mathsf{h})) - k^0, \tag{6.186}$$

  where $\mathbf{T}^0(\mathsf{h})$ is a tensor-valued function such that $\mathbf{T}^0(0) = \mathbf{O}$, representing the translation of the yield surface in the stress space.

6.11 Incremental Variational Principles

In this section the concept of mechanical equilibrium for a body $\mathscr{B}$ made of elasto-plastic material will be addressed. This implies that displacements and stresses in the body can be described as a program of external loads evolves. Thus, the problem consists of determining the displacement field $\mathbf{u}$ and the stress field $\mathbf{T}$ for a known load program $f(t)$. With this goal in mind, consider that at a given time instant, say t_0, the body is at equilibrium and, for such an instant, loads are incremented as dictated by the rate $\dot{f}$. Consider also that the body is kinematically constrained and the admissible velocity fields are equal to the field $\dot{\bar{\mathbf{u}}}$ over the portion of the boundary $\partial\mathscr{B}_v$. Then, we have to find the kinematically admissible velocity field $\dot{\mathbf{u}}_0 \in \mathrm{Kin}_v$, the stress rate field $\dot{\mathbf{T}}_0$, related to the strain field $\mathscr{D}(\dot{\mathbf{u}}_0)$ through a certain elasto-plastic constitutive equation such that equilibrates the load rate $\dot{f}$. Within the variational framework discussed in previous chapters, this problem is fully characterized by the Principle of Virtual Power, which states that

$$(\dot{\mathbf{T}}_0, \mathscr{D}\hat{\mathbf{v}}) = \langle \dot{f}, \hat{\mathbf{v}} \rangle \qquad \forall \hat{\mathbf{v}} \in \mathrm{Var}_v, \tag{6.187}$$

where $\mathscr{D}$ is the strain rate operator, Var_v is the space of kinematically admissible velocity fields, and

$$\langle \dot{f}, \hat{\mathbf{v}} \rangle = \int_{\mathscr{B}} \dot{\mathbf{b}} \cdot \hat{\mathbf{v}} dV + \int_{\partial\mathscr{B}_a} \dot{\mathbf{a}} \cdot \hat{\mathbf{v}} dS. \tag{6.188}$$

As we shall see, this problem is equivalent to a minimum problem in kinematical variables and also equivalent to a minimum principle in (dual) stress variables.

6.11.1 Principle of Minimum for the Velocity

Let us enunciate the variational equation (6.187) as a minimization problem in terms of the velocity field. This minimum principle is known as the Greenberg Principle.

The ideal elasto-plastic problem corresponding to the mechanical equilibrium is equivalent to finding the field $\dot{\mathbf{u}}_0 \in \mathrm{Kin}_v$ that is a minimum of the functional

$$\Pi_1(\dot{\mathbf{u}}_0) = \min_{\mathbf{v} \in \mathrm{Kin}_v} \Pi_1(\mathbf{v}) = \min_{\mathbf{v} \in \mathrm{Kin}_v} \left[\frac{1}{2} \int_{\mathscr{B}} \dot{\mathbf{T}}(\mathbf{D}, \mathbf{T}_0) \cdot \mathbf{D} dV - \langle \dot{f}, \mathbf{v} \rangle \right], \tag{6.189}$$

where $\mathbf{D} = \mathscr{D}\mathbf{v}$ and the notation $\dot{\mathbf{T}}(\mathbf{D}, \mathbf{T}_0)$ stands for the constitutive relation. Using the identity $\dot{\mathbf{T}} = \mathbb{D}(\mathbf{D} - \dot{\lambda}\mathsf{f}_\mathbf{T})$, problem (6.189) can be reshaped as

$$\Pi_1(\dot{\mathbf{u}}_0) = \min_{\mathbf{v} \in \mathrm{Kin}_v} \Pi_1(\mathbf{v}) = \min_{\mathbf{v} \in \mathrm{Kin}_v} \left[\frac{1}{2} \int_{\mathscr{B}} (\mathbb{D}\mathbf{D} \cdot \mathbf{D} - \mathbb{D}\mathbf{D} \cdot \mathsf{f}_\mathbf{T}\dot{\lambda}) dV - \langle \dot{f}, \mathbf{v} \rangle \right], \tag{6.190}$$

with $\dot{\lambda}$ verifying

- if $\mathsf{f}(\mathbf{T}^0) = 0$ then $\dot{\lambda} \geq 0, \dot{f} \leq 0$, $\dot{\lambda} \cdot \dot{\mathsf{f}} = 0$,
- if $\mathsf{f}(\mathbf{T}^0) < 0$ then $\dot{\lambda} = 0$,

for $\dot{\mathsf{f}} = \mathsf{f}_\mathbf{T} \cdot \dot{\mathbf{T}}$ given in terms of $\mathbf{D}$ as

$$\dot{\mathsf{f}} = \mathbb{D}\mathbf{D} \cdot \mathsf{f}_\mathbf{T}(\mathbf{T}_0) - \dot{\lambda}\mathbb{D}\mathsf{f}_\mathbf{T}(\mathbf{T}_0) \cdot \mathsf{f}_\mathbf{T}(\mathbf{T}_0). \tag{6.191}$$

We shall now see that the fields $\dot{\mathbf{u}}_0$, $\mathbf{D}_0$ and $\dot{\mathbf{T}}_0$, which are the solution of the variational problem (6.187), minimize the Greenberg functional. Consider the difference

$$\Delta = \Pi_1(\mathbf{v}) - \Pi_1(\dot{\mathbf{u}}_0) = \frac{1}{2} \int_{\mathscr{B}} (\dot{\mathbf{T}} \cdot \mathbf{D} - \dot{\mathbf{T}}_0 \cdot \mathbf{D}_0) dV - \langle \dot{f}, \mathbf{v} - \dot{\mathbf{u}}_0 \rangle, \tag{6.192}$$

for $\mathbf{v} \in \mathrm{Kin}_v$, $\mathbf{D} = \mathscr{D}\mathbf{v}$ and $\dot{\mathbf{T}} = \dot{\mathbf{T}}(\mathbf{D}, \mathbf{T}_0)$. So, we have $\mathbf{v} - \dot{\mathbf{u}}_0 \in \mathrm{Var}_v$ and $\dot{\mathbf{T}}_0$ is at equilibrium with $\dot{f}$, then

$$\langle \dot{f}, \mathbf{v} - \dot{\mathbf{u}}_0 \rangle = \int_{\mathscr{B}} \dot{\mathbf{T}}_0 \cdot (\mathbf{D} - \mathbf{D}_0) dV, \tag{6.193}$$

and so with (6.192) into (6.193) we obtain

$$\Delta = \frac{1}{2} \int_{\mathscr{B}} (\dot{\mathbf{T}} \cdot \mathbf{D} + \dot{\mathbf{T}}_0 \cdot \mathbf{D}_0 - 2\dot{\mathbf{T}}_0 \cdot \mathbf{D}) dV. \tag{6.194}$$

This difference can be conveniently written utilizing the constitutive relations

$$\dot{\mathbf{T}} \cdot \mathbf{D}^p = 0 \Rightarrow \dot{\mathbf{T}} \cdot \mathbf{D} = \mathbb{D}^{-1}\dot{\mathbf{T}} \cdot \dot{\mathbf{T}}, \tag{6.195}$$

$$\dot{\mathbf{T}}_0 \cdot \mathbf{D}_0^p = 0 \Rightarrow \dot{\mathbf{T}}_0 \cdot \mathbf{D}_0 = \mathbb{D}^{-1}\dot{\mathbf{T}}_0 \cdot \dot{\mathbf{T}}_0, \tag{6.196}$$

hence it yields

$$\Delta = \frac{1}{2}\int_{\mathscr{B}} [\mathbb{D}^{-1}(\dot{\mathbf{T}} - \dot{\mathbf{T}}_0) \cdot (\dot{\mathbf{T}} - \dot{\mathbf{T}}_0) - 2\dot{\mathbf{T}}_0 \cdot (\mathbf{D} - \mathbb{D}^{-1}\dot{\mathbf{T}})]dV. \tag{6.197}$$

As we can appreciate, the first term in the integrand is strictly positive if $\dot{\mathbf{T}} \neq \dot{\mathbf{T}}_0$. The second term is rewritten as

$$-2\dot{\mathbf{T}}_0 \cdot \mathbf{D}^p = -2\dot{\lambda}[\dot{\mathbf{T}}_0 \cdot \mathsf{f}_{\mathbf{T}}(\mathbf{T}_0)]. \tag{6.198}$$

On the one hand, if $\mathbf{T}_0$ is an elastic stress state, that is, $\mathsf{f}(\mathbf{T}_0) < 0$, then $\dot{\lambda} = 0$ and the second term is nullified. On the other hand, if $\mathsf{f}(\mathbf{T}_0) = 0$, the rate $\dot{\mathbf{T}}_0$ will be plastically admissible if $\dot{\mathbf{T}}_0 \cdot \mathsf{f}_{\mathbf{T}}(\mathbf{T}_0) \leq 0$ because the material is perfectly plastic and, in turn, $\dot{\lambda} \geq 0$. Then, the second term is also positive or null. As a consequence, we see that $\Pi_1(\mathbf{v})$ will always be larger than or equal to $\Pi_1(\dot{\mathbf{u}}_0)$, given that the equality is verified if and only if $\dot{\mathbf{T}} = \dot{\mathbf{T}}_0$. Hence, the proof is completed.

It is easy to note that, notwithstanding that the uniqueness of the stress rate field solution for the problem is guaranteed, the same does not hold for the velocity field. In effect, if $\dot{\mathbf{u}}_0$ is a solution, then $\dot{\mathbf{u}}_0 + \mathbf{v}^*$, for all $\mathbf{v}^*$ associated to a purely plastic strain, also minimizes the functional. The most frequent case of lack of uniqueness in the velocities occurs when plastic collapse develops. This is a phenomenon in which unlimited plastic strains take place under constant loading.

6.11.2 Principle of Minimum for the Stress Rate

We will explore now the minimum principle dual to the Greenberg Principle. This dual principle is expressed in terms of the stress rate and is known as the Prager Principle.

The ideal elasto-plastic problem corresponding to the mechanical equilibrium is equivalent to determining $\dot{\mathbf{T}}_0 \in \mathrm{Est}_{\dot{T}}$ such that it is a constrained minimum of the functional

$$\Pi_2(\dot{\mathbf{T}}_0) = \min_{\dot{\mathbf{T}} \in \mathrm{Est}_{\dot{T}}} \left[\Pi_2(\dot{\mathbf{T}}),\ \dot{\mathbf{T}} \cdot \mathsf{f}_{\mathbf{T}}(\dot{\mathbf{T}}_0) \leq 0,\ \forall \mathbf{x} \in \mathscr{B} \text{ where } \mathsf{f}(\mathbf{T}) = 0\right], \tag{6.199}$$

where

$$\Pi_2(\dot{\mathbf{T}}) = \frac{1}{2}\int_{\mathscr{B}} \mathbb{D}^{-1}\dot{\mathbf{T}} \cdot \dot{\mathbf{T}} dV - \int_{\partial\mathscr{B}_v} \dot{\mathbf{T}}\mathbf{n} \cdot \dot{\bar{\mathbf{u}}} dS. \tag{6.200}$$

This formulation poses the problem of finding the critical point of a differentiable quadratic functional (the complementary elastic strain energy) with linear equality constraints, $\dot{\mathbf{T}} \in \mathrm{Est}_{\dot{T}}$, and inequality constraints, $\dot{\mathbf{T}} \cdot \mathsf{f}_{\mathbf{T}}(\dot{\mathbf{T}}_0) \leq 0$. Next, we shall show that the solution of the variational problem (6.187), given by $\dot{\mathbf{u}}_0$, $\mathbf{D}_0$ and $\dot{\mathbf{T}}_0$, also minimizes the functional in (6.199). In order to do this, we proceed similarly to the proof given in the previous section. Consider the difference

$$\Delta = \Pi_2(\dot{\mathbf{T}}) - \Pi_2(\dot{\mathbf{T}}_0) = \frac{1}{2}\int_{\mathscr{B}} (\mathbb{D}^{-1}\dot{\mathbf{T}} \cdot \dot{\mathbf{T}} - \mathbb{D}^{-1}\dot{\mathbf{T}}_0 \cdot \dot{\mathbf{T}}_0) dV - \int_{\partial\mathscr{B}_v} (\dot{\mathbf{T}} - \dot{\mathbf{T}}_0)\mathbf{n} \cdot \dot{\bar{\mathbf{u}}} dS. \tag{6.201}$$

Since $\dot{\mathbf{T}}$ and $\dot{\mathbf{T}}_0$ belong to $\mathrm{Est}_{\dot{\mathbf{T}}}$, we have that $\dot{\mathbf{T}} - \dot{\mathbf{T}}_0$ is a self-equilibrated stress state. Because $\mathbf{D}_0$ is kinematically admissible, from the PCVP (see Chapter 3) it holds that

$$\int_{\mathscr{B}} (\dot{\mathbf{T}} - \dot{\mathbf{T}}_0) \cdot \mathbf{D}_0 dV = \int_{\partial\mathscr{B}_v} (\dot{\mathbf{T}} - \dot{\mathbf{T}}_0)\mathbf{n} \cdot \bar{\dot{\mathbf{u}}} dS. \tag{6.202}$$

Recalling that $\mathbf{D}_0 = \mathbb{D}^{-1}\dot{\mathbf{T}}_0 + \mathbf{D}_0^p$, and replacing these results into (6.201) we get

$$\Delta = \frac{1}{2}\int_{\mathscr{B}} [\mathbb{D}^{-1}(\dot{\mathbf{T}} - \dot{\mathbf{T}}_0) \cdot (\dot{\mathbf{T}} - \dot{\mathbf{T}}_0) + \mathbb{D}^{-1}\dot{\mathbf{T}}_0 \cdot \dot{\mathbf{T}}_0 + 2\dot{\mathbf{T}}_0 \cdot \mathbf{D}_0^p - 2\dot{\mathbf{T}} \cdot \mathbf{D}_0^p] dV. \tag{6.203}$$

Now we analyze the different terms of the right-hand side integrand in the above expression. Straightforwardly, the first two terms are strictly positive, and the first is nullified only if $\dot{\mathbf{T}} = \dot{\mathbf{T}}_0$. The third term is always null because $\mathbf{D}_0^p$ is associated with $\dot{\mathbf{T}}_0$ by the constitutive equation for perfectly plastic materials, ensuring $\dot{\mathbf{T}}_0 \cdot \mathbf{D}_0^p = 0$. The remaining fourth term can be written as

$$-2\dot{\mathbf{T}} \cdot \mathbf{D}_0^p = -2\dot{\lambda}(\dot{\mathbf{T}}_0, \mathbf{T}_0)[\dot{\mathbf{T}} \cdot \mathrm{f}_{\mathbf{T}}(\mathbf{T}_0)], \tag{6.204}$$

which turns always to be positive or null. In fact, if $\mathrm{f}(\mathbf{T}_0) < 0$ then $\dot{\lambda}(\dot{\mathbf{T}}_0, \mathbf{T}_0) = 0$. In turn, if $\mathrm{f}(\mathbf{T}_0) = 0$, the constraint prescribed over the virtual rates $\dot{\mathbf{T}}$ guarantees that $\dot{\mathbf{T}} \cdot \mathrm{f}_{\mathbf{T}}(\mathbf{T}_0) \leq 0$, yielding $\dot{\lambda} \geq 0$. Hence, the result follows and, once again, the uniqueness of the stress rate field is demonstrated.

6.11.3 Uniqueness of the Stress Field

In the previous sections we have shown the uniqueness of the stress rate field, which is solution of the ideal elasto-plastic problem. Accordingly, the stress field will also be unique provided the whole loading history is specified from the initial time for which the material was in its virgin state (no loads applied) up to the current loading level. This is a structural difference with elasticity, where to know the stress distribution in a static problem it is only required to have knowledge of the current loading state, as well as of the displacements over the boundary.

When the material is plastic (non-ideally plastic), the uniqueness of the stress is guaranteed if, in addition to the external loads and boundary velocity, the plastic strains are specified. This uniqueness result for the stresses can be transformed into an equivalent result through the concept of residual stresses. The residual stress field is the difference between the real stresses and the elastic stresses obtained in an identical body, with the same elastic constants, but now unlimitedly elastic. Thus, the residual stresses are at equilibrium with the null load. The uniqueness result is then enunciated stating that the self-equilibrated residual stress field is univocally determined by the plastic strain field.

6.11.4 Variational Inequality for the Stress

An alternative characterization for the problem under study consists of the formulation of an evolution variational inequality which relates the stress state $\mathbf{T}(t)$ and its corresponding stress rate.

The evolution of the stress state in the ideal elasto-plastic problem is described by a stress field $\mathbf{T} = \mathbf{T}(\mathbf{x}, t)$ which satisfies the initial conditions for the stress and, at every time t, is such that $\mathbf{T} \in \mathscr{K}$ and verifies

$$\int_{\mathscr{B}} \mathbb{D}^{-1}(\mathbf{T} - \mathbf{T}^*) \cdot \dot{\mathbf{T}} dV - \int_{\partial\mathscr{B}_v} (\mathbf{T} - \mathbf{T}^*)\mathbf{n} \cdot \dot{\bar{\mathbf{u}}} dS \leq 0 \quad \forall \mathbf{T}^* \in \mathscr{K}, \tag{6.205}$$

where $\mathscr{K}$ is the set of all statically admissible stresses, that is, at equilibrium with the load f at time t, and plastically admissible, implying that $\mathsf{f}(\mathbf{T}^*) \leq 0$ for every point $\mathbf{x}$ in the body $\mathscr{B}$ at that time t. The set $\mathscr{K}$ is formally defined by

$$\mathscr{K} = \mathrm{Est}_T \cap \mathscr{P}, \tag{6.206}$$

where

$$\mathscr{P} = \{\mathbf{T};\ \mathsf{f}(\mathbf{T}) \leq 0 \ \ \forall \mathbf{x} \in \mathscr{B}\}. \tag{6.207}$$

This variational inequality can be obtained from the PCVP as explained next. Consider an arbitrary virtual field $\mathbf{T}^*$, at equilibrium with the load f at time t. Then, the difference between this field and the real stress field $\mathbf{T}$, denoted by $\mathbf{T} - \mathbf{T}^*$ is self-equilibrated. In addition, the real strain rate, which is $\mathbb{D}^{-1}\dot{\mathbf{T}} + \mathbf{D}^p$, is kinematically compatible. Hence, the PCVP determines that

$$\int_{\mathscr{B}} (\mathbf{T} - \mathbf{T}^*) \cdot (\mathbb{D}^{-1}\dot{\mathbf{T}} + \mathbf{D}^p) dV = \int_{\partial\mathscr{B}_v} (\mathbf{T} - \mathbf{T}^*)\mathbf{n} \cdot \dot{\bar{\mathbf{u}}} dS \qquad \forall \mathbf{T}^* \in \mathrm{Est}_T. \tag{6.208}$$

Now, exploiting the PMP for $\mathbf{T}$ and $\mathbf{D}^p$, which are related through the constitutive equation, yields

$$\int_{\mathscr{B}} (\mathbf{T} - \mathbf{T}^*) \cdot \mathbf{D}^p dV \geq 0 \quad \forall \mathbf{T}^* \in \mathscr{P}. \tag{6.209}$$

Combining (6.208) and (6.209) for all $\mathbf{T}^* \in \mathscr{K} = \mathrm{Est}_T \cap \mathscr{P}$, and using the symmetry of $\mathbb{D}$, leads us to the proposed evolution variational inequality (6.205).

6.11.5 Principle of Minimum with Two Fields

The minimum principle for the velocity field (Greenberg Principle) formulated as the minimization of Greenberg functional (6.190) presents a nondifferentiable component given by the term

$$\Pi_1^{nd}(\mathbf{v}) = -\frac{1}{2} \int_{\mathscr{B}} \mathbb{D}^{-1}\mathbf{D} \cdot \mathsf{f}_{\mathbf{T}} \dot{\lambda} dV, \tag{6.210}$$

where $\mathbf{D} = \mathscr{D}\mathbf{v}$ and $\dot{\lambda}$ is the nondifferentiable function of $\mathbf{v}$ which corresponds to the ideal elasto-plastic equation

$$\dot{\lambda} = \frac{\mathscr{H}(\mathsf{f})(\mathbb{D}\mathbf{D} \cdot \mathsf{f}_{\mathbf{T}})^+}{\mathbb{D}\mathsf{f}_{\mathbf{T}} \cdot \mathsf{f}_{\mathbf{T}}}. \tag{6.211}$$

The differentiable component of the Greenberg functional (6.190) is

$$\Pi_1^d(\mathbf{v}) = \int_{\mathscr{B}} \mathbb{D}\mathbf{D} \cdot \mathbf{D} dV - \langle f, \mathbf{v} \rangle. \tag{6.212}$$

A strategy to circumvent this obstacle consists of replacing this term by another functional, now a function of two fields $\mathbf{v}$ and $\dot{\lambda}$, which are independent from each other. In this sense, the functional

$$\Psi(\mathbf{D}, \dot{\lambda}) = \int_{\mathscr{B}} \left(\frac{1}{2} \mathbb{D}\mathsf{f}_{\mathrm{T}} \cdot \mathsf{f}_{\mathrm{T}} \dot{\lambda}^2 - \mathbb{D}\mathbf{D} \cdot \mathsf{f}_{\mathrm{T}} \dot{\lambda} \right) dV, \tag{6.213}$$

is suitable to this end because

$$\min_{\dot{\lambda} \in \dot{\Lambda}} \Psi(\mathscr{D}\mathbf{v}^*, \dot{\lambda}) = \Pi_1^{nd}(\mathbf{v}^*), \tag{6.214}$$

where (remembering that $\mathsf{f}(\mathbf{T}_0) = 0 \; \forall \mathbf{x} \in \mathscr{B}$)

$$\dot{\Lambda} = \{\dot{\lambda};\ \dot{\lambda} \geq 0,\ \dot{\lambda} \cdot \mathsf{f}(\mathbf{T}_0) = 0 \ \ \forall \mathbf{x} \in \mathscr{B}\}. \tag{6.215}$$

The identity (6.214), whose proof is left as an exercise (see [310]), allows us to formulate the so-called Capurso Principle. This principle states that all the pairs $(\dot{\mathbf{u}}_0, \dot{\lambda}_0)$ which are solution of the elasto-plastic problem written in rates (which implies that $\dot{\mathbf{u}}_0$ is a critical point of the Greenberg functional and $\dot{\lambda}_0$ is associated with $\dot{\mathbf{u}}_0$ through the constitutive equation), constitute the set of fields which minimize the following functional

$$\begin{aligned} \Pi_3(\dot{\mathbf{u}}_0, \dot{\lambda}_0) &= \min_{\mathbf{v} \in \mathrm{Kin}_v} \min_{\dot{\lambda} \in \dot{\Lambda}} \Pi_3(\mathbf{v}, \dot{\lambda}) = \min_{\mathbf{v} \in \mathrm{Kin}_v} \min_{\dot{\lambda} \in \dot{\Lambda}} \left[\Pi_1^d(\mathbf{v}) + \Psi(\mathbf{D}, \dot{\lambda}) \right] \\ &= \min_{\mathbf{v} \in \mathrm{Kin}_v} \min_{\dot{\lambda} \in \dot{\Lambda}} \left[\int_{\mathscr{B}} \left(\frac{1}{2} \mathbb{D}\mathbf{D} \cdot \mathbf{D} + \frac{1}{2} \mathbb{D}\mathsf{f}_{\mathrm{T}} \cdot \mathsf{f}_{\mathrm{T}} \dot{\lambda}^2 - \mathbb{D}\mathbf{D} \cdot \mathsf{f}_{\mathrm{T}} \dot{\lambda} \right) dV - \langle \dot{f}, \mathbf{v} \rangle \right], \end{aligned} \tag{6.216}$$

where $\mathbf{D} = \mathscr{D}\mathbf{v}$.

6.12 Incremental Constitutive Equations

Throughout this chapter we have approached the elasto-plastic material behavior employing a phenomenological standpoint. This has led us to the derivation of constitutive equations written in terms of displacement rates, strain rates, plastic strain rates, and stress rates, which satisfy certain plastic admissibility criteria, allowing for each pseudo-time of the evolution process, either loading processes (which may or may not induce plastic strain rates) or unloading processes (which induce purely elastic material response), to occur. By using these phenomenological constitutive equations, we have also shown the construction of functionals whose minimization provided the characterization of such rates for each pseudo-time in the evolution problem both by fulfilling the elasto-plastic equations and by equilibrating the load rate $\dot{f}$.

These formulations are exact when expressed in terms of rates. In practice, there is the need to discretize this evolution problem and advance in the time variable through finite increments of size Δt. This implies that much care is needed in order to avoid the violation of the plastic admissibility, as well as the existence of local unloading elastic processes that may occur within such finite increment. This can be tackled through proper a posteriori corrections once a finite step has been given.

Hence, the need for a finite increment elasto-plastic formulation materializes. This formulation should be expressed in terms of finite increments in displacements, strains, plastic strains and stresses, and at the same time it must be capable of fulfilling, as in the continuous evolution problem, all the constraints posed by the elasto-plastic behavior and by the loading process.

Moved by the previous discussion, the goal of this section is to construct incremental constitutive equations and to formulate variational principles which enable the characterization of these increments incorporating the whole phenomenological machinery of elasto-plasticity, which accounts for the mechanical equilibrium, compatibility and admissibility of plastic strains, and local unloading elastic processes. In this way, these principles are particularly appropriate to develop consistent numerical methods which eliminate the need for the a posteriori corrections mentioned above.

In order to feature a compact presentation, we will resort to a slightly different approach to the one taken in previous sections. This approach consists of the derivation of generalized potentials for stress and strain increments rooted in thermodynamic concepts, and where the nondifferentiability will be fulfilled through the subdifferential concept.

6.12.1 Constitutive Equations for Rates

In this section we will revisit the elasto-plastic constitutive equations from previous sections within the formalism provided by the so-called generalized potentials. The basis for this approach lies in the thermodynamic formalism. To this end, and as done from the beginning of the chapter, we will continue to limit the presentation to the case of infinitesimal strains. We will also admit the existence of a thermodynamic potential given by the free energy functional $\Psi = \Psi(\mathbf{E}, \mathbf{E}^p, \alpha)$, which is a function of the total strain, $\mathbf{E}$, of the internal variable described by the accumulated plastic strain, $\mathbf{E}^p$, and of the internal variable associated with the hardening phenomenon, α, which could be a scalar-valued, a vector-valued or a tensor-valued entity, in keeping with the hardening mechanism being modeled [136]. In particular, let us assume that this free energy functional is differentiable and is given by the sum of two convex functionals, namely, Ψ^e, depending upon the difference $\mathbf{E} - \mathbf{E}^p$ (that is depending on the purely elastic strain at the corresponding time), and Ψ^h, responsible for modeling material hardening. Then

$$\Psi(\mathbf{E}, \mathbf{E}^p, \alpha) = \Psi^e(\mathbf{E} - \mathbf{E}^e) + \Psi^h(\alpha). \tag{6.217}$$

According to classical thermodynamic principles applied to material behavior, we have that the generalized internal thermodynamic forces, characterized by the stress state $\mathbf{T}$ and by the generalized internal force A associated with the hardening phenomenon, are given by

$$\mathbf{T} = (\Psi^e)_{\mathbf{E}^e}(\mathbf{E} - \mathbf{E}^p), \tag{6.218}$$

$$A = (\Psi^h)_\alpha(\alpha). \tag{6.219}$$

Hence, it follows that A can be characterized by a field which depends on the nature of the internal variable α associated with material hardening.

From the convexity of these functionals, we can construct the conjugate (complementary) potentials by means of the Legendre (also Fenchel–Legendre) transformation, that is

$$\Psi_c^e(\mathbf{T}) = \max_{\mathbf{E}^e \in \mathscr{W}} \left[\mathbf{T} \cdot \mathbf{E}^e - \Psi^e(\mathbf{E}^e)\right], \tag{6.220}$$

$$\Psi_c^h(A) = \max_{\alpha^* \in \mathfrak{H}} \left[A \cdot \alpha^* - \Psi^h(\alpha^*) \right], \tag{6.221}$$

where $\mathscr{W}$ is the strain space, and $\mathfrak{H}$ is the space of hardening internal variables. These complementary functionals lead us to the inverse constitutive equations to (6.218) and (6.219), which give

$$\mathbf{E} - \mathbf{E}^p = (\Psi_c^e)_{\mathbf{T}}(\mathbf{T}), \tag{6.222}$$

$$\alpha = (\Psi_c^h)_A(A). \tag{6.223}$$

On the other hand, for these inverse constitutive relations the second law of thermodynamics entails the following constraint

$$\mathbf{T} \cdot \mathbf{D}^p - A \cdot \dot{\alpha} \leq 0. \tag{6.224}$$

The sufficient condition for the material to fulfill the second law of thermodynamics is to fulfill the Principle of Maximum Dissipation (see the Principle of Maximal Power (6.114)). As a result, there exists a Dissipation Potential, which is denoted by $\mathfrak{D} = \mathfrak{D}(\mathbf{D}^p, -\dot{\alpha})$, for the thermodynamic forces $(\mathbf{T}, A)$ which is a nondifferentiable convex functional in terms of the rates of internal variables, that is, $(\mathbf{D}^p, -\dot{\alpha})$, and which also depends upon the current state of $\mathbf{E}$, $\mathbf{E}^p$ and α. Hence, the constitutive equations (6.218) and (6.219) and (6.222) and (6.223) are complemented by the normality condition

$$(\mathbf{T}, A) \in \partial\mathfrak{D}(\mathbf{D}^p, -\dot{\alpha}), \tag{6.225}$$

where the notation $\partial\mathfrak{D}(\mathbf{D}^p, -\dot{\alpha})$ stands for the subdifferential of $\mathfrak{D}$ at point $(\mathbf{D}^p, -\dot{\alpha})$, that is to say the set of all subgradients $(\mathbf{T}, A)$ which satisfy the variational condition [78]

$$\mathfrak{D}(\mathbf{D}^{p*}, -\dot{\alpha}^*) - \mathfrak{D}(\mathbf{D}^p, -\dot{\alpha}) \geq \mathbf{T} \cdot (\mathbf{D}^{p*} - \mathbf{D}^p) - A \cdot (\dot{\alpha}^* - \dot{\alpha}) \quad \forall (\mathbf{D}^{p*}, \dot{\alpha}^*) \in \mathscr{W} \times \dot{\mathfrak{H}}, \tag{6.226}$$

where $\dot{\mathfrak{H}}$ is the space of rates of hardening internal variables. Thus, it is enough for $\mathfrak{D}$ to have a minimum at $(\mathbf{O}, 0)$ in order to satisfy the second law of thermodynamics (6.224). In fact, under these hypotheses, we can take $(\mathbf{D}^{p*}, -\dot{\alpha}^*) = (\mathbf{O}, 0)$ in (6.226) to obtain (6.224).

The inverse expression to the normality condition (6.225) can be obtained by introducing the complementary dissipation potential $\mathfrak{D}_c(\mathbf{T}, A)$ using, again, the Fenchel–Legendre transformation

$$\mathfrak{D}_c(\mathbf{T}, A) = \max_{(\mathbf{D}^{p*}, -\alpha^*) \in \mathscr{W} \times \mathfrak{H}} \left[\mathbf{T} \cdot \mathbf{D}^{p*} - A \cdot \dot{\alpha}^* - \mathfrak{D}(\mathbf{D}^{p*}, -\dot{\alpha}^*) \right]. \tag{6.227}$$

Then, the inverse to equation (6.225) can be written as

$$(\mathbf{D}^p, -\dot{\alpha}) \in \partial\mathfrak{D}_c(\mathbf{T}, A). \tag{6.228}$$

In particular, the dissipation potential that characterizes a standard plastic material is given by the support function of the set $\mathscr{P}$ [3] which contains all the thermodynamically

3 Given the convex set $C = \{u;\ u \in \mathscr{U}\}$ the function $S_C : \mathscr{U}' \to \mathbb{R}$ defined by $S_C(d) = \sup_{u \in C} \langle d, u \rangle$, where $\langle \cdot, \cdot \rangle$ is the duality product between the elements in the space $\mathscr{U}$ and those in its dual space $\mathscr{U}'$, is called support function of set C.

admissible internal forces $(\mathbf{T}, A)$ and that, obviously, includes the origin $(\mathbf{O}, 0)$. Then, in this case we have

$$\mathfrak{D}(\mathbf{D}^p, -\dot{\alpha}) = \max_{(\mathbf{T},A)\in\mathscr{P}} \left[\mathbf{T} \cdot \mathbf{D}^p - A \cdot \dot{\alpha}\right]. \tag{6.229}$$

It is important to note that $\mathfrak{D}$ is positively homogeneous of the first degree, and it is not strictly convex with minimum at the origin $(\mathbf{O}, 0)$. In turn, it is not differentiable at the origin at least, and $\partial\mathfrak{D}(\mathbf{O}, 0) = \mathscr{P}$. Moreover, given $(\mathbf{D}^p, -\dot{\alpha})$ the maximum in (6.229) is reached for $(\mathbf{T}, A)$ such that

$$(\mathbf{T}^* - \mathbf{T}) \cdot \mathbf{D}^p - (A^* - A) \cdot \dot{\alpha} \leq 0 \qquad \forall(\mathbf{T}^*, A^*) \in \mathscr{P}. \tag{6.230}$$

As seen in previous sections, for practical reasons, it is usual to define $\mathscr{P}$ through m smooth surfaces $\mathsf{f}_i(\mathbf{T}, A)$ called yield modes [181]. Then, defining f as the vector-valued function whose components are these yield modes, the set $\mathscr{P}$ is characterized as follows

$$\mathscr{P} = \{(\mathbf{T}, A);\ \mathsf{f}(\mathbf{T}, A) \leq 0\}. \tag{6.231}$$

If the transformation indicated in (6.227) is applied to the definition of the dissipation potential $\mathfrak{D}$ given by (6.229), we obtain

$$\mathfrak{D}_c(\mathbf{T}, A) = \mathscr{I}_{\mathscr{P}}(\mathbf{T}, A) = \begin{cases} 0 & \text{if } (\mathbf{T}, A) \in \mathscr{P}, \\ +\infty & \text{if } (\mathbf{T}, A) \notin \mathscr{P}, \end{cases} \tag{6.232}$$

where $\mathscr{I}_{\mathscr{P}}$ is the indicator function for the set $\mathscr{P}$. Observe that the constraint $(\mathbf{T}, A) \in \mathscr{P}$ is implicitly prescribed by (6.228) since $\partial\mathscr{I}_{\mathscr{P}}(\mathbf{T}, A)$ is an empty set if $(\mathbf{T}, A) \notin \mathscr{P}$. With this, the plastic flow rule (6.228) can be geometrically interpreted imposing that plastic rates $(\mathbf{D}^p, -\dot{\alpha})$ belong to the cone defined by normals to $\mathscr{P}$ at $(\mathbf{T}, A)$, characterized by (6.230). In terms of the yield modes this becomes

$$\mathbf{D}^p = \mathsf{f}_{\mathbf{T}}(\mathbf{T}, A)\dot{\lambda} = \sum_{i=1}^{m} (\mathsf{f}_i)_{\mathbf{T}}(\mathbf{T}, A)\dot{\lambda}_i, \tag{6.233}$$

$$-\alpha = \mathsf{f}_A(\mathbf{T}, A) \cdot \dot{\lambda} = \sum_{i=1}^{m} (\mathsf{f}_i)_A(\mathbf{T}, A)\dot{\lambda}_i, \tag{6.234}$$

where the (vector field) yield factor $\dot{\lambda}$ fulfills the following linear complementarity relations

$$\mathsf{f}_{\mathbf{T}}(\mathbf{T}, A)\dot{\lambda} = 0 \qquad \mathsf{f}_{\mathbf{T}}(\mathbf{T}, A) \leq 0 \qquad \dot{\lambda} \geq 0. \tag{6.235}$$

Finally, according to (6.230), the following yield consistency condition is reached

$$\dot{\mathbf{T}} \cdot \mathbf{D}^p - \dot{A} \cdot \dot{\alpha} = 0. \tag{6.236}$$

6.12.2 Constitutive Equations for Increments

The relations derived in the previous section represent the exact constitutive behavior within the framework of the so-called standard plastic materials. Notwithstanding this, these expressions are not sufficient when we, in the time-discrete realm, have to leap in time through finite increments of stresses and strains. Indeed, such increments cannot be simply obtained from the corresponding rates. Should we use the rates to move on through time, there exists the possibility of violating the constraints underlying the elasto-plastic behavior. Hence, the goal of this section is to construct consistent incremental constitutive equations. By consistent equations we mean that these equations will not require a posteriori numerical corrections in order to fix eventual violations of unloading elastic processes or yield admissibility which may have been provoked at the end of the incremental step. The proposed equations will be approximate in some sense, given that there is no exact formulation when a path-dependent problem (which requires the knowledge of the entire process history) is written in incremental form.

With the previous considerations, suppose that at the instant t we know the state of the strain variables $\mathbf{E}, \mathbf{E}^p, \mathbf{E}^e = \mathbf{E} - \mathbf{E}^p$ in the problem. We also know the state of the hardening variable α and the internal thermodynamic forces $(\mathbf{T}, A)$. The finite increments in these variables, when jumping from t to $t + \Delta t$, are represented by the same variables preceded by the symbol Δ.

In what follows we will propose the following incremental version of the set of expressions given by the plastic flow equation (6.228) and the potential relation between variables α and A established by (6.219)

$$(\Delta\mathbf{E}^p, -\Delta\alpha) \in \partial\mathcal{I}_{\mathcal{P}}(\mathbf{T} + \Delta\mathbf{T}, A + \Delta A), \tag{6.237}$$

with

$$\Delta\alpha = (J_c^h)_{\Delta A}(\Delta A), \tag{6.238}$$

where

$$J_c^h(\Delta A) = \Psi_c^h(A + \Delta A) - [\Psi_c^h(A) + (\Psi_c^h)_{\Delta A}(A) \cdot \Delta A]. \tag{6.239}$$

The following arguments justify this approximate strategy.

- Changes in the plastic strains $\mathbf{E}^p$ and α as established by these incremental expressions are assumed to be related to the value of the thermodynamic forces at the end of the increment. This is a good approximation for the elastic processes and for local unloading elastic processes which take place at the beginning of the increment as well as for the case of plastic processes, provided there is no elastic unloading during the increment.
- The yield admissibility of internal forces at the end of the increment $(\mathbf{T} + \Delta\mathbf{T}, A + \Delta A)$ is prescribed by equation (6.238) because $\partial\mathcal{I}_{\mathcal{P}}(\mathbf{T} + \Delta\mathbf{T}, A + \Delta A)$ would be an empty set otherwise.
- Equation (6.238) is the incremental version of equation (6.219), involving no approximation whatsoever. In addition, since Ψ_c^h is convex and differentiable then J_c^h inherits these properties.

Before going ahead, let us introduce a result which will be important for the subsequent developments. Consider

$$j(s) = \inf_a \{h(a);\ (s, a) \in G\} = \inf_a \gamma(s, a), \tag{6.240}$$

where

$$\gamma(s, a) = h(a) + \mathcal{I}_G(s, a), \tag{6.241}$$

with the following hypotheses

i) function h is differentiable and strictly convex;
ii) the set G is closed and (not strictly) convex, and contains the origin, that is the pair $(s, a) = (0, 0)$.

Then, the following statement holds [78]: $j(s)$ is a continuous convex function defined in the domain

$$\text{Dom}(j) = \{s;\ \exists a \text{ such that } (s, a) \in G\}.$$

In addition, $h(a) \geq h(0)\ \forall a$, then, for all s such that $(s, 0) \in G$ results $j(s) = 0$.

Hence, it follows that, for $s \in \text{Dom}(j)$, with j as defined in (6.240), the following relations are equivalent

- $p \in \partial j(s)$
- for some a this is $(p, -\nabla h(a)) \in \partial \mathcal{I}_G(s, a)$.

The proof for these results follows from classical convex analysis, as in [78].

Using the previous result, equations (6.237) and (6.238) are equivalent to

$$\Delta E^p \in \partial J_c^p(\Delta \mathrm{T}), \tag{6.242}$$

where

$$J_c^p(\Delta \mathrm{T}) = \min_{\Delta A^*} \left[J_c^h(\Delta A^*),\ (\mathrm{T} + \Delta \mathrm{T}, A + \Delta A^*) \in \mathscr{P} \right], \tag{6.243}$$

or equivalently

$$J_c^p(\Delta \mathrm{T}) = \min_{\Delta A^*} \left[J_c^h(\Delta A^*) + \mathcal{I}_{\mathscr{P}}(\mathrm{T} + \Delta \mathrm{T}, A + \Delta A^*) \right]. \tag{6.244}$$

Importantly, notice that for perfect plasticity this functional is reduced to $\mathcal{I}_{\mathscr{P}}(\mathrm{T} + \Delta \mathrm{T})$.

The pseudo-potencial (6.242) corresponding to the plastic flow is the basis for the formal derivation of the incremental elasto-plastic relations. To accomplish that task let us rewrite (6.218) in its exact form in terms of increments

$$\Delta \mathrm{E} = \Delta \mathrm{E}^e + \Delta \mathrm{E}^p, \tag{6.245}$$

$$\Delta \mathrm{E}^e = (J_c^e)_{\Delta \mathrm{T}}(\Delta \mathrm{T}), \tag{6.246}$$

where

$$J_c^e(\Delta \mathrm{T}) = \Psi_c^e(\mathrm{T} + \Delta \mathrm{T}) - [\Psi_c^e(\mathrm{T}) + \Delta \mathrm{T} \cdot (\Psi_c^e)_{\mathrm{T}}(\mathrm{T})]. \tag{6.247}$$

Now, combining (6.245), (6.246) and (6.242), we get

$$\Delta \mathrm{E} \in \partial J_c^{ep}(\Delta \mathrm{T}), \tag{6.248}$$

where

$$\begin{aligned} J_c^{ep}(\Delta \mathrm{T}) &= J_c^e(\Delta \mathrm{T}) + J_c^p(\Delta \mathrm{T}) = \min_{\Delta A^*} \left[J_c^e(\Delta \mathrm{T}) + J_c^h(\Delta A^*),\ (\mathrm{T} + \Delta \mathrm{T}, A + \Delta A^*) \in \mathscr{P} \right] \\ &= \min_{\Delta A^*} \left[J_c^e(\Delta \mathrm{T}) + J_c^h(\Delta A^*) + \mathcal{I}_{\mathscr{P}}(\mathrm{T} + \Delta \mathrm{T}, A + \Delta A^*) \right]. \end{aligned} \tag{6.249}$$

This functional is lower semi-continuous, strictly convex, and coercive by virtue of the elastic component being convex and coercive, and of the plastic term being convex. Such complementary elasto-plastic incremental potential is nondifferentiable whenever J_c^p is. The example for such a situation is in perfect plasticity.

The inverse constitutive equation can be attained by deriving the potential J^{ep} conjugate to J_c^{ep} through the Fenchel–Legendre transformation

$$J^{ep}(\Delta \mathbf{E}) = \max_{\Delta \mathbf{T}^*} \left[\Delta \mathbf{T}^* \cdot \Delta \mathbf{E} - J_c^{ep}\right]. \tag{6.250}$$

Introducing (6.249) in the expression above leads to

$$\begin{aligned} J^{ep}(\Delta \mathbf{E}) &= \max_{(\Delta \mathbf{T}^*, \Delta A^*)} \left[\Delta \mathbf{T}^* \cdot \Delta \mathbf{E} - J_c^e(\Delta \mathbf{T}^*) - J_c^h(\Delta A^*),\ (\mathbf{T} + \Delta \mathbf{T}^*, A + \Delta A^*) \in \mathscr{P}\right] \\ &= \max_{(\Delta \mathbf{T}^*, \Delta A^*)} \left[\Delta \mathbf{T}^* \cdot \Delta \mathbf{E} - J_c^e(\Delta \mathbf{T}^*) - J_c^h(\Delta A^*) - \mathcal{I}_{\mathscr{P}}(\mathbf{T} + \Delta \mathbf{T}^*, A + \Delta A^*)\right]. \end{aligned} \tag{6.251}$$

As a result, we get

$$\Delta \mathbf{T} \in \partial J^{ep}(\Delta \mathbf{E}) = (J^{ep})_{\Delta \mathbf{E}}(\Delta \mathbf{E}) \qquad \Leftrightarrow \qquad \Delta \mathbf{E} \in \partial J_c^{ep}(\Delta \mathbf{T}). \tag{6.252}$$

Indeed, classical results of convex analysis (see [78, 236]) establish that the set $\partial J^{ep}(\Delta \mathbf{E})$ coincides with the set of solutions to the problem (6.250). In particular, this problem admits a unique solution because J_c^{ep} is coercive and strictly convex. Hence, the subdifferential of J^{ep} only contains a single element $\Delta \mathbf{T}$ for each $\Delta \mathbf{E}$. In other words, J^{ep} is differentiable.

6.12.3 Variational Principle in Finite Increments

For ease of presentation, we will assume that the kinematical boundary constraints do not depend on time, and that the state of the body is fully known at time t. Then, the PVP at time $t + \Delta t$ states that

$$\int_{\mathscr{B}} (\mathbf{T} + \Delta \mathbf{T}) \cdot \mathscr{D}(\Delta \mathbf{u}^* - \Delta \mathbf{u}) dV = \langle f_{t+\Delta t}, \Delta \mathbf{u}^* - \Delta \mathbf{u} \rangle \qquad \forall \Delta \mathbf{u}^* \in \mathrm{Var}_u, \tag{6.253}$$

where $\Delta \mathbf{u}$ and $\Delta \mathbf{T}$ are related through the incremental elasto-plastic constitutive equation $\Delta \mathbf{T} = (J^{ep})_{\Delta \mathbf{E}}(\mathscr{D}(\Delta \mathbf{u}))$.

As seen, this constitutive equation has the property

$$J^{ep}(\Delta \mathbf{E}^*) - J^{ep}(\mathscr{D}(\Delta \mathbf{u})) \geq \Delta \mathbf{T} \cdot (\Delta \mathbf{E}^* - \mathscr{D}(\Delta \mathbf{u})) \qquad \forall \Delta \mathbf{E}^* \in \mathscr{W}. \tag{6.254}$$

In particular, for any compatible $\Delta \mathbf{E}^*$, the previous expression takes the form

$$J^{ep}(\mathscr{D}(\Delta \mathbf{u}^*)) - J^{ep}(\mathscr{D}(\Delta \mathbf{u})) \geq \Delta \mathbf{T} \cdot \mathscr{D}(\Delta \mathbf{u}^* - \Delta \mathbf{u}) \qquad \forall \Delta \mathbf{u}^* \in \mathrm{Var}_u. \tag{6.255}$$

Integrating all over the domain of the body $\mathscr{B}$ yields

$$\int_{\mathscr{B}} J^{ep}(\mathscr{D}(\Delta \mathbf{u}^*)) dV - \int_{\mathscr{B}} J^{ep}(\mathscr{D}(\Delta \mathbf{u})) dV \geq \int_{\mathscr{B}} \Delta \mathbf{T} \cdot \mathscr{D}(\Delta \mathbf{u}^* - \Delta \mathbf{u}) dV \quad \forall \Delta \mathbf{u}^* \in \mathrm{Var}_u. \tag{6.256}$$

Introducing the total elasto-plastic incremental potential, denoted by $\mathsf{J}^{ep}(\mathscr{D}(\Delta \mathbf{u})) = \int_{\mathscr{B}} J^{ep}(\mathscr{D}(\Delta \mathbf{u})) dV$, the expression above can be rewritten as

$$\mathsf{J}^{ep}(\mathscr{D}(\Delta \mathbf{u}^*)) - \mathsf{J}^{ep}(\mathscr{D}(\Delta \mathbf{u})) \geq \int_{\mathscr{B}} \Delta \mathbf{T} \cdot \mathscr{D}(\Delta \mathbf{u}^* - \Delta \mathbf{u}) dV \quad \forall \Delta \mathbf{u}^* \in \mathrm{Var}_u. \tag{6.257}$$

Finally, replacing this result into (6.253) we get

$$\begin{aligned} J^{ep}(\mathcal{D}(\Delta\mathbf{u}^*)) - J^{ep}(\mathcal{D}(\Delta\mathbf{u})) &\geq \langle f_{t+\Delta t}, \Delta\mathbf{u}^* - \Delta\mathbf{u}\rangle \\ &- \int_{\mathscr{B}} \mathbf{T}\cdot\mathcal{D}(\Delta\mathbf{u}^* - \Delta\mathbf{u})dV \quad \forall\Delta\mathbf{u}^* \in \mathrm{Var}_u, \end{aligned} \tag{6.258}$$

and rearranging terms, and calling

$$\Pi^{ep}(\Delta\mathbf{u}) = J^{ep}(\mathcal{D}(\Delta\mathbf{u})) + \int_{\mathscr{B}} \mathbf{T}\cdot\mathcal{D}(\Delta\mathbf{u})dV - \langle f_{t+\Delta t}, \Delta\mathbf{u}\rangle, \tag{6.259}$$

we obtain an incremental minimum principle which consists of finding $\Delta\mathbf{u} \in \mathrm{Var}_u$ such that

$$\begin{aligned} \Pi^{ep}(\Delta\mathbf{u}) &= \min_{\Delta\mathbf{u}^*\in\mathrm{Var}_u} \Pi^{ep}(\Delta\mathbf{u}^*) \\ &= \min_{\Delta\mathbf{u}^*\in\mathrm{Var}_u} \left[\int_{\mathscr{B}} J^{ep}(\mathcal{D}(\Delta\mathbf{u}^*))dV + \int_{\mathscr{B}} \mathbf{T}\cdot\mathcal{D}(\Delta\mathbf{u}^*)dV - \langle f_{t+\Delta t}, \Delta\mathbf{u}^*\rangle\right]. \end{aligned} \tag{6.260}$$

We have arrived at a variational minimum principle in terms of finite increments whose solution yields the increments $\Delta\mathbf{u}$, $\Delta\mathbf{E}^e$, $\Delta\mathbf{T}$, ΔA, $\Delta\alpha$, $\Delta\mathbf{E}^p$ and $\Delta\lambda$ that satisfy the proposed incremental elasto-plastic constitutive equations, and which guarantee that at the end of the incremental step all constraints relate to plastic admissibility, unloading elastic processes or loading plastic processes. It is important to keep in mind that it is the potential J^{ep} that provides the mean to characterize these incremental variables.

At last, in a completely analogous manner, we could have made use of the complementary functional J_c^{ep} which, in conjunction with the Principle of Complementary Virtual Power, would lead to a minimum principle in terms of the incremental stress. This procedure, as well as the developments of other alternative mixed variational principles involving different incremental variables, can be directly pursued, as done in the domain of elasticity (see Chapter 4).

6.13 Complementary Reading

The reader who wishes to go further into the study of the topics addressed in this chapter is directed to the following contributions:

1) **Books**: J. B. Martin [187], R. Hill [140], L. M. Kachanov [153], W. T. Koiter [159], J. Mandel [184], M. A. Cohn and G. Maier [51], C. Johnson [150], R. A. Feijóo and E. Taroco [92].
2) **Scientific articles**: H. J. Greenberg [128], G. Maier [180, 181], G. Maier and J. Munzo [182], N. Zouain, R. A. Feijóo and E. Taroco [311, 312], N. Zouain and R. A. Feijóo [309, 310].

Part III

Modeling of Structural Components

Throughout the development of humanity, and each time compelled by the need for the design and construction of increasingly intricate structural elements inherent to this development, engineers, architects, physicists, and mathematicians have realized the necessity for the idealization of accurate mechanical models capable of adequately representing the functional response of such components, while maintaining a level of simplicity such that the calculation of analytical solutions is feasible. In increasing degree of complexity we cite arches, vaults, beams, plates, and shells. The facilities provided by such simplified models, together with the capacity of these components to provide proper support when subjected to diverse loading conditions, have been fundamental pillars for such applications to spread to diverse and different areas of the engineering domain.

Unfortunately, it is rather common to find, in the specialized scientific literature, the presentation of such mechanical models within a theoretical framework commingling the foundational mechanical aspects of the model with the complexity of the material behavior. By employing such an approach, the reader can easily get confounded in separating basic concepts/principles, which are valid for any kind of material, and the aspects underlying the description of the material response. Usually, the former go unnoticed given the intricacy posited by the mathematical representation utilized for the material model.

The reasons for this entanglement lie at the heart of the initial theoretical developments in the field. In fact, most of the embryonic theories were brought forward for the first time in conjunction with elastic material models. In 18th and 19th centuries, an extremely productive period was encountered, where simple models for elastic structural components were fully and clearly instituted. In particular, the beam theory initiated by Euler, by different members of the Bernoulli family, and by Coulomb remains among the most widely used groundwork for structural mechanics, primarily because of the simplicity and utility in structural engineering in which beams (and columns) play a fundamental role.

There is therefore a crucial need to provide a unified incarnation from which these mechanical models naturally emerge by employing a certain set of hypotheses, which must be valid for any kind of material behavior. In this spirit, the variational formulation, and its foundational structure based on the Method of Virtual Power, through the enunciation of the Principle of Virtual Power presented in Chapter 3, successfully faces this goal by providing the mechanical and mathematical understructure governing the kinematics and the concept of equilibrium for the problem under study. In 1705, James Bernoulli proposed that the curvature in (elastic) beams was proportional to the bending moment applied. In turn, in 1751, Euler, jointly with Daniel Bernoulli, used this theory to incorporate transversal vibrations, and in 1744 Euler finished his landmark work on beam (and columns) buckling when subjected to compressional loads.

Also, and following the piece of advice given by Daniel Bernoulli in 1742, Euler, in 1744, introduced for the first time the notion of strain energy per unit length in the (elastic) beam, as a functional proportional to the square of the curvature. Furthermore, Euler adopted total strain energy as the quantity that should render stationariness, and by making use of the tools of variational calculus (which were under development by these same researchers), he obtained the equilibrium equations and was able to obtain the corresponding displacement field solution to the problem. These very same variational concepts were of great significance in the first part of the 19th century for the contributions of French mathematicians within the context of small transversal displacements and vibrations in elastic plates. Such theory was developed by Sophie Germain, and posteriorly by Simeon Denis Poisson, who considered the plate as an elastic plane featuring resistance to curvature. The ultimate progress was achieved by Navier some years later, introducing the correct expressions to describe the strain energy, as well as the partial differential equations which govern the problem. At this point, it is important to observe that some difficulties emerged related to the conditions prescribed over the boundary of these structural components, particularly in the case of plates. In fact, plates cannot be prescribed simultaneously with bending moments and shear transversal stresses. This issue was solved by Gustav Robert Kirchhoff who exploited the Principle of Virtual Work (recall its equivalence with the Principle of Virtual Power) under the hypothesis that originally orthogonal fibers to the midplane of the plate continue to remain orthogonal after the deformation.

As a first step in the conception of the shells theory, around 1770, Euler studied the deformation in beams subjected to initial curvatures (in contrast with the usual study of straight beams) carrying out a simplified analysis of the vibration of elastic bells, which were approximated by a set of beams arranged in a ring shape. Notwithstanding this, the first works in shell theory arose almost 100 years later. English mathematicians Augustus Edward Hough Love, in 1888, and Horace Lamb, in 1890, were the first to develop a theory adequate to represent the behavior of shells under small strains. The theories that came after were compared and placed within a larger theoretical structure in 1945 by the Dutch mechanicist Warner Tjardus Koiter and by the Russian mechanicist Valentin Valentinovich Novozhilov. Moreover, the whole formalism presented by Koiter remained unperceived because of the Second World War. This shell theory continues to be of the utmost importance and interest even now because of the large number of applications of shell-like components in civil, mechanical, aeronautical, and aerospace engineering, among others.

The chapters that constitute Part III of the book are written with the intention to deliver to the reader a unified variational framework from which the different beam and bar (Chapter 7 and Chapter 8), and plates and shells (Chapter 9) models naturally arise as a consequence of clear and unambiguous hypotheses. To this end, we will fully utilize a compact notation based on intrinsic coordinates to describe the kinematics adopted for the different structural components. Adding the concept of duality and finally introducing the Principle of Virtual Power will enable us to characterize the loads which are admitted by the specific models, as well as the corresponding equilibrium equations. Remarkably, this will be achieved regardless of the specific material behavior. In other words, all the developments are valid for any kind of material used in the construction of the component. The consideration of specific material behaviors will only be regarded when addressing particular aspects of the theory. The development of complex constitutive laws for structural components was briefly addressed in early chapters and is out of the scope of the present book.

7

Bending of Beams

7.1 Introduction

This chapter is devoted to the presentation of the simplest beam model (studied by Bernoulli, Euler, and others), which is characterized by a very particular geometry, as illustrated in Figure 7.1. First, a beam is a structural component which has one of its characteristic dimensions, say L and called the beam length, much larger than the other characteristic dimensions, which are the height, h_0, and the width, b_0. That is $L \gg h_0$ and $L \gg b_0$. Second, a beam has a longitudinal plane of symmetry, implying that transversal sections have symmetry with respect to this plane. Thirdly, in a beam, when it is deflected as in Figure 7.1, the superior longitudinal fibers are shortened, while inferior longitudinal fibers are elongated. In the middle of the beam, there are fibers that maintain their length. When such a condition is satisfied, fibers are said to be in a neutral state. The set of these neutral fibers forms the beam neutral surface. It will be considered that the intersection of the longitudinal symmetry plane and the neutral surface defines, in the initial load-free configuration, a straight line that will be referred to as the longitudinal beam axis (x-axis in Figure 7.1).

As we have seen in Chapter 3, the construction of the mechanical model for this structural component is firstly conducted by the definition of the kinematics which will serve to describe the motion of the particles that constitute the beam. This implies in the definition of the set/space of admissible displacement fields Kin_u, the space of virtual motion actions Var_v, the strain rate operator $\mathcal{D}$ and the space where these strain rates belong to $\mathcal{W}$. Only after these elements have been properly and fully sketched does the concept of duality arise, enabling the characterization of power-conjugate internal and external generalized forces associated with the kinematical model. Finally, the Principle of Virtual Power (PVP) posits the concept of mechanical equilibrium between these dual entities. This is the path that is followed in the remaining sections of this chapter.

7.2 Kinematics

Prior to the characterization of the kinematics to be adopted for this structural component let us go back to the definition of strain rate for bodies in the three-dimensional space reworked for the case of the beam geometry. Let us suppose that motion actions featured by the beam are such that they maintain the symmetry with respect to the longitudinal symmetry plane. In addition, and because of the small width of the beam, it is

Introduction to the Variational Formulation in Mechanics: Fundamentals and Applications, First Edition.
Edgardo O. Taroco, Pablo J. Blanco and Raúl A. Feijóo.

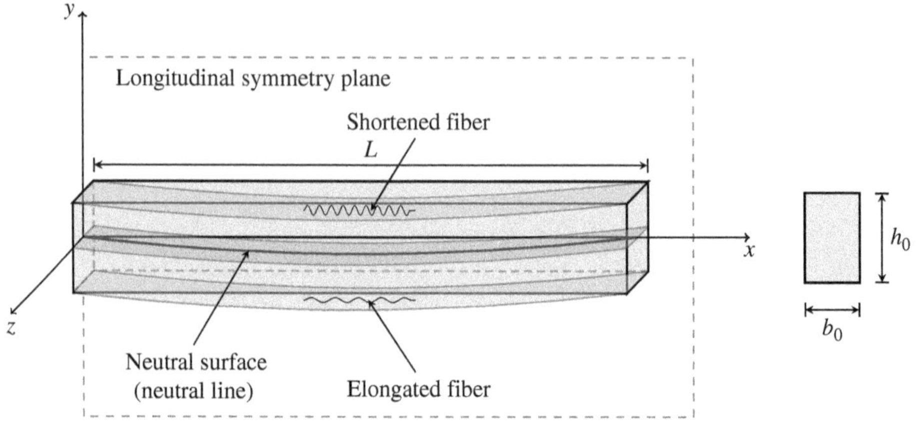

Figure 7.1 Geometry of a simple straight beam. Definition of neutral surface and neutral line.

reasonable to consider both that motion actions are independent from the z direction (this means constancy with respect to that coordinate) and that the velocity component v_z is identically zero.

With these hypotheses, and by virtue of the large difference among the characteristic dimensions posed by the beam geometry, we transform the coordinates as follows

$$X = \frac{1}{L}x, \qquad Y = \frac{1}{h}y \quad (x, y) \in [0, L] \times \left[\frac{1}{h}, -\frac{1}{h}\right], \tag{7.1}$$

where, for simplicity, but without loss of generality, we are considering the transformation for a beam of rectangular cross-section with height $2h$ and width b. Hence, the derivatives transform accordingly

$$\frac{\partial(\cdot)}{\partial x} = \frac{1}{L}\frac{\partial(\cdot)}{\partial X}, \qquad \frac{\partial(\cdot)}{\partial y} = \frac{1}{h}\frac{\partial(\cdot)}{\partial Y}. \tag{7.2}$$

Finally, let us place the mechanical regime within the context of infinitesimal strains, and also consider that motion actions take place over the straight initial (reference) configuration. Thus, and recalling that in Cartesian coordinates it is

$$[\mathscr{D}\mathbf{v}]_{ab} = [\nabla^s \mathbf{v}]_{ab} = \frac{1}{2}\left[\frac{\partial v_a}{\partial b} + \frac{\partial v_b}{\partial a}\right] \qquad a, b = x, y, z, \tag{7.3}$$

we can obtain the $L^2(\Omega)$ norm of this strain rate (strain action) as

$$\begin{aligned} \|\mathscr{D}\mathbf{v}\|_\Omega^2 &= Lhb \int_0^1 \int_{-1}^1 \left[\frac{1}{L^2}\left(\frac{\partial v_x}{\partial X}\right)^2 + \frac{1}{h^2}\left(\frac{\partial v_y}{\partial Y}\right)^2 + \frac{1}{2}\left(\frac{1}{h}\frac{\partial v_x}{\partial Y} + \frac{1}{L}\frac{\partial v_y}{\partial X}\right)^2\right] dXdY \\ &= \frac{bh}{L}\int_0^1 \int_{-1}^1 \left[\left(\frac{\partial v_x}{\partial X}\right)^2 + \left(\frac{L}{h}\right)^2\left(\frac{\partial v_y}{\partial Y}\right)^2 + \frac{1}{2}\left(\frac{L}{h}\frac{\partial v_x}{\partial Y} + \frac{\partial v_y}{\partial X}\right)^2\right] dXdY. \end{aligned} \tag{7.4}$$

Now, in order to avoid the norm being unbounded when $h \to 0$ (elastic line bearing bending, according to Bernoulli and Euler) the following expressions must be satisfied

$$\frac{\partial v_y}{\partial Y} = 0, \tag{7.5}$$

$$\frac{L}{h}\frac{\partial v_x}{\partial Y} + \frac{\partial v_y}{\partial X} = 0. \tag{7.6}$$

First, (7.5) leads to

$$\frac{\partial v_y}{\partial Y} = 0 \quad \Rightarrow \quad v_y = v_y(x), \tag{7.7}$$

meaning that the transversal action must be constant in the y direction. Therefore, the functional form $v_y = v_y(x)$ characterizes the transversal action over the neutral beam axis (neutral line). Second, (7.6) implies

$$\frac{\partial v_x}{\partial Y} = -\frac{h}{L}\frac{\partial v_y}{\partial X} \quad \Rightarrow \quad v_x(x,y) = \overline{v}_x(x) - \frac{h}{L}\frac{\partial v_y}{\partial X}Y = \overline{v}_x(x) - y\frac{\partial v_y}{\partial x}, \tag{7.8}$$

where $\overline{v}_x(x)$ stands for the motion action in the longitudinal direction of the beam, which may vary with coordinate x, but which is transversally constant (it does not depend on y).

Thus, for this class of motion actions, the only component of the strain rate tensor which is different from zero gives

$$[\mathscr{D}\mathbf{v}]_{xx} = \frac{\partial v_x}{\partial x} = \frac{\partial \overline{v}_x}{\partial x} - y\frac{\partial^2 v_y}{\partial x^2}. \tag{7.9}$$

In this manner, we have arrived at the description of the strain actions for simple straight beams. This class of motion actions is known as Kirchhoff–Love in the domain of plates and shells. We can then introduce a formal definition for this class of motion. In fact, it is said that the vector field $\mathbf{v}$ defined in $\Omega = [0, L] \times [-h, h] \times [-\frac{b}{2}, \frac{b}{2}]$ is a Bernoulli–Euler motion action if it satisfies the conditions

$$\mathbf{v}(x,y,z) = \mathbf{v}(x,y), \qquad [\mathscr{D}\mathbf{v}]_{yy} = [\mathscr{D}\mathbf{v}]_{xy} = [\mathscr{D}\mathbf{v}]_{xz} = [\mathscr{D}\mathbf{v}]_{yz} = [\mathscr{D}\mathbf{v}]_{zz} = 0, \tag{7.10}$$

for all material points in the beam. In particular, the space of all Bernoulli–Euler motion actions is represented with the notation $\mathscr{V}_{BE}$. This space $\mathscr{V}_{BE}$ can be characterized in an entirely equivalent way as described next.

Given the unit vector $\mathbf{e}_x$ in the direction of the neutral (straight) beam axis, and the orthogonal unit vector $\mathbf{e}_y$, the field $\mathbf{v}$ belongs to $\mathscr{V}_{BE}$ if and only if it can be represented as

$$\mathbf{v}(x,y,z) = \mathbf{v}(x,y) = v_x(x,y)\mathbf{e}_x + v_y(x)\mathbf{e}_y, \tag{7.11}$$

where

$$v_x(x,y) = \overline{v}_x(x) - y\frac{\partial v_y}{\partial x}(x). \tag{7.12}$$

Let us now unveil the physical meaning of these Bernoulli–Euler fields. First, it is observed that all transversal sections remain orthogonal after the realization of a Bernoulli–Euler motion action. Consider fibers which are contained in the transversal plane, and which are orthogonal to the neutral line, like fiber p–q in Figure 7.2. After the Bernoulli–Euler motion action, these fibers maintain its length (see fiber P–Q in the same figure). Finally, a point over the neutral line, with coordinate $(x, 0)$ before the motion action, corresponds the coordinate (x^*, y^*) after the Bernoulli–Euler motion action, which is given by

$$x^* = x + \overline{v}_x(x), \qquad y^* = v_y(x). \tag{7.13}$$

Calculating the differentials we obtain

$$dx^* = \left(1 + \frac{\partial \overline{v}_x}{\partial x}\right) dx, \qquad dy^* = \frac{\partial v_y}{\partial x} dx. \tag{7.14}$$

Hence, the tangent to the neutral line after the realization of a Bernoulli–Euler motion action forms an angle ϕ with the x-axis given by

$$\tan \phi = \frac{\frac{\partial v_y}{\partial x}}{(1 + \frac{\partial \overline{v}_x}{\partial x})}, \tag{7.15}$$

as seen in Figure 7.2.

Since we are assuming that $\frac{\partial \overline{v}_x}{\partial x} \ll 1$ (or in the case where only transversal motion actions are considered, leading to $\overline{v}_x = 0$), it yields

$$\phi = \frac{\partial v_y}{\partial x}, \tag{7.16}$$

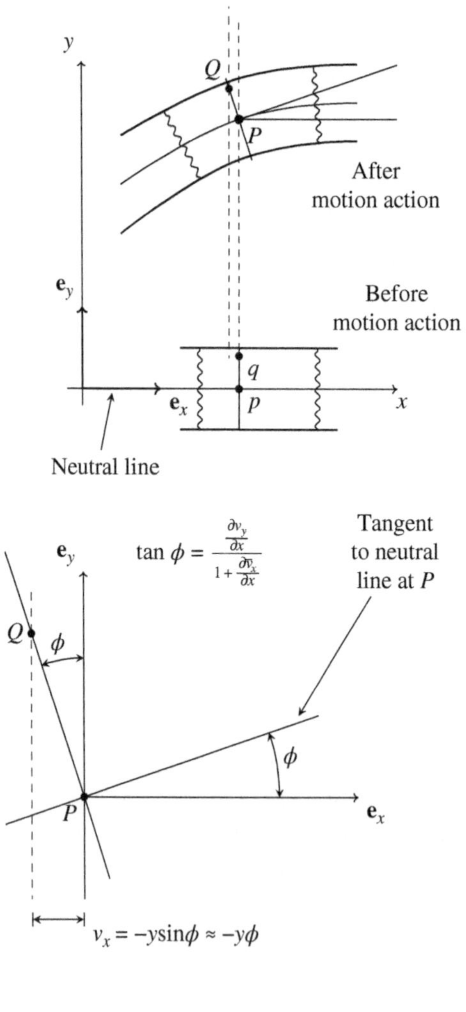

Figure 7.2 Physical interpretation of Bernoulli–Euler motion actions.

that is, the transversal plane, which was orthogonal to the neutral line prior to the motion action, continues to be orthogonal to this line after the Bernoulli–Euler motion action. In short, we obtain the original statement given by Bernoulli and Euler for this theory: transversal planes remain planes and orthogonal, with no strain occurring in them.

We have thus established the elements required to define the space of motion actions Var_v for the beam model under study. Then, calling $\mathscr{V}_{\mathrm{BE}}$ the space of Bernoulli–Euler motion actions, we have

$$\mathrm{Var}_v = \{\mathbf{v} \in \mathscr{V}_{\mathrm{BE}}, + \text{homogeneous essential constraints}\}, \tag{7.17}$$

or, equivalently, in terms of fields $\bar{v}_x$ and v_y

$$\mathrm{Var}_v = \{(\bar{v}_x, v_y) \in \mathscr{V}, + \text{homogeneous essential constraints}\}, \tag{7.18}$$

where $\mathscr{V}$ is a space of functions with enough regularity for all the mathematical operations to be well-posed, and where homogeneous essential boundary conditions are prescribed in a finite number of points in the domain $[0, L]$ either over $\bar{v}_x$, v_y and/or $\frac{\partial v_y}{\partial x}$ fields. This will be discussed in detail in Section 7.4. The strain rate operator is reduced to the operator $\frac{\partial}{\partial x}$, and the space of strain rates, $\mathscr{W}$, is reduced to a space whose elements are scalar fields. As we saw in Chapter 3, and as will be seen in the next section, these two spaces together with the duality play a fundamental role in the characterization of internal and external generalized forces which are consistent with the so-constructed kinematical model.

7.3 Generalized Forces

The first step towards the characterization of generalized forces lies in the definition of the linear (and continuous) functional, corresponding to the internal power, defined over the space of strain actions. The following internal power functional is adopted here

$$\begin{aligned}
P_i &= -(\mathbf{T}, \mathscr{D}\hat{\mathbf{v}}) = -\int_0^L \int_A \sigma \frac{\partial \hat{v}_x}{\partial x} dA dx = -\int_0^L \int_A \sigma \left(\frac{\partial \hat{\bar{v}}_x}{\partial x} - y \frac{\partial^2 \hat{v}_y}{\partial x^2} \right) dA dx \\
&= -\int_0^L \left(\int_A \sigma dA \right) \frac{\partial \hat{\bar{v}}_x}{\partial x} dx + \int_0^L \left(\int_A \sigma y dA \right) \frac{\partial^2 \hat{v}_y}{\partial x^2} dx \\
&= -\int_0^L N \frac{\partial \hat{\bar{v}}_x}{\partial x} dx + \int_0^L M \frac{\partial^2 \hat{v}_y}{\partial x^2} dx \qquad \hat{\mathbf{v}} \in \mathrm{Var}_v,
\end{aligned} \tag{7.19}$$

where $A(x)$ stands for the beam transversal area at point x of the neutral line.

Clearly, the duality enables us to perform the characterization of the internal stress, say σ, which stands for the stress acting over the longitudinal fibers, in duality with the longitudinal strain action $[\mathscr{D}\hat{\mathbf{v}}]_{xx}$, which in turn, by integrating in the transversal section, $A(x)$, gives place to two internal generalized stresses $(N, M) \in \mathscr{W}'$, designated, respectively, as normal stress and bending stress, which act in the transversal section at each point x over the neutral beam axis. These two generalized internal stresses are, correspondingly, power-conjugates to the longitudinal strain action, $\frac{\partial \hat{\bar{v}}_x}{\partial x}$, and to the transversal section rotation rate, $\frac{\partial}{\partial x}(\frac{\partial \hat{v}_y}{\partial x})$.

It is worthwhile noting that if motion actions are constrained to purely transversal actions (that is, $\hat{\hat{v}}_x = 0$), the only internal stress in the beam model is the bending stress M. Moreover, if the beam is regarded as a three-dimensional structural object, the internal power is actually reduced to a line integral over the interval $[0, L]$, whatever the specific form of the cross-sectional area $A(x)$ (provided the initially considered symmetries in the geometrical setting are verified). This is the reason why Bernoulli and Euler, in 1750, considered a beam as a line with resistance to flexion.

Having defined the internal power, there are no further difficulties in finding a representation for the external forces in duality with the elements from Var_v. In order to do this, we have to explicitly find the dual (equilibrium) operator $\mathcal{D}^*$. Then, it is

$$
\begin{aligned}
(\mathbf{T}, \mathcal{D}\hat{\mathbf{v}}) &= \int_0^L N\frac{\partial \hat{\hat{v}}_x}{\partial x}dx - \int_0^L M\frac{\partial^2 \hat{v}_y}{\partial x^2}dx = N\hat{\hat{v}}_x|_0^L - \int_0^L \frac{\partial N}{\partial x}\hat{\hat{v}}_x dx - M\frac{\partial \hat{v}_y}{\partial x}\bigg|_0^L \\
&\quad + \frac{\partial M}{\partial x}\hat{v}_y\bigg|_0^L - \int_0^L \frac{\partial^2 M}{\partial x^2}\hat{v}_y dx = \langle \mathcal{D}^*\mathbf{T}, \hat{\mathbf{v}}\rangle,
\end{aligned} \tag{7.20}
$$

and, consequently

$$
\begin{aligned}
\langle f, \hat{\mathbf{v}}\rangle &= F_x\hat{\hat{v}}_x\Big|_0^L + \int_0^L q_x\hat{\hat{v}}_x dx + m\frac{\partial \hat{v}_y}{\partial x}\bigg|_0^L + F_y\hat{v}_y\Big|_0^L + \int_0^L q_y\hat{v}_y dx \\
&= (F_x\hat{\hat{v}}_x + F_y\hat{v}_y)\Big|_0^L + m\frac{\partial \hat{v}_y}{\partial x}\bigg|_0^L + \int_0^L (q_x\hat{\hat{v}}_x + q_y\hat{v}_y)dx \qquad \hat{\mathbf{v}} \in \text{Var}_v.
\end{aligned} \tag{7.21}
$$

In this fashion, generalized external forces, generically denoted by f, which are admissible to the kinematical beam model under consideration, are represented by forces F_x and F_y, and by the bending moment m acting at points $x = 0$ and $x = L$ (or in a finite set of internal points in the interval $[0, L]$), and by forces per unit length characterized by q_x and q_y. Evidently, at points where constraints over the motion actions are considered, the corresponding force (or moment) will exert null power, becoming a reactive generalized force.

7.4 Mechanical Equilibrium

Once the internal and external power functionals, dictated by the kinematical model, are fully established, the concept of mechanical equilibrium is ruled by the PVP. In the context of the Bernoulli–Euler beam model under study, the PVP states that given the system of loads to which the beam is subjected, which is defined by the concentrated forces F_x^i, F_y^i (applied at point $x_i \in [0, L]$), for $i = 1, \ldots, n$, by the bending concentrated moment m_j (applied at point $x_j \in [0, L]$) for $j = 1, \ldots, r$, and by the distributed loads q_x, q_y, the state of generalized internal stress given by the pair (N, M), equilibrates the system of loads if the following variational equation is satisfied

$$
\begin{aligned}
\int_0^L \left(N\frac{\partial \hat{\hat{v}}_x}{\partial x} - M\frac{\partial^2 \hat{v}_y}{\partial x^2}\right)dx &= \sum_{i=1}^n (F_x^i\hat{\hat{v}}_x(x_i) + F_y^i\hat{v}_y(x_i)) + \sum_{j=1}^r m_j\frac{\partial \hat{v}_y}{\partial x}(x_j) \\
&\quad + \int_0^L (q_x\hat{\hat{v}}_x + q_y\hat{v}_y)dx \qquad \forall(\hat{\hat{v}}_x, \hat{v}_y) \in \text{Var}_v.
\end{aligned} \tag{7.22}
$$

Variational equation (7.22) defines the fields N and M which are equilibrated with the system of loads characterized by the left-hand side in the equation. Clearly, there is no uniqueness result for these fields, as for a certain system of loads $(\{F^i_x, F^i_y\}^n_{i=1}, \{m_j\}^r_{j=1}, q_x, q_y)$, there is an infinite number of internal stress states (N, M) which satisfy the equilibrium, all of them being different in a self-equilibrated internal stress state, that is, up to an internal stress state at equilibrium with a system of loads producing null power. The set of all internal stress states at equilibrium with the system of external loads is denoted by Est_T (see Chapter 3), while self-equilibrated fields define the subspace Var_T, being

$$\mathrm{Est}_T = \{(N,M);\ (N,M) \text{ satisfy } (7.22)\}, \tag{7.23}$$

$$\mathrm{Var}_T = \left\{ (N,M);\ \int_0^L \left(N\frac{\partial \hat{\hat{v}}_x}{\partial x} - M\frac{\partial^2 \hat{v}_y}{\partial x^2} \right) dx = 0 \quad \forall(\hat{\hat{v}}_x, \hat{v}_y) \in \mathrm{Var}_v \right\}. \tag{7.24}$$

Let us derive the Euler–Lagrange equations associated with the variational equation (7.22). In order to achieve this goal, and to make the presentation simple, let us assume that there is a single point $a \in [0, L]$ in which forces and moments are applied, and that the fields in the integrands are smooth enough so that the mathematical operations to be carried out next make perfect sense. Then, repeatedly integrating by parts yields

$$\begin{aligned}
&-\int_{0,a}^{a,L} \left(N\frac{\partial \hat{\hat{v}}_x}{\partial x} - M\frac{\partial^2 \hat{v}_y}{\partial x^2} \right) dx \\
&\quad + F^a_x \hat{\hat{v}}_x(a) + F^a_y \hat{v}_y(a) + m_a \frac{\partial \hat{v}_y}{\partial x}(a) + \int_{0,a}^{a,L} (q_x \hat{\hat{v}}_x + q_y \hat{v}_y) dx \\
&= -N\hat{\hat{v}}_x\Big|_{0,a}^{a,L} + M\frac{\partial \hat{v}_y}{\partial x}\Bigg|_{0,a}^{a,L} - \frac{\partial M}{\partial x}\hat{v}_y\Bigg|_{0,a}^{a,L} + \int_{0,a}^{a,L} \left(\frac{\partial N}{\partial x}\hat{\hat{v}}_x + \frac{\partial^2 M}{\partial x^2}\hat{v}_y \right) dx \\
&\quad + F^a_x \hat{\hat{v}}_x(a) + F^a_y \hat{v}_y(a) + m_a \frac{\partial \hat{v}_y}{\partial x}(a) + \int_{0,a}^{a,L} (q_x \hat{\hat{v}}_x + q_y \hat{v}_y) dx = 0 \quad \forall(\hat{\hat{v}}_x, \hat{v}_y) \in \mathrm{Var}_v.
\end{aligned} \tag{7.25}$$

Rearranging the terms in the expression above leads us to the following

$$\begin{aligned}
&\int_{0,a}^{a,L} \left(\frac{\partial N}{\partial x} + q_x \right) \hat{\hat{v}}_x dx + \int_{0,a}^{a,L} \left(\frac{\partial^2 M}{\partial x^2} + q_y \right) \hat{v}_y dx \\
&\quad - N\hat{\hat{v}}_x|_L + N\hat{\hat{v}}_x|_0 - \left([\![N]\!]_a - F^a_x \right) \hat{\hat{v}}_x(a) \\
&\quad + M\frac{\partial \hat{v}_y}{\partial x}\Bigg|_L - M\frac{\partial \hat{v}_y}{\partial x}\Bigg|_0 + \left([\![M]\!]_a + m_a \right) \frac{\partial \hat{v}_y}{\partial x}(a) \\
&\quad - \frac{\partial M}{\partial x}\hat{v}_y\Bigg|_L + \frac{\partial M}{\partial x}\hat{v}_y\Bigg|_0 - \left(\left[\!\!\left[\frac{\partial M}{\partial x} \right]\!\!\right]_a - F^a_y \right) \hat{v}_y(a) = 0 \qquad \forall(\hat{\hat{v}}_x, \hat{v}_y) \in \mathrm{Var}_v,
\end{aligned} \tag{7.26}$$

where notation $[\![\cdot]\!]_a$ stands for $(\cdot)(a^-) - (\cdot)(a^+)$, that is, it represents the jump in the variable $(\cdot)$ when moving from a^- to a^+.

Equation (7.26) contains all the information required about the problem. Indeed, closer inspection will quickly allow us to uncover the types of kinematical constraints (essential boundary conditions) which can be considered for the model, the natural boundary conditions which, in duality to the essential boundary conditions, can be prescribed, the partial differential equations which hold in the beam domain $[0, L]$ and, finally, the (natural) jump conditions which are verified at point a where some discontinuity is expected to occur. Thus, from the fundamental theorem of the calculus of variations we arrive at the following results.

- Essential/natural boundary conditions. If at $x = 0$ the motion in the x direction is prescribed (essential boundary condition), it implies that $\hat{v}_x(0) = 0$ holds. However, if no motion is prescribed at that point, we obtain the natural boundary condition which states that it must necessarily be

$$N(0) = 0. \tag{7.27}$$

Notice here the duality machinery working for a rather simple product between a kinematical entity and its corresponding generalized force (actually two real numbers). These two quantities cannot be prescribed simultaneously.

If at $x = 0$ the motion in the y direction is prescribed (essential boundary condition), then we must have $\hat{v}_y(0) = 0$. However, if the motion in that direction is free, the natural boundary condition

$$\frac{\partial M}{\partial x}(0) = 0, \tag{7.28}$$

must necessarily hold. Again, the duality between the kinematical and force entities establishes that either one or the other can be prescribed, but never in a simultaneous manner.

If at $x = 0$ the rotation of the transversal section is prescribed (essential boundary condition), this implies that $\frac{\partial \hat{v}_y}{\partial x}(0) = 0$. Nevertheless, if no constraint is given to the rotation at that point, the variational equation requires that

$$M(0) = 0, \tag{7.29}$$

resulting in the corresponding natural boundary condition. Once again, notice the dual game between corresponding kinematical and force entitites, both are not allowed to be prescribed simultaneously.

For the other end point $x = L$ we have the same reasoning as for $x = 0$. That is, either the kinematics is essentially constrained or a boundary condition for the generalized internal stresses naturally emerges.

- Partial differential equations. From the variational equation we conclude that the following so-called Euler–Lagrange equations for the Bernoulli–Euler beam model are verified in the beam domains $(0, a)$ and (a, L)

$$\frac{\partial N}{\partial x} + q_x = 0 \qquad \text{in } (0, a) \cup (a, L), \tag{7.30}$$

$$\frac{\partial^2 M}{\partial x^2} + q_y = 0 \qquad \text{in } (0, a) \cup (a, L). \tag{7.31}$$

- Jump equations. At point $x = a \in [0, L]$ three additional equalitites naturally arise as a consequence of the variational equation. These are regarded as jump equations, and for the beam model under study they are

$$[\![N]\!]_a - F_x^a = 0, \tag{7.32}$$

Space of motion actions

Space of external forces

$$P_e = \int_0^L (q_x \bar{v}_x + q_y v_y)\, dx$$
$$+ \sum_i (F_x^i \bar{v}_x(x_i) + F_y^i v_y(x_i)) + \sum_j m_j \frac{\partial v_y}{\partial x}(x_j)$$

$(q_x, q_y,$
$\{F_x^i\}, \{F_y^i\}, \{m_j\})$

$$-P_i = \int_0^L \left(N \frac{\partial \bar{v}_x}{\partial x} - M \frac{\partial^2 v_y}{\partial x^2} \right) dx$$

(N, M)

Space of deformation actions

Space of internal stresses

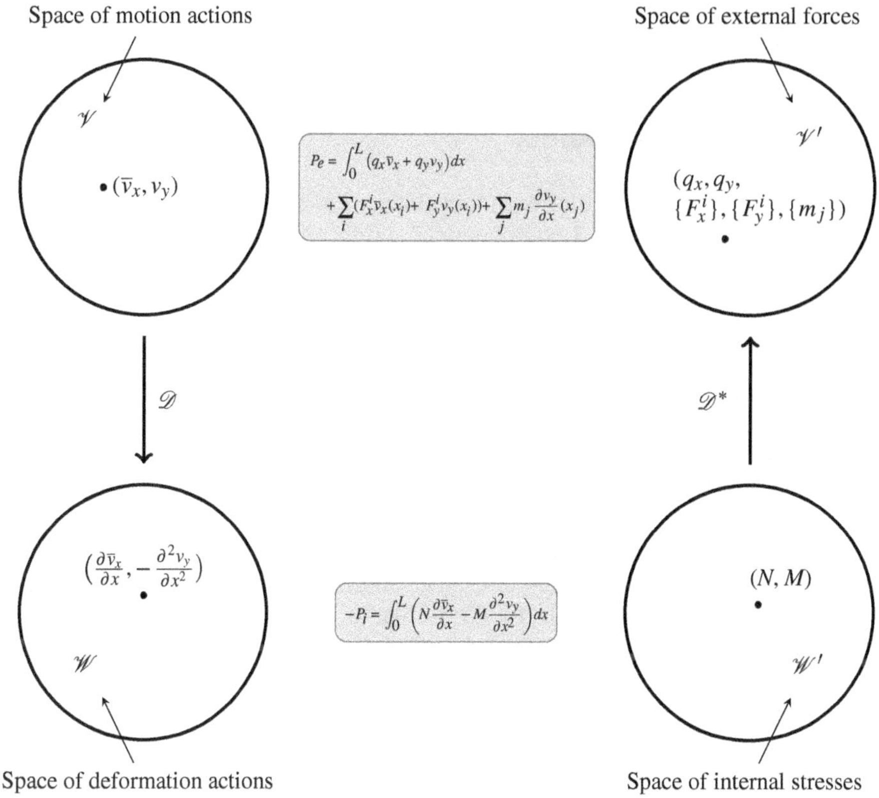

Figure 7.3 Ingredients in the Bernoulli–Euler theory of beams.

$$\left[\!\!\left[\frac{\partial M}{\partial x} \right]\!\!\right]_a - F_y^a = 0, \tag{7.33}$$

$$\left[\!\!\left[M \right]\!\!\right]_a + m_a = 0. \tag{7.34}$$

The schematic summary presented in Figure 7.3 features the ingredients in the Bernoulli–Euler theory for the bending of beams.

The set of equations presented above constitutes the classical boundary value problem formulation for the beam deflection problem. It is important to note that we have arrived at this local form of the governing equations by considering the hypothesis that the fields were regular enough for the integration by parts to make sense. Hence, if such regularity is not verified, the equations of the boundary value problem are not valid any more. In contrast, the mechanical equilibrium remains perfectly defined even when using the PVP, through the variational equation (7.22).

A further relevant aspect that deserves to be mentioned is that the beam model was conceived regardless of the material which constitutes the beam. That is, the principles and conclusions exposed above hold regardless of the material behavior. The material model will be embedded within the constitutive equations, which are inserted into the formulation so variational equation (7.22) becomes a field problem. Such constitutive equations establish a relation between the generalized internal stress (N, M) and the history of the real strain experienced by the beam. See Chapters 5 and 6, and bibliographic references cited therein.

7.5 Timoshenko Beam Model

In previous sections we studied the Bernoulli–Euler beam theory, which is obtained by considering that the structural component is significantly slender, that is $L \gg h_0$. In contrast to this model, there is an alternative mathematical model for beams proposed early in the 20th century by Timoshenko. Within this theory, the hypothesis of slenderness is somewhat relaxed (the beam is not that slender), but it keeps important aspects of the Bernoulli–Euler framework. In fact, the Timoshenko theory considers that transversal planes remain planes after the motion, the transversal planes can rotate but, in general, they are not orthogonal with respect to the neutral line (angle ψ in Figure 7.4 is not equal to the angle formed by the tangent to the neutral line and the x axis), and, finally, fibers in the transversal plane do not undergo deformation processes.

Based on these considerations of kinematical nature, and taking into account the characterization performed for Bernoulli–Euler motion actions, it is now easy to provide a kinematical representation for Timoshenko motion actions. In particular, the space of such motion actions is denoted by $\mathscr{V}_{\mathrm{T}}$. As a matter of fact, and given the unit vectors $\mathbf{e}_x$, in the direction of the (straight) neutral line, and the orthogonal unit vector $\mathbf{e}_y$ over the longitudinal symmetry plane, the field $\mathbf{v}$ belongs to $\mathscr{V}_T$ if it can be written as

$$\mathbf{v}(x, y, z) = \mathbf{v}(x, y) = v_x(x, y)\mathbf{e}_x + v_y(x)\mathbf{e}_y, \tag{7.35}$$

where

$$v_x(x, y) = \overline{v}_x(x) - y\psi(x), \tag{7.36}$$

where ψ is the motion action field which stands for the rotation of the transversal section at point x.

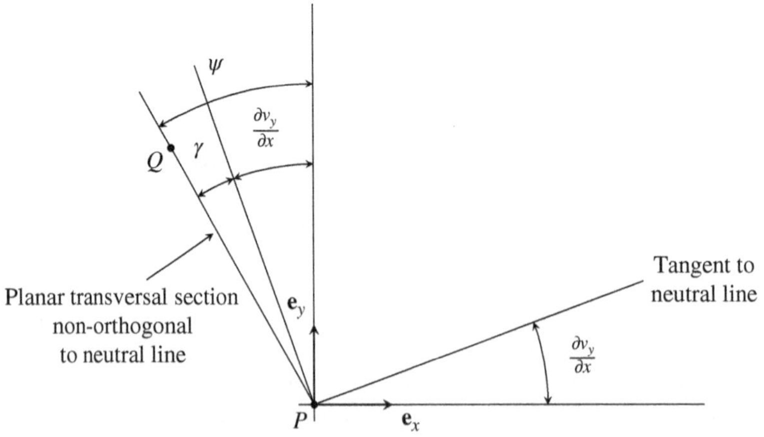

Figure 7.4 Physical interpretation of beam displacement fields in the Timoshenko model.

With the same strain action operator as used in the Bernoulli–Euler formalism, that is $\mathscr{D}(\cdot) = (\nabla(\cdot))^s$, it is easy to verify that for elements $\mathbf{v} \in \mathscr{V}_T$ we get

$$[\mathscr{D}\mathbf{v}]_{xz} = [\mathscr{D}\mathbf{v}]_{zx} = [\mathscr{D}\mathbf{v}]_{yz} = [\mathscr{D}\mathbf{v}]_{zy} = [\mathscr{D}\mathbf{v}]_{zz} = [\mathscr{D}\mathbf{v}]_{yy} = 0,$$
$$[\mathscr{D}\mathbf{v}]_{xx} = \left(\frac{\partial \overline{v}_x}{\partial x} - y\frac{\partial \psi}{\partial x}\right)(x),$$
$$[\mathscr{D}\mathbf{v}]_{xy} = [\mathscr{D}\mathbf{v}]_{yx} = \frac{1}{2}\left(\frac{\partial v_y}{\partial x} - \psi\right)(x). \tag{7.37}$$

With a similar procedure to the one developed for Bernoulli–Euler beams, we have that the space of virtual motion actions, Var_v, is defined by

$$\mathrm{Var}_v = \{\mathbf{v} \in \mathscr{V}_T, + \text{homogeneous essential constraints}\}, \tag{7.38}$$

where the kinematical (essential) boundary conditions are prescribed over all or part of the kinematical variables $\overline{v}_x$, v_y and ψ. Equivalently, it is possible to provide a representation for Var_v as follows

$$\mathrm{Var}_v = \{(\overline{v}_x, v_y, \psi) \in \mathscr{V}, + \text{homogeneous essential constraints}\}, \tag{7.39}$$

where $\mathscr{V}$ is a space of functions which are smooth enough for the forthcoming mathematical operations to make sense.

Analogously to the Bernoulli–Euler beam model, the internal power functional is shaped as follows

$$\begin{aligned}
P_i = -(\mathbf{T}, \mathscr{D}\hat{\mathbf{v}}) &= -\int_0^L \int_A \sigma\left(\frac{\partial \hat{\overline{v}}_x}{\partial x} - y\frac{\partial \hat{\psi}}{\partial x}\right) dAdx - \int_0^L \int_A \tau\frac{1}{2}\left(\frac{\partial \hat{v}_y}{\partial x} - \hat{\psi}\right) dAdx \\
&= -\int_0^L \left(\int_A \sigma dA\right)\frac{\partial \hat{\overline{v}}_x}{\partial x}dx + \int_0^L \left(\int_A \sigma y dA\right)\frac{\partial \hat{\psi}}{\partial x}dx \\
&\quad - \int_0^L \left(\int_A \frac{1}{2}\tau dA\right)\left(\frac{\partial \hat{v}_y}{\partial x} - \hat{\psi}\right) dx \\
&= -\int_0^L N\frac{\partial \hat{\overline{v}}_x}{\partial x}dx + \int_0^L M\frac{\partial \hat{\psi}}{\partial x}dx - \int_0^L Q\left(\frac{\partial \hat{v}_y}{\partial x} - \hat{\psi}\right) dx \\
&= -\int_0^L N\frac{\partial \hat{\overline{v}}_x}{\partial x}dx + \int_0^L M\frac{\partial \hat{\psi}}{\partial x}dx + \int_0^L Q\hat{\gamma}dx \qquad \hat{\mathbf{v}} \in \mathrm{Var}_v.
\end{aligned} \tag{7.40}$$

It can be noted that, in addition to the generalized internal stresses N and M already present in the Bernoulli–Euler beam model, there is an additional generalized stress, denoted by Q, power-conjugate to the rotation $\hat{\gamma} = \hat{\psi} - \frac{\partial \hat{v}_y}{\partial x}$ (which stands for the loss of orthogonality between the transversal fibers and the neutral line, see Figure 7.4), and which is known as the shear stress at point x. This is a clear demonstration of the versatility provided by the duality concept. In other words, once the Timoshenko motion action space has been enriched by adding a further independent kinematical descriptor (observe that $\mathscr{V}_{BE} \subset \mathscr{V}_T$ and therefore the same happens with the corresponding spaces Var_v defined in (7.18) and (7.39)) the dual space of internal stresses $\mathscr{W}'$ is also enriched.

Having established the internal power functional for the Timoshenko beam model, it is possible to determine the external generalized loads which are admissible for this kinematical model. Similarly to that made with the Bernoulli–Euler beam model, after integration by parts we arrive at the following representation for the external forces

$$(\mathbf{T}, \mathcal{D}\hat{\mathbf{v}}) = N\hat{\bar{v}}_x|_0^L - \int_0^L \frac{\partial N}{\partial x}\hat{\bar{v}}_x dx - M\hat{\psi}|_0^L + \int_0^L \left(\frac{\partial M}{\partial x} - Q\right)\hat{\psi} dx$$

$$+ Q\hat{v}_y|_0^L - \int_0^L \frac{\partial Q}{\partial x}\hat{v}_y dx = \langle \mathcal{D}^*\mathbf{T}, \hat{\mathbf{v}}\rangle, \quad (7.41)$$

and therefore

$$\langle f, \hat{\mathbf{v}}\rangle = (F_x\hat{\bar{v}}_x + F_y\hat{v}_y)|_0^L + m\hat{\psi}|_0^L + \int_0^L (q_x\hat{\bar{v}}_x + q_y\hat{v}_y)dx + \int_0^L m_d\hat{\psi} dx \quad \hat{\mathbf{v}} \in \mathrm{Var}_v. \quad (7.42)$$

This procedure naturally yields the external generalized forces compatible with the Timoshenko kinematics, being characterized by the forces F_x and F_y, by the bending moment m, which are concentrated in a finite number points over the neutral line, and, in addition, by the distributed forces q_x and q_y, and by the distributed moment m_d. Compared with the generalized forces admitted by the Bernoulli–Euler beam model, we see that now the Timoshenko admits a moment distributed in the whole beam domain (which was not admitted by the Bernoulli–Euler theory). In this sense, it is important to mention that there are some authors who present the Bernoulli–Euler theory including these distributed moments, something that has no mathematical support and which is not consistent with the mechanical theory.

At this point we have defined the internal power functional, and the external power has been derived from the procedure followed above. Thus, the characterization of internal generalized stresses and external generalized forces is fully accomplished for the Timoshenko beam model. The mechanical equilibrium is then defined by the PVP, which is written in the following variational form. Consider the system of forces applied over the beam which is characterized by the concentrated loads F_x^i, F_y^i (applied at point $x_i \in [0, L]$), for $i = 1, \dots, n$, by the bending concentrated moment m_j (applied at point $x_j \in [0, L]$), for $j = 1, \dots, r$, by the forces per unit length q_x and q_y, and by the bending moment per unit length m_d. Then, the generalized internal stress given by the triple (N, M, Q) is at equilibrium with that system of generalized forces if the following variational equation is satisfied

$$\int_0^L \left[N\frac{\partial \hat{\bar{v}}_x}{\partial x} - M\frac{\partial \hat{\psi}}{\partial x} - Q\left(\hat{\psi} - \frac{\partial \hat{v}_y}{\partial x}\right)\right] dx = \sum_{i=1}^{n}(F_x^i\hat{\bar{v}}_x(x_i) + F_y^i\hat{v}_y(x_i)) + \sum_{j=1}^{r} m_j\hat{\psi}(x_j)$$

$$+ \int_0^L (q_x\hat{\bar{v}}_x + q_y\hat{v}_y)dx + \int_0^L m_d\hat{\psi} dx \qquad \forall(\hat{\bar{v}}_x, \hat{v}_y, \hat{\psi}) \in \mathrm{Var}_v. \quad (7.43)$$

Let us now derive the Euler–Lagrange equations associated with the variational equation (7.43). For simplicity in the presentation, we will consider that we have a single point $a \in [0, L]$ where the concentrated forces and the moment are applied, and that the functions in the integrands are regular enough so that integration by parts can be safely performed. Then, integration by parts yields

$$
\begin{aligned}
&-\int_{0,a}^{a,L}\left(N\frac{\partial\hat{\bar{v}}_x}{\partial x}-M\frac{\partial\hat{\psi}}{\partial x}-Q\left(\hat{\psi}-\frac{\partial\hat{v}_y}{\partial x}\right)\right)dx\\
&\quad+F_x^a\hat{\bar{v}}_x(a)+F_y^a\hat{v}_y(a)+m_a\hat{\psi}(a)+\int_{0,a}^{a,L}(q_x\hat{\bar{v}}_x+q_y\hat{v}_y+m_d\hat{\psi})dx\\
&=-N\hat{\bar{v}}_x|_{0,a}^{a,L}+M\hat{\psi}|_{0,a}^{a,L}-Q\hat{v}_y|_{0,a}^{a,L}+\int_{0,a}^{a,L}\left(\frac{\partial N}{\partial x}\hat{\bar{v}}_x-\left(\frac{\partial M}{\partial x}-Q\right)\hat{\psi}+\frac{\partial Q}{\partial x}\hat{v}_y\right)dx\\
&\quad+F_x^a\hat{\bar{v}}_x(a)+F_y^a\hat{v}_y(a)+m_a\hat{\psi}(a)\\
&\quad+\int_{0,a}^{a,L}(q_x\hat{\bar{v}}_x+q_y\hat{v}_y+m_d\hat{\psi})dx=0\quad\forall(\hat{\bar{v}}_x,\hat{v}_y,\hat{\psi})\in\mathrm{Var}_v.
\end{aligned}
\tag{7.44}
$$

Rearranging terms leads to

$$
\begin{aligned}
&\int_{0,a}^{a,L}\left(\frac{\partial N}{\partial x}+q_x\right)\hat{\bar{v}}_xdx+\int_{0,a}^{a,L}\left(Q-\frac{\partial M}{\partial x}+m_d\right)\hat{\psi}dx+\int_{0,a}^{a,L}\left(\frac{\partial Q}{\partial x}+q_y\right)\hat{v}_ydx\\
&-N\hat{\bar{v}}_x|_L+N\hat{\bar{v}}_x|_0-\left([\![N]\!]_a-F_x^a\right)\hat{\bar{v}}_x(a)+M\hat{\psi}|_L-M\hat{\psi}|_0+\left([\![M]\!]_a+m_a\right)\hat{\psi}(a)\\
&-Q\hat{v}_y|_L+Q\hat{v}_y|_0-\left([\![Q]\!]_a-F_y^a\right)\hat{v}_y(a)=0\qquad\forall(\hat{\bar{v}}_x,\hat{v}_y,\hat{\psi})\in\mathrm{Var}_v.
\end{aligned}
\tag{7.45}
$$

This variational equation in conjunction with the fundamental lema of the calculus of variations enable us to conclude the following.

- Essential/natural boundary conditions. If at $x = 0$ the motion in the x direction is prescribed (essential boundary condition), then we have $\hat{\bar{v}}_x(0) = 0$. However, if no motion is prescribed at that point, we obtain the following natural boundary condition

  $$N(0) = 0. \tag{7.46}$$

  If at $x = 0$ the motion in the y direction is prescribed (essential boundary condition), then $\hat{v}_y(0) = 0$. However, if the motion in that direction is not prescribed, the following natural boundary condition holds

  $$Q(0) = 0. \tag{7.47}$$

  If at $x = 0$ the rotation action of the transversal section is prescribed (essential boundary condition), this implies $\hat{\psi}(0) = 0$. However, if the rotation action is not constrained, the variational equation naturally leads to

  $$M(0) = 0. \tag{7.48}$$

  The same reasoning holds for the other end point $x = L$. Either the kinematics is essentially constrained or a natural boundary condition for the generalized internal stresses emerge.
- Partial differential equations. The variational equation implies the following Euler–Lagrange equations for the Timoshenko beam in the domains $(0, a)$ and (a, L)

  $$\frac{\partial N}{\partial x}+q_x=0\qquad\text{in }(0,a)\cup(a,L), \tag{7.49}$$

  $$\frac{\partial Q}{\partial x}+q_y=0\qquad\text{in }(0,a)\cup(a,L), \tag{7.50}$$

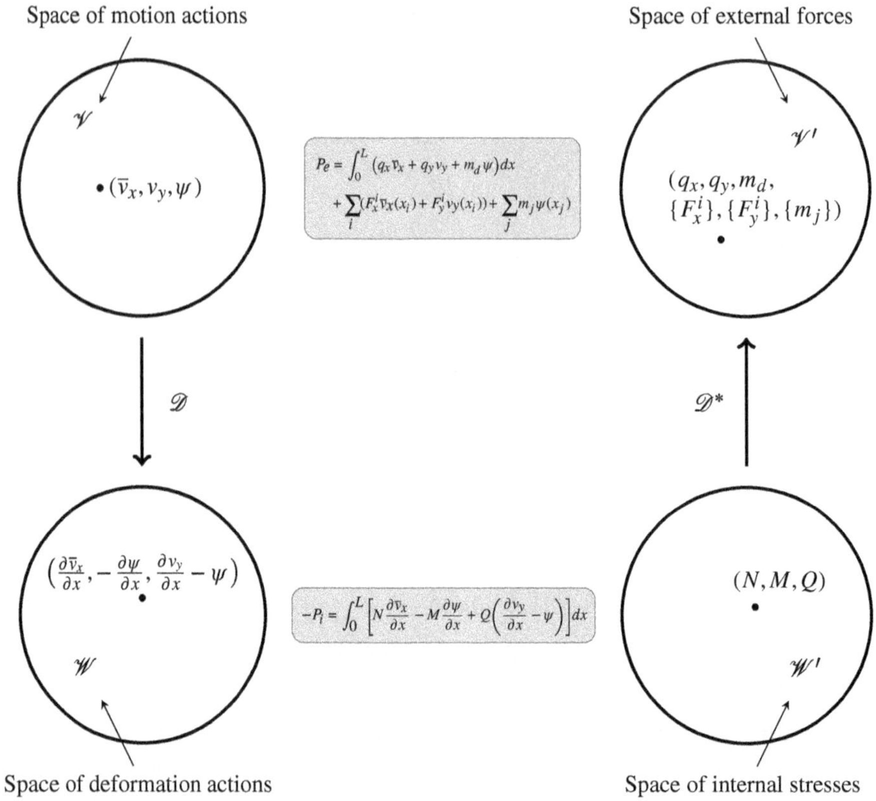

Figure 7.5 Ingredients in the Timoshenko theory of beams.

$$Q - \frac{\partial M}{\partial x} + m_d = 0 \qquad \text{in } (0, a) \cup (a, L). \tag{7.51}$$

- Jump equations. At point $x = a \in [0, L]$ we have three additional equations that are naturally satisfied when the mechanical equilibrium is verified. These equations are known as jump equations, and for the Timoshenko beam they read

$$[\![N]\!]_a - F_x^a = 0, \tag{7.52}$$

$$[\![Q]\!]_a - F_y^a = 0, \tag{7.53}$$

$$[\![M]\!]_a + m_a = 0. \tag{7.54}$$

The main elements that play a mechanical role in the modeling of beam bending under Timoshenko assumptions are presented in Figure 7.5.

7.6 Final Remarks

As can be appreciated throughout this chapter, we have introduced, with a minimal set of hypotheses, the theoretical foundations for the Bernoulli–Euler beam model and the Timoshenko beam model. The combination of adequate kinematical hypotheses and

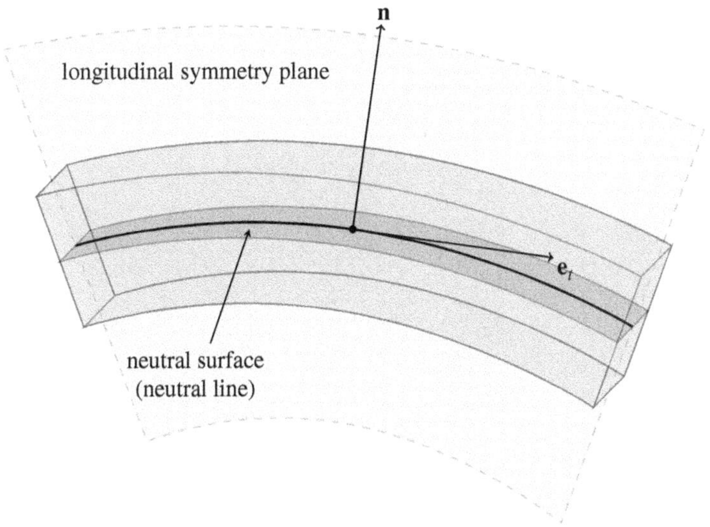

Figure 7.6 Neutral axis in a curved beam along the longitudinal symmetry plane.

mathematical duality are the basic pillars required to establish, clearly and unequivocally, the generalized internal stresses and external forces which are consistent with each kinematical model. As a final ingredient, the Principle of Virtual Power materializes the concept of mechanical equilibrium between these generalized efforts. Note that we have not been forced to resort, at any step in the development of the models, to the material behavior of the matter that composes the structural component. The material model will arise when there is a need to define a relation between the generalized internal stresses and the history of the strain the beam was subjected to. With this relation, the PVP turns into a variational field problem which allows the (history of the) displacement field along the loading process to be determined.

An additional remark is in order concerning the beam geometry. In this chapter, for both beam models studied, we have assumed that, in the current configuration, the beam featured a straight neutral axis. The formulation developed here cannot be directly applied to beams in which this neutral axis is curved, as illustrated in Figure 7.6. The kinematical hypotheses remain the same than in the case of a straight beam. This implies that transversal sections remain planes and either orthogonal (Bernoulli–Euler theory) or non-orthogonal (Timoshenko theory) when a motion action is imprinted. Notwithstanding this, an additional technical difficulty emerges by virtue of the transversal sections having different orientations as we move along the beam axis. In fact, while we have employed a universal Cartesian frame of reference defined by the classical unit vectors $(\mathbf{e}_x, \mathbf{e}_y, \mathbf{e}_z)$, that is, constant for the whole beam neutral axis, in a curved beam this triple of vectors is formed by the unit vectors intrinsically attached to the beam neutral axis, denoted by $(\mathbf{e}_t, \mathbf{n}, \mathbf{e}_z)$, as seen in Figure 7.6. The orientation of these vectors, particularly $\mathbf{e}_t$ and $\mathbf{n}$, changes along the beam axial coordinate. Since the construction of curved beam models is a particular case of the theory for shells, the geometric tools utilized to model curved beams will be similar to those that arise in the modeling of shells. This topic is addressed in detail in Chapter 9, and the interested reader can refer to that chapter to gain a flavor of the geometry for curved structural components.

8

Torsion of Bars

8.1 Introduction

In this chapter the mechanical theory for the modeling of torsion of regular prismatic bars under the kinematical hypotheses of Saint-Venant is presented. The variational framework developed in Chapter 3 is manipulated to achieve this goal. We will see in detail the construction of kinematically admissible sets and spaces of motion actions, and the strain (rate) operator will be defined. From this Saint-Venant kinematical model, compatible internal stresses and external forces are then characterized by mathematical duality and, through the concept of mechanical equilibrium and compatibility provided, respectively, by the Principle of Virtual Power (PVP) and by the Principle of Complementary Virtual Power (PCVP) the variational formulations will be exposed regardless of the material behavior. Then, as an illustrative example, the case of elastic materials within the range of infinitesimal strains is addressed.

8.2 Kinematics

The kinematical model is governed by the so-called Saint-Venant kinematical hypotheses (see Figure 8.1), which are the following

- regular prismatic bars (equally called bars) are considered, that is, the transversal section, denoted by Ω, with a smooth delimiting boundary Γ, is constant along the longitudinal bar axis, of length L
- each transversal section can undergo a rigid rotation around the normal axis to Ω, such that the rate of change along the longitudinal bar axis is constant
- the motion action in the direction of the longitudinal bar axis is constant for all transversal sections, so taking an orthogonal system of coordinates as displayed in Figure 8.1, where $\mathbf{e}_z$ is oriented in the direction of the longitudinal bar axis, the previous statement implies that the motion action $\mathbf{v}$ is such that the component in the $\mathbf{e}_z$ direction gives $v_z = v_z(x, y)$, where (x, y) are the coordinates in the transversal plane aligned with Ω.

Thus, the motion action $\mathbf{v} = v_a \mathbf{e}_a$, $a = x, y, z$, compatible with the previous hypotheses is given by

$$v_x = -r\dot{\theta} \sin \beta = -y\dot{\theta}, \tag{8.1}$$

Introduction to the Variational Formulation in Mechanics: Fundamentals and Applications, First Edition.
Edgardo O. Taroco, Pablo J. Blanco and Raúl A. Feijóo.

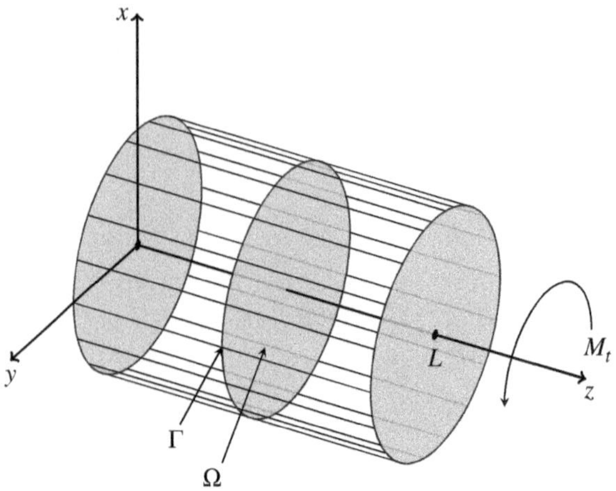

Figure 8.1 Prismatic bar subjected to torsion.

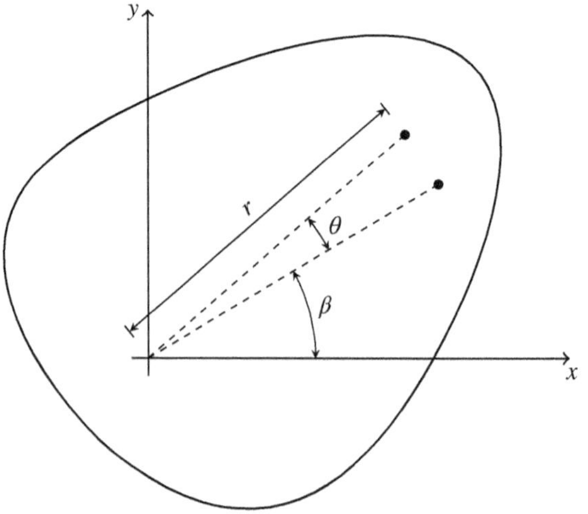

Figure 8.2 Kinematics of a prismatic bar subjected to torsion within the Saint-Venant theory.

$$v_y = r\dot{\theta}\cos\beta = x\dot{\theta}, \tag{8.2}$$

$$v_z = v_z(x, y) = \kappa(x, y), \tag{8.3}$$

where $\dot{\theta} = \dot{\theta}(z)$ is the motion action representing the angular rotation around the axis $\mathbf{e}_z$ corresponding to coordinate z (see Figure 8.2).

The previous functional relations imply that the non-null components of the strain rate tensor $\mathbf{D} = \mathscr{D}\mathbf{v} = (\nabla\mathbf{v})^s$ are

$$[\mathscr{D}\mathbf{v}]_{xz} = [\mathscr{D}\mathbf{v}]_{zx} = \frac{1}{2}\left[-y\frac{\partial\dot{\theta}}{\partial z} + \frac{\partial\kappa}{\partial x}\right], \tag{8.4}$$

$$[\mathscr{D}\mathbf{v}]_{yz} = [\mathscr{D}\mathbf{v}]_{zy} = \frac{1}{2}\left[x\frac{\partial\dot{\theta}}{\partial z} + \frac{\partial\kappa}{\partial y}\right]. \tag{8.5}$$

Then, from the previous kinematical hypotheses it turns out that $\alpha = \frac{\partial \theta}{\partial z} \in \mathbb{R}$ (constant), and, finally, the kinematics (motion action) takes the form

$$v_x = -\alpha yz, \tag{8.6}$$

$$v_y = \alpha xz, \tag{8.7}$$

$$v_z = \kappa(x, y), \tag{8.8}$$

while the strain rate assumes the form

$$[\mathscr{D}\mathbf{v}]_{xz} = [\mathscr{D}\mathbf{v}]_{zx} = \frac{1}{2}\left[-\alpha y + \frac{\partial \kappa}{\partial x}\right], \tag{8.9}$$

$$[\mathscr{D}\mathbf{v}]_{yz} = [\mathscr{D}\mathbf{v}]_{zy} = \frac{1}{2}\left[\alpha x + \frac{\partial \kappa}{\partial y}\right]. \tag{8.10}$$

To simplify the mathematical operations, we introduce the vector field

$$\gamma(\mathbf{v}) = \nabla_\Omega \kappa + \alpha \chi, \tag{8.11}$$

where κ is a scalar field and χ is a vector field defined by

$$\nabla_\Omega \kappa = \frac{\partial \kappa}{\partial x}\mathbf{e}_x + \frac{\partial \kappa}{\partial y}\mathbf{e}_y, \tag{8.12}$$

$$\chi = -y\mathbf{e}_x + x\mathbf{e}_y. \tag{8.13}$$

We can, therefore, define the space of kinematically admissible motion actions Kin_v as

$$\text{Kin}_v = \{(v_x, v_y, v_z) \in \mathscr{V}\,;\; v_x = -\alpha yz,\; v_y = \alpha xz,\; v_z = \kappa(x, y)\}. \tag{8.14}$$

In order to characterize the space of kinematically admissible virtual motion actions Var_v we resort to the definition

$$\text{Kin}_v = \overline{\mathbf{v}} + \text{Var}_v \qquad \overline{\mathbf{v}} \in \text{Kin}_v \text{ arbitrary}. \tag{8.15}$$

Hence, given arbitrary $\mathbf{v}^1, \mathbf{v}^2 \in \text{Kin}_v$, we get

$$\mathbf{v}^1 = (-\alpha_1 zy, \alpha_1 zx, \kappa_1), \tag{8.16}$$

$$\mathbf{v}^2 = (-\alpha_2 zy, \alpha_2 zx, \kappa_2). \tag{8.17}$$

In this manner $\mathbf{w} = \mathbf{v}^2 - \mathbf{v}^1 \in \text{Var}_v$, and so

$$\mathbf{w} = (\alpha_2 - \alpha_1)\left(-zy, zx, \frac{\kappa_2}{\alpha_2 - \alpha_1} - \frac{\kappa_1}{\alpha_2 - \alpha_1}\right) = (-\hat{\alpha} zy, \hat{\alpha} zx, \hat{\kappa}). \tag{8.18}$$

From this, we readily conclude that $\text{Kin}_v = \text{Var}_v$. Equivalently, we can characterize the space Var_v in terms of (κ, α) as

$$\text{Var}_v = \{(\kappa, \alpha) \in \mathscr{V}_\Omega \times \mathbb{R}\}, \tag{8.19}$$

where $\mathscr{V}_\Omega$ is a space of sufficiently smooth functions defined in the bar transversal section Ω.

To finalize the study of the kinematics we have to characterize rigid motion actions. Note that we have $\mathbf{v} \in \mathscr{N}(\mathscr{D})$ if and only if $(\nabla \mathbf{v})^s = \mathbf{O}$, that in this case it becomes $\gamma = \mathbf{0}$, and thus (see (8.11))

$$\frac{\partial \kappa}{\partial x} - \alpha y = 0, \tag{8.20}$$

$$\frac{\partial \kappa}{\partial y} + \alpha x = 0, \tag{8.21}$$

that is

$$\kappa = \alpha xy + f(y), \tag{8.22}$$

$$\kappa = -\alpha xy + g(x), \tag{8.23}$$

which are satisfied whenever

$$\alpha = 0, \tag{8.24}$$

$$f(y) = g(x) = C, \tag{8.25}$$

where C is a constant. Hence

$$\mathcal{N}(\mathcal{D}) = \{\mathbf{v} \in \mathcal{V}\,;\; v_x = v_y = 0,\; v_z = C\}, \tag{8.26}$$

or equivalently

$$\mathcal{N}(\mathcal{D}) = \{(0, 0, C) \in \mathcal{V}\,;\; C \in \mathbb{R}\}. \tag{8.27}$$

8.3 Generalized Forces

Having defined the kinematics and the sets Kin_v, Var_v and $\mathcal{N}(\mathcal{D})$, we can advance to the next step, which consists of the characterization of the external loads compatible with the adopted kinematical model. We have seen in Chapter 3 that the PVP tells us that the body $\mathcal{B}$ in the configuration $\mathcal{B}_t$ is at equilibrium under the action of the load f if

$$\langle f, \hat{\mathbf{v}} \rangle = 0 \quad \forall \hat{\mathbf{v}} \in \text{Var}_v \cap \mathcal{N}(\mathcal{D}) = \{\mathbf{v} \in \mathcal{V}\,;\; v_x = v_y = 0,\; v_z = \kappa = C\}, \tag{8.28}$$

$$(\mathbf{T}, \mathcal{D}(\hat{\mathbf{v}})) = \langle f, \hat{\mathbf{v}} \rangle \quad \forall \hat{\mathbf{v}} \in \text{Var}_v. \tag{8.29}$$

The first expression states that the component of f dual to the kinematical variable κ, say f_z, has to be orthogonal to $\mathbb{R}$, that is, it must be a function with zero mean. The second expression allows us to identify the type of load compatible with the model. In fact, introducing the vector

$$\sigma = T_{xz}\mathbf{e}_x + T_{yz}\mathbf{e}_y, \tag{8.30}$$

and recalling that, by hypothesis, the strain rate is the same for all transversal sections, it follows that the corresponding stress state also satisfies this property. In other words, the stress state does not depend on variable z. We then have

$$\begin{aligned} -P_i = (\mathbf{T}, \mathcal{D}\hat{\mathbf{v}}) &= \int_{\mathcal{B}} \mathbf{T} \cdot (\nabla \hat{\mathbf{v}})^s dV = \int_{\mathcal{B}} \sigma \cdot \hat{\gamma}\, dV \\ &= L \int_{\Omega} \sigma \cdot \hat{\gamma}\, d\Omega = L \int_{\Omega} [\sigma \cdot \nabla_{\Omega} \hat{\kappa} + (\sigma \cdot \chi)\hat{\alpha}]\, d\Omega \qquad \hat{\mathbf{v}} \in \text{Var}_v. \end{aligned} \tag{8.31}$$

Now, we have to put in evidence the equilibrium operator $\mathscr{D}^*$ as follows

$$\begin{aligned}(\mathbf{T}, \mathscr{D}\hat{\mathbf{v}}) &= L\int_\Omega [\sigma \cdot \nabla_\Omega \hat{\kappa} + (\sigma \cdot \chi)\hat{\alpha}]d\Omega \\ &= L\int_\Omega [\mathrm{div}_\Omega(\hat{\kappa}\sigma) - \hat{\kappa}\mathrm{div}_\Omega\sigma + (\sigma \cdot \chi)\hat{\alpha}]d\Omega \\ &= L\left[\int_\Gamma (\sigma \cdot \mathbf{n})\hat{\kappa}d\Gamma - \int_\Omega \hat{\kappa}\mathrm{div}_\Omega\sigma d\Omega + \int_\Omega \hat{\alpha}(\sigma \cdot \chi)d\Omega\right] = \langle \mathscr{D}^*\mathbf{T}, \hat{\mathbf{v}}\rangle,\end{aligned} \tag{8.32}$$

and, as a consequence, this gives

$$\langle f, \hat{\mathbf{v}}\rangle = L\left[\int_\Gamma f_n\hat{\kappa}d\Gamma + \int_\Omega f_b\hat{\kappa}d\Omega + \int_\Omega m_t\hat{\alpha}d\Omega\right] \qquad \hat{\mathbf{v}} \in \mathrm{Var}_v. \tag{8.33}$$

From the constraint over f we get $\int_\Gamma f_n d\Gamma + \int_\Omega f_b d\Omega = 0$, and since $\hat{\alpha}$ is constant, finally

$$P_e = \langle f, \hat{\mathbf{v}}\rangle = L\int_\Gamma f_n\hat{\kappa}d\Gamma + L\int_\Omega f_b\hat{\kappa}d\Omega + M_t L\hat{\alpha} \qquad \hat{\mathbf{v}} \in \mathrm{Var}_v, \tag{8.34}$$

where

$$M_t = \int_\Omega m_t d\Omega, \tag{8.35}$$

is the torsional moment applied over the bar end in which the motions are not prescribed. Hereafter, we limit the presentation to the case of pure torsion in which only torsional moment is allowed to act over the body. In other words we take $f_n = f_b = 0$ and $M_t \neq 0$, so we finally have

$$P_e = \langle f, \hat{\mathbf{v}}\rangle = M_t L\hat{\alpha} \qquad \hat{\mathbf{v}} \in \mathrm{Var}_v. \tag{8.36}$$

The main elements that form the mechanical model for the case of torsion of prismatic bars can be appreciated in Figure 8.3.

8.4 Mechanical Equilibrium

The external forces compatible with the kinematical model were characterized in the previous section. Now we can formulate the definition of the mechanical equilibrium for this structural component through the PVP.

First, note that the balance of internal and external powers establishes that

$$\int_\Omega \sigma \cdot \gamma(\hat{\mathbf{v}})d\Omega = M_t\hat{\alpha} \qquad \forall\hat{\mathbf{v}} \in \mathrm{Var}_v. \tag{8.37}$$

From the defintion in (8.11), we get

$$\int_\Omega (\sigma \cdot \nabla_\Omega \hat{\kappa} + (\sigma \cdot \chi)\hat{\alpha})d\Omega = M_t\hat{\alpha} \qquad \forall(\hat{\kappa}, \hat{\alpha}) \in \mathscr{V}_\Omega \times \mathbb{R}, \tag{8.38}$$

and since $\hat{\alpha} \in \mathbb{R}$, we have

$$\int_\Omega \sigma \cdot \nabla_\Omega \hat{\kappa} d\Omega = \hat{\alpha}\left[M_t - \int_\Omega \sigma \cdot \chi d\Omega\right] \qquad \forall(\hat{\kappa}, \hat{\alpha}) \in \mathscr{V}_\Omega \times \mathbb{R}. \tag{8.39}$$

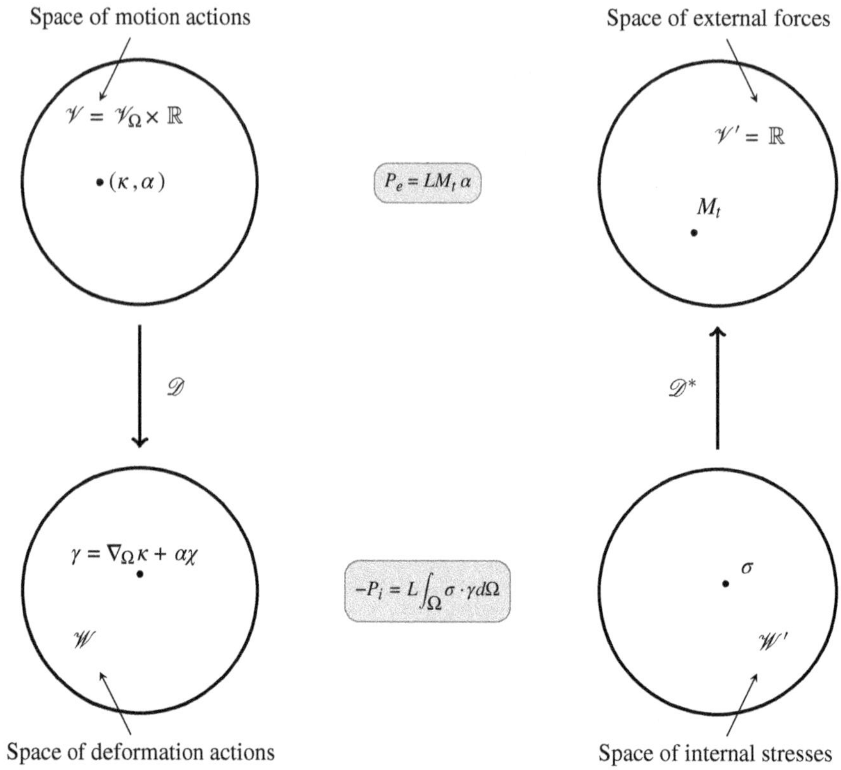

Figure 8.3 Ingredients in the Saint-Venant theory for the torsion of prismatic bars.

Hence, we arrive at the variational formulation of the mechanical equilibrium for the problem of torsion of prismatic bars in the Saint-Venant theory. It states that, given a load applied over the bar, characterized by the torsional moment M_t, the internal generalized stress, given by σ, is at equilibrium with such load if

$$\int_\Omega \sigma \cdot \nabla_\Omega \hat{\kappa} d\Omega = 0 \qquad \forall \hat{\kappa} \in \mathscr{V}_\Omega, \tag{8.40}$$

$$\int_\Omega \sigma \cdot \chi d\Omega = M_t. \tag{8.41}$$

Note that the variational equation (8.41) is valid for any material that constitutes the bar.

Expressions (8.40) and (8.41) provide the variational formulation, which was achieved through the manipulation of the kinematics. This is why such formulation is also known as primal (or kinematical) variational formulation of the torsion problem. It is important to highlight that the PVP allows us to characterize σ, that is, the components T_{xz} and T_{yz} of $\mathbf{T}$. The other components of the stress tensor $\mathbf{T}$ are reactive internal stresses which correspond to the constraints prescribed by the kinematical hypotheses in the model. The derivation of the exact (or approximate) expressions for these reactive stresses will be tackled within the variational framework of the shell theory (see Chapter 9). The interested reader will be able to readily obtain these reactive stresses for the torsion model presented here.

As we know from Chapter 3, the generalized stress σ belongs to the linear manifold Est_T whose elements are those fields which are at equilibrium with the applied load M_t. Recalling that

$$\text{Est}_T = \overline{\sigma} + \text{Var}_T \qquad \overline{\sigma} \in \text{Est}_T \text{ arbitrary,} \tag{8.42}$$

we have that all these fields are determined through the PVP up to a self-equilibrated stress field. The indetermination is removed when the material behavior of the bar is specified through a constitutive equation.

The presentation will now be constrained to the case of bars featuring a linear elastic isotropic behavior within the infinitesimal strain regime. We then have

$$\mathbf{T} = 2\mu\mathbf{E} + \lambda(\text{tr}\mathbf{E})\mathbf{I}, \tag{8.43}$$

where μ and λ are the Lamé coefficients (hereafter considered constant and strictly positive numbers in $\mathscr{B}$), $\mathbf{E}$ is the (infinitesimal) strain tensor, and $\mathbf{I}$ is the identity tensor. According to the kinematics of the model

$$\text{tr}\mathbf{E} = 0, \tag{8.44}$$

and therefore

$$\mathbf{T} = 2\mu\mathbf{E}, \tag{8.45}$$

which is equivalent, in generalized variables, to

$$\sigma = \mu\gamma. \tag{8.46}$$

Going back to the variational equation (8.40) with the constitutive equation (8.46) we obtain

$$\int_\Omega \mu\gamma \cdot \nabla_\Omega \hat{\kappa} d\Omega = 0 \qquad \forall \hat{\kappa} \in \mathscr{V}_\Omega. \tag{8.47}$$

Introducing the definition of γ and taking into account that μ is constant, the variational equation becomes

$$\int_\Omega (\nabla_\Omega \kappa + \alpha\chi) \cdot \nabla_\Omega \hat{\kappa} d\Omega = 0 \qquad \forall \hat{\kappa} \in \mathscr{V}_\Omega. \tag{8.48}$$

Writing $\kappa = \alpha\varphi$, where $\varphi = \varphi(x, y)$ is known in the literature as the warping function, the previous expression gives

$$\begin{aligned} \int_\Omega \nabla_\Omega \varphi \cdot \nabla_\Omega \hat{\kappa} d\Omega &= -\int_\Omega \chi \cdot \nabla_\Omega \hat{\kappa} d\Omega = \int_\Omega \hat{\kappa} \text{div}_\Omega \chi d\Omega - \int_\Omega \text{div}_\Omega (\hat{\kappa}\chi) d\Omega \\ &= -\int_\Gamma \hat{\kappa}\chi \cdot \mathbf{n} d\Gamma \qquad \hat{\kappa} \in \mathscr{V}_\Omega. \end{aligned} \tag{8.49}$$

In this way, we have

$$\int_\Omega \nabla_\Omega \varphi \cdot \nabla \hat{\kappa} d\Omega = -\int_\Gamma \hat{\kappa}\chi \cdot \mathbf{n} d\Gamma \qquad \forall \hat{\kappa} \in \mathscr{V}_\Omega, \tag{8.50}$$

$$\int_\Omega \mu(\nabla\varphi \cdot \chi + |\chi|^2) d\Omega = \frac{M_t}{\alpha}, \tag{8.51}$$

where $|\chi|^2 = \chi \cdot \chi$ and the ratio $\frac{M_t}{\alpha}$ is known as the torsional stiffness. The two equations above define the variational formulation for the problem of torsion of elastic isotropic prismatic bars under Saint-Venant hypotheses.

In order to obtain the local form of the equilibrium, that is the Euler–Lagrange equations associated with the variational equation (8.50), we make use of the equilibrium operator $\mathcal{D}^*$ and of the PVP, as follows

$$\begin{aligned}\int_\Omega \mu\gamma \cdot (\nabla_\Omega \hat{\kappa} + \hat{\alpha}\chi) d\Omega &= \int_\Omega (\text{div}_\Omega(\hat{\kappa}\mu\gamma) - \hat{\kappa}\text{div}_\Omega(\mu\gamma)) d\Omega \\ &+ \hat{\alpha}\int_\Omega \mu\gamma \cdot \chi d\Omega = \int_\Gamma (\mu\gamma \cdot \mathbf{n})\hat{\kappa} d\Gamma - \int_\Omega \text{div}_\Omega(\mu\gamma)\hat{\kappa} d\Omega \\ &+ \hat{\alpha}\int_\Omega \mu\gamma \cdot \chi d\Omega = M_t\hat{\alpha} \quad \forall(\hat{\kappa}, \hat{\alpha}) \in \mathcal{V}_\Omega \times \mathbb{R}. \end{aligned} \tag{8.52}$$

After rearranging terms we find

$$\int_\Gamma (\mu\gamma \cdot \mathbf{n})\hat{\kappa} d\Gamma - \int_\Omega \text{div}_\Omega(\mu\gamma)\hat{\kappa} d\Omega - \hat{\alpha}\left[M_t - \int_\Omega \mu\gamma \cdot \chi d\Omega\right] = 0 \quad \forall(\hat{\kappa}, \hat{\alpha}) \in \mathcal{V}_\Omega \times \mathbb{R}, \tag{8.53}$$

which is equivalent to

$$\text{div}_\Omega(\mu\gamma) = 0 \quad \text{in } \Omega, \tag{8.54}$$

$$\mu\gamma \cdot \mathbf{n} = 0 \quad \text{on } \Gamma, \tag{8.55}$$

$$\int_\Omega \alpha\mu(\nabla_\Omega\varphi \cdot \chi + |\chi|^2) d\Omega = M_t. \tag{8.56}$$

Since $\gamma = \kappa + \alpha\chi$, and noting that μ and α are constants, then, observing that $\text{div}_\Omega\chi = 0$ and utilizing function φ, we get

$$\mu\alpha\text{div}_\Omega(\nabla_\Omega\varphi) = 0 \quad \text{in } \Omega, \tag{8.57}$$

$$\mu\alpha(\nabla_\Omega\varphi + \chi) \cdot \mathbf{n} = 0 \quad \text{on } \Gamma, \tag{8.58}$$

$$\int_\Omega \alpha\mu(\nabla_\Omega\varphi \cdot \chi + |\chi|^2) d\Omega = M_t. \tag{8.59}$$

Integrating by parts the first term on the left-hand side of equation (8.59) we find

$$\begin{aligned}\int_\Omega \alpha\mu(\nabla_\Omega\varphi \cdot \chi + |\chi|^2) d\Omega &= -\alpha\mu\int_\Omega \varphi\text{div}_\Omega\chi d\Omega + \mu\alpha\int_\Gamma \varphi(\chi \cdot \mathbf{n}) d\Gamma \\ &+ \mu\alpha\int_\Omega |\chi|^2 d\Omega = \mu\alpha\int_\Gamma \varphi(\chi \cdot \mathbf{n}) d\Gamma + \mu\alpha\int_\Omega |\chi|^2 d\Omega. \end{aligned} \tag{8.60}$$

As a result, the local form of the variational problems (8.50) and (8.51) is characterized by the following boundary value problem

$$\triangle_\Omega\varphi = 0 \quad \text{in } \Omega, \tag{8.61}$$

$$\nabla_\Omega\varphi \cdot \mathbf{n} = -\chi \cdot \mathbf{n} \quad \text{on } \Gamma, \tag{8.62}$$

$$\int_\Gamma \varphi(\chi \cdot \mathbf{n}) d\Gamma = \frac{M_t}{\alpha\mu} - \int_\Omega |\chi|^2 d\Omega. \tag{8.63}$$

Thus, the local form of the torsion problem for elastic bars is reduced to the solution of a Laplace partial differential equation in φ with Neumann-like boundary conditions and a constraint over the weighted mean value of the field φ over the boundary of the transversal section in the bar. It is worth mentioning that, despite the relative practical simplicity of the variational problem (8.50), in most cases the use of the dual formulation is preferred, which is described in the next section.

8.5 Dual Formulation

The so-called dual variational formulation is derived by using the PCVP, which establishes that the strain rate $\mathbf{D}$, defined in the actual equilibrium configuration, is compatible if and only if

$$(\hat{\mathbf{T}}, \mathbf{D}) = (\hat{\mathbf{T}}, \mathcal{D}\mathbf{w}) \qquad \forall \hat{\mathbf{T}} \in \mathrm{Var}_T, \tag{8.64}$$

where $\mathbf{w} \in \mathrm{Kin}_v$ is arbitrary.

In extended form, the previous expression reads

$$\int_{\mathcal{B}_t} \hat{\mathbf{T}} \cdot \mathbf{D}\, dV = \int_{\partial\mathcal{B}_{tv}} \hat{\mathbf{T}}\mathbf{n} \cdot \overline{\mathbf{v}}\, dS \qquad \forall \hat{\mathbf{T}} \in \mathrm{Var}_T, \tag{8.65}$$

where $\overline{\mathbf{v}}$ is the velocity field prescribed over the boundary $\partial\mathcal{B}_{tv}$.

Keeping in mind all the hypotheses introduced at the beginning of this chapter, and with the aim of facilitating the presentation, we will assume that the value of α is known. That is, α is regarded as a datum for the problem. From a mechanical point of view, this implies that we are prescribing the rotation of any transversal section of the bar. As a consequence, the components v_x and v_y of the velocity field (or displacement in the infinitesimal strain regime) are prescribed.

Now, we have to define the set Est_T and the space Var_T for the kinematics under consideration. This is achieved first by writing

$$\mathrm{Var}_T = \{\mathbf{T} \in \mathcal{W}'; \ (\mathbf{T}, \mathcal{D}\hat{\mathbf{v}}) = 0 \quad \forall \hat{\mathbf{v}} \in \mathrm{Var}_v\}, \tag{8.66}$$

which for the torsion problem takes the form

$$\begin{aligned}(\mathbf{T}, \mathcal{D}\hat{\mathbf{v}}) &= \int_{\mathcal{B}_t} \mathbf{T} \cdot (\nabla\hat{\mathbf{v}})^s\, dV = L\int_\Omega \sigma \cdot \hat{\gamma} d\Omega = L\int_\Omega \sigma \cdot \nabla_\Omega \hat{\kappa} d\Omega \\ &= L\int_\Gamma (\sigma \cdot \mathbf{n})\hat{\kappa} d\Gamma - L\int_\Omega \hat{\kappa}\,\mathrm{div}_\Omega \sigma d\Omega = 0 \qquad \forall \hat{\kappa} \in \mathcal{V}_\Omega.\end{aligned} \tag{8.67}$$

This entails

$$\mathrm{div}_\Omega \sigma = \frac{\partial \sigma_x}{\partial x} + \frac{\partial \sigma_y}{\partial y} = 0 \qquad \text{in } \Omega. \tag{8.68}$$

At this point, we conviniently introduce the function ψ such that

$$\sigma_x = \frac{\partial \psi}{\partial y} \qquad \sigma_y = -\frac{\partial \psi}{\partial x}, \tag{8.69}$$

resulting in

$$\sigma \cdot \mathbf{n} = \frac{\partial \psi}{\partial y} n_x - \frac{\partial \psi}{\partial x} n_y = 0 \qquad \text{on } \Gamma, \tag{8.70}$$

which implies

$$\nabla_{\Omega}\psi \cdot \mathbf{t} = \frac{\partial \psi}{\partial t} = 0, \tag{8.71}$$

and so the function ψ remains unchanged along Γ, that is

$$\psi = \overline{\psi} \qquad \text{on } \Gamma, \tag{8.72}$$

where $\mathbf{t}$ is the unit vector tangent to the boundary Γ and $\overline{\psi}$ is a constant value.

Since α is known (it is prescribed) we have $\text{Est}_T = \text{Var}_T$. Hence, we obtain

$$\text{Est}_T = \text{Var}_T = \left\{ \sigma \in \mathscr{W}'_{\Omega};\ \sigma_x = \frac{\partial \psi}{\partial y},\ \sigma_y = -\frac{\partial \psi}{\partial x},\ \psi = \overline{\psi} \text{ on } \Gamma \right\}. \tag{8.73}$$

As seen in Chapter 3, it is important to notice that the dual problem (8.64) does not guarantee uniqueness because, given a compatible strain rate $\mathbf{D}$, for $\hat{\mathbf{v}} \in \text{Var}_v$ we have

$$(\hat{\mathbf{T}}, \mathbf{D} + \mathscr{D}\hat{\mathbf{v}}) = (\hat{\mathbf{T}}, \mathbf{D}) + (\hat{\mathbf{T}}, \mathscr{D}\hat{\mathbf{v}}) = (\hat{\mathbf{T}}, \mathscr{D}\mathbf{w}) + (\hat{\mathbf{T}}, \mathscr{D}\hat{\mathbf{v}}) = (\hat{\mathbf{T}}, \mathscr{D}(\mathbf{w} + \hat{\mathbf{v}})), \tag{8.74}$$

where $\mathscr{D}(\mathbf{w} + \hat{\mathbf{v}})$ is, clearly, compatible by construction.

Again, the indetermination is removed by introducing the material behavior of the bar. Thus, from the constitutive equation for linear elastic isotropic materials described in the previous section, from the properties of Var_T and Est_T, and from the kinematics defined for the bar, the PCVP becomes

$$\begin{aligned}\int_{\mathscr{B}_t} \hat{\mathbf{T}} \cdot \mathbf{D} dV &= L \int_{\Omega} \hat{\sigma} \cdot \gamma(\sigma) d\Omega = L \int_{\Omega} \hat{\sigma} \cdot \frac{\sigma}{\mu} d\Omega = \int_{\partial \mathscr{B}_{tv}} \hat{\mathbf{T}}\mathbf{n} \cdot \overline{\mathbf{v}} dS \\ &= L \int_{\Omega} \hat{\sigma} \cdot \alpha \chi d\Omega = L \int_{\Omega} \hat{\sigma} \cdot \alpha \chi d\Omega \qquad \forall \hat{\sigma} \in \text{Var}_T.\end{aligned} \tag{8.75}$$

In this fashion, the expression of the PCVP can be rewritten as

$$\int_{\Omega} \frac{1}{\mu} \nabla_{\Omega}\psi \cdot \nabla_{\Omega}\hat{\psi} d\Omega = -\int_{\Omega} \alpha \nabla_{\Omega}\hat{\psi} \cdot \mathbf{r} d\Omega \qquad \forall \hat{\psi} \in \text{Var}_{\psi}, \tag{8.76}$$

where $\mathbf{r} = x\mathbf{e}_x + y\mathbf{e}_y$ is the position vector and

$$\text{Var}_{\psi} = \{\psi \in H^1(\Omega);\ \psi = \text{constant in } \Gamma\}. \tag{8.77}$$

The right-hand side in (8.76) can be reworked to give

$$\begin{aligned}-\int_{\Omega} \alpha \nabla_{\Omega}\hat{\psi} \cdot \mathbf{r} d\Omega &= \int_{\Omega} \alpha \hat{\psi} \text{div}_{\Omega}\mathbf{r} d\Omega - \int_{\Omega} \alpha \text{div}_{\Omega}(\hat{\psi}\mathbf{r}) d\Omega \\ &= \alpha \left[\int_{\Omega} 2\hat{\psi} d\Omega - \sum_{i=1}^{nb} \overline{\psi}_i \int_{\Gamma_i} \mathbf{r} \cdot \mathbf{n} d\Gamma \right],\end{aligned} \tag{8.78}$$

where we have made use of the property that ψ is constant along the nb boundaries (see Figure 8.4).

For multiply connected domains, as in the case of Figure 8.4, ψ takes the value $\overline{\psi}_i$ at Γ_i. We must also note that the normal vectors $\mathbf{n}$ are always oriented outward from the domain in the sense dictated by the integration over the boundary which, related to the normal vector, is counterclockwise for the external boundary and clockwise for the

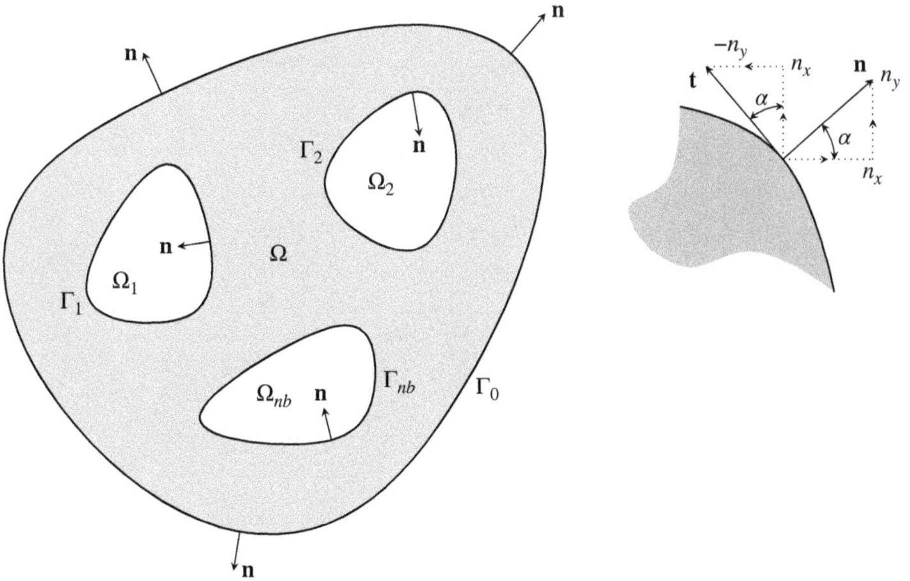

Figure 8.4 Domain of analysis for the torsion problem.

internal boundaries. Then, to solve the last integral on the right-hand side in expression (8.78), we proceed with

$$\int_{\Gamma_0} \mathbf{r} \cdot \mathbf{n} d\Gamma = \int_{\Omega_0} \mathrm{div}_{\Omega} \mathbf{r} d\Omega = 2\Omega_0 = 2\left(\Omega + \sum_{i=1}^{nb} \Omega_i\right), \tag{8.79}$$

$$\int_{\Gamma_i} \mathbf{r} \cdot \mathbf{n} d\Gamma = -\int_{\Omega_i} \mathrm{div}_{\Omega} \mathbf{r} d\Omega = -2\Omega_i \qquad i = 1, \ldots, nb, \tag{8.80}$$

where Ω_i stands for the area limited by Γ_i, $i = 1, \ldots, nb$. Analogously, Ω represents the value of the area of the domain of analysis and nb is the number of orifices in the transversal section.

Replacing expressions (8.79) and (8.80) into (8.78), and subsequently introducing that result into (8.76), we reach the variational form of the problem, which is given by

$$\int_{\Omega} \frac{1}{\mu} \nabla_{\Omega} \psi \cdot \nabla_{\Omega} \hat{\psi} d\Omega = 2\alpha \left[\int_{\Omega} \hat{\psi} d\Omega - \overline{\psi}_0 \Omega_0 + \sum_{i=1}^{nb} \overline{\psi}_i \Omega_i \right] \qquad \forall \hat{\psi} \in \mathrm{Var}_{\psi}. \tag{8.81}$$

It is easy to appreciate that, if ψ solves the variational equation, then $\psi + C$ also does, where C is a constant. Therefore, putting $\overline{\psi}_0 = 0$ and dividing the equation by α, we get

$$\int_{\Omega} \frac{1}{\mu} \nabla_{\Omega} \Psi \cdot \nabla_{\Omega} \hat{\psi} d\Omega = 2 \left[\int_{\Omega} \hat{\psi} d\Omega + \sum_{i=1}^{nb} \overline{\psi}_i \Omega_i \right] \qquad \forall \hat{\psi} \in \mathrm{Var}_{\psi}, \tag{8.82}$$

where $\Psi = \frac{\psi}{\alpha}$. This is the final variational form for the torsion problem obtained from the PCVP.

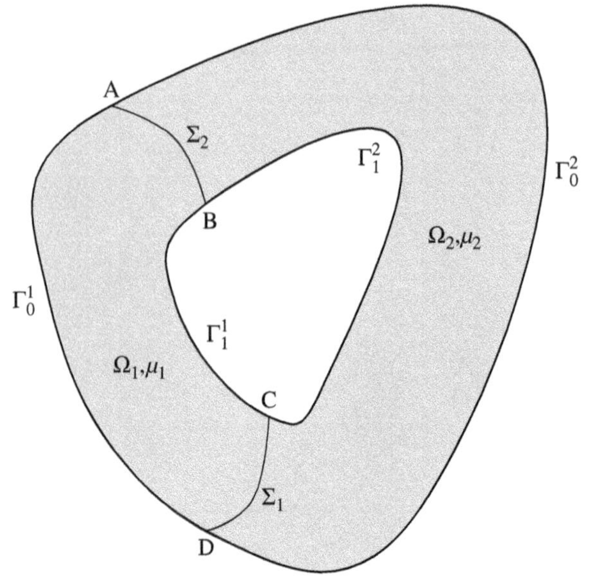

Figure 8.5 Torsion of a bar constituted by two different materials.

Now we derive the local (strong) form of the problem, that is the Euler–Lagrange equations associated with (8.82). This is done for the case of two materials with different properties. Consider the domain of analysis shown in Figure 8.5. When integrating by parts the left-hand side of (8.82), we obtain

$$\begin{aligned}\int_\Omega \frac{1}{\mu}\nabla_\Omega \Psi \cdot \nabla_\Omega \hat{\psi}\, d\Omega &= \int_\Omega \frac{1}{\mu}\mathrm{div}_\Omega(\hat{\psi}\nabla_\Omega \Psi)d\Omega - \int_\Omega \frac{1}{\mu}\hat{\psi}\triangle_\Omega \Psi d\Omega \\ &= \int_\Gamma \frac{1}{\mu}\hat{\psi}\nabla_\Omega \Psi \cdot \mathbf{n} d\Gamma - \int_\Omega \frac{1}{\mu}\hat{\psi}\triangle_\Omega \Psi d\Omega. \end{aligned} \tag{8.83}$$

Given that (see Figure 8.5)

$$\begin{aligned}\Gamma &= \Gamma_0 \cup \Gamma_1 = \Gamma_0^1 \cup \Gamma_0^2 \cup \Gamma_1^1 \cup \Gamma_1^2 \\ &= \Gamma_0^1 \cup \Gamma_1^1 \cup \Sigma_1^{DC} \cup \Sigma_2^{BA} \cup \Gamma_0^2 \cup \Gamma_1^2 \cup \Sigma_2^{AB} \cup \Sigma_1^{CD}.\end{aligned} \tag{8.84}$$

By virtue of the orientation of the boundary, we have that the integrals over Σ_1^{DC} and Σ_1^{CD} and over Σ_2^{AB} and Σ_2^{BA} cancel each other, respectively. Then, by doing

$$\Gamma^1 = \Gamma_0^1 \cup \Sigma_1^{DC} \cup \Gamma_1^1 \cup \Sigma_2^{BA}, \tag{8.85}$$

$$\Gamma^2 = \Gamma_0^2 \cup \Sigma_2^{AB} \cup \Gamma_1^2 \cup \Sigma_1^{CD}, \tag{8.86}$$

we obtain $\Gamma = \Gamma^1 \cup \Gamma^2$ and thus

$$\int_\Gamma \frac{1}{\mu}\hat{\psi}\nabla_\Omega \Psi \cdot \mathbf{n} d\Gamma - \int_\Omega \frac{1}{\mu}\hat{\psi}\triangle_\Omega \Psi d\Omega = \sum_{i=1}^{2}\left[\int_{\Gamma^i} \frac{1}{\mu_i}\hat{\psi}\nabla_\Omega \Psi \cdot \mathbf{n} d\Gamma - \int_{\Omega_i} \frac{1}{\mu_i}\hat{\psi}\triangle_\Omega \Psi d\Omega\right]. \tag{8.87}$$

Working with the right-hand side of equation (8.82), and remembering that $\hat{\psi}$ is constant in Γ and $\hat{\psi} = 0$ on Γ_0, we have

$$2\left[\int_\Omega \hat{\psi} d\Omega + \sum_{i=1}^{nb} \overline{\psi}_i \Omega_i\right] \quad \rightarrow \quad \sum_{i=1}^{2}\int_{\Omega_i} 2\hat{\psi}\, d\Omega - \int_{\Gamma_1} \hat{\psi}\mathbf{r}\cdot\mathbf{n}\, d\Gamma. \tag{8.88}$$

Equating expressions (8.87) and (8.88), taking into account that $\hat{\psi} = 0$ on Γ_0, $\hat{\psi} = C$ on Γ_1 and arranging terms, we finally have

$$\sum_{i=1}^{2} \left[\int_{\Omega_i} \left(\frac{1}{\mu_i} \triangle_\Omega \Psi + 2 \right) \hat{\psi} d\Omega \right] - \sum_{i=1}^{2} \int_{\Gamma^i} \frac{1}{\mu_i} \hat{\psi} \nabla_\Omega \Psi \cdot \mathbf{n} d\Gamma$$
$$- \int_{\Gamma_1} \hat{\psi} \mathbf{r} \cdot \mathbf{n} d\Gamma = 0 \quad \forall \hat{\psi} \in \mathrm{Var}_\psi, \tag{8.89}$$

and, if we make use of (8.85) and (8.86) to expand the boundary integrals this yields

$$\sum_{i=1}^{2} \int_{\Omega_i} \left(\frac{1}{\mu_i} \triangle_\Omega \Psi + 2 \right) \hat{\psi} d\Omega - \sum_{i=1}^{2} \int_{\Sigma_i} \left[\!\left[\frac{1}{\mu} \nabla_\Omega \Psi \right]\!\right] \cdot \mathbf{n} \hat{\psi} d\Gamma$$
$$- \int_{\Gamma_1} \left(\frac{1}{\mu} \nabla_\Omega \Psi + \mathbf{r} \right) \cdot \mathbf{n} d\Gamma \hat{\psi} = 0 \qquad \forall \hat{\psi} \in \mathrm{Var}_\psi. \tag{8.90}$$

The previous expression, together with the fundamental theorem of the calculus of variations, provides the local form of the dual torsion problem, which reads

$$\triangle_\Omega \Psi = -2\mu_i \qquad \text{in } \Omega_i, \tag{8.91}$$

$$\left[\!\left[\frac{1}{\mu} \nabla \Psi \right]\!\right] \cdot \mathbf{n} = 0 \qquad \text{on } \Sigma_i,\ i = 1, 2, \tag{8.92}$$

$$\int_{\Gamma_1} \left(\frac{1}{\mu} \nabla_\Omega \Psi + \mathbf{r} \right) \cdot \mathbf{n} d\Gamma = 0. \tag{8.93}$$

To further understand the mechanical meaning of expressions (8.92) and (8.93), we have to rewrite them in terms of the angular distortion γ and the warping function φ. Thus, we have

$$\left[\!\left[\frac{1}{\mu} \nabla \Psi \cdot \mathbf{n} \right]\!\right] = \left[\!\left[\frac{1}{\mu\alpha} \nabla \psi \cdot \mathbf{n} \right]\!\right] = \left[\!\left[\frac{1}{\mu\alpha} \begin{pmatrix} \partial\psi/\partial x \\ \partial\psi/\partial y \end{pmatrix} \cdot \begin{pmatrix} n_x \\ n_y \end{pmatrix} \right]\!\right]$$
$$= \left[\!\left[\frac{1}{\mu\alpha} \mu \begin{pmatrix} -\gamma_y \\ \gamma_x \end{pmatrix} \cdot \begin{pmatrix} n_x \\ n_y \end{pmatrix} \right]\!\right] = \left[\!\left[-\frac{1}{\alpha} \gamma \cdot \mathbf{t} \right]\!\right], \tag{8.94}$$

and, therefore, from (8.92) we conclude

$$[\![\gamma \cdot \mathbf{t}]\!] = 0 \qquad \text{on } \Sigma_i, \quad i = 1, 2. \tag{8.95}$$

Recalling that only the components $[\mathbf{D}]_{xz}$, $[\mathbf{D}]_{yz}$, $[\mathbf{D}]_{zx}$ and $[\mathbf{D}]_{zy}$ of the strain rate tensor are non-null, and applying this strain rate $\mathbf{D}$ over the tangent direction $\mathbf{t}$ we obtain

$$\mathbf{D}\mathbf{t} = \frac{1}{2} \begin{pmatrix} 0 \\ 0 \\ \gamma_x t_x + \gamma_y t_y \end{pmatrix}, \tag{8.96}$$

and, consequently

$$\mathbf{t} \cdot \mathbf{D}\mathbf{t} = 0, \tag{8.97}$$

$$\mathbf{n} \cdot \mathbf{D}\mathbf{t} = 0, \tag{8.98}$$

$$\mathbf{e}_z \cdot \mathbf{D}\mathbf{t} = \gamma \cdot \mathbf{t}. \tag{8.99}$$

Hence, from (8.95) we get

$$[\![\mathbf{e}_z \cdot \mathbf{Dt}]\!] = 0. \tag{8.100}$$

In this manner we arrive at three consequences. First, the fibers in the tangent direction $\mathbf{t}$ do not undergo deformation. Second, there is no angular distortion between the pair of vectors $(\mathbf{n}, \mathbf{t})$. Third, there is no jump in the angular distortion between $\mathbf{e}_z$ and $\mathbf{t}$.

In addition, expression (8.93) can be rewritten as follows

$$\begin{aligned}
\int_{\Gamma_1} \left(\frac{1}{\mu}\nabla_\Omega \Psi + \mathbf{r}\right) \cdot \mathbf{n} d\Gamma &= \int_{\Gamma_1} \left(\frac{1}{\mu\alpha}\nabla_\Omega \psi + \mathbf{r}\right) \cdot \mathbf{n} d\Gamma \\
&= \int_{\Gamma_1} \left[\frac{1}{\mu\alpha}\begin{pmatrix} -\mu(\partial\kappa/\partial y + \alpha x) \\ \mu(\partial\kappa/\partial x - \alpha y) \end{pmatrix} + \mathbf{r}\right] \cdot \mathbf{n} d\Gamma \\
&= \int_{\Gamma_1} \left[\begin{pmatrix} -\partial\varphi/\partial y \\ \partial\varphi/\partial x \end{pmatrix}\right] \cdot \mathbf{n} d\Gamma = \int_{\Gamma_1} -\nabla_\Omega \varphi \cdot \mathbf{t} d\Gamma = -\int_{\Gamma_1} d\varphi = 0,
\end{aligned} \tag{8.101}$$

and then

$$[\![\varphi]\!] = 0 \quad \text{on } \Gamma_1. \tag{8.102}$$

This equation is called the compatibility condition, and, mechanically, entails that it is not possible to have displacements in the direction of $\mathbf{e}_z$ anywhere over Γ_1.

Finally, let us see the expression of the torsional moment. From (8.41), we have that

$$\begin{aligned}
M_t &= \int_\Omega \sigma \cdot \chi d\Omega = \int_\Omega \begin{pmatrix} \partial\psi/\partial y \\ -\partial\psi/\partial x \end{pmatrix} \cdot \begin{pmatrix} -y \\ x \end{pmatrix} d\Omega = -\int_\Omega \alpha \nabla_\Omega \Psi \cdot \mathbf{r} d\Omega \\
&= -\alpha \left[\int_\Omega \text{div}_\Omega(\Psi\mathbf{r}) d\Omega - \int_\Omega 2\Psi d\Omega\right] = \alpha\left[\int_\Omega 2\Psi d\Omega - \int_\Gamma \Psi \mathbf{r} \cdot \mathbf{n} d\Gamma\right] \\
&= \alpha\left[\int_\Omega 2\Psi \, d\Omega + \sum_{i=1}^{nb} 2\Psi_i \Omega_i\right],
\end{aligned} \tag{8.103}$$

where Ψ_i is the value attained by Ψ on the boundary Γ_i, nb is the number of orifices in the domain, and Ω_i is the area limited by Γ_i. Defining $\frac{M_t}{\alpha}$ as the torsional stiffness of the bar, representing the torsional moment required to exert a one-radian angular rotation, we get

$$\frac{M_t}{\alpha} = 2\left[\int_\Omega \Psi d\Omega + \sum_{i=1}^{nb} \Psi_i \Omega_i\right]. \tag{8.104}$$

9

Plates and Shells

9.1 Introduction

As highlighted in the introduction to Part III, Roman architects built aqueducts to convey water from remote locations covering large distances over rivers and valleys through arches. This arrangement enabled the loads over the structure to be transferred to the foundations through a fully compressive stress state in the whole structure, which was particularly appropriate for the material available at the time. An example of this structure is the Pont du Gard in the south of France. This structure was built before the Christian age to allow the Nîmes aqueduct (approximately 50km in length) to cross the Gard river. The architects and hydraulic engineers who designed this aqueduct in the ancient Roman empire created an unprecedented piece of technical art.

In contemporaneous civilization (20th and 21st centuries) shells (considering beams and plates as particular cases) are thin (slender) structural components when compared to the gargantuan areas they are capable of covering thanks to their shape (curvature), and can be adapted to the different demands from civil, mechanical, nuclear, naval, and aerospace engineering, among others.

These slender structural components become marvelous artworks combining architectonic needs, light weight, and beauty. Many famous European and Latin American architects made use of these components in the last century and at the beginning of the present century. As examples of these pioneers we can mention the works of the Uruguayan architect Eladio Dieste, the works of the Spaniard architects Félix Candela (who developed his professional career in Mexico), Santiago Calatrava, and Eduardo Torroja, as well as the contributions made by the Italian architect Pier Luigi Nervi. A particularly outstanding structure is the Baha'i Lotus temple in New Delhi, designed by the Iranian architect Fariborz Sahba.

All these examples demonstrate the magnificence of these structures in civil construction. However, such components are also commonplace in industrial applications, for example cooling towers in nuclear plants. Similarly, in the mechanical engineering field, pressure vessels constitute a clear example, among many others, of the application of shell components.

These examples are but a small number within a myriad of applications of shells. At the same time, they demonstrate that shells have been ubiquitously present in human development. This is a fundamental motivation to present the theoretical bases for the variational formalism that provides the groundwork for the construction of shell models. In this sense, we will pursue a similar path to that employed in the presentation of the

Introduction to the Variational Formulation in Mechanics: Fundamentals and Applications, First Edition.
Edgardo O. Taroco, Pablo J. Blanco and Raúl A. Feijóo.

theory of the bending of beams and torsion of bars. This means that we will first address the kinematics associated with the motion actions, which ultimately characterizes the space Var_v, the strain rate action operator, which allows us to define the space $\mathscr{W}$ and, by using duality notions, we will proceed to determine the generalized forces, both internal and external, associated with the different theories. Finally, the concept of mechanical equilibrium between these generalized forces will be reached through the Principle of Virtual Power (PVP).

It is important to point out that the treatment given to this structural component in the present chapter, from a geometrical perspective, will be general in the sense that the curvature of the structure will always be considered in the theoretical developments. This is one of the main characteristics of shells. The case of plates is then intrinsically addressed as a particular case of shells featuring null curvature. The reader interested in plate modeling should remove the terms contributed by the curvature of the shell, easily finding the corresponding formulation for plates. Throughout the chapter this is left as an exercise for the reader.

9.2 Geometric Description

As briefly discussed in Section 9.1, a shell is a three-dimensional solid body that has one characteristic dimension, the shell thickness, much smaller than the other two. This property allows the mechanics of shells to be analyzed as a particular case of three-dimensional deformable bodies.

The (purely geometric) property that the thickness is smaller than the other two characteristic dimensions enables, when combined with proper simplifications underlying the kinematics of the shell, the analysis of shells to be reduced to that of surfaces, similarly to the reduction from three-dimensional bodies to analysis of lines seen in Chapter 7 (bending of beams) and Chapter 8 (torsion of bars). Therefore, the shell theory we will develop in this chapter aims to translate the three-dimensional conception of this class of structural component into a conception in which fields are exclusively defined over the representative surface of the shell. In this regard, Section 9.9 addresses basic notions of differential geometry applicable to surfaces defined in three-dimensional space.

The theory of shells was born with the landmark work of Love in 1888 [176, 177], and has been imprinted by remarkable contributions provided by the Russian, Dutch, German, and American schools, among others. We can, for instance, mention the analogy established between statics and geometry by Gol'denveizer [123] and Lur'e [179], and its applications by means of complex variable analysis; the introduction of symmetric stresses by Sanders [271] and Leonard [173], the theory of plane stress states by Koiter [160] and its comparison with the theory proposed by Love; the method of asymptotic expansions by Gol'denveizer; or the theory of Green and Naghdi based on Cosserat surfaces, just to mention a few.

The development of these theories has always been distinguished by an elevated degree of complexity. Specifically, complications inherent to the geometrical framework posed by surfaces are implied in the conception of physical fields which are defined in nonlinear manifolds, whose adequate treatment is found within the realm of differential geometry. Another source of intricacy (this also occurs in the analysis of beams) is related to the constitutive equation of the material that composes the shell.

As mentioned before, this chapter presents the basic principles and concepts underlying different classical shell theories. As in previous chapters in this part of the book, the variational approach provides a mechanical framework that holds regardless of the material behavior of the component. This will be accomplished by making use of an intrinsic system of coordinates defined over the shell, which is independent from the specific system of coordinates adopted when performing numerical calculations. By proceeding this way, and framed by the variational formulation, it is possible to develop the kinematical concepts, unequivocally describe the generalized (internal and external) forces, and define the corresponding mechanical equilibrium in a compact, unambiguous, and elegant manner, thus avoiding that the complexity brought by differential calculus and by the specificities of the material behavior which can obscure the conceptual bases of the different shell models.

Then, based on previous considerations, we observe that the current configuration $\mathscr{B}_t$ of the shell, hereafter simply denoted by Ω, is placed in the three-dimensional Euclidean space $\mathscr{E}$, and can be understood as being generated by a line segment of variable (small) length that spans a certain region of space, and whose mid point defines a surface, say Σ_o, to which the line segment is orthogonal. The length of such a line segment is called the thickness of the shell, and Σ_o is called the middle surface of the shell. Thus, if $\mathbf{x}_o \in \Sigma_o$ is an arbitrary point in the middle surface and $\mathbf{n} = \mathbf{n}(\mathbf{x}_o)$ is the unit normal vector to this surface at point $\mathbf{x}_o$, we can formalize this description as follows

$$\Omega = \{\mathbf{x} \in \mathscr{E};\ \mathbf{x} = \mathbf{x}_o + \xi\mathbf{n},\ \mathbf{x}_o \in \Sigma_o,\ \xi \in H\}, \tag{9.1}$$

where $H = (-\frac{h(\mathbf{x}_o)}{2}, \frac{h(\mathbf{x}_o)}{2})$ and $h(\mathbf{x}_o)$ is the shell thickness at point $\mathbf{x}_o$. For the application $\mathbf{x} \in \Omega \leftrightarrow (\mathbf{x}_o, \xi)$ to be well-posed[1] we are going to assume that the following constraints are verified

- $\mathbf{n}(\mathbf{x}_o)$ is uniquely defined for any point $\mathbf{x}_o \in \Sigma_o$. That means there are no points (or curves) over Σ_o in which the normal is not defined. In other words, we are postulating that the middle surface is smooth in the sense that the normal vector and the tangent plane (orthogonal to the normal vector) are univocally defined for each point over the middle surface. As stated before, the case of plates constitutes a special situation of shells in which the middle surface Σ_o is simply characterized by a plane. Then, for plates, the vector $\mathbf{n}$ is the same (a single normal vector) whatever the point $\mathbf{x}_o \in \Sigma_o$.
- $|\xi| \leq \frac{h(\mathbf{x}_o)}{2} \leq R_m$ where R_m is the minimum radius of curvature of the shell at point $\mathbf{x}_o$.

Taking into consideration the description established by (9.1), the domain of the shell Ω is limited by the superior surface, Σ^+, and the inferior surface, Σ^-, which are, respectively, defined by

$$\Sigma^+ = \left\{\mathbf{x} \in \mathscr{E};\ \mathbf{x} = \mathbf{x}_o + \frac{h(\mathbf{x}_o)}{2}\mathbf{n},\ \mathbf{x}_o \in \Sigma_o\right\}, \tag{9.2}$$

$$\Sigma^- = \left\{\mathbf{x} \in \mathscr{E};\ \mathbf{x} = \mathbf{x}_o - \frac{h(\mathbf{x}_o)}{2}\mathbf{n},\ \mathbf{x}_o \in \Sigma_o\right\}, \tag{9.3}$$

1 This implies that, for a given $\mathbf{x} \in \Omega$, there is a unique pair $(\mathbf{x}_o, \xi)$, and vice versa.

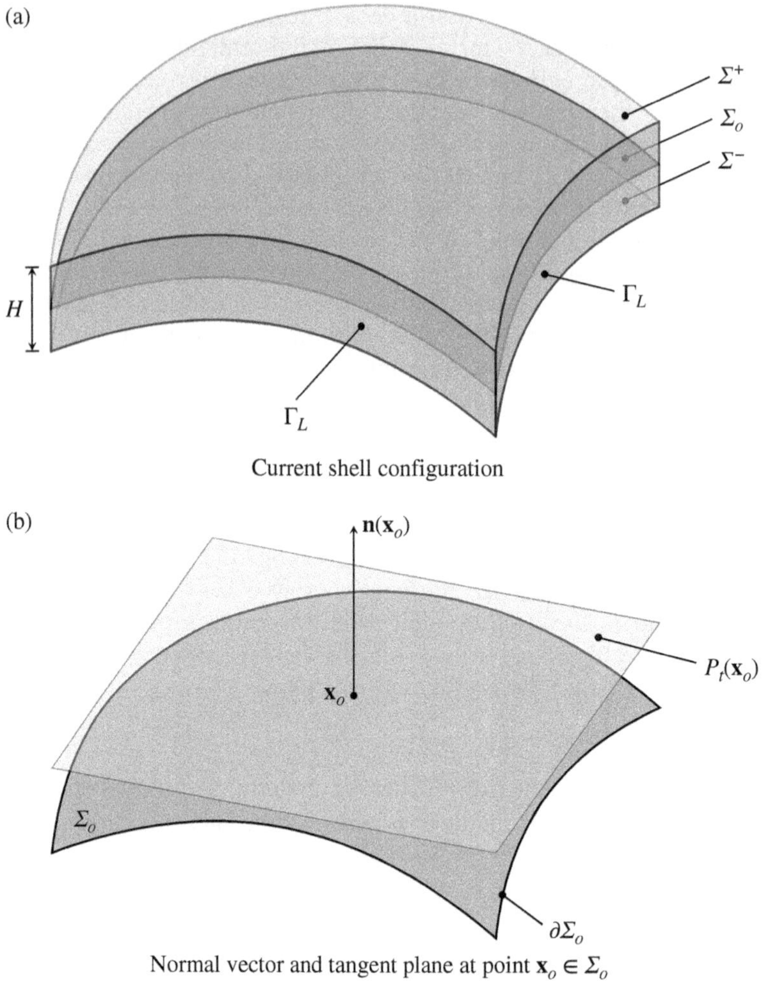

Figure 9.1 Elements for the geometric description of a shell.

and by the lateral surface of the shell, Γ_L, generated by the line segment that goes through all the points $\mathbf{x}_o$ placed over the boundary of Σ_o, that is, when $\mathbf{x}_o$ is located over $\partial\Sigma_o$. This lateral surface is defined by

$$\Gamma_L = \{\mathbf{x} \in \mathscr{E};\ \mathbf{x} = \mathbf{x}_o + \xi\mathbf{n},\ \mathbf{x}_o \in \partial\Sigma_o,\ \xi \in H\}. \tag{9.4}$$

Figure 9.1 shows the elements that participate in the geometric description of the current shell configuration.

Our main goal in this chapter is to provide a theoretical formalism by employing a compact notation that is independent of the specific system of coordinates adopted at the time of the calculations. Hence, we observe that the middle surface of the shell has an intrinsic system of coordinates which meets our requirements. Given the conditions we have placed on the shell geometry, for each point $\mathbf{x}_o \in \Sigma_o$ we have the intrinsic system denoted by $\{\mathbf{n}(\mathbf{x}_o), P_t(\mathbf{x}_o)\}$, which is given by the normal vector $\mathbf{n} = \mathbf{n}(\mathbf{x}_o)$ and by the tangent plane $P_t(\mathbf{x}_o)$, which is orthogonal to $\mathbf{n}$. Thus, given $\mathbf{x}_o \in \Sigma_o$, the tangent plane to the middle surface at that point is defined by

$$P_t(\mathbf{x}_o) = \{\mathbf{y} \in \mathscr{E}; (\mathbf{y} - \mathbf{x}_o) \cdot \mathbf{n}(\mathbf{x}_o) = 0\}. \tag{9.5}$$

This plane changes continuously across the middle surface of the shell (because of the postulated regularity), and in the case of plates it turns into a unique tangent plane, which is the very middle surface.

With these definitions, we see that a vector, say $\mathbf{v}$, can be decomposed as the sum of its projection over the tangent plane and its projection over the normal vector. That is

$$\mathbf{v} = \mathbf{v}_t + \mathbf{v}_n = \mathbf{v}_t + (\mathbf{v} \cdot \mathbf{n})\mathbf{n}, \tag{9.6}$$

from where

$$\mathbf{v}_t = \mathbf{v} - (\mathbf{v} \cdot \mathbf{n})\mathbf{n} = \mathbf{v} - (\mathbf{n} \otimes \mathbf{n})\mathbf{v} = [\mathbf{I} - (\mathbf{n} \otimes \mathbf{n})]\mathbf{v} = \Pi_t\mathbf{v}. \tag{9.7}$$

This expression defines the orthogonal projection operator over the tangent plane P_t, which is given by

$$\Pi_t = \mathbf{I} - (\mathbf{n} \otimes \mathbf{n}), \tag{9.8}$$

and that evidently results in a symmetric second-order tensor. With this definition, we can write the identity second-order tensor as follows

$$\mathbf{I} = \Pi_t(\mathbf{x}_o) + \mathbf{n}(\mathbf{x}_o) \otimes \mathbf{n}(\mathbf{x}_o). \tag{9.9}$$

Then, for an arbitrary vector $\mathbf{v}$ we obtain the following decomposition in the intrinsic basis

$$\mathbf{v} = \mathbf{I}\mathbf{v} = \Pi_t(\mathbf{x}_o)\mathbf{v} + (\mathbf{n}(\mathbf{x}_o) \otimes \mathbf{n}(\mathbf{x}_o))\mathbf{v} = \mathbf{v}_t + (\mathbf{v} \cdot \mathbf{n})\mathbf{n} = \mathbf{v}_t + v_n\mathbf{n} = \mathbf{v}_t + \mathbf{v}_n, \tag{9.10}$$

where $\mathbf{v}_t$ is a vector lying over the tangent plane and v_n is the component of $\mathbf{v}$ in the direction of $\mathbf{n}$.

In turn, given a second-order tensor, say $\mathbf{S}$, the decomposition in this intrinsic basis is easily derived as follows

$$\begin{aligned}
\mathbf{S} &= \mathbf{I}\mathbf{S}\mathbf{I} = (\Pi_t + \mathbf{n} \otimes \mathbf{n})\mathbf{S}(\Pi_t + \mathbf{n} \otimes \mathbf{n}) \\
&= \Pi_t\mathbf{S}\Pi_t + \Pi_t\mathbf{S}\mathbf{n} \otimes \mathbf{n} + \mathbf{n} \otimes \Pi_t\mathbf{S}^T\mathbf{n} + (\mathbf{n} \cdot \mathbf{S}\mathbf{n})(\mathbf{n} \otimes \mathbf{n}) \\
&= \mathbf{S}_t + \mathbf{S}_s \otimes \mathbf{n} + \mathbf{n} \otimes \mathbf{S}_s^* + S_n(\mathbf{n} \otimes \mathbf{n}),
\end{aligned} \tag{9.11}$$

where we have removed the functional dependence on $\mathbf{x}_o$ for notational simplicity.

In the previous decomposition we can see that $\mathbf{S}_t = \Pi_t\mathbf{S}\Pi_t$ is a second-order tensor that transforms vectors in the tangent plane into vectors in the same tangent plane, and vectors parallel to the normal $\mathbf{n}$ into the null vector. As for $\mathbf{S}_s = \Pi_t\mathbf{S}\mathbf{n}$ and $\mathbf{S}_s^* = \Pi_t\mathbf{S}^T\mathbf{n}$, these are vectors in the tangent plane, which are the same if $\mathbf{S}$ is symmetric. Moreover, the tensor $\mathbf{S}_s \otimes \mathbf{n}$ transforms vectors in the normal direction into vectors over the tangent plane, while vectors in this tangent plane are mapped into the null vector. Similarly, the tensor $\mathbf{n} \otimes \mathbf{S}_s^*$ transforms vectors in the tangent plane into vectors pointing in the normal direction, such that normal vectors to the middle surface are mapped into the null vector. Finally, tensor $S_n(\mathbf{n} \otimes \mathbf{n})$ transforms normal vectors into normal vectors, while vectors in the tangent plane are mapped into the null vector.

In this manner, as expected, this intrinsic and orthogonal system of reference enables the decomposition of vectors and second-order tensors into components which are mutually orthogonal. Hence, the inner products between two arbitrary vectors, say **u** and **v**, and between two arbitrary tensors, say **R** and **S**, take, respectively, the following forms

$$\mathbf{u} \cdot \mathbf{v} = \mathbf{u}_t \cdot \mathbf{v}_t + u_n v_n, \tag{9.12}$$

$$\mathbf{R} \cdot \mathbf{S} = \mathbf{R}_t \cdot \mathbf{S}_t + \mathbf{R}_s \cdot \mathbf{S}_s + \mathbf{R}_s^* \cdot \mathbf{S}_s^* + R_n\, S_n. \tag{9.13}$$

Furthermore, the application of a second-order tensor **S** over a vector **v** gives

$$\mathbf{S}\mathbf{v} = (\mathbf{S}\mathbf{v})_t + (\mathbf{S}\mathbf{v})_n \mathbf{n} = \mathbf{S}_t \mathbf{v}_t + v_n \mathbf{S}_s + ((\mathbf{v}_t \cdot \mathbf{S}_s^*) + v_n S_n)\mathbf{n}. \tag{9.14}$$

9.3 Differentiation and Integration

According to the geometric concepts presented in the previous section, we have that an arbitrary point in the shell, say $\mathbf{x} \in \Omega$, is univocally related to the pair $(\mathbf{x}_o, \xi)$ through the mapping

$$\mathbf{x} = \mathbf{x}_o + \xi \mathbf{n}(\mathbf{x}_o), \tag{9.15}$$

where $\mathbf{x}_o$ is the orthogonal projection (minimum distance) of $\mathbf{x}$ over the middle surface of the shell Σ_o. With this transformation, any field (scalar, vector, second-order tensor, etc.) defined in Ω can be transformed into a field defined in $\Sigma_o \times H$. This, together with suitable kinematical hypotheses, will allow us to transform the integrals in Ω and over $\partial\Omega$, which appear in the PVP, into integrals in Σ_o and over its boundary $\partial\Sigma_o$. In other words, from a variational problem defined in a three-dimensional domain we will switch to a variational problem written in a surface. This procedure is entirely similar to the problem involving beams, when reducing the three-dimensional to a one-dimensional set.

According to the transformation (9.15), we have that an infinitesimal fiber $d\mathbf{x}$ in Ω can be expressed as

$$\begin{aligned} d\mathbf{x} &= d\mathbf{x}_o + \xi d\mathbf{n}(\mathbf{x}_o) + d\xi \mathbf{n}(\mathbf{x}_o) = d\mathbf{x}_o + \xi \nabla_{\mathbf{x}_o} \mathbf{n}\, d\mathbf{x}_o + d\xi \mathbf{n} \\ &= (\mathbf{I} + \xi \nabla_{\mathbf{x}_o} \mathbf{n}(\mathbf{x}_o)) d\mathbf{x}_o + d\xi \mathbf{n} = \Lambda(\mathbf{x}_o, \xi) d\mathbf{x}_o + d\xi \mathbf{n}, \end{aligned} \tag{9.16}$$

where

$$\Lambda(\mathbf{x}_o, \xi) = \mathbf{I} + \xi \nabla_{\mathbf{x}_o} \mathbf{n}(\mathbf{x}_o). \tag{9.17}$$

Expression (9.17) defines the operator denoted by Λ in which $\nabla_{\mathbf{x}_o}(\cdot)$ represents the gradient of $(\cdot)$ with respect to coordinates $\mathbf{x}_o$ defined over the tangent plane P_t to the middle surface Σ_o at point $\mathbf{x}_o$. Likewise, $\nabla_{\mathbf{x}_o}\mathbf{n}$ is the middle surface curvature tensor which informs us about the change of the tangent plane in the neighborhood of $\mathbf{x}_o$. Here it is important to note that Λ is a symmetric tensor ($\Lambda = \Lambda^T$), and that it is invertible, that is there exists Λ^{-1} such that $\Lambda\Lambda^{-1} = \mathbf{I}$, by virtue of the biunivocal correspondence between $\mathbf{x} \in \Omega$ and the pair $(\mathbf{x}_o, \xi)$. Moreover, it is straightforward to observe that

$$\Pi_t(\mathbf{x}_o) d\mathbf{x} = \Lambda(\mathbf{x}_o, \xi) d\mathbf{x}_o, \tag{9.18}$$

$$d\mathbf{x} \cdot \mathbf{n}(\mathbf{x}_o) = d\xi. \tag{9.19}$$

In the case of plates, the reader will readily note that operator Λ reduces to the identity tensor **I**, and becomes independent from $(\mathbf{x}_o, \xi)$.

Consider a scalar field v defined in Ω. Then, thanks to the mapping (9.15), this field can be understood as a field defined in $\Sigma_o \times H$, that is

$$v(\mathbf{x}) = \tilde{v}(\mathbf{x}_o, \xi). \tag{9.20}$$

The differentiation of this field yields

$$dv = \nabla v \cdot d\mathbf{x} = \nabla_{\mathbf{x}_o} \tilde{v} \cdot d\mathbf{x}_o + \frac{\partial \tilde{v}}{\partial \xi} d\xi. \tag{9.21}$$

In turn, making use of expression (9.16) we have

$$\nabla v \cdot d\mathbf{x} = \nabla v \cdot \Lambda d\mathbf{x}_o + \nabla v \cdot \mathbf{n} d\xi. \tag{9.22}$$

Comparing (9.21) and (9.22) and exploiting the symmetry of operator Λ leads us to the identities

$$\Lambda \nabla v = \nabla_{\mathbf{x}_o} \tilde{v}, \tag{9.23}$$

$$\nabla v \cdot \mathbf{n} = \frac{\partial \tilde{v}}{\partial \xi}. \tag{9.24}$$

Therefore, the decomposition of the vector ∇v into tangent and normal components

$$\nabla v = (\nabla v)_t + (\nabla v)_n \mathbf{n}, \tag{9.25}$$

gives

$$(\nabla v)_t = \Pi_t \nabla v = \Pi_t \Lambda^{-1} \nabla_{\mathbf{x}_o} \tilde{v}, \tag{9.26}$$

$$(\nabla v)_n = \nabla v \cdot \mathbf{n} = \frac{\partial \tilde{v}}{\partial \xi}. \tag{9.27}$$

Let us see now the differentiation of a vector field $\mathbf{v}$ originally defined in Ω and posteriorly understood as a field $\tilde{\mathbf{v}}$ defined in $\Sigma_o \times H$, that is

$$\mathbf{v}(\mathbf{x}) = \tilde{\mathbf{v}}(\mathbf{x}_o, \xi). \tag{9.28}$$

Analogously to the previous development, we find

$$d\mathbf{v} = \nabla \mathbf{v} d\mathbf{x} = \nabla_{\mathbf{x}_o} \tilde{\mathbf{v}}\, d\mathbf{x}_o + \frac{\partial \tilde{\mathbf{v}}}{\partial \xi} d\xi. \tag{9.29}$$

By recalling (9.16), the previous expression takes the form

$$\nabla \mathbf{v} d\mathbf{x} = (\nabla_{\mathbf{x}_o} \tilde{\mathbf{v}}) \Lambda^{-1} \Pi_t d\mathbf{x} + \frac{\partial \tilde{\mathbf{v}}}{\partial \xi} \mathbf{n} \cdot d\mathbf{x} = \left((\nabla_{\mathbf{x}_o} \tilde{\mathbf{v}}) \Lambda^{-1} \Pi_t + \frac{\partial \tilde{\mathbf{v}}}{\partial \xi} \otimes \mathbf{n} \right) d\mathbf{x}, \tag{9.30}$$

which renders

$$\nabla \mathbf{v} = (\nabla_{\mathbf{x}_o} \tilde{\mathbf{v}}) \Lambda^{-1} \Pi_t + \frac{\partial \tilde{\mathbf{v}}}{\partial \xi} \otimes \mathbf{n}. \tag{9.31}$$

Relation (9.31) provides a connection between the gradient with respect to coordinates $\mathbf{x} \in \Omega$ of an arbitrary vector field $\mathbf{v}$ defined in Ω and the gradient with respect to coordinates $(\mathbf{x}_o, \xi)$ of the same field, but now understood as a field defined in $\Sigma_o \times H$ and denoted by $\tilde{\mathbf{v}}$. Resorting to decomposition (9.10), when the gradient operator is applied we obtain

$$\nabla \mathbf{v} = \nabla \mathbf{v}_t + v_n \nabla \mathbf{n} + \mathbf{n} \otimes \nabla v_n. \tag{9.32}$$

For the fields $\mathbf{v}_t$ and $\mathbf{n}$ we can then apply the result obtained in (9.31), and for the gradient of the scalar field v_n, ∇v_n, we utilize the relations obtained in (9.26) and (9.27). Thus, we have

$$\nabla\mathbf{v} = (\nabla_{\mathbf{x}_o}\tilde{\mathbf{v}}_t)\Lambda^{-1}\Pi_t + \frac{\partial\tilde{\mathbf{v}}_t}{\partial\xi}\otimes\mathbf{n} + \tilde{v}_n(\nabla_{\mathbf{x}_o}\mathbf{n})\Lambda^{-1}\Pi_t + \mathbf{n}\otimes\Pi_t\Lambda^{-1}\nabla_{\mathbf{x}_o}\tilde{v}_n + \frac{\partial\tilde{v}_n}{\partial\xi}\mathbf{n}\otimes\mathbf{n}. \tag{9.33}$$

In the previous expression all the terms on the right-hand side, except for the first, are written in the intrinsic system of coordinates. It is only required now to decompose the first term in order to have the full decomposition of tensor $\nabla\mathbf{v}$. This is achieved as follows

$$(\Pi_t + \mathbf{n}\otimes\mathbf{n})(\nabla_{\mathbf{x}_o}\tilde{\mathbf{v}}_t)\Lambda^{-1}\Pi_t(\Pi_t + \mathbf{n}\otimes\mathbf{n}) = \Pi_t(\nabla_{\mathbf{x}_o}\tilde{\mathbf{v}}_t)\Lambda^{-1}\Pi_t + \mathbf{n}\otimes\Pi_t^T\Lambda^{-1}(\nabla_{\mathbf{x}_o}\tilde{\mathbf{v}}_t)^T\mathbf{n}. \tag{9.34}$$

From the orthogonality between $\mathbf{n}$ and $\tilde{\mathbf{v}}_t$, the term $(\nabla_{\mathbf{x}_o}\tilde{\mathbf{v}}_t)^T$ can be rewritten in a more convenient manner. In effect, since $\mathbf{n}\cdot\tilde{\mathbf{v}}_t = 0$, for any point $\mathbf{x}_o$ in Σ_o, we get

$$\nabla_{\mathbf{x}_o}(\mathbf{n}\cdot\tilde{\mathbf{v}}_t) = (\nabla_{\mathbf{x}_o}\mathbf{n})\tilde{\mathbf{v}}_t + (\nabla_{\mathbf{x}_o}\tilde{\mathbf{v}}_t)^T\mathbf{n} = \mathbf{0} \quad\Rightarrow\quad (\nabla_{\mathbf{x}_o}\tilde{\mathbf{v}}_t)^T\mathbf{n} = -(\nabla_{\mathbf{x}_o}\mathbf{n})\tilde{\mathbf{v}}_t. \tag{9.35}$$

Replacing (9.35) into (9.34) we get

$$(\nabla_{\mathbf{x}_o}\tilde{\mathbf{v}}_t)\Lambda^{-1}\Pi_t = \Pi_t(\nabla_{\mathbf{x}_o}\tilde{\mathbf{v}}_t)\Lambda^{-1}\Pi_t - \mathbf{n}\otimes\Pi_t^T\Lambda^{-1}(\nabla_{\mathbf{x}_o}\mathbf{n})\tilde{\mathbf{v}}_t. \tag{9.36}$$

From (9.36) and using (9.33) we finally attain the required expression

$$\begin{aligned}\nabla\mathbf{v} = {}& \Pi_t(\nabla_{\mathbf{x}_o}\tilde{\mathbf{v}}_t)\Lambda^{-1}\Pi_t + \tilde{v}_n(\nabla_{\mathbf{x}_o}\mathbf{n})\Lambda^{-1}\Pi_t + \frac{\partial\tilde{\mathbf{v}}_t}{\partial\xi}\otimes\mathbf{n}\\ &+ \mathbf{n}\otimes\Pi_t\Lambda^{-1}(\nabla_{\mathbf{x}_o}\tilde{v}_n - (\nabla_{\mathbf{x}_o}\mathbf{n})\tilde{\mathbf{v}}_t) + \frac{\partial\tilde{v}_n}{\partial\xi}\mathbf{n}\otimes\mathbf{n},\end{aligned} \tag{9.37}$$

where we can distinguish the components of $\nabla\mathbf{v}$ in the intrinsic basis. In particular,

$$(\nabla\mathbf{v})_t = \Pi_t(\nabla_{\mathbf{x}_o}\tilde{\mathbf{v}}_t)\Lambda^{-1}\Pi_t + \tilde{v}_n(\nabla_{\mathbf{x}_o}\mathbf{n})\Lambda^{-1}\Pi_t, \tag{9.38}$$

$$(\nabla\mathbf{v})_s = \frac{\partial\tilde{\mathbf{v}}_t}{\partial\xi}, \tag{9.39}$$

$$(\nabla\mathbf{v})_s^* = \Pi_t\Lambda^{-1}(\nabla_{\mathbf{x}_o}\tilde{v}_n - (\nabla_{\mathbf{x}_o}\mathbf{n})\tilde{\mathbf{v}}_t), \tag{9.40}$$

$$(\nabla\mathbf{v})_n = \frac{\partial\tilde{v}_n}{\partial\xi}. \tag{9.41}$$

Exercise 9.1 Derive expressions (9.38)–(9.41) for the particular case of plates.

So far we have seen how the geometric description of the shell, by using the mapping (9.15), molds the expressions of the gradients of scalar and vector fields. Now let us see how this description shapes the expressions of integrals defined in Ω and of integrals over the boundaries Σ^+, Σ^- and Γ_L. With this in mind, let φ and $\mathbf{w}$ be arbitrary scalar and vector fields, respectively. Then

$$\int_\Omega \varphi\, d\Omega = \int_{\Sigma_o}\int_H \varphi \det\Lambda d\xi d\Sigma_o, \tag{9.42}$$

$$\int_{\Sigma^+} \varphi\, d\Sigma^+ = \int_{\Sigma_o} (\varphi\det\Lambda)|_{\xi=\frac{h}{2}} d\Sigma_o, \tag{9.43}$$

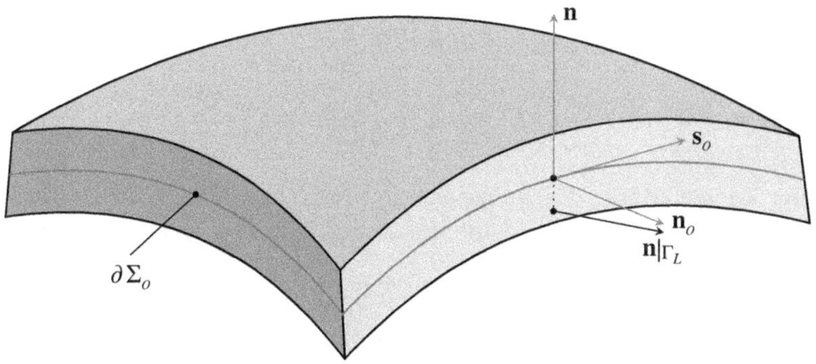

Figure 9.2 Geometric description of the normal vectors to $\partial\Sigma_o$ and Γ_L.

$$\int_{\Sigma^-} \varphi \, d\Sigma^- = \int_{\Sigma_o} (\varphi \det \Lambda)|_{\xi=-\frac{h}{2}} d\Sigma_o, \tag{9.44}$$

$$\int_{\Gamma_L} \mathbf{w} \cdot \mathbf{n}|_{\Gamma} \, d\Gamma_L = \int_{\partial\Sigma_o} \int_H \mathbf{w} \cdot \Lambda^{-1}\mathbf{n}_o \det \Lambda d\xi d\partial\Sigma_o, \tag{9.45}$$

where $\mathbf{n}|_{\Gamma}$ and $\mathbf{n}_o$ are, respectively, the unit outward vectors to Γ_L and to $\partial\Sigma_o$ at point $\mathbf{x}_o$, as seen in Figure 9.2.

Exercise 9.2 Obtain expressions (9.42)–(9.45) for the particular case of plates.

The roadmap to be followed from this point onwards comprises the use of the concept of mathematical duality and the PVP. These concepts involve integrals that will be expressed through the transformations indicated in (9.42)–(9.45), which in combination with expressions such as those derived in (9.38)–(9.41) to describe the tensor $\nabla\mathbf{v}$, will provide a complete characterization of the generalized internal stresses and generalized external forces which emerge when employing such geometric description. For ease of notation, in what follows we will remove the $\tilde{(\cdot)}$ in the representation of the field $(\cdot)$ when defined in terms of $\Sigma_o \times H$.

9.4 Principle of Virtual Power

A shell, defined in the current configuration Ω, can be seen as a three-dimensional body. For this body, we define the space of kinematically admissible motion actions Var_v as follows

$$\text{Var}_v = \{\mathbf{v} \in \mathscr{V} \, ; \, \mathbf{v}|_{\partial\Omega_\mathbf{u}} = \mathbf{0}\}, \tag{9.46}$$

where $\mathscr{V}$ is a space of sufficiently regular vector functions defined in Ω, and $\partial\Omega_\mathbf{u}$ is the part of the boundary $\partial\Omega$ where displacements are prescribed. To simplify the presentation, let us assume that this body (the shell) in its current configuration Ω is subjected to forces per unit volume $\mathbf{b}$, defined in Ω, and to forces per unit area $\mathbf{f}$ defined in $\partial\Omega_\mathbf{f}$, with $\partial\Omega_\mathbf{f}$ and $\partial\Omega_\mathbf{u}$ being complementary with respect to $\partial\Omega$, that is $\partial\Omega = \partial\Omega_\mathbf{u} \cup \partial\Omega_\mathbf{f}$ and

$\partial\Omega_{\mathbf{u}} \cap \partial\Omega_{\mathbf{f}} = \emptyset$. We have seen in Chapter 3 that the mechanical equilibrium is introduced by the PVP through the variational equation

$$\int_{\Omega} \mathbf{T} \cdot (\nabla \mathbf{v})^s d\Omega = \int_{\Omega} \mathbf{b} \cdot \mathbf{v} d\Omega + \int_{\partial\Omega_{\mathbf{f}}} \mathbf{f} \cdot \mathbf{v} d\partial\Omega_{\mathbf{f}} \qquad \forall \mathbf{v} \in \mathrm{Var}_v. \tag{9.47}$$

Since in this chapter we will address the variational formulation for shell components always in the spatial configuration, the gradient operator of field $(\cdot)$ with respect to the spatial coordinates $\mathbf{x}$ is denoted by $\nabla(\cdot)$.

In what follows we will rewrite the PVP (9.47) exploiting the geometric description of the shell, namely introducing differentiation and integration in terms of intrinsic coordinates and decomposing vector and tensor fields into the components defined in the intrinsic system of reference. Thus, the space of kinematically admissible virtual motion actions becomes

$$\mathrm{Var}_v = \{(\mathbf{v}_t, v_n) \in \mathscr{V}_o;\ (\mathbf{v}_t, v_n)|_{\partial\Sigma^+_{\mathbf{u}} \cup \partial\Sigma^-_{\mathbf{u}} \cup \Gamma_{L\mathbf{u}}} = (\mathbf{0}, 0)\}, \tag{9.48}$$

where $\mathscr{V}_o$ is a space of pair functions $(\mathbf{v}_t, v_n)$ sufficiently regular in $\Sigma_o \times H$, and $\partial\Sigma^+_{\mathbf{u}}, \partial\Sigma^-_{\mathbf{u}}$ and $\Gamma_{L\mathbf{u}}$ are the parts of the corresponding boundaries $\partial\Sigma^+, \partial\Sigma^-$ and Γ_L where motion actions are prescribed. In addition, on the corresponding complementary parts $\partial\Sigma^+_{\mathbf{f}}$, $\partial\Sigma^-_{\mathbf{f}}$ and $\Gamma_{L\mathbf{f}}$ forces per unit area are exerting some action. Now, from (9.38)–(9.41), we conclude that the virtual strain (rate) action is given by

$$(\nabla \mathbf{v})^s = (\nabla \mathbf{v})^s_t + (\nabla \mathbf{v})^s_s \otimes \mathbf{n} + \mathbf{n} \otimes (\nabla \mathbf{v})^s_s + (\nabla \mathbf{v})^s_n \mathbf{n} \otimes \mathbf{n}, \tag{9.49}$$

where the components are defined through

$$(\nabla \mathbf{v})^s_t = \left(\Pi_t(\nabla_{\mathbf{x}_o} \mathbf{v}_t)\, \Lambda^{-1}\Pi_t + v_n(\nabla_{\mathbf{x}_o} \mathbf{n})\, \Lambda^{-1}\Pi_t\right)^s, \tag{9.50}$$

$$(\nabla \mathbf{v})^s_s = \frac{1}{2}\left(\frac{\partial \mathbf{v}_t}{\partial \xi} + \Pi_t \Lambda^{-1}(\nabla_{\mathbf{x}_o} v_n - (\nabla_{\mathbf{x}_o} \mathbf{n})\, \mathbf{v}_t)\right), \tag{9.51}$$

$$(\nabla \mathbf{v})^s_n = \frac{\partial v_n}{\partial \xi}. \tag{9.52}$$

Combining (9.13) with (9.42) and (9.50)–(9.52), the left-hand side in the PVP (9.47) (internal virtual power) takes the form

$$\int_{\Omega} \mathbf{T} \cdot (\nabla \mathbf{v})^s d\Omega = \int_{\Sigma_o}\int_H \Bigg[\mathbf{T}_t \cdot \left((\Pi_t(\nabla_{\mathbf{x}_o} \mathbf{v}_t) + v_n \nabla_{\mathbf{x}_o} \mathbf{n})\Lambda^{-1}\Pi_t\right)^s + \mathbf{T}_s \cdot \left(\frac{\partial \mathbf{v}_t}{\partial \xi} - \Pi_t \Lambda^{-1}((\nabla_{\mathbf{x}_o} \mathbf{n})\mathbf{v}_t - \nabla_{\mathbf{x}_o} v_n)\right) + T_n \frac{\partial v_n}{\partial \xi}\Bigg] \det \Lambda\, d\xi\, d\Sigma_o, \tag{9.53}$$

where the tensor field $\mathbf{T}_t$, the vector field $\mathbf{T}_s$, and the scalar field T_n entirely characterize the components (see (9.11)) of the Cauchy stress tensor $\mathbf{T}$ in the intrinsic system of coordinates defined over the shell, as shown in Figure 9.3.

Analogously, we decompose the forces per unit volume $\mathbf{b} = \mathbf{b}_t + b_n\mathbf{n}$ and the forces per unit area $\mathbf{f}^+ = \mathbf{f}^+_t + f^+_n\mathbf{n}$, $\mathbf{f}^- = \mathbf{f}^-_t + f^-_n\mathbf{n}$ and $\bar{\mathbf{f}} = \bar{\mathbf{f}}_t + \bar{f}_n\mathbf{n}$ (forces applied on $\Sigma^+_{\mathbf{f}}$, $\Sigma^-_{\mathbf{f}}$ and $\Gamma_{L\mathbf{f}}$, respectively) into the components of the intrinsic system of reference. In this manner, and utilizing expressions (9.12) and (9.43)–(9.45), the right-hand side in the

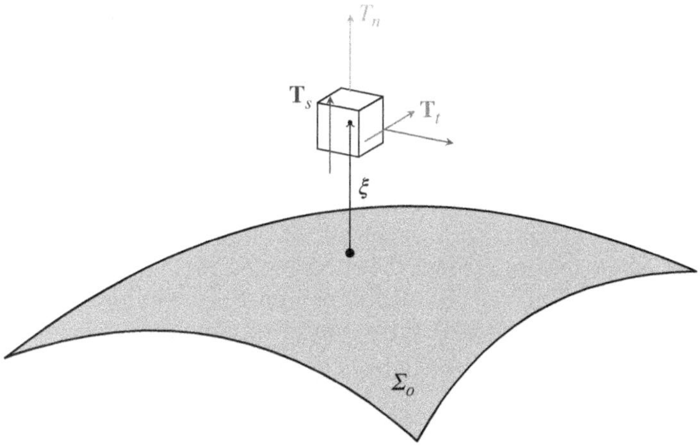

Figure 9.3 Decomposition of the Cauchy stress tensor according to the intrinsic system of reference of the shell.

PVP (external virtual power) is shaped as follows

$$\int_{\Omega} \mathbf{b} \cdot \mathbf{v} d\Omega + \int_{\partial\Omega_{\mathrm{f}}} \mathbf{f} \cdot \mathbf{v} d\partial\Omega_{\mathrm{f}} = \int_{\Sigma_o}\int_H (\mathbf{b}_t \cdot \mathbf{v}_t + b_n v_n) \det \Lambda d\xi d\Sigma_o$$
$$+ \int_{\Sigma_{\mathrm{f}}^+} (\mathbf{f}_t^+ \cdot \mathbf{v}_t + f_n^+ v_n) d\Sigma_{\mathrm{f}}^+ + \int_{\Sigma_{\mathrm{f}}^-} (\mathbf{f}_t^- \cdot \mathbf{v}_t + f_n^- v_n) d\Sigma_{\mathrm{f}}^- + \int_{\Gamma_{L\mathrm{f}}} (\bar{\mathbf{f}}_t \cdot \mathbf{v}_t + \bar{f}_n v_n) d\Gamma_{L\mathrm{f}}. \tag{9.54}$$

By putting the expressions (9.53) and (9.54) into (9.47) we can rework the PVP for the case of shells described in an intrinsic system of reference and seen as three-dimensional bodies for which no kinematical hypotheses aimed at characterizing the motion actions of this structural component have yet been postulated. This implies that the kinematics, so far, remains complete, that is with full three-dimensional nature. Hence, the mechanical equilibrium through the PVP states that the stress state characterized by $(\mathbf{T}_t, \mathbf{T}_s, T_n)$ is at equilibrium with the system of external forces defined by $(\mathbf{b}_t, b_n, \mathbf{f}_t^+, f_n^+, \mathbf{f}_t^-, f_n^-, \bar{\mathbf{f}}_t, \bar{f}_n)$ if and only if the following variational equation is satisfied

$$\int_{\Sigma_o}\int_H \Bigg[\mathbf{T}_t \cdot \left((\Pi_t (\nabla_{\mathbf{x}_o} \mathbf{v}_t) + v_n \nabla_{\mathbf{x}_o} \mathbf{n}) \Lambda^{-1} \Pi_t \right)^s$$
$$+ \mathbf{T}_s \cdot \left(\frac{\partial \mathbf{v}_t}{\partial \xi} - \Pi_t \Lambda^{-1} ((\nabla_{\mathbf{x}_o} \mathbf{n}) \mathbf{v}_t - \nabla_{\mathbf{x}_o} v_n) \right) + T_n \frac{\partial v_n}{\partial \xi} \Bigg] \det \Lambda d\xi d\Sigma_o$$
$$= \int_{\Sigma_o}\int_H (\mathbf{b}_t \cdot \mathbf{v}_t + b_n v_n) \det \Lambda d\xi d\Sigma_o + \int_{\Sigma_{\mathrm{f}}^+} (\mathbf{f}_t^+ \cdot \mathbf{v}_t + f_n^+ v_n) d\Sigma_{\mathrm{f}}^+$$
$$+ \int_{\Sigma_{\mathrm{f}}^-} (\mathbf{f}_t^- \cdot \mathbf{v}_t + f_n^- v_n) d\Sigma_{\mathrm{f}}^- + \int_{\Gamma_{L\mathrm{f}}} (\bar{\mathbf{f}}_t \cdot \mathbf{v}_t + \bar{f}_n v_n) d\Gamma_{L\mathrm{f}} \quad \forall (\mathbf{v}_t, v_n) \in \mathrm{Var}_v. \tag{9.55}$$

Exercise 9.3 Specify the variational equation (9.55) for the case of plates and provide an interpretation for the different terms.

It is important to note here that the PVP ruled according to (9.55) is as general and valid as the original expression of the PVP stated in (9.47). Notwithstanding this, the convenience and benefits in working with (9.55) will remain clear in the forthcoming sections, when kinematical hypotheses, which will be introduced to describe specific shell kinematics associated with the different shell models proposed in the literature, will enable the reduction of the three-dimensional mechanical problem to a problem described over the middle surface of the shell. As a matter of fact, existing shell models consist of postulating specific, a priori defined, expressions for the fields $\mathbf{v}_t(\mathbf{x}_o, \xi)$ and $v_n(\mathbf{x}_o, \xi)$ in terms of the variable ξ, making possible the integration across the shell thickness and reshaping the integrals in Ω (and its boundary) now on Σ_o (and its boundary). Briefly, the PVP for the three-dimensional body turns into a two-dimensional PVP for fields defined on the middle surface.

9.5 Unified Framework for Shell Models

The mechanical theory developed in the previous section is, kinematically speaking, as rich as the standard formalism for three-dimensional continua. This is manifested in the variational equation of the PVP given by (9.55). Even if such a balance between internal and external virtual powers has no practical utility, the unified groundwork articulated in the previous section greatly paves the way for the development of the diverse and different shell models available in the literature. In fact, expression (9.55) constitutes the starting point for the systematic construction of shell models with different mechanical features.

By inspecting equation (9.55), it is evident that the reduction of the domain of analysis to the middle surface is required to adopt kinematical hypotheses capable of explicitly providing a full characterization of the motion actions $\mathbf{v}_t(\mathbf{x}_o, \xi)$ and $v_n(\mathbf{x}_o, \xi)$ in terms of the variable ξ. This procedure yields two major consequences. The first one is that the integrals in Ω (and its boundary) are reduced to integrals in the middle surface Σ_o (and its boundary). The second outcome is that, by performing such integration across the thickness, generalized stresses and forces naturally come into existence as entities which are dual, in the sense that they are power-conjugate, to the objects that govern the behavior of the fields $\mathbf{v}_t(\mathbf{x}_o, \xi)$ and $v_n(\mathbf{x}_o, \xi)$.

To see this in full detail let us consider that we can expand the field $\mathbf{v}(\mathbf{x}_o, \xi)$, which is described in terms of the intrinsic components as $\mathbf{v}(\mathbf{x}_o, \xi) = \mathbf{v}_t(\mathbf{x}_o, \xi) + v_n(\mathbf{x}_o, \xi)\mathbf{n}(\mathbf{x}_o)$, by polynomial functions in the variable ξ. In particular, we adopt the expansions

$$\mathbf{v}_t(\mathbf{x}_o, \xi) = \mathbf{v}_t^o(\mathbf{x}_o) + \xi\omega_t(\mathbf{x}_o), \tag{9.56}$$

$$v_n(\mathbf{x}_o, \xi) = v_n^o(\mathbf{x}_o) + \xi\omega_n(\mathbf{x}_o) + \frac{\xi^2}{2}\alpha_n(\mathbf{x}_o), \tag{9.57}$$

where $\mathbf{v}_t^o(\mathbf{x}_o)$ and $\omega_t(\mathbf{x}_o)$ are vector fields and $v_n^o(\mathbf{x}_o)$, $\omega_n(\mathbf{x}_o)$ and $\alpha_n(\mathbf{x}_o)$ are scalar fields, all of them defined in Σ_o. More specifically, $\mathbf{v}_t^o(\mathbf{x}_o)$ and $\omega_t(\mathbf{x}_o)$ are vectors belonging to the tangent plane at point $\mathbf{x}_o$ in the shell. In turn, as $\mathbf{v}_t^o(\mathbf{x}_o)$ and $v_n^o(\mathbf{x}_o)$ are standard motion actions (velocities), $\omega_t(\mathbf{x}_o)$ and $\omega_n(\mathbf{x}_o)$ represent, respectively, the rotation action in the tangent plane around the axes contained in such a plane and the rotation action of the tangent plane around the normal vector $\mathbf{n}$. Finally, the scalar field α_n stands for the rate of change of the rotation with respect to ξ. These elements are displayed in Figure 9.4.

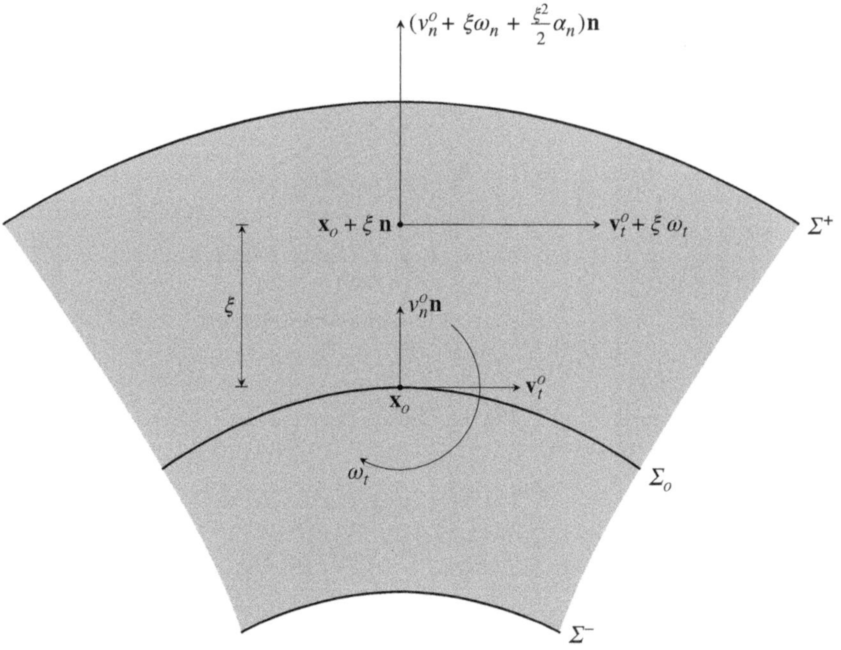

Figure 9.4 Physical interpretation for the kinematical representation of motion actions in the shell.

The expansions presented in (9.56) and (9.57) correspond to the shell model reported in [139]. Introducing (9.56) and (9.57) into (9.50)–(9.52), we obtain

$$
\begin{aligned}
(\nabla\mathbf{v})_t^s &= \left(\Pi_t(\nabla_{\mathbf{x}_o}\mathbf{v}_t^o)\Lambda^{-1}\Pi_t + v_n^o(\nabla_{\mathbf{x}_o}\mathbf{n})\Lambda^{-1}\Pi_t\right)^s \\
&\quad + \xi\left(\Pi_t(\nabla_{\mathbf{x}_o}\omega_t)\Lambda^{-1}\Pi_t + \omega_n(\nabla_{\mathbf{x}_o}\mathbf{n})\Lambda^{-1}\Pi_t\right)^s + \frac{\xi^2}{2}\left(\alpha_n(\nabla_{\mathbf{x}_o}\mathbf{n})\Lambda^{-1}\Pi_t\right)^s, \quad (9.58)
\end{aligned}
$$

$$
\begin{aligned}
(\nabla\mathbf{v})_s^s &= \frac{1}{2}\left(\Pi_t\Lambda^{-1}(\omega_t + \nabla_{\mathbf{x}_o}v_n^o - (\nabla_{\mathbf{x}_o}\mathbf{n})\,\mathbf{v}_t^o)\right) \\
&\quad + \frac{\xi}{2}\left(\Pi_t\Lambda^{-1}(\nabla_{\mathbf{x}_o}\omega_n)\right) + \frac{\xi^2}{4}\left(\Pi_t\Lambda^{-1}(\nabla_{\mathbf{x}_o}\alpha_n)\right), \quad (9.59)
\end{aligned}
$$

$$
(\nabla\mathbf{v})_n^s = \omega_n + \xi\alpha_n. \quad (9.60)
$$

Putting these results into the internal virtual power functional (see (9.53)) yields

$$
\begin{aligned}
\int_\Omega \mathbf{T}\cdot(\nabla\mathbf{v})^s d\Omega &= \int_{\Sigma_o}\int_H \mathbf{T}_t\cdot\left(\Pi_t(\nabla_{\mathbf{x}_o}\mathbf{v}_t^o)\Lambda^{-1}\Pi_t + v_n^o(\nabla_{\mathbf{x}_o}\mathbf{n})\Lambda^{-1}\Pi_t\right)^s \det\Lambda d\xi d\Sigma_o \\
&\quad + \int_{\Sigma_o}\int_H \mathbf{T}_t\cdot\xi\left(\Pi_t(\nabla_{\mathbf{x}_o}\omega_t)\Lambda^{-1}\Pi_t + \omega_n(\nabla_{\mathbf{x}_o}\mathbf{n})\Lambda^{-1}\Pi_t\right)^s \det\Lambda d\xi d\Sigma_o \\
&\quad + \int_{\Sigma_o}\int_H \mathbf{T}_t\cdot\frac{\xi^2}{2}(\alpha_n(\nabla_{\mathbf{x}_o}\mathbf{n})\Lambda^{-1}\Pi_t)^s \det\Lambda d\xi d\Sigma_o
\end{aligned}
$$

$$+\int_{\Sigma_o}\int_H \mathbf{T}_s \cdot \Pi_t \Lambda^{-1}\left(\omega_t + \nabla_{\mathbf{x}_o} v_n^o - (\nabla_{\mathbf{x}_o}\mathbf{n})\mathbf{v}_t^o\right) \det \Lambda d\xi d\Sigma_o$$
$$+\int_{\Sigma_o}\int_H \mathbf{T}_s \cdot \xi\left(\Pi_t \Lambda^{-1}(\nabla_{\mathbf{x}_o}\omega_n)\right) \det \Lambda d\xi d\Sigma_o$$
$$+\int_{\Sigma_o}\int_H \mathbf{T}_s \cdot \frac{\xi^2}{2}\left(\Pi_t \Lambda^{-1}(\nabla_{\mathbf{x}_o}\alpha_n)\right) \det \Lambda d\xi d\Sigma_o$$
$$+\int_{\Sigma_o}\int_H T_n \omega_n \det \Lambda d\xi d\Sigma_o + \int_{\Sigma_o}\int_H T_n \xi \alpha_n \det \Lambda d\xi d\Sigma_o. \tag{9.61}$$

By virtue of the symmetry of the Cauchy stress tensor and of the tensor Λ, and from the properties of the projection operator Π_t, expression (9.61) is rewritten as

$$\int_\Omega \mathbf{T} \cdot (\nabla \mathbf{v})^s d\Omega = \int_{\Sigma_o}\left(\int_H \mathbf{T}_t \Lambda^{-1} \det \Lambda d\xi\right) \cdot \left(\Pi_t(\nabla_{\mathbf{x}_o}\mathbf{v}_t^o) + v_n^o(\nabla_{\mathbf{x}_o}\mathbf{n})\right) d\Sigma_o$$
$$+\int_{\Sigma_o}\left(\int_H \mathbf{T}_t \Lambda^{-1}\xi \det \Lambda d\xi\right) \cdot \left(\Pi_t(\nabla_{\mathbf{x}_o}\omega_t) + \omega_n(\nabla_{\mathbf{x}_o}\mathbf{n})\right) d\Sigma_o$$
$$+\int_{\Sigma_o}\left(\int_H \mathbf{T}_t \Lambda^{-1}\frac{\xi^2}{2} \det \Lambda d\xi\right) \cdot (\alpha_n(\nabla_{\mathbf{x}_o}\mathbf{n}))d\Sigma_o$$
$$+\int_{\Sigma_o}\left(\int_H \Lambda^{-1}\mathbf{T}_s \det \Lambda d\xi\right) \cdot (\omega_t + \nabla_{\mathbf{x}_o} v_n^o - (\nabla_{\mathbf{x}_o}\mathbf{n})\mathbf{v}_t^o)d\Sigma_o$$
$$+\int_{\Sigma_o}\left(\int_H \Lambda^{-1}\mathbf{T}_s \xi \det \Lambda d\xi\right) \cdot (\nabla_{\mathbf{x}_o}\omega_n)d\Sigma_o$$
$$+\int_{\Sigma_o}\left(\int_H \Lambda^{-1}\mathbf{T}_s \frac{\xi^2}{2} \det \Lambda d\xi\right) \cdot (\nabla_{\mathbf{x}_o}\alpha_n)d\Sigma_o$$
$$+\int_{\Sigma_o}\left(\int_H T_n \det \Lambda d\xi\right)\omega_n d\Sigma_o + \int_{\Sigma_o}\left(\int_H T_n \xi \det \Lambda d\xi\right)\alpha_n d\Sigma_o. \tag{9.62}$$

The integrals across the thickness in the internal virtual power functional given by (9.62) can be now lumped together, giving rise to the following generalized internal stresses defined in the middle surface of the shell Σ_o

$$\begin{aligned}
\mathbf{N}_t &= \textstyle\int_H \mathbf{T}_t \Lambda^{-1} \det \Lambda d\xi, & \mathbf{M}_t &= \textstyle\int_H \mathbf{T}_t \Lambda^{-1}\xi \det \Lambda d\xi,\\
\mathbf{P}_t &= \textstyle\int_H \mathbf{T}_t \Lambda^{-1}\frac{\xi^2}{2} \det \Lambda d\xi, & \mathbf{Q} &= \textstyle\int_H \Lambda^{-1}\mathbf{T}_s \det \Lambda d\xi,\\
\mathbf{S} &= \textstyle\int_H \Lambda^{-1}\mathbf{T}_s\xi \det \Lambda d\xi, & \mathbf{H} &= \textstyle\int_H \Lambda^{-1}\mathbf{T}_s\frac{\xi^2}{2} \det \Lambda d\xi,\\
A &= \textstyle\int_H T_n \det \Lambda d\xi, & B &= \textstyle\int_H T_n\xi \det \Lambda d\xi.
\end{aligned} \tag{9.63}$$

From a physical viewpoint, the generalized internal stresses can be interpreted as follows: $\mathbf{N}_t$ is the membrane stress tensor, $\mathbf{M}_t$ is the membrane moment tensor, $\mathbf{Q}$ is the shear stress vector, and A is the normal stress (scalar field). In addition, tensor $\mathbf{P}_t$, vectors $\mathbf{S}$ and $\mathbf{H}$, and the scalar field B are, respectively, the moment produced by tangent stresses, by shear stresses, and by the normal stress. Here it is noteworthy that even if the $\mathbf{T}_t$ is symmetric, tensors $\mathbf{N}_t$, $\mathbf{M}_t$ and $\mathbf{P}_t$ are not symmetric in general. Below we will see that, thanks to further hypotheses over the behavior of tensor Λ, and therefore over

Λ^{-1}, these generalized stresses recover the symmetry property. Also, notice that even if **Q**, **S** and **H** are vector fields, we choose capital letters to avoid confusion with other objects that will arise in the forthcoming developments.

With the definition of the generalized stresses (9.63), and given the expansion of kinematic entities in terms of ξ, it turns out that the internal power in the shell is written through an integral defined over the middle surface, as follows

$$\begin{aligned}\int_{\Omega} \mathbf{T} \cdot (\nabla \mathbf{v})^s d\Omega = & \int_{\Sigma_o} \mathbf{N}_t \cdot \left(\Pi_t(\nabla_{\mathbf{x}_o} \mathbf{v}_t^o) + v_n^o(\nabla_{\mathbf{x}_o} \mathbf{n})\right) d\Sigma_o \\ & + \int_{\Sigma_o} \mathbf{M}_t \cdot \left(\Pi_t(\nabla_{\mathbf{x}_o} \omega_t) + \omega_n(\nabla_{\mathbf{x}_o} \mathbf{n})\right) d\Sigma_o + \int_{\Sigma_o} (\mathbf{P}_t \cdot \nabla_{\mathbf{x}_o} \mathbf{n}) \alpha_n d\Sigma_o \\ & + \int_{\Sigma_o} \mathbf{Q} \cdot (\omega_t + \nabla_{\mathbf{x}_o} v_n^o - (\nabla_{\mathbf{x}_o} \mathbf{n}) \mathbf{v}_t^o) d\Sigma_o + \int_{\Sigma_o} \mathbf{S} \cdot \nabla_{\mathbf{x}_o} \omega_n d\Sigma_o \\ & + \int_{\Sigma_o} \mathbf{H} \cdot \nabla_{\mathbf{x}_o} \alpha_n d\Sigma_o + \int_{\Sigma_o} A \omega_n d\Sigma_o + \int_{\Sigma_o} B \alpha_n d\Sigma_o. \end{aligned} \tag{9.64}$$

Exercise 9.4 Derive the expressions of the generalized stresses in (9.63) and of the internal power (9.64) for the case of plates. Discuss the properties of the stress tensors $\mathbf{N}_t$, $\mathbf{M}_t$ and $\mathbf{P}_t$.

Analogously, we will now manipulate the expression of the external virtual power (9.54) to reduce the dimension of the domain of integration. First, consider the following relation between the differentials $d\Gamma$ and $d\xi d\partial\Sigma_o$:

$$d\Gamma = [\Lambda^2 \mathbf{s}_o \cdot \mathbf{s}_o]^{1/2} d\xi \, d\partial\Sigma_o, \tag{9.65}$$

where $\mathbf{s}_o$ is the unit vector tangent to the boundary $\partial\Sigma_o$ at point $\mathbf{x}_o$ defined by $\mathbf{s}_o = \mathbf{n} \times \mathbf{n}_o$, as shown in Figure 9.2. Now, we insert expansions (9.56) and (9.57) into (9.54), and the external virtual power functional gives

$$\begin{aligned}\int_{\Omega} \mathbf{b} \cdot \mathbf{v} d\Omega + \int_{\partial\Omega_{\mathrm{f}}} \mathbf{f} \cdot \mathbf{v} d\partial\Omega_{\mathrm{f}} = & \int_{\Sigma_o} \int_H (\mathbf{b}_t \cdot \mathbf{v}_t + b_n v_n) \det \Lambda d\xi d\Sigma_o \\ & + \int_{\Sigma_{\mathrm{f}}^+} (\mathbf{f}_t^+ \cdot \mathbf{v}_t + f_n^+ v_n) d\Sigma_{\mathrm{f}}^+ + \int_{\Sigma_{\mathrm{f}}^-} (\mathbf{f}_t^- \cdot \mathbf{v}_t + f_n^- v_n) d\Sigma_{\mathrm{f}}^- + \int_{\Gamma_{L\mathrm{f}}} (\bar{\mathbf{f}}_t \cdot \mathbf{v}_t + \bar{f}_n v_n) d\Gamma_{L\mathrm{f}} \\ = & \int_{\Sigma_o} \int_H \mathbf{b}_t \cdot (\mathbf{v}_t^o + \xi \omega_t) \det \Lambda d\xi d\Sigma_o + \int_{\Sigma_o} \int_H b_n \left(v_n^o + \xi \omega_n + \frac{\xi^2}{2} \alpha_n\right) \det \Lambda d\xi d\Sigma_o \\ & + \int_{\Sigma_o} \mathbf{f}_t^+ \cdot \left(\mathbf{v}_t^o + \frac{h}{2} \omega_t\right) \det \Lambda^+ d\Sigma_o + \int_{\Sigma_o} f_n^+ \left(v_n^o + \frac{h}{2} \omega_n + \frac{h^2}{8} \alpha_n\right) \det \Lambda^+ d\Sigma_o \\ & + \int_{\Sigma_o} \mathbf{f}_t^- \cdot \left(\mathbf{v}_t^o - \frac{h}{2} \omega_t\right) \det \Lambda^- d\Sigma_o + \int_{\Sigma_o} f_n^- \left(v_n^o - \frac{h}{2} \omega_n + \frac{h^2}{8} \alpha_n\right) \det \Lambda^- d\Sigma_o \\ & + \int_{\partial\Sigma_{o\mathrm{f}}} \int_H \bar{\mathbf{f}}_t \cdot (\mathbf{v}_t^o + \xi \omega_t) [\Lambda^2 \mathbf{s}_o \cdot \mathbf{s}_o]^{1/2} d\xi d\partial\Sigma_{o\mathrm{f}} \\ & + \int_{\partial\Sigma_{o\mathrm{f}}} \int_H \bar{f}_n \left(v_n^o + \xi \omega_n + \frac{\xi^2}{2} \alpha_n\right) [\Lambda^2 \mathbf{s}_o \cdot \mathbf{s}_o]^{1/2} d\xi d\partial\Sigma_{o\mathrm{f}}. \end{aligned} \tag{9.66}$$

Carrying out the integrals across the shell thickness, and introducing the following definitions for the so-called generalized external forces gives

$$
\begin{aligned}
\mathbf{p}_{bt} &= \int_H \mathbf{b}_t \det \Lambda d\xi, & p_{bn} &= \int_H b_n \det \Lambda d\xi, \\
\mathbf{m}_{bt} &= \int_H \mathbf{b}_t \xi \det \Lambda d\xi, & m_{bn} &= \int_H b_n \xi \det \Lambda d\xi, \\
\mathbf{p}_t &= \mathbf{f}_t^+ \det \Lambda^+ + \mathbf{f}_t^- \det \Lambda^-, & p_n &= f_n^+ \det \Lambda^+ + f_n^- \det \Lambda^-, \\
\mathbf{m}_t &= \tfrac{h}{2}(\mathbf{f}_t^+ \det \Lambda^+ - \mathbf{f}_t^- \det \Lambda^-), & m_n &= \tfrac{h}{2}(f_n^+ \det \Lambda^+ - f_n^- \det \Lambda^-), \\
\overline{\mathbf{p}}_t &= \int_H \overline{\mathbf{f}}_t [\Lambda^2 \mathbf{s}_o \cdot \mathbf{s}_o]^{1/2} d\xi, & \overline{p}_n &= \int_H \overline{f}_n [\Lambda^2 \mathbf{s}_o \cdot \mathbf{s}_o]^{1/2} d\xi, \\
\overline{\mathbf{m}}_t &= \int_H \overline{\mathbf{f}}_t \xi [\Lambda^2 \mathbf{s}_o \cdot \mathbf{s}_o]^{1/2} d\xi, & \overline{m}_n &= \int_H \overline{f}_n \xi [\Lambda^2 \mathbf{s}_o \cdot \mathbf{s}_o]^{1/2} d\xi, \\
q_{bn} &= \int_H b_n \tfrac{\xi^2}{2} \det \Lambda d\xi, & q_n &= \tfrac{h^2}{8}(f_n^+ \det \Lambda^+ + f_n^- \det \Lambda^-), \\
\overline{q}_n &= \int_H \overline{f}_n \tfrac{\xi^2}{2} [\Lambda^2 \mathbf{s}_o \cdot \mathbf{s}_o]^{1/2} d\xi.
\end{aligned}
\tag{9.67}
$$

Then we obtain the final form of the external virtual power

$$
\begin{aligned}
\int_\Omega \mathbf{b} \cdot \mathbf{v} d\Omega + \int_{\partial\Omega_\mathrm{f}} \mathbf{f} \cdot \mathbf{v} d\partial\Omega_\mathrm{f} = \int_{\Sigma_o} &\big[(\mathbf{p}_{bt} + \mathbf{p}_t) \cdot \mathbf{v}_t^o + (\mathbf{m}_{bt} + \mathbf{m}_t) \cdot \omega_t + (p_{bn} + p_n) v_n^o \\
&+ (m_{bn} + m_n)\omega_n + (q_{bn} + q_n)\alpha_n\big] \, d\Sigma_o \\
&+ \int_{\partial\Sigma_{o\mathbf{f}}} \big[\overline{\mathbf{p}}_t \cdot \mathbf{v}_t^o + \overline{\mathbf{m}}_t \cdot \omega_t + \overline{p}_n v_n^o + \overline{m}_n \omega_n + \overline{q}_n \alpha_n\big] \, d\partial\Sigma_{o\mathbf{f}}.
\end{aligned}
\tag{9.68}
$$

In the expressions above, $\mathbf{p}_{bt}$, p_{bn}, $\mathbf{p}_t$, p_n, $\overline{\mathbf{p}}_t$ and $\overline{p}_n$ are the additive decompositions of the forces which are applied to the body over the surfaces Σ^+, Σ^- and Γ_L, while $\mathbf{m}_{bt}$, $\mathbf{m}_t$, and $\overline{\mathbf{m}}_t$ represent the moments with respect to the middle surface Σ_o produced by the forces applied over Σ^+, Σ^- and Γ_L. The other generalized forces, m_{bn}, m_n, $\overline{m}_n$, q_{bn}, q_n and $\overline{q}_n$, can be understood using duality arguments with respect to the motion actions ω_n and α_n, defined, respectively, in Σ_o and $\partial\Sigma_o$. It is important to highlight here that the generalized forces (9.67) compatible with the general shell model adopted in this section have been naturally derived, and not defined a priori, thanks to the identification of the duality pairings with each of the entities that characterize the kinematics. Finally, $\partial\Sigma_{o\mathbf{f}}$ corresponds to the part of the boundary $\partial\Sigma_o$ where there are applied forces (natural boundary conditions in the PVP or Neumann-like boundary conditions in the context of Euler–Lagrange equations).

Exercise 9.5 Characterize the generalized external forces (9.67) and the external virtual power functional (9.68) for the case of plates.

After having deduced the expressions of the internal virtual power (9.64) and the external virtual power (9.68) for the present general shell model it is possible to formalize the mechanical equilibrium through the PVP cast in the two-dimensional domain of definition of the middle surface. Thus, we have that for the system of generalized forces, defined in (9.67), the generalized stress state in the shell, as characterized by (9.63), is at equilibrium if the following variational equation is satisfied

$$
\begin{aligned}
&\int_{\Sigma_o} \mathbf{N}_t \cdot \big(\Pi_t(\nabla_{\mathbf{x}_o} \mathbf{v}_t^o) + v_n^o(\nabla_{\mathbf{x}_o} \mathbf{n})\big) \, d\Sigma_o + \int_{\Sigma_o} \mathbf{M}_t \cdot \big(\Pi_t(\nabla_{\mathbf{x}_o} \omega_t) + \omega_n(\nabla_{\mathbf{x}_o} \mathbf{n})\big) \, d\Sigma_o \\
&+ \int_{\Sigma_o} (\mathbf{P}_t \cdot \nabla_{\mathbf{x}_o} \mathbf{n})\alpha_n d\Sigma_o + \int_{\Sigma_o} \mathbf{Q} \cdot (\omega_t + \nabla_{\mathbf{x}_o} v_n^o - (\nabla_{\mathbf{x}_o} \mathbf{n})\mathbf{v}_t^o) d\Sigma_o
\end{aligned}
$$

$$
\begin{aligned}
&+ \int_{\Sigma_o} \mathbf{S} \cdot \nabla_{\mathbf{x}_o} \omega_n d\Sigma_o + \int_{\Sigma_o} \mathbf{H} \cdot \nabla_{\mathbf{x}_o} \alpha_n d\Sigma_o + \int_{\Sigma_o} A\omega_n d\Sigma_o + \int_{\Sigma_o} B\alpha_n d\Sigma_o \\
&= \int_{\Sigma_o} \left[(\mathbf{p}_{bt} + \mathbf{p}_t) \cdot \mathbf{v}_t^o + (\mathbf{m}_{bt} + \mathbf{m}_t) \cdot \omega_t + (p_{bn} + p_n) v_n^o \right. \\
&\left. + (m_{bn} + m_n)\omega_n + (q_{bn} + q_n)\alpha_n \right] d\Sigma_o \\
&+ \int_{\partial\Sigma_{of}} \left[\overline{\mathbf{p}}_t \cdot \mathbf{v}_t^o + \overline{\mathbf{m}}_t \cdot \omega_t + \overline{p}_n v_n^o + \overline{m}_n \omega_n + \overline{q}_n \alpha_n \right] d\partial\Sigma_{of} \\
&\forall (\mathbf{v}_t^o, \omega_t, v_n^o, \omega_n, \alpha_n) \in \mathrm{Var}_v,
\end{aligned} \tag{9.69}
$$

where Var_v is the functional space of admissible virtual motion actions in the shell, which is defined by

$$
\mathrm{Var}_v = \{(\mathbf{v}_t^o, \omega_t, v_n^o, \omega_n, \alpha_n) \in \mathscr{V}_o^G;\ (\mathbf{v}_t^o, \omega_t, v_n^o, \omega_n, \alpha_n)|_{\partial\Sigma_{ou}} = (\mathbf{0}, \mathbf{0}, 0, 0, 0)\}, \tag{9.70}
$$

where $\mathscr{V}_o^G$ is a space of functions defined over the middle surface of the shell that are regular enough for the mathematical operations to make sense, while $\partial\Sigma_{o\mathbf{u}}$ is the portion of $\partial\Sigma_o$ where motion actions are prescribed.

After integration by parts, the Euler–Lagrange equations rendered by the variational statement (9.69) are

$$
\begin{cases}
\mathrm{div}_{\mathbf{x}_o} \mathbf{N}_t + (\nabla_{\mathbf{x}_o} \mathbf{n})\mathbf{Q} + (\mathbf{p}_{bt} + \mathbf{p}_t) = \mathbf{0} & \text{on } \Sigma_o, \\
\mathrm{div}_{\mathbf{x}_o} \mathbf{Q} - \mathbf{N}_t \cdot \nabla_{\mathbf{x}_o} \mathbf{n} + (p_{bn} + p_n) = 0 & \text{on } \Sigma_o, \\
\mathrm{div}_{\mathbf{x}_o} \mathbf{M}_t - \mathbf{Q} + (\mathbf{m}_{bt} + \mathbf{m}_t) = \mathbf{0} & \text{on } \Sigma_o, \\
\mathrm{div}_{\mathbf{x}_o} \mathbf{S} - \mathbf{M}_t \cdot \nabla_{\mathbf{x}_o} \mathbf{n} - A + (m_{bn} + m_n) = 0 & \text{on } \Sigma_o, \\
\mathrm{div}_{\mathbf{x}_o} \mathbf{H} - \mathbf{P}_t \cdot \nabla_{\mathbf{x}_o} \mathbf{n} - B + (q_{bn} + q_n) = 0 & \text{on } \Sigma_o, \\
\mathbf{N}_t \mathbf{n}_o = \overline{\mathbf{p}}_t & \text{on } \partial\Sigma_{of}, \\
\mathbf{Q} \cdot \mathbf{n}_o = \overline{p}_n & \text{on } \partial\Sigma_{of}, \\
\mathbf{M}_t \mathbf{n}_o = \overline{\mathbf{m}}_t & \text{on } \partial\Sigma_{of}, \\
\mathbf{S} \cdot \mathbf{n}_o = \overline{m}_n & \text{on } \partial\Sigma_{of}, \\
\mathbf{H} \cdot \mathbf{n}_o = \overline{q}_n & \text{on } \partial\Sigma_{of}.
\end{cases} \tag{9.71}
$$

Once again, we can observe that each of the previous equations, including the natural conditions over the portion of the boundary where no motion actions are prescribed, is associated with each one of the variables that describe the kinematics in this general shell model, originally proposed in [139].

Exercise 9.6 Integrate by parts in (9.69) to effectively arrive at (9.71).

Exercise 9.7 Obtain the PVP for the general model of plates which is derived from the variational expression (9.69), and then obtain the corresponding Euler–Lagrange equations.

Exercise 9.8 Compare the Euler–Lagrange equations obtained in the previous exercise with those obtained in (9.71), and discuss the role of the curvature tensor in the coupling of the different generalized stresses in the equilibrium equations.

9.6 Classical Shell Models

In this section we will develop the diverse shell models available in the literature. As will be seen, we can perform this unified derivation based on the geometric construct provided in previous sections. Indeed, the shell models developed next correspond to different kinematical representations established by constraints on the behavior of the fields $\mathbf{v}_t(\mathbf{x}_o, \xi)$ and $v_n(\mathbf{x}_o, \xi)$ which enabled, as seen in the previous section, the integration across the shell thickness, reducing the problem from a three-dimensional domain to the two-dimensional middle surface domain.

9.6.1 Naghdi Model

In the shell theory presented in Section 9.4 all possible motion actions for shells were allowed, yielding the richest kinematical description possible (that of three-dimensional continua), and therefore putting in evidence all components (tangent, transversal, and normal) of the strain action operator. As a consequence of the duality, this resulted in the appearance of all kinds of internal stresses. In turn, in Section 9.5 the kinematics was constrained according to expansions (9.56) and (9.57), as proposed by [139]. Such a kinematical framework promoted the appearance of certain internal generalized stresses as defined in (9.63), which play a role in the PVP. In this section, we will present another shell model known in the literature as the Naghdi model, proposed in [207, 208]. To arrive at this model, further simplifications will be considered in the expansions (9.56) and (9.57). In essence, in this theory it is postulated that the transversal components to the motion associated with the expansion in ξ are null. This hypothesis entails that $\omega_n(\mathbf{x}_o) = 0$ and $\alpha_n(\mathbf{x}_o) = 0$, and therefore we obtain

$$\mathbf{v}_t(\mathbf{x}_o, \xi) = \mathbf{v}_t^o(\mathbf{x}_o) + \xi\omega_t(\mathbf{x}_o), \tag{9.72}$$

$$v_n(\mathbf{x}_o, \xi) = v_n^o(\mathbf{x}_o). \tag{9.73}$$

With these considerations, the strain actions (9.58)–(9.60) are reduced to the following expressions

$$(\nabla\mathbf{v})_t^s = \left(\Pi_t(\nabla_{\mathbf{x}_o}\mathbf{v}_t^o)\Lambda^{-1}\Pi_t + v_n^o(\nabla_{\mathbf{x}_o}\mathbf{n})\Lambda^{-1}\Pi_t\right)^s + \xi\left(\Pi_t(\nabla_{\mathbf{x}_o}\omega_t)\Lambda^{-1}\Pi_t\right)^s, \tag{9.74}$$

$$(\nabla\mathbf{v})_s^s = \frac{1}{2}\left(\Pi_t\Lambda^{-1}(\omega_t + \nabla_{\mathbf{x}_o}v_n^o - (\nabla_{\mathbf{x}_o}\mathbf{n})\mathbf{v}_t^o)\right), \tag{9.75}$$

$$(\nabla\mathbf{v})_n^s = 0. \tag{9.76}$$

Clearly, and due to the kinematical hypotheses assumed in (9.72) and (9.73), the strain action component $(\nabla\mathbf{v})_n^s$ is null. This implies that the stress state in the shell obtained from duality arguments to this direction (which actually exists to ensure the kinematical constraint is satisfied) cannot be quantified through the PVP. This is easy to see because the power exerted by this stress component in duality with kinematically admissible motion actions (i.e. $(\nabla\mathbf{v})_n^s = 0$) will trivially be zero. In other words, the generalized stress state in that direction must be interpreted as a reactive stress (the constraint here is the restriction imposed on the motion by the Naghdi theory).

Analogously to the procedure followed in the previous section, with the strain actions (9.74)–(9.76), we arrive at the internal power functional for the shell within this theory. Briefly, removing the null terms from (9.64) yields

$$\int_{\Omega} \mathbf{T} \cdot (\nabla \mathbf{v})^s d\Omega = \int_{\Sigma_o} \mathbf{N}_t \cdot \left(\Pi_t(\nabla_{\mathbf{x}_o} \mathbf{v}_t^o) + v_n^o(\nabla_{\mathbf{x}_o} \mathbf{n}) \right) d\Sigma_o$$
$$+ \int_{\Sigma_o} \mathbf{M}_t \cdot \left(\Pi_t(\nabla_{\mathbf{x}_o} \omega_t) \right) d\Sigma_o + \int_{\Sigma_o} \mathbf{Q} \cdot (\omega_t + \nabla_{\mathbf{x}_o} v_n^o - (\nabla_{\mathbf{x}_o} \mathbf{n}) \mathbf{v}_t^o) d\Sigma_o. \tag{9.77}$$

Defining now

$$\varepsilon = \Pi_t(\nabla_{\mathbf{x}_o} \mathbf{v}_t^o) + v_n^o(\nabla_{\mathbf{x}_o} \mathbf{n}), \tag{9.78}$$
$$\chi = \Pi_t(\nabla_{\mathbf{x}_o} \omega_t), \tag{9.79}$$
$$\psi = \omega_t + \nabla_{\mathbf{x}_o} v_n^o - (\nabla_{\mathbf{x}_o} \mathbf{n}) \mathbf{v}_t^o, \tag{9.80}$$

where ε is the membrane strain (rate) action tensor, χ is the bending strain (rate) action tensor, and ψ is the shear strain (rate) action vector, we can rewrite the internal power in more compact form

$$\int_{\Omega} \mathbf{T} \cdot (\nabla \mathbf{v})^s d\Omega = \int_{\Sigma_o} (\mathbf{N}_t \cdot \varepsilon + \mathbf{M}_t \cdot \chi + \mathbf{Q} \cdot \psi) d\Sigma_o. \tag{9.81}$$

In turn, taking into account once again the kinematics associated with the Naghdi model as stated by (9.72) and (9.73), the external virtual power (9.68) is simplified into the following form

$$\int_{\Omega} \mathbf{b} \cdot \mathbf{v} d\Omega + \int_{\partial\Omega_{\mathrm{f}}} \mathbf{f} \cdot \mathbf{v} d\partial\Omega_{\mathrm{f}}$$
$$= \int_{\Sigma_o} \left[(\mathbf{p}_{bt} + \mathbf{p}_t) \cdot \mathbf{v}_t^o + (\mathbf{m}_{bt} + \mathbf{m}_t) \cdot \omega_t + (p_{bn} + p_n) v_n^o \right] d\Sigma_o$$
$$+ \int_{\partial\Sigma_{o\mathrm{f}}} \left[\overline{\mathbf{p}}_t \cdot \mathbf{v}_t^o + \overline{\mathbf{m}}_t \cdot \omega_t + \overline{p}_n v_n^o \right] d\partial\Sigma_{o\mathrm{f}}. \tag{9.82}$$

Once we have arrived at the characterization of the internal virtual power and external virtual power functionals within the Naghdi theory, we are able to enunciate the mechanical equilibrium by means of the PVP, which results in the following variational statement

$$\int_{\Sigma_o} \left[\mathbf{N}_t \cdot \left(\Pi_t(\nabla_{\mathbf{x}_o} \mathbf{v}_t^o) + v_n^o(\nabla_{\mathbf{x}_o} \mathbf{n}) \right) + \mathbf{M}_t \cdot \left(\Pi_t(\nabla_{\mathbf{x}_o} \omega_t) \right) \right] d\Sigma_o$$
$$+ \int_{\Sigma_o} \mathbf{Q} \cdot (\omega_t + \nabla_{\mathbf{x}_o} v_n^o - (\nabla_{\mathbf{x}_o} \mathbf{n}) \mathbf{v}_t^o) d\Sigma_o$$
$$= \int_{\Sigma_o} \left[(\mathbf{p}_{bt} + \mathbf{p}_t) \cdot \mathbf{v}_t^o + (\mathbf{m}_{bt} + \mathbf{m}_t) \cdot \omega_t + (p_{bn} + p_n) v_n^o \right] d\Sigma_o$$
$$+ \int_{\partial\Sigma_{o\mathrm{f}}} \left[\overline{\mathbf{p}}_t \cdot \mathbf{v}_t^o + \overline{\mathbf{m}}_t \cdot \omega_t + \overline{p}_n v_n^o \right] d\partial\Sigma_{o\mathrm{f}} \qquad \forall (\mathbf{v}_t^o, \omega_t, v_n^o) \in \mathrm{Var}_v, \tag{9.83}$$

where Var_v is the space of kinematically admissible virtual motion actions in the shell defined by

$$\mathrm{Var}_v = \{ (\mathbf{v}_t^o, \omega_t, v_n^o) \in \mathscr{V}_o^N, \ (\mathbf{v}_t^o, \omega_t, v_n^o)|_{\partial\Sigma_{o\mathrm{u}}} = (\mathbf{0}, \mathbf{0}, 0) \}, \tag{9.84}$$

with $\mathscr{V}_o^N$ being a space of sufficiently regular functions.

The Euler–Lagrange equations resulting from the variational equation (9.83) are given by

$$\begin{cases} \operatorname{div}_{\mathbf{x}_o} \mathbf{N}_t + (\nabla_{\mathbf{x}_o} \mathbf{n})\mathbf{Q} + (\mathbf{p}_{bt} + \mathbf{p}_t) = \mathbf{0} & \text{on } \Sigma_o, \\ \operatorname{div}_{\mathbf{x}_o} \mathbf{Q} - \mathbf{N}_t \cdot \nabla_{\mathbf{x}_o} \mathbf{n} + (p_{bn} + p_n) = 0 & \text{on } \Sigma_o, \\ \operatorname{div}_{\mathbf{x}_o} \mathbf{M}_t - \mathbf{Q} + (\mathbf{m}_{bt} + \mathbf{m}_t) = \mathbf{0} & \text{on } \Sigma_o, \\ \mathbf{N}_t \mathbf{n}_o = \overline{\mathbf{p}}_t & \text{on } \partial\Sigma_{o\mathrm{f}}, \\ \mathbf{Q} \cdot \mathbf{n}_o = \overline{p}_n & \text{on } \partial\Sigma_{o\mathrm{f}}, \\ \mathbf{M}_t \mathbf{n}_o = \overline{\mathbf{m}}_t & \text{on } \partial\Sigma_{o\mathrm{f}}. \end{cases} \tag{9.85}$$

Exercise 9.9 Construct the Naghdi model for the particular case of plates and compare the results with the theory for shells.

At this point, we must recall that the construction of this Naghdi model was performed on top of the more general shell model, which is kinematically richer, developed in Section 9.5. Finally, all the ingredients in the theory proposed by Naghdi, framed in the unified variational framework of this book, are presented in Figure 9.5. In that figure, note that we denote $\tilde{\mathbf{p}}_t = \mathbf{p}_{bt} + \mathbf{p}_t$, $\tilde{\mathbf{m}}_t = \mathbf{m}_{bt} + \mathbf{m}_t$ and $\tilde{p}_n = p_{bn} + p_n$.

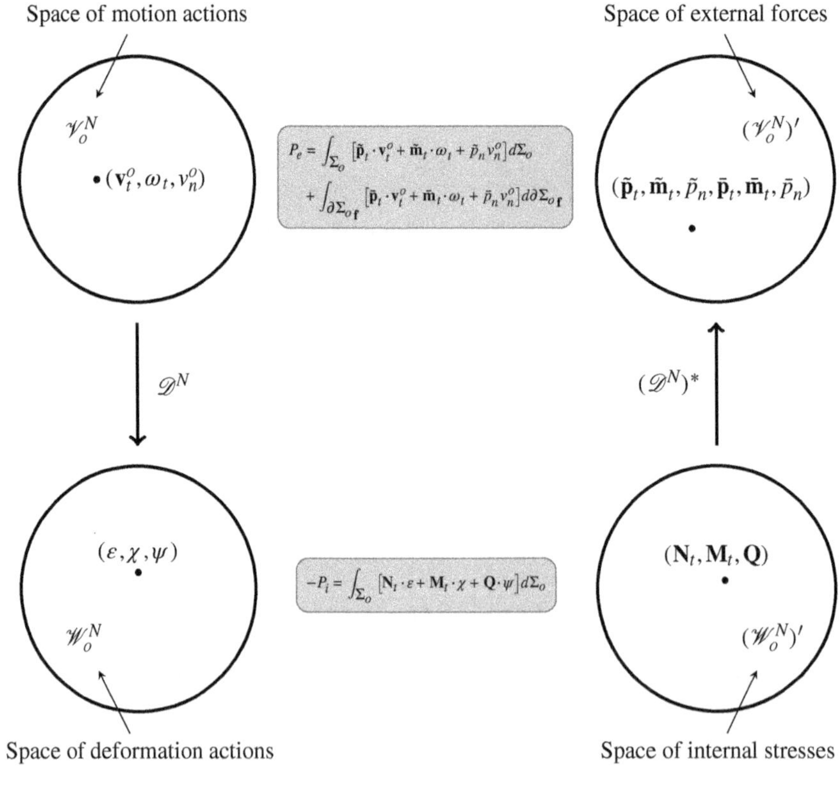

Figure 9.5 Ingredients in the Naghdi theory for shells and plates.

9.6.2 Kirchhoff–Love Model

The shell theory presented in this section (acknowledged as the exact Kirchhoff–Love theory) is constructed after applying the kinematical hypotheses proposed by Kirchhoff and Love (see [285] and [286]) to the structural component presented in the previous section. This theory is said to be exact because the hypotheses introduced are exclusively of kinematical nature. This means that the theory does not comprise simplifying hypotheses related to the relation between the thickness and the curvature of the shell, that is $\frac{h}{R}$, as we will see in forthcoming sections.

Such kinematical constraints can be cast in the following format: we say that the motion actions in the shell, $\mathbf{v}$, are of Kirchhoff–Love type if the strain action components $(\nabla\mathbf{v})^s_s$ and $(\nabla\mathbf{v})^s_n$ (see (9.51) and (9.52)) are null for all points $(\mathbf{x}_o, \xi)$, which implies

$$(\nabla\mathbf{v})^s_s = \frac{1}{2}\left(\frac{\partial \mathbf{v}_t}{\partial \xi} + \Pi_t \Lambda^{-1}(\nabla_{\mathbf{x}_o} v_n - (\nabla_{\mathbf{x}_o}\mathbf{n})\,\mathbf{v}_t)\right) = \mathbf{0}, \tag{9.86}$$

$$(\nabla\mathbf{v})^s_n = \frac{\partial v_n}{\partial \xi} = 0. \tag{9.87}$$

Since these constraints are linear, the motion actions which satisfy these equations define the space of Kirchhoff–Love motion actions given by

$$\mathscr{V}_{KL} = \{\mathbf{v}(\mathbf{x}_o, \xi) \in \mathscr{V}\,;\; (\nabla\mathbf{v})^s_s = \mathbf{0} \text{ and } (\nabla\mathbf{v})^s_n = 0 \quad \forall(\mathbf{x}_o, \xi) \in \Sigma_o \times H\}. \tag{9.88}$$

This space $\mathscr{V}_{KL}$ can be characterized in a completely equivalent manner. In fact, the motion action $\mathbf{v}$ defined in the shell is said to be a Kirchhoff–Love field, that is, $\mathbf{v} \in \mathscr{V}_{KL}$, if and only if the following representation holds

$$\mathbf{v}_t(\mathbf{x}_o, \xi) = \mathbf{v}^o_t(\mathbf{x}_o) + \xi \omega^{KL}_t(\mathbf{x}_o), \tag{9.89}$$

$$v_n(\mathbf{x}_o, \xi) = v^o_n(\mathbf{x}_o), \tag{9.90}$$

where the rotation ω^{KL}_t is particularly given by

$$\omega^{KL}_t(\mathbf{x}_o) = (\nabla_{\mathbf{x}_o}\mathbf{n}(\mathbf{x}_o))\mathbf{v}^o_t(\mathbf{x}_o) - \nabla_{\mathbf{x}_o} v^o_n(\mathbf{x}_o). \tag{9.91}$$

To understand the derivation of this characterization, observe first that if the field possesses the representation given by (9.89) and (9.90) then the kinematical constraints (9.86) and (9.87) are immediately verified, as readily seen when replacing (9.89) and (9.90) into (9.86) and (9.87), considering (9.91) and recalling that $\Lambda = \mathbf{I} + \xi\nabla_{\mathbf{x}_o}\mathbf{n}$

$$\begin{aligned} 2(\nabla\mathbf{v})^s_s &= \frac{\partial \mathbf{v}_t}{\partial \xi} + \Pi_t \Lambda^{-1}(\nabla_{\mathbf{x}_o} v_n - (\nabla_{\mathbf{x}_o}\mathbf{n})\,\mathbf{v}_t) \\ &= \Pi_t \Lambda^{-1}\left(\Lambda\omega^{KL}_t + \nabla_{\mathbf{x}_o} v^o_n - (\nabla_{\mathbf{x}_o}\mathbf{n})\mathbf{v}^o_t - \xi(\nabla_{\mathbf{x}_o}\mathbf{n})\omega^{KL}_t\right) \\ &= \Pi_t \Lambda^{-1}\left(\omega^{KL}_t + \nabla_{\mathbf{x}_o} v^o_n - (\nabla_{\mathbf{x}_o}\mathbf{n})\mathbf{v}^o_t\right) = \mathbf{0}, \end{aligned} \tag{9.92}$$

$$(\nabla\mathbf{v})^s_n = \frac{\partial v_n}{\partial \xi} = \frac{\partial v^o_n}{\partial \xi} = 0. \tag{9.93}$$

To actually construct a Kirchhoff–Love field it is necessary to see that the solution to (9.86) and (9.87) has the form established in (9.89) and (9.90), with (9.91). Particularly, from

$$(\nabla\mathbf{v})^s_n = \frac{\partial v_n}{\partial \xi} = 0 \quad \Rightarrow \quad v_n = v^o_n(\mathbf{x}_o), \tag{9.94}$$

we have that the field v_n can only be a function of $\mathbf{x}_o$, that means $v_n = v_n^o(\mathbf{x}_o)$, and then the representation (9.90) is proved. In addition, and with this result in mind, equation (9.86) can be rewritten as

$$\begin{aligned}2\Lambda^{-1}\Pi_t(\nabla\mathbf{v})_s^s &= \Lambda^{-1}\frac{\partial\mathbf{v}_t}{\partial\xi} + \Lambda^{-1}\Pi_t\Lambda^{-1}(\nabla_{\mathbf{x}_o}v_n^o) - \Lambda^{-1}\Pi_t\Lambda^{-1}(\nabla_{\mathbf{x}_o}\mathbf{n})\,\mathbf{v}_t\\ &= \Lambda^{-1}\frac{\partial\mathbf{v}_t}{\partial\xi} + \Lambda^{-1}\Pi_t\Lambda^{-1}(\nabla_{\mathbf{x}_o}v_n^o) - \Lambda^{-2}(\nabla_{\mathbf{x}_o}\mathbf{n})\,\mathbf{v}_t\\ &= \Lambda^{-1}\frac{\partial\mathbf{v}_t}{\partial\xi} + \Lambda^{-1}\Pi_t\Lambda^{-1}(\nabla_{\mathbf{x}_o}v_n^o) - \Lambda^{-2}\frac{\partial\Lambda}{\partial\xi}\,\mathbf{v}_t = \mathbf{0}.\end{aligned} \tag{9.95}$$

Noting now the following properties of the tensor Λ we have

$$\Lambda = \Lambda^2\Lambda^{-1} \quad\Rightarrow\quad \frac{\partial\Lambda}{\partial\xi} = 2\Lambda\frac{\partial\Lambda}{\partial\xi}\Lambda^{-1} + \Lambda^2\frac{\partial\Lambda^{-1}}{\partial\xi}, \tag{9.96}$$

$$\mathbf{I} = \Lambda\Lambda^{-1} \quad\Rightarrow\quad \frac{\partial\Lambda}{\partial\xi}\Lambda^{-1} + \Lambda\frac{\partial\Lambda^{-1}}{\partial\xi} = \mathbf{O}. \tag{9.97}$$

From the previous expressions we then obtain

$$\frac{\partial\Lambda}{\partial\xi} = -2\Lambda^2\frac{\partial\Lambda^{-1}}{\partial\xi} + \Lambda^2\frac{\partial\Lambda^{-1}}{\partial\xi} = -\Lambda^2\frac{\partial\Lambda^{-1}}{\partial\xi}. \tag{9.98}$$

Replacing (9.98) into (9.95) leads us to the following ordinary differential equation in the variable ξ which provides the form of $\mathbf{v}_t$

$$2\Lambda^{-1}\Pi_t(\nabla\mathbf{v})_s^s = \frac{\partial}{\partial\xi}(\Lambda^{-1}\mathbf{v}_t) + \Lambda^{-1}\Pi_t\Lambda^{-1}(\nabla_{\mathbf{x}_o}v_n^o) = \mathbf{0}. \tag{9.99}$$

The solution of this equation consists of a homogeneous solution, called $\mathbf{v}_t^h$, and a particular solution, called $\mathbf{v}_t^p$. Therefore we have $\mathbf{v}_t = \mathbf{v}_t^h + \mathbf{v}_t^p$, where

$$\frac{\partial}{\partial\xi}(\Lambda^{-1}\mathbf{v}_t^h) = \mathbf{0} \qquad\Rightarrow\qquad \Lambda^{-1}\mathbf{v}_t^h = \mathbf{v}_t^o(\mathbf{x}_o), \tag{9.100}$$

that is,

$$\mathbf{v}_t^h = \Lambda\mathbf{v}_t^o(\mathbf{x}_o) = \mathbf{v}_t^o(\mathbf{x}_o) + \xi(\nabla_{\mathbf{x}_o}\mathbf{n})\mathbf{v}_t^o(\mathbf{x}_o). \tag{9.101}$$

The particular solution is such that

$$\frac{\partial}{\partial\xi}(\Lambda^{-1}\mathbf{v}_t^p) + \Lambda^{-1}\Pi_t\Lambda^{-1}(\nabla_{\mathbf{x}_o}v_n^o) = \mathbf{0}, \tag{9.102}$$

which results in

$$\Lambda^{-1}\mathbf{v}_t^p = \Lambda^{-1}\Pi_t\mathbf{v}_t^p = -\xi\Lambda^{-1}\Pi_t(\nabla_{\mathbf{x}_o}v_n^o) \qquad\Rightarrow\qquad \mathbf{v}_t^p = -\xi(\nabla_{\mathbf{x}_o}v_n^o). \tag{9.103}$$

Utilizing again (9.98), and from the fact that Λ^{-1} and Π_t are commutative tensors ($\Lambda^{-1}\Pi_t = \Pi_t\Lambda^{-1}$), we arrive at

$$\begin{aligned}&\frac{\partial}{\partial\xi}(-\xi\Lambda^{-1}\Pi_t(\nabla_{\mathbf{x}_o}v_n^o)) + \Lambda^{-1}\Pi_t\Lambda^{-1}(\nabla_{\mathbf{x}_o}v_n^o)\\ &\quad = -\Lambda^{-1}\Pi_t(\nabla_{\mathbf{x}_o}v_n^o) - \xi\frac{\partial\Lambda^{-1}}{\partial\xi}\Pi_t(\nabla_{\mathbf{x}_o}v_n^o) + \Lambda^{-1}\Pi_t\Lambda^{-1}(\nabla_{\mathbf{x}_o}v_n^o)\end{aligned}$$

$$
\begin{aligned}
&= -\Lambda^{-1}\Pi_t(\nabla_{\mathbf{x}_o} v_n^o) + \xi\Lambda^{-2}\frac{\partial \Lambda}{\partial \xi}\Pi_t(\nabla_{\mathbf{x}_o} v_n^o) + \Lambda^{-1}\Pi_t\Lambda^{-1}(\nabla_{\mathbf{x}_o} v_n^o) \\
&= \Lambda^{-2}\left[-\Lambda\Pi_t(\nabla_{\mathbf{x}_o} v_n^o) + \xi(\nabla_{\mathbf{x}_o}\mathbf{n})\Pi_t(\nabla_{\mathbf{x}_o} v_n^o)\right] + \Lambda^{-1}\Pi_t\Lambda^{-1}(\nabla_{\mathbf{x}_o} v_n^o) \\
&= -\Lambda^{-2}\Pi_t(\nabla_{\mathbf{x}_o} v_n^o) + \Lambda^{-1}\Pi_t\Lambda^{-1}(\nabla_{\mathbf{x}_o} v_n^o) \\
&= -\Lambda^{-1}\Lambda^{-1}\Pi_t(\nabla_{\mathbf{x}_o} v_n^o) + \Lambda^{-1}\Pi_t\Lambda^{-1}(\nabla_{\mathbf{x}_o} v_n^o) \\
&= -\Lambda^{-1}\Pi_t\Lambda^{-1}(\nabla_{\mathbf{x}_o} v_n^o) + \Lambda^{-1}\Pi_t\Lambda^{-1}(\nabla_{\mathbf{x}_o} v_n^o) = \mathbf{0}. \qquad (9.104)
\end{aligned}
$$

Thus, from (9.94), (9.101), and (9.103), we have that a Kirchhoff–Love field is given by

$$
\begin{aligned}
\mathbf{v}_t &= \mathbf{v}_t^h + \mathbf{v}_t^p = \mathbf{v}_t^o(\mathbf{x}_o) + \xi(\nabla_{\mathbf{x}_o}\mathbf{n})\mathbf{v}_t^o(\mathbf{x}_o) - \xi\nabla_{\mathbf{x}_o} v_n^o \\
&= \mathbf{v}_t^o(\mathbf{x}_o) + \xi\left((\nabla_{\mathbf{x}_o}\mathbf{n})\mathbf{v}_t^o(\mathbf{x}_o) - \nabla_{\mathbf{x}_o} v_n^o\right) = \mathbf{v}_t^o(\mathbf{x}_o) + \xi\omega_t^{KL}(\mathbf{x}_o), \qquad (9.105)\\
v_n &= v_n^o(\mathbf{x}_o). \qquad (9.106)
\end{aligned}
$$

In this manner, we have characterized the space $\mathscr{V}_{KL}$, originally defined in (9.88), but now in terms of the tangent and normal motion actions, as

$$
\begin{aligned}
\mathscr{V}_{KL} = \{&\mathbf{v}(\mathbf{x}_o,\xi) \in \mathscr{V}\ ;\ \mathbf{v} = \mathbf{v}_t + v_n\mathbf{n};\ \mathbf{v}_t \text{ given by (9.105) and} \\
&v_n \text{ given by (9.106)} \quad \forall(\mathbf{x}_o,\xi) \in \Sigma_o \times H\}.
\end{aligned}
$$

Exercise 9.10 Characterize the Kirchhoff–Love motion actions for plates.

It is worthwhile underlining that the kinematical hypotheses in this Kirchhoff–Love model correspond to assuming that fibers normal to the middle surface of the shell (and therefore of length $h(\mathbf{x}_o)$) are subjected to virtual motion actions, $\mathbf{v}_t^o$ and v_n^o, such that the fibers, after the deformation, remain normal to the middle surface and do not modify its length ($(\nabla\mathbf{v})_s^s = \mathbf{0}$ and $(\nabla\mathbf{v})_n^s = 0$).

Now, it is possible to establish the characterization of the internal virtual power functional for all virtual motion actions of Kirchhoff–Love type. Since for these fields the normal component of the strain (rate) action tensor is null, and from the definition of ω_t^{KL} (see (9.91)) we get (see (9.80))

$$
\psi = \omega_t^{KL} + \nabla_{\mathbf{x}_o} v_n^o - (\nabla_{\mathbf{x}_o}\mathbf{n})\mathbf{v}_t^o = \mathbf{0}, \qquad (9.107)
$$

the internal virtual power becomes

$$
\begin{aligned}
\int_\Omega \mathbf{T}\cdot(\nabla\mathbf{v})^s d\Omega = \int_{\Sigma_o} &\left[\mathbf{N}_t\cdot(\Pi_t(\nabla_{\mathbf{x}_o}\mathbf{v}_t^o) + v_n^o\nabla_{\mathbf{x}_o}\mathbf{n})\right. \\
&\left.+\mathbf{M}_t\cdot\Pi_t(\nabla_{\mathbf{x}_o}((\nabla_{\mathbf{x}_o}\mathbf{n})\mathbf{v}_t^o - \nabla_{\mathbf{x}_o} v_n^o))\right] d\Sigma_o. \qquad (9.108)
\end{aligned}
$$

Introducing the notation

$$
\begin{aligned}
\varepsilon^{KL} &= \Pi_t(\nabla_{\mathbf{x}_o}\mathbf{v}_t^o) + v_n^o\nabla_{\mathbf{x}_o}\mathbf{n} = \varepsilon, \qquad (9.109)\\
\chi^{KL} &= \Pi_t(\nabla_{\mathbf{x}_o}((\nabla_{\mathbf{x}_o}\mathbf{n})\mathbf{v}_t^o - \nabla_{\mathbf{x}_o} v_n^o)) = \chi|_{\omega_t=\omega_t^{KL}}, \qquad (9.110)
\end{aligned}
$$

where ε and χ have been defined in the presentation of the Naghdi model (see (9.78) and (9.79)), the internal virtual power for all Kirchhoff–Love motion actions can be expressed in a more compact form as

$$
\int_\Omega \mathbf{T}\cdot(\nabla\mathbf{v})^s d\Omega = \int_{\Sigma_o}\left[\mathbf{N}_t\cdot\varepsilon^{KL} + \mathbf{M}_t\cdot\chi^{KL}\right] d\Sigma_o. \qquad (9.111)
$$

Exercise 9.11 Particularize expressions (9.109) and (9.110) for the case of Kirchhoff–Love plates.

To construct the external virtual power functional for Kirchhoff–Love motion actions we resort to the definitions (9.89) and (9.90), with (9.91), and introduce them into the external virtual power from the Naghdi model (9.82). Because now the field ω_t^{KL} is not an independent kinematic variable (as in the Naghdi model), but now depends on $\mathbf{v}_t^o$ and v_n^o (and particularly on $\nabla_{\mathbf{x}_o} v_n^o$), integration by parts is now required. To do this, and without loss of generality, we will assume that the shell is smooth, that is the normal $\mathbf{n}(\mathbf{x}_o)$ is uniquely defined for all points over the middle surface Σ_o, as well as the shell boundary (there are no edges) and therefore $\mathbf{n}_o$ is uniquely defined in $\partial\Sigma_o$. Hence, we obtain

$$\begin{aligned}
&\int_\Omega \mathbf{b}\cdot\mathbf{v}d\Omega + \int_{\partial\Omega_\mathbf{f}} \mathbf{f}\cdot\mathbf{v}d\partial\Omega_\mathbf{f}\\
&= \int_{\Sigma_o} \left[(\tilde{\mathbf{p}}_{bt}^{KL} + \tilde{\mathbf{p}}_t^{KL})\cdot\mathbf{v}_t^o + (p_{bn} + p_n + \mathrm{div}_{\mathbf{x}_o}(\mathbf{m}_{bt} + \mathbf{m}_t))v_n^o\right] d\Sigma_o\\
&\quad + \int_{\partial\Sigma_{o\mathrm{f}}} \Big[\overline{\mathbf{p}}_t^{KL}\cdot\mathbf{v}_t^o + (\overline{p}_n + (\nabla_{\mathbf{x}_o}(\overline{\mathbf{m}}_t\cdot\mathbf{s}_o))\cdot\mathbf{s}_o - (\mathbf{m}_{bt} + \mathbf{m}_t)\cdot\mathbf{n}_o)v_n^o\\
&\quad + (\overline{\mathbf{m}}_t\cdot\mathbf{n}_o)(\nabla_{\mathbf{x}_o}v_n^o)\cdot\mathbf{n}_o\Big] d\partial\Sigma_{o\mathrm{f}},
\end{aligned} \tag{9.112}$$

where the generalized forces appearing now in the Kirchhoff–Love shell model, $\tilde{\mathbf{p}}_{bt}^{KL}$, $\tilde{\mathbf{p}}_t^{KL}$ and $\overline{\mathbf{p}}_t^{KL}$, are defined similarly to those presented in (9.67), and are specifically given by

$$\begin{aligned}
\tilde{\mathbf{p}}_{bt}^{KL} &= \int_H \Lambda\mathbf{b}_t \det\Lambda d\xi,\\
\tilde{\mathbf{p}}_t^{KL} &= \Lambda^+\mathbf{f}_t^+ \det\Lambda^+ + \Lambda^-\mathbf{f}_t^- \det\Lambda^-,\\
\overline{\mathbf{p}}_t^{KL} &= \int_H \Lambda\overline{\mathbf{f}}_t[\Lambda^2\mathbf{s}_o\cdot\mathbf{s}_o]^{1/2}d\xi.
\end{aligned} \tag{9.113}$$

Also, in (9.112) we have introduced some of the generalized forces already defined in (9.67).

Now, taking into consideration the derived forms of the internal virtual power and external virtual power functionals, we can formalize the concept of mechanical equilibrium for the (exact) Kirchhoff–Love model through the variational equation

$$\begin{aligned}
&\int_{\Sigma_o} \left[\mathbf{N}_t\cdot(\Pi_t(\nabla_{\mathbf{x}_o}\mathbf{v}_t^o) + v_n^o\nabla_{\mathbf{x}_o}\mathbf{n}) + \mathbf{M}_t\cdot\Pi_t(\nabla_{\mathbf{x}_o}((\nabla_{\mathbf{x}_o}\mathbf{n})\mathbf{v}_t^o - \nabla_{\mathbf{x}_o}v_n^o))\right] d\Sigma_o\\
&= \int_{\Sigma_o} \left[(\tilde{\mathbf{p}}_{bt}^{KL} + \tilde{\mathbf{p}}_t^{KL})\cdot\mathbf{v}_t^o + (p_{bn} + p_n + \mathrm{div}_{\mathbf{x}_o}(\mathbf{m}_{bt} + \mathbf{m}_t))v_n^o\right] d\Sigma_o\\
&\quad + \int_{\partial\Sigma_{o\mathrm{f}}} \Big[\overline{\mathbf{p}}_t^{KL}\cdot\mathbf{v}_t^o + (\overline{p}_n + (\nabla_{\mathbf{x}_o}(\overline{\mathbf{m}}_t\cdot\mathbf{s}_o))\cdot\mathbf{s}_o - (\mathbf{m}_{bt} + \mathbf{m}_t)\cdot\mathbf{n}_o)v_n^o\\
&\quad + (\overline{\mathbf{m}}_t\cdot\mathbf{n}_o)(\nabla_{\mathbf{x}_o}v_n^o)\cdot\mathbf{n}_o\Big] d\partial\Sigma_{o\mathrm{f}} \qquad \forall(\mathbf{v}_t^o, v_n^o)\in \mathrm{Var}_v,
\end{aligned} \tag{9.114}$$

where

$$\mathrm{Var}_v = \{(\mathbf{v}_t^o, v_n^o)\in\mathscr{V}_o^{KL};\ (\mathbf{v}_t^o, v_n^o)|_{\partial\Sigma_{o\mathbf{u}}} = (\mathbf{0}, 0),\ (\nabla_{\mathbf{x}_o}v_n^o)\cdot\mathbf{n}_o|_{\partial\Sigma_{o\mathbf{u}}} = 0\}. \tag{9.115}$$

with $\mathscr{V}_o^{KL}$ being a space of functions with the regularity required by the Kirchhoff–Love hypotheses.

Integrating by parts we are led to the Euler–Lagrange equations associated with the variational formulation (9.114), which are

$$\begin{cases} \operatorname{div}_{\mathbf{x}_o}\mathbf{N}_t + (\nabla_{\mathbf{x}_o}\mathbf{n})\operatorname{div}_{\mathbf{x}_o}\mathbf{M}_t + \mathbf{p}_t^{KL} = \mathbf{0} & \text{on } \Sigma_o, \\ \operatorname{div}_{\mathbf{x}_o}\operatorname{div}_{\mathbf{x}_o}\mathbf{M}_t - \mathbf{N}_t \cdot \nabla_{\mathbf{x}_o}\mathbf{n} + p_n^{KL} = 0 & \text{on } \Sigma_o, \\ (\mathbf{N}_t + (\nabla_{\mathbf{x}_o}\mathbf{n})\mathbf{M}_t)\mathbf{n}_o = \overline{\mathbf{p}}_t^{KL} & \text{on } \partial\Sigma_{of}, \\ (\operatorname{div}_{\mathbf{x}_o}\mathbf{M}_t) \cdot \mathbf{n}_o + (\nabla_{\mathbf{x}_o}(\mathbf{M}_t\mathbf{n}_o \cdot \mathbf{s}_o)) \cdot \mathbf{s}_o = \overline{p}_n^{KL} & \text{on } \partial\Sigma_{of}, \\ \mathbf{M}_t\mathbf{n}_o \cdot \mathbf{n}_o = -\overline{m}_n^{KL} & \text{on } \partial\Sigma_{of}, \end{cases} \tag{9.116}$$

where we have introduced the following notations for the generalized external forces acting over the shell:

$$\mathbf{p}_t^{KL} = \tilde{\mathbf{p}}_{bt}^{KL} + \tilde{\mathbf{p}}_t^{KL}, \tag{9.117}$$

$$p_n^{KL} = p_{bn} + p_n + \operatorname{div}_{\mathbf{x}_o}(\mathbf{m}_{bt} + \mathbf{m}_t), \tag{9.118}$$

$$\overline{p}_n^{KL} = \overline{p}_n + (\nabla_{\mathbf{x}_o}(\overline{\mathbf{m}}_t \cdot \mathbf{s}_o)) \cdot \mathbf{s}_o - (\mathbf{m}_{bt} + \mathbf{m}_t) \cdot \mathbf{n}_o, \tag{9.119}$$

$$\overline{m}_n^{KL} = \overline{\mathbf{m}}_t \cdot \mathbf{n}_o. \tag{9.120}$$

Exercise 9.12 Obtain the PVP and the corresponding Euler–Lagrange equations for the Kirchhoff–Love plate model. Discuss the role of the curvature tensor in the coupling of the generalized internal stresses arising in the equilibrium equation for shells.

Before ending the presentation of the so-called exact Kirchhoff–Love shell model it is worth noting once again the difference between this theory and that underlying the Naghdi model. Even if both models apparently adopt the same kinematics for the motion actions (compare the pair (9.72) and (9.73) with the pair (9.89) and (9.90)), these are fundamentally different. As a matter of fact, in the Naghdi model the fields $\mathbf{v}_t^o(\mathbf{x}_o)$, $\omega_t(\mathbf{x}_o)$, and $v_n^o(\mathbf{x}_o)$ are independent of each other. This does not occur in the Kirchhoff–Love model, since the field $\omega_t^{KL}(\mathbf{x}_o)$ becomes a field which is a function of the fields $\mathbf{v}_t^o(\mathbf{x}_o)$ and $v_n^o(\mathbf{x}_o)$ (through $\nabla_{\mathbf{x}_o} v_n^o$). This is a consequence of the kinematical constraint (9.91). Therefore, in the Kirchhoff–Love model, the kinematics describing the motion actions depends upon $\mathbf{v}_t^o(\mathbf{x}_o)$, $v_n^o(\mathbf{x}_o)$ and $\nabla_{\mathbf{x}_o} v_n^o(\mathbf{x}_o)$.

The condensation of all the ingredients in the Kirchhoff–Love theory for shells, according to our unified variational framework, is illustrated in Figure 9.6. In that figure, note that we denoted $\tilde{\mathbf{p}}_t = \mathbf{p}_{bt} + \mathbf{p}_t$, $\tilde{\mathbf{m}}_t = \mathbf{m}_{bt} + \mathbf{m}_t$ and $\tilde{p}_n = p_{bn} + p_n$.

In the sections that follow we will address other shell models known in the specialized literature. These models will be built on top of the exact Kirchhoff–Love model, by assuming further simplifying geometric hypotheses, in terms of the relation between the shell thickness $h(\mathbf{x}_o)$ and its smaller radius of curvature, denoted by $R_m(\mathbf{x}_o)$. For these hypotheses to hold they must be satisfied everywhere in the middle surface of the shell. As will be shown, these hypotheses reshape the geometric operators $\Lambda(\mathbf{x}_o)$ and $\Lambda^{-1}(\mathbf{x}_o)$. Such approximations, on the one hand, allow us to rework the generalized internal stresses, recovering, for example, the property of symmetry which had been

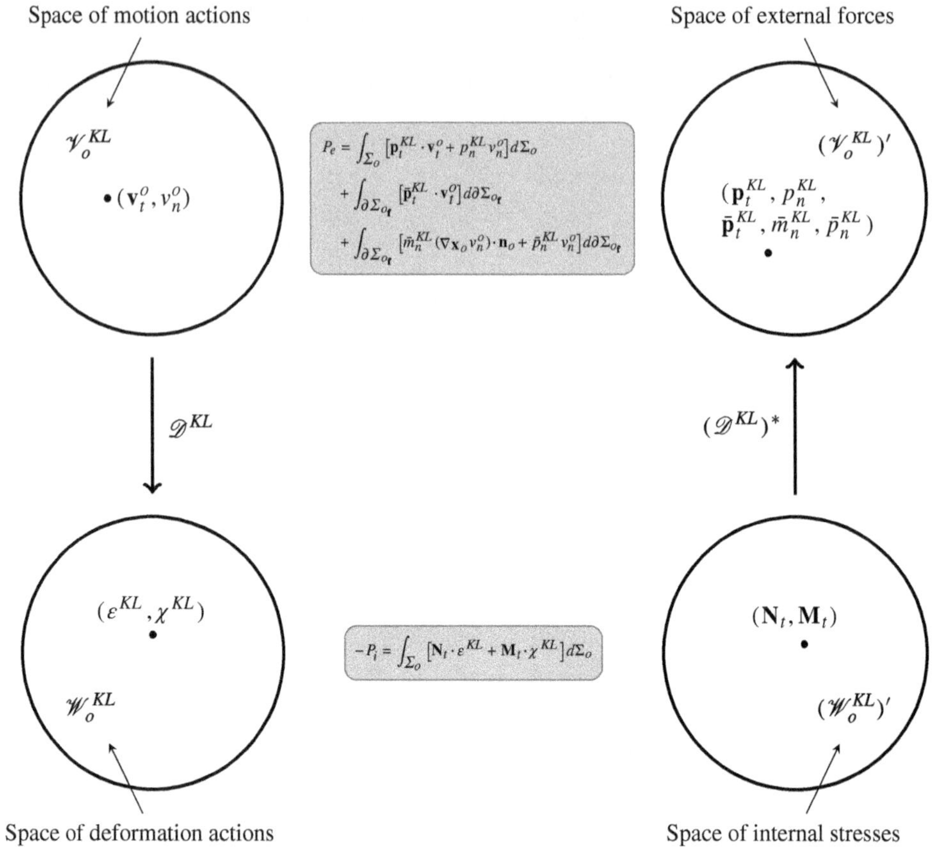

Figure 9.6 Ingredients in the Kirchhoff–Love theory for shells and plates.

lost in the models studied so far. On the other hand, the approximations also introduce modifications in the expressions of the strain (rate) actions.

9.6.3 Love Model

The model addressed in this section was originally proposed by Love in [177], and it is known as the Love's first approximation. This model is based on the Kirchhoff–Love kinematical hypotheses (9.89) and (9.90), while, in addition, it is assumed that the shell is thin such that it verifies the geometric constraint

$$\frac{h(\mathbf{x}_o)}{R_m(\mathbf{x}_o)} \ll 1 \qquad \forall \mathbf{x}_o \in \Sigma_o. \tag{9.121}$$

From the definition of the tensor $\Lambda = \mathbf{I} + \xi \nabla_{\mathbf{x}_o} \mathbf{n}(\mathbf{x}_o)$, the previous geometric constraint implies that

$$\Lambda = \mathbf{I} + \mathcal{O}\left(\frac{h}{R_m}\right) \qquad \Rightarrow \qquad \Lambda \approx \mathbf{I} \qquad \Rightarrow \qquad \det \Lambda \approx 1. \tag{9.122}$$

With such simplifications the generalized (non-symmetric) internal stresses $\mathbf{N}_t$ and $\mathbf{M}_t$, defined in (9.63), now denoted by $\mathbf{N}_t^L$ and $\mathbf{M}_t^L$, become symmetric, in fact

$$\mathbf{N}_t^L = \int_H \mathbf{T}_t d\xi,$$
$$\mathbf{M}_t^L = \int_H \mathbf{T}_t \xi d\xi. \tag{9.123}$$

Introducing these approximations into the internal virtual power of the exact Kirchhoff–Love model (see (9.111)) we get the internal power for the Love model

$$\int_\Omega \mathbf{T}\cdot(\nabla\mathbf{v})^s d\Omega \approx \int_{\Sigma_o} \left[\mathbf{N}_t^L\cdot\varepsilon^L + \mathbf{M}_t^L\cdot\chi^L\right] d\Sigma_o, \tag{9.124}$$

where strain actions dual to the generalized stresses $\mathbf{N}_t^L$ and $\mathbf{M}_t^L$ are given, respectively, by ε^L and χ^L defined as

$$\varepsilon^L = (\varepsilon^{KL})^s, \tag{9.125}$$
$$\chi^L = (\chi^{KL})^s, \tag{9.126}$$

with ε^{KL} and χ^{KL} defined in (9.109) and (9.110).

The external power for the Love model is easily obtained from the external power functional constructed in the Kirchhoff–Love model. Now, the generalized forces are modified due to the geometric constraint (9.122), giving rise to the following generalized forces compatible with the Love model

$$\begin{aligned}
\tilde{\mathbf{p}}_{bt}^L &= \textstyle\int_H \mathbf{b}_t d\xi, & p_{bn}^L &= \textstyle\int_H b_n d\xi,\\
\tilde{\mathbf{p}}_t^L &= \mathbf{f}_t^+ + \mathbf{f}_t^-, & p_n^L &= f_n^+ + f_n^-,\\
\overline{\mathbf{p}}_t^L &= \textstyle\int_H \overline{\mathbf{f}}_t[\mathbf{s}_o\cdot\mathbf{s}_o]^{1/2} d\xi, & \overline{p}_n^L &= \textstyle\int_H \overline{f}_n[\mathbf{s}_o\cdot\mathbf{s}_o]^{1/2} d\xi,\\
\mathbf{m}_{bt}^L &= \textstyle\int_H \mathbf{b}_t \xi d\xi, & \mathbf{m}_t^L &= \tfrac{h}{2}(\mathbf{f}_t^+ - \mathbf{f}_t^-),\\
\overline{\mathbf{m}}_t^L &= \textstyle\int_H \overline{\mathbf{f}}_t \xi[\mathbf{s}_o\cdot\mathbf{s}_o]^{1/2} d\xi,
\end{aligned} \tag{9.127}$$

yielding the following external virtual power functional

$$\begin{aligned}
&\int_\Omega \mathbf{b}\cdot\mathbf{v}d\Omega + \int_{\partial\Omega_\mathrm{f}} \mathbf{f}\cdot\mathbf{v}d\partial\Omega_\mathrm{f}\\
&\approx \int_{\Sigma_o} \left[(\tilde{\mathbf{p}}_{bt}^L + \tilde{\mathbf{p}}_t^L)\cdot\mathbf{v}_t^o + (p_{bn}^L + p_n^L + \mathrm{div}_{\mathbf{x}_o}(\mathbf{m}_{bt}^L + \mathbf{m}_t^L))v_n^o\right] d\Sigma_o\\
&+ \int_{\partial\Sigma_{o\mathrm{f}}} \Big[\overline{\mathbf{p}}_t^L\cdot\mathbf{v}_t^o + (\overline{p}_n^L + (\nabla_{\mathbf{x}_o}(\overline{\mathbf{m}}_t^L\cdot\mathbf{s}_o))\cdot\mathbf{s}_o - (\mathbf{m}_{bt}^L + \mathbf{m}_t^L)\cdot\mathbf{n}_o)v_n^o\\
&+(\overline{\mathbf{m}}_t^L\cdot\mathbf{n}_o)(\nabla_{\mathbf{x}_o}v_n^o)\cdot\mathbf{n}_o\Big] d\partial\Sigma_{o\mathrm{f}}.
\end{aligned} \tag{9.128}$$

Then, it is straightforward to write the PVP in the Love model through the corresponding variational expression

$$\begin{aligned}
&\int_{\Sigma_o} \left[\mathbf{N}_t^L\cdot\varepsilon^L + \mathbf{M}_t^L\cdot\chi^L\right] d\Sigma_o\\
&= \int_{\Sigma_o} \left[(\tilde{\mathbf{p}}_{bt}^L + \tilde{\mathbf{p}}_t^L)\cdot\mathbf{v}_t^o + (p_{bn}^L + p_n^L + \mathrm{div}_{\mathbf{x}_o}(\mathbf{m}_{bt}^L + \mathbf{m}_t^L))v_n^o\right] d\Sigma_o
\end{aligned}$$

$$
+ \int_{\partial\Sigma_{of}} \Big[\overline{\mathbf{p}}_t^L \cdot \mathbf{v}_t^o + (\overline{p}_n^L + (\nabla_{\mathbf{x}_o}(\overline{\mathbf{m}}_t^L \cdot \mathbf{s}_o)) \cdot \mathbf{s}_o - (\mathbf{m}_{bt}^L + \mathbf{m}_t^L) \cdot \mathbf{n}_o) v_n^o
$$

$$
+ (\overline{\mathbf{m}}_t^L \cdot \mathbf{n}_o)(\nabla_{\mathbf{x}_o} v_n^o) \cdot \mathbf{n}_o \Big] \, d\partial\Sigma_{of} \qquad \forall (\mathbf{v}_t^o, v_n^o) \in \mathrm{Var}_v, \tag{9.129}
$$

with Var_v defined as in (9.115). Finally, the Euler–Lagrange equations associated with the variational statement (9.129) are derived from integrating by parts as in the exact Kirchhoff–Love model, with the corresponding modifications in the generalized internal stresses and external forces produced by the geometric approximation imposed by the Love model.

Exercise 9.13 Derive the Euler–Lagrange equations for the Love model. Then discuss the similarities and differences between the Love model and the Kirchhoff–Love model.

9.6.4 Koiter Model

We address now the model proposed by Koiter in [158]. Let us recall the internal power from the exact Kirchhoff–Love theory, given by expression (9.108), and which is repeated here for convenience

$$
\int_\Omega \mathbf{T} \cdot (\nabla \mathbf{v})^s d\Omega
$$

$$
= \int_{\Sigma_o} \int_H \mathbf{T}_t \Lambda^{-1} \cdot \left[(\Pi_t (\nabla_{\mathbf{x}_o} \mathbf{v}_t^o) + v_n^o \nabla_{\mathbf{x}_o} \mathbf{n} + \xi \Pi_t \nabla_{\mathbf{x}_o} \omega_t^{KL}) \right] \det \Lambda d\xi d\Sigma_o. \tag{9.130}
$$

Exploiting the relation that establishes that $\mathbf{R} \cdot \mathbf{ST} = \mathbf{RT}^T \cdot \mathbf{S} = \mathbf{S}^T\mathbf{R} \cdot \mathbf{T}$, which holds for arbirtary second-order tensor fields $\mathbf{R}$, $\mathbf{S}$ and $\mathbf{T}$, and taking into account the symmetry of tensor Λ^{-1}, the internal power can be rewritten as

$$
\int_\Omega \mathbf{T} \cdot (\nabla \mathbf{v})^s d\Omega
$$

$$
= \int_{\Sigma_o} \int_H \Lambda^{-1} \mathbf{T}_t \Lambda^{-1} \cdot \left[\Lambda (\Pi_t (\nabla_{\mathbf{x}_o} \mathbf{v}_t^o) + v_n^o \nabla_{\mathbf{x}_o} \mathbf{n} + \xi \Pi_t \nabla_{\mathbf{x}_o} \omega_t^{KL}) \right]^s \det \Lambda d\xi d\Sigma_o. \tag{9.131}
$$

Since $\Lambda = \mathbf{I} + \xi \nabla_{\mathbf{x}_o} \mathbf{n}$, the term in brackets is expressed as

$$
\left[\Lambda (\Pi_t (\nabla_{\mathbf{x}_o} \mathbf{v}_t^o) + v_n^o \nabla_{\mathbf{x}_o} \mathbf{n} + \xi \Pi_t \nabla_{\mathbf{x}_o} \omega_t^{KL}) \right]^s = (\Pi_t (\nabla_{\mathbf{x}_o} \mathbf{v}_t^o))^s + v_n^o \nabla_{\mathbf{x}_o} \mathbf{n}
$$

$$
+ \xi \left[(\Lambda \Pi_t \nabla_{\mathbf{x}_o} \omega_t^{KL}) + (\nabla_{\mathbf{x}_o} \mathbf{n}) \Pi_t (\nabla_{\mathbf{x}_o} \mathbf{v}_t^o) + v_n^o (\nabla_{\mathbf{x}_o} \mathbf{n})^2) \right]^s. \tag{9.132}
$$

By neglecting terms of order $\frac{h}{R_m}$ when compared to the identity, we obtain

$$
\left[\Lambda (\Pi_t (\nabla_{\mathbf{x}_o} \mathbf{v}_t^o) + v_n^o \nabla_{\mathbf{x}_o} \mathbf{n} + \xi \Pi_t \nabla_{\mathbf{x}_o} \omega_t^{KL}) \right]^s \approx (\Pi_t (\nabla_{\mathbf{x}_o} \mathbf{v}_t^o))^s + v_n^o \nabla_{\mathbf{x}_o} \mathbf{n}
$$

$$
+ \xi \left[(\Pi_t \nabla_{\mathbf{x}_o} \omega_t^{KL})^s + \left[(\nabla_{\mathbf{x}_o} \mathbf{n}) \Pi_t (\nabla_{\mathbf{x}_o} \mathbf{v}_t^o) \right]^s + v_n^o (\nabla_{\mathbf{x}_o} \mathbf{n})^2 \right]. \tag{9.133}
$$

Remembering the definitions of the strain actions in the Love model given by (9.125) and (9.126) (see also the expressions of the strain actions in the Kirchhoff–Love model defined in (9.109) and (9.110)), and introducing the definitions

$$
\varepsilon^K = \varepsilon^L, \tag{9.134}
$$

$$\chi^K = \chi^L + ((\nabla_{\mathbf{x}_o}\mathbf{n})\Pi_t(\nabla_{\mathbf{x}_o}\mathbf{v}_t^o))^s + v_n^o(\nabla_{\mathbf{x}_o}\mathbf{n})^2, \tag{9.135}$$

we note that expression (9.133) can be written as

$$\left[\Lambda(\Pi_t(\nabla_{\mathbf{x}_o}\mathbf{v}_t^o) + v_n^o\nabla_{\mathbf{x}_o}\mathbf{n} + \xi\Pi_t\nabla_{\mathbf{x}_o}\omega_t^{KL})\right]^s \approx \varepsilon^K + \xi\chi^K. \tag{9.136}$$

Inserting this result into the definition of the internal virtual power functional (9.131) yields

$$\int_\Omega \mathbf{T}\cdot(\nabla\mathbf{v})^s d\Omega \approx \int_{\Sigma_o}\int_H \Lambda^{-1}\mathbf{T}_t\Lambda^{-1}\cdot\left[\varepsilon^K + \xi\chi^K\right]\det\Lambda d\xi d\Sigma_o. \tag{9.137}$$

This expression allows us to deduce the generalized internal stresses which are power-conjugate to the strain actions defined in the Koiter model. These generalized stresses are

$$\begin{aligned}\mathbf{N}_t^K &= \int_H \Lambda^{-1}\mathbf{T}_t\Lambda^{-1}\det\Lambda d\xi,\\ \mathbf{M}_t^K &= \int_H \Lambda^{-1}\mathbf{T}_t\Lambda^{-1}\xi\det\Lambda d\xi.\end{aligned} \tag{9.138}$$

With these results we arrive at the definition of the internal power functional in the Koiter model

$$\int_\Omega \mathbf{T}\cdot(\nabla\mathbf{v})^s d\Omega \approx \int_{\Sigma_o}\left[\mathbf{N}_t\cdot\varepsilon^K + \mathbf{M}_t\cdot\chi^K\right]d\Sigma_o. \tag{9.139}$$

It is important to note that the approximation introduced in the expression of the internal virtual power from the (exact) Kirchhoff–Love theory (see Section 9.6.2) to obtain the internal power functional in the Koiter model maintains the same order of approximation as that utilized in the Love model (see Section 9.6.3). In other words, both models (Love and Koiter) drop all terms which are of order h/R_m when compared to the identity. However, the term

$$\begin{aligned}\xi\left((\nabla_{\mathbf{x}_o}\mathbf{n})\Pi_t(\nabla_{\mathbf{x}_o}\mathbf{v}_t^o) + v_n^o(\nabla_{\mathbf{x}_o}\mathbf{n})^2\right)^s &= \mathcal{O}\left(\frac{h}{R_m}\right)\left(\Pi_t(\nabla_{\mathbf{x}_o}\mathbf{v}_t^o) + v_n^o(\nabla_{\mathbf{x}_o}\mathbf{n})\right)^s\\ &= \mathcal{O}\left(\frac{h}{R_m}\right)\varepsilon^L,\end{aligned} \tag{9.140}$$

in (9.133) remains. That is to say, all terms which are linear in ξ have been used in the deduction of the internal virtual power for the Koiter model. This strategy enables some inconsistencies that appear in the Love model related to the equilibrium around the normal vector and to rigid modes to be resolved. This will be addressed in more detail in Section 9.8.

The Euler–Lagrange equations associated with the PVP in the Koiter model (with the internal power given by (9.139)) are obtained following the standard procedure of integration by parts, and are given by

$$\begin{cases}\operatorname{div}_{\mathbf{x}_o}\mathbf{N}_t^K + (\nabla_{\mathbf{x}_o}\mathbf{n})\operatorname{div}_{\mathbf{x}_o}\mathbf{M}_t^K + \operatorname{div}_{\mathbf{x}_o}((\nabla_{\mathbf{x}_o}\mathbf{n})\mathbf{M}_t^K) + \mathbf{p}_t^{KL} = \mathbf{0} & \text{on } \Sigma_o,\\ \operatorname{div}_{\mathbf{x}_o}\operatorname{div}_{\mathbf{x}_o}\mathbf{M}_t^K - \mathbf{M}_t^K\cdot(\nabla_{\mathbf{x}_o}\mathbf{n})^2 - \mathbf{N}_t^K\cdot(\nabla_{\mathbf{x}_o}\mathbf{n}) + p_n^{KL} = 0 & \text{on } \Sigma_o,\\ (\mathbf{N}_t^K + 2(\nabla_{\mathbf{x}_o}\mathbf{n})\mathbf{M}_t^K)\mathbf{n}_o = \overline{\mathbf{p}}_t^{KL} & \text{on } \partial\Sigma_{o\mathbf{f}},\\ \operatorname{div}_{\mathbf{x}_o}\mathbf{M}_t^K\cdot\mathbf{n}_o + (\nabla_{\mathbf{x}_o}(\mathbf{M}_t^K\mathbf{n}_o\cdot\mathbf{s}_o))\cdot\mathbf{s}_o = \overline{p}_n^{KL} & \text{on } \partial\Sigma_{o\mathbf{f}},\\ \mathbf{M}_t^K\mathbf{n}_o\cdot\mathbf{n}_o = -\overline{m}_n^{KL} & \text{on } \partial\Sigma_{o\mathbf{f}}.\end{cases} \tag{9.141}$$

Notice that two additional terms appear in the Koiter model, given by $\mathrm{div}_{\mathbf{x}_o}((\nabla_{\mathbf{x}_o}\mathbf{n})\mathbf{M}_t^K)$ and $-\mathbf{M}_t^K \cdot (\nabla_{\mathbf{x}_o}\mathbf{n})^2$ in the equations corresponding to the membrane equilibrium and bending equilibrium, respectively. Finally, the natural boundary conditions are the same as in the Kirchhoff–Love model, see $(9.116)_3$ and $(9.116)_5$, we have just to replace $\mathbf{N}_t$ and $\mathbf{M}_t$ by $\mathbf{N}_t^K$ and $\mathbf{M}_t^K$.

Exercise 9.14 Derive the Koiter model for the case of plates and compare with previous plate models. Discuss the results.

9.6.5 Sanders Model

The model discussed in this section was formerly proposed in [270]. The construction of this model follows the same guidelines as previous sections. Thus, let us take the internal power functional from the exact Kirchhoff–Love theory, established by expression (9.108), and rewrite it as

$$\int_\Omega \mathbf{T}\cdot(\nabla\mathbf{v})^s d\Omega$$
$$= \int_{\Sigma_o}\int_H \mathbf{T}_t \cdot \left[\left(\Pi_t(\nabla_{\mathbf{x}_o}\mathbf{v}_t^o) + v_n^o\nabla_{\mathbf{x}_o}\mathbf{n} + \xi\Pi_t(\nabla_{\mathbf{x}_o}\omega^{KL})\right)\Lambda^{-1}\right]^s \det\Lambda d\xi d\Sigma_o. \quad (9.142)$$

This internal power can equivalently be written in the format

$$\int_\Omega \mathbf{T}\cdot(\nabla\mathbf{v})^s d\Omega$$
$$= \int_{\Sigma_o}\int_H \Lambda^{-1}\mathbf{T}_t \cdot \Lambda\left[\left(\Pi_t(\nabla_{\mathbf{x}_o}\mathbf{v}_t^o) + v_n^o\nabla_{\mathbf{x}_o}\mathbf{n} + \xi\Pi_t(\nabla_{\mathbf{x}_o}\omega^{KL})\right)\Lambda^{-1}\right]^s \det\Lambda d\xi d\Sigma_o.$$
$$(9.143)$$

Now, exploiting the expansion presented in (9.239), the term $\Lambda[\dots]^s$ in (9.143) can be described as

$$\Lambda\left[\left(\Pi_t(\nabla_{\mathbf{x}_o}\mathbf{v}_t^o) + v_n^o\nabla_{\mathbf{x}_o}\mathbf{n} + \xi(\Pi_t(\nabla_{\mathbf{x}_o}\omega^{KL}))\right)\Lambda^{-1}\right]^s$$
$$= \left(\Pi_t(\nabla_{\mathbf{x}_o}\mathbf{v}_t^o)\right)^s + v_n^o\nabla_{\mathbf{x}_o}\mathbf{n} + \xi\Bigg[\left(\mathbf{I} + \mathcal{O}\left(\frac{h}{R_m}\right)\right)\left(\Pi_t\left(\nabla_{\mathbf{x}_o}\omega^{KL}\right)\right)^s$$
$$-\left(\Pi_t\left(\nabla_{\mathbf{x}_o}\mathbf{v}_t^o\right)\left(\nabla_{\mathbf{x}_o}\mathbf{n}\right)\right)^s + \left(\nabla_{\mathbf{x}_o}\mathbf{n}\right)\left(\Pi_t\left(\nabla_{\mathbf{x}_o}\mathbf{v}_t^o\right)\right)^s\Bigg]. \quad (9.144)$$

Dropping the terms of order $\mathcal{O}\left(\frac{h}{R_m}\right)$ gives

$$\Lambda\left[\left(\Pi_t(\nabla_{\mathbf{x}_o}\mathbf{v}_t^o) + v_n^o\nabla_{\mathbf{x}_o}\mathbf{n} + \xi(\Pi_t(\nabla_{\mathbf{x}_o}\omega^{KL}))\right)\Lambda^{-1}\right]^s$$
$$\approx (\Pi_t(\nabla_{\mathbf{x}_o}\mathbf{v}_t^o))^s + v_n^o\nabla_{\mathbf{x}_o}\mathbf{n} + \xi(\Pi_t(\nabla_{\mathbf{x}_o}\omega^{KL}))^s$$
$$-\xi(\Pi_t(\nabla_{\mathbf{x}_o}\mathbf{v}_t^o)(\nabla_{\mathbf{x}_o}\mathbf{n}))^s + \xi(\nabla_{\mathbf{x}_o}\mathbf{n})(\Pi_t(\nabla_{\mathbf{x}_o}\mathbf{v}_t^o))^s. \quad (9.145)$$

Further manipulation of the last two terms yields

$$-\xi(\Pi_t(\nabla_{\mathbf{x}_o}\mathbf{v}_t^o)(\nabla_{\mathbf{x}_o}\mathbf{n}))^s + \xi(\nabla_{\mathbf{x}_o}\mathbf{n})(\Pi_t(\nabla_{\mathbf{x}_o}\mathbf{v}_t^o))^s$$
$$= -\xi\frac{1}{2}\left[\Pi_t(\nabla_{\mathbf{x}_o}\mathbf{v}_t^o)(\nabla_{\mathbf{x}_o}\mathbf{n}) + (\nabla_{\mathbf{x}_o}\mathbf{n})(\Pi_t(\nabla_{\mathbf{x}_o}\mathbf{v}_t^o))^T\right]$$
$$+\xi(\nabla_{\mathbf{x}_o}\mathbf{n})\frac{1}{2}\left[(\Pi_t(\nabla_{\mathbf{x}_o}\mathbf{v}_t^o)) + (\Pi_t(\nabla_{\mathbf{x}_o}\mathbf{v}_t^o))^T\right]$$

$$= \xi(\nabla_{\mathbf{x}_o}\mathbf{n})\frac{1}{2}\left[(\Pi_t(\nabla_{\mathbf{x}_o}\mathbf{v}_t^o)) - (\Pi_t(\nabla_{\mathbf{x}_o}\mathbf{v}_t^o))^T\right]$$
$$+ \xi\frac{1}{2}\left[(\nabla_{\mathbf{x}_o}\mathbf{n})(\Pi_t(\nabla_{\mathbf{x}_o}\mathbf{v}_t^o))^T - (\Pi_t(\nabla_{\mathbf{x}_o}\mathbf{v}_t^o))(\nabla_{\mathbf{x}_o}\mathbf{n})\right]$$
$$= \xi(\nabla_{\mathbf{x}_o}\mathbf{n})\Omega_t^o + \xi\left[(\nabla_{\mathbf{x}_o}\mathbf{n})(\Pi_t(\nabla_{\mathbf{x}_o}\mathbf{v}_t^o))^T\right]^a, \tag{9.146}$$

where

$$\Omega_t^o = \frac{1}{2}\left[(\Pi_t(\nabla_{\mathbf{x}_o}\mathbf{v}_t^o)) - (\Pi_t(\nabla_{\mathbf{x}_o}\mathbf{v}_t^o))^T\right] = \left[\Pi_t(\nabla_{\mathbf{x}_o}\mathbf{v}_t^o)\right]^a, \tag{9.147}$$

represents the rotation around the normal vector of the shell middle surface and $(\cdot)^a$ is the skew-symmetric part of tensor $(\cdot)$.

With these results, the internal virtual power is approximated as

$$\int_\Omega \mathbf{T}\cdot(\nabla\mathbf{v})^s d\Omega \approx \int_{\Sigma_o}\int_H \Lambda^{-1}\mathbf{T}_t\cdot\left[(\Pi_t(\nabla_{\mathbf{x}_o}\mathbf{v}_t^o))^s + v_n^o\nabla_{\mathbf{x}_o}\mathbf{n} + \xi(\Pi_t(\nabla_{\mathbf{x}_o}\omega^{KL}))^s\right.$$
$$\left. +\xi(\nabla_{\mathbf{x}_o}\mathbf{n})\Omega_t^o + \xi\left[(\nabla_{\mathbf{x}_o}\mathbf{n})(\Pi_t(\nabla_{\mathbf{x}_o}\mathbf{v}_t^o))^T\right]^a\right]\det\Lambda d\xi d\Sigma_o. \tag{9.148}$$

The next step consists of finding an approximation for the tensor $\Lambda^{-1}\mathbf{T}_t$, in which terms of order $\mathcal{O}\left(\frac{h}{R_m}\right)$ are to be neglected. To do this, we will make use of the expansion for Λ^{-1}, as

$$\Lambda^{-1}\mathbf{T}_t = (\Lambda^{-1}\mathbf{T}_t)^s + (\Lambda^{-1}\mathbf{T}_t)^a$$
$$= \frac{1}{2}(\Lambda^{-1}\mathbf{T}_t + \mathbf{T}_t\Lambda^{-1}) + \frac{1}{2}(\Lambda^{-1}\mathbf{T}_t - (\Lambda^{-1}\mathbf{T}_t)^T)$$
$$= \frac{1}{2}\left[\Lambda^{-1}\mathbf{T}_t + \mathbf{T}_t\Lambda^{-1} + \mathbf{T}_t - \xi(\nabla_{\mathbf{x}_o}\mathbf{n})\Lambda^{-1}\mathbf{T}_t - \mathbf{T}_t + \xi\mathbf{T}_t\Lambda^{-1}(\nabla_{\mathbf{x}_o}\mathbf{n})\right]$$
$$= \frac{1}{2}\left[\Lambda^{-1}\mathbf{T}_t\left(\mathbf{I} - \mathcal{O}\left(\frac{h}{R_m}\right)\right) + \mathbf{T}_t\Lambda^{-1}\left(\mathbf{I} + \mathcal{O}\left(\frac{h}{R_m}\right)\right)\right]$$
$$\approx \frac{1}{2}\left[\Lambda^{-1}\mathbf{T}_t + \mathbf{T}_t\Lambda^{-1}\right] = (\Lambda^{-1}\mathbf{T}_t)^s. \tag{9.149}$$

Introducing this result into the internal virtual power (9.148), and recalling that $\mathbf{S}\cdot\mathbf{A} = 0$ where $\mathbf{S}$ is a symmetric tensor and $\mathbf{A}$ is skew-symmetric, we obtain

$$\int_\Omega \mathbf{T}\cdot(\nabla\mathbf{v})^s d\Omega \approx \int_{\Sigma_o}\int_H (\Lambda^{-1}\mathbf{T}_t)^s\cdot\left[(\Pi_t(\nabla_{\mathbf{x}_o}\mathbf{v}_t^o))^s + v_n^o\nabla_{\mathbf{x}_o}\mathbf{n}\right.$$
$$\left. + \xi(\Pi_t(\nabla_{\mathbf{x}_o}\omega^{KL}))^s + \xi((\nabla_{\mathbf{x}_o}\mathbf{n})\Omega_t^o)^s\right]\det\Lambda d\xi d\Sigma_o. \tag{9.150}$$

This approximated expression clearly leads us to the generalized strain actions and generalized stresses in the Sanders model. Indeed, consider the following definitions for the generalized stresses

$$\mathbf{N}_t^S = \int_H (\Lambda^{-1}\mathbf{T}_t)^s \det\Lambda\, d\xi = (\mathbf{N}_t)^s,$$
$$\mathbf{M}_t^S = \int_H (\Lambda^{-1}\mathbf{T}_t)^s \xi \det\Lambda\, d\xi = (\mathbf{M}_t)^s, \tag{9.151}$$

and the definitions for the corresponding generalized strain actions

$$\varepsilon^S = \varepsilon^L,$$
$$\chi^S = \chi^L + ((\nabla_{\mathbf{x}_o}\mathbf{n})\Omega_t^o)^s, \tag{9.152}$$

with ε^L and χ^L being the corresponding strain actions from the Love model defined in (9.125) and (9.126) (see also the strain actions in the Kirchhoff–Love model, (9.109) and (9.110)). Hence, we arrive at the approximated expression for the internal virtual power in the Sanders model

$$\int_{\Omega} \mathbf{T} \cdot (\nabla \mathbf{v})^s d\Omega \approx \int_{\Sigma_o} \left[\mathbf{N}_t^S \cdot \varepsilon^S + \mathbf{M}_t^S \cdot \chi^S \right] d\Sigma_o. \tag{9.153}$$

As with the Koiter model, the Sanders model aims to resolve the inconsistencies which appear in the Love model by maintaining some specific terms of order $\mathcal{O}\left(\frac{h}{R_m}\right)$.

Similarly to previous sections, the Euler–Lagrange equations associated with the PVP in the Sanders model are

$$\begin{cases} \mathrm{div}_{\mathbf{x}_o} \left[\mathbf{N}_t^S + ((\nabla_{\mathbf{x}_o} \mathbf{n}) \mathbf{M}_t^S)^a) \right] + (\nabla_{\mathbf{x}_o} \mathbf{n}) \mathrm{div}_{\mathbf{x}_o} \mathbf{M}_t^S + \mathbf{p}_t^{KL} = \mathbf{0} & \text{on } \Sigma_o, \\ \mathrm{div}_{\mathbf{x}_o} \mathrm{div}_{\mathbf{x}_o} \mathbf{M}_t^S - \mathbf{N}_t^S \cdot (\nabla_{\mathbf{x}_o} \mathbf{n}) + p_n^{KL} = 0 & \text{on } \Sigma_o, \\ \left[\mathbf{N}_t^K + ((\nabla_{\mathbf{x}_o} \mathbf{n}) \mathbf{M}_t^S))^s \right] \mathbf{n}_o = \overline{\mathbf{p}}_t^{KL} & \text{on } \partial\Sigma_{o\mathbf{f}}, \\ \mathrm{div}_{\mathbf{x}_o} \mathbf{M}_t^S \cdot \mathbf{n}_o + (\nabla_{\mathbf{x}_o} (\mathbf{M}_t^S \mathbf{n}_o \cdot \mathbf{s}_o)) \cdot \mathbf{s}_o = \overline{p}_n^{KL} & \text{on } \partial\Sigma_{o\mathbf{f}}, \\ \mathbf{M}_t^S \mathbf{n}_o \cdot \mathbf{n}_o = -\overline{m}_n^{KL} & \text{on } \partial\Sigma_{o\mathbf{f}}. \end{cases} \tag{9.154}$$

Exercise 9.15 Obtain the corresponding Sanders model for plates and compare the formulation with those obtained in previous sections.

9.6.6 Donnell–Mushtari–Vlasov Model

The last shell model presented in this chapter was proposed by Donnell for cylindrical shells, and by Mushtari and Vlasov for thin shells (see [213]).

This model introduces a different type of approximation. Here, the strain (rate) tensor ε^L is considered, while the contribution of the gradient of the tangent velocity $\mathbf{v}_t^o$ is disregarded in the definition of the strain (rate) tensor χ_t^L (see (9.126)). In this way, there is a new tensor that measures the curvature variation, which characterizes the Donnell–Mushtari–Vlasov model. Hence, the strain (rate) action for this model is

$$\varepsilon^{DMV} = \varepsilon^L, \tag{9.155}$$

$$\chi^{DMV} = -\Pi_t \nabla_{\mathbf{x}_o} \nabla_{\mathbf{x}_o} v_n^o. \tag{9.156}$$

In turn, the generalized stresses are similar to those given in (9.123), that is

$$\mathbf{N}_t^{DMV} = \mathbf{N}_t^L, \tag{9.157}$$

$$\mathbf{M}_t^{DMV} = \mathbf{M}_t^L. \tag{9.158}$$

With these elements, the internal virtual power takes the form

$$\int_{\Omega} \mathbf{T} \cdot (\nabla \mathbf{v})^s d\Omega \approx \int_{\Sigma_o} \left[\mathbf{N}_t^{DMV} \cdot \varepsilon^{DMV} + \mathbf{M}_t^{DMV} \cdot \chi^{DMV} \right] d\Sigma_o. \tag{9.159}$$

The Euler–Lagrange equations associated with the Donnell–Mushtari–Vlasov model can directly be derived using the standard arguments, but taking into account the internal power given in (9.159). These equations are

$$\begin{cases} \operatorname{div}_{\mathbf{x}_o} \mathbf{N}_t^{DMV} + \mathbf{p}_t^{KL} = \mathbf{0} & \text{on } \Sigma_o, \\ \operatorname{div}_{\mathbf{x}_o} \operatorname{div}_{\mathbf{x}_o} \mathbf{M}_t^{DMV} - \mathbf{N}_t^{DMV} \cdot (\nabla_{\mathbf{x}_o} \mathbf{n}) + p_n^{KL} = 0 & \text{on } \Sigma_o, \\ \mathbf{N}_t^{DMV} \mathbf{n}_o = \overline{\mathbf{p}}_t^{KL} & \text{on } \partial\Sigma_{of}, \\ \operatorname{div}_{\mathbf{x}_o} \mathbf{M}_t^{DMV} \cdot \mathbf{n}_o + (\nabla_{\mathbf{x}_o} (\mathbf{M}_t^{DMV} \mathbf{n}_o \cdot \mathbf{s}_o)) \cdot \mathbf{s}_o = \overline{p}_n^{KL} & \text{on } \partial\Sigma_{of}, \\ \mathbf{M}_t^{DMV} \mathbf{n}_o \cdot \mathbf{n}_o = -\overline{m}_n^{KL} & \text{on } \partial\Sigma_{of}. \end{cases} \tag{9.160}$$

Exercise 9.16 Obtain the Donnell–Mushtari–Vlasov model for plates. Compare and discuss the results in the light of the theories developed in previous sections.

9.7 Constitutive Equations and Internal Constraints

In previous sections we have made use of the PVP to define the equilibrium for the different shell models addressed. Remarkably, we have shown that many models available in the literature could be covered by this unified variational tapestry, while clearly and unambiguously deriving the generalized internal stresses and generalized external forces that are compatible according to the corresponding kinematical hypotheses.

As already stressed when presenting the basic concepts underlying the PVP, the variational construct is defined in the current spatial (deformed) configuration of the body, where equilibrium actually and physically occurs. Thus, the provided variational equations are valid for any level of strains occurring within the body during the deformation process. Moreover, these equations remain valid for any material behavior of the shell. In fact, the material model was never explicitly defined within the variational framework developed so far. This material model, expressed through a constitutive equation, relates the generalized stresses and the generalized strains, and possibly the generalized strain rates.

With these general considerations, in this section we will illustrate the construction of admissible constitutive equations for some of the shell models. For simplicity, the presentation will be limited to the case of bodies undergoing infinitesimal strains and small displacements. This limitation is for simplicity and to avoid obscuring the concepts with unnecessary details. However, the interested reader will have the elements to make the extension to finite strains. In this context, the reference (material) and the current (spatial) configurations can be confounded, and then the displacements and the velocities can be treated indistinctly, and written in the same manner. Hence, the generalized strains in each model are defined in an identical manner to the generalized strain rates for the different shell models.

It is important to point out that some inconsistencies in thin shell models can be avoided if the kinematical hypotheses considered in the construction of the model are also incorporated as internal constraints in the definition of the material behavior. A detailed account of such a class of inconsistencies in the theory of thin plates can be found in [241]. Specifically, the goal of this section is to pursue a variational strategy that circumvents this kind of inconsistency for the shell models developed earlier in this chapter. The interested reader can refer also to [100, 241, 285, 286].

9.7.1 Preliminary Concepts

In the mechanics of continuum media, a material internal constraint consists of a constraint on the deformations to which the material can be subjected [132]. An example of this is found in the mechanics of incompressible materials. These materials can only be subjected to strain fields which, locally, do not change the volume of the material. In the shell models studied in this chapter, fibers orthogonal to the middle surface of the shell remain unaltered. Then, given the duality we have established in the understanding of the mechanics, we know that for each constraint over the motion, or equally over the strain, we will have, correspondingly, reactive forces and stresses which will be orthogonal to the respective sets of admissible virtual variations (see Chapter 3).

In what follows, and since we are in the context of infinitesimal strains, notice that the generalized strain measures can be confounded with (can be regarded equally to) the generalized strain rate measures appearing in each shell model. Hence, we will denote by $\mathbf{E}$ the (symmetric) strain tensor and by $\mathscr{C}(\mathbf{E}) = 0$ the constraint that must be satisfied to be considered an admissible constitutive strain. Then, the set of admissible strains in the space of strains $\mathscr{W}$ formed by all second-order symmetric tensor fields can be characterized as

$$\mathscr{W}_{\mathscr{C}} = \{\mathbf{E} \in \mathscr{W};\ \mathscr{C}(\mathbf{E}) = 0\}. \tag{9.161}$$

However, and as a consequence of mathematical duality, the previous set can be represented in a completely equivalent manner by means of the orthogonality between the reactive stresses $\mathbf{T}^{(r)} \in \mathscr{W}'$ and the constraint $\mathscr{C}$. Therefore, if we denote

$$\mathscr{W}'_{\mathscr{C}} = \{\mathbf{T}^{(r)} \in \mathscr{W}';\ \mathbf{T}^{(r)} \in (\mathscr{W}_{\mathscr{C}})^{\perp}\}, \tag{9.162}$$

then we have

$$\mathscr{W}_{\mathscr{C}} = \{\mathbf{E} \in \mathscr{W};\ \mathbf{E} \cdot \mathbf{T}^{(r)} = 0 \quad \forall \mathbf{T}^{(r)} \in \mathscr{W}'_{\mathscr{C}}\}. \tag{9.163}$$

From a purely mechanical point of view, the previous expressions entail that reactive stresses $\mathbf{T}^{(r)}$ produce no internal power, that is, they do not contribute to the internal power developed in the material for all materially admissible strains. Using the same argument, we see that it is not possible to characterize the stress $\mathbf{T}^{(r)}$ through the PVP (or its equivalent, the PVW). In other words, the stress state to which the body is subjected is given by

$$\mathbf{T} = \mathbf{T}^{(a)} + \mathbf{T}^{(r)}, \tag{9.164}$$

where $\mathbf{T}^{(a)} \in \mathscr{W}'$ is the active stress state associated with the strain $\mathbf{E} \in \mathscr{W}_{\mathscr{C}}$ by means of the constitutive equations which characterize the material behavior. This stress state $\mathbf{T}^{(a)}$ remains fully and uniquely defined through the PVP placed in the set of kinematically admissible motion actions (or virtual displacements in the context of infinitesimal strains). Recall that this set is defined by constraints over the boundary as well as by the same constraints imposed internally that define the material admissibility of the strains. In other words, the set Kin_u is given by

$$\begin{aligned}\mathrm{Kin}_u &= \{\mathbf{u} \in \mathscr{U};\ \mathbf{u} \text{ satisfies essential boundary conditions and } \mathbf{E}(\mathbf{u}) \in \mathscr{W}_{\mathscr{C}}\} \\ &= \overline{\mathbf{u}} + \mathrm{Var}_u,\end{aligned} \tag{9.165}$$

with $\overline{\mathbf{u}}$ an arbitrary element in Kin_u and Var_u the subspace that generates the linear manifold Kin_u.

Once the active stresses have been calculated through the PVP, the body is then released from all constraints (particularly the internal constitutive constraint $\mathscr{C}$), and the PVP is applied again with the known active stress $\mathbf{T}^{(a)}$ and the reactive stresses $\mathbf{T}^{(r)}$ introduced, which now play a role in the internal power because the constraints have been removed. This provides the variational equations to calculate the reactive stress state $\mathbf{T}^{(r)}$. As a matter of fact, we have the following sequence of variational problems in order to determine $\mathbf{T}^{(a)}$ and $\mathbf{T}^{(r)}$. The first variational problem consists of determining $\mathbf{u}_o \in \mathrm{Kin}_u$, such that it produces a strain $\mathbf{E}(\mathbf{u}_o) \in \mathscr{W}_{\mathscr{C}}$ that, through the constitutive equations, yields an active stress $\mathbf{T}^{(a)}(\mathbf{E}(\mathbf{u}_o))$ which satisfies the PVP (or its equivalent the PVW)

$$(\mathbf{T}^{(a)}(\mathbf{E}(\mathbf{u}_o)), \mathbf{E}(\mathbf{u} - \mathbf{u}_o)) = \langle f, \mathbf{u} - \mathbf{u}_o\rangle \qquad \forall \mathbf{u} \in \mathrm{Kin}_u. \tag{9.166}$$

With $\mathbf{T}^{(a)}$ defined in this manner, we remove the constitutive constraint $\mathbf{E}(\mathbf{u}) \in \mathscr{W}_{\mathscr{C}}$, in other words, the set Kin_u is now extended to be defined as

$$\mathrm{Kin}_u^* = \{\mathbf{u} \in \mathscr{U};\ \mathbf{u} \text{ satisfies essential boundary conditions}\} = \overline{\mathbf{u}}^* + \mathrm{Var}_u^*, \tag{9.167}$$

where $\overline{\mathbf{u}}^*$ is an arbitrary element in Kin_u^* and Var_u^* is the subspace that generates this linear manifold, and then we proceed again by utilizing the PVP using this admissible space, where now $\mathbf{T}^{(r)}$ produces internal power. This variational problem consists of determining $\mathbf{T}^{(r)} \in \mathscr{W}_{\mathscr{C}}'$ such that

$$(\mathbf{T}^{(a)}(\mathbf{E}(\mathbf{u}_o)) + \mathbf{T}^{(r)}, \mathbf{E}(\mathbf{u})) = \langle f, \mathbf{u}\rangle \qquad \forall \mathbf{u} \in \mathrm{Var}_u^*. \tag{9.168}$$

Because in the previous expression the stress $\mathbf{T}^{(a)}(\mathbf{E}(\mathbf{u}_o))$ is known, the variational equation is arranged as

$$(\mathbf{T}^{(r)}, \mathbf{E}(\mathbf{u})) = -(\mathbf{T}^{(a)}(\mathbf{E}(\mathbf{u}_o)), \mathbf{E}(\mathbf{u})) + \langle f, \mathbf{u}\rangle \qquad \forall \mathbf{u} \in \mathrm{Var}_u^*. \tag{9.169}$$

Importantly, the right-hand side above is different from zero because $\mathbf{u} \in \mathrm{Var}_u^*$, that is, it does not belong to Var_u anymore. For the case of shell models seen here, the previous procedure will be demonstrated in detail in Section 9.8.3, where, instead of employing expression (9.169), we will obtain approximate solutions to the associated Euler–Lagrange equations.

Evidently, if the material is such that there are m constitutive constraints, $\mathscr{C}_i$, $i = 1, \dots, m$, the kinematical admissibility from the point of view of the material behavior is associated with displacement fields $\mathbf{u}$ which yield strains $\mathbf{E}(\mathbf{u})$ that belong to the intersection of all the sets $\mathscr{W}_{\mathscr{C}_i}$. That is, in such a case we have

$$\mathbf{E}(\mathbf{u}) \in \mathscr{W}_\cap = \bigcap_{i=1}^{m} \mathscr{W}_{\mathscr{C}_i}, \tag{9.170}$$

and, in this manner, the set of reactive stresses associated with this set of constraints becomes

$$\mathscr{W}'_{\cap} = \{\mathbf{T}^{(r)} \in \mathscr{W}', \ \mathbf{T}^{(r)} \in (\mathscr{W}_{\cap})^{\perp}\}. \tag{9.171}$$

Finally, the set of kinematically admissible displacements is now defined by

$$\text{Kin}_u = \left\{\mathbf{u} \in \mathscr{U}; \ \mathbf{u} \text{ satisfies essential boundary conditions and } \mathbf{E}(\mathbf{u}) \in \mathscr{W}_{\cap}\right\}. \tag{9.172}$$

9.7.2 Model with Naghdi Hypothesis

This section is devoted to the shell model presented in Section 9.6.1 for which, under the infinitesimal strains hypothesis, the strain tensors have the same representation as the strain actions defined in (9.78)–(9.80). Thus, from the constitutive point of view, the generalized stresses $\mathbf{N}_t$, $\mathbf{M}_t$ and $\mathbf{Q}$ will be expressed in terms of the following strain tensor measures

$$\begin{aligned}
\varepsilon &= \Pi_t(\nabla_{\mathbf{x}_o}\mathbf{u}_t^o) + u_n^o\nabla_{\mathbf{x}_o}\mathbf{n}, \\
\chi &= \Pi_t(\nabla_{\mathbf{x}_o}\theta_t), \\
\psi &= \theta_t - (\nabla_{\mathbf{x}_o}\mathbf{n})\mathbf{u}_t^o + \nabla_{\mathbf{x}_o}u_n^o,
\end{aligned} \tag{9.173}$$

where $\mathbf{u}_t^o$ and u_n^o stand for the tangent (vector) and normal (scalar) displacements fields, while θ_t defines the rotation of the tangent plane, according to the kinematical hypotheses (9.72) and (9.73). For such assumptions, it is clear that the strain tensor $\mathbf{E}$, corresponding to the three-dimensional body, is such that $E_n = \mathbf{En}\cdot\mathbf{n} = \mathbf{E}\cdot(\mathbf{n}\otimes\mathbf{n}) = 0$. As a result, for this kinematics we have the internal constraint

$$\mathscr{C}(\mathbf{E}) = \mathbf{En}\cdot\mathbf{n} = 0, \tag{9.174}$$

and the set (subspace in this case) of the stresses which become reactive to this constraint is characterized by

$$\mathscr{W}'_{\mathscr{C}} = \{\mathbf{T}^{(r)} \in \mathscr{W}'; \ \mathbf{T}^{(r)} = \alpha\mathbf{n}\otimes\mathbf{n}, \ \alpha \in \mathbb{R}\}. \tag{9.175}$$

Then, the set of admissible strains becomes

$$\mathscr{W}_{\mathscr{C}} = \{\mathbf{E} \in \mathscr{W}; \ \mathbf{E}\cdot\mathbf{T}^{(r)} = 0 \quad \forall\mathbf{T}^{(r)} \in \mathscr{W}'_{\mathscr{C}}\}. \tag{9.176}$$

In turn, the stress state in the shell is now given by $\mathbf{T} = \mathbf{T}^{(a)} + \mathbf{T}^{(r)}$, where each of these tensors can be written in terms of the components in the intrinsic system of coordinates in the middle surface of the shell

$$\mathbf{T}^{(a)} = \mathbf{T}_t^{(a)} + \mathbf{T}_s^{(a)}\otimes\mathbf{n} + \mathbf{n}\otimes\mathbf{T}_s^{(a)}, \tag{9.177}$$

$$\mathbf{T}^{(r)} = T_n^{(r)}(\mathbf{n}\otimes\mathbf{n}), \tag{9.178}$$

where it is evident that $\mathbf{T}_t = \mathbf{T}_t^{(a)}$, $\mathbf{T}_s = \mathbf{T}_s^{(a)}$ and $T_n = T_n^{(r)}$.

Let us now denote, respectively, $W = W(\mathbf{E})$ and $W^* = W^*(\mathbf{T})$, the strain energy density function (per unit volume) and the complementary strain energy density function. As we saw in Chapter 4, these functions are such that for all stress states $\mathbf{T}$ and all strain states $\mathbf{E}$, both symmetric tensors, the following relation holds

$$\int_\Omega W^*(\mathbf{T})d\Omega + \int_\Omega W(\mathbf{E})d\Omega \geq \int_\Omega \mathbf{T}\cdot\mathbf{E}d\Omega \qquad \forall\mathbf{T} \in \mathscr{W}', \ \forall\mathbf{E} \in \mathscr{W}, \tag{9.179}$$

with the equality being valid when **T** and **E** are related through the constitutive equation. Moreover, we have seen that the Legendre transformation allows us, by means of a maximization problem, to write the strain energy function in terms of the complementary function, and vice versa. For example, once the complementary strain energy density function $W^*(\mathbf{T})$ is known, and for a given **E**, we have that

$$\int_\Omega W(\mathbf{E})d\Omega = \max_{\mathbf{T}} \Pi^*(\mathbf{T},\mathbf{E}), \tag{9.180}$$

where

$$\Pi^*(\mathbf{T},\mathbf{E}) = \int_\Omega (\mathbf{T}\cdot\mathbf{E} - W^*(\mathbf{T}))d\Omega. \tag{9.181}$$

Then, for a given strain state **E**, the problem of finding the stress state **T** associated with **E** through the constitutive equation consists of solving the maximization problem stated above. We know from Chapter 4 that the solution to this problem verifies

$$\mathbf{E} = \frac{\partial W^*}{\partial \mathbf{T}}. \tag{9.182}$$

Assuming the material to be linear elastic, and with a complementary strain energy function given by

$$W^*(\mathbf{T}) = \frac{1}{2}\mathbb{D}^{-1}\mathbf{T}\cdot\mathbf{T}, \tag{9.183}$$

with $\mathbb{D}$ being a fourth-order tensor, the solution of the maximization problem yields

$$W(\mathbf{E}) = \frac{1}{2}\mathbb{D}\mathbf{E}\cdot\mathbf{E}. \tag{9.184}$$

A similar procedure can be performed to calculate the complementary strain energy function when the strain energy function $W(\mathbf{E})$ is given.

Briefly, from the previous results we get

$$\mathbf{E} = \frac{\partial W^*}{\partial \mathbf{T}} = \mathbb{D}^{-1}\mathbf{T} \qquad \text{and} \qquad \mathbf{T} = \frac{\partial W}{\partial \mathbf{E}} = \mathbb{D}\mathbf{E}. \tag{9.185}$$

In a material without internal constraints, the tensor $\mathbb{D}$ is a linear mapping between $\mathscr{W}$ and $\mathscr{W}$ (here we identify $\mathscr{W}'$ with the same space $\mathscr{W}$). Nevertheless, for a material with internal constraints given by $\mathscr{C}$ we have

$$\mathbb{D} : \mathscr{W}_{\mathscr{C}} \to \mathscr{W}_{\mathscr{C}}, \tag{9.186}$$

with the set $\mathscr{W}_{\mathscr{C}}$ defined in (9.176) because, as we have already stated, only the active component $\mathbf{T}^{(a)}$ of the stress state **T** can be evaluated through constitutive equations.

Let us introduce now the concept of symmetry group. Recall first that *Orth* is the set of all orthogonal tensors with positive determinant. Then, the symmetry group denoted by $\mathcal{Q}$, of a material described by the constitutive tensor $\mathbb{D}$, constrained to $\mathscr{W}_{\mathscr{C}}$, is the collection of all tensors $\mathbf{Q} \in Orth$ such that

$$\mathbf{QEQ}^T \in \mathscr{W}_{\mathscr{C}} \qquad \text{and} \qquad \mathbf{QT}^{(a)}\mathbf{Q}^T = \mathbb{D}[\mathbf{QEQ}^T] \qquad \forall \mathbf{E} \in \mathscr{W}_{\mathscr{C}}. \tag{9.187}$$

In a material with internal constraints, the group $\mathcal{Q}$ cannot be arbitrary because it must verify (9.187). For example, in the case of isotropic materials, the material can be incompressible, which corresponds to a constraint given by $\mathscr{C}(\mathbf{E}) = \mathrm{tr}(\nabla\mathbf{u}) = \nabla\mathbf{u}\cdot\mathbf{I} =$

$\mathbf{E} \cdot \mathbf{I} = 0$, implying that $\mathscr{W}_{\mathscr{C}} = \{\mathbf{E} \in \mathscr{W}\ ;\ \mathbf{E} \cdot \alpha\mathbf{I} = 0\quad \forall\alpha \in \mathbb{R}\}$, but it cannot be rigid in the normal direction, $E_n = 0$. Within this context, the maximal symmetry group compatible with the internal constraint is formed by materials with transversal isotropy [241]. This implies that we have to consider, in what follows, orthotropic materials with rhombic symmetry.

Consider an orthonormal basis $(\mathbf{e}_{(1)}, \mathbf{e}_{(2)}, \mathbf{n})$, where $\mathbf{e}_{(1)}$ and $\mathbf{e}_{(2)}$ are two unit vectors orthogonal between them and lying on the tangent plane of the middle surface of the shell. The symmetry group, by taking into consideration this basis, can be generated by means of rotations characterized by an angle ϕ around $\mathbf{e}_{(1)}$ and $\mathbf{e}_{(2)}$.

Let us term the components of the fourth-order constitutive tensor $\mathbb{D}$ by D_{ijkl} $(i, j, k, l = 1, 2, n)$ with respect to the orthonormal basis. For an orthotropic material, the components for which the indices are repeated once or three times must be null. Furthermore, considering the symmetry of the constitutive tensor we have $D_{klij} = D_{klji} = D_{lkij} = D_{ijkl}$, and the material will ultimately be characterized only by nine material coefficients. In particular, using the classical vector representation for $\mathbf{T}^{(a)}$ and $\mathbf{E}$, we get

$$\begin{pmatrix} T^{(a)}_{11} \\ T^{(a)}_{22} \\ T^{(a)}_{12} \\ T^{(a)}_{1n} \\ T^{(a)}_{2n} \\ T^{(a)}_{n} \end{pmatrix} = \begin{pmatrix} D_{1111} & D_{1122} & 0 & 0 & 0 & D_{11nn} \\ D_{1122} & D_{2222} & 0 & 0 & 0 & D_{22nn} \\ 0 & 0 & 2D_{1212} & 0 & 0 & 0 \\ 0 & 0 & 0 & 2D_{1n1n} & 0 & 0 \\ 0 & 0 & 0 & 0 & 2D_{2n2n} & 0 \\ D_{11nn} & D_{22nn} & 0 & 0 & 0 & D_{nnnn} \end{pmatrix} \begin{pmatrix} E_{11} \\ E_{22} \\ E_{12} \\ E_{1n} \\ E_{2n} \\ E_{n} \end{pmatrix}. \tag{9.188}$$

In this expression, $T^{(a)}_{11}, \ldots, T^{(a)}_{n}$ and $E_{11}, \ldots, E_n$ are, respectively, the components of $\mathbf{T}^{(a)}$ and $\mathbf{E}$ with respect to the adopted basis $(\mathbf{e}_{(1)}, \mathbf{e}_{(2)}, \mathbf{n})$.

Remember now that in the Naghdi shell model $E_n = 0$ and, therefore, $T^{(a)}_n = 0$. Thus, for this material the following coefficients are nullified

$$D_{11nn} = D_{22nn} = 0. \tag{9.189}$$

Moreover, for the chosen basis, any symmetric second-order tensor $\mathbf{S}$ satisfies

$$\mathbf{S}_t = \begin{pmatrix} S_{11} & S_{12} \\ S_{21} & S_{22} \end{pmatrix} \qquad \mathbf{S}_s = \mathbf{S}_s^* = \begin{pmatrix} S_{1n} \\ S_{2n} \end{pmatrix}. \tag{9.190}$$

Using (9.188), (9.189) and (9.190) we then obtain the relation

$$\mathbf{T}^{(a)}_t = \mathbb{D}_t \mathbf{E}_t, \tag{9.191}$$

for the tangent component, where $\mathbb{D}_t$ is the fourth-order tensor defined by

$$\begin{aligned} \mathbb{D}_t = {} & D_{1111}(\mathbf{e}_{(1)} \otimes \mathbf{e}_{(1)} \otimes \mathbf{e}_{(1)} \otimes \mathbf{e}_{(1)}) + D_{2222}(\mathbf{e}_{(2)} \otimes \mathbf{e}_{(2)} \otimes \mathbf{e}_{(2)} \otimes \mathbf{e}_{(2)}) \\ & + D_{1122}(\mathbf{e}_{(1)} \otimes \mathbf{e}_{(1)} \otimes \mathbf{e}_{(2)} \otimes \mathbf{e}_{(2)}) + D_{1122}(\mathbf{e}_{(2)} \otimes \mathbf{e}_{(2)} \otimes \mathbf{e}_{(1)} \otimes \mathbf{e}_{(1)}) \\ & + D_{1212}(\mathbf{e}_{(1)} \otimes \mathbf{e}_{(2)} \otimes \mathbf{e}_{(1)} \otimes \mathbf{e}_{(2)}) + D_{1212}(\mathbf{e}_{(1)} \otimes \mathbf{e}_{(2)} \otimes \mathbf{e}_{(2)} \otimes \mathbf{e}_{(1)}) \\ & + D_{1212}(\mathbf{e}_{(2)} \otimes \mathbf{e}_{(1)} \otimes \mathbf{e}_{(2)} \otimes \mathbf{e}_{(1)}) + D_{1212}(\mathbf{e}_{(2)} \otimes \mathbf{e}_{(1)} \otimes \mathbf{e}_{(1)} \otimes \mathbf{e}_{(2)}), \end{aligned} \tag{9.192}$$

and also the relation

$$\mathbf{T}^{(a)}_s \otimes \mathbf{n} + \mathbf{n} \otimes \mathbf{T}^{(a)}_s = 2(\mathbb{D}_s \mathbf{E}_s \otimes \mathbf{n} + \mathbf{n} \otimes \mathbb{D}_s \mathbf{E}_s), \tag{9.193}$$

for the transversal components, where $\mathbb{D}_s$ is a second-order tensor defined by

$$\mathbb{D}_s = D_{1n1n}(\mathbf{e}_{(1)} \otimes \mathbf{e}_{(1)}) + D_{2n2n}(\mathbf{e}_{(2)} \otimes \mathbf{e}_{(2)}). \tag{9.194}$$

Adding expressions (9.191) and (9.193) we arrive at the constitutive relation $\mathbf{T}^{(a)} = \mathbb{D}\mathbf{E}$ for an orthotropic material, which is compatible with the kinematical hypotheses of the Naghdi shell model.

To obtain the inverse constitutive equation, $\mathbf{E} = \mathbb{D}^{-1}\mathbf{T}^{(a)}$, it is necessary to invert (9.191) and (9.193), yielding

$$\mathbf{E}_t = \mathbb{A}_t \mathbf{T}_t^{(a)}, \tag{9.195}$$

where $\mathbb{A}_t = \mathbb{D}_t^{-1}$ and

$$\mathbf{E}_s \otimes \mathbf{n} + \mathbf{n} \otimes \mathbf{E}_s = \frac{1}{2}(\mathbb{A}_s \mathbf{T}_s^{(a)} \otimes \mathbf{n} + \mathbf{n} \otimes \mathbb{A}_s \mathbf{T}_s^{(a)}), \tag{9.196}$$

with $\mathbb{A}_s = \mathbb{D}_s^{-1}$.

For an orthotropic material, the components of tensors $\mathbb{A}_t$ and $\mathbb{A}_s$ are such that the following relations must be satisfied

$$\begin{pmatrix} E_{11} \\ E_{22} \\ E_{12} \\ E_{1n} \\ E_{2n} \end{pmatrix} = \begin{pmatrix} \frac{1}{\mathit{IE}_1} & -\frac{\nu_{12}}{\mathit{IE}_2} & 0 & 0 & 0 \\ -\frac{\nu_{21}}{\mathit{IE}_1} & \frac{1}{\mathit{IE}_2} & 0 & 0 & 0 \\ 0 & 0 & \frac{1}{\mathit{IG}_{12}} & 0 & 0 \\ 0 & 0 & 0 & \frac{1}{\mathit{IG}_{1n}} & 0 \\ 0 & 0 & 0 & 0 & \frac{1}{\mathit{IG}_{2n}} \end{pmatrix} \begin{pmatrix} T_{11}^{(a)} \\ T_{22}^{(a)} \\ T_{12}^{(a)} \\ T_{1n}^{(a)} \\ T_{2n}^{(a)} \end{pmatrix}, \tag{9.197}$$

where we have used the fact that $D_{11nn} = D_{22nn} = 0$ (see (9.189)). From (9.197) the components of $\mathbb{A}_t$ and $\mathbb{A}_s$ can directly be identified. Here, IE_1 and IE_2 are, respectively, the Young moduli in the directions $\mathbf{e}_{(1)}$ and $\mathbf{e}_{(2)}$. Also, IG_{12}, IG_{1n} and IG_{2n} are the transversal stiffness moduli, and ν_{12} and ν_{21} are the Poisson coefficients. Here, we have to keep in mind that the identity $\frac{\nu_{12}}{\mathit{IE}_2} = \frac{\nu_{21}}{\mathit{IE}_1}$ must be verified to meet the symmetry condition for $\mathbf{T}^{(a)}$. In effect, we have the following symmetries for $\mathbb{A}$, $A_{1212} = A_{2112} = A_{2121} = A_{1221}$.

Now, introducing (9.195) and (9.196) into (9.183), the complementary strain energy density function is given by

$$W^*(\mathbf{T}^{(a)}) = \frac{1}{2}(\mathbb{A}_t \mathbf{T}_t^{(a)} \cdot \mathbf{T}_t^{(a)} + \mathbb{A}_s \mathbf{T}_s^{(a)} \cdot \mathbf{T}_s^{(a)}), \tag{9.198}$$

and replacing this expression into (9.181) we obtain the functional $\Pi^*(\mathbf{T}^{(a)}, \mathbf{E})$ (which will allow the calculation of the strain energy density function (9.180)) for an orthotropic material with internal constraints such that $\mathbf{E} \in \mathscr{W}_{\mathscr{C}}$, with $\mathscr{W}_{\mathscr{C}}$ as in (9.176), and which is given by

$$\Pi^*(\mathbf{T}^{(a)}, \mathbf{E}) = \int_{\Sigma_o} \int_H \left[\mathbf{T}^{(a)} \cdot \mathbf{E} - \frac{1}{2}(\mathbb{A}_t \mathbf{T}_t^{(a)} \cdot \mathbf{T}_t^{(a)} + \mathbb{A}_s \mathbf{T}_s^{(a)} \cdot \mathbf{T}_s^{(a)})\right] \det \Lambda \, d\xi \, d\Sigma_o. \tag{9.199}$$

This functional establishes (through the maximization problem) the constitutive relation between the stress state $\mathbf{T}^{(a)}$ and the strain state $\mathbf{E}$. However, for the shell model we are studying here, it is necessary that the previous functional is written in terms of the generalized stresses $\mathbf{N}_t$, $\mathbf{M}_t$ and $\mathbf{Q}$ and in terms of generalized strains ε, χ and ψ. To do this, we need to write $\mathbf{T}_t$ and $\mathbf{T}_s$ as functions of $\mathbf{N}_t$, $\mathbf{M}_t$ and $\mathbf{Q}$, and the problem is that these functions are not known a priori. According to (9.63) we only know the inverse relation among these variables. This forces us to introduce approximations for $\mathbf{T}_t$ and $\mathbf{T}_s$ such that definitions (9.63) are satisfied by construction.

To this end, the first hypothesis to be considered is the assumption that $\mathbf{T}_t\Lambda^{-1}\det\Lambda$ is a linear function in the variable ξ across the shell thickness $[-\frac{h}{2},\frac{h}{2}]$. Then, in this case we have the following form for $\mathbf{T}_t$

$$\mathbf{T}_t = \frac{\Lambda}{\det\Lambda}\left[\frac{1}{h}\mathbf{N}_t + \varsigma\frac{6}{h^2}\mathbf{M}_t\right] \qquad \text{where} \qquad \varsigma = \frac{\xi}{h/2}. \tag{9.200}$$

The second hypothesis consists of postulating that $\Lambda^{-1}\mathbf{T}_s\det\Lambda$ is a parabolic function of ξ across the shell thickness, leading to

$$\begin{aligned}\mathbf{T}_s = \frac{\Lambda}{\det\Lambda}\Big[&\frac{3}{2h}(1-\varsigma^2)\mathbf{Q} - \frac{1}{4}(\Lambda^{-1}\det\Lambda)^+(1-2\varsigma-3\varsigma^2)\mathbf{f}_t^+ \\ &+\frac{1}{4}(\Lambda^{-1}\det\Lambda)^-(1+2\varsigma-3\varsigma^2)\mathbf{f}_t^-\Big],\end{aligned} \tag{9.201}$$

where $(\cdot)^\pm$ is employed to symbolize the value of field $(\cdot)$ at $\xi = \pm\frac{h}{2}$.

It is worth of note that expression (9.201), for $\varsigma = \pm 1$, takes the value $\mathbf{T}_s|_{\varsigma=\pm 1} = \mathbf{f}_t^\pm$, verifying the boundary conditions in the superior and inferior boundaries of the shell.

A further important remark is that, within this Naghdi theory, the tensor $\mathbf{T}_t$ is not symmetric and, therefore, in general, the solution to the Euler–Lagrange equations written for the approximations presented above $\mathbf{T}_t$ and $\mathbf{T}_s$ may not exist (see [72]).

Now, from (9.173) the internal virtual power acquires the form

$$\mathbf{T}^{(a)}\cdot\mathbf{E} = \mathbf{T}_t\cdot(\varepsilon+\xi\chi)\Lambda^{-1} + \mathbf{T}_s\cdot\frac{1}{2}\Lambda^{-1}\psi. \tag{9.202}$$

Putting (9.200), (9.201) and (9.202) into (9.199) we obtain the functional Π^* in terms of the generalized stresses and strains for this shell model, that is, we arrive at $\Pi^* = \Pi^\star(\mathbf{N}_t,\mathbf{M}_t,\mathbf{Q},\varepsilon,\chi,\psi)$. The necessary condition (and for the present case also the sufficient condition) for the critical point to be a maximum (see (9.180)) corresponds to nullifying the first Gâteaux variation of functional Π^*, denoted by $\delta\Pi^*$, at a generalized strain measure ε,χ,ψ and in the direction given by $\delta\mathbf{N}_t$, $\delta\mathbf{M}_t$ and $\delta\mathbf{Q}$. Hence, this Gâteaux variation gives

$$\begin{aligned}\delta\Pi^* = -\int_{\Sigma_o}\Big[&\left(\frac{1}{h^2}\overline{\mathbb{A}}_t^0\mathbf{N}_t + \frac{12}{h^4}\overline{\mathbb{A}}_t^1\mathbf{M}_t - \varepsilon\right)\cdot\delta\mathbf{N}_t \\ &+\left(\frac{12}{h^4}\overline{\mathbb{A}}_t^1\mathbf{N}_t + \frac{144}{h^6}\overline{\mathbb{A}}_t^2\mathbf{M}_t - \chi\right)\cdot\delta\mathbf{M}_t \\ &+\left(\overline{\mathbb{A}}_s\mathbf{Q} - \overline{\mathbb{A}}_s^+\mathbf{f}_t^+ + \overline{\mathbb{A}}_s^-\mathbf{f}_t^- - \psi\right)\cdot\delta\mathbf{Q}\Big]\, d\Sigma_o,\end{aligned} \tag{9.203}$$

where tensors $\overline{\mathbb{A}}_t^k$, $k = 0,1,2$, $\overline{\mathbb{A}}_s$, $\overline{\mathbb{A}}_s^+$ and $\overline{\mathbb{A}}_s^-$ are termed inverse generalized elasticity tensors.

The previous tensors $\overline{\mathbb{A}}_t^k$, $k = 0,1,2$ are defined in intrinsic (compact) notation through the corresponding application over second-order tensors, say $\mathbf{S}$, defined in the middle surface of the shell

$$\overline{\mathbb{A}}_t^k\mathbf{S} = \int_H [\mathbb{A}_t(\mathbf{S}\Lambda)]\Lambda\frac{\xi^k}{\det\Lambda}\, d\xi \qquad k = 0,1,2. \tag{9.204}$$

With this definition, it is not difficult to obtain the components of these tensors

$$[\overline{\mathbb{A}}_t^k]_{\alpha\mu\gamma\rho} = \int_H [\mathbb{A}_t]_{\alpha\beta\gamma\delta}[\Lambda]_{\rho\delta}[\Lambda]_{\beta\mu}\frac{\xi^k}{\det\Lambda}\,d\xi \qquad k = 0, 1, 2, \tag{9.205}$$

where we have to remember that $[\Lambda]_{\rho\delta} = \delta_{\rho\delta} + \xi[\nabla_{\mathbf{x}_o}\mathbf{n}]_{\rho\delta}$, where $\delta_{\rho\delta}$ is the Kronecker delta.

In turn, tensors $\overline{\mathbb{A}}_s$, $\overline{\mathbb{A}}_s^+$ and $\overline{\mathbb{A}}_s^-$ are characterized by

$$\begin{aligned}
\overline{\mathbb{A}}_s &= \frac{9}{4h^2}\int_H (1-\varsigma^2)^2\frac{\Lambda\mathbb{A}_s\Lambda}{\det\Lambda}\,d\xi,\\
\overline{\mathbb{A}}_s^+ &= \frac{3}{8h}(\Lambda^{-1}\det\Lambda)^+\int_H (1-\varsigma^2)(1-2\varsigma-3\varsigma^2)\frac{\Lambda\mathbb{A}_s\Lambda}{\det\Lambda}\,d\xi,\\
\overline{\mathbb{A}}_s^- &= \frac{3}{8h}(\Lambda^{-1}\det\Lambda)^-\int_H (1-\varsigma^2)(1+2\varsigma-3\varsigma^2)\frac{\Lambda\mathbb{A}_s\Lambda}{\det\Lambda}\,d\xi.
\end{aligned} \tag{9.206}$$

Clearly, the functional Π^* reaches a maximum value when the first Gâteaux variation (9.203) is null for all $\delta\mathbf{N}_t$, $\delta\mathbf{M}_t$ and $\delta\mathbf{Q}$. Thus, we are led to the constitutive equations that relate the generalized strains as functions of the generalized stresses, which are given by

$$\begin{aligned}
\varepsilon &= \frac{1}{h^2}\overline{\mathbb{A}}_t^0\mathbf{N}_t + \frac{12}{h^4}\overline{\mathbb{A}}_t^1\mathbf{M}_t,\\
\chi &= \frac{12}{h^4}\overline{\mathbb{A}}_t^1\mathbf{N}_t + \frac{144}{h^6}\overline{\mathbb{A}}_t^2\mathbf{M}_t,\\
\psi &= \overline{\mathbb{A}}_s\mathbf{Q} - \overline{\mathbb{A}}_s^+\mathbf{f}_t^+ + \overline{\mathbb{A}}_s^-\mathbf{f}_t^-.
\end{aligned} \tag{9.207}$$

The constitutive equations obtained in (9.207) are called exact in the sense that no hypothesis related to the ratio $\frac{h}{R_m}$ (with R_m the minor curvature radius of the shell) has been introduced.

The next step consists of assuming that $\frac{h}{R_m} \ll 1$ in order to derive approximated constitutive equations. To this end we will make use of the identity

$$\det\Lambda = 1 + \xi\operatorname{tr}(\nabla_{\mathbf{x}_o}\mathbf{n}) + \xi^2\det(\nabla_{\mathbf{x}_o}\mathbf{n}). \tag{9.208}$$

Since $\xi \in [-\frac{h}{2}, \frac{h}{2}]$ and with $R_m = \min\{R_1, R_2\}$ being the minor of the principal radii of curvature of the shell (R_1 and R_2 are the eigenvalues of tensor $\nabla_{\mathbf{x}_o}\mathbf{n}$) the following inequality holds

$$|\xi\operatorname{tr}(\nabla_{\mathbf{x}_o}\mathbf{n}) + \xi^2\det(\nabla_{\mathbf{x}_o}\mathbf{n})| < \frac{h}{R_m} + \frac{h^2}{4R_m^2}. \tag{9.209}$$

Considering thin shells, we have that $\frac{h}{R_m} \ll 1$ and so $|\xi\operatorname{tr}(\nabla_{\mathbf{x}_o}\mathbf{n}) + \xi^2\det(\nabla_{\mathbf{x}_o}\mathbf{n})| \ll 1$. Therefore, we can expand the inverse of the determinant of tensor Λ as follows

$$(\det\Lambda)^{-1} = 1 - \xi\operatorname{tr}(\nabla_{\mathbf{x}_o}\mathbf{n}) - \xi^2\det(\nabla_{\mathbf{x}_o}\mathbf{n}) + \xi^2(\operatorname{tr}(\nabla_{\mathbf{x}_o}\mathbf{n}))^2 + \mathcal{O}\left(\frac{h^3}{R_m^3}\right). \tag{9.210}$$

Hence, it is

$$\xi\nabla_{\mathbf{x}_o}\mathbf{n} = \mathcal{O}\left(\frac{h}{R_m}\right), \quad \xi\operatorname{tr}(\nabla_{\mathbf{x}_o}\mathbf{n}) = \mathcal{O}\left(\frac{h}{R_m}\right), \quad \xi^2\det(\nabla_{\mathbf{x}_o}\mathbf{n}) = \mathcal{O}\left(\frac{h^2}{R_m^2}\right). \tag{9.211}$$

Putting (9.210) into (9.206), considering that the properties of the material are uniform along the thickness of the shell (that implies $\mathbb{D} = \mathbb{D}(\mathbf{x}_o)$ and $\mathbb{A} = \mathbb{A}(\mathbf{x}_o)$), neglecting the terms of order $\mathcal{O}\left(\frac{h^3}{R_m^3}\right)$ and, finally, integrating across the thickness, yields the following approximate expressions for the generalized tensors $\overline{\mathbb{A}}_s$, $\overline{\mathbb{A}}_s^+$ and $\overline{\mathbb{A}}_s^-$

$$\overline{\mathbb{A}}_s = \frac{6}{5h}\left[\mathbb{A}_s + \frac{h^2}{28}\left((\nabla_{\mathbf{x}_o}\mathbf{n})\mathbb{A}_s(\nabla_{\mathbf{x}_o}\mathbf{n}) - \mathrm{tr}(\nabla_{\mathbf{x}_o}\mathbf{n})(\mathbb{A}_s(\nabla_{\mathbf{x}_o}\mathbf{n}) + (\nabla_{\mathbf{x}_o}\mathbf{n})\mathbb{A}_s)\right.\right.$$
$$\left.\left. + ((\mathrm{tr}(\nabla_{\mathbf{x}_o}\mathbf{n}))^2 - \det(\nabla_{\mathbf{x}_o}\mathbf{n}))\mathbb{A}_s\right)\right], \tag{9.212}$$

$$\overline{\mathbb{A}}_s^+ = \frac{3}{8h}(\Lambda^{-1}\det\Lambda)^+\left[\frac{4h}{15}\mathbb{A}_s - \frac{2h^2}{15}\left(\mathbb{A}_s(\nabla_{\mathbf{x}_o}\mathbf{n}) + (\nabla_{\mathbf{x}_o}\mathbf{n})\mathbb{A}_s - \mathrm{tr}(\nabla_{\mathbf{x}_o}\mathbf{n})\mathbb{A}_s\right)\right.$$
$$- \frac{h^3}{105}\left((\nabla_{\mathbf{x}_o}\mathbf{n})\mathbb{A}_s(\nabla_{\mathbf{x}_o}\mathbf{n}) - \mathrm{tr}(\nabla_{\mathbf{x}_o}\mathbf{n})(\mathbb{A}_s(\nabla_{\mathbf{x}_o}\mathbf{n}) + (\nabla_{\mathbf{x}_o}\mathbf{n})\mathbb{A}_s)\right.$$
$$\left.\left. + ((\mathrm{tr}(\nabla_{\mathbf{x}_o}\mathbf{n}))^2 - \det(\nabla_{\mathbf{x}_o}\mathbf{n}))\mathbb{A}_s\right)\right], \tag{9.213}$$

$$\overline{\mathbb{A}}_s^- = \frac{3}{8h}(\Lambda^{-1}\det\Lambda)^-\left[\frac{4h}{15}\mathbb{A}_s + \frac{2h^2}{15}\left(\mathbb{A}_s(\nabla_{\mathbf{x}_o}\mathbf{n}) + (\nabla_{\mathbf{x}_o}\mathbf{n})\mathbb{A}_s - \mathrm{tr}(\nabla_{\mathbf{x}_o}\mathbf{n})\mathbb{A}_s\right)\right.$$
$$- \frac{h^3}{105}\left((\nabla_{\mathbf{x}_o}\mathbf{n})\mathbb{A}_s(\nabla_{\mathbf{x}_o}\mathbf{n}) - \mathrm{tr}(\nabla_{\mathbf{x}_o}\mathbf{n})(\mathbb{A}_s(\nabla_{\mathbf{x}_o}\mathbf{n}) + (\nabla_{\mathbf{x}_o}\mathbf{n})\mathbb{A}_s)\right.$$
$$\left.\left. + ((\mathrm{tr}(\nabla_{\mathbf{x}_o}\mathbf{n}))^2 - \det(\nabla_{\mathbf{x}_o}\mathbf{n}))\mathbb{A}_s\right)\right]. \tag{9.214}$$

Similarly, putting (9.210) into expression (9.205), we reach the following approximated expressions for the tensors $\overline{\mathbb{A}}_t^k$, $k = 0, 1, 2$

$$[\overline{\mathbb{A}}_t^0]_{\alpha\mu\gamma\rho} = [\mathbb{A}_t]_{\alpha\beta\gamma\delta}\left[h\delta_{\rho\delta}\delta_{\mu\beta} + \frac{h^3}{12}\left([\nabla_{\mathbf{x}_o}\mathbf{n}]_{\rho\delta}[\nabla_{\mathbf{x}_o}\mathbf{n}]_{\beta\mu}\right.\right.$$
$$- \mathrm{tr}(\nabla_{\mathbf{x}_o}\mathbf{n})(\delta_{\rho\delta}[\nabla_{\mathbf{x}_o}\mathbf{n}]_{\beta\mu} + [\nabla_{\mathbf{x}_o}\mathbf{n}]_{\rho\delta}\delta_{\beta\mu})$$
$$\left.\left. - (\det(\nabla_{\mathbf{x}_o}\mathbf{n}) + (\mathrm{tr}(\nabla_{\mathbf{x}_o}\mathbf{n}))^2)\delta_{\rho\delta}\delta_{\mu\beta}\right)\right], \tag{9.215}$$

$$[\overline{\mathbb{A}}_t^1]_{\alpha\mu\gamma\rho} = [\mathbb{A}_t]_{\alpha\beta\gamma\delta}\frac{h^3}{12}\left[\delta_{\rho\delta}[\nabla_{\mathbf{x}_o}\mathbf{n}]_{\beta\mu} + [\nabla_{\mathbf{x}_o}\mathbf{n}]_{\rho\delta}\delta_{\beta\mu} - \mathrm{tr}(\nabla_{\mathbf{x}_o}\mathbf{n})\delta_{\rho\delta}\delta_{\mu\beta}\right], \tag{9.216}$$

$$[\overline{\mathbb{A}}_t^2]_{\alpha\mu\gamma\rho} = [\mathbb{A}_t]_{\alpha\beta\gamma\delta}\left[\frac{h^3}{12}\delta_{\rho\delta}\delta_{\mu\beta} + \frac{h^5}{80}\left([\nabla_{\mathbf{x}_o}\mathbf{n}]_{\rho\delta}[\nabla_{\mathbf{x}_o}\mathbf{n}]_{\beta\mu}\right.\right.$$
$$- \mathrm{tr}(\nabla_{\mathbf{x}_o}\mathbf{n})(\delta_{\rho\delta}[\nabla_{\mathbf{x}_o}\mathbf{n}]_{\beta\mu} + [\nabla_{\mathbf{x}_o}\mathbf{n}]_{\rho\delta}\delta_{\beta\mu})$$
$$\left.\left. - (\det(\nabla_{\mathbf{x}_o}\mathbf{n}) + (\mathrm{tr}(\nabla_{\mathbf{x}_o}\mathbf{n}))^2)\delta_{\rho\delta}\delta_{\mu\beta}\right)\right]. \tag{9.217}$$

Expressions (9.212)–(9.214) and (9.215)–(9.217) characterize the constitutive behavior of an orthotropic material with internal constraints consistent with the kinematical constraints of the Naghdi shell model. Such formulae can be explicitly given if we consider that the basis $(\mathbf{e}_{(1)}, \mathbf{e}_{(2)}, \mathbf{n})$ employed in the derivation of the inverse generalized elasticity tensors is such that the vectors $\mathbf{e}_{(1)}$ and $\mathbf{e}_{(2)}$ are parallel to the principal directions of the curvature of the shell. In such case we get

$$\nabla_{\mathbf{x}_o}\mathbf{n} = \begin{pmatrix} \frac{1}{R_1} & 0 \\ 0 & \frac{1}{R_2} \end{pmatrix}, \tag{9.218}$$

where, as before, R_1 and R_2 are, respectively, the principal radii of curvature of the shell associated with the directions $\mathbf{e}_{(1)}$ and $\mathbf{e}_{(2)}$. In this case we have $\det(\nabla_{\mathbf{x}_o}\mathbf{n}) = \frac{1}{R_1 R_2}$ and $\mathrm{tr}(\nabla_{\mathbf{x}_o}\mathbf{n}) = \frac{1}{R_1} + \frac{1}{R_2}$. With these particular forms in (9.212)–(9.214) and (9.215)–(9.217), and then in (9.207) we get

$$
\begin{aligned}
[\varepsilon]_{11} &= \frac{1}{\mathbb{E}_1 h}\left[\left(1+\frac{h^2}{12R_2}\left(\frac{1}{R_2}-\frac{1}{R_1}\right)\right)[\mathbf{N}_t]_{11} - \nu_{21}[\mathbf{N}_t]_{22} - \left(\frac{1}{R_2}-\frac{1}{R_1}\right)[\mathbf{M}_t]_{11}\right],\\
[\varepsilon]_{22} &= \frac{1}{\mathbb{E}_2 h}\left[\left(1+\frac{h^2}{12R_1}\left(\frac{1}{R_1}-\frac{1}{R_2}\right)\right)[\mathbf{N}_t]_{22} - \nu_{12}[\mathbf{N}_t]_{11} - \left(\frac{1}{R_1}-\frac{1}{R_2}\right)[\mathbf{M}_t]_{22}\right],\\
[\varepsilon]_{12} &= \frac{1}{\mathbb{G}_{12} h}\left[\left(1+\frac{h^2}{12R_1}\left(\frac{1}{R_1}-\frac{1}{R_2}\right)\right)[\mathbf{N}_t]_{12} + [\mathbf{N}_t]_{21} - \left(\frac{1}{R_1}-\frac{1}{R_2}\right)[\mathbf{M}_t]_{12}\right],\\
[\varepsilon]_{21} &= \frac{1}{\mathbb{G}_{12} h}\left[\left(1+\frac{h^2}{12R_2}\left(\frac{1}{R_2}-\frac{1}{R_1}\right)\right)[\mathbf{N}_t]_{21} + [\mathbf{N}_t]_{12} - \left(\frac{1}{R_2}-\frac{1}{R_1}\right)[\mathbf{M}_t]_{21}\right],
\end{aligned}
\tag{9.219}
$$

$$
\begin{aligned}
[\chi]_{11} &= \frac{12}{\mathbb{E}_1 h^3}\left[\left(1+\frac{3h^2}{20R_2}\left(\frac{1}{R_2}-\frac{1}{R_1}\right)\right)[\mathbf{M}_t]_{11} - \nu_{21}[\mathbf{M}_t]_{22} - \frac{h^2}{12}\left(\frac{1}{R_2}-\frac{1}{R_1}\right)[\mathbf{N}_t]_{11}\right],\\
[\chi]_{22} &= \frac{12}{\mathbb{E}_2 h^3}\left[\left(1+\frac{3h^2}{20R_1}\left(\frac{1}{R_1}-\frac{1}{R_2}\right)\right)[\mathbf{M}_t]_{22} - \nu_{12}[\mathbf{M}_t]_{11} - \frac{h^2}{12}\left(\frac{1}{R_1}-\frac{1}{R_2}\right)[\mathbf{N}_t]_{22}\right],\\
[\chi]_{12} &= \frac{12}{\mathbb{G}_{12} h^3}\left[\left(1+\frac{3h^2}{20R_1}\left(\frac{1}{R_1}-\frac{1}{R_2}\right)\right)[\mathbf{M}_t]_{12} + [\mathbf{M}_t]_{21} - \frac{h^2}{12}\left(\frac{1}{R_1}-\frac{1}{R_2}\right)[\mathbf{N}_t]_{12}\right],\\
[\chi]_{21} &= \frac{12}{\mathbb{G}_{12} h^3}\left[\left(1+\frac{3h^2}{20R_2}\left(\frac{1}{R_2}-\frac{1}{R_1}\right)\right)[\mathbf{M}_t]_{21} + [\mathbf{M}_t]_{12} - \frac{h^2}{12}\left(\frac{1}{R_2}-\frac{1}{R_1}\right)[\mathbf{N}_t]_{21}\right],
\end{aligned}
\tag{9.220}
$$

$$
\begin{aligned}
[\psi]_1 = \frac{1}{\mathbb{G}_{1n} h}&\left[\frac{6}{5}[\mathbf{Q}]_1\left(1+\frac{h^2}{28R_2}\left(\frac{1}{R_2}-\frac{1}{R_1}\right)\right) - \frac{h^2}{20}[\mathbf{p}_t]_1\left(\frac{1}{R_2}-\frac{1}{R_1}\right)\right.\\
&\left. -\frac{1}{5}[\mathbf{m}_t]_1\left(1-\frac{h^2}{28R_2}\left(\frac{1}{R_2}-\frac{1}{R_1}\right)\right)\right],\\
[\psi]_2 = \frac{1}{\mathbb{G}_{2n} h}&\left[\frac{6}{5}[\mathbf{Q}]_2\left(1+\frac{h^2}{28R_1}\left(\frac{1}{R_1}-\frac{1}{R_2}\right)\right) - \frac{h^2}{20}[\mathbf{p}_t]_2\left(\frac{1}{R_1}-\frac{1}{R_2}\right)\right.\\
&\left. -\frac{1}{5}[\mathbf{m}_t]_2\left(1-\frac{h^2}{28R_1}\left(\frac{1}{R_1}-\frac{1}{R_2}\right)\right)\right].
\end{aligned}
\tag{9.221}
$$

In the previous expressions the components $[\varepsilon]_{ij}$, $[\chi]_{ij}$, $[\psi]_i$, $[\mathbf{N}_t]_{ij}$, $[\mathbf{M}_t]_{ij}$, $[\mathbf{Q}]_i$, $[\mathbf{p}_t]_i$, $[\mathbf{m}_t]_i$, $i,j = 1,2$ are the components corresponding to the directions $\mathbf{e}_{(1)}$ and $\mathbf{e}_{(2)}$.

9.7.3 Model with Kirchhoff–Love Hypothesis

A theory similar to that developed in the previous section will now be presented for the case of shell models based on the Kirchhoff–Love model. For the kinematical hypotheses established by (9.89) and (9.90), and under the hypothesis of infinitesimal strains and displacements, the generalized strains for this class of models is

$$
\begin{aligned}
\varepsilon^L &= (\Pi_t(\nabla_{\mathbf{x}_o}\mathbf{u}_t^o))^s + u_n^o \nabla_{\mathbf{x}_o}\mathbf{n},\\
\chi^L &= (\Pi_t(\nabla_{\mathbf{x}_o}\theta_t^{KL}))^s,
\end{aligned}
$$

$$\chi^K = (\Pi_t(\nabla_{\mathbf{x}_o}\theta_t^{KL}))^s + ((\nabla_{\mathbf{x}_o}\mathbf{n})\Pi_t(\nabla_{\mathbf{x}_o}\mathbf{u}_t^o))^s + u_n^o(\nabla_{\mathbf{x}_o}\mathbf{n})^2,$$
$$\chi^S = (\Pi_t(\nabla_{\mathbf{x}_o}\theta_t^{KL}))^s + ((\nabla_{\mathbf{x}_o}\mathbf{n})\Xi_t^o)^s, \tag{9.222}$$

where $\Xi_t^o = (\Pi_t(\nabla_{\mathbf{x}_o}\mathbf{u}_t^o))^a$. Recalling the components of $\mathbf{E}$, for the models based on the Kirchhoff–Love hypotheses we have that $\mathbf{E}_s = \Pi_t\mathbf{E}\mathbf{n} = \mathbf{0}$ (that is $E_{1n} = E_{2n} = 0$) and $E_n = \mathbf{E}\mathbf{n}\cdot\mathbf{n} = 0$. Analogously to the previous section, using the basis $(\mathbf{e}_{(1)}, \mathbf{e}_{(2)}, \mathbf{n})$, we identify three internal constraints associated with the following subspaces

$$\begin{aligned}\mathscr{W}_{\mathscr{C}_1} &= \{\mathbf{E}\in\mathscr{W},\ \mathbf{E}\cdot(\mathbf{e}_{(1)}\otimes\mathbf{n}) = 0\},\\ \mathscr{W}_{\mathscr{C}_2} &= \{\mathbf{E}\in\mathscr{W},\ \mathbf{E}\cdot(\mathbf{e}_{(2)}\otimes\mathbf{n}) = 0\},\\ \mathscr{W}_{\mathscr{C}_3} &= \{\mathbf{E}\in\mathscr{W},\ \mathbf{E}\cdot(\mathbf{n}\otimes\mathbf{n}) = 0\}.\end{aligned} \tag{9.223}$$

In this manner, for all strains admissible with the Kirchhoff–Love model, tensor $\mathbf{E}$ must satisfy

$$\mathbf{E}\in\mathscr{W}_\cap \quad \text{with} \quad \mathscr{W}_\cap = \mathscr{W}_{\mathscr{C}_1}\cap\mathscr{W}_{\mathscr{C}_2}\cap\mathscr{W}_{\mathscr{C}_3}. \tag{9.224}$$

Once again, if we identify the elements $\mathbf{T}\in\mathscr{W}'$ with the elements in the space of symmetric tensor fields $\mathscr{W}$, we similarly have that the active stress component is understood as $\mathbf{T}^{(a)}\in\mathscr{W}_\cap$, and so

$$\begin{aligned}\mathbf{T}^{(a)} &= \mathbf{T}_t^{(a)},\\ \mathbf{T}^{(r)} &= \mathbf{T}_s^{(r)}\otimes\mathbf{n} + \mathbf{n}\otimes\mathbf{T}_s^{(r)} + T_n^{(r)}(\mathbf{n}\otimes\mathbf{n}),\end{aligned} \tag{9.225}$$

where $\mathbf{T}_t = \mathbf{T}_t^{(a)}$, $\mathbf{T}_s = \mathbf{T}_s^{(r)}$ and $T_n = T_n^{(r)}$. Therefore, in this case we have $\mathbf{T}^{(a)} = \mathbb{D}\mathbf{E}$, with $\mathbf{E}\in\mathscr{W}_\cap$ and $\mathbb{D} : \mathscr{W}_\cap \to \mathscr{W}_\cap$. In addition, the symmetry group for this material is given by the collection of all tensors $\mathbf{Q}\in Orth$ such that

$$\mathbf{Q}\mathbf{E}\mathbf{Q}^T\in\mathscr{W}_\cap \qquad \text{and} \qquad \mathbf{Q}\mathbf{T}^{(a)}\mathbf{Q}^T = \mathbb{D}[\mathbf{Q}\mathbf{E}\mathbf{Q}^T] \qquad \forall\mathbf{E}\in\mathscr{W}_\cap. \tag{9.226}$$

The forthcoming steps follow the developments already shown in Section 9.7.2 to construct generalized constitutive equations for the Kirchhoff–Love model. Then, we have

$$\mathbf{T}_t^{(a)} = \mathbb{D}_t\mathbf{E}_t, \tag{9.227}$$

and its inverse relation

$$\mathbf{E}_t = \mathbb{A}_t\mathbf{T}_t^{(a)}. \tag{9.228}$$

As in the previous section, these relations are given by

$$\begin{pmatrix} E_{11}\\ E_{22}\\ E_{12}\end{pmatrix} = \begin{pmatrix} \frac{1}{\mathbb{E}_1} & -\frac{\nu_{12}}{\mathbb{E}_2} & 0\\ -\frac{\nu_{21}}{\mathbb{E}_1} & \frac{1}{\mathbb{E}_2} & 0\\ 0 & 0 & \frac{1}{\mathbb{G}_{12}}\end{pmatrix}\begin{pmatrix} T_{11}^{(a)}\\ T_{22}^{(a)}\\ T_{12}^{(a)}\end{pmatrix}, \tag{9.229}$$

while the functional $\Pi^*(\mathbf{T}^{(a)},\mathbf{E})$, as defined in Section 9.7.2, gives

$$\Pi^*(\mathbf{T}^{(a)},\mathbf{E}) = \int_{\Sigma_o}\int_H \left[\mathbf{T}_t^{(a)}\cdot\mathbf{E}_t - \frac{1}{2}(\mathbb{A}_t\mathbf{T}_t^{(a)}\cdot\mathbf{T}_t^{(a)})\right]\det\Lambda\ d\xi d\Sigma_o. \tag{9.230}$$

Here we will assume that the ratio $\frac{h}{R_m}$ is small. Our goal is to find approximate expressions that characterize the stress tensor $\mathbf{T}_t^A$ in terms of the generalized internal stresses

$\mathbf{N}_t^A$ and $\mathbf{M}_t^A$, where the index A refers, respectively, to $A = L, K, S$, the Love, Koiter and Sanders shell models. As before, we consider that the stress tensor $\mathbf{T}_t^A$ depends linearly on ξ. In order to satisfy definitions (9.123), (9.138) and (9.151) it is necessary that

$$\begin{aligned}
\mathbf{T}_t^L &= \left[\frac{1}{h}\mathbf{N}_t^L + \frac{6}{h^2}\varsigma\mathbf{M}_t^L\right], \\
\mathbf{T}_t^K &= \Lambda\left[\frac{1}{h}\mathbf{N}_t^K + \frac{6}{h^2}\varsigma\mathbf{M}_t^K\right]\frac{\Lambda}{\det\Lambda}, \\
\mathbf{T}_t^S &= \frac{\Lambda}{\det\Lambda}\left[\frac{1}{h}\mathbf{N}_t^S + \frac{6}{h^2}\varsigma\mathbf{M}_t^S\right],
\end{aligned} \tag{9.231}$$

where $\varsigma = \frac{\xi}{h/2}$. Now, introducing (9.231) into (9.230) we solve the problem of finding the element that maximizes functional $\Pi^*(\mathbf{N}_t^A, \mathbf{M}_t^A, \varepsilon^L, \chi^A)$ by requiring $\delta\Pi^* = 0$, $\forall\delta\mathbf{N}_t^A, \delta\mathbf{M}_t^A$. Notice that the tensor ε^L is the same for the three models (Love, Koiter and Sanders). The solution to this problem leads to

$$\begin{aligned}
\varepsilon^L &= \frac{1}{h^2}\tilde{\mathbb{A}}_t^0\mathbf{N}_t^A + \frac{12}{h^4}\tilde{\mathbb{A}}_t^1\mathbf{M}_t^A, \\
\chi^A &= \frac{12}{h^4}\tilde{\mathbb{A}}_t^1\mathbf{N}_t^A + \frac{144}{h^6}\tilde{\mathbb{A}}_t^2\mathbf{M}_t^A,
\end{aligned} \tag{9.232}$$

with $A = L, K, S$, and where tensors $\tilde{\mathbb{A}}_t^k$, $k = 0, 1, 2$ are defined by

$$\tilde{\mathbb{A}}_t^k = \int_H \mathbb{A}_t\xi^k\, d\xi \qquad k = 0, 1, 2. \tag{9.233}$$

Considering that the properties of the material that composes the shell remain invariant across the thickness, we can integrate (9.233) to give

$$\tilde{\mathbb{A}}_t^0 = h\mathbb{A}_t \quad \tilde{\mathbb{A}}_t^1 = \mathbf{O} \quad \tilde{\mathbb{A}}_t^2 = \frac{h^3}{12}\mathbb{A}_t. \tag{9.234}$$

Putting this result into (9.232) we obtain the generalized constitutive equations, which are

$$\begin{aligned}
\varepsilon^L &= \frac{1}{h}\mathbb{A}_t\mathbf{N}_t^A, \\
\chi^A &= \frac{12}{h^3}\mathbb{A}_t\mathbf{M}_t^A,
\end{aligned} \tag{9.235}$$

with $A = L, K, S$. Importantly, note that by virtue of the hypotheses introduced during the derivation of these constitutive equations, the membrane and bending stresses are decoupled in the constitutive equations. This is in contrast to the constitutive equations derived in the Naghdi model. According to Koiter [158], the decoupling of these membrane and bending stresses at the constitutive level is compatible with the first-order approximation that underlies the theory of thin shells.

Finally, using the basis $(\mathbf{e}_{(1)}, \mathbf{e}_{(2)}, \mathbf{n})$, the constitutive equations (9.235) turn into

$$\begin{aligned}
[\varepsilon^L]_{11} &= \frac{1}{\mathbb{E}_1 h}([\mathbf{N}_t^A]_{11} - \nu_{21}[\mathbf{N}_t^A]_{22}), \\
[\varepsilon^L]_{22} &= \frac{1}{\mathbb{E}_2 h}([\mathbf{N}_t^A]_{22} - \nu_{12}[\mathbf{N}_t^A]_{11}), \\
[\varepsilon^L]_{12} &= \frac{2}{\mathbb{G}_{12} h}[\mathbf{N}_t^A]_{12} = [\varepsilon^L]_{21},
\end{aligned} \tag{9.236}$$

$$
\begin{aligned}
[\chi^A]_{11} &= \frac{12}{\mathbb{E}_1 h^3}([\mathbf{M}_t^A]_{11} - \nu_{21}[\mathbf{M}_t^A]_{22}),\\
[\chi^A]_{22} &= \frac{12}{\mathbb{E}_2 h^3}([\mathbf{M}_t^A]_{22} - \nu_{12}[\mathbf{M}_t^A]_{11}),\\
[\chi^A]_{12} &= \frac{24}{\mathbb{G}_{12} h^3}[\mathbf{M}_t^A]_{12} = [\chi^A]_{21},
\end{aligned}
\tag{9.237}
$$

with $A = L, K, S$.

9.8 Characteristics of Shell Models

In this section we will visit three mechanically relevant aspects related to the shell models studied in this chapter. These issues are

- the relation between generalized stresses corresponding to the different models
- consistency of the models with respect to the so-called concept of equilibrium around the normal
- the determination of reactive stresses in each model.

9.8.1 Relation Between Generalized Stresses

Let us expose the relation between the generalized stresses $\mathbf{N}_t$, $\mathbf{M}_t$, $\mathbf{N}_t^K$, $\mathbf{M}_t^K$, $\mathbf{N}_t^S$ and $\mathbf{M}_t^S$ described in previous sections. This goal is attained by considering first that the inverse of tensor Λ can be written as

$$
\Lambda^{-1} = \mathbf{I} - \xi(\nabla_{\mathbf{x}_o}\mathbf{n})\Lambda^{-1},
\tag{9.238}
$$

which leads to the following recurrent expansion for this inverse tensor (see also (9.328))

$$
\Lambda^{-1} = \mathbf{I} - \xi\nabla_{\mathbf{x}_o}\mathbf{n} + \xi^2(\nabla_{\mathbf{x}_o}\mathbf{n})^2 - \dots + (-1)^n\xi^n(\nabla_{\mathbf{x}_o}\mathbf{n})^n\Lambda^{-1}.
\tag{9.239}
$$

Replacing (9.238) into (9.138), and making use of (9.239) yields

$$
\begin{aligned}
\mathbf{N}_t^K &= \int_H \mathbf{T}_t\Lambda^{-1}\det\Lambda\, d\xi - \int_H (\nabla_{\mathbf{x}_o}\mathbf{n})\Lambda^{-1}\mathbf{T}_t\xi\left(\mathbf{I} - \mathcal{O}\left(\frac{h}{R_m}\right)\right)\det\Lambda d\xi,\\
\mathbf{M}_t^K &= \int_H \left(\mathbf{I} - \mathcal{O}\left(\frac{h}{R_m}\right)\right)\mathbf{T}_t\xi\Lambda^{-1}\det\Lambda d\xi.
\end{aligned}
\tag{9.240}
$$

Dropping the terms of order $\mathcal{O}\left(\frac{h}{R_m}\right)$ and noting that, by definition, only the symmetric component of tensor $\mathbf{M}_t$ plays a role in the Koiter model, we obtain

$$
\begin{aligned}
\mathbf{N}_t^K &= \mathbf{N}_t - (\nabla_{\mathbf{x}_o}\mathbf{n})(\mathbf{M}_t)^T,\\
\mathbf{M}_t^K &= (\mathbf{M}_t)^s.
\end{aligned}
\tag{9.241}
$$

We have seen in Section 9.6.5 that

$$
\begin{aligned}
\mathbf{N}_t^S &= (\mathbf{N}_t)^s,\\
\mathbf{M}_t^S &= (\mathbf{M}_t)^s.
\end{aligned}
\tag{9.242}
$$

To relate $\mathbf{N}_t^K$ and $\mathbf{M}_t^K$ with $\mathbf{N}_t^S$ and $\mathbf{M}_t^S$ note the symmetry of $\mathbf{N}_t^K$ and also notice that in (9.240), without dropping terms of order $\mathcal{O}\left(\frac{h}{R_m}\right)$ it can be written as

$$\mathbf{N}_t^K = \int_H (\mathbf{T}_t \Lambda^{-1})^s \det \Lambda d\xi - \int_H ((\nabla_{\mathbf{x}_o}\mathbf{n})\Lambda^{-1}\mathbf{T}_t\Lambda^{-1})^s \xi \det \Lambda d\xi. \tag{9.243}$$

Finally, combining (9.138), (9.151) and (9.241) with (9.243), the generalized stress tensors in the Sanders model give

$$\begin{aligned}\mathbf{N}_t^S &= \mathbf{N}_t^K + ((\nabla_{\mathbf{x}_o}\mathbf{n})\mathbf{M}_t^K)^s,\\ \mathbf{M}_t^S &= \mathbf{M}_t^K.\end{aligned} \tag{9.244}$$

9.8.2 Equilibrium Around the Normal

In this section we analyze an important aspect in the mechanical theory of shells which is related to the equilibrium around the normal. This issue emerges once hypotheses are introduced in order to constrain the admissible shell kinematics. From a three-dimensional perspective, when no kinematical hypothesis has yet been considered, the equilibrium around the normal vector is inherently verified by virtue of the symmetry of the Cauchy stress tensor, which ensures the equilibrium around orthonormal axes. Nevertheless, and because of the introduction of kinematicalhypotheses and further geometric approximations in the shell models studied, this equilibrium must be carefully screened. This is carried out for the different shell models already presented.

9.8.2.1 Kirchhoff–Love Model

In the first place, and as already observed, $\mathbf{v}_t^o$ is a vector belonging to the tangent plane and $\mathbf{n}$ is the normal vector to the middle surface of the shell. Then we have

$$(\nabla_{\mathbf{x}_o}\mathbf{n})\mathbf{v}_t^o = -(\nabla_{\mathbf{x}_o}\mathbf{v}_t^o)^T\mathbf{n}, \tag{9.245}$$

and ω_t^{KL} given by (9.91) can be rewritten as

$$\omega_t^{KL} = -(\nabla_{\mathbf{x}_o}\mathbf{v}_t^o)^T\mathbf{n} - \nabla_{\mathbf{x}_o} v_n^o. \tag{9.246}$$

Inserting (9.246) into the expression of the internal virtual power (9.108) we get

$$\begin{aligned}\int_\Omega \mathbf{T}\cdot(\nabla\mathbf{v})^s d\Omega = \int_{\Sigma_o} &\left[\mathbf{N}_t\cdot(\Pi_t(\nabla_{\mathbf{x}_o}\mathbf{v}_t^o) + v_n^o\nabla_{\mathbf{x}_o}\mathbf{n})\right.\\ &\left.-\mathbf{M}_t\cdot((\nabla_{\mathbf{x}_o}\mathbf{v}_t^o)^T(\nabla_{\mathbf{x}_o}\mathbf{n}) + (\Pi_t\nabla_{\mathbf{x}_o}\nabla_{\mathbf{x}_o}\mathbf{v}_t^o)\mathbf{n} + \nabla_{\mathbf{x}_o}\nabla_{\mathbf{x}_o}v_n^o)\right] d\Sigma_o.\end{aligned} \tag{9.247}$$

Hence, considering that $\nabla_{\mathbf{x}_o}\mathbf{n} = \Pi_t(\nabla_{\mathbf{x}_o}\mathbf{n})$, we get

$$(\nabla_{\mathbf{x}_o}\mathbf{v}_t^o)^T(\nabla_{\mathbf{x}_o}\mathbf{n}) = (\nabla_{\mathbf{x}_o}\mathbf{v}_t^o)^T\Pi_t(\nabla_{\mathbf{x}_o}\mathbf{n}) = (\Pi_t(\nabla_{\mathbf{x}_o}\mathbf{v}_t^o))^T(\nabla_{\mathbf{x}_o}\mathbf{n}). \tag{9.248}$$

Decomposing $\Pi_t(\nabla_{\mathbf{x}_o}\mathbf{v}_t^o)$ into the symmetric and skew-symmetric components gives

$$\Pi_t(\nabla_{\mathbf{x}_o}\mathbf{v}_t^o) = (\Pi_t(\nabla_{\mathbf{x}_o}\mathbf{v}_t^o))^s + \Omega_t^o, \tag{9.249}$$

where Ω_t^o has been defined in (9.147). This tensor is associated with the surface rotational of the velocity field $\mathbf{v}_t^o$, and represents the rotation around the normal vector [238]. Using (9.248) and (9.249) in (9.247) we obtain

$$\int_\Omega \mathbf{T}\cdot(\nabla\mathbf{v})^s d\Omega = \int_{\Sigma_o} \Big[\mathbf{N}_t \cdot ((\Pi_t(\nabla_{\mathbf{x}_o}\mathbf{v}_t^o))^S + \Omega_t^o + v_n^o\nabla_{\mathbf{x}_o}\mathbf{n}) \\ - \mathbf{M}_t \cdot \big((\Pi_t(\nabla_{\mathbf{x}_o}\mathbf{v}_t^o))^s(\nabla_{\mathbf{x}_o}\mathbf{n}) - \Omega_t^o(\nabla_{\mathbf{x}_o}\mathbf{n}) \\ +(\Pi_t\nabla_{\mathbf{x}_o}\nabla_{\mathbf{x}_o}\mathbf{v}_t^o)\mathbf{n} + \nabla_{\mathbf{x}_o}\nabla_{\mathbf{x}_o}v_n^o\big)\Big]\, d\Sigma_o. \tag{9.250}$$

Taking the part of the internal power associated with Ω_t^o in expression (9.250), and noting that the external power associated with Ω_t^o is null, we have the equation for the equilibrium around the normal vector

$$\int_{\Sigma_o} (\mathbf{N}_t + \mathbf{M}_t(\nabla_{\mathbf{x}_o}\mathbf{n}))\cdot\Omega_t^o d\Sigma_o = 0 \qquad \forall\Omega_t^o \in \mathscr{W}^a, \tag{9.251}$$

where

$$\mathscr{W}^a = \{\Omega_t^o \in \mathscr{W};\ \Omega_t^o = (\Pi_t(\nabla_{\mathbf{x}_o}\mathbf{v}_t^o))^a,\ \ (\mathbf{v}_t^o, 0) \in \mathrm{Var}_v\}. \tag{9.252}$$

Notably, since $\Pi_t(\nabla_{\mathbf{x}_o}\mathbf{v}_t^o)$ is a tensor field over the tangent plane, $\mathscr{W}^a$ is nothing but the subspace of skew-symmetric tensors defined in such a plane.

Finally, from (9.251) we conclude that the equilibrium around the normal vector can be written in the following manner

$$(\mathbf{N}_t + \mathbf{M}_t(\nabla_{\mathbf{x}_o}\mathbf{n}))^a = \mathbf{O}. \tag{9.253}$$

To verify this let us recall the definitions of $\mathbf{N}_t$ and $\mathbf{M}_t$ given in (9.63), and also keep in mind that $(\nabla_{\mathbf{x}_o}\mathbf{n})\Lambda^{-1} = \Lambda^{-1}(\nabla_{\mathbf{x}_o}\mathbf{n})$. Then, we can verify that

$$(\mathbf{N}_t + \mathbf{M}_t(\nabla_{\mathbf{x}_o}\mathbf{n}))^a = \int_H (\mathbf{T}_t\Lambda^{-1} + \xi\mathbf{T}_t\Lambda^{-1}(\nabla_{\mathbf{x}_o}\mathbf{n}))^a \det\Lambda d\xi \\ = \int_H (\mathbf{T}_t(\mathbf{I} + \xi\nabla_{\mathbf{x}_o}\mathbf{n})\Lambda^{-1})^a \det\Lambda d\xi = \int_H (\mathbf{T}_t)^a \det\Lambda d\xi = \mathbf{O}. \tag{9.254}$$

This result clearly states that the exact theory behind the Kirchhoff–Love model intrinsically verifies the equilibrium around the normal.

9.8.2.2 Love Model

Following the same path as the previous section, we can straightforwardly see that the equilibrium around the normal in the Love model takes a similar form to (9.253), but now with $\mathbf{N}_t^L$ and $\mathbf{M}_t^L$ instead of $\mathbf{N}_t$ and $\mathbf{M}_t$

$$(\mathbf{N}_t^L + \mathbf{M}_t^L(\nabla_{\mathbf{x}_o}\mathbf{n}))^a \stackrel{?}{=} \mathbf{O}. \tag{9.255}$$

To check if this equation is actually verified let us consider the generalized stresses from the Love model defined in (9.123). Then, from (9.255) we have

$$(\mathbf{N}_t^L + \mathbf{M}_t^L(\nabla_{\mathbf{x}_o}\mathbf{n}))^a = \int_H (\mathbf{T}_t + \xi\mathbf{T}_t(\nabla_{\mathbf{x}_o}\mathbf{n}))^a d\xi = \int_H (\mathbf{T}_t\Lambda)^a d\xi \neq \mathbf{O}. \tag{9.256}$$

As can be appreciated from (9.256), the equation for the equilibrium around the normal for the Love model given by (9.255) is not satisfied in general. Nevertheless, we should note that in the following particular cases this equilibrium is satisfied: (i) plates, in which $\nabla_{\mathbf{x}_o}\mathbf{n} = \mathbf{O}$, (ii) spherical shells, for which $\nabla_{\mathbf{x}_o}\mathbf{n} = \frac{1}{R}\Pi_t$ (R is the curvature radius of the sphere), and (iii) shells with symmetry of revolution in both the definition of the geometry and the definition of the applied loads.

9.8.2.3 Koiter Model

In this case let us decompose the tensor $\Pi_t(\nabla_{\mathbf{x}_o}\mathbf{v}_t^o)$ into the symmetric component $(\Pi_t(\nabla_{\mathbf{x}_o}\mathbf{v}_t^o))^s$ and the skew-symmetric component Ω_t^o, as indicated in (9.249). Then, the internal virtual power given by (9.139) can be rewritten as

$$\begin{aligned}\int_\Omega \mathbf{T}\cdot(\nabla\mathbf{v})^s d\Omega = \int_{\Sigma_o} &\left[\mathbf{N}_t^K\cdot((\Pi_t(\nabla_{\mathbf{x}_o}\mathbf{v}_t^o))^s + \Omega_t^o + v_n^o\nabla_{\mathbf{x}_o}\mathbf{n})\right.\\ &- \mathbf{M}_t^K\cdot((\Pi_t(\nabla_{\mathbf{x}_o}\mathbf{v}_t^o))^s(\nabla_{\mathbf{x}_o}\mathbf{n}) - \Omega_t^o(\nabla_{\mathbf{x}_o}\mathbf{n}) + (\Pi_t\nabla_{\mathbf{x}_o}\nabla_{\mathbf{x}_o}\mathbf{v}_t^o)\mathbf{n}\\ &\left.+\nabla_{\mathbf{x}_o}\nabla_{\mathbf{x}_o}v_n^o - (\nabla_{\mathbf{x}_o}\mathbf{n})\Omega_t^o - (\nabla_{\mathbf{x}_o}\mathbf{n})(\Pi_t(\nabla_{\mathbf{x}_o}\mathbf{v}_t^o))^s - v_n^o(\nabla_{\mathbf{x}_o}\mathbf{n})^2)\right] d\Sigma_o.\end{aligned} \tag{9.257}$$

Analogously to previous sections, we will analyze the internal virtual power associated with the rotation around the normal, that is, for motion actions $\Omega_t^o \in \mathscr{W}^a$. The counterpart expression to (9.251), but now in the context of the Koiter model, leads to

$$\int_{\Sigma_o} (\mathbf{N}_t^K + \mathbf{M}_t^K(\nabla_{\mathbf{x}_o}\mathbf{n}) + (\nabla_{\mathbf{x}_o}\mathbf{n})\mathbf{M}_t^K)\cdot\Omega_t^o d\Sigma_o = 0 \quad \forall\Omega_t^o \in \mathscr{W}^a, \tag{9.258}$$

where $\mathscr{W}^a$ is as indicated in (9.252). Then, the Euler–Lagrange equations which manifest the equilibrium around the normal take the form

$$(\mathbf{N}_t^K + \mathbf{M}_t^K(\nabla_{\mathbf{x}_o}\mathbf{n}) + (\nabla_{\mathbf{x}_o}\mathbf{n})\mathbf{M}_t^K)^a = \mathbf{O}, \tag{9.259}$$

which is automatically verified once the tensors $\mathbf{N}_t^K$, $\mathbf{M}_t^K$ and $\nabla_{\mathbf{x}_o}\mathbf{n}$ are all symmetric. Therefore, we conclude that the Koiter shell model (like the Kirchhoff–Love model) also satisfies the equation that defines the equilibrium around the normal.

9.8.2.4 Sanders Model

Let us now decompose the tensor $\Pi_t(\nabla_{\mathbf{x}_o}\mathbf{v}_t^o)$ into the symmetric and skew-symmetric components (see (9.249)). Using the internal virtual power functional from the Sanders model given by (9.150), introducing in this expression the definition (9.151) for the generalized stresses and considering the previous decomposition for $\Pi_t(\nabla_{\mathbf{x}_o}\mathbf{v}_t^o)$ yields

$$\begin{aligned}\int_\Omega \mathbf{T}\cdot(\nabla\mathbf{v})^s d\Omega = \int_{\Sigma_o} &\left[\mathbf{N}_t^S\cdot((\Pi_t(\nabla_{\mathbf{x}_o}\mathbf{v}_t^o))^s + \Omega_t^o + v_n^o\nabla_{\mathbf{x}_o}\mathbf{n})\right.\\ &- \mathbf{M}_t^S\cdot\big((\Pi_t(\nabla_{\mathbf{x}_o}\mathbf{v}_t^o))^S(\nabla_{\mathbf{x}_o}\mathbf{n}) - \Omega_t^o(\nabla_{\mathbf{x}_o}\mathbf{n}) + (\Pi_t\nabla_{\mathbf{x}_o}\nabla_{\mathbf{x}_o}\mathbf{v}_t^o)\mathbf{n}\\ &\left.+\nabla_{\mathbf{x}_o}\nabla_{\mathbf{x}_o}v_n^o - (\nabla_{\mathbf{x}_o}\mathbf{n})\Omega_t^o\big)\right] d\Sigma_o.\end{aligned} \tag{9.260}$$

The contribution to the internal virtual power associated with $\Omega_t^o \in \mathscr{W}^a$ is given by

$$\int_{\Sigma_o} (\mathbf{N}_t^S + \mathbf{M}_t^S(\nabla_{\mathbf{x}_o}\mathbf{n}) + (\nabla_{\mathbf{x}_o}\mathbf{n})\mathbf{M}_t^S)\cdot\Omega_t^o d\Sigma_o = 0 \qquad \forall\Omega_t^o \in \mathscr{W}^a, \tag{9.261}$$

where $\mathscr{W}^a$ was defined in (9.252). Finally, the Euler–Lagrange equations corresponding to the equilibrium around the normal can be written in the following form

$$(\mathbf{N}_t^S + \mathbf{M}_t^S(\nabla_{\mathbf{x}_o}\mathbf{n}) + (\nabla_{\mathbf{x}_o}\mathbf{n})\mathbf{M}_t^S)^a = \mathbf{O}. \tag{9.262}$$

From the definitions of $\mathbf{N}_t^S$ and $\mathbf{M}_t^S$ given in (9.151), and noting the symmetry of $\nabla_{\mathbf{x}_o}\mathbf{n}$, we conclude that (9.262) is automatically satisfied in the Sanders model. In this manner, as in the Koiter model, we have demonstrated that the Sanders model is consistent in the sense that, in spite of the kinematical and geometrical hypotheses introduced to drop terms of order $\mathcal{O}(\frac{h}{R_m})$, the equilibrium around the normal is satisfied.

9.8.3 Reactive Generalized Stresses

The shell models studied in this chapter all are based on kinematical hypotheses which enable the reduction of the dimension of the domain of analysis where the problem is defined. In effect, these models started with the consideration of a mechanical problem posed in a three-dimensional body and turned into a mechanical problem cast over the two-dimensional domain defined by the middle surface of the shell.

In the previous context a natural question that emerges is concerned with the suitability of a given shell model when facing a real problem. Evidently, the answer to this question depends upon the geometry of the body as well as on the nature of the applied loads to which the body is subjected.

Moreover, once a shell model has been chosen to represent the mechanics of a three-dimensional body, the next natural question that arises is concerned with the quantification of the noncomformity between the chosen shell model and the reality posed by the original three-dimensional model. In order to provide tools to address this question we have, at least, to evaluate the terms that have been disregarded in the expression of the internal virtual power as a consequence of the kinematical and geometrical hypotheses adopted. Thus, an important step to establish the degree of confidence in the selected shell model consists of constructing expressions to measure the magnitude of the reactive generalized stresses (which play no role in the mechanical equilibrium because of the null power for all admissible generalized strain actions) when compared to the active generalized stresses. If these reactive stresses feature significant values relative to the active stresses, then we have an indicator that the shell model is not adequate to approximate the original three-dimensional problem.

In the following sections we will derive expressions which allow us to assess the reactive generalized stresses for the shell models based on the Naghdi and Kirchhoff–Love hypotheses. This will be achieved by establishing approximate expressions for such reactive stresses through the Euler–Lagrange equations associated with the variational formulation defined in (9.169) and adapted to the different shell models under study.

9.8.3.1 Reactions in the Naghdi Model

The theory behind the Naghdi model developed in Section 9.6.1 together with the derivation of the constitutive equations for this model (see Section 9.7.2) fully characterize the corresponding PVP (see (9.83)). As we already saw, once this equilibrium problem is solved we obtain the generalized stresses $\mathbf{N}_t$, $\mathbf{M}_t$ and $\mathbf{Q}$.

The calculation of reactive stresses in the Naghdi model consists of obtaining a closed form for the stress component T_n. Since this is a reactive stress, it must be calculated after having solved the variational equation for the mechanical equilibrium in the shell, and therefore after knowing the corresponding generalized stresses $\mathbf{N}_t$, $\mathbf{M}_t$ and $\mathbf{Q}$ that make the equilibrium effective. For this, as said above, it will be necessary to solve a new variational problem corresponding to the mechanical equilibrium wherein the normal component of the strain is allowed. Releasing this constraint the power exerted by T_n is not null and can be quantified. This new problem is governed by the variational equation seen in (9.55), which corresponds to a three-dimensional body, but having exploited the geometric description for shells. With such a variational equation, where now the only field to be determined is the unknown reactive stress T_n, we obtain the following Euler–Lagrange equations for T_n

$$\begin{cases} \frac{\partial}{\partial \xi}(T_n \det \Lambda) = (\mathbf{T}_t \Lambda^{-1} \det \Lambda) \cdot (\nabla_{\mathbf{x}_o} \mathbf{n}) \\ \qquad\qquad -\mathrm{div}_{\mathbf{x}_o}(\Lambda^{-1}\mathbf{T}_s \det \Lambda) - b_n \det \Lambda & \text{on } \Sigma_o, \\ T_n = f_n^+ & \text{at } \xi = \frac{h}{2}, \\ T_n = -f_n^- & \text{at } \xi = -\frac{h}{2}. \end{cases} \tag{9.263}$$

This boundary value problem, written in terms of the transversal variable ξ, is a system of first-order differential equations which admits a unique solution if the stresses $\mathbf{T}_t$ and $\mathbf{T}_s$ are the ones that provide the solution to the variational equation of the three-dimensional body and considering the kinematical hypotheses (9.72) and (9.73) (see [70]). Integrating $(9.263)_1$ and using $(9.263)_3$, the following characterization is obtained for the reactive stress T_n

$$T_n = -f_n^- + \frac{1}{\det \Lambda} \int_{-\frac{h}{2}}^{\xi} \left[(\mathbf{T}_t \Lambda^{-1} \cdot (\nabla_{\mathbf{x}_o} \mathbf{n}) - b_n) \det \Lambda - \mathrm{div}_{\mathbf{x}_o}(\Lambda^{-1}\mathbf{T}_s \det \Lambda) \right] d\zeta. \tag{9.264}$$

This is the exact expression for T_n corresponding to the Naghdi kinematical hypotheses. Clearly, this expression naturally satisfies the boundary condition at $\xi = -\frac{h}{2}$. Notwithstanding this, the solution of the equilibrium in the shell does not provide the stresses $\mathbf{T}_t$ and $\mathbf{T}_s$, but it delivers the generalized stresses which, for the Naghdi model, are given by $\mathbf{N}_t$, $\mathbf{M}_t$ and $\mathbf{Q}$. Then, it is interesting to obtain the approximation for T_n as a function of these generalized stresses. This approximation can be obtained by inserting in (9.264) the approximate expressions we have constructed for $\mathbf{T}_t$ and $\mathbf{T}_s$ in terms of the generalized stresses (see (9.200) and (9.201)). Proceeding this way, and integrating across the thickness, we have the resulting approximation for T_n as a function of $\mathbf{N}_t$, $\mathbf{M}_t$ and $\mathbf{Q}$, and of the generalized forces f_n^-, $\mathbf{f}_t^+$, $\mathbf{f}_t^-$ and b_n

$$\begin{aligned} T_n = -f_n^- - \frac{1}{\det \Lambda} \Bigg[& \mathrm{div}_{\mathbf{x}_o}\left(\frac{1}{4}\mathbf{Q}(2 + 3\varsigma - \varsigma^3)\right) \\ & + \mathrm{div}_{\mathbf{x}_o}\left(\frac{h}{8}(\Lambda^{-1}\det \Lambda)^+ \mathbf{f}_t^+ (\varsigma^3 + \varsigma^2 - \varsigma - 1)\right) \\ & - \mathrm{div}_{\mathbf{x}_o}\left(\frac{h}{8}(\Lambda^{-1}\det \Lambda)^- \mathbf{f}_t^- (\varsigma^3 - \varsigma^2 - \varsigma + 1)\right) \\ & - \left(\frac{1}{2}\mathbf{N}_t(\varsigma + 1) + \frac{3}{2h}\mathbf{M}_t(\varsigma^2 - 1)\right) \cdot (\nabla_{\mathbf{x}_o}\mathbf{n}) + \int_{-\frac{h}{2}}^{\xi} b_n \det \Lambda d\zeta \Bigg], \end{aligned} \tag{9.265}$$

where $\varsigma = \frac{\xi}{h/2}$. This approximation for T_n satisfies the boundary condition at $\xi = -\frac{h}{2}$. Let us now verify if the boundary condition at $\xi = \frac{h}{2}$ is satisfied. To calculate the value of T_n at $\xi = \frac{h}{2}$ we recall the definition of the generalized force p_{bn}, defined in (9.67). Thus, we obtain

$$T_n|_{\xi=\frac{h}{2}} = -f_n^- - \frac{1}{(\det \Lambda)^+}\left[\mathrm{div}_{\mathbf{x}_o}\mathbf{Q} - \mathbf{N}_t \cdot (\nabla_{\mathbf{x}_o}\mathbf{n}) + p_{bn}\right]. \tag{9.266}$$

Since $\mathbf{N}_t$ and $\mathbf{Q}$ solve the mechanical equilibrium for the Naghdi model, we can replace $(9.85)_2$ in (9.266) to yield

$$T_n|_{\xi=\frac{h}{2}} = f_n^+ + \left[\frac{(\det \Lambda)^-}{(\det \Lambda)^+} - 1\right] f_n^-, \tag{9.267}$$

where we have made use of the definition of the generalized force p_n given in (9.67).

Analyzing expression (9.267) we observe that, since approximate expressions were involved in its derivation, the reactive stress T_n does not satisfy, in general, the boundary condition $(9.263)_2$. However, it is important to notice that in the case when $f_n^- = 0$, also for the case of plates (where $\Lambda = \mathbf{I}$), and for the case of thin shells (when we can neglect the terms of order $\mathcal{O}(\frac{h}{R_m})$), the expression we have arrived at is consistent with the boundary condition $(9.263)_2$.

9.8.3.2 Reactions in the Kirchhoff–Love Model

The theoretical formulation of the Kirchhoff–Love model developed in Sections 9.6.2 and 9.7.3 took us to the variational equation (9.114) and to the generalized constitutive equations (9.232). As we already know, the solution of the equilibrium problem (9.114) enables us to determine the generalized stresses $\mathbf{N}_t$ and $\mathbf{M}_t$.

Considering the kinematical hypotheses corresponding to the Kirchhoff–Love theory (and also to the models that branch from this theory), the calculation of the reactive generalized stresses consists of recovering the stresses $\mathbf{T}_s$ and T_n which, within this model, exert null internal power in the variational formulation (9.114). In this way, the procedure to evaluate these reactive stresses will follow, basically, the same strategy as the previous section. This implies that we will have to solve a new problem of mechanical equilibrium in which the normal and shear strains are admissible strain actions, rendering a non-null internal power that will allow us to characterize the reactions T_n and $\mathbf{T}_s$. This new mechanical problem consists of the variational equation for a three-dimensional body and using the shell descriptive geometry, as seen in (9.55). With this variational equation, where the only unknown fields to be determined are T_n and $\mathbf{T}_s$, it is possible to obtain the Euler–Lagrange equations for each field. First, for $\mathbf{T}_s$ these equations are

$$\begin{cases} \frac{\partial}{\partial \xi}(\mathbf{T}_s \det \Lambda) + (\nabla_{\mathbf{x}_o}\mathbf{n})\Lambda^{-1}\mathbf{T}_s \det \Lambda = \\ \qquad -\mathrm{div}_{\mathbf{x}_o}(\mathbf{T}_t \Lambda^{-1} \det \Lambda) - \mathbf{b}_t \det \Lambda & \text{on } \Sigma_o, \\ \mathbf{T}_s = \mathbf{f}_t^+ & \text{at } \xi = \frac{h}{2}, \\ \mathbf{T}_s = -\mathbf{f}_t^- & \text{at } \xi = -\frac{h}{2}. \end{cases} \tag{9.268}$$

As in the previous section, we have arrived at a system of first-order differential equations in ξ with two boundary conditions. The solution exists provided the field

$\mathbf{T}_t$ is the solution of the variational problem in the three-dimensional body under the Kirchhoff–Love kinematical hypotheses (9.89) and (9.90) (see [70]). Now, given $\Lambda = \mathbf{I} + \xi\nabla_{\mathbf{x}_o}\mathbf{n}$, we can rewrite equation $(9.268)_1$, as

$$\Lambda^{-1}\frac{\partial}{\partial\xi}(\Lambda\mathbf{T}_s \det\Lambda) = -\mathrm{div}_{\mathbf{x}_o}(\mathbf{T}_t\Lambda^{-1}\det\Lambda) - \mathbf{b}_t\det\Lambda \qquad \text{on } \Sigma_o. \tag{9.269}$$

Integrating this expression across the thickness and using $(9.268)_3$ we obtain the exact expression for the generalized reactive stress $\mathbf{T}_s$

$$\mathbf{T}_s = -\mathbf{f}_t^- - \frac{\Lambda^{-1}}{\det\Lambda}\int_{-\frac{h}{2}}^{\xi}\Lambda\left[\mathrm{div}_{\mathbf{x}_o}(\mathbf{T}_t\Lambda^{-1}\det\Lambda) + \mathbf{b}_t\det\Lambda\right]\,d\zeta. \tag{9.270}$$

Similarly, the Euler–Lagrange equations for T_n are exactly the same as derived in (9.263). These equations provide the exact expression already given in (9.264). In this case, $\mathbf{T}_t$ must correspond to the solution of the three-dimensional problem wherein the Kirchhoff–Love hypotheses were introduced, while $\mathbf{T}_s$ is the reactive stress obtained in (9.270).

Let us now replace $\mathbf{T}_t$ by its approximate expression as a function of the generalized stresses from the models under analysis. In particular, consider the Love model as a representative model which features an approximation of order $\frac{h}{R_m}$. Then the results derived next can be easily extended to Koiter and Sanders models.

To make the presentation simpler, consider that $\mathbf{b}_t = b_n\mathbf{n} = \mathbf{0}$. Also, since in the Love model (as in Koiter and Sanders models) the approximation is related to the assumption $\frac{h}{R_m} \ll 1$, we therefore take the approximations $\Lambda \simeq \mathbf{I}$ and $\det\Lambda \simeq 1$ as valid.

In what follows, we will first derive an approximation for $\mathbf{T}_s$. From the previous considerations, expression (9.270) is simplified as

$$\mathbf{T}_s = -\mathbf{f}_t^- - \int_{-\frac{h}{2}}^{\xi}\left[\mathrm{div}_{\mathbf{x}_o}\mathbf{T}_t + \zeta(\nabla_{\mathbf{x}_o}\mathbf{n})\mathrm{div}_{\mathbf{x}_o}\mathbf{T}_t\right]\,d\zeta. \tag{9.271}$$

Inserting the approximation for $\mathbf{T}_t$ given by $(9.231)_1$ (corresponding to the Love model), and integrating (9.271) in the shell thickness, we obtain the characterization for $\mathbf{T}_s$ in terms of the generalized stresses $\mathbf{N}_t^L$ and $\mathbf{M}_t^L$ and of the generalized force $\mathbf{f}_t^-$

$$\begin{aligned}\mathbf{T}_s = &-\mathbf{f}_t^- - \mathrm{div}_{\mathbf{x}_o}\left(\frac{1}{2}\mathbf{N}_t^L(\varsigma+1) + \frac{3}{2h}\mathbf{M}_t^L(\varsigma^2-1)\right)\\ &-(\nabla_{\mathbf{x}_o}\mathbf{n})\mathrm{div}_{\mathbf{x}_o}\left(\frac{h}{8}\mathbf{N}_t^L(\varsigma^2-1) + \frac{1}{2}\mathbf{M}_t^L(\varsigma^3+1)\right),\end{aligned} \tag{9.272}$$

which automatically satisfies the boundary condition at $\xi = -\frac{h}{2}$ (i.e. $\varsigma = -1$). Moreover, for $\xi = \frac{h}{2}$ we have

$$\mathbf{T}_s|_{\xi=\frac{h}{2}} = -\mathbf{f}_t^- - \mathrm{div}_{\mathbf{x}_o}\mathbf{N}_t^L - (\nabla_{\mathbf{x}_o}\mathbf{n})\mathrm{div}_{\mathbf{x}_o}\mathbf{M}_t, \tag{9.273}$$

and using $(9.116)_1$ this yields

$$\mathbf{T}_s|_{\xi=\frac{h}{2}} = -\mathbf{f}_t^- + \mathbf{p}_t^{KL}. \tag{9.274}$$

Hence, with the definition of the generalized force $\mathbf{p}_t^{KL}$, recalling that $\mathbf{b}_t = \mathbf{0}$ and $b_n = 0$, and assuming $\Lambda\det\Lambda \simeq \mathbf{I}$, expression (9.274) gives

$$\mathbf{T}_s|_{\xi=\frac{h}{2}} = \mathbf{f}_t^+. \tag{9.275}$$

We conclude that the approximate expression for the reactive stress $\mathbf{T}_s$ also verifies the boundary condition $(9.268)_2$ over the other surface of the shell.

Let us now see the approximation for T_n. With the considerations presented above, equation (9.264) becomes

$$T_n = -f_n^- + \int_{-\frac{h}{2}}^{\xi} \left[\mathbf{T}_t \cdot (\nabla_{\mathbf{x}_o}\mathbf{n}) - \mathrm{div}_{\mathbf{x}_o}\mathbf{T}_s\right] d\zeta. \tag{9.276}$$

Introducing $(9.231)_1$ and (9.272) into (9.276) yields

$$\begin{aligned} T_n = -f_n^- + \int_{-\frac{h}{2}}^{\xi} \Bigg[& \left(\frac{1}{h}\mathbf{N}_t^L + \frac{6}{h^2}\mathbf{M}_t^L\varpi\right) \cdot (\nabla_{\mathbf{x}_o}\mathbf{n}) + \mathrm{div}_{\mathbf{x}_o}\mathbf{f}_t^- \\ & + \mathrm{div}_{\mathbf{x}_o}\mathrm{div}_{\mathbf{x}_o}\left(\frac{1}{2}\mathbf{N}_t^L(\varpi+1) + \frac{3}{2h}\mathbf{M}_t^L(\varpi^2-1)\right) \\ & + \mathrm{div}_{\mathbf{x}_o}\left((\nabla_{\mathbf{x}_o}\mathbf{n})\mathrm{div}_{\mathbf{x}_o}\left(\frac{h}{8}\mathbf{N}_t^L(\varpi^2-1) + \frac{1}{2}\mathbf{M}_t^L(\varpi^3+1)\right)\right)\Bigg] d\zeta, \end{aligned} \tag{9.277}$$

where $\varpi = \frac{\zeta}{h/2}$. Taking into account the order of approximation of this model, the following expression also holds

$$\mathrm{div}_{\mathbf{x}_o}\left(\frac{1}{2}\mathbf{N}_t^L(\varpi+1)\right) + (\nabla_{\mathbf{x}_o}\mathbf{n})\mathrm{div}_{\mathbf{x}_o}\left(\frac{h}{8}\mathbf{N}_t^L(\varpi^2-1)\right) \simeq \frac{1}{2}(\varpi+1)\mathrm{div}_{\mathbf{x}_o}\mathbf{N}_t^L. \tag{9.278}$$

With this result in (9.277), after arranging terms, and integrating across the shell thickness, we obtain the characterization for T_n as a function of the generalized stresses $\mathbf{N}_t^L$ and $\mathbf{M}_t^L$ and of the generalized forces f_n^- and $\mathbf{f}_t^-$

$$\begin{aligned} T_n = & -f_n^- + \left(\frac{1}{2}\mathbf{N}_t^L(\varsigma+1) + \frac{3}{2h}\mathbf{M}_t^L(\varsigma^2-1)\right) \cdot (\nabla_{\mathbf{x}_o}\mathbf{n}) + \mathrm{div}_{\mathbf{x}_o}\left(\frac{h}{2}\mathbf{f}_t^-(\varsigma+1)\right) \\ & + \mathrm{div}_{\mathbf{x}_o}\mathrm{div}_{\mathbf{x}_o}\left(\frac{h}{8}\mathbf{N}_t^L(\varsigma^2+2\varsigma+1) + \frac{1}{4}\mathbf{M}_t^L(\varsigma^3-3\varsigma-2)\right) \\ & + \mathrm{div}_{\mathbf{x}_o}\left((\nabla_{\mathbf{x}_o}\mathbf{n})\mathrm{div}_{\mathbf{x}_o}\left(\frac{h}{16}(\varsigma^4+4\varsigma+3)\mathbf{M}_t^L\right)\right). \end{aligned} \tag{9.279}$$

Clearly, this equation satisfies the boundary condition at $\xi = -\frac{h}{2}$. For the surface corresponding to $\xi = \frac{h}{2}$, the previous expression leads to

$$\begin{aligned} T_n|_{\xi=\frac{h}{2}} = & -f_n^- + \mathbf{N}_t^L \cdot (\nabla_{\mathbf{x}_o}\mathbf{n}) - \mathrm{div}_{\mathbf{x}_o}\mathrm{div}_{\mathbf{x}_o}\mathbf{M}_t^L + h\,\mathrm{div}_{\mathbf{x}_o}\mathbf{f}_t^- \\ & + \frac{h}{2}\mathrm{div}_{\mathbf{x}_o}\left(\mathrm{div}_{\mathbf{x}_o}\mathbf{N}_t^L + (\nabla_{\mathbf{x}_o}\mathbf{n})\mathrm{div}_{\mathbf{x}_o}\mathbf{M}_t^L\right). \end{aligned} \tag{9.280}$$

Noting that $\mathbf{N}_t^L$ and $\mathbf{M}_t^L$ satisfy $(9.116)_1$ and $(9.116)_2$ we get

$$T_n|_{\xi=\frac{h}{2}} = -f_n^- + p_n^{KL} + h\,\mathrm{div}_{\mathbf{x}_o}\mathbf{f}_t^- - \frac{h}{2}\mathrm{div}_{\mathbf{x}_o}\mathbf{p}_t^{KL}. \tag{9.281}$$

Finally, from the definitions of $\mathbf{p}_t^{KL}$ and p_n^{KL} we get

$$T_n|_{\xi=\frac{h}{2}} = f_n^+. \tag{9.282}$$

We therefore see that, in this theory, the reactive stress satisfies the boundary condition at the superior shell surface. This tells us that the proposed approximation is consistent with the boundary condition $(9.263)_2$, valid for the exact problem which defines T_n.

Finally, observe that the approximate characterizations for the reactive stresses in terms of the generalized stresses $\mathbf{N}_t^L$ and $\mathbf{M}_t^L$ resulted from the solution of the equilibrium equations associated with Love's first approximation. Notwithstanding this, and given that the Koiter and Sanders models rely on the same approximation established by relation $\frac{h}{R_m}$, in the expressions for $\mathbf{T}_s$ and T_n (see (9.272) and (9.279)), $\mathbf{N}_s^L$ and $\mathbf{M}_s^L$ can directly be replaced by the counterparts corresponding to either the Koiter or Sanders models, that is, by $(\mathbf{N}_s^K, \mathbf{M}_s^K)$ or $(\mathbf{N}_s^S, \mathbf{M}_s^S)$.

9.9 Basics Notions of Surfaces

9.9.1 Preliminaries

Let Ω be an open set of $\mathbb{R}^2$. For simplicity, we will consider that in this set Ω we have adopted an orthogonal system of coordinates $\{0, \xi^\alpha, \alpha = 1, 2\}$ with a basis defined by the unit vectors $\mathbf{e}_\alpha$, $\alpha = 1, 2$. Then, each point $\xi \in \Omega$ can be characterized by $\xi = \xi^\alpha \mathbf{e}_\alpha$. We say that the vector function (application) $\phi : \xi \in \Omega \mapsto \mathbf{X} \in \mathbb{R}^3$ is of class $\mathscr{C}^1$ if ϕ and the partial derivatives $\phi_\alpha = \frac{\partial \phi}{\partial \xi^\alpha}$, $\alpha = 1, 2$, are all continuous. Similarly, we say that ϕ is of class $\mathscr{C}^k$, $k \geq 2$, if ϕ and all its partial derivatives up to order k are all continuous functions. In particular, we say that ϕ is smooth if it is of class $\mathscr{C}^k$ for all positive integers k. A fundamental aspect in differential geometry is that if ϕ is of class $\mathscr{C}^k$, $k \geq 2$, then one can verify that $\frac{\partial^2 \phi}{\partial \xi^1 \partial \xi^2} = \frac{\partial^2 \phi}{\partial \xi^2 \partial \xi^1}$ (and something similar holds for higher-order partial derivatives). With these elements we can introduce the following definition.

We say that a smooth vector function $\phi : \xi \in \Omega \subset \mathbb{R}^2 \mapsto \mathbf{X} \in \Sigma \subset \mathbb{R}^3$ is a parameterization of the surface $\Sigma \subset \mathbb{R}^3$ constituted by all points $\mathbf{X}$ such that $\mathbf{X} = \phi(\xi)$ with $\xi \in \Omega$ (see Figure 9.7) if

1) ϕ is a one-to-one mapping (injective), that is, for every point in Σ there is a unique point in Ω
2) the partial derivatives $\phi_{\xi^1}(\xi_o) = \frac{\partial \phi}{\partial \xi^1}|_{\xi_o}$ and $\phi_{\xi^2}(\xi_o) = \frac{\partial \phi}{\partial \xi^2}|_{\xi_o}$ are linearly independent for every point $\xi_o = (\xi_o^1, \xi_o^2) \in \Omega$.

In this manner, a family of straight orthogonal lines parallel to directions $\mathbf{e}_1, \mathbf{e}_2$ in Ω are transformed through the parameterization ϕ in the ξ^1-curve and ξ^2-curve over the surface Σ, as illustrated in Figure 9.8, defining what is usually termed as curvilinear coordinates over the surface. Since for each point $\mathbf{X} = \phi(\xi)$ the vectors ϕ_{ξ^1} and ϕ_{ξ^2} are linearly independent, we can define at point $\mathbf{X}$ from Σ the (unique) tangent plane $P_t(\mathbf{X})$ characterized by the unit normal vector $\mathbf{n}(\mathbf{X})$ which, for the parameterization ϕ, is defined by

$$\mathbf{n}(\mathbf{X}) = \frac{\phi_{\xi^1} \times \phi_{\xi^2}}{\|\phi_{\xi^1} \times \phi_{\xi^2}\|}. \tag{9.283}$$

Notice that in doing this we are further assigning an orientation to the surface Σ.

With the elements presented so far, we are able to define, for any regular (smooth) surface, an intrinsic system of coordinates defined at each point of the surface by its

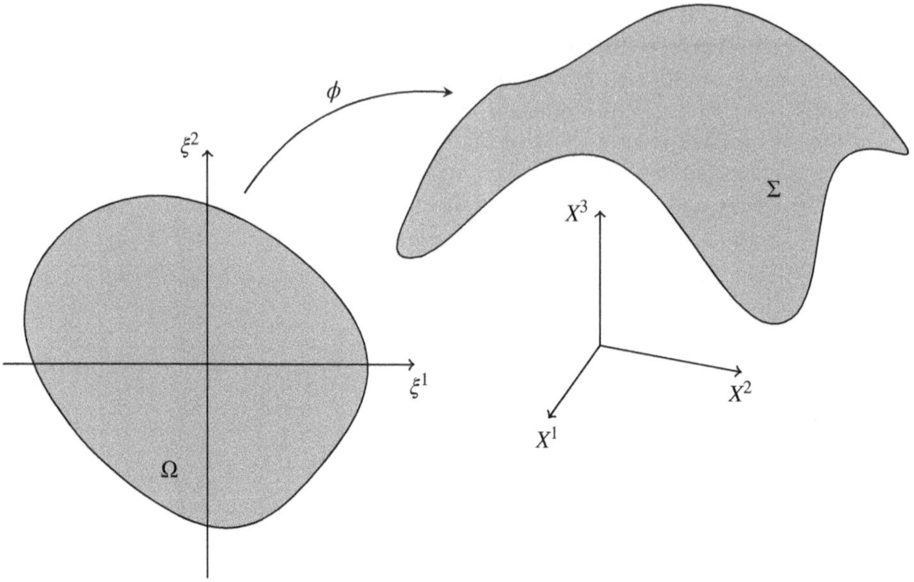

Figure 9.7 Parameterization of a surface.

tangent plane and its normal vector, which is independent from the parameterization utilized in defining the surface. Specifically, the basis $\{P_t(\mathbf{X}), \mathbf{n}(\mathbf{X})\}$ (which for a certain parameterization ϕ is given by the triple $\{\phi_{\xi^1}, \phi_{\xi^2}, \mathbf{n}\}$), is a basis for the vector space $\mathscr{V}$ associated with the Euclidean point space $\mathbb{R}^3$.

9.9.2 First Fundamental Form

In this section we will study how the inner product between a differential vector lying on the tangent plane and a differential vector in the direction normal to the surface behaves in terms of the differential element $d\xi \in \Omega$.

Observe that for a given parameterization ϕ of a surface Σ, even if the vectors ϕ_{ξ^1} and ϕ_{ξ^2} are linearly independent, we can define a basis for the tangent plane given by $\{\phi_{\xi^1}, \phi_{\xi^2}\}$, in general, they are neither orthogonal vectors nor unit vectors. Therefore, the length (norm) of any vector in this tangent plane written as a function of a linear combination of the elements in such a basis cannot be computed using Pythagoras' theorem (which holds for a basis formed by orthogonal unit vectors).

To see this in detail we can, from the aforementioned basis, define another basis, $\{\phi^{\xi^1}, \phi^{\xi^2}\}$, which in Chapter 1 was termed the dual basis, and which is characterized by

$$\phi^{\xi^\alpha} \cdot \phi_{\xi^\beta} = \delta^\alpha_\beta, \qquad \delta^\alpha_\beta = \begin{cases} 1 & \text{if } \alpha = \beta, \\ 0 & \text{if } \alpha \neq \beta. \end{cases} \tag{9.284}$$

In this way, a vector $\mathbf{v}$ belonging to the tangent plane ($\mathbf{v} \in P_t(\mathbf{X})$, $\mathbf{X} \in \Sigma$) can be written as a linear combination of vectors in the original basis, or of dual vectors, that is

$$\mathbf{v} = v^\alpha \phi_{\xi^\alpha} = v_\alpha \phi^{\xi^\alpha}, \tag{9.285}$$

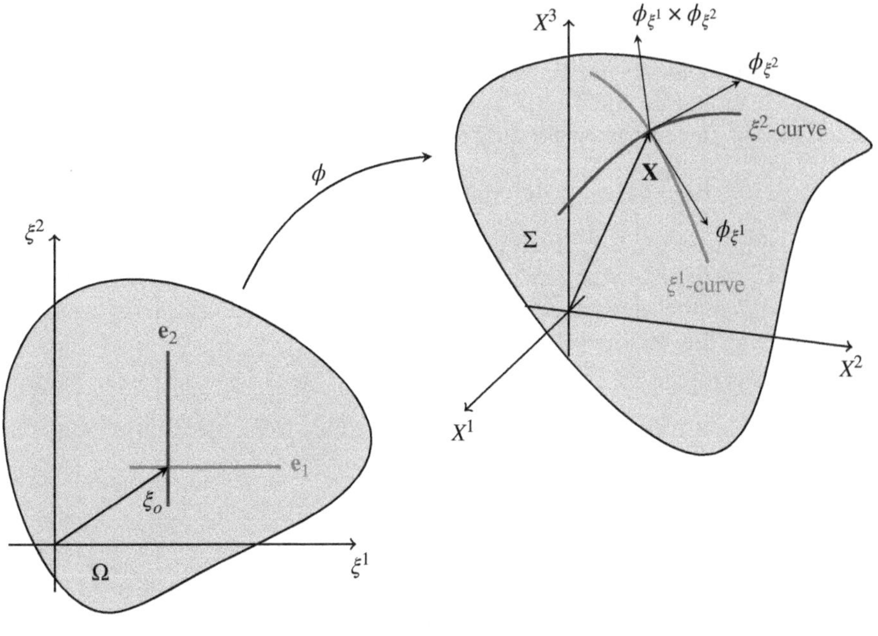

Figure 9.8 Tangent plane and normal vector. Intrinsic system of coordinates for a surface.

where the repeated index denotes the implicit sum spanning all possible values the index can take (that is $\alpha = 1, 2$), and where $v_\alpha = \mathbf{v} \cdot \phi_{\xi^\alpha}$ and $v^\alpha = \mathbf{v} \cdot \phi^{\xi^\alpha}$ are called covariant components and contravariant components, respectively. In turn, we can easily verify that the inner product of vectors in the tangent plane, say $\mathbf{v}$ and $\mathbf{u}$, is given by

$$\mathbf{v} \cdot \mathbf{u} = g_{\alpha\beta} v^\alpha v^\beta = g^{\alpha\beta} v_\alpha v_\beta = v^\alpha u_\alpha = v_\alpha u^\alpha, \tag{9.286}$$

with

$$g_{\alpha\beta} = \phi_{\xi^\alpha} \cdot \phi_{\xi^\beta} \qquad g^{\alpha\beta} = \phi^{\xi^\alpha} \cdot \phi^{\xi^\beta}, \tag{9.287}$$

and where, given the symmetry of the inner product, we get $g_{\alpha\beta} = g_{\beta\alpha}$ and $g^{\alpha\beta} = g^{\beta\alpha}$. Since the basis vectors are vector fields defined over the surface, it is important to notice that the coefficients $g_{\alpha\beta}$ (and so $g^{\alpha\beta}$) are scalar fields defined over the surface, usually denoted as

$$\begin{aligned} E(\xi) &= g_{11}(\xi) = \phi_{\xi^1}(\xi) \cdot \phi_{\xi^1}(\xi), \\ F(\xi) &= g_{12}(\xi) = \phi_{\xi^1}(\xi) \cdot \phi_{\xi^2}(\xi) = g_{21}(\xi), \\ G(\xi) &= g_{22}(\xi) = \phi_{\xi^2}(\xi) \cdot \phi_{\xi^2}(\xi). \end{aligned} \tag{9.288}$$

Evidently, $E > 0$ and $G > 0$, and in general they are different from the unit value, which means that the vectors of the basis for the tangent plane are not unit vectors. Also, the field F indicates how the curvilinear coordinates intersect each other at each point over the surface. In fact, if $F = 0$ we have that the curvilinear coordinates are orthogonal. If now we adopt $\mathbf{u} = \mathbf{v}$, we obtain

$$\|\mathbf{v}\|^2 = g_{11}(v^1)^2 + 2g_{12}v^1v^2 + g_{22}(v^2)^2 = E(v^1)^2 + 2Fv^1v^2 + G(v^2)^2. \tag{9.289}$$

Thus, the coefficients (fields) E, F and G can be interpreted as the correction factors of the classical Pythagoras' theorem.

Repeating the previous development, let us study the transformation of differential vectors in Ω ($d\xi \in \Omega$) into differential vectors on Σ ($d\mathbf{X} \in P_t(\mathbf{X})$). Again, given the parameterization ϕ, this transformation takes the form

$$d\mathbf{X} = \frac{\partial \phi}{\partial \xi^1} d\xi^1 + \frac{\partial \phi}{\partial \xi^2} d\xi^2 = \phi_{\xi^1} d\xi^1 + \phi_{\xi^2} d\xi^2. \tag{9.290}$$

The previous expression can be written in compact form (independent from the coordinate system) by introducing the tensor field $\nabla_\xi \phi(\xi)$ which puts in evidence the relation between a differential element $d\xi$ in $\xi \in \Omega$ and its corresponding differential $d\mathbf{X} = d\mathbf{X}(\xi)$ on Σ through the parameterization ϕ

$$d\mathbf{X} = (\nabla_\xi \phi) d\xi. \tag{9.291}$$

For this system of coordinates, Ω gives $d\xi = d\xi^1 \mathbf{e}_1 + d\xi^2 \mathbf{e}_2$. Replacing this in (9.291) it is easy to verify the following relations

$$\begin{aligned} \phi_{\xi^1} &= (\nabla_\xi \phi)\mathbf{e}_1, \\ \phi_{\xi^2} &= (\nabla_\xi \phi)\mathbf{e}_2. \end{aligned} \tag{9.292}$$

From these relations we also establish the connection between the length of an infinitesimal element on the tangent plane and the corresponding element in the domain Ω. In fact, we have

$$\begin{aligned} \|d\mathbf{X}\|^2 &= d\mathbf{X} \cdot d\mathbf{X} = (\nabla_\xi \phi) d\xi \cdot (\nabla_\xi \phi) d\xi \\ &= (\nabla_\xi \phi)^T (\nabla_\xi \phi) d\xi \cdot d\xi = \mathbf{g} d\xi \cdot d\xi, \end{aligned} \tag{9.293}$$

which is known in the literature as the First Fundamental Form of the surface Σ, and where $\mathbf{g} = (\nabla_\xi \phi)^T (\nabla_\xi \phi)$ is called the metric tensor associated with this surface. By definition, the metric tensor is symmetric and its components in the basis $\{\mathbf{e}_1, \mathbf{e}_2\}$ are given by

$$g_{\alpha\beta} = \mathbf{g}\mathbf{e}_\alpha \cdot \mathbf{e}_\beta = \mathbf{g}\mathbf{e}_\beta \cdot \mathbf{e}_\alpha = g_{\beta\alpha}. \tag{9.294}$$

Finally, taking into consideration that a differential area element $d\Sigma$ at each point $\mathbf{X} \in \Sigma$ is related to the corresponding area element in $d\Omega$ through

$$d\Sigma = \sqrt{\det \mathbf{g}}\, d\Omega, \tag{9.295}$$

the integral of a field ψ in Σ, either scalar, vector or tensor, can be obtained through the form

$$\int_\Sigma \psi(\mathbf{X}) d\Sigma = \int_\Omega \psi(\phi(\xi)) \sqrt{\det \mathbf{g}}\, d\Omega. \tag{9.296}$$

9.9.3 Second Fundamental Form

For a smooth surface Σ, we have seen that at every point $\mathbf{X} \in \Sigma$ the tangent plane $P_t(\mathbf{X})$ and the corresponding normal vector $\mathbf{n}(\mathbf{X})$ are uniquely defined. Since $\mathbf{n}(\mathbf{X})$ is a unit vector by definition, we have that the scalar field $\mathbf{n} \cdot \mathbf{n}$ is constant (with value equal to one) all over the surface. Then

$$d(\mathbf{n} \cdot \mathbf{n}) = 2d\mathbf{n} \cdot \mathbf{n} = 0 \qquad \Rightarrow \qquad d\mathbf{n} \perp \mathbf{n} \quad \text{everywhere in } \Sigma. \tag{9.297}$$

In other words, the vector $d\mathbf{n}(\mathbf{X})$ belongs to the tangent plane $P_t(\mathbf{X})$.

In addition, the definition of the gradient of a vector field is

$$d\mathbf{n} = (\nabla\mathbf{n})d\mathbf{X}, \tag{9.298}$$

where $\nabla\mathbf{n}$ is the gradient (with respect to coordinates $\mathbf{X}$) of the vector field $\mathbf{n}$ at $\mathbf{X} \in \Sigma$.[2] In particular, the tensor field $\nabla\mathbf{n}$ is called the curvature tensor of the surface and has the following properties. First, since $d\mathbf{X}$ is a vector belonging to the tangent plane at $\mathbf{X}$, $\nabla\mathbf{n}$ is an endomorphism of the tangent plane because it transforms vectors in the tangent plane into vectors also lying over the same tangent plane. Second, the curvature tensor is symmetric, that is $\nabla\mathbf{n} = (\nabla\mathbf{n})^T$, or equivalently

$$(\nabla\mathbf{n})\mathbf{u}\cdot\mathbf{v} = \mathbf{u}\cdot(\nabla\mathbf{n})\mathbf{v} \quad \forall\mathbf{u},\mathbf{v} \in P_t(\mathbf{X}). \tag{9.299}$$

To show this, consider a smooth parameterization $\phi : \xi \in \Omega \mapsto \mathbf{X} \in \Sigma$. Next, consider that $\mathbf{n}$ is not a function of $\mathbf{X}$, but a function of ξ, that is, $\mathbf{n}(\xi) = \mathbf{n}(\mathbf{X}(\xi))$. Then, the partial derivative of $\mathbf{n}$ with respect to ξ^α, $\alpha = 1, 2$, can be calculated through the chain rule

$$\mathbf{n}_{\xi^\alpha} = \frac{\partial\mathbf{n}}{\partial\xi^\alpha} = (\nabla\mathbf{n})\frac{\partial\phi}{\partial\xi^\alpha} = (\nabla\mathbf{n})\phi_{\xi^\alpha}. \tag{9.300}$$

Moreover, and given that ϕ_{ξ^α}, $\alpha = 1, 2$, belongs to the tangent plane $P_t(\mathbf{X})$, this gives

$$\begin{aligned} \mathbf{n}\cdot\phi_{\xi^\alpha} &= 0,\\ \mathbf{n}\cdot\phi_{\xi^\beta} &= 0. \end{aligned} \tag{9.301}$$

Deriving the first with respect to ξ^β and the second with respect to ξ^α we obtain, respectively

$$\mathbf{n}_{\xi^\beta}\cdot\phi_{\xi^\alpha} + \mathbf{n}\cdot\phi_{\xi^\alpha\xi^\beta} = 0, \tag{9.302}$$

$$\mathbf{n}_{\xi^\alpha}\cdot\phi_{\xi^\beta} + \mathbf{n}\cdot\phi_{\xi^\beta\xi^\alpha} = 0. \tag{9.303}$$

On the other hand, because ϕ is smooth, we have $\phi_{\xi^\alpha\xi^\beta} = \phi_{\xi^\beta\xi^\alpha}$. This, together with the previous expressions, leads us to

$$\mathbf{n}_{\xi^\beta}\cdot\phi_{\xi^\alpha} = \mathbf{n}_{\xi^\alpha}\cdot\phi_{\xi^\beta}, \tag{9.304}$$

which, by making use of (9.300), can be rewritten as

$$(\nabla\mathbf{n})\phi_{\xi^\beta}\cdot\phi_{\xi^\alpha} = (\nabla\mathbf{n})\phi_{\xi^\alpha}\cdot\phi_{\xi^\beta}. \tag{9.305}$$

Therefore, given arbitrary vectors in the tangent plane at $\mathbf{X}$, say $\mathbf{u}$ and $\mathbf{v}$, we arrive at

$$\begin{aligned} (\nabla\mathbf{n})\mathbf{u}\cdot\mathbf{v} &= (\nabla\mathbf{n})(u^\alpha\phi_{\xi^\alpha})\cdot(v^\beta\phi_{\xi^\beta}) = (\nabla\mathbf{n})\phi_{\xi^\alpha}\cdot\phi_{\xi^\beta}u^\alpha v^\beta\\ &= (\nabla\mathbf{n})\phi_{\xi^\beta}\cdot\phi_{\xi^\alpha}u^\alpha v^\beta = (\nabla\mathbf{n})(v^\beta\phi_{\xi^\beta})\cdot(u^\alpha\phi_{\xi^\alpha}) = (\nabla\mathbf{n})\mathbf{v}\cdot\mathbf{u}, \end{aligned} \tag{9.306}$$

and so the symmetry property is demonstrated.

In addition, from expression (9.300) we can write the curvature tensor in terms of its components in the original basis ($\{\phi_{\xi^1}, \phi_{\xi^2}, \mathbf{n}\}$) or in its dual basis ($\{\phi^{\xi^1}, \phi^{\xi^2}, \mathbf{n}\}$). As a

2 Observe that this gradient $\nabla\mathbf{n}$ with respect to coordinates $\mathbf{X}$ corresponds to the tensor $\nabla_{\mathbf{x}_o}\mathbf{n}$ that we have introduced throughout this chapter when studying the mechanical models for shell components. Such a tensor expresses the gradient of the normal vector of the middle surface of the shell with respect to coordinates defined over the middle surface Σ_o. In this section, we have chosen to avoid overloading with notation unnecessarily.

matter of fact, keeping in mind that this tensor is an endomorphism of the tangent plane, we obtain

$$\nabla \mathbf{n} = \mathbf{n}_{\xi^\alpha} \otimes \boldsymbol{\phi}^{\xi^\alpha}, \tag{9.307}$$

and its components are

$$[\nabla \mathbf{n}]^{\beta}_{\cdot\alpha} = -b^{\beta}_{\cdot\alpha} = (\nabla \mathbf{n})\boldsymbol{\phi}_{\xi^\alpha} \cdot \boldsymbol{\phi}^{\xi^\beta} = \mathbf{n}_{\xi^\alpha} \cdot \boldsymbol{\phi}^{\xi^\beta}, \tag{9.308}$$

$$[\nabla \mathbf{n}]_{\beta\alpha} = -b_{\beta\alpha} = (\nabla \mathbf{n})\boldsymbol{\phi}_{\xi^\alpha} \cdot \boldsymbol{\phi}_{\xi^\beta} = \mathbf{n}_{\xi^\alpha} \cdot \boldsymbol{\phi}_{\xi^\beta}, \tag{9.309}$$

where the minus sign has been introduced to adopt the convention established in the literature. Then

$$\nabla \mathbf{n} = -b^{\beta}_{\cdot\alpha}(\boldsymbol{\phi}_{\xi^\beta} \otimes \boldsymbol{\phi}^{\xi^\alpha}) = -b_{\beta\alpha}(\boldsymbol{\phi}^{\xi^\beta} \otimes \boldsymbol{\phi}^{\xi^\alpha}), \tag{9.310}$$

and recalling (9.300), this gives

$$\mathbf{n}_{\xi^\alpha} = -b^{\beta}_{\cdot\alpha}(\boldsymbol{\phi}_{\xi^\beta} \otimes \boldsymbol{\phi}^{\xi^\alpha})\boldsymbol{\phi}_{\xi^\alpha} = -b^{\beta}_{\cdot\alpha}\boldsymbol{\phi}_{\xi^\beta}, \tag{9.311}$$

$$\mathbf{n}_{\xi^\alpha} = -b_{\beta\alpha}(\boldsymbol{\phi}^{\xi^\beta} \otimes \boldsymbol{\phi}^{\xi^\alpha})\boldsymbol{\phi}_{\xi^\alpha} = -b_{\beta\alpha}\boldsymbol{\phi}^{\xi^\beta}. \tag{9.312}$$

Now, the property of tensor $\nabla \mathbf{n}$ we want to highlight is its relation with the Second Fundamental Form of the surface which determines the inner product between the vectors $d\mathbf{n}$ and $d\mathbf{X}$. Indeed, from the expression for $d\mathbf{X}$ as a function of $d\xi$ given by (9.290), the Second Fundamental Form acquires the form

$$d\mathbf{n} \cdot d\mathbf{X} = (\nabla \mathbf{n})d\mathbf{X} \cdot d\mathbf{X} = \nabla \mathbf{n}(\boldsymbol{\phi}_{\xi^\alpha} d\xi^\alpha) \cdot (\boldsymbol{\phi}_{\xi^\beta} d\xi^\beta) = -b_{\beta\alpha} d\xi^\alpha d\xi^\beta. \tag{9.313}$$

Finally, given the symmetry of the curvature tensor, there exists an orthonormal basis in the tangent plane, denoted by $\{\mathbf{a}_1, \mathbf{a}_2\}$, such that the curvature tensor can be characterized as

$$\nabla \mathbf{n} = \frac{1}{R_m}\mathbf{a}_1 \otimes \mathbf{a}_1 + \frac{1}{R_M}\mathbf{a}_2 \otimes \mathbf{a}_2, \tag{9.314}$$

where $\frac{1}{R_m}$ and $\frac{1}{R_M}$ are known as the principal curvatures, minimum and maximum respectively, associated with the principal coordinate (orthogonal) curves which go through point $\mathbf{X}$ and whose tangent lines are given by the vectors (eigenvectors) $\mathbf{a}_1$ and $\mathbf{a}_2$. In turn, it is easy to verify that the invariants of the curvature tensor are given by

$$\text{tr}(\nabla \mathbf{n}) = \frac{1}{R_m} + \frac{1}{R_M} = 2\mathsf{H}, \tag{9.315}$$

$$\det(\nabla \mathbf{n}) = \frac{1}{R_m R_M} = \mathsf{K}, \tag{9.316}$$

where H is the mean curvature and K is the Gaussian curvature (or total curvature) of the surface Σ at point $\mathbf{X}$.

9.9.4 Third Fundamental Form

We have seen that the First Fundamental Form of a surface is related to the inner product $d\mathbf{X} \cdot d\mathbf{X}$. The Second Fundamental Form is, in turn, related to the inner product $d\mathbf{n} \cdot d\mathbf{X}$. In this section, we analyze the remaining inner product $d\mathbf{n} \cdot d\mathbf{n}$, called Third Fundamental Form, which is given by

$$d\mathbf{n} \cdot d\mathbf{n} = (\nabla\mathbf{n})d\mathbf{X} \cdot (\nabla\mathbf{n})d\mathbf{X} = (\nabla\mathbf{n})(\nabla\mathbf{n})d\mathbf{X} \cdot d\mathbf{X} = \mathbf{C}d\mathbf{X} \cdot d\mathbf{X}. \tag{9.317}$$

From the definition it is verified that $\mathbf{C} = (\nabla\mathbf{n})\nabla\mathbf{n} = (\nabla\mathbf{n})^2$ is a symmetric second-order tensor whose components in the original basis of the tangent plane $\{\boldsymbol{\phi}_{\xi^1}, \boldsymbol{\phi}_{\xi^2}\}$ are related to the components of the curvature tensor as follows

$$\begin{aligned} d\mathbf{n} \cdot d\mathbf{n} &= (\nabla\mathbf{n})(\nabla\mathbf{n})(d\xi^\beta \boldsymbol{\phi}_{\xi^\beta}) \cdot (d\xi^\alpha \boldsymbol{\phi}_{\xi^\alpha}) \\ &= (\nabla\mathbf{n})^2 \boldsymbol{\phi}_{\xi^\beta} \cdot \boldsymbol{\phi}_{\xi^\alpha} d\xi^\beta d\xi^\alpha = [\mathbf{C}]_{\alpha\beta} d\xi^\beta d\xi^\alpha. \end{aligned} \tag{9.318}$$

Then, from (9.310) we have

$$\begin{aligned} \mathbf{C} = (\nabla\mathbf{n})^2 &= (-b_{\alpha\gamma}(\boldsymbol{\phi}^{\xi^\alpha} \otimes \boldsymbol{\phi}^{\xi^\gamma}))(-b^{\gamma}_{\cdot\beta}(\boldsymbol{\phi}_{\xi^\gamma} \otimes \boldsymbol{\phi}^{\xi^\beta})) \\ &= b_{\alpha\gamma}\, b^{\gamma}_{\cdot\beta}(\boldsymbol{\phi}^{\xi^\alpha} \otimes \boldsymbol{\phi}^{\xi^\beta}), \end{aligned} \tag{9.319}$$

and then we obtain

$$[\mathbf{C}]_{\alpha\beta} = b_{\alpha\gamma} b^{\gamma}_{\cdot\beta} = b^{\cdot\gamma}_{\alpha} b_{\gamma\beta}, \tag{9.320}$$

which follows from the following identity

$$(\boldsymbol{\phi}^{\xi^\alpha} \otimes \boldsymbol{\phi}^{\xi^\gamma})(\boldsymbol{\phi}_{\xi^\gamma} \otimes \boldsymbol{\phi}^{\xi^\beta}) = (\boldsymbol{\phi}^{\xi^\alpha} \otimes \boldsymbol{\phi}_{\xi^\gamma})(\boldsymbol{\phi}^{\xi^\gamma} \otimes \boldsymbol{\phi}^{\xi^\beta}). \tag{9.321}$$

9.9.5 Complementary Properties

When the geometric description of a shell component was established, we defined any point in the three-dimensional body, $\mathbf{x}$, as a function of the orthogonal projection over the middle surface, yielding $\mathbf{x}_o$, and of the distance between these two points, called ξ, resulting in the relation $\mathbf{x} = \mathbf{x}_o + \xi\mathbf{n}(\mathbf{x}_o)$ that takes the pair $(\mathbf{x}_o, \xi) \in \Sigma_o \times H$, with $H = [-\frac{h(\mathbf{x}_o)}{2}, \frac{h(\mathbf{x}_o)}{2}]$, to the point $\mathbf{x} \in \Omega$ (see Section 9.2). At that time, we admitted that this was a one-to-one (that is invertible) transformation. The necessary condition for the existence of the inverse transformation is that $\det \Lambda \neq 0$, where $\Lambda = \mathbf{I} + \xi\nabla_{\mathbf{x}_o}\mathbf{n}(\mathbf{x}_o)$. Hence, we have

$$\det \Lambda = \det(\mathbf{I} + \xi\nabla_{\mathbf{x}_o}\mathbf{n}(\mathbf{x}_o)) = 1 + \xi \mathrm{tr}(\nabla_{\mathbf{x}_o}\mathbf{n}) + \xi^2 \det(\nabla_{\mathbf{x}_o}\mathbf{n}). \tag{9.322}$$

Remembering the expressions for $\mathrm{tr}(\nabla_{\mathbf{x}_o}\mathbf{n})$ and $\det(\nabla_{\mathbf{x}_o}\mathbf{n})$ in terms of the minimum and maximum curvatures given in (9.315) and (9.316), the limit condition for the inverse to exist can be cast as a function of ξ as

$$\det \Lambda = 1 + \xi\left(\frac{1}{R_m} + \frac{1}{R_M}\right) + \xi^2 \frac{1}{R_m R_M} = 0. \tag{9.323}$$

The roots for this quadratic equation are $\xi_1 = -R_m$ and $\xi_2 = -R_M$. Therefore, the condition

$$|\xi| \leq \frac{h(\mathbf{x}_o)}{2} \leq |R_m|, \tag{9.324}$$

guarantees the invertibility of the application $(\mathbf{x}_o, \xi) \mapsto \mathbf{x}$.

Moreover, from $\mathbf{x} = \mathbf{x}_o + \xi\mathbf{n}(\mathbf{x}_o)$ we have that the differential is

$$d\mathbf{x} = d\mathbf{x}_o + \xi(\nabla_{\mathbf{x}_o}\mathbf{n})d\mathbf{x}_o + d\xi\mathbf{n} = (\mathbf{I} + \xi\nabla_{\mathbf{x}_o}\mathbf{n})d\mathbf{x}_o + d\xi\mathbf{n} = \Lambda d\mathbf{x}_o + d\xi\mathbf{n}, \tag{9.325}$$

which projected onto the tangent plane $P_t(\mathbf{X}_o)$ yields

$$\Pi_t d\mathbf{x} = \Pi_t \Lambda d\mathbf{x}_o = \Lambda d\mathbf{x}_o, \tag{9.326}$$

resulting in

$$d\mathbf{x}_o = \Lambda^{-1}\Pi_t d\mathbf{x}. \tag{9.327}$$

These expressions illustrate the role the tensors Λ and Λ^{-1} play in the transformations between the differential element $\Pi_t d\mathbf{x}$ parallel to the tangent plane and the corresponding element $d\mathbf{x}_o$ in that plane.

From the expression $\Lambda = \mathbf{I} + \xi\nabla_{\mathbf{x}_o}\mathbf{n}$, we can obtain an exact form for the inverse tensor Λ^{-1}. In effect, keeping in mind that $\Lambda^{-1}\Lambda = \mathbf{I}$, we easily obtain the following recurrent expansion for this inverse tensor

$$\begin{aligned}\Lambda^{-1} &= \mathbf{I} - \xi(\nabla_{\mathbf{x}_o}\mathbf{n})\Lambda^{-1} \\ &= \mathbf{I} - \xi\nabla_{\mathbf{x}_o}\mathbf{n} + \xi^2(\nabla_{\mathbf{x}_o}\mathbf{n})^2 - \ldots + (-1)^n\xi^n(\nabla_{\mathbf{x}_o}\mathbf{n})^n\Lambda^{-1}.\end{aligned} \tag{9.328}$$

Let us see the properties of the derivatives of Λ with respect to ξ, which are denoted by $\dot{(\cdot)} = \frac{\partial \cdot}{\partial \xi}$. First, it is straightforward to verify that tensors Λ and $\dot{\Lambda}$ are commutative. In fact, knowing that $\Lambda = \mathbf{I} + \xi\nabla_{\mathbf{x}_o}\mathbf{n}$, we have $\dot{\Lambda} = \nabla_{\mathbf{x}_o}\mathbf{n}$ and

$$\Lambda\dot{\Lambda} = \dot{\Lambda}\Lambda = \nabla_{\mathbf{x}_o}\mathbf{n} + \xi(\nabla_{\mathbf{x}_o}\mathbf{n})^2. \tag{9.329}$$

Deriving the relation $\Lambda\Lambda^{-1} = \mathbf{I}$ yields

$$\dot{\Lambda}\Lambda^{-1} + \Lambda(\dot{\Lambda^{-1}}) = \mathbf{O}. \tag{9.330}$$

Left-multiplying the previous identity by Λ^{-1} and right-multiplying by Λ we have the following relations between the derivatives of Λ and Λ^{-1}

$$(\dot{\Lambda^{-1}}) = -\Lambda^{-1}\dot{\Lambda}\Lambda^{-1}, \tag{9.331}$$

$$\dot{\Lambda} = \Lambda(\dot{\Lambda^{-1}})\Lambda. \tag{9.332}$$

Now, deriving the identity $\Lambda = \Lambda^{-1}\Lambda^2 = \Lambda^2\Lambda^{-1}$ gives

$$\dot{\Lambda} = (\dot{\Lambda^{-1}})\Lambda^2 + \Lambda^{-1}2\Lambda\dot{\Lambda} = (\dot{\Lambda^{-1}})\Lambda^2 + 2\dot{\Lambda}, \tag{9.333}$$

$$\dot{\Lambda} = 2\Lambda\dot{\Lambda}\Lambda^{-1} + \Lambda^2(\dot{\Lambda^{-1}}) = 2\dot{\Lambda} + \Lambda^2(\dot{\Lambda^{-1}}), \tag{9.334}$$

which finally yields

$$\dot{\Lambda} = -(\dot{\Lambda^{-1}})\Lambda^2 = -\Lambda^2(\dot{\Lambda^{-1}}) = -\Lambda(\dot{\Lambda^{-1}})\Lambda, \tag{9.335}$$

$$(\dot{\Lambda^{-1}}) = -\dot{\Lambda}\Lambda^{-2} = -\Lambda^{-2}\dot{\Lambda} = -\Lambda^{-1}\dot{\Lambda}\Lambda^{-1}. \tag{9.336}$$

Part IV

Other Problems in Physics

In this part of the book we will address the modeling of other problems typically encountered in the physics of continua through the use of the variational framework. In Chapter 10 we will concentrate attention on steady-state heat transfer in rigid bodies. This domain of physics has usually been tackled by employing modeling techniques based on local (strong) forms, that is, making use of partial differential equations (Euler–Lagrange equations). By exploiting the general concepts developed in Chapter 3, we will build a variational model for this problem, making possible the analysis of several aspects such as constitutive equations, reactive forces, and kinematical constraints. Moreover, the connection to traditional minimization problems will be established, in line with that presented in Chapter 4. In particular, regarding the heat transfer problem in rigid bodies, the work by P. Podio-Guidugli deserves special mention [242], where the Principle of Virtual Power is developed in parallel for both the duality between strain rate and stress, and the duality between temperature and entropy. These duality pairings in fact deliver power as outcome. In our case, we will utilize a more classical variational approach in which the duality is understood as a result of pairing temperature and heat flux, which renders power per unit temperature.

In Chapter 11 we will examine the modeling of incompressible fluid flow. As with the heat transfer problem, this area of research is usually studied in the context of conservation laws, and the strategy to derive the governing equations from basic principles has, therefore, been based on balance equations posed in control volumes. Such an approach, like the case of the energy conservation law (which yields the heat transfer problem), requires that the existence of the entity stress is assumed a priori. We will see that, using a completely analogous modeling approach to that of solid mechanics (i.e. using a purely kinematical model and using the concept of virtual power) is enough to develop classical models from fluid mechanics, while we retain the advantage featured by the variational formulation which allows us to work with a minimum set of hypotheses in the construction of the model, and enables us to understand the consequence of kinematical constraints in this class of (incompressible) flows. Moreover, the variational form also provides a direct path for the formulation of the fluid flow problem as a minimization problem, similarly to Chapter 4, provided some additional hypotheses are introduced.

In Chapter 12 we will address a topic that is closely related to the domain of multiscale modeling (whose presentation in detail is left for Part V). The model to be developed in this chapter features an enriched kinematical model when compared to what we have seen in previous chapters. In this context, we will address the development of high-order models, also known as models of high-order continua. Within this class of kinematically enriched models are Cosserat type models [57, 125, 126], which gave rise to the emersion of micromorphic models [80, 81], among which second gradient models can be interpreted as particular cases [243, 244]. We will see that these models can be systematically constructed, making use of the theoretical variational formalism proposed in Chapter 3, similar (but not entirely equal) to the work by P. Germain in [116].

The problems discussed in this part of the book are nothing but a few representative examples of a rather larger class of problems in the physical realm for which the variational formulation can be applied in an entirely analogous manner. These other problems are usually and generically referred to as field problems, among which we can mention, for instance, the study of the distribution of electric and/or magnetic fields, the transport of species, lubrication problems, and porous media flow, among others. The former two can find a parallel treatment to the heat transfer problem, while the later two are particular cases of fluid flow formulation.

10

Heat Transfer

10.1 Introduction

In this chapter we develop the variational formulation to model the steady-state heat transfer in rigid continuum media, that is, in bodies where the spatial position of particles remains invariant.

According to the roadmap established in Chapter 3, the first step in the construction of a variational model consists of defining the kinematics for the model, that is, the motion actions that particles can execute. In the heat transfer problem, the temperature is the primary scalar field that characterizes the average kinetic energy of molecules, and so it characterizes the kinematics in this problem. The temperature in a body is constrained to satisfy certain conditions, and therefore it is possible to define the concept of admissible variations of the temperature fields. With this, it is possible to introduce the generalized strain action operator, which in this context is denoted by $\mathscr{G}$ (in Chapter 3 it was denoted by $\mathscr{D}$), which leads us to the conception of the virtual internal power. From there, the characterization of generalized internal and external forces will follow, and then the application of the Principle of Virtual Power will be exploited to define the concept of equilibrium for the system.

Let us consider a body occupying a bounded and regular region $\mathscr{B}$[1], with boundary $\partial\mathscr{B}$, in the three-dimensional Euclidean space $\mathscr{E}$. As a consequence, spatial coordinates $\mathbf{x}$ coincide with the material (reference) coordinates $\mathbf{X}$ admitted for the particles. That is, the spatial configuration and the material configuration of the body agree. For the heat transfer problem in bodies whose deformations are substantial, the developments of this chapter remain valid, but the reader has to keep in mind that the configuration of the body referred to in what follows is the spatial (actual, deformed) configuration the body occupies in space once the mechanical equilibrium has been achieved.

10.2 Kinematics

Making the analogy with the velocity field in the mechanics of continuum media, the scalar field called temperature, and denoted by θ, becomes the primal (also primary) variable in the heat transfer problem.[2]

---

1 An additional hypothesis is that this region is spatially fixed.

2 In [242] the concept of thermal displacement α is utilized in the development of the theory, whose relation with the temperature here is $\theta = \dot{\alpha}$.

Introduction to the Variational Formulation in Mechanics: Fundamentals and Applications, First Edition.
Edgardo O. Taroco, Pablo J. Blanco and Raúl A. Feijóo.

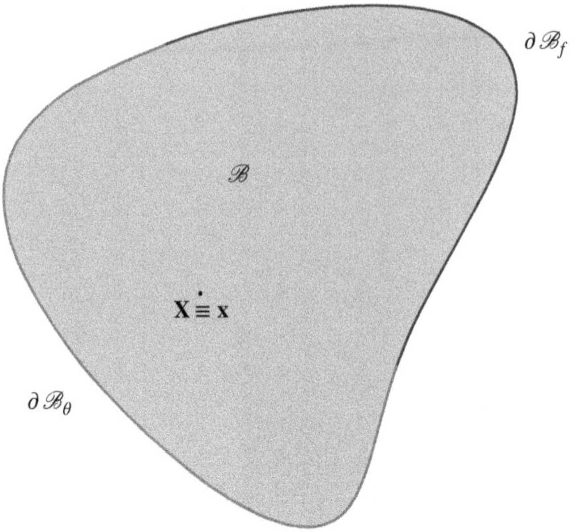

Figure 10.1 Spatial configuration of the body for the heat transfer problem.

We call $\mathscr{T}$ the set of all possible temperature scalar fields that are sufficiently regular and can be defined over the body $\mathscr{B}$. Here, the set $\mathscr{T}$, endowed with the usual operations of addition and multiplication by a real number, becomes a vector space. By regular it is understood that θ is smooth enough such that all operations to be executed over this field are well-defined.

Consider that there exists a portion of the boundary, called $\partial\mathscr{B}_\theta$, where the temperature is prescribed. Then, we define the set

$$\mathrm{Kin}_\theta = \{\theta \in \mathscr{T}\ ;\ \theta|_{\partial\mathscr{B}_\theta} = \overline{\theta}\}, \tag{10.1}$$

where $\overline{\theta}$ is the value of the prescribed temperature on $\partial\mathscr{B}_\theta$, as illustrated in Figure 10.1.

The set Kin_θ contains all admissible temperature fields for the problem under study. Moreover, we define the vector subspace associated with $\mathscr{T}$ with fields satisfying homogeneous boundary conditions on $\partial\mathscr{B}_\theta$, that is

$$\mathrm{Var}_\theta = \{\theta \in \mathscr{T}\ ;\ \theta|_{\partial\mathscr{B}_\theta} = 0\}. \tag{10.2}$$

It is easy to verify that Kin_θ is the translation of the subspace Var_θ. The elements $\hat{\theta} \in \mathrm{Var}_\theta$ are called virtual thermal variations, or admissible temperature variations. With the previous elements we can write

$$\mathrm{Kin}_\theta = \theta_0 + \mathrm{Var}_\theta, \tag{10.3}$$

where $\theta_0 \in \mathrm{Kin}_\theta$ is arbitrary. Now, consider that the regularity of the temperature fields is such that we can express the temperature at a point $\mathbf{x} \in \mathscr{B}$ close to $\mathbf{x}_o \in \mathscr{B}$ using a Taylor expansion as follows

$$\theta(\mathbf{x}) = \theta(\mathbf{x}_o) + \nabla\theta(\mathbf{x}_o)\cdot(\mathbf{x}-\mathbf{x}_o) + \mathscr{O}(\mathbf{x}-\mathbf{x}_o). \tag{10.4}$$

Calling $\mathbf{g}$ to the vector field $\nabla\theta$ and for all points $\mathbf{x}$ in a sufficiently small neighborhood of $\mathbf{x}_o$ we can admit that the following holds

$$\theta(\mathbf{x}) = \theta(\mathbf{x}_o) + \mathbf{g}(\mathbf{x}_o)\cdot(\mathbf{x}-\mathbf{x}_o). \tag{10.5}$$

In addition, the field θ is called constant, or rigid, if it verifies

$$\theta(\mathbf{x}) = \theta(\mathbf{x}_o) \qquad \forall \mathbf{x}_o \in \mathscr{B}, \tag{10.6}$$

or equivalently

$$\mathbf{g} = \mathbf{0} \qquad \forall \mathbf{x} \in \mathscr{B}. \tag{10.7}$$

Thus, another fundamental ingredient in the present model is the vector field $\mathbf{g}$ which represents the temperature spatial gradient (or thermal gradient). In particular, the set of all sufficiently regular fields $\mathbf{g}$ is denoted by $\mathscr{S}$.

We will say that $\mathbf{g} \in \mathscr{S}$ is a thermal gradient, or compatible gradient, if it is possible to determine $\theta \in \mathscr{T}$ such that

$$\mathbf{g} = \nabla\theta. \tag{10.8}$$

Thus, we can define the operator $\mathscr{G} = \nabla(\cdot) : \mathscr{T} \to \mathscr{S}$ which for each $\theta \in \mathscr{T}$ assigns a thermal gradient $\mathbf{g} = \mathscr{G}\theta \in \mathscr{S}$.

The kernel of operator $\mathscr{G}$, denoted by $\mathscr{N}(\mathscr{G})$, is then characterized by

$$\mathscr{N}(\mathscr{G}) = \{\theta \in \mathscr{T};\ \mathscr{G}\theta = \mathbf{0} \quad \forall \mathbf{x} \in \mathscr{B}\}, \tag{10.9}$$

and consists of all rigid thermal actions. All the elements and kinematical concepts introduced so far have a full parallel with the fundamental ingredients introduced in Chapter 3.

10.3 Principle of Thermal Virtual Power

As in the mechanics of continuum media, let us admit that external thermal loads that exert some action over the body $\mathscr{B}$ are characterized by linear (and continuous) functionals $t : \mathscr{T} \to \mathbb{R}$. Hence, the external thermal loads are elements of the space dual to $\mathscr{T}$, here represented by $\mathscr{T}'$. We will call the value this functional takes in $\mathbb{R}$ the external thermal power, which is then given by

$$P_e = \langle t, \theta \rangle, \tag{10.10}$$

where $\langle \cdot, \cdot \rangle$ is the duality product $\langle \cdot, \cdot \rangle : \mathscr{T}' \times \mathscr{T} \to \mathbb{R}$.

In this section we will see that these external thermal loads can be explicitly characterized by making use of the Principle of Thermal Virtual Power, which will be enunciated next.

First, let us define the internal thermal stresses through proper linear (and continuous) functionals defined over the space of temperatures $\mathscr{T}$ and of thermal gradients $\mathscr{S}$. The value taken by such functionals at point $(\theta, \mathbf{g})$ is termed the internal thermal power, or simply P_i. Once again, as in the mechanics domain, we will introduce a series of hypotheses that will allow us to find a representation for such a functional.

First, let us assume that the functional has the following general representation

$$P_i = \int_{\mathscr{B}} (p\theta + \mathbf{q} \cdot \mathbf{g}) dV. \tag{10.11}$$

The second hypothesis corresponds to admitting that P_i is null for all rigid thermal actions (uniform θ), that is

$$P_i = \int_{\mathscr{B}} (p\theta + \mathbf{q} \cdot \mathbf{g}) dV = 0 \qquad \forall \theta \in \mathscr{N}(\mathscr{G}), \tag{10.12}$$

which yields

$$P_i = \int_{\mathscr{B}} p\theta \, dV = 0 \qquad \forall \theta \in \mathscr{N}(\mathscr{G}). \tag{10.13}$$

Since the expression above must hold for every part $\mathscr{P}$ of the body $\mathscr{B}$, it implies

$$p = 0 \qquad \forall \mathbf{x} \in \mathscr{B}. \tag{10.14}$$

Thus, we have

$$P_i = \int_{\mathscr{B}} \mathbf{q} \cdot \mathbf{g} dV. \tag{10.15}$$

With this procedure, we have arrived at a duality between the thermal gradient and the heat flux.[3]

We then conclude that internal thermal stresses are linear (and continuous) functionals defined over $\mathscr{S}$, that means they are elements in the space dual to $\mathscr{S}$, here designated by $\mathscr{S}'$. By virtue of the Riesz representation theorem we can put these functionals in correspondence with a vector field, say $\mathbf{q}$, which is named as the heat flux vector field. Therefore, the value of this functional at $\mathbf{g} = \mathscr{G}\theta = \nabla\theta \in \mathscr{S}$ is given by

$$P_i = (\mathbf{q}, \mathbf{g}) = \int_{\mathscr{B}} \mathbf{q} \cdot \mathbf{g} dV. \tag{10.16}$$

We have then defined the spaces $\mathscr{T}$, $\mathscr{T}'$, $\mathscr{S}$ and $\mathscr{S}'$ and the dual products (linear forms) $\langle \cdot, \cdot \rangle$ and $(\cdot, \cdot)$ which provide a correspondence between these spaces. Now, we define the adjoint operator $\mathscr{G}^* : \mathscr{S}' \to \mathscr{T}'$, called the thermal equilibrium operator, as

$$(\mathbf{q}, \mathscr{G}\theta) = \langle \mathscr{G}^*\mathbf{q}, \theta \rangle \qquad \theta \in \mathscr{S}. \tag{10.17}$$

At this stage we have all the elements required to formalize the variational principle which rules the problem, and which we call the Principle of Thermal Virtual Power (PTVP). Then, we say that a body $\mathscr{B}$, under the action of the external thermal loads $t \in \mathscr{T}'$, is at static[4] thermal equilibrium if

$$P_e = \langle t, \hat{\theta} \rangle = 0 \qquad \forall \hat{\theta} \in \mathrm{Var}_\theta \cap \mathscr{N}(\mathscr{G}), \tag{10.18}$$

which implies that the external thermal virtual power is nullified for all rigid thermal actions and, in addition, satisfies

$$P_i + P_e = (\mathbf{q}, \mathscr{G}\hat{\theta}) + \langle t, \hat{\theta} \rangle = 0 \qquad \forall \hat{\theta} \in \mathrm{Var}_\theta, \tag{10.19}$$

3 Actually, the dual variable to the temperature, in the sense of mechanical power, is the so-called entropy flux, denoted by $\mathbf{h}$. In the present development, since we are working with mechanical power per unit temperature, here simply called thermal power, the dual variable to the thermal gradient is, in fact, the heat flux, whose relation to the entropy flux is $\mathbf{h} = \frac{\mathbf{q}}{\theta}$.

4 We added this characterization because at the beginning of this chapter we tackled the steady-state heat transfer problem. Extension to the unsteady case is straightforward and is also carried out in this section.

that is, the sum of the internal thermal virtual power and of the external thermal virtual power equals zero for all admissible virtual thermal actions.

The first part of the PTVP enables us to establish which are the external thermal loads compatible with the thermal equilibrium. Note that this first part makes sense provided the existence of $\mathbf{q}$ for a given t at equilibrium is demonstrated. This is, in fact, the reason to show the following result.

Theorem 10.1 Representation Theorem. Let $t \in \mathcal{T}'$ be at thermal equilibrium, then there exist $\mathbf{q} \in \mathcal{S}'$ and $v \in \mathcal{T}'$ such that

$$t = -\mathcal{G}^*\mathbf{q} - v, \tag{10.20}$$

$$\langle v, \hat{\theta} \rangle = 0 \qquad \forall \hat{\theta} \in \text{Var}_\theta, \tag{10.21}$$

or equivalently $v \in \text{Var}_\theta^\perp$, where v is called the reactive flux. Moreover, it is said that $\mathbf{q}$ and v equilibrate t.

Proof. If t is at equilibrium we have

$$t \in (\text{Var}_\theta \cap \mathcal{N}(\mathcal{G}))^\perp \quad \Leftrightarrow \quad t \in \text{Var}_\theta^\perp + \mathcal{N}(\mathcal{G})^\perp. \tag{10.22}$$

Since $\theta \in \mathcal{N}(\mathcal{G})$, we get $\mathcal{G}\theta = \mathbf{0}$, which implies, in such a case, that

$$\langle \mathcal{G}^*\mathbf{q}, \theta \rangle = (\mathbf{q}, \mathcal{G}\theta) = 0 \qquad \forall \mathbf{q} \in \mathcal{S}', \tag{10.23}$$

that is, $\theta \in R(\mathcal{G}^*)^\perp$, and therefore $\mathcal{N}(\mathcal{G}) = R(\mathcal{G}^*)^\perp$, or equivalently

$$\mathcal{N}(\mathcal{G})^\perp = R(\mathcal{G}^*). \tag{10.24}$$

Putting this result into (10.22) yields

$$t \in \text{Var}_\theta^\perp + R(\mathcal{G}^*), \tag{10.25}$$

and then we conclude that there exist $-\mathbf{q} \in \mathcal{S}'$ and $-v \in \text{Var}_\theta^\perp$ such that

$$t = -\mathcal{G}^*\mathbf{q} - v, \tag{10.26}$$

and furthermore since $-v \in \text{Var}_\theta^\perp$ we get

$$\langle v, \hat{\theta} \rangle = 0 \qquad \forall \hat{\theta} \in \text{Var}_\theta. \tag{10.27}$$

■

According to the previous results we conclude that

$$\langle t, \hat{\theta} \rangle = \langle -\mathcal{G}^*\mathbf{q} - v, \hat{\theta} \rangle = -\langle \mathcal{G}^*\mathbf{q}, \hat{\theta} \rangle = -(\mathbf{q}, \mathcal{G}\hat{\theta}) \qquad \forall \hat{\theta} \in \text{Var}_\theta, \tag{10.28}$$

which means

$$-(\mathbf{q}, \mathcal{G}\hat{\theta}) = \langle t, \hat{\theta} \rangle \qquad \forall \hat{\theta} \in \text{Var}_\theta. \tag{10.29}$$

Hence, the theorem 10.1 and the concept of equilibrium for rigid thermal actions allow us to obtain the second part of the PTVP.

In this manner, we arrived at the point in which we have been given all the necessary elements to formulate the variational equations and retrieve from them the associated Euler–Lagrange equations which govern the steady-state heat transfer problem in strong form. Hereafter, we will provide a more concrete notation to these ingredients, specifying the context of three-dimensional heat transfer with $\mathscr{G}(\cdot) = \nabla(\cdot)$. Therefore, from the PTVP we have

$$-\int_{\mathscr{B}} \mathbf{q} \cdot \nabla\hat{\theta} dV = \langle t, \hat{\theta} \rangle \qquad \forall \hat{\theta} \in \mathrm{Var}_{\theta}. \tag{10.30}$$

If $\mathbf{q}$ and θ are sufficiently regular fields, we can integrate by parts as follows

$$\int_{\mathscr{B}} \mathbf{q} \cdot \nabla\hat{\theta} dV = \int_{\mathscr{B}} \left[\mathrm{div}(\mathbf{q}\hat{\theta}) - \mathrm{div}\mathbf{q}\hat{\theta}\right] dV = \int_{\partial\mathscr{B}_f} (\mathbf{q} \cdot \mathbf{n})\hat{\theta} dS - \int_{\mathscr{B}} \mathrm{div}\mathbf{q}\hat{\theta} dV, \tag{10.31}$$

where we have used the fact that $\hat{\theta} = 0$ over $\partial\mathscr{B}_{\theta}$, where $\partial\mathscr{B} = \partial\mathscr{B}_{\theta} \cup \partial\mathscr{B}_f$ with $\partial\mathscr{B}_{\theta} \cap \partial\mathscr{B}_f = \emptyset$ (see Figure 10.1).

Placing (10.31) into expression (10.30) provides the characterization for the element $t \in \mathscr{T}'$

$$\langle t, \hat{\theta} \rangle = -\int_{\partial\mathscr{B}_f} (\mathbf{q} \cdot \mathbf{n})\hat{\theta} dS + \int_{\mathscr{B}} \mathrm{div}\mathbf{q}\hat{\theta} dV. \tag{10.32}$$

This result tells us that the external thermal loads compatible with the kinematical model consist of a thermal load per unit volume, denoted by r, and a thermal load per unit area, denoted by $\overline{q}$.

In diverse applications of practical interest, where the material surrounding the solid body is a flowing fluid, the load $\overline{q}$ depends upon the temperature of the solid boundary, θ, of the fluid temperature at remote locations, say θ_{∞}, and of the properties of the fluid and of the type of surface, lumped into a parameter h, through a relation $\overline{q} = h(\theta_{\infty} - \theta)$. We call this physical phenomenon the convective heat exchange, and the surface where this exchange takes place is denoted $\partial\mathscr{B}_c$. If we consider $\partial\mathscr{B}_f = \partial\mathscr{B}_q \cup \partial\mathscr{B}_c$ we obtain the following representation for $t \in \mathscr{T}'$

$$\langle t, \hat{\theta} \rangle = \int_{\partial\mathscr{B}_q} \overline{q}\hat{\theta} dS + \int_{\partial\mathscr{B}_c} h(\theta_{\infty} - \theta)\hat{\theta} dS + \int_{\mathscr{B}} r\hat{\theta} dV. \tag{10.33}$$

Then, the PTVP consists of the variational equation

$$-\int_{\mathscr{B}} \mathbf{q} \cdot \nabla\hat{\theta} dV = \int_{\partial\mathscr{B}_q} \overline{q}\hat{\theta} dS + \int_{\partial\mathscr{B}_c} h(\theta_{\infty} - \theta)\hat{\theta} dS + \int_{\mathscr{B}} r\hat{\theta} dV \quad \forall \hat{\theta} \in \mathrm{Var}_{\theta}. \tag{10.34}$$

To extend the PTVP to the case of unsteady heat transfer it is enough to replace r by $r^* = r - \rho c\dot{\theta}$ where ρ is the density for the material in $\mathscr{B}$ and $c > 0$ is the specific heat, being $c = c_p = c_v$ in the case of solids.

Thus, in order to find the local form of the variational problem (10.34) we have to find the associated Euler–Lagrange equations, whose abstract form is $-\mathscr{G}^*\mathbf{q} = t + v$ as we have seen in (10.20). We proceed with the integration by parts for the left-hand side of equation (10.34), yielding

$$\begin{aligned} -\int_{\partial\mathscr{B}_f} (\mathbf{q} \cdot \mathbf{n})\hat{\theta} dS + \int_{\mathscr{B}} \mathrm{div}\mathbf{q}\hat{\theta} dV \\ = \int_{\partial\mathscr{B}_q} \overline{q}\hat{\theta} dS + \int_{\partial\mathscr{B}_c} h(\theta_{\infty} - \theta)\hat{\theta} dV + \int_{\mathscr{B}} r\hat{\theta} dV \qquad \forall \hat{\theta} \in \mathrm{Var}_{\theta}. \end{aligned} \tag{10.35}$$

Using now standard variational arguments (fundamental theorem of the calculus of variations), we find

$$\text{div}\mathbf{q} = r \qquad \text{in } \mathscr{B}, \tag{10.36}$$

$$\mathbf{q} \cdot \mathbf{n} = -\overline{q} \qquad \text{on } \partial\mathscr{B}_q, \tag{10.37}$$

$$\mathbf{q} \cdot \mathbf{n} = -h(\theta_\infty - \theta) \qquad \text{on } \partial\mathscr{B}_c, \tag{10.38}$$

which are the Euler–Lagrange equations sought for the variational equation (10.34).

Notice that, even in situations in which the fields are not regular enough so that integration by parts can be safely performed as we did, the PTVP still holds. In this regard, consider, for example, that there exists an internal surface, say S, which divides the body into two parts, each of which features smooth fields so that integration by parts can be pursued. In this case, from the integration by parts procedure emerges an additional term of the form $\int_S([\![\mathbf{q}]\!] \cdot \mathbf{n})\hat{\theta} dS$. This indicates that the system of external thermal loads t could account for a load per unit area defined over the internal surface S, say $\overline{q}_S$. Hence, the equilibrium in such case implies

$$[\![\mathbf{q}]\!] \cdot \mathbf{n} = -\overline{q}_S \qquad \text{on } S. \tag{10.39}$$

Let us now characterize the reactive flux $\nu \in \mathscr{T}'$ by releasing the kinematical constraints from space Var_θ. To do this, note that (10.20) is equivalent to the following

$$\langle t, \hat{\theta}\rangle = \langle -\mathscr{G}^*\mathbf{q} - \nu, \hat{\theta}\rangle = -\langle \mathscr{G}^*\mathbf{q}, \hat{\theta}\rangle - \langle \nu, \hat{\theta}\rangle \qquad \forall \hat{\theta} \in \mathscr{T}. \tag{10.40}$$

Using the PTVP yields

$$\begin{aligned}\langle \nu, \hat{\theta}\rangle = &-\int_{\mathscr{B}} \mathbf{q} \cdot \nabla\hat{\theta} dV - \int_{\partial\mathscr{B}_q} \overline{q}\hat{\theta} dS - \int_{\partial\mathscr{B}_c} h(\theta_\infty - \theta)\hat{\theta} dS \\ &- \int_{\mathscr{B}} r\hat{\theta} dV \quad \forall \hat{\theta} \in \mathscr{T},\end{aligned} \tag{10.41}$$

and, after integration by parts, it becomes

$$\begin{aligned}\langle \nu, \hat{\theta}\rangle &= -\int_{\partial\mathscr{B}} (\mathbf{q} \cdot \mathbf{n})\hat{\theta} dS + \int_{\mathscr{B}} \text{div}\mathbf{q}\hat{\theta} dV \\ &\quad - \int_{\partial\mathscr{B}_q} \overline{q}\hat{\theta} dS - \int_{\partial\mathscr{B}_c} h(\theta_\infty - \theta)\hat{\theta} dS - \int_{\mathscr{B}} r\hat{\theta} dV \\ &= -\int_{\partial\mathscr{B}_\theta} (\mathbf{q} \cdot \mathbf{n})\hat{\theta} dS \quad \forall \hat{\theta} \in \mathscr{T}.\end{aligned} \tag{10.42}$$

As a direct consequence, we have that

$$\int_{\partial\mathscr{B}_\theta} [\nu + (\mathbf{q} \cdot \mathbf{n})]\hat{\theta} dS = 0 \quad \forall \hat{\theta} \in \mathscr{T}, \tag{10.43}$$

which, in local form, implies

$$\nu = -\mathbf{q} \cdot \mathbf{n} \quad \text{on } \partial\mathscr{B}_\theta. \tag{10.44}$$

The reactive flux ν is the normal component of the heat flux vector field over the part of the boundary in which the temperature has been constrained.

Finally, the reader will find in Figure 10.2 the main concepts and results which characterize the variational formulation.

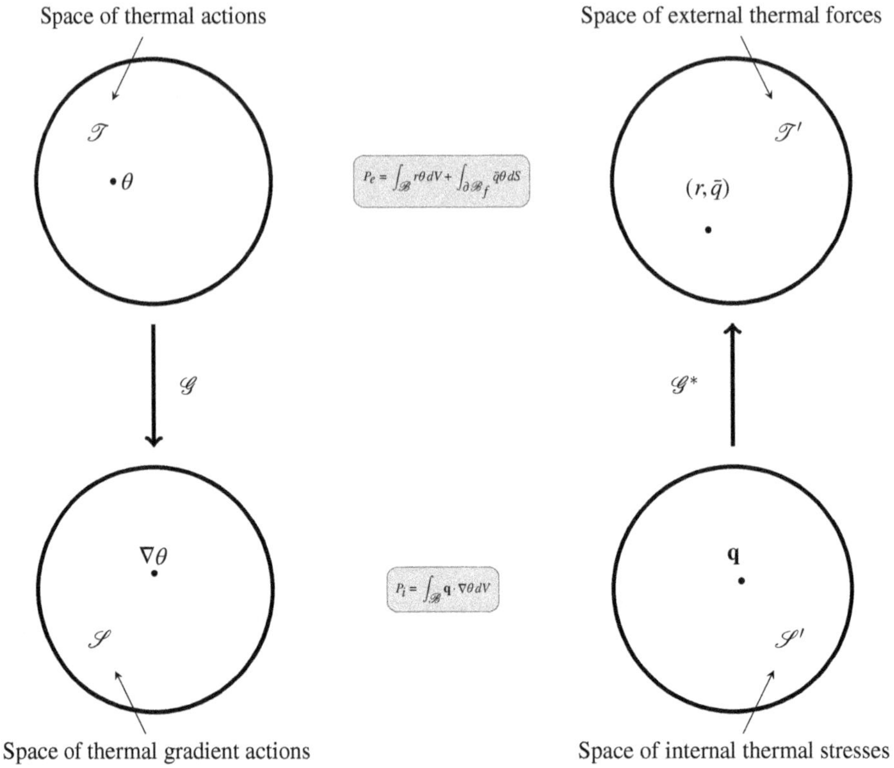

Figure 10.2 Ingredients in the variational theory for the heat transfer model.

10.4 Principle of Complementary Thermal Virtual Power

The previous section dealt with the primal variational form of the steady-state heat transfer problem, characterized through the PTVP, which establishes a connection between the internal thermal stress $\mathbf{q}$ and the external thermal load t so that (steady-state) equilibrium is achieved.

As in the mechanics of deformable bodies, for the heat transfer problem it is also possible to introduce a problem which can be regarded as dual to the primal problem, in the sense that it provides a characterization of the compatibility of the thermal gradient vector field $\mathbf{g}$, that is, to determine whether there exists $\theta \in \mathrm{Kin}_\theta$ such that $\mathscr{G}\theta = \mathbf{g}$. To formulate this problem we first define the following sets

$$\mathrm{Est}_q = \{\mathbf{q} \in \mathcal{S}';\ (\mathbf{q}, \mathscr{G}\hat{\theta}) + \langle t, \hat{\theta}\rangle = 0 \quad \forall \hat{\theta} \in \mathrm{Var}_\theta\}, \tag{10.45}$$

which is the set of all internal thermal stresses that are equilibrated with the external thermal load $t \in \mathscr{T}'$, and

$$\mathrm{Var}_q = \{\mathbf{q} \in \mathcal{S}';\ (\mathbf{q}, \mathscr{G}\hat{\theta}) = 0 \quad \forall \hat{\theta} \in \mathrm{Var}_\theta\}, \tag{10.46}$$

which is the vector subspace of $\mathcal{S}'$ whose elements are internal thermal stresses (heat flux fields) $\mathbf{q}$ at thermal equilibrium with the null thermal load.

As before, it is important to note that if $\mathbf{q}_0 \in \text{Est}_q$ and $\hat{\mathbf{q}} \in \text{Var}_q$ then

$$\mathbf{q} = \mathbf{q}_0 + \hat{\mathbf{q}} \in \text{Est}_q, \tag{10.47}$$

because

$$-(\mathbf{q}, \mathscr{G}\hat{\theta}) = -(\mathbf{q}_0, \mathscr{G}\hat{\theta}) - (\hat{\mathbf{q}}, \mathscr{G}\hat{\theta}) = -(\mathbf{q}_0, \mathscr{G}\hat{\theta}) = \langle t, \hat{\theta}\rangle \qquad \forall\hat{\theta} \in \text{Var}_\theta. \tag{10.48}$$

This result shows us that Est_q is a translation of Var_q, that is

$$\text{Est}_q = \mathbf{q}_0 + \text{Var}_q \quad \mathbf{q}_0 \in \text{Est}_q. \tag{10.49}$$

With the previous definitions, we can now enunciate the Principle of Complementary Thermal Virtual Power (PCTVP). We say that the vector field $\mathbf{g} \in \mathscr{S}$ is thermally compatible for an external thermal load $t \in \mathscr{T}'$ at equilibrium if and only if

$$(\hat{\mathbf{q}}, \mathbf{g}) = (\hat{\mathbf{q}}, \mathscr{G}\theta_0) \qquad \forall\hat{\mathbf{q}} \in \text{Var}_q, \tag{10.50}$$

for some $\theta_0 \in \text{Kin}_\theta$.

To see this result, consider first that $\mathbf{g}$ is compatible, then $\exists\theta_1 \in \mathscr{T}$ such that $\mathbf{g} = \mathscr{G}\theta_1$. Since Kin_θ is a translation of Var_θ, that is, $\text{Kin}_\theta = \theta_0 + \text{Var}_\theta$, with θ_0 arbitrary, we have $\theta_1 - \theta_0 \in \text{Var}_\theta$. Then

$$(\hat{\mathbf{q}}, \mathscr{G}(\theta_1 - \theta_0)) = 0 \qquad \forall\hat{\mathbf{q}} \in \text{Var}_q, \tag{10.51}$$

which results in the very definition of the space Var_q.

Suppose now that

$$(\hat{\mathbf{q}}, \mathbf{g}) = (\hat{\mathbf{q}}, \mathscr{G}\theta_0) \qquad \forall\hat{\mathbf{q}} \in \text{Var}_q, \tag{10.52}$$

where $\theta_0 \in \text{Kin}_\theta$. Hence

$$(\hat{\mathbf{q}}, \mathbf{g} - \mathscr{G}\theta_0) = 0 \qquad \forall\hat{\mathbf{q}} \in \text{Var}_q, \tag{10.53}$$

and we conclude that

$$\mathbf{g} - \mathscr{G}\theta_0 \in \text{Var}_q^\perp \quad \Leftrightarrow \quad \mathbf{g} \in \mathscr{G}\theta_0 + \text{Var}_q^\perp. \tag{10.54}$$

Moreover, from (10.46) we also know that

$$\text{Var}_q = (\mathscr{G}(\text{Var}_\theta))^\perp, \tag{10.55}$$

and then

$$\text{Var}_q^\perp = \mathscr{G}(\text{Var}_\theta). \tag{10.56}$$

Replacing (10.56) into (10.54) yields

$$\mathbf{g} \in \mathscr{G}\theta_0 + \mathscr{G}(\text{Var}_\theta) = \mathscr{G}(\theta_0 + \text{Var}_\theta) = \mathscr{G}(\text{Kin}_\theta), \tag{10.57}$$

which means $\mathbf{g} \in \mathscr{G}(\text{Kin}_\theta)$ or, equivalently, $\exists\theta_0 \in \text{Kin}_\theta$ such that $\mathscr{G}\theta_0 = \mathbf{g}$.

For the case under analysis, expression (10.50) corresponds to

$$\int_{\mathscr{B}} \hat{\mathbf{q}} \cdot \mathbf{g}\, dV = \int_{\mathscr{B}} \hat{\mathbf{q}} \cdot \nabla\theta_0\, dV \qquad \forall\hat{\mathbf{q}} \in \text{Var}_q, \tag{10.58}$$

so by making use of integration by parts we find

$$\int_{\mathscr{B}} \hat{\mathbf{q}} \cdot \nabla\theta_0 dV = \int_{\mathscr{B}} \mathrm{div}(\theta_0\hat{\mathbf{q}})dV - \int_{\mathscr{B}} \mathrm{div}\hat{\mathbf{q}}\theta_0 dV$$
$$= \int_{\partial\mathscr{B}} (\hat{\mathbf{q}} \cdot \mathbf{n})\theta_0 dS - \int_{\mathscr{B}} \mathrm{div}\hat{\mathbf{q}}\theta_0 dV = \int_{\partial\mathscr{B}_\theta} \hat{\mathbf{q}} \cdot \mathbf{n}\overline{\theta} dS, \quad (10.59)$$

where we make use of the fact that $\mathrm{div}\hat{\mathbf{q}} = 0$ in $\mathscr{B}$, $\hat{\mathbf{q}} \cdot \mathbf{n} = 0$ on $\partial\mathscr{B}_q$ and on $\partial\mathscr{B}_c$, and $\theta_0 = \overline{\theta}$ on $\partial\mathscr{B}_\theta$.

Therefore, we conclude that the PCTVP acquires the following variational form

$$\int_{\mathscr{B}} \hat{\mathbf{q}} \cdot \mathbf{g} dV = \int_{\partial\mathscr{B}_\theta} (\hat{\mathbf{q}} \cdot \mathbf{n})\overline{\theta} dS \qquad \forall \hat{\mathbf{q}} \in \mathrm{Var}_q. \quad (10.60)$$

10.5 Constitutive Equations

In our road towards the formulation of a field problem for the heat transfer problem there is still a missing aspect concerning the material behavior in the present context. This constitutive material response determines the way in which the heat flux $\mathbf{q}$ is related to the thermal gradient $\mathbf{g} = \nabla\theta$. That is, we have to look for an operator that maps the space $\mathscr{S}$ into its dual $\mathscr{S}'$. A typical constitutive model is that given by a linear relation, which states that

$$\mathbf{q} = -\mathbf{K}\mathbf{g}, \quad (10.61)$$

where $\mathbf{K}$ is a second-order tensor and the sign indicates that the heat flux opposes the thermal gradient. Next, we further assume that the tensor $\mathbf{K}$ satisfies the following properties

$$\mathbf{K}\mathbf{a} \cdot \mathbf{b} = \mathbf{K}\mathbf{b} \cdot \mathbf{a} \qquad \forall \mathbf{a}, \mathbf{b} \in \mathscr{S} \Rightarrow \mathbf{K} = \mathbf{K}^T, \quad (10.62)$$

$$\mathbf{K}\mathbf{g} \cdot \mathbf{g} \geq 0 \qquad \forall \mathbf{g} \in \mathscr{S}, \quad (10.63)$$

$$\mathbf{K}\mathbf{g} \cdot \mathbf{g} = 0 \qquad \Leftrightarrow \mathbf{g} = \mathbf{0}. \quad (10.64)$$

Let us now define the thermal gradient energy density function $\varpi = \varpi(\mathbf{g})$ as

$$\varpi(\overline{\mathbf{g}}) = \int_{\mathbf{0}}^{\overline{\mathbf{g}}} \mathbf{q}(\mathbf{g}) \cdot d\mathbf{g} = -\int_{\mathbf{0}}^{\overline{\mathbf{g}}} \mathbf{K}\mathbf{g} \cdot d\mathbf{g} = -\frac{1}{2}\mathbf{K}\overline{\mathbf{g}} \cdot \overline{\mathbf{g}}. \quad (10.65)$$

Notice that from the properties of $\mathbf{K}$ it follows that

$$\varpi(\overline{\mathbf{g}}) \leq 0, \quad (10.66)$$

where the equality is verified for $\overline{\mathbf{g}} = \mathbf{0}$.

Thus, the heat flux $\overline{\mathbf{q}}$ related to the thermal gradient $\overline{\mathbf{g}}$ through the constitutive equation is such that

$$\overline{\mathbf{q}} = \left.\frac{\partial\varpi}{\partial\mathbf{g}}\right|_{\mathbf{g}=\overline{\mathbf{g}}} = -\mathbf{K}\overline{\mathbf{g}}. \quad (10.67)$$

In turn, a Taylor's expansion ϖ around the point $\mathbf{g}_0$ provides

$$\begin{aligned}\varpi(\mathbf{g}) &= \varpi(\mathbf{g}_0) + \left.\frac{\partial \varpi}{\partial \mathbf{g}}\right|_{\mathbf{g}_0} \cdot (\mathbf{g} - \mathbf{g}_0) + \left.\frac{1}{2}\frac{\partial^2 \varpi}{\partial \mathbf{g}^2}\right|_{\mathbf{g}_0} (\mathbf{g} - \mathbf{g}_0) \cdot (\mathbf{g} - \mathbf{g}_0) \\ &= \varpi(\mathbf{g}_0) - \mathbf{K}\mathbf{g}_0 \cdot (\mathbf{g} - \mathbf{g}_0) - \frac{1}{2}\mathbf{K}(\mathbf{g} - \mathbf{g}_0) \cdot (\mathbf{g} - \mathbf{g}_0),\end{aligned} \tag{10.68}$$

which gives

$$\varpi(\mathbf{g}) - \varpi(\mathbf{g}_0) \leq -\mathbf{K}\mathbf{g}_0 \cdot (\mathbf{g} - \mathbf{g}_0) = \mathbf{q}_0 \cdot (\mathbf{g} - \mathbf{g}_0), \tag{10.69}$$

while the equality is verified for $\mathbf{g} = \mathbf{g}_0$. In this case we conclude that ϖ is a strictly concave function.

In the case of materials featuring nonlinear constitutive behavior we assume the following property for ϖ holds

$$(\mathbf{q}^B - \mathbf{q}^A) \cdot (\mathbf{g}^B - \mathbf{g}^A) \leq 0 \qquad \forall \mathbf{g}^B, \mathbf{g}^A \in \mathscr{S}, \tag{10.70}$$

with the equality being valid for $\mathbf{g}^B = \mathbf{g}^A$, and $\mathbf{q}^B = \mathbf{q}(\mathbf{g}^B)$, $\mathbf{q}^A = \mathbf{q}(\mathbf{g}^A)$. Let $\mathbf{g} = \mathbf{g}^A + d\mathbf{g}$ be a thermal gradient whose heat flux is $\mathbf{q} = \mathbf{q}(\mathbf{g})$. Then

$$(\mathbf{q} - \mathbf{q}^A) \cdot d\mathbf{g} \leq 0, \tag{10.71}$$

and integrating the process that takes us from $\mathbf{g}^A$ to $\mathbf{g}^B$ we arrive at

$$\int_{\mathbf{g}^A}^{\mathbf{g}^B} (\mathbf{q} - \mathbf{q}^A) \cdot d\mathbf{g} \leq 0, \tag{10.72}$$

and so we get

$$\varpi(\mathbf{g}^B) - \varpi(\mathbf{g}^A) \leq \mathbf{q}^A \cdot (\mathbf{g}^B - \mathbf{g}^A), \tag{10.73}$$

where we have recovered the concavity of function ϖ from the monotonicity property stated by (10.70).

By virtue of the positive-definiteness of tensor $\mathbf{K}$, we can calculate the inverse $\mathbf{K}^{-1}$, and we can find the inverse constitutive equation as follows

$$\mathbf{g} = \mathbf{g}(\mathbf{q}) = -\mathbf{K}^{-1}\mathbf{q}, \tag{10.74}$$

therefore it is possible to introduce the complementary thermal gradient energy density function as

$$\varpi^*(\overline{\mathbf{q}}) = \int_{\mathbf{0}}^{\overline{\mathbf{q}}} \mathbf{g}(\mathbf{q}) \cdot d\mathbf{q} = -\int_{\mathbf{0}}^{\overline{\mathbf{q}}} \mathbf{K}^{-1}\mathbf{q} \cdot d\mathbf{q} = -\frac{1}{2}\mathbf{K}^{-1}\overline{\mathbf{q}} \cdot \overline{\mathbf{q}} \leq 0, \tag{10.75}$$

where the equality is verified for $\overline{\mathbf{q}} = \mathbf{0}$. Furthermore,

$$\overline{\mathbf{g}} = \left.\frac{\partial \varpi^*}{\partial \mathbf{q}}\right|_{\mathbf{q}=\overline{\mathbf{q}}}, \tag{10.76}$$

and also

$$\varpi^*(\mathbf{q}) - \varpi^*(\mathbf{q}_0) \leq \mathbf{g}_0 \cdot (\mathbf{q} - \mathbf{q}_0), \tag{10.77}$$

which allows us to conclude that function ϖ^* is also concave in the space $\mathscr{S}'$.

From the properties of ϖ and ϖ^* we verify that

$$\varpi(\mathbf{g}) + \varpi^*(\mathbf{q}) \leq \mathbf{g} \cdot \mathbf{q}, \tag{10.78}$$

where the equality is satisfied if $\mathbf{g}$ and $\mathbf{q}$ are related by means of the constitutive equation.

10.6 Principle of Minimum Total Thermal Energy

The goal of this section is to transform the steady-state heat transfer problem into a minimization problem with a corresponding cost functional. Therefore, we proceed analogously to the steps presented in Chapter 4.

Since function ϖ is concave, from (10.73) we can write

$$\varpi(\mathscr{G}\theta) - \varpi(\mathscr{G}\theta_0) \leq \left.\frac{\partial \varpi}{\partial \mathbf{g}}\right|_{\mathscr{G}\theta_0} \cdot \mathscr{G}(\theta - \theta_0), \tag{10.79}$$

where the equality holds for all $\theta - \theta_0 \in \mathscr{T} \setminus \mathscr{N}(\mathscr{G})$. Integrating (10.79) in the domain occupied by the body $\mathscr{B}$, we obtain the following

$$\int_{\mathscr{B}} \left(\varpi(\mathscr{G}\theta) - \varpi(\mathscr{G}\theta_0)\right) dV \leq \int_{\mathscr{B}} \left.\frac{\partial \varpi}{\partial \mathbf{g}}\right|_{\mathscr{G}\theta_0} \cdot \mathscr{G}(\theta - \theta_0) dV \qquad \forall \theta \in \text{Kin}_\theta. \tag{10.80}$$

Replacing this result into the PTVP expressed by equation (10.30) gives

$$-\int_{\mathscr{B}} \left(\varpi(\mathscr{G}\theta) - \varpi(\mathscr{G}\theta_0)\right) dV \geq \langle t, (\theta - \theta_0)\rangle \qquad \forall \theta \in \text{Kin}_\theta. \tag{10.81}$$

Let us define the Total Thermal Energy as follows

$$\Theta(\theta) = -\int_{\mathscr{B}} \varpi(\mathscr{G}\theta) dV - \langle t, \theta\rangle, \tag{10.82}$$

then after arranging terms we get

$$\Theta(\theta) = -\int_{\mathscr{B}} \varpi(\mathscr{G}\theta) dV - \langle t, \theta\rangle \geq -\int_{\mathscr{B}} \varpi(\mathscr{G}\theta_0) dV - \langle t, \theta_0\rangle = \Theta(\theta_0) \qquad \forall \theta \in \text{Kin}_\theta. \tag{10.83}$$

In this way, we have arrived at the Principle of Minimum Total Thermal Energy, which consists of finding $\theta_0 \in \text{Kin}_\theta$ that minimizes the Total Thermal Energy functional, that is

$$\Theta(\theta_0) = \min_{\theta \in \text{Kin}_\theta} \Theta(\theta) = \min_{\theta \in \text{Kin}_\theta} \left[-\int_{\mathscr{B}} \varpi(\mathscr{G}\theta) dV - \langle t, \theta\rangle\right]. \tag{10.84}$$

10.7 Poisson and Laplace Equations

Consider the constitutive model most widely used in practice for this kind of problem. Such a model is known as the Fourier law and it establishes that the heat flux $\mathbf{q}$ and the thermal gradient $\mathbf{g} = \nabla\theta$ are related as follows

$$\mathbf{q} = -\mathbf{K}\nabla\theta = -k\nabla\theta, \tag{10.85}$$

where $\mathbf{K} = k\mathbf{I}$ is characterized by a single (positive) scalar field k, called the material thermal conductivity, and $\mathbf{I}$ is the second-order identity tensor.

Then, we rewrite the PTVP given by (10.34) for the particular choice of a material ruled by constitutive equation (10.85)

$$\int_{\mathscr{B}} k\nabla\theta \cdot \nabla\hat{\theta} dV = \int_{\partial\mathscr{B}_q} \overline{q}\hat{\theta} dS + \int_{\partial\mathscr{B}_c} h(\theta_\infty - \theta)\hat{\theta} dS + \int_{\mathscr{B}} r\hat{\theta} dV \qquad \forall\hat{\theta} \in \mathrm{Var}_\theta. \tag{10.86}$$

The Euler–Lagrange equations are obtained by integrating by parts the left-hand side in the above expression and using standard arguments from the calculus of variations

$$-\mathrm{div}(k\nabla\theta) = r \qquad \text{in } \mathscr{B}, \tag{10.87}$$

$$k\nabla\theta \cdot \mathbf{n} = \overline{q} \qquad \text{on } \partial\mathscr{B}_q, \tag{10.88}$$

$$k\nabla\theta \cdot \mathbf{n} = h(\theta_\infty - \theta) \qquad \text{on } \partial\mathscr{B}_c. \tag{10.89}$$

Let us further simplify the presentation for the case of thermally homogeneous materials, that is, for materials featuring a constant conductivity k, and also for homogeneous essential boundary conditions over the whole boundary, that is, $\overline{\theta} = 0$ over $\partial\mathscr{B}_\theta = \partial\mathscr{B}$. Then, the problem is reduced to the widely known Poisson problem

$$\Delta\theta = -\frac{r}{k} \qquad \text{in } \mathscr{B}, \tag{10.90}$$

$$\theta = 0 \qquad \text{on } \partial\mathscr{B}. \tag{10.91}$$

If now we consider a given $\overline{\theta}$ over the boundary $\partial\mathscr{B}$, and that $r = 0$, we arrive at the Dirichlet problem for the Laplace equation

$$\Delta\theta = 0 \qquad \text{in } \mathscr{B}, \tag{10.92}$$

$$\theta = \overline{\theta} \qquad \text{on } \partial\mathscr{B}, \tag{10.93}$$

which characterizes the so-called harmonic functions.

Next, we present the variational problem (10.86) in the format of a minimization problem. Thus, and noting that for the Fourier law the thermal energy density function is given by

$$\varpi(\nabla\theta) = -\frac{1}{2}k|\nabla\theta|^2, \tag{10.94}$$

the total thermal energy functional (10.82) gives

$$\Theta(\theta) = \int_{\mathscr{B}} \frac{1}{2}k|\nabla\theta|^2 dV - \int_{\partial\mathscr{B}_q} \overline{q}\theta dS - \int_{\partial\mathscr{B}_c} h(\theta_\infty - \theta)\theta dS - \int_{\mathscr{B}} r\theta dV. \tag{10.95}$$

Finally, after rearranging terms, the minimization problem standing from (10.84) amounts to finding $\theta_0 \in \mathrm{Kin}_\theta$ such that

$$\Theta(\theta_0) = \min_{\theta\in\mathrm{Kin}_\theta} \Theta(\theta) = \min_{\theta\in\mathrm{Kin}_\theta} \left[\frac{1}{2}\int_{\mathscr{B}} k|\nabla\theta|^2 dV + \frac{1}{2}\int_{\partial\mathscr{B}_c} h\theta^2 dS - \int_{\partial\mathscr{B}_q} \overline{q}\theta dS - \int_{\partial\mathscr{B}_c} h\theta_\infty\theta dS - \int_{\mathscr{B}} r\theta dV\right]. \tag{10.96}$$

Note that the term $h(\theta_\infty - \theta)$ becomes quadratic when paired with θ, this gives rise to the term of the form $\frac{1}{2}h\theta^2$ in the expression above. Here we have simplified the presentation by directly modifying the cost functional to account for this fact, and to ensure that the first Gâteaux derivative of functional Θ in (10.95) actually corresponds to the variational equation (10.86). An alternative more systematic way to do this involves postulating that the internal energy is not only composed by ϖ integrated over the body domain $\mathscr{B}$, but also by a surface energy ψ of the form $\frac{1}{2}h\theta^2$, defined over the boundary $\partial\mathscr{B}_c$. In turn, the external system of body forces must include a heat flux of the form $h\theta_\infty$.

11

Incompressible Fluid Flow

11.1 Introduction

In this chapter we will tackle the construction of a model for incompressible fluid flow phenomena exploiting the variational framework presented earlier. This implies that we will make use of the Principle of Virtual Power (PVP) presented in Chapter 3 to find the variational equation that governs the problem of mechanical equilibrium. It is worth mentioning that, notwithstanding that the fluid flow problem is addressed here as a separated chapter, the variational structure comprises the very same ingredients already discussed in the mechanics of solid continua also seen in Chapter 3. In fact, from the kinematical perspective, the same kinematical descriptors will be utilized to characterize the motion actions that can be executed over particles composing the (fluid) continuum. A peculiarity to be discussed in detail here is that of the incompressibility constraint, which characteristically emerges in fluid mechanics problems at low Mach numbers.[1] We will underline the way in which this constraint enters into the variational formulation and the consequences of such a constraint.

Thus, this chapter differs from most of the literature in the fluid mechanics domain in that the derivation of the governing equations is realized through a purely variational formalism, instead of the standard approach based on (mass and momentum) conservation laws. As said before, the PVP provides a unified framework to model this problem and, unlike the standard approach based on conservation laws, it is neither required to assume the existence of the stress tensor beforehand nor to postulate the existence of the mechanical pressure field in the fluid. These entities naturally emerge from the PVP and from the kinematical constraints considered for the problem. Moreover, through the standard procedure of integration by parts it will be possible to find the classical equations in the format of partial differential equations, which for the case of Newtonian fluids are nothing but the Navier–Stokes equations. As a corollary, we will examine some particular cases of interest such as the case of irrotational incompressible flows and the case of Stokes flows. Moreover, in the latter case we will see how to express the problem as a minimization problem.

For a more detailed exposition of the fluid flow problem see [45, 56, 62, 120, 290, 294].

---

1 The Mach number is the non-dimensional number defined as the ratio between a characteristic velocity in the problem and the speed of sound in the continuum. For most engineering applications, the Mach number remains below 0.3 and the incompressibility hypothesis is reasonable.

Introduction to the Variational Formulation in Mechanics: Fundamentals and Applications, First Edition.
Edgardo O. Taroco, Pablo J. Blanco and Raúl A. Feijóo.

11.2 Kinematics

For continuum media whose constitutive behavior is that of a fluid, we will consider the same kinematics as introduced in the case of solid continua. That is, let us consider a set of particles that compose the continuum within the Euclidean space $\mathscr{E}$. To describe the motion of the particles through the Euclidean space in a given time interval $t \in [t_0, t_f]$ we make use of motion mapping (see Chapter 3, where it was termed simply deformation)

$$\begin{aligned} \mathscr{X} \; : \mathscr{B} \times [t_0, t_f] &\rightarrow \mathscr{E}, \\ (\mathbf{X}, t) &\mapsto \mathbf{x} = \mathscr{X}(\mathbf{X}, t), \end{aligned} \tag{11.1}$$

where $\mathscr{B}$ denotes the set of particles that compose the physical system under analysis and $\mathbf{X}$ represents the position of the particles at a certain reference instant (reference configuration), for example at $t = t_0$. In this manner, $\mathbf{x}$ stands, through the motion mapping $\mathscr{X}$, for the position occupied by the particle $\mathbf{X}$ at time t.

As we have seen in Chapter 3, the velocity field in the material and spatial descriptions is defined, respectively, by

$$\dot{\mathbf{x}}(\mathbf{X}, t) = \frac{\partial \mathscr{X}}{\partial t}(\mathbf{X}, t)\bigg|_{\mathbf{X}\text{ fixed}}, \tag{11.2}$$

$$\mathbf{v}(\mathbf{x}, t) = \frac{\partial \mathscr{X}}{\partial t}(\mathscr{X}^{-1}(\mathbf{x}, t), t), \tag{11.3}$$

where $\mathscr{X}^{-1}$ is the inverse mapping, that is, $\mathbf{x} = \mathscr{X}(\mathscr{X}^{-1}(\mathbf{x}, t), t)$, which is assumed to exist and to be differentiable.

We have also proved that the acceleration of particles, in the material and spatial descriptions, respectively, was given by

$$\ddot{\mathbf{x}}(\mathbf{X}, t) = \frac{\partial^2 \mathscr{X}}{\partial t^2}(\mathbf{X}, t)\bigg|_{\mathbf{X}\text{ fixed}}, \tag{11.4}$$

$$\dot{\mathbf{v}}(\mathbf{x}, t) = \frac{\partial \mathbf{v}}{\partial t}(\mathbf{x}, t) + [(\nabla \mathbf{v})(\mathbf{x}, t)]\mathbf{v}(\mathbf{x}, t), \tag{11.5}$$

where $\nabla(\cdot)$ denotes the gradient operator with respect to spatial coordinates $\mathbf{x}$.

According to the developments of Chapter 3, the deformation (strain) rate, denoted by ε, experienced by a particle is characterized by

$$\varepsilon = \frac{1}{2}(\nabla \mathbf{v} + (\nabla \mathbf{v})^T) = (\nabla \mathbf{v})^s, \tag{11.6}$$

that is, it is characterized by the symmetric component of the velocity gradient tensor field. Furthermore, we have seen that for isochoric (volume preserving) motions, which are of fundamental interest in this chapter, it is verified that

$$\operatorname{div} \mathbf{v} = 0. \tag{11.7}$$

Let us consider the set $\mathscr{V}$ of all velocity fields that can characterize the motion of particles in the fluid medium. Since we will work with incompressible flows and we will further consider kinematical constraints over some portion of the boundary that defines the system (domain) under analysis $\mathscr{B}_0$, which we call $\partial\mathscr{B}_{0v}$, we have the following set of kinematically admissible velocity fields

$$\text{Kin}_v = \{\mathbf{v} \in \mathscr{V}\,;\; \mathbf{v}|_{\partial\mathscr{B}_{0v}} = \bar{\mathbf{v}},\; \operatorname{div} \mathbf{v} = 0 \quad \forall \mathbf{x} \in \mathscr{B}_0\}. \tag{11.8}$$

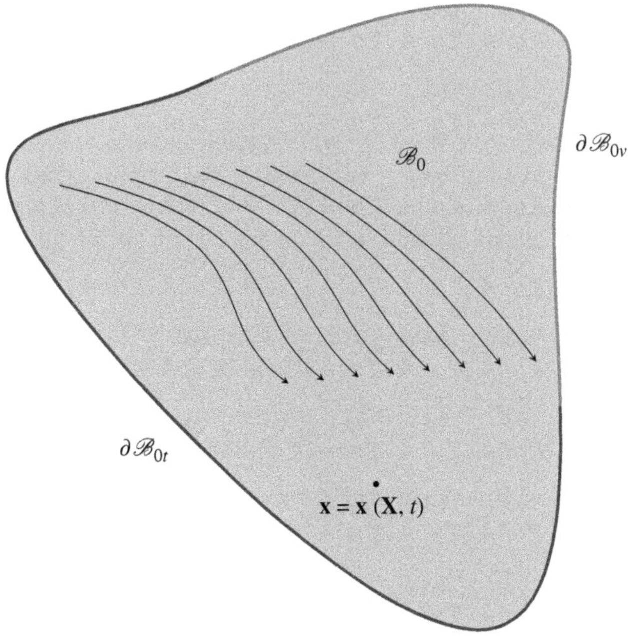

Figure 11.1 Fixed Eulerian domain for the setting of the incompressible fluid flow problem.

It is important to note here that, given $\mathbf{v}(\mathbf{x}, t)$, we will adopt an Eulerian description to frame the motion of particles. Therefore, the fields are defined in the domain $\mathscr{B}_0$ which is arbitrarily chosen according to the needs of the problem under study. This domain can be, for example, the domain occupied by particles $\mathbf{X}$ at time $t = t_0$. Nevertheless, as said, its definition is completely arbitrary (such as in the mechanics solid continua). In this chapter we assume that $\mathscr{B}_0$ remains invariant along time.[2] Finally, observe that we have considered that the velocity of particles over the portion of the boundary $\partial\mathscr{B}_{0v}$ is prescribed to be equal to $\overline{\mathbf{v}}$. Figure 11.1 displays the domain of analysis and the portion of the boundary where essential boundary conditions are imposed.

Note that Kin_v defined in (11.8) is a linear manifold and therefore is a translation of a vector space. This vector space is defined by

$$\mathrm{Var}_v = \{\mathbf{v} \in \mathscr{V}\ ;\ \mathbf{v}|_{\partial\mathscr{B}_{0v}} = \mathbf{0},\ \operatorname{div}\mathbf{v} = 0 \quad \forall \mathbf{x} \in \mathscr{B}_0\}, \tag{11.9}$$

yielding

$$\mathrm{Kin}_v = \mathbf{v}_0 + \mathrm{Var}_v, \tag{11.10}$$

with $\mathbf{v}_0 \in \mathrm{Kin}_v$ arbitrary.

2 In problems where the fluid flows through deforming domains, this hypothesis is removed and the domain of definition of the problem is time-dependent. What is more, the definition of the domain may depend upon the very solution of the problem, such as in the mechanics domain within finite strain regimes. Typical examples are those encountered in fluid structure interaction problems, where structures undergo large displacements within the fluid media.

We say that the strain rate $\varepsilon \in \mathscr{W}$ is compatible if it is possible to find $\mathbf{v} \in \mathscr{V}$ such that

$$\varepsilon = (\nabla \mathbf{v})^s. \tag{11.11}$$

Then, we define the operator $\mathscr{D}(\cdot) = (\nabla(\cdot))^s : \mathscr{V} \rightarrow \mathscr{W}$ which establishes a correspondence between every field $\mathbf{v} \in \mathscr{V}$ and strain rate tensor fields of the form $\varepsilon = (\nabla \mathbf{v})^s$. Moreover, since $\mathbf{v}$ is divergence-free, the strain rate operator $\mathscr{D}$ yields traceless strain rate tensor fields, that is, for $\mathbf{v} \in \mathrm{Var}_v$ we have that $\mathscr{D}\mathbf{v} = \varepsilon = (\nabla \mathbf{v})^s$, which verifies

$$\mathrm{tr}\varepsilon = \mathrm{tr}(\nabla \mathbf{v})^s = \mathrm{div}\,\mathbf{v} = 0. \tag{11.12}$$

Therefore, we say that the strain actions belong to the subspace $\mathscr{W}^D \subset \mathscr{W}$ of second-order traceless tensor fields.

As in Chapter 3, the null space $\mathscr{N}(\mathscr{D})$ is given by

$$\mathscr{N}(\mathscr{D}) = \{\mathbf{v} \in \mathscr{V}\,;\ \mathscr{D}\mathbf{v} = \mathbf{O} \ \ \forall \mathbf{x} \in \mathscr{B}_0\}, \tag{11.13}$$

that is, by all those velocity fields whose functional form is

$$\mathbf{v} = \mathbf{v}_O + \mathbf{W}_O(\mathbf{x} - O), \tag{11.14}$$

where $\mathbf{v}_O$ is a uniform velocity field in $\mathscr{B}_0$, O is a constant vector, and $\mathbf{W}_O$ is a uniform skew-symmetric tensor in $\mathscr{B}_0$.

11.3 Principle of Virtual Power

The previous section introduced all the basic elements that describe the kinematics in the present problem. Hence, we note that the motion actions which are able to generate power are vector fields which stand for the flow velocity field. Let us proceed to characterize the external forces through linear and continuous functionals $f : \mathscr{V} \rightarrow \mathbb{R}$, that is, $f \in \mathscr{V}'$. Then, the external power functional is characterized by

$$P_e = \langle f, \mathbf{v} \rangle, \tag{11.15}$$

where $\langle \cdot, \cdot \rangle$ is the duality product between pairs of $\mathscr{V}'$ and $\mathscr{V}$. We will see that the PVP will allow us to explore in detail the functional form of the elements $f \in \mathscr{V}'$.

Concerning the characterization of internal stresses, we will follow the same procedure utilized in Chapter 3, which will be repeated here to make the present chapter self-contained. Let us start by introducing the internal power functional, denoted by P_i, as a linear and continuous functional defined over the space of motion actions and over the space of first gradient motion actions tensor fields. That means

$$P_i = -\int_{\mathscr{B}_0} (\mathbf{f} \cdot \mathbf{v} + \sigma \cdot \nabla \mathbf{v}) dV, \tag{11.16}$$

where $\mathbf{f}$ is a vector field and σ is a second-order tensor field. However, we know that rigid motion actions do not allow us to evaluate the internal power, that is, $P_i = 0$, $\forall \mathbf{v} \in \mathscr{N}(\mathscr{D})$. Hence, considering a uniform velocity field $\mathbf{v} = \mathbf{c}$ in $\mathscr{B}_0$, from (11.16) we have

$$P_i = -\int_{\mathscr{B}_0} \mathbf{f} \cdot \mathbf{c} dV = 0 \qquad \forall \mathbf{c} \in \mathbb{R}^3, \tag{11.17}$$

and since this must hold for every part of the domain $\mathscr{B}_0$, we arrive at the following

$$\mathbf{f} = \mathbf{0} \qquad \forall \mathbf{x} \in \mathscr{B}_0. \tag{11.18}$$

Considering a rigid motion action of the form $\mathbf{v} = \mathbf{v}_O + \mathbf{W}_O(\mathbf{x} - O)$, with $\mathbf{W}_O$ skew-symmetric and uniform in $\mathscr{B}_0$, expression (11.16) gives

$$P_i = -\int_{\mathscr{B}_0} \sigma \cdot \mathbf{W}_O dV = 0 \qquad \forall \mathbf{W}_O \in Skw, \tag{11.19}$$

and since this must be verified for every part of $\mathscr{B}_0$ we obtain

$$\sigma \in Sym \qquad \forall \mathbf{x} \in \mathscr{B}_0. \tag{11.20}$$

These results indicate that the internal power must be shaped as follows

$$P_i = -\int_{\mathscr{B}_0} \sigma \cdot \varepsilon dV = -(\sigma, \varepsilon), \tag{11.21}$$

that is, the internal stress in the fluid medium is characterized by a linear and continuous functional defined over the space of strain rates $\mathscr{W}$. Similarly to what was done in Chapter 3, the symmetric second-order tensor field σ is called the Cauchy stress field.

Before going ahead with the formulation of the PVP let us see a particularization of the expression of the internal power for incompressible motion actions, which are of interest for the framework posed by the PVP. Thus, for incompressible motion actions we have that (11.7) is satisfied, which implies

$$\text{tr}\varepsilon = 0. \tag{11.22}$$

Let us recall now the decomposition of a second-order tensor $\mathbf{A}$ into the hydrostatic $\mathbf{A}^H$ and deviatoric $\mathbf{A}^D$ components

$$\mathbf{A} = \mathbf{A}^H + \mathbf{A}^D, \tag{11.23}$$

$$\mathbf{A}^H = \frac{1}{3}(\text{tr}\mathbf{A})\mathbf{I}, \tag{11.24}$$

$$\mathbf{A}^D = \mathbf{A} - \frac{1}{3}(\text{tr}\mathbf{A})\mathbf{I}, \tag{11.25}$$

and so clearly $\text{tr}\mathbf{A}^D = 0$.

Taking (11.21) and decomposing each tensor field into the corresponding hydrostatic and deviatoric components, and making use of (11.22) we get

$$\begin{aligned} P_i &= -\int_{\mathscr{B}_0} \sigma \cdot \varepsilon dV = -\int_{\mathscr{B}_0} \left(\frac{1}{3}(\text{tr}\sigma)\mathbf{I} + \sigma^D\right) \cdot \left(\frac{1}{3}(\text{tr}\varepsilon)\mathbf{I} + \varepsilon^D\right) dV \\ &= -\int_{\mathscr{B}_0} \left(\frac{1}{3}(\text{tr}\sigma)\mathbf{I} + \sigma^D\right) \cdot \varepsilon^D dV = -\int_{\mathscr{B}_0} \left(\frac{1}{3}(\text{tr}\sigma)(\text{tr}\varepsilon^D) + \sigma^D \cdot \varepsilon^D\right) dV \\ &= -\int_{\mathscr{B}_0} \sigma^D \cdot \varepsilon^D dV = -(\sigma^D, \varepsilon^D). \end{aligned} \tag{11.26}$$

Expression (11.26) highlights that the internal power is a linear and continuous functional over the subspace $\mathscr{W}^D \subset \mathscr{W}$ formed by all incompressible motion actions. Dually, the internal power is characterized by a deviatoric Cauchy stress, in other words, the hydrostatic component σ^H of the full Cauchy stress tensor does not contribute to the internal power because it represents the reaction associated, by duality, with the incompressibility kinematical restriction.

Let us enunciate the PVP. We say that the system, whose domain of analysis is $\mathscr{B}_0$, is at mechanical equilibrium with the external force $f \in \mathscr{V}'$, if

$$P_e = \langle f, \hat{\mathbf{v}} \rangle = 0 \qquad \forall \hat{\mathbf{v}} \in \mathrm{Var}_v \cap \mathscr{N}(\mathscr{D}), \tag{11.27}$$

which implies that the external power must be nullified for all kinematically admissible rigid motion actions and, in addition, it is verified that

$$P_i + P_e = -(\sigma^D, (\nabla\hat{\mathbf{v}})^s) + \langle f, \hat{\mathbf{v}} \rangle = 0 \qquad \forall \hat{\mathbf{v}} \in \mathrm{Var}_v, \tag{11.28}$$

that is, the sum of the internal and external powers equals zero for all kinematically admissible motion actions.

Notice that in (11.28) we have exploited the following identity for isochoric motions

$$\begin{aligned} \varepsilon^D = ((\nabla\mathbf{v})^s)^D &= (\nabla\mathbf{v})^s - \frac{1}{3}(\mathrm{tr}[(\nabla\mathbf{v})^s])\mathbf{I} = (\nabla\mathbf{v})^s - \frac{1}{3}(\mathrm{tr}(\nabla\mathbf{v}))\mathbf{I} \\ &= (\nabla\mathbf{v})^s - \frac{1}{3}(\mathrm{div}\,\mathbf{v})\mathbf{I} = (\nabla\mathbf{v})^s. \end{aligned} \tag{11.29}$$

In the following we establish the existence of the field σ^D for a given equilibrated external force f, that is, f satisfies the first part of the PVP.

Theorem 11.1 Representation Theorem. If $f \in \mathscr{V}'$ is at equilibrium, there exists $\sigma^D \in \mathscr{W}'$ and $r \in \mathscr{V}'$ such that

$$f = \mathscr{D}^*\sigma^D - r, \tag{11.30}$$

$$\langle r, \hat{\mathbf{v}} \rangle = 0 \qquad \forall \hat{\mathbf{v}} \in \mathrm{Var}_v, \tag{11.31}$$

where $r \in \mathrm{Var}_v^\perp$, which is the reactive force. In this way, we say that σ^D and r equilibrate f.

Proof. Since f is at equilibrium we have

$$f \in (\mathrm{Var}_v \cap \mathscr{N}(\mathscr{D}))^\perp \quad \Leftrightarrow \quad f \in \mathrm{Var}_v^\perp + \mathscr{N}(\mathscr{D})^\perp. \tag{11.32}$$

Because $\mathbf{v} \in \mathscr{N}(\mathscr{D})$, then $\mathscr{D}\mathbf{v} = \mathbf{O}$, and consequently

$$\langle \mathscr{D}^*\sigma^D, \mathbf{v} \rangle = (\sigma^D, \mathscr{D}\mathbf{v}) = 0 \qquad \forall \sigma^D \in \mathscr{W}', \tag{11.33}$$

which implies $\mathbf{v} \in R(\mathscr{D}^*)^\perp$, then $\mathscr{N}(\mathscr{D}) = R(\mathscr{D}^*)^\perp$, that is

$$\mathscr{N}(\mathscr{D})^\perp = R(\mathscr{D}^*). \tag{11.34}$$

Putting this into (11.32) allows us to conclude that

$$f \in \mathrm{Var}_v^\perp + R(\mathscr{D}^*), \tag{11.35}$$

from where we obtain that there exist $\sigma^D \in \mathscr{W}'$ and $-r \in \mathrm{Var}_v^\perp$ such that

$$f = \mathscr{D}^*\sigma^D - r, \tag{11.36}$$

and since $-r \in \mathrm{Var}_v^\perp$, we straightforwardly have

$$\langle r, \hat{\mathbf{v}} \rangle = 0 \quad \forall \hat{\mathbf{v}} \in \mathrm{Var}_v. \tag{11.37}$$

■

The previous representation result guarantees that the field σ^D exists. However, this characterization is particularly important in this incompressible flow problem because it provides the structure of the reactive force r.

From the representation theorem we conclude the following

$$\langle f, \hat{\mathbf{v}} \rangle = \langle \mathcal{D}^* \sigma^D - r, \hat{\mathbf{v}} \rangle = \langle \mathcal{D}^* \sigma^D, \hat{\mathbf{v}} \rangle = (\sigma^D, \mathcal{D} \hat{\mathbf{v}}) \qquad \forall \hat{\mathbf{v}} \in \mathrm{Var}_v, \tag{11.38}$$

that is

$$(\sigma^D, \mathcal{D} \hat{\mathbf{v}}) = \langle f, \hat{\mathbf{v}} \rangle \qquad \forall \hat{\mathbf{v}} \in \mathrm{Var}_v, \tag{11.39}$$

arriving, in this manner, at the second part of the PVP.

Let us explicitly write the functional form of the internal power functional, which will enable the full characterization of external forces compatible with the model. Therefore, from the PVP we have

$$\int_{\mathcal{B}_0} \sigma^D \cdot (\nabla \hat{\mathbf{v}})^s dV = \langle f, \hat{\mathbf{v}} \rangle \qquad \forall \hat{\mathbf{v}} \in \mathrm{Var}_v. \tag{11.40}$$

Assuming that the fields are regular enough so that integration by parts can be performed, and recalling that σ^D is a symmetric second-order tensor, we proceed in the following manner

$$\begin{aligned} \int_{\mathcal{B}_0} \sigma^D \cdot (\nabla \hat{\mathbf{v}})^s dV &= \int_{\mathcal{B}_0} \sigma^D \cdot \nabla \hat{\mathbf{v}} dV = \int_{\mathcal{B}_0} \left[\mathrm{div}\,(\sigma^D \hat{\mathbf{v}}) - (\mathrm{div}\,\sigma^D) \cdot \hat{\mathbf{v}} \right] dV \\ &= \int_{\partial \mathcal{B}_{0t}} (\sigma^D \mathbf{n}) \cdot \hat{\mathbf{v}} dS - \int_{\mathcal{B}_0} (\mathrm{div}\,\sigma^D) \cdot \hat{\mathbf{v}} dV, \end{aligned} \tag{11.41}$$

where we have taken into consideration the fact that $\hat{\mathbf{v}} = \mathbf{0}$ over $\partial \mathcal{B}_{0v}$, and where we have put $\partial \mathcal{B}_0 = \partial \mathcal{B}_{0v} \cup \partial \mathcal{B}_{0t}$ with $\partial \mathcal{B}_{0v} \cap \partial \mathcal{B}_{0t} = \emptyset$. The vector $\mathbf{n}$ denotes the outward unit normal vector to these boundaries. Hence, we can characterize the element $f \in \mathcal{V}'$ as

$$\langle f, \hat{\mathbf{v}} \rangle = \int_{\partial \mathcal{B}_{0t}} (\sigma^D \mathbf{n}) \cdot \hat{\mathbf{v}} dS - \int_{\mathcal{B}_0} (\mathrm{div}\,\sigma^D) \cdot \hat{\mathbf{v}} dV, \tag{11.42}$$

that is, f is defined by means of two elements, a force per unit volume, here denoted by $\mathbf{b}^*$, defined in $\mathcal{B}_0$, and a force per unit area, called $\mathbf{t}$, defined over the portion of the boundary $\partial \mathcal{B}_{0t}$. Then

$$\langle f, \hat{\mathbf{v}} \rangle = \int_{\partial \mathcal{B}_{0t}} \mathbf{t} \cdot \hat{\mathbf{v}} dS + \int_{\mathcal{B}_0} \mathbf{b}^* \cdot \hat{\mathbf{v}} dV. \tag{11.43}$$

In the case of fluid flow problems, the accelerations experienced by particles provide an important contribution to the phenomenology under study. This is manifested through the corresponding contribution to the external power, and remains characterized within the element $\mathbf{b}^*$. In effect, introducing the forces caused by particle accelerations, denoted by $-\rho \dot{\mathbf{v}}$, where ρ is the fluid density (which is constant for each particle because the fluid is incompressible), and defining $\mathbf{b}^* = \mathbf{b} - \rho \dot{\mathbf{v}}$, expression (11.43) becomes

$$\langle f, \hat{\mathbf{v}} \rangle = \int_{\partial \mathcal{B}_{0t}} \mathbf{t} \cdot \hat{\mathbf{v}} dS + \int_{\mathcal{B}_0} \mathbf{b} \cdot \hat{\mathbf{v}} dV - \int_{\mathcal{B}_0} \rho \dot{\mathbf{v}} \cdot \hat{\mathbf{v}} dV. \tag{11.44}$$

Therefore, the PVP acquires the following variational form

$$\int_{\mathscr{B}_0} \left[\rho\dot{\mathbf{v}}\cdot\hat{\mathbf{v}} + \sigma^D\cdot(\nabla\hat{\mathbf{v}})^s\right] dV = \int_{\partial\mathscr{B}_{0t}} \mathbf{t}\cdot\hat{\mathbf{v}}dS + \int_{\mathscr{B}_0} \mathbf{b}\cdot\hat{\mathbf{v}}dV \qquad \forall\hat{\mathbf{v}}\in \mathrm{Var}_v, \tag{11.45}$$

where, from (11.5), we recall that

$$\dot{\mathbf{v}} = \frac{\partial\mathbf{v}}{\partial t} + (\nabla\mathbf{v})\mathbf{v}, \tag{11.46}$$

where the first term is the contribution to the accelerations given by time-dependent processes and the second term incorporates the convective accelerations, necessary in this Eulerian framework to properly account for the variation of the velocity field across space.

Observe that there exists a complete analogy between the development of the PVP for fluid media treated here and the PVP for solid media presented in Chapter 3. Actually, both PVP are exactly the same, but for the incompressibility constraint, and the differences arise in two fundamental aspects. First, while the kinematics of solid particles is described using a Lagrangian framework, the kinematics of fluid particles is described using an Eulerian framework. This gives rise to a different functional form for the acceleration terms. Second, the constitutive behavior is essentially different in nature, which is the consequence of different constitution of the matter in both types of media. While in solid media we have relations for the stress and the (history of) strain (and eventually strain rate), in simple fluid media the stress is related to the strain rate.

In order to find the Euler–Lagrange equations associated with the variational problem (11.45) we assume, once again, that the fields are regular enough so that the integration by parts procedure is mathematically well posed. Thus, we have

$$\begin{aligned}&\int_{\mathscr{B}_0} \left[\rho\dot{\mathbf{v}}\cdot\hat{\mathbf{v}} - (\operatorname{div}\sigma^D)\cdot\hat{\mathbf{v}}\right] dV + \int_{\partial\mathscr{B}_{0t}} (\sigma^D\mathbf{n})\cdot\hat{\mathbf{v}}dS \\ &= \int_{\partial\mathscr{B}_{0t}} \mathbf{t}\cdot\hat{\mathbf{v}}dS + \int_{\mathscr{B}_0} \mathbf{b}\cdot\hat{\mathbf{v}}dV \qquad \forall\hat{\mathbf{v}}\in\mathrm{Var}_v,\end{aligned} \tag{11.47}$$

and rearranging terms gives

$$\int_{\mathscr{B}_0} \left[\rho\dot{\mathbf{v}} - \operatorname{div}\sigma^D - \mathbf{b}\right]\cdot\hat{\mathbf{v}}dV + \int_{\partial\mathscr{B}_{0t}} \left[\sigma^D\mathbf{n} - \mathbf{t}\right]\cdot\hat{\mathbf{v}}dS = 0 \qquad \forall\hat{\mathbf{v}}\in\mathrm{Var}_v. \tag{11.48}$$

The identification of the Euler–Lagrange equations in this case requires caution by virtue of the peculiarity of the kinematical constraint, introduced by the incompressibility assumption, and which is present in the space Var_v. Let us first consider $\mathrm{Var}_v^0 \subset \mathrm{Var}_v$ defined by

$$\mathrm{Var}_v^0 = \{\mathbf{v}\in\mathscr{V}\ ;\ \mathbf{v}|_{\partial\mathscr{B}_0} = \mathbf{0},\ \operatorname{div}\mathbf{v} = 0\ \ \forall\mathbf{x}\in\mathscr{B}_0\}. \tag{11.49}$$

Therefore, from (11.48) we have

$$\int_{\mathscr{B}_0} \left[\rho\dot{\mathbf{v}} - \operatorname{div}\sigma^D - \mathbf{b}\right]\cdot\hat{\mathbf{v}}dV = 0 \qquad \forall\hat{\mathbf{v}}\in\mathrm{Var}_v^0. \tag{11.50}$$

This implies that the first element between brackets belongs to the orthogonal complement of Var_v^0, that is

$$\rho\dot{\mathbf{v}} - \operatorname{div}\sigma^D - \mathbf{b} \in (\mathrm{Var}_v^0)^\perp. \tag{11.51}$$

This means that this element is a vector field written as the gradient of a scalar field, say $-p$, that is

$$\rho\dot{\mathbf{v}} - \operatorname{div}\sigma^D - \mathbf{b} = -\nabla p \qquad \text{in } \mathscr{B}_0. \tag{11.52}$$

As a matter of fact, to understand this conclusion let us take the field $\mathbf{w} \in \mathrm{Var}_v^0$ and consider also a vector field $\mathbf{s}$. Now, let us write

$$\int_{\mathscr{B}_0} \mathbf{s} \cdot \mathbf{w} dV = 0 \qquad \forall \mathbf{w} \in \mathrm{Var}_v^0. \tag{11.53}$$

Putting $\mathbf{s}$ as the gradient of a scalar field, that is, $\mathbf{s} = \nabla\phi$, we have

$$\int_{\mathscr{B}_0} \nabla\phi \cdot \mathbf{w} dV = 0 \qquad \forall \mathbf{w} \in \mathrm{Var}_v^0, \tag{11.54}$$

and integrating by parts (assuming that the fields are regular enough) gives

$$-\int_{\mathscr{B}_0} \phi \operatorname{div} \mathbf{w} dV + \int_{\partial\mathscr{B}_0} \phi \mathbf{n} \cdot \mathbf{w} dS = -\int_{\mathscr{B}_0} \phi \operatorname{div} \mathbf{w} dV = 0 \qquad \forall \mathbf{w} \in \mathrm{Var}_v^0, \tag{11.55}$$

where we have used the fact that $\mathbf{w} = \mathbf{0}$ on $\partial\mathscr{B}_0$. This identity is verified trivially because $\mathbf{w} \in \mathrm{Var}_v^0$. Then we conclude that the vector fields that originate as the gradient of a scalar field characterize the orthogonal complement to Var_v^0. That is, for $\mathbf{s} = \nabla\phi$, then $\mathbf{s} \in (\mathrm{Var}_v^0)^\perp$. In this way, we confirm that by using (11.52) we are implicitly verifying the orthogonality condition (11.51). An extended discussion about this topic with a more rigorous mathematical analysis can be found in [120, 290].

With the result found in (11.52) we go back to expression (11.48), yielding

$$-\int_{\mathscr{B}_0} \nabla p \cdot \hat{\mathbf{v}} dV + \int_{\partial\mathscr{B}_{0t}} (\sigma^D \mathbf{n} - \mathbf{t}) \cdot \hat{\mathbf{v}} dS = 0 \qquad \forall \hat{\mathbf{v}} \in \mathrm{Var}_v. \tag{11.56}$$

Integrating by parts the first term (assuming fields are regular enough), we get

$$\int_{\mathscr{B}_0} p \operatorname{div} \hat{\mathbf{v}} dV - \int_{\partial\mathscr{B}_{0t}} p\mathbf{n} \cdot \hat{\mathbf{v}} dV + \int_{\partial\mathscr{B}_{0t}} (\sigma^D \mathbf{n} - \mathbf{t}) \cdot \hat{\mathbf{v}} dS = 0 \qquad \forall \hat{\mathbf{v}} \in \mathrm{Var}_v. \tag{11.57}$$

Using the fact that $\operatorname{div}\hat{\mathbf{v}} = 0$ for fields in Var_v, and rearranging terms we obtain

$$\int_{\partial\mathscr{B}_{0t}} \big((-p\mathbf{I} + \sigma^D)\mathbf{n} - \mathbf{t}\big) \cdot \hat{\mathbf{v}} dS = 0 \qquad \forall \hat{\mathbf{v}} \in \mathrm{Var}_v. \tag{11.58}$$

This equation leads us to the conclusion that

$$-p\mathbf{n} + \sigma^D \mathbf{n} - \mathbf{t} = \mathbf{0} \qquad \text{on } \partial\mathscr{B}_{0t}. \tag{11.59}$$

Finally, we arrive at the following Euler–Lagrange equations

$$\rho\dot{\mathbf{v}} + \nabla p - \operatorname{div}\sigma^D = \mathbf{b} \qquad \text{in } \mathscr{B}_0, \tag{11.60}$$

$$-p\mathbf{n} + \sigma^D \mathbf{n} = \mathbf{t} \qquad \text{on } \partial\mathscr{B}_{0t}, \tag{11.61}$$

which, together with the kinematical constraints

$$\operatorname{div}\mathbf{v} = 0 \qquad \text{in } \mathscr{B}_0, \tag{11.62}$$

$$\mathbf{v} = \bar{\mathbf{v}} \qquad \text{on } \partial\mathscr{B}_{0v}, \tag{11.63}$$

allow us to write the PVP (11.45) in the local, or strong, form.

Notice that a scalar field p emerges in the equilibrium equations. This scalar field is called the mechanical pressure, or simply the fluid pressure. In particular, the pressure field contributes to the complete characterization of the Cauchy stress tensor through the hydrostatic component σ^H. Indeed, we can write

$$\sigma = -p\mathbf{I} + \sigma^D, \tag{11.64}$$

and in this manner equations (11.60) and (11.61) take the following format, which resembles that from solid mechanics

$$\rho\dot{\mathbf{v}} - \operatorname{div}\sigma = \mathbf{b} \qquad \text{in } \mathscr{B}_0, \tag{11.65}$$

$$\sigma\mathbf{n} = \mathbf{t} \qquad \text{on } \partial\mathscr{B}_{0t}. \tag{11.66}$$

Even having provided an explicit characterization for the orthogonal complement to the space of divergence-free vector fields, we will now unveil the nature of the reactive force $r \in \mathscr{V}'$. To achieve this, we must remove all the kinematical constraints from the space Var_v, that is

$$\langle f, \hat{\mathbf{v}}\rangle = \langle \mathscr{D}^*\sigma^D, \hat{\mathbf{v}}\rangle - \langle r, \hat{\mathbf{v}}\rangle \qquad \forall \hat{\mathbf{v}} \in \mathscr{V}, \tag{11.67}$$

and employing the PVP through expression (11.45) we have

$$\langle r, \hat{\mathbf{v}}\rangle = \int_{\mathscr{B}_0} \left[\rho\dot{\mathbf{v}}\cdot\hat{\mathbf{v}} + \sigma^D\cdot(\nabla\hat{\mathbf{v}})^s\right] dV - \int_{\partial\mathscr{B}_{0t}} \mathbf{t}\cdot\hat{\mathbf{v}}dS - \int_{\mathscr{B}_0} \mathbf{b}\cdot\hat{\mathbf{v}}dV \quad \forall\hat{\mathbf{v}} \in \mathscr{V}, \tag{11.68}$$

which, after integration by parts, yields

$$\langle r, \hat{\mathbf{v}}\rangle = \int_{\mathscr{B}_0} \left[\rho\dot{\mathbf{v}} - \operatorname{div}\sigma^D - \mathbf{b}\right]\cdot\hat{\mathbf{v}}dV + \int_{\partial\mathscr{B}_{0t}} \left[\sigma^D\mathbf{n} - \mathbf{t}\right]\cdot\hat{\mathbf{v}}dS + \int_{\partial\mathscr{B}_{0v}} (\sigma^D\mathbf{n})\cdot\hat{\mathbf{v}}dS \quad \forall\hat{\mathbf{v}} \in \mathscr{V}. \tag{11.69}$$

Therefore, by using (11.60) and (11.61) we arrive at the following

$$\langle r, \hat{\mathbf{v}}\rangle = -\int_{\mathscr{B}_0} \nabla p\cdot\hat{\mathbf{v}}dV + \int_{\partial\mathscr{B}_{0t}} p\mathbf{n}\cdot\hat{\mathbf{v}}dS + \int_{\partial\mathscr{B}_{0v}} (\sigma^D\mathbf{n})\cdot\hat{\mathbf{v}}dS \qquad \forall\hat{\mathbf{v}} \in \mathscr{V}. \tag{11.70}$$

Now, integrating by parts the first term on the right-hand side gives

$$\langle r, \hat{\mathbf{v}}\rangle = \int_{\mathscr{B}_0} p\operatorname{div}\hat{\mathbf{v}}dV + \int_{\partial\mathscr{B}_{0v}} \left[-p\mathbf{n} + \sigma^D\mathbf{n}\right]\cdot\hat{\mathbf{v}}dS \qquad \forall\hat{\mathbf{v}} \in \mathscr{V}. \tag{11.71}$$

Clearly, the reactive force r is fully described by two elements in correspondence with the kinematical constraints, that of incompressibility $\operatorname{div}\mathbf{v} = 0$ all over $\mathscr{B}_0$ and that of prescribed velocity over the portion of the boundary $\partial\mathscr{B}_{0v}$. If we name these two elements r_1 (a scalar field defined in $\mathscr{B}_0$) and $\mathbf{r}_2$ (a vector field defined on $\partial\mathscr{B}_{0v}$), respectively, we have

$$\langle r, \hat{\mathbf{v}}\rangle = \int_{\mathscr{B}_0} r_1\operatorname{div}\hat{\mathbf{v}}dV + \int_{\partial\mathscr{B}_{0v}} \mathbf{r}_2\cdot\hat{\mathbf{v}}dS. \tag{11.72}$$

Comparing (11.72) with (11.71), we arrive at the local form which characterizes the reactive forces r_1 and $\mathbf{r}_2$, which gives

$$r_1 = p \qquad \text{in } \mathscr{B}_0, \tag{11.73}$$

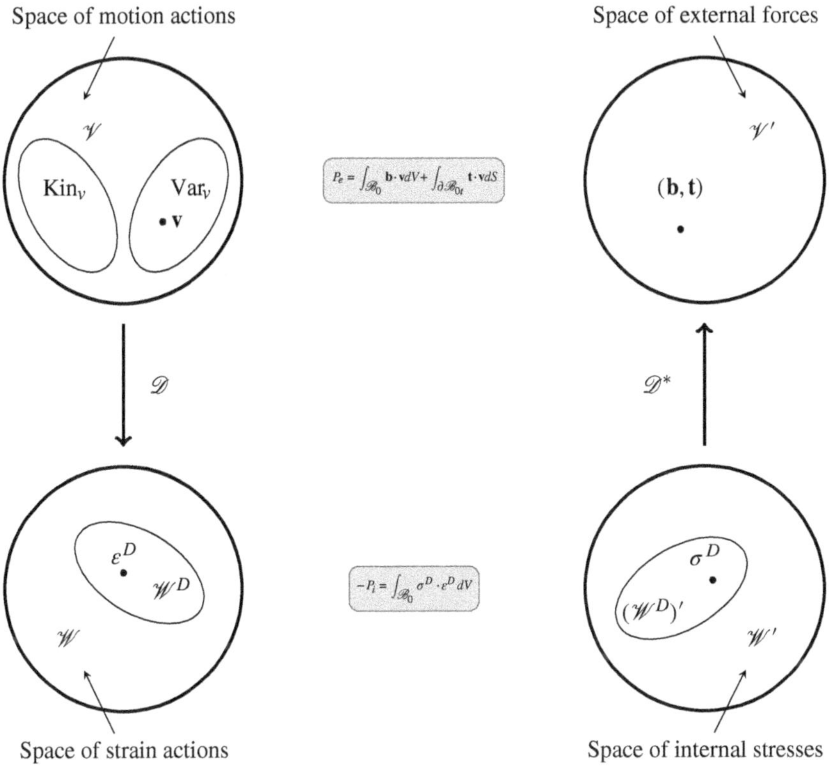

Figure 11.2 Ingredients in the variational theory for the modeling of incompressible fluid flow.

$$\mathbf{r}_2 = -p\mathbf{n} + \sigma^D\mathbf{n} \qquad \text{on } \partial\mathscr{B}_{0v}. \tag{11.74}$$

The reactive scalar field r_1 is the pressure field. That is, the existence of the pressure field in this case was not a hypothesis for the model, but it came to the surface as a natural consequence of the incompressibility constraint. Thus, in the current version of the variational problem for fluid flow phenomena, the pressure cannot be characterized through the equilibrium, but only after the equilibrium was found and the constraints were removed.

Finaly, all the ingredients that play a role in the variational formulation for the incompressible fluid flow problem are displayed in Figure 11.2.

11.4 Navier–Stokes Equations

Let us consider in this section the particular case in which the fluid features a Newtonian constitutive behavior. This behavior is characterized by a linear relation between the deviatoric component of the Cauchy stress tensor and the strain rate as

$$\sigma^D = 2\mu\varepsilon^D = \mu(\nabla\mathbf{v} + \nabla\mathbf{v}^T)^D, \tag{11.75}$$

where μ is the dynamic viscosity of the fluid and is a strictly positive property of the fluid medium, that is, $\mu \geq \mu_o > 0$. Since the velocity field is divergence free, it is equivalent to write

$$\sigma^D = \mu(\nabla\mathbf{v} + \nabla\mathbf{v}^T). \tag{11.76}$$

Applying the divergence operation to the deviatoric component of the Cauchy stress tensor σ^D given by (11.76), and assuming that μ is a uniform property in the whole domain of analysis, gives

$$\operatorname{div}\sigma^D = \operatorname{div}[\mu(\nabla\mathbf{v} + \nabla\mathbf{v}^T)] = \mu\operatorname{div}(\nabla\mathbf{v}) + \mu\operatorname{div}(\nabla\mathbf{v}^T) = \mu\Delta\mathbf{v}, \tag{11.77}$$

where we have used the fact that $\operatorname{div}(\nabla\mathbf{v}^T) = \nabla(\operatorname{div}\mathbf{v}) = \mathbf{0}$ (see Chapter 2).

Considering the constitutive behavior given by (11.76), particularly expression (11.77), and taking into account the definition of the material derivative (11.46), from the Euler–Lagrange equations (11.60) and (11.61), and considering the kinematical constraints (11.62) and (11.63), we arrive at the following system of equations

$$\rho\frac{\partial\mathbf{v}}{\partial t} + \rho(\nabla\mathbf{v})\mathbf{v} + \nabla p - \mu\Delta\mathbf{v} = \mathbf{b} \qquad \text{in } \mathscr{B}_0, \tag{11.78}$$

$$\operatorname{div}\mathbf{v} = 0 \qquad \text{in } \mathscr{B}_0, \tag{11.79}$$

$$-p\mathbf{n} + \mu(\nabla\mathbf{v} + (\nabla\mathbf{v})^T)\mathbf{n} = \mathbf{t} \qquad \text{on } \partial\mathscr{B}_{0t}, \tag{11.80}$$

$$\mathbf{v} = \overline{\mathbf{v}} \qquad \text{on } \partial\mathscr{B}_{0v}. \tag{11.81}$$

This system of equations describes the local (or strong) form of the problem involving the incompressible flow of a homogeneous Newtonian fluid. These equations are known in the literature as the Navier–Stokes equations.

The typical approach to derive these equations is to make use of the so-called conservation laws. In particular, the differential form of the conservation law for the mass yields (11.79), while the differential form of the conservation law for the momentum results in (11.78). Differently, we have arrived at these equations employing a variational approach, in total analogy to the governing equations in solid mechanics. As stated, the differences lie in two fundamental aspects: the Eulerian description utilized for fluid flow problems and the constitutive behavior. A subtle difference appeared by virtue of the incompressibility constraint introduced in the analysis of fluid problems (which was not considered in the study of solid mechanics). However, this small difference vanishes if, in the solid mechanics domain, we also consider the hypothesis of incompressible kinematics.

Before closing this section let us particularize the PVP given by (11.45) for the case of Newtonian fluids as follows

$$\int_{\mathscr{B}_0}\left[\rho\frac{\partial\mathbf{v}}{\partial t}\cdot\hat{\mathbf{v}} + \rho(\nabla\mathbf{v})\mathbf{v}\cdot\hat{\mathbf{v}} + \mu(\nabla\mathbf{v} + (\nabla\mathbf{v})^T)\cdot\nabla\hat{\mathbf{v}}\right]dV$$
$$= \int_{\partial\mathscr{B}_{0t}}\mathbf{t}\cdot\hat{\mathbf{v}}dS + \int_{\mathscr{B}_0}\mathbf{b}\cdot\hat{\mathbf{v}}dV \qquad \forall\hat{\mathbf{v}}\in\mathrm{Var}_v. \tag{11.82}$$

As we have seen in (11.73), the fluid pressure possesses the physical meaning of being the force which is a reaction to the imposition of the divergence-free motion actions. Therefore, we can relax this constraint by removing it from the linear manifold Kin_v and consequently from the linear space Var_v, and we can force this constraint through a Lagrange multiplier which, evidently, becomes the pressure field p. To this, consider

$$\mathrm{Kin}_v^* = \{\mathbf{v}\in\mathscr{V}\ ;\ \mathbf{v}|_{\partial\mathscr{B}_{0v}} = \overline{\mathbf{v}}\}, \tag{11.83}$$

$$\mathrm{Var}_v^* = \{\mathbf{v} \in \mathscr{V}\,;\ \mathbf{v}|_{\partial\mathscr{B}_{0v}} = \mathbf{0}\}. \tag{11.84}$$

Then, we must add to the left-hand side of (11.82) the following terms which contribute to the power exerted in duality by the pair given by the Lagrange multiplier (the pressure p, previously a reactive force) and the corresponding incompressibility constraint

$$-\int_{\mathscr{B}_0} \left[p\,\mathrm{div}\,\hat{\mathbf{v}} + \hat{p}\,\mathrm{div}\,\mathbf{v}\right] dV, \tag{11.85}$$

where p and $\hat{p}$ belong to a space of functions, say $\mathscr{P}$, which are regular enough for mathematical operations to make sense. In this case, such a space is the space of square integrable scalar functions, that is, $\mathscr{P} = L^2(\mathscr{B}_0)$. As a result, from (11.82) we are led to the variational problem which consists of finding $(\mathbf{v}, p) \in \mathrm{Kin}_v^* \times \mathscr{P}$ such that

$$\int_{\mathscr{B}_0} \left[\rho\frac{\partial \mathbf{v}}{\partial t}\cdot\hat{\mathbf{v}} + \rho(\nabla\mathbf{v})\mathbf{v}\cdot\hat{\mathbf{v}} + \mu(\nabla\mathbf{v} + (\nabla\mathbf{v})^T)\cdot\nabla\hat{\mathbf{v}} - p\,\mathrm{div}\,\hat{\mathbf{v}} - \hat{p}\,\mathrm{div}\,\mathbf{v}\right] dV$$
$$= \int_{\partial\mathscr{B}_{0t}} \mathbf{t}\cdot\hat{\mathbf{v}}dS + \int_{\mathscr{B}_0} \mathbf{b}\cdot\hat{\mathbf{v}}dV \qquad \forall(\hat{\mathbf{v}},\hat{p}) \in \mathrm{Var}_v^* \times \mathscr{P}. \tag{11.86}$$

It can readily be verified that the Euler–Lagrange equations associated with the variational formulation (11.86) are those given by (11.78)–(11.81).

11.5 Stokes Flow

A particular situation that usually appears in several fluid mechanics problems is that in which the Navier–Stokes equations can be substantially simplified because of the dominance of shear stress diffusion processes over convective accelerations. Such cases are characterized by the so-called Reynolds number, defined by

$$\mathrm{Re} = \frac{\rho V L}{\mu}, \tag{11.87}$$

where V is a characteristic fluid velocity in the problem and L is a characteristic length in the problem. When $\mathrm{Re} \ll 1$ convective phenomena are negligible compared to the contribution of shear stresses caused by viscosity, and, therefore, the acceleration forces can be neglected in the variational equation (11.86), or, equivalently, in the balance of forces given by the local form (11.78). In such cases, the variational formulation (11.86) consists of finding $(\mathbf{v}, p) \in \mathrm{Kin}_v^* \times \mathscr{P}$ such that

$$\int_{\mathscr{B}_0} \left[\mu(\nabla\mathbf{v} + (\nabla\mathbf{v})^T)\cdot\nabla\hat{\mathbf{v}} - p\,\mathrm{div}\,\hat{\mathbf{v}} - \hat{p}\,\mathrm{div}\,\mathbf{v}\right] dV$$
$$= \int_{\partial\mathscr{B}_{0t}} \mathbf{t}\cdot\hat{\mathbf{v}}dS + \int_{\mathscr{B}_0} \mathbf{b}\cdot\hat{\mathbf{v}}dV \quad \forall(\hat{\mathbf{v}},\hat{p}) \in \mathrm{Var}_v^* \times \mathscr{P}. \tag{11.88}$$

Analogously, we can work with Kin_v and Var_v where the incompressibility constraint has been included in the definition of these sets. Hence, the problem (11.88) implies finding $\mathbf{v} \in \mathrm{Kin}_v$ such that

$$\int_{\mathscr{B}_0} \mu(\nabla\mathbf{v} + (\nabla\mathbf{v})^T)\cdot\nabla\hat{\mathbf{v}}dV = \int_{\partial\mathscr{B}_{0t}} \mathbf{t}\cdot\hat{\mathbf{v}}dS + \int_{\mathscr{B}_0} \mathbf{b}\cdot\hat{\mathbf{v}}dV \qquad \forall\hat{\mathbf{v}} \in \mathrm{Var}_v. \tag{11.89}$$

Clearly, the Euler–Lagrange equations associated with problem (11.88) (or equivalently with (11.89)) are

$$\nabla p - \mu \Delta \mathbf{v} = \mathbf{b} \qquad \text{in } \mathscr{B}_0, \tag{11.90}$$

$$\operatorname{div} \mathbf{v} = 0 \qquad \text{in } \mathscr{B}_0, \tag{11.91}$$

$$-p\mathbf{n} + \mu(\nabla \mathbf{v} + (\nabla \mathbf{v})^T)\mathbf{n} = \mathbf{t} \qquad \text{on } \partial\mathscr{B}_{0t}, \tag{11.92}$$

$$\mathbf{v} = \overline{\mathbf{v}} \qquad \text{on } \partial\mathscr{B}_{0v}. \tag{11.93}$$

In particular, the variational problem (11.89) can be understood as the necessary (and in this case sufficient) condition for a certain cost functional (to be proposed next) to reach a minimum value. We will proceed with the case of Newtonian fluids, but the procedure is the same for more general situations. Note that the constitutive law (11.75) can be written as

$$\overline{\sigma}^D = 2\mu\overline{\varepsilon}^D = \left.\frac{\partial \pi}{\partial \varepsilon}\right|_{\varepsilon=\overline{\varepsilon}^D}, \tag{11.94}$$

where π is a constitutive functional given, for the Newtonian case, by

$$\pi(\varepsilon) = \mu\varepsilon \cdot \varepsilon = \mu|\varepsilon|^2, \tag{11.95}$$

which is a quadratic functional in the argument ε, and therefore, as happened in the case of linear elasticity in Chapter 4, it verifies

$$\begin{aligned}\pi(\varepsilon) &= \pi(\varepsilon_0) + \left.\frac{\partial \pi}{\partial \varepsilon}\right|_{\varepsilon_0} \cdot (\varepsilon - \varepsilon_0) + \frac{1}{2}\left.\frac{\partial^2 \pi}{\partial \varepsilon^2}\right|_{\varepsilon_0} (\varepsilon - \varepsilon_0) \cdot (\varepsilon - \varepsilon_0) \\ &= \pi(\varepsilon_0) + 2\mu\varepsilon_0 \cdot (\varepsilon - \varepsilon_0) + \mu(\varepsilon - \varepsilon_0) \cdot (\varepsilon - \varepsilon_0),\end{aligned} \tag{11.96}$$

and so we obtain

$$\pi(\varepsilon) - \pi(\varepsilon_0) \geq 2\mu\varepsilon_0 \cdot (\varepsilon - \varepsilon_0), \tag{11.97}$$

where the equality is verified for $\varepsilon = \varepsilon_0$. For compatible strain rates, that is, $\varepsilon = (\nabla \mathbf{v})^s$ and $\varepsilon_0 = (\nabla \mathbf{v}_0)^s$, the expression above is equivalent to

$$\pi((\nabla \mathbf{v})^s) - \pi((\nabla \mathbf{v}_0)^s) \geq 2\mu(\nabla \mathbf{v}_0)^s \cdot ((\nabla \mathbf{v})^s - (\nabla \mathbf{v}_0)^s). \tag{11.98}$$

Integrating now over the whole domain of analysis $\mathscr{B}_0$ yields

$$\int_{\mathscr{B}_0} \left(\pi((\nabla \mathbf{v})^s) - \pi((\nabla \mathbf{v}_0)^s)\right) dV \geq \int_{\mathscr{B}_0} 2\mu(\nabla \mathbf{v}_0)^s \cdot ((\nabla \mathbf{v})^s - (\nabla \mathbf{v}_0)^s) dV \quad \forall \mathbf{v} \in \text{Kin}_v. \tag{11.99}$$

The right-hand side in the expression above is nothing but the first term in the variational equation (11.89), which gives

$$\begin{aligned}&\int_{\mathscr{B}_0} \left(\pi((\nabla \mathbf{v})^s) - \pi((\nabla \mathbf{v}_0)^s)\right) dV \\ &\geq \int_{\partial\mathscr{B}_{0t}} \mathbf{t} \cdot (\mathbf{v} - \mathbf{v}_0) dS + \int_{\mathscr{B}_0} \mathbf{b} \cdot (\mathbf{v} - \mathbf{v}_0) dV \quad \forall \mathbf{v} \in \text{Kin}_v.\end{aligned} \tag{11.100}$$

Now, arranging terms on both sides of the inequality gives

$$\begin{aligned}\Pi(\mathbf{v}) &= \int_{\mathscr{B}_0} \pi((\nabla\mathbf{v})^s)dV - \int_{\partial\mathscr{B}_{0t}} \mathbf{t}\cdot\mathbf{v}dS - \int_{\mathscr{B}_0} \mathbf{b}\cdot\mathbf{v}dV \\ &\geq \int_{\mathscr{B}_0} \pi((\nabla\mathbf{v}_0)^s)dV - \int_{\partial\mathscr{B}_{0t}} \mathbf{t}\cdot\mathbf{v}_0dS - \int_{\mathscr{B}_0} \mathbf{b}\cdot\mathbf{v}_0dV \\ &= \Pi(\mathbf{v}_0) \quad \forall\mathbf{v}\in \mathrm{Kin}_v.\end{aligned} \tag{11.101}$$

The equality above is verified if $\mathbf{v}$ and $\mathbf{v}_0$ are the same up to a rigid motion. In this manner, we have arrived at the formulation of a minimum principle equivalent to the PVP given by (11.89). In analogy to the solid mechanics domain, this minimum principle is called the Principle of Minimum Total Potential Energy, and consists of finding $\mathbf{v}_0 \in \mathrm{Kin}_v$ such that the Total Potential Energy functional is minimized, that is

$$\Pi(\mathbf{v}_0) = \min_{\mathbf{v}\in\mathrm{Kin}_v} \Pi(\mathbf{v}) = \min_{\mathbf{v}\in\mathrm{Kin}_v}\left[\int_{\mathscr{B}_0} \mu|(\nabla\mathbf{v})^s|^2 dV - \int_{\partial\mathscr{B}_{0t}} \mathbf{t}\cdot\mathbf{v}dS - \int_{\mathscr{B}_0} \mathbf{b}\cdot\mathbf{v}dV\right]. \tag{11.102}$$

Alternatively, we can relax the constraint $\mathrm{div}\,\mathbf{v} = 0$ present in Kin_v through a proper Lagrange multiplier. Hence, the minimization problem given by (11.102) turns into the problem of finding a stationary (saddle) point $(\mathbf{v}_0, p_0) \in \mathrm{Kin}_v^* \times \mathscr{P}$ of the following Lagrangian functional

$$\Lambda(\mathbf{v}, p) = \int_{\mathscr{B}_0} \mu|(\nabla\mathbf{v})^s|^2 dV - \int_{\mathscr{B}_0} p\mathrm{div}\,\mathbf{v}dV - \int_{\partial\mathscr{B}_{0t}} \mathbf{t}\cdot\mathbf{v}dS - \int_{\mathscr{B}_0} \mathbf{b}\cdot\mathbf{v}dV. \tag{11.103}$$

It is straightforward to show that the variational formulation (11.88) emerges as the necessary condition satisfied by the stationary point of the functional (11.103).

11.6 Irrotational Flow

Let us incorporate a further constraint of kinematical nature to analyze another class of problem arising in the fluid mechanics realm. With this additional constraint we admit that the velocity field is irrotational, that means that $\mathrm{curl}\mathbf{v} = \mathbf{0}$ in $\mathscr{B}_0$. With this constraint, and knowing that $\mathrm{div}\,\mathbf{v} = 0$, we have that the velocity field is harmonic. Indeed, from Chapter 2 let us recall the identity

$$\Delta\mathbf{v} = \mathrm{curl}\mathrm{curl}\mathbf{v} - \nabla(\mathrm{div}\,\mathbf{v}), \tag{11.104}$$

from which the harmonic property follows directly, that is

$$\Delta\mathbf{v} = \mathbf{0}. \tag{11.105}$$

Let us define the following sets

$$\mathrm{Kin}_v^i = \{\mathbf{v}\in\mathscr{V}\,;\ \mathbf{v}|_{\partial\mathscr{B}_0} = \bar{\mathbf{v}},\ \mathrm{div}\,\mathbf{v} = 0,\ \mathrm{curl}\mathbf{v} = \mathbf{0}\ \ \forall\mathbf{x}\in\mathscr{B}_0\}, \tag{11.106}$$

$$\mathrm{Var}_v^i = \{\mathbf{v}\in\mathscr{V}\,;\ \mathbf{v}|_{\partial\mathscr{B}_0} = \mathbf{0},\ \mathrm{div}\,\mathbf{v} = 0,\ \mathrm{curl}\mathbf{v} = \mathbf{0}\ \ \forall\mathbf{x}\in\mathscr{B}_0\}. \tag{11.107}$$

Therefore, the PVP (11.82) for the particular case of irrotational flow becomes

$$\int_{\mathcal{B}_0} \left[\rho \frac{\partial \mathbf{v}}{\partial t} \cdot \hat{\mathbf{v}} + \rho(\nabla \mathbf{v})\mathbf{v} \cdot \hat{\mathbf{v}} + \mu(\nabla \mathbf{v} + (\nabla \mathbf{v})^T) \cdot \nabla \hat{\mathbf{v}}\right] dV = \int_{\mathcal{B}_0} \mathbf{b} \cdot \hat{\mathbf{v}} dV \quad \forall \hat{\mathbf{v}} \in \mathrm{Var}_v^i. \tag{11.108}$$

To find the Euler-Lagrange equations we relax the incompressible and irrotational kinematical constraints, leading to

$$\int_{\mathcal{B}_0} \left[\rho \frac{\partial \mathbf{v}}{\partial t} \cdot \hat{\mathbf{v}} + \rho(\nabla \mathbf{v})\mathbf{v} \cdot \hat{\mathbf{v}} + \mu(\nabla \mathbf{v} + (\nabla \mathbf{v})^T) \cdot \nabla \hat{\mathbf{v}}\right] dV$$
$$- \int_{\mathcal{B}_0} \left[p \operatorname{div} \hat{\mathbf{v}} + \mathbf{h} \cdot \operatorname{curl} \hat{\mathbf{v}}\right] dV = \int_{\mathcal{B}_0} \mathbf{b} \cdot \hat{\mathbf{v}} dV \quad \forall \hat{\mathbf{v}} \in \mathcal{V}_0, \tag{11.109}$$

where

$$\mathcal{V}_0 = \{\mathbf{v} \in \mathcal{V}\ ;\ \mathbf{v}|_{\partial \mathcal{B}_0} = \mathbf{0}\}. \tag{11.110}$$

Knowing that

$$\int_{\mathcal{B}_0} p \operatorname{div} \hat{\mathbf{v}} dV = \int_{\partial \mathcal{B}_0} p\mathbf{n} \cdot \hat{\mathbf{v}} dS - \int_{\mathcal{B}_0} \nabla p \cdot \hat{\mathbf{v}} dV, \tag{11.111}$$

$$\int_{\mathcal{B}_0} \mathbf{h} \cdot \operatorname{curl} \hat{\mathbf{v}} dV = \int_{\partial \mathcal{B}_0} (\mathbf{n} \times \mathbf{h}) \cdot \hat{\mathbf{v}} dS - \int_{\mathcal{B}_0} (\operatorname{curl} \mathbf{h}) \cdot \hat{\mathbf{v}} dV, \tag{11.112}$$

after integration by parts in (11.109) and exploiting the harmonic property of the velocity field we obtain

$$\int_{\mathcal{B}_0} \left[\rho \frac{\partial \mathbf{v}}{\partial t} + \rho(\nabla \mathbf{v})\mathbf{v} + \nabla p + \operatorname{curl} \mathbf{h}\right] \cdot \hat{\mathbf{v}} dV = \int_{\mathcal{B}_0} \mathbf{b} \cdot \hat{\mathbf{v}} dV \qquad \forall \hat{\mathbf{v}} \in \mathcal{V}_0. \tag{11.113}$$

Hence, the Euler–Lagrange equations together with the kinematical constraints give

$$\rho \frac{\partial \mathbf{v}}{\partial t} + \rho(\nabla \mathbf{v})\mathbf{v} + \nabla p + \operatorname{curl} \mathbf{h} = \mathbf{b} \qquad \text{in } \mathcal{B}_0, \tag{11.114}$$

$$\operatorname{div} \mathbf{v} = 0 \qquad \text{in } \mathcal{B}_0, \tag{11.115}$$

$$\operatorname{curl} \mathbf{v} = \mathbf{0} \qquad \text{in } \mathcal{B}_0, \tag{11.116}$$

$$\mathbf{v} = \overline{\mathbf{v}} \qquad \text{on } \partial \mathcal{B}_0. \tag{11.117}$$

Let us admit that the external force $\mathbf{b}$ derives from a potential function Ψ, that is

$$\mathbf{b} = -\nabla \Psi. \tag{11.118}$$

Since the flow is irrotational it is $\operatorname{curl} \mathbf{v} \times \mathbf{a} = (\nabla \mathbf{v} - (\nabla \mathbf{v})^T)\mathbf{a} = \mathbf{0}$ for arbitrary $\mathbf{a}$, and so $(\nabla \mathbf{v})\mathbf{a} = (\nabla \mathbf{v})^T \mathbf{a}$, and therefore

$$(\nabla \mathbf{v})\mathbf{v} = (\nabla \mathbf{v})^T \mathbf{v} = \frac{1}{2} \nabla(\mathbf{v} \cdot \mathbf{v}) = \frac{1}{2} \nabla |\mathbf{v}|^2. \tag{11.119}$$

Consequently, expression (11.114) gives

$$\rho \frac{\partial \mathbf{v}}{\partial t} + \nabla \left(p + \frac{1}{2} \rho |\mathbf{v}|^2 + \Psi\right) + \operatorname{curl} \mathbf{h} = \mathbf{0} \qquad \text{in } \mathcal{B}_0. \tag{11.120}$$

Moreover, since $\operatorname{curl}\mathbf{v} = \mathbf{0}$, the vector field $\mathbf{v}$ can be written as the gradient of a scalar field, say Φ, that is

$$\mathbf{v} = \nabla\Phi, \tag{11.121}$$

then, expression (11.120) turns into

$$\nabla\left(\rho\frac{\partial\Phi}{\partial t} + p + \frac{1}{2}\rho|\mathbf{v}|^2 + \Psi\right) + \operatorname{curl}\mathbf{h} = \mathbf{0} \qquad \text{in } \mathscr{B}_0. \tag{11.122}$$

From the Helmholtz decomposition we conclude that each of the terms in (11.122) must be independently null. That means

$$\nabla\left(\rho\frac{\partial\Phi}{\partial t} + p + \frac{1}{2}\rho|\mathbf{v}|^2 + \Psi\right) = \mathbf{0} \qquad \text{in } \mathscr{B}_0, \tag{11.123}$$

$$\operatorname{curl}\mathbf{h} = \mathbf{0} \qquad \text{in } \mathscr{B}_0. \tag{11.124}$$

For steady-state problems we get $\frac{\partial\Phi}{\partial t} = 0$, and therefore

$$\nabla\left(p + \frac{1}{2}\rho|\mathbf{v}|^2 + \Psi\right) = \mathbf{0} \qquad \text{in } \mathscr{B}_0. \tag{11.125}$$

Let us introduce the quantity

$$H = p + \frac{1}{2}\rho|\mathbf{v}|^2 + \Psi, \tag{11.126}$$

then (11.125) tells us that, under the flow regime considered here, the quantity H is constant in space, that is

$$H = \text{constant} \quad \text{in } \mathscr{B}_0, \tag{11.127}$$

which implies

$$p + \frac{1}{2}\rho|\mathbf{v}|^2 + \Psi = H_o \quad \text{in } \mathscr{B}_0, \tag{11.128}$$

where H_o is the constant.

Making $\Psi = \rho g h$, where g is the magnitude of the acceleration due to gravity and h is the height measured from a given reference, we find

$$p + \frac{1}{2}\rho|\mathbf{v}|^2 + \rho g h = H_o \quad \text{in } \mathscr{B}_0, \tag{11.129}$$

which is the celebrated Bernoulli equation that holds for any streamline in the irrotational and incompressible velocity field.

The Bernoulli equation enables us to calculate, in flows that satisfy the considered hypotheses, either the value of the pressure or the velocity magnitude for which we must know, correspondingly, the velocity magnitude or the pressure and, in addition, the value of H_o at some reference point in the system.

We are concerned now with the problem of finding $\mathbf{v}$ for the present type of flow. Given the constraints taken for the flow, that is $\operatorname{div}\mathbf{v} = 0$ and $\operatorname{curl}\mathbf{v} = \mathbf{0}$ in $\mathscr{B}_0$, and $\mathbf{v} = \overline{\mathbf{v}}$ on $\partial\mathscr{B}_0$, we have that $\mathbf{v}$ remains uniquely determined. Thus, the constraints over the kinematics provide the characterization of the velocity field for the calculation of p.

In fact, from $\operatorname{curl}\mathbf{v} = \mathbf{0}$ we have that expression (11.121) must be verified, and knowing that $\operatorname{div}\mathbf{v} = 0$ in $\mathscr{B}_0$ this leads to

$$\Delta\Phi = 0 \qquad \text{in } \mathscr{B}_0, \tag{11.130}$$

with the boundary condition

$$\nabla \Phi \cdot \mathbf{n} = \overline{\mathbf{v}} \cdot \mathbf{n} = \overline{v}_n \qquad \text{in } \partial \mathscr{B}_0. \tag{11.131}$$

The reader will readily identify that the boundary value problem described by (11.130) and (11.131) stands for the Euler–Lagrange equations of the variational problem that consists of determining $\Phi \in \mathscr{Z}$ such that

$$\int_{\mathscr{B}_0} \nabla \Phi \cdot \nabla \hat{\Phi} dV = \int_{\partial \mathscr{B}_0} \overline{v}_n \hat{\Phi} dS \qquad \forall \hat{\Phi} \in \mathscr{Z}, \tag{11.132}$$

where, in this case, $\mathscr{Z} = H^1(\mathscr{B}_0)$.

Variational problem (11.132) is known in the literature as the potential formulation for the irrotational and incompressible flow problem. This variational equation allows us to determine the potential Φ which, ultimately, fully characterizes the velocity field $\mathbf{v}$.

Note that problem (11.132) comprises natural (Neumann type) boundary conditions in the whole boundary $\partial \mathscr{B}_0$. This implies that the boundary condition must obey a compatibility condition. In fact, considering in (11.132) that $\hat{\Phi} = \hat{C}$ where $\hat{C}$ is a constant, then

$$\int_{\partial \mathscr{B}_0} \overline{v}_n \hat{C} dS = \hat{C} \int_{\partial \mathscr{B}_0} \overline{v}_n dS = 0, \tag{11.133}$$

which means that $\overline{v}_n$ must verify

$$\int_{\partial \mathscr{B}_0} \overline{v}_n dS = 0. \tag{11.134}$$

However, this condition is already being satisfied by the velocity field by virtue of the incompressibility constraint. In fact, notice that

$$\int_{\mathscr{B}_0} \operatorname{div} \mathbf{v} dV = \int_{\partial \mathscr{B}_0} \overline{\mathbf{v}} \cdot \mathbf{n} dS = \int_{\partial \mathscr{B}_0} \overline{v}_n dS = 0. \tag{11.135}$$

Therefore, we have that the variational problem (11.132) is well posed in this sense, and the solution Φ is defined up to a constant.

12

High-Order Continua

12.1 Introduction

The theory underlying the mechanics of the so-called micromorphic materials was systematically developed in the mid the 20th century by A.C. Eringen and was compiled in [80, 81]. Within the framework of this theory the continuum body is regarded as composed of deformable particles, which are characterized by an inner structure with finite size. The motivation to provide a richer description than that given by classical continuum mechanics is rooted in the possibility of taking into account the mechanical phenomenology occurring at a smaller scale. This can be accomplished, within the context of continuum mechanics, by exploiting high-order theories of continua. By high-order it must be understood that the kinematical description of the particles that form the body possess additional entities which account for high-order motion modes than a simple velocity vector field.

The main drawback of this kind of theory lies in the construction of adequate constitutive laws. From the experimental point of view, the development of experimental settings to quantify the relation between generalized strain measures and generalized internal stresses results in a more than arduous task.

In this chapter we will once again make use of the variational structure provided by the Method of Virtual Power through the formulation of the Principle of Virtual Power (PVP) presented in Chapter 3. Hence, by exploiting duality arguments between generalized strain rates and generalized internal stresses, and between motion actions and external forces, the PVP will allow us to systematically construct the variational formulation for the theory of micromorphic media. As a complementary goal we will analyze two particular cases: the second-order theory and the micropolar theory. Once more, the reader will find similarities between the developments in this chapter and the theory reported in [116] for microstructured materials. However, it will also be possible to appreciate the differences with that theoretical framework, mainly in related to the characterization of external forces which are compatible with the kinematical model.

It is worth stressing that the theory developed here holds for both solid and fluid media, provided they can be cast within the high-order kinematical framework proposed in forthcoming sections. As in previous chapters, this equivalence holds as long

Introduction to the Variational Formulation in Mechanics: Fundamentals and Applications, First Edition.
Edgardo O. Taroco, Pablo J. Blanco and Raúl A. Feijóo.

as the solid body is regarded in its spatial (current or deformed) configuration. Moreover, the reader will be ready to develop, from the ideas exposed here and in analogy to that seen in Chapter 10, high-order heat transfer variational models.

Thus, we will first establish the kinematical description for the continuum and then we will characterize the generalized strain rates that will be involved in the assessment of the internal power, and that will serve to characterize the internal stresses through a proper linear and continuous functional. Similarly, we will deduce the generalized external forces which are compatible with the adopted kinematical model. With these elements and the PVP we will then present the variational equations that govern the problem, as well as the local equilibrium equations (strong formulation in terms of partial differential equations) for the static and dynamic cases. Finally, we will particularize the theory for the cases of micropolar media and second gradient theory.

12.2 Kinematics

The first step for the construction of a variational formulation for micromorphic media consists of the definition of the kinematics for the problem. To this end, let $\mathscr{B}_t$ be the domain occupied by the macro-particles of the continuum at time instant t, and let $\partial\mathscr{B}_t$ be the boundary, whose outward unit normal vector is $\mathbf{n}$. Let us call $\mathscr{B}$ to the domain occupied by particles at a reference time instant t_0. In this context we consider that each macro-particle, denoted by P, is formed by a set of micro-particles.

That is, from the point of view of the internal structure of macro-particles (micro-structure), each macro-particle P is associated with a micro-continuum of finite size, denoted by Ω_μ, which consists of a set of micro-particles, as seen in Figure 12.1. These micro-particles occupy the position $\mathbf{X}' = \mathbf{X} + \mathbf{Y}$, where $\mathbf{X}$ represents the reference position vector of the macro-particle and $\mathbf{Y}$ is the relative position vector of micro-particles in the micro-structure of the macro-particle.

At time t, micro-particles that form each one of the macro-particles occupy the position $\mathscr{X}'(\mathbf{X},\mathbf{Y},t) = \mathscr{X}(\mathbf{X},t) + \mathscr{R}(\mathbf{X},\mathbf{Y},t)$, where $\mathbf{x} = \mathscr{X}(\mathbf{X},t)$ is the position vector of the macro-particle and $\mathbf{y} = \mathscr{R}(\mathbf{X},\mathbf{Y},t)$ is the relative position vector of micro-particles in the micro-scale.

Hence, we see that the motion of particles in the continuum remains characterized by two families of motion mappings, for macro-particles

$$\begin{aligned} \mathscr{X} \,:\, &\mathscr{B} \times [t_0, t_f] \to \mathscr{E}, \\ &(\mathbf{X}, t) \mapsto \mathbf{x} = \mathscr{X}(\mathbf{X}, t), \end{aligned} \tag{12.1}$$

and for micro-particles

$$\begin{aligned} \mathscr{R} \,:\, &\mathscr{B} \times \Omega_\mu \times [t_0, t_f] \to \mathscr{E}, \\ &(\mathbf{X}, \mathbf{Y}, t) \mapsto \mathbf{y} = \mathscr{R}(\mathbf{X}, \mathbf{Y}, t), \end{aligned} \tag{12.2}$$

where $\mathscr{X}$ and $\mathscr{R}$ are, respectively, macro-motion and micro-motion [80].

The first hypothesis is to assume that the mapping $\mathscr{R}$ is linear in the argument $\mathbf{Y}$, that is to say

$$\mathscr{R}(\mathbf{X}, \mathbf{Y}, t) = \Theta(\mathbf{X}, t)\mathbf{Y}, \tag{12.3}$$

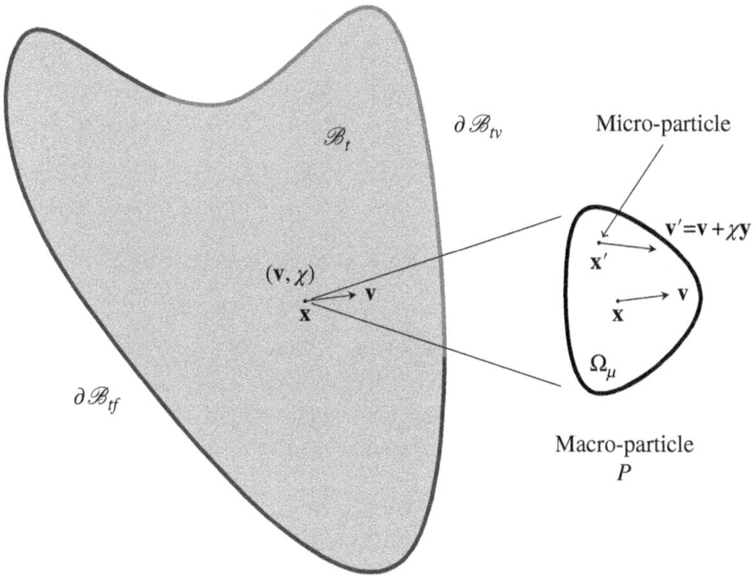

Figure 12.1 Micromorphic continuum and micro-region associated with each macro-particle P in the continuum body. The observable kinematics at the scale of the problem under study is defined by the vector field **v** and by the second-order tensor field χ (see (12.55)).

where Θ is a second-order tensor which represents the internal motion taking place in the macro-particle P. Therefore, the position of micro-particles is fully characterized by

$$\mathbf{x}' = \mathscr{X}'(\mathbf{X}, \mathbf{Y}, t) = \mathbf{x} + \mathbf{y} = \mathscr{X}(\mathbf{X}, t) + \Theta(\mathbf{X}, t)\mathbf{Y}. \tag{12.4}$$

Thus, macro-particles that form the body $\mathscr{B}$, characterized solely by the position vector $\mathbf{X}$ in Chapter 3, are now described by the pair $(\mathbf{X}, \Theta)$. In particular, we consider that $\mathbf{X}$ is the center of mass of a set of micro-particles,[1] being therefore $\mathbf{y} = \mathbf{0}$ for $\mathbf{Y} = \mathbf{0}$.

In the present context, the displacement field is then given by $\mathbf{U}'(\mathbf{X}, \mathbf{Y}, t) = \mathbf{x}' - \mathbf{X}'$, that is

$$\mathbf{U}'(\mathbf{X}, \mathbf{Y}, t) = \mathscr{X}(\mathbf{X}, t) + \Theta(\mathbf{X}, t)\mathbf{Y} - \mathbf{X} - \mathbf{Y}, \tag{12.5}$$

or equivalently

$$\mathbf{x} + \mathbf{y} = \mathbf{X} + \mathbf{Y} + \mathbf{U}'(\mathbf{X}, \mathbf{Y}, t), \tag{12.6}$$

where $\mathbf{U}'$ can be written as

$$\mathbf{U}'(\mathbf{X}, \mathbf{Y}, t) = \mathbf{U}(\mathbf{X}, t) + \mathbf{u}(\mathbf{X}, \mathbf{Y}, t), \tag{12.7}$$

where

$$\mathbf{U}(\mathbf{X}, t) = \mathbf{x} - \mathbf{X} = \mathscr{X}(\mathbf{X}, t), \tag{12.8}$$

$$\mathbf{u}(\mathbf{X}, \mathbf{Y}, t) = \mathbf{y} - \mathbf{Y} = \Theta(\mathbf{X}, t)\mathbf{Y} - \mathbf{Y}. \tag{12.9}$$

1 For such hypothesis it is enough to assume that $\int_{\Omega_\mu} \rho' \mathbf{y} dV_\mu = \mathbf{0}$, where ρ' is the density of the medium comprising the macro-particle P.

In this case, for a sufficiently small neighborhood of $(\mathbf{X}_o, \mathbf{Y}_o) \in \mathscr{B}$, called $\mathcal{S}(\mathbf{X}_o, \mathbf{Y}_o)$, we have that the following expression holds

$$\begin{aligned}\mathcal{X}'(\mathbf{X}, \mathbf{Y}, t) = {} & \mathcal{X}'(\mathbf{X}_o, \mathbf{Y}_o, t) + \nabla_{\mathbf{X}}\mathcal{X}'(\mathbf{X}_o, \mathbf{Y}_o, t)(\mathbf{X} - \mathbf{X}_o)\\ & + \nabla_{\mathbf{Y}}\mathcal{X}'(\mathbf{X}_o, \mathbf{Y}_o, t)(\mathbf{Y} - \mathbf{Y}_o)\\ & + o(\mathbf{X} - \mathbf{X}_o, \mathbf{Y} - \mathbf{Y}_o) \qquad \forall (\mathbf{X}, \mathbf{Y}) \in \mathcal{S}(\mathbf{X}_o, \mathbf{Y}_o),\end{aligned} \tag{12.10}$$

which, for our case, gives

$$\begin{aligned}\mathcal{X}'(\mathbf{X}, \mathbf{Y}, t) = {} & \mathcal{X}'(\mathbf{X}_o, \mathbf{Y}_o, t) + \mathbf{F}(\mathbf{X}_o, t)(\mathbf{X} - \mathbf{X}_o) + \mathsf{H}(\mathbf{X}_o, t)(\mathbf{Y} \otimes (\mathbf{X} - \mathbf{X}_o))\\ & + \boldsymbol{\Theta}(\mathbf{X}_o, t)(\mathbf{Y} - \mathbf{Y}_o) + o(\mathbf{X} - \mathbf{X}_o, \mathbf{Y} - \mathbf{Y}_o) \qquad \forall (\mathbf{X}, \mathbf{Y}) \in \mathcal{S}(\mathbf{X}_o, \mathbf{Y}_o),\end{aligned} \tag{12.11}$$

where $\mathbf{F} = \nabla_{\mathbf{X}}\mathcal{X}$ and $\mathsf{H} = \nabla_{\mathbf{X}}\boldsymbol{\Theta}$, and also the operation $\nabla_{\mathbf{X}}\boldsymbol{\Theta}(\mathbf{X}, t)(\mathbf{Y} \otimes (\mathbf{X} - \mathbf{X}_o))$ must be understood in the following sense: $(\mathbf{a}_1 \otimes \mathbf{a}_2 \otimes \mathbf{a}_3)(\mathbf{b} \otimes \mathbf{c}) = (\mathbf{b} \cdot \mathbf{a}_2)(\mathbf{c} \cdot \mathbf{a}_3)\mathbf{a}_1$. With the elements $\mathbf{F}$, H and $\boldsymbol{\Theta}$, we can now seek the strain rates associated with this class of continuum. Let us put $(\mathbf{X}, \mathbf{Y}) = (\mathbf{X}_o + d\mathbf{X}, \mathbf{Y}_o + d\mathbf{Y})$ into (12.11), then

$$d\mathbf{x}' = d\mathbf{x} + d\mathbf{y} = \mathbf{F}d\mathbf{X} + \mathsf{H}(\mathbf{Y} \otimes d\mathbf{X}) + \boldsymbol{\Theta}d\mathbf{Y}. \tag{12.12}$$

Defining $(\cdot)^{\mathsf{t}}$ as the operation satisfying $(\mathbf{a}_1 \otimes \mathbf{a}_2 \otimes \mathbf{a}_3)^{\mathsf{t}} = \mathbf{a}_1 \otimes \mathbf{a}_3 \otimes \mathbf{a}_2$ (see Chapter 1), we can write

$$d\mathbf{x}' = d\mathbf{x} + d\mathbf{y} = (\mathbf{F} + \mathsf{H}^{\mathsf{t}}\mathbf{Y})d\mathbf{X} + \boldsymbol{\Theta}d\mathbf{Y}. \tag{12.13}$$

For a single fiber, its squared length gives

$$\begin{aligned}d\mathbf{x}' \cdot d\mathbf{x}' & = \left[(\mathbf{F} + \mathsf{H}^{\mathsf{t}}\mathbf{Y})d\mathbf{X} + \boldsymbol{\Theta}d\mathbf{Y}\right] \cdot \left[(\mathbf{F} + \mathsf{H}^{\mathsf{t}}\mathbf{Y})d\mathbf{X} + \boldsymbol{\Theta}d\mathbf{Y}\right]\\ & = (\mathbf{F} + \mathsf{H}^{\mathsf{t}}\mathbf{Y})^T(\mathbf{F} + \mathsf{H}^{\mathsf{t}}\mathbf{Y})d\mathbf{X} \cdot d\mathbf{X} + 2\boldsymbol{\Theta}^T(\mathbf{F} + \mathsf{H}^{\mathsf{t}}\mathbf{Y})d\mathbf{X} \cdot d\mathbf{Y} + \boldsymbol{\Theta}^T\boldsymbol{\Theta}d\mathbf{Y} \cdot d\mathbf{Y}.\end{aligned} \tag{12.14}$$

Therefore, the strain measures for a differential element are given by

$$\begin{aligned}& d\mathbf{x}' \cdot d\mathbf{x}' - [d\mathbf{X} \cdot d\mathbf{X} + 2d\mathbf{Y} \cdot d\mathbf{X} + d\mathbf{Y} \cdot d\mathbf{Y}]\\ & \quad = 2\mathbf{A}_1 d\mathbf{X} \cdot d\mathbf{X} + 2\mathbf{A}_2 d\mathbf{X} \cdot d\mathbf{Y} + 2\mathbf{A}_3 d\mathbf{Y} \cdot d\mathbf{Y},\end{aligned} \tag{12.15}$$

with

$$\mathbf{A}_1 = \frac{1}{2}\left[(\mathbf{F} + \mathsf{H}^{\mathsf{t}}\mathbf{Y})^T(\mathbf{F} + \mathsf{H}^{\mathsf{t}}\mathbf{Y}) - \mathbf{I}\right], \tag{12.16}$$

$$\mathbf{A}_2 = \boldsymbol{\Theta}^T(\mathbf{F} + \mathsf{H}^{\mathsf{t}}\mathbf{Y}) - \mathbf{I}, \tag{12.17}$$

$$\mathbf{A}_3 = \frac{1}{2}\left[\boldsymbol{\Theta}^T\boldsymbol{\Theta} - \mathbf{I}\right]. \tag{12.18}$$

These tensors enable us to define feasible strain measures. For example, a viable set of strain measures is

$$\mathsf{C} = \mathbf{F}^T\mathbf{F}, \tag{12.19}$$

$$\mathbf{K} = \mathbf{F}^T\boldsymbol{\Theta}, \tag{12.20}$$

$$\mathsf{G}^{\mathsf{t}}\mathbf{a} = \mathbf{F}^T(\mathsf{H}^{\mathsf{t}}\mathbf{a}) \quad \forall \mathbf{a}. \tag{12.21}$$

The third-order tensor G can be explicitly written as

$$\mathsf{G} = \left[\mathsf{H}^T \circ \mathbf{F}\right]^{\frac{1}{T}}, \tag{12.22}$$

where $\mathsf{H}^T \circ \mathbf{F}$ is an operation that must be understood in the sense $(\mathbf{a}_1 \otimes \mathbf{a}_2 \otimes \mathbf{a}_3)\circ(\mathbf{b}_1 \otimes \mathbf{b}_2) = (\mathbf{a}_3 \cdot \mathbf{b}_1)(\mathbf{a}_1 \otimes \mathbf{a}_2 \otimes \mathbf{b}_2)$ (see Chapter 1).

In effect, working with expressions (12.16)–(12.18) we obtain

$$\mathbf{A}_1 = \frac{1}{2}[\mathbf{C} - \mathbf{I}] + \frac{1}{2}\left[\mathsf{G}^t\mathbf{Y} + (\mathsf{G}^t\mathbf{Y})^T + (\mathsf{G}^t\mathbf{Y})^T\mathbf{C}^{-1}(\mathsf{G}^t\mathbf{Y})\right], \tag{12.23}$$

$$\mathbf{A}_2 = \mathbf{K}^T + \mathbf{K}^T\mathbf{C}^{-1}(\mathsf{G}^t\mathbf{Y}) - \mathbf{I}, \tag{12.24}$$

$$\mathbf{A}_3 = \frac{1}{2}\left[\mathbf{K}^T\mathbf{C}^{-1}\mathbf{K} - \mathbf{I}\right]. \tag{12.25}$$

In this way, when $\mathbf{C} = \mathbf{I}$, $\mathbf{K} = \mathbf{I}$ and $\mathsf{G} = \mathbf{0}$, we get $\mathbf{A}_1 = \mathbf{A}_2 = \mathbf{A}_3 = \mathbf{O}$, and therefore $d\mathbf{x}' \cdot d\mathbf{x}' = d\mathbf{X}' \cdot d\mathbf{X}'$, which is the case of the fiber without deformation.

To analyze the strain rate of a fiber we observe

$$\overline{d\mathbf{x}' \cdot d\mathbf{x}'}^{\cdot} = 2d\mathbf{x}' \cdot \overline{d\mathbf{x}'}^{\cdot}, \tag{12.26}$$

where $\overline{d\mathbf{x}'}^{\cdot}$ is calculated by deriving (12.13)

$$\overline{d\mathbf{x}'}^{\cdot} = (\dot{\mathbf{F}} + \dot{\mathsf{H}}^t\mathbf{Y})d\mathbf{X} + \dot{\Theta}d\mathbf{Y}. \tag{12.27}$$

Note also that

$$d\mathbf{x} = \mathbf{F}d\mathbf{X}, \tag{12.28}$$

$$d\mathbf{y} = (\mathsf{H}^t\mathbf{Y})d\mathbf{X} + \Theta d\mathbf{Y}, \tag{12.29}$$

which can be inverted, yielding

$$d\mathbf{X} = \mathbf{F}^{-1}d\mathbf{x}, \tag{12.30}$$

$$d\mathbf{Y} = -\Theta^{-1}(\mathsf{H}^t\mathbf{Y})\mathbf{F}^{-1}d\mathbf{x} + \Theta^{-1}d\mathbf{y}. \tag{12.31}$$

In turn, we know that

$$\dot{\mathbf{F}} = \nabla_{\mathbf{X}}\dot{\mathbf{U}} = (\nabla_{\mathbf{x}}\mathbf{v})_m\mathbf{F}, \tag{12.32}$$

$$\dot{\mathsf{H}} = \nabla_{\mathbf{X}}\dot{\Theta} = [(\nabla_{\mathbf{x}}\omega)_m \circ \mathbf{F}], \tag{12.33}$$

$$\dot{\Theta} = (\omega)_m, \tag{12.34}$$

where $(\cdot)_m$ denotes the material description of the field, $\mathbf{v}$ is the spatial description of the velocity field of the center of mass of the macro-particle, and ω is the micro-strain rate tensor relative to the particle. Replacing the previous expressions in (12.27) we have

$$\begin{aligned}\overline{d\mathbf{x}'}^{\cdot} &= ((\nabla_{\mathbf{x}}\mathbf{v})_m\mathbf{F} + [(\nabla_{\mathbf{x}}\omega)_m \circ \mathbf{F}]^t\mathbf{Y})\mathbf{F}^{-1}d\mathbf{x} - (\omega)_m[\Theta^{-1}(\mathsf{H}^t\mathbf{Y})\mathbf{F}^{-1}d\mathbf{x} - \Theta^{-1}d\mathbf{y}] \\ &= (\nabla_{\mathbf{x}}\mathbf{v})_m d\mathbf{x} + [(\nabla_{\mathbf{x}}\omega)^t_m \circ (\theta)^{-1}_m]\mathbf{y}d\mathbf{x} - (\omega)_m(\theta)^{-1}_m((\nabla_{\mathbf{x}}\theta)^t_m \circ (\theta)^{-1}_m)\mathbf{y}d\mathbf{x} \\ &\quad + (\omega)_m(\theta)^{-1}_m d\mathbf{y},\end{aligned} \tag{12.35}$$

where θ is the spatial description of the field Θ, and we have also used the fact that $\mathbf{Y} = \Theta^{-1}\mathbf{y}$. We can rewrite this expression in the spatial configuration, and so after arranging terms we obtain

$$\overline{d\mathbf{x}'}^{\cdot} = \left[\nabla_{\mathbf{x}}\mathbf{v} + [(\nabla_{\mathbf{x}}\omega)^t \circ \theta^{-1}]\mathbf{y} - \omega\theta^{-1}[(\nabla_{\mathbf{x}}\theta)^t \circ \theta^{-1}]\mathbf{y}\right]d\mathbf{x} + \omega\theta^{-1}d\mathbf{y}. \tag{12.36}$$

Let us define now the tensor

$$\chi = \omega\theta^{-1}, \tag{12.37}$$

called the microgyration tensor in [80, 81]. This tensor χ possesses information not only about the rotation, but also about the micro-strain rate experienced by the macro-particle, as we shall see later. First, observe that the following identity holds

$$[\nabla_{\mathbf{x}}\chi]^{\mathbf{t}}\mathbf{y} = [\nabla_{\mathbf{x}}(\omega\theta^{-1})]^{\mathbf{t}}\mathbf{y} = [(\nabla_{\mathbf{x}}\omega)^{\mathbf{t}}\circ\theta^{-1}]\mathbf{y} - \omega\theta^{-1}[(\nabla_{\mathbf{x}}\theta)^{\mathbf{t}}\circ\theta^{-1}]\mathbf{y}, \tag{12.38}$$

which follows from noting that the gradient of the inverse of a tensor $\mathbf{S}$ is

$$\nabla_{\mathbf{x}}(\mathbf{S}^{-1}) = -\mathbf{S}^{-1}((\nabla_{\mathbf{x}}\mathbf{S})^{\mathbf{t}}\circ\mathbf{S}^{-1}). \tag{12.39}$$

Then, introducing (12.37) and (12.38) into (12.36) gives

$$\dot{\overline{d\mathbf{x}'}} = \left[\nabla_{\mathbf{x}}\mathbf{v} + (\nabla_{\mathbf{x}}\chi)^{\mathbf{t}}\mathbf{y}\right] d\mathbf{x} + \chi d\mathbf{y}. \tag{12.40}$$

Using this expression in (12.26) allows us to arrive at

$$\begin{aligned}\dot{\overline{d\mathbf{x}'\cdot d\mathbf{x}'}} &= 2(d\mathbf{x}+d\mathbf{y})\cdot\left[\nabla_{\mathbf{x}}\mathbf{v} + (\nabla_{\mathbf{x}}\chi)^{\mathbf{t}}\mathbf{y}\right] d\mathbf{x} + 2(d\mathbf{x}+d\mathbf{y})\cdot(\chi d\mathbf{y})\\ &= 2\left[\nabla_{\mathbf{x}}\mathbf{v} + (\nabla_{\mathbf{x}}\chi)^{\mathbf{t}}\mathbf{y}\right] d\mathbf{x}\cdot d\mathbf{x} + 2\left[\nabla_{\mathbf{x}}\mathbf{v} + \chi^T + (\nabla_{\mathbf{x}}\chi)^{\mathbf{t}}\mathbf{y}\right] d\mathbf{x}\cdot d\mathbf{y} + 2\chi d\mathbf{y}\cdot d\mathbf{y},\end{aligned} \tag{12.41}$$

and since $\mathbf{A}\mathbf{b}\cdot\mathbf{b} = \mathbf{A}^s\mathbf{b}\cdot\mathbf{b}$, we have that the material rate of the length of the fiber is

$$\begin{aligned}\dot{\overline{d\mathbf{x}'\cdot d\mathbf{x}'}} &= 2\left[(\nabla_{\mathbf{x}}\mathbf{v})^s + ((\nabla_{\mathbf{x}}\chi)^{\mathbf{t}}\mathbf{y})^s\right] d\mathbf{x}\cdot d\mathbf{x}\\ &\quad + 2\left[\nabla_{\mathbf{x}}\mathbf{v} + \chi^T + (\nabla_{\mathbf{x}}\chi)^{\mathbf{t}}\mathbf{y}\right] d\mathbf{x}\cdot d\mathbf{y} + 2\chi^s d\mathbf{y}\cdot d\mathbf{y}.\end{aligned} \tag{12.42}$$

Moreover, notice that for $\dot{\overline{d\mathbf{x}'\cdot d\mathbf{x}'}} = 0$ the following must hold

$$(\nabla_{\mathbf{x}}\mathbf{v})^s + ((\nabla_{\mathbf{x}}\chi)^{\mathbf{t}}\mathbf{y})^s = \mathbf{O}, \tag{12.43}$$

$$\nabla_{\mathbf{x}}\mathbf{v} + \chi^T + (\nabla_{\mathbf{x}}\chi)^{\mathbf{t}}\mathbf{y} = \mathbf{O}, \tag{12.44}$$

$$\chi^s = \mathbf{O}, \tag{12.45}$$

which is verified if and only if

$$\nabla_{\mathbf{x}}\mathbf{v} - \chi = \mathbf{O}, \tag{12.46}$$

$$\nabla_{\mathbf{x}}\chi = \mathbf{0}, \tag{12.47}$$

$$\chi^s = \mathbf{O}. \tag{12.48}$$

In fact, taking the symmetric and skew-symmetric components of (12.46) and exploiting (12.48) yields

$$(\nabla_{\mathbf{x}}\mathbf{v})^s = \mathbf{O}, \tag{12.49}$$

$$(\nabla_{\mathbf{x}}\mathbf{v})^a - \chi^a = \mathbf{O}. \tag{12.50}$$

Therefore, we have arrived at a set of entities which characterize the strain rate measures in the continuum under analysis

$$\varepsilon = (\nabla_{\mathbf{x}}\mathbf{v})^s, \tag{12.51}$$

$$\eta = \nabla_{\mathbf{x}}\mathbf{v} - \chi, \tag{12.52}$$

$$\zeta = \nabla_{\mathbf{x}}\chi, \tag{12.53}$$

where, given the equivalence between (12.49) and (12.48) we have decided to replace the strain rate measure given by χ^s by that established by $(\nabla_{\mathbf{x}}\mathbf{v})^s$.

Thus, with these strain rates, expression (12.42) is written as

$$\begin{aligned}\dot{\overline{d\mathbf{x}' \cdot d\mathbf{x}'}} &= 2\left[\varepsilon + (\zeta^{\mathbf{t}}\mathbf{y})^s\right] d\mathbf{x} \cdot d\mathbf{x} \\ &\quad + 2\left[2\varepsilon - \eta^T + \zeta^{\mathbf{t}}\mathbf{y}\right] d\mathbf{x} \cdot d\mathbf{y} + 2[\varepsilon - \eta^s] d\mathbf{y} \cdot d\mathbf{y}.\end{aligned} \tag{12.54}$$

Now, notice that from the description found in (12.40) we can interpret that the field of motion actions the continuum can experience is characterized by velocity fields that have the form

$$\mathbf{v}' = \mathbf{v} + \chi\mathbf{y}, \tag{12.55}$$

where $\mathbf{v}$ is the velocity of the center of mass of Ω_μ and χ is the second-order tensor already seen, which is uniform all over Ω_μ and represents the gradient of the velocity field (consequently uniform) in Ω_μ. Therefore, χ is associated with the micro-strain and micro-rotation in such a micro-domain.

With the adopted kinematics we see that, at the macroscopic level, the kinematics of each particle is fully characterized by these two kinematical variables $\mathbf{v}$ and χ. With this, we can define for all time instants t the space of kinematically possible motion actions $\mathscr{V}$, the linear manifold Kin_v of kinematically admissible motion actions, and the associated linear space Var_v of kinematically admissible virtual motion actions

$$\mathscr{V} = \{(\mathbf{v}, \chi);\ \mathbf{v} \in \mathbf{H}^1(\mathscr{B}_t), \chi \in \mathsf{H}^1(\mathscr{B}_t)\}, \tag{12.56}$$

$$\mathrm{Kin}_v = \{(\mathbf{v}, \chi) \in \mathscr{V};\ (\mathbf{v}, \chi)|_{\partial\mathscr{B}_{tv}} = (\overline{\mathbf{v}}, \overline{\chi})\}, \tag{12.57}$$

$$\mathrm{Var}_v = \{(\mathbf{v}, \chi) \in \mathscr{V};\ (\mathbf{v}, \chi)|_{\partial\mathscr{B}_{tv}} = (\mathbf{0}, \mathbf{O})\}, \tag{12.58}$$

where $\mathbf{H}^1(\mathscr{B}_t)$ is the space of vector functions such that each component belongs to $H^1(\mathscr{B}_t)$, $\mathsf{H}^1(\mathscr{B}_t)$ is the space of second-order tensor functions where each component belongs to $H^1(\mathscr{B}_t)$, and $H^1(\mathscr{B}_t)$ is the space of continuous functions with squared integrable first derivative. Moreover, $\partial\mathscr{B}_t = \partial\mathscr{B}_{tv} \cup \partial\mathscr{B}_{tf}$, where $\partial\mathscr{B}_{tv} \cap \partial\mathscr{B}_{tf} = \emptyset$. For simplicity, we have assumed that both variables $\mathbf{v}$ and χ are constrained over the same portion of the boundary, denoted by $\partial\mathscr{B}_{tv}$.

Having found a set of variables that describe the strain rate in the continuum, we can now identify the strain rate operator $\mathscr{D}$ as

$$\begin{aligned}&\mathscr{D} : \mathscr{V} \to \mathscr{W}, \\ &(\mathbf{v}, \chi) \mapsto \mathscr{D}(\mathbf{v}, \chi) = ((\nabla_{\mathbf{x}}\mathbf{v})^s, \nabla_{\mathbf{x}}\mathbf{v} - \chi, \nabla_{\mathbf{x}}\chi),\end{aligned} \tag{12.59}$$

that is, the space $\mathscr{W}$ of strain rates is formed by three fields, two second-order tensors $(\nabla_{\mathbf{x}}\mathbf{v})^s$ and $\nabla_{\mathbf{x}}\mathbf{v} - \chi$ and a third-order tensor $\nabla_{\mathbf{x}}\chi$.

Hence, we can define the rigid motion actions. It is easy to verify that the velocity fields of the form

$$\mathbf{v}'_R = \mathbf{v}_R + \chi_R \mathbf{y} = \mathbf{v}_O + \mathbf{W}_O\mathbf{x} + \mathbf{W}_O\mathbf{y}, \tag{12.60}$$

with $\mathbf{v}_O$ a uniform vector field in $\mathscr{B}_t$ and $\mathbf{W}_O \in Skw$ a skew-symmetric uniform tensor field in $\mathscr{B}_t$, constitute rigid motion actions. As a matter of fact

$$\nabla_{\mathbf{x}}\mathbf{v}_R = \mathbf{W}_O, \tag{12.61}$$

$$\chi_R = \mathbf{W}_O, \tag{12.62}$$

and therefore (12.46)–(12.48) are satisfied. We can then introduce the null space of the operator $\mathscr{D}$

$$\mathscr{N}(\mathscr{D}) = \{(\mathbf{v}, \chi) \in \mathscr{V}\ ;\ (\mathbf{v}, \chi) = (\mathbf{v}_O + \mathbf{W}_O\mathbf{x}, \mathbf{W}_O) \\ \text{with } \mathbf{v}_O \in \mathbb{R}^3,\ \mathbf{W}_O \in (\mathbb{R}^{3\times 3})^a\}. \tag{12.63}$$

Before ending the section, it is important to add some further comments. The micromorphic model that we have constructed is characterized by

- a classical strain rate measure given by $(\nabla_{\mathbf{x}}\mathbf{v})^s$
- a measure of the distortion existing between the macro-kinematics and the micro-kinematics given by $\nabla_{\mathbf{x}}\mathbf{v} - \chi$
- and, finally, a high-order generalized strain given by $\nabla_{\mathbf{x}}\chi$.

Now, if we incorporate the internal kinematical constraint $\nabla_{\mathbf{x}}\mathbf{v} = \chi$ the new model thus obtained will be characterized by a second gradient kinematics because it would be governed by strain rates given by $(\nabla_{\mathbf{x}}\mathbf{v})^s$ and $\nabla_{\mathbf{x}}\chi = \nabla_{\mathbf{x}}\nabla_{\mathbf{x}}\mathbf{v}$. This model will be discussed in more detail in the following sections.

12.3 Principle of Virtual Power

As in previous chapters, especially in Chapter 3, we will define the external forces through the concept of external power. This notion is mathematically shaped by the external power functional, which is a linear and continuous functional defined over the space $\mathscr{V}$ of kinematical pairs $(\mathbf{v}, \chi)$. Introducing the dual space $\mathscr{V}'$ we can define the external power as the operation

$$\begin{aligned} \langle \cdot, \cdot \rangle : \mathscr{V}' \times \mathscr{V} &\to \mathbb{R}, \\ (f, (\mathbf{v}, \chi)) &\mapsto P_e(\mathbf{v}, \chi) = \langle f, (\mathbf{v}, \chi) \rangle. \end{aligned} \tag{12.64}$$

What is more, since the functional is linear and continuous, and because the kinematical entities $\mathbf{v}$ and χ are mutually independent, we can provide an explicit form of this external power

$$\begin{aligned} \langle \cdot, \cdot \rangle : \mathscr{V}' \times \mathscr{V} &\to \mathbb{R}, \\ ((f_v, f_\chi), (\mathbf{v}, \chi)) &\mapsto P_e(\mathbf{v}, \chi) = \langle f_v, \mathbf{v} \rangle + \langle f_\chi, \chi \rangle. \end{aligned} \tag{12.65}$$

The characterization of these duality pairings between f_v and $\mathbf{v}$, and between f_χ and χ will be realized through the PVP, as is seen next.

At this point, we have already characterized the kinematical objects that will be utilized as strain rate measures (see definitions of ε, η and ζ in (12.51), (12.52) and (12.53), respectively). The next step consists of postulating, as done in Chapter 3, that the internal power that characterizes the internal stresses compatible with the kinematical model is given by a linear and continuous functional defined over the space of kinematical objects $(\mathbf{v}, \chi)$ and their first-order gradients, that is

$$P_i = -\int_{\mathscr{B}_t} (\mathbf{a}\cdot\mathbf{v} + \mathbf{B}\cdot\nabla_{\mathbf{x}}\mathbf{v} + \mathbf{C}\cdot\chi + \mathsf{D}\cdot\nabla_{\mathbf{x}}\chi)dV. \tag{12.66}$$

The internal power must be an objective quantity, which implies that it must be null when rigid motion actions are executed over the body. Then

$$P_i = -\int_{\mathscr{B}_t} (\mathbf{a}\cdot\mathbf{v} + \mathbf{B}\cdot\nabla_{\mathbf{x}}\mathbf{v} + \mathbf{C}\cdot\chi + \mathsf{D}\cdot\nabla_{\mathbf{x}}\chi)dV = 0 \qquad \forall(\mathbf{v},\chi)\in\mathcal{N}(\mathscr{D}), \tag{12.67}$$

and from the form of the elements in $\mathcal{N}(\mathscr{D})$, we have

$$\begin{aligned} P_i = &-\left(\int_{\mathscr{B}_t}\mathbf{a}dV\right)\cdot\mathbf{v}_O - \left(\int_{\mathscr{B}_t}\mathbf{B}dV\right)\cdot\mathbf{W}_O \\ &-\left(\int_{\mathscr{B}_t}\mathbf{C}dV\right)\cdot\mathbf{W}_O = 0 \quad \forall(\mathbf{v}_O,\mathbf{W}_O)\in\mathbb{R}^3\times(\mathbb{R}^{3\times3})^a. \end{aligned} \tag{12.68}$$

Since this must hold for every part of the body, we conclude that

$$\mathbf{a} = \mathbf{0}, \tag{12.69}$$

$$\mathbf{B} + \mathbf{C} \in Sym. \tag{12.70}$$

That is, for the present model, the internal stresses cannot be characterized by vector fields, and, in addition, whatever the type of second-order tensor field, $\mathbf{B}$ and $\mathbf{C}$, which characterizes the internal stresses, its sum must yield a symmetric tensor. Thus, introducing the notation $\sigma = (\mathbf{B}+\mathbf{C})$ and $\gamma = -\mathbf{C}$, and from the symmetry of σ, the internal power takes the final form

$$\begin{aligned} P_i &= -\int_{\mathscr{B}_t} (\sigma\cdot(\nabla_{\mathbf{x}}\mathbf{v})^s + \gamma\cdot(\nabla_{\mathbf{x}}\mathbf{v}-\chi) + \mathsf{D}\cdot\nabla_{\mathbf{x}}\chi)dV \\ &= -\int_{\mathscr{B}_t} (\sigma\cdot\varepsilon + \gamma\cdot\eta + \mathsf{D}\cdot\zeta)dV = -((\sigma,\gamma,\mathsf{D}),(\varepsilon,\eta,\zeta)). \end{aligned} \tag{12.71}$$

We have, as throughout the previous chapters, that the internal power results in a linear and continuous functional defined over the space of strain rate actions.

Let us now enunciate the PVP. We say that the body $\mathscr{B}_t$ is at mechanical equilibrium, for a given system of external forces $f \in \mathscr{V}'$, or equivalently $(f_v, f_\chi) \in \mathscr{V}'$, if the following holds

$$P_e = \langle f, (\hat{\mathbf{v}}, \hat{\chi})\rangle = 0 \qquad \forall(\hat{\mathbf{v}},\hat{\chi})\in \mathrm{Var}_v \cap \mathcal{N}(\mathscr{D}), \tag{12.72}$$

or equivalently

$$P_e = \langle f_v, \hat{\mathbf{v}} \rangle + \langle f_\chi, \hat{\chi} \rangle = 0 \qquad \forall(\hat{\mathbf{v}}, \hat{\chi}) \in \mathrm{Var}_v \cap \mathcal{N}(\mathcal{D}). \tag{12.73}$$

It is also verified that

$$P_i + P_e = -((\sigma, \gamma, \mathbf{D}), ((\nabla_{\mathbf{x}} \hat{\mathbf{v}})^s, \nabla_{\mathbf{x}} \hat{\mathbf{v}} - \hat{\chi}, \nabla_{\mathbf{x}} \hat{\chi})) + \langle f, (\hat{\mathbf{v}}, \hat{\chi}) \rangle = 0 \quad \forall(\hat{\mathbf{v}}, \hat{\chi}) \in \mathrm{Var}_v. \tag{12.74}$$

As before, the following representation theorem guarantees that there exists the triple $(\sigma, \gamma, \mathbf{D})$ that equilibrates the system of forces (f_v, f_χ).

Theorem 12.1 Representation Theorem. If $(f_v, f_\chi) \in \mathscr{V}'$ is at equilibrium, then there exists $(\sigma, \gamma, \mathbf{D}) \in \mathscr{W}'$ and $(r_v, r_\chi) \in \mathscr{V}'$ such that

$$f_v = [\mathcal{D}^*(\sigma, \gamma, \mathbf{D})]_v - r_v, \tag{12.75}$$

$$f_\chi = [\mathcal{D}^*(\sigma, \gamma, \mathbf{D})]_\chi - r_\chi, \tag{12.76}$$

and also satisfying

$$\langle (r_v, r_\chi), (\hat{\mathbf{v}}, \hat{\chi}) \rangle = 0 \qquad \forall(\hat{\mathbf{v}}, \hat{\chi}) \in \mathrm{Var}_v, \tag{12.77}$$

which implies that $(r_v, r_\chi) \in \mathrm{Var}_v^\perp$. Here, the pair (r_v, r_χ) is the system of reactive forces. Therefore, we say that $(\sigma, \gamma, \mathbf{D})$ and (r_v, r_χ) equilibrate (f_v, f_χ).

Proof. We know that $f = (f_v, f_\chi)$ is at equilibrium, then

$$(f_v, f_\chi) \in (\mathrm{Var}_v \cap \mathcal{N}(\mathcal{D}))^\perp \quad \Leftrightarrow \quad (f_v, f_\chi) \in \mathrm{Var}_v^\perp + \mathcal{N}(\mathcal{D})^\perp. \tag{12.78}$$

Since $(\mathbf{v}, \chi) \in \mathcal{N}(\mathcal{D})$ we have $\mathcal{D}\mathbf{v} = (\mathbf{O}, \mathbf{O}, \mathbf{0})$, and then

$$\langle \mathcal{D}^*(\sigma, \gamma, \mathbf{D}), (\mathbf{v}, \chi) \rangle = ((\sigma, \gamma, \mathbf{D}), \mathcal{D}(\mathbf{v}, \chi)) = 0 \qquad \forall(\sigma, \gamma, \mathbf{D}) \in \mathscr{W}', \tag{12.79}$$

that means $(\mathbf{v}, \chi) \in R(\mathcal{D}^*)^\perp$, and so we conclude $\mathcal{N}(\mathcal{D}) = R(\mathcal{D}^*)^\perp$, that is

$$\mathcal{N}(\mathcal{D})^\perp = R(\mathcal{D}^*). \tag{12.80}$$

Putting this into (12.78) gives

$$(f_v, f_\chi) \in \mathrm{Var}_v^\perp + R(\mathcal{D}^*), \tag{12.81}$$

consequently, there exist $(\sigma, \gamma, \mathbf{D}) \in \mathscr{W}'$ and $(-r_v, -r_\chi) \in \mathrm{Var}_v^\perp$ such that

$$f_v = [\mathcal{D}^*(\sigma, \gamma, \mathbf{D})]_v - r_v, \tag{12.82}$$

$$f_\chi = [\mathcal{D}^*(\sigma, \gamma, \mathbf{D})]_\chi - r_\chi, \tag{12.83}$$

and since $(-r_v, -r_\chi) \in \mathrm{Var}_v^\perp$, it is verified that

$$\langle (r_v, r_\chi), (\hat{\mathbf{v}}, \hat{\chi}) \rangle = 0 \qquad \forall(\hat{\mathbf{v}}, \hat{\chi}) \in \mathrm{Var}_v. \tag{12.84}$$

■

The representation theorem allows us to conclude the second part of the PVP. In other words, it enables us to write

$$\begin{aligned}\langle (f_v, f_\chi), (\hat{\mathbf{v}}, \hat{\chi})\rangle &= \langle \mathscr{D}^*(\sigma, \gamma, \mathbf{D}) - (r_v, r_\chi), (\hat{\mathbf{v}}, \hat{\chi})\rangle \\ &= \langle \mathscr{D}^*(\sigma, \gamma, \mathbf{D}), (\hat{\mathbf{v}}, \hat{\chi})\rangle = ((\sigma, \gamma, \mathbf{D}), \mathscr{D}(\hat{\mathbf{v}}, \hat{\chi})) \qquad \forall (\hat{\mathbf{v}}, \hat{\chi}) \in \mathrm{Var}_v,\end{aligned} \tag{12.85}$$

or equivalently

$$((\sigma, \gamma, \mathbf{D}), \mathscr{D}(\hat{\mathbf{v}}, \hat{\chi})) = \langle (f_v, f_\chi), (\hat{\mathbf{v}}, \hat{\chi})\rangle \qquad \forall (\hat{\mathbf{v}}, \hat{\chi}) \in \mathrm{Var}_v. \tag{12.86}$$

Then, replacing the definition of the internal power (12.71) into the second part of the PVP yields

$$\begin{aligned}&\int_{\mathscr{B}_t} \left(\sigma \cdot (\nabla_{\mathbf{x}}\hat{\mathbf{v}})^s + \gamma \cdot (\nabla_{\mathbf{x}}\hat{\mathbf{v}} - \hat{\chi}) + \mathbf{D} \cdot \nabla_{\mathbf{x}}\hat{\chi}\right) dV \\ &\quad = \langle (f_v, f_\chi), (\hat{\mathbf{v}}, \hat{\chi})\rangle \quad \forall (\hat{\mathbf{v}}, \hat{\chi}) \in \mathrm{Var}_v.\end{aligned} \tag{12.87}$$

Let us now characterize the external forces compatible with the model, that is, the elements $f \in \mathscr{V}'$, or equivalently $(f_v, f_\chi) \in \mathscr{V}'$. Assuming the fields are regular enough, and knowing that σ is a symmetric tensor, integration by parts in expression (12.87) renders

$$\begin{aligned}&\int_{\mathscr{B}_t} \left(\sigma \cdot (\nabla_{\mathbf{x}}\hat{\mathbf{v}})^s + \gamma \cdot (\nabla_{\mathbf{x}}\hat{\mathbf{v}} - \hat{\chi}) + \mathbf{D} \cdot \nabla_{\mathbf{x}}\hat{\chi}\right) dV \\ &\quad = \int_{\partial\mathscr{B}_{tf}} ((\sigma + \gamma)\mathbf{n} \cdot \hat{\mathbf{v}} + \mathbf{D}\mathbf{n} \cdot \hat{\chi})\, dS \\ &\quad\quad - \int_{\mathscr{B}_t} ((\mathrm{div}(\sigma + \gamma)) \cdot \hat{\mathbf{v}} + (\gamma + \mathrm{div}\,\mathbf{D}) \cdot \hat{\chi})\, dV,\end{aligned} \tag{12.88}$$

where we have taken into consideration that $(\hat{\mathbf{v}}, \hat{\chi}) = (\mathbf{0}, \mathbf{O})$ on $\partial\mathscr{B}_{tv}$. In this manner, from (12.87) we observe that

$$\begin{aligned}\langle (f_v, f_\chi), (\hat{\mathbf{v}}, \hat{\chi})\rangle &= \int_{\partial\mathscr{B}_{tf}} ((\sigma + \gamma)\mathbf{n} \cdot \hat{\mathbf{v}} + \mathbf{D}\mathbf{n} \cdot \hat{\chi})\, dS \\ &\quad - \int_{\mathscr{B}_t} ((\mathrm{div}(\sigma + \gamma)) \cdot \hat{\mathbf{v}} + (\gamma + \mathrm{div}\,\mathbf{D}) \cdot \hat{\chi})\, dV \\ &= \int_{\partial\mathscr{B}_{tf}} (\sigma + \gamma)\mathbf{n} \cdot \hat{\mathbf{v}} dS - \int_{\mathscr{B}_t} (\mathrm{div}(\sigma + \gamma)) \cdot \hat{\mathbf{v}} dV \\ &\quad + \int_{\partial\mathscr{B}_{tf}} \mathbf{D}\mathbf{n} \cdot \hat{\chi} dS - \int_{\mathscr{B}_t} (\gamma + \mathrm{div}\,\mathbf{D}) \cdot \hat{\chi} dV.\end{aligned} \tag{12.89}$$

Hence, we obtain explicit functional forms for both elements f_v and f_χ

$$\langle f_v, \hat{\mathbf{v}}\rangle = \int_{\partial\mathscr{B}_{tf}} (\sigma + \gamma)\mathbf{n} \cdot \hat{\mathbf{v}} dS - \int_{\mathscr{B}_t} (\mathrm{div}(\sigma + \gamma)) \cdot \hat{\mathbf{v}} dV, \tag{12.90}$$

$$\langle f_\chi, \hat{\chi}\rangle = \int_{\partial\mathscr{B}_{tf}} \mathbf{D}\mathbf{n} \cdot \hat{\chi} dS - \int_{\mathscr{B}_t} (\gamma + \mathrm{div}\,\mathbf{D}) \cdot \hat{\chi} dV. \tag{12.91}$$

Note that f_v is characterized by a vector field per unit area defined in $\partial\mathscr{B}_{tf}$ and by a vector field per unit volume defined in $\mathscr{B}_t$, while the element f_χ is characterized by a second-order tensor field per unit area on $\partial\mathcal{B}_{tf}$ and by a second-order tensor field per unit volume in $\mathcal{B}_t$, that is

$$\langle f_v, \hat{\mathbf{v}}\rangle = \int_{\partial\mathcal{B}_{tf}} \mathbf{t}\cdot\hat{\mathbf{v}}dS + \int_{\mathcal{B}_t} \mathbf{b}\cdot\hat{\mathbf{v}}dV, \tag{12.92}$$

$$\langle f_\chi, \hat{\chi}\rangle = \int_{\partial\mathcal{B}_{tf}} \mathbf{N}\cdot\hat{\chi}dS + \int_{\mathcal{B}_t} \mathbf{M}\cdot\hat{\chi}dV. \tag{12.93}$$

Therefore, we have described the components of the equilibrium operator $\mathcal{D}^*$, $[\mathcal{D}^*(\cdot)]_v$ and $[\mathcal{D}^*(\cdot)]_\chi$ because $f_v = [\mathcal{D}^*(\sigma, \gamma, \mathsf{D})]_v$ and $f_\chi = [\mathcal{D}^*(\sigma, \gamma, \mathsf{D})]_\chi$.

Going back to the PVP, it acquires the final form

$$\begin{aligned}\int_{\mathcal{B}_t} & \left(\sigma\cdot(\nabla_{\mathbf{x}}\hat{\mathbf{v}})^s + \gamma\cdot(\nabla_{\mathbf{x}}\hat{\mathbf{v}} - \hat{\chi}) + \mathsf{D}\cdot\nabla_{\mathbf{x}}\hat{\chi}\right) dV \\ &= \int_{\partial\mathcal{B}_{tf}} \mathbf{t}\cdot\hat{\mathbf{v}}dS + \int_{\mathcal{B}_t} \mathbf{b}\cdot\hat{\mathbf{v}}dV + \int_{\partial\mathcal{B}_{tf}} \mathbf{N}\cdot\hat{\chi}dS \\ &+ \int_{\mathcal{B}_t} \mathbf{M}\cdot\hat{\chi}dV \quad \forall(\hat{\mathbf{v}}, \hat{\chi}) \in \mathrm{Var}_v.\end{aligned} \tag{12.94}$$

The Euler–Lagrange equations associated with the variational statement of the PVP (12.94) follow from integrating by parts as before. Then

$$\begin{aligned}\int_{\partial\mathcal{B}_{tf}} & ((\sigma+\gamma)\mathbf{n}\cdot\hat{\mathbf{v}} + \mathsf{D}\mathbf{n}\cdot\hat{\chi})\, dS - \int_{\mathcal{B}_t} ((\mathrm{div}(\sigma+\gamma))\cdot\hat{\mathbf{v}} + (\gamma + \mathrm{div}\,\mathsf{D})\cdot\hat{\chi})\, dV \\ &= \int_{\partial\mathcal{B}_{tf}} \mathbf{t}\cdot\hat{\mathbf{v}}dS + \int_{\mathcal{B}_t} \mathbf{b}\cdot\hat{\mathbf{v}}dV + \int_{\partial\mathcal{B}_{tf}} \mathbf{N}\cdot\hat{\chi}dS \\ &+ \int_{\mathcal{B}_t} \mathbf{M}\cdot\hat{\chi}dV \quad \forall(\hat{\mathbf{v}}, \hat{\chi}) \in \mathrm{Var}_v,\end{aligned} \tag{12.95}$$

and rearranging terms gives

$$\begin{aligned}\int_{\partial\mathcal{B}_{tf}} & ((\sigma+\gamma)\mathbf{n} - \mathbf{t})\cdot\hat{\mathbf{v}}dS \\ &+ \int_{\partial\mathcal{B}_{tf}} (\mathsf{D}\mathbf{n} - \mathbf{N})\cdot\hat{\chi}dS - \int_{\mathcal{B}_t} (\mathrm{div}(\sigma+\gamma) + \mathbf{b})\cdot\hat{\mathbf{v}}dV \\ &- \int_{\mathcal{B}_t} (\gamma + \mathrm{div}\,\mathsf{D} + \mathbf{M})\cdot\hat{\chi}dV = 0 \quad \forall(\hat{\mathbf{v}}, \hat{\chi}) \in \mathrm{Var}_v.\end{aligned} \tag{12.96}$$

In this way, we have arrived at the Euler–Lagrange equations

$$\mathrm{div}(\sigma+\gamma) + \mathbf{b} = \mathbf{0} \quad \text{in } \mathcal{B}_t, \tag{12.97}$$

$$\gamma + \mathrm{div}\,\mathsf{D} + \mathbf{M} = \mathbf{O} \quad \text{in } \mathcal{B}_t, \tag{12.98}$$

$$(\sigma+\gamma)\mathbf{n} = \mathbf{t} \quad \text{on } \partial\mathcal{B}_{tf}, \tag{12.99}$$

$$\mathsf{D}\mathbf{n} = \mathbf{N} \quad \text{on } \partial\mathcal{B}_{tf}, \tag{12.100}$$

which, together with the kinematical constraints

$$\mathbf{v} = \overline{\mathbf{v}} \quad \text{on } \partial\mathscr{B}_{tv}, \tag{12.101}$$

$$\chi = \overline{\chi} \quad \text{on } \partial\mathscr{B}_{tv}, \tag{12.102}$$

provide a complete description of the strong formulation of the mechanical problem for micromorphic continua.

Let us assume now that there exists an internal surface S in which the conditions of regularity of the fields is not verified so that integration by parts cannot be performed. However, we can integrate by parts in the portions of the domain $\mathscr{B}_t$ where fields are actually regular enough. This procedure results in terms over the internal surface S of the form $\int_S [\![\sigma + \gamma]\!]\mathbf{n} \cdot \hat{\mathbf{v}} dS + \int_S [\![\mathbf{D}]\!]\mathbf{n} \cdot \hat{\chi} dS$, and in this case the external forces f_v and f_χ could include, respectively, a vector field $\mathbf{t}_S$ and a tensor field $\mathbf{N}_S$ per unit area defined over S. In such a situation, the Euler–Lagrange equations provide the following jump conditions

$$[\![\sigma + \gamma]\!]\mathbf{n} = \mathbf{t}_S \quad \text{on } S, \tag{12.103}$$

$$[\![\mathbf{D}]\!]\mathbf{n} = \mathbf{N}_S \quad \text{on } S. \tag{12.104}$$

Before ending this section we calculate the reactive forces $(r_v, r_\chi) \in \mathscr{V}'$ which, we recall, do not contribute with power in the statement of the PVP and can only be assessed by removing the kinematical constraints from the space Var_v, that is

$$\begin{aligned} \langle (f_v, f_\chi), (\hat{\mathbf{v}}, \hat{\chi}) \rangle = {} & \langle \mathscr{D}^*(\sigma, \gamma, \mathbf{D}), (\hat{\mathbf{v}}, \hat{\chi}) \rangle \\ & - \langle (r_v, r_\chi), (\hat{\mathbf{v}}, \hat{\chi}) \rangle \quad \forall (\hat{\mathbf{v}}, \hat{\chi}) \in \mathscr{V}. \end{aligned} \tag{12.105}$$

Putting the PVP (12.94) into (12.105) we have

$$\begin{aligned} \langle (r_v, r_\chi), (\hat{\mathbf{v}}, \hat{\chi}) \rangle = {} & \int_{\mathscr{B}_t} \left(\sigma \cdot (\nabla_{\mathbf{x}} \hat{\mathbf{v}})^s + \gamma \cdot (\nabla_{\mathbf{x}} \hat{\mathbf{v}} - \hat{\chi}) + \mathbf{D} \cdot \nabla_{\mathbf{x}} \hat{\chi} \right) dV \\ & - \int_{\partial\mathscr{B}_{tf}} \mathbf{t} \cdot \hat{\mathbf{v}} dS + \int_{\mathscr{B}_t} \mathbf{b} \cdot \hat{\mathbf{v}} dV + \int_{\partial\mathscr{B}_{tf}} \mathbf{N} \cdot \hat{\chi} dS \\ & + \int_{\mathscr{B}_t} \mathbf{M} \cdot \hat{\chi} dV \quad \forall (\hat{\mathbf{v}}, \hat{\chi}) \in \mathscr{V}, \end{aligned} \tag{12.106}$$

and thus, integrating by parts, grouping terms, and making use of the equilibrium, we have

$$\begin{aligned} \langle (r_v, r_\chi), (\hat{\mathbf{v}}, \hat{\chi}) \rangle = {} & \int_{\partial\mathscr{B}_{tf}} ((\sigma + \gamma)\mathbf{n} - \mathbf{t}) \cdot \hat{\mathbf{v}} dS + \int_{\partial\mathscr{B}_{tf}} (\mathbf{D}\mathbf{n} - \mathbf{N}) \cdot \hat{\chi} dS \\ & + \int_{\partial\mathscr{B}_{tv}} (\sigma + \gamma)\mathbf{n} \cdot \hat{\mathbf{v}} dS + \int_{\partial\mathscr{B}_{tv}} \mathbf{D}\mathbf{n} \cdot \hat{\chi} dS \\ & - \int_{\mathscr{B}_t} (\operatorname{div}(\sigma + \gamma) + \mathbf{b}) \cdot \hat{\mathbf{v}} dV - \int_{\mathscr{B}_t} (\gamma + \operatorname{div} \mathbf{D} + \mathbf{M}) \cdot \hat{\chi} dV \\ = {} & \int_{\partial\mathscr{B}_{tv}} (\sigma + \gamma)\mathbf{n} \cdot \hat{\mathbf{v}} dS + \int_{\partial\mathscr{B}_{tv}} \mathbf{D}\mathbf{n} \cdot \hat{\chi} dS \qquad \forall (\hat{\mathbf{v}}, \hat{\chi}) \in \mathscr{V}. \end{aligned} \tag{12.107}$$

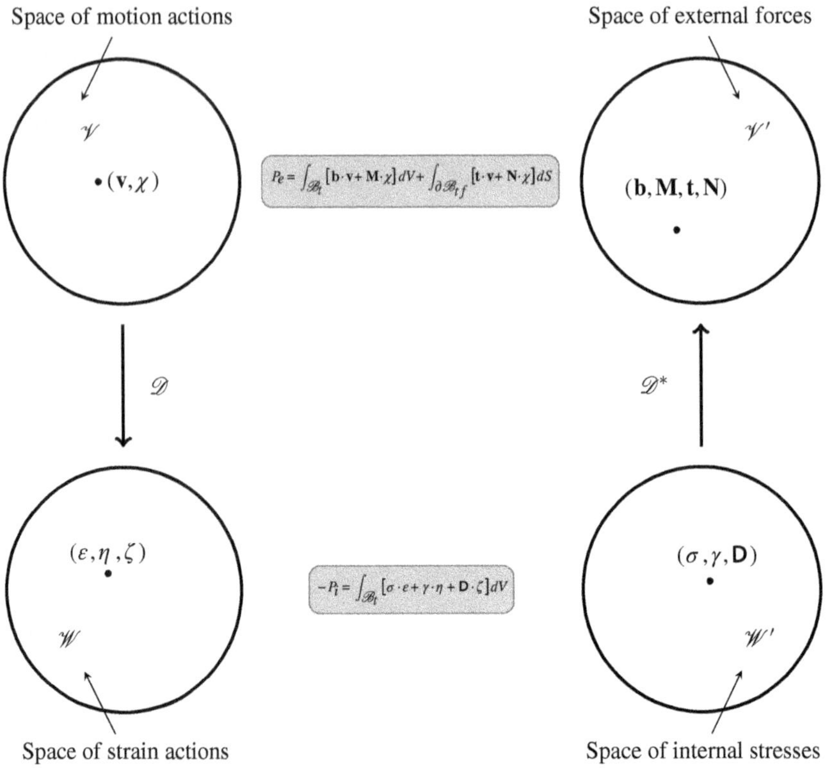

Figure 12.2 Ingredients in the variational theory for the modeling of high-order continua.

Consequently, we managed to characterize the reactive forces r_v and r_χ as

$$\langle r_v, \hat{\mathbf{v}}\rangle + \langle r_\chi, \hat{\chi}\rangle = \int_{\partial\mathscr{B}_{tv}} (\sigma+\gamma)\mathbf{n}\cdot\hat{\mathbf{v}}dS + \int_{\partial\mathscr{B}_{tv}} \mathbf{Dn}\cdot\hat{\chi}dS \quad \forall(\hat{\mathbf{v}},\hat{\chi})\in\mathscr{V}, \tag{12.108}$$

or equivalently

$$\int_{\partial\mathscr{B}_{tv}} (\mathbf{r}_v-(\sigma+\gamma)\mathbf{n})\cdot\hat{\mathbf{v}}dS + \int_{\partial\mathscr{B}_{tv}} (\mathbf{R}_\chi-\mathbf{Dn})\cdot\hat{\chi}dS = 0 \quad \forall(\hat{\mathbf{v}},\hat{\chi})\in\mathscr{V}. \tag{12.109}$$

In local form, the reactive forces take the form

$$\mathbf{r}_v = (\sigma+\gamma)\mathbf{n} \quad \text{on } \partial\mathscr{B}_{tv}, \tag{12.110}$$

$$\mathbf{R}_\chi = \mathbf{Dn} \quad \text{on } \partial\mathscr{B}_{tv}, \tag{12.111}$$

where $\mathbf{r}_v$ is the vector that represents the reaction r_v and $\mathbf{R}_\chi$ is the second-order tensor that represents the reaction r_χ.

Finally, Figure 12.2 displays the relevant ingredients in the variational theory of high-order continuum media.

12.4 Dynamics

This section tackles the dynamic problem for which it is necessary to incorporate the power exerted by forces caused by inertia of particles. This is of the utmost importance in time-dependent problems.

As we have seen in Chapter 3, the PVP also allows the dynamic equilibrium to be characterized. To this end, it is necessary to add the virtual power associated with the inertial forces, which is called in this context $\mathbf{b}^i$. To determine this force in the present problem let us recall that in the micromorphic continuum we have associated with each macro-particle P in $\mathscr{B}_t$ a micro-region Ω_μ, whose local origin of coordinates is located at the center of mass of that region.

With this in mind, we adopt the following hypothesis: the virtual power associated with the inertial forces of the macro-particle P must be equal to the average virtual power of the inertial forces in the corresponding micro-region, that is

$$\mathbf{b}^i|_P \cdot (\hat{\mathbf{v}}, \hat{\chi})|_P = \frac{1}{|\Omega_\mu|} \int_{\Omega_\mu} \rho' \mathbf{a}' \cdot \hat{\mathbf{v}}' dV_\mu, \tag{12.112}$$

where the notation $(\cdot)|_P$ indicates the association with the macro-particle P. In the previous expression ρ' stands for the density (not necessarily homogeneous) of the material in the micro-region Ω_μ and $\mathbf{a}'$ is the acceleration of micro-particles which constitute the macro-particle P.

Before going ahead with the derivation of $\mathbf{b}^i|_P$ let us introduce some useful definitions. The average density of the micro-domain is the density of the macro-particle P, that is

$$\rho = \frac{1}{|\Omega_\mu|} \int_{\Omega_\mu} \rho' dV_\mu. \tag{12.113}$$

Since P occupies the center of mass of the micro-region, and this center is chosen as the origin for the system of coordinates $\mathbf{y} \in \Omega_\mu$, we get

$$\int_{\Omega_\mu} \rho' \mathbf{y} dV_\mu = \mathbf{0}. \tag{12.114}$$

Consider also the symmetric second-order tensor field $\mathbf{J}$ corresponding to the polar moment of inertia of the micro-domain, that is

$$\mathbf{J} = \frac{1}{\rho|\Omega_\mu|} \int_{\Omega_\mu} \rho' \mathbf{y} \otimes \mathbf{y} d\Omega_\mu. \tag{12.115}$$

With these definitions we can now obtain the form for the inertial force $\mathbf{b}^i|_P$. To this end, let us remember our first hypothesis over the velocity field in the micro-domain stated by (12.55). Then, the acceleration of any point of Ω_μ is established by

$$\mathbf{a}' = \dot{\mathbf{v}}' = \dot{\mathbf{v}} + \dot{\chi}\mathbf{y} + \chi\dot{\mathbf{y}}, \tag{12.116}$$

but $\dot{\mathbf{y}}$ is the velocity of point $\mathbf{y}$ relative to the velocity of the origin of coordinates (center of mass), that is

$$\dot{\mathbf{y}} = \mathbf{v}' - \mathbf{v} = \chi\mathbf{y}, \tag{12.117}$$

which, when replaced into (12.116), provides us with the final expression for the acceleration experienced by any point of Ω_μ

$$\mathbf{a}' = \dot{\mathbf{v}} + (\dot{\chi} + \chi\chi)\mathbf{y}. \tag{12.118}$$

Analogously, going back to (12.55), virtual motion actions in the micro-domain are given by

$$\hat{\mathbf{v}}' = \hat{\mathbf{v}} + \hat{\chi}\mathbf{y}. \tag{12.119}$$

Substituting (12.118) and (12.119) into (12.112) yields

$$\begin{aligned}
\mathbf{b}^i|_P \cdot (\hat{\mathbf{v}}, \hat{\chi})|_P &= \frac{1}{|\Omega_\mu|}\int_{\Omega_\mu} \rho' \mathbf{a}' \cdot \hat{\mathbf{v}}' dV_\mu = \frac{1}{|\Omega_\mu|}\int_{\Omega_\mu} \rho'(\dot{\mathbf{v}} + (\dot{\chi} + \chi\chi)\mathbf{y}) \cdot (\hat{\mathbf{v}} + \hat{\chi}\mathbf{y}) dV_\mu \\
&= (\dot{\mathbf{v}} \cdot \hat{\mathbf{v}})\left(\frac{1}{|\Omega_\mu|}\int_{\Omega_\mu} \rho' dV_\mu\right) + \hat{\chi}^T \dot{\mathbf{v}} \cdot \left(\frac{1}{|\Omega_\mu|}\int_{\Omega_\mu} \rho' \mathbf{y} dV_\mu\right) \\
&\quad + \dot{\chi}^T \hat{\mathbf{v}} \cdot \left(\frac{1}{|\Omega_\mu|}\int_{\Omega_\mu} \rho' \mathbf{y} dV_\mu\right) + \dot{\chi}^T \hat{\chi} \cdot \left(\frac{1}{|\Omega_\mu|}\int_{\Omega_\mu} \rho' \mathbf{y} \otimes \mathbf{y} dV_\mu\right) \\
&\quad + (\chi\chi)^T \hat{\mathbf{v}} \cdot \left(\frac{1}{|\Omega_\mu|}\int_{\Omega_\mu} \rho' \mathbf{y} dV_\mu\right) + (\chi\chi)^T \hat{\chi} \cdot \left(\frac{1}{|\Omega_\mu|}\int_{\Omega_\mu} \rho' \mathbf{y} \otimes \mathbf{y} dV_\mu\right).
\end{aligned} \tag{12.120}$$

From the definitions introduced previously, this expression reduces to

$$\mathbf{b}^i|_P \cdot (\hat{\mathbf{v}}, \hat{\chi})|_P = \rho\dot{\mathbf{v}} \cdot \hat{\mathbf{v}} + \rho(\dot{\chi} + \chi\chi)^T \hat{\chi} \cdot \mathbf{J} = \rho\dot{\mathbf{v}} \cdot \hat{\mathbf{v}} + \rho(\dot{\chi} + \chi\chi)\mathbf{J} \cdot \hat{\chi}. \tag{12.121}$$

We thus obtain the representation of the inertial force associated with each macro-particle P from $\mathscr{B}_t$. The PVP, including the virtual power exerted by this inertial force, provides the statement of the mechanical equilibrium for the micromorphic medium in the dynamic case.

The system of forces including inertia phenomena becomes $f - f^i = (f_v - f_v^i, f_\chi - f_\chi^i) \in \mathscr{V}'$, where f_v^i and f_χ^i are

$$\langle f_v^i, \hat{\mathbf{v}} \rangle = \int_{\mathscr{B}_t} \rho\dot{\mathbf{v}} \cdot \hat{\mathbf{v}} dV, \tag{12.122}$$

$$\langle f_\chi^i, \hat{\chi} \rangle = \int_{\mathscr{B}_t} \rho(\dot{\chi} + \chi\chi)\mathbf{J} \cdot \hat{\chi} dV, \tag{12.123}$$

and the state of internal stresses $(\sigma, \gamma, \mathsf{D}) \in \mathscr{W}'$ is at (dynamic) equilibrium in $\mathscr{B}_t$ if the PVP is satisfied at each time instant $t \in (t_0, T]$, that is

$$((\sigma, \gamma, \mathsf{D}), \mathscr{D}(\hat{\mathbf{v}}, \hat{\chi})) = \langle (f_v - f_v^i, f_\chi - f_\chi^i), (\hat{\mathbf{v}}, \hat{\chi}) \rangle \qquad \forall (\hat{\mathbf{v}}, \hat{\chi}) \in \mathrm{Var}_v, \tag{12.124}$$

or, in extended form

$$\begin{aligned}
&\int_{\mathscr{B}_t} \rho(\dot{\mathbf{v}} \cdot \hat{\mathbf{v}} + (\dot{\chi} + \chi\chi)\mathbf{J} \cdot \hat{\chi}) dV + \int_{\mathscr{B}_t} \left(\sigma \cdot (\nabla_{\mathbf{x}}\hat{\mathbf{v}})^s + \gamma \cdot (\nabla_{\mathbf{x}}\hat{\mathbf{v}} - \hat{\chi}) + \mathsf{D} \cdot \nabla_{\mathbf{x}}\hat{\chi}\right) dV \\
&\quad = \int_{\partial\mathscr{B}_{tf}} \mathbf{t} \cdot \hat{\mathbf{v}} dS + \int_{\mathscr{B}_t} \mathbf{b} \cdot \hat{\mathbf{v}} dV + \int_{\partial\mathscr{B}_{tf}} \mathbf{N} \cdot \hat{\chi} dS \\
&\qquad + \int_{\mathscr{B}_t} \mathbf{M} \cdot \hat{\chi} dV \quad \forall (\hat{\mathbf{v}}, \hat{\chi}) \in \mathrm{Var}_v.
\end{aligned} \tag{12.125}$$

Integrating by parts as before we reach the Euler–Lagrange equations for the dynamic case, which are

$$\rho\dot{\mathbf{v}} - \operatorname{div}(\sigma + \gamma) = \mathbf{b} \quad \text{in } \mathscr{B}_t, \tag{12.126}$$

$$\rho(\dot{\chi} + \chi\chi)\mathbf{J} - \gamma - \operatorname{div} \mathbf{D} = \mathbf{M} \quad \text{in } \mathscr{B}_t, \tag{12.127}$$

$$(\sigma + \gamma)\mathbf{n} = \mathbf{t} \quad \text{on } \partial\mathscr{B}_{tf}, \tag{12.128}$$

$$\mathbf{D}\mathbf{n} = \mathbf{N} \quad \text{on } \partial\mathscr{B}_{tf}, \tag{12.129}$$

which now, together with the kinematical constraints

$$\mathbf{v} = \overline{\mathbf{v}} \quad \text{on } \partial\mathscr{B}_{tv}, \tag{12.130}$$

$$\chi = \overline{\chi} \quad \text{on } \partial\mathscr{B}_{tv}, \tag{12.131}$$

and the initial conditions

$$\mathbf{v}(t_0) = \mathbf{v}_0 \quad \text{in } \mathscr{B}_t, \tag{12.132}$$

$$\chi(t_0) = \chi_0 \quad \text{in } \mathscr{B}_t, \tag{12.133}$$

define the initial boundary value problem associated with the mechanical equilibrium in its strong form for the micromorphic continuum model under analysis.

Before ending this section we highlight once again the fundamental concept when building a variational model. Notably, the variational formulation encapsulates in a single integral expression, given by the PVP (12.125), all the information required to characterize the model. Moreover, generalized internal stresses in the model and compatible external forces are forged by the very kinematics for the particles that has been previously defined for the continuum. Consequently, the classical (strong) formulation given by the set of Euler–Lagrange equations (12.126)–(12.129) is only valid if there exists enough regularity for operations to make sense. Whenever this regularity requirement is not met, the local form requires modifications, which are naturally accounted for by the variational formulation through the appearance of natural jump conditions.

12.5 Micropolar Media

The case of micropolar (fluid or solid) media can be derived from the theory developed in previous sections as a particular case by assuming that each particle behaves as a rigid body. As a consequence of this additional hypothesis, the tensor field χ, now called χ^p, which represents the gradient of the relative velocity field of micro-particles, becomes a skew-symmetric second-order tensor, that is $\chi^p \in Skw\,((\chi^p)^T = -\chi^p)$, which represents the rotation of the macro-particle P. Let us then introduce the axial vector λ associated with the tensor χ^p, then we have $\chi^p\mathbf{a} = \lambda^p \times \mathbf{a}$ for an arbitrary vector $\mathbf{a}$. In this manner, the velocity field (12.55) is described as

$$\mathbf{v}' = \mathbf{v} + \lambda \times \mathbf{y}. \tag{12.134}$$

With this constraint, the sets $\mathscr{V}$, Kin_v and Var_v (see (12.56), (12.57) and (12.58)) are defined for the micropolar case in the following way

$$\mathscr{V}^p = \{(\mathbf{v}, \lambda);\ \mathbf{v} \in \mathbf{H}^1(\mathscr{B}_t), \lambda \in \mathbf{H}^1(\mathscr{B}_t)\}, \tag{12.135}$$

$$\text{Kin}_v^p = \{(\mathbf{v}, \lambda) \in \mathscr{V}^p;\ (\mathbf{v}, \lambda) = (\overline{\mathbf{v}}, \overline{\lambda}) \text{ on } \partial\mathscr{B}_{tv}\}, \tag{12.136}$$

$$\text{Var}_v^p = \{(\mathbf{v}, \lambda) \in \mathscr{V}^p;\ (\mathbf{v}, \lambda) = (\mathbf{0}, \mathbf{0}) \text{ on } \partial\mathscr{B}_{tv}\}. \tag{12.137}$$

In order to find the strain rate measures corresponding to this model, first notice that in this case $\varepsilon - \eta^s = \mathbf{O}$, and therefore, from (12.54), we obtain

$$\begin{aligned}\overline{d\mathbf{x}' \cdot d\mathbf{x}'}^{\,\cdot} = {} & 2\left[(\nabla_{\mathbf{x}}\mathbf{v})^s + ((\nabla_{\mathbf{x}}\chi^p)^{\mathbf{t}}\mathbf{y})^s\right] d\mathbf{x} \cdot d\mathbf{x} \\ & + 2\left[\nabla_{\mathbf{x}}\mathbf{v} + (\chi^p)^T + (\nabla_{\mathbf{x}}\chi^p)^{\mathbf{t}}\mathbf{y}\right] d\mathbf{x} \cdot d\mathbf{y}.\end{aligned} \tag{12.138}$$

Let us write the following tensor

$$\varepsilon^p = \nabla_{\mathbf{x}}\mathbf{v} + (\chi^p)^T = \nabla_{\mathbf{x}}\mathbf{v} - \chi^p, \tag{12.139}$$

where we have used the skew-symmetry of χ^p. Then, we put $(\nabla_{\mathbf{x}}\chi^p)^{\mathbf{t}}\mathbf{y}$ as follows

$$(\nabla_{\mathbf{x}}\chi^p)^{\mathbf{t}}\mathbf{y} = \nabla_{\mathbf{x}}(\chi^p\mathbf{y}) = \nabla_{\mathbf{x}}(\lambda \times \mathbf{y}), \tag{12.140}$$

where $\nabla_{\mathbf{x}}(\lambda \times \mathbf{a}_1)$ is the second-order tensor such that, for $\mathbf{a}_1$ and $\mathbf{a}_2$ independent from $\mathbf{x}$, it is verified that

$$[\nabla_{\mathbf{x}}(\lambda \times \mathbf{a}_1)]^T\mathbf{a}_2 = (\nabla_{\mathbf{x}}\lambda)^T(\mathbf{a}_1 \times \mathbf{a}_2). \tag{12.141}$$

An alternative way to write this operation is through the cross product between a tensor and a vector (see Chapter 1)

$$\nabla_{\mathbf{x}}(\lambda \times \mathbf{a}_1) = -\mathbf{a}_1 \times \nabla_{\mathbf{x}}\lambda, \tag{12.142}$$

and calling

$$\zeta^p = \nabla_{\mathbf{x}}\lambda, \tag{12.143}$$

we have

$$(\nabla_{\mathbf{x}}\chi^p)^{\mathbf{t}}\mathbf{y} = -\mathbf{y} \times \zeta^p. \tag{12.144}$$

Hence, expression (12.138) can be written as

$$\overline{d\mathbf{x}' \cdot d\mathbf{x}'}^{\,\cdot} = 2\left[\varepsilon^p - (\mathbf{y} \times \zeta^p)^s\right] d\mathbf{x} \cdot d\mathbf{x} + 2\left[\varepsilon^p - \mathbf{y} \times \zeta^p\right] d\mathbf{x} \cdot d\mathbf{y}, \tag{12.145}$$

that is, it is characterized by the following entities that measure the strain rate

$$\varepsilon^p = \nabla_{\mathbf{x}}\mathbf{v} - \chi^p(\lambda), \tag{12.146}$$

$$\zeta^p = \nabla_{\mathbf{x}}\lambda, \tag{12.147}$$

where ζ^p is a second-order tensor.

The internal power functional P_i is assumed, as before, to be a linear and continuous functional defined over the space of strain rates, which for this case gives

$$P_i = -\int_{\mathscr{B}_t} \left(\tau \cdot (\nabla_{\mathbf{x}}\mathbf{v} - \chi^p(\lambda)) + \mathbf{D} \cdot \nabla_{\mathbf{x}}\lambda\right) dV = -\int_{\mathscr{B}_t} (\tau \cdot \varepsilon^p + \mathbf{D} \cdot \zeta^p) dV, \tag{12.148}$$

where τ and $\mathbf{D}$ are second-order tensors. Remembering that $\tau \cdot \chi^p(\lambda) = -\lambda \cdot \tau_\times$ (see Chapter 1), we rewrite the internal power as

$$P_i = -\int_{\mathscr{B}_t} (\tau \cdot \nabla_{\mathbf{x}}\mathbf{v} + \tau_\times \cdot \lambda + \mathbf{D} \cdot \nabla_{\mathbf{x}}\lambda) dV. \tag{12.149}$$

Then, the PVP in this case becomes

$$\begin{aligned}\int_{\mathscr{B}_t} (\tau \cdot \nabla_{\mathbf{x}}\hat{\mathbf{v}} + \tau_\times \cdot \hat{\lambda} + \mathbf{D} \cdot \nabla_{\mathbf{x}}\hat{\lambda}) dV \\ = \int_{\partial\mathscr{B}_{tf}} \mathbf{t} \cdot \hat{\mathbf{v}} dS + \int_{\mathscr{B}_t} \mathbf{b} \cdot \hat{\mathbf{v}} dV + \int_{\partial\mathscr{B}_{tf}} \mathbf{m} \cdot \hat{\lambda} dS \\ + \int_{\mathscr{B}_t} \mathbf{g} \cdot \hat{\lambda} dV \quad \forall (\hat{\mathbf{v}}, \hat{\lambda}) \in \mathrm{Var}_v^p.\end{aligned} \tag{12.150}$$

The Euler–Lagrange equations are found by using the same arguments as employed before. This means that we have to integrate by parts and arrange terms to give

$$\begin{aligned}\int_{\partial\mathscr{B}_{tf}} (\tau\mathbf{n} - \mathbf{t}) \cdot \hat{\mathbf{v}} dS + \int_{\partial\mathscr{B}_{tf}} (\mathbf{D}\mathbf{n} - \mathbf{m}) \cdot \hat{\lambda} dS - \int_{\mathscr{B}_t} (\operatorname{div}\tau + \mathbf{b}) \cdot \hat{\mathbf{v}} dV \\ - \int_{\mathscr{B}_t} (-\tau_\times + \operatorname{div}\mathbf{D} + \mathbf{g}) \cdot \hat{\lambda} dV = 0 \quad \forall (\hat{\mathbf{v}}, \hat{\lambda}) \in \mathrm{Var}_v^p.\end{aligned} \tag{12.151}$$

Therefore, the Euler–Lagrange equations are

$$\operatorname{div}\tau + \mathbf{b} = \mathbf{0} \quad \text{in } \mathscr{B}_t, \tag{12.152}$$

$$-\tau_\times + \operatorname{div}\mathbf{D} + \mathbf{g} = \mathbf{0} \quad \text{in } \mathscr{B}_t, \tag{12.153}$$

$$\tau\mathbf{n} = \mathbf{t} \quad \text{on } \partial\mathscr{B}_{tf}, \tag{12.154}$$

$$\mathbf{D}\mathbf{n} = \mathbf{m} \quad \text{on } \partial\mathscr{B}_{tf}, \tag{12.155}$$

with essential boundary conditions given by

$$\mathbf{v} = \overline{\mathbf{v}} \quad \text{on } \partial\mathscr{B}_{tv}, \tag{12.156}$$

$$\lambda = \overline{\lambda} \quad \text{on } \partial\mathscr{B}_{tv}. \tag{12.157}$$

Figure 12.3 presents the basic elements that are part of the variational formulation for micropolar media.

12.6 Second Gradient Theory

This section presents in full detail the construction of the theory of second-order continua, which can be obtained as a particular case of the micromorphic theory developed before. In fact, to this end it is just necessary to postulate that micro-particles are subjected to the same strain rate of the body (i.e. we remove the micro-distortion).

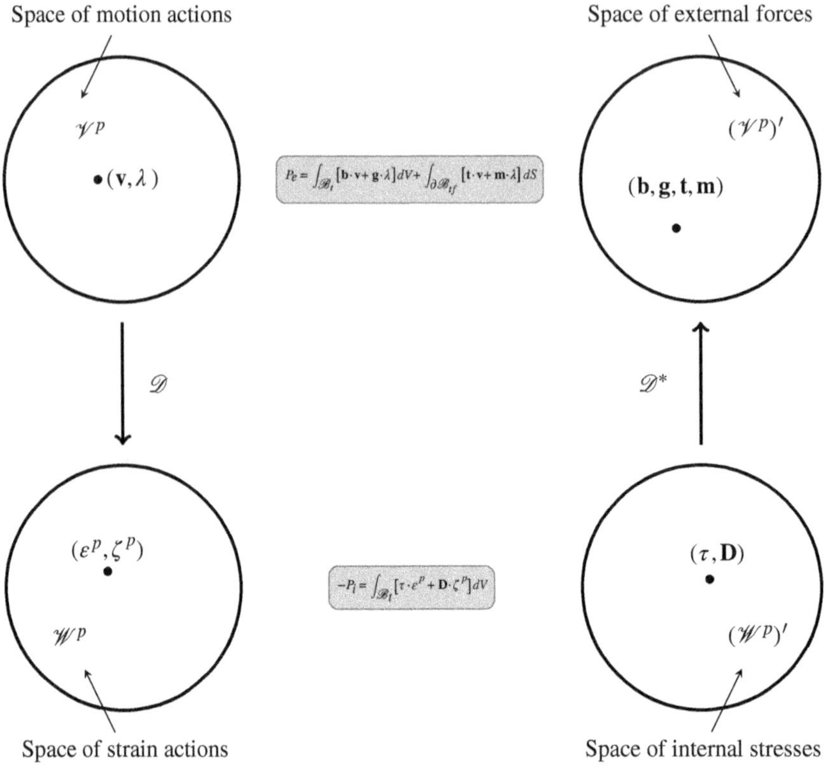

Figure 12.3 Ingredients in the variational theory for the modeling of micropolar continua.

In mathematical terms, this amounts to considering in the kinematical description given by (12.55), the constraint

$$\chi^g = \nabla_{\mathbf{x}}\mathbf{v}, \tag{12.158}$$

that is, the velocity field acquires the form

$$\mathbf{v}' = \mathbf{v} + (\nabla_{\mathbf{x}}\mathbf{v})\mathbf{y}. \tag{12.159}$$

This means that variables $\mathbf{v}$ and χ are not independent any longer, and since $\nabla_{\mathbf{x}}\chi^g = \nabla_{\mathbf{x}}\nabla_{\mathbf{x}}\mathbf{v}$, we have to require higher regularity in the model for the kinematical variable $\mathbf{v}$. Under these conditions, the space of possible motion actions gives

$$\mathscr{V}^g = \{\mathbf{v};\ \mathbf{v} \in \mathbf{H}^2(\mathscr{B}_t)\}, \tag{12.160}$$

where $\mathbf{H}^2(\mathscr{B}_t)$ is the space of continuous vector functions with continuous first derivatives and square integrable second derivatives. In turn, kinematically admissible motion actions belong to the linear manifold

$$\mathrm{Kin}_v^g = \left\{ \mathbf{v} \in \mathscr{V}^g;\ \mathbf{v}|_{\partial\mathscr{B}_{tv}} = \overline{\mathbf{v}},\ \left.\frac{\partial\mathbf{v}}{\partial\xi}\right|_{\partial\mathscr{B}_{tv}} = \overline{\mathbf{w}},\ \mathbf{v}|_{\Gamma_{tv}} = \tilde{\mathbf{v}} \right\}, \tag{12.161}$$

where Γ_{tv} represents a curve over the boundary $\partial\mathscr{B}_t$ through which there exists a discontinuity in the normal vector $\mathbf{n}$. In addition, the boundary $\partial\mathscr{B}_t$ has coordinate in the direction of the normal vector given by ξ. Moreover, the normal vector to the curve Γ_{tv} which is tangent to $\partial\mathscr{B}_t$ is called $\mathbf{m}$. The space associated with Kin_v^g is

$$\mathrm{Var}_v^g = \left\{ \mathbf{v} \in \mathscr{V}^g;\ \mathbf{v}|_{\partial\mathscr{B}_{tv}} = \mathbf{0},\ \left.\frac{\partial \mathbf{v}}{\partial \xi}\right|_{\partial\mathscr{B}_{tv}} = \mathbf{0},\ \mathbf{v}|_{\Gamma_{tv}} = \mathbf{0} \right\}. \tag{12.162}$$

For this particular case, observing that

$$((\nabla_{\mathbf{x}}\chi^g)^{\mathbf{t}}\mathbf{y})^s = (\nabla_{\mathbf{x}}((\nabla_{\mathbf{x}}\mathbf{v})^s))\mathbf{y}, \tag{12.163}$$

$$(\nabla_{\mathbf{x}}\chi^g)^{\mathbf{t}}\mathbf{y} = (\nabla_{\mathbf{x}}\nabla_{\mathbf{x}}\mathbf{v})\mathbf{y}, \tag{12.164}$$

the strain rate experienced by a fiber (12.42) becomes

$$\begin{aligned}\overline{d\mathbf{x}' \cdot d\mathbf{x}'}^{\cdot} &= 2\left[(\nabla_{\mathbf{x}}\mathbf{v})^s + (\nabla_{\mathbf{x}}((\nabla_{\mathbf{x}}\mathbf{v})^s))\mathbf{y}\right] d\mathbf{x} \cdot d\mathbf{x} \\ &\quad + 2\left[\nabla_{\mathbf{x}}\mathbf{v} + (\nabla_{\mathbf{x}}\mathbf{v})^T + (\nabla_{\mathbf{x}}\nabla_{\mathbf{x}}\mathbf{v})\mathbf{y}\right] d\mathbf{x} \cdot d\mathbf{y} \\ &\quad + 2(\nabla_{\mathbf{x}}\mathbf{v})^s d\mathbf{y} \cdot d\mathbf{y},\end{aligned} \tag{12.165}$$

and introducing the strain rate measures

$$\varepsilon = (\nabla_{\mathbf{x}}\mathbf{v})^s, \tag{12.166}$$

$$\Lambda = \nabla_{\mathbf{x}}\nabla_{\mathbf{x}}\mathbf{v}, \tag{12.167}$$

we have

$$\overline{d\mathbf{x}' \cdot d\mathbf{x}'}^{\cdot} = 2\left[\varepsilon + (\nabla_{\mathbf{x}}\varepsilon)\mathbf{y}\right] d\mathbf{x} \cdot d\mathbf{x} + 2\left[2\varepsilon + \Lambda\mathbf{y}\right] d\mathbf{x} \cdot d\mathbf{y} + 2\varepsilon\mathbf{y} \cdot d\mathbf{y}. \tag{12.168}$$

Thus, the internal power P_i takes the form

$$P_i = -\int_{\mathscr{B}_t} \left(\sigma \cdot (\nabla_{\mathbf{x}}\mathbf{v})^s + \mathsf{M} \cdot \nabla_{\mathbf{x}}\nabla_{\mathbf{x}}\mathbf{v}\right) dV = -\int_{\mathscr{B}_t} (\sigma \cdot \varepsilon + \mathsf{M} \cdot \Lambda) dV, \tag{12.169}$$

where σ is a symmetric second-order tensor and M is a third-order tensor with the symmetry property

$$\mathsf{M}^{\mathbf{t}} = \mathsf{M}. \tag{12.170}$$

It is worthwhile highlighting that notwithstanding we have used the same notation, both tensors σ and M are essentially different from those introduced in the micromorphic theory.

To obtain the characterization of external forces compatible with the kinematical model we proceed to integrate by parts the expression of the internal power

$$\begin{aligned}P_i &= \int_{\mathscr{B}_t} (\mathrm{div}\sigma \cdot \mathbf{v} + \mathrm{div}\mathsf{M} \cdot \nabla_{\mathbf{x}}\mathbf{v}) dV - \int_{\partial\mathscr{B}_{tf}} \left(\sigma\mathbf{n} \cdot \mathbf{v} + \mathsf{M}\mathbf{n} \cdot (\nabla_{\mathbf{x}}\mathbf{v})\right) dS \\ &= \int_{\mathscr{B}_t} \mathrm{div}(\sigma - \mathrm{div}\,\mathsf{M}) \cdot \mathbf{v} dV - \int_{\partial\mathscr{B}_{tf}} \left((\sigma - \mathrm{div}\,\mathsf{M})\mathbf{n} \cdot \mathbf{v} + \mathsf{M}\mathbf{n} \cdot (\nabla_{\mathbf{x}}\mathbf{v})\right) dS.\end{aligned} \tag{12.171}$$

The last term in (12.171) can be further integrated by parts only in the direction tangent to $\partial\mathscr{B}_{tf}$. To acomplish this, we have to decompose the tensor $\nabla_{\mathbf{x}}\mathbf{v}$ as follows (see Chapter 9)

$$\nabla_{\mathbf{x}}\mathbf{v} = (\nabla_{\mathbf{x}}^{\partial}\mathbf{v})\Pi + \frac{\partial\mathbf{v}}{\partial\xi} \otimes \mathbf{n}, \tag{12.172}$$

where $\nabla_{\mathbf{x}}^{\partial}(\cdot)$ is the surface gradient operator and $\Pi = \mathbf{I} - \mathbf{n}\otimes\mathbf{n}$ is the projection operator over the tangent plane to $\partial\mathscr{B}_{tf}$. With this, we write the last term in (12.171) as follows, where we integrate by parts the surface gradient

$$\begin{aligned}\int_{\partial\mathscr{B}_{tf}} \mathbf{Mn}\cdot(\nabla_{\mathbf{x}}\mathbf{v})dS &= \int_{\partial\mathscr{B}_{tf}} \mathbf{Mn}\cdot\left((\nabla_{\mathbf{x}}^{\partial}\mathbf{v})\Pi + \frac{\partial\mathbf{v}}{\partial\xi}\otimes\mathbf{n}\right)dS\\ &= \int_{\partial\mathscr{B}_{tf}} \left((\mathbf{Mn})\Pi\cdot\nabla_{\mathbf{x}}^{\partial}\mathbf{v} + (\mathbf{Mn})\mathbf{n}\cdot\frac{\partial\mathbf{v}}{\partial\xi}\right)dS\\ &= -\int_{\partial\mathscr{B}_{tf}} \mathrm{div}^{\partial}\left[(\mathbf{Mn})\Pi\right]\cdot\mathbf{v}dS + \int_{\Gamma_{tf}} [\![(\mathbf{Mn})\mathbf{m}]\!]\cdot\mathbf{v}dc\\ &\quad + \int_{\partial\mathscr{B}_{tf}} (\mathbf{Mn})\mathbf{n}\cdot\frac{\partial\mathbf{v}}{\partial\xi}dS,\end{aligned} \tag{12.173}$$

where Γ_{tf} represents the curve over the boundary where there exists a discontinuity in the normal vector $\mathbf{n}$ (and therefore in the tangent plane), such that the vector $\mathbf{m}$ represents the vector normal to the curve Γ_{tf} which is over the tangent plane, and can be written as $\mathbf{m} = \mathbf{l}\times\mathbf{n}$, with $\mathbf{l}$ the tangent unit vector to the curve Γ_{tf}. Also, div^{∂} is the surface divergence operator. Putting (12.173) into (12.171) yields

$$\begin{aligned}P_i = &\int_{\mathscr{B}_t} \mathrm{div}(\sigma - \mathrm{div}\mathbf{M})\cdot\mathbf{v}dV - \int_{\partial\mathscr{B}_{tf}} (\sigma - \mathrm{div}\mathbf{M})\mathbf{n}\cdot\mathbf{v}dS\\ &+ \int_{\partial\mathscr{B}_{tf}} \mathrm{div}^{\partial}\left[(\mathbf{Mn})\Pi\right]\cdot\mathbf{v}dS - \int_{\Gamma_{tf}} [\![(\mathbf{Mn})\mathbf{m}]\!]\cdot\mathbf{v}dc\\ &- \int_{\partial\mathscr{B}_{tf}} (\mathbf{Mn})\mathbf{n}\cdot\frac{\partial\mathbf{v}}{\partial\xi}dS,\end{aligned} \tag{12.174}$$

and rearranging terms we get

$$\begin{aligned}P_i = &\int_{\mathscr{B}_t} \mathrm{div}(\sigma - \mathrm{div}\mathbf{M})\cdot\mathbf{v}dV - \int_{\partial\mathscr{B}_{tf}} \left[(\sigma - \mathrm{div}\mathbf{M})\mathbf{n} - \mathrm{div}^{\partial}((\mathbf{Mn})\Pi)\right]\cdot\mathbf{v}dS\\ &- \int_{\Gamma_{tf}} [\![(\mathbf{Mn})]\!]\mathbf{m}\cdot\mathbf{v}dc - \int_{\partial\mathscr{B}_{tf}} (\mathbf{Mn})\mathbf{n}\cdot\frac{\partial\mathbf{v}}{\partial\xi}dS.\end{aligned} \tag{12.175}$$

In this way, we see that the external forces compatible with the kinematical model have the following characterization

$$\langle f, \mathbf{v}\rangle = \int_{\mathscr{B}_t} \mathbf{b}\cdot\mathbf{v}dV + \int_{\partial\mathscr{B}_{tf}} \mathbf{t}\cdot\mathbf{v}dS + \int_{\partial\mathscr{B}_{tf}} \mathbf{r}\cdot\frac{\partial\mathbf{v}}{\partial\xi}dS + \int_{\Gamma_{tf}} \mathbf{s}\cdot\mathbf{v}dc. \tag{12.176}$$

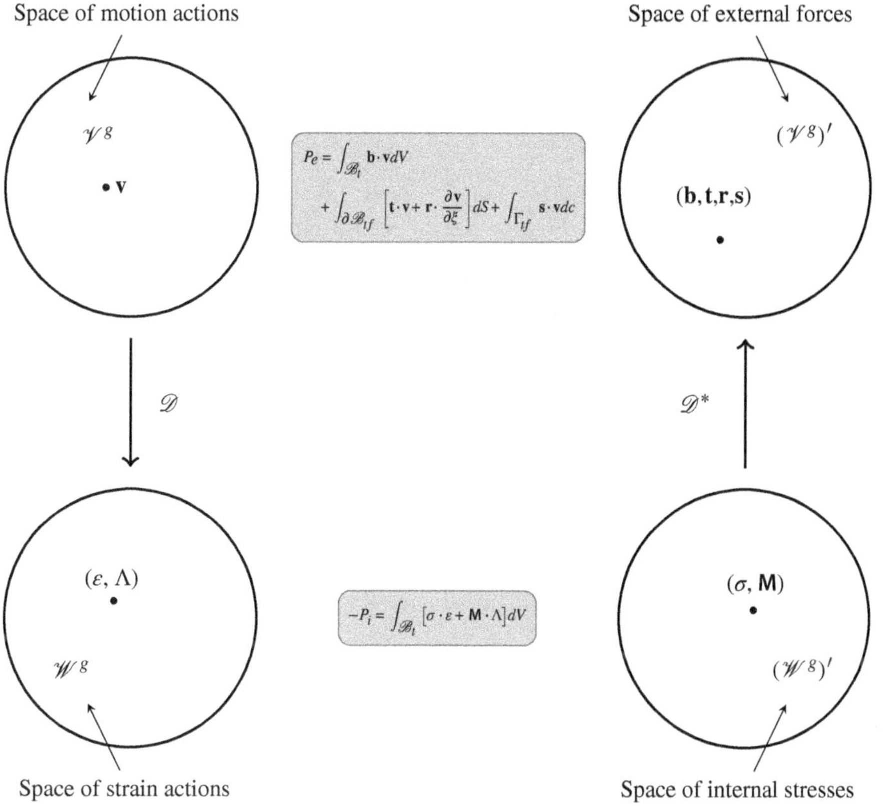

Figure 12.4 Ingredients in the variational theory for the modeling of second gradient continua.

We now provide the statement of the PVP for the second gradient model through the following variational equation

$$\begin{aligned}
&\int_{\mathscr{B}_t} \left(\sigma \cdot (\nabla_{\mathbf{x}}\hat{\mathbf{v}})^s + \mathbf{M} \cdot \nabla_{\mathbf{x}}\nabla_{\mathbf{x}}\hat{\mathbf{v}}\right) dV \\
&= \int_{\mathscr{B}_t} \mathbf{b} \cdot \hat{\mathbf{v}} dV + \int_{\partial\mathscr{B}_{tf}} \mathbf{t} \cdot \hat{\mathbf{v}} dS + \int_{\partial\mathscr{B}_{tf}} \mathbf{r} \cdot \frac{\partial \hat{\mathbf{v}}}{\partial \xi} dS \\
&\quad + \int_{\Gamma_{tf}} \mathbf{s} \cdot \hat{\mathbf{v}} dc \quad \forall \hat{\mathbf{v}} \in \mathrm{Var}_v^g. \qquad (12.177)
\end{aligned}$$

Finally, and following the same integration by part steps realized before when the external forces were characterized, we obtain the Euler–Lagrange equations associated with the PVP of the second gradient theory developed here

$$\mathrm{div}\sigma - \mathrm{div}\mathrm{div}\mathbf{M} + \mathbf{b} = \mathbf{0} \quad \text{in } \mathscr{B}_t, \qquad (12.178)$$

$$(\sigma - \mathrm{div}\mathbf{M})\mathbf{n} - \mathrm{div}^{\partial}((\mathbf{M}\mathbf{n})\Pi) = \mathbf{t} \quad \text{on } \partial\mathscr{B}_{tf}, \qquad (12.179)$$

$$(\mathbf{M}\mathbf{n})\mathbf{n} = \mathbf{r} \quad \text{on } \partial\mathscr{B}_{tf}, \qquad (12.180)$$

$$[\![(\mathbf{M}\mathbf{n})\mathbf{m}]\!] = \mathbf{s} \quad \text{on } \Gamma_{tf}, \qquad (12.181)$$

which, together with the kinematical constraints

$$\mathbf{v} = \overline{\mathbf{v}} \quad \text{on } \partial\mathscr{B}_{tv}, \tag{12.182}$$

$$\frac{\partial \mathbf{v}}{\partial \xi} = \overline{\mathbf{w}} \quad \text{on } \partial\mathscr{B}_{tv}, \tag{12.183}$$

$$\mathbf{v} = \tilde{\mathbf{v}} \quad \text{on } \mathit{\Gamma}_{tv}, \tag{12.184}$$

provide us with the description of the mechanical problem in its strong form.

Figure 12.4 shows the elements that are part of the variational formulation for second gradient media.

Part V

Multiscale Modeling

Theories relating the behavior of continuum media at a macroscopic (or large) scale with the physical interactions occurring at microscopic (small) scales are termed *multiscale theories*. These theories originated in the second half of the last century, where the landmark contributions of Kirkwood and collaborators placed the groundwork to assemble the governing equations of transport phenomena in continuum media starting from statistical mechanics arguments within the molecular dynamics realm [148, 154–156]. Posteriorly, in the field of solid mechanics, substantial theoretical developments towards the estimation of macroscopic properties of heterogeneous materials began with the pioneer work of Hashin [137], Hill [141–144], Budiansky [43], Mandel [183], and Gurson [130], among others. A parallel stream of developments was supported by the asymptotic analysis of partial differential equations with periodic coefficients for the modeling of continuum media with periodic microstructure, which was initiated in the 1970s. Fundamental contributions in this specific field are the books by Bensoussan and collaborators [27] and by Sanchez-Palencia [268, 269]. A common aspect in all these theories is the fact that variables at the macroscopic level, usually named homogenized variables, are invariably related to some kind of averaging process of the fields defined at the microscopic level.

Recent decades have witnessed a large increase in the appearance of multiscale theories within the context of computational mechanics. Attention has been mainly centered on formulations that establish the concept of the Representative Volume Element (RVE) where stresses and strains in the macroscale are obtained as volumetric averages of the fields corresponding to the RVE. Although the RVE is also generally modeled as a continuum, it can be considered as being formed by discrete interactions. However, a common characteristic among all these contributions resides in the fact that they are, almost exclusively, rooted in computational homogenization techniques which make use of finite element techniques as the numerical strategy for the approximation of the problem [102, 161–163, 167, 192–196, 210, 235, 281, 291].

Nowadays, applications of this kind of approach in solid mechanics embrace a wide scope of phenomenology including, among others, plasticity, thermomechanical coupling, dynamics and vibrations, and problems involving material damage and failure. For example, in the field of plasticity, the review reported by McDowell presents a compendium of the application of multiscale theories not only for continuum scales, but also at the level of molecular and atomistic physical interactions [189]. The literature in this area offers clear evidence of the capability of multiscale modeling, which allows several challenges to be surmounted when modeling plastic behavior resulting from complex phenomena such as dislocation dynamics, crystal plasticity, and phase transformation, all of them under intricate histories of strain. Nonetheless, it is also important to emphasize that many fundamental issues still remain open. That is, many questions related to the understanding and modeling of the mechanisms underlying the microscale as well as the development of suitable multiscale theories are yet to be answered (see [189] and references therein).

Multiscale formulations have also been demonstrated to be useful in the assembly of high-order constitutive models [162, 163, 168, 169, 280]. These formulations are adequate to model the material behavior when scales are not sufficiently separated, and the material behavior depends upon the size of microscale material structures. An appealing aspect of RVE multiscale strategies is that they are capable of delivering high-order behavior at the large scale in spite of being considering conventional mechanical models at the small scale posed by the RVE.

A detailed description of the vast universe of applications relying upon these RVE-based multiscale theories is out of the scope of the present material. The interested reader is referred to the work in [32–35, 121, 266, 267, 275–278, 296–298]. In these contributions, specific bibliographical references regarding the different areas of application of multiscale modeling can be found.

From the previous exposition, it is clear that the scope of applicability of multiscale theories based on the RVE concept is extensive. Moreover, it is important to add that the adoption of such theories is gaining momentum in the present context. This is easily confirmed by the large number of scientific contributions published in the field in recent years, as well as by the number of conferences and workshops organized around the multiscale topic. One of the main causes of the advancement of such multiscale strategies is the compelling need to employ computational tools capable of predicting with a high degree of precision the material response in situations where complex physical mechanisms develop at small scales, but which cannot be captured using conventional phenomenological constitutive modeling. Another reason for the steady growth in interest for multiscale modeling is the need to gain essential insight into the underlying mechanisms occurring at the microscale level and how these mechanisms are manifested at the macroscale (observable) level [292, 304]. Such understanding, together with the potential capabilities of predicting the impact on the observable material behavior, is crucial to optimize the use of existing materials, as well as to project new materials in a rational and scientific manner.

Notwithstanding the widespread use of RVE-based multiscale theories, a general and unified theoretical framework for the development and treatment of such theories was lacking until recently [34]. As a matter of fact, the multiscale approach exploiting the concept of the RVE in classical solid mechanics, where both macroscopic and microscopic kinematics are described in the traditional framework of continuum media, has reached solidly formulated theoretical bases through the contributions of Hill [144] and Mandel [183]. However, it is quite probable that any attempt to extend this approach beyond the classical arena faces severe challenges. This is because classical theory (and available extensions) was developed without a clear distinction between the fundamentals and the corresponding consequences. Hence, even now most of the literature in the field lacks a clear and precise presentation of the extensions required in classical theory so other physical phenomenology can be incorporated in the different scales.

Therefore, a proper extension of the classical principles which play a role in the modeling across the scales requires careful consideration which cannot be solely established on top of physical intuition, and even less can be based on computational techniques (as occurs in most of the publications in the field). In fact, this transfer between scales, which has a fundamental role in the definition of the boundary conditions at the domain which stands for the RVE, as well as in the development of homogenization formulae for variables such as, for example, stress, is in general postulated without an underlying fundamental principle [25, 26, 174, 274].

As a response to the questions exposed above, that is, to the lack of a general and well-fundamented multiscale modeling structure, and to the pressing need for the development of sophisticated multiscale models, in this part of the book, making use of the concepts of mathematical duality and the Principle of Virtual Power (PVP) studied throughout the book, we present a unified variational theory for a wide range of multiscale models based on the concept of RVE. Our main goal here is to present/create a sufficiently general substructure within which novel multiscale models, capable of incorporating new physics in the microscale, can be developed on top of solid mathematical (variational) bases by following clear and precise systematic steps. Our experience is that, the theory presented in the following chapters is capable of dealing with multiphysics problems, including material failure due to microscale strain localization, solid dynamics, thermomechanics, and fluid mechanics, among others.

The unified structure presented has been christened the *Method of Multiscale Virtual Power*, and it is rooted in three fundamental principles

- *kinematical admissibility*[1]
- *duality*[2]
- *Principle of Multiscale Virtual Power (PMVP).*[3]

To simplify the presentation we will limit it to the case of two scales: (i) the macroscale characterized by the characteristic length of the domain where the macroscopic problem is defined and (ii) the microscale characterized by the typical length of the RVE adopted.

1 Within the general context of the proposed method, the expression *kinematics* must be understood in a wide (generalized) sense, which is the one established by the primal physical variables in a given formulation. Kinematical variables are those quantities whose rates produce power in duality with corresponding fluxes (stress-like variables). In mechanical problems these correspond to generalized displacements and the associated strains and rates. In thermal problems, they correspond to the temperature and its gradient, and the same for other kinds of problems.

2 Duality was the mathematical foundation for the edifice of the PVP and of the PCVP in Chapter 3. As we saw, this duality allows us to determine, consistently with the kinematical model adopted, the class of generalized (internal and external) forces which are admissible for the problem under study. In other words, with the *duality* we avoid the *a priori* definitions of internal and external generalized forces that are developed in the body as a consequence of the chosen kinematical description. More precisely, and as we have seen throughout the previous chapters, when admitting a certain kinematics in a given problem we are intrinsically morphing the internal and external forces by means of the characterization provided by linear and continuous functionals defined over the spaces of kinematically admissible virtual motion actions.

3 This principle is nothing but the generalization of the PVP but now applied within the context of multiple scales where it is possible to exert power. According to what we have seen, such a principle, jointly with the concepts of kinematical admissibility and duality, defines a solid mathematical basis and a systematic structure for the modeling of innumerous problems.

Multiscale modeling in time will not be addressed here, but the concepts presented can be extended to also consider such a scenario.

With this in mind, the concept of kinematical admissibility establishes the connection between the adopted kinematics in the macro and micro (RVE) levels. For this, two operators will be introduced, the insertion and homogenization operators. These operators are responsible for the transfer of kinematical information between the two scales, which, in general, can be modeled by utilizing different kinematical descriptions. Therefore, these operators establish the constraints that kinematical fields have to satisfy to guarantee the kinematical admissibility of the multiscale model. That is, kinematical admissibility consists of the preservation of the magnitude of these kinematical variables during the macro–micro transfer process. In turn, these restrictions are the ones that allow us to characterize the spaces and sets the sought solutions belong to, and over which the power defined by the linear functionals will be calculated. The duality plays a crucial role in the establishment of the correct definition of internal and external forces for the model. As we already have mentioned, these variables cannot be defined independently from the kinematics. Rather, these variables are mere consequences of the chosen kinematics, and so they emerge in an unambiguous manner via mathematical duality (power) with respect to the entities that describe the kinematics. Finally, the PMVP can be seen as the generalization of the (classical) Principle of Macrohomogenization proposed by Hill and Mandel [144, 183], which is cast in a variational structure and extended to comprehend not only internal virtual power but also virtual power of an external nature, accounting thus for the total virtual power developed at both macroscale and microscale. In this regard, the reader interested in seeing how these ideas were shaped and from which this unified formulation was originated can inspect our work in [32–35, 121, 266, 267, 275–278, 296–298]. In particular, in [34] the reader will find the present unified variational formulation for RVE-based multiscale modeling.

When developing a model employing the rational structure proposed in the following chapters we have, at the core of the theory, that the only degree of freedom during the modeling process resides (i) in the definition of the variables that describe the chosen kinematics in each of the spatial scales and (ii) in the definition of how these variables are related with each other in a way such that the kinematical admissibility is guaranteed. Once this has been set, the remaining part of the model is fully characterized in an unequivocal manner over the basis provided by the mathematical duality and by the PMVP. As a result, the equilibrium equations at both scales as well as the homogenization operators, which establish the relation between the generalized (internal and external) forces at both scales, are postulated through variational arguments which, as discussed before, also provide a proper framework for the posterior numerical treatment in the search for approximate solutions. With this, the proposed formulation provides a clear separation between the theoretical concepts and the subsequent numerical approximation, something that, in most of the publications in the field, continues to be confusingly and incorrectly reported.

Finally, we understand that the proposed framework examined in this part of the book contains a logic structure with a degree of flexibility which not only provides a rational ground for existing multiscale models, but, more importantly, significantly facilitates the rigorous development, through systematic and well-defined steps, of new and refined multiscale theories.

13

Method of Multiscale Virtual Power

13.1 Introduction

In this chapter we will present the derivation of the proposed multiscale modeling framework. This task is accomplished by using an abstract notation which provides the theory with a general unified framework able to meet the needs for modeling a wide variety of physical problems, such as classical formulations in solid and fluid mechanics, strain localization at macro and micro scales, damage and failure mechanics, high-order continua, multiphysics problems, and multiscaling between scales featuring different physical phenomena, among others.

For ease of reading, each time a given piece of theory has been established, the adopted abstract notation will be specified for the classical scenario encountered in solid mechanics within the infinitesimal strains regime throughout the chapter.

The multiscale modeling strategy adopted in this chapter follows the derivation/discussion of the basic foundational elements on top of which the proposed formulation is sustained. That is to say, we will sequentially address the concepts of kinematical admissibility, mathematical duality, and the Principle of Multiscale Virtual Power (PMVP). However, and since all this procedure is extensively supported by the Principle of Virtual Power (PVP), we initiate the chapter with a concise revision of the Method of Virtual Power introduced in Chapter 3. Furthermore, this methodical remembrance will serve to set the basic notation, as well as to make this part self-contained.

13.2 Method of Virtual Power

13.2.1 Kinematics

This section recalls the main concepts involved in the kinematical description necessary to model physical problems within the context of the Method of Virtual Power developed in Chapter 3. To this end, and for simplicity, the presentation is limited to the case of continuum media. Let $\mathscr{B}$ be the body occupying the spatial domain Ω with regular boundary Γ, and let us denote by $\mathbf{x} \in \Omega$ an arbitrary point in this region of space.

Introduction to the Variational Formulation in Mechanics: Fundamentals and Applications, First Edition.
Edgardo O. Taroco, Pablo J. Blanco and Raúl A. Feijóo.

The set of generalized displacements[1] that characterizes the kinematics associated with $\mathscr{B}$ belongs to the space $\mathscr{V}$ whose elements $u \in \mathscr{V}$ are n-tuples of tensor fields regular enough such that the mathematical operations are well-posed. The components of an arbitrary element $u \in \mathscr{V}$ are denoted by u^i, $i = 1, \dots, n$, that is, $u = (u^1, \dots, u^n)$. Moreover, each component u^i can, in turn, be a zero-order tensor field (scalar), a first-order tensor field (vector), a second-order tensor field, and so on, depending upon the physics represented by the model. In this way, each component u^i will be described by r^i scalar fields. Thus, the number of scalar fields which fully describe the element u is $R = \sum_{i=1}^{n} r^i$. Finally, each u^i has the domain of definition $\Omega^{\mathscr{V}_i} := \mathrm{Dom}(\mathscr{V}_i)$, $i = 1, \dots, n$, that is

$$
\begin{aligned}
u^i : \Omega^{\mathscr{V}_i} &\to \mathscr{V}_i, \\
\mathbf{x} &\mapsto u^i(\mathbf{x}),
\end{aligned}
\tag{13.1}
$$

where $\Omega^{\mathscr{V}_i} \subseteq \Omega$. In turn, each domain $\Omega^{\mathscr{V}_i}$ can be a set of points, lines, surfaces or volumes. This can be expressed in compact form in the following way

$$
\begin{aligned}
u : \Omega^{\mathscr{V}} &\to \mathscr{V}, \\
\mathbf{x} &\mapsto u(\mathbf{x}),
\end{aligned}
\tag{13.2}
$$

where $\Omega^{\mathscr{V}} := \mathrm{Dom}(\mathscr{V}) = (\mathrm{Dom}(\mathscr{V}_1), \dots, \mathrm{Dom}(\mathscr{V}_n))$, or, equivalently, we can put $\Omega^{\mathscr{V}} = (\Omega^{\mathscr{V}_1}, \dots, \Omega^{\mathscr{V}_n})$. Let us see now how this structure/notation accounts for classical examples in solid mechanics.

Classical solid mechanics. In this case, the domain Ω is a region of the Euclidean space and the generalized displacements are reduced to a single field, which is given by the displacement field $u = \mathbf{u}$ (i.e. it is a first-order tensor field), $\mathscr{V}$ is a proper Sobolev functional space in Ω. Usually, this space is adopted to be $\mathscr{V} = \mathbf{H}^1(\Omega)$, that is, a space of square integrable functions with square integrable derivatives in Ω. Hereafter we will maintain this standard assumption. Also, we have $\Omega^{\mathscr{V}} = \Omega$. A more elaborate case can be found in multiphysics problems. For example, when studying electro-mechanical interactions, the kinematics is given by $u = (\mathbf{u}, \phi)$, where $\mathbf{u}$ is the displacement in the solid body (a first-order tensor, that is, a vector field) and ϕ is a electrostatic potential field (a zero-order tensor, that is, a scalar field) (see [73]). For the case of micromorphic fluid media studied in Chapter 12, the generalized displacement is characterized by $u = (\mathbf{v}, \chi)$ including a velocity field $\mathbf{v}$, which is a first-order tensor field, and a strain rate field χ, which is a second-order tensor field (see [80, 81, 115, 116]). ■

Now, we need to define the set $\mathrm{Kin}_u \subset \mathscr{V}$ of kinematically admissible generalized displacements. That is, the elements $u \in \mathrm{Kin}_u$ satisfy some kind of kinematical constraint, for example essential boundary conditions prescribed over a portion of the boundary or distributed constraints over the whole domain. A solution to the equilibrium problem associated with the physical system under consideration (yet to be defined) will be sought in this set Kin_u. For ease of presentation, we will consider that Kin_u is a linear manifold generated by the subspace Var_u of kinematically admissible generalized motion actions

$$
\mathrm{Var}_u = \{\hat{v} \in \mathscr{V};\ \hat{v} = v_1 - v_2,\ v_1, v_2 \in \mathrm{Kin}_u\}. \tag{13.3}
$$

---

1 In this chapter, and making abuse of notation, we use the term generalized displacements also for the case of generalized motion actions. In this manner, $\mathscr{V}$ will be the space where all these elements belong to.

Classical solid mechanics. For this case we have, for example, $\text{Kin}_u = \{\mathbf{u} \in \mathbf{H}^1(\Omega);\ \mathbf{u}|_{\Gamma_u} = \mathbf{u}^*\}$ and $\text{Var}_u = \{\mathbf{u} \in \mathbf{H}^1(\Omega);\ \mathbf{u}|_{\Gamma_u} = \mathbf{0}\}$, where Γ_u is the part of the boundary Γ where the displacements are essentially prescribed. ■

A further fundamental concept we have established for the variational formulation of the PVP is that of generalized strain (or deformation) action, which belongs to the space denoted by $\mathscr{W}$. In general, any element $D \in \mathscr{W}$ is characterized by an m-tuple of tensor fields. That is, $D = (D^1, \dots, D^m)$, where each component D^i can be a zero-order tensor field (scalar), a first-order tensor field (vector), or any arbitrary order tensor field depending upon the physics of the problem. Each component D^i, $i = 1, \dots, m$, is described by s^i scalar fields, and so the total number of scalar fields which characterize D is given by $S = \sum_{i=1}^m s^i$. Also, each component D^i is defined in its corresponding domain of definition $\Omega^{\mathscr{W}_i} := \text{Dom}(\mathscr{W}_i)$, that is

$$\begin{aligned} D^i : \Omega^{\mathscr{W}_i} &\to \mathscr{W}_i, \\ \mathbf{x} &\mapsto D^i(\mathbf{x}), \end{aligned} \tag{13.4}$$

where, as before, each $\Omega^{\mathscr{W}_i} \subseteq \Omega$ can be a set of points, lines, surfaces or volumes. This can be put in compact form as follows

$$\begin{aligned} D : \Omega^{\mathscr{W}} &\to \mathscr{W}, \\ \mathbf{x} &\mapsto D(\mathbf{x}), \end{aligned} \tag{13.5}$$

where we write $\Omega^{\mathscr{W}} := \text{Dom}(\mathscr{W}) = (\text{Dom}(\mathscr{W}_1), \dots, \text{Dom}(\mathscr{W}_m))$, that is to say, $\Omega^{\mathscr{W}} = (\Omega^{\mathscr{W}_1}, \dots, \Omega^{\mathscr{W}_m})$.

Finally, we define the linear operator $\mathscr{D}$ which takes elements from the space $\mathscr{V}$ to the space $\mathscr{W}$, and so this operator is called the generalized strain (or deformation) action operator, that is

$$\begin{aligned} \mathscr{D} : \mathscr{V} &\to \mathscr{W}, \\ u &\mapsto D = \mathscr{D}u. \end{aligned} \tag{13.6}$$

Classical solid mechanics. In this case, $\mathscr{D} = (\nabla(\cdot))^s$, and so we have $\mathscr{D}\mathbf{u} = (\nabla\mathbf{u})^s$, $n = 1$, $R = 3$, $m = 1$ and $S = 6$. The field $(\nabla\mathbf{u})^s$ is defined over the whole body, and therefore $\Omega^{\mathscr{W}} = \Omega$ and belongs to the space of functions $\mathscr{W} = \mathsf{L}^2_{\text{sym}}(\Omega) = \{\varepsilon \in \mathsf{L}^2(\Omega);\ \varepsilon = \varepsilon^T\}$ (recall that we have adopted a standard mathematical setting). Another example, in the case of electro-mechanical coupling, is that for which the generalized strain action operator is such that $\mathscr{D}(\mathbf{u}, \phi) = ((\nabla\mathbf{u})^s, \nabla\phi)$, where $\mathbf{u}$ and ϕ have been introduced in a previous remark. Thus, in this case

$$\mathscr{D} = \begin{pmatrix} (\nabla(\cdot))^s & 0 \\ 0 & \nabla(\cdot) \end{pmatrix}, \tag{13.7}$$

where $n = 2$, $R = 4$, $m = 2$ and $S = 9$. For the micromorphic fluid studied in Chapter 12, the generalized strain action operator characterizes the triple $\mathscr{D}(\mathbf{v}, \chi) = ((\nabla\mathbf{v})^s, \nabla\mathbf{v} - \chi, \nabla\chi)$, to give

$$\mathscr{D} = \begin{pmatrix} (\nabla(\cdot))^s & 0 \\ \nabla(\cdot) & -\mathbf{I} \\ 0 & \nabla(\cdot) \end{pmatrix}, \tag{13.8}$$

where $n = 2$, $R = 12$, $m = 3$ and $S = 45$. It is important to note that the sense of the gradient operator $\nabla(\cdot)$ will depend on the configuration in which we are describing the problem. Thus, in solid mechanics it will usually be the material gradient (it could also be the spatial gradient), while in the case of fluid mechanics, including micromorphic media, it is the spatial gradient operator. ∎

Let us revisit the concept of kinematical compatibility within the context of the Method of Virtual Power. An element $D \in \mathscr{W}$ is said to be a kinematically compatible generalized deformation action if there exists an element $u \in \mathscr{V}$ such that $D = \mathscr{D} u$. The domain where a kinematically compatible generalized deformation action is defined will be denoted by $\Omega^{\mathscr{W}}$ and can be written as $\Omega^{\mathscr{W}} = \mathrm{Dom}(\mathscr{D}(\mathscr{V}))$.

Observe that since $\mathscr{D}$ is a linear operator it has a well-defined (rectangular) matrix representation given by

$$\mathscr{D} = \begin{pmatrix} \mathscr{D}^{11} & \mathscr{D}^{12} & \dots & \mathscr{D}^{1n} \\ \mathscr{D}^{21} & \mathscr{D}^{22} & \dots & \mathscr{D}^{2n} \\ \vdots & \vdots & \ddots & \vdots \\ \mathscr{D}^{m1} & \mathscr{D}^{m2} & \dots & \mathscr{D}^{mn} \end{pmatrix}. \tag{13.9}$$

In this representation we have $\Omega^{\mathscr{W}_i} = \mathrm{Dom}(\mathscr{D}^{i1}(u^1)) = \dots = \mathrm{Dom}(\mathscr{D}^{in}(u^n))$, $i = 1, \dots, m$.

Another important subspace of $\mathscr{V}$, which plays an important role in the Method of Virtual Power, is given by the null space of operator $\mathscr{D}$, denoted by $\mathscr{N}(\mathscr{D}) \subset \mathscr{V}$, and which is defined in the following way

$$\mathscr{N}(\mathscr{D}) = \{u \in \mathscr{V}\,;\ \mathscr{D} u = 0\}. \tag{13.10}$$

This is the subspace of $\mathscr{V}$ whose elements are motion actions resulting in null generalized deformation actions. This is why these generalized motion actions are also called generalized rigid motion actions.

Classical solid mechanics. In this case, the null space of operator $\mathscr{D} = (\nabla(\cdot))^s$ is the space of all infinitesimal rigid motion actions which, therefore, admit the following representation $\mathbf{u}(\mathbf{x}) = \mathbf{u}_o + \mathbf{W}(\mathbf{x} - \mathbf{x}_o)$, where $\mathbf{u}_o$ is a uniform field, $\mathbf{W}$ is a uniform second-order skew-symmetric tensor field, and $\mathbf{x}_o$ is an arbitrary reference point. ∎

It is also important to highlight the subspace $\mathscr{D}(\mathrm{Var}_u) \subset \mathscr{W}$, which is the image of Var_u under the transformation given by the operator $\mathscr{D}$. Naturally, the elements in this space are termed kinematically compatible generalized deformation actions.

13.2.2 Duality

As we have seen throughout the present book, the foundational first hypothesis in the modeling based on the Method of Virtual Power is to admit that the internal generalized forces, denoted by Σ, and the generalized external forces, denoted by f, which are admissible for the physical system we are modeling, are dual (in the sense that they are linear and continuous functionals) to the adopted kinematical variables we have chosen to describe the system. That is, these internal and external forces cannot be defined a priori, but they are a consequence of the kinematics and of the duality concept. Hence, denoting by $\mathscr{W}'$ and $\mathscr{V}'$ the dual spaces of $\mathscr{W}$ and $\mathscr{V}$, respectively, this first hypothesis can be formalized in the following manner

- The nature of the generalized internal forces admissible to the system, which will be called generalized stresses, $\Sigma \in \mathscr{W}'$, is characterized by linear and continuous functionals over $\mathscr{W}$, defined by the duality product denoted by $\langle \Sigma, D\rangle_{\mathscr{W}'\times\mathscr{W}}$.
- Similarly, the nature of the generalized external forces $f \in \mathscr{V}'$ is characterized by linear and continuous functionals in $\mathscr{V}$, defined by the duality pairing denoted by $\langle f, u\rangle_{\mathscr{V}'\times\mathscr{V}}$.

These duality products satisfy well-known properties

- $\langle \Sigma, D\rangle_{\mathscr{W}'\times\mathscr{W}} = 0 \quad \forall D \in \mathscr{W} \Rightarrow \Sigma = 0$
- $\langle \Sigma, D\rangle_{\mathscr{W}'\times\mathscr{W}} = 0 \quad \forall \Sigma \in \mathscr{W}' \Rightarrow D = 0$
- $\langle f, u\rangle_{\mathscr{V}'\times\mathscr{V}} = 0 \quad \forall u \in \mathscr{V} \Rightarrow f = 0$
- $\langle f, u\rangle_{\mathscr{V}'\times\mathscr{V}} = 0 \quad \forall f \in \mathscr{V}' \Rightarrow u = 0.$

To characterize these generalized stresses and forces for the kinematical model under development, first we have to choose a proper duality product $\langle \cdot, \cdot\rangle_{\mathscr{W}'\times\mathscr{W}}$. Evidently, the definition of this duality product will depend on the nature of the phyisical phenomena to be described by the model. Also, as it will be seen next, this definition will enable the characterization of the duality pairing $\langle \cdot, \cdot\rangle_{\mathscr{V}'\times\mathscr{V}}$.

Therefore, with the notation introduced in the previous section, the duality product between generalized stresses and generalized deformation actions takes the form

$$\langle \Sigma, \mathscr{D} u\rangle_{\mathscr{W}'\times\mathscr{W}} = \sum_{i=1}^{m} \langle \Sigma^i, (\mathscr{D} u)^i\rangle_{\mathscr{W}_i'\times\mathscr{W}_i}, \tag{13.11}$$

or, equivalently, using (13.9) we have

$$\langle \Sigma, \mathscr{D} u\rangle_{\mathscr{W}'\times\mathscr{W}} = \sum_{i=1}^{m} \sum_{j=1}^{n} \langle \Sigma^i, \mathscr{D}^{ij} u^j\rangle_{\mathscr{W}_i'\times\mathscr{W}_i}, \tag{13.12}$$

where $\langle \cdot, \cdot\rangle_{\mathscr{W}_i'\times\mathscr{W}_i}$ denotes the generalized inner product over the domain of definition of component i of the generalized deformation. For example, if $\Omega^{\mathscr{W}_i}$ is a surface or a volume in the Euclidean space, this duality product could take the form

$$\langle \Sigma^i, (\mathscr{D} u)^i\rangle_{\mathscr{W}_i'\times\mathscr{W}_i} = (\Sigma^i, (\mathscr{D} u)^i)_{\Omega^{\mathscr{W}_i}} = \int_{\Omega^{\mathscr{W}_i}} \Sigma^i \cdot (\mathscr{D} u)^i d\Omega^{\mathscr{W}_i}. \tag{13.13}$$

In turn, if $\Omega^{\mathscr{W}_i}$ is a set of points, the product could be given by

$$\langle \Sigma^i, (\mathscr{D} u)^i\rangle_{\mathscr{W}_i'\times\mathscr{W}_i} = \sum_{i=1}^{N^{\mathscr{W}_i}} \Sigma^i \cdot (\mathscr{D} u)^i, \tag{13.14}$$

where $N^{\mathscr{W}_i}$ represents the cardinality of the set $\Omega^{\mathscr{W}_i}$. In this manner, and once the duality product has been chosen, we are able to establish a representation (actually an identification) for Σ. The next example shows this in detail.

Classical solid mechanics. The duality product adopted for this model consists of the expression $\langle \Sigma, \mathscr{D} u\rangle_{\mathscr{W}'\times\mathscr{W}} = \int_\Omega \sigma \cdot (\nabla \mathbf{u})^s \, d\Omega$. Hence, the stress σ is dual to the strain rate, which is given by the symmetric gradient of the motion action field **u**. In this manner, we can identify (or represent) σ as being a second-order symmetric tensor field, known in this model as the Cauchy stress tensor. ■

It is worthwhile underlining that in the PVP the product $\langle \cdot, \cdot \rangle_{\mathscr{W}' \times \mathscr{W}}$ is evaluated in a smaller space composed by kinematically compatible generalized deformation actions $\mathscr{D}(\text{Var}_u)$. Moreover, this product is called the Internal Virtual Power and is denoted by P^i [2]. That is,

$$P^i(\mathscr{D}\hat{u}) = \langle \Sigma, \mathscr{D}\hat{u} \rangle_{\mathscr{W}' \times \mathscr{W}} \qquad \hat{u} \in \text{Var}_u. \tag{13.15}$$

Summing up, when the kinematics for a model corresponding to a particular physical system is defined together with the associated duality product for the internal virtual power, the representation of generalized stresses which are admissible for the model remains univocally determined.

Before going ahead, it is important to note that the definition of the internal virtual power must be such that it is invariant to observer changes, that is, invariant to the superposition of rigid motion actions. In other words, it must satisfy the condition

$$P^i(\mathscr{D}\hat{u}) = 0 \qquad \forall \hat{u} \in \text{Var}_u \cap \mathscr{N}(\mathscr{D}). \tag{13.16}$$

Let us now see that the selection of a specific form for the duality product between generalized stresses and deformation actions enables the, also univocal, characterization of the generalized external forces $f \in \mathscr{V}'$ admissible to the kinematical model, and also of the corresponding functional associated with the External Virtual Power. To observe this, recall the definition of the adjoint operator (also called the equilibrium operator), $\mathscr{D}^*$ (see [220]), given by

$$\begin{aligned} \mathscr{D}^* : \mathscr{W}' &\rightarrow \mathscr{V}', \\ \Sigma &\mapsto f = \mathscr{D}^*(\Sigma), \end{aligned} \tag{13.17}$$

where $\mathscr{D}^*$ is the adjoint operator to operator $\mathscr{D}$, in the sense that it is the operator that satisfies

$$\langle \Sigma, \mathscr{D}u \rangle_{\mathscr{W}' \times \mathscr{W}} = \langle \mathscr{D}^*\Sigma, u \rangle_{\mathscr{V}' \times \mathscr{V}}. \tag{13.18}$$

From the previous definition we notice that, once a specific form for the duality product $\langle \Sigma, \mathscr{D}u \rangle_{\mathscr{W}' \times \mathscr{W}}$ has been adopted, a characterization for f intrinsically admissible to the chosen kinematical model is retrieved. In other words, (13.18) implies that

$$f \text{ must have the structure established by } \mathscr{D}^*\Sigma \in \mathscr{V}'. \tag{13.19}$$

In turn, $\langle \mathscr{D}^*\Sigma, u \rangle_{\mathscr{V}' \times \mathscr{V}}$ in expanded form corresponds to

$$\langle \mathscr{D}^*\Sigma, u \rangle_{\mathscr{V}' \times \mathscr{V}} = \sum_{i=1}^{n} \langle (\mathscr{D}^*\Sigma)^i, u^i \rangle_{\mathscr{V}_i' \times \mathscr{V}_i}, \tag{13.20}$$

2 Frequently, the internal virtual power is defined with a negative sign as we have done in Chapter 3. However, this is immaterial as it depends in the way the PVP is written, that is, the sum of internal and external virtual powers equal to zero (here the negative sign is required) or the internal virtual power equal to the external virtual power (in this case the negative sign is dropped).

or, equivalently, using (13.9), it is

$$\langle \mathscr{D}^* \Sigma, u \rangle_{\mathscr{V}' \times \mathscr{V}} = \sum_{i=1}^{n} \sum_{j=1}^{m} \langle \mathscr{D}^{*ij} \Sigma^j, u^i \rangle_{\mathscr{V}_i' \times \mathscr{V}_i}. \tag{13.21}$$

In this manner, just as happened with the representation for Σ, the characterization of admissible forces f for the kinematical model remains inevitably identified once the adjoint operator has been obtained for the model under study. The following example shows this in detail.

Classical solid mechanics. In this case the adjoint operator $\mathscr{D}^*$ is obtained by integrating by parts the expression of the internal virtual power, that is

$$\begin{aligned} \langle \Sigma, \mathscr{D} u \rangle_{\mathscr{W}' \times \mathscr{W}} &= \int_\Omega \sigma \cdot (\nabla \mathbf{u})^s \, d\Omega = -\int_\Omega \operatorname{div}\sigma \cdot \mathbf{u} \, d\Omega + \int_{\Gamma_t} \sigma\mathbf{n} \cdot \mathbf{u} \, d\Gamma \\ &= \int_\Omega \mathbf{b} \cdot \mathbf{u} \, d\Omega + \int_{\Gamma_t} \mathbf{t} \cdot \mathbf{u} \, d\Gamma = \langle \mathscr{D}^* \Sigma, u \rangle_{\mathscr{V}' \times \mathscr{V}}, \end{aligned} \tag{13.22}$$

where $\mathbf{n}$ is the unit outward normal vector to the boundary Γ_t, which is such that $\Gamma = \Gamma_u \cup \Gamma_t$, $\Gamma_u \cap \Gamma_t = \emptyset$. With this we clearly see that the external forces $f \in \mathscr{V}'$ can be represented: through a force per unit volume vector field, $\mathbf{b}$, defined in Ω and by a force per unit area vector field $\mathbf{t}$, defined over the portion of the boundary Γ_t where no kinematical constraints are prescribed.[3] ■

Given that the nature of the external forces $f \in \mathscr{V}'$ has been identified, we can introduce the External Virtual Power functional by constraining the evaluation of this duality product to the space Var_u, that is

$$P^e(\hat{u}) = \langle f, \hat{u} \rangle_{\mathscr{V}' \times \mathscr{V}} \qquad \hat{u} \in \mathrm{Var}_u, \tag{13.23}$$

or equivalently

$$P^e(\hat{u}) = \sum_{i=1}^{n} \langle f^i, \hat{u}^i \rangle_{\mathscr{V}_i' \times \mathscr{V}_i} \qquad \hat{u} \in \mathrm{Var}_u. \tag{13.24}$$

With these definitions we define the Total Virtual Power functional as

$$P^t(\hat{u}, \mathscr{D}\hat{u}) = P^i(\mathscr{D}\hat{u}) - P^e(\hat{u}) \qquad \hat{u} \in \mathrm{Var}_u. \tag{13.25}$$

13.2.3 Principle of Virtual Power

The second foundational hypothesis for the Method of Virtual Power consists of admitting the validity of the PVP which, as we have already seen, establishes the conditions for which a system of generalized internal and external forces is at equilibrium. More specifically, the PVP states that the generalized stresses $\Sigma \in \mathscr{W}'$ and the

---

3 If we admit that σ is not sufficiently regular in the whole domain Ω featuring, for example, a discontinuity in the inner surface S from Ω, the integration by parts automatically leads to the characterization of a further class of admissible force for the model, which is given by the vector $\mathbf{s}$, which stands for a force per unit area acting over the surface S.

generalized forces $f \in \mathscr{V}'$ are at equilibrium if and only if the following variational equation is satisfied[4]

$$P^t(\hat{u}, \mathscr{D}\hat{u}) = 0 \qquad \forall \hat{u} \in \mathrm{Var}_u. \tag{13.26}$$

This equation can be rewritten as

$$P^i(\mathscr{D}\hat{u}) = P^e(\hat{u}) \qquad \forall \hat{u} \in \mathrm{Var}_u, \tag{13.27}$$

or

$$\langle \Sigma, \mathscr{D}\hat{u}\rangle_{\mathscr{W}'\times\mathscr{W}} = \langle f, \hat{u}\rangle_{\mathscr{V}'\times\mathscr{V}} \qquad \forall \hat{u} \in \mathrm{Var}_u. \tag{13.28}$$

Moreover, from (13.26), f has to satisfy

$$\langle f, \hat{u}\rangle_{\mathscr{V}'\times\mathscr{V}} = 0 \qquad \forall \hat{u} \in \mathrm{Var}_u \cap \mathscr{N}(\mathscr{D}). \tag{13.29}$$

In the following example we see the application of the PVP.

Classical solid mechanics. The Cauchy stress field σ and the system of external forces $(\mathbf{b}, \mathbf{t})$ are at equilibrium if and only if

$$\int_\Omega \sigma \cdot (\nabla \hat{\mathbf{u}})^s \, d\Omega = \int_\Omega \mathbf{b} \cdot \hat{\mathbf{u}} \, d\Omega + \int_{\Gamma_t} \mathbf{t} \cdot \hat{\mathbf{u}} \, d\Gamma \qquad \forall \hat{\mathbf{u}} \in \mathrm{Var}_u. \tag{13.30}$$

Obviously, this variational equation also accounts for the dynamic equilibrium if, for example, $\mathbf{b}$ is the inertia force field $\mathbf{b} = -\rho\ddot{\mathbf{u}}$, where ρ is the density and $\ddot{\mathbf{u}}$ is the acceleration of the particles in the body. It is important to notice that (13.29) implies that there is no system of forces (surface forces, body forces, and inertia forces if any) that is able to produce power in duality with kinematically admissible rigid motion actions. ■

Finally, Figure 13.1 illustrates a conceptual diagram involving the basic ingredients that play a role in the Method of Virtual Power.

13.2.4 Equilibrium Problem

To provide a complete description of the model for the physical problem under analysis it is still necessary to incorporate the behavior of the material that comprises the continuum. Hence, constitutive equations must be introduced to characterize the generalized stress state Σ in terms of the history of the kinematical variables adopted for the model. If we denote by u^t the history of the generalized displacements experienced by the body up to time t, the constitutive relation can be expressed in the following manner

$$\Sigma = \Sigma(u^t). \tag{13.31}$$

Also, note that for all time instants $\tau \in [0, t]$ the corresponding displacement $u(\tau)$ is kinematically admissible, that is, $u(\tau) \in \mathrm{Kin}_u$. Hence, we use the notation $u^t \in \mathrm{Kin}_u$ to indicate that the displacements in the body $\mathscr{B}$ are kinematically admissible for all time instants $\tau \in [0, t]$. With these definitions we can formulate the equilibrium problem for the body taking into consideration the material behavior of the continuum.

4 The expression *equilibrium* is not limited to static equilibrium. If the system of forces f includes generalized inertia forces associated with the physical problem under consideration, the dynamic equilibrium is automatically considered by the PVP.

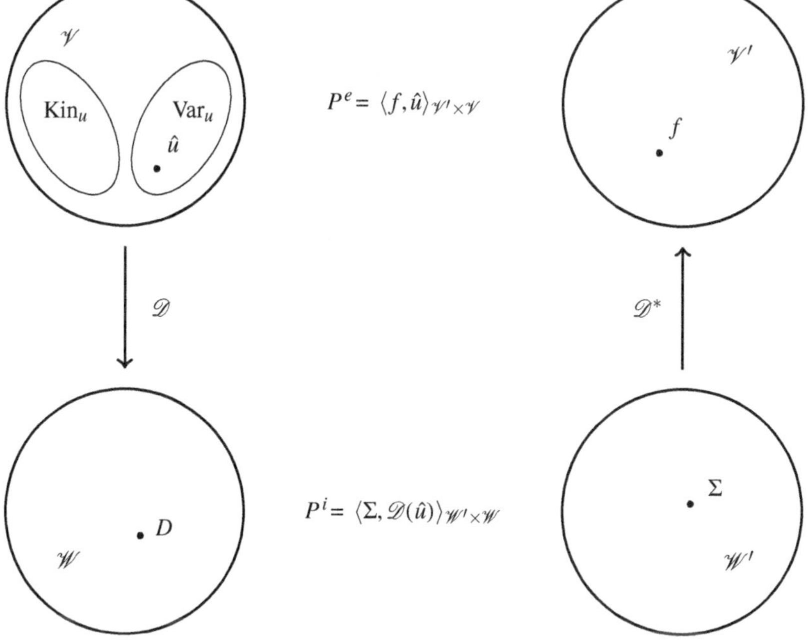

Figure 13.1 Method of Virtual Power. Conceptual diagram displaying sets, spaces, functionals, and operators of the theoretical framework.

For a given constitutive equation, such as that from (13.31), and given the history of the loading process exerted over the body, denoted by f^t, find the history $u^t \in \text{Kin}_u$ of kinematically admissible generalized displacements such that

$$\langle \Sigma(u^\tau), \mathscr{D}\hat{u} \rangle_{\mathscr{W}' \times \mathscr{W}} = \langle f(\tau), \hat{u} \rangle_{\mathscr{V}' \times \mathscr{V}} \qquad \forall \hat{u} \in \text{Var}_u, \text{for each } \tau \in [0, t]. \tag{13.32}$$

13.3 Fundamentals of the Multiscale Theory

From this section onwards we present the unified variational multiscale theory that we have named the Method of Multiscale Virtual Power, and, as described in the introduction to this part of the book, will make possible the construction of multiscale physical models based on the concept of Representative Volume Element (RVE).

The family of multiscale theories that are embraced by the proposed formulation is based on the idea that each point at the macroscale characterized by the body, occupying the domain Ω_M, is connected with an RVE that occupies the domain Ω_μ whose characteristic size is ℓ_μ, and which is much smaller than the characteristic size ℓ_M of Ω_M (see Figure 13.2). Domains Ω_M and Ω_μ are also designated, respectively, by macroscale and microscale. Points (or coordinates) in the macroscale are denoted by $\mathbf{x} \in \Omega_M$, while points (or coordinates) in the microscale are denoted by $\mathbf{y} \in \Omega_\mu$. In addition, from here onwards we make use of the subscripts M and μ to refer, respectively, to entities (variables) defined in the macroscale and microscale, respectively.

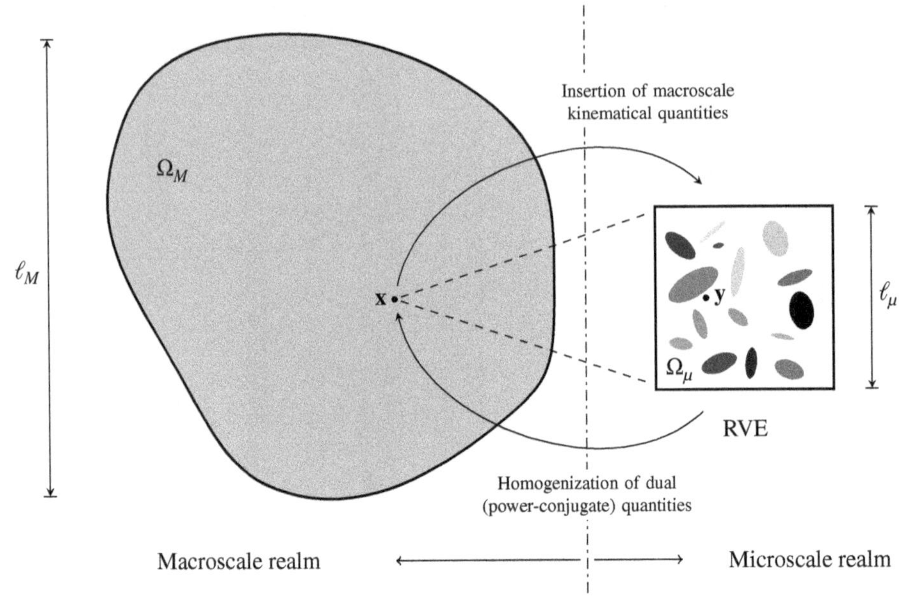

Figure 13.2 RVE-based multiscale modeling. The observable scale of the body is denoted macroscale and the substructure that forms the medium is denoted by microscale.

The path to be pursued along the development of the Method of Multiscale Virtual Power will be the same as that followed in Section 13.2 to describe the Method of Virtual Power which, within the current context, corresponds to a monoscale model of a problem. In particular, the concepts of kinematical admissibility, mathematical duality, and virtual power will be extended in order to set the groundwork to formulate the Principle of Multiscale Virtual Power. Hence, the method is based on three fundamental principles

i) *Principle of Kinematical Admissibility*, where the macroscale and microscale kinematics are defined and related through the insertion and homogenization operators, ensuring a coherent physical transfer (conservation of transferred quantities) between the kinematical variables connected through the scales.
ii) *Mathematical Duality*, where the generalized stresses and forces at macro and micro levels are defined in duality with their kinematical variables adopted in the corresponding scales.
iii) *Principle of Multiscale Virtual Power*, which generalizes the well-known Hill–Mandel Principle of Macrohomogeneity, through which the equilibrium equations and homogenization formulae for generalized stresses and forces are obtained in an unambiguous manner by using solid variational arguments.

As it will be seen, the proposed theory provides a clear and logical structure, a framework in which existing variational formulations can be rationally justified and novel multiscale models can be rigorously derived following well-defined steps.

The sense of the rationality of the proposed approach is totally equivalent to that of the Method of Virtual Power, which was revisited in Section 13.2 and was used throughout this book for the development of monoscale variational models of physical systems. In particular, we will see that once the kinematical variables at both scales are postulated, the kinematical admissibility is established for the physical system under study, and the corresponding generalized stresses and forces at both scales are identified by means of duality arguments, all the equations of the multiscale model are derived from the Principle of Multiscale Virtual Power by using standard variational arguments.

13.4 Kinematical Admissibility between Scales

When considering a physical system as composed of two scales, we assume from the beginning that the kinematics that describes the relevant phenomena at the macroscale can, in general, differ from the kinematics used to describe the microscale. However, since the ideas and definitions presented in Section 13.2.1 can individually be applied to each of the two scales, analogous procedures will be followed.

When postulating the kinematics for a physical system described through two scales, ultimately one seeks to define the space of admissible generalized displacements in the microscale, which will be denoted by Kin_{u_μ}, and that, evidently, will depend on the kinematical variables selected for the macroscale.

In particular, this process of characterizing Kin_{u_μ} will be referred to as the Principle of Kinematical Admissibility and consists of the following four steps

i) Definition of the kinematics at macroscale and microscale.
ii) Definition of the insertion operators which describe the way in which the macroscale kinematical variables are inserted (mapped) into the microscale.
iii) Definition of the homogenization operators which establish the way in which the microscale kinematical variables are averaged to render the macroscale kinematical variables. In this manner, the homogenization process for the microscale kinematical variables is, to some extent but not entirely, the inverse process to that given by the insertion operation. Therefore, these homogenization operators must guarantee that the kinematical variables that play a role in the micro–macro transfer process are, somehow, preserved (in the sense of conservation).
iv) Statement of the kinematical admissibility, which allows the definition of the set Kin_{u_μ} that contains all the microscale generalized displacements which are admissible within the context of the multiscale theory to be obtained.

In particular, item (i), which amounts to establishing the definition of kinematical variables, depends fundamentally on the physical phenomena (and level of detail) which one wants to capture with the model and, consequently, is strongly influenced by the experience and knowledge of the person in charge of the construction of the model. Thus, it is at this specific stage where lies one of the pinpoints of arbitrariness in the conception of the multiscale model within the proposed framework.

13.4.1 Macroscale Kinematics

Following in the footsteps of Section 13.2, the kinematics in the macroscale is characterized by the generalized displacement $u_M \in \mathscr{V}_M$, defined by an n_M-tuple of tensor fields, where each component u_M^i is described by r_M^i scalar fields. Hence, the total number of scalar fields that describe u_M gives $R_M = \sum_{i=1}^{n_M} r_M^i$, and the components have a domain of definition $\Omega_M^{\mathscr{V}_i} := \text{Dom}(\mathscr{V}_{Mi})$, $i = 1, \dots, n_M$, that is

$$\begin{aligned} u_M^i : \Omega_M^{\mathscr{V}_i} &\to \mathscr{V}_{Mi}, \\ \mathbf{x} &\mapsto u_M^i(\mathbf{x}), \end{aligned} \tag{13.33}$$

where $\Omega_M^{\mathscr{V}_i} \subseteq \Omega_M$, and where $\Omega_M^{\mathscr{V}_i}$ can be a set of points, lines, surfaces or volumes. In compact form, this can be written as

$$\begin{aligned} u_M : \Omega_M^{\mathscr{V}} &\to \mathscr{V}_M, \\ \mathbf{x} &\mapsto u_M(\mathbf{x}), \end{aligned} \tag{13.34}$$

and, as in Section 13.2, we have that the domain $\Omega_M^{\mathscr{V}}$ is defined as $\Omega_M^{\mathscr{V}} := \text{Dom}(\mathscr{V}_M) = (\text{Dom}(\mathscr{V}_{M1}), \dots, \text{Dom}(\mathscr{V}_{Mn_M}))$, or, equivalently, $\Omega_M^{\mathscr{V}} = (\Omega_M^{\mathscr{V}_1}, \dots, \Omega_M^{\mathscr{V}_{n_M}})$.

The set of kinematically admissible generalized displacements is denoted by Kin_{u_M} and the associated space of kinematically admissible generalized motion actions is denoted by Var_{u_M}. The space of macroscale generalized deformation actions is denoted by $\mathscr{W}_M$. Each element $D_M \in \mathscr{W}_M$ is an m_M-tuple of tensor fields, where each component D_M^i is described by s_M^i scalar fields (the total number of scalar fields describing D_M is $S_M = \sum_{i=1}^{m_M} s_M^i$). In addition, the domain of definition of these components is designated by $\Omega_M^{\mathscr{W}_i} := \text{Dom}(\mathscr{W}_{Mi})$, $i = 1, \dots, m_M$, that is

$$\begin{aligned} D_M^i : \Omega_M^{\mathscr{W}_i} &\to \mathscr{W}_{Mi}, \\ \mathbf{x} &\mapsto D_M^i(\mathbf{x}), \end{aligned} \tag{13.35}$$

where $\Omega_M^{\mathscr{W}_i} \subseteq \Omega_M$ can be a set of points, lines, surfaces or volumes, and in compact form it can be written as

$$\begin{aligned} D_M : \Omega_M^{\mathscr{W}} &\to \mathscr{W}_M, \\ \mathbf{x} &\mapsto D_M(\mathbf{x}), \end{aligned} \tag{13.36}$$

where $\Omega_M^{\mathscr{W}} := \text{Dom}(\mathscr{W}_M) = (\text{Dom}(\mathscr{W}_{M1}), \dots, \text{Dom}(\mathscr{W}_{Mm_M}))$, or equivalently $\Omega_M^{\mathscr{W}} = (\Omega_M^{\mathscr{W}_1}, \dots, \Omega_M^{\mathscr{W}_{m_M}})$.

In the macroscale, the kinematically admissible generalized deformation actions are characterized by the linear operator

$$\begin{aligned} \mathscr{D}_M : \mathscr{V}_M &\to \mathscr{W}_M, \\ u_M &\mapsto D_M = \mathscr{D}_M(u_M). \end{aligned} \tag{13.37}$$

In turn, the domain of definition of these generalized deformation actions is denoted by $\Omega_M^{\mathscr{W}} = \text{Dom}(\mathscr{D}_M(u_M))$.

Notice that, as in (13.9), because $\mathscr{D}_M$ is a linear operator, we have that the following representation holds

$$\mathscr{D}_M = \begin{pmatrix} \mathscr{D}_M^{11} & \mathscr{D}_M^{12} & \cdots & \mathscr{D}_M^{1n_M} \\ \mathscr{D}_M^{21} & \mathscr{D}_M^{22} & \cdots & \mathscr{D}_M^{2n_M} \\ \vdots & \vdots & \ddots & \vdots \\ \mathscr{D}_M^{m_M 1} & \mathscr{D}_M^{m_M 2} & \cdots & \mathscr{D}_M^{m_M n_M} \end{pmatrix}, \tag{13.38}$$

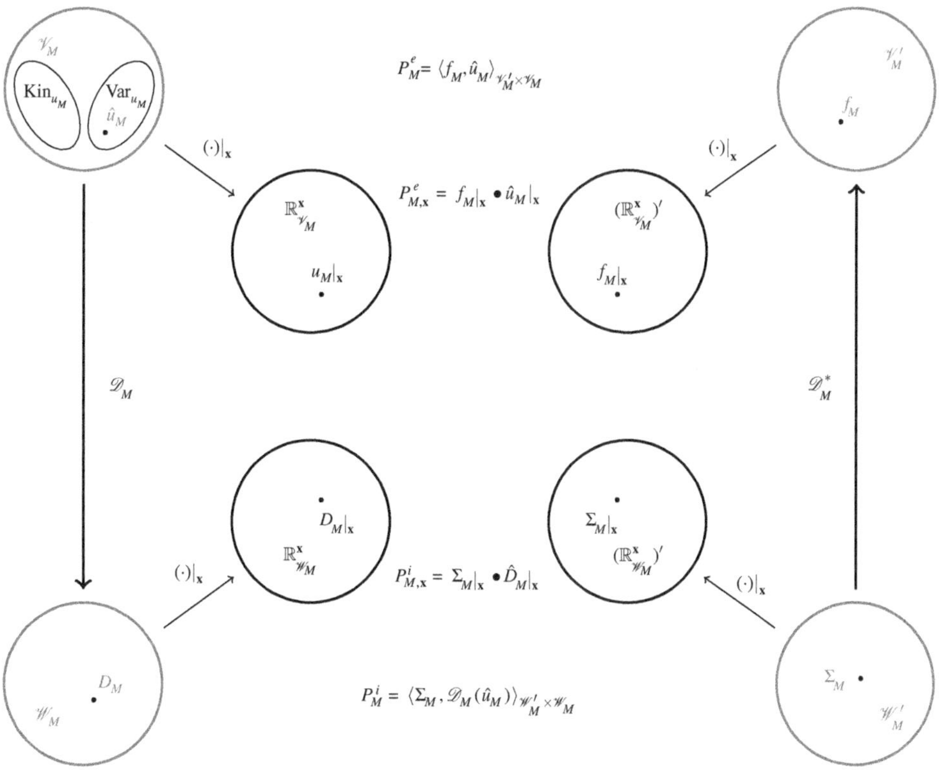

Figure 13.3 Method of Multiscale Virtual Power. Diagram of sets, spaces, functionals and basic operations at the macroscale. Gray circles denote the ingredients when the variables are to be understood as macroscale fields, while black circles denote these ingredients when variables are evaluated at a point in the macroscale.

where $\Omega_M^{\mathscr{W}_i} = \mathrm{Dom}(\mathscr{D}_M^{i1}(u_M^1)) = \dots = \mathrm{Dom}(\mathscr{D}_M^{in_M}(u_M^{n_M})),\ i = 1, \dots, m_M$.

Figure 13.3 shows the functional macroscale structure (equivalent to Figure 13.1 but in the multiscale context). In this figure, we have introduced sets and additional operations related to individual points in the macroscale domain. In particular, the value of the variable $(\cdot)$ at an arbitrary macroscale point $\mathbf{x}$ is denoted by $(\cdot)|_{\mathbf{x}}$. As will be seen next, these values are related to the kinematics adopted at the microscale.

13.4.2 Microscale Kinematics

As already emphasized, the primordial concept within the RVE-based variational multiscale theoretical framework addressed here is that each macroscale point $\mathbf{x}$ is related to a domain which stands for the microscale realm, and is briefly denoted by RVE. Now we address the kinematics defined at a general RVE domain by following a completely analogous scheme to the one pursued for the macroscale kinematics. That is to say, the RVE occupies the spatial domain Ω_μ, whose points are denoted by $\mathbf{y}$. The space of generalized displacements in the RVE is called $\mathscr{V}_\mu$, where each element $u_\mu \in \mathscr{V}_\mu$ is an n_μ-tuple of tensor fields. Each component u_μ^i is described by r_μ^i scalar fields (the total number of fields that characterize u_μ is $R_\mu = \sum_{i=1}^{n_\mu} r_\mu^i$), and the domain of definition of each field is $\Omega_\mu^{\mathscr{V}_i} := \mathrm{Dom}(\mathscr{V}_{\mu_i})$, $i = 1, \dots, n_\mu$, that is

$$\begin{aligned} u_\mu^i : \Omega_\mu^{\mathscr{V}_i} &\to \mathscr{V}_{\mu_i}, \\ \mathbf{y} &\mapsto u_\mu^i(\mathbf{y}), \end{aligned} \tag{13.39}$$

where each $\Omega_\mu^{\mathscr{V}_i} \subseteq \Omega_\mu$ is a set of points, lines, surfaces or volumes. In compact notation we can write

$$\begin{aligned} u_\mu : \Omega_\mu^{\mathscr{V}} &\to \mathscr{V}_\mu, \\ \mathbf{y} &\mapsto u_\mu(\mathbf{y}), \end{aligned} \tag{13.40}$$

where $\Omega_\mu^{\mathscr{V}} := \mathrm{Dom}(\mathscr{V}_\mu) = (\mathrm{Dom}(\mathscr{V}_{\mu_1}), \dots, \mathrm{Dom}(\mathscr{V}_{\mu_{n_\mu}}))$, or equivalently $\Omega_\mu^{\mathscr{V}} = (\Omega_\mu^{\mathscr{V}_1}, \dots, \Omega_\mu^{\mathscr{V}_{n_\mu}})$.

Without loss of generality, it is convenient to express the microscale generalized displacement $u_\mu \in \mathscr{V}_\mu$ as the sum

$$u_\mu = \overline{u}_\mu + \tilde{u}_\mu, \tag{13.41}$$

where the first term, $\overline{u}_\mu$, depends on the macroscale kinematics at the point $\mathbf{x}$ associated with the RVE, and the second term, $\tilde{u}_\mu$, is called the generalized displacement fluctuation. In general, the field $\overline{u}_\mu$ is not uniform in the microscale domain, that is, it may depend on the coordinate $\mathbf{y}$. The set of all the microscale generalized displacements $\overline{u}_\mu$ forms the subspace denoted by $\overline{\mathscr{V}}_\mu$, and the collection of all fields $\tilde{u}_\mu$ gives rise to the subspace $\tilde{\mathscr{V}}_\mu$ of generalized displacement fluctuations.

Next, we define the space $\mathscr{W}_\mu$ of microscale generalized deformation actions. Each element $D_\mu \in \mathscr{W}_\mu$ is an m_μ-tuple of tensor fields whose components are denoted by D_μ^i. Each component is described by s_μ^i scalar fields (the total number of fields which describe D_μ is $S_\mu = \sum_{i=1}^{m_\mu} s_\mu^i$), and these components have the domain of definition $\Omega_\mu^{\mathscr{W}_i} := \mathrm{Dom}(\mathscr{W}_{\mu_i})$, $i = 1, \dots, m_\mu$, that is

$$\begin{aligned} D_\mu^i : \Omega_\mu^{\mathscr{W}_i} &\to \mathscr{W}_{\mu_i}, \\ \mathbf{y} &\mapsto D_\mu^i(\mathbf{y}), \end{aligned} \tag{13.42}$$

where $\Omega_\mu^{\mathscr{W}_i} \subseteq \Omega_\mu$ can be a set of points, lines, surfaces or volumes. In compact form this gives

$$\begin{aligned} D_\mu : \Omega_\mu^{\mathscr{W}} &\to \mathscr{W}_\mu, \\ \mathbf{y} &\mapsto D_\mu(\mathbf{y}), \end{aligned} \tag{13.43}$$

where $\Omega_\mu^{\mathscr{W}} := \mathrm{Dom}(\mathscr{W}_\mu) = (\mathrm{Dom}(\mathscr{W}_{\mu_1}), \dots, \mathrm{Dom}(\mathscr{W}_{\mu_{m_\mu}}))$, or in an equivalent manner $\Omega_\mu^{\mathscr{W}} = (\Omega_\mu^{\mathscr{W}_1}, \dots, \Omega_\mu^{\mathscr{W}_{m_\mu}})$.

In turn, we define the linear operator of microscale generalized deformation actions

$$\begin{aligned} \mathscr{D}_\mu : \mathscr{V}_\mu &\to \mathscr{W}_\mu, \\ u_\mu &\mapsto D_\mu = \mathscr{D}_\mu(u_\mu), \end{aligned} \tag{13.44}$$

where the domain of definition is given by $\Omega_\mu^{\mathscr{W}} = \mathrm{Dom}(\mathscr{D}_\mu(u_\mu))$.

Analogously to (13.38), we have the following representation for the operator $\mathscr{D}_\mu$

$$\mathscr{D}_\mu = \begin{pmatrix} \mathscr{D}_\mu^{11} & \mathscr{D}_\mu^{12} & \dots & \mathscr{D}_\mu^{1n_\mu} \\ \mathscr{D}_\mu^{21} & \mathscr{D}_\mu^{22} & \dots & \mathscr{D}_\mu^{2n_\mu} \\ \vdots & \vdots & \ddots & \vdots \\ \mathscr{D}_\mu^{m_\mu 1} & \mathscr{D}_\mu^{m_\mu 2} & \dots & \mathscr{D}_\mu^{m_\mu n_\mu} \end{pmatrix}, \tag{13.45}$$

where $\Omega_\mu^{\mathscr{W}^i} = \mathrm{Dom}(\mathscr{D}_\mu^{i1}(u_\mu^1)) = \dots = \mathrm{Dom}(\mathscr{D}_\mu^{in_\mu}(u_\mu^{n_\mu}))$, $i = 1, \dots, m_\mu$.

13.4.3 Insertion Operators

As was mentioned at the beginning of this section, the microscale RVE kinematics associated with an arbitrary macroscale point $\mathbf{x} \in \Omega_M$ is connected with the kinematics defined at the macroscale by virtue of (i) the insertion operators, which define a macro–micro kinematical transfer, and (ii) the homogenization operators, which define a micro–macro kinematical transfer. These two operators are linear in their arguments and must be adequately fabricated in order to proffer a consistent mechanical/physical transfer between scales for data encoded into generalized displacements and deformations. Evidently, the definition of these operators depends on the physical system that is being modeled.

For simplicity, hereafter we will assume that all the kinematical variables defined at the macroscale participate in the transfer between scales. In a more general scenario, we could have that only a subset of the macroscale kinematical variables plays a role in this multiscale kinematical transfer.

Let us define the set $\mathbb{R}^{\mathbf{x}}_{\mathscr{V}_M}$ of elements belonging to $\mathscr{V}_M$ when they are evaluated at the point $\mathbf{x} \in \Omega_M$

$$\mathbb{R}^{\mathbf{x}}_{\mathscr{V}_M} = \{w = (w^1, \dots, w^{n_M});\ w^i \in \mathbb{R}^{r_M^i}, i = 1, \dots, n_M, w = u|_{\mathbf{x}},\ u \in \mathscr{V}_M\}, \tag{13.46}$$

where $\mathbb{R}^{r_M^i} = \overbrace{\mathbb{R} \times \cdots \times \mathbb{R}}^{r_M^i}$ and r_M^i is the number of scalars that describe the ith tensor field from the n_M-tuple. Note that $\dim(\mathbb{R}^{\mathbf{x}}_{\mathscr{V}_M}) = R_M$. We refer to $\mathbb{R}^{\mathbf{x}}_{\mathscr{V}_M}$ as the set of point-valued macroscale generalized displacements.

Similarly, we define the set of point-valued macroscale generalized deformation actions, $\mathbb{R}^{\mathbf{x}}_{\mathscr{W}_M}$, formed by all the elements from $\mathscr{W}_M$ when evaluated at point $\mathbf{x} \in \Omega_M$, that is

$$\mathbb{R}^{\mathbf{x}}_{\mathscr{W}_M} = \{V = (V^1, \dots, V^{m_M});\ V^i \in \mathbb{R}^{s_M^i}, i = 1, \dots, m_M, V = D|_{\mathbf{x}},\ D \in \mathscr{W}_M\}, \tag{13.47}$$

where $\mathbb{R}^{s_M^i} = \overbrace{\mathbb{R} \times \cdots \times \mathbb{R}}^{s_M^i}$ and s_M^i is the number of scalar fields required to describe the ith tensor field in the m_M-tuple. Observe that $\dim(\mathbb{R}^{\mathbf{x}}_{\mathscr{W}_M}) = S_M$.

Within these two sets of point-valued generalized displacements and generalized deformation actions in the macroscale we are going to distinguish the set of point-valued macroscale generalized motion actions, $\hat{w}$, and macroscale generalized deformation actions, $\hat{V}$, denoting these two sets by $\widehat{\mathbb{R}^{\mathbf{x}}_{\mathscr{V}_M}}$ and $\widehat{\mathbb{R}^{\mathbf{x}}_{\mathscr{W}_M}}$, respectively. It is interesting to note here that depending on the specific kinematical multiscale model under development, some components of these point-valued macroscale generalized displacements and deformation actions can be null elements, even when the real kinematics at that point is not necessarily null.

Classical multiscale solid mechanics. For this case, the set $\mathbb{R}^{\mathbf{x}}_{\mathscr{V}_M}$ is defined in the following way: $\mathbb{R}^{\mathbf{x}}_{\mathscr{V}_M} = \{\mathbf{w} \in \mathbb{R}^3;\ \mathbf{w} = \mathbf{u}_M|_{\mathbf{x}},\ \mathbf{u}_M \in \mathbf{H}^1(\Omega_M)\}$. In addition, we have $\mathbb{R}^{\mathbf{x}}_{\mathscr{W}_M} = \{\varepsilon \in \mathbb{R}^{3\times 3};\ \varepsilon = \varepsilon_M|_{\mathbf{x}},\ \varepsilon_M \in \mathbf{L}^2_{\text{sym}}(\Omega_M)\}$. ■

At this point, we are ready to introduce the concept of insertion operator. Within the present theory, the insertion operators are essential to define the way in which the macroscale kinematics contributes to the microscale kinematics. In other words, these operators define how the macroscale kinematics is inserted in the microscale. Hence, the insertion operators are defined next

- The u_M-insertion operator,

$$\begin{aligned} \mathcal{I}^{\mathscr{V}}_{\mu} &: \mathbb{R}^{\mathbf{x}}_{\mathscr{V}_M} \to \overline{\mathscr{V}}_{\mu}, \\ & u_M|_{\mathbf{x}} \mapsto \overline{u}_{\mu} = \mathcal{I}^{\mathscr{V}}_{\mu}(u_M|_{\mathbf{x}}), \end{aligned} \tag{13.48}$$

  maps the point-valued macroscale generalized displacement $u_M|_{\mathbf{x}}$ into a field $\overline{u}_{\mu}$ that contributes to the microscale generalized displacement with the expression (13.41).
- The D_M-insertion operator,

$$\begin{aligned} \mathcal{I}^{\mathscr{W}}_{\mu} &: \mathbb{R}^{\mathbf{x}}_{\mathscr{W}_M} \to \overline{\mathscr{V}}_{\mu}, \\ & D_M|_{\mathbf{x}} \mapsto \overline{u}_{\mu} = \mathcal{I}^{\mathscr{W}}_{\mu}(D_M|_{\mathbf{x}}), \end{aligned} \tag{13.49}$$

  maps the point-valued macroscale generalized deformation $D_M|_{\mathbf{x}}$ into a field that contributes to the microscale generalized displacement according to expression (13.41).

Importantly, these two operators are linear in their respective arguments.

Classical multiscale solid mechanics. The u_M-insertion operator is defined in this case as

$$\mathcal{I}^{\mathscr{V}}_{\mu}(\mathbf{u}_M|_{\mathbf{x}}) = \mathbf{u}_M|_{\mathbf{x}}, \tag{13.50}$$

that is, it maps $\mathbf{u}_M|_{\mathbf{x}}$ into a uniform field in Ω_μ whose value is equal to the point-wise value of the macroscale generalized displacement at point $\mathbf{x}$ connected to the microscale domain. In turn, the D_M-insertion operator is postulated as

$$\mathcal{I}^{\mathscr{W}}_{\mu}(\varepsilon_M|_{\mathbf{x}}) = \varepsilon_M|_{\mathbf{x}}(\mathbf{y} - \mathbf{y}_o), \tag{13.51}$$

where $\mathbf{y}_o = \frac{1}{|\Omega_\mu|}\int_{\Omega_\mu} \mathbf{y}\, d\Omega_\mu$, that is, it maps the point-wise value of the macroscale generalized deformation action (in this case infinitesimal deformation) into a displacement field which is linearly distributed in the microscale domain. ■

The proposition of these two insertion operators is not arbitrary. On the contrary, they must ensure that point-wise values of macroscale generalized motion actions and generalized deformation actions are preserved when mapped to the microscale domain.

This will be addressed together with the enforcement of additional constraints over these operators that will be presented in due course (see, for example, (13.70) and (13.71)).

From (13.48) and (13.49), the kinematical variables $u_M|_{\mathbf{x}}$ and $D_M|_{\mathbf{x}}$ can be combined in such a way that results a non-uniform generalized displacement field (i.e. it depends on $\mathbf{y}$) at the microscale. In particular, we point out that the domain $\Omega_\mu^{\mathscr{V}_i}$, $i = 1, \dots, n_\mu$ corresponds to the domain of insertion where component i of the image of the insertion operators $\mathcal{J}_\mu^{\mathscr{V}}$ and $\mathcal{J}_\mu^{\mathscr{W}}$ is defined.

Classical multiscale solid mechanics. For this case, the point values of macroscale kinematical variables, $\mathbf{u}_M|_{\mathbf{x}}$ and $\varepsilon_M|_{\mathbf{x}}$ contribute with the microscale displacement through $\overline{\mathbf{u}}_\mu$ in the following manner

$$\overline{\mathbf{u}}_\mu = \mathcal{J}_\mu^{\mathscr{V}}(\mathbf{u}_M|_{\mathbf{x}}) + \mathcal{J}_\mu^{\mathscr{W}}(\varepsilon_M|_{\mathbf{x}}) = \mathbf{u}_M|_{\mathbf{x}} + \varepsilon_M|_{\mathbf{x}}(\mathbf{y} - \mathbf{y}_o). \tag{13.52}$$

Evidently, the domain of insertion in the microscale corresponds in this case to the whole RVE domain Ω_μ. ■

The theory developed so far allows much more general insertion operators than those found in typical multiscale formulations. Most of these formulations only consider mappings of the macroscale deformation that render affine displacement fields at the microscale (this is exactly what is used in the classical multiscale theory). The kind of operators used in the literature are not adequate for problems that require more complex insertions, such as those necessary to deal with mechanical problems featuring progressive strain localization, nucleation and evolution of fractures, shear bands, damage, and other classes of failure mechanisms taking place at the microscale domain. Moreover, as will be shown in Section 13.7, the insertion operators play a fundamental role in the characterization of the homogenization of macroscale generalized stresses and forces, something that is not considered in current formulations.

Note that since $\mathcal{J}_\mu^{\mathscr{V}}$ and $\mathcal{J}_\mu^{\mathscr{W}}$ are, by definition, linear operators, the following representations hold

$$\mathcal{J}_\mu^{\mathscr{V}} = \begin{pmatrix} \mathcal{J}_\mu^{\mathscr{V}\,11} & \mathcal{J}_\mu^{\mathscr{V}\,12} & \cdots & \mathcal{J}_\mu^{\mathscr{V}\,1n_M} \\ \mathcal{J}_\mu^{\mathscr{V}\,21} & \mathcal{J}_\mu^{\mathscr{V}\,22} & \cdots & \mathcal{J}_\mu^{\mathscr{V}\,2n_M} \\ \vdots & \vdots & \ddots & \vdots \\ \mathcal{J}_\mu^{\mathscr{V}\,n_\mu 1} & \mathcal{J}_\mu^{\mathscr{V}\,n_\mu 2} & \cdots & \mathcal{J}_\mu^{\mathscr{V}\,n_\mu n_M} \end{pmatrix}, \tag{13.53}$$

and

$$\mathcal{J}_\mu^{\mathscr{W}} = \begin{pmatrix} \mathcal{J}_\mu^{\mathscr{W}\,11} & \mathcal{J}_\mu^{\mathscr{W}\,12} & \cdots & \mathcal{J}_\mu^{\mathscr{W}\,1m_M} \\ \mathcal{J}_\mu^{\mathscr{W}\,21} & \mathcal{J}_\mu^{\mathscr{W}\,22} & \cdots & \mathcal{J}_\mu^{\mathscr{W}\,2m_M} \\ \vdots & \vdots & \ddots & \vdots \\ \mathcal{J}_\mu^{\mathscr{W}\,n_\mu 1} & \mathcal{J}_\mu^{\mathscr{W}\,n_\mu 2} & \cdots & \mathcal{J}_\mu^{\mathscr{W}\,n_\mu m_M} \end{pmatrix}. \tag{13.54}$$

Now we introduce the following definition. We say that $u_\mu \in \mathscr{V}_\mu$ is linked to the macroscale kinematics at point $\mathbf{x} \in \Omega_M$ if there exists $u_M|_{\mathbf{x}} \in \mathbb{R}^{\mathbf{x}}_{\mathscr{V}_M}$ and $D_M|_{\mathbf{x}} \in \mathbb{R}^{\mathbf{x}}_{\mathscr{W}_M}$ such that

$$u_\mu = \overline{u}_\mu + \tilde{u}_\mu = \underbrace{\mathcal{J}_\mu^{\mathscr{V}}(u_M|_{\mathbf{x}}) + \mathcal{J}_\mu^{\mathscr{W}}(D_M|_{\mathbf{x}})}_{=\overline{u}_\mu} + \tilde{u}_\mu. \tag{13.55}$$

Hence, for all microscale generalized displacements, $u_\mu \in \mathscr{V}_\mu$, linked to the macroscale kinematics at point $\mathbf{x}$, the corresponding kinematically admissible microscale generalized deformation action is given by

$$\begin{aligned} D_\mu &= \mathscr{D}_\mu(u_\mu) = \mathscr{D}_\mu(\overline{u}_\mu) + \mathscr{D}_\mu(\tilde{u}_\mu) \\ &= \mathscr{D}_\mu(\mathscr{I}_\mu^{\mathscr{V}}(u_M|_\mathbf{x})) + \mathscr{D}_\mu(\mathscr{I}_\mu^{\mathscr{W}}(D_M|_\mathbf{x})) + \mathscr{D}_\mu(\tilde{u}_\mu). \end{aligned} \tag{13.56}$$

For physical reasons, we impose the following restriction to operator $\mathscr{I}_\mu^{\mathscr{V}}$ in order to avoid the insertion of u_M inducing microscale generalized deformation actions

$$\mathscr{D}_\mu(\mathscr{I}_\mu^{\mathscr{V}}(u_M|_\mathbf{x})) = 0 \qquad \forall u_M|_\mathbf{x} \in \mathbb{R}^\mathbf{x}_{\mathscr{V}_M}. \tag{13.57}$$

From a mechanical standpoint, this restriction establishes that the generalized displacement inserted from the macroscale must belong to the null space of $\mathscr{D}_\mu$, $\mathscr{N}(\mathscr{D}_\mu)$, that is, the image of operator $\mathscr{I}_\mu^{\mathscr{V}}$ belongs to the space of microscale generalized rigid displacements.

Classical multiscale solid mechanics. An example of the previous restriction is evident in the context of the classical multiscale problem in solid mechanics. In fact, writing for this case $\mathbf{u}_\mu = \mathbf{u}_M|_\mathbf{x} + \varepsilon_M|_\mathbf{x}(\mathbf{y} - \mathbf{y}_o) + \tilde{\mathbf{u}}_\mu$, we ensure that the microscale displacement, $\mathbf{u}_\mu$, is linked to the macroscale displacement at point $\mathbf{x}$. Hence, given that $\mathscr{I}_\mu^{\mathscr{V}}(\mathbf{u}_M|_\mathbf{x}) = \mathbf{u}_M|_\mathbf{x}$ is a uniform field in the microscale, the corresponding deformation action is given by $\varepsilon_\mu = (\nabla_\mathbf{y}(\mathbf{u}_M|_\mathbf{x} + \varepsilon_M|_\mathbf{x}(\mathbf{y} - \mathbf{y}_o) + \tilde{\mathbf{u}}_\mu))^s = \varepsilon_M|_\mathbf{x} + (\nabla_\mathbf{y}\tilde{\mathbf{u}}_\mu)^s$, satisfying by construction restriction (13.57). ■

From (13.56), the microscale generalized deformation action can be written as

$$D_\mu = \overline{D}_\mu + \tilde{D}_\mu, \tag{13.58}$$

where $\overline{D}_\mu$ is the contribution brought by the macroscale kinematics into the microscale generalized deformation and $\tilde{D}_\mu$, the fluctuation component in the microscale generalized deformation action, depends solely on entities defined at the microscale, that is to say

$$\overline{D}_\mu = \mathscr{D}_\mu(\mathscr{I}_\mu^{\mathscr{W}}(D_M|_\mathbf{x})), \tag{13.59}$$

$$\tilde{D}_\mu = \mathscr{D}_\mu(\tilde{u}_\mu). \tag{13.60}$$

From the previous exposition it follows that the contribution of the macroscale generalized deformation action into its microscale counterpart can be directly obtained by applying the composed insertion operator defined by $\mathcal{I}_\mu = \mathscr{D}_\mu \mathscr{I}_\mu^{\mathscr{W}}$, that is

$$\begin{aligned} \mathcal{I}_\mu &: \mathbb{R}^\mathbf{x}_{\mathscr{W}_M} \to \mathscr{W}_\mu, \\ &D_M|_\mathbf{x} \mapsto D_\mu = \mathcal{I}_\mu(D_M|_\mathbf{x}). \end{aligned} \tag{13.61}$$

13.4.4 Homogenization Operators

The homogenization operators also have a fundamental character in the proposed multiscale theory. Evidently, in the first place, these operators depend upon the physical model that is being developed. Independently, here we are interested in characterizing essential properties/aspects for these operators. Thus, the function of these operators is to define the way in which the microscale kinematics is homogenized (averaged) to yield the corresponding point-wise values (at the macroscale point linked to the RVE) of the macroscale kinematics. In other words, these are operators that map microscale kinematical information into the macroscale kinematics. There exist two homogenization operators

- The u_μ-homogenization operator maps the microscale generalized displacement into the point-wise value of the macroscale generalized displacement at point $\mathbf{x}$

$$\begin{aligned} \mathscr{H}_\mu^{\mathscr{V}} &: \mathscr{V}_\mu \to \mathbb{R}^{\mathbf{x}}_{\mathscr{V}_M}, \\ u_\mu &\mapsto \mathscr{H}_\mu^{\mathscr{V}}(u_\mu) \in \mathbb{R}^{\mathbf{x}}_{\mathscr{V}_M}. \end{aligned} \tag{13.62}$$

- The D_μ-homogenization operator maps the microscale generalized deformation action into the point-wise value of the macroscale generalized deformation action at point $\mathbf{x}$

$$\begin{aligned} \mathscr{H}_\mu^{\mathscr{W}} &: \mathscr{W}_\mu \to \mathbb{R}^{\mathbf{x}}_{\mathscr{W}_M}, \\ D_\mu &\mapsto \mathscr{H}_\mu^{\mathscr{W}}(D_\mu) \in \mathbb{R}^{\mathbf{x}}_{\mathscr{W}_M}. \end{aligned} \tag{13.63}$$

Both operators are linear and involve averaging operations over the domains of definition established by the corresponding insertion operations, $\Omega_\mu^{\mathscr{V}_i}, i = 1, \dots, n_\mu$, and $\Omega_\mu^{\mathscr{W}_i}, i = 1, \dots, m_\mu$.

Classical multiscale solid mechanics. In this case, the u_μ-homogenization operator is defined by

$$\mathscr{H}_\mu^{\mathscr{V}}(\mathbf{u}_\mu) = \frac{1}{|\Omega_\mu|}\int_{\Omega_\mu} \mathbf{u}_\mu \, d\Omega_\mu, \tag{13.64}$$

while the D_μ-homogenization operator is defined by

$$\mathscr{H}_\mu^{\mathscr{W}}(\varepsilon_\mu) = \frac{1}{|\Omega_\mu|}\int_{\Omega_\mu} \varepsilon_\mu \, d\Omega_\mu. \tag{13.65}$$

That is, $\mathbf{u}_M|_{\mathbf{x}}$ and $\varepsilon_M|_{\mathbf{x}}$ are simply volume averages of the corresponding kinematical fields defined at the RVE. ■

From the linearity of the homogenization operators, the following matrix representation can be established

$$\mathscr{H}_\mu^{\mathscr{V}} = \begin{pmatrix} \mathscr{H}_\mu^{\mathscr{V}\,11} & \mathscr{H}_\mu^{\mathscr{V}\,12} & \dots & \mathscr{H}_\mu^{\mathscr{V}\,1n_\mu} \\ \mathscr{H}_\mu^{\mathscr{V}\,21} & \mathscr{H}_\mu^{\mathscr{V}\,22} & \dots & \mathscr{H}_\mu^{\mathscr{V}\,2n_\mu} \\ \vdots & \vdots & \ddots & \vdots \\ \mathscr{H}_\mu^{\mathscr{V}\,n_M1} & \mathscr{H}_\mu^{\mathscr{V}\,n_M2} & \dots & \mathscr{H}_\mu^{\mathscr{V}\,n_Mn_\mu} \end{pmatrix}, \tag{13.66}$$

and

$$\mathscr{H}_\mu^{\mathscr{W}} = \begin{pmatrix} \mathscr{H}_\mu^{\mathscr{W}\,11} & \mathscr{H}_\mu^{\mathscr{W}\,12} & \dots & \mathscr{H}_\mu^{\mathscr{W}\,1m_\mu} \\ \mathscr{H}_\mu^{\mathscr{W}\,21} & \mathscr{H}_\mu^{\mathscr{W}\,22} & \dots & \mathscr{H}_\mu^{\mathscr{W}\,2m_\mu} \\ \vdots & \vdots & \ddots & \vdots \\ \mathscr{H}_\mu^{\mathscr{W}\,m_M1} & \mathscr{H}_\mu^{\mathscr{W}\,m_M2} & \dots & \mathscr{H}_\mu^{\mathscr{W}\,m_Mm_\mu} \end{pmatrix}. \tag{13.67}$$

Let us go back to the question formulated before regarding the restrictions the insertion operators $\mathcal{J}_\mu^{\mathscr{V}}$ and $\mathcal{J}_\mu^{\mathscr{W}}$, defined by (13.48) and (13.49), are required to satisfy. Over physical grounds, we demanded that these operators should be defined in such a way that the magnitude of the kinematical variables that participate in the transfer between scales are, in some sense, conserved. This must be understood as the principle of conservation of macroscale generalized displacements and conservation of macroscale generalized deformation actions, or simply the Principle of Kinematical Conservation. That is, we want to ensure that the homogenization of the insertion of each component $u_M^i|_{\mathbf{x}}$ from $u_M|_{\mathbf{x}}$ effectively yields $u_M^i|_{\mathbf{x}}$. The same criterion is applied for the components $D_M^i|_{\mathbf{x}}$. To formalize this requirement, we define $u_M^{\{i\}} \in \mathbb{R}^{\mathbf{x}}_{\mathscr{V}_M}$ and $D_M^{\{i\}} \in \mathbb{R}^{\mathbf{x}}_{\mathscr{W}_M}$ such that

$$(u_M^{\{i\}})^j = \begin{cases} u_M^i|_{\mathbf{x}} & \text{if } j = i, \\ 0 & \text{if } j \neq i, \end{cases} \tag{13.68}$$

$$(D_M^{\{i\}})^j = \begin{cases} D_M^i|_{\mathbf{x}} & \text{if } j = i, \\ 0 & \text{if } j \neq i. \end{cases} \tag{13.69}$$

Then, the principle of kinematical conservation is satisfied if $\mathcal{J}_\mu^{\mathscr{V}}$ and $\mathcal{J}_\mu^{\mathscr{W}}$ verify the following restrictions

$$\mathscr{H}_\mu^{\mathscr{V}}(\mathcal{J}_\mu^{\mathscr{V}}(u_M^{\{i\}})) = u_M^{\{i\}} \quad i = 1, \dots, n_M, \tag{13.70}$$

$$\mathscr{H}_\mu^{\mathscr{W}}(\mathscr{D}_\mu(\mathcal{J}_\mu^{\mathscr{W}}(D_M^{\{i\}}))) = D_M^{\{i\}} \quad i = 1, \dots, m_M. \tag{13.71}$$

Clearly, in the case where not all the kinematical variables are inserted into the microscale, restrictions (13.70) and (13.71) must be understood for the variables that are effectively transferred to the microscale.

Classical multiscale solid mechanics. In this case we have

$$\mathscr{H}_\mu^{\mathscr{V}}(\mathcal{J}_\mu^{\mathscr{V}}(\mathbf{u}_M|_{\mathbf{x}})) = \frac{1}{|\Omega_\mu|}\int_{\Omega_\mu} \mathbf{u}_M|_{\mathbf{x}}\, d\Omega_\mu = \mathbf{u}_M|_{\mathbf{x}}. \tag{13.72}$$

In the same manner

$$\mathscr{H}_\mu^{\mathscr{W}}(\mathscr{D}_\mu(\mathcal{J}_\mu^{\mathscr{W}}(\varepsilon_M|_{\mathbf{x}}))) = \frac{1}{|\Omega_\mu|}\int_{\Omega_\mu} (\nabla_{\mathbf{y}}(\varepsilon_M|_{\mathbf{x}}(\mathbf{y} - \mathbf{y}_o)))^s\, d\Omega_\mu = \varepsilon_M|_{\mathbf{x}}. \tag{13.73}$$

We thus see that restrictions (13.70) and (13.71) are verified in the context of the classical multiscale theory. ∎

From (13.70) it is easy to verify that the composition $\mathscr{H}_\mu^{\mathscr{V}}\mathcal{J}_\mu^{\mathscr{V}}$ defines the identity operator in $\mathbb{R}^{\mathbf{x}}_{\mathscr{V}_M}$. Similarly, from (13.71), the composition $\mathscr{H}_\mu^{\mathscr{W}}\mathscr{D}_\mu\mathcal{J}_\mu^{\mathscr{W}}$ defines the identity operator in $\mathbb{R}^{\mathbf{x}}_{\mathscr{W}_M}$.

13.4.5 Kinematical Admissibility

Let us now introduce the central concept of kinematical admissibility in the kinematical transfer between scales. We say that the microscale generalized displacement $u_\mu \in \mathscr{V}_\mu$, linked to the macroscale kinematics, and the associated generalized deformation action

$\mathscr{D}_\mu(u_\mu) \in \mathscr{W}_\mu$ are kinematically admissible with respect to $u_M|_\mathbf{x} \in \mathbb{R}^\mathbf{x}_{\mathscr{V}_M}$ and $D_M|_\mathbf{x} \in \mathbb{R}^\mathbf{x}_{\mathscr{W}_M}$, if the following relations are verified

$$\mathscr{H}^\mathscr{V}_\mu(u_\mu) = \mathscr{H}^\mathscr{V}_\mu(\mathscr{J}^\mathscr{V}_\mu(u_M|_\mathbf{x})), \tag{13.74}$$

$$\mathscr{H}^\mathscr{W}_\mu(\mathscr{D}_\mu(u_\mu)) = \mathscr{H}^\mathscr{W}_\mu(\mathscr{D}_\mu(\mathscr{J}^\mathscr{W}_\mu(D_M|_\mathbf{x}))). \tag{13.75}$$

The definition of kinematical admissibility implies additional constraints. As a matter of fact, since $u_\mu \in \mathscr{V}_\mu$ is linked to the macroscale kinematics, (13.55) is satisfied, therefore the left-hand side from (13.74) gives

$$\mathscr{H}^\mathscr{V}_\mu(u_\mu) = \mathscr{H}^\mathscr{V}_\mu(\mathscr{J}^\mathscr{V}_\mu(u_M|_\mathbf{x})) + \mathscr{H}^\mathscr{V}_\mu(\mathscr{J}^\mathscr{W}_\mu(D_M|_\mathbf{x})) + \mathscr{H}^\mathscr{V}_\mu(\tilde{u}_\mu). \tag{13.76}$$

Imposing the additional restriction in the operator $\mathscr{J}^\mathscr{W}_\mu$ given by

$$\mathscr{H}^\mathscr{V}_\mu(\mathscr{J}^\mathscr{W}_\mu(D_M|_\mathbf{x})) = 0, \tag{13.77}$$

we have, as a consequence of (13.74), (13.76) and (13.77), that $\tilde{u}_\mu$ must satisfy the following kinematical constraint

$$\mathscr{H}^\mathscr{V}_\mu(\tilde{u}_\mu) = 0. \tag{13.78}$$

Because $\mathscr{H}^\mathscr{V}_\mu$ represents a mean operation (average) involving the measure of the domain in which each component of the insertion operator is defined, equations (13.70) and (13.74) establish a relation between $u_M|_\mathbf{x}$ and the homogenization of the microscale generalized displacement u_μ. In turn, equation (13.78) introduces n_M tensor constraints (i.e. R_M scalar constraints) which must be satisfied by the generalized displacement fluctuation at the microscale, $\tilde{u}_\mu$, to link the microscale kinematics to the macroscale kinematics, hence verifying the kinematical admissibility.

Classical multiscale solid mechanics. In this classical example, and recalling the definition of the u_μ-homogenization operator, the kinematical admissibility of $\mathbf{u}_\mu$ implies that

$$\mathscr{H}^\mathscr{V}_\mu(\mathbf{u}_\mu) = \mathscr{H}^\mathscr{V}_\mu(\mathscr{J}^\mathscr{V}_\mu(\mathbf{u}_M|_\mathbf{x})) = \mathbf{u}_M|_\mathbf{x}. \tag{13.79}$$

In turn, by construction, operator $\mathscr{J}^\mathscr{W}_\mu$ is such that

$$\frac{1}{|\Omega_\mu|}\int_{\Omega_\mu} \varepsilon_M|_\mathbf{x}(\mathbf{y}-\mathbf{y}_o)\, d\Omega_\mu = \varepsilon_M|_\mathbf{x}\left(\frac{1}{|\Omega_\mu|}\int_{\Omega_\mu}(\mathbf{y}-\mathbf{y}_o)\, d\Omega_\mu\right) = \mathbf{0}, \tag{13.80}$$

and so we have

$$\frac{1}{|\Omega_\mu|}\int_{\Omega_\mu}(\mathbf{u}_M|_\mathbf{x} + \varepsilon_M|_\mathbf{x}(\mathbf{y}-\mathbf{y}_o) + \tilde{\mathbf{u}}_\mu)\, d\Omega_\mu$$

$$= \mathbf{u}_M|_\mathbf{x} + \frac{1}{|\Omega_\mu|}\int_{\Omega_\mu}\tilde{\mathbf{u}}_\mu\, d\Omega_\mu = \mathbf{u}_M|_\mathbf{x}, \tag{13.81}$$

which is verified only if

$$\int_{\Omega_\mu}\tilde{\mathbf{u}}_\mu\, d\Omega_\mu = \mathbf{0}. \tag{13.82}$$

■

We can proceed in an analogous way with equation (13.75). Taking into account equation (13.57), we have

$$
\begin{aligned}
&\mathscr{H}_{\mu}^{\mathscr{W}}(\mathscr{D}_{\mu}(u_{\mu}))\\
&= \underbrace{\mathscr{H}_{\mu}^{\mathscr{W}}(\mathscr{D}_{\mu}(\mathscr{J}_{\mu}^{\mathscr{V}}(u_M|_{\mathbf{x}})))}_{=0} + \mathscr{H}_{\mu}^{\mathscr{W}}(\mathscr{D}_{\mu}(\mathscr{J}_{\mu}^{\mathscr{W}}(D_M|_{\mathbf{x}}))) + \mathscr{H}_{\mu}^{\mathscr{W}}(\mathscr{D}_{\mu}(\tilde{u}_{\mu}))\\
&= \mathscr{H}_{\mu}^{\mathscr{W}}(\mathscr{D}_{\mu}(\mathscr{J}_{\mu}^{\mathscr{W}}(D_M|_{\mathbf{x}}))).
\end{aligned} \tag{13.83}
$$

This entails that the following must be verified

$$
\mathscr{H}_{\mu}^{\mathscr{W}}(\mathscr{D}_{\mu}(\tilde{u}_{\mu})) = 0. \tag{13.84}
$$

Hence, equations (13.71) and (13.75) determine the relation between $D_M|_{\mathbf{x}}$ and the homogenization of the microscale generalized deformation action D_{μ}.

To sum up, we have determined that all kinematically admissible microscale generalized displacement field u_{μ} is such that the associated fluctuation field, $\tilde{u}_{\mu}$, satisfies the kinematical constraints (13.78) and (13.84). Consequently, we can rationally define the space of kinematically admissible microscale generalized displacement fluctuations as follows

$$
\mathrm{Kin}_{\tilde{u}_{\mu}} = \{\tilde{u}_{\mu} \in \mathscr{V}_{\mu};\ \mathscr{H}_{\mu}^{\mathscr{V}}(\tilde{u}_{\mu}) = 0,\ \mathscr{H}_{\mu}^{\mathscr{W}}(\mathscr{D}_{\mu}(\tilde{u}_{\mu})) = 0\}. \tag{13.85}
$$

The elements $\tilde{u}_{\mu} \in \mathrm{Kin}_{\tilde{u}_{\mu}}$ satisfy the minimum kinematical constraints which make the kinematical transfer between the scales admissible. Evidently, it is possible to consider additional kinematical constraints. In that case, the resulting space is such that $\mathrm{Kin}^{*}_{\tilde{u}_{\mu}} \subset \mathrm{Kin}_{\tilde{u}_{\mu}}$, yielding different multiscale models, in particular, multiscale models which are stiffer than the one characterized by the space $\mathrm{Kin}_{\tilde{u}_{\mu}}$ due to the simple fact that the latter features the minimal set of admissible kinematical constraints.

In turn, since the kinematical constraints considered over $\tilde{u}_{\mu}$ are linear and homogeneous, it follows that the space of kinematically admissible microscale generalized motion action fluctuations coincides with the same space $\mathrm{Kin}_{\tilde{u}_{\mu}}$, that is

$$
\mathrm{Var}_{\tilde{u}_{\mu}} = \mathrm{Kin}_{\tilde{u}_{\mu}}. \tag{13.86}
$$

Classical multiscale solid mechanics. In this classical multiscale example, and with the definition of the D_{μ}-homogenization operator, the kinematical admissibility of D_{μ} implies

$$
\begin{aligned}
\mathscr{H}_{\mu}^{\mathscr{W}}((\nabla_{\mathbf{y}}\mathbf{u}_{\mu})^{s}) &= \mathscr{H}_{\mu}^{\mathscr{W}}((\nabla_{\mathbf{y}}(\mathbf{u}_M|_{\mathbf{x}} + \varepsilon_M|_{\mathbf{x}}(\mathbf{y}-\mathbf{y}_{o}) + \tilde{\mathbf{u}}_{\mu}))^{s})\\
&= \mathscr{H}_{\mu}^{\mathscr{W}}(\varepsilon_M|_{\mathbf{x}} + (\nabla_{\mathbf{y}}\tilde{\mathbf{u}}_{\mu})^{s}) = \varepsilon_M|_{\mathbf{x}}.
\end{aligned} \tag{13.87}
$$

Hence,

$$
\begin{aligned}
&\frac{1}{|\Omega_{\mu}|}\int_{\Omega_{\mu}} (\varepsilon_M|_{\mathbf{x}} + (\nabla_{\mathbf{y}}\tilde{\mathbf{u}}_{\mu})^{s})\, d\Omega_{\mu}\\
&= \varepsilon_M|_{\mathbf{x}} + \frac{1}{|\Omega_{\mu}|}\int_{\Omega_{\mu}} (\nabla_{\mathbf{y}}\tilde{\mathbf{u}}_{\mu})^{s}\, d\Omega_{\mu} = \varepsilon_M|_{\mathbf{x}},
\end{aligned} \tag{13.88}
$$

which is satisfied if

$$
\int_{\Omega_{\mu}} (\nabla_{\mathbf{y}}\tilde{\mathbf{u}}_{\mu})^{s}\, d\Omega_{\mu} = \mathbf{O}. \tag{13.89}
$$

Equivalently, after using the Green formula we obtain

$$\int_{\partial\Omega_\mu} (\tilde{\mathbf{u}}_\mu \otimes \mathbf{n}_\mu)^s \, d\partial\Omega_\mu = \mathbf{O}, \tag{13.90}$$

where $\mathbf{n}_\mu$ is the unit outward normal vector to the boundary $\partial\Omega_\mu$ of Ω_μ. Therefore, the space of kinematically admissible microscale generalized displacement fluctuations and the associated space of admissible motion actions are

$$\begin{aligned}\mathrm{Kin}_{\tilde{u}_\mu} &= \mathrm{Var}_{\tilde{u}_\mu}\\ &= \left\{ \tilde{\mathbf{u}}_\mu \in \mathbf{H}^1(\Omega_\mu); \int_{\Omega_\mu} \tilde{\mathbf{u}}_\mu \, d\Omega_\mu = \mathbf{0}, \int_{\partial\Omega_\mu} (\tilde{\mathbf{u}}_\mu \otimes \mathbf{n}_\mu)^s \, d\partial\Omega_\mu = \mathbf{O} \right\}.\end{aligned} \tag{13.91}$$

■

Notice that the kinematical constraints imposed by the kinematical admissibility between scales is reduced to a set of n_M tensor constraints given by (13.78) in addition to the m_M tensor constraints given by (13.84). Moreover, because in the present framework the macroscale and microscale kinematics can in general be different, the microscale kinematical fields may not be properly controled. That is, some microscale kinematical descriptors may not be observable from the macroscale. In these cases, it will be necessary to add further constraints over $\tilde{u}_\mu$ in order to guarantee that the microscale problem is a mathematically well-posed problem (it has a solution). These restrictions are homogeneous and depend on physical aspects associated with the microscale kinematical model.

As we have already highlighted, and as it will be exposed soon in detail, the space $\mathrm{Kin}_{\tilde{u}_\mu}$ plays a fundamental role in the definition of the equilibrium concept at the microscale. Hence, if additional constraints are added to those minimally required by the space $\mathrm{Kin}_{\tilde{u}_\mu}$, the solution of the multiscale problem will evidently be different. A strategy to consider additional constraints is to enforce $\tilde{u} \equiv 0$ in the whole microscale domain. This corresponds to what in the classical multiscale literature in the field of solid mechanics is known as the Taylor model (or mixture model). Here, within a more general multiscale context, we refer to this space as the Taylor fluctuations space. This space is formed by a single element, which is the null element in $\mathrm{Kin}_{\tilde{u}_\mu}$

$$\mathrm{Kin}_{\tilde{u}_\mu}^{Taylor} = \{\tilde{u}_\mu \in \mathscr{V}_\mu;\ \tilde{u}_\mu = 0 \text{ in } \Omega_\mu^{\mathscr{V}}\} = \{0\}. \tag{13.92}$$

Obviously, this space corresponds to the maximally constrained space of fluctuations that can be established. In turn, and as mentioned before, we can adopt alternative not-so-constrained spaces compared to $\mathrm{Kin}_{\tilde{u}_\mu}^{Taylor}$.

Classical multiscale solid mechanics. In this context, additionally to the Taylor space $\mathrm{Kin}_{\tilde{\mathbf{u}}_\mu}^{Taylor}$, we could adopt a constraint stating that the fluctuations are null over the boundary, yielding the space $\mathrm{Kin}_{\tilde{\mathbf{u}}_\mu}^{nbc}$, obtained by prescribing $\tilde{\mathbf{u}}_\mu = \mathbf{0}$, $\forall \mathbf{y} \in \partial\Omega_\mu$. Another space, called $\mathrm{Kin}_{\tilde{\mathbf{u}}_\mu}^{pbc}$, can easily be constructed for RVE domains with periodic geometry (typically found in periodic media). The microscale domains in periodic RVEs feature normal vectors $\mathbf{n}_\mu$ which are antiperiodic over the boundary $\partial\Omega_\mu$. In such cases, for every $\mathbf{y} \in \partial\Omega_\mu$ there exists a binunivocal correspondence with the point $\mathbf{y}^* \in \partial\Omega_\mu$ over the opposite boundary to $\partial\Omega_\mu$ such that $\mathbf{n}_\mu(\mathbf{y}) = -\mathbf{n}_\mu(\mathbf{y}^*)$. Thus, the kinematical admissibility is ensured if the displacement fluctuation field $\tilde{\mathbf{u}}_\mu$ is periodic on $\partial\Omega_\mu$, that is, $\tilde{\mathbf{u}}_\mu(\mathbf{y}) = \tilde{\mathbf{u}}_\mu(\mathbf{y}^*)$. For these cases it is also easy to verify that $\mathrm{Kin}_{\tilde{\mathbf{u}}_\mu}^{Taylor} \subset \mathrm{Kin}_{\tilde{\mathbf{u}}_\mu}^{nbc} \subset \mathrm{Kin}_{\tilde{\mathbf{u}}_\mu}^{pbc} \subset \mathrm{Kin}_{\tilde{\mathbf{u}}_\mu}$. From a purely mechanical perspective, this means that the RVE-based multiscale model will render the stiffest behavior for the maximally constrained space $\mathrm{Kin}_{\tilde{\mathbf{u}}_\mu}^{Taylor}$ and the most flexible model will be obtained with the minimally constrained space $\mathrm{Kin}_{\tilde{\mathbf{u}}_\mu}$. ■

The next step consists of the characterization of the subspace Kin_{u_μ} of kinematically admissible microscale generalized displacements. This subspace is formed by all the generalized displacements, $u_\mu \in \mathscr{V}_\mu$, linked to the macroscale kinematics at point $\mathbf{x} \in \Omega_M$ and kinematically admissible with respect to $u_M|_\mathbf{x} \in \mathbb{R}^\mathbf{x}_{\mathscr{V}_M}$ and $D_M|_\mathbf{x} \in \mathbb{R}^\mathbf{x}_{\mathscr{W}_M}$, that is

$$\mathrm{Kin}_{u_\mu} = \Big\{ u_\mu \in \mathscr{V}_\mu;\ u_\mu = \mathscr{J}_\mu^{\mathscr{V}}(u_M|_\mathbf{x}) + \mathscr{J}_\mu^{\mathscr{W}}(D_M|_\mathbf{x}) + \tilde{u}_\mu, \\ u_M|_\mathbf{x} \in \mathbb{R}^\mathbf{x}_{\mathscr{V}_M},\ D_M|_\mathbf{x} \in \mathbb{R}^\mathbf{x}_{\mathscr{W}_M}, \tilde{u}_\mu \in \mathrm{Kin}_{\tilde{u}_\mu} \Big\}. \tag{13.93}$$

The corresponding space of kinematically admissible microscale generalized motion actions, denoted by Var_{u_μ}, is given by

$$\mathrm{Var}_{u_\mu} = \{\hat{u}_\mu \in \mathscr{V}_\mu;\ \hat{u}_\mu = u_\mu^1 - u_\mu^2,\ u_\mu^1, u_\mu^2 \in \mathrm{Kin}_{u_\mu}\}, \tag{13.94}$$

or, taking into consideration (13.86),

$$\mathrm{Var}_{u_\mu} = \mathrm{Kin}_{u_\mu}. \tag{13.95}$$

13.5 Duality in Multiscale Modeling

The goal of this section is to apply the concept of duality presented in Section 13.2.2 for both scales. Within the multiscale framework proposed here, special attention must be given to the definition of the virtual power at a generic macroscale point $\mathbf{x}$. Regarding the microscale, special attention is required for the identification of generalized stresses and generalized forces compatible with the microscale kinematics.

13.5.1 Macroscale Virtual Power

Following the developments of Section 13.2, the macroscale internal virtual power is defined by

$$P_M^i(\mathscr{D}_M(\hat{u}_M)) = \langle \Sigma_M, \mathscr{D}_M(\hat{u}_M)\rangle_{\mathscr{W}_M' \times \mathscr{W}_M} \qquad \hat{u}_M \in \mathrm{Var}_{u_M}, \tag{13.96}$$

or, equivalently, by

$$P_M^i(\mathscr{D}_M(\hat{u}_M)) = \sum_{k=1}^{m_M} \langle \Sigma_M^k, (\mathscr{D}_M(\hat{u}_M))^k\rangle_{\mathscr{W}_{Mk}' \times \mathscr{W}_{Mk}} \qquad \hat{u}_M \in \mathrm{Var}_{u_M}. \tag{13.97}$$

In this context, we are interested in assessing the internal virtual power associated with a single generic macroscale point $\mathbf{x}$ to be, posteriorly, related to the microscale internal virtual power through what will be called the Principle of Multiscale Virtual Power, which will be introduced soon.

In this sense, note that at the point $\mathbf{x} \in \Omega_M$, the kinematical entity that allows the internal power at that point to be evaluated is given by $\mathscr{D}_M(\hat{u}_M)|_{\mathbf{x}}$. Hence, with the notation $\hat{D}_M|_{\mathbf{x}} = \mathscr{D}_M(\hat{u}_M)|_{\mathbf{x}}$, the macroscale internal virtual power, called $P^i_{M,\mathbf{x}}(\hat{D}_M|_{\mathbf{x}})$, at point $\mathbf{x}$ can be expressed in the following manner

$$P^i_{M,\mathbf{x}}(\hat{D}_M|_{\mathbf{x}}) = \sum_{k=1}^{m_M} \omega_k (\Sigma_M|_{\mathbf{x}})^k \cdot (\hat{D}_M|_{\mathbf{x}})^k =: \Sigma_M|_{\mathbf{x}} \bullet \hat{D}_M|_{\mathbf{x}} \qquad \hat{D}_M|_{\mathbf{x}} \in \widehat{\mathbb{R}^{\mathbf{x}}_{\mathscr{W}_M}}, \tag{13.98}$$

where ω_k, $k = 1, \dots, m_M$, are dimensional scalars (see explanation at the end of this section) which ensure the dimensional compatibility of the products $(\Sigma_M|_{\mathbf{x}})^k \cdot (\hat{D}_M|_{\mathbf{x}})^k$, $k = 1, \dots, m_M$, in each contribution in the internal virtual power. It is important to observe that (13.98) has units of density of power, where each product $(\Sigma_M|_{\mathbf{x}})^k \cdot (\hat{D}_M|_{\mathbf{x}})^k$ is a density of power, that is power per unit measure ω_k of the corresponding subset of the macroscale domain. Note that each of these subsets can have different dimensionality (e.g. point, line, surface, volume). We underline that this level of generality is crucial to model physical systems which feature simultaneous phenomena of different natures defined in different subdomains, such as localization phenomena cohesive fracturing or discrete mechanical interactions, among others.

In turn, the operation $(\cdot) \bullet (\cdot)$ represents the duality product between the variables

$$\begin{aligned}
&(\cdot) \bullet (\cdot) : (\mathbb{R}^{\mathbf{x}}_{\mathscr{W}_M})' \times \mathbb{R}^{\mathbf{x}}_{\mathscr{W}_M} \to \mathbb{R}, \\
&\qquad (\Sigma_M|_{\mathbf{x}}, D_M|_{\mathbf{x}}) \mapsto \Sigma_M|_{\mathbf{x}} \bullet D_M|_{\mathbf{x}} = \sum_{k=1}^{m_M} \omega_k \, (\Sigma_M|_{\mathbf{x}})^k \cdot (D_M|_{\mathbf{x}})^k.
\end{aligned} \tag{13.99}$$

Classical multiscale solid mechanics. In this case the macroscale internal virtual power is given by the product between the deformation virtual action (rate of deformation) and the Cauchy stress field,

$$P^i_M = \int_{\Omega_M} \sigma_M \cdot (\nabla_{\mathbf{x}} \hat{\mathbf{u}}_M)^s \, d\Omega_M. \tag{13.100}$$

The internal virtual power evaluated at point $\mathbf{x}$ (connected to the RVE) is

$$P^i_{M,\mathbf{x}} = \sigma_M|_{\mathbf{x}} \bullet \hat{\varepsilon}_M|_{\mathbf{x}} = \omega_1 \sigma_M|_{\mathbf{x}} \cdot \hat{\varepsilon}_M|_{\mathbf{x}}. \tag{13.101}$$

■

In a similar way, we define the macroscale external virtual power

$$P^e_M(\hat{u}_M) = \langle f_M, \hat{u}_M \rangle_{\mathscr{V}'_M \times \mathscr{V}_M} \qquad \hat{u}_M \in \mathrm{Var}_{u_M}, \tag{13.102}$$

where f_M has the structure of $\mathscr{D}^*_M(\Sigma_M)$. The external virtual power at point $\mathbf{x}$ takes the form

$$P^e_{M,\mathbf{x}}(\hat{u}_M|_{\mathbf{x}}) = \sum_{k=1}^{n_M} \gamma_k \, (f_M|_{\mathbf{x}})^k \cdot (\hat{u}_M|_{\mathbf{x}})^k =: f_M|_{\mathbf{x}} \bullet \hat{u}_M|_{\mathbf{x}} \qquad \hat{u}_M|_{\mathbf{x}} \in \widehat{\mathbb{R}^{\mathbf{x}}_{\mathscr{V}_M}}. \tag{13.103}$$

Dimensional parameters γ_k, $k = 1, \dots, n_M$, are entirely analogous to parameters ω_k appearing in (13.98). Notice that in (13.103) we have used the same notation for the duality product, $(\cdot) \bullet (\cdot)$ as in (13.98), but the real sense of this product is given by the context in which it is established. In this case, the duality product $(\cdot) \bullet (\cdot)$ is defined by

$$
\begin{aligned}
&(\cdot) \bullet (\cdot) : (\mathbb{R}^{\mathbf{x}}_{\mathscr{V}_M})' \times \mathbb{R}^{\mathbf{x}}_{\mathscr{V}_M} \to \mathbb{R}, \\
&(f_M|_{\mathbf{x}}, u_M|_{\mathbf{x}}) \mapsto f_M|_{\mathbf{x}} \bullet u_M|_{\mathbf{x}} = \sum_{k=1}^{n_M} \gamma_k \, (f_M|_{\mathbf{x}})^k \cdot (u_M|_{\mathbf{x}})^k.
\end{aligned}
\tag{13.104}
$$

Classical multiscale solid mechanics. In this case the macroscale external virtual power is given by

$$
P^e_M = \int_{\Omega_M} \mathbf{f}_M \cdot \hat{\mathbf{u}}_M \, d\Omega_M + \int_{\Gamma_{Mt}} \mathbf{t}_M \cdot \hat{\mathbf{u}}_M \, d\Gamma_M. \tag{13.105}
$$

The corresponding power at the macroscale point $\mathbf{x} \in \Omega_M$ is

$$
P^e_{M,\mathbf{x}} = \mathbf{f}_M|_{\mathbf{x}} \bullet \hat{\mathbf{u}}_M|_{\mathbf{x}} = \gamma_1 \mathbf{f}_M|_{\mathbf{x}} \cdot \hat{\mathbf{u}}_M|_{\mathbf{x}}. \tag{13.106}
$$

■

Now we have established the macroscale internal and external virtual powers, we define the macroscale total virtual power at point $\mathbf{x}$ simply by using (13.98) and (13.103), giving

$$
\begin{aligned}
P^t_{M,\mathbf{x}}(\hat{u}_M|_{\mathbf{x}}, \hat{D}_M|_{\mathbf{x}}) &= \Sigma_M|_{\mathbf{x}} \bullet \hat{D}_M|_{\mathbf{x}} - f_M|_{\mathbf{x}} \bullet \hat{u}_M|_{\mathbf{x}} \\
\hat{u}_M|_{\mathbf{x}} &\in \widehat{\mathbb{R}^{\mathbf{x}}_{\mathscr{V}_M}}, \hat{D}_M|_{\mathbf{x}} \in \widehat{\mathbb{R}^{\mathbf{x}}_{\mathscr{W}_M}}.
\end{aligned}
\tag{13.107}
$$

Figure 13.3 shows a schematic representation of the different concepts utilized in the macroscale segment of the proposed multiscale variational framework.

At this point it is important to note that the macroscale point $\mathbf{x}$ is a point that belongs to a certain geometrical object from the macroscale domain. In general, it could be a point in a volume, a surface, a line or simply a point itself. In this manner, the external virtual power, $P^e_{M,\mathbf{x}}$, at that point is associated with external generalized forces defined over the geometric object to which the point belongs and that are properly characterized through duality arguments. For example, in the case of a point in a three-dimensional solid continuum we will have, from the duality, that the characterization of the generalized forces in the body includes notions of passive forces (e.g. that caused by gravity in classical solid mechanics) as well as inertia forces (those associated with accelerations in classical solid mechanics). Hence, when we refer to generalized body forces we mean these two classes of generalized forces (passive and inertia). In this way, dynamics phenomena are automatically taken into account by the proposed formulation. Importantly, we underline the fact that both macroscale and microscale possess the same time scale.

13.5.2 Microscale Virtual Power

Exploiting the concept of duality from Section 13.2, the microscale internal virtual power can be expressed in the following manner

$$
P^i_\mu(\mathscr{D}_\mu(\hat{u}_\mu)) = \langle \Sigma_\mu, \mathscr{D}_\mu(\hat{u}_\mu) \rangle_{\mathscr{W}'_\mu \times \mathscr{W}_\mu} \qquad \hat{u}_\mu \in \mathrm{Var}_{u_\mu}. \tag{13.108}
$$

Considering (13.55) and (13.57), and with a certain abuse of notation, we obtain the following equivalent representation

$$P^i_\mu(\hat{D}_M|_{\mathbf{x}}, \mathscr{D}_\mu(\hat{u}_\mu)) = \langle \Sigma_\mu, \mathscr{D}_\mu(\mathscr{I}^{\mathscr{W}}_\mu(\hat{D}_M|_{\mathbf{x}}) + \hat{u}_\mu)\rangle_{\mathscr{W}'_\mu \times \mathscr{W}_\mu}$$
$$\hat{D}_M|_{\mathbf{x}} \in \widehat{\mathbb{R}^{\mathbf{x}}_{\mathscr{W}_M}},\ \hat{u}_\mu \in \mathrm{Var}_{\tilde{u}_\mu}, \tag{13.109}$$

in terms of the macroscale deformation action, $\hat{D}_M|_{\mathbf{x}}$, and of the microscale motion action fluctuations, $\hat{u}_\mu$. Observe that the macroscale kinematics is mapped into the microscale through the insertion operator $\mathscr{I}^{\mathscr{W}}_\mu$. Figure 13.4 illustrates the basic concepts of the microscale realm which set the basis for the proposed unified multiscale variational formulation.

Classical multiscale solid mechanics. Taking into account the terms that participate in the definition of the microscale deformation action, the internal virtual power in this example is shaped as follows

$$P^i_\mu = \int_{\Omega_\mu} \sigma_\mu \cdot (\nabla_{\mathbf{y}}\hat{\mathbf{u}}_\mu)^s \, d\Omega_\mu = \int_{\Omega_\mu} \sigma_\mu \cdot (\hat{\varepsilon}_M|_{\mathbf{x}} + (\nabla_{\mathbf{y}}\hat{\tilde{\mathbf{u}}}_\mu)^s) \, d\Omega_\mu$$
$$= \int_{\Omega_\mu} \sigma_\mu \cdot \hat{\varepsilon}_M|_{\mathbf{x}} \, d\Omega_\mu + \int_{\Omega_\mu} \sigma_\mu \cdot (\nabla_{\mathbf{y}}\hat{\tilde{\mathbf{u}}}_\mu)^s \, d\Omega_\mu. \tag{13.110}$$

■

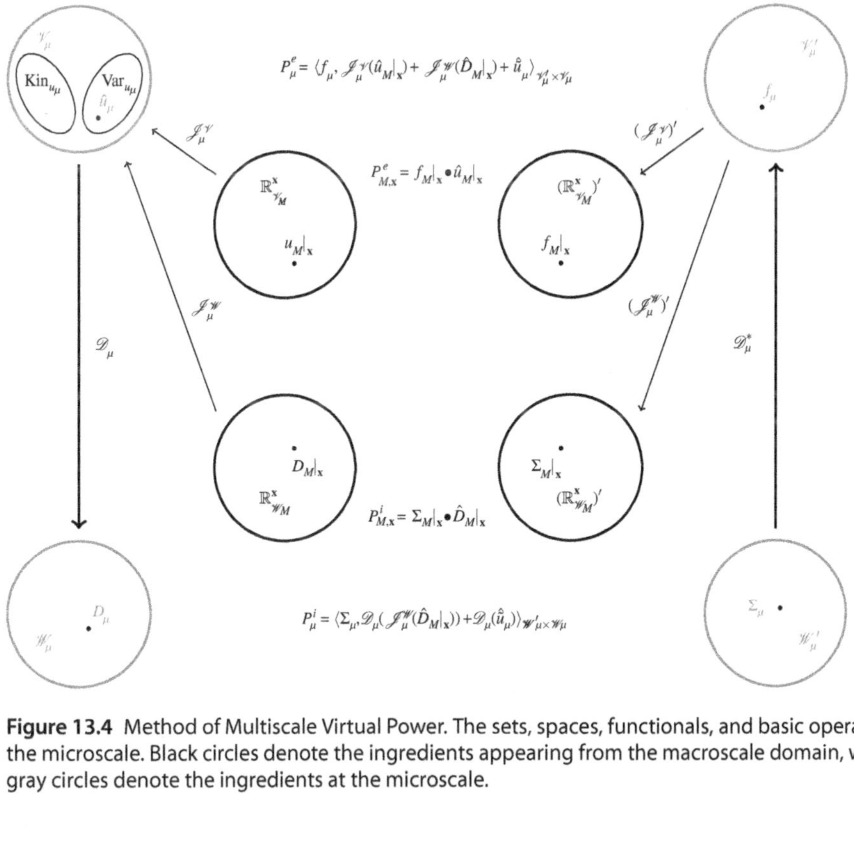

Figure 13.4 Method of Multiscale Virtual Power. The sets, spaces, functionals, and basic operations at the microscale. Black circles denote the ingredients appearing from the macroscale domain, while gray circles denote the ingredients at the microscale.

Let us now present the microscale external virtual power. As we have seen, this external power is defined as a linear functional over the space $\mathrm{Var}_{u_\mu} = \mathrm{Kin}_{u_\mu}$, then

$$P^e_\mu(\hat{u}_\mu) = \langle f_\mu, \hat{u}_\mu \rangle_{\mathscr{V}'_\mu \times \mathscr{V}_\mu} \qquad \hat{u}_\mu \in \mathrm{Var}_{u_\mu}. \tag{13.111}$$

From (13.108) and from the definition of the adjoint operator $\mathscr{D}^*_\mu$ we can fully characterize the nature of generalized external forces $f_\mu \in \mathscr{V}'_\mu$ admissible at the microscale. In fact, we have

$$\begin{aligned}\langle \Sigma_\mu, \mathscr{D}_\mu(\hat{u}_\mu) \rangle_{\mathscr{W}'_\mu \times \mathscr{W}_\mu} &= \langle \mathscr{D}^*_\mu(\Sigma_\mu), \hat{u}_\mu \rangle_{\mathscr{V}'_\mu \times \mathscr{V}_\mu} \\ &= \langle f_\mu, \mathscr{J}^{\mathscr{V}}_\mu(\hat{u}_M|_{\mathbf{x}}) + \mathscr{J}^{\mathscr{W}}_\mu(\hat{D}_M|_{\mathbf{x}}) + \hat{\tilde{u}}_\mu \rangle_{\mathscr{V}'_\mu \times \mathscr{V}_\mu} \qquad \hat{u}_\mu \in \mathrm{Var}_{u_\mu},\end{aligned} \tag{13.112}$$

that is

$$\begin{aligned}P^e_\mu(\hat{u}_M|_{\mathbf{x}}, \hat{D}_M|_{\mathbf{x}}, \hat{\tilde{u}}_\mu) &= \langle f_\mu, \mathscr{J}^{\mathscr{V}}_\mu(\hat{u}_M|_{\mathbf{x}}) + \mathscr{J}^{\mathscr{W}}_\mu(\hat{D}_M|_{\mathbf{x}}) + \hat{\tilde{u}}_\mu \rangle_{\mathscr{V}'_\mu \times \mathscr{V}_\mu} \\ &\qquad \hat{u}_M|_{\mathbf{x}} \in \widehat{\mathbb{R}^{\mathbf{x}}_{\mathscr{V}_M}},\ \hat{D}_M|_{\mathbf{x}} \in \widehat{\mathbb{R}^{\mathbf{x}}_{\mathscr{W}_M}},\ \hat{\tilde{u}}_\mu \in \mathrm{Var}_{\tilde{u}_\mu}.\end{aligned} \tag{13.113}$$

Note that the evaluation of the microscale external virtual power relies on kinematical entities defined at both the microscale and the macroscale.

Classical multiscale solid mechanics. For this example, the microscale external virtual power is given by

$$\begin{aligned}P^e_\mu &= \int_{\Omega_\mu} \mathbf{f}_\mu \cdot \hat{\mathbf{u}}_\mu \, d\Omega_\mu = \int_{\Omega_\mu} \mathbf{f}_\mu \cdot (\hat{\mathbf{u}}_M|_{\mathbf{x}} + \hat{\varepsilon}_M|_{\mathbf{x}}(\mathbf{y} - \mathbf{y}_o) + \hat{\tilde{\mathbf{u}}}_\mu) \, d\Omega_\mu \\ &= \int_{\Omega_\mu} \mathbf{f}_\mu \cdot \hat{\mathbf{u}}_M|_{\mathbf{x}} \, d\Omega_\mu + \int_{\Omega_\mu} (\mathbf{f}_\mu \otimes (\mathbf{y} - \mathbf{y}_o))^s \cdot \hat{\varepsilon}_M|_{\mathbf{x}} \, d\Omega_\mu \\ &\quad + \int_{\Omega_\mu} \mathbf{f}_\mu \cdot \hat{\tilde{\mathbf{u}}}_\mu \, d\Omega_\mu.\end{aligned} \tag{13.114}$$

■

With the definition of the internal and external virtual powers, we then introduce the microscale total virtual power as follows

$$\begin{aligned}P^t_\mu(\hat{u}_M|_{\mathbf{x}}, \hat{D}_M|_{\mathbf{x}}, \hat{\tilde{u}}_\mu) &= P^i_\mu(\hat{D}_M|_{\mathbf{x}}, \mathscr{D}_\mu(\hat{\tilde{u}}_\mu)) - P^e_\mu(\hat{u}_M|_{\mathbf{x}}, \hat{D}_M|_{\mathbf{x}}, \hat{\tilde{u}}_\mu) \\ &\qquad \hat{u}_M|_{\mathbf{x}} \in \widehat{\mathbb{R}^{\mathbf{x}}_{\mathscr{V}_M}},\ \hat{D}_M|_{\mathbf{x}} \in \widehat{\mathbb{R}^{\mathbf{x}}_{\mathscr{W}_M}},\ \hat{\tilde{u}}_\mu \in \mathrm{Var}_{\tilde{u}_\mu},\end{aligned} \tag{13.115}$$

characterized by the sum of linear and continuous functionals defined in Var_{u_μ}.

The contribution of kinematical entities from the macroscale, $\hat{u}_M|_{\mathbf{x}}$ and $\hat{D}_M|_{\mathbf{x}}$, in the evaluation of the microscale total virtual power, given by (13.115), has fundamental implications in the theory. To see this, let us assess the power P^t_μ for $\hat{\tilde{u}}_\mu = 0$,

$$\begin{aligned}P^t_\mu(\hat{u}_M|_{\mathbf{x}}, \hat{D}_M|_{\mathbf{x}}, 0) &= P^i_\mu(\hat{D}_M|_{\mathbf{x}}, 0) - P^e_\mu(\hat{u}_M|_{\mathbf{x}}, \hat{D}_M|_{\mathbf{x}}, 0) \\ &= \langle \Sigma_\mu, \mathscr{D}_\mu(\mathscr{J}^{\mathscr{W}}_\mu(\hat{D}_M|_{\mathbf{x}})) \rangle_{\mathscr{W}'_\mu \times \mathscr{W}_\mu} - \langle f_\mu, \mathscr{J}^{\mathscr{V}}_\mu(\hat{u}_M|_{\mathbf{x}}) + \mathscr{J}^{\mathscr{W}}_\mu(\hat{D}_M|_{\mathbf{x}}) \rangle_{\mathscr{V}'_\mu \times \mathscr{V}_\mu} \\ &= \langle (\mathscr{D}^*_\mu \Sigma_\mu - f_\mu), \mathscr{J}^{\mathscr{W}}_\mu(\hat{D}_M|_{\mathbf{x}}) \rangle_{\mathscr{V}'_\mu \times \mathscr{V}_\mu} - \langle f_\mu, \mathscr{J}^{\mathscr{V}}_\mu(\hat{u}_M|_{\mathbf{x}}) \rangle_{\mathscr{V}'_\mu \times \mathscr{V}_\mu}.\end{aligned} \tag{13.116}$$

From this expression, entities of the kind of forces and stresses, which we denote by $f^\mu_M|_{\mathbf{x}} \in (\mathbb{R}^{\mathbf{x}}_{\mathscr{V}_M})'$ and $\Sigma^\mu_M|_{\mathbf{x}} \in (\mathbb{R}^{\mathbf{x}}_{\mathscr{W}_M})'$, in duality, respectively, with the macroscale virtual motion actions $\hat{u}_M|_{\mathbf{x}}$ and $\hat{D}_M|_{\mathbf{x}}$, can be promptly identified, as we show next. Making use of the adjoint operators $(\mathcal{J}_\mu^{\mathcal{W}})' : \mathcal{V}_\mu' \to (\mathbb{R}_{\mathcal{W}_M}^{\mathbf{x}})'$ and $(\mathcal{J}_\mu^{\mathcal{V}})' : \mathcal{V}_\mu' \to (\mathbb{R}_{\mathcal{V}_M}^{\mathbf{x}})'$, from (13.116) we obtain

$$\begin{aligned}\Sigma_M^\mu|_{\mathbf{x}} \bullet \hat{D}_M|_{\mathbf{x}} &= \sum_{k=1}^{m_M} \omega_k\, (\Sigma_M^\mu|_{\mathbf{x}})^k \cdot (\hat{D}_M|_{\mathbf{x}})^k\\ &:= \langle (\mathcal{D}_\mu^* \Sigma_\mu - f_\mu), \mathcal{J}_\mu^{\mathcal{W}} (\hat{D}_M|_{\mathbf{x}})\rangle_{\mathcal{V}_\mu' \times \mathcal{V}_\mu}\\ &= \langle (\mathcal{J}_\mu^{\mathcal{W}})'(\mathcal{D}_\mu^* \Sigma_\mu - f_\mu), \hat{D}_M|_{\mathbf{x}}\rangle_{(\mathbb{R}_{\mathcal{W}_M}^{\mathbf{x}})' \times \mathbb{R}_{\mathcal{W}_M}^{\mathbf{x}}},\end{aligned} \tag{13.117}$$

$$\begin{aligned}f_M^\mu|_{\mathbf{x}} \bullet \hat{u}_M|_{\mathbf{x}} &= \sum_{k=1}^{n_M} \gamma_k\, (f_M^\mu|_{\mathbf{x}})^k \cdot (\hat{u}_M|_{\mathbf{x}})^k\\ &:= \langle f_\mu, \mathcal{J}_\mu^{\mathcal{V}} (\hat{u}_M|_{\mathbf{x}})\rangle_{\mathcal{V}_\mu' \times \mathcal{V}_\mu}\\ &= \langle (\mathcal{J}_\mu^{\mathcal{V}})' f_\mu, \hat{u}_M|_{\mathbf{x}}\rangle_{(\mathbb{R}_{\mathcal{V}_M}^{\mathbf{x}})' \times \mathbb{R}_{\mathcal{V}_M}^{\mathbf{x}}}.\end{aligned} \tag{13.118}$$

Thus, putting the definitions introduced in (13.117) and (13.118) into (13.116) we readily obtain

$$P_\mu^t(\hat{u}_M|_{\mathbf{x}}, \hat{D}_M|_{\mathbf{x}}, 0) = \Sigma_M^\mu|_{\mathbf{x}} \bullet \hat{D}_M|_{\mathbf{x}} - f_M^\mu|_{\mathbf{x}} \bullet \hat{u}_M|_{\mathbf{x}}. \tag{13.119}$$

In short, by using duality concepts we have demonstrated that, as a consequence of the kinematical admissibility (connecting the macroscale and microscale kinematics), the microscale total virtual power has contributions which are evaluated by means of macroscale generalized virtual motion actions. The comparison between (13.119) and (13.107) suggests that an additional connection can be postulated between the macroscale and microscale virtual powers. This is addressed in the next section.

13.6 Principle of Multiscale Virtual Power

The Principle of Multiscale Virtual Power (PMVP) introduced in this section provides the physical connection between the scales by postulating the interplay between the virtual powers defined in each scale (macro and micro) that we have defined in the previous section. Since the physics is encoded in the formulation of the PMVP, we will see that as subproducts of the PMVP we have that the following consequences are intrinsic to the formulation

- Characterization of the n_μ variational equilibrium equations at the microscale.
- Characterization of the m_M homogenization formulae for the macroscale generalized stresses. In other words, the expressions that allow us to calculate the macroscale generalized stresses in terms of the microscale generalized stresses and forces occurring at the microscale are delivered.
- Characterization of the n_M homogenization formulae for the macroscale generalized forces. That is to say, the expressions that allow us to calculate the macroscale generalized forces in terms of the microscale generalized forces are delivered.

This principle can be acknowledged as an extended variational version of the Hill–Mandel Principle of Macrohomogeneity [144, 184]. Next, we formulate the Principle of Multiscale Virtual Power.

The total virtual power at the macroscale point $\mathbf{x}$ must be equal to the total virtual power at the corresponding microscale domain (RVE associated with point $\mathbf{x}$) for all kinematically admissible virtual motion actions from macroscale and microscale that is

$$P^t_{M,\mathbf{x}}(\hat{u}_M|_{\mathbf{x}}, \hat{D}_M|_{\mathbf{x}}) = P^t_\mu(\hat{u}_M|_{\mathbf{x}}, \hat{D}_M|_{\mathbf{x}}, \hat{\tilde{u}}_\mu)$$
$$\forall(\hat{u}_M|_{\mathbf{x}}, \hat{D}_M|_{\mathbf{x}}, \hat{\tilde{u}}_\mu) \text{ kinematically admissible,} \tag{13.120}$$

or in a more explicit form

$$\Sigma_M|_{\mathbf{x}} \bullet \hat{D}_M|_{\mathbf{x}} - f_M|_{\mathbf{x}} \bullet \hat{u}_M|_{\mathbf{x}}$$
$$= \langle \Sigma_\mu, \mathscr{D}_\mu(\mathcal{J}^{\mathscr{W}}_\mu(\hat{D}_M|_{\mathbf{x}}) + \hat{\tilde{u}}_\mu)\rangle_{\mathscr{W}'_\mu \times \mathscr{W}_\mu} - \langle f_\mu, \mathcal{J}^{\mathscr{V}}_\mu(\hat{u}_M|_{\mathbf{x}})$$
$$+ \mathcal{J}^{\mathscr{W}}_\mu(\hat{D}_M|_{\mathbf{x}}) + \hat{\tilde{u}}_\mu\rangle_{\mathscr{V}'_\mu \times \mathscr{V}_\mu} \quad \forall(\hat{u}_M|_{\mathbf{x}}, \hat{D}_M|_{\mathbf{x}}, \hat{\tilde{u}}_\mu) \in \widehat{\mathbb{R}^{\mathbf{x}}_{\mathscr{V}_M}} \times \widehat{\mathbb{R}^{\mathbf{x}}_{\mathscr{W}_M}} \times \mathrm{Var}_{\tilde{u}_\mu}. \tag{13.121}$$

As an extension of the classical Hill–Mandel Principle, in which only the balance of macroscale and microscale internal virtual powers intervene, the PMVP implies the balance between the total virtual power (i.e. internal plus external powers) at the macroscale point $\mathbf{x}$ and the total virtual power in the corresponding microscale domain (RVE).

Classical multiscale solid mechanics. In this example, the macroscale body forces and stresses $(\sigma_M|_{\mathbf{x}}, \mathbf{f}_M|_{\mathbf{x}})$ and the corresponding fields living in the microscale $(\sigma_\mu, \mathbf{f}_\mu)$ satisfy the PMVP if and only if the following variational equation is satisfied

$$\sigma_M|_{\mathbf{x}} \bullet \hat{\varepsilon}_M|_{\mathbf{x}} - \mathbf{f}_M|_{\mathbf{x}} \bullet \hat{\mathbf{u}}_M|_{\mathbf{x}} = \int_{\Omega_\mu} \sigma_\mu \cdot \hat{\varepsilon}_M|_{\mathbf{x}}\, d\Omega_\mu + \int_{\Omega_\mu} \sigma_\mu \cdot (\nabla_{\mathbf{y}} \hat{\tilde{\mathbf{u}}}_\mu)^s\, d\Omega_\mu$$
$$- \int_{\Omega_\mu} \mathbf{f}_\mu \cdot \hat{\mathbf{u}}_M|_{\mathbf{x}}\, d\Omega_\mu - \int_{\Omega_\mu} (\mathbf{f}_\mu \otimes (\mathbf{y} - \mathbf{y}_o))^s \cdot \hat{\varepsilon}_M|_{\mathbf{x}}\, d\Omega_\mu$$
$$- \int_{\Omega_\mu} \mathbf{f}_\mu \cdot \hat{\tilde{\mathbf{u}}}_\mu\, d\Omega_\mu \quad \forall(\hat{\mathbf{u}}_M|_{\mathbf{x}}, \hat{\varepsilon}_M|_{\mathbf{x}}, \hat{\tilde{\mathbf{u}}}_\mu) \in \widehat{\mathbb{R}^{\mathbf{x}}_{\mathscr{V}_M}} \times \widehat{\mathbb{R}^{\mathbf{x}}_{\mathscr{W}_M}} \times \mathrm{Var}_{\tilde{\mathbf{u}}_\mu}. \tag{13.122}$$

■

It is also important to remark that the Principle of Multiscale Virtual Power also provides the definition for the dimensional scalars ω_k, $k = 1, \dots, m_M$ and γ_k, $k = 1, \dots, n_M$, which appear on the left-hand side of (13.121) (following identities (13.99) and (13.104)).

13.7 Dual Operators

The dual homogenization operators for the macroscale generalized stresses and forces as well as the equilibrium equations governing the microscale phenomenology emerge from the PMVP as natural consequences. As it will be seen in what follows, these consequences stand for the Euler–Lagrange equations associated with the variational equation (13.121) corresponding to the PMVP.

13.7.1 Microscale Equilibrium

Taking $\hat{D}_M|_{\mathbf{x}} = 0$ and $\hat{u}_M|_{\mathbf{x}} = 0$ in (13.121), we obtain the variational form of the equilibrium equations at the microscale

$$\langle \Sigma_\mu, \mathscr{D}_\mu(\hat{\tilde{u}}_\mu)\rangle_{\mathscr{W}'_\mu \times \mathscr{W}_\mu} - \langle f_\mu, \hat{\tilde{u}}_\mu\rangle_{\mathscr{V}'_\mu \times \mathscr{V}_\mu} = 0 \qquad \forall \hat{\tilde{u}}_\mu \in \mathrm{Var}_{\tilde{u}_\mu}. \tag{13.123}$$

Evidently, these equations have the same format as the equilibrium equations corresponding to the formulation of the PVP for the physical problem at the microscale, which is described by generalized stresses and forces, and which is essentially affected by the kinematical constraints implicit in the definition of the space $\mathrm{Var}_{\tilde{u}_\mu}$.

Classical multiscale solid mechanics. In this case, when we adopt $\hat{\mathbf{u}}_M|_\mathbf{x} = \mathbf{0}$ and $\hat{\varepsilon}_M|_\mathbf{x} = \mathbf{O}$ in the expression of the PMVP, the equilibrium at the RVE is obtained through the variational equation

$$\int_{\Omega_\mu} \sigma_\mu \cdot (\nabla_\mathbf{y} \hat{\mathbf{u}}_\mu)^s \, d\Omega_\mu - \int_{\Omega_\mu} \mathbf{f}_\mu \cdot \hat{\mathbf{u}}_\mu \, d\Omega_\mu = 0 \qquad \forall \hat{\mathbf{u}}_\mu \in \mathrm{Var}_{\tilde{u}_\mu}. \tag{13.124}$$

■

Utilizing the adjoint operator $\mathscr{D}_\mu^*$ in (13.123), we have the following variational variant

$$\langle \mathscr{D}_\mu^*(\Sigma_\mu) - f_\mu, \hat{u}_\mu \rangle_{\mathscr{V}_\mu' \times \mathscr{V}_\mu} = 0 \qquad \forall \hat{u}_\mu \in \mathrm{Var}_{\tilde{u}_\mu}, \tag{13.125}$$

or equivalently

$$\mathscr{D}_\mu^*(\Sigma_\mu) - f_\mu \in (\mathrm{Var}_{\tilde{u}_\mu})^\perp \subset \mathscr{V}_\mu'. \tag{13.126}$$

Moreover, recalling that f_μ must satisfy (13.29), that is,

$$\langle f_\mu, \hat{u}_\mu \rangle_{\mathscr{V}_\mu' \times \mathscr{V}_\mu} = 0 \qquad \forall \hat{u}_\mu \in \mathrm{Var}_{\tilde{u}_\mu} \cap \mathscr{N}(\mathscr{D}_\mu), \tag{13.127}$$

we arrive at

$$f_\mu \in (\mathrm{Var}_{\tilde{u}_\mu} \cap \mathscr{N}(\mathscr{D}_\mu))^\perp, \tag{13.128}$$

where $(\cdot)^\perp$ represents the orthogonal complement of $(\cdot)$. That is, the system of generalized forces f_μ is orthogonal to the rigid virtual motion action fluctuations which are kinematically admissible at the microscale.

Classical multiscale solid mechanics. By introducing the following decomposition $\mathbf{f}_\mu = \bar{\mathbf{f}}_\mu + \tilde{\mathbf{f}}_\mu$, where $\bar{\mathbf{f}}_\mu = \frac{1}{|\Omega_\mu|}\int_{\Omega_\mu} \mathbf{f}_\mu \, d\Omega_\mu$, and $\tilde{\mathbf{f}}_\mu = \mathbf{f}_\mu - \bar{\mathbf{f}}_\mu$, we obtain that $\int_{\Omega_\mu} \mathbf{f}_\mu \cdot \hat{\mathbf{u}}_\mu \, d\Omega_\mu = \int_{\Omega_\mu} \tilde{\mathbf{f}}_\mu \cdot \hat{\mathbf{u}}_\mu \, d\Omega_\mu$. Therefore, only the fluctuations of the body force, $\tilde{\mathbf{f}}_\mu$, which are not orthogonal to $\mathrm{Var}_{\tilde{\mathbf{u}}_\mu}$, are relevant to the microscale equilibrium problem.

Now, to reach Euler–Lagrange equations associated with the microscale variational equilibrium equation we integrate by parts the term corresponding to the internal virtual power and take into consideration the remark made in the previous paragraph concerning the fluctuations of the microscale force field. Thus, we obtain

$$\int_{\partial\Omega_\mu} \sigma_\mu \mathbf{n}_\mu \cdot \hat{\mathbf{u}}_\mu \, d\partial\Omega_\mu - \int_{\Omega_\mu} \left(\mathrm{div}_\mathbf{y} \sigma_\mu + \tilde{\mathbf{f}}_\mu\right) \cdot \hat{\mathbf{u}}_\mu \, d\Omega_\mu = 0 \quad \forall \hat{\mathbf{u}}_\mu \in \mathrm{Var}_{\tilde{u}_\mu}. \tag{13.129}$$

Hence, by using standard variational arguments and noting the constraint satisfied by $\hat{\mathbf{u}}_\mu$ over $\partial\Omega_\mu$, we conclude that

$$\mathrm{div}_\mathbf{y} \sigma_\mu + \tilde{\mathbf{f}}_\mu = \mathbf{0} \quad \text{in } \Omega_\mu, \tag{13.130}$$

$$\sigma_\mu \mathbf{n}_\mu = \mathbf{T}\mathbf{n}_\mu \quad \text{on } \partial\Omega_\mu. \tag{13.131}$$

We highlight two important issues in the previous equations. First, note that the constant part of the force field plays no role in the equilibrium equation. Second, the stress field is uniform, with value equal to $\mathbf{T}$, over the whole boundary $\partial\Omega_\mu$, as a consequence of the orthogonality condition with respect to the constraint brought by the deformation homogenization when it is expressed in terms of variables defined over the RVE boundary. This is why this minimally constrained model is also known in the literature as the uniform traction model. Clearly, kinematically more constrained models yield stress fields which are not uniform across the boundary. ■

Finally, we also need to note that the microscale equilibrium problem remains fully defined through the variational equation (13.123) when the generalized external force f_μ is furnished (a datum for the problem), and when the constitutive law $\Sigma_\mu = \Sigma_\mu(u_\mu^t)$ is defined as a function of the history of the field u_μ at each point of the RVE. Consequently, the microscale equilibrium problem can be stated as follows.

Given the constitutive equation $\Sigma_\mu = \Sigma_\mu(u_\mu^t)$, given the histories of generalized displacements and deformation actions, $u_M^t|_\mathbf{x}$ and $D_M^t|_\mathbf{x}$, and given the history of the generalized external forces admissible at the microscale, f_μ^t, determine the history $u_\mu^t \in \mathrm{Kin}_{u_\mu}$ of kinematically admissible microscale generalized displacements such that

$$\langle \Sigma_\mu(u_\mu^\tau), \mathscr{D}_\mu(\hat{u}_\mu)\rangle_{\mathscr{W}'\times\mathscr{W}} = \langle f(\tau), \hat{u}_\mu\rangle_{\mathscr{V}'\times\mathscr{V}} \quad \forall \hat{u}_\mu \in \mathrm{Var}_{u_\mu}, \text{ for each } \tau \in [0,t]. \tag{13.132}$$

13.7.2 Homogenization of Generalized Stresses

By taking $\hat{u}_M|_\mathbf{x} = 0$ and $\tilde{\hat{u}}_\mu = 0$ in (13.121) (see also (13.117)) we get

$$\begin{aligned}
\Sigma_M|_\mathbf{x} \bullet \hat{D}_M|_\mathbf{x} &= \langle \Sigma_\mu, \mathscr{D}_\mu(\mathscr{I}_\mu^{\mathscr{W}}(\hat{D}_M|_\mathbf{x}))\rangle_{\mathscr{W}_\mu'\times\mathscr{W}_\mu} - \langle f_\mu, \mathscr{I}_\mu^{\mathscr{W}}(\hat{D}_M|_\mathbf{x})\rangle_{\mathscr{V}_\mu'\times\mathscr{V}_\mu} \\
&= \langle \mathscr{D}_\mu^*\Sigma_\mu - f_\mu, \mathscr{I}_\mu^{\mathscr{W}}(\hat{D}_M|_\mathbf{x})\rangle_{\mathscr{V}_\mu'\times\mathscr{V}_\mu} \\
&= \langle (\mathscr{I}_\mu^{\mathscr{W}})'(\mathscr{D}_\mu^*\Sigma_\mu - f_\mu), \hat{D}_M|_\mathbf{x}\rangle_{(\mathbb{R}^\mathbf{x}_{\mathscr{W}_M})'\times\mathbb{R}^\mathbf{x}_{\mathscr{W}_M}} \\
&= \Sigma_M^\mu|_\mathbf{x} \bullet \hat{D}_M|_\mathbf{x} \qquad \forall \hat{D}_M|_\mathbf{x} \in \widehat{\mathbb{R}^\mathbf{x}_{\mathscr{W}_M}}.
\end{aligned} \tag{13.133}$$

This expression enables us to readily identify the Σ_M-homogenization operator, which is a linear operator defined as

$$\begin{aligned}
&\mathfrak{H}_\Sigma : \mathscr{V}_\mu' \to (\mathbb{R}^\mathbf{x}_{\mathscr{W}_M})', \\
&(\mathscr{D}_\mu^*\Sigma_\mu - f_\mu) \mapsto \Sigma_M^\mu|_\mathbf{x} = \mathfrak{H}_\Sigma(\mathscr{D}_\mu^*\Sigma_\mu - f_\mu),
\end{aligned} \tag{13.134}$$

such that

$$\begin{aligned}
&\langle (\mathscr{I}_\mu^{\mathscr{W}})'(\mathscr{D}_\mu^*\Sigma_\mu - f_\mu), \hat{D}_M|_\mathbf{x}\rangle_{(\mathbb{R}^\mathbf{x}_{\mathscr{W}_M})'\times\mathbb{R}^\mathbf{x}_{\mathscr{W}_M}} \\
&= \mathfrak{H}_\Sigma(\mathscr{D}_\mu^*\Sigma_\mu - f_\mu) \bullet \hat{D}_M|_\mathbf{x} \quad \forall \hat{D}_M|_\mathbf{x} \in \widehat{\mathbb{R}^\mathbf{x}_{\mathscr{W}_M}}.
\end{aligned} \tag{13.135}$$

From (13.133) and from the previous definition we arrive at

$$(\Sigma_M|_\mathbf{x} - \mathfrak{H}_\Sigma(\mathscr{D}_\mu^*\Sigma_\mu - f_\mu)) \bullet \hat{D}_M|_\mathbf{x} = 0 \qquad \forall \hat{D}_M|_\mathbf{x} \in \widehat{\mathbb{R}^\mathbf{x}_{\mathscr{W}_M}}. \tag{13.136}$$

Therefore, we obtain the homogenization formula for the macroscale generalized stresses

$$\Sigma_M|_\mathbf{x} - \mathfrak{H}_\Sigma(\mathscr{D}_\mu^*\Sigma_\mu - f_\mu) \in (\widehat{\mathbb{R}^\mathbf{x}_{\mathscr{W}_M}})^\perp \subseteq (\mathbb{R}^\mathbf{x}_{\mathscr{W}_M})'. \tag{13.137}$$

The Σ_M-homogenization operator, that is, the corresponding homogenization formulae for the stress-like variables, has been here deduced as a natural consequences of the PMVP. This is a meaningful difference between the proposed formulation and what is usually found in the specialized literature, where, typically, stress homogenization formulae are given *a priori*, and without theoretical grounds.

Classical multiscale solid mechanics. In this classical scenario, putting $\hat{\mathbf{u}}_M|_{\mathbf{x}} = \mathbf{0}$ and $\tilde{\hat{\mathbf{u}}}_\mu = \mathbf{0}$ in the PMVP, we obtain

$$\begin{aligned}\sigma_M|_{\mathbf{x}} \bullet \hat{\varepsilon}_M|_{\mathbf{x}} &= \int_{\Omega_\mu} \sigma_\mu \cdot \hat{\varepsilon}_M|_{\mathbf{x}}\, d\Omega_\mu \\ &- \int_{\Omega_\mu} (\mathbf{f}_\mu \otimes (\mathbf{y} - \mathbf{y}_o))^s \cdot \hat{\varepsilon}_M|_{\mathbf{x}}\, d\Omega_\mu \quad \forall \hat{\varepsilon}_M|_{\mathbf{x}} \in \widehat{\mathbb{R}^{\mathbf{x}}_{\mathscr{W}_M}}.\end{aligned} \tag{13.138}$$

The homogenization formula is attained by first noticing that

$$\sigma_M|_{\mathbf{x}} \bullet \hat{\varepsilon}_M|_{\mathbf{x}} = |\Omega_\mu|\, \sigma_M|_{\mathbf{x}} \cdot \hat{\varepsilon}_M|_{\mathbf{x}}, \tag{13.139}$$

which results in

$$\sigma_M|_{\mathbf{x}} = \frac{1}{|\Omega_\mu|} \int_{\Omega_\mu} \left(\sigma_\mu - (\mathbf{f}_\mu \otimes (\mathbf{y} - \mathbf{y}_o))^s\right)\, d\Omega_\mu. \tag{13.140}$$

In this case we have $\omega_1 = |\Omega_\mu|$ and so the formulation naturally renders the formula characterizing the Σ_M-homogenization operator, and that in this case coincides with the physically intuitive stress homogenization formula one can postulate for the present case. This was achieved through a shortcut. Let us see the same result using a more detailed analysis. Consider the operator $\mathscr{D}_\mu(\cdot) = (\nabla_{\mathbf{y}}(\cdot))^s$ explicitly as follows

$$\begin{aligned}\sigma_M|_{\mathbf{x}} \bullet \hat{\varepsilon}_M|_{\mathbf{x}} &= \int_{\Omega_\mu} \sigma_\mu \cdot (\nabla_{\mathbf{y}}(\hat{\varepsilon}_M|_{\mathbf{x}}(\mathbf{y} - \mathbf{y}_o)))^s\, d\Omega_\mu \\ &- \int_{\Omega_\mu} (\mathbf{f}_\mu \otimes (\mathbf{y} - \mathbf{y}_o))^s \cdot \hat{\varepsilon}_M|_{\mathbf{x}}\, d\Omega_\mu \quad \forall \hat{\varepsilon}_M|_{\mathbf{x}} \in \widehat{\mathbb{R}^{\mathbf{x}}_{\mathscr{W}_M}}.\end{aligned} \tag{13.141}$$

Then, integrating by parts the first term on the right-hand side gives

$$\begin{aligned}\sigma_M|_{\mathbf{x}} \bullet \hat{\varepsilon}_M|_{\mathbf{x}} &= \int_{\Omega_\mu} (-\mathrm{div}_{\mathbf{y}} \sigma_\mu \otimes (\mathbf{y} - \mathbf{y}_o))^s \cdot \hat{\varepsilon}_M|_{\mathbf{x}}\, d\Omega_\mu \\ &+ \int_{\partial\Omega_\mu} (\sigma_\mu \mathbf{n}_\mu \otimes (\mathbf{y} - \mathbf{y}_o))^s \cdot \hat{\varepsilon}_M|_{\mathbf{x}}\, d\partial\Omega_\mu \\ &- \int_{\Omega_\mu} (\mathbf{f}_\mu \otimes (\mathbf{y} - \mathbf{y}_o))^s \cdot \hat{\varepsilon}_M|_{\mathbf{x}}\, d\Omega_\mu \quad \forall \hat{\varepsilon}_M|_{\mathbf{x}} \in \widehat{\mathbb{R}^{\mathbf{x}}_{\mathscr{W}_M}}.\end{aligned} \tag{13.142}$$

Now, making use of the strong form of the microscale equilibrium leads to

$$\begin{aligned}&\sigma_M|_{\mathbf{x}} \bullet \hat{\varepsilon}_M|_{\mathbf{x}} \\ &= \left(\int_{\partial\Omega_\mu} (\sigma_\mu \mathbf{n}_\mu \otimes (\mathbf{y} - \mathbf{y}_o))^s\, d\partial\Omega_\mu\right) \cdot \hat{\varepsilon}_M|_{\mathbf{x}} \quad \forall \hat{\varepsilon}_M|_{\mathbf{x}} \in \widehat{\mathbb{R}^{\mathbf{x}}_{\mathscr{W}_M}}.\end{aligned} \tag{13.143}$$

Hence, using similar arguments to those employed previously, we conclude that the formula for the Σ_M-homogenization operator is given by

$$\sigma_M|_{\mathbf{x}} = \frac{1}{|\Omega_\mu|}\int_{\partial\Omega_\mu} (\sigma_\mu \mathbf{n}_\mu \otimes (\mathbf{y} - \mathbf{y}_o))^s \, d\partial\Omega_\mu. \tag{13.144}$$

This expression is completely analogous to that reached in (13.140). The advantage in the last form over (13.140) is that the homogenization solely depends on variables defined over the boundary of the RVE, which is something that had been stated by Hill as a crucial aspect of theoretical and practical relevance in multiscale theories that are built on top of the RVE concept [144]. Notwithstanding this, it is important to note that the first formula requires less regularity than the second one regarding the fields involved in the calculations. ■

According to expression (13.133), coefficients ω_k, $k = 1, \ldots, m_M$, which appear on the left-hand side (see also (13.99)), are recognized through the homogenization procedure given by the operator $\mathfrak{H}_\Sigma$ in order for the homogenization operation to be physically consistent (see the previous example in the domain of classical multiscale modeling in solid mechanics).

13.7.3 Homogenization of Generalized Forces

Consider the particular case of (13.121) when we take $\hat{D}_M|_{\mathbf{x}} = 0$ and $\hat{\tilde{u}}_\mu = 0$ (see also (13.118)). This leads to

$$\begin{aligned} f_M|_{\mathbf{x}} \bullet \hat{u}_M|_{\mathbf{x}} &= \langle f_\mu, \mathcal{I}_\mu^{\mathcal{V}}(\hat{u}_M|_{\mathbf{x}})\rangle_{\mathcal{V}_\mu' \times \mathcal{V}_\mu} \\ &= \langle (\mathcal{I}_\mu^{\mathcal{V}})'(f_\mu), \hat{u}_M|_{\mathbf{x}}\rangle_{(\mathbb{R}^{\mathbf{x}}_{\mathcal{V}_M})' \times \mathbb{R}^{\mathbf{x}}_{\mathcal{V}_M}} \\ &= f_M^\mu|_{\mathbf{x}} \bullet \hat{u}_M|_{\mathbf{x}} \qquad \forall \hat{u}_M|_{\mathbf{x}} \in \widehat{\mathbb{R}^{\mathbf{x}}_{\mathcal{V}_M}}. \end{aligned} \tag{13.145}$$

Similarly to the procedure followed to obtain the characterization of the stress homogenization operator, from the previous variational equation we easily identify the f_M-homogenization operator defined as

$$\begin{aligned} \mathfrak{H}_f &: \mathcal{V}_\mu' \to (\mathbb{R}^{\mathbf{x}}_{\mathcal{V}_M})', \\ & f_\mu \mapsto f_M^\mu|_{\mathbf{x}} = \mathfrak{H}_f(f_\mu), \end{aligned} \tag{13.146}$$

as explicitly given by

$$\langle (\mathcal{I}_\mu^{\mathcal{V}})'(f_\mu), \hat{u}_M|_{\mathbf{x}}\rangle_{(\mathbb{R}^{\mathbf{x}}_{\mathcal{V}_M})' \times \mathbb{R}^{\mathbf{x}}_{\mathcal{V}_M}} = \mathfrak{H}_f(f_\mu) \bullet \hat{u}_M|_{\mathbf{x}} \qquad \forall \hat{u}_M|_{\mathbf{x}} \in \widehat{\mathbb{R}^{\mathbf{x}}_{\mathcal{V}_M}}. \tag{13.147}$$

With the definition of $\mathfrak{H}_f$, (13.145) gives

$$(f_M|_{\mathbf{x}} - \mathfrak{H}_f(f_\mu)) \bullet \hat{u}_M|_{\mathbf{x}} = 0 \qquad \forall \hat{u}_M|_{\mathbf{x}} \in \widehat{\mathbb{R}^{\mathbf{x}}_{\mathcal{V}_M}}. \tag{13.148}$$

Thus, we obtain the homogenization formula that enables us to characterize the macroscale generalized forces in terms of the microscale generalized forces

$$f_M|_{\mathbf{x}} - \mathfrak{H}_f(f_\mu) \in (\widehat{\mathbb{R}^{\mathbf{x}}_{\mathcal{V}_M}})^\perp \subseteq (\mathbb{R}^{\mathbf{x}}_{\mathcal{V}_M})'. \tag{13.149}$$

Classical multiscale solid mechanics. Considering $\hat{\varepsilon}_M|_{\mathbf{x}} = \mathbf{O}$ and $\hat{\mathbf{u}}_\mu = \mathbf{0}$ in the PMVP we obtain

$$\mathbf{f}_M|_{\mathbf{x}} \bullet \hat{\mathbf{u}}_M|_{\mathbf{x}} = \int_{\Omega_\mu} \mathbf{f}_\mu \cdot \hat{\mathbf{u}}_M|_{\mathbf{x}} \, d\Omega_\mu \qquad \forall \hat{\mathbf{u}}_M|_{\mathbf{x}} \in \widehat{\mathbb{R}^{\mathbf{x}}_{\mathscr{V}_M}}. \tag{13.150}$$

Here, we identify

$$\mathbf{f}_M|_{\mathbf{x}} \bullet \hat{\mathbf{u}}_M|_{\mathbf{x}} = |\Omega_\mu| \, \mathbf{f}_M|_{\mathbf{x}} \cdot \hat{\mathbf{u}}_M|_{\mathbf{x}}, \tag{13.151}$$

from where we get

$$\mathbf{f}_M|_{\mathbf{x}} = \frac{1}{|\Omega_\mu|} \int_{\Omega_\mu} \mathbf{f}_\mu \, d\Omega_\mu. \tag{13.152}$$

This defines the f_M-homogenization operator. Observe that γ_1 is here identified as $\gamma_1 = |\Omega_\mu|$, ensuring the physical consistency of the homogenization operation. ■

It is important to appreciate that, according to (13.145), the coefficients γ_k, $k = 1, \dots, n_M$, present on the left-hand side of this expression (see also (13.104)), remain identified through the homogenization process defined by the operator $\mathfrak{H}_f$ in such a way that the operation is physically consistent (see the example in the context of classical multiscale modeling in solid mechanics).

13.8 Final Remarks

To recap, in this chapter we have established a full theory for multiscale models of physical systems based on the RVE concept and rooted in variational grounds by exploiting a minimal set of hypotheses and removing several spots of arbitrariness frequently encountered in multiscale formulations available in the literature. Within the proposed theoretical framework, the RVE-based models are created through a set of systematic and well-defined logical steps, embraced by the structure provided by the Method of Multiscale Virtual Power. Using this method allows us in the first place to adopt proper kinematical descriptions in both scales, as well as a way of connecting these kinematics. In addition, and by exploiting duality arguments, the method provides the identification of the nature of stress- and force-like variables in each of the scales. Finally, the equilibrium equations and the homogenization relations for the dual variables such as stress-like variables and force-like variables are univocally deduced from standard variational arguments as consequences of the Principle of Multiscale Virtual Power.

An interesting point in the multiscale theory developed here is that the concepts of macroscale internal and external powers are not completely independent from each other, as occurs in the conventional variational theory for problems characterized by a single scale. As a matter of fact, and in general, the macroscale generalized internal stress, Σ_M, which, by duality, is characterized by the macroscale internal power, is morphed by a contribution of the microscale generalized stress, Σ_μ (and therefore affected by the microscale internal power), and by a contribution of the microscale generalized force, f_μ (and thus affected by the microscale external power). Furthermore, the effects of Σ_μ and f_μ in the macroscale generalized stress Σ_M are inextricably combined through the microscale equilibrium problem defined by (13.120).

The phenomenological interactions described in the previous paragraph clearly alter the standard notion of constitutiveness that we have when we refer to the material behavior because the microscale external forces (e.g. body forces and inertia forces in solid mechanics) can contribute to the definition of the macroscale stresses. In turn, a situation in which this constitutive concept is still preserved occurs when the transfer between scales only involves the balance between macroscale and microscale internal virtual powers. That is, only when $P^i_{M,\mathbf{x}}$ and P^i_μ are considered in the Principle of Multiscale Virtual Power, or, in other words, when the external virtual powers at both scales are not taken into account in the multiscale model. In such scenario, the structure of a purely constitutive multiscale model is recovered. This is the situation mostly found in the literature, where the models are theoretically supported by the Hill–Mandel Principle of Macrohomogeneity [144, 184]. The interesting aspect in this case is that the macroscale generalized stress, $\Sigma_M|_{\mathbf{x}}$, only depends on the constitutive behavior in the microscale and on the phenomenological interactions among the different structural elements that characterize the RVE domain and the microscale physics, and for which the microscale generalized stress Σ_μ is solely responsible. In fact, even if f_μ can exist, this field is orthogonal to the space of generalized virtual displacements and, therefore, does not exert power. In other words, f_μ is in that case the reactive force associated with the kinematical constraints that are incorporated in the definition of the space $\mathrm{Var}_{\tilde{u}_\mu}$ [278] and, accordingly, f_μ cannot be arbitrarily chosen.

14

Applications of Multiscale Modeling

14.1 Introduction

This chapter aims to present the application of the Method of Multiscale Virtual Power in two problems that appear in solid mechanics. For these two examples we will assume that the bodies are constituted by highly heterogeneous and stable materials at the microscale in the sense that, during the whole loading process, no strain localization, failure or damage phenomena, leading to loss of structural stability, is observed. In turn, we will consider the effects of dynamic forces (inertia) in the microscale in order to study the impact that these forces have in the definition of the macroscale model.

In the first example we consider both macroscale and microscale kinematics to be the same, and that we are within the range of finite deformations. The second example differs in the sense that different types of constitutive models are considered at both scales. In particular, we consider a given constitutive material response at the macroscale, but at the same time the continuum is considered to behave as incompressible. At the microscale, the incompressibility constraint will not be considered. These examples are sufficient to show the versatility of the proposed method when approaching different situations usually encountered in multiscale modeling. In this sense, reading the scientific contributions published by the authors, where we report the application of this technique in diverse problems including, among others, fracture and failure, strain localization, thermomechanics, and fluid mechanics, can be of further help in understanding the usefulness of the multiscale methodology.

14.2 Solid Mechanics with External Forces

In this section we will illustrate the development of a multiscale model in the field of solid mechanics in the finite strain regime, where we will include the effect of external forces of a passive as well as of an active nature (inertia). In order to deal with the dynamics, we assume that both spatial scales share the same phenomenological time scale.[1] As we will

1 The tools presented within the proposed Method of Multiscale Virtual Power can easily be extended in a way that incorporates different time scales. An example of the application of a multiscale problem in space and time can be found in the modeling of biological systems, for example the modeling of the cardiovascular system, where aging phenomena in biological tissues occurs along the human life span (10^9 seconds) while the heart beat scale affecting the development of local stresses driving remodeling and growth phenomena is in the range of 1 second.

Introduction to the Variational Formulation in Mechanics: Fundamentals and Applications, First Edition.
Edgardo O. Taroco, Pablo J. Blanco and Raúl A. Feijóo.

see in this part of the chapter, the incorporation of this effect (dynamic forces) is quite simple and straightforward. In effect, to achieve this goal it is enough to include the virtual power associated with these forces in the formulation of the Principle of Multiscale Virtual Power.

14.2.1 Multiscale Kinematics

Since we will work within the regime of finite deformations at both scales, the model will be developed in the corresponding reference (material) configurations occupied by the solid body (which will be adopted also as the initial configuration), and which will be assumed to be free of loads, initial stresses, and strains. That is, these reference configurations correspond to the virgin state of the material.

With this in mind, and following the method proposed in Chapter 13, we denote by $\Omega_M \subset \mathbb{R}^3$ the macroscale reference configuration, by $\partial\Omega_M$ its boundary, which we consider to be smooth enough, and $\mathbf{n}_M$ is the outward unit normal vector to that boundary. In turn, we denote by $\mathbf{x} \in \Omega_M$ an arbitrary material point in Ω_M.

The kinematics adopted at the macroscale is the classical one corresponding to solid mechanics. Then the kinematics is characterized by the displacement field $\mathbf{u}_M \in \text{Kin}_{u_M}$, where

$$\text{Kin}_{u_M} = \{\mathbf{u} \in \mathbf{H}^1(\Omega_M);\ \mathbf{u}|_{\partial\Omega_{Mu}} = \overline{\mathbf{u}}\}, \tag{14.1}$$

is the linear manifold formed by all kinematically admissible displacements with the essential boundary condition prescribed over $\partial\Omega_{Mu}$. In addition, the deformation action operator is given by the gradient operator in the reference configuration, that is, $\mathcal{D}_M(\cdot) = \nabla_\mathbf{x}(\cdot)$. Therefore, a deformation action is compatible if it is given by the gradient of the displacement field, that is, $D_M = \nabla_\mathbf{x}\mathbf{u}_M = \mathbf{G}_M$. With this, the space of deformation actions is characterized by all second-order tensor fields (not necessarily symmetric) such that are, for example, square integrable in Ω_M, that is, $\mathscr{W}_M = \mathbf{L}^2(\Omega_M)$. From the previous definitions it is easy to see that the value taken by displacements and by deformation actions at each point $\mathbf{x} \in \Omega_M$ are given, respectively, by vectors $\mathbf{w}$ and by second-order tensors $\mathbf{H}$ belonging to the following spaces

$$\mathbb{R}^\mathbf{x}_{\mathscr{V}_M} = \{\mathbf{w} \in \mathbb{R}^3;\ \mathbf{w} = \mathbf{u}|_\mathbf{x}, \mathbf{u} \in \mathscr{V}_M\}, \tag{14.2}$$

$$\mathbb{R}^\mathbf{x}_{\mathscr{W}_M} = \{\mathbf{H} \in \mathbb{R}^{3\times3};\ \mathbf{H} = \mathbf{G}_M|_\mathbf{x}, \mathbf{G}_M \in \mathscr{W}_M\}. \tag{14.3}$$

As established in the multiscale context, let us consider that every macroscale point $\mathbf{x} \in \Omega_M$ is in correspondence with an RVE domain, whose reference configuration is denoted by Ω_μ, and is such that its geometric center is chosen as the origin of the system of coordinates at the microscale, where generic points are denoted by $\mathbf{y}$. Then, it is verified that

$$\int_{\Omega_\mu} \mathbf{y}\, d\Omega_\mu = \mathbf{0}. \tag{14.4}$$

Analogously, at the microscale the boundary Ω_μ is assumed to be smooth, and it is denoted by $\partial\Omega_\mu$ with outward unit normal vector $\mathbf{n}_\mu$.

As mentioned before, the microscale kinematics is of the same kind as the macroscale kinematics. This microscale kinematics is described by displacement fields $\mathbf{u}_\mu \in \mathscr{V}_\mu$, where $\mathscr{V}_\mu = \mathbf{H}^1(\Omega_\mu)$. Moreover, the deformation action operator is the gradient operator $\mathscr{D}_\mu(\cdot) = \nabla_\mathbf{y}(\cdot)$ and compatible deformation actions are given by $D_\mu = \nabla_\mathbf{y}\mathbf{u}_\mu = \mathbf{G}_\mu \in \mathscr{W}_\mu = \mathbf{L}^2(\Omega_\mu)$.

Since at the microscale we are considering materials that are not expected to feature during the whole loading process, any strain localization phenomena, and/or other kinds of failure mechanisms, let us suppose that the insertion of the displacement field is uniform, that is

$$\mathscr{I}_\mu^{\mathscr{V}}(\mathbf{u}_M|_\mathbf{x}) = \mathbf{u}_M|_\mathbf{x} \qquad \forall \mathbf{y} \in \Omega_\mu, \tag{14.5}$$

while the insertion operator for the macroscale deformation actions is defined such that

$$\mathscr{I}_\mu^{\mathscr{W}}(\mathbf{G}_M|_\mathbf{x}) = \mathbf{G}_M|_\mathbf{x}\mathbf{y} \qquad \mathbf{y} \in \Omega_\mu, \tag{14.6}$$

where we are considering that the geometric center in Ω_μ coincides with the origin of the microscale coordinate system.

With the introduction of these elements, the microscale displacement takes the form

$$\mathbf{u}_\mu = \mathbf{u}_M|_\mathbf{x} + \mathbf{G}_M|_\mathbf{x}\mathbf{y} + \tilde{\mathbf{u}}_\mu \qquad \mathbf{y} \in \Omega_\mu. \tag{14.7}$$

Since $\mathscr{D}_\mu(\cdot) = \nabla_\mathbf{y}(\cdot)$, it is directly verified that $\mathscr{D}_\mu(\mathscr{I}_\mu^{\mathscr{V}}(\mathbf{u}_M|_\mathbf{x})) = \mathbf{O}$ and also that $\mathscr{D}_\mu(\mathscr{I}_\mu^{\mathscr{W}}(\mathbf{G}_M|_\mathbf{x})) = \mathbf{G}_M|_\mathbf{x}$. Therefore, we obtain

$$\mathbf{G}_\mu = \mathscr{D}_\mu(\mathbf{u}_\mu) = \mathbf{G}_M|_\mathbf{x} + \nabla_\mathbf{y}\tilde{\mathbf{u}}_\mu. \tag{14.8}$$

Let us now choose the homogenization operators for the kinematical variables. To relate the microscale displacements to the macroscale we select the operation

$$\mathscr{H}_\mu^{\mathscr{V}}(\mathbf{u}_\mu) = \frac{1}{|\Omega_\mu|}\int_{\Omega_\mu} \mathbf{u}_\mu \, d\Omega_\mu, \tag{14.9}$$

while the operator mapping the microscale deformation action to the macroscale is defined as

$$\mathscr{H}_\mu^{\mathscr{W}}(\mathbf{G}_\mu) = \frac{1}{|\Omega_\mu|}\int_{\Omega_\mu} \mathbf{G}_\mu \, d\Omega_\mu. \tag{14.10}$$

It is interesting to note at this point that, by construction, the Principle of Kinematical Conservation, that is preservation of displacement and deformation given, respectively, by (13.70) and by (13.71), is automatically satisfied, that is, the following identities are trivially verified

$$\mathscr{H}_\mu^{\mathscr{V}}(\mathscr{I}_\mu^{\mathscr{V}}(\mathbf{u}_M|_\mathbf{x})) = \frac{1}{|\Omega_\mu|}\int_{\Omega_\mu} \mathscr{I}_\mu^{\mathscr{V}}(\mathbf{u}_M|_\mathbf{x}) \, d\Omega_\mu = \mathbf{u}_M|_\mathbf{x}, \tag{14.11}$$

$$\mathscr{H}_\mu^{\mathscr{W}}(\mathscr{D}_\mu(\mathscr{I}_\mu^{\mathscr{W}}(\mathbf{G}_M|_\mathbf{x}))) = \frac{1}{|\Omega_\mu|}\int_{\Omega_\mu} \mathscr{D}_\mu(\mathscr{I}_\mu^{\mathscr{W}}(\mathbf{G}_M|_\mathbf{x})) \, d\Omega_\mu = \mathbf{G}_M|_\mathbf{x}. \tag{14.12}$$

To satisfy the kinematical admissibility, see expressions (13.74) and (13.75), additional constraints are required to be prescribed to the fluctuation field $\tilde{\mathbf{u}}_\mu$. In effect, for the present application example, the expression (13.74) takes the form

$$\frac{1}{|\Omega_\mu|}\int_{\Omega_\mu} \mathbf{u}_\mu \, d\Omega_\mu = \frac{1}{|\Omega_\mu|}\int_{\Omega_\mu} \mathscr{I}_\mu^{\mathscr{V}}(\mathbf{u}_M|_\mathbf{x}) \, d\Omega_\mu = \mathbf{u}_M|_\mathbf{x}. \tag{14.13}$$

Keeping in mind the decomposition of the displacement field in the microscale given by (14.7), the adopted definition for the insertion operator $\mathcal{J}_\mu^{\mathscr{V}}$, see (14.5), and since

$$\int_{\Omega_\mu} \mathbf{G}_M|_{\mathbf{x}}\mathbf{y}\, d\Omega_\mu = \mathbf{0}, \tag{14.14}$$

we see that the microscale displacements satisfy the kinematical admissibility required by expression (14.13) if and only if the fluctuation field satisfies the constraint

$$\int_{\Omega_\mu} \tilde{\mathbf{u}}_\mu\, d\Omega_\mu = \mathbf{0}. \tag{14.15}$$

In the same way, the kinematical admissibility required by expression (13.75) takes, for the present example, the form

$$\frac{1}{|\Omega_\mu|}\int_{\Omega_\mu} \mathbf{G}_\mu\, d\Omega_\mu = \frac{1}{|\Omega_\mu|}\int_{\Omega_\mu} \mathscr{D}_\mu(\mathcal{J}_\mu^{\mathscr{V}}(\mathbf{u}_M|_{\mathbf{x}}))\, d\Omega_\mu = \mathbf{G}_M|_{\mathbf{x}}. \tag{14.16}$$

Once again, from (14.6) and from (14.8), the kinematical admissibility demanded by (14.16) is satisfied if and only if the microscale gradient of the displacement fluctuation verifies the additional constraint

$$\int_{\Omega_\mu} \nabla_{\mathbf{y}}\tilde{\mathbf{u}}_\mu\, d\Omega_\mu = \mathbf{O}, \tag{14.17}$$

or its equivalent form after using the Green formula

$$\int_{\partial\Omega_\mu} \tilde{\mathbf{u}}_\mu \otimes \mathbf{n}_\mu\, d\partial\Omega_\mu = \mathbf{O}. \tag{14.18}$$

The kinematical admissibility is thus guaranteed, allowing a connection between the macroscale kinematics, characterized at each point $\mathbf{x} \in \Omega_M$ by $\mathbf{u}_M|_{\mathbf{x}}$ and $\mathbf{G}_M|_{\mathbf{x}}$, and the kinematics defined in the associated RVE, given by the expansion (14.7), to be established. This yields corresponding constraints over the fluctuation field given by (14.15) and (14.17), or equivalently (14.18).

As a consequence of this, the kinematically admissible microscale motion actions are completely characterized, as well as the corresponding space of virtual actions

$$\begin{aligned}\mathrm{Kin}_{u_\mu} = \mathrm{Var}_{u_\mu} = \Bigg\{ &\mathbf{u}_\mu \in \mathbf{H}^1(\Omega_\mu);\ \mathbf{u}_\mu = \mathbf{u}_M|_{\mathbf{x}} + \mathbf{G}_M|_{\mathbf{x}}\mathbf{y} + \tilde{\mathbf{u}}_\mu,\\ &\mathbf{u}_M|_{\mathbf{x}} \in \mathbb{R}^{\mathbf{x}}_{\mathscr{V}_M}, \mathbf{G}_M|_{\mathbf{x}} \in \mathbb{R}^{\mathbf{x}}_{\mathscr{W}_M}, \int_{\Omega_\mu} \tilde{\mathbf{u}}_\mu\, d\Omega_\mu = \mathbf{0}, \int_{\Omega_\mu} \nabla_{\mathbf{y}}\tilde{\mathbf{u}}_\mu\, d\Omega_\mu = \mathbf{O}\Bigg\},\end{aligned} \tag{14.19}$$

or, alternatively

$$\begin{aligned}\mathrm{Kin}_{u_\mu} = \mathrm{Var}_{u_\mu} = \Bigg\{ &\mathbf{u}_\mu \in \mathbf{H}^1(\Omega_\mu);\ \mathbf{u}_\mu = \mathbf{u}_M|_{\mathbf{x}} + \mathbf{G}_M|_{\mathbf{x}}\mathbf{y} + \tilde{\mathbf{u}}_\mu,\\ &\mathbf{u}_M|_{\mathbf{x}} \in \mathbb{R}^{\mathbf{x}}_{\mathscr{V}_M}, \mathbf{G}_M|_{\mathbf{x}} \in \mathbb{R}^{\mathbf{x}}_{\mathscr{W}_M}, \int_{\Omega_\mu} \tilde{\mathbf{u}}_\mu\, d\Omega_\mu = \mathbf{0}, \int_{\partial\Omega_\mu} \tilde{\mathbf{u}}_\mu \otimes \mathbf{n}_\mu\, d\partial\Omega_\mu = \mathbf{O}\Bigg\}.\end{aligned} \tag{14.20}$$

We can appreciate that the space Kin_{u_μ} can be rewritten, putting in evidence the space of kinematically admissible microscale fluctuations, denoted by $\text{Kin}_{\tilde{u}_\mu}$ (and therefore the space of virtual fluctuations $\text{Var}_{\tilde{u}_\mu}$), which gives

$$\begin{aligned}\text{Kin}_{\tilde{u}_\mu} &= \text{Var}_{\tilde{u}_\mu}\\ &= \left\{ \tilde{\mathbf{u}}_\mu \in \mathbf{H}^1(\Omega_\mu);\ \int_{\Omega_\mu} \tilde{\mathbf{u}}_\mu \, d\Omega_\mu = \mathbf{0},\ \int_{\Omega_\mu} \nabla_{\mathbf{y}} \tilde{\mathbf{u}}_\mu \, d\Omega_\mu = \mathbf{O} \right\},\end{aligned} \tag{14.21}$$

or equivalently

$$\begin{aligned}\text{Kin}_{\tilde{u}_\mu} &= \text{Var}_{\tilde{u}_\mu}\\ &= \left\{ \tilde{\mathbf{u}}_\mu \in \mathbf{H}^1(\Omega_\mu);\ \int_{\Omega_\mu} \tilde{\mathbf{u}}_\mu \, d\Omega_\mu = \mathbf{0},\ \int_{\partial\Omega_\mu} \tilde{\mathbf{u}}_\mu \otimes \mathbf{n}_\mu \, d\partial\Omega_\mu = \mathbf{O} \right\}.\end{aligned} \tag{14.22}$$

With this definition, we now have

$$\begin{aligned}\text{Kin}_{u_\mu} = \Big\{ & \mathbf{u}_\mu = \mathbf{u}_M|_{\mathbf{x}} + \mathbf{G}_M|_{\mathbf{x}}\mathbf{y} + \tilde{\mathbf{u}}_\mu;\\ & \mathbf{u}_M|_{\mathbf{x}} \in \mathbb{R}^{\mathbf{x}}_{\mathscr{V}_M}, \mathbf{G}_M|_{\mathbf{x}} \in \mathbb{R}^{\mathbf{x}}_{\mathscr{W}_M}, \tilde{\mathbf{u}}_\mu \in \text{Kin}_{\tilde{u}_\mu} \Big\}.\end{aligned} \tag{14.23}$$

An equivalent alternative to characterize the microscale displacement, $\mathbf{u}_\mu$, that will ease the derivation of Euler–Lagrange equations associated with the Principle of Multiscale Virtual Power, consists of explicitly introducing the homogenization operators for displacements and deformations implied by the kinematical admissibility (see (14.13) and (14.16)), instead of fluctuations. In other words, this alternative seeks to not explicitly write the expansion (14.7), which is now an implicit consequence of this new characterization. In fact, we have

$$\begin{aligned}\text{Kin}^*_{u_\mu} = \Big\{ & \mathbf{u}_\mu \in \mathbf{H}^1(\Omega_\mu);\ \int_{\Omega_\mu} \mathbf{u}_\mu \, d\Omega_\mu = \mathbf{u}_M|_{\mathbf{x}},\ \int_{\Omega_\mu} \nabla_{\mathbf{y}} \mathbf{u}_\mu \, d\Omega_\mu = \mathbf{G}_M|_{\mathbf{x}},\\ & \mathbf{u}_M|_{\mathbf{x}} \in \mathbb{R}^{\mathbf{x}}_{\mathscr{V}_M}, \mathbf{G}_M|_{\mathbf{x}} \in \mathbb{R}^{\mathbf{x}}_{\mathscr{W}_M} \Big\},\end{aligned} \tag{14.24}$$

or equivalently

$$\begin{aligned}\text{Kin}^*_{u_\mu} = \Big\{ & \mathbf{u}_\mu \in \mathbf{H}^1(\Omega_\mu);\ \int_{\Omega_\mu} \mathbf{u}_\mu \, d\Omega_\mu = \mathbf{u}_M|_{\mathbf{x}},\ \int_{\partial\Omega_\mu} \mathbf{u}_\mu \otimes \mathbf{n}_\mu \, d\partial\Omega_\mu = \mathbf{G}_M|_{\mathbf{x}},\\ & \mathbf{u}_M|_{\mathbf{x}} \in \mathbb{R}^{\mathbf{x}}_{\mathscr{V}_M}, \mathbf{G}_M|_{\mathbf{x}} \in \mathbb{R}^{\mathbf{x}}_{\mathscr{W}_M} \Big\}.\end{aligned} \tag{14.25}$$

Note that it is easy to verify that all elements such that $\mathbf{u}_\mu \in \text{Kin}_{u_\mu}$ also belong to $\text{Kin}^*_{u_\mu}$ and, in turn, all elements satisfying the constraints in $\text{Kin}^*_{u_\mu}$ are also elements of Kin_{u_μ}. That is, both definitions provided above are equivalent, so

$$\text{Kin}^*_{u_\mu} = \text{Kin}_{u_\mu}. \tag{14.26}$$

14.2.2 Characterization of Virtual Power

The next step in the construction of the multiscale model consists of the characterization of the internal and external powers at both scales. In this sense, and since we are working in the context of finite deformations utilizing a classical kinematics, the macroscale total power is given by

$$P^t_M(\hat{\mathbf{u}}_M, \nabla_{\mathbf{x}}\hat{\mathbf{u}}_M) = \underbrace{\int_{\Omega_M} \mathbf{P}_M \cdot \nabla_{\mathbf{x}}\hat{\mathbf{u}}_M \, d\Omega_M}_{=P^i_M} - \underbrace{\int_{\Omega_M} \mathbf{f}_M \cdot \hat{\mathbf{u}}_M \, d\Omega_M}_{=P^e_M}, \tag{14.27}$$

where $\mathbf{P}_M$ is the Piola–Kirchhoff stress tensor of the first kind. Here we should point out that $\mathbf{f}_M$ can be integrated by a passive force, say $\mathbf{f}^p_M$ (e.g. generated by the gravity), and/or by an active, or dynamic, force, represented by $\mathbf{f}^a_M$, that is, $\mathbf{f}_M = \mathbf{f}^p_M - \mathbf{f}^a_M$.

From this, we see that for each macroscale point $\mathbf{x} \in \Omega_M$ the density of internal and external powers (that are to be connected to the power developed at the RVE through the PMVP) are, respectively, given by

$$P^i_{M,\mathbf{x}} = \mathbf{P}_M|_{\mathbf{x}} \bullet \nabla_{\mathbf{x}}\hat{\mathbf{u}}_M|_{\mathbf{x}} = \mathbf{P}_M|_{\mathbf{x}} \bullet \hat{\mathbf{G}}_M|_{\mathbf{x}} \quad \hat{\mathbf{G}}_M|_{\mathbf{x}} \in \mathbb{R}^{\mathbf{x}}_{\mathscr{W}_M}, \tag{14.28}$$

$$P^e_{M,\mathbf{x}} = \mathbf{f}_M|_{\mathbf{x}} \bullet \hat{\mathbf{u}}_M|_{\mathbf{x}} \quad \hat{\mathbf{u}}_M|_{\mathbf{x}} \in \mathbb{R}^{\mathbf{x}}_{\mathscr{V}_M}, \tag{14.29}$$

where the operations $(\cdot) \bullet (\cdot)$ will posteriorly be characterized via the PMVP.

Concerning the microscale, the internal power is given by

$$P^i_\mu(\nabla_{\mathbf{y}}\hat{\mathbf{u}}_\mu) = \int_{\Omega_\mu} \mathbf{P}_\mu \cdot \nabla_{\mathbf{y}}\hat{\mathbf{u}}_\mu \, d\Omega_\mu \qquad \hat{\mathbf{u}}_\mu \in \mathrm{Var}_{u_\mu}, \tag{14.30}$$

which in extended form reads

$$\begin{aligned} P^i_\mu &= \int_{\Omega_\mu} \mathbf{P}_\mu \cdot (\hat{\mathbf{G}}_M|_{\mathbf{x}} + \nabla_{\mathbf{y}}\hat{\tilde{\mathbf{u}}}_\mu) \, d\Omega_\mu \\ &= \int_{\Omega_\mu} \mathbf{P}_\mu \cdot \hat{\mathbf{G}}_M|_{\mathbf{x}} \, d\Omega_\mu + \int_{\Omega_\mu} \mathbf{P}_\mu \cdot \nabla_{\mathbf{y}}\hat{\tilde{\mathbf{u}}}_\mu \, d\Omega_\mu \\ &\hat{\mathbf{G}}_M|_{\mathbf{x}} \in \mathbb{R}^{\mathbf{x}}_{\mathscr{W}_M}, \ \hat{\tilde{\mathbf{u}}}_\mu \in \mathrm{Var}_{\tilde{u}_\mu}. \end{aligned} \tag{14.31}$$

The incorporation of the effects carried by the presence of body forces (either passive or dynamic) in the microscale is carried out in a simple manner, just by admitting their existence in the definition of the external virtual power at that scale. As a matter of fact, as occurred at the macroscale, the force field in the microscale is designated by $\mathbf{f}_\mu$, and will be formed by a passive component (e.g. gravity), denoted by $\mathbf{f}^p_\mu$, and by an active component associated with the acceleration of microscale particles, represented by $\mathbf{f}^a_\mu$. Hence, we write

$$\mathbf{f}_\mu = \mathbf{f}^p_\mu - \mathbf{f}^a_\mu, \tag{14.32}$$

$$\mathbf{f}^a_\mu = \rho_\mu \ddot{\mathbf{u}}_\mu = \rho_\mu(\ddot{\mathbf{u}}_M|_{\mathbf{x}} + \ddot{\mathbf{G}}_M|_{\mathbf{x}}\mathbf{y} + \ddot{\tilde{\mathbf{u}}}_\mu). \tag{14.33}$$

Consequently, and taking into account that $\hat{\mathbf{u}}_\mu = \hat{\mathbf{u}}_M|_\mathbf{x} + \hat{\mathbf{G}}_M|_\mathbf{x}\mathbf{y} + \hat{\tilde{\mathbf{u}}}_\mu \in \mathrm{Var}_{u_\mu}$, the microscale external power is given by

$$
\begin{aligned}
P^e_\mu(\hat{\mathbf{u}}_\mu) &= \int_{\Omega_\mu} \mathbf{f}^p_\mu \cdot \hat{\mathbf{u}}_\mu \, d\Omega_\mu - \int_{\Omega_\mu} \mathbf{f}^a_\mu \cdot \hat{\mathbf{u}}_\mu \, d\Omega_\mu \\
&= \int_{\Omega_\mu} \mathbf{f}^p_\mu \cdot \hat{\mathbf{u}}_M|_\mathbf{x} \, d\Omega_\mu + \int_{\Omega_\mu} (\mathbf{f}^p_\mu \otimes \mathbf{y}) \cdot \hat{\mathbf{G}}_M|_\mathbf{x} \, d\Omega_\mu \\
&\quad + \int_{\Omega_\mu} \mathbf{f}^p_\mu \cdot \hat{\tilde{\mathbf{u}}}_\mu \, d\Omega_\mu - \int_{\Omega_\mu} \rho_\mu \ddot{\mathbf{u}}_\mu \cdot \hat{\mathbf{u}}_M|_\mathbf{x} \, d\Omega_\mu \\
&\quad - \int_{\Omega_\mu} \rho_\mu (\ddot{\mathbf{u}}_\mu \otimes \mathbf{y}) \cdot \hat{\mathbf{G}}_M|_\mathbf{x} \, d\Omega_\mu - \int_{\Omega_\mu} \rho_\mu \ddot{\mathbf{u}}_\mu \cdot \hat{\tilde{\mathbf{u}}}_\mu \, d\Omega_\mu \\
&\hat{\mathbf{u}}_M|_\mathbf{x} \in \mathbb{R}^\mathbf{x}_{\mathscr{V}_M}, \quad \hat{\mathbf{G}}_M|_\mathbf{x} \in \mathbb{R}^\mathbf{x}_{\mathscr{W}_M}, \quad \hat{\tilde{\mathbf{u}}}_\mu \in \mathrm{Var}_{\tilde{u}_\mu}.
\end{aligned}
\tag{14.34}
$$

14.2.3 Principle of Multiscale Virtual Power

The groundwork is now ready to formulate the Principle of Multiscale Virtual Power which, as we saw in Chapter 13, will enable the final set up of the multiscale model that is being developed. In fact, by using standard concepts from variational calculus, from the PMVP we are granted: (i) the homogenization formulae (and corresponding operators) that enable us to describe the macroscale stresses in terms of the microscale stresses and (passive and active) forces, (ii) the homogenization formulae (and corresponding operators) that allow the characterization of macroscale forces as a function of the microscale counterparts, (iii) the variational equations describing the mechanical equilibrium that microscale stresses and forces must satisfy, and, finally, (iv) given the kinematical constraints considered in the space $\mathrm{Var}_{\tilde{u}_\mu}$ as a result of the kinematical admissibility, we will be able to obtain the reactive forces associated, through duality, with these restrictions. Once again, we highlight the fact that these expressions are consequences of the kinematical hypotheses introduced at both scales, of the kinematical admissibility connecting the scales, and of the duality, materialized through the PMVP. In other words, these consequences are naturally extracted from the PMVP, neither being arbitrarily established nor *ad hoc* considered, as happens in many RVE-based multiscale theories available in the literature of the field.

Thus, we say that $(\mathbf{P}_M|_\mathbf{x}, \mathbf{f}_M|_\mathbf{x}) \in (\mathbb{R}^\mathbf{x}_{\mathscr{W}_M})' \times (\mathbb{R}^\mathbf{x}_{\mathscr{V}_M})'$ and $(\mathbf{P}_\mu, \mathbf{f}_\mu) \in \mathscr{W}'_\mu \times \mathscr{V}'_\mu$ satisfy the PMVP if and only if the following variational equation is satisfied

$$
\begin{aligned}
\mathbf{P}_M|_\mathbf{x} \bullet \hat{\mathbf{G}}_M|_\mathbf{x} - \mathbf{f}_M|_\mathbf{x} \bullet \hat{\mathbf{u}}_M|_\mathbf{x} &= \int_{\Omega_\mu} \mathbf{P}_\mu \cdot \hat{\mathbf{G}}_M|_\mathbf{x} \, d\Omega_\mu + \int_{\Omega_\mu} \mathbf{P}_\mu \cdot \nabla_\mathbf{y} \hat{\tilde{\mathbf{u}}}_\mu \, d\Omega_\mu \\
&- \int_{\Omega_\mu} \mathbf{f}^p_\mu \cdot \hat{\mathbf{u}}_M|_\mathbf{x} \, d\Omega_\mu - \int_{\Omega_\mu} (\mathbf{f}^p_\mu \otimes \mathbf{y}) \cdot \hat{\mathbf{G}}_M|_\mathbf{x} \, d\Omega_\mu \\
&- \int_{\Omega_\mu} \mathbf{f}^p_\mu \cdot \hat{\tilde{\mathbf{u}}}_\mu \, d\Omega_\mu + \int_{\Omega_\mu} \rho_\mu \ddot{\mathbf{u}}_\mu \cdot \hat{\mathbf{u}}_M|_\mathbf{x} \, d\Omega_\mu \\
&+ \int_{\Omega_\mu} \rho_\mu (\ddot{\mathbf{u}}_\mu \otimes \mathbf{y}) \cdot \hat{\mathbf{G}}_M|_\mathbf{x} \, d\Omega_\mu + \int_{\Omega_\mu} \rho_\mu \ddot{\mathbf{u}}_\mu \cdot \hat{\tilde{\mathbf{u}}}_\mu \, d\Omega_\mu \\
&\forall (\hat{\mathbf{u}}_M|_\mathbf{x}, \hat{\mathbf{G}}_M|_\mathbf{x}, \hat{\tilde{\mathbf{u}}}_\mu) \in \mathbb{R}^\mathbf{x}_{\mathscr{V}_M} \times \mathbb{R}^\mathbf{x}_{\mathscr{W}_M} \times \mathrm{Var}_{\tilde{u}_\mu}.
\end{aligned}
\tag{14.35}
$$

In order to study all the consequences that emerge from the PMVP we will rewrite it but now relaxing the constraints of kinematical admissibility existing in the definition of the space $\mathrm{Var}_{\tilde{u}_\mu}$, which are given by $\int_{\Omega_\mu} \tilde{\mathbf{u}}_\mu \, d\Omega_\mu = \mathbf{0}$ and $\int_{\partial\Omega_\mu} \tilde{\mathbf{u}}_\mu \otimes \mathbf{n}_\mu \, d\partial\Omega_\mu = \mathbf{O}$. In this way, we are characterizing the reactive forces associated with the kinematical constraints. To accomplish this task we resort to the theory of Lagrange multipliers, which are given, respectively, by $\lambda \in (\mathbb{R}^{\mathbf{x}}_{\mathscr{V}_M})'$ and $\Lambda \in (\mathbb{R}^{\mathbf{x}}_{\mathscr{W}_M})'$. Hence, we obtain the following variational equation equivalent to the PMVP formulated in (14.35)

$$
\begin{aligned}
&\mathbf{P}_M \bullet \hat{\mathbf{G}}_M|_{\mathbf{x}} - \mathbf{f}_M \bullet \hat{\mathbf{u}}_M|_{\mathbf{x}} \\
&= \int_{\Omega_\mu} \mathbf{P}_\mu \cdot \hat{\mathbf{G}}_M|_{\mathbf{x}} \, d\Omega_\mu + \int_{\Omega_\mu} \mathbf{P}_\mu \cdot \nabla_{\mathbf{y}} \hat{\tilde{\mathbf{u}}}_\mu \, d\Omega_\mu - \int_{\Omega_\mu} \mathbf{f}^p_\mu \cdot \hat{\mathbf{u}}_M|_{\mathbf{x}} \, d\Omega_\mu \\
&\quad - \int_{\Omega_\mu} (\mathbf{f}^p_\mu \otimes \mathbf{y}) \cdot \hat{\mathbf{G}}_M|_{\mathbf{x}} \, d\Omega_\mu - \int_{\Omega_\mu} \mathbf{f}^p_\mu \cdot \hat{\tilde{\mathbf{u}}}_\mu \, d\Omega_\mu + \int_{\Omega_\mu} \rho_\mu \ddot{\mathbf{u}}_\mu \cdot \hat{\mathbf{u}}_M|_{\mathbf{x}} \, d\Omega_\mu \\
&\quad + \int_{\Omega_\mu} \rho_\mu (\ddot{\mathbf{u}}_\mu \otimes \mathbf{y}) \cdot \hat{\mathbf{G}}_M|_{\mathbf{x}} \, d\Omega_\mu + \int_{\Omega_\mu} \rho_\mu \ddot{\mathbf{u}}_\mu \cdot \hat{\tilde{\mathbf{u}}}_\mu \, d\Omega_\mu \\
&\quad + \lambda \cdot \int_{\Omega_\mu} \hat{\tilde{\mathbf{u}}}_\mu \, d\Omega_\mu + \hat{\lambda} \cdot \int_{\Omega_\mu} \tilde{\mathbf{u}}_\mu \, d\Omega_\mu \\
&\quad - \Lambda \cdot \int_{\partial\Omega_\mu} \hat{\tilde{\mathbf{u}}}_\mu \otimes \mathbf{n}_\mu \, d\partial\Omega_\mu - \hat{\Lambda} \cdot \int_{\partial\Omega_\mu} \tilde{\mathbf{u}}_\mu \otimes \mathbf{n}_\mu \, d\partial\Omega_\mu \\
&\forall (\hat{\mathbf{u}}_M|_{\mathbf{x}}, \hat{\mathbf{G}}_M|_{\mathbf{x}}, \hat{\tilde{\mathbf{u}}}_\mu, \hat{\lambda}, \hat{\Lambda}) \in \mathbb{R}^{\mathbf{x}}_{\mathscr{V}_M} \times \mathbb{R}^{\mathbf{x}}_{\mathscr{W}_M} \times \mathbf{H}^1(\Omega_\mu) \times (\mathbb{R}^{\mathbf{x}}_{\mathscr{V}_M})' \times (\mathbb{R}^{\mathbf{x}}_{\mathscr{W}_M})'.
\end{aligned}
\tag{14.36}
$$

Clearly, there exists complete equivalence between variational expressions (14.35) and (14.36). As will be seen, it is more practical and conceptually simpler to retrieve all the Euler–Lagrange equations encoded in the variational equation of the PMVP from (14.36) because the space of microscale fluctuations is free of constraints.

14.2.4 Equilibrium Problem and Homogenization

Let us now proceed to obtain the Euler–Lagrange equations associated with the PMVP given by the variational equation (14.36), where the kinematical constraints have been relaxed through proper Lagrange multipliers, which, ultimately, represent the reactive forces of the system to the enforcement of such prescriptions.

Hence, from the variational equation (14.36), and using standard concepts from variational calculus, we find that the following identities are satisfied.

- **Kinematical admissibility**. Nullifying all the variations except $\hat{\lambda} \in (\mathbb{R}^{\mathbf{x}}_{\mathscr{V}_M})'$ and $\hat{\Lambda} \in (\mathbb{R}^{\mathbf{x}}_{\mathscr{W}_M})'$, respectively, yields the two constraints that establish the kinematical admissibility of the fluctuation fields, that is

$$\int_{\Omega_\mu} \tilde{\mathbf{u}}_\mu \, d\Omega_\mu = \mathbf{0}, \tag{14.37}$$

$$\int_{\partial\Omega_\mu} \tilde{\mathbf{u}}_\mu \otimes \mathbf{n}_\mu \, d\partial\Omega_\mu = \mathbf{O}. \tag{14.38}$$

- **Macroscale stress homogenization.** Nullifying all the variations except $\hat{\mathbf{G}}_M|_{\mathbf{x}} \in \mathbb{R}^{\mathbf{x}}_{\mathscr{W}_M}$ leads to

$$\mathbf{P}_M|_{\mathbf{x}} \bullet \hat{\mathbf{G}}_M|_{\mathbf{x}} = \int_{\Omega_\mu} \mathbf{P}_\mu \, d\Omega_\mu \cdot \hat{\mathbf{G}}_M|_{\mathbf{x}} - \int_{\Omega_\mu} (\mathbf{f}^p_\mu \otimes \mathbf{y}) \, d\Omega_\mu \cdot \hat{\mathbf{G}}_M|_{\mathbf{x}}$$
$$+ \int_{\Omega_\mu} \rho_\mu (\ddot{\mathbf{u}}_\mu \otimes \mathbf{y}) \, d\Omega_\mu \cdot \hat{\mathbf{G}}_M|_{\mathbf{x}} \qquad \forall \hat{\mathbf{G}}_M|_{\mathbf{x}} \in \mathbb{R}^{\mathbf{x}}_{\mathscr{W}_M}. \tag{14.39}$$

From this expression we conclude that the operation $\mathbf{P}_M|_{\mathbf{x}} \bullet \hat{\mathbf{G}}_M|_{\mathbf{x}}$ is identified as $\mathbf{P}_M|_{\mathbf{x}} \bullet \hat{\mathbf{G}}_M|_{\mathbf{x}} = |\Omega_\mu| \mathbf{P}_M \cdot \hat{\mathbf{G}}_M|_{\mathbf{x}}$, and so we obtain the homogenization formula for the macroscale stress in terms of the microscale stresses and (passive and active) forces

$$\mathbf{P}_M|_{\mathbf{x}} = \frac{1}{|\Omega_\mu|} \int_{\Omega_\mu} (\mathbf{P}_\mu - \mathbf{f}^p_\mu \otimes \mathbf{y} + \rho_\mu (\ddot{\mathbf{u}}_\mu \otimes \mathbf{y})) \, d\Omega_\mu$$
$$= \frac{1}{|\Omega_\mu|} \int_{\Omega_\mu} (\mathbf{P}_\mu - \mathbf{f}_\mu \otimes \mathbf{y}) \, d\Omega_\mu. \tag{14.40}$$

As can be appreciated, from the previous expression the macroscale stress explicitly and implicitly depends on $\ddot{\mathbf{u}}_\mu$. As a matter of fact, this expression can be recast in the following way

$$\mathbf{P}_M|_{\mathbf{x}} = \mathbf{A}_M + \mathbf{S}_M$$
$$= \frac{1}{|\Omega_\mu|} \int_{\Omega_\mu} \rho_\mu (\ddot{\mathbf{u}}_\mu \otimes \mathbf{y}) \, d\Omega_\mu$$
$$+ \frac{1}{|\Omega_\mu|} \int_{\Omega_\mu} (\mathbf{P}_\mu - \mathbf{f}^p_\mu \otimes \mathbf{y}) \, d\Omega_\mu, \tag{14.41}$$

where $\mathbf{A}_M$ takes into account the explicit dependence, while $\mathbf{S}_M$ carries the implicit dependence through the microscale stress state, $\mathbf{P}_\mu$, which we will see depends on $\ddot{\mathbf{u}}_\mu$ by means of the microscale equilibrium problem.
- **Macroscale force homogenization.** In a similar way to before, considering all null variations except $\hat{\mathbf{u}}_M|_{\mathbf{x}} \in \mathbb{R}^{\mathbf{x}}_{\mathscr{V}_M}$ we get

$$\mathbf{f}_M|_{\mathbf{x}} \bullet \hat{\mathbf{u}}_M|_{\mathbf{x}} = \left(\int_{\Omega_\mu} \mathbf{f}^p_\mu \, d\Omega_\mu - \int_{\Omega_\mu} \rho_\mu \ddot{\mathbf{u}}_\mu \, d\Omega_\mu \right) \cdot \hat{\mathbf{u}}_M|_{\mathbf{x}} \quad \forall \hat{\mathbf{u}}_M|_{\mathbf{x}} \in \mathbb{R}^{\mathbf{x}}_{\mathscr{V}_M}. \tag{14.42}$$

This expression enables us to identify $\mathbf{f}_M|_{\mathbf{x}} \bullet \hat{\mathbf{u}}_M|_{\mathbf{x}}$ as $\mathbf{f}_M|_{\mathbf{x}} \bullet \hat{\mathbf{u}}_M|_{\mathbf{x}} = |\Omega_\mu| \mathbf{f}_M|_{\mathbf{x}} \cdot \hat{\mathbf{u}}_M|_{\mathbf{x}}$, and therefore it provides the homogenization formula for the macroscale force in terms of the microscale forces

$$\mathbf{f}_M|_{\mathbf{x}} = \frac{1}{|\Omega_\mu|} \int_{\Omega_\mu} (\mathbf{f}^p_\mu - \rho_\mu \ddot{\mathbf{u}}_\mu) \, d\Omega_\mu. \tag{14.43}$$

Before going on, let us see some consequences of this homogenization formula. Consider first the case in which we neglect the active forces in the problem, and suppose that the microscale passive forces are caused by the action of gravity. Then $\mathbf{f}_\mu = \rho_\mu \mathbf{g}|_{\mathbf{x}}$ because the microscale gravity field can reasonably be admitted to be uniform, where we have denoted by $\mathbf{g}|_{\mathbf{x}}$ the value of the acceleration induced by the force of gravitation at point $\mathbf{x}$. With this hypothesis, from (14.43) we obtain

$$\mathbf{f}_M^p|_{\mathbf{x}} = \frac{1}{|\Omega_\mu|}\int_{\Omega_\mu} \rho_\mu \mathbf{g}|_{\mathbf{x}}\, d\Omega_\mu = \left(\frac{1}{|\Omega_\mu|}\int_{\Omega_\mu} \rho_\mu\, d\Omega_\mu\right)\mathbf{g}|_{\mathbf{x}}, \tag{14.44}$$

identifying in this manner the density at the macroscale given by

$$\rho_M|_{\mathbf{x}} = \frac{1}{|\Omega_\mu|}\int_{\Omega_\mu} \rho_\mu\, d\Omega_\mu. \tag{14.45}$$

Consider now the presence of active (inertia) forces. Then, from the previous development we write

$$\mathbf{f}_M^a|_{\mathbf{x}} = \frac{1}{|\Omega_\mu|}\int_{\Omega_\mu} \rho_\mu \ddot{\mathbf{u}}_\mu\, d\Omega_\mu = \frac{1}{|\Omega_\mu|}\int_{\Omega_\mu} \mathbf{f}_\mu^a\, d\Omega_\mu. \tag{14.46}$$

Recalling (14.33), this expression takes the form

$$\begin{aligned}
\mathbf{f}_M^a|_{\mathbf{x}} &= \frac{1}{|\Omega_\mu|}\int_{\Omega_\mu} \rho_\mu(\ddot{\mathbf{u}}_M|_{\mathbf{x}} + \ddot{\mathbf{G}}_M|_{\mathbf{x}}\mathbf{y} + \ddot{\tilde{\mathbf{u}}}_\mu)\, d\Omega_\mu \\
&= \rho_M|_{\mathbf{x}}\ddot{\mathbf{u}}_M|_{\mathbf{x}} + \ddot{\mathbf{G}}_M|_{\mathbf{x}}\frac{1}{|\Omega_\mu|}\int_{\Omega_\mu} \rho_\mu \mathbf{y}\, d\Omega_\mu \\
&\quad + \frac{1}{|\Omega_\mu|}\int_{\Omega_\mu} \rho_\mu \ddot{\tilde{\mathbf{u}}}_\mu\, d\Omega_\mu.
\end{aligned} \tag{14.47}$$

That is to say, according to the proposed multiscale theory, the active force at the macroscale point $\mathbf{x}$ is explicitly given by the contribution of the classical term arising in mechanics, $\rho_M|_{\mathbf{x}}\ddot{\mathbf{u}}_M|_{\mathbf{x}}$, jointly with two other nonstandard terms associated with accelerations $\ddot{\mathbf{G}}_M|_{\mathbf{x}}$ and $\ddot{\tilde{\mathbf{u}}}_\mu$. In addition, note that an implicit dependence is hidden in the formulation through the equilibrium problem.

- **Microscale mechanical equilibrium and Lagrange multipliers.** In order to deduce the equations that govern the microscale equilibrium we have to nullify all the variations except $\hat{\tilde{\mathbf{u}}}_\mu \in \mathbf{H}^1(\Omega_\mu)$ in the variational equation (14.36) of the PMVP. Therefore, we are led to

$$\begin{aligned}
&\int_{\Omega_\mu} \mathbf{P}_\mu \cdot \nabla_{\mathbf{y}}\hat{\tilde{\mathbf{u}}}_\mu\, d\Omega_\mu - \int_{\Omega_\mu} \mathbf{f}_\mu^p \cdot \hat{\tilde{\mathbf{u}}}_\mu\, d\Omega_\mu + \int_{\Omega_\mu} \rho_\mu \ddot{\mathbf{u}}_\mu \cdot \hat{\tilde{\mathbf{u}}}_\mu\, d\Omega_\mu \\
&+ \lambda \cdot \int_{\Omega_\mu} \hat{\tilde{\mathbf{u}}}_\mu\, d\Omega_\mu - \Lambda \cdot \int_{\partial\Omega_\mu} \hat{\tilde{\mathbf{u}}}_\mu \otimes \mathbf{n}_\mu\, d\partial\Omega_\mu = 0 \quad \forall \hat{\tilde{\mathbf{u}}}_\mu \in \mathbf{H}^1(\Omega_\mu),
\end{aligned} \tag{14.48}$$

and making use of the property $\int_{\Omega_\mu} \nabla_{\mathbf{y}}\hat{\tilde{\mathbf{u}}}_\mu\, d\Omega_\mu = \int_{\partial\Omega_\mu} \hat{\tilde{\mathbf{u}}}_\mu \otimes \mathbf{n}_\mu\, d\partial\Omega_\mu$ (see (14.17) and (14.18)), we obtain

$$\begin{aligned}
&\int_{\Omega_\mu} (\mathbf{P}_\mu - \Lambda) \cdot \nabla_{\mathbf{y}}\hat{\tilde{\mathbf{u}}}_\mu\, d\Omega_\mu - \int_{\Omega_\mu} \mathbf{f}_\mu^p \cdot \hat{\tilde{\mathbf{u}}}_\mu\, d\Omega_\mu + \int_{\Omega_\mu} \rho_\mu \ddot{\mathbf{u}}_\mu \cdot \hat{\tilde{\mathbf{u}}}_\mu\, d\Omega_\mu \\
&+ \int_{\Omega_\mu} \lambda \cdot \hat{\tilde{\mathbf{u}}}_\mu\, d\Omega_\mu = 0 \qquad \forall \hat{\tilde{\mathbf{u}}}_\mu \in \mathbf{H}^1(\Omega_\mu).
\end{aligned} \tag{14.49}$$

  - **Characterization of Lagrange multiplier λ.** Since $\hat{\tilde{\mathbf{u}}}_\mu \in \mathbf{H}^1(\Omega_\mu)$, we can adopt $\hat{\tilde{\mathbf{u}}}_\mu$ as an arbitrary constant within the variational equation (14.49), and recalling (14.43) we get

$$\lambda = \frac{1}{|\Omega_\mu|}\int_{\Omega_\mu} (\mathbf{f}_\mu^p - \rho_\mu \ddot{\mathbf{u}}_\mu)\, d\Omega_\mu = \mathbf{f}_M|_\mathbf{x}. \tag{14.50}$$

That is, the reactive force identified in duality to the kinematical constraint given by the kinematical admissibility $\int_{\Omega_\mu} \tilde{\mathbf{u}}_\mu d\Omega_\mu = \mathbf{0}$ is given by the macroscale external force $\mathbf{f}_M|_\mathbf{x}$.

- **Microscale equilibrium.** Putting the previous result (14.50) into (14.49) we obtain the variational equation characterizing the microscale equilibrium problem

$$\int_{\Omega_\mu} (\mathbf{P}_\mu - \Lambda) \cdot \nabla_\mathbf{y} \hat{\mathbf{u}}_\mu\, d\Omega_\mu$$
$$- \int_{\Omega_\mu} (\mathbf{f}_\mu - \mathbf{f}_M|_\mathbf{x}) \cdot \hat{\mathbf{u}}_\mu\, d\Omega_\mu = 0 \quad \forall \hat{\mathbf{u}}_\mu \in \mathbf{H}^1(\Omega_\mu). \tag{14.51}$$

The Euler–Lagrange equations associated with the previous variational equation are easily deduced by integrating by parts, and become

$$-\mathrm{div}(\mathbf{P}_\mu - \Lambda) = \mathbf{f}_\mu - \mathbf{f}_M|_\mathbf{x} \quad \text{in } \Omega_\mu, \tag{14.52}$$

$$\mathbf{P}_\mu \mathbf{n}_\mu = \Lambda \mathbf{n}_\mu \quad \text{on } \partial\Omega_\mu. \tag{14.53}$$

Since Λ is a uniform second-order tensor, and, moreover, it will be shown (see (14.63) below) that the Lagrange multiplier Λ is such that $\Lambda = \mathbf{P}_M|_\mathbf{x}$, we readily obtain

$$-\mathrm{div}\mathbf{P}_\mu = \mathbf{f}_\mu - \mathbf{f}_M|_\mathbf{x} \quad \text{in } \Omega_\mu, \tag{14.54}$$

$$\mathbf{P}_\mu \mathbf{n}_\mu = \mathbf{P}_M|_\mathbf{x} \mathbf{n}_\mu \quad \text{on } \partial\Omega_\mu. \tag{14.55}$$

These equations, together with the kinematical constraints stating the kinematical admissibility, given by (14.37) and (14.38), and with the constitutive equation for the microscale stress represented by the Piola–Kirchhoff tensor of the first kind, that is, $\mathbf{P}_\mu = \mathscr{P}_\mu(\nabla_\mathbf{y} \mathbf{u}_\mu)$, define the set of equations that characterizes the microscale equilibrium in its strong form. We notice that these equilibrium equations correspond to a problem with Neumann boundary conditions, where the force per unit area over the boundary $\partial\Omega_\mu$ is given by $\Lambda \mathbf{n}_\mu = \mathbf{P}_M|_\mathbf{x} \mathbf{n}_\mu$, and with subsidiary restrictions established by the kinematical admissibility.

- **Characterization of Lagrange multiplier Λ.** To obtain this characterization, let us first see the following equivalence

$$\int_{\Omega_\mu} \mathbf{P}_\mu\, d\Omega_\mu = \int_{\Omega_\mu} \mathbf{P}_\mu \nabla_\mathbf{y} \mathbf{y}\, d\Omega_\mu$$
$$= -\int_{\Omega_\mu} \mathrm{div}\mathbf{P}_\mu \otimes \mathbf{y}\, d\Omega_\mu + \int_{\partial\Omega_\mu} \mathbf{P}_\mu \mathbf{n}_\mu \otimes \mathbf{y}\, d\partial\Omega_\mu. \tag{14.56}$$

With this identity, and making use of the equilibrium equations (14.52) and (14.53) (knowing that Λ is uniform), we obtain

$$\int_{\Omega_\mu} \mathbf{P}_\mu\, d\Omega_\mu = \int_{\Omega_\mu} (\mathbf{f}_\mu - \mathbf{f}_M|_\mathbf{x}) \otimes \mathbf{y}\, d\Omega_\mu + \int_{\partial\Omega_\mu} \Lambda \mathbf{n}_\mu \otimes \mathbf{y}\, d\partial\Omega_\mu. \tag{14.57}$$

Thus, since $\mathbf{f}_M|_\mathbf{x}$ is a uniform vector, and since we have picked the origin of the coordinate system as the geometric center of the domain Ω_μ (see (14.4)), we get

$$\int_{\Omega_\mu} \mathbf{f}_M|_\mathbf{x} \otimes \mathbf{y}\, d\Omega_\mu = \mathbf{f}_M|_\mathbf{x} \otimes \int_{\Omega_\mu} \mathbf{y}\, d\Omega_\mu = \mathbf{O}. \tag{14.58}$$

Putting this into (14.57) and rearranging terms gives

$$\int_{\Omega_\mu} \mathbf{P}_\mu\, d\Omega_\mu - \int_{\Omega_\mu} \mathbf{f}_\mu \otimes \mathbf{y}\, d\Omega_\mu = \int_{\partial\Omega_\mu} \Lambda \mathbf{n}_\mu \otimes \mathbf{y}\, d\partial\Omega_\mu. \tag{14.59}$$

Since Λ is a constant tensor, and recalling that

$$\begin{aligned}\int_{\partial\Omega_\mu} \mathbf{n}_\mu \otimes \mathbf{y}\, d\partial\Omega_\mu &= \left(\int_{\partial\Omega_\mu} \mathbf{y} \otimes \mathbf{n}_\mu\, d\partial\Omega_\mu\right)^T \\ &= \left(\int_{\Omega_\mu} \nabla_\mathbf{y}\mathbf{y}\, d\Omega_\mu\right)^T = \int_{\Omega_\mu} \mathbf{I}\, d\Omega_\mu = |\Omega_\mu|\mathbf{I},\end{aligned} \tag{14.60}$$

we get

$$\int_{\partial\Omega_\mu} \Lambda \mathbf{n}_\mu \otimes \mathbf{y}\, d\partial\Omega_\mu = |\Omega_\mu|\Lambda. \tag{14.61}$$

Putting this result into (14.59) we have

$$\frac{1}{|\Omega_\mu|}\int_{\Omega_\mu} \mathbf{P}_\mu\, d\Omega_\mu - \frac{1}{|\Omega_\mu|}\int_{\Omega_\mu} \mathbf{f}_\mu \otimes \mathbf{y}\, d\Omega_\mu = \Lambda. \tag{14.62}$$

This last identity, together with the homogenization formula for the macroscale stress tensor given by (14.40), allows us to finally obtain the full characterization of the Lagrange multiplier Λ, as follows

$$\Lambda = \frac{1}{|\Omega_\mu|}\int_{\Omega_\mu} (\mathbf{P}_\mu - \mathbf{f}_\mu \otimes \mathbf{y})\, d\Omega_\mu = \mathbf{P}_M|_\mathbf{x}. \tag{14.63}$$

Clearly, the reactive force identified in duality to the kinematical constraint corresponding to the kinematical admissibility given by $\int_{\Omega_\mu} \nabla_\mathbf{y}\tilde{\mathbf{u}}_\mu\, d\Omega_\mu = \int_{\partial\Omega_\mu} \tilde{\mathbf{u}}_\mu \otimes \mathbf{n}_\mu\, d\partial\Omega_\mu = \mathbf{O}$ is the stress state $\mathbf{P}_M|_\mathbf{x}$ at the macroscale point $\mathbf{x}$.

- **Macroscale stress homogenization as a boundary integral formula**. Let us now develop an alternative, fully equivalent, homogenization formula for the macroscale stress tensor. In fact, from (14.59) and from the equilibrium equation over the boundary given by (14.55) we get

$$\int_{\Omega_\mu} \mathbf{P}_\mu\, d\Omega_\mu - \int_{\Omega_\mu} \mathbf{f}_\mu \otimes \mathbf{y}\, d\Omega_\mu = \int_{\partial\Omega_\mu} \mathbf{P}_\mu \mathbf{n}_\mu \otimes \mathbf{y}\, d\partial\Omega_\mu, \tag{14.64}$$

which, together with the macroscale stress homogenization formula given by (14.40), enables us to establish a new way to homogenize the macroscale stress, as follows

$$\mathbf{P}_M|_\mathbf{x} = \frac{1}{|\Omega_\mu|}\int_{\partial\Omega_\mu} \mathbf{P}_\mu \mathbf{n}_\mu \otimes \mathbf{y}\, d\partial\Omega_\mu. \tag{14.65}$$

It is interesting to highlight that this expression is general even if the explicit dependence in terms of the microscale forces (passive or active) involved in the problem has been dropped. As a matter of fact, these forces are implicitly carried by the microscale stress tensor, $\mathbf{P}_\mu$, which equilibrates such loads. In other words, either we can use (14.40) to deliver the homogenization of the macroscale stress, for which we have to explicitly consider the microscale force $\mathbf{f}_\mu$ or, alternatively, we can use the expression (14.65) that, as we have seen, is explicitly independent from such forces. From a theoretical point of view both expressions are identical. However, this equivalence does not hold at the discrete (computational) level, where both expressions possess advantages and disadvantages. As is well known, the fields that intervene in both expressions are better approximated in the bulk of the domain than over the boundary. Then, and from this perspective, expression (14.40) is advantageous over (14.65).

Exercise 14.1 Redefine the Principle of Multiscale Virtual Power given in (14.35) but now with the space $\mathrm{Kin}^*_{u_\mu}$ (defined in (14.24)), putting in evidence the microscale displacement $\mathbf{u}_\mu$. After that, extend the principle by incorporating the corresponding Lagrange multipliers relaxing now the subsidiary restrictions existing in the definition of this space, which correspond to the homogenization of displacements and deformations given by (14.13) and (14.16). Finally, obtain the stress homogenization formula, the force homogenization formula, and the microscale mechanical equilibrium, in addition to the characterization of the Lagrange multipliers. Compare the results with those obtained in this section and comment on the role of the Lagrange multipliers in both variational formulations.

14.2.5 Tangent Operators

This section is devoted to studying the macroscale tangent operator associated with the macroscale stress homogenization formula, which is given by expression (14.40). This is necessary because the nonlinearity of the macroscale equilibrium problem poses the question of what is the macroscale stress increment for a given macroscale deformation increment. This is useful in Newton-like numerical algorithms which aim to seek the equilibrium solution by exploiting the tangent operator.

In particular, and to simplify the presentation, we limit the material developed in the present section to static problems where dynamics can be neglected. With this hypothesis, the stress homogenization (14.40) becomes

$$\mathbf{P}_M|_\mathbf{x} = \frac{1}{|\Omega_\mu|}\int_{\Omega_\mu} (\mathbf{P}_\mu - \mathbf{f}^p_\mu \otimes \mathbf{y})d\Omega_\mu. \tag{14.66}$$

Remember also the microscale equilibrium problem given by the variational equation (14.35) for $\hat{\mathbf{u}}_M|_\mathbf{x} = \mathbf{0}$ and $\hat{\mathbf{G}}_M|_\mathbf{x} = \mathbf{O}$. This problem is well-posed once a proper microscale constitutive model has been chosen, that is, once the constitutive equation for the material present in Ω_μ is specified. In particular, let us suppose that the material response depends on the history of the process, and is not associated with fracturing, failure or strain localization phenomena. With this, the constitutive equation that characterizes the stress at the microscale at time t as a function of the deformation up to that instant, represented here by $(\mathbf{F}_\mu)^t = \mathbf{I} + (\nabla_\mathbf{y}\mathbf{u}_\mu)^t$, can be described through the constitutive functional

$$\mathbf{P}_\mu = \mathfrak{F}_\mu((\mathbf{F}_\mu)^t) = \mathfrak{F}_\mu((\mathbf{I} + \mathbf{G}_M|_\mathbf{x} + \nabla_\mathbf{y}\tilde{\mathbf{u}}_\mu)^t). \tag{14.67}$$

With this definition at hand, the microscale equilibrium problem consists of determining for each time instant t the field $\tilde{\mathbf{u}}_\mu^t \in \mathrm{Kin}_{\tilde{u}_\mu}$ such that

$$\int_{\Omega_\mu} \mathfrak{F}_\mu((\mathbf{I} + \mathbf{G}_M|_\mathbf{x} + \nabla_\mathbf{y}\tilde{\mathbf{u}}_\mu)^t) \cdot \nabla_\mathbf{y}\hat{\mathbf{u}}_\mu \, d\Omega_\mu - \int_{\Omega_\mu} \mathbf{f}_\mu^p(\mathbf{y}, t) \cdot \hat{\mathbf{u}}_\mu \, d\Omega_\mu = 0 \quad \forall \hat{\mathbf{u}}_\mu \in \mathrm{Var}_{\tilde{u}_\mu}, \tag{14.68}$$

which shows us clearly the dependence of the history of the fluctuation field with the history of the macroscale deformation. This relation between the history of the displacement fluctuation $\tilde{\mathbf{u}}_\mu^t$ (and therefore its gradient) and the history of macroscale deformation, $(\mathbf{G}_M|_\mathbf{x})^t$, is nonlinear and is implicitly defined by (14.68). Moreover, we can express this relation through the functional

$$(\nabla_\mathbf{y}\tilde{\mathbf{u}}_\mu)^t = \mathfrak{C}((\mathbf{G}_M|_\mathbf{x})^t). \tag{14.69}$$

Replacing (14.69) into (14.67), and making abuse of notation, we obtain

$$\mathbf{P}_\mu = \mathfrak{F}_\mu((\mathbf{G}_M|_\mathbf{x})^t) = \mathfrak{F}_\mu((\mathbf{I} + \mathbf{G}_M|_\mathbf{x} + \mathfrak{C}(\mathbf{G}_M|_\mathbf{x}))^t). \tag{14.70}$$

That is, the microscale phenomenology explicitly depends upon the history of the macroscale deformation and implicitly on the same macroscale deformation through the fluctuation field. Furthermore, substituting this result in the stress homogenization formula (14.66) we finally obtain

$$\begin{aligned}\mathbf{P}_M|_\mathbf{x} &= \mathfrak{P}_M|_\mathbf{x}((\mathbf{G}_M|_\mathbf{x})^t)\\ &= \frac{1}{|\Omega_\mu|}\int_{\Omega_\mu} (\mathfrak{F}_\mu((\mathbf{I} + \mathbf{G}_M|_\mathbf{x} + \mathfrak{C}(\mathbf{G}_M|_\mathbf{x}))^t) - \mathbf{f}_\mu^p \otimes \mathbf{y}) d\Omega_\mu.\end{aligned} \tag{14.71}$$

At this point, we have all the elements to adequately characterize the macroscale tangent constitutive operator we are looking for. Such an operator corresponds to the directional derivative of operator $\mathfrak{P}_M|_\mathbf{x}$ calculated at point $(\mathbf{G}_M|_\mathbf{x})^t$ in the direction given by $\delta\mathbf{G}_M|_\mathbf{x}$, that is

$$\begin{aligned}D\mathfrak{P}_M|_\mathbf{x}((\mathbf{G}_M|_\mathbf{x})^t)\delta\mathbf{G}_M|_\mathbf{x} &= \frac{d}{d\varepsilon}\mathfrak{P}_M|_\mathbf{x}((\mathbf{G}_M|_\mathbf{x})^t + \varepsilon\delta\mathbf{G}_M|_\mathbf{x})\Big|_{\varepsilon=0}\\ &= \frac{1}{|\Omega_\mu|}\int_{\Omega_\mu} D\mathfrak{F}_\mu((\mathbf{G}_M|_\mathbf{x})^t)[\delta\mathbf{G}_M|_\mathbf{x} + D\mathfrak{C}((\mathbf{G}_M|_\mathbf{x})^t)\delta\mathbf{G}_M|_\mathbf{x}] d\Omega_\mu\\ &= \left[\frac{1}{|\Omega_\mu|}\int_{\Omega_\mu} \left(D\mathfrak{F}_\mu((\mathbf{G}_M|_\mathbf{x})^t) + D\mathfrak{F}_\mu((\mathbf{G}_M|_\mathbf{x})^t)D\mathfrak{C}((\mathbf{G}_M|_\mathbf{x})^t)\right) d\Omega_\mu\right]\delta\mathbf{G}_M|_\mathbf{x},\end{aligned} \tag{14.72}$$

where, for a generic microscale point, $D\mathfrak{F}_\mu((\mathbf{G}_M|_\mathbf{x})^t)\delta\mathbf{G}_M|_\mathbf{x}$ represents the directional derivative of operator $\mathfrak{F}_\mu$ evaluated at point $(\mathbf{G}_M|_\mathbf{x})^t$ in the direction of $\delta\mathbf{G}_M|_\mathbf{x}$, and

$D\mathfrak{C}((\mathbf{G}_M|_\mathbf{x})^t)\delta\mathbf{G}_M|_\mathbf{x}$ represents the directional derivative of operator $\mathfrak{C}$ at point $(\mathbf{G}_M|_\mathbf{x})^t$ according to direction $\delta\mathbf{G}_M|_\mathbf{x}$, that is

$$\nabla_\mathbf{y}\delta\tilde{\mathbf{u}}_\mu = D\mathfrak{C}((\mathbf{G}_M|_\mathbf{x})^t)\delta\mathbf{G}_M|_\mathbf{x}. \tag{14.73}$$

Here it is important to observe (and this observation remains valid for any kind of multiscale model constructed within the present framework) that the inspection of expression (14.72) shows us that the macroscale tangent constitutive operator is formed by the contribution of two terms. The first term corresponds to the homogenization of the directional derivative of the explicit component of the microscale constitutive functional, and is equivalent to assuming that the fluctuation field is the null function. That is, this contribution corresponds to adopting the kinematical model acknowledged as the Taylor model. The second term is associated with the contribution brought by the fluctuation field. To put in evidence these contributions, the macroscale constitutive equation is written in the following manner

$$D\mathfrak{P}_M|_\mathbf{x}((\mathbf{G}_M|_\mathbf{x})^t) = D^{Taylor}\mathfrak{P}_M|_\mathbf{x}((\mathbf{G}_M|_\mathbf{x})^t) + \tilde{D}\mathfrak{P}_M|_\mathbf{x}((\mathbf{G}_M|_\mathbf{x})^t), \tag{14.74}$$

where

$$D^{Taylor}\mathfrak{P}_M|_\mathbf{x}((\mathbf{G}_M|_\mathbf{x})^t) = \frac{1}{|\Omega_\mu|}\int_{\Omega_\mu} D\mathfrak{F}_\mu((\mathbf{G}_M|_\mathbf{x})^t)d\Omega_\mu, \tag{14.75}$$

and

$$\tilde{D}\mathfrak{P}_M|_\mathbf{x}((\mathbf{G}_M|_\mathbf{x})^t) = \frac{1}{|\Omega_\mu|}\int_{\Omega_\mu} D\mathfrak{F}_\mu((\mathbf{G}_M|_\mathbf{x})^t)D\mathfrak{C}((\mathbf{G}_M|_\mathbf{x})^t)d\Omega_\mu. \tag{14.76}$$

Finally, note that all these tangent operators are fourth-order tensor entities. In particular, the contribution $D^{Taylor}\mathfrak{P}_M|_\mathbf{x}$ exclusively depends on the microscale material behavior and, therefore, can only be characterized once the material has been adopted. In fact, $D\mathfrak{F}_\mu((\mathbf{G}_M|_\mathbf{x})^t)$ is the fourth-order tensor field corresponding to the classical constitutive tangent operator for the material at the microscale. Now, let us illustrate the way in which the tangent operator $D\mathfrak{C}((\mathbf{G}_M|_\mathbf{x})^t)$ can be explicitly characterized. To accomplish this task notice that, from the definition, this operator represents the tangent relation between $(\nabla_\mathbf{y}\tilde{\mathbf{u}}_\mu)^t$ and $(\mathbf{G}_M|_\mathbf{x})^t$ given by expression (14.69) and formalized by expression (14.73). To characterize this operator we proceed to linearize the variational equation that defines the microscale equilibrium problem and, therefore, the relation (14.69). This linearization procedure consists of determining, for each time instant t, the field $\delta\tilde{\mathbf{u}}_\mu \in \mathrm{Kin}_{\tilde{u}_\mu}$ such that

$$\begin{aligned}&\int_{\Omega_\mu} D\mathfrak{F}_\mu((\mathbf{G}_M|_\mathbf{x})^t)\nabla_\mathbf{y}\delta\tilde{\mathbf{u}}_\mu \cdot \nabla_\mathbf{y}\hat{\mathbf{u}}_\mu \, d\Omega_\mu \\ &= -\int_{\Omega_\mu} D\mathfrak{F}_\mu((\mathbf{G}_M|_\mathbf{x})^t)\delta\mathbf{G}_M|_\mathbf{x} \cdot \nabla_\mathbf{y}\hat{\mathbf{u}}_\mu \, d\Omega_\mu \quad \forall\hat{\mathbf{u}}_\mu \in \mathrm{Var}_{\tilde{u}_\mu}.\end{aligned} \tag{14.77}$$

For $\delta\mathbf{G}_M|_\mathbf{x}$ written in Cartesian coordinates, that is, $\delta\mathbf{G}_M|_\mathbf{x} = [\delta\mathbf{G}_M|_\mathbf{x}]_{ij}(\mathbf{e}_i \otimes \mathbf{e}_j)$, we can introduce the vector field $\delta\tilde{\mathbf{u}}_\mu^{ij} \in \mathrm{Kin}_{\tilde{u}_\mu}$ defined by the following variational equation

$$\int_{\Omega_\mu} D\mathfrak{F}_\mu((\mathbf{G}_M|_\mathbf{x})^t)\nabla_\mathbf{y}\delta\tilde{\mathbf{u}}_\mu^{ij} \cdot \nabla_\mathbf{y}\hat{\mathbf{u}}_\mu \, d\Omega_\mu$$

$$= -\int_{\Omega_\mu} D\mathfrak{F}_\mu((\mathbf{G}_M|_\mathbf{x})^t)(\mathbf{e}_i \otimes \mathbf{e}_j) \cdot \nabla_\mathbf{y}\hat{\mathbf{u}}_\mu \, d\Omega_\mu \quad \forall \hat{\mathbf{u}}_\mu \in \mathrm{Var}_{\tilde{u}_\mu}, \tag{14.78}$$

from where

$$\delta\tilde{\mathbf{u}}_\mu = [\delta\mathbf{G}_M|_\mathbf{x}]_{ij}\delta\tilde{\mathbf{u}}_\mu^{ij}, \tag{14.79}$$

leading to

$$\nabla_\mathbf{y}\delta\tilde{\mathbf{u}}_\mu = [\delta\mathbf{G}_M|_\mathbf{x}]_{ij}\nabla_\mathbf{y}\delta\tilde{\mathbf{u}}_\mu^{ij}. \tag{14.80}$$

In this manner, noting that we are using Cartesian coordinates, the previous expression is rewritten as

$$\nabla_\mathbf{y}\delta\tilde{\mathbf{u}}_\mu = [\nabla_\mathbf{y}\delta\tilde{\mathbf{u}}_\mu]_{pq}(\mathbf{e}_p \otimes \mathbf{e}_q) = \frac{\partial(\delta\tilde{\mathbf{u}}_\mu)_p}{\partial x_q}(\mathbf{e}_p \otimes \mathbf{e}_q)$$

$$= [\delta\mathbf{G}_M|_\mathbf{x}]_{ij}\frac{\partial(\delta\tilde{\mathbf{u}}_\mu^{ij})_p}{\partial x_q}(\mathbf{e}_p \otimes \mathbf{e}_q). \tag{14.81}$$

Comparing (14.73) with this last expression we finally obtain the characterization of the tangent operator $D\mathfrak{C}((\mathbf{G}_M|_\mathbf{x})^t)$ in terms of the Cartesian components

$$[D\mathfrak{C}((\mathbf{G}_M|_\mathbf{x})^t)]_{ijkl} = \frac{\partial(\delta\tilde{\mathbf{u}}_\mu^{kl})_i}{\partial x_j}. \tag{14.82}$$

This result, in conjunction with the classical constitutive tangent operator for the material at the microscale, enables the full characterization of the Cartesian components of the macroscale tangent constitutive operator associated with the contributions of fluctuations (14.76), whose Cartesian components are

$$[\tilde{D}\mathfrak{P}_M|_\mathbf{x}((\mathbf{G}_M|_\mathbf{x})^t)]_{ijkl}$$

$$= \frac{1}{|\Omega_\mu|}\int_{\Omega_\mu} [D\mathfrak{F}_\mu((\mathbf{G}_M|_\mathbf{x})^t)]_{ijpq}[D\mathfrak{C}((\mathbf{G}_M|_\mathbf{x})^t)]_{pqkl} d\Omega_\mu$$

$$= \frac{1}{|\Omega_\mu|}\int_{\Omega_\mu} [D\mathfrak{F}_\mu((\mathbf{G}_M|_\mathbf{x})^t)]_{ijpq}\frac{\partial(\delta\tilde{\mathbf{u}}_\mu^{kl})_p}{\partial x_q} d\Omega_\mu. \tag{14.83}$$

14.3 Mechanics of Incompressible Solid Media

As anticipated at the beginning of this chapter, the goal of this section is to look into the application of the PMVP to model macroscale incompressible media, even if at the microscale the medium features either compressible or incompressible behavior. In fact, as will be shown in the forthcoming sections, the PMVP provides a tool to study in a clear and concise manner an aspect that usually arises in the multiscale modeling of rubber-like materials, and even in the context of biological tissues. Specifically, the next section addresses the problem of coupling a macroscale incompressible medium with a microscale in which the constituent material features, in general, a compressible behavior. We exclude from the analysis cases of strain localization, damage, and related phenomena. The mechanical problem is described in the actual (spatial or Eulerian) description. We then present the model where both scales feature incompressible behavior, also by employing a spatial description. Finally, we present the same cases but using a material (Lagrangian) description.

In all that follows, and to ease the presentation, we suppose that the kinematical descriptions for both scales are the same, and are given by the classical kinematics used in the context of finite strains.

14.3.1 Principle of Virtual Power

Before jumping into multiscale modeling, let us revisit some concepts employed in the description of the macroscale problem. In the present context, we denote by Ω^s_M the spatial configuration of the body, with coordinates $\mathbf{x}_s$ (index s indicates spatial description). Also, we introduce the material description Ω^m_M with coordinates $\mathbf{x}_m$ (index m stands for the material description). The displacement field between material and spatial (deformed) configurations is denoted by $\mathbf{u}_M$ regardless of the type of description (spatial or material), provided the context makes clear the description to be considered. In the first part of the section we describe the problem using the spatial configuration. Hence, the field $\mathbf{u}_M$ is the spatial description of the displacement field. When we describe the problem using the material configuration, the field $\mathbf{u}_M$ stands for the material description of the displacement field.

For incompressible media, let us recall that the tensor $\mathbf{F} = \mathbf{I} + \nabla_{\mathbf{x}_m} \mathbf{u}_M$ (here $\mathbf{u}_M$ is a material field) has to satisfy

$$\det \mathbf{F} = \det(\mathbf{I} + \nabla_{\mathbf{x}_m} \mathbf{u}_M) = 1. \tag{14.84}$$

This is equivalent to requiring that

$$(\det \mathbf{F})_s^{-1} = (\det \mathbf{F}^{-1})_s = \det \mathbf{F}_s^{-1} = 1 = \det(\mathbf{I} - \nabla_{\mathbf{x}_s} \mathbf{u}_M) = 1, \tag{14.85}$$

where now $\mathbf{u}_M$ is a spatial field. Then, the equilibrium problem, without relaxing the kinematical constraint associated with the incompressibility, is given by the variational problem that consists of determining the spatial field $\mathbf{u}_M \in \mathrm{Kin}_{u_M}$ such that

$$\int_{\Omega^s_M} \sigma^D_M \cdot (\nabla_{\mathbf{x}_s} \hat{\mathbf{v}}_M)^s \, d\Omega^s_M = \langle f, \hat{\mathbf{v}}_M \rangle \qquad \forall \hat{\mathbf{v}}_M \in \mathrm{Var}_{u_M}, \tag{14.86}$$

where

$$\mathrm{Kin}_{u_M} = \{\mathbf{u} \in \mathbf{H}^1(\Omega^s_M);\ (\det \mathbf{F}^{-1})_s = 1 \text{ in } \Omega^s_M, \mathbf{u}|_{\partial\Omega^s_{Mu}} = \overline{\mathbf{u}}\}, \tag{14.87}$$

$$\mathrm{Var}_{u_M} = \{\mathbf{u} \in \mathbf{H}^1(\Omega^s_M);\ \mathrm{div}_{\mathbf{x}_s} \mathbf{u} = 0 \text{ in } \Omega^s_M, \mathbf{u}|_{\partial\Omega^s_{Mu}} = \mathbf{0}\}. \tag{14.88}$$

In addition, $\nabla_{\mathbf{x}_s}(\cdot)$ represents the gradient operator with respect to the spatial coordinates $\mathbf{x}_s$, $\mathrm{div}_{\mathbf{x}_s}$ is the divergence operator in these coordinates, and $(\mathbf{A})^D$ stands for the deviatoric component of the second-order tensor $\mathbf{A}$. Observe the parallel between this formulation and that developed in the field of fluid mechanics in Chapter 11. In fact, the internal power is associated with the stress state σ_M, characterized through the deviatoric component, σ^D_M, because its dual variable is the field $\nabla_{\mathbf{x}_s} \hat{\mathbf{v}}_M$, which possesses null trace. That is, the pressure (which exists in the solid body) cannot be calculated through the equilibrium because it exerts null power in duality with the divergence free virtual motion actions.

It is important to notice here that the set Kin_{u_M} defined in (14.87) is a nonlinear manifold. Hence, the space Var_{u_M} is not obtained from taking the difference between elements belonging to Kin_{u_M}, but is obtained as the tangent space to Kin_{u_M}. This explains the fact that the real displacements of the body have to satisfy the constraint (14.85), while the virtual motion actions (velocities) have to satisfy the tangent constraint, which yields motion actions featuring divergence-free behavior.

Let us now proceed to relax the incompressibility constraint via the corresponding Lagrange multiplier. Therefore, the equilibrium problem now consists of determining $(\mathbf{u}_M, \lambda_M) \in \mathrm{Kin}^*_{u_M} \times L^2(\Omega^s_M)$ such that

$$\int_{\Omega^s_M} \left(\sigma^D_M \cdot (\nabla_{\mathbf{x}_s}\hat{\mathbf{v}}_M)^s + \lambda_M \mathrm{div}_{\mathbf{x}_s}\hat{\mathbf{v}}_M + \hat{\lambda}_M(1 - \det(\mathbf{I} - \nabla_{\mathbf{x}_s}\mathbf{u}_M))\right) d\Omega^s_M = \langle f, \hat{\mathbf{v}}_M \rangle$$
$$\forall (\hat{\mathbf{v}}_M, \hat{\lambda}_M) \in \mathrm{Var}^*_{u_M} \times L^2(\Omega^s_M), \tag{14.89}$$

where

$$\mathrm{Kin}^*_{u_M} = \{\mathbf{u} \in \mathbf{H}^1(\Omega^s_M);\ \mathbf{u}|_{\partial\Omega^s_{Mu}} = \overline{\mathbf{u}}\}, \tag{14.90}$$

$$\mathrm{Var}^*_{u_M} = \{\mathbf{u} \in \mathbf{H}^1(\Omega^s_M);\ \mathbf{u}|_{\partial\Omega^s_{Mu}} = \mathbf{0}\}. \tag{14.91}$$

It is important to note that the stress field σ^D_M is a deviatoric and symmetric second-order tensor, therefore we have to always keep in mind that the internal power functional can be written in the following alternative ways

$$\int_{\Omega^s_M} \sigma^D_M \cdot (\nabla_{\mathbf{x}_s}\hat{\mathbf{v}}_M)^s \, d\Omega^s_M = \int_{\Omega^s_M} \sigma^D_M \cdot (\nabla_{\mathbf{x}_s}\hat{\mathbf{v}}_M)^D \, d\Omega^s_M$$
$$= \int_{\Omega^s_M} \sigma^D_M \cdot \nabla_{\mathbf{x}_s}\hat{\mathbf{v}}_M \, d\Omega^s_M, \tag{14.92}$$

in spite of relaxing, or not, the incompressibility kinematical constraint. In addition, λ_M is the reactive force associated with the macroscale kinematical constraint, and that is not determined through a constitutive functional, but through the equilibrium (in this case the macroscale equilibrium). What is more, as for the variable $\mathbf{u}_M$ the solution of the variational problem that features the relaxed condition, (14.89), is the same as that we would obtain from the original variational equation (14.86). That is, $\mathbf{u}_M$ is such that $\det(\mathbf{I} - \nabla_{\mathbf{x}_s}\mathbf{u}_M) = 1$ and σ^D_M is the stress state that we obtain by means of the constitutive equation adopted for the deviatoric component of the stress in the incompressible material.

These aspects related to the internal power and the incompressibility constraint are worth emphasizing. On the one hand, the internal power is always given by (14.92), and since the solution is always a field whose motion actions are divergence free, we have, from the multiscale viewpoint and for both cases (original and relaxed formulations), that the deformation actions will be given by the deviatoric component of the gradient $((\nabla_{\mathbf{x}_s}\mathbf{v}_M)^s)^D|_{\mathbf{x}_s} = \mathbf{G}^{sD}_M|_{\mathbf{x}_s}$. On the other hand, the power associated with the kinematical constraint (incompressibility) is not understood either as an internal power or as an external power. Such power is associated with the relaxation of an internal constraint.

That is, the release of the restriction enables the generalized force to be shaped, as a dual variable, which is required to be applied over the body in order to enforce the constraint.

In the sections to come we focus on the formulation of the multiscale model seeking to determine the constitutive material behavior, that is, to determine the tensor σ_M^D. For simplicity, in this section we neglect the effect of body forces in the theory.

14.3.2 Multiscale Kinematics

First, consider the case in which the microscale is formed by materials that, in general, feature compressible behavior. Despite this, we want to recover, at the macroscale, the constitutive behavior of an incompressible material.

In the previous section we established the PVP for the macroscale problem, and we also defined the kinematical description for the problem. That is, we introduced the displacement field $\mathbf{u}_M$ and the space $\mathscr{V}_M$ where they belong, as well as the set Kin_{u_M} of candidates to be the solution of the problem (see (14.87)) and the kinematically admissible motion actions in the space Var_{u_M} (see (14.88)). We also wrote the PVP in two equivalent forms, without releasing the incompressibility constraint (see (14.86)) and freeing the constraint through the introduction of a proper Lagrange multiplier (see (14.89)).

The deformation rate operator has been defined as $\mathscr{D}_M(\cdot) = (\nabla_{\mathbf{x}_s}(\cdot))^s$, then $\mathscr{W}_M = \mathbf{L}^2(\Omega_M^s)$. Observe that $\mathscr{D}_M$ acts over the space Var_{u_M}, which are divergence free motion actions. Thus, the compatible deformation rates are such that $D_M = (\nabla_{\mathbf{x}_s}(\cdot))^{sD} = \mathbf{G}_M^{sD}$, that is, second-order trace-free, and symmetric tensor fields (i.e. deviatoric symmetric tensors). In this manner, motion actions and deformation rates at a point $\mathbf{x}_s \in \Omega_M^s$ belong to the following spaces

$$\mathbb{R}^{\mathbf{x}}_{\mathscr{V}_M} = \{\mathbf{w} \in \mathbb{R}^3;\ \mathbf{w} = \mathbf{u}|_{\mathbf{x}_s}, \mathbf{u} \in \mathscr{V}_M\}, \tag{14.93}$$

$$\mathbb{S}^{\mathbf{x}}_{\mathscr{W}_M} = \{\mathbf{H} \in \mathbb{R}^{\mathbf{x}}_{\mathscr{W}_M};\ \mathbf{H} = \mathbf{H}^T,\ \mathrm{tr}\mathbf{H} = 0\}, \tag{14.94}$$

with

$$\mathbb{R}^{\mathbf{x}}_{\mathscr{W}_M} = \{\mathbf{H} \in \mathbb{R}^{3\times3};\ \mathbf{H} = \mathbf{G}_M|_{\mathbf{x}_s}, \mathbf{G}_M \in \mathscr{W}_M\}. \tag{14.95}$$

As for the microscale, whose spatial domain is Ω_μ^s, with coordinates $\mathbf{y}_s$, the kinematics is the classical one, and so we have microscale displacement vector fields $\mathbf{u}_\mu$ (in its spatial description) which belong to the space $\mathscr{V}_\mu = \mathbf{H}^1(\Omega_\mu^s)$, and for which the deformation rate operator is $\mathscr{D}_\mu(\cdot) = (\nabla_{\mathbf{y}_s}(\cdot))^s = \mathbf{G}_\mu \in \mathscr{W}_\mu = \mathbf{L}^2(\Omega_\mu^s)$.

The insertion operators in this case are defined as

$$\mathscr{I}_\mu^{\mathscr{V}}(\mathbf{u}_M|_{\mathbf{x}_s}) = \mathbf{u}_M|_{\mathbf{x}_s} \quad \forall \mathbf{y}_s \in \Omega_\mu^s, \tag{14.96}$$

$$\mathscr{I}_\mu^{\mathscr{W}}(\mathbf{G}_M|_{\mathbf{x}_s}) = \mathbf{G}_M|_{\mathbf{x}_s}\mathbf{y}_s \quad \forall \mathbf{y}_s \in \Omega_\mu^s, \tag{14.97}$$

where we have considered that the geometric center of Ω_μ^s coincides with the origin of coordinates at the microscale. With this, we obtain the following representation for the motion actions and deformation rates at the microscale

$$\mathbf{u}_\mu = \mathbf{u}_M|_{\mathbf{x}_s} + \mathbf{G}_M|_{\mathbf{x}_s}\mathbf{y}_s + \tilde{\mathbf{u}}_\mu \quad \mathbf{y}_s \in \Omega_\mu^s, \tag{14.98}$$

$$\mathbf{G}_\mu = (\mathbf{G}_M|_{\mathbf{x}_s})^s + (\nabla_{\mathbf{x}_s}\tilde{\mathbf{u}}_\mu)^s \quad \mathbf{y}_s \in \Omega_\mu^s. \tag{14.99}$$

Next, we introduce the homogenization operators for the kinematics

$$\mathscr{H}_{\mu}^{\mathscr{V}}(\mathbf{u}_{\mu}) = \frac{1}{|\Omega_{\mu}^{s}|}\int_{\Omega_{\mu}^{s}} \mathbf{u}_{\mu}\, d\Omega_{\mu}^{s}, \tag{14.100}$$

$$\mathscr{H}_{\mu}^{\mathscr{W}}(\mathbf{G}_{\mu}) = \frac{1}{|\Omega_{\mu}^{s}|}\int_{\Omega_{\mu}^{s}} \mathbf{G}_{\mu}\, d\Omega_{\mu}^{s}. \tag{14.101}$$

Hence, it is straightforward to verify that the fluctuations must satisfy the kinematical constraints

$$\int_{\Omega_{\mu}^{s}} \tilde{\mathbf{u}}_{\mu}\, d\Omega_{\mu}^{s} = \mathbf{0}, \tag{14.102}$$

$$\int_{\partial\Omega_{\mu}^{s}} (\tilde{\mathbf{u}}_{\mu} \otimes \mathbf{n}_{\mu}^{s})^{s}\, d\partial\Omega_{\mu}^{s} = \mathbf{O}. \tag{14.103}$$

As a consequence of this we obtain

$$\begin{aligned}\mathrm{Var}_{u_{\mu}} = \Big\{ & \mathbf{u}_{\mu} = \mathbf{u}_{M}|_{\mathbf{x}_s} + \mathbf{G}_{M}|_{\mathbf{x}_s}\mathbf{y}_s + \tilde{\mathbf{u}}_{\mu}; \\ & \mathbf{u}_{M}|_{\mathbf{x}_s} \in \mathbb{R}^{\mathbf{x}}_{\mathscr{V}_M}, \mathbf{G}_{M}|_{\mathbf{x}_s} \in \mathbb{S}^{\mathbf{x}}_{\mathscr{W}_M}, \tilde{\mathbf{u}}_{\mu} \in \mathrm{Var}_{\tilde{u}_{\mu}} \Big\},\end{aligned} \tag{14.104}$$

where

$$\mathrm{Var}_{\tilde{u}_{\mu}} = \left\{ \tilde{\mathbf{u}}_{\mu} \in \mathbf{H}^{1}(\Omega_{\mu}^{s});\ \int_{\Omega_{\mu}^{s}} \tilde{\mathbf{u}}_{\mu}\, d\Omega_{\mu}^{s} = \mathbf{0},\ \int_{\partial\Omega_{\mu}^{s}} (\tilde{\mathbf{u}}_{\mu} \otimes \mathbf{n}_{\mu}^{s})^{s}\, d\partial\Omega_{\mu}^{s} = \mathbf{O} \right\}. \tag{14.105}$$

It is worthwhile emphasizing here that the expansions (14.93) and (14.94), as well as the constraints (14.102) and (14.103), and the space of kinematically admissible motion actions $\mathrm{Var}_{u_{\mu}}$ defined in (14.104) provide a characterization of the motion actions and deformation actions. That is to say, the multiscale theory developed here has been conceived to provide a sense of consistency to the information transferred between the scales, and to achieve this it is of the utmost importance to correctly specify the spaces in which the motion actions and the deformation actions belong. Based on such a crucial aspect, the concept of duality morphs these quantities into dual variables, whose pairing yields power. Recall that we have postulated that the power functionals as well as the insertion and homogenization operators are linear operators.

When addressing the problem of an incompressible microscale material, all the elements presented in this section remain invariant, except for the space $\mathrm{Var}_{\tilde{u}_{\mu}}$ defined in (14.105), which, for the incompressible microscale becomes

$$\begin{aligned}\mathrm{Var}^{+}_{\tilde{u}_{\mu}} = \Big\{ & \tilde{\mathbf{u}}_{\mu} \in \mathbf{H}^{1}(\Omega_{\mu}^{s});\ \mathrm{div}_{\mathbf{y}_s}\tilde{\mathbf{u}}_{\mu} = 0,\ \int_{\Omega_{\mu}^{s}} \tilde{\mathbf{u}}_{\mu}\, d\Omega_{\mu}^{s} = \mathbf{0}, \\ & \int_{\partial\Omega_{\mu}^{s}} (\tilde{\mathbf{u}}_{\mu} \otimes \mathbf{n}_{\mu}^{s})^{s}\, d\partial\Omega_{\mu}^{s} = \mathbf{O} \Big\}.\end{aligned} \tag{14.106}$$

14.3.3 Principle of Multiscale Virtual Power

At the beginning of this section we saw that the internal power (recall we are neglecting external forces in the multiscale model) for this problem is

$$P^i_M((\nabla_{\mathbf{x}_s}\hat{\mathbf{u}}_M)^{sD}) = \int_{\Omega^s_M} \sigma^D_M \cdot (\nabla_{\mathbf{x}_s}\hat{\mathbf{v}}_M)^{sD} d\Omega^s_M, \tag{14.107}$$

from which we readily verify that

$$P^i_{M,\mathbf{x}_s} = \sigma^D_M|_{\mathbf{x}_s} \bullet (\nabla_{\mathbf{x}_s}\hat{\mathbf{u}}_M)^{sD}|_{\mathbf{x}_s} = \sigma^D_M|_{\mathbf{x}_s} \bullet \mathbf{G}^{sD}_M|_{\mathbf{x}_s} \qquad \mathbf{G}^{sD}_M|_{\mathbf{x}_s} \in \mathbb{S}^{\mathbf{x}}_{\mathscr{W}_M}, \tag{14.108}$$

and in explicit form, anticipating the characterization of the operation $(\cdot) \bullet (\cdot)$, we have

$$P^i_{M,\mathbf{x}_s} = |\Omega^s_\mu| \sigma^D_M|_{\mathbf{x}_s} \cdot \mathbf{G}^{sD}_M|_{\mathbf{x}_s} \qquad \mathbf{G}^{sD}_M|_{\mathbf{x}_s} \in \mathbb{S}^{\mathbf{x}}_{\mathscr{W}_M}. \tag{14.109}$$

In turn, the microscale internal power is written as

$$P^i_\mu((\nabla_{\mathbf{y}_s}\hat{\mathbf{u}}_\mu)^s) = \int_{\Omega^s_\mu} \sigma_\mu \cdot (\nabla_{\mathbf{y}_s}\hat{\mathbf{u}}_\mu)^s d\Omega^s_\mu \qquad \hat{\mathbf{u}}_\mu \in \mathrm{Var}_{u_\mu}, \tag{14.110}$$

and in extended form it is

$$\begin{aligned} P^i_\mu &= \int_{\Omega^s_\mu} \sigma_\mu \cdot (\hat{\mathbf{G}}^{sD}_M|_{\mathbf{x}_s} + (\nabla_{\mathbf{y}_s}\hat{\mathbf{u}}_\mu)^s) d\Omega^s_\mu \\ &= \int_{\Omega^s_\mu} \sigma_\mu \cdot \hat{\mathbf{G}}^{sD}_M|_{\mathbf{x}_s} d\Omega^s_\mu + \int_{\Omega^s_\mu} \sigma_\mu \cdot (\nabla_{\mathbf{y}_s}\hat{\mathbf{u}}_\mu)^s d\Omega^s_\mu \\ &\hat{\mathbf{G}}^{sD}_M|_{\mathbf{x}_s} \in \mathbb{S}^{\mathbf{x}}_{\mathscr{W}_M}, \ \hat{\mathbf{u}}_\mu \in \mathrm{Var}_{\tilde{u}_\mu}. \end{aligned} \tag{14.111}$$

We now formulate the PMVP for this problem. We say that the macroscale stress $\sigma^D_M|_{\mathbf{x}} \in (\mathbb{S}^{\mathbf{x}}_{\mathscr{W}_M})'$ and the microscale stress $\sigma_\mu \in \mathscr{W}'_\mu$ satisfy the PMVP if and only if the following variational equation is verified

$$\begin{aligned} \sigma^D_M|_{\mathbf{x}_s} \bullet \hat{\mathbf{G}}^{sD}_M|_{\mathbf{x}_s} &= \int_{\Omega^s_\mu} \sigma_\mu \cdot \hat{\mathbf{G}}^{sD}_M|_{\mathbf{x}_s} d\Omega^s_\mu \\ &+ \int_{\Omega^s_\mu} \sigma_\mu \cdot (\nabla_{\mathbf{y}_s}\hat{\mathbf{u}}_\mu)^s d\Omega^s_\mu \quad \forall(\hat{\mathbf{G}}^{sD}_M|_{\mathbf{x}_s}, \hat{\mathbf{u}}_\mu) \in \mathbb{S}^{\mathbf{x}}_{\mathscr{W}_M} \times \mathrm{Var}_{\tilde{u}_\mu}, \end{aligned} \tag{14.112}$$

or equivalently

$$\begin{aligned} \sigma^D_M|_{\mathbf{x}_s} \cdot \hat{\mathbf{G}}^{sD}_M|_{\mathbf{x}_s} &= \frac{1}{|\Omega^s_\mu|} \int_{\Omega^s_\mu} \sigma_\mu \cdot \hat{\mathbf{G}}^{sD}_M|_{\mathbf{x}_s} d\Omega^s_\mu \\ &+ \frac{1}{|\Omega^s_\mu|} \int_{\Omega^s_\mu} \sigma_\mu \cdot (\nabla_{\mathbf{y}_s}\hat{\mathbf{u}}_\mu)^s d\Omega^s_\mu \quad \forall(\hat{\mathbf{G}}^{sD}_M|_{\mathbf{x}_s}, \hat{\mathbf{u}}_\mu) \in \mathbb{S}^{\mathbf{x}}_{\mathscr{W}_M} \times \mathrm{Var}_{\tilde{u}_\mu}. \end{aligned} \tag{14.113}$$

Hence, as a first consequence extracted from the PMVP, by taking $\hat{\mathbf{u}}_\mu = \mathbf{0}$ in (14.113), we get the stress homogenization formula

$$\left(\sigma^D_M|_{\mathbf{x}_s} - \frac{1}{|\Omega^s_\mu|} \int_{\Omega^s_\mu} \sigma_\mu \, d\Omega^s_\mu\right) \cdot \hat{\mathbf{G}}^{sD}_M|_{\mathbf{x}_s} = 0 \quad \forall \hat{\mathbf{G}}^{sD}_M|_{\mathbf{x}_s} \in \mathbb{S}^{\mathbf{x}}_{\mathscr{W}_M}, \tag{14.114}$$

that is

$$\sigma_M^D|_{\mathbf{x}_s} - \frac{1}{|\Omega_\mu^s|}\int_{\Omega_\mu^s} \sigma_\mu \, d\Omega_\mu^s \in (\mathbb{S}_{\mathscr{W}_M}^{\mathbf{x}})^\perp \subset (\mathbb{R}_{\mathscr{W}_M}^{\mathbf{x}})', \tag{14.115}$$

and from the definition of the space $\mathbb{S}_{\mathscr{W}_M}^{\mathbf{x}}$ given in (14.94) we can conclude that $\sigma_M^D|_{\mathbf{x}}$ is, in fact, a symmetric second-order tensor with null trace (deviatoric) that is characterized through

$$\sigma_M^D|_{\mathbf{x}_s} = \frac{1}{|\Omega_\mu^s|}\int_{\Omega_\mu^s} \sigma_\mu^D \, d\Omega_\mu^s, \tag{14.116}$$

implying that it consists of the homogenization of the deviatoric component of the microscale stress tensor.

Now, considering $\hat{\mathbf{G}}_M^{sD}|_{\mathbf{x}_s} = \mathbf{O}$ in (14.113), the microscale equilibrium problem in variational form is recovered

$$\int_{\Omega_\mu^s} \sigma_\mu \cdot (\nabla_{\mathbf{y}_s} \hat{\mathbf{u}}_\mu)^s d\Omega_\mu^s = 0 \qquad \forall \hat{\mathbf{u}}_\mu \in \mathrm{Var}_{\tilde{u}_\mu}. \tag{14.117}$$

For the case of microscale incompressible media, the PMVP given by expression (14.113) turns into the following

$$\begin{aligned} \sigma_M^D|_{\mathbf{x}_s} \cdot \hat{\mathbf{G}}_M^{sD}|_{\mathbf{x}_s} &= \frac{1}{|\Omega_\mu^s|}\int_{\Omega_\mu^s} \sigma_\mu^D \cdot \hat{\mathbf{G}}_M^{sD}|_{\mathbf{x}_s} \, d\Omega_\mu^s \\ &+ \frac{1}{|\Omega_\mu^s|}\int_{\Omega_\mu^s} \sigma_\mu^D \cdot (\nabla_{\mathbf{y}_s} \hat{\mathbf{u}}_\mu)^s d\Omega_\mu^s \quad \forall (\hat{\mathbf{G}}_M^{sD}|_{\mathbf{x}_s}, \hat{\mathbf{u}}_\mu) \in \mathbb{S}_{\mathscr{W}_M}^{\mathbf{x}} \times \mathrm{Var}_{\tilde{u}_\mu}^+, \end{aligned} \tag{14.118}$$

where $\mathrm{Var}_{\tilde{u}_\mu}^+$ is defined in (14.106). The consequences of the PMVP are then the stress homogenization formula, which is identical to (14.116), that is

$$\sigma_M^D|_{\mathbf{x}_s} = \frac{1}{|\Omega_\mu^s|}\int_{\Omega_\mu^s} \sigma_\mu^D \, d\Omega_\mu^s, \tag{14.119}$$

and the microscale equilibrium problem, which is

$$\int_{\Omega_\mu^s} \sigma_\mu^D \cdot (\nabla_{\mathbf{y}_s} \hat{\mathbf{u}}_\mu)^s d\Omega_\mu^s = 0 \qquad \forall \hat{\mathbf{u}}_\mu \in \mathrm{Var}_{\tilde{u}_\mu}^+. \tag{14.120}$$

Note that in this case we can relax the kinematical constraint over the microscale displacement fluctuation field, in a similar way to that carried out in the PVP (14.89). We must underline that this does not change the course of the development of the proposed multiscale model.

It is important to note that we have not introduced the relaxation of the kinematical constraint within the power balance established by the PMVP. Freeing this constraint can only be done once the equilibrium problem at each scale is to be tackled. In other words, the PMVP takes us to the microscale equilibrium problem, and this problem can be solved in two alternative ways, with or without freeing the incompressibility constraint. That entails that the release of this kinematical constraint does not play a role in the concept of the PMVP, since it only intervenes in the manner to find the microscale equilibrium.

14.3.4 Incompressibility and Material Configuration

In this section we skip the treatment of the equilibrium problem and will exclusively focus on the stress homogenization formula.

Let $\mathbf{P}_M$ and $\mathbf{P}_\mu$ be the Piola–Kirchhoff stress tensors of the first kind at the macroscale and microscale, respectively. We utilize the notation $\hat{\mathbf{G}}_M|_{\mathbf{x}_m}$ to describe the gradient (material) of the motion action described in the material configuration, and we recall that the following identity holds for every macroscale point $\mathbf{x}_m$ in the reference configuration

$$[\hat{\mathbf{G}}_M|_{\mathbf{x}_s}]_m = \hat{\mathbf{G}}_M|_{\mathbf{x}_m}(\mathbf{F}_M|_{\mathbf{x}_m})^{-1}, \tag{14.121}$$

that is

$$[\hat{\mathbf{G}}_M|_{\mathbf{x}_s}]_m \mathbf{F}_M|_{\mathbf{x}_m} = \hat{\mathbf{G}}_M|_{\mathbf{x}_m}, \tag{14.122}$$

where $\hat{\mathbf{G}}_M|_{\mathbf{x}_s}$ is the macroscale gradient in its spatial description and $\mathbf{F}_M|_{\mathbf{x}_m} = \mathbf{I} + \mathbf{G}_M|_{\mathbf{x}_m}$ is the deformation gradient at the macroscale point $\mathbf{x}_m$.

By virtue of the microscale incompressibility we have

$$\begin{aligned} \mathrm{tr}\hat{\mathbf{G}}_M|_{\mathbf{x}_s} &= \mathbf{I}\cdot\hat{\mathbf{G}}_M|_{\mathbf{x}_s} = \mathbf{I}\cdot[\hat{\mathbf{G}}_M|_{\mathbf{x}_s}]_m = \mathbf{I}\cdot\hat{\mathbf{G}}_M|_{\mathbf{x}_m}(\mathbf{F}_M|_{\mathbf{x}_m})^{-1} \\ &= (\mathbf{F}_M|_{\mathbf{x}_m})^{-T}\cdot\hat{\mathbf{G}}_M|_{\mathbf{x}_m} = 0. \end{aligned} \tag{14.123}$$

Because we want to characterize the stress homogenization formula in the reference configuration, we take the stress homogenization formula delivered by the PMVP written in the material configuration (14.36) (disregarding volume forces). In this case, the PMVP yields

$$\begin{aligned} &\mathbf{P}_M|_{\mathbf{x}_m}\cdot\hat{\mathbf{G}}_M|_{\mathbf{x}_m} \\ &= \frac{1}{|\Omega_\mu^m|}\int_{\Omega_\mu^m}\mathbf{P}_\mu\cdot(\hat{\mathbf{G}}_M|_{\mathbf{x}_m} + \nabla_{\mathbf{x}_m}\hat{\mathbf{u}}_\mu)\,d\Omega_\mu^m \quad \forall(\hat{\mathbf{G}}_M|_{\mathbf{x}_m}, \hat{\mathbf{u}}_\mu) \in \mathbb{T}^{\mathbf{x}}_{\mathscr{W}_M}(\mathbf{G}_M|_{\mathbf{x}_m}) \times \mathrm{Var}_{\tilde{u}_\mu}, \end{aligned} \tag{14.124}$$

where

$$\mathbb{T}^{\mathbf{x}}_{\mathscr{W}_M}(\mathbf{G}_M|_{\mathbf{x}_m}) = \{\mathbf{H} \in \mathbb{R}^{\mathbf{x}}_{\mathscr{W}_M};\ (\mathbf{I} + \mathbf{G}_M|_{\mathbf{x}_m})^{-T}\cdot\mathbf{H} = 0\}, \tag{14.125}$$

with

$$\mathbb{R}^{\mathbf{x}}_{\mathscr{W}_M} = \{\mathbf{H} \in \mathbb{R}^{3\times3};\ \mathbf{H} = \mathbf{G}_M|_{\mathbf{x}_m},\ \mathbf{G}_M \in \mathscr{W}_M\}. \tag{14.126}$$

Observe that $\mathbb{T}^{\mathbf{x}}_{\mathscr{W}_M}$ is a space depending on $\mathbf{G}_M|_{\mathbf{x}_m}$, which is totally consistent because it is the tangent space of a nonlinear manifold.

Then, concentrating attention only in the derivation of the homogenization formula, and using (14.122), leads to

$$\int_{\Omega_\mu^m} \mathbf{P}_\mu \cdot \hat{\mathbf{G}}_M|_{\mathbf{x}_m} \, d\Omega_\mu^m = \left(\int_{\Omega_\mu^m} \mathbf{P}_\mu \, d\Omega_\mu^m \right) \cdot [\hat{\mathbf{G}}_M|_{\mathbf{x}_s}]_m \mathbf{F}_M|_{\mathbf{x}_m}$$
$$= \left(\int_{\Omega_\mu^m} \mathbf{P}_\mu \, d\Omega_\mu^m \right) (\mathbf{F}_M|_{\mathbf{x}_m})^T \cdot [\hat{\mathbf{G}}_M|_{\mathbf{x}_s}]_m$$
$$= \left[\left(\int_{\Omega_\mu^m} \mathbf{P}_\mu \, d\Omega_\mu^m \right) (\mathbf{F}_M|_{\mathbf{x}_m})^T \right]^D \cdot [\hat{\mathbf{G}}_M|_{\mathbf{x}_s}]_m, \tag{14.127}$$

where the last equality is attained by making use of expression (14.123), which establishes that tensor $[\hat{\mathbf{G}}_M|_{\mathbf{x}_s}]_m$ has null trace. Nevertheless, in (14.124) we have the tensor $\hat{\mathbf{G}}_M|_{\mathbf{x}_m}$, and so, using again (14.122), we manipulate the previous expression in the following manner

$$\int_{\Omega_\mu^m} \mathbf{P}_\mu \cdot \hat{\mathbf{G}}_M|_{\mathbf{x}_m} \, d\Omega_\mu^m$$
$$= \left[\left(\int_{\Omega_\mu^m} \mathbf{P}_\mu \, d\Omega_\mu^m \right) (\mathbf{F}_M|_{\mathbf{x}_m})^T \right]^D \cdot [\hat{\mathbf{G}}_M|_{\mathbf{x}_s}]_m$$
$$= \left[\left(\int_{\Omega_\mu^m} \mathbf{P}_\mu \, d\Omega_\mu^m \right) (\mathbf{F}_M|_{\mathbf{x}_m})^T \right]^D \cdot \hat{\mathbf{G}}_M|_{\mathbf{x}_m} (\mathbf{F}_M|_{\mathbf{x}_m})^{-1}$$
$$= \left[\left(\int_{\Omega_\mu^m} \mathbf{P}_\mu \, d\Omega_\mu^m \right) (\mathbf{F}_M|_{\mathbf{x}_m})^T \right]^D (\mathbf{F}_M|_{\mathbf{x}_m})^{-T} \cdot \hat{\mathbf{G}}_M|_{\mathbf{x}_m}. \tag{14.128}$$

Putting this result into the PMVP given by (14.124), and considering $\hat{\mathbf{u}}_\mu = \mathbf{0}$, we are led to

$$\left(\mathbf{P}_M|_{\mathbf{x}_m} - \frac{1}{|\Omega_\mu^m|} \left[\left(\int_{\Omega_\mu^m} \mathbf{P}_\mu \, d\Omega_\mu^m \right) (\mathbf{F}_M|_{\mathbf{x}_m})^T \right]^D (\mathbf{F}_M|_{\mathbf{x}_m})^{-T} \right) \cdot \hat{\mathbf{G}}_M|_{\mathbf{x}_m} = 0$$
$$\forall \hat{\mathbf{G}}_M|_{\mathbf{x}_m} \in \mathbb{T}^{\mathbf{x}}_{\mathscr{W}_M}(\mathbf{G}_M|_{\mathbf{x}_m}). \tag{14.129}$$

Correspondingly, since the macroscale medium is regarded as incompressible, by employing (14.121) the following relation is verified

$$\sigma_M^D|_{\mathbf{x}_s} \cdot \hat{\mathbf{G}}_M^{sD}|_{\mathbf{x}_s} = \sigma_M^D|_{\mathbf{x}_s} \cdot \hat{\mathbf{G}}_M|_{\mathbf{x}_s} = \sigma_M^D|_{\mathbf{x}_s} \cdot [\hat{\mathbf{G}}_M|_{\mathbf{x}_m} (\mathbf{F}_M|_{\mathbf{x}_m})^{-1}]_s, \tag{14.130}$$

and in the material configuration this is

$$[\sigma_M^D|_{\mathbf{x}_s} \cdot \hat{\mathbf{G}}_M^{sD}|_{\mathbf{x}_s}]_m = [\sigma_M^D|_{\mathbf{x}_s}]_m \cdot \hat{\mathbf{G}}_M|_{\mathbf{x}_m} (\mathbf{F}_M|_{\mathbf{x}_m})^{-1}$$
$$= [\sigma_M^D|_{\mathbf{x}_s}]_m (\mathbf{F}_M|_{\mathbf{x}_m})^{-T} \cdot \hat{\mathbf{G}}_M|_{\mathbf{x}_m}. \tag{14.131}$$

Consequently, introducing the Piola–Kirchhoff stress tensor of the first kind which is associated with $\sigma_M^D|_{\mathbf{x}_s}$, which is denoted by $\mathbf{P}_M^d|_{\mathbf{x}_m}$, we have

$$\det \mathbf{F}_M|_{\mathbf{x}_m} [\sigma_M^D|_{\mathbf{x}_s}]_m (\mathbf{F}_M|_{\mathbf{x}_m})^{-T} \cdot \hat{\mathbf{G}}_M|_{\mathbf{x}_m} = \mathbf{P}_M^d|_{\mathbf{x}_m} \cdot \hat{\mathbf{G}}_M|_{\mathbf{x}_m}. \tag{14.132}$$

Putting this result into (14.129) gives

$$\left[\det \mathbf{F}_M|_{\mathbf{x}_m}[\sigma_M|_{\mathbf{x}_s}]_m - \frac{1}{|\Omega_\mu^m|}\left(\int_{\Omega_\mu^m}\mathbf{P}_\mu\, d\Omega_\mu^m\right)(\mathbf{F}_M|_{\mathbf{x}_m})^T\right]^D \cdot \hat{\mathbf{G}}_M|_{\mathbf{x}_m}(\mathbf{F}_M|_{\mathbf{x}_m})^{-1} = 0$$
$$\forall \hat{\mathbf{G}}_M|_{\mathbf{x}_m} \in \mathbb{T}^{\mathbf{x}}_{\mathscr{W}_M}(\mathbf{G}_M|_{\mathbf{x}_m}). \tag{14.133}$$

This implies

$$\det \mathbf{F}_M|_{\mathbf{x}_m}[\sigma^D_M|_{\mathbf{x}_s}]_m - \frac{1}{|\Omega_\mu^m|}\left[\left(\int_{\Omega_\mu^m}\mathbf{P}_\mu\, d\Omega_\mu^m\right)(\mathbf{F}_M|_{\mathbf{x}_m})^T\right]^D = \alpha\mathbf{I}. \tag{14.134}$$

Nevertheless, since the left-hand side in the previous expression is the sum of two deviatoric tensors, this means that it has to necessarily be $\alpha = 0$. Thus, we arrive at the stress homogenization formula in the material configuration for the case of an incompressible macroscale medium

$$\begin{aligned}\mathbf{P}^d_M|_{\mathbf{x}_m} &= \det \mathbf{F}_M|_{\mathbf{x}_m}[\sigma^D_M|_{\mathbf{x}_s}]_m(\mathbf{F}_M|_{\mathbf{x}_m})^{-T}\\ &= \frac{1}{|\Omega_\mu^m|}\left[\left(\int_{\Omega_\mu^m}\mathbf{P}_\mu\, d\Omega_\mu^m\right)(\mathbf{F}_M|_{\mathbf{x}_m})^T\right]^D (\mathbf{F}_M|_{\mathbf{x}_m})^{-T}.\end{aligned} \tag{14.135}$$

Let us make a quick comparison between the previous formula and that of (14.116). In order to do this, from the previous expression, and considering the relation between the Cauchy and Piola–Kirchhoff stresses at the microscale, we obtain the following

$$\begin{aligned}[\sigma^D_M|_{\mathbf{x}_s}]_m &= \frac{1}{\det \mathbf{F}_M|_{\mathbf{x}_m}|\Omega_\mu^m|}\left[\left(\int_{\Omega_\mu^m}\mathbf{P}_\mu\, d\Omega_\mu^m\right)(\mathbf{F}_M|_{\mathbf{x}_m})^T\right]^D\\ &= \frac{1}{\det \mathbf{F}_M|_{\mathbf{x}_m}|\Omega_\mu^m|}\left[\left(\int_{\Omega_\mu^m}\det \mathbf{F}_\mu[\sigma_\mu]_m\mathbf{F}_\mu^{-T}\, d\Omega_\mu^m\right)(\mathbf{F}_M|_{\mathbf{x}_m})^T\right]^D.\end{aligned} \tag{14.136}$$

It can be appreciated that, because (14.136) differs from (14.116), both models deliver different stress homogenization formulae in the spatial and material configurations. This is coherent with the fact that the transformation between these configurations entails a nonlinear mapping. Hence, the application of the multiscale approach in the material configuration subtly differs from the application in the spatial configuration.

For the particular case in which the microscale is also an incompressible medium, by utilizing an analogous reasoning it is possible to arrive at the following homogenization formula

$$\mathbf{P}^d_M|_{\mathbf{x}_m} = \frac{1}{|\Omega_\mu^m|}\left[\left(\int_{\Omega_\mu^m}\mathbf{P}^d_\mu\, d\Omega_\mu^m\right)(\mathbf{F}_M|_{\mathbf{x}_m})^T\right]^D (\mathbf{F}_M|_{\mathbf{x}_m})^{-T}, \tag{14.137}$$

where we have used the definitions $\mathbf{P}^d_M|_{\mathbf{x}_m} = \det \mathbf{F}_M|_{\mathbf{x}_m}[\sigma^D_M|_{\mathbf{x}_s}]_m(\mathbf{F}_M|_{\mathbf{x}_m})^{-T}$ and $\mathbf{P}^d_\mu = \det \mathbf{F}_\mu[\sigma^D_\mu]_m\mathbf{F}_\mu^{-T}$.

14.4 Final Remarks

As was mentioned at the beginning of this chapter, and also in the previous chapter, we strongly recommend reading the scientific papers [32–35, 121, 266, 267, 275–278, 296–298], in which the application of the Method of Multiscale Virtual Power is reported for diverse problems arising in solid and fluid mechanics. By doing this, the reader will have a complete vision of the potential and versatility of the RVE-based multiscale approach for the modeling of complex physical systems. As pointed out throughout this part of the book, the Method of Multiscale Virtual Power is rooted in variational grounds which enable the modeler to clearly define the basic hypotheses in the model, and to be unambiguously led to the consequences of such hypotheses, removing any sort of arbitrary and/or unnecessary assumptions.

Part VI

Appendices

A

Definitions and Notations

A.1 Introduction

In order to make the presentation of this book more self-contained, this appendix briefly presents some concepts and results from linear algebra, real analysis, and convex analysis, as well as some notations that are utilized throughout this book. We also provide bibliographic references so that the reader can get a more rigorous grasp of the mathematical concepts involved in the different topics, which are relevant for a better understanding of the variational formulation in mechanics.

A.2 Sets

A set $\mathscr{B}$ is a collection of elements b. If b is an element of the set $\mathscr{B}$ we say that b belongs to $\mathscr{B}$, and is represented by $b \in \mathscr{B}$. With $b \notin \mathscr{B}$ we indicate that the element b does not belong to the set $\mathscr{B}$.

An alternative manner to define a set $\mathscr{B}$ is through the introduction of a referential set $\mathscr{C}$, whose elements belong to the set $\mathscr{B}$ if they satisfy a certain property $\mathscr{P}$. This is represented in the following way

$$\mathscr{B} = \{c \in \mathscr{C};\ c \text{ satisfies } \mathscr{P}\}, \tag{A.1}$$

which is read as follows: the set $\mathscr{B}$ formed by all the elements from $\mathscr{C}$ such that satisfy the property $\mathscr{P}$.

We say that a set is finite if it contains a finite number of elements, otherwise we say that the set is infinite. A set $\mathscr{B}$ is enumerable if it is possible to establish a biunivocal correspondence between $\mathscr{B}$ and any subset from $\mathbb{N}$ ($\mathbb{N}$ is the set of natural numbers). Any enumerable set can be finite or infinite. A non-enumerable set is an infinite set which is not enumerable.

We say that $\mathscr{A}$ is a subset from $\mathscr{B}$, $\mathscr{A} \subseteq \mathscr{B}$, if all the elements from $\mathscr{A}$ belong to $\mathscr{B}$. In particular, if $\mathscr{B}$ has elements that do not belong to $\mathscr{A}$ we say that $\mathscr{A}$ is a proper subset, and in this case it is represented by $\mathscr{A} \subset \mathscr{B}$. In turn, two sets $\mathscr{A}$ and $\mathscr{B}$ are said to be equal if it is verified that

$$\mathscr{A} \subset \mathscr{B}, \qquad \mathscr{B} \subset \mathscr{A}. \tag{A.2}$$

Introduction to the Variational Formulation in Mechanics: Fundamentals and Applications, First Edition.
Edgardo O. Taroco, Pablo J. Blanco and Raúl A. Feijóo.

The union, intersection and difference of two sets $\mathscr{A}$ and $\mathscr{B}$ is represented, respectively, by $\mathscr{A} \cup \mathscr{B}$, $\mathscr{A} \cap \mathscr{B}$ and $\mathscr{A} \setminus \mathscr{B}$, where

$$\mathscr{A} \cup \mathscr{B} = \{x;\ x \in \mathscr{A} \text{ or } x \in \mathscr{B}\}, \tag{A.3}$$

$$\mathscr{A} \cap \mathscr{B} = \{x;\ x \in \mathscr{A} \text{ and } x \in \mathscr{B}\}, \tag{A.4}$$

$$\mathscr{A} \setminus \mathscr{B} = \{x;\ x \in \mathscr{A} \text{ and } x \notin \mathscr{B}\}. \tag{A.5}$$

Another concept used is that of Cartesian product. Let $\mathscr{A}$ and $\mathscr{B}$ be two non-empty sets (they have at least one element), the Cartesian product of $\mathscr{A}$ and $\mathscr{B}$, which is represented by $\mathscr{A} \times \mathscr{B}$, is the set of ordered pairs (a, b) such that $a \in \mathscr{A}$ and $b \in \mathscr{B}$, that is

$$\mathscr{A} \times \mathscr{B} = \{(a, b);\ a \in \mathscr{A} \text{ and } b \in \mathscr{B}\}. \tag{A.6}$$

By ordered pair we mean that (a, b) and (a', b') are equal if and only if $a = a'$ and $b = b'$.

A.3 Functions and Transformations

A function, application or transformation, consists of

i) a set $\mathscr{A}$
ii) a set $\mathscr{B}$, not necessarily different from $\mathscr{A}$
iii) a rule, or correspondence, f that allows each element from $\mathscr{A}$ to be associated with a single element from $\mathscr{B}$.

In this case we say that f is a $\mathscr{B}$-valued function in $\mathscr{A}$, or that f is a function defined in $\mathscr{A}$ with value in $\mathscr{B}$. The terms transformation, or operator, are employed as synonyms for the term function. We also say that f is a function that transforms, or applies, $\mathscr{A}$ in $\mathscr{B}$. This is indicated in the following manner

$$f : \mathscr{A} \to \mathscr{B}. \tag{A.7}$$

The set of all the elements of $\mathscr{A}$ for which f is defined is called the domain of f, and it is denoted by $D(f)$.

Let $f : \mathscr{A} \to \mathscr{B}$ then, given $x \in D(f)$, the element $y \in \mathscr{B}$ associated with x through the function f is denoted indistinctly by

$$y = f(x) \qquad \text{or} \qquad f : x \mapsto y, \tag{A.8}$$

and is termed the value of f at x, or the image of x under f. In turn, x is also called a preimage of y under f. Note that we put *a preimage*, instead of *the preimage.*

The set of all values that f is able to take in $\mathscr{B}$ is called the image (or codomain) of $\mathscr{B}$ through f. This is represented as follows

$$R(f) = \{y \in \mathscr{B};\ \exists x \in D(f) \text{ such that } f(x) = y\}. \tag{A.9}$$

Let $f : \mathscr{A} \to \mathscr{B}$, then f is

- injective: if for different elements in $\mathscr{A}$, we have different elements in $\mathscr{B}$, that is,

  $$f(x_1) = f(x_2) \qquad \Longrightarrow \qquad x_1 = x_2, \tag{A.10}$$

  or, alternatively, f is injective if and only if every element $y \in R(f)$ has one and only one preimage x in $D(f)$

- surjective: or also that f applies $\mathscr{A}$ over $\mathscr{B}$, if $R(f) = \mathscr{B}$
- bijective: if it is both injective and surjective.

For example, let $\mathbb{R}$ be the set of all real numbers and let $f : \mathbb{R} \to \mathbb{R}^+$ ($\mathbb{R}^+$ is the set of all non-negative real numbers) be a function such that $f : x \mapsto x^2$. As can be seen, this function is not injective because the number 4 has two preimages (2 and −2). Let now $f : \mathbb{R}^+ \mapsto \mathbb{R}^+$ be the same function defined before, then f is bijective.

Another interesting example in mechanics is the application that allows us to go from a configuration of the body to another different configuration for the same body, we denote this by $\mathscr{X}_t : \mathscr{B} \to \mathscr{B}_t$, where $\mathscr{B}$ is the set that can be identified with a reference configuration of the body, while $\mathscr{B}_t$ is the set which is identified with the configuration of the body at time t. In this case, $\mathscr{X}_t$ describes the motion of the body. As seen in Chapter 3, this application must be bijective.

The surjective application $I : \mathscr{A} \to \mathscr{A}$ defined by

$$\begin{aligned} I : \mathscr{A} &\to \mathscr{A}, \\ x &\mapsto I(x) = x, \end{aligned} \tag{A.11}$$

is called the identity function in $\mathscr{A}$.

If $f : \mathscr{A} \to \mathscr{B}$ and $g : \mathscr{A} \to \mathscr{B}$ are such that $f(x) = g(x)$ for all $x \in \mathscr{A}$ then we say that both functions are the same ($f = g$). Here we have to recall that function must not be confounded with representation of the function. Hence, for example, $f(x) = |x|$ and $g(x) = \exp^{\log |x|}$ are representations of the same function.

Let $f : \mathscr{X} \to \mathscr{Y}$ be a function with domain $D(f)$. Function g such that $D(f) \subset D(g)$ and $f(x) = g(x)$ for all $x \in D(f)$ is an extension of f. In turn, if $D(f) = \mathscr{X}$ and $\mathscr{A}$ is a subset of $\mathscr{X}$, function h such that $D(h) = \mathscr{A}$ and $h(x) = f(x)$ for all $x \in \mathscr{A}$ is a restriction of f in $\mathscr{A}$. This last operation is denoted by $h = f|_{\mathscr{A}}$.

Let $f : \mathscr{A} \to \mathscr{B}$ and $g : \mathscr{B} \to \mathscr{C}$ such that $R(f) \cap D(g) \neq \emptyset$, then we can introduce the function $g \circ f$ defined by

$$\begin{aligned} g \circ f : \mathscr{A} &\to \mathscr{C}, \\ a &\mapsto (g \circ f)(a) = g(f(a)), \end{aligned} \tag{A.12}$$

with domain given by

$$D(g \circ f) = \{x \in D(f);\ f(x) \in D(g)\}, \tag{A.13}$$

and image given by

$$R(g \circ f) = \{y \in R(g);\ y = g(f(x)),\ x \in D(g \circ f)\}. \tag{A.14}$$

Now, the function $g \circ f$ is termed the composition of f with g (it is usually said that f is composed with g).

If g and f are bijective, and there exists $g \circ f$, then $g \circ f$ is also bijective. From the definition it follows that the composition is not commutative because in general $f \circ g$ is not necessarily defined. Composition is also associative, that is

$$r \circ (g \circ f) = (r \circ g) \circ f. \tag{A.15}$$

In what follows we interchangeably employ the notation $g \circ f$ and gf to represent the composition of f with g.

The transformation $f : \mathcal{X} \to \mathcal{Y}$ is invertible if there exists an application $g : \mathcal{Y} \to \mathcal{X}$ such that gf and fg are the identity functions in $\mathcal{X}$ and $\mathcal{Y}$, respectively. In this case we say that g is the inverse of f.

In the following we provide relevant consequences involving the inverse function.

If the function $f : \mathcal{X} \to \mathcal{Y}$ is invertible then the inverse is unique.

This result can easily be proven assuming that there exist different functions g_1 and g_2 ($g_1 \neq g_2$) which are both inverses of f. Let y be an arbitrary element in $\mathcal{Y}$. Since fg_1 and g_2f represent the identity function in $\mathcal{Y}$ and in $\mathcal{X}$, respectively, we get

$$g_2(y) = g_2(fg_1(y)) = g_2f(g_1(y)) = g_1(y), \tag{A.16}$$

then

$$g_2 = g_1. \tag{A.17}$$

We have arrived at a contradiction, as a consequence of the assumption that there are two different inverse functions. Therefore, the inverse is unique.

Hereafter, the inverse of function f is simply denoted by f^{-1}.

If the function $f : \mathcal{X} \to \mathcal{Y}$ is invertible, then f^{-1} is also invertible, and $(f^{-1})^{-1} = f$. The validity of this theorem is straightforward.

If the function $f : \mathcal{X} \to \mathcal{Y}$ is invertible, then it is injective.

Suppose that f is invertible. We have to prove that f is injective, that is, for every $y \in \mathcal{Y}$ there is a unique preimage $x \in \mathcal{X}$. Consider the opposite, that is, there exist two preimages x_1 and x_2, that is

$$f(x_1) = f(x_2) = y. \tag{A.18}$$

Since f is invertible

$$x_1 = f^{-1}f(x_1) = f^{-1}f(x_2) = x_2. \tag{A.19}$$

That is, $x_1 = x_2$. Then f is injective.

If the function $f : \mathcal{X} \to \mathcal{Y}$ is invertible, then $R(f) = \mathcal{Y}$.

Let $y \in \mathcal{Y}$ be an arbitrary element, then

$$y = ff^{-1}(y) = f(x) \implies y \in R(f) \quad \forall y \in \mathcal{Y}, \tag{A.20}$$

which implies

$$R(f) = \mathcal{Y}, \tag{A.21}$$

and the result follows.

Now note that the necessary and sufficient condition for $f : \mathcal{X} \to \mathcal{Y}$ to be invertible is that f is bijective.

The results presented above provide the proof for the necessary condition. To demonstrate the sufficient condition suppose that f is bijective and let $y \in \mathcal{Y}$ be an arbitrary element. This corresponds to the case $R(f) = \mathcal{Y}$ and to assume that f is injective. Then, for every $y \in \mathcal{Y}$ there exists a unique preimage $x \in \mathcal{X}$. Let us call g the function that relates every $y \in \mathcal{Y}$ with its unique preimage $x \in \mathcal{X}$, that is, $g(y) = x$. Then, gf and fg are the identity functions in $\mathcal{X}$ and $\mathcal{Y}$, respectively. Therefore, f is invertible.

From the previous result we conclude that if g and f are two bijective functions such that the composition gf is defined, then

$$(g \circ f)^{-1} = f^{-1} \circ g^{-1}. \tag{A.22}$$

A.4 Groups

Let $\mathscr{P}$ be a set in which we define an internal operation (represented by $*$) which satisfies the properties

i) for $x, y \in \mathscr{P}$ then, $x * y \in \mathscr{P}$
ii) for $x, v, z \in \mathscr{P}$ then the associative property is verified, that is

$$x * v * z = x * (v * z) = (x * v) * z, \tag{A.23}$$

iii) there exists an element $i \in \mathscr{P}$, called the identity element, such that

$$i * x = x \quad \forall x \in \mathscr{P}, \tag{A.24}$$

iv) for an arbitrary $x \in \mathscr{P}$, there exists $x^{-1} \in \mathscr{P}$, called the inverse element of x, such that

$$x^{-1} * x = i, \tag{A.25}$$

then, we say that $\mathbf{P} = (\mathscr{P}, *)$ is a group.

For example, consider the set of natural numbers $\mathbb{N}$ and the internal operation $+$, where $+$ represents the usual addition operation between natural numbers, then $(\mathbb{N}, +)$ is a group. In contrast, $(\mathbb{N}, \cdot)$, where $\cdot$ is the usual multiplication operation between natural numbers, is not a group.

Let us present another example. The set of all bijective functions that apply a set over the very same set with the internal operation $\circ$ (composition operation) form a group.

A group $\mathscr{X}$ is said to be Abelian if the internal operation satisfies the symmetry, or commutativity, property, that is

$$x * y = y * x \qquad \forall x, y \in \mathscr{X}. \tag{A.26}$$

In an Abelian group, the internal operation is usually denoted with the symbol $+$. In this manner, the properties of this internal operation are

$$x + y + z = x + (y + z) = (x + y) + z \quad \text{associativity}, \tag{A.27}$$

$$x + y = y + x \quad \text{commutativity}. \tag{A.28}$$

In general, in Abelian groups we denote the identity element by 0 and the inverse element of x by $-x$, that is

$$x + 0 = x, \tag{A.29}$$

$$x + (-x) = 0. \tag{A.30}$$

In this case we also say that the group is additive.

The binary operation that allows us to define the algebraic structure of a group within a set must satisfy the four properties formulated as axioms at the beginning of this section. In particular, representing this operation as the usual operations of multiplication, addition, etc. characterizes the nature of the elements of the set.

A group can contain a finite number or an infinite number of elements. If the group is finite, the number of elements is called the order of the group, or the cardinality of the group.

A subgroup of a group is formed by a subset of the set that defines the group and inherits the same internal operation from the group. The concept of a proper subgroup is similar to that of a proper subset.

Consider now the relation

$$\bar{z} = w * z * w^{-1}, \tag{A.31}$$

where $z, w \in \mathbf{P} = (\mathscr{P}, *)$, then $\bar{z}$ also belongs to $\mathbf{P}$. If we consider w as a fixed element in $\mathbf{P}$ and z an arbitrary element in $\mathbf{P}$, the previous expression univocally defines $\bar{z}$ in terms of z. From this point of view, the expression (A.31) defines a bijective application from the group $\mathbf{P}$ onto itself, and the application preserves the relations of the group.

In fact, consider for example

$$\bar{c} * \bar{z} = (w * c * w^{-1}) * (w * z * w^{-1}) = w * c * z * w^{-1} = \overline{c * z}. \tag{A.32}$$

In turn, given a subgroup $\mathbf{S}$ from $\mathbf{P}$, the application (A.31) transforms $\mathbf{S}$ in another subgroup $\bar{\mathbf{S}}$ called the conjugate to $\mathbf{S}$. This transformation is symbolically indicated by

$$\bar{\mathbf{S}} = w\mathbf{S}w^{-1} \quad \mathbf{S} \subset \mathbf{P} \text{ and } w \in \mathbf{P}. \tag{A.33}$$

If a group $\mathbf{S}$ is equal to its conjugate group $\bar{\mathbf{S}}$, that is

$$\mathbf{S} = w\mathbf{S}w^{-1} \qquad \forall w \in \mathbf{P}, \tag{A.34}$$

or its equivalent

$$\mathbf{S}w = w\mathbf{S}, \tag{A.35}$$

we say that $\mathbf{S}$ is an invariant subgroup, or a self-conjugate subgroup. Notice that (A.34) does not imply that

$$z = w * z * w^{-1} \quad \forall z \in \mathbf{S}, \tag{A.36}$$

but simply that

$$\bar{z} = w * z * w^{-1} \in \mathbf{S} \quad \forall z \in \mathbf{S}. \tag{A.37}$$

If the elements $f_1, f_2, \dots, f_n$, with n a finite number belong to a group $\mathbf{P}$, the group of all the elements formed through the internal operation realized among all these elements and with the corresponding inverses is a subgroup of $\mathbf{P}$ called the subgroup generated by $\{f_i\}_{i \leq n}$.

A subgroup generated with a single element f (in this case $n = 1$) is called cyclic. In this case the elements

$$i, f, f * f, f * f * f, \dots, f^{-1}, f^{-1} * f^{-1}, f^{-1} * f^{-1} * f^{-1}, \dots, \tag{A.38}$$

are called the powers of f, and we can formally write

$$f^0 = i, f^1 = f, f^2 = f * f, \dots, f^{-2} = f^{-1} * f^{-1}, \dots. \tag{A.39}$$

From the very definition, in a cyclic group it is verified that

$$f^m * f^n = f^n * f^m = f^{n+m}, \tag{A.40}$$

$$(f^m)^n = (f^n)^m = f^{nm}. \tag{A.41}$$

As a consequence, we have that every cyclic group is Abelian.

A.5 Morphisms

Let $(\mathscr{A}, \bullet)$ and $(\mathscr{B}, *)$ be two groups. The function

$$f : \mathscr{A} \to \mathscr{B}, \tag{A.42}$$

that does not alter these structures is called morphism, or homomorphism.

A bijective morphism is called isomorphism. In this case we say that $\mathscr{A}$ and $\mathscr{B}$ are isomorphic.

A morphism whose domain and codomain are the same set ($\mathscr{A} = \mathscr{B}$) is called as endomorphism. Finally, a bijective endomorphism is called automorphism.

A.6 Vector Spaces

A real vector space (or linear real space) is a set V formed by elements called vectors and is provided by two operations

i) an internal operation

$$A : V \times V \to V, \tag{A.43}$$

called addition, and usually represented by

$$A(\mathbf{u}, \mathbf{v}) = \mathbf{u} + \mathbf{v}, \tag{A.44}$$

ii) an external operation

$$M : \mathbb{R} \times V \to V, \tag{A.45}$$

called multiplication by scalar, and generally represented by

$$M(\alpha, \mathbf{u}) = \alpha\mathbf{u}. \tag{A.46}$$

These two operations satisfy the following axioms

i) the addition is commutative

$$\mathbf{u} + \mathbf{v} = \mathbf{v} + \mathbf{u} \qquad \forall \mathbf{u}, \mathbf{v} \in V, \tag{A.47}$$

ii) the addition is associative

$$\mathbf{u} + (\mathbf{v} + \mathbf{w}) = (\mathbf{u} + \mathbf{v}) + \mathbf{w} \qquad \forall \mathbf{u}, \mathbf{v}, \mathbf{w} \in V, \tag{A.48}$$

iii) there exists the null element in V, denoted by $\mathbf{0}$, called the null vector of V, such that

$$\mathbf{u} + \mathbf{0} = \mathbf{u} \qquad \forall \mathbf{u} \in V, \tag{A.49}$$

iv) for each $\mathbf{u} \in V$, there is a vector, $-\mathbf{u}$, called the opposite vector, or the negative of $\mathbf{u}$, such that

$$\mathbf{u} + (-\mathbf{u}) = \mathbf{0} \qquad \forall \mathbf{u} \in V, \tag{A.50}$$

v) the multiplication by scalar is associative

$$\alpha(\beta\mathbf{u}) = (\alpha\beta)\mathbf{u} \qquad \forall \mathbf{u} \in V, \tag{A.51}$$

vi) the multiplication by scalar is distributive with respect to the addition by scalar

$$(\alpha + \beta)\mathbf{u} = \alpha\mathbf{u} + \beta\mathbf{u} \qquad \forall \alpha, \beta \in \mathbb{R},\ \forall \mathbf{u} \in V, \tag{A.52}$$

vii) the multiplication by scalar is distributive with respect to the addition of vectors

$$\alpha(\mathbf{u} + \mathbf{v}) = \alpha\mathbf{u} + \alpha\mathbf{v} \qquad \forall \alpha \in \mathbb{R},\ \forall \mathbf{u}, \mathbf{v} \in V. \tag{A.53}$$

Notice that the seven properties enumerated above are not independent from each other.

The addition of vectors together with the four axioms introduced at the beginning of the section defines an additive structure.

With the previous elements we can introduce a third operation, called subtraction, which is defined in the following manner

$$\begin{aligned} S &: V \times V \to V, \\ &(\mathbf{u}, \mathbf{v}) \mapsto S(\mathbf{u}, \mathbf{v}) = \mathbf{u} - \mathbf{v} = \mathbf{u} + (-\mathbf{v}). \end{aligned} \tag{A.54}$$

All these operations (addition, multiplication by scalar, and subtraction) are also called linear operations.

Finally, given the vectors $\mathbf{v}_1, \mathbf{v}_2, \dots \mathbf{v}_n \in V$ and the scalars $a_1, a_2, \dots a_n \in \mathbb{R}$ we introduce the vector $\mathbf{v} = a_1\mathbf{v}_1 + a_2\mathbf{v}_2 + \cdots + a_n\mathbf{v}_n$ which is called the linear combination of vectors $\mathbf{v}_i$, $i = 1, \dots n$.

An important example is the vector space (also called the n-dimensional Euclidean space) denoted by $\mathbb{R}^n = \{\mathbf{v};\ \mathbf{v} = (v_1, v_2, \dots, v_n),\ v_i \in \mathbb{R}\}$ where the operations of addition and multiplication by scalar are defined

$$\mathbf{u} + \mathbf{v} = (u_1 + v_1, u_2 + v_2, \dots, u_n + v_n), \tag{A.55}$$

$$a\mathbf{u} = (au_1, au_2, \dots, au_n). \tag{A.56}$$

A subset U of V is said to be a subspace of V if it is closed with respect to the linear combination. In other words, U is a subspace of V if every linear combination of vectors from U also belongs to U. From this definition it follows that if U is a subspace, then $\mathbf{0} \in U$. A subspace of $\mathbb{R}^n$ is, for example, the set of all vectors of the form $\mathbf{v} = (0, v_2, v_3, \dots, v_n)$.

From the previous definitions, the following results are direct easy-to-see consequences.

First, note that in every vector space there exists a unique null element. Indeed, the third axiom at the beginning of the section tells us that there exists at least one null vector element. Suppose that there are two null elements, called $\mathbf{0}_1$ and $\mathbf{0}_2$, which are different ($\mathbf{0}_1 \neq \mathbf{0}_2$). By taking $\mathbf{u} = \mathbf{0}_1$ and $\mathbf{0} = \mathbf{0}_2$, we obtain

$$\mathbf{0}_1 + \mathbf{0}_2 = \mathbf{0}_1. \tag{A.57}$$

Similarly, taking $\mathbf{u} = \mathbf{0}_2$ and $\mathbf{0} = \mathbf{0}_1$, gives

$$\mathbf{0}_2 + \mathbf{0}_1 = \mathbf{0}_2. \tag{A.58}$$

From the first axiom we have that addition is commutative, so

$$\mathbf{0}_1 + \mathbf{0}_2 = \mathbf{0}_2 + \mathbf{0}_1, \tag{A.59}$$

and so it follows that $\mathbf{0}_1 = \mathbf{0}_2$.

As a second result we have that if V is a vector space, then for all $\mathbf{u} \in V$ there exists a unique negative vector $\mathbf{v} = -\mathbf{u} \in V$ such that $\mathbf{u} + \mathbf{v} = \mathbf{0}$. In fact, from the fourth axiom we have that there exists at least one negative element. Assume that $\mathbf{u}$ possesses two negative elements, say $\mathbf{v}_1$ and $\mathbf{v}_2$, which are different ($\mathbf{v}_1 \neq \mathbf{v}_2$), then

$$\mathbf{u} + \mathbf{v}_1 = \mathbf{0}, \tag{A.60}$$

$$\mathbf{u} + \mathbf{v}_2 = \mathbf{0}. \tag{A.61}$$

Adding $\mathbf{v}_2$ to both sides of the first equation we get

$$\mathbf{v}_2 + (\mathbf{u} + \mathbf{v}_1) = \mathbf{v}_2 + \mathbf{0} = \mathbf{v}_2, \tag{A.62}$$

but, from the second axiom we have

$$\mathbf{v}_2 + (\mathbf{u} + \mathbf{v}_1) = (\mathbf{v}_2 + \mathbf{u}) + \mathbf{v}_1 = \mathbf{0} + \mathbf{v}_1 = \mathbf{v}_1, \tag{A.63}$$

and so $\mathbf{v}_1 = \mathbf{v}_2$.

Now, let V be a vector space. Then, for any $\mathbf{u}, \mathbf{v} \in V$ and $a, b \in \mathbb{R}$, it is easy to verify that

i) $0\mathbf{u} = \mathbf{0}$
ii) $a\mathbf{0} = \mathbf{0}$
iii) $(-a)\mathbf{u} = (-a\mathbf{u}) = a(-\mathbf{u})$
iv) if $a\mathbf{u} = \mathbf{0}$, then it can be either $a = 0$ or $\mathbf{u} = \mathbf{0}$
v) if $a\mathbf{u} = a\mathbf{v}$ and $a \neq 0$ then $\mathbf{u} = \mathbf{v}$
vi) if $a\mathbf{u} = b\mathbf{u}$ and $\mathbf{u} \neq \mathbf{0}$ then $a = b$
vii) $\mathbf{u} + \mathbf{u} = 2\mathbf{u}$, $\mathbf{u} + \mathbf{u} + \mathbf{u} = 3\mathbf{u}$ and in general $\sum_{i=1}^{n} \mathbf{u} = n\mathbf{u}$.

Let us quickly see the proof of results i)–iii). The rest is left as an exercise for the reader.

i) Consider $\mathbf{w} = 0\mathbf{u}$, we have to prove that $\mathbf{w} = \mathbf{0}$. To this end, let us add $\mathbf{w}$ to itself

$$\mathbf{w} + \mathbf{w} = 0\mathbf{u} + 0\mathbf{u} = (0 + 0)\mathbf{u} = 0\mathbf{u} = \mathbf{w}. \tag{A.64}$$

Subtracting $\mathbf{w}$ to both sides of the previous expression gives $\mathbf{w} = \mathbf{0}$.

ii) Consider $\mathbf{w} = a\mathbf{0}$, adding $\mathbf{w}$ to itself gives

$$\mathbf{w} + \mathbf{w} = a\mathbf{0} + a\mathbf{0} = a(\mathbf{0} + \mathbf{0}) = a\mathbf{0} = \mathbf{w}. \tag{A.65}$$

Subtracting $\mathbf{w}$ to both sides of the previous expression gives $\mathbf{w} = \mathbf{0}$.

iii) Consider $\mathbf{w} = (-a)\mathbf{u}$, adding $a\mathbf{u}$ to both sides of the equation yields

$$\mathbf{w} + a\mathbf{u} = (-a)\mathbf{u} + a\mathbf{u} = (-a + a)\mathbf{u} = 0\mathbf{u} = \mathbf{0}. \tag{A.66}$$

Hence, $\mathbf{w}$ is the negative vector of $a\mathbf{u}$, that is, $\mathbf{w} = -(a\mathbf{u})$. Therefore, adding $a(-\mathbf{u})$ to $a\mathbf{u}$ gives

$$a(-\mathbf{u}) + a\mathbf{u} = a(-\mathbf{u} + \mathbf{u}) = \mathbf{0}, \tag{A.67}$$

and therefore $(-a)\mathbf{u} = (-a\mathbf{u}) = a(-\mathbf{u})$.

Let B be a non-empty subset of a vector space V. The set of all combinations of elements in B is a subspace of V. We call this subspace spanned, or generated, subspace, and it is represented by span$B$. Many questions may appear at this point. For example, what kind of spaces can be generated by finite sets of elements? If a space can be generated by a finite set of elements, which is the smallest number of elements required? To answer these questions we will see in next sections the concepts of dependence, independence, bases, and dimension.

A.7 Sets and Dependence in Vector Spaces

Let $S = \{\mathbf{v}_1, \mathbf{v}_2, \dots, \mathbf{v}_k\}$ be a set of k vectors. We say that S is linearly independent if the equation

$$a_1\mathbf{v}_1 + a_2\mathbf{v}_2 + \dots + a_k\mathbf{v}_k = \mathbf{0} \qquad a_i \in \mathbb{R}, i = 1, \dots, k, \tag{A.68}$$

is satisfied only for $a_1 = a_2 = \dots = a_k = 0$. The set that is not linearly independent is called linearly dependent.

Let S be a set of vectors in the vector space V. This set S is linearly independent if and only if every subset of S is linearly independent. Otherwise it is linearly dependent.

The following result establishes an important property of a generated space. Let $S_k = \{\mathbf{v}_1, \mathbf{v}_2, \dots, \mathbf{v}_k\}$ be a linearly independent set containing k vectors from the vector space V. Also, let span(S_k) be the space generated by S_k. Therefore, every set containing $k + 1$ elements from span(S_k) is linearly dependent.

To see this result we proceed by induction. Suppose first that $k = 1$. Then, S_1 consists of a single element $\mathbf{v}_1 \neq \mathbf{0}$ and therefore S_1 is linearly independent. Consider two arbitrary elements $(k + 1 = 2)$ from span(S_1). Let $\mathbf{w}_1$ and $\mathbf{w}_2$ be these elements. Since they belong to span(S_1), they take the form

$$\mathbf{w}_1 = a_1\mathbf{v}_1 \quad \mathbf{w}_2 = a_2\mathbf{v}_1, \tag{A.69}$$

where $a_1, a_2 \in \mathbb{R}$ are not simultaneously zero. By multiplying both sides of the first expression by a_2 and the second expression by a_1, and subtracting these vectors we get

$$a_2\mathbf{w}_1 - a_1\mathbf{w}_2 = \mathbf{0}, \tag{A.70}$$

for all a_1, a_2 which are not simultaneously null. Then, $\{\mathbf{w}_1, \mathbf{w}_2\}$ is linearly dependent. Since the chosen elements $\mathbf{w}_i$ were arbitrary it follows that any set with $k + 1 = 2$ elements from span(S_1) is linearly dependent. Thus, we have demonstrated the result for $k = 1$. Assume that the result holds for $k - 1$, and let us see that the same result remains valid for k. So, consider an arbitrary set of $k + 1$ different elements from span(S_k), designated by $T = \{\mathbf{w}_1, \mathbf{w}_2, \dots, \mathbf{w}_{k+1}\}$. The task is to prove that T is linearly dependent. Because $\mathbf{w}_i \in$ span(S_k), $\forall i = 1, \dots, k + 1$, we have

$$\mathbf{w}_i = \sum_{j=1}^{k} a_{ij}\mathbf{v}_j \quad i = 1, \dots, k + 1. \tag{A.71}$$

Here, we have two possibilities.

i) Suppose that $a_{i1} = 0, i = 1, \dots, k+1$ then, every element in T is a linear combination of the elements of the set $S' = \{\mathbf{v}_2, \mathbf{v}_3, \dots, \mathbf{v}_k\}$. However, S' is linearly independent and consists of $k-1$ elements. Since the result was assumed to hold for $k-1$, it follows that T is linearly dependent.

ii) Consider the case in which not all the scalars a_{i1} are null. In particular, supppose $a_{11} \neq 0$, if required, we can renumber the elements $\mathbf{w}_i$ to achieve this. By making the expression (A.71) particular for $i = 1$, and multiplying both sides by $c_i = a_{i1}/a_{11}$, yields

$$c_i \mathbf{w}_1 = a_{i1}\mathbf{v}_1 + \sum_{j=2}^{k} c_i a_{1j} \mathbf{v}_j. \tag{A.72}$$

Subtracting (A.71) from (A.72) gives

$$c_i \mathbf{w}_1 - \mathbf{w}_i = \sum_{j=2}^{k} (c_i a_{1j} - a_{ij}) \mathbf{v}_j, \tag{A.73}$$

for $i = 2, 3, \dots, k+1$. This last equation tells us that the k vectors $c_i\mathbf{w}_1 - \mathbf{w}_i$ are linear combinations of the set $S' = \{\mathbf{v}_2, \mathbf{v}_3, \dots, \mathbf{v}_k\}$ which is linearly independent. Again, since the result holds for $k-1$, it follows that the set of k vectors $c_i\mathbf{w}_1 - \mathbf{w}_i$ is linearly dependent. From this, and for real numbers $b_2, b_3, \cdots, b_{k+1}$ not all simultaneously zero, we get

$$\sum_{j=2}^{k+1} b_j (c_j \mathbf{w}_1 - \mathbf{w}_j) = \mathbf{0}, \tag{A.74}$$

from where

$$\left(\sum_{j=2}^{k+1} b_j c_j \right) \mathbf{w}_1 - \sum_{j=2}^{k+1} b_j \mathbf{w}_j = \mathbf{0}, \tag{A.75}$$

which implies that the set of $k+1$ vectors $\mathbf{w}_1, \mathbf{w}_2, \dots, \mathbf{w}_{k+1}$ is linearly dependent.

Finally, using the induction argument we conclude the proof of this result.

A.8 Bases and Dimension

We say that a finite set S of elements belonging to the vector space V is a (finite) basis for V if S is linearly independent and if $\text{span}(S) = V$.

The vector space V is finite dimensional if it has a finite basis. If $V = \{\mathbf{0}\}$, then $\dim V = 0$, if $V \neq \{\mathbf{0}\}$ and if it has no finite basis then we say that V is infinite dimensional.

For example, let $X = \{\mathbf{x};\ \mathbf{x} = (x_1, x_2, \dots, x_n, \dots)\}$ be the set of all real successions. Then, X is an infinite dimensional vector space over $\mathbb{R}$, with the operations of addition and multiplication by scalar defined component-wise.

We now present an important result. Let V be a finite dimensional vector space. Then, every basis for V has the same number of elements. In particular, the number of elements in any basis of V is the dimension of V.

Let S and T be two bases for V. Since V is finite dimensional, then S and T have a finite number of elements. Let s be the number of elements in S and t the number of elements in T. Since S is a linearly independent set of s vectors, from the previous result we have that every set of $s+1$ elements of V is linearly dependent. Because T is, by hypothesis, linearly independent, it gives $t \leq s$. Changing S by T in the previous reasoning we arrive at $s \leq t$. From these results we conclude that $s = t$ and the result follows.

For example, consider $\mathbb{R}^n = \{\mathbf{x};\ \mathbf{x} = (x_1, x_2, \dots, x_n),\ x_i \in \mathbb{R}\}$. The set $S = \{\mathbf{e}_i, i = 1, 2, \dots, n\}$, where $\mathbf{e}_i = (0, 0, \dots, 1, 0, \dots, 0)$ with 1 in the ith position, is a basis for $\mathbb{R}^n$ and $\dim \mathbb{R}^n = n$.

Another important result establishes that for a finite-dimensional vector space V with $\dim V = n$ (n-dimensional space) we have that every set with n linearly independent elements from V is a basis for V, and also that every set of linearly independent elements from V is a subset of some basis for V.

Given that $\dim V = n$, there exists in V a linearly independent subset B with n such that $\text{span}(B) = V$. Let $T = \{\mathbf{v}_1, \mathbf{v}_2, \dots, \mathbf{v}_n\}$ be a linearly independent set of n elements in V. Then, $\text{span}(T) \subseteq V$. If $\text{span}(T) \neq V$, there exists $\mathbf{w} \in V$ such that $\mathbf{w} \notin \text{span}(T)$. Thus, $T' = \{\mathbf{v}_1, \mathbf{v}_2, \dots, \mathbf{v}_n, \mathbf{w}\}$ is a linearly independent set and also $T' \subset V$. From the previous result we know that T' is linearly dependent, therefore we arrive at a contradiction originated from assuming that $\text{span}(T) \neq V$. As a consequence, T is a basis for V and we have shown the first statement.

To prove the second statement let us consider an arbitrary linearly independent set of elements in V, say $S = \{\mathbf{v}_1, \mathbf{v}_2, \dots, \mathbf{v}_k\}$. If $k = n$, S is a basis for V. Otherwise, there exists an element $\mathbf{w} \in V$ which does not belong to $\text{span}(S)$. Incorporating this element into S we obtain the set $S' = \{\mathbf{v}_1, \mathbf{v}_2, \dots, \mathbf{v}_k, \mathbf{w}\}$. If this set is linearly dependent there exist real numbers $a_1, a_2, \dots, a_{k+1}$ not all simultaneously zero for which

$$\sum_{i=1}^{k} a_i \mathbf{v}_i + a_{k+1}\mathbf{w} = \mathbf{0}. \tag{A.76}$$

Since $\{\mathbf{v}_1, \mathbf{v}_2, \dots, \mathbf{v}_k\}$ is linearly independent, it follows that $a_{k+1} \neq 0$. We can, thus, write $\mathbf{w}$ in terms of a linear combination of the elements $\{\mathbf{v}_1, \mathbf{v}_2, \dots, \mathbf{v}_k\}$, that is $\mathbf{w} \in \text{span}(S)$, which contradicts the original hypothesis that $\mathbf{w} \notin \text{span}(S)$. Therefore, S' is a linearly independent set. Once again, if $k + 1 = n$ then S' is a basis for V, and because $S \subset S'$ we have that the second statement above holds. In the case that S' is not a basis for V, we repeat the previous reasoning until we arrive at a set $\overline{S}$ that is a basis for V. The number of steps to arrive at this set $\overline{S}$ is clearly finite, otherwise we would eventually obtain a linearly independent set with $n + 1$ elements, which is in conflict with the results presented previously.

Moreover, now we can easily see that the following is valid. If V is a finite-dimensional space and W is a subspace of V, then $\dim W \leq \dim V$. Also, if V is an infinite-dimensional space, whichever the value of n considered, there exists a linearly independent subset $\{\mathbf{v}_1, \mathbf{v}_2, \dots, \mathbf{v}_n\}$ of elements from V.

A.9 Components

Let V be an n-dimensional vector space, and let $\{\mathbf{e}_1, \mathbf{e}_2, \dots, \mathbf{e}_n\} = \{\mathbf{e}_i\}_{i \leq n}$ be a basis for V. Then

$$V = \text{span}(\mathbf{e}_1, \mathbf{e}_2, \dots, \mathbf{e}_n), \tag{A.77}$$

and, therefore, every vector $\mathbf{v} \in V$ is a linear combination of the vectors in the basis, that is to say

$$\mathbf{v} = v^1\mathbf{e}_1 + v^2\mathbf{e}_2 + \ldots + v^n\mathbf{e}_n \qquad v^i \in \mathbb{R}. \tag{A.78}$$

Coefficients v^i are called the components of $\mathbf{v}$ with respect to the basis $\{\mathbf{e}_i\}_{i\leq n}$. These components are unique because the elements in the basis are linearly independent. In fact, suppose that $\mathbf{v}$ admits a second representation of the kind given by (A.78), and designate with d^i the components of such a representation. Subtracting both representations we obtain

$$\sum_{i=1}^{n}(v^i - d^i)\mathbf{e}_i = \mathbf{0}, \tag{A.79}$$

and since $\{\mathbf{e}_1, \mathbf{e}_2, \ldots, \mathbf{e}_n\}$ is linearly independent, it follows that (A.79) implies $v^i - d^i = 0$ for all components i, then $v^i = d^i$.

Usually, throughout this book we will find expressions of the kind given by (A.78) (also (A.79)), and in many other cases the number of indexes is larger. For notational simplicity we introduce the convention that establishes that any expression containing repeated indexes has a summation implied that covers the whole range of validity of the referred index (or indexes). With this convention, (A.78) is now written simply as

$$\mathbf{v} = v^i\mathbf{e}_i. \tag{A.80}$$

In particular

$$\mathbf{e}_i = \delta^j_i\mathbf{e}_j \quad \text{where} \quad \delta^j_i = \begin{cases} 1 & i = j, \\ 0 & i \neq j. \end{cases} \tag{A.81}$$

Let $\{\mathbf{e}_i\}_{i\leq n}$ and $\{\overline{\mathbf{e}}_i\}_{i\leq n}$ be two different bases for the n-dimensional vector space V. We call the real numbers e^j_k and $\overline{e}^j_k$, respectively, jth component of the kth element in these bases with respect to each other, that is

$$\mathbf{e}_k = e^j_k\overline{\mathbf{e}}_j \qquad \overline{\mathbf{e}}_k = \overline{e}^j_k\mathbf{e}_j. \tag{A.82}$$

Thus, we obtain

$$\mathbf{e}_i = e^j_i\overline{e}^k_j\mathbf{e}_k \qquad \overline{\mathbf{e}}_i = \overline{e}^j_i e^k_j\overline{\mathbf{e}}_k. \tag{A.83}$$

Comparing (A.83) with (A.81), gives

$$e^j_i\overline{e}^k_j = \overline{e}^j_i e^k_j = \delta^j_i. \tag{A.84}$$

The previous expression also tells us that the matrixes with components $[e^j_i]$ and $[\overline{e}^j_i]$ are, correspondingly, the inverse of each other.

Moreover, notice that if v^i and $\overline{v}^i$ are the components of $\mathbf{v} \in V$ with respect to the bases $\{\mathbf{e}_i\}_{i\leq n}$ and $\{\overline{\mathbf{e}}_i\}_{i\leq n}$, respectively, then

$$v^i = \overline{e}^i_j\overline{v}^j, \tag{A.85}$$

$$\overline{v}^i = e^i_j v^j, \tag{A.86}$$

where e_j^i and $\overline{e}_j^i$ are defined in (A.82). As a matter of fact, by taking the representation of $\mathbf{v}$ in the basis $\{\overline{\mathbf{e}}_i\}_{i\leq n}$, and recalling the representation of the basis in terms of the basis $\{\mathbf{e}_i\}_{i\leq n}$ given by (A.82), we obtain

$$\mathbf{v} = \overline{v}^j\overline{\mathbf{e}}_j = \overline{v}^j\overline{e}_j^i\mathbf{e}_i = v^i\mathbf{e}_i. \tag{A.87}$$

From this expression we conclude that

$$v^i = \overline{e}_j^i\overline{v}^j. \tag{A.88}$$

To show that (A.86) holds we follow an analogous reasoning to the previous one, but now expressing $\mathbf{v}$ as a linear combination of the elements of the basis $\{\mathbf{e}_i\}_{i\leq n}$, and the result follows straightforwardly.

A.10 Sum of Sets and Subspaces

Let X and Y be two subsets in the vector space V. The sum of these subsets, represented by $X + Y$, consists of the set of all vectors of the form $\mathbf{x} + \mathbf{y}$, where $\mathbf{x} \in X$ and $\mathbf{y} \in Y$. If X and Y are two subspaces of V, then we have that $X + Y$ is also a subspace of V.

A.11 Linear Manifolds

Let S be a subspace of the vector space U and let $\mathbf{u}_0 \in U$ be a given element in this space. The set of all the elements $\mathbf{u} \in U$ such that

$$\mathbf{u} = \mathbf{u}_0 + \eta \quad \eta \in S, \tag{A.89}$$

is called a linear manifold of U. Observe that if $\mathbf{u}_0 \in S$ we have that the linear manifold is itself a subspace that coincides with S. The linear manifold is also designated as translation of S in $\mathbf{u}_0$, and is represented by

$$X = \mathbf{u}_0 + S. \tag{A.90}$$

This expression emphasizes that for all elements $\mathbf{u} \in X$, there always exists an element $\eta \in S$ such that $\mathbf{u} = \mathbf{u}_0 + \eta$.

A.12 Convex Sets and Cones

The set K of the vector space V is said to be a convex set if for all $\mathbf{u}, \mathbf{v} \in K$ and $\alpha \in [0, 1]$ we get

$$\alpha\mathbf{u} + (1 - \alpha)\mathbf{v} \in K. \tag{A.91}$$

From a geometrical perspective, the previous expression tells us that if two elements belong to K, every element lying over the straight line (in an abstract sense) defined by $\mathbf{u}$ and $\mathbf{v}$ also belongs to K. In addition, the empty set is considered to be convex.

It is easy to note that vector spaces, subspaces, and linear manifolds are all convex sets. In turn, if K and G are convex sets from the vector space V it follows that the set

$$\alpha K = \{\mathbf{x};\ \mathbf{x} = \alpha\mathbf{v},\ \mathbf{v} \in K\}, \tag{A.92}$$

is also convex for all $\alpha \in \mathbb{R}$, and also that the set $K + G$ is convex.

Let C be an arbitrary collection of convex sets (which can be infinite), then $\cap_{K\in C} K$ is convex.

The set C of the vector space V is said to be a convex cone centered at the origin if

$$\mathbf{v} \in C \quad \Rightarrow \quad \alpha\mathbf{v} \in C \qquad \forall\alpha \geq 0. \tag{A.93}$$

Observe that the null element of V belongs to the cone C. With this definition we can introduce the following concepts. The set C which is both a cone and a convex set is called convex cone. In turn, let C_0 be a cone and let $\mathbf{v}_0$ be an element of the vector space V, then the set $C = \mathbf{v}_0 + C_0 = \{\mathbf{v} \in C;\ \mathbf{v} = \mathbf{v}_0 + \alpha\mathbf{x},\ \mathbf{x} \in C_0,\ \alpha \geq 0\}$ is defined as the translation in $\mathbf{v}_0$ of the cone C_0. In particular, the element $\mathbf{v}_0$ is the vertex of the cone. For this reason, C_0 is also called a cone with a vertex at the origin.

Clearly, the concept of cone is more general than that of a space and linear manifold, which are both particular classes of convex cones.

A.13 Direct Sum of Subspaces

Let X and Y be two subspaces of the vector space V. We say that V is the direct sum of X and Y if and only if

$$V = X + Y, \tag{A.94}$$

$$X \cap Y = \mathbf{0}, \tag{A.95}$$

where $\mathbf{0} \in \mathbf{V}$ is the null element. The direct sum is represented by $V = X \oplus Y$.

Notice that we can write $V = X \oplus Y$ if and only if $\mathbf{v} \in V$ has a unique representation $\mathbf{v} = \mathbf{x} + \mathbf{y}$ with $\mathbf{x} \in X$ and $\mathbf{y} \in Y$.

A.14 Linear Transformations

Let V and W be two vector spaces, the function $\mathbf{T} : V \to W$ is a linear transformation from V into W if the following properties are satisfied

i) $\mathbf{T}(\mathbf{u} + \mathbf{v}) = \mathbf{T}(\mathbf{u}) + \mathbf{T}(\mathbf{v}), \forall\mathbf{u}, \mathbf{v} \in V$,
ii) $\mathbf{T}(a\mathbf{u}) = a\mathbf{T}(\mathbf{u}), \forall\mathbf{u} \in V$ and $\forall a \in \mathbb{R}$.

In other words, the function $\mathbf{T}$ preserves the addition and the multiplication by scalar. This tells us that any linear transformation is a morphism. The two properties a linear transformation has to satisfy can be condensed into a single property, which is

$$\mathbf{T}(a\mathbf{u} + b\mathbf{v}) = a\mathbf{T}(\mathbf{u}) + b\mathbf{T}(\mathbf{v}) \qquad \forall\mathbf{u}, \mathbf{v} \in V \quad \forall a, b \in \mathbb{R}. \tag{A.96}$$

Observe that the identity transformation $\mathbf{I} : V \to V$ where $\mathbf{I}(\mathbf{v}) = \mathbf{v}$ for all $\mathbf{v} \in V$, is a linear transformation. The multiplication by a given real number a, $\mathbf{T} : V \to V$, where

$\mathbf{T}(\mathbf{v}) = a\mathbf{v}$ for all $\mathbf{v} \in V$ is also a linear transformation. Let V be the vector space containing all continuous functions in $[a, b]$, then $\mathbf{T} : V \to V$ defined by $\mathbf{T}(\mathbf{v}) = \int_a^x \mathbf{v}(t)dt$, $a \leq x \leq b$, is a linear transformation.

Consider a linear transformation $\mathbf{T} : V \to W$, then the set $\mathbf{T}(V)$ (that is, $R(\mathbf{T})$) is a subspace of W. Moreover, $\mathbf{T}$ transforms the null vector of V into the null vector of W.

To see that $\mathbf{T}(V)$ is a subspace of W it is enough to show that $\mathbf{T}(V)$ is closed with respect to the addition and multiplication by scalar. Let us take arbitrary elements of $\mathbf{T}(V)$, as $\mathbf{T}(\mathbf{u})$ and $\mathbf{T}(\mathbf{v})$. By adding these vectors, and since $\mathbf{T}$ is a linear transformation, we get

$$\mathbf{T}(\mathbf{u}) + \mathbf{T}(\mathbf{v}) = \mathbf{T}(\mathbf{u} + \mathbf{v}) \quad \Rightarrow \quad \mathbf{T}(\mathbf{u}) + \mathbf{T}(\mathbf{v}) \in \mathbf{T}(V). \tag{A.97}$$

Also, let a be an arbitrary real number and let $\mathbf{T}(\mathbf{u})$ be an arbitrary element in $\mathbf{T}(V)$. By considering the product of such an element with the real number a, and recalling that $\mathbf{T}$ is a linear transformation, we get

$$a\mathbf{T}(\mathbf{u}) = \mathbf{T}(a\mathbf{u}) \quad \Rightarrow \quad a\mathbf{T}(\mathbf{u}) \in \mathbf{T}(V). \tag{A.98}$$

Therefore, $\mathbf{T}(V)$ is a subspace of W. On the other hand, if in the last expression we consider $a = 0$, we get

$$\mathbf{0} = 0\mathbf{T}(\mathbf{u}) = \mathbf{T}(0\mathbf{u}) = \mathbf{T}(\mathbf{0}). \tag{A.99}$$

The set of all the elements of V such that $\mathbf{T}$ maps them into the null element of W is called the null space of $\mathbf{T}$, or the kernel of $\mathbf{T}$. This space is denoted by $\mathcal{N}(\mathbf{T})$. Then

$$\mathcal{N}(\mathbf{T}) = \{\mathbf{v} \in V; \ \mathbf{T}(\mathbf{v}) = \mathbf{0}\}. \tag{A.100}$$

Let $\mathbf{T} : V \to W$ be a linear transformation between two vector spaces V and W. Then, $\mathcal{N}(\mathbf{T})$ is a subspace of V. To see this result, it is enough to show that $\mathcal{N}(\mathbf{T})$ is closed with respect to the addition and multiplication by scalar. Consider the arbitrary elements $\mathbf{u}, \mathbf{v} \in \mathcal{N}(\mathbf{T})$ and $a \in \mathbb{R}$, then

$$\mathbf{T}(\mathbf{u} + \mathbf{v}) = \mathbf{T}(\mathbf{u}) + \mathbf{T}(\mathbf{v}) = \mathbf{0}, \tag{A.101}$$

$$\mathbf{T}(a\mathbf{u}) = a\mathbf{T}(\mathbf{u}) = \mathbf{0}. \tag{A.102}$$

Let us see another important result. Consider the linear transformation $\mathbf{T} : V \to W$ between two vector spaces V and W, where V is finite-dimensional. Then, $\mathbf{T}(V)$ (that is, $R(\mathbf{T})$) is also finite-dimensional, such that

$$\dim \mathcal{N}(\mathbf{T}) + \dim \mathbf{T}(V) = \dim V. \tag{A.103}$$

Moreover, if V is infinite-dimensional, at least one of the spaces $\mathcal{N}(\mathbf{T})$ or $\mathbf{T}(V)$ is infinite dimensional.

To prove the first statement consider $n = \dim V$ and let $\{\mathbf{e}_k\}_{k \leq n}$ be a basis of $\mathcal{N}(\mathbf{T})$. Since $\mathcal{N}(\mathbf{T})$ is a subspace of V it follows that $k \leq n$. However, we know that the set $\{\mathbf{e}_k\}_{k \leq n}$ is a subset of some basis of V. Let

$$\mathbf{e}_1, \mathbf{e}_2, \ldots, \mathbf{e}_k, \ldots, \mathbf{e}_{k+r}, \tag{A.104}$$

be this basis, where

$$\mathbf{T}(\mathbf{e}_{k+1}), \mathbf{T}(\mathbf{e}_{k+2}), \ldots, \mathbf{T}(\mathbf{e}_{k+r}), \tag{A.105}$$

forms a basis for $\mathbf{T}(V)$. With this, we show that $\dim \mathbf{T}(V) = r$ and the statement is proven. In order to do this, let us see first that (A.105) spans $\mathbf{T}(V)$. Consider an arbitrary element $\mathbf{w} \in \mathbf{T}(V)$. Then, there exists $\mathbf{v} \in V$ such that $\mathbf{w} = \mathbf{T}(\mathbf{v})$. Because (A.104) is a basis for V, we have

$$\mathbf{w} = \mathbf{T}(v^i\mathbf{e}_i) = v^i\mathbf{T}(\mathbf{e}_i) = \sum_{i=1}^{k} v^i\mathbf{T}(\mathbf{e}_i) + \sum_{i=k+1}^{k+r} v^i\mathbf{T}(\mathbf{e}_i) = \sum_{i=k+1}^{k+r} v^i\mathbf{T}(\mathbf{e}_i), \tag{A.106}$$

since $\mathbf{T}(\mathbf{e}_1) = \mathbf{T}(\mathbf{e}_2), \dots, \mathbf{T}(\mathbf{e}_k) = \mathbf{0}$ by virtue of being a basis for $\mathcal{N}(\mathbf{T})$. Expression (A.106) tells us that (A.105) spans $\mathbf{T}(V)$. The next step consists of showing that these elements form a linearly independent set, that is, they constitute a basis for $\mathbf{T}(V)$. To do this suppose that there exist scalars, not all of them null, $c^{k+1}, \dots, c^{k+r}$ such that

$$\sum_{i=k+1}^{k+r} c^i\mathbf{T}(\mathbf{e}_i) = \mathbf{0}, \tag{A.107}$$

that is, we are assuming that (A.105) is a linearly dependent set. Expression (A.107), because $\mathbf{T}$ is a linear transformation, implies

$$\mathbf{T}\left(\sum_{i=k+1}^{k+r} c^i\mathbf{e}_i\right) = \mathbf{0}, \tag{A.108}$$

and so

$$\mathbf{c} = \sum_{i=k+1}^{k+r} c^i\mathbf{e}_i \in \mathcal{N}(\mathbf{T}). \tag{A.109}$$

From this expression, we see that $\mathbf{c}$ can also be written as a linear combination of the elements that form the basis for $\mathcal{N}(\mathbf{T})$ that is

$$\mathbf{c} = \sum_{i=1}^{k} c^i\mathbf{e}_i \in \mathcal{N}(\mathbf{T}). \tag{A.110}$$

Subtracting (A.109) from (A.110), gives

$$\sum_{i=1}^{k} c^i\mathbf{e}_i - \sum_{i=k+1}^{k+r} c^i\mathbf{e}_i = \mathbf{0}. \tag{A.111}$$

Therefore, the set $\{\mathbf{e}_i\}_{i\leq k+r}$ is linearly dependent, which is a contradiction with our original hypothesis, since we have assumed that this set was a basis for V. Hence, the set (A.105) is linearly independent. We have therefore demonstrated the first statement.

To prove the second statement we have to show that one of the spaces, $\mathcal{N}(\mathbf{T})$ or $\mathbf{T}(V)$, is infinite dimensional. In particular, if $\dim \mathcal{N}(\mathbf{T})$ is infinite then the result follows trivially. Consider that $\dim \mathcal{N}(\mathbf{T}) = k$, with k finite, and let us prove that $\dim \mathbf{T}(V)$ is infinite. To this, it is sufficient to demonstrate that, whichever the value of $n \in \mathbb{N}$, there exists a linearly independent set with n elements from $\mathbf{T}(V)$. So, let $\{\mathbf{e}_i\}_{i\leq k}$ be a basis for $\mathcal{N}(\mathbf{T}) \subset V$. Because V is infinite dimensional, whichever the value of $n \in \mathbb{N}$, we can construct

$$S = \{\mathbf{e}_1, \mathbf{e}_2, \dots, \mathbf{e}_k, \dots, \mathbf{e}_{k+n}\}, \tag{A.112}$$

where $S \subset V$ and S is linearly independent. Now, since $\mathbf{T}$ is a linear transformation and $\{\mathbf{e}_i\}_{i\leq k}$ is a basis for $\mathcal{N}(\mathbf{T})$, it follows that the set $\{\mathbf{Te}_i\}_{k+1\leq i\leq k+n}$ is a linearly independent set containing n vectors from $\mathbf{T}(V)$ (if it is linearly dependent, then S would be linearly dependent). Since n is arbitrary it is demonstrated that dim $\mathbf{T}(V)$ is infinite.

According to the morphism concept and related topics, we see that the linear transformation $\mathbf{T} : U \to V$ is an isomorphism if it is bijective. If such a transformation exists we say that the spaces U and V are isomorphic.

A further useful result says that two finite-dimensional vector spaces U and V are isomorphic if and only if they have the same dimension. Consider that U and V are isomorphic, then there exists $\mathbf{T} : U \to V$, which is bijective. Let $\{\mathbf{e}_i\}_{i\leq n}$ be a basis for U. We want to show that the elements $\{\mathbf{Te}_i\}_{i\leq n}$ form a basis for V and therefore $\dim U = \dim V$. In order to do this, suppose that the set $\{\mathbf{Te}_i\}_{i\leq n}$ is linearly dependent, so there exist scalars $a_i \in \mathbb{R}$ not all of them zero, such that

$$a_i\mathbf{Te}_i = \mathbf{0}. \tag{A.113}$$

Since $\mathbf{T}$ is linear we have

$$\mathbf{T}(a_i\mathbf{e}_i) = \mathbf{0} \quad \Rightarrow \quad a_i\mathbf{e}_i = \mathbf{0}, \tag{A.114}$$

which implies that $\{\mathbf{e}_i\}_{i\leq n}$ is linearly dependent, which is a contradiction because it is a basis for U. Therefore, all the numbers a_i are zero. Hence, and from (A.113), we conclude that the set $\{\mathbf{Te}_i\}_{i\leq n}$ is linearly independent. Because $\mathbf{T}$ is bijective, for an arbitrary element $\mathbf{v} \in V$, there exists a unique $\mathbf{u} \in U$ such that

$$\mathbf{v} = \mathbf{T}(\mathbf{u}) = \mathbf{T}(u^i\mathbf{e}_i) = u^i\mathbf{T}(\mathbf{e}_i). \tag{A.115}$$

That is, any element $\mathbf{v} \in V$ can be uniquely written as a linear combination of the elements of the set $\{\mathbf{Te}_i\}_{i\leq n}$. Then span$(\{\mathbf{Te}_i\}_{i\leq n}) = V$ and $\dim U = \dim V$.

To end the proof, we suppose that both spaces have the same dimension, $\dim U = \dim V$. We denote by $\{\mathbf{e}_i\}_{i\leq n}$ and $\{\mathbf{d}_i\}_{i\leq n}$ the bases for U and V, respectively. Since these bases have the same number of elements, we can construct a bijective linear transformation $\hat{f}$ such that

$$\hat{f} : \{\mathbf{e}_i\}_{i\leq n} \mapsto \{\mathbf{d}_i\}_{i\leq n}. \tag{A.116}$$

Let $\mathbf{u} \in U$ be an arbitrary element, then the transformation $\mathbf{T}$ defined by

$$\mathbf{T}(\mathbf{u}) = u^i\mathbf{T}(\mathbf{e}_i), \tag{A.117}$$

where u^i, $i \leq n$, are the components of $\mathbf{u}$ with respect to the basis $\{\mathbf{e}_i\}_{i\leq n}$, is clearly linear, injective, and $R(\mathbf{T}) = V$. Thus, $\mathbf{T}$ is bijective. Consequently, U and V are isomorphic.

As a corollary, we have that any n-dimensional vector space is isomorphic to the Euclidean space $\mathbb{R}^n$.

Now, let U and V be two isomorphic vector spaces, then $\mathbf{T} : U \to V$ is an isomorphism if and only if the following conditions are verified

i) $\mathcal{N}(\mathbf{T}) = \mathbf{0}$
ii) $R(\mathbf{T}) = V$
iii) $\mathbf{T}$ transforms the bases for U into bases for V.

Consider two vector spaces with possibly different dimensions. The set of all linear transformations from U into V is designated by $\mathscr{L}(U;V)$

$$\mathscr{L}(U;V)=\{\mathbf{T};\ \mathbf{T}:U\to V,\mathbf{T}\text{ is linear}\}. \tag{A.118}$$

If in $\mathscr{L}(U;V)$ we introduce the operations of addition and multiplication by scalar, defined in the following manner

$$(\mathbf{T}_1+\mathbf{T}_2)(\mathbf{u})=\mathbf{T}_1(\mathbf{u})+\mathbf{T}_2(\mathbf{u}),\quad \forall\mathbf{u}\in U, \tag{A.119}$$

$$(a\mathbf{T})(\mathbf{u})=a\mathbf{T}(\mathbf{u}),\quad \forall\mathbf{u}\in U,\forall a\in\mathbb{R}, \tag{A.120}$$

we have that the set $\mathscr{L}(U;V)$ is a vector space. The null element in $\mathscr{L}(U;V)$ is represented by the symbol $\mathbf{O}$.

With the previous definition we can see that $\dim\mathscr{L}(U;V)=\dim U\dim V$. In fact, let $\{\mathbf{f}_i\}_{i\leq m}$ $\{\mathbf{e}_i\}_{i\leq n}$ be the bases for U and V, respectively, and define the linear transformation

$$(\mathbf{e}_i\otimes\mathbf{f}^\alpha):U\to V, \tag{A.121}$$

in the following manner

$$(\mathbf{e}_i\otimes\mathbf{f}^\alpha)(\mathbf{f}_\beta)=\delta^\alpha_\beta\mathbf{e}_i, \tag{A.122}$$

where $i=1,\dots,n$ and $\alpha,\beta=1,\dots,m$. At this point, we notice that $(\mathbf{e}_i\otimes\mathbf{f}^\alpha)$ must be interpreted as a symbol to indicate the transformation defined in (A.122). The operations $\otimes$ and $\mathbf{f}^\alpha$ are defined next. Let us first show that the set $\{(\mathbf{e}_i\otimes\mathbf{f}^\alpha),i=1,\dots,n,\ \alpha=1,\dots,m\}$ forms a basis for $\mathscr{L}(U;V)$. To this end, we will first see that it is a linearly independent set. Suppose that the set is linearly dependent, then there exist scalars $a^{i}_{.\alpha}$ $i=1,\dots,n$, $\alpha=1,\dots m$, not all of them zero such that

$$a^{i}_{.\alpha}(\mathbf{e}_i\otimes\mathbf{f}^\alpha)=\mathbf{O}. \tag{A.123}$$

From the addition and multiplication by scalar results

$$a^{i}_{.\alpha}(\mathbf{e}_i\otimes\mathbf{f}^\alpha)(\mathbf{u})=\mathbf{0}\qquad\forall\mathbf{u}\in U. \tag{A.124}$$

In particular, for $\mathbf{u}=\mathbf{f}_\beta$, $\beta=1,\dots,m$,

$$a^{i}_{.\alpha}(\mathbf{e}_i\otimes\mathbf{f}^\alpha)(\mathbf{f}_\beta)=\mathbf{0}\qquad\beta=1,\dots,m. \tag{A.125}$$

From (A.122) we have

$$a^{i}_{.\alpha}\delta^\alpha_\beta\mathbf{e}_i=a^{i}_{.\beta}\mathbf{e}_i=\mathbf{0}\qquad\beta=1,\dots,m. \tag{A.126}$$

Given that $\{\mathbf{e}_i,\ i=1,\dots,m\}$ is a basis for V, (A.126) is equivalent to

$$a^{i}_{.\beta}=0\qquad i=1,\dots,n,\ \ \beta=1,\dots,m. \tag{A.127}$$

Hence, the set $\{(\mathbf{e}_i\otimes\mathbf{f}^\alpha),i=1,\dots,n,\ \alpha=1,\dots,m\}$ is linearly independent. Now let us show that this set spans $\mathscr{L}(U;V)$. Let $\mathbf{T}$ be an arbitrary element from $\mathscr{L}(U;V)$. Thus, $\mathbf{T}(\mathbf{f}_\beta)$ is a vector of V and therefore we can write it as a linear combination of the elements of the basis, that is

$$\mathbf{T}(\mathbf{f}_\beta)=T^{i}_{.\beta}\mathbf{e}_i. \tag{A.128}$$

From this expression, and from (A.122), we obtain

$$\mathbf{T}(\mathbf{f}_\beta)=T^{i}_{.\alpha}(\mathbf{e}_i\otimes\mathbf{f}^\alpha)(\mathbf{f}_\beta), \tag{A.129}$$

from where

$$\mathbf{T} = T^{i}_{.\alpha}(\mathbf{e}_i \otimes \mathbf{f}^{\alpha}). \tag{A.130}$$

In turn, since we have already seen that $\{(\mathbf{e}_i \otimes \mathbf{f}^{\alpha})\}$ is linearly independent, we conclude that the set is a basis for $\mathscr{L}(U; V)$.

The basis $\{(\mathbf{e}_i \otimes \mathbf{f}^{\alpha})\}$ for $\mathscr{L}(U; V)$ described above is called the product basis with respect to the basis $\{\mathbf{e}_i\}$ and $\{\mathbf{f}^{\alpha}\}$. The components $T^{i}_{.\alpha}$ of $\mathbf{T}$ relative to $\{(\mathbf{e}_i \otimes \mathbf{f}^{\alpha})\}$ are called the components of $\mathbf{T}$ with respect to the product basis. Expression (A.128) characterizes these components. Specifically, if we take an arbitrary vector $\mathbf{u} \in U$, whose components are u^{α} with respect to the basis $\{\mathbf{f}_{\alpha}\}$ for U, the image $\mathbf{T}(\mathbf{u})$ of $\mathbf{u}$ under the transformation $\mathbf{T}$ takes the following form as a function of the components with respect to the basis $\{\mathbf{e}_i\}$ for V

$$\mathbf{T}(\mathbf{u}) = T^{i}_{.\alpha}(\mathbf{e}_i \otimes \mathbf{f}^{\alpha})u^{\alpha}\mathbf{f}_{\alpha} = T^{i}_{.\alpha}u^{\alpha}\mathbf{e}_i. \tag{A.131}$$

As a consequence of (A.121) and (A.122), the basis $\{(\mathbf{e}_i \otimes \mathbf{f}^{\alpha})\}$ for $\mathscr{L}(U; V)$ provides an isomorphism between $\mathscr{L}(U; V)$ and $\mathbb{R}^{n\times m}$

$$K : \mathscr{L}(U; V) \to \mathbb{R}^{n\times m}, \tag{A.132}$$

defined by

$$K(\mathbf{T}) = [T^{i}_{.\alpha}], \tag{A.133}$$

where $[T^{i}_{.\alpha}]$ represents a matrix where the coefficient placed at the ith row and αth column is $T^{i}_{.\alpha}$. In particular, $K(\mathbf{e}_i \otimes \mathbf{f}^{\alpha})$ is a matrix with all the coefficients zero except for that located at row i, column α, whose value is 1.

Consider now that we transform the bases, and let $\{\overline{\mathbf{e}}_i\}$ and $\{\overline{\mathbf{f}}_{\alpha}\}$ be the new bases for V and U, respectively. The components of $\mathbf{T}$ satisfy the following transformation rule (change of bases)

$$T^{i}_{.\alpha} = \overline{e}^{i}_{k}\overline{T}^{k}_{.\beta}f^{\beta}_{\alpha}, \tag{A.134}$$

where $\overline{e}^{i}_{j}$ is defined in $(A.82)_2$ and f^{β}_{α} is defined by a similar expression to $(A.82)_1$. To show (A.134) consider $\mathbf{T}(\mathbf{f}_{\alpha})$, from (A.128) and (A.82) yields

$$T^{i}_{.\alpha}\mathbf{e}_i = \mathbf{T}(\mathbf{f}_{\alpha}) = \mathbf{T}(f^{\beta}_{\alpha}\overline{\mathbf{f}}_{\beta}) = f^{\beta}_{\alpha}\overline{T}^{k}_{.\beta}\overline{\mathbf{e}}_k = f^{\beta}_{\alpha}\overline{T}^{k}_{.\beta}\overline{e}^{i}_{k}\mathbf{e}_i. \tag{A.135}$$

Next, we will consider some examples of the space $\mathscr{L}(U; V)$ that appear when adopting particular spaces for U and V.

A.15 Canonical Isomorphism

Take $U = \mathbb{R}$. In this case the vector space $\mathscr{L}(\mathbb{R}; V)$ is a copy of V. In fact, consider $\mathbf{T} \in \mathscr{L}(\mathbb{R}; V)$ and let us construct the following application

$$\mathbf{I} : \mathscr{L}(\mathbb{R}; V) \to V, \tag{A.136}$$

defined by

$$\mathbf{I}(\mathbf{T}) = \mathbf{T}(1), \tag{A.137}$$

which, as we notice, is an isomorphism. Since this isomorphism can be established without resorting to any specific basis, it receives the name of canonical isomorphism. This is written as

$$\mathscr{L}(\mathbb{R}; V) \simeq V. \tag{A.138}$$

Two vector spaces between which there exists a canonical isomorphism are said to be canonically isomorphic.

A.16 Algebraic Dual Space

Consider the space $\mathscr{L}(V; \mathbb{R})$, which is acknowledged as the algebraic dual space of V. This space is denoted by V^*. The elements of V^* are real-valued linear functionals defined in V. If $\{\mathbf{e}_i\}$ is a basis of V then the product basis of V^* is

$$(1 \otimes \mathbf{e}^i), \tag{A.139}$$

where 1 is the normal basis for $\mathbb{R}$. In this case we simplify the notation of operator (A.139), removing the 1 and the symbol $\otimes$. In this manner $\{\mathbf{e}^i\}$ is called the dual basis of $\{\mathbf{e}_i\}$. From (A.122), the bases $\{\mathbf{e}_i\}$ and $\{\mathbf{e}^i\}$ are related through

$$\mathbf{e}^i(\mathbf{e}_j) = \delta^i_j. \tag{A.140}$$

In order to distinguish the elements of V and V^* we call them, respectively, vectors and covectors. In turn, if the components of the vectors are taken with respect to the basis $\{\mathbf{e}_i\}$, those of the covectors are taken with respect to the dual basis, that is, the basis $\{\mathbf{e}^i\}$.

Let now $\{\mathbf{e}_i\}$ and $\{\overline{\mathbf{e}}_i\}$ be two bases for V related as in expression (A.82), then the corresponding dual bases $\{\mathbf{e}^i\}$ and $\{\overline{\mathbf{e}}^i\}$ are related in the following way

$$\mathbf{e}^i(\mathbf{e}_i) = \overline{\mathbf{e}}^k(\overline{\mathbf{e}}_k) = \overline{\mathbf{e}}^k(\overline{e}^i_k\mathbf{e}_i) = (\overline{e}^i_k\overline{\mathbf{e}}^k)(\mathbf{e}_i), \tag{A.141}$$

$$\overline{\mathbf{e}}^i(\overline{\mathbf{e}}_i) = \mathbf{e}^k(\mathbf{e}_k) = \mathbf{e}^k(e^i_k\overline{\mathbf{e}}_i) = (e^i_k\mathbf{e}^k)(\overline{\mathbf{e}}_i), \tag{A.142}$$

from where

$$\mathbf{e}^i = \overline{e}^i_k\overline{\mathbf{e}}^k \quad \overline{\mathbf{e}}^i = e^i_k\mathbf{e}^k. \tag{A.143}$$

Moreover, the transformation rules for the components of the covector $u^* = u_i\mathbf{e}^i = \overline{u}_i\overline{\mathbf{e}}^i$ readily become

$$u_i = e^k_i\overline{u}_k \quad \overline{u}_i = \overline{e}^k_i u_k, \tag{A.144}$$

as can be easily verified.

Consider now the dual space of V^*, that is $\mathscr{L}(V^*; \mathbb{R})$, which we denote by V^{**}. Let us show that V^{**} is canonically isomorphic to V. To this end, take an arbitrary, but fixed, element $\mathbf{u} \in V$ and define the real-valued function $\mathbf{u}^{**}$ in V^* given by

$$\mathbf{u}^{**}(\mathbf{w}) = \mathbf{w}(\mathbf{u}) \qquad \forall \mathbf{w} \in V^*. \tag{A.145}$$

Hence, the transformation

$$(\cdot)^{**} : \mathscr{L}(V^*; \mathbb{R}) \to V, \tag{A.146}$$

is a canonical isomorphism. Therefore, we say that the relation between V and V^* is reflexive, as already suggested by the word dual.

Expression (A.145) can be expressed through a more symmetric notation. Take $\mathbf{u} \in V$ and $\mathbf{w} \in V^*$, and designate (A.145) in the following manner

$$\langle \mathbf{u}, \mathbf{w} \rangle = \mathbf{w}(\mathbf{u}) = \mathbf{u}^{**}(\mathbf{w}). \tag{A.147}$$

The application

$$\langle \cdot, \cdot \rangle : V \times V^* \to \mathbb{R}, \tag{A.148}$$

has the following properties

i) $\langle \cdot, \cdot \rangle$ is bilinear, meaning that it is a linear application with respect to each one of the arguments
ii) $\langle \cdot, \cdot \rangle$ is defined, meaning that if $\mathbf{u} \in V$ is chosen and $\langle \mathbf{u}, \mathbf{w} \rangle = 0$, $\forall \mathbf{w} \in V^*$ then we have that $\mathbf{u} = \mathbf{0}$, and, reciprocally, if $\mathbf{w} \in V^*$ is chosen and $\langle \mathbf{u}, \mathbf{w} \rangle = 0$, $\forall \mathbf{u} \in V$, one has $\mathbf{w} = \mathbf{0}$.

This definition enables us to extend the concept of orthogonality. Thus, if $\langle \mathbf{u}, \mathbf{w} \rangle = 0$, then we say that the vector $\mathbf{u}$ is orthogonal to the covector $\mathbf{w}$, and vice versa.

A.16.1 Orthogonal Complement

Let X be a subspace of the vector space V and let V^* be the dual (algebraic) space of V. The set of all linear functionals $\mathbf{f} \in \mathbf{V}^*$ such that

$$\langle \mathbf{f}, \mathbf{x} \rangle = 0 \qquad \forall \mathbf{x} \in X, \tag{A.149}$$

is called the orthogonal complement of X and is represented by $X^\perp$. As can be appreciated, $X^\perp$ is a vector subspace of the dual space V^* ($X^\perp \subset V^*$).

Let us present an important result. Let X be a finite-dimensional subspace of the vector space V. Then the following holds

$$(X^\perp)^\perp = X^{\perp\perp} = X. \tag{A.150}$$

In fact, an implicit representation of the subspace X is given by

$$X = \{\mathbf{x} \in V;\ \langle \mathbf{f}, \mathbf{x} \rangle = 0 \quad \forall \mathbf{f} \in X^\perp\}. \tag{A.151}$$

So, if $\mathbf{x} \in X^{\perp\perp}$ we know that $\langle \mathbf{f}, \mathbf{x} \rangle = 0$, $\forall \mathbf{f} \in X^\perp$, which implies $\mathbf{x} \in X$ and therefore $X^{\perp\perp} \subset X$. The proof that $X \subset X^{\perp\perp}$ is trivial and the result follows.

Now, let X and Y be two finite-dimensional subspaces of V. Then

$$(X \cap Y)^\perp = X^\perp + Y^\perp. \tag{A.152}$$

To see this, take $\mathbf{w} \in (X^\perp + Y^\perp)^\perp$, then

$$\langle \mathbf{f}, \mathbf{w} \rangle = 0 \quad \forall \mathbf{f} \in \mathbf{X}^\perp \quad \Rightarrow \mathbf{w} \in X^{\perp\perp}, \tag{A.153}$$

$$\langle \mathbf{g}, \mathbf{w} \rangle = 0 \quad \forall \mathbf{g} \in \mathbf{Y}^\perp \quad \Rightarrow \mathbf{w} \in Y^{\perp\perp}, \tag{A.154}$$

and as a consequence

$$\mathbf{w} \in (X^\perp + Y^\perp)^\perp \Rightarrow \mathbf{w} \in X^{\perp\perp} \cap Y^{\perp\perp} = X \cap Y, \tag{A.155}$$

which results in

$$(X^{\perp} + Y^{\perp})^{\perp} \subset X \cap Y. \tag{A.156}$$

Taking the orthogonal complement we have

$$(X \cap Y)^{\perp} \subset (X^{\perp} + Y^{\perp}). \tag{A.157}$$

Reciprocally, if $\mathbf{w} \in X \cap Y$, we have

$$\langle \mathbf{f}, \mathbf{w} \rangle = 0 \quad \forall \mathbf{f} \in \mathbf{X}^{\perp}, \tag{A.158}$$

$$\langle \mathbf{g}, \mathbf{w} \rangle = 0 \quad \forall \mathbf{g} \in \mathbf{Y}^{\perp}, \tag{A.159}$$

and so we obtain that $\mathbf{w} \in (X^{\perp} + Y^{\perp})^{\perp}$, which implies $X \cap Y \subset (X^{\perp} + Y^{\perp})^{\perp}$, and taking again the orthogonal complement

$$X^{\perp} + Y^{\perp} \subset (X \cap Y)^{\perp}, \tag{A.160}$$

arriving thus at the result.

Notice that the previous two results begin with the hypothesis that the subspaces are finite-dimensional. Only in these cases, and by exploiting the algebraic structure of the vector space, was it possible to provide the corresponding proofs. In the case of infinite-dimensional subspaces, the results stated above are not verified in general. For this to occur in the infinite-dimensional case, it will be necessary to endow the space V with a suitable topological structure.

A.16.2 Positive and Negative Conjugate Cones

The positive conjugate cone of the set S in the vector space V is the subset S^+ of V^* defined by

$$S^+ = \{\mathbf{f} \in V^*;\ \langle \mathbf{f}, \mathbf{v} \rangle \geq 0 \quad \forall \mathbf{v} \in S\}. \tag{A.161}$$

Analogously, the negative conjugate cone of the set S in the vector space V is the subset S^- of V^* defined by

$$S^- = \{\mathbf{f} \in V^*;\ \langle \mathbf{f}, \mathbf{v} \rangle \leq 0 \quad \forall \mathbf{v} \in S\}. \tag{A.162}$$

Observe that S^+ and S^- are non-empty convex cones because they always contain at least the null functional. Furthermore, if S is a subspace of V we get

$$S^+ = S^- = S^{\perp}, \tag{A.163}$$

and so the concept of a conjugate cone can be understood as a generalization of the concept of the orthogonal complement.

Likewise, the following results constitute corresponding generalizations of the results stated in the previous section. Moreover, the proofs follow exactly the same course as the proofs sketched in the previous section.

Let C be a convex cone with vertex at the origin and assume that there exists a finite number of elements $\mathbf{f}_\alpha \in \mathbf{V}^*$, $\alpha \in J = \{1, \dots, n\}$ such that

$$C = \{\mathbf{v} \in V;\ \langle \mathbf{f}_\alpha, \mathbf{v} \rangle \geq 0 \quad \forall \alpha \in J\}, \tag{A.164}$$

then, the following holds

$$(C^+)^+ = C, \tag{A.165}$$

$$(C^-)^- = C. \tag{A.166}$$

Consider two convex cones C_1 and C_2 with vertexes at the origin, then

$$(C_1 \cap C_2)^+ = C_1^+ + C_2^+, \tag{A.167}$$

$$(C_1 \cap C_2)^- = C_1^- + C_2^-. \tag{A.168}$$

A.17 Algebra in *V*

The vector space $\mathscr{L}(V; V)$ consists of all the linear transformations from V into itself. These transformations are called tensors.[1] If $\{\mathbf{e}_i\}$ is a basis for V, the product basis for $\mathscr{L}(V; V)$ is $\{(\mathbf{e}_i \otimes \mathbf{e}^j,\ i,j = 1, \dots, n\}$ defined as in (A.122) by replacing $\mathbf{f}^\alpha$ by $\mathbf{e}^j$. In this manner, the components of a tensor $\mathbf{T} \in \mathscr{L}(V; V)$ with respect to this basis are

$$\mathbf{T} = T^i_{.j}(\mathbf{e}_i \otimes \mathbf{e}^j). \tag{A.169}$$

When we handle tensors, that is linear transformations belonging to $\mathscr{L}(V; V)$, it is usual to remove the parentheses in the vector argument, so hereafter we employ the notation

$$\mathbf{T}(\mathbf{u}) = \mathbf{T}\mathbf{u}. \tag{A.170}$$

Therefore, the components of $\mathbf{T}$ are defined by (A.128), that is

$$\mathbf{T}\mathbf{e}_i = T^j_{.i}\mathbf{e}_j \quad i = 1, \dots, n. \tag{A.171}$$

From (A.171), if $\mathbf{v} \in V$ and $\mathbf{v} = v^i\mathbf{e}_i$, we get

$$\mathbf{T}\mathbf{v} = v^j T^i_{.j}\mathbf{e}_i. \tag{A.172}$$

In other words, the components of the image of $\mathbf{v}$ under the transformation (the application of the tensor) $\mathbf{T}$ with respect to the basis for V are given by

$$T^i_{.j}v^j \quad i = 1, \dots, n. \tag{A.173}$$

From (A.171) we conclude that the identity transformation $\mathbf{I}$ has, with respect to any basis for V, the components δ^i_j.

In $\mathscr{L}(V; V)$ we can define the composition among its elements in a natural manner. To do this, given $\mathbf{T}$ and $\mathbf{L} \in \mathscr{L}(V; V)$ we introduce the multiplication $\mathbf{TL}$ as the composition of tensors $\mathbf{T}$ and $\mathbf{L}$, that is

$$\mathbf{TL}\mathbf{v} = (\mathbf{T} \circ \mathbf{L})\mathbf{v} = \mathbf{T}(\mathbf{L}\mathbf{v}) \qquad \mathbf{v} \in V. \tag{A.174}$$

As already seen when studying the composition of functions, the multiplication is associative, distributive but, in general, is not commutative. A vector space in which such multiplication operation is defined is termed algebra. From the previous definitions, we notice that there exists an isomorphism between the algebra $\mathscr{L}(V; V)$ and the matrix algebra $\mathbb{R}^{n\times n}$. Indeed, recalling (A.132), the tensor $\mathbf{T}$ is in correspondence with the matrix whose coefficients are $[T^i_{.j}]$ with respect to any basis $\{\mathbf{e}_i\}$ for V. From (A.174) we can easily show that the multiplication of tensors corresponds to the product of the

---

1 More precisely, we would have to say second-order tensors, but we simplify the denomination in the present context.

respective matrixes. In effect, the components of **TL** are defined through (A.128), that is

$$\mathbf{TLe}_j = (\mathbf{TL})^i_{.j}\mathbf{e}_i, \tag{A.175}$$

while from (A.174) we get

$$\mathbf{TLe}_j = \mathbf{T}(\mathbf{Le}_j) = \mathbf{T}(L^k_{.j}\mathbf{e}_k) = L^k_{.j}\mathbf{Te}_k = L^k_{.j}T^i_{.k}\mathbf{e}_i. \tag{A.176}$$

From both expressions above we finally obtain

$$(\mathbf{TL})^i_{.j} = T^i_{.k}L^k_{.j}. \tag{A.177}$$

Hence, the matrix $[(\mathbf{TL})^i_{.j}]$, is the matrix product between the matrixes $[T^i_{.j}]$ and $[L^i_{.j}]$.

Suppose now a change of basis from $\{\mathbf{e}_i\}$ to $\{\overline{\mathbf{e}}_i\}$. Then the components of the tensor $\mathbf{T} \in \mathscr{L}(V; V)$ satisfy the following transformation rule

$$T^i_{.j} = \overline{e}^i_r \overline{T}^r_{.k} e^k_j, \tag{A.178}$$

where $\overline{e}^i_j$ and e^i_j are defined as in (A.82).

According to the transformation rule (A.178), it follows that the following functions can be defined

$$\mathrm{tr}\mathbf{T} = T^i_{.i} = \overline{T}^i_{.i}, \tag{A.179}$$

$$\det \mathbf{T} = \det[T^i_{.j}] = \det[\overline{T}^i_{.j}], \tag{A.180}$$

called, respectively, the trace and the determinant of the tensor **T**. These functions are well-defined in the sense that for each tensor we have a unique trace and a unique determinant. That is, these functions admit a representation which is independent from the component representation given to the tensor. In addition, we can note that

i) while the trace is a linear function, the determinant is not, except for the case $\dim V = 1$
ii) it is verified that

$$\mathrm{tr}(\mathbf{TL}) = \mathrm{tr}(\mathbf{LT}), \tag{A.181}$$

in fact

$$\mathrm{tr}(\mathbf{TL}) = (\mathbf{TL})^i_{.i} = T^i_{.k}L^k_{.i} = L^k_{.i}T^i_{.k} = (\mathbf{LT})^k_{.k} = \mathrm{tr}(\mathbf{LT}), \tag{A.182}$$

iii) and also it is verified that

$$\det(\mathbf{TL}) = (\det \mathbf{T})(\det \mathbf{L}) = \det(\mathbf{LT}), \tag{A.183}$$

where to see this we have to recall that

$$\det[T^i_{.j}] = \frac{1}{n!}\delta^{i_1\cdots i_n}_{j_1\cdots j_n} T^{j_1}_{.i_1} \cdots T^{j_n}_{.i_n}, \tag{A.184}$$

where

$$\delta^{i_1\cdots i_n}_{j_1\cdots j_n} = \begin{cases} 1 & \text{if } (i_1\cdots i_n) \text{ is an even permutation of } (j_1\cdots j_n), \\ -1 & \text{if } (i_1\cdots i_n) \text{ is an odd permutation of } (j_1\cdots j_n), \\ 0 & \text{otherwise,} \end{cases} \tag{A.185}$$

then, from (A.180), we arrive at the result

$$\begin{aligned}\det(\mathbf{TL}) &= \det[(TL)^{i}_{.j}]\\ &= \left(\frac{1}{n!}\delta^{k_1\cdots k_n}_{j_1\cdots j_n}(T^{j_1}_{.k_1}\cdots T^{j_n}_{.k_n})\right)\left(\frac{1}{n!}\delta^{i_1\cdots i_n}_{k_1\cdots k_n}(L^{k_1}_{.i_1}\cdots L^{k_n}_{.i_n}\right)\\ &= (\det\mathbf{T})(\det\mathbf{L}) = (\det\mathbf{L})(\det\mathbf{T}) = \det(\mathbf{LT}).\end{aligned} \tag{A.186}$$

The transformation $\mathbf{L}$ is non-singular if it is an isomorphism in V, otherwise we say that it is singular. If $\mathbf{L}$ is non-singular then the inverse $\mathbf{L}^{-1}$ is defined. Given the isomorphism between the algebra $\mathscr{L}(V;V)$ and the matrix algebra $\mathbb{R}^{n\times n}$ it follows that $\mathbf{L}$ is invertible, or non-singular, if and only if $\det\mathbf{L}\neq 0$. It is not difficult to show that the following relation holds

$$\det\mathbf{L}^{-1} = (\det\mathbf{L})^{-1} \qquad \text{for all non-singular } \mathbf{L}. \tag{A.187}$$

In addition, the set containing all non-singular linear transformations from V form a group with respect to the multiplication in $\mathscr{L}(V;V)$. We represent this group with the notation $\mathscr{G}\mathscr{L}\mathscr{G}(V)$, and we call it the general linear group of V. The set of all transformations for which the determinant is ± 1 is a subgroup of $\mathscr{G}\mathscr{L}\mathscr{G}(V)$ called unimodular group and represented by $\mathscr{U}\mathscr{G}(V)$. If we consider only those linear transformations whose determinant equals 1, we have another subgroup, termed the special linear group, which is represented by $\mathscr{S}\mathscr{L}\mathscr{G}(V)$. The relation between $\mathscr{L}(V;V)$ and these groups can be appreciated in the following

$$\underbrace{\mathscr{L}(V;V)\supset\mathscr{G}\mathscr{L}\mathscr{G}(V)\supset\mathscr{U}\mathscr{G}(V)\supset\mathscr{S}\mathscr{L}\mathscr{G}(V).}_{\substack{\Downarrow\ \det\\ \mathbb{R}\supset\mathbb{R}\backslash\{0\}\supset\pm1\supset1}} \tag{A.188}$$

A.18 Adjoint Operators

The space of all linear transformations from V^* into itself, $\mathscr{L}(V^*;V^*)$, is isomorphic to the space $\mathscr{L}(V;V)$. The isomorphism is determined through the transformation

$$\begin{aligned}(\cdot)^A : \mathscr{L}(V;V) &\rightarrow \mathscr{L}(V^*;V^*),\\ \mathbf{L} &\mapsto (\mathbf{L})^A = \mathbf{L}^*,\end{aligned} \tag{A.189}$$

characterized by

$$\langle\mathbf{w}^*,\mathbf{Lv}\rangle = \langle\mathbf{L}^*\mathbf{w}^*,\mathbf{v}\rangle \quad \mathbf{v}\in V, \mathbf{w}^*\in V^*. \tag{A.190}$$

Operator $\mathbf{L}^*$ is called the adjoint operator of $\mathbf{L}$. In turn, from the definition it follows that $\mathbf{L}^*\mathbf{w}^*(\mathbf{v}) = \mathbf{w}^*(\mathbf{Lv})$.

This concept can be further developed. As a matter of fact, let X and Y be two vector spaces, and let us consider X^* and Y^* as the respective (algebraic) dual spaces. That is, if $\mathbf{x}^*\in X^*$ we have $\mathbf{x}^*\in\mathscr{L}(X;\mathbb{R})$ and the duality operation is represented by

$$\mathbf{x}^*(\mathbf{x}) = \langle\mathbf{x}^*,\mathbf{x}\rangle_{X^*\times X}\in\mathbb{R}. \tag{A.191}$$

Similarly, for $\mathbf{y}^*\in Y^*$ we have $\mathbf{y}^*\in\mathscr{L}(Y;\mathbb{R})$ represented by

$$\mathbf{y}^*(\mathbf{y}) = \langle\mathbf{y}^*,\mathbf{y}\rangle_{Y^*\times Y}\in\mathbb{R}. \tag{A.192}$$

Hence, we can establish the following correspondence

$$\begin{aligned}(\cdot)^A : \mathscr{L}(X;Y) \to \mathscr{L}(Y^*;X^*),\\ \mathbf{L} \mapsto (\mathbf{L})^A = \mathbf{L}^*,\end{aligned} \tag{A.193}$$

characterized by

$$\langle \mathbf{y}^*, \mathbf{L}\mathbf{x}\rangle_{Y^*\times Y} = \langle \mathbf{L}^*\mathbf{y}^*, \mathbf{x}\rangle_{X^*\times X} \qquad \mathbf{x} \in X, \mathbf{y}^* \in Y^*. \tag{A.194}$$

As before, operator $\mathbf{L}^*$ is called the adjoint operator of $\mathbf{L}$, and from the definition it is $\mathbf{L}^*\mathbf{y}^*(\mathbf{x}) = \mathbf{y}^*(\mathbf{L}\mathbf{x})$.

Finally, it is easy to see that the adjoint operator is unique.

Now, if the spaces X and Y are finite-dimensional we can verify the following relations between the operator and its adjoint operator

i) $\mathscr{N}(\mathbf{T}) = R(\mathbf{T}^*)^\perp$
ii) $R(\mathbf{T}) = \mathscr{N}(\mathbf{T}^*)^\perp$
iii) $\mathscr{N}(\mathbf{T}^*) = R(\mathbf{T})^\perp$
iv) $R(\mathbf{T}^*) = \mathscr{N}(\mathbf{T})^\perp$.

Let us focus on the first two statements.

i) $\forall \mathbf{x} \in \mathscr{N}(\mathbf{T})$, we have $\mathbf{T}\mathbf{x} = \mathbf{0}$ and therefore $\langle \mathbf{y}^*, \mathbf{T}\mathbf{x}\rangle_{Y^*\times Y} = 0$, $\forall \mathbf{y}^* \in Y^*$, that is $\langle \mathbf{T}^*\mathbf{y}^*, \mathbf{x}\rangle_{X^*\times X} = 0$, $\forall \mathbf{y}^* \in Y^*$, which is equivalent to $\mathbf{x} \in R(\mathbf{T}^*)^\perp$.
ii) $\forall \mathbf{y}^* \in \mathscr{N}(\mathbf{T}^*)$, it is verified $\mathbf{T}^*\mathbf{y}^* = \mathbf{0}$, and consequently $\langle \mathbf{T}^*\mathbf{y}^*, \mathbf{x}\rangle_{X^*\times X} = 0$, $\forall \mathbf{x} \in X$ which implies $\langle \mathbf{y}^*, \mathbf{T}\mathbf{x}\rangle_{Y^*\times Y} = 0$, $\forall \mathbf{x} \in X$ and therefore $\mathbf{y}^* \in R(\mathbf{T})^\perp$. Hence, $\mathscr{N}(\mathbf{T}^*) = R(\mathbf{T})^\perp$. Taking now the orthogonal complement at both sides of this expression the result follows.

The last two statements are easily proven by interchanging the role of operator $\mathbf{T}$ with that of operator $\mathbf{T}^*$.

A.19 Transposition and Bilinear Functions

Consider the space $\mathscr{L}(V;V^*)$. With respect to the basis $\{\mathbf{e}^i\}$ for V^*, every element of $\mathscr{L}(V;V^*)$ can be written as a function of the components, in the following manner

$$\mathbf{A} = A_{ij}(\mathbf{e}^i \otimes \mathbf{e}^j), \tag{A.195}$$

where components A_{ij} are given by

$$\mathbf{A}\mathbf{e}_j = A_{ij}\mathbf{e}^i. \tag{A.196}$$

Notation $(\mathbf{e}^i \otimes \mathbf{e}^j)$ is nothing but a particular case of the more general notation adopted in (A.121) and (A.122). In fact, if $\mathbf{v} \in V$ and v^i is the ith component of $\mathbf{v}$ with respect to the basis $\{\mathbf{e}_i\}$ for V, from (A.196) it follows that

$$\mathbf{A}\mathbf{v} = A_{ij}(\mathbf{e}^i \otimes \mathbf{e}^j)v^k\mathbf{e}_k = A_{ij}v^j\mathbf{e}^i. \tag{A.197}$$

Thus, the components of the covector $\mathbf{w} = \mathbf{A}\mathbf{v} \in V^*$ (image vector of $\mathbf{v}$ under the linear transformation $\mathbf{A}$), with respect to the basis $\mathbf{e}^i$ for V^*, are the scalars $A_{ij}v^j$. The transformation rule for the components of $\mathbf{A}$ for a change of basis from $\{\mathbf{e}^i\}$ to $\{\overline{\mathbf{e}}^i\}$ reads

$$A_{ij} = e_i^r\overline{A}_{rk}e_j^k. \tag{A.198}$$

In effect, from (A.196), (A.82) and (A.143), we get

$$A_{ij}\mathbf{e}^i = \mathbf{A}\mathbf{e}_j = \mathbf{A}(e_j^k\overline{\mathbf{e}}_k) = e_j^k\mathbf{A}\overline{\mathbf{e}}_k = e_j^k\overline{A}_{rk}\overline{\mathbf{e}}^r = e_j^k\overline{A}_{rk}e_i^r\mathbf{e}^i = e_i^r\overline{A}_{rk}e_j^k\mathbf{e}^i. \tag{A.199}$$

The space $\mathscr{L}(V; V^*)$ is isomorphic to itself through the operation called transposition, which is defined in the following way

$$(\cdot)^T : \mathscr{L}(V; V^*) \to \mathscr{L}(V; V^*), \tag{A.200}$$

such that

$$\langle \mathbf{u}, \mathbf{A}\mathbf{v}\rangle = \langle \mathbf{A}^T\mathbf{u}, \mathbf{v}\rangle \quad \mathbf{u}, \mathbf{v} \in V, \mathbf{A} \in \mathscr{L}(V; V^*). \tag{A.201}$$

With respect to any basis $\{\mathbf{e}_i\}$ for V, the matrix of the components of $\mathbf{A}^T$ is the transpose of the matrix corresponding to $\mathbf{A}$

$$A_{ji}^T = A_{ij} \qquad i, j = 1, \dots, n. \tag{A.202}$$

To show this, it is enough to recall (A.138), (A.147) and (A.196). Indeed

$$\langle \mathbf{e}_i, \mathbf{A}\mathbf{e}_j\rangle = \langle \mathbf{e}_i, A_{rj}\mathbf{e}^r\rangle = A_{rj}\langle \mathbf{e}_i, \mathbf{e}^r\rangle = A_{rj}\mathbf{e}^r(\mathbf{e}_i) = A_{rj}\delta_i^r = A_{ij}, \tag{A.203}$$

and also

$$\langle \mathbf{e}_i, \mathbf{A}\mathbf{e}_j\rangle = \langle \mathbf{A}^T\mathbf{e}_i, \mathbf{e}_j\rangle = \langle A_{ri}^T\mathbf{e}^r, \mathbf{e}_j\rangle = A_{ri}^T\mathbf{e}^r(\mathbf{e}_j) = A_{ri}^T\delta_j^r = A_{ji}^T. \tag{A.204}$$

We say that $\mathbf{A} \in \mathscr{L}(V; V^*)$ is symmetric if it is equal to its transpose, that is if $\mathbf{A} = \mathbf{A}^T$. From the definition of transposition (A.200) we have that $\mathbf{A}$ is symmetric if and only if

$$\langle \mathbf{u}, \mathbf{A}\mathbf{v}\rangle = \langle \mathbf{A}^T\mathbf{u}, \mathbf{v}\rangle = \langle \mathbf{A}\mathbf{u}, \mathbf{v}\rangle \qquad \forall \mathbf{u}, \mathbf{v} \in V. \tag{A.205}$$

We say that $\mathbf{A} \in \mathscr{L}(V; V^*)$ is skew-symmetric if it is equal to the negative of its transpose, that is $\mathbf{A} = -\mathbf{A}^T$, implying

$$\langle \mathbf{u}, \mathbf{A}\mathbf{v}\rangle = \langle \mathbf{A}^T\mathbf{u}, \mathbf{v}\rangle = \langle -\mathbf{A}\mathbf{u}, \mathbf{v}\rangle = -\langle \mathbf{A}\mathbf{u}, \mathbf{v}\rangle \qquad \forall \mathbf{u}, \mathbf{v} \in V. \tag{A.206}$$

The function (also form or functional) given by

$$f : V \times V \to \mathbb{R}, \tag{A.207}$$

is called bilinear if it is linear with respect to each one of its arguments. The set $\mathscr{L}(V \times V; \mathbb{R})$ of all bilinear forms in $V \times V$ is a vector space. In particular, $\mathscr{L}(V \times V; \mathbb{R})$ is canonically isomorphic to $\mathscr{L}(V; V^*)$, and the canonical isomorphism

$$\begin{aligned} (\hat{\cdot}) : \mathscr{L}(V; V^*) &\to \mathscr{L}(V \times V; \mathbb{R}), \\ \mathbf{A} &\mapsto \hat{\mathbf{A}}(\mathbf{u}, \mathbf{v}), \end{aligned} \tag{A.208}$$

is defined by

$$\hat{\mathbf{A}}(\mathbf{u}, \mathbf{v}) = \mathbf{A}\mathbf{v}(\mathbf{u}) = \langle \mathbf{u}, \mathbf{A}\mathbf{v}\rangle \qquad \mathbf{A} \in L(V; V^*), \mathbf{u}, \mathbf{v} \in V. \tag{A.209}$$

In terms of the components with respect to the basis $\{\mathbf{e}_i\}$ for V, the previous expression is given by

$$\begin{aligned} \hat{\mathbf{A}}(\mathbf{u}, \mathbf{v}) &= \langle \mathbf{u}, \mathbf{A}\mathbf{v}\rangle = \langle u^i\mathbf{e}_i, \mathbf{A}v^j\mathbf{e}_j\rangle = u^iv^j\langle \mathbf{e}_i, \mathbf{A}\mathbf{e}_j\rangle \\ &= u^iv^jA_{ij}\langle \mathbf{e}_i, \mathbf{e}^i\rangle = u^iv^jA_{ij}. \end{aligned} \tag{A.210}$$

The bilinear function $\hat{\mathbf{A}}$ is said to be

i) positive definite, if

$$\hat{\mathbf{A}}(\mathbf{u},\mathbf{u}) > 0 \qquad \forall \mathbf{u} \in V, \mathbf{u} \neq \mathbf{0}, \tag{A.211}$$

and in this case we also say that the associated operator $\mathbf{A}$ is positive definite

ii) positive semi-definite, if

$$\hat{\mathbf{A}}(\mathbf{u},\mathbf{u}) \geq 0 \qquad \forall \mathbf{u} \in V, \mathbf{u} \neq \mathbf{0}, \tag{A.212}$$

and in this case we also say that the associated operator $\mathbf{A}$ is positive semi-definite

iii) symmetric, if the associated operator $\mathbf{A}$ is symmetric

iv) $\hat{\mathbf{A}}(\mathbf{u},\mathbf{v})$ is called quadratic form associated to the operator $\mathbf{A}$, see (A.210).

The spaces seen above are canonically isomorphic to other spaces as we will see next. Let U and V be two arbitrary vector spaces and let, for example, $\{\mathbf{f}_\alpha,\ \alpha = 1,\dots,m\}$ and $\{\mathbf{e}_i, i = 1,\cdots,n\}$ be the corresponding bases.

The dual space of the space of all bilinear transformations in $U \times V$ is called the tensor product space and is represented by $U \otimes V$, then

$$U \otimes V = (\mathscr{L}(U \times V; \mathbb{R}))^*. \tag{A.213}$$

We call the tensor product to the linear function $(\mathbf{u} \otimes \mathbf{v}) \in (\mathscr{L}(U \times V; \mathbb{R}))^*$ defined by the ordered pair $\mathbf{u} \in U$ and $\mathbf{v} \in V$ such that

$$(\mathbf{u} \otimes \mathbf{v})(\hat{\mathbf{A}}) = \hat{\mathbf{A}}(\mathbf{u},\mathbf{v}) \qquad \hat{\mathbf{A}} \in \mathscr{L}(U \times V; \mathbb{R}). \tag{A.214}$$

In particular, $(\mathbf{f}_\alpha \otimes \mathbf{e}_i)$ is the tensor product between the elements of the bases for U and V. As before, we can show that $\{(\mathbf{f}_\alpha \otimes \mathbf{e}_i),\ \alpha = 1,\dots,m,\ i = 1,\dots,n\}$ is a basis for $U \otimes V$.

Moreover, we can define a canonical isomorphism between $V \otimes U^*$ and $\mathscr{L}(U, V)$ in such a way that for each element from $V \otimes U^*$, $(\mathbf{v} \otimes \mathbf{w})$ corresponds the linear transformation from U into V defined in the following manner

$$(\mathbf{v} \otimes \mathbf{w})\mathbf{u} = \langle \mathbf{w},\mathbf{u} \rangle \mathbf{v} \qquad \forall \mathbf{u} \in U. \tag{A.215}$$

This isomorphism allows us to reinterpret the notation $(\mathbf{e}_i \otimes \mathbf{f}^\alpha)$, introduced for the first time in (A.121), as the tensor product of $\mathbf{e}_i$ with $\mathbf{f}^\alpha$.

We can now generalize (A.213) to tensor products involving more than two vector spaces, say $U, V, \dots, W$. The space of all multilinear transformations in $U \times V \times \cdots \times W$ is designated $\mathscr{L}(U \times V \times \cdots \times W; \mathbb{R})$ and the tensor product space is

$$U \otimes V \otimes \cdots \otimes W = (\mathscr{L}(U \times V \times \cdots \times W; \mathbb{R}))^*. \tag{A.216}$$

In the same way, if $\mathbf{u} \in U$, $\mathbf{v} \in V$, ..., $\mathbf{w} \in W$, we define the tensor product $\mathbf{u} \otimes \mathbf{v} \otimes \cdots \otimes \mathbf{w}$ as

$$[\mathbf{u} \otimes \mathbf{v} \otimes \cdots \otimes \mathbf{w}](\Psi) \equiv \Psi(\mathbf{u},\mathbf{v},\dots,\mathbf{w}) \qquad \forall \Psi \in \mathscr{L}(U \times V \times \cdots \times W; \mathbb{R}). \tag{A.217}$$

Having introduced the general tensor product, we now introduce the tensor space τ_s^r of order (r,s) over V as

$$\tau_s^r = \underbrace{V \otimes V \otimes \cdots \otimes V}_{r \text{ times}} \otimes \underbrace{V \otimes V \otimes \cdots \otimes V}_{s \text{ times}}. \tag{A.218}$$

In terms of the components with respect to the basis $\{\mathbf{e}_i\}$ of V, a tensor $\Psi \in \tau_s^r$ can be rewritten as

$$\Psi = \Psi^{i_1 \cdots i_r}_{j_1 \cdots j_s} \mathbf{e}_{i_1} \otimes \mathbf{e}_{i_2} \otimes \cdots \otimes \mathbf{e}_{i_r} \otimes \mathbf{e}^{j_1} \otimes \mathbf{e}^{j_2} \otimes \cdots \otimes \mathbf{e}^{j_s}. \tag{A.219}$$

For a change of basis from $\{\mathbf{e}_i\}$ to $\{\overline{\mathbf{e}}_i\}$ the components satisfy the following transformation rule

$$\overline{\Psi}^{i_1 \cdots i_r}_{j_1 \cdots j_s} = \overline{\mathbf{e}}^{i_1}_{k_1} \cdots \overline{\mathbf{e}}^{i_r}_{k_r} \overline{\mathbf{e}}^{t_1}_{j_1} \cdots \overline{\mathbf{e}}^{t_s}_{j_s} \Psi^{k_1 \cdots k_r}_{t_1 \cdots t_s}. \tag{A.220}$$

In (A.219) and (A.220) the numbers r and s are, respectively, the contravariant and covariant orders of the tensor under consideration. If one of these numbers is zero, then it must be omitted. Thus, the vector $\mathbf{v}$ in expression (A.85) is a contravariant tensor of order 1 (see (A.85)), and the covector $\mathbf{u}$ in (A.144) is a covariant tensor of order 1 (see $(A.144)_1$).

In addition, we can define different types of contractions. In particular, we can construct the operation

$$\mathbf{C}_1^1 : \tau_s^r \rightarrow \tau_{s-1}^{r-1}, \tag{A.221}$$

characterized by

$$\mathbf{C}_1^1(\mathbf{v}_1 \otimes \cdots \otimes \mathbf{v}_r \otimes \mathbf{u}^1 \otimes \cdots \otimes \mathbf{u}^s) = \langle \mathbf{v}_1, \mathbf{u}^1 \rangle (\mathbf{v}_2 \otimes \cdots \otimes \mathbf{v}_r \otimes \mathbf{u}^2 \otimes \cdots \otimes \mathbf{u}^s), \tag{A.222}$$

and putting the components in evidence yields

$$[\mathbf{C}_1^1 \Psi]^{i_2 \cdots i_r}_{j_2 \cdots j_s} = \Psi^{i_1 \cdots i_r}_{j_1 \cdots j_s} \delta^{j_1}_{i_1}. \tag{A.223}$$

A.20 Inner Product Spaces

In the previous section we saw that the spaces V and its dual V^* are isomorphic, but not canonically isomorphic. However, we can identify V with V^* through a particular linear application

$$\mathbf{G} : V \rightarrow V^*, \tag{A.224}$$

and, in this way, we will not have to distinguish between vectors and covectors. As was explained in Section A.19, $\mathbf{G}$ corresponds to a unique bilinear function $\hat{\mathbf{G}}$ defined in $V \times V$

$$\hat{\mathbf{G}} \in \mathscr{L}(V \times V; \mathbb{R}). \tag{A.225}$$

In order to individualize this function, we will call it bilinear product in V. A bilinear function is a bilinear product if and only if satisfies

$$\hat{\mathbf{G}}(\mathbf{u}, \mathbf{v}) = 0 \qquad \forall \mathbf{u} \quad \Rightarrow \mathbf{v} = \mathbf{0}, \tag{A.226}$$

$$\hat{\mathbf{G}}(\mathbf{u}, \mathbf{v}) = 0 \qquad \forall \mathbf{v} \quad \Rightarrow \mathbf{u} = \mathbf{0}. \tag{A.227}$$

Therefore, when building $\mathbf{G}$ we have to take into account these conditions.

Also, the inverse of $\mathbf{G}$, denoted by $\mathbf{G}^{-1}$, is an isomorphism between V and V^*, and corresponds to a unique bilinear function $\hat{\mathbf{G}}^*$, dual to $\hat{\mathbf{G}}$, defined in $V^* \times V^*$

$$\hat{\mathbf{G}}^* : V^* \times V^* \to \mathbb{R} \quad \mathbf{G}^{-1} : V^* \to V. \tag{A.228}$$

The operations $\mathbf{G}$ and $\mathbf{G}^{-1}$ allow us to go from vectors to covectors and vice versa. This implies that it is possible to modify the position of the indexes. In fact, from (A.196) it follows that the components of $\mathbf{G}$ with respect to the basis $\{\mathbf{e}^i\}$ are defined by

$$\mathbf{G}\mathbf{e}_i = g_{ji}\mathbf{e}^j, \tag{A.229}$$

and those of $\mathbf{G}^{-1}$ with respect to the basis $\{\mathbf{e}_i\}$ are given by

$$\mathbf{G}^{-1}\mathbf{e}^i = g^{ji}\mathbf{e}_j. \tag{A.230}$$

From these expressions we obtain

$$\mathbf{v}^* = v_j\mathbf{e}^j = \mathbf{G}\mathbf{v} = v^i\mathbf{G}\mathbf{e}_i = v^i g_{ji}\mathbf{e}^j \quad \Rightarrow v_j = g_{ji}v^i, \tag{A.231}$$

$$\mathbf{v} = v^j\mathbf{e}_j = \mathbf{G}^{-1}\mathbf{v}^* = v_i\mathbf{G}^{-1}\mathbf{e}^i = v_i g^{ji}\mathbf{e}_j \quad \Rightarrow v^j = g^{ji}v_i, \tag{A.232}$$

where v^i and v_i are the ith components of $\mathbf{v}$ and $\mathbf{v}^* = \mathbf{G}\mathbf{v}$ with respect to the bases $\{\mathbf{e}_i\}$ for V and $\{\mathbf{e}^i\}$ for V^*, respectively.

We say that the covector $\mathbf{G}\mathbf{v}$ is associated with $\mathbf{v}$ through $\mathbf{G}$. Once $\mathbf{G}$ has been established and kept fixed, we can see $\mathbf{v}$ and $\mathbf{G}\mathbf{v}$ as the same object because the former fully determines the latter, and vice versa. We can thus systematically replace all the concepts and operations that, in general, involve the dual space V^* by the corresponding operations in the very same space V. In other words, contractions are functions that take elements in $V \times V$, tensors defined in V^* can be replaced by the corresponding in V, and so on.

Through these operations involving $\mathbf{G}$ and $\mathbf{G}^{-1}$ it is possible, given a tensor $\mathbf{A}$, to define other tensors associated with it. In effect, suppose that $\mathbf{A}$ has the components $A_{ij}{}^{l}$, then

$$A_{ij}{}^{l} = g_{jk}A_i{}^{kl} \qquad A^{ik}_j = g^{ir}g_{js}A_r{}^{sk}. \tag{A.233}$$

As can be seen from these operations, the relative position of the indexes turns out to be an important aspect to take into consideration.

Now, a bilinear product is an inner product if it is symmetric and positive definite. So, a bilinear function $\hat{\mathbf{A}}$ is an inner product if it satisfies:

i) $\hat{\mathbf{A}}(\mathbf{u},\mathbf{v}) = \hat{\mathbf{A}}(\mathbf{v},\mathbf{u})$, that is, operator $\mathbf{A}$ associated with $\hat{\mathbf{A}}$ must be symmetric, which means $\mathbf{A} = \mathbf{A}^T$
ii) $\hat{\mathbf{A}}(\mathbf{u},\mathbf{u}) > 0, \forall \mathbf{u} \in V, \mathbf{u} \neq \mathbf{0}$.

A vector space in which a certain inner product has been defined is an inner product space, or Euclidean space. To define this inner product, consider a basis, say $\{\mathbf{e}_i\}$, for the vector space, then we can take the following linear application $\mathbf{G}$

$$\mathbf{G} = \delta_{ij}(\mathbf{e}^i \otimes \mathbf{e}^j) = (\mathbf{e}^1 \otimes \mathbf{e}^1)\cdots(\mathbf{e}^n \otimes \mathbf{e}^n). \tag{A.234}$$

Let us see that $\mathbf{G}$ defined above is symmetric and positive definite. While the symmetry is straightforward, for the positive-definiteness we have

$$\hat{\mathbf{G}}(\mathbf{u},\mathbf{u}) = \langle \mathbf{u}, \mathbf{G}\mathbf{u}\rangle = \langle \mathbf{u}, u^i\delta_{ji}\mathbf{e}^j\rangle = \delta_{ji}u^k u^i\langle \mathbf{e}_k, \mathbf{e}^j\rangle = \delta_{ji}\delta_{kj}u^k u^i = u^i u^i, \tag{A.235}$$

and then

$$\hat{\mathbf{G}}(\mathbf{u},\mathbf{u}) > 0 \qquad \forall \mathbf{u} \neq \mathbf{0}. \tag{A.236}$$

In particular, the basis $\{\mathbf{e}_i\}$ for V verifying the property $\hat{\mathbf{G}}(\mathbf{e}_i, \mathbf{e}_j) = \delta_{ij}$ is called the orthonormal basis. This designation is motivated by the fact that

$$\langle \mathbf{e}_k, \mathbf{G}\mathbf{e}_i \rangle = \delta_{ji} \langle \mathbf{e}_k, \mathbf{e}^j \rangle = \delta_{ji}\delta^j_k = \delta_{ki} \quad \Rightarrow \quad \langle \mathbf{e}_k, \mathbf{G}\mathbf{e}_i \rangle = \begin{cases} 1 & \text{if } k = i, \\ 0 & \text{if } k \neq i. \end{cases} \tag{A.237}$$

Hence, we can enunciate the following two results.

First, given an arbitrary basis $\{\mathbf{f}_i\}$, there exists an orthonormal basis $\{\mathbf{e}_i\}$ such that

$$\text{span}\{\mathbf{f}_1, \dots, \mathbf{f}_n\} = \text{span}\{\mathbf{e}_1, \dots, \mathbf{e}_n\}, \tag{A.238}$$

where $n = \dim(\text{span}\{\mathbf{f}_i\})$.

The construction of this basis is performed by an inductive process. First take

$$\mathbf{e}_1 = \frac{1}{\sqrt{\langle \mathbf{f}_1, \mathbf{G}\mathbf{f}_1 \rangle}} \mathbf{f}_1, \tag{A.239}$$

this vector $\mathbf{e}_1$ is unitary with respect to the inner product (A.234). Suppose that we have already built $\{\mathbf{e}_1, \dots, \mathbf{e}_r\}$ that satisfies (A.237) for $k = 1, \dots, r$, that means that these are orthonormal vectors in the sense of the inner product. Let us construct $\mathbf{e}_{r+1}$. In order to do so consider

$$\tilde{\mathbf{e}}_{r+1} = \mathbf{f}_{r+1} - \sum_{i=1}^{r} a_i \mathbf{e}_i, \tag{A.240}$$

where the scalars a_i are to be defined such that (A.237) is verified. Thus, the inner product between $\tilde{\mathbf{e}}_{r+1}$ and $\mathbf{e}_j$ is

$$\begin{aligned} \langle \tilde{\mathbf{e}}_{r+1}, \mathbf{G}\mathbf{e}_j \rangle &= \langle \mathbf{f}_{r+1}, \mathbf{G}\mathbf{e}_j \rangle - \sum_{i=1}^{r} a_i \langle \mathbf{e}_i, \mathbf{G}\mathbf{e}_j \rangle \\ &= \langle \mathbf{f}_{r+1}, \mathbf{G}\mathbf{e}_j \rangle - a_j \qquad j = 1, \dots, r, \end{aligned} \tag{A.241}$$

given that, by construction, all the elements $\{\mathbf{e}_1, \dots, \mathbf{e}_r\}$ are orthonormal among them. Then, since we want $\mathbf{e}_{r+1}$ to be orthogonal to $\mathbf{e}_1, \dots, \mathbf{e}_r$, we nullify the previous expression by selecting

$$a_j = \langle \mathbf{f}_{r+1}, \mathbf{G}\mathbf{e}_j \rangle \qquad j = 1, \dots, r. \tag{A.242}$$

In this manner

$$\mathbf{e}_{r+1} = \frac{1}{\sqrt{\langle \tilde{\mathbf{e}}_{r+1}, \mathbf{G}\tilde{\mathbf{e}}_{r+1} \rangle}} \tilde{\mathbf{e}}_{r+1}. \tag{A.243}$$

This serves to prove that it is possible to build an orthonormal set. We have to prove now that (A.238) holds. We consider that the result was valid for $k = r$, then it is $\text{span}\{\mathbf{f}_1, \dots, \mathbf{f}_r\} = \text{span}\{\mathbf{e}_1, \dots, \mathbf{e}_r\}$, and so

$$\text{span}\{\mathbf{e}_1, \dots, \mathbf{e}_r\} \subset \text{span}\{\mathbf{f}_1, \dots, \mathbf{f}_{r+1}\}. \tag{A.244}$$

From (A.240) and (A.243) it follows that

$$\mathbf{e}_{r+1} \in \text{span}\{\mathbf{f}_1, \dots, \mathbf{f}_{r+1}\}, \tag{A.245}$$

consequently

$$\operatorname{span}\{\mathbf{e}_1, \dots, \mathbf{e}_{r+1}\} \subset \operatorname{span}\{\mathbf{f}_1, \dots, \mathbf{f}_{r+1}\}. \tag{A.246}$$

Reciprocally, these expressions show us that

$$\mathbf{f}_{r+1} \in \operatorname{span}\{\mathbf{e}_1, \dots, \mathbf{e}_{r+1}\}, \tag{A.247}$$

so

$$\operatorname{span}\{\mathbf{f}_1, \dots, \mathbf{f}_{r+1}\} \subset \operatorname{span}\{\mathbf{e}_1, \dots, \mathbf{e}_{r+1}\}, \tag{A.248}$$

and the result follows.

Second, note that the basis $\{\mathbf{e}_i\}$ is orthonormal if and only if

$$\mathbf{e}^i = \mathbf{G}\mathbf{e}_i. \tag{A.249}$$

Indeed, if the basis is orthonormal, from (A.249) and (A.234) we get

$$\mathbf{G}\mathbf{e}_i = \delta_{ji}\mathbf{e}^j = \mathbf{e}^i. \tag{A.250}$$

If (A.249) is verified we get

$$\mathbf{G}\mathbf{e}_i = \delta_{ji}\mathbf{e}^j, \tag{A.251}$$

then

$$\langle \mathbf{e}_k, \mathbf{G}\mathbf{e}_i \rangle = \langle \mathbf{e}_k, \delta_{ji}\mathbf{e}^j \rangle = \delta_{ji}\delta^j_k = \delta_{ki}. \tag{A.252}$$

It is interesting to observe that, given that the components of $\mathbf{G}$ with respect to an orthonormal basis $\{\mathbf{e}_i\}$ are the scalars δ_{ij}, it is not necessary to distinguish the relative position of the indexes. In effect, from (A.233), we have that for an arbitrary relative positioning of the indexes

$$A^{\ l}_{ij} = \delta_{jk}A^{kl}_{i} = A^{jl}_{i}. \tag{A.253}$$

When the inner product defined by a certain (fixed) $\mathbf{G}$ is adopted, it is not necessary to keep $\mathbf{G}$ in the notation of the inner product, thus arriving at the classical notation

$$\mathbf{u} \cdot \mathbf{v} = \hat{\mathbf{G}}(\mathbf{u}, \mathbf{v}) = \langle \mathbf{u}, \mathbf{G}\mathbf{v} \rangle. \tag{A.254}$$

Recalling that $\mathbf{G}$ with respect to any basis $\{\mathbf{e}_i\}$ is such that

$$\mathbf{G}\mathbf{e}_i = g_{ji}\mathbf{e}^j, \tag{A.255}$$

and that

$$\mathbf{e}^i(\mathbf{e}_j) = \langle \mathbf{e}^i, \mathbf{e}_j \rangle = \delta_{ij}, \tag{A.256}$$

as a consequence we have that (A.254) can be written in terms of the components as follows

$$\mathbf{u} \cdot \mathbf{v} = \langle u^i\mathbf{e}_i, v^j g_{kj}\mathbf{e}^k \rangle = g_{kj}u^i v^j \langle \mathbf{e}_i, \mathbf{e}^k \rangle = g_{kj}u^i v^j \delta_{ik} = g_{ij}u^i v^j = u_j v^j = u^i v_i, \tag{A.257}$$

where we have made use of (A.231) and (A.232). In particular, if $\{\mathbf{e}_i\}$ is an orthonormal basis we get

$$\mathbf{u} \cdot \mathbf{v} = u^i v^i. \tag{A.258}$$

Let us introduce some definitions. We say that a linear transformation $\mathbf{Q}$ in the inner product vector space V is orthogonal if it preserves the inner product, that is,

$$\mathbf{u} \cdot \mathbf{v} = \mathbf{Qu} \cdot \mathbf{Qv} \qquad \forall \mathbf{u}, \mathbf{v} \in V. \tag{A.259}$$

Using (A.259) for the particular case of an orthonormal basis yields

$$\mathbf{e}_i \cdot \mathbf{e}_j = \delta_{ij} = \mathbf{Qe}_i \cdot \mathbf{Qe}_j = Q_{ki}Q_{kj}, \tag{A.260}$$

which tells us that the transformation is orthogonal if the associated matrix $[Q_{ij}]$ is an orthogonal matrix.

The set of all orthogonal transformations in V is a group, denoted by $\mathscr{O}(V)$, which is a subgroup in $\mathscr{GLG}(V)$ (see Section A.17). From (A.260) it follows that all orthogonal transformations have determinant equal to ± 1. Therefore, $\mathscr{O}(V)$ is a subgroup of the unimodular group $\mathscr{U}(V)$ (see Section A.17).

An orthogonal transformation is said to be a proper orthogonal transformation if the determinant equals 1. The set of all proper orthogonal transformations is a group, denoted by $\mathscr{P}\,\mathscr{O}(V)$, which is a subgroup in $\mathscr{SLG}(V)$ (see Section A.17).

As already presented in Section A.18, for all linear transformations $\mathbf{L} \in \mathscr{L}(V; V)$ there exists an adjoint linear transformation $\mathbf{L}^A \in \mathscr{L}(V^*, V^*)$. When an inner product $\hat{\mathbf{G}}$ has been established, we can define a new transformation, say $\tilde{\mathbf{L}}^A \in \mathscr{L}(V; V)$, in the following manner

$$\tilde{\mathbf{L}}^A = \mathbf{G}^{-1} \circ \mathbf{L}^A \circ \mathbf{G}, \tag{A.261}$$

as can be deduced from

$$\begin{array}{ccc} & V^* \to^{\mathbf{L}^A} V^* & \\ \mathbf{G} \uparrow & & \downarrow \mathbf{G}^{-1} \\ & V \to^{\tilde{\mathbf{L}}^A} V & \end{array} \tag{A.262}$$

As a function of the components, we have

$$(\tilde{L}^A)^i_j = g^{ik}(L^A)^l_k g_{lj}, \tag{A.263}$$

but since $\mathbf{L}^A$ is adjoint to $\mathbf{L}$, we get

$$(L^A)^l_k = (L)^l_k, \tag{A.264}$$

and therefore it follows that

$$(\tilde{L}^A)^i_j = g^{ik}(L)^l_k g_{lj}. \tag{A.265}$$

Next, we show the derivation of (A.263) and (A.264)

$$\mathbf{L}^A : V^* \to V^* \quad \Rightarrow \mathbf{L}^A = (L^A)^j_i(\mathbf{e}^i \otimes \mathbf{e}_j), \tag{A.266}$$

$$\tilde{\mathbf{L}}^A : V \to V \quad \Rightarrow \tilde{\mathbf{L}}^A = (\tilde{L}^A)^i_j(\mathbf{e}_i \otimes \mathbf{e}^j), \tag{A.267}$$

in turn

$$(\mathbf{e}^i \otimes \mathbf{e}_j)\mathbf{e}^k = \delta^k_j \mathbf{e}^i, \tag{A.268}$$

$$(\mathbf{e}_i \otimes \mathbf{e}^j)\mathbf{e}_k = \delta^j_k \mathbf{e}_i, \tag{A.269}$$

and also

$$\tilde{\mathbf{L}}^A \mathbf{e}_j = (\tilde{L}^A)^i_j \mathbf{e}_i, \tag{A.270}$$

$$\tilde{\mathbf{L}}^A \mathbf{e}_j = \mathbf{G}^{-1} \circ \mathbf{L}^A(\mathbf{G}\mathbf{e}_j) = g_{lj}\mathbf{G}^{-1}(\mathbf{L}^A \mathbf{e}^l)$$
$$= g_{lj}(L^A)^l_k \mathbf{G}^{-1}\mathbf{e}^k = g_{lj}(L^A)^l_k g^{ik} e_i. \tag{A.271}$$

From (A.270) and (A.271), (A.263) follows. To show that (A.264) is valid, let us recall the definition (A.189) and (A.190)

$$\langle \mathbf{e}^i, \mathbf{L}\mathbf{e}_j \rangle = \langle \mathbf{L}^A \mathbf{e}^i, \mathbf{e}_j \rangle. \tag{A.272}$$

Thus, the left-hand side gives

$$\langle \mathbf{e}^i, \mathbf{L}\mathbf{e}_j \rangle = L^i_j, \tag{A.273}$$

and the right-hand side becomes

$$\langle \mathbf{L}^A \mathbf{e}^i, \mathbf{e}_j \rangle = (L^A)^i_j. \tag{A.274}$$

From (A.263) we note that, if the bases are orthonormal, the matrix of the components of $\tilde{\mathbf{L}}^A$ is equal to the matrix of the components of $\mathbf{L}^A$ and, as a consequence, it is the transpose of the matrix associated with the components of $\mathbf{L}$. Moreover, the bilinear function corresponding to $\tilde{\mathbf{L}}^A$ is the transpose of that corresponding to $\mathbf{L}$ (see (A.254)). As suggested by the previous statement, we can also characterize the operator $\tilde{\mathbf{L}}^A$ from the following condition

$$\mathbf{u} \cdot \tilde{\mathbf{L}}^A \mathbf{v} = \mathbf{L}\mathbf{u} \cdot \mathbf{v} \qquad \forall \mathbf{u}, \mathbf{v} \in V. \tag{A.275}$$

In this manner, the operator $\tilde{\mathbf{L}}^A \in \mathscr{L}(V; V)$ is also named as the transpose operator of the operator $\mathbf{L} \in \mathscr{L}(V; V)$ and, for this reason, it is simply represented by $\mathbf{L}^T$.

Comparing this condition with (A.259) we have that the tensor $\mathbf{Q}$ is orthogonal if and only if

$$\mathbf{Q}\mathbf{Q}^T = \mathbf{Q}^T\mathbf{Q} = \mathbf{I}, \tag{A.276}$$

or equivalently

$$\mathbf{Q}^T = \mathbf{Q}^{-1}. \tag{A.277}$$

From the previous definitions, in an inner product space, the dual space need not be specified, or even used or mentioned.

The inner product $\mathbf{G}$ in V allows us to define the inner products in the tensor spaces defined over V. For example, we can define an inner product in τ^1_2 through the condition

$$(\mathbf{u} \otimes \mathbf{v} \otimes \mathbf{w}) \cdot (\mathbf{a} \otimes \mathbf{b} \otimes \mathbf{c}) = (\mathbf{u} \cdot \mathbf{a})(\mathbf{v} \cdot \mathbf{b})(\mathbf{w} \cdot \mathbf{c}), \tag{A.278}$$

valid for all tensors $(\mathbf{u} \otimes \mathbf{v} \otimes \mathbf{w}) \in \tau^1_2$. In particular, for the space of linear transformations in V, $\mathscr{L}(V; V)$, the inner product is given by

$$\mathbf{L} \cdot \mathbf{N} = \mathrm{tr}(\mathbf{L}^T\mathbf{N}), \tag{A.279}$$

which is consistent with the condition $(\mathbf{u} \otimes \mathbf{v}) \cdot (\mathbf{a} \otimes \mathbf{b}) = (\mathbf{u} \cdot \mathbf{a})(\mathbf{v} \cdot \mathbf{b})$ for all tensors $(\mathbf{u} \otimes \mathbf{v}) \in \tau^1_1 \simeq \mathscr{L}(V; V)$.

Equivalently, if V is an inner product space, a linear transformation in $\mathscr{L}(V; V)$ is said to be symmetric if coincides with its transpose, that is, $\mathbf{L} = \mathbf{L}^T$, and it is said to be skew-symmetric if $\mathbf{L} = -\mathbf{L}^T$. Moreover, a linear transformation $\mathbf{L}$ is positive definite if the bilinear function associated with it is positive definite, that is

$$\mathbf{u} \cdot \mathbf{L}\mathbf{u} > 0 \qquad \forall \mathbf{u} \in V, \ \mathbf{u} \neq \mathbf{0}. \tag{A.280}$$

The symmetric part of $\mathbf{L}$, also denoted by $\mathbf{L}^s$, is given by

$$\mathbf{L}^s = \frac{1}{2}(\mathbf{L} + \mathbf{L}^T), \tag{A.281}$$

and the skew-symmetric part is given by

$$\mathbf{L}^a = \frac{1}{2}(\mathbf{L} - \mathbf{L}^T). \tag{A.282}$$

From the previous definitions it follows that all linear transformations $\mathbf{L}$ can be written as $\mathbf{L} = \mathbf{L}^s + \mathbf{L}^a$, and these components also satisfy

$$\mathbf{u} \cdot \mathbf{L}\mathbf{u} = \mathbf{u} \cdot \mathbf{L}^s\mathbf{u}, \tag{A.283}$$

$$\mathbf{u} \cdot \mathbf{L}^a\mathbf{u} = 0 \qquad \forall \mathbf{u} \in V. \tag{A.284}$$

Hence, we conclude that $\mathbf{L}$ is positive definite if and only if the symmetric component is positive definite.

B

Elements of Real and Functional Analysis

B.1 Introduction

In the previous appendix we introduced a series of concepts and definitions which are required from Chapter 3 onwards. In particular, the definitions of the translation of a subpsace, of a cone, the concept of dual space, orthogonal complement, and conjugate cone, among others, were useful in the conception of the Principle of Virtual Power and the Principle of Complementary Virtual Power. The same happens with the concept of the adjoint operator, which in the mechanical realm corresponds to the concept of the equilibrium operator ($\mathscr{D}$ and $\mathscr{D}^*$).

However, in the previous appendix as well as, to some extent, in Chapter 3, we employed a purely algebraic vision that, for the case of finite-dimensional vector spaces, was enough to scrutinize the underlying theoretical concepts. Strictly, this is not enough if we are dealing with infinite-dimensional spaces (as is the case of the spaces usually employed in the mechanics of continua) because, for example, if $\mathcal{M}$ is a vector subspace of X, in general, $(\mathcal{M}^{\perp})^{\perp} \subset \mathcal{M}$. Thus, it is necessary to introduce further concepts such as those of open and closed sets, convergence and continuity, among others, which are associated with the topological structure that is required by such vector spaces for the mathematical treatment to be correct.

In order to do this, in this appendix we will briefly address some concepts of the real number system for which we consider the usual addition and multiplication operations, as well as the natural ordering of the real numbers (if $x \leq y$ then $x + z \leq z + y$, $\forall z \in \mathbb{R}$; if $x \geq 0$ and $y \geq 0$ then $xy \geq 0$ and, therefore, "$\leq$" is a complete ordering!).

If $x \in \mathbb{R}$, the absolute value of x, denoted by $|x|$, is defined by

$$|x| = \begin{cases} x & \text{if } x \geq 0, \\ -x & \text{if } x \leq 0. \end{cases} \tag{B.1}$$

Consider $x, y \in \mathbb{R}$, then it is verified the triangle inequality is

$$|x + y| \leq |x| + |y|. \tag{B.2}$$

Also, observe that, for $x, y \in \mathbb{R}$, the absolute value satisfies $|x - y| \geq 0$, $|x - y| = 0$ if and only if $x = y$ and $|x - y| = |y - x|$.

Moreover, the following difference between absolute values is also verified, establishing, for $x, y \in \mathbb{R}$, that

$$||x| - |y|| \leq |x - y|. \tag{B.3}$$

Introduction to the Variational Formulation in Mechanics: Fundamentals and Applications, First Edition.
Edgardo O. Taroco, Pablo J. Blanco and Raúl A. Feijóo.

Let $a_1, a_2, \dots, a_n$ and $b_1, b_2, \dots, b_n$ be real numbers, the Cauchy–Schwarz inequality establishes that

$$\left(\sum_{i=1}^{n} a_i b_i\right)^2 \leq \left(\sum_{i=1}^{n} a_i^2\right)\left(\sum_{i=1}^{n} b_i^2\right). \tag{B.4}$$

Let $\mathscr{A} \subset \mathbb{R}$ be a subset of $\mathbb{R}$ such that for all $x \in \mathscr{A}$ it is verified that $x \leq a$. Then a is an upper bound for the subset $\mathscr{A}$ and, in this case, we say that the subset $\mathscr{A}$ is bounded from above.

Let $\mathscr{A} \subset \mathbb{R}$ be bounded from above, and suppose that there exists $a \in \mathbb{R}$ such that a is an upper bound for $\mathscr{A}$, and $a \leq b$ for all upper bound b for $\mathscr{A}$. Then, a is the smallest upper bound, or the supremum of $\mathscr{A}$. Analogously, we can define sets bounded from below, lower bound, and infimum of a subset.

Here it is important to note that the complete ordering simply establishes that every subset bounded from above (below) has a supremum (an infimum). In turn, it is not difficult to prove that the supremum (infimum) is unique.

Let us now see some properties of the sets of points that can be extended to other, more general, mathematical systems. The topology of the system of real numbers refers to concepts of open sets, neighborhood, and specific classifications of the points in the sets.

Let $a \in \mathbb{R}$, the neighborhood $N_\delta(a)$ is the set of points $x \in \mathbb{R}$ such that $|x - a| < \delta$ where δ is a given positive number.

Let $\mathscr{A} \in \mathbb{R}$ be non-empty. The point $a \in \mathbb{R}$ is an interior point if there exists $\delta > 0$ such that $N_\delta(a) \subset \mathscr{A}$, that is, all the points $x \in N_\delta(a) \subset \mathscr{A}$.

The set $\mathscr{A} \subset \mathbb{R}$ is an open set if all $x \in \mathscr{A}$ is an interior point. Let us see some examples. Consider $a < b$, then

$$(a, b) = \{x \in \mathbb{R};\ a < x < b\}, \tag{B.5}$$

$$[a, b) = \{x \in \mathbb{R};\ a \leq x < b\}, \tag{B.6}$$

$$(a, b] = \{x \in \mathbb{R};\ a < x \leq b\}, \tag{B.7}$$

$$(a, \infty), (-\infty, a), (-\infty, \infty) = \mathbb{R}, \emptyset, \tag{B.8}$$

are examples of open sets.

Let $\mathscr{A} \subset \mathbb{R}$. The point a, not necessarily belonging to $\mathscr{A}$, is an accumulation point in $\mathscr{A}$ if and only if every neighborhood $N_\delta(a)$ contains at least a point of $\mathscr{A}$ different from a. Thus, $N_\delta(a)$ contains infinite points (see the result presented below).

Let $\mathscr{A} \subset \mathbb{R}$, the closure of this set is the set that contains all the points of $\mathscr{A}$ and all the accumulation points. Then, if $\hat{\mathscr{A}}$ is the subset of the accumulation points, $\overline{\mathscr{A}} = \mathscr{A} \cup \hat{\mathscr{A}}$ is the closure of $\mathscr{A}$.

The set $\mathscr{A}$ is closed if and only if $\mathscr{A} = \overline{\mathscr{A}}$. Also, this is equivalent to saying that $\mathscr{A}$ is closed if its complement, $\mathbb{R} \setminus \mathscr{A}$, is open. Observe that $\mathbb{R}$ is closed because its complement, the empty set $\emptyset$, is an open set.

Let us enunciate a first result. Let a be an acumulation point in $\mathscr{A} \subseteq \mathbb{R}$, then, every neighborhood of a contains an infinite number of points from $\mathscr{A}$ (proof follows by contradiction!).

The previous result tells us that every set containing a finite number of points cannot contain accumulation points. The question therefore is concerned with the equivalent result in infinite sets.

The following result is known as the Bolzano–Weierstrass theorem. Let $\mathscr{A} \subset \mathbb{R}$ be a bounded infinite set, then there exists at least one point in $\mathbb{R}$ which is an accumulation point.

Let $\mathscr{A} = \{1, \frac{1}{2}, \frac{1}{3}, \dots, \frac{1}{n}, \dots\}$ be an infinite set where all elements are such that $a \in \mathscr{A}$ for $0 < a \leq 1$, Therefore, the number 0 is an accumulation point.

The set $\mathscr{A} \subset \mathbb{R}$ is a dense set in $\mathscr{B} \subset \mathbb{R}$ if $\mathscr{B} \subseteq \overline{\mathscr{A}}$. Moreover, if $\mathscr{B} = \overline{\mathscr{A}}$, we say that $\mathscr{A}$ is dense almost everywhere in $\mathscr{B}$.

The following result is known as the Heine–Borel theorem. The set $\mathscr{A} \subset \mathbb{R}$ is compact if and only if $\mathscr{A}$ is closed and bounded. In turn, if $\mathscr{A} \subset \mathbb{R}$ is compact, every infinite subset of $\mathscr{A}$ has a point of accumulation in $\mathscr{A}$.

B.2 Sequences

If at each positive integer number n we associate a point $a_n \in \mathbb{R}$, the set $a_1, a_2, \dots, a_n, \dots$ forms a sequence that is represented by $\{a_n\}$. For example, if we adopt the following expressions for a_n

$$a_n = \frac{1}{n} \quad a_n = \frac{1}{2^n} \quad a_n = \left(\frac{n+2}{n+1}\right)^n, \tag{B.9}$$

we can describe the following sequences in $\mathbb{R}$

$$1, \frac{1}{2}, \frac{1}{3}, \dots, \tag{B.10}$$

$$1, \frac{1}{4}, \frac{1}{8}, \dots, \tag{B.11}$$

$$\frac{3}{2}, \left(\frac{4}{3}\right)^2, \left(\frac{5}{4}\right)^3, \dots. \tag{B.12}$$

We say that the sequence $\{a_n\} \subset \mathbb{R}$ has a limit $a^* \in \mathbb{R}$ if for all $\epsilon > 0$ there exists an integer number N such that $|a_n - a^*| < \epsilon$ for all $n > N$. If the sequence has a limit, we say that the sequence converges, or that it is a convergent sequence. This is denoted by

$$\lim_{n\to\infty} a_n = a^*, \tag{B.13}$$

otherwise, we say that the sequence diverges, or that it is a divergent sequence.

We say that a sequence is monotonically increasing if $a_{n+1} > a_n$ for all n (decreasing if $a_{n+1} < a_n$). We say that the sequence is bounded from above if there exists a number b such that $a_n < b$ for all n. Likewise, the sequence is bounded from below if there exists b such that $b < a_n$ for all n.

Given a sequence $\{u_1, u_2, \dots, u_n, \dots\}$, then the sequences $\{u_1, u_4, u_7, \dots\}$ or $\{u_2, u_4, u_6, \dots\}$, etc., are called subsequences of $\{u_n\}$.

An important result states that all bounded monotonic sequences (increasing or decreasing) are convergent. The so-called Bolzano–Weierstrass theorem for sequences states that all bounded sequences have a subsequence that converges.

In fact, bounded sequences which are non-monotonic can either converge or not. For example, the sequence

$$\{1, -1, 1, -1, \dots, (-1)^{n+1}, \dots\}, \tag{B.14}$$

is bounded, but it is not convergent. However, if we take the subsequence

$$\{1, 1, \ldots\}, \tag{B.15}$$

it is clearly convergent.

The set $\mathscr{A} \subset \mathbb{R}$ is said to be compact if every infinite sequence of points in $\mathscr{A}$ has a subsequence that converges.

We say that the sequence $\{a_n\}$ is a Cauchy sequence if for all $\epsilon > 0$ there exists an integer number N such that

$$|a_n - a_m| < \epsilon \qquad \forall n, m > N. \tag{B.16}$$

Observe that, if the sequence is convergent, then it is also a Cauchy sequence. In fact, let a^* be the limit of the sequence, then

$$|a_n - a_m| = |a_n - a^* - (a_m - a^*)| \leq |a_n - a^*| + |a_m - a^*|, \tag{B.17}$$

since $\{a_n\}$ is convergent, we can always pick N such that for $n, m > N$ we get

$$|a_n - a^*| < \epsilon_1 \quad |a_m - a^*| < \epsilon_2 \qquad \text{with } \epsilon_1 + \epsilon_2 = \epsilon, \tag{B.18}$$

then (B.16) is verified.

The set $\mathscr{A} \subset \mathbb{R}$ is said to be complete if every Cauchy sequence of points in $\mathscr{A}$ converges for some point in $\mathscr{A}$.

Let us present two examples. The set of rational numbers is not complete because, for example, the sequence $\{1, 1.4, 1.41, 1.412, \ldots\}$ has as a limit the irrational number $\sqrt{2}$. The set $\mathbb{R}$ is complete.

B.3 Limit and Continuity of Functions

Consider the function

$$\begin{aligned} f : \mathbb{R} &\to \mathbb{R}, \\ x &\mapsto f(x), \end{aligned} \tag{B.19}$$

we say that this function has the limit a at point x_0 if for all $\epsilon > 0$ there exists a number $\delta > 0$ such that if $0 < |x - x_0| < \delta$, then $|f(x_0) - a| < \epsilon$. This is represented by

$$\lim_{x \to x_0} f(x) = a. \tag{B.20}$$

The function $f : \mathscr{A} \subset \mathbb{R} \to \mathbb{R}$ is continuous at the accumulation point $x_0 \in \mathscr{A}$ if and only if $f(x_0)$ exists, and $\lim_{x \to x_0} f(x) = f(x_0)$. Clearly, this definition is achieved via the limit definition.

The function $f : \mathscr{A} \subset \mathbb{R} \to \mathbb{R}$ is continuous at point $x_0 \in \mathscr{A}$ if for all $\epsilon > 0$ there exists $\delta > 0$ such that

$$|f(x) - f(x_0)| < \epsilon \text{ whenever } |x - x_0| < \delta. \tag{B.21}$$

This definition is called the ϵ-δ definition.

The notion of continuity is closer to the concept of open sets than to the concept of limit or to the concept given by the ϵ-δ definition. In fact, the definition $|f(x) - f(x_0)| < \epsilon$ defines a neighborhood $N_\epsilon(f(x_0))$ in $R(f)$. In particular

$$N_\epsilon(f(x_0)) = \{f(x);\ |f(x) - f(x_0)| < \epsilon\}, \tag{B.22}$$

defines an open neighborhood with radius ϵ in $f(x_0)$. If f is continuous at x_0, then the points x mapped by f to this neighborhood can be found in another neighborhood of x_0 whose radius is δ. In other words, for each $\epsilon > 0$ there exists $\delta > 0$ such that the inverse image of all open neighborhoods in $f(x_0)$ are included in an open neighborhood of x_0.

The previous observation leads to the following definition of continuity, which is introduced through the concept of open neighborhood. Let $\mathscr{A}, \mathscr{B} \subset \mathbb{R}$ be two sets, the function

$$f : \mathscr{A} \to \mathscr{B}, \tag{B.23}$$

is continuous at $x_0 \in \mathscr{A}$ if the inverse image of every open neighborhood of $f(x_0)$ in $\mathscr{B}$ is contained in an open neighborhood of x_0 in $\mathscr{A}$. If the continuity is verified for all $x_0 \in \mathscr{A}$ we say that f is continuous in $\mathscr{A}$.

An important result is established in the so-called Weierstrass theorem for a continuous function. Let $f : \mathbb{R} \to \mathbb{R}$ be a continuous function in the closed interval $[a, b]$. Then, $f(x)$ assumes all the intermediate values between $f(a)$ and $f(b)$.

Finally, let us recall some concepts that we have used in the chapters of this text. Consider $f : \mathscr{A} \subset \mathbb{R} \to \mathbb{R}$, and let x_0 be an accumulation point in $\mathscr{A}$. We say that $f'(x_0)$ is the derivative of f at x_0 if the following limit

$$\lim_{x \to x_0} \frac{f(x) - f(x_0)}{x - x_0} = f'(x_0), \tag{B.24}$$

exists.

If $f'(x_0)$ exists, we also say that f is differentiable at x_0. If f is differentiable for all $x_0 \in \mathscr{A}$ we say that f is differentiable in $\mathscr{A}$.

Let $J : \mathbb{R} \to \mathbb{R}$ be a function defined by

$$J(\epsilon) = \int_{x_1(\epsilon)}^{x_2(\epsilon)} f(x, \epsilon) dx, \tag{B.25}$$

then

$$\frac{dJ}{d\epsilon} = f(x_2, \epsilon)\frac{dx_2}{d\epsilon} - f(x_1, \epsilon)\frac{dx_1}{d\epsilon} + \int_{x_1}^{x_2} \frac{\partial f}{\partial \epsilon} dx, \tag{B.26}$$

is the so-called Leibniz rule to calculate the derivative of an integral expression.

In turn, the integration by parts rule is repeatedly utilized throughout this text, and expresses the following

$$\int_{x_1}^{x_2} g\frac{df}{dx} dx = gf\Big|_{x_1}^{x_2} - \int_{x_1}^{x_2} \frac{dg}{dx} f dx. \tag{B.27}$$

B.4 Metric Spaces

Let us now generalize the concept of absolute value from real numbers to more complex objects. To this we introduce the concept of metric.

Let X be a non-empty space, the function $d : X \times X \to \mathbb{R}$ is called a metric if the following properties are satisfied

i) $d(x, y) \geq 0$ for all $x, y \in X$
ii) $d(x, y) = 0$ if and only if $x = y$ (strictly positive)
iii) $d(x, y) = d(y, x)$ for all $x, y \in X$ (symmetric)
iv) $d(x, z) \leq d(x, y) + d(y, z)$ for all $x, y, z \in X$ (triangle inequality).

In general, in the literature we see that when a metric is introduced in a set the elements are called points and $d(x, y)$ is the distance between these points. Observe that properties (i)–(iv) are a generalization of the properties of absolute value, and of the geometric concept we have about distance between points in the Euclidean space.

Remember at this point that the presentation is being limited to real vector spaces. Consider the real vector space X and define a metric d, then (X, d) is called a metric space.

Let us present some examples of metric spaces.

- The classical metric space is the real vector space $\mathbb{R}$ where the metric is given by
$$d(x, y) = |x - y|. \tag{B.28}$$
- Consider $X = \mathbb{R} \times \mathbb{R}$, and given arbitrary $x = (x_1, x_2)$ and $y = (y_1, y_2)$ and the metric defined by
$$d(x, y) = \sqrt{|x_1 - y_1|^2 + |x_2 - y_2|^2}, \tag{B.29}$$
then (X, d) corresponds to the two-dimensional Euclidean space.
- For the space $X = \mathbb{R} \times \mathbb{R}$, the metric $d(x, y)$ given by
$$d_1(x, y) = |x_1 - y_1| + |x_2 - y_2|, \tag{B.30}$$
satisfies all the properties to be considered a metric.

We thus see that for the same set of elements we can choose different metrics, giving rise to different metric spaces.

We introduce now the so-called Hölder's inequality, which states that

$$\sum_{i=1}^{n} |x_i y_i| \leq \left(\sum_{i=1}^{n} |x_i|^p \right)^{\frac{1}{p}} \left(\sum_{i=1}^{n} |y_i|^q \right)^{\frac{1}{q}} \quad 1 < p < \infty, \ \frac{1}{p} + \frac{1}{q} = 1. \tag{B.31}$$

This property permits Minkowski's inequality to be demonstrated, which establishes that, for $1 \leq p < \infty$, the following holds

$$\left(\sum_{i=1}^{n} |x_i \pm y_i|^p \right)^{\frac{1}{p}} \leq \left(\sum_{i=1}^{n} |x_i|^p \right)^{\frac{1}{p}} + \left(\sum_{i=1}^{n} |y_i|^p \right)^{\frac{1}{p}}. \tag{B.32}$$

With these elements, we can prove that (X, d_p) is a metric space, where

$$X = \underbrace{\mathbb{R} \times \mathbb{R} \times \cdots \times \mathbb{R}}_{n \text{ times}}, \tag{B.33}$$

and the metric $d_p : X \to \mathbb{R}$ is given by

$$d_p(x,y) = \left(\sum_{i=1}^{n} |x_i - y_i|^p \right)^{\frac{1}{p}}. \tag{B.34}$$

With this result, we have the following metrics in $\mathbb{R} \times \mathbb{R}$

$$d_1(x,y) = |x_1 - y_1| + |x_2 - y_2|, \tag{B.35}$$

$$d_2(x,y) = (|x_1 - y_1|^2 + |x_2 - y_2|^2)^{\frac{1}{2}}, \tag{B.36}$$

$$d_3(x,y) = (|x_1 - y_1|^3 + |x_2 - y_2|^3)^{\frac{1}{3}}, \tag{B.37}$$

$$\vdots$$

$$d_p(x,y) = (|x_1 - y_1|^p + |x_2 - y_2|^p)^{\frac{1}{p}}, \tag{B.38}$$

$$d_\infty(x,y) = \max_{1 \le i \le n} |x_i - y_i|. \tag{B.39}$$

Let us present other examples of metric spaces but now using function spaces.

- The space $C(a,b)$ of real-valued functions which are continuous in the interval (a,b) with the Chebyshev metric

$$d_\infty(f,g) = \max_{x \in (a,b)} |f(x) - g(x)|, \tag{B.40}$$

is a metric space.

- For the spaces $L^p(a,b)$, for $1 \le p < \infty$, where the functions (equivalence classes) are such that (measures in the sense of the Lebesgue measure) they satisfy

$$\int_a^b |f(x)|^p dx < \infty, \tag{B.41}$$

the following metric can be defined

$$d_p(f,g) = \left(\int_a^b |f - g|^p dx \right)^{\frac{1}{p}}, \tag{B.42}$$

where the triangle inequality is obtained from Minkowski's inequality for integrals

$$\left(\int_a^b |f(x) \pm g(x)|^p dx \right)^{\frac{1}{p}} \le \left(\int_a^b |f(x)|^p dx \right)^{\frac{1}{p}} + \left(\int_a^b |g(x)|^p dx \right)^{\frac{1}{p}}. \tag{B.43}$$

Making use of the metric, the concepts of continuity and convergence that we saw previously for functions and sequences in $\mathbb{R}$ can be extended to functions and sequences in metric spaces.

Consider $\mathcal{A} \subset X$ and let (X,d) be a metric space. Then, $\mathcal{A}$ is dense everywhere in X if and only if for all $\epsilon > 0$ and all $x \in X$ there exists an element $x_A \in \mathcal{A}$ such that

$$d(x, x_A) < \epsilon. \tag{B.44}$$

The space (X, d) is said to be separable if there exists a numerable set of elements which is dense everywhere in X. In other words, if there exists $\{x_1, x_2, \dots, x_n, \dots\}$, $x_i \in X$, $\forall i$, such that for each $\epsilon > 0$ and each $x \in X$ there exists an element x_n in this set such that

$$d(x, x_n) < \epsilon. \tag{B.45}$$

Let (X, d) be a metric space. The sequence $\{x_n\}$ is a Cauchy sequence if

$$d(x_n, x_m) \to 0 \qquad \text{for } n, m \to \infty. \tag{B.46}$$

A sequence $\{x_n\}$ in (X, d) is convergent if there exists $x \in X$ such that

$$d(x_n, x) \to 0 \qquad \text{for } n \to \infty. \tag{B.47}$$

From this definition we note that every convergent sequence is a Cauchy sequence. However, a Cauchy sequence is not necessarily convergent.

Consider, for example, the sequence $\{\frac{1}{n}\}$ in (X, d), where

$$X = (0, 1] \qquad d(x, y) = |x - y|. \tag{B.48}$$

Clearly

$$\lim_{n\to\infty} \frac{1}{n} = 0 \notin X. \tag{B.49}$$

If we add the zero element to the set X (we complete the set), the sequence $\{\frac{1}{n}\}$ is convergent.

The metric space (X, d) is said to be complete if every Cauchy sequence in (X, d) is a converging sequence in (X, d).

For example, we have that the metric space $(C(a, b), d_\infty)$ is complete, and the metric space $(L^p(a, b), d_p)$ is also complete.

B.5 Normed Spaces

Let V be a real vector space, the function $\|\cdot\| : V \to \mathbb{R}$ is a norm in V if the following properties are satisfied

i) $\|u\| \geq 0$, $\forall u \in V$, and $\|u\| = 0$ if and only if $u = 0$
ii) $\|\alpha u\| = |\alpha|\|u\|$, $\forall u \in V$, $\forall \alpha \in \mathbb{R}$
iii) $\|u + v\| \leq \|u\| + \|v\|$.

With this definition we have that $(V, \|\cdot\|)$ is a normed vector space.

Let us present some examples. The space of continuous functions in the interval (a, b) with the norm $\|f\| = \max\limits_{x\in(a,b)} |f(x)|$, that is $(C(a, b), \max\limits_{x\in(a,b)} |f(x)|)$, is a normed space. If we consider other norms such as $\|f\| = (\int_a^b |f(x)|^2 dx)^{\frac{1}{2}}$ and $\|f\| = \int_a^b |f(x)| dx$, then the spaces given by $(C(a, b), (\int_a^b |f(x)|^2 dx)^{\frac{1}{2}})$ and $(C(a, b), \int_a^b |f(x)| dx)$ are normed spaces.

From the properties of a norm, we can observe that the norm naturally induces the metric

$$d(f, g) = \|f - g\|, \tag{B.50}$$

from which we conclude that every normed space is a metric space. The reciprocal statement is not true. In this manner, and from the previous examples, we see that, notwithstanding the real vector space $V = C(a,b)$ remains the same for all cases, the normed spaces are different. In particular, the cases $(C(a,b), (\int_a^b |f(x)|^2 dx)^{\frac{1}{2}})$, and $(C(a,b), \int_a^b |f(x)|dx)$ are not complete with respect to the metric (topology) induced by the norm. The first case in which we have $(C(a,b), \max_{x\in(a,b)} |f(x)|)$ is a complete normed space.

The sequence $\{x_n\}$ is a Cauchy sequence in $(V, \|\cdot\|)$ if for all $\epsilon > 0$ there exists N such that

$$\|u_n - u_m\| < \epsilon \quad \forall n, m > N, \tag{B.51}$$

or equivalently

$$\|u_n - u_m\| \to 0 \qquad \text{for } n, m \to \infty. \tag{B.52}$$

The space $(V, \|\cdot\|)$ is complete if every Cauchy sequence converges to some element in the same space. In this case the space is a Banach space. Normed vector spaces where all Cauchy sequences are convergent (i.e., Banach spaces) are particularly important in the realm of mechanics. In fact, in the spaces $\mathscr{V}$ and $\mathscr{W}$ of motion actions and deformation actions, respectively, adequate norms are chosen in order to ensure that these are Banach spaces. This has profound implications when in the analysis of the problem we are interested in proving the existence and uniqueness of solutions. Also the approximation of the problem depends upon the norms selected because the construction of a sequence of approximate solutions must be such that it converges to the solution of the problem.

Let us present some examples of Banach spaces.

i) Consider the n-dimensional Euclidean space $\mathbb{R}^n$. Let $\{x_i\}$ be a Cauchy sequence, then for all $\epsilon > 0$ there exists N such that

$$\|x_m - x_n\|^2 = \sum_{i=1}^{n} (x_m^{(i)} - x_n^{(i)})^2 < \epsilon^2 \qquad n, m > N. \tag{B.53}$$

For each component i it is verified that $|x_m^{(i)} - x_n^{(i)}| < \epsilon$, for $m, n > N$, and so we have that $\{x_m^{(i)}\}$ is a Cauchy sequence of real numbers, which converges, for example, to $x^{(i)} = \lim_{m\to\infty} \{x_m^{(i)}\}$. We can thus construct $x = \{x^{(i)}, \dots, x^{(n)}\}$, and get $\|x_m - x\| \to 0$, for $m \to \infty$, concluding that $\mathbb{R}^n$ is complete.

ii) Every finite-dimensional normed space is complete. This result has been utilized throughout this text.

iii) Many infinite-dimensional normed spaces are not complete. The space $C(0,1)$ with the norm

$$\|f\| = \left(\int_0^1 \|f(x)\|^2 dx \right)^{\frac{1}{2}}, \tag{B.54}$$

is not complete. Indeed, the sequence of continuous functions given by

$$u_n(x) = \begin{cases} 2^n x^{n+1} & 0 \le x \le \frac{1}{2}, \\ 1 - 2^n(1-x)^{n+1} & \frac{1}{2} \le x \le 1, \end{cases} \tag{B.55}$$

is a Cauchy sequence with respect to the normed defined above. Nevertheless, the sequence converges to the function $u \notin C(0,1)$, in fact $\|u_n - u\| \to 0$ for $n \to \infty$ (see

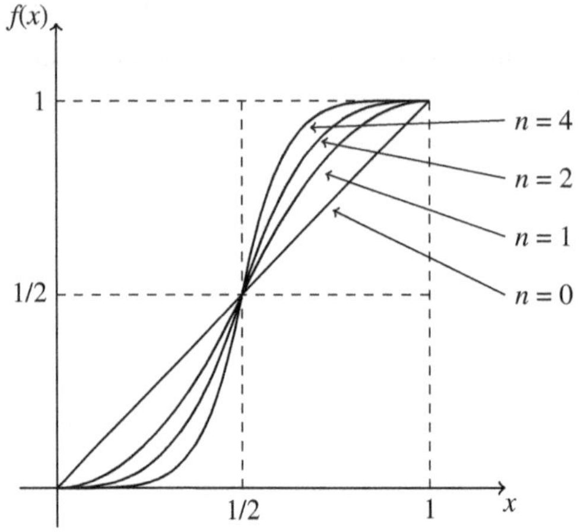

Figure B.1 The space C(0, 1) with the norm defined in (B.54) is not complete.

Figure B.1) where u is given by

$$u(x) = \begin{cases} 0 & 0 \leq x < \frac{1}{2}, \\ \frac{1}{2} & x = \frac{1}{2}, \\ 1 & \frac{1}{2} < x \leq 1. \end{cases} \tag{B.56}$$

iv) The space $C(0, 1)$ with the norm

$$\|f\| = \int_0^1 |f(x)| dx, \tag{B.57}$$

is not complete. In effect, the Cauchy sequence of continuous functions

$$f_n(x) = \begin{cases} 0 & x \leq \frac{1}{2} - \frac{1}{n}, \\ nx - \frac{n}{2} + 1 & \frac{1}{2} - \frac{1}{n} \leq x \leq \frac{1}{2}, \\ 1 & \frac{1}{2} \leq x < 1, \end{cases} \tag{B.58}$$

converges to the discontinuous function shown in Figure B.2.

v) The space $C(0, 1)$ is complete when it is endowed with the norm

$$\|f\| = \sup_{x \in (0,1)} |f(x)|. \tag{B.59}$$

In fact, for an arbitrary $x_0 \in (0, 1)$, the Cauchy sequence $\{f_n(x)\}$ leads to a Cauchy sequence of real numbers

$$|f_n(x_0) - f_m(x_0)| < \sup_{x \in (0,1)} |f_n(x) - f_m(x)| \to 0. \tag{B.60}$$

Hence, there exists a number, say $f(x_0)$, for which $f_n(x_0)$ converges for all $x_0 \in (0, 1)$. In this way, the Cauchy sequence $\{f_n(x)\}$ uniformly converges in $(0, 1)$ to $f(x)$. We have yet to prove that $f(x) \in C(0, 1)$. To do this, let $\{x_m\}$ be a sequence that converges to x in $(0, 1)$. Then

$$|f(x) - f(x_m)| \leq |f(x) - f_n(x_m)| + |f_n(x_m) - f(x_m)|. \tag{B.61}$$

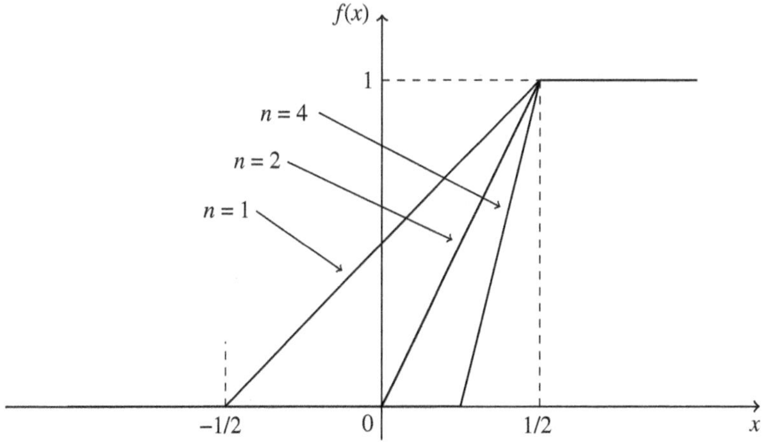

Figure B.2 The space $C(0, 1)$ with the norm defined in (B.57) is not complete.

Since f_n is continuous $f_n(x_m) \rightarrow f_n(x)$ for $m \rightarrow \infty$. In addition, $f_n(x) \rightarrow f(x)$. Similarly, $f_n(x_m) \rightarrow f(x_m)$ for $n \rightarrow \infty$. Hence, the right-hand side can be made as small as we want, that is

$$|f(x) - f(x_m)| \rightarrow 0 \qquad m \rightarrow \infty. \tag{B.62}$$

In other words

$$f(x_m) \rightarrow_{m\rightarrow\infty} f(x), \tag{B.63}$$

then f is continuous. Then $(C(0, 1), \sup_{x\in(0,1)} |f(x)|)$ is a complete space, that is, it is a Banach space.

vi) Let us go back to the example in which $C(0, 1)$ was an incomplete space in relation to the norm

$$||f|| = \left(\int_0^1 |f(x)|^2\right)^{\frac{1}{2}}. \tag{B.64}$$

Notwithstanding this, we can complete this space. Such completion is termed the space of $L^2(0, 1)$ functions, or more precisely equivalence classes, which are integrable in the sense given by the Lebesgue measure. That is, in this space we cannot distinguish the difference between functions that differ solely at a set of null measure. In such a case we say that f and g are equal almost everywhere if

$$\int_0^1 (f - g)^2 dx = 0. \tag{B.65}$$

B.6 Quotient Space

A further important concept in mechanics is that of the quotient space. To illustrate this concept let us consider the subspace M of the vector space X, and let us generate in X linear manifolds V through translations of M. The linear manifolds obtained in this manner can be viewed as elements of a new vector space that we will call the quotient space of X with modulus M, and is represented by X/M. For example, let M be a plane that goes through the origin of the space $\mathscr{V}$ associated with the three-dimensional Euclidean point space $\mathscr{E}$. Then, $\mathscr{V}/M$ results in the vector space whose elements are all the planes parallel to M. Next, we formalize this idea.

Let M be a subspace of the vector space X. Two elements $x_1, x_2 \in X$ are said to be modulus M equivalent if $x_1 - x_2 \in M$. In particular, all the elements in X modulus M equivalents to $x \in X$ are represented simply by $[x]$.

This equivalence relation enables us to partition X into equivalence classes and, in this way, the quotient space X/M consists of all the equivalence classes thus defined, and where the addition and multiplication by scalar are defined by

$$[x_1] + [x_2] = [x_1 + x_2], \tag{B.66}$$

$$\alpha[x] = [\alpha x]. \tag{B.67}$$

In several problems in the domain of mechanics the uniqueness of the solution can be established up to a rigid body motion. In particular, the set of all the rigid body motions, given by the kernel of the deformation rate operator, $\mathscr{N}(\mathscr{D})$ (see Chapter 3), is a subspace of the vector space $\mathscr{V}$. Then, working with the quotient space $\mathscr{V}/\mathscr{N}(\mathscr{D})$ we recover the uniqueness of the solution. Finally, it is interesting to observe that if X is a Banach space and M is a closed subspace, then X/M is also a Banach space.

B.7 Linear Transformations in Normed Spaces

Consider now the normed vector spaces U and V, and let T be the linear (operator) transformation $T : U \to V$. We define:

- the null space (or kernel) to $\mathscr{N}(T) = \{u \in U;\ Tu = 0\}$
- the image to $R(T) = \{v \in V;\ \exists u \in U \text{ such that } v = Tu\}$.

Observe that T is injective if and only if $\mathscr{N}(T) = \{0\}$.

We say that T is a bounded (or bounded from above) linear transformation if and only if there exists a non-negative number K such that

$$\|Tu\|_V \le K\|u\|_U \qquad \forall u \in U, \tag{B.68}$$

where $\|\cdot\|_V$ and $\|\cdot\|_U$ are, respectively, the norms in V and U.

From the previous definition we can always associate with each bounded linear transformation the collection of positive numbers K such that $\|Tu\|_V \le K\|u\|_U$. If, from this set, we take the smallest of all the values K we are establishing a correspondence between the linear transformation T and that number that we will call the norm of T

$$\|T\| = \inf\{K;\ \|Tu\|_V \le K\|u\|_U,\ \ \forall u \in U\}. \tag{B.69}$$

Clearly, the previous definition is equivalent to

$$\|T\| = \sup_{\substack{u\in U\\ u\neq 0}} \frac{\|Tu\|_V}{\|u\|_U}, \tag{B.70}$$

$$||T|| = \sup_{\substack{u \in U \\ ||u||_U \leq 1}} ||Tu||_V, \tag{B.71}$$

$$||T|| = \sup_{\substack{u \in U \\ ||u||_U = 1}} ||Tu||_V. \tag{B.72}$$

Let us analyze the following example. Consider the derivative operator

$$\mathscr{D} u = \frac{du}{dx}, \tag{B.73}$$

defined in the set of differentiable functions with norm given by

$$||u|| = \sup_{x \in (0,1)} |u(x)|. \tag{B.74}$$

Evidently, $\mathscr{D} : C^1(0,1) \to C(0,1)$ is linear but is not bounded in the previous norm. In fact, let

$$u_n(x) = \sin(nx), \tag{B.75}$$

then

$$|u_n| = \sup_{x \in (0,1)} |\sin(nx)| = 1 \qquad \forall n. \tag{B.76}$$

Hence, we have

$$\mathscr{D} u_n = n\cos(nx) \quad \Rightarrow \quad ||\mathscr{D} u_n|| = n, \tag{B.77}$$

given that $||u_n|| = 1$ and $||\mathscr{D} u_n||$ grows indefinitely as $n \to \infty$, then we cannot find a number, say K, such that

$$||\mathscr{D} u|| < K||u|| \qquad \forall u \in C^1(0,1). \tag{B.78}$$

However, if we introduce the norm

$$||u|| = \max\left\{ \sup_{x \in (0,1)} |u(x)|, \sup_{x \in (0,1)} |(\mathscr{D} u)(x)| \right\}, \tag{B.79}$$

we have that $\mathscr{D}$ now becomes a bounded linear operator. Here it is interesting to note that, in general, the differential operators of order m can be considered bounded operators provided that an adequate norm is adopted.

In the case of nonlinear operators $P : U \to V$, the concept of boundedness in $S \subset U$ takes the form

$$||P(u) - P(v)||_V \leq K||u - v||_U \qquad \forall u, v \in S \subset U, \tag{B.80}$$

and we say that the operator satisfies the Lipschitz condition in S.

The concept of continuity for transformations P (either linear or not) in normed spaces is now given as follows. Let $P : U \to V$ be a transformation where U and V are normed vector spaces. Then, P is continuous at $u_0 \in U$ if for all $\epsilon > 0$ there exists $\delta > 0$ such that

$$||P(u) - P(u_0)||_V < \epsilon, \tag{B.81}$$

provided that

$$||u - u_0||_U < \delta. \tag{B.82}$$

Equivalently, P is continuous in $u_0 \in U$ if for every sequence $\{u_n\} \in U$ converging to u_0 (that is $\lim_{n\to\infty} \|u - u_0\|_U = 0$), the sequence $\{P(u_n)\}$ converges to $P(u_0)$ in the norm of space V.

For linear operators, there exists a correspondence between boundedness and continuity. Let us present this result. A linear transformation $T : U \to V$ is continuous if and only if it is bounded.

Observe that if P is a nonlinear operator that satisfies the Lipschitz condition, then it is also a continuous operator.

Many times in the field of mechanics we find operators that satisfy an inequality that is exactly the opposite to the previous one (boundedness from above). These operators are said to be bounded from below. In particular, the linear operator $T : U \to V$ is limited from below if and only if there exists a constant $C > 0$ such that

$$\|Tu\|_V \geq C\|u\|_U \qquad \forall u \in U. \tag{B.83}$$

As an example, let us consider the operator $\mathscr{D} u = \frac{du}{dx} : C^1(0,1) \to C(0,1)$ which, as we have already seen, is not bounded for the norm $\|u\| = \sup_{x\in(0,1)} |u(x)|$. This operator is still not bounded in the case we take the space V to be $C_0(0,1)$, where the index "0" implies that the elements are continuous functions in $(0,1)$ that are zero at $x = 0$ and at $x = 1$. Now, if $u \in C_0(0,1)$ we have

$$u(x) = \int_0^x \frac{du}{dt} dt \leq \sup_{x\in(0,1)} \left|\frac{du}{dx}\right|, \tag{B.84}$$

thus

$$\|\mathscr{D} u\| \geq \sup_{x\in(0,1)} \left|\frac{du}{dx}\right| = \|u\|, \tag{B.85}$$

and so we conclude that this operator is bounded from below.

An important aspect of this kind of operator resides in the fact that, although they are not continuous, they always have a continuous inverse operator defined in the image. This result is presented in the following statement.

Let $A : U \to V$ be an operator bounded from below and let U and V be normed vector spaces. Then, A has an continuous inverse operator $A^{-1} : R(A) \to U$. Reciprocally, if A has a continuous inverse operator, then it is limited from below.

B.8 Topological Dual Space

In Section A.16 we defined the algebraic dual space V^* of the vector space V as the vector space formed by all linear functionals (not necessarily continuous) in V. Let us consider now the space formed by all linear and continuous functionals (therefore bounded) in V. This space is known as the topological dual space of V, and is represented by V'

$$V' = \mathscr{L}(V, \mathbb{R}). \tag{B.86}$$

From the definition it follows that the topological dual space is a smaller space than the algebraic dual space. In turn, and as we have seen in the previous section, this space is a normed space where the norm of an element $\ell \in V'$ is defined by

$$\|\ell\|_{V'} = \sup_{\substack{v\in V\\ v\neq 0}} \frac{|\langle \ell, v\rangle|}{\|v\|_V}, \tag{B.87}$$

then

$$|\langle \ell, v\rangle| \leq \|\ell\|_{V'}\|v\|_V \qquad \forall \ell \in V',\ v \in V. \tag{B.88}$$

It can be shown that even in the case that V is not a Banach space, the topological dual space is always a Banach space. Thus, we can talk about the space formed by all the linear and continuous functionals in V', called the bidual space, which is a complete space and is represented by V''. In general, V cannot be identified with its bidual, but when this identification is possible we say that V is a reflexive space.

Let M be a subspace of the normed vector space U. Then, given a linear and continuous functional in M, it can be extended to the whole space U as established by the Hahn–Banach theorem, which states the following. Let M be a subspace of the separable normed vector space U and let ℓ be a linear and continuous functional in M. Then, there exists a linear and continuous functional, denoted by $\tilde{\ell}$, defined in U, such that $\tilde{\ell}$ is an extension of ℓ, that is $\tilde{\ell}(u) = \ell(u)$, $\forall u \in M$, and $\|\tilde{\ell}\|_{U'} = \|\ell\|_{M'}$.

As a corollary of the previous result we have that if $\alpha_0 \in \mathbb{R}^+$ and $u_0 \in U$ is an arbitrary element, then there exists a linear functional $\ell \in U'$ such that

$$\|\ell\|_{U'} = \alpha_0 \quad \ell(u_0) = \|\ell\|_{U'}\|u_0\|_U. \tag{B.89}$$

These two concepts allow us to prove the following result, which was employed in Chapter 3.

Let M be a closed subspace of the normed vector space U, then

$$(M^{\perp})^{\perp} = M. \tag{B.90}$$

Finally, the following results were also used in Chapter 3.

Let U and V be two normed vector spaces and let $T : U \to V$ be a linear and continuous transformation. Then, $\mathcal{N}(T)$ is a closed subspace of U. Moreover, observe that, from the previous result we have that $(\mathcal{N}(T)^{\perp})^{\perp} = \mathcal{N}(T)$.

Let U and V be two Banach spaces and let $T : U \to V$ be an injective linear and continuous transformation. Then, the image $R(T)$ of T is closed if and only if T is bounded from below.

B.9 Weak and Strong Convergence

Let U be a Banach space and U' the topological dual space. Then, given the sequence $\{u_n\} \in U$ we say that it converges strongly, or that converges in the norm of U, to the element $u_0 \in U$ if for all $\epsilon > 0$ there exists N such that

$$\|u_n - u_0\|_U < \epsilon \qquad \forall n > N, \tag{B.91}$$

or equivalently

$$\lim_{n\to\infty} \|u_n - u_0\|_U = 0. \tag{B.92}$$

With the definition of the dual space we can therefore introduce a new sense of convergence, known as weak convergence. Thus, we say that u_n converges weakly to u_0 if the following is verified

$$|\ell(u_n - u_0)| = |\langle \ell, u_n - u_0 \rangle| < \epsilon \qquad \forall n > N, \quad \forall \ell \in U', \tag{B.93}$$

which is equivalent to

$$\lim_{n\to\infty} |\ell(u_n - u_0)| = 0 \qquad \forall \ell \in U'. \tag{B.94}$$

In turn, since the dual space is also a normed space (moreover, it is a Banach space), we can talk about strong convergence and weak convergence of linear functionals. So, given the sequence $\{\ell_n\} \in U'$, we say that it converges strongly* to $\ell \in U'$ if

$$\lim_{n\to\infty} \|\ell_n - \ell\|_{U'} = 0. \tag{B.95}$$

In turn, we say that this sequence converges weakly* if

$$\lim_{n\to\infty} \langle \ell_n - \ell, u \rangle = 0 \qquad \forall u \in U. \tag{B.96}$$

C

Functionals and the Gâteaux Derivative

C.1 Introduction

In this appendix we study in more detail several results that were utilized in Chapter 4 when dealing with the stiffness mechanical operators, with the Gâteaux derivatives of functionals, and with convex functions.

C.2 Properties of Operator $\mathscr{K}$

Let $\mathscr{D}$ be the deformation rate operator satisfying the properties already established in Chapter 3 and let $\mathscr{K} = \mathscr{D}^*\mathbb{D}\mathscr{D}$ be the stiffness operator corresponding to the analyzed hyperelastic materials (recall that $\mathbb{D}$ is a positive definite symmetric fourth-order tensor in the space of deformations). Next, we show a series of fundamental results.

The first result establishes that

$$\mathscr{N}(\mathscr{D}) = \mathscr{N}(\mathscr{K}). \tag{C.1}$$

To prove (C.1) let us recall that $\mathbb{D}$ is positive definite, and then it follows that $\mathscr{D}u = 0$, $\forall u \in \mathscr{N}(\mathscr{D})$, so $\mathbb{D}\mathscr{D}u = 0$, $\forall u \in \mathscr{N}(\mathscr{D})$. Therefore, we conclude that $\mathscr{D}^*\mathbb{D}\mathscr{D}u = 0$, $\forall u \in \mathscr{N}(\mathscr{D})$, that is $\mathscr{K}u = 0$, $\forall u \in \mathscr{N}(\mathscr{D})$. In this manner we obtain $\mathscr{N}(\mathscr{D}) \subset \mathscr{N}(\mathscr{K})$. In the same manner, we have that $\mathscr{K}u = 0$, $\forall u \in \mathscr{N}(\mathscr{K})$, therefore $\langle \mathscr{K}u, u\rangle = 0$, $\forall u \in \mathscr{N}(\mathscr{K})$, or equivalently $\langle \mathscr{K}u, u\rangle = (\mathbb{D}\mathscr{D}u, \mathscr{D}u) = 0$, $\forall u \in \mathscr{N}(\mathscr{K})$. From the positive definiteness of $\mathbb{D}$ we get $\mathscr{D}u = 0$, $\forall u \in \mathscr{N}(\mathscr{K})$, resulting in $\mathscr{N}(\mathscr{K}) \subset \mathscr{N}(\mathscr{D})$. We have thus demonstrated that $\mathscr{N}(\mathscr{D}) \subset \mathscr{N}(\mathscr{K}) \subset \mathscr{N}(\mathscr{D})$, from which (C.1) follows.

The second result establishes that

$$\mathscr{N}(\mathscr{K}) = R(\mathscr{K})^{\perp}. \tag{C.2}$$

To prove (C.2) we know, on the one hand, that $\mathscr{K}u = 0$, $\forall u \in \mathscr{N}(\mathscr{K})$, and so we get $\langle \mathscr{K}u, w\rangle = 0$, $\forall w \in \mathscr{U}$. Then, $\langle \mathscr{D}^*\mathbb{D}\mathscr{D}u, w\rangle = (\mathbb{D}\mathscr{D}u, \mathscr{D}w) = (\mathbb{D}\mathscr{D}w, \mathscr{D}u) = \langle \mathscr{D}^*\mathbb{D}\mathscr{D}w, u\rangle = \langle \mathscr{K}w, u\rangle = 0$, $\forall w \in \mathscr{U}$, which implies $u \in R(\mathscr{K})^{\perp}$, that is to say, $\mathscr{N}(\mathscr{K}) \subset R(\mathscr{K})^{\perp}$. On the other hand, if $u \in R(\mathscr{K})^{\perp}$ then $\langle \mathscr{K}w, u\rangle = 0$, $\forall w \in \mathscr{U}$, and in particular $\langle \mathscr{K}u, u\rangle = 0$, that is $(\mathbb{D}\mathscr{D}u, \mathscr{D}u) = 0$, and so $\mathscr{D}u = 0$, implying that $u \in \mathscr{N}(\mathscr{D})$, and finally $R(\mathscr{K})^{\perp} \subset \mathscr{N}(\mathscr{D})$. Now, from (C.1) we have that $R(\mathscr{K})^{\perp} \subset \mathscr{N}(\mathscr{K})$, then $\mathscr{N}(\mathscr{K}) \subset R(\mathscr{K})^{\perp} \subset \mathscr{N}(\mathscr{K}) \Leftrightarrow \mathscr{N}(\mathscr{K}) = R(\mathscr{K})^{\perp}$, and then (C.2) follows.

Introduction to the Variational Formulation in Mechanics: Fundamentals and Applications, First Edition.
Edgardo O. Taroco, Pablo J. Blanco and Raúl A. Feijóo.

Under additional hypotheses for the space $\mathscr{U}$, for example

- for a three-dimensional body it is enough to consider $\mathscr{U} = [H^1(\mathscr{B})]^n$, where $[H^1(\mathscr{B})]^n$ is the Sobolev (actually a Hilbert) space whose elements u are vector functions such that are, together with their first derivatives, square integrable in the sense of Lebesgue in the region $\mathscr{B}$ of the n-dimensional Euclidean space
- removing the rigid motion, that is, by working with the quotient space $\mathscr{U}/\mathscr{N}(\mathscr{D})$
- considering that the region $\mathscr{B}$ is regular

it is possible to show that the operator $\mathscr{K} : \mathscr{U} \to \mathscr{U}'$ is bounded from below, and so $R(\mathscr{K})$ is a closed vector subspace of $\mathscr{U}'$. With this, when taking the orthogonal complement to expression (C.2) we obtain the relation

$$\mathscr{N}(\mathscr{K})^{\perp} = R(\mathscr{K}). \tag{C.3}$$

The third result follows from the previous ones, and states that

$$\mathscr{N}(\mathscr{D}) = \mathscr{N}(\mathscr{K}) = R(\mathscr{K})^{\perp}, \tag{C.4}$$

$$\mathscr{N}(\mathscr{D})^{\perp} = \mathscr{N}(\mathscr{K})^{\perp} = R(\mathscr{K}). \tag{C.5}$$

The fourth result establishes that

$$\text{Var}_u \cap \mathscr{N}(\mathscr{K}) = \text{Var}_u \cap \mathscr{K}(\text{Var}_u)^{\perp}. \tag{C.6}$$

In fact, for $u \in \text{Var}_u \cap \mathscr{N}(\mathscr{K})$ we have $u \in \text{Var}_u$ and $u \in \mathscr{N}(\mathscr{K})$, that is $u \in \text{Var}_u$ and $\langle \mathscr{K} u, w\rangle = 0, \forall w \in \text{Var}_u$, which implies that $u \in \text{Var}_u$ and $\langle \mathscr{K} w, u\rangle = 0, \forall w \in \text{Var}_u$, and therefore $u \in \text{Var}_u$ and $u \in \mathscr{K}(\text{Var}_u)^{\perp}$. In this manner, $u \in \text{Var}_u \cap \mathscr{K}(\text{Var}_u)^{\perp}$, and so we have demonstrated that (C.6) holds.

The fifth result establishes that $[\text{Var}_u \cap \mathscr{K}(\text{Var}_u)^{\perp}]^{\perp} = \text{Var}_u^{\perp} \oplus \mathscr{K}(\text{Var}_u)$, where the operation $\oplus$ indicates direct sum.

To prove the result above note that

$$[\text{Var}_u \cap \mathscr{K}(\text{Var}_u)^{\perp}]^{\perp} = \text{Var}_u^{\perp} + [\mathscr{K}(\text{Var}_u)^{\perp}]^{\perp}, \tag{C.7}$$

and from the properties of $\mathscr{K}$ we have

$$[\text{Var}_u \cap \mathscr{K}(\text{Var}_u)^{\perp}]^{\perp} = \text{Var}_u^{\perp} + \mathscr{K}(\text{Var}_u). \tag{C.8}$$

We need to prove now that the addition can be replaced by the direct sum. To see this we have to prove that $\text{Var}_u^{\perp} \cap \mathscr{K}(\text{Var}_u) = 0$, that is the null element in $\mathscr{U}'$. Consider $f \in \text{Var}_u^{\perp} \cap \mathscr{K}(\text{Var}_u)$, then $f \in \text{Var}_u^{\perp}$ and $f \in \mathscr{K}(\text{Var}_u)$, that is $f \in \text{Var}_u^{\perp}$ and $\exists u \in \text{Var}_u$ such that $\mathscr{K} u = f$, which implies $\langle \mathscr{K} u, w\rangle = 0, \forall w \in \text{Var}_u$ and $f = \mathscr{K} u$, or equivalently $\langle \mathscr{K} u, u\rangle = 0, u \in \text{Var}_u$ and $f = \mathscr{K} u$, which means $u \in \mathscr{N}(\mathscr{K})$ and $f = \mathscr{K} u$, resulting in $f = 0$. Therefore, the only element in the intersection is the zero element, and the result follows.

C.3 Convexity and Semi-Continuity

Let X be a real vector space. We say that the set $\mathscr{C} \subset X$ is convex if

$$\theta x + (1-\theta)y \in \mathscr{C} \qquad \forall x, y \in \mathscr{C}, \quad \theta \in (0,1). \tag{C.9}$$

The functional $\mathscr{F} : \mathscr{C} \to \mathbb{R}$ is convex if

$$\mathscr{F}(\theta x + (1-\theta)y) \leq \theta\mathscr{F}(x) + (1-\theta)\mathscr{F}(y) \qquad \forall x, y \in \mathscr{C}, \quad \theta \in (0,1). \tag{C.10}$$

The functional $\mathscr{F}$ is said to be strictly convex if the strict inequality is satisfied in (C.10).

Given a functional $\mathscr{F}$ defined in the convex $\mathscr{C} \subset X$, we can always define the functional $\tilde{\mathscr{F}}$ in the whole X given by

$$\tilde{\mathscr{F}} : X \to \mathbb{R},$$
$$x \mapsto \tilde{\mathscr{F}}(x) = \begin{cases} \mathscr{F}(x) & \text{if } x \in \mathscr{C}, \\ +\infty & \text{if } x \notin \mathscr{C}. \end{cases} \tag{C.11}$$

From the expression above we conclude that $\mathscr{F}$ is convex if and only if $\tilde{\mathscr{F}}$ is convex.

Let now V be a Banach space, the functional $\mathscr{F} : V \to \overline{\mathbb{R}} = \mathbb{R} \cup \{-\infty, +\infty\}$ is lower semi-continuous if the following equivalent conditions are verified

- $\forall \alpha \in \mathbb{R}$ the set $\{x \in V;\ \mathscr{F}(x) \leq \alpha\}$ is closed
- $\forall \alpha \in \mathbb{R}$ the set $\{x \in V;\ \mathscr{F}(x) > \alpha\}$ is open.

C.4 Gâteaux Differential

We say that the functional $\mathscr{F} : V \to \mathbb{R}$ is Gâteaux differentiable at point $v_0 \in V$ in the direction $v \in V$ if the quotient

$$\frac{\mathscr{F}(v_0 + \epsilon v) - \mathscr{F}(v_0)}{\epsilon}, \tag{C.12}$$

has a limit when $\epsilon \to 0$. In this case, this is represented by

$$\delta\mathscr{F}(v_0; v) = \lim_{\epsilon \to 0} \frac{1}{\epsilon}[\mathscr{F}(v_0 + \epsilon v) - \mathscr{F}(v_0)] = \frac{\partial}{\partial \epsilon}\mathscr{F}(v_0 + \epsilon v)\bigg|_{\epsilon=0}. \tag{C.13}$$

In particular $\delta\mathscr{F}(v_0; v)$ is termed the Gâteaux differential of the functional $\mathscr{F}$ at point v_0 according to the direction v. This differential is also known as the first variation of $\mathscr{F}$ at point v_0 in the direction given by v.

Moreover, if there exists $D\mathscr{F}(v_0) \in V'$ such that

$$\delta\mathscr{F}(v_0; v) = \langle D\mathscr{F}(v_0), v \rangle, \tag{C.14}$$

we say that $D\mathscr{F}(v_0)$ is the Gâteaux derivative of the functional $\mathscr{F}$ at point v_0.

Let $\mathscr{F} : V \to \mathbb{R}$ be a Gâteaux differentiable functional. Then, the functional $\mathscr{F}$ is convex if and only if the mapping $D\mathscr{F}(v) \in V'$ is monotone, which implies

$$\langle D\mathscr{F}(v_2) - D\mathscr{F}(v_1), v_2 - v_1 \rangle \geq 0 \qquad \forall v_1, v_2 \in V. \tag{C.15}$$

C.5 Minimization of Convex Functionals

Let us consider the functional $\mathscr{J} : V \to \mathbb{R}$ and let $\mathscr{C} \subset V$ be a set of V, where V is a Banach space. We say that $\mathscr{J}(u_0)$ is a relative minimum of $\mathscr{J}$ in $\mathscr{C}$ if $u_0 \in \mathscr{C}$ and there exists a neighborhood $\mathscr{N}_\delta(u_0)$ such that

$$\mathscr{J}(u) \geq \mathscr{J}(u_0) \qquad \forall u \in \mathscr{C} \cap \mathscr{N}_\delta(u_0). \tag{C.16}$$

In turn, we say that $\mathcal{J}(u_0)$ is an absolute minimum of $\mathcal{J}$ in $\mathcal{C}$ if $u_0 \in \mathcal{C}$ and if

$$\mathcal{J}(u) \geq \mathcal{J}(u_0) \qquad \forall u \in \mathcal{C}. \tag{C.17}$$

Let V be a reflexive Banach space. If $\mathcal{J}$ is lower semi-continuous and if $\mathcal{C}$ is a bounded closed set of V, then there exists at least an absolute minimum in $\mathcal{C}$.

We must observe that when V is a finite-dimensional space the previous result tells us that every (real-valued) function which is continuous over a bounded closed set reaches its minimum value at least once.

If V is a reflexive Banach space and $\mathcal{J}$ is lower semi-continuous and coercive, that means if $\mathcal{J}$ is such that

$$\lim_{\|u\|_V \to \infty} \mathcal{J}(u) = +\infty, \tag{C.18}$$

then there exists at least an absolute minimum.

Let $\mathcal{C}$ be an open set in V, then:

i) if $u_0 \in \mathcal{C}$ is a relative minimum of $\mathcal{J}$ in $\mathcal{C}$ and if $\mathcal{J}$ is Gâteaux differentiable, then

$$\delta\mathcal{J}(u_0; v) = 0 \qquad \forall v \in V, \tag{C.19}$$

ii) if $\mathcal{J}$ is convex and Gâteaux differentiable, then every local minimum is an absolute minimum, and also

$$\mathcal{J}(u) \geq \mathcal{J}(u_0) \qquad \forall u \in \mathcal{C}, \tag{C.20}$$

is equivalent to the following variational expression

$$\delta\mathcal{J}(u_0; u) = 0 \qquad \forall u \in V, \tag{C.21}$$

iii) if $\mathcal{J}$ is strictly convex, then u_0 is unique.

If $\mathcal{J}$ is Gâteaux differentiable in $\mathcal{C}$, and $\mathcal{C}$ is a convex subset in the reflexive Banach space V, and if $u_0 \in \mathcal{C}$ is such that

$$\mathcal{J}(u) \geq \mathcal{J}(u_0) \qquad \forall u \in \mathcal{C}, \tag{C.22}$$

then

$$\delta\mathcal{J}(u_0; u - u_0) \geq 0 \qquad \forall u \in \mathcal{C}. \tag{C.23}$$

Moreover, if $\mathcal{J}$ is convex expressions (C.22) and (C.23) are equivalent.

References

1 Akhiezer, N.I.: *The Calculus of Variations*. Blaisdell Pub. (1962)
2 Ames, W.F.: *Nonlinear Partial Differential Equations in Engineering*, vol. 1-2. Academic Press (1972)
3 Ames, W.F.: *Numerical Methods for Partial Differential Equations*. Academic Press (1977)
4 Andrade, E.N.: On the viscous flow in metals and allied phenomena. *Proc. of the Royal Society A* 84(1) (1910)
5 Andrade, E.N.: The flow in metals under large constant stresses. *Proc. of the Royal Society A* 90(329) (1914)
6 Argyris, J.H.: Energy theorems and structural analysis. *Aircraft Eng.* 26, 347–356 (1954)
7 Argyris, J.H.: Energy theorems and structural analysis. *Aircraft Eng.* 27, 42–58 (1955)
8 Argyris, J.H.: Matrix analysis of three dimensional elastic media, small and large deflections. *AIAA J.* 3, 45–51 (1965)
9 Argyris, J.H.: Tetrahedron elements with linearly varying strain for the matrix displacement method. *J. Roy. Aeron. Soc.* 69, 877–880 (1965)
10 Argyris, J.H.: Triangular elements with linearly varying strain for the matrix displacement method. *J. Roy. Aeron. Soc.* 69, 711–713 (1965)
11 Argyris, J.H.: Matrix displacement analysis of plates and shells. *Engr. Arch* 35, 102–142 (1966)
12 Argyris, J.H., Fried, I., Scharpf, D.W.: The TUBA family of plate elements for the matrix displacement method. *Aeron. J. Roy. Aeron. Soc.* 72, 701–709 (1968)
13 Argyris, J.H., Kelsey, S.: *Energy Theorems and Structural Analysis*. Butterworth and Co., London (1960)
14 Argyris, J.H., Scharpf, D.W.: The SHEBA family of shell elements for the displacement method. *Aeron. J. Roy. Aeron. Soc.* 72, 873–883 (1968)
15 Argyris, J.H., Scharpf, D.W.: Finite element in space and time. *Nucl. Eng. Design* 10(4), 456–464 (1969)
16 Ashby, M.F.: A first report on deformation mechanism maps. *Acta Metallurgical* 20, 887–898 (1972)
17 Ashwell, D.G., Gallagher, R.H.: Finite elements for thin shells and curved members. In: *Conference on Finite Elements Applied to Thin Shells and Curved Members*. John Wiley, Cardiff, 1974 (1976)
18 Aubin, J.P.: *Approximation of Elliptic Boundary Value Problems*. John Wiley (1972)
19 Bachman, G., Narici, L.: *Functional Analysis*. Academic Press (1966)

Introduction to the Variational Formulation in Mechanics: Fundamentals and Applications, First Edition.
Edgardo O. Taroco, Pablo J. Blanco and Raúl A. Feijóo.

20 Bailey, R.W.: Creep of steel under simple and coumpound stresses and the use of high initial temperature in steam power plants. In: *Trans. World Power Conference* (Vol. 3), Tokyo, p. 1089 (1929). Also published in Engineering, 129, 265, 1930
21 Bailey, R.W.: The utilization of creep test data in engineering design. *Proc. of the Inst. of Mech. Engineers* 131, 260 (1936)
22 Barbosa, H.J.C., Feijóo, R.A.: Numerical algorithms for contact problems in linear elastostatic. In: *Conference on Stress Analysis in Nuclear Power Plants.* Porto Alegre, Brazil (1984)
23 Bartle, R.G.: *The Elements of Real Analysis*, 2nd edn. Wiley (1976)
24 Bazaraa, M.S., Shetty, C.M.: *Nonlinear Programming: Theory and Algorithms.* John Wiley (1979)
25 Belytschko, T., Loehnert, S., Song, J.H.: Multiscale aggregating discontinuities: A method for circumventing loss of material stability. *Int. J. Numer. Meth. Eng.* 73, 869–894 (2008)
26 Belytschko, T., Song, J.H.: Coarse-graining of multiscale crack propagation. *Int. J. Numer. Meth. Eng.* 81, 537–563 (2010)
27 Bensoussan, A., Lions, J., Papanicolaou, G.: *Asymptotic analysis for periodic structures.* Elsevier, North-Holland (1978)
28 Bernasconi, G., Piatti, G.: Creep of Engineering Materials and Structures. Applied Science Publishers (1979)
29 Besseling, J.F.: A theory of elastic, plastic and creep deformations of an initially isotropic material, showing anisotropic strain-hardening, creep recovery and secondary creep. *ASME J. Appl. Mech.* 25, 529–536 (1958)
30 Bevilacqua, L., Feijóo, R.A., Rojas, L.: Torsión de barras seccionalmente homogéneas. Technical Report 22.73, COPPE/UFRJ, Rio de Janeiro, Brazil (1973)
31 Bevilacqua, L., Feijóo, R.A., Rojas, L.: A variational principle for the Laplace's operator with application in the torsion of composite rods. *Int. J. Solids Struct.* 10, 1091–1102 (1974)
32 Blanco, P., Clausse, A., Feijóo, R.: Homogenization of the Navier-Stokes equations by means of the Multi-scale Virtual Power Principle. *Comput. Meth. App. Mech. Eng.* 315, 760–779 (2017)
33 Blanco, P., Giusti, S.: Thermomechanical multiscale constitutive modeling: accounting for microstructural thermal effects. *J. Elast.* 115, 27–46 (2014)
34 Blanco, P., Sánchez, P., de Souza Neto, E., Feijóo, R.: Variational foundations and generalized unified theory of RVE-based multiscale models. *Arch. Comput. Methods Eng.* 23, 191–253 (2016)
35 Blanco, P., Sánchez, P., de Souza Neto, E., Feijóo, R.: The Method of Multiscale Virtual Power for the derivation of a second order mechanical model. *Mech. Mater.* 99, 53–67 (2016)
36 Bliss, G.A.: *Lectures of the Calculus of Variations.* University of Chicago Press (1945)
37 Bolza, O.: *Lectures of the Calculus of Variations.* Dover Pub. (1961)
38 Boyle, J.T., Spence, J.: Generalized structural model in creep mechanics. In: *Creep in Structures, 3rd Symposium.* A.R.S. Ponter and D.R. Hayhurst (eds.). Springer-Verlag (1981)
39 Boyle, J.T., Spence, J.: *Stress Analysis for Creep.* Butterworths (1983)
40 Bramble, J.H., Hilbert, S.R.: Bounds for a class of linear functionals with applications to Hermite interpolation. *Num. Math.* 16, 362–369 (1971)

41 Brebbia, C.A., Tottenham, H. (eds.): *Variational Methods in Engineering*. Southampton University Press (1973)

42 Brézis, H.: Problèmes unilatéraux. *J. Math. Pures et Appl.* 51, 1–168 (1972)

43 Budiansky, B.: On the elastic moduli of some heterogeneous materials. *J. Mech. Phys. Solids* 13, 223–227 (1965)

44 Calladine, C.R., Drucker, D.C.: Nesting surfaces of constant rate of energy dissipation in creep. *Quart. Appl. Math.* 20(1), 79–84 (1962)

45 Carey, G.F., Oden, J.T.: *Finite Elements: Fluid Mechanics, The Texas Finite Element Series*, vol. 6. Prentice-Hall (1986)

46 Carnahan, B., Luther, H.A., Wilkes, J.: *Applied Numerical Methods*. John Wiley (1969)

47 Chadwic, P.: *Continuum Mechanics*. John Wiley (1976)

48 Cheney, E.W.: *Introduction to Approximation Theory*. McGraw Hill (1966)

49 Ciarlet, P.G.: *Orders of Convergence in Finite Element Methods*. Mathematics of Finite Elements and Applications. Academic Press (1973)

50 Ciarlet, P.G.: *The Finite Element Method for Elliptic Problems, Studies in Mathematics and its Applications*, vol. 4. North–Holland Publishing Company (1978)

51 Cohn, M.A., Maier, G.: *Engineering Plasticity by Mathematical Programming*. Pergamon Press (1979)

52 Collatz, L.: The Numerical Treatment of Differential Equations, 3rd. edn. Springer Verlag (1966)

53 Comitê sobre Comportamento Inelástico de Materiais da ABCM: Influência dos parâmetros da lei de Norton na determinação de tensões e deformações em materiais que experimentam fenômenos de creep. In: I SIBRAT. Salvador, Brazil. Published also as Technical Report 011/80, Laboratório de Computação Científica, LCC/CNPq, Rio de Janeiro, Brazil (1980)

54 Comitê sobre Comportamento Inelástico de Materiais da ABCM: Comparação entre as equações de Norton e Mukherjee na determinação de tensões e deformações em materiais que experimentan fenômenos de fluência. In: Inter-American Conf. on Materiais Technology, pp. 87–90. Mexico (1981)

55 Comitê sobre Comportamento Inelástico de Materiais da ABCM: Modelagem mecânica do desempenho estrutural de um vaso de pressão submetido a fluência. In: II SIBRAT. Salvador, Brazil. Published also as Technical Report 014/82, Laboratório de Computação Científica, LCC/CNPq, Rio de Janeiro, Brazil (1982)

56 Connor, J.J., Brebbia, C.A.: *Finite Element Techniques for Fluid Flows*. Newnes–Butterworths (1977)

57 Cosserat, E., Cosserat, F.: *Théorie des corps déformables*. Hermann, Paris (1909)

58 Cottle, R., Pang, J.S., Stone, R.: *The Linear Complementarity Problem (Computer Science and Scientific Computing)*. Academic Press, Inc, Boston (1992)

59 Courant, B.: Variational methods for the solutions of problems of equilibrium and vibrations. *Bull. American Math. Soc.* 49, 1–23 (1943)

60 Courant, B.: *Differential and Integral Calculus*, vol. 2. Wiley–Interscience Publishers (1968)

61 Courant, B., Hilbert, D.: *Methods of Mathematical Physics*, vol. 1-2. Interscience Publishers (1953)

62 Cuvelier, C., Segal, A., van Stenhoven, A.A.: *Finite Element Methods and Navier–Stokes Equations*. Mathematics and Its Applications. D. Reidel Publishing Company (1986)

63 Da Fonseca, Z.: Soluciones numéricas via métodos variacionales en problemas de termoelasticidad dinámica acoplada. Master's thesis, Prog. Eng. Mecânica, COPPE/UFRJ, Rio de Janeiro, Brazil (1977)

64 Dahlquist, G., Bjorck, A.: *Numerical Methods*. Prentice Hall (1974)

65 Davis, P.J.: *Interpolation and Approximation*. Dover Pub. (1963)

66 Fraeijs de Veubeke, B.: Bending and stretching of plates. In: Conf. on Matrix Methods in Structural Mechanics. Wright-Patterson Air Force, Ohio, USA (1965)

67 Fraeijs de Veubeke, B.: Displacements and equilibrium models in the finite element method. In: Stress Analysis. O.C. Zienkiewicz and G.S. Holister (eds.), pp. 145–197. John Wiley (1965)

68 Fraeijs de Veubeke, B.: A conforming finite element for plate bending. *Int. J. Solids Struct.* 4, 95–108 (1968)

69 Fraeijs de Veubeke, B.: Variational principles and the patch test. *Int. J. Numer. Meth. Eng.* 8, 783–801 (1974)

70 Destuynder, P.: Introduction à la théorie des coques minces èlastiques. Analyse numérique de problemes de coques minces - description des résultats et exemples. Tech. rep., *INRIA Report* (1981)

71 Detteman, J.W.: *Mathematical Methods in Physics and Engineering*. McGraw Hill (1962)

72 Dikmen, M.: *Theory of thin elastic shells*. Pitman Advanced Publ. Program (1982)

73 Dorfmann, L., Ogden, R.: *Nonlinear Theory of Electroelastic and Magnetoelastic Interactions*. Springer, New York (2014)

74 Dorn, J.E.: Progress in understanding high-temperature creep. In: H.W. Gilette Memorial Lecture. Am. Soc. for Testing and Mat., Philadelphia (1962)

75 Dorn, J.E.: Some fundamental experiments on high temperature creep. *J. Mech. Phys. Solids* 8, 85–116 (1954)

76 Duvaut, G., Lions, J.L.: *Les Inéquations en Méchanique et en Physiques*. Dunod (1972)

77 Dym, C., Shames, I.H.: *Solid Mechanics: A Variational Approach*. McGraw Hill (1973)

78 Ekeland, I., Temam, R.: *Analyse Convexe et Problémes Variationnels*. Dunod (1974)

79 Elsgoltz, L.: *Ecuaciones Diferenciales y Cálculo Variacional*. Editora Mir, Moscú (1969)

80 Eringen, A.: *Microcontinuum Field Theories. I: Foundations and Solids*. Springer-Verlag, New York (1999)

81 Eringen, A.: *Microcontinuum Field Theories. II: Fluent Media*. Springer-Verlag, New York (1999)

82 Faddeeva, F.: *Computational Methods of Linear Algebra*. Freeman, San Francisco (1963)

83 Feijóo, R.A.: Aplicación del método de Ritz a funcionales relajados en mecánica de los sólidos. Master Thesis, Prog. de Eng. Civil, COPPE/UFRJ, Rio de Janeiro, Brazil (1973)

84 Feijóo, R.A.: Introducción a Mecánica del Contínuo. In: I Escola de Matemática Aplicada. Laboratório de Computação Científica, LCC/CNPq, Rio de Janeiro, Brazil (1978)

85 Feijóo, R.A.: Algunos aspectos de la matemática aplicada en mecánica del continuo. *Boletim da Soc. Bras. Mat. Aplic. e Computacional* 2, 33–62 (1982)

86 Feijóo, R.A., Barbosa, H.J.C.: Static analysis of piping systems with unilateral supports. In: VII COBEM-Congresso Brasileiro de Engenharia Mecânica, pp. D:635–638. Brazil (1983)

87 Feijóo, R.A., Bevilacqua, L.: Sequential interpolation fuctions. *Int. J. Numer. Meth. Eng.* 10, 133–144 (1976)

88 Feijóo, R.A., Rojas, L., Bevilacqua, L.: Application of variational principles to extended functional in discontinuous fields for the beam problem. In: *I Simpósio Brasileiro de Ciências Mecânicas.* Rio de Janeiro, Brazil (1973)

89 Feijóo, R.A., Rojas, L., Bevilacqua, L.: Generalización del método de Ritz vía funcionales relajados. In: *XVI Jornadas Sudamericanas de Ingeniería Estructural.* Buenos Aires, Argentina (1974)

90 Feijóo, R.A., Rojas, L., Taroco, E., Bevilacqua, L.: Métodos variacionales en cáscaras axisimétricas sometidas a cargas arbitrarias. In: *XVIII Jornadas Sudamericanas de Ingeniería Estructural.* Salvador, Bahia, Brazil (1976)

91 Feijóo, R.A., Taroco, E.: Algoritmos numéricos en creep secundario. In: II Congreso Latinoamericano sobre Métodos Computacionales. Curitiba, Brazil. Published also as Technical Report 010/80, Laboratório de Computação Científica, LCC/CNPq, Rio de Janeiro, Brazil (1980)

92 Feijóo, R.A., Taroco, E.: Introducción a Plasticidad y su Formulación Variacional. In: II Escola de Matemática Aplicada, pp. 1–156. Laboratório de Computação Científica, LCC/CNPq, Rio de Janeiro, Brazil (1980)

93 Feijóo, R.A., Taroco, E.: Métodos Variacionais em Mecânica dos Sólidos. In: II Escola de Matemática Aplicada. Laboratório de Computação Científica, LCC/CNPq, Rio de Janeiro, Brazil (1980)

94 Feijóo, R.A., Taroco, E.: Principios y Métodos Variacionales en Mecánica. In: I Curso de Mecânica Teórica e Aplicada, Módulo 1, pp. 117–195. Laboratório de Computação Científica, LCC/CNPq, Rio de Janeiro, Brazil (1982)

95 Feijóo, R.A., Taroco, E.: Principios Variacionales y el Método de los Elementos Finitos en la Teoría de Placas y Cáscaras. In: I Curso de Mecânica Teórica e Aplicada, Módulo 2, pp. 1–116. Laboratório de Computação Científica, LCC/CNPq, Rio de Janeiro, Brazil (1983)

96 Feijóo, R.A., Taroco, E., Guerreiro, J.N.C.: Análisis de tensiones y deformaciones en problemas de creep no estacionario. In: *I SIBRAT*. Salvador, Brazil (1980)

97 Feijóo, R.A., Taroco, E., Guerreiro, J.N.C.: Introducción a Fluencia. In: I Curso de Mecânica Teórica e Aplicada, Módulo 2, pp. 251–298. Laboratório de Computação Científica, LCC/CNPq, Rio de Janeiro, Brazil (1983)

98 Feijóo, R.A., Taroco, E., Guerreiro, J.N.C.: Métodos numéricos en el análisis de tensiones. In: ACETE - Análise de Componentes Estruturais em Temperaturas Elevadas. Laboratório de Computação Científica, LCC/CNPq, Rio de Janeiro, Brazil (1983)

99 Feijóo, R.A., Taroco, E., Zouain, N.: Nuevos resultados en el análisis limite vía creep secundario modificado. Revista Brasileira de Ciências Mecânicas, 2 (IV) 25–34, ABCM - Associação Brasileira de Ciências Mecânicas. Published also as Technical Report 004/82, Laboratório de Computação Científica, LCC/CNPq, Rio de Janeiro, Brazil (1982)

100 Feijóo, R.A., Taroco, E., Zouain, N.: Princípios Variacionais em Mecânica, In: II Curso de Mecânica Teórica e Aplicada. Módulo I, pp. 1–200. Laboratório de Computação Científica, LCC/CNPq, Rio de Janeiro, Brazil (1984)

101 Feijóo, R.A., Zouain, N.: Análisis elastoplástica via optimización. Technical Report 004/83, Laboratório de Computação Científica, LCC/CNPq, Rio de Janeiro, Brazil (1983)

102 Feyel, F., Chaboche, J-L.: FE^2 multiscale approach for modelling the elastoviscoplastic behaviour of long fibre SiC/Ti composite materials. *Comput. Meth. App. Mech. Eng.* 183, 309–330 (2000)

103 Fichera, G.: Boundary Value Problems of Elasticity with Unilateral Constraints, *Handbuch der Physics*, vol. Band via/2, pp. 391–424. Springer Verlag (1972)

104 Fichera, G.: Existence Theorems in Elasticity, *Handbuch der Physics*, vol. Band via/2, pp. 347–389. Springer Verlag (1972)

105 Finlayson, B.A.: *The Method of Weighted Residuals and Variational Principles.* Academic Press (1972)

106 Flugge, W.: *Tensor Analysis and Continuum Mechanics.* Springer Verlag (1972)

107 Forsythe, G.E., Moler, C.: *Computer Solutions of Linear Algebraic Systems.* Prentice Hall (1967)

108 Forsythe, G.E., Wason, W.R.: *Finite Difference Methods for Partial Differential Equations.* John Wiley (1960)

109 Fremond, M.: Méthodes Variationnelles en Calcul des Structures. *École Nationale des Ponts et Chaussées, Paris* (1980)

110 Frost, J.J., Ashby, M.F.: Deformation mechanism maps for pure iron, two austenitic stainless steels, and a low-alloy ferritic steel. Tech. rep., Univ. Cambridge, Department of Engng. (1975)

111 Fung, Y.C.: *Foundations of Solid Mechanics.* Prentice Hall (1965)

112 Gelfand, I.M., Fomin, S.V.: *Calculus of Variations.* Prentice Hall (1963)

113 Geradin, M.: On the variational method in the direct integration of the transient structural response. *J. of Sound and Vibration* 34(4), 479–487 (1974)

114 Germain, P.: Sur l'application de la méthode des puissances virtuelles en mécanique des milieux continus. C. R. Acad. Sci., *Série A* 274, 1051–1055 (1972)

115 Germain, P.: La méthode des puissances virtuelles en méchanique des milieux continus. Première partie: Théorie du second gradient. *Journal de Méchanique* 12(2), 235–274 (1973)

116 Germain, P.: The method of virtual power in continuum mechanics. Part 2: Microstructure. *SIAM J. Appl. Math.* 25(3), 556–575 (1973)

117 Germain, P.: Course de Mécanique, vol. 1, 2. École Polytechnique (1980)

118 Germain, P.: Four Lectures on the Foundation of Shell Theory. In: I Curso de Mecânica Teórica e Aplicada, Módulo 1, Laboratório de Computação Científica, LCC/CNPq, Rio de Janeiro, Brazil (1982)

119 Germain, P., Nguyen, Q.S., Suquet, P.: Continuum thermodynamics. Personal Communication

120 Girault, V., Raviart, P.: Finite Element Methods for Navier-Stokes Equations: Theory and Algorithms. Springer-Verlag (1986)

121 Giusti, S., Blanco, P., de Souza Neto, E., Feijóo, R.: An assessment of the Gurson yield criterion by a computational multi-scale approach. *Engineering Computations: International Journal for Computer-Aided Engineering and Software* 26(3), 281–301 (2009)

122 Glowinsky, R., Lions, J.L., Tremoliers, R.: *Analyse Numérique des Inéquations Variationnelles.* Dunod (1976)

123 Gol'denveizer, A.: *Theory of Elastic Shells*. Pergamon Press, New York (1961)

124 Gonzales Guirnaldes, C.: Métodos variacionales en termoelasticidad. Master Thesis, Prog. de Eng. Mecânica, COPPE/UFRJ, Rio de Janeiro, Brazil (1977)

125 Green, A.: Micro-materials and multipolar continuum mechanics. *Int. J. Eng. Sci.* 3, 533–537 (1965)

126 Green, A., Rivlin, R.: Multipolar continuum mechanics: functional theory. *Proc. of the Royal Society A* 284, 303–324 (1965)

127 Greenbaum, G.A., Rubinstein, M.F.: Creep analysis of axisymmetric bodies using finite elements. *Nuclear Engng. Design* 7, 379–397 (1968)

128 Greenberg, H.J.: Complementary minimum principles for an elastic-plastic material. *Quart. Appl. Math.* 7, 85 (1949)

129 Guerreiro, J.N.C.: Análise de Casos Particulares - Fluência. In: I Curso de Mecânica Teórica e Aplicada. Módulo 2, pp. 415–448. Laboratório de Computação Científica, LCC/CNPq, Rio de Janeiro, Brazil (1983)

130 Gurson, A.: Continuum theory of ductile rupture by void nucleation and growth. Part I: Yield criteria and flow rule for porous media. *J. Engng. Mat. Techn.* 99, 2–15 (1949)

131 Gurtin, M.E.: *The Linear Theory of Elasticity*. In: Truesdell C. (ed.) Handbuch der Physik. Springer-Verlag, Berlin (1972)

132 Gurtin, M.E.: *An Introduction to Continuum Mechanics*. Academic Press (1981)

133 Gurtin, M.E.: *Configurational Forces as Basic Concepts of Continuum Physics*. Springer (1991)

134 Gurtin, M.E., Martins, L.C.: Cauchy's theory in classical physics. *Arch. Rat. Mech. Anal.* 60, 305–324 (1976)

135 Halmos, P.R.: *Finite-Dimensional Vector Spaces*, 2nd edn. Van Nostrand-Reinhold (1958)

136 Halphen, B., Nguyen, Q.S.: Sur les matériaux standards généralisés. *Journal de Mécanique* 14, 39–63 (1975)

137 Hashin, Z., Shtrikman, S.: A variational approach to the theory of elastic behaviour of multiphase materials. *J. Mech. Phys. Solids* 11, 127–140 (1963)

138 Hault, P.: Some comments on thermomechanical constitutive equations for inelastic analysis of LMFBR components. In: 4th Int. Conf. Struct. Mech. Reactor Technology. San Francisco (1977)

139 Hildebrand, F.B., Reissner, E., Thomas, G.: Notes on the foundations of the theory of small displacements of orthotropic shells. NACA Technical Note 1833, NACA (1949)

140 Hill, R.: *The Mathematical Theory of Plasticity*. Oxford University Press (1950)

141 Hill, R.: Elastic properties of reinforced solids: Some theoretical principles. *J. Mech. Phys. Solids* 11, 357–372 (1963)

142 Hill, R.: Continuum micro-mechanics of elastoplastic polycrystals. *J. Mech. Phys. Solids* 13, 89–101 (1965)

143 Hill, R.: A self-consistent mechanics of composite materials. *J. Mech. Phys. Solids* 13, 213–222 (1965)

144 Hill, R.: On constitutive macro-variables for heterogeneous solids at finite strain. *Proc. of the Royal Society A* 326, 131–147 (1972)

145 Hodge, P.G.: *Plastic Analysis of Structures*. McGraw Hill (1959)

146 Hodge, P.G.: *Limit Analysis of Rotationally Symmetric Plates and Shells*. Prentice-Hall (1963)

147 Hult, J.A.H.: *Creep in Engineering Structures*. Blaisdell Pub. Com. (1966)

148 Irving, J., Kirkwood, J.: The statistical mechanical theory of transport processes. IV. The equations of hydrodynamics. *J. Chem. Phys.* 18, 817–829 (1950)

149 Isaacson, E., Keller, H.B.: *Analysis of Numerical Methods*. John Wiley and Sons (1966)

150 Johnson, C.: On finite element methods for plasticity problems. *Numerische Mathematik* 26, 79–84 (1976)

151 Johnson, C.: *Numerical Solution of Partial Differential Equations by the Finite Element Method*. Cambridge University Press (1987)

152 Kachanov, L.M.: *The Theory of Creep*. Lending Library (1967)

153 Kachanov, L.M.: *Fundamentals of The Theory of Plasticity*. MIR Publishers (1974)

154 Kanouté, P., Boso, D.P., Chaboche, J.L. and Schrefler, B.A.: Multiscale methods for composites: A review. *Arch. Comput. Methods Eng.* 16, 31–75 (2009)

155 Kirkwood, J.: The statistical mechanical theory of transport processes. I. General theory. *J. Chem. Phys.* 14, 180–201 (1946)

156 Kirkwood, J.: The statistical mechanical theory of transport processes. II. Transport in gases. *J. Chem. Phys.* 15, 72–76 (1947)

157 Kirkwood, J., Buff, F., Greenn, M.: The statistical mechanical theory of transport processes. III. The coefficients of shear and bulk viscosity of liquids. *J. Chem. Phys.* 17, 988–994 (1949)

158 Koiter, W.T.: A consistent first approximation in the general theory of thin elastic shells. In: IUTAM Symposium on the theory of thin elastic shells. North-Holland, Amsterdam, pages 12-32, (1960)

159 Koiter, W.T.: General theorems for elastic-plastic solids. In: I.N. Sneddon, R. Hill (eds.) *Progress in Solid Mechanics*. North-Holland Press (1960)

160 Koiter, W.T.: On the nonlinear theory of thin elastic shells. *Proceedings of the Koninklijke Nederlandse Akademie van Wetenschappen, Series B* 69, 1–54 (1966)

161 Kolmogorov, A.N., Fomin, S.V.: Elementos de la Teoría de Funciones y del Análisis Funcional. Editora Mir, Moscú (1972)

162 Kouznetsova, V., Brekelmans, W., Baaijens, F.: An approach to micro-macro modeling of heterogeneous materials. *Comp. Mech.* 27, 37–48 (2001)

163 Kouznetsova, V., Geers, M., Brekelmans, W.: Multiscale constitutive modelling of heterogeneous materials with a gradient-enhanced computational homogenization scheme. *Int. J. Numer. Meth. Eng.* 54, 1235–1260 (2002)

164 Kouznetsova, V., Geers, M., Brekelmans, W.: Multiscale second order computational homogenization of multi-phase materials: A nested finite element solution strategy. *Comput. Meth. App. Mech. Eng.* 193, 5525–5550 (2004)

165 Kraus, H.: *Creep Analysis*. John Wiley (1980)

166 Kreyszig, E.: *Introductory Functional Analysis with Applications*. John Wiley (1978)

167 Lanczos, C.: *The Variational Principles of Mechanics*, 4th. edn. University Toronto Press (1970)

168 Larsson, F., Runesson, K., Su, F.: Variationally consistent computational homogenization of transient heat flow. *Int. J. Numer. Meth. Eng.* 81, 1659–1686 (2010)

169 Larsson, R., Diebels, S.: A second-order homogenization procedure for multi-scale analysis based on micropolar kinematics. *Int. J. Numer. Meth. Eng.* 69, 2485–2512 (2007)

170 Larsson, R., Zhang, Y.: Homogenization of microsystem interconnects based on micropolar theory and discontinuous kinematics. *J. Mech. Phys. Solids* 55, 819–841 (2007)

171 Lebedev, L., Cloud, M., Eremeyev, V.: *Tensor Analysis with Applications in Mechanics.* World Scientific, Singapore (2010)

172 Leipholz, H.: Direct Variational Methods and Eigenvalue Problems in Engineering. Noordhoff Int. Pub. (1975)

173 Leonard, R.: Nonlinear first approximation thin shell and membrane theory. Ph.D. thesis, Virginia Polytechnique Institute, Virginia, United States (1961)

174 Liu, I.S.: Introduction to continuum mechanics. In: Textos de Métodos Matemáticos. UFRJ, Rio de Janeiro, Brazil (1988)

175 Loehnert, S., Belytschko, T.: A multiscale projection method for macro/microcrack simulations. *Int. J. Numer. Meth. Eng.* 71, 1466–1482 (2007)

176 Love, A.E.H.: The small free vibrations and deformation of a thin elastic shell. *Proc. of the Royal Society A* 179, 491–546 (1888)

177 Love, A.E.H.: *A treatise on the mathematical theory of elasticity.* Cambridge University Press, 2 Edition (1906)

178 Luenberger, D.G.: *Optimization by Vector Space Methods.* John Wiley (1969)

179 Lur'e, A.: General theory of elastic shells. *Prikl. Mat. Mekh.* 4, 7–34 (1940)

180 Maier, G.: Quadratic programming and theory of elastic perfectly plastic structures. *Meccanica* 3, 265–273 (1968)

181 Maier, G.: Piecewise linearization of yield criteria in structural plasticity. *S. M. Archives* 213, 239–281 (1976)

182 Maier, G., Munzo, J.: Mathematical programming applications to engineering plastic analysis. *Appl. Mech. Rev.* 35, 1631–1643 (1982)

183 Malvern, L.E.: *Introduction to the Mechanics of a Continuous Medium.* Series in Engineering of the Physical Sciences. Prentice–Hall (1969)

184 Mandel, J.: Plasticité Classique et Viscoplasticité. CISM Lecture Notes N^o97. Springer-Verlag (1971)

185 Marchuk, G.I.: *Methods of Numerical Mathematics.* Springer Verlag (1975)

186 Marin, J., Pao, Y.N.: An analitical theory of the creep deformation of materials. *J. Appl. Mech.* 20, 245–252 (1953)

187 Martin, J.B.: *Plasticity, Fundamentals and General Results.* Mit Press (1975)

188 Maugin, G.A: *Material Inhomogeneities in Elasticity.* Chapman & Hall (1993)

189 Maugin, G.A.: The method of virtual power in continuum mechanics: Application to coupled fields. *Acta Mechanica* 35, 1–70 (1980)

190 McDowell, D.: A perspective on trends in multiscale plasticity. *Int. J. Plasticity* 26, 1280–1309 (2010)

191 McVetty, P.: Creep of metals at elevated temperatures. The Hyperbolic-Sine relation between stress and creep rate. *Transactions of the ASME* 65, 761–767 (1943)

192 Meyer, H.D.: The numerical solution of nonlinear parabolic problems by variational methods. *SIAM J. Num. Anal.* 10, 700–722 (1973)

193 Michel, J., Moulinec, H., Suquet, P.: Effective properties of composite materials with periodic microstructure: A computational approach. *Comput. Meth. App. Mech. Eng.* 172, 109–143 (1999)

194 Miehe, C., Koch, A.: Computational micro-to-macro transition of discretized microstructures undergoing small strain. *Arch. Appl. Mech.* 72, 300–317 (2002)

195 Miehe, C., Schotte, J., Lambrecht, J.: Homogenization of inelastic solid materials at finite strains based on incremental minimization principles. Application to the texture analysis of polycrystals. *J. Mech. Phys. Solids* 50, 2123–2167 (2002)

196 Miehe, C., Schotte, J., Schröder, J.: Computational micro-macro transitions and overall moduli in the analysis of polycrystals at large strains. *Computational Materials Science* 16, 372–382 (1999)

197 Miehe, C., Schröder, J., Becker, M.: Computational homogenization analysis in finite elasticity: material and structural instabilities on the micro- and macro-scales of periodic composites and their interaction. *Comput. Meth. App. Mech. Eng.* 191, 4971–5005 (2002)

198 Mikhlin, S.G.: *Variational Methods in Mathematical Physics*. Pergamon Press (1964)

199 Mikhlin, S.G.: *The Problem of the Minimum of Quadratic Functionals*. Holden Day (1965)

200 Mikhlin, S.G.: *Mathematical Physics, an Advanced Course*. North Holland (1970)

201 Mikhlin, S.G.: *The Numerical Performance of Variational Methods*. Nordhoff Pub. (1971)

202 Mikhlin, S.G.: *Approximation on a Rectangular Grid*. Sijthoff-Noordhoff (1979)

203 Mikhlin, S.G., Smolitskiy, K.L.: *Approximate Methods of Solution for Differential and Integral Equations*. Elsevier (1967)

204 Morse, M.: *Variational Analysis, Critical Extremals and Sturmian Extensions*. John Wiley (1973)

205 Mukherjee, A.K., Bird, J.E., Dorn, J.E.: Experimental correlations for high temperature creep. *Transactions of the ASME* 62, 155–179 (1969)

206 Nadai, A.: The Influence of Time upon Creep, The Hyperbolic Sine Creep Law. S. Timoshenko 60th Anniversary Volume, pp. 155–170, Macmillan, New York (1938)

207 Naghdi, P.M.: On the theory of thin elastic shells. *Quart. Appl. Math.* 14, 369–380 (1957)

208 Naghdi, P.M.: Foundations of elastic shell theory. *Prog. Solid Mech.* 4, 1–90 (1963)

209 Naylor, A.W., Sell, G.R.: *Linear Operator Theory in Engineering and Sciences*. Hold, Rinehart, Winston (1971)

210 Necas, J.: *Les Méthodes Directes en Théorie des Équations Elliptiques*. Elsevier Masson (1967)

211 Nemat-Nasser, S.: Averaging theorems in finite deformation plasticity. *Mech. Mater.* 31, 493–523 (1999)

212 Norton, F.H.: *Creep of Steel at High Temperatures*. McGraw Hill (1929)

213 Novozhilov, V.V.: *Thin shell theory*. Wolters-Noordhoff, Groningem (1970)

214 Oden, J.T.: Calculation of stiffness matrices for finite elements of thin shells of arbitrary shape. *AIAA J.* 6, 969–972 (1968)

215 Oden, J.T.: Finite element analysis of nonlinear problems in the dynamic theory of coupled thermoelasticity. *Nuclear Engn. Design* 10, 465–475 (1969)

216 Oden, J.T.: Finite element formulation of problems of finite deformation and irreversible thermodynamics of nonlinear continua - a survey and extension of recent developments. In: *Japan - U. S. Seminar on Matrix Methods of Structural Analysis and Design*. Tokio, Japan (1969)

217 Oden, J.T.: Finite element applications in mathematical physics. In: J.R. Whiteman (ed.) *The Mathematics of Finite Elements and Applications*, pp. 239–282. Academic Press, London (1972)

218 Oden, J.T.: *Finite Elements of Nonlinear Continua*. McGraw Hill (1972)

219 Oden, J.T.: Imbedding technique for the generation of weak-weak finite approximations of linear and nonlinear operators. Journal of the Engineering

Mechanics Division, Proceedings of the American Society of Civil Engineers, pp. 1327–1330 (1972)
220 Oden, J.T.: *Applied Functional Analysis*. Prentice Hall (1979)
221 Oden, J.T. (ed.): Computational Methods in Nonlinear Mechanics. Proceedings of the TICOM Second International Conference, North-Holland, Amsterdam-New York (1980)
222 Oden, J.T.: Exterior penalty methods for contact problems in elasticity. In: W. Wunderlich (ed.) *Nonlinear Finite Element Analysis in Structural Mechanics*. Springer Verlag (1981)
223 Oden, J.T., Aguirre-Ramirez, G.: Formulation of general discrete models of thermomechanical behavior of materials with memory. *Int. J. Solids Struct.* 5, 1077–1093 (1969)
224 Oden, J.T., Armstrong, W.H.: Analysis of nonlinear dynamic coupled thermoviscoelasticity problems by the finite element methods. *Comput. Struct.* 1, 603–621 (1971)
225 Oden, J.T., Carey, G.F.: *Finite Elements: A Second Course, The Texas Finite Element Series*, vol. 2. Prentice–Hall (1983)
226 Oden, J.T., Carey, G.F.: *Finite Elements: Mathematical Aspects, The Texas Finite Element Series*, vol. 4. Prentice–Hall (1983)
227 Oden, J.T., Carey, G.F., Becker, E.B.: *Finite Elements: An Introduction, The Texas Finite Element Series*, vol. 1. Prentice–Hall (1981)
228 Oden, J.T., Kelley, B.E.: Finite element formulation of general electrothermoelasticity problems. *Int. J. Numer. Meth. Eng.* 3, 161–179 (1971)
229 Oden, J.T., Reddy, J.N.: *An Introduction to the Mathematical Theory of Finite Elements*. John Wiley (1976)
230 Oden, J.T., Reddy, J.N.: *Variational Methods in Theoretical Mechanics*. Springer Verlag (1976)
231 Oden, J.T., Somogyi, D.: Finite element applications in fluid dynamics. *J. Engng. Mech. Division ASCE* 5(EM3), 821–826 (1969)
232 Odqvist, F.K.: From Stanford 1960 to Gothenburg 1970. In: J. Hult (ed.) *IUTAM Symposium on Creep in Structures*. Springer Verlag (1972)
233 Odqvist, F.K.: *Mathematical Theory of Creep and Creep Rupture*. Clarendon Press, Oxford (1974)
234 Oikawa, H.: Three-dimensional presentation of deformation mechanism diagrams. *Scripta Metallurgica* 13, 701–705 (1979)
235 Olszak, W., Sawczuk, A.: *Inelastic Behaviour in Shells*. Nordhoff (1967)
236 Panagiotopoulos, P.D.: *Inequality Problems in Mechanics and Applications*. Birkhauser (1990)
237 Penny, R.K., Marriot, D.L.: *Design for Creep*. McGraw Hill (1971)
238 Phillips, H.B.: Vector analysis. John Wiley & Sons (1933)
239 del Piero, G.: Variational methods in Limit Analysis. In: R.A. Feijóo and E.A. Taroco (Eds.) 2nd. Applied Mathematic Summer School, Laboratório de Computação Científica, LCC/CNPq, Rio de Janeiro, Brazil (1980)
240 del Piero, G., Podio-Guidugli, P.: Seminario di Elasticitá Lineare. In: 1a. Scuola Estiva di Fisica Matematica, Ed. Tecnico Scientifica, Italy (1978)
241 Podio-Guidugli, P.: An exact derivation of the thin plate equation. *J. Elast.* 22(2–3), 121–133 (1989)

242 Podio-Guidugli, P.: A virtual power format for thermomechanics. *Continuum Mech. Thermodyn.* 20, 479–487 (2009)

243 Polizzotto, C.: A gradient elasticity theory for second-grade materials and higher order inertia. *Int. J. Solids Struct.* 49, 2121–2137 (2012)

244 Polizzotto, C.: A second strain gradient elasticity theory with second velocity gradient inertia – Part II: Dynamic behavior. *Int. J. Solids Struct.* 50, 3766–3777 (2013)

245 Prager, W.: Strain hardening under combined stresses. *J. Appl. Physics.* 16, 837–840 (1945)

246 Prager, W.: *An Introduction to Plasticity.* Wesley (1959)

247 Prenter, P.M.: *Splines and Variational Methods.* John Wiley (1975)

248 Prezemieniecki, J.S.: *Theory of Matrix Structural Analysis.* McGraw Hill (1968)

249 Rabotnov, Y.N.: *Creep Problems in Structural Members.* North Holland (1969)

250 Rektorys, K.: *Variational Methods in Mathematical Sciences and Engineering.* D. Reidel Publishing Company (1975)

251 Richtmyer, R.D., Morton, K.W.: *Difference Methods for Initial–Value Problems,* second edn. No. 4 in Interscience Tracts in Pure and Applied Mathematics. Interscience Publishers (1967)

252 Robinson, E.L.: 100.000 hours creep test. *Mech. Eng.* 65, 166–168 (1943)

253 Romano, G.: Variational inequalities and extremum principles in incremental elastoplasticity, Part I and II Tech. Rep. 307, Univ. Calabria, Cosenza, Italy (1975)

254 Romano, G.: A general variational theory of incremental elastic-plastic boundary value problems. Tech. Rep. 306, Univ. Napoli, Ist. Scienza delle Costruzion, Italy (1979)

255 Romano, G.: Duality and variational principles in structural mechanics under bilateral and unilateral constrainsts. Tech. Rep. 13, Laboratório de Computação Científica, LCC-CNPq, Rio de Janeiro, Brazil (1982)

256 Romano, G.: Principi e metodi variazionali nella meccanica delle struture e dei solidi, Parte I: Principi. Tech. rep., Facoltà di Ingegneria dell'Università di Napoli, Italy (1984)

257 Romano, G., Alfano, G.: Variational principles and discrete formulations in plasticity. Personal communication

258 Romano, G., Romano, M.: On the foundation of variational principles in linear structural mechanics. Tech. Rep. 307, Univ. Napoli, Ist. Scienza delle Costruzioni, Italy (1979)

259 Romano, G., Rosati, L., Marotti de Sciarra, F.: Variational formulations of non-linear and non-smooth structural problems. *Int. J. Non-Linear Mechanics* 28(2), 195–208 (1993)

260 Romano, G., Rosati, L., Marotti de Sciarra, F.: Variational principles for a class of finite-step elasto-plastic problems with non-linear mixed hardening. *Comp. Meth. App. Mech Engng.* 109, 293–314 (1993)

261 Romano, G., Rosati, L., Marotti de Sciarra, F.: A variational theory for finite-step elasto-plastic problems. *Int. J. Solids Struct.* 30(17), 2317–2334 (1993)

262 Romano, G., Rosati, L., Marotti de Sciarra, F., Bisegna, P.: A potential theory for monotone multivalued operators. *Quart. Appl. Math.* 51(4), 613–631 (1993)

263 Rozenblium, V.I.: Approximate equations of creep in thin shells. *PMM* 27(1), 154–159 (1963)

264 Sagan, H.: *Introduction to the Calculus of Variations.* McGraw Hill (1969)

265 Sagan, H.: *Advanced Calculus of Real-Valued Functions of a Real Variable and Vector-Valued Functions of a Vector Variable.* Houghton Mifflin (1974)

266 Sánchez, P., Blanco, P., Huespe, A., Feijóo, R.: Failure-oriented multi-scale variational formulation for softening materials. Tech. Rep. P&D N°6, LNCC-MCTI Laboratório Nacional de Computação Científica, Brazil (2011)

267 Sánchez, P., Blanco, P., Huespe, A., Feijóo, R.: Failure-oriented multi-scale variational formulation: micro-structures with nucleation and evolution of softening bands. *Comput. Meth. App. Mech. Eng.* 257, 221–247 (2013)

268 Sánchez-Palencia, E.: Non-Homogeneous Media and Vibration Theory, *Lecture Notes in Physics*, vol. 127. Springer-Verlag, Berlin (1980)

269 Sánchez-Palencia, E.: Homogenization method for the study of composite media. In: F. Verhulst (ed.) *Asymptotic Analysis II. Surveys and New Trends. Lecture Notes in Mathematics*, vol. 985, pp. 192–214. Springer Verlag (1981)

270 Sanders, J.L.: An improved first approximation theory for thin shells. Tech. Rep. R-24, NASA Technical Report (1959)

271 Sanders, J.L.: Nonlinear theories for thin shells. *Quart. Appl. Math.* 21, 21–36 (1963)

272 Save, M.A., Massonnet, C.E.: *Plastic Analysis and Design of Beams and Frames.* Blaisdell (1965)

273 Save, M.A., Massonnet, C.E.: *Plastic Analysis and Design of Plates, Shells and Disks.* North-Holland (1972)

274 Song, J.H., Belytschko, T.: Multiscale aggregating discontinuities method for micro-macro failure of composites. *Composites Part B* 40, 417–426 (2009)

275 de Souza Neto, E., Blanco, P., Sánchez, P., Feijóo, R.: An RVE-based multiscale theory of solids with micro-scale inertia and body force effects. *Mech. Mater.* 80, 136–144 (2015)

276 de Souza Neto, E., Feijóo, R.: Variational foundations of multi-scale constitutive models of solids: small and large strain kinematical formulation. Tech Rep. P&D No. 16, LNCC, Laboratório Nacional de Computação Científica, Brazil (2006)

277 de Souza Neto, E., Feijóo, R.: On the equivalence between spatial and material volume averaging of stress in large strain multi-scale constitutive models. *Mech. Mater.* 40, 803–811 (2008)

278 de Souza Neto, E., Feijóo, R.: Variational foundations of large strain multiscale solid constitutive models: Kinematical formulation. In: M. Vaz Jr., E. de Souza Neto, P. Muñoz Rojas (eds.) *Computational Materials Modelling: From Classical to Multi-Scale Techniques*, pp. 341–378. Wiley, Chichester (2010)

279 Strang, G., Fix, G.: *An Analysis of the Finite Element Method.* Prentice Hall (1973)

280 Sunyk, R., Steinmann, P.: On higher gradients in continuum-atomistic modelling. *Int. J. Solids Struct.* 40, 6877–6896 (2003)

281 Swan, C.: Techniques for stress- and strain-controlled homogenization of inelastic periodic composites. *Comput. Meth. App. Mech. Eng.* 117, 249–267 (1994)

282 Taroco, E., Feijóo, R.A.: Problema de creep en discos que giran a velocidad constante. In: V COBEM. Brazil (1979)

283 Taroco, E., Feijóo, R.A.: Viscoplasticidad y su Formulación Variacional. In: Segunda Escola de Matemática Aplicada, vol. 2, pp. 157–331. Laboratório de Computação Científica, LCC/CNPq, Rio de Janeiro, Brazil (1980)

284 Taroco, E., Feijóo, R.A.: Análisis limite via creep. In: VI COBEM. Published also as Technical Report 013/81, Laboratório de Computação Científica, LCC/CNPq, Rio de Janeiro, Brazil (1981)

285 Taroco, E., Feijóo, R.A.: Introducción a la Teoría de Cáscaras. In: I Curso de Mecânica Teórica e Aplicada. Módulo 1, pp. 197–248. Laboratório de Computação Científica, LCC/CNPq, Rio de Janeiro, Brazil (1982)

286 Taroco, E., Feijóo, R.A.: Teoría de Placas y Cáscaras. In: I Curso de Mecânica Teórica e Aplicada. Módulo 2, pp. 117–176. Laboratório de Computação Científica, LCC/CNPq, Rio de Janeiro, Brazil (1983)

287 Taroco, E., Feijóo, R.A., Martins, L.C.: Forma incremental del teorema de los trabajos virtuales aplicados a grandes deformaciones. In: G. Marshal (ed.) *Simposio sobre Métodos Numéricos en la Mecánica del Continuo*, pp. 199–206. Editorial Universitaria de Buenos Aires (1978)

288 Tauchert, T.R.: *Energy Principles in Structural Mechanics*. McGraw Hill (1974)

289 Taylor, A.E.: *Introduction to Functional Analysis*. John Wiley (1958)

290 Temam, R.: *Navier-Stokes Equations: Theory and Numerical Analysis*. North-Holland (1977)

291 Terada, K., Kikuchi, N.: A class of general algorithms for multi-scale analysis of heterogeneous media. *Comput. Meth. App. Mech. Eng.* 190, 5427–5464 (2001)

292 Terada, K., Watanabe, I., Akiyama, M.: Effects of shape and size of crystal grains on the strengths of polycrystalline metals. *Int. J. Multiscale Computational Engineering* 4, 445–460 (2006)

293 Thomas, J.M.: Sur l'analyse numérique des méthodes d'elements finis hybrides et mixes. Ph.D. thesis, Université Pierre et Marie Curie (1977)

294 Thomasset, F.: *Implementation of Finite Element Methods for Navier–Stokes Equations*. Springer Series in Computational Physics. Springer–Verlag (1989)

295 Timoshenko, S.P., Goodier, J.N.: *Theory of Elasticity*. McGraw Hill (1968)

296 Toro, S., Sánchez, P., Blanco, P., de Souza Neto, E., Huespe, A., Feijóo, R.: Multiscale formulation for material failure accounting for cohesive cracks at the macro and micro scales. *Int. J. Plasticity* 76, 75–110 (2016)

297 Toro, S., Sánchez, P., Huespe, A., Giusti, S., Blanco, P., Feijóo, R.: A two-scale failure model for heterogeneous materials: numerical implementation based on the finite element method. *Int. J. Numer. Meth. Eng.* 97, 313–351 (2014)

298 Toro, S., Sánchez, P., Podestá, J., Blanco, P., Huespe, A., Feijóo, R.: Cohesive surface model for fracture based on a two-scale formulation: computational implementation aspects. *Comput. Mech.* 58, 549–585 (2016)

299 Truesdell, C.: *A First Course in Rational Continuum Mechanics*, vol. 1. Academic Press (1981)

300 Truesdell, C., Toupin, R.A.: The Classical Field Theories, *Handbuch der Physik*, vol. 3. Springer Verlag (1960)

301 Vainberg, M.M.: *Variational Methods for the Study of Nonlinear Operators*. Holden-Day, San Francisco (1964)

302 Vicat, L.T.: Note sur l'allongement progressif du fil de fer soumis à diverses tensions. In: Annales Ponts et Chaussées, Mémoires et Documents, N CLXV, pp. 40–44 (1834)

303 Washizu, K.: *Variational Methods in Elasticity and Plasticity*, 3rd. edn. Pergamon (1982)

304 Watanabe, I., Terada, K., de Souza Neto, E., Perić, D.: Characterization of macroscopic tensile strength of polycrystalline metals with two-scale finite element analysis. *J. Mech. Phys. Solids* 56, 1105–1125 (2006)

305 Weinstock, R.: *Calculus of Variations with Applications to Physics and Engineering*. Dover Publications (1974)

306 Zienkiewicz, O.C., Taylor, R.L.: *The Finite Element Method*, vol. 1, fourth edn. McGraw Hill (1989)

307 Zienkiewicz, O.C., Taylor, R.L.: *The Finite Element Method*, vol. 2, fourth edn. McGraw Hill (1991)

308 Zienkiewicz, O.C., Zhu, J.Z.: A simple error estimator and adaptive procedure for practical engineering analysis. *International Journal for Numerical Methods in Engineering* 24, 337–357 (1987)

309 Zouain, N., Feijóo, R.A.: Análisis elasto-plástico vía optimización. *Revista Brasileira de Ciências Mecânicas* 5(3), 1–15 (1983)

310 Zouain, N., Feijóo, R.A.: On kinematical minimum principles for rates and increments in plasticity. Technical Report 014/84, Laboratório de Computação Científica, LCC/CNPq, Rio de Janeiro, Brazil (1984)

311 Zouain, N., Feijóo, R.A., Taroco, E.: Análise de Cascas Plásticas. In: Curso de Mecânica Teórica e Aplicada sobre Teoria das Cascas e suas Aplicações na Engenharia. Módulo 2, pp. 177–250. Laboratório de Computação Científica, LCC/CNPq, Rio de Janeiro, Brazil (1983)

312 Zouain, N., Feijóo, R.A., Taroco, E.: Pressão limite de becais. Technical Report 007/83, Laboratório de Computação Científica, LCC/CNPq, Rio de Janeiro, Brazil (1983)

Index

Introduction to the Variational Formulation in Mechanics: Fundamentals and Applications, First Edition.
Edgardo O. Taroco, Pablo J. Blanco and Raúl A. Feijóo.

g

n

o

p

q

r

t